ELEMENTS OF THE NATURE AND PROPERTIES OF SOILS

FOURTH EDITION

Ray R. Weil
Professor of Soil Science
University of Maryland

Nyle C. Brady (late)
Professor of Soil Science, Emeritus
Cornell University

330 Hudson Street, NY NY 10013

Vice President, Portfolio Management: Andrew Gilfillan
Portfolio Manager: Pamela Chirls
Editorial Assistant: Lara Dimmick
Field Marketing Manager: Bob Nisbet
Product Marketing Manager: Elizabeth Mackenzie-Lamb
Director, Digital Studio and Content Production: Brian Hyland
Managing Producer: Jennifer Sargunar
Content Producer: Rinki Kaur
Manager, Rights Management: Johanna Burke
Manufacturing Buyer: Deidra Smith
Creative Digital Lead: Mary Siener
Full-Service Management and Composition: Integra Software Services Pvt. Ltd.
Full-Service Project Manager: Gowthaman Sadhanandham
Cover Design: Studio Montage
Cover Photos: Ray Weil
Printer/Binder: LSC Communications, Inc.
Cover Printer: Phoenix Color
Text Font: Garamond3LTPro

Copyright © 2019, 2010, 2004 by Pearson Education, Inc. or its affiliates. All Rights Reserved. Manufactured in the United States of America. This publication is protected by Copyright, and permission should be obtained from the publisher prior to any prohibited reproduction, storage in a retrieval system, or transmission in any form or by any means, electronic, mechanical, photocopying, recording, or likewise. For information regarding permissions, request forms, and the appropriate contacts within the Pearson Education Global Rights and Permissions department, please visit www.pearsoned.com/permissions/.

Acknowledgments of third-party content appear on the appropriate page within the text.

PEARSON and ALWAYS LEARNING are exclusive trademarks owned by Pearson Education, Inc. or its affiliates in the U.S. and/or other countries.

Unless otherwise indicated herein, any third-party trademarks, logos, or icons that may appear in this work are the property of their respective owners, and any references to third-party trademarks, logos, icons, or other trade dress are for demonstrative or descriptive purposes only. Such references are not intended to imply any sponsorship, endorsement, authorization, or promotion of Pearson's products by the owners of such marks, or any relationship between the owner and Pearson Education, Inc., authors, licensees, or distributors.

Library of Congress Cataloging-in-Publication Data

Library of Congress Cataloging in Publication Control Number: 2018015179

ISBN-10:   0-13-325459-3
ISBN-13: 978-0-13-325459-4

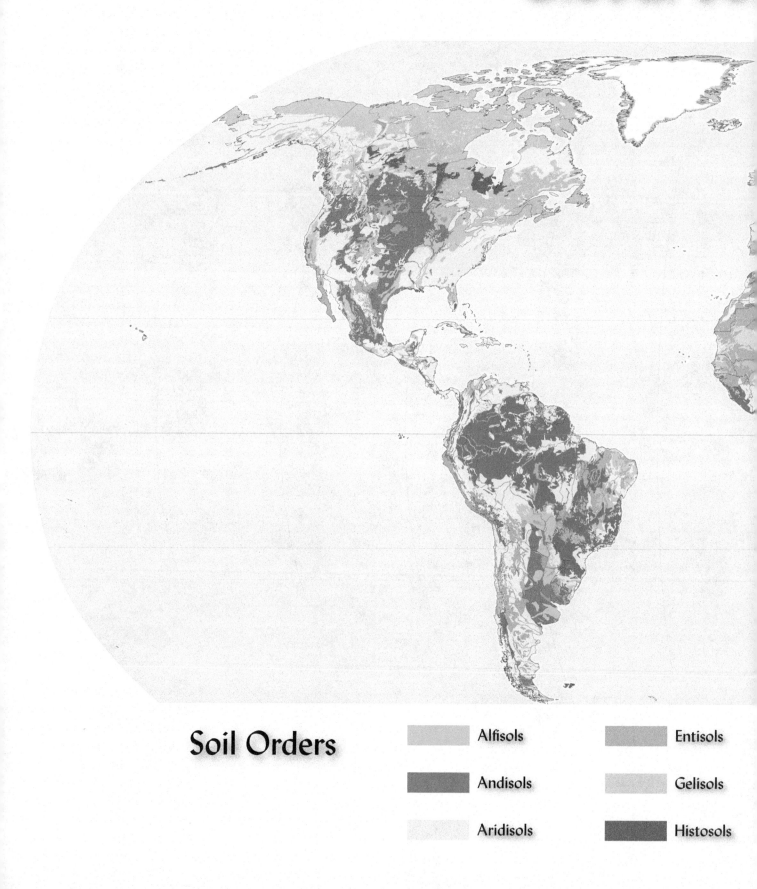

Map Courtesy of Natural Resources Conservation Service, USDA

Global So

Soil Orders

- Alfisols
- Andisols
- Aridisols
- Entisols
- Gelisols
- Histosols

Regions

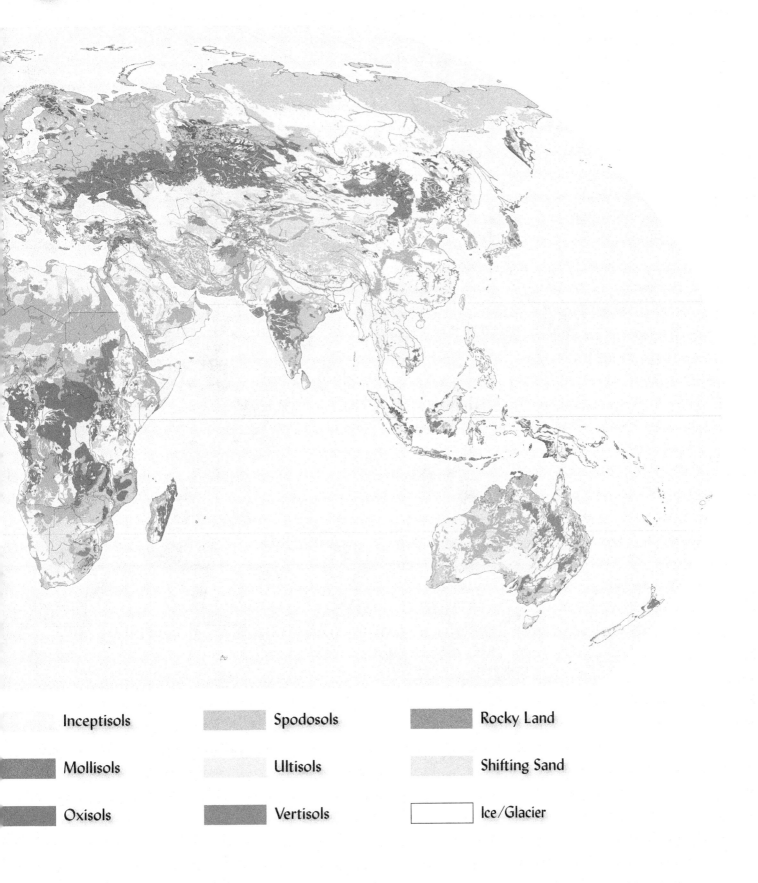

Inceptisols	Spodosols	Rocky Land
Mollisols	Ultisols	Shifting Sand
Oxisols	Vertisols	Ice/Glacier

Map Courtesy of Natural Resources Conservation Service, USDA

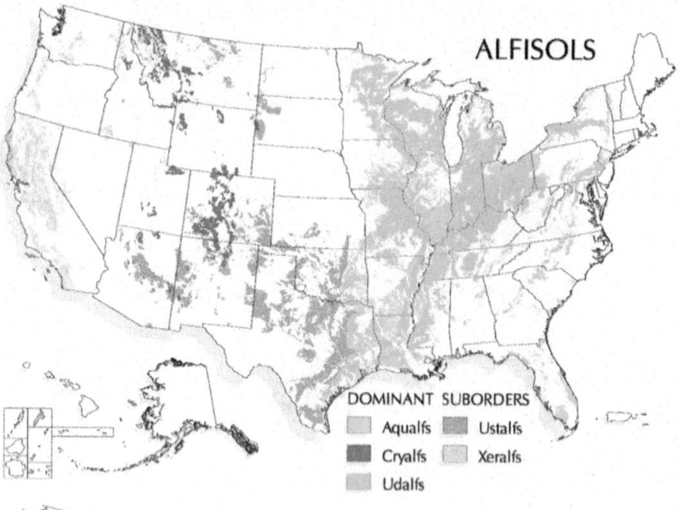

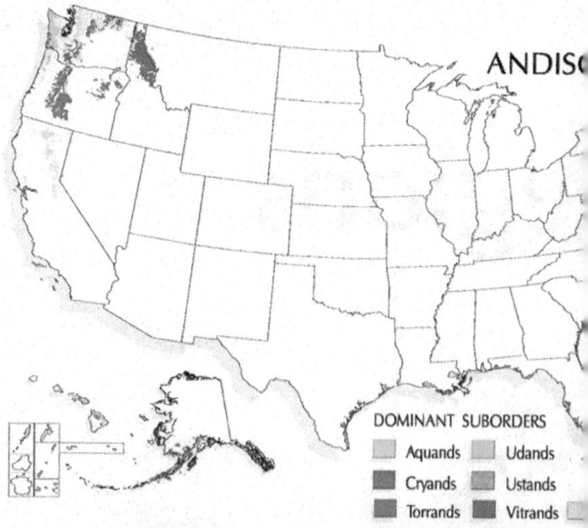

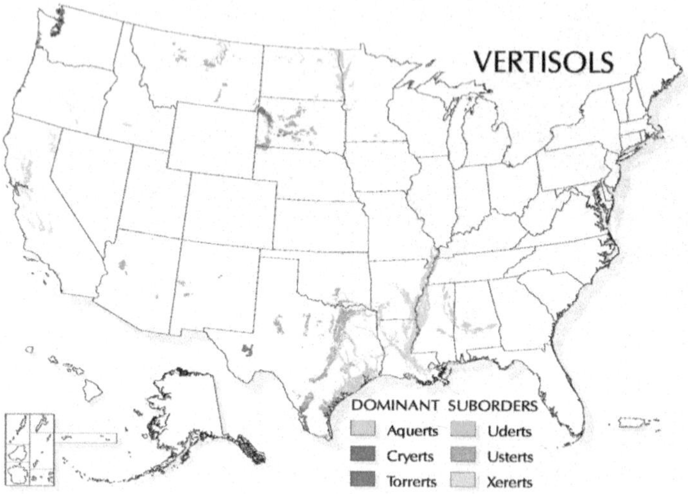

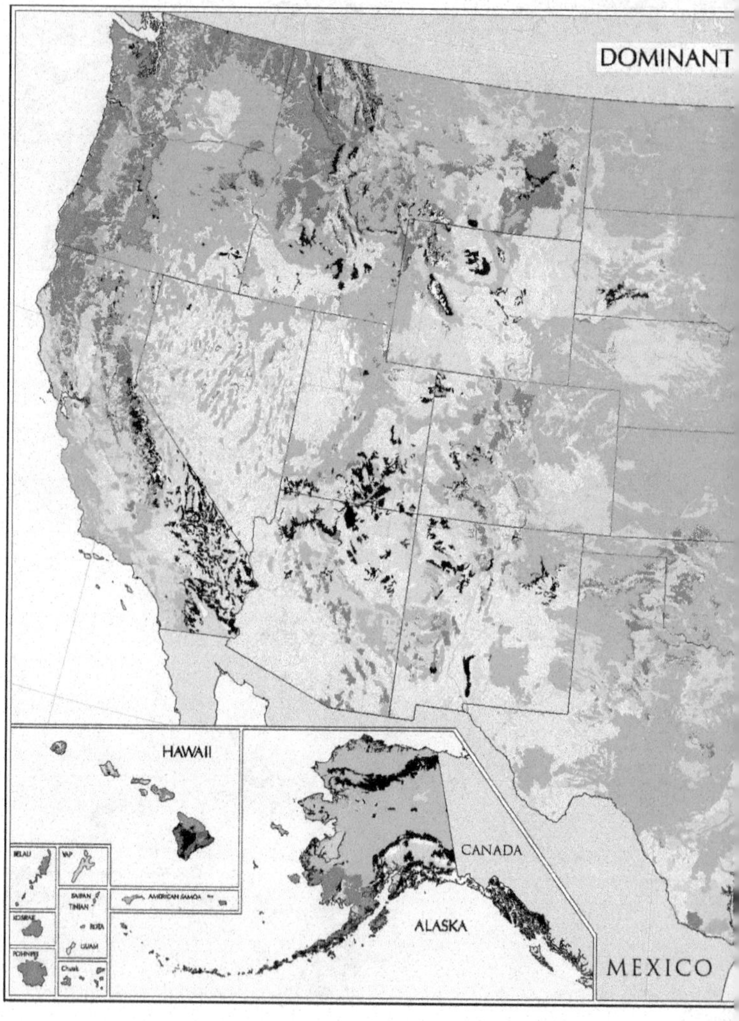

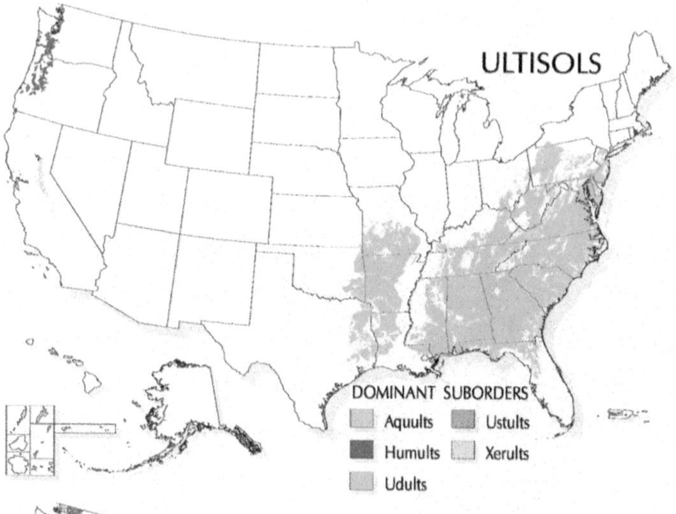

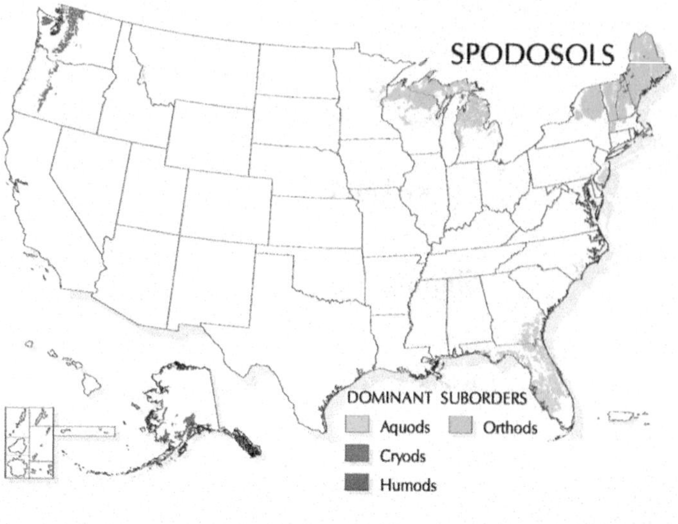

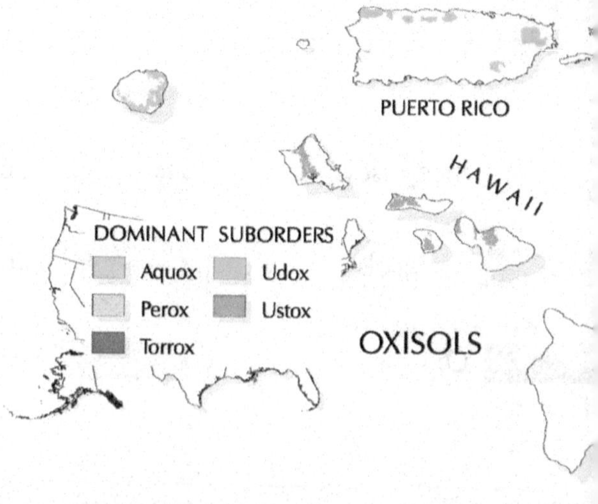

ALFISOLS

DOMINANT SUBORDERS
- Aqualfs
- Cryalfs
- Udalfs
- Ustalfs
- Xeralfs

ANDISOLS

DOMINANT SUBORDERS
- Aquands
- Cryands
- Torrands
- Udands
- Ustands
- Vitrands

VERTISOLS

DOMINANT SUBORDERS
- Aquerts
- Cryerts
- Torrerts
- Uderts
- Usterts
- Xererts

ULTISOLS

DOMINANT SUBORDERS
- Aquults
- Humults
- Udults
- Ustults
- Xerults

SPODOSOLS

DOMINANT SUBORDERS
- Aquods
- Cryods
- Humods
- Orthods

DOMINANT

HAWAII
CANADA
ALASKA
MEXICO
PUERTO RICO
HAWAII

OXISOLS

DOMINANT SUBORDERS
- Aquox
- Perox
- Torrox
- Udox
- Ustox

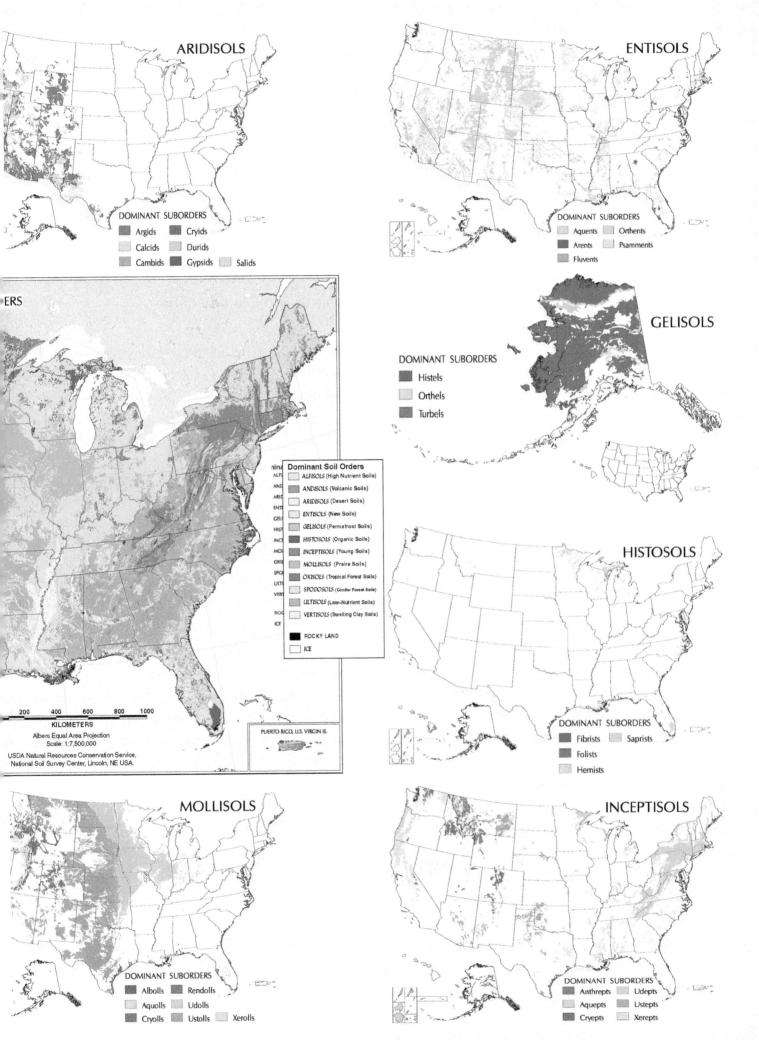

Brief Contents

1
The Soils Around Us 1

2
Formation of Soils from Parent Materials 29

3
Soil Classification 72

4
Soil Architecture and Physical Properties 117

5
Soil Water: Characteristics and Behavior 164

6
Soil and the Hydrologic Cycle 197

7
Soil Aeration and Temperature 239

8
The Colloidal Fraction: Seat of Soil Chemical and Physical Activity 275

9
Soil Acidity, Alkalinity, Salinity, and Sodicity 312

10
Organisms and Ecology of the Soil 369

11
Soil Organic Matter 419

12
Nutrient Cycles and Soil Fertility 466

13
Practical Nutrient Management 548

14
Soil Erosion and Its Control 606

15
Soils and Chemical Pollution 652

Contents

Preface xxi
About the Authors xxiv

1 The Soils Around Us 1

1.1 What Ecosystem Services Do Soils Perform? 2
1.2 How Do Soils Support Plant Growth? 3
1.3 How Do Soils Regulate Water Supplies? 6
1.4 How Do Soils Recycle Raw Materials? 7
1.5 How Do Soils Modify the Atmosphere? 7
1.6 What Lives in the Soil Habitat? 7
1.7 How are Soils Used in Building and Engineering? 9
1.8 The Pedosphere and the Critical Zone? 10
1.9 Soils as Natural Bodies 11
1.10 The Soil Profile and Its Layers (Horizons) 13
1.11 How Does Topsoil Differ from Subsoil? 15
1.12 Soil—Interface of Air, Minerals, Water, and Life 17
1.13 What Are the Mineral (Inorganic) Constituents of Soils? 17
   Soil Texture 18
   Soil Minerals 18
   Soil Structure 19
1.14 What Is Soil Organic Matter Like? 19
1.15 Why Is Soil Water So Dynamic and Complex? 21
   Soil Solution 21
1.16 Soil Air: A Changing Mixture of Gases 21
1.17 How Do Soil Components Interact to Supply Nutrients to Plants? 22
   Essential Element Availability 22
1.18 How Do Plant Roots Obtain Nutrients? 24
1.19 Soil Health, Degradation, and Resilience 25
   Soil Quality and Health 26
   Soil Degradation and Resilience 26
1.20 Conclusion 27
Study Questions 27
References 28

2 Formation of Soils from Parent Materials 29

2.1 Weathering of Rocks and Minerals 29
   Characteristics of Rocks and Minerals 30
   Weathering: A General Case 32
   Physical Weathering (Disintegration) 33
   Biogeochemical Weathering 34
2.2 What Environmental Factors Influence Soil Formation? 36
2.3 Parent Materials 36
   Classification of Parent Materials 37
   Residual Parent Material 37
   Colluvial Debris 38
   Alluvial Stream Deposits 39
   Coastal Sediments 41
   Parent Materials Transported by Glacial Ice and Meltwaters 42
   Parent Materials Transported by Wind 43
   Organic Deposits 44
2.4 How Does Climate Affect Soil Formation? 47
   Effective Precipitation 48
   Temperature 48
2.5 How Do Living Organisms (Including People) Affect Soil Formation? 49
   Role of Natural Vegetation 50
   Role of Animals 51
2.6 How Does Topography Affect Soil Formation? 53
2.7 How Does Time Affect Soil Formation 55
2.8 Four Basic Processes of Soil Formation 58
   Soil-Forming Processes in Action: A Simplified Example 60
2.9 The Soil Profile 62
   The Master Horizons and Layers 62
   Subdivisions Within Master Horizons 64
   Transition Horizons 65

Subhorizon Distinctions 65
Horizons in a Given Profile 66
Soil Genesis in Nature 67
2.10 Urban Soils 67
Pedological Properties Unique to Urban Soils 67
Physical Properties Unique to Urban Soils 68
Chemical Properties Unique to Urban Soils 68
Biological Properties Unique to Urban Soils 69
2.11 Conclusion 70
Study Questions 70
References 71

3
Soil Classification 72

3.1 Concept of Individual Soils 72
Pedon and Polypedon 73
Groupings of Soil Individuals 74
3.2 *Soil Taxonomy*: A Comprehensive Classification System 75
Bases of Soil Classification 75
Diagnostic Surface Horizons of Mineral Soils 75
Diagnostic Subsurface Horizons 75
3.3 Categories and Nomenclature of *Soil Taxonomy* 79
Nomenclature of *Soil Taxonomy* 79
3.4 Soil Orders 81
3.5 Entisols (Recent: Little If Any Profile Development) 83
Distribution and Use 83
3.6 Inceptisols (Few Diagnostic Features: Inception of B Horizon) 85
Distribution and Use 86
3.7 Andisols (Volcanic Ash Soils) 86
Distribution and Use 87
3.8 Gelisols (Permafrost and Frost Churning) 87
Distribution and Use 88
3.9 Histosols (Organic Soils Without Permafrost) 89
Distribution and Use 91
3.10 Aridisols (Dry Soils) 92
Distribution and Use 93
3.11 Vertisols (Dark, Swelling, and Cracking Clays) 94
Distribution and Use 95

3.12 Mollisols (Dark, Soft Soils of Grasslands) 96
Distribution and Use 97
3.13 Alfisols (Argillic or Natric Horizon, Moderately Leached) 98
Distribution and Use 99
3.14 Ultisols (Argillic Horizon, Highly Leached) 99
Distribution and Use 100
3.15 Spodosols (Acid, Sandy, Forest Soils, Highly Leached) 101
Distribution and Use 102
3.16 Oxisols (Oxic Horizon, Highly Weathered) 102
Distribution and Use 102
3.17 Lower-Level Categories in *Soil Taxonomy* 104
Suborders 104
Soil Moisture Regimes (SMRs) 104
Great Groups 104
Subgroups 105
Families 108
Soil Temperature Regimes 108
Series 108
3.18 Mapping the Different Soils in a Landscape 111
Soil Description 111
Delineating Soil Boundaries 112
Online Interactive Soil Survey 113
How to Use Web Soil Survey 113
"There's an App for That" 114
3.19 Conclusion 115
Study Questions 115
References 116

4
Soil Architecture and Physical Properties 117

4.1 Soil Color 117
Causes and Interpretation of Soil Colors 118
4.2 Soil Texture (Size Distribution of Soil Particles) 120
Nature of Soil Separates 120
Influence of Surface Area on Other Soil Properties 122
4.3 Soil Textural Classes 123
Alteration of Soil Textural Class 124

Determination of Textural Class by the "Feel" Method 124
Laboratory Particle-Size Analyses 125
4.4 Structure of Mineral Soils 127
Types of Soil Structure 128
Description of Soil Structure in the Field 131
4.5 Formation and Stabilization of Soil Aggregates 131
Hierarchical Organization of Soil Aggregates 131
Factors Influencing Aggregate Formation and Stability in Soils 132
Physical–Chemical Processes 132
Biological Processes 134
4.6 Tillage and Structural Management of Soils 137
Tillage and Soil Tilth 137
Conventional Tillage and Crop Production 138
Conservation Tillage and Soil Tilth 139
Soil Crusting 139
Soil Conditioners 140
General Guidelines for Managing Soil Tilth 140
4.7 Soil Density 141
Particle Density 141
Bulk Density 141
Factors Affecting Bulk Density 141
Useful Density Figures 144
Management Practices Affecting Bulk Density 145
Influence of Bulk Density on Soil Strength and Root Growth 149
4.8 Pore Space of Mineral Soils 150
Factors Influencing Total Pore Space 150
Size of Pores 150
Cultivation and Pore Size 152
4.9 Soil Properties Relevant to Engineering Uses 153
Field Rating of Soil Consistence and Consistency 153
Soil Strength and Sudden Failure 154
Settlement—Gradual Compression 156
Expansive Soils 157
Atterberg Limits 157
Unified Classification System for Soil Materials 158

4.10 Conclusion 161
Study Questions 161
References 162

5
Soil Water: Characteristics and Behavior 164

5.1 Structure and Related Properties of Water 165
Cohesion Versus Adhesion 166
Surface Tension 166
5.2 Capillary Fundamentals and Soil Water 166
Capillary Mechanism 166
Height of Rise in Soils 166
5.3 Soil Water Energy Concepts 168
Forces Affecting Potential Energy 168
Soil Water Potential 168
Gravitational Potential 169
Pressure Potential (Including Hydrostatic and Matric Potentials) 169
Osmotic Potential 170
Units Used to Quantify Water Potentials 171
Combined Potentials 171
5.4 Soil Water Content and Soil Water Potential 171
Soil Water Versus Energy Curves 171
Measurement of Soil Water Status 172
Volumetric Water Content 173
Measuring Soil Water Status 174
5.5 How Does Water Move in Soil? 177
Saturated Flow Through Soils 177
Factors Influencing the Hydraulic Conductivity of Saturated Soils 179
Unsaturated Flow in Soils 180
5.6 Infiltration and Percolation 181
Infiltration 182
Percolation 182
Water Movement in Stratified Soils 183
Water Movement in Stratified Soils 185
5.7 Qualitative Description of Soil Wetness 185
Maximum Retentive Capacity 186
Field Capacity 186

Permanent Wilting Percentage or Wilting Coefficient 187

Hygroscopic Coefficient 188

5.8 Factors Affecting Amount of Plant-Available Soil Water 189

Water Content–Potential Relationship 189

Compaction Effects on Matric Potential, Aeration, and Root Growth 190

Osmotic Potential 190

Soil Depth and Layering 190

5.9 Mechanisms by Which Plants Are Supplied with Water 193

Rate of Capillary Movement 193

Rate of Root Extension 193

Root Distribution 193

Root–Soil Contact 194

5.10 Conclusion 194

Study Questions 195

References 196

6
Soil and the Hydrologic Cycle 197

6.1 The Global Hydrologic Cycle 198

Global Stocks of Water 198

The Hydrologic Cycle 198

Water Balance Equation 199

6.2 Fate of Incoming Water 200

Effects of Vegetation and Soils on Infiltration 200

6.3 The Soil–Plant–Atmosphere Continuum (SPAC) 205

Evapotranspiration 205

6.4 Control of ET 209

Control of Transpiration 209

Control of Surface Evaporation 211

6.5 Liquid Losses of Water from the Soil 213

Percolation and Leaching 213

Percolation–Evaporation Balance 214

6.6 Percolation and Groundwater 216

Groundwater Resources 216

Shallow Groundwater 217

Movement of Chemicals in the Drainage Water 218

Chemical Movement Through Macropores 218

6.7 Enhancing Soil Drainage 220

Reasons for Enhancing Soil Drainage 220

Surface Drainage Systems 222

Subsurface (Internal) Drainage 222

6.8 Septic Tank Drain Fields 226

Operation of a Septic System 227

Soil Properties Influencing Suitability for a Septic Drain Field 228

6.9 Irrigation Principles and Practices 229

Importance of Irrigation Today 230

Water-Use Efficiency 231

Surface Irrigation 232

Sprinkler Systems 233

Microirrigation 235

6.10 Conclusion 236

Study Questions 237

References 238

7
Soil Aeration and Temperature 239

7.1 Soil Aeration—The Process 239

Soil Aeration in the Field 240

Excess Moisture 240

Gaseous Interchange 240

7.2 Means of Characterizing Soil Aeration 241

Gaseous Composition of the Soil Air 241

Air-Filled Porosity 242

7.3 Oxidation–Reduction (Redox) Potential 242

Redox Reactions 242

Role of Oxygen Gas 243

Other Electron Acceptors 243

7.4 Factors Affecting Soil Aeration and E_H 245

Rates of Respiration in the Soil 245

Depth in the Soil Profile 245

Drainage of Excess Water 246

Small-Scale Soil Heterogeneity 246

Seasonal Differences 247

Effects of Vegetation 247

7.5 Ecological Effects of Soil Aeration 247

Effects on Organic Residue Degradation 247

Oxidation–Reduction of Elements 248

Effects on Activities of Higher Plants 249

- 7.6 Soil Aeration in Urban Landscapes 251
  - Container-Grown Plants 251
  - Tree and Lawn Management 252
- 7.7 Wetlands and Their Poorly Aerated Soils 253
  - Defining a Wetland 253
  - Wetland Hydrology 254
  - Hydric Soils 255
  - Hydrophytic Vegetation 255
  - Wetland Chemistry 257
  - Constructed Wetlands 257
- 7.8 Processes Affected by Soil Temperature 259
  - Plant Processes 259
  - Microbial Processes 261
  - Freezing and Thawing 262
  - Permafrost 262
  - Soil Heating by Fire 263
  - Contaminant Removal 263
- 7.9 Absorption and Loss of Solar Energy 264
  - Slope Angle and Aspect 265
- 7.10 Thermal Properties of Soils 266
  - Specific Heat of Soils 266
  - Heat of Vaporization 267
  - Thermal Conductivity of Soils 268
  - Variation with Time and Depth 269
- 7.11 Soil Temperature Control 270
  - Organic Mulches and Plant-Residue Management 270
  - Plastic Mulches 271
  - Moisture Control 272
- 7.12 Conclusion 273

Study Questions 273

References 273

8
The Colloidal Fraction: Seat of Soil Chemical and Physical Activity 275

- 8.1 General Properties and Types of Soil Colloids 276
  - Size 276
  - Surface Area 276
  - Surface Charges 276
  - Adsorption of Cations and Anions 276
  - Adsorption of Water 277
  - Types of Soil Colloids 278
- 8.2 Fundamentals of Layer Silicate Clay Structure 279
  - Tetrahedral and Octahedral Sheets 279
  - Source of Charges 281
- 8.3 Mineralogical Organization of Silicate Clays 281
  - 1:1-Type Silicate Clays 281
  - Expanding 2:1-Type Silicate Clays 283
  - Nonexpanding 2:1 Silicate Minerals 285
- 8.4 Structural Characteristics of Nonsilicate Colloids 286
  - Iron and Aluminum Oxides 286
  - Humus 287
- 8.5 Genesis and Geographic Distribution of Soil Colloids 288
  - Genesis of Colloids 288
  - Distribution of Clays by Geography and Soil Order 289
- 8.6 Sources of Charges on Soil Colloids 290
  - Constant Charges on Silicate Clays 290
  - pH-Dependent Charges 291
- 8.7 Adsorption of Cations and Anions 293
  - Outer- and Inner-Sphere Complexes 294
- 8.8 Cation Exchange Reactions 295
  - Principles Governing Cation Exchange Reactions 296
- 8.9 Cation Exchange Capacity (CEC) 298
  - Methods of Determining CEC 298
  - Cation Exchange Capacities of Soils 300
  - pH and Cation Exchange Capacity 301
- 8.10 Exchangeable Cations in Field Soils 302
  - Cation Saturation and Nutrient Availability 303
  - Influence of Complementary Cations 303
  - Effect of Type of Colloid 304
- 8.11 Anion Exchange 304
  - Inner-Sphere Complexes 304
  - Weathering and CEC/AEC Levels 305
- 8.12 Sorption of Organic Molecules to Soil Colloids 306
  - Distribution Coefficients 307
  - Binding of Biomolecules to Clay and Humus 308
- 8.13 Conclusion 309

Study Questions 310

References 311

9
Soil Acidity, Alkalinity, Salinity, and Sodicity 312

- 9.1 What Processes Cause Soil Acidity and Alkalinity? 313
  - Acidifying Processes That Produce Hydrogen Ions 314
  - Alkalizing Processes That Consume Hydrogen Ions or Produce Hydroxyl Ions 315
- 9.2 What Role Does Aluminum Play in Soil Acidity? 317
- 9.3 Pools of Soil Acidity 318
  - Principal Pools of Soil Acidity 318
  - Cation Saturation Percentages 320
  - Acid (or Nonacid) Cation Saturation and pH 320
- 9.4 Buffering of pH in Soils 321
  - Why Is Soil pH Buffering Important? 322
- 9.5 How Can We Measure Soil pH? 323
  - Potentiometric Methods 323
  - Variability in the Field 324
- 9.6 How Do Humans Acidify Soils? 325
  - Nitrogen Fertilization 325
  - Acid Deposition from the Atmosphere 326
  - Exposure of Potential Acid Sulfate Materials 328
- 9.7 How Does Soil pH Affect Living Things? 330
  - Aluminum Toxicity 330
  - Manganese, Hydrogen, and Iron Toxicity to Plants 331
  - Nutrient Availability to Plants 332
  - Microbial Effects 333
  - Optimal pH Conditions for Plant Growth 333
  - Soil pH and Organic Molecules 335
- 9.8 Raising Soil pH by Liming 336
  - Agricultural Liming Materials 336
  - How Do Liming Materials React to Raise Soil pH? 337
  - Lime Requirement: How Much Lime Is Needed to Do the Job? 338
  - How Lime Is Applied 340
  - Special Liming Situations 340
- 9.9 Ameliorating Acidity Without Lime 341
  - Using Gypsum 341
  - Using Organic Matter 341
- 9.10 Lowering Soil pH 343
  - Acid Organic Matter 343
  - Inorganic Chemicals 343
- 9.11 Development of Salt-Affected Soils 346
  - Accumulation of Salts in Nonirrigated Soils 346
  - Irrigation-Induced Salinity and Alkalinity 347
- 9.12 Measuring Salinity and Sodicity 348
  - Salinity 348
  - Sodium Status (Sodicity) 350
- 9.13 Classes of Salt-Affected Soils 350
  - Saline Soils 350
  - Saline–Sodic Soils 352
  - Sodic Soils 352
- 9.14 Physical Degradation in Soil–Sodic Soils 352
  - Slaking, Swelling, and Dispersion 353
  - Two Causes of Soil Dispersion 353
- 9.15 Biological Impacts of Salt–Affected Soils 354
  - How Salts Affect Plants 354
  - Selective Tolerance of Higher Plants to Saline and Sodic Soils 355
  - Salt Problems Not Related to Arid Climates 355
- 9.16 Water-Quality Considerations for Irrigation 356
- 9.17 Reclamation of Saline Soils 358
  - Leaching Requirement (LR) 358
  - Management of Soil Salinity 359
  - Some Limitations of the Leaching Requirement Approach 361
- 9.18 Reclamation of Saline–Sodic and Sodic Soils 362
  - Gypsum 362
  - Sulfur and Sulfuric Acid 362
  - Physical Condition 363
  - Management of Reclaimed Soils 363
- 9.19 Conclusion 364

Study Questions 366
References 367

10
Organisms and Ecology of the Soil 369

- 10.1 The Diversity of Organisms in the Soil 370
- 10.2 Organisms in Action 373
  - Trophic Levels and the Soil Food Web 373
  - Sources of Energy and Carbon 373

Primary Producers 375
Primary Consumers 375
Secondary Consumers 375
Tertiary Consumers 376
Ecosystem Engineers 377

10.3 Abundance, Biomass, and Metabolic Activity 378
Comparative Organism Activity 378

10.4 Earthworms 379
Influence on Soil Fertility, Productivity, and Environmental Quality 380
Deleterious Effects of Earthworms 381
Factors Affecting Earthworm Activity 382

10.5 Ants and Termites 384
Ants 384
Termites 384

10.6 Soil Microanimals 386
Nematodes 386
Protozoa 388
Other Fascinating Soil Microcreatures 389

10.7 Plant Roots 390
Root Morphology 390
How Roots Alter Soil Conditions 391
Rhizosphere 391
Rhizodeposition 391

10.8 Soil Algae 393

10.9 Soil Fungi 393
Molds 394
Mushroom Fungi 395
Activities of Fungi 395
Mycorrhizae 397

10.10 Soil Prokaryotes: Bacteria and Archaea 400
Characteristics 400
Prokaryote Diversity in Soils 400
Sources of Energy 402
Importance of Prokaryotes 402
Soil Actinomycetes 403

10.11 Conditions Affecting the Growth and Activity of Soil Microorganisms 405
Organic Resources 405
Oxygen Requirements 405
Moisture and Temperature 405
Exchangeable Calcium and pH 405

10.12 Beneficial Effects of Soil Organisms on Plant Communities 406
Soil Organic Matter Formation and Nutrient Cycling 406
Breakdown of Toxic Compounds 406
Inorganic Transformations 406
Nitrogen Fixation 406
Rhizobacteria 407
Plant Protection 407

10.13 Soil Organisms and Plant Damage 407
Plant Pests and Parasites 408
Plant Disease Control by Soil Management 408
Disease-Suppressive Soils 409

10.14 Ecological Relationships Among Soil Organisms 412
Mutualistic Associations 412
Biocrusts 412
Effects of Agricultural Practices on Soil Organisms 413

10.15 Conclusion 415

Study Questions 415
References 416

11
Soil Organic Matter 419

11.1 The Global Carbon Cycle 419
Basic Processes 420
Carbon Sources 421

11.2 Organic Decomposition in Soils 423
Composition of Plant Residues 423
Decomposition of Organic Compounds in Aerobic Soils 424
Example of Organic Decay 425
Production of Simple Inorganic Products 426
Decomposition in Anaerobic Soils 426

11.3 Factors Controlling Rates of Residue Decomposition and Mineralization 427
Physical Factors Influencing Residue Quality 427
Carbon/Nitrogen Ratio of Organic Materials and Soils 428
Influence of Carbon/Nitrogen Ratio on Residue Decomposition 429

Examples of Inorganic Nitrogen Release During Decay 430
Influence of Soil Ecology 431
Influence of Lignin and Polyphenol Content 431
11.4 Genesis and Nature of Soil Organic Matter and Humus 432
Microbial Transformations 434
Examples of Biomolecules in Soil Organic Matter 436
Colloid Characteristics of Humus 436
Stability of Humus 438
11.5 Influences of Organic Matter on Plant Growth and Soil Function 438
Direct Influence of Humus on Plant Growth 438
Allelochemical Effects 439
Influence of Organic Matter on Soil Properties and Indirectly on Plants 440
11.6 Amounts and Quality of Organic Matter in Soils 442
Labile Organic Matter 442
Protected or Stable Organic Matter (Humus) 442
Changes in Labile and Humus Pools with Soil Management 443
11.7 Carbon Balance in the Soil-Plant-Atmosphere System 444
Agroecosystems 444
Natural Ecosystems 447
11.8 Environmental Factors Influencing Soil Organic Carbon Levels 447
Differences Among Soil Orders 448
Influence of Climate 449
Influence of Natural Vegetation 450
Effects of Texture and Drainage 450
11.9 Soil Organic Matter Management 451
Influence of Agricultural Management and Tillage 451
Influence of Rotations, Residues, and Plant Nutrients 452
The Conundrum of Soil Organic Matter Management 453
General Guidelines for Managing Soil Organic Matter 453
11.10 Soils and Climate Change 454
Global Climate Change 454
Carbon Dioxide 455

11.11 Composts and Composting 461
11.12 Conclusion 463
Study Questions 464
References 464

12
Nutrient Cycles and Soil Fertility 466

12.1 Nitrogen in the Soil System 467
Nitrogen and Plant Growth and Development 467
Distribution and Cycling of Nitrogen 468
Nitrogen Immobilization and Mineralization 470
Ammonium Fixation by Clay Minerals 471
Dissolved Organic Nitrogen 471
Ammonia Volatilization 472
Nitrification 473
Soil Conditions Affecting Nitrification 474
Gaseous Losses by Denitrification 475
Anammox 476
Atmospheric Pollution and Greenhouse Gas Emissions 476
Symbiotic Nitrogen Fixation with Legumes 479
N Fixation in Nodule-Forming Nonlegumes 482
Symbiotic Nitrogen Fixation Without Nodules 483
Nonsymbiotic Fixation by Heterotrophs 484
Fixation by Autotrophs 484
Nitrogen Deposition from the Atmosphere 484
12.2 Sulfur in the Soil System 487
Roles of Sulfur in Plants and Animals 487
Natural Sources of Sulfur in Soils 488
Cycling of Sulfur in Soils 491
Sulfur Retention and Exchange 494
Sulfur Fertility Maintenance 495
12.3 Phosphorus in Plant Nutrition and Soil Fertility 496
Phosphorus and Plant Growth 496
The Phosphorus Problem in Soil Fertility and Environmental Quality 497
The Phosphorus Cycle 498
Organic Phosphorus in Soils 501

Inorganic Phosphorus in Soils 503

Phosphorus-Fixation Capacity of Soils 507

How Do Plants Obtain Adequate Phosphorus? 509

Management Strategies for Meeting Plant Phosphorus Needs in Low-P Soils 510

Management Strategies for Controlling Over-Enrichment of Soils and Water Pollution 512

12.4 Potassium: Nature and Ecological Roles 514

Potassium in Plant and Animal Nutrition 514

The Potassium Cycle in Soil–Plant Systems 516

The Potassium Problem in Soil Fertility 518

Forms and Availability of Potassium in Soils 518

Factors Affecting Potassium Fixation in Soils 520

Some Practical Aspects of Potassium Management 521

12.5 Calcium as an Essential Nutrient 522

Calcium in Plants 522

Soil Forms and Processes 523

12.6 Magnesium as a Plant Nutrient 524

Magnesium in Plants 524

Magnesium in Soil 524

Ratio of Calcium to Magnesium 525

12.7 Silicon in Soil–Plant Ecology 526

Silicon in Plants 526

Silicon in Soils 527

12.8 Micronutrients in the Soil–Plant System 528

Deficiency Versus Toxicity 528

Micronutrient Cycles, Forms, and Reactions in the Soil 532

Organic 534

Influence of Soil pH 534

Oxidation State 535

Organic Matter and Clay 536

Role of Mycorrhizae 537

Organic Chelating Agents 538

Soil Management and Trace Element Needs 539

Fertilizer Applications 540

Fighting Micronutrient Hunger 540

12.9 Conclusion 542

Study Questions 543

References 545

13 Practical Nutrient Management 548

13.1 Goals of Nutrient Management 548

Plant Production 549

Soil Health and Productivity 549

Conservation of Nutrient Resources 549

Environmental Impact: Nutrient Budgets and Balances 550

13.2 Nutrients as Pollutants 551

Nutrient Damage to Aquatic Ecosystems 551

Nutrient Management Plans 552

Best Management Practices (BMPs) 554

Buffer Strips 554

Cover Crops for Nutrient Management 556

Conservation Tillage 559

Combining Practices on the Landscape 560

13.3 Ecosystem Nutrient Cycles 560

Nutrient Cycling in Grasslands 562

13.4 Recycling Nutrients Through Animal Manures 564

Nutrient Composition of Animal Manures 564

Concentrated Animal-Feeding Operations (CAFOs) 564

Storage, Treatment, and Management of Animal Manures 567

Methods of Manure Application 569

13.5 Industrial and Municipal By-Products 570

Garbage and Yard Wastes 570

Food-Processing Wastes 570

Wood Wastes 570

Wastewater Treatment By-Products 571

Sewage Effluent 571

Sewage Sludge or Biosolids 572

Integrated Recycling of Wastes 572

13.6 Practical Utilization of Organic Nutrient Sources 573

13.7 Inorganic Commercial Fertilizers 576

Origin and Processing of Inorganic Fertilizers 577

Properties and Use of Inorganic Fertilizers 577

Fertilizer Grade 578

Fate of Fertilizer Nutrients 578

The Concept of the Limiting Factor 581

13.8 Fertilizer Application Methods 581
  Broadcasting 581
  Localized Placement 584
  Foliar Application 585
13.9 Timing of Nutrient Application 585
  Availability When the Plants Need It 585
  Environmentally Sensitive Periods 586
  Physiologically Appropriate Timing 586
  Practical Field Limitations 586
13.10 Diagnostic Tools and Methods 586
  Plant Symptoms and Field Observations 587
  Plant Analysis 587
13.11 Soil Analysis 591
  Sampling the Soil 591
  Chemical Analysis of the Sample 593
  Interpreting the Results to Make a Recommendation 594
  Merits of Soil Testing 595
13.12 Site-Index Approach to Phosphorus Management 596
  Overenrichment of Soils 596
  Transport of Phosphorus from Land to Water 597
  Phosphorus Soil Test Level as Indicator of Potential Losses 597
  Phosphorus Site Index 599
13.13 Some Advances and Challenges in Nitrogen Management 599
13.14 Conclusion 602
Study Questions 603
References 603

14 Soil Erosion and Its Control 606

14.1 Significance of Soil Erosion and Land Degradation 607
  Land Degradation 607
  Soil-Vegetation Interdependency 607
  Geological Versus Accelerated Erosion 608
14.2 On-Site and Off-Site Impacts of Accelerated Soil Erosion 610
  Types of On-Site Damages 610
  Types of Off-Site Damages 610

14.3 Mechanics of Water Erosion 612
  Influence of Raindrops 612
  Transportation of Soil 613
  Types of Water Erosion 613
  Deposition of Eroded Soil 613
14.4 Models to Predict Water-Induced Erosion 614
  The Universal Soil-Loss Equation (USLE) 615
  The Revised Universal Soil-Loss Equation (RUSLE) 616
14.5 Factors Affecting Interrill and Rill Erosion 616
  Rainfall Erosivity Factor, R 616
  Soil Erodibility Factor, K 616
  Topographic Factor, LS 616
  Cover and Management Factor, C 618
  Support Practice Factor, P 619
14.6 Conservation Tillage 622
  Conservation Tillage Systems 624
  Adaptation by Farmers 625
  Erosion Control by Conservation Tillage 626
  Effect on Soil Properties 627
14.7 Vegetative Barriers 628
14.8 Control of Gully Erosion and Mass Wasting 629
  Remedial Treatment of Gullies 629
  Mass Wasting on Unstable Slopes 630
14.9 Control of Accelerated Erosion on Range and Forestland 631
  Rangeland Problems 631
  Erosion on Forestlands 631
  Practices to Reduce Soil Loss Caused by Timber Production 632
14.10 Erosion and Sediment Control on Construction Sites 634
  Principles of Erosion Control on Construction Sites 634
  Keeping the Disturbed Soil Covered 634
  Controlling the Runoff 636
  Trapping the Sediment 637
14.11 Wind Erosion: Importance and Factors Affecting It 638
  Mechanics of Wind Erosion 640
  Factors Affecting Wind Erosion 641

14.12 Predicting and Controlling Wind Erosion 641
   Control of Wind Erosion 642
14.13 Tillage Erosion 644
   Movement of Soil by Tillage 644
   Quantification of Tillage Erosion 645
14.14 Land Capability Classification and Progress in Soil Conservation 647
   Conservation Management to Enhance Soil Health 648
   Finding Soil Conservation Win-Win Systems 648
14.15 Summary and Conclusion 649
Study Questions 650
References 650

15
Soils and Chemical Pollution 652

15.1 Toxic Organic Chemicals 652
   Environmental Damage from Organic Chemicals 653
   The Nature of the Pesticide Problem 653
15.2 Kinds of Organic Contaminants 657
   Industrial Organics 657
   Pesticides 657
15.3 Behavior of Organic Chemicals in Soil 659
   Contamination of Groundwater 662
   Chemical Reactions 662
   Microbial Metabolism 662
   Plant Absorption and Breakdown 664
   Persistence in Soils 664
15.4 Effects of Pesticides on Soil Organisms 665
   Fumigants 665
   Effects on Soil Fauna 665
   Effects on Soil Microorganisms 666
15.5 Remediation of Soils Contaminated with Organic Chemicals 667
   Physical and Chemical Methods 667
   Bioremediation 669
   Phytoremediation 672

15.6 Soil Contamination with Toxic Inorganic Substances 674
   Sources of the Contaminants 674
   Accumulation in Soils 675
   Concentration in Living Tissues 675
   Some Inorganic Contaminants and Their Reactions in Soils 678
15.7 Potential Hazards of Chemicals in Sewage Sludge 678
   Heavy Metals in Sewage Sludge 679
15.8 Prevention and Remediation of Inorganic Soil Contamination 681
   Reducing Soil Application 681
   Immobilizing the Toxins 681
   Bioremediation by Metal Hyperaccumulating Plants 683
   Management to Enhance Phytoremediation 684
15.9 Landfills 685
   The Municipal Solid Waste Problem 685
   Two Basic Types of Landfill Design 686
   Natural Attenuation Landfills 686
   Containment or Secured Landfills 688
   Environmental Impacts of Landfills 688
   Land Use After Completion 690
15.10 Radionuclides in Soil 690
   Radioactivity from Nuclear Fission 690
   Nuclear Accident at Chernobyl 691
   Nuclear Accident at Fukushima 692
   Radioactive Wastes 692
15.11 Radon Gas from Soils 693
   The Health Hazard 693
   How Radon Accumulates in Buildings 693
   Radon Testing and Remediation 694
15.12 Conclusion 695
Study Questions 695
References 696

APPENDIX A. World Reference Base, Canadian, and Australian Soil Classification Systems 698

APPENDIX B. SI Units, Conversion Factors, Periodic Table of the Elements, and Plant Names 703

Glossary of Soil Science Terms 709

INDEX 728

Nyle C. Brady 1920–2015

On November 24, 2015, soil science lost one of its giants. Nyle C. Brady passed away at the age of 95. Dr. Brady was a global leader in soil science, in agriculture, and in humanity. He was born in 1920 in the tiny rural town of Manassa, Colorado, USA. He earned a BS degree in chemistry from Brigham Young University in 1941 and went on to complete his PhD in soil science at North Carolina State University in 1947. Dr. Brady then served as a member of the faculty at Cornell University in Ithaca, NY, USA, for 26 years, rising from assistant professor to professor and chair of the agronomy department and finally to assistant dean of the College of Agriculture. During this period, he was elected president of both the American Society of Agronomy and the Soil Science Society of America.

Soon after arriving at Cornell University, he was recruited by Professor Harry O. Buckman to assist in co-authoring the then already classic soil science textbook, *The Nature and Properties of Soils*. The first edition of this textbook to bear Nyle Brady's name as co-author was published in 1952. Under Nyle's hand, this book rose to prominence throughout the world and several generations of soil scientists got their introduction to the field through its pages. He was the sole author of editions published between 1974 and 1990. He continued to work on revised editions of this book with co-author Ray Weil until 2004.

Dr. Brady was of that generation of American soil scientists that contributed so much to the original green revolution. He conducted research into the chemistry of phosphorus and the management of fertilizers, and he was an early researcher on minimum tillage. Known for his active interest in international development and for his administrative skills, he was recruited in 1973 to be the third Director General of the International Rice Research Institute (IRRI) in the Philippines. Dr. Brady pioneered new cooperative relationships between IRRI and the national agricultural research institutions in many Asian countries, including a breakthrough visit to China at a time when that country was still quite closed to the outside world. He oversaw the transition to a second-generation of green revolution soil management and plant breeding designed to overcome some of the shortcomings of the first generation.

After leaving IRRI, he served as senior assistant administrator for Science and Technology at the U.S. Agency for International Development from 1981 to 1989. He was a fierce champion of international scientific cooperation to promote sustainable resource use and agricultural development. During the 1990s Dr. Brady, then in his 70s, served as senior international development consultant for the United Nations Development Programme (UNDP) and for the World Bank, in which capacity he continued to promote scientific collaboration in advances in environmental stewardship and agricultural development.

Dr. Brady was always open-minded and ready to accept new truths supported by scientific evidence, as can be seen by the evolution of the discussion of such topics as pesticide use, fertilizer management, manure utilization, tillage, soil organic matter, and soil acidity management in *The Nature and Properties of Soils* under his guidance. Nyle Brady had a larger-than-life personality, a deep sense of empathy, and an incredible understanding of how to work with people to get positive results. He was the kind of person that friends, associates, and even strangers would go to for advice when they found themselves in a perplexing position as a scientist, administrator, or even in their personal life. He will be very much missed for a long time to come by his family and by all who knew him or were touched by his work.

Preface

By opening this fourth edition of *The Elements of the Nature and Properties of Soils*, you are tapping into a narrative that has been at the forefront of soil science for more than a century. The first version of the parent book from which this book has been abridged, was published in 1909. It was largely a guide to good soil management for farmers in the glaciated regions of New York State in the northeastern United States. Since then, the books have evolved to provide a globally relevant framework for an integrated understanding of the diversity of soils, the soil system, and its role in the ecology of planet Earth.

If you are a student reading this, you have chosen a truly auspicious time to take up the study of soil science. Scientists and managers well versed in soil science are in short supply and becoming increasingly sought after. Much of what you learn from these pages will be of enormous practical value in equipping you to meet the many natural-resource challenges of the twenty-first century. You will soon find that the soil system provides many opportunities to see practical applications for principles from such sciences as biology, chemistry, physics, and geology.

The importance of soils and soil science is increasingly recognized by business and political leaders, by the scientific community, and by those who work with the land. Soils are now widely recognized as the underpinning of terrestrial ecosystems and the source of a wide range of essential ecosystem services. An understanding of the soil system is therefore critical for the success and environmental harmony of almost any human endeavor on the land.

This latest edition of *Elements of the Nature and Properties of Soils* is the first to feature *full color illustrations* throughout. As is the case for its parent book, *The Nature and Properties of Soils*, 15th edition, this newest edition of *Elements of the Nature and Properties of Soils* strives to explain the fundamental principles of soil science in a manner that you will find relevant to your interests. The text emphasizes the soil as a natural resource and soils as ecosystems. It highlights the many interactions between soils and other components of the larger forest, range, agricultural, wetland, and constructed ecosystems. This book is designed to serve you well, whether you expect this to be your only formal exposure to soil science or you are embarking on a comprehensive soil science education. It is meant to provide both an exciting, accessible introduction to the world of soils and a reliable reference that you will want to keep for your professional bookshelf.

Every chapter has been thoroughly updated with the latest advances, concepts, and applications. This edition includes new or updated discussions on soils and human health, organic farming, engineering properties of soils, colloids and CEC, humus and organic matter, the proton-balance approach to soil acidity, soil salinity and alkalinity, irrigation techniques, soil food-web ecology, disease suppressive soils, soil archaea, soil contamination and bioremediation, nutrient management, soil health, soil ecosystem services, soil interactions with global climate change, and many other topics of current interest in soil science. At the same time, this abridgement of the original book omits or simplifies some of the more technical details, presents fewer chemical equations and calculations, and focuses the text more clearly on the basics of soil science such that a survey of the field is be accomplished in 15 instead of 20 chapters, comprising about 700 instead of nearly 1,100 pages.

If you are an instructor or a soil scientist, you will benefit from changes in this latest edition. Most noticeable is the use of full-color throughout, which improves the new and refined figures and illustrations to help make the study of soils more efficient, engaging, and intellectually satisfying. Every topic, from soil classification to soil carbon, has been updated to reflect current thinking in the discipline. Hundreds of new key references have been added. This edition includes in-depth discussions on such topics of cutting edge soil science as carbon sequestration, subaqueous soils, urban and human engineered soils, cycling and plant use of silicon, inner- and outer-sphere complexes, radioactive soil contamination, new understandings of the nitrogen cycle, cation saturation and ratios, acid sulfate soils, water-saving irrigation techniques, hydraulic redistribution, cover crop effects on soil health, soil invertebrate ecology, disease suppressive soils, soil microbial genomics, soil ecosystem services, biochar, soil interactions with global climate change, digital soil maps, and many others.

In response to their popularity in recent editions, I have also added or improved "boxes" that present either fascinating examples and applications or technical details and calculations. These boxes both *highlight* material of special interest and allow the logical thread of the regular text to flow smoothly without digression or interruption. Examples of applications boxes or case study vignettes include the following:

- "Dirt for Dinner"
- "Subaqueous Soils—Underwater Pedogenesis"
- "Practical Applications of Unsaturated Water Flow in Contrasting Layers"
- "Char: Is Black the New Gold?"
- "Where have All the Humics Gone?"
- "Tragedy in the Big Easy—A Levee Doomed to Fail"
- "Costly and Embarrassing Soil pH Mystery"
- "Gardeners' Friend Not Always So Friendly
- "Soil Microbiology in the Molecular Age"
- "The Law of Return Made Easy: Using Human Urine"

Boxes also are provided to explain detailed calculations and practical numerical problems. Examples include the following:

- "Estimating CEC and Clay Mineralogy"
- "Calculating Lime Needs Based on pH Buffering"
- "Leaching Requirement for Saline Soils"
- "Calculation of Percent Pore Space in Soils"
- "Calculating Soil CEC from Lab Data"
- "Concentrations and Toxicity of Contaminants"
- "Calculation of Nitrogen Mineralization"

As the global economy expands, exponentially societies face new challenges with managing their natural resources. Soil as a fundamental natural resource is critical to sustained economic growth and the prosperity of people in all parts of the world. To achieve balanced growth with a sustainable economy while improving environmental quality, it will be necessary to have a deep understanding of soils, including their properties, functions, ecological roles, and management. I have written this textbook in a way designed to engage inquisitive minds and challenge them to understand soils and actively do their part as environmental and agricultural scientists, in the interest of ensuring a prosperous and healthy future for humanity on planet Earth.

In this textbook, I have tried to take a broad view of soils in the environment and in relation to human society. In so doing, the book focuses on six major ecological roles of soil. Soils provide for the growth of plants, which, in turn, provide wildlife habitat, food for people and animals, bio-energy, clothing, pharmaceuticals, and building materials. In addition to plant production, soils also dramatically influence the Earth's atmosphere and therefore the direction of future climate change. Soils serve a recycling function that, if taken advantage of, can help societies to conserve and reuse valuable and finite resources. Soils harbor a large proportion of the Earth's biodiversity—a resource which modern technology has allowed us to harness for any number of purposes. Water, like soil, will be a critical resource for the future generations. Soils functions largely determine both the amount of water that is supplied for various uses and also the quality and purification of that water. Finally, knowledge of soil physical properties and behavior, as well as an understanding of how different soils relate to each other in the landscape, will be critical for successful and sustainable engineering projects aimed at effective and safe land development.

For all these reasons it will be essential for the next generation of scientists, business people, teachers, and other professionals to learn enough about soils to appreciate their importance and to take them into full consideration for development projects and all activities on the land. It is my sincere hope that this book, early editions of which have served so many generations of soil students and scientists, will allow new generations of future soil scientists to benefit from the global ecological view of soils that this textbook expounds.

Dr. Nyle Brady, although long in retirement and recently deceased, remains as co-author in recognition of the fact that his vision, wisdom, and inspiration continue to permeate the book. Although the responsibility for writing this edition was solely mine, I certainly could not have made all of the many improvements without innumerable suggestions, ideas, and corrections contributed by soil scientists, instructors, and students from around the world. This 4th edition of *Elements of the Nature and Properties of Soils*, like preceding editions, has greatly benefited from the high level of professional devotion and camaraderie that characterizes the global soil science community.

Special thanks go to Dr. Rachel Gilker for her invaluable editorial and research assistance. I also thank the following colleagues (listed alphabetically by institution) for their especially, valuable suggestions, contributions, reviews, and inspiration: Alan Bayless (Mineral Area College); Doug Malo (South Dakota State University); Gobena Huluka (Auburn University); James Crum (Michigan State University); Kamalesh Panthi (East Carolina University); Vivek Tandon (University of Texas at El Paso); William Moore (Abraham Baldwin Agricultural College); Pichu Rengasamy (The University of Adelaide); Pedro Sanchez and Cheryl Palm (University of Florida); Johannes Lehmann (Cornell University); Eric Brevik (Dickinson State University); Dan Richter (Duke University); Robert Darmody, Laura Flint Gentry, Colin Thorn, and Michelle Wander (University of Illinois); Lee Burras (Iowa State University); Aurore Kaisermann (Laboratoire Bioemco); Daniel Hillel (University of Massachusetts, Emeritus); Rafiq Islam and Rattan Lal (The Ohio State University); Darrell Schultze (Purdue University); Joel Gruver (Western Illinois University); Ivan Fernandez (University of Maine); David Lobb (University of Manitoba); Mark Carroll, Glade Dlott, Delvin Fanning, Nicole Fiorellino, Robert Hill, Bruce James, Natalie Lounsbury, Brian Needelman, Martin Rabenhorst, Patricia Steinhilber, Barret Wessel, and Stephane Yarwood (University of Maryland); Martha Mamo (University of Nebraska); Jose Amador (University of Rhode Island); Allen Franzluebbers, Jeff Herrick, Scott Lesch, and Jim Rhoades (USDA/Agricultural Research Service); Bob Ahrens, Bob Engel, Maxine Levine, Paul Reich, Kenneth Scheffe, and Sharon Waltman (USDA/Natural Resources Conservation Service); Markus Kleber (Oregon State University); Henry Lin and Charlie White (The Pennsylvania State University); Joseph Heckman (Rutgers, The State University of New Jersey); Fred Magdoff (University of Vermont); W. Lee Daniels and John Galbraith (Virginia Tech); and Peter Abrahams (University of Wales).

Last, but not least, I deeply appreciate the good humor, forbearance, and patience of my wife, Trish, and those students and colleagues who may have felt some degree of neglect as I focused so much of my energy, time, and attention on this labor of love.

About the Authors

Ray R. Weil is Professor of Soil Science. He has earned degrees at Michigan State University, Purdue University, and Virginia Tech. Before coming to Maryland, he served in the Peace Corps in Ethiopia, managed a 500 acre organic farm in North Carolina, and was a lecturer at the University of Malawi. He has become an international leader in sustainable agricultural systems in both developed and developing countries. Published in over 95 scientific journal articles and 8 books, his research focuses on cover crops and organic matter management for enhanced soil quality and nutrient cycling for water quality and sustainability. His research lab developed analytical methods for soil microbial biomass and active soil C that have been adopted by the USDA/NRCS and are used in ecosystem studies worldwide. His contributions to improved cropping systems and soil management have been put into practice on farms large and small.

As a University of Maryland professor, Dr. Weil teaches undergraduate and graduate classes in soil science and sustainable agriculture. He has taught over 7,000 students, addressed over 5,000 farmers and farm advisors at meetings and field days, and helped train hundreds of researchers and managers in various companies and organizations. He has been the major advisor for 42 MS and PhD students. Weil is a fellow of both the Soil Science Society of America and the American Society of Agronomy. He has twice been awarded a Fulbright Fellowship to support his work in developing countries. The synergism between Dr. Weil's teaching and research, and his ecological approach to soil science have found expression in various editions of this textbook since 1995.

Nyle C. Brady (late), a native of Colorado, graduated from Brigham Young University in 1941 with a bachelor's degree in Chemistry. In 1947 he received his PhD degree in soil science from the University of North Carolina. He served on the Cornell faculty from 1947 to 1973, and in 1952 became co-author of the world's most widely used college textbook on Soil Science. He was head of the Department of Agronomy from July 1955 to December 1963 and served as the director of the Cornell University Agricultural Experiment Station from September 1965 to July 1973. He was associate dean of the New York State College of Agriculture and Life Sciences from October 1970 to July 1973.

Dr. Brady served as the director of Science and Education for the U.S. Department of Agriculture in Washington, D.C., from December 1963 to September 1965. From July 1973 to July 1981, he was director general of the International Rice Research Institute in the Philippines. From 1981 to 1989, he served as senior assistant administrator for Science and Technology of the United States Agency for International Development in Washington, D.C. From 1990 to 1994 he was a full-time senior consultant for collaborative research and development programs of the World Bank in Washington, D.C., and the United Nations Development Program in New York. Dr. Brady passed away in 2015 at the age of 95.

1
The Soils Around Us

Earth, our unique soil- and water-covered planet. (NASA)

For in the end we will conserve only what we love.
We will love only what we understand.
And we will understand only what we are taught.
—BABA DIOUM, AFRICAN CONSERVATIONIST

The Earth, our only home in the vastness of the universe, is unique for living systems sustained by its air, water, and soil resources. Among the millions of life forms on our planet, one species, the human species, has become so dominant that the quality of those resources now depends on that species learning to exercise a whole new level of stewardship.

With about half of the land surface now appropriated for human use, our activities are changing the very nature of the Earth's ecology. Depletion of stratospheric ozone is threatening to overload us with ultraviolet radiation. Emissions of carbon dioxide, nitrogen oxides, and methane gases are warming the planet and destabilizing the global climate. Tropical rain forests are disappearing. Groundwater supplies are being contaminated and depleted. In parts of the world, the capacity of soils to produce food is being degraded, even as the number of people needing food is increasing. Bringing the global environment back into balance may well be the defining challenge for the current generation of students studying soils.

Soils[1] are crucial to life on Earth. The quality of the soil present largely determines the capacity of land to support plants, animals, and society. Soils also play a central role in many of today's environmental challenges. From water pollution and climate change to biodiversity loss and human food supply, the world's ecosystems are impacted in far-reaching ways by processes carried out in the soil. As human societies become increasingly urbanized, fewer people have intimate contact with the soil, and individuals tend to lose sight of the many ways in which they depend upon soils. Indeed, our dependence on soils is likely to increase, not decrease, in the future.

Soils will continue to supply us with nearly all of our food, yet how many of us remember, as we eat a slice of pizza, that the pizza's crust began in a field of wheat and its cheese began with grass, clover, and corn rooted in the soils of a dairy farm? Most of the fiber we use for lumber, paper, and clothing has its roots in the soils of forests and farmland. Although we sometimes use plastics and fiber synthesized from fossil petroleum as substitutes, in the long term we will continue to depend on terrestrial ecosystems for these needs.

In addition, biomass grown on soils is likely to become an increasingly important feedstock for fuels and manufacturing as the world's finite supplies of petroleum are depleted during the course of this century. The early marketplace signs of this trend can be

---

[1] Throughout this text, bold type indicates key terms whose definitions can be found in the glossary.

Figure 1.1 *The many functions and ecosystem services performed by soil can be grouped into six crucial ecological roles.* (Diagram courtesy of Ray R. Weil)

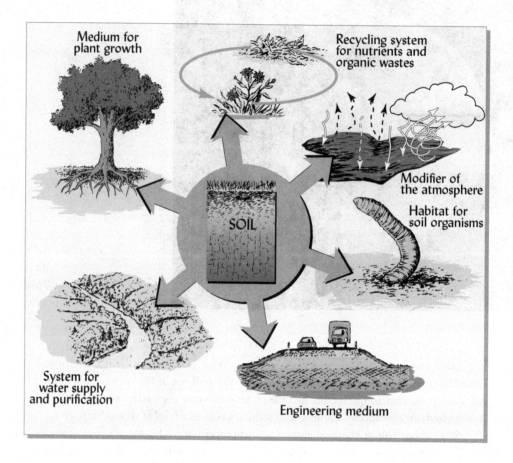

seen in the form of ethanol and biodegradable plastics synthesized from corn or biodiesel fuels and printers' inks made from soybean oil.

One of the stark realities of the twenty-first century is that the population of humans that demands all of these products will increase by several billion (population is expected to stabilize later this century at 9 to 10 billion). Unfortunately, the amount of soil available to meet these demands will not increase at all. In fact, the resource base is actually *shrinking* because of soil degradation and urbanization. Understanding how to better manage the soil resource is essential to our survival and to the maintenance of sufficient habitat for the other creatures that share this planet with us. In short, the study of soil science has never been more important than it is today.

1.1 WHAT ECOSYSTEM SERVICES DO SOILS PERFORM?[2]

Scientists now recognize that the world's ecosystems provide goods and services estimated to be worth about $60 trillion (in 2018 dollars) every year—as much as the gross national products (GNP) of all the world's economies. Over half of global **ecosystem services** arise on land, and soils are thought to contribute about $12 trillion worth, nearly equivalent to the entire US economy. Ecosystem services can be thought of as:

- *provisioning* (providing goods such as water, food, medicines, and lumber),
- *regulating* (processes that purify water, decompose wastes, control pests, or modify atmospheric gases),

---

[2] For a review of global value of ecosystem services, see Costanza et al. (2014); for the proportion attributable to soils, see McBratney et al. (2017).

- *supportive* (assisting with nutrient cycling, seed dispersal, primary biomass production, etc.), and
- *cultural/informational* (archeological preservation, spiritual uplift, outdoor recreation opportunities).

Whether occurring in your backyard, a farm, a forest, or a regional watershed, soils have six key roles to play (Figure 1.1) in the provision of ecosystem services. *First*, soils support plant growth and thereby often determine the nature of the vegetation present and, indirectly, the number and types of animals (including people) that the vegetation can support. *Second*, soils regulate water supplies. Water loss, utilization, contamination, and purification are all affected by the soil. *Third*, the soil functions as nature's recycling system. Within the soil, waste products and dead bodies of plants, animals, and people are assimilated, and their basic elements are made available for reuse by the next generation of life. In addition to recycling, soil can serve as a protective covering of human artifacts for centuries before they are unearthed by archeologists. *Fourth*, soils markedly influence the composition and physical condition of the atmosphere by taking up and releasing large quantities of carbon dioxide, oxygen, and other gases and by contributing dust and re-radiated heat energy to the air. *Fifth*, soils are alive and are home to creatures from small mammals and reptiles to tiny insects to microorganisms of unimaginable numbers and diversity. *Finally*, soil plays an important role as an engineering medium. Soil not only is an important building material in the form of earth fill and bricks (baked soil material) but provides the foundation for virtually every road, airport, and house we build.

1.2 HOW DO SOILS SUPPORT PLANT GROWTH?

When we think of the forests, prairies, gardens, and fields that surround us, we usually envision the **shoots**—the aboveground parts of plants. Because plant roots are usually hidden from our view and difficult to study, we know much less about plant–environment interactions belowground than aboveground, but we must understand both to truly understand either. Here is what plants obtain from the soil in which their roots proliferate:

- Physical support
- Air
- Water
- Temperature moderation
- Protection from toxins
- Nutrient elements

First, the soil mass provides physical support, anchoring the root system so that the plant does not fall over or blow away. Occasionally, strong wind or heavy snow does topple a plant whose root system has been restricted by shallow or inhospitable soil conditions.

Because root respiration, like our own respiration, produces carbon dioxide (CO_2) and uses oxygen (O_2), an important function of the soil is *ventilation*—maintaining the quantity and quality of air by allowing CO_2 to escape and fresh O_2 to enter the root zone. This ventilation is accomplished via networks of soil pores.

An equally important function of soil pores is to absorb water and hold it where it can be used by plant roots. As long as plant leaves are exposed to sunlight, the plant requires a continuous stream of water to use in cooling, nutrient transport, turgor maintenance, and photosynthesis. Since plants use water continuously, but in most places it rains only occasionally, the water-holding capacity of soils is essential for plant survival. A deep soil may store enough water to allow plants to survive long periods without rain (Figure 1.2).

The soil also moderates temperature fluctuations. Perhaps you can recall digging in garden soil (or even beach sand) on a summer afternoon and feeling how hot the soil was at the surface and how much cooler just a few centimeters below. The insulating properties of soil protect the deeper portion of the root system from extremes of hot and cold that often occur at the soil surface. For example, it is not unusual for the mid-afternoon temperature at the surface of bare soil to reach 40°C, a condition lethal to most plant roots. Just a few centimeters deeper, however, the temperature may be 10°C cooler, allowing roots to function normally.

Phytotoxic substances in soils may result from human pollution, or they may be produced by plant roots, by microorganisms, or by natural chemical reactions. The soil may protect plants from such substances by ventilating gases, by decomposing or adsorbing organic

Figure 1.2 *A family of African elephants finds welcome shade under the leafy canopy of a huge acacia tree in this East African savanna. The photo was taken in a long dry season; no rain had fallen for almost five months. The tree roots are still using water stored several meters deep in the soil. The light-colored grasses are more shallow-rooted and have either set seed and died or gone into a dried-up, dormant condition.* (Photo courtesy of Ray R. Weil)

toxins, or by suppressing toxin-producing organisms. Also, some soil microorganisms produce organic, growth-stimulating compounds that may improve plant vigor.

A fertile soil will provide a continuing supply of dissolved **mineral nutrients** in amounts and relative proportions appropriate for optimal plant growth. These nutrients include such metallic elements as potassium, calcium, iron, and copper, as well as such nonmetallic elements as nitrogen, sulfur, phosphorus, and boron. Roots take these elements out of the soil solution and the plant incorporates most of them into the organic compounds that constitute its tissues. Animals usually obtain their mineral nutrients from the soil, indirectly, by eating plants. Under some circumstances, animals (including humans) satisfy their craving for minerals by ingesting soil directly (Box 1.1). Plants also take up some elements that they do not appear to use, which is fortunate as animals do require several elements that plants do not (see periodic table, Appendix B).

BOX 1.1
DIRT FOR DINNER?[a]

You are probably thinking, "dirt (excuse me, *soil*) for dinner? Yuck!" Of course, various birds, reptiles, and mammals are well known to consume soil at special "licks," and involuntary, inadvertent ingestion of soil by humans (especially children) is widely recognized as a pathway for exposure to environmental toxins (see Chapter 15, Box 15.2), but many people, anthropologists and nutritionists included, find it hard to believe that anyone would *purposefully* eat soil. Yet, research on the subject shows that many people do routinely eat soil, often in amounts of 20 to 100 g (up to $1/4$ pound) daily. **Geophagy** (deliberate "soil eating") is practiced in societies as disparate as those in Thailand, Turkey, rural Alabama, and urban Tanzania (Figure 1.3). Immigrants have brought the practice of soil eating to such cities as London and New York. In fact, scientists studying the practice suggest that geophagy is a widespread and normal human behavior. Children and women (especially when pregnant) appear more likely than men to be geophagists. Poor people eat soil more commonly than the relatively well-to-do.

People usually do not eat just any soil, but seek out particular soils, generally high in clay and low in sand, be it the hardened clay of a termite nest, the soft, white clay exposed in a riverbank, or the dark red clay from a certain deep soil layer. People in different places and circumstances seek to consume different types of soils—some seek sodium- or calcium-rich soils, others soil with high amounts of certain clays, still others seek soils rich in iron. Interestingly, unlike many other animals, humans rarely appear to eat soil to obtain salt. Possible benefits from eating soil may include mineral nutrient supplementation, although only iron appears to be sufficiently bioavailable to actually improve nutrition. While other mammals seem to obtain significant amounts of

(continued)

BOX 1.1
DIRT FOR DINNER?[a] (CONTINUED)

Figure 1.3 *Bars of reddish clay soil sold for human consumption in a market in Morogoro, Tanzania. The soil bars (stacked neatly on the circular tray in foreground) are sold individually or by the bagful mainly to pregnant women, who commonly consume about 10 bars per day.*
(Photo courtesy of Ray R. Weil)

mineral nutrients from eating soil, the main benefit that humans receive is probably detoxification of ingested poisons and parasites (e.g., by adsorption to clay—see Chapter 8), relief from stomachaches, survival in times of famine, and psychological comfort. Geophagists have been known to go to great lengths to satisfy their cravings for soil. But before you run out and add some local soil to your menu, consider the potential downsides to geophagy. Aside from the possibly difficult task of developing a taste for the stuff, the drawbacks to eating soil (especially surface soils) can include parasitic worm infection, lead poisoning, and mineral nutrient imbalances (because of adsorption of some mineral nutrients and release of others)—as well as premature tooth wear!

[a]This box is largely based on a fascinating review by Young et al. (2011) and Abrahams (2012), (2005).

Of the 92 naturally occurring chemical elements, 17 have been shown to be **essential elements**, meaning that plants cannot grow and complete their life cycles without them. Table 1.1 lists these and several additional elements that appear to be quasi-essential (needed by some but not all plants). Essential elements used by plants in relatively large amounts are called **macronutrients**; those used in smaller amounts are known as **micronutrients**.

In addition to the mineral nutrients listed, plants may also use minute quantities of organic compounds from soils. However, uptake of these substances is not necessary for normal plant growth. The organic metabolites, enzymes, and structural compounds making up a plant's dry matter consist mainly of carbon, hydrogen, and oxygen, which the plant obtains by photosynthesis from air and water, not from the soil.

Plants *can* be grown in nutrient solutions without any soil (a method termed **hydroponics**), but then the plant-support functions of soils must be engineered into the system and maintained at a high cost of time, energy, and management. In fact, imagining the expense

Table 1.1
ELEMENTS NEEDED FOR PLANT GROWTH AND THEIR SOURCES[a]
The chemical forms most commonly taken in by plants are shown in parentheses, with the chemical symbol for the element in bold type.

Macronutrients: Used in relatively large amounts (>0.1% of dry plant tissue)		Micronutrients: Used in relatively small amounts (<0.1% of dry plant tissue)
Mostly from air and water	Mostly from soil solids	From soil solids
Carbon (CO_2)	*Cations:*	*Cations:*
Hydrogen (H_2O)	Calcium (Ca^{2+})	Copper (Cu^{2+})
Oxygen (O_2, H_2O)	Magnesium (Mg^{2+})	*Cobalt (Co^{2+})[b]
	Nitrogen (NH_4^+)	Iron (Fe^{2+})
	Potassium (K^+)	Manganese (Mn^{2+})
	Anions:	Nickel (Ni^{2+})
	Nitrogen (NO_3^-)	*Sodium (Na^+)[b]
	Phosphorus ($H_2PO_4^-$, HPO_4^{2-})	Zinc (Zn^{2+})
	Sulfur (SO_4^{2-})	*Anions:*
	*Silicon (H_4SiO_4, $H_3SiO_4^-$)[b]	Boron (H_3BO_3, $H_4BO_4^-$)
		Chlorine (Cl^-)
		Molybdenum (MoO_4^{2-})

[a]Many other elements are taken up from soils by plants but are not *essential* for plant growth. See periodic table in Appendix B.
[b]Elements marked by (*) are quasi-essential elements (*sensu*, Epstein and Bloom (2005)), required for some, but not for all, plants. Silicon is used in large amounts to play important roles in most plants, so is considered a plant-beneficial element, but has been proved essential only for diatoms and plants in the *Equisetaceae* family. Cobalt has been proved essential for only *Leguminosae* when in symbiosis with nitrogen-fixing bacteria (see Section 13.10). Sodium can partially substitute for potassium in many plants and is essential in small amounts for plants using the C4 photosynthesis pathway (mainly tropical grasses).

of attempting to grow enough food to support 7 billion people in hydroponic greenhouses is a good way to comprehend the economic value of the soil's food provisioning ecosystem service. Thus, although hydroponic production is feasible for high-value plants on a small scale, production of the world's food and fiber and maintenance of natural ecosystems will always depend on millions of square kilometers of productive soils.

1.3 HOW DO SOILS REGULATE WATER SUPPLIES?

To maintain or improve water quality, we must recognize that nearly every drop of water in our rivers, lakes, estuaries, and aquifers has either traveled through the soil or flowed over its surface (excluding the relatively minor quantity of precipitation that falls directly into bodies of fresh surface water). Imagine, for example, a heavy rain falling on the hills surrounding a river. If the soil allows the rain to soak in, some of the water will be stored in the soil, some used by the trees, and some will seep slowly down through the soil layers to the groundwater, eventually entering the river over a period of months or years as **base flow**. As it soaks through the upper layers of soil, contaminated water is purified and cleansed by soil processes that remove many impurities and kill potential disease organisms.

Contrast the preceding scenario with what would occur if the soil were so shallow or impermeable that most of the rain could not penetrate the soil, but ran off the land surface, scouring surface soil and debris as it sped toward the river. The result would be a destructive flash flood of muddy contaminated water. This comparison highlights how the nature and management of soils in a watershed will influence the purity and amount of water finding its way to aquatic systems. For those who live in rural homes, the purifying action of the soil (in a septic drain field as described in Section 6.8) is the main barrier that stands between what flushes down the toilet and the water running into the kitchen sink!

1.4 HOW DO SOILS RECYCLE RAW MATERIALS?

What would a world be like without the recycling functions performed by soils? Without reuse of nutrients, plants and animals would have run out of nourishment long ago. The world would be covered with a layer, possibly hundreds of meters high, of plant and animal wastes and corpses. Obviously, recycling is vital to ecosystems, whether forests, farms, or cities. The soil system plays a pivotal role in the major geochemical cycles. Soils have the capacity to assimilate great quantities of organic waste, turning it into beneficial **soil organic matter**, converting the mineral nutrients in the waste to forms that can be utilized by plants and animals, and returning the carbon to the atmosphere as carbon dioxide, where it again will become a part of living organisms through plant photosynthesis. Some soils can accumulate large amounts of carbon as soil organic matter, thus reducing the concentration of atmospheric carbon dioxide and potentially mitigating global climate change (see Sections 1.5, 1.14, and 11.10).

1.5 HOW DO SOILS MODIFY THE ATMOSPHERE?

As the soil "breathes" in and out it interacts in many ways with the Earth's blanket of air. That is, soils absorb oxygen and other gases such as methane, while they release gases such as carbon dioxide and nitrous oxide. These gas exchanges between the soil and the atmosphere have a significant influence on atmospheric composition and global climate change. The evaporation of soil moisture is a major source of water vapor in the atmosphere, altering air temperature, composition, and weather patterns.

In places where the soil is dry, poorly structured, and unvegetated, soil particles can be picked up by winds and contribute great quantities of dust to the atmosphere, reducing visibility, increasing human health hazards from breathing dirty air, and altering the temperature of the air and of the Earth itself. Moist, well-vegetated, and structured soils can prevent such dust-laden air.

1.6 WHAT LIVES IN THE SOIL HABITAT?

When speaking of ecosystems needing protection, most people envision a stand of old-growth forest with its abundant wildlife, or an estuary with oyster beds and fisheries. Perhaps, once you have read this book, you will envision a handful of soil when someone speaks of an ecosystem. Soil is not a mere pile of broken rock and dead debris. A handful of soil may be home to *billions* of organisms, belonging to thousands of species that act as predators, prey, producers, consumers, and parasites (Figure 1.4). This complex community of organisms influences human well-being through many ecosystem functions, but soils also influence human health directly, for good or for ill (see Box 1.2).

Figure 1.4 *The soil is home to a wide variety of organisms, both relatively large and very small. Here, a relatively large predator, a centipede (shown at about actual size), hunts for its next meal—which is likely to be one of the many smaller animals that feed on dead plant debris.* (Photo courtesy of Ray R. Weil)

BOX 1.2
SOILS AND HUMAN HEALTH[a]

Although human health impacts of soils often go unrecognized, soils affect us all for better and for worse. Soils impact our health indirectly via all six of the ecological soil functions described in Sections 1.2–1.7. Soils and soil components (such soil particles, mineral elements, and microorganisms) also directly affect our health when we come in contact with them by handling soil or in the food we eat, the water we drink, and the air we breathe.

THE FOOD WE EAT

The composition of our food reflects the nature of the soil in which it was grown. Because some soils fail to make iron easily available for plant uptake, the iron contents of vegetables and meats are commonly insufficient to prevent human iron deficiency anemia. Zinc, which is involved in the function of hundreds of our body's enzymes, is another case in point; with insufficient dietary intake, we may suffer such symptoms as hair loss and impairment of immune system function, fertility, and sex drive. About half of the world's agricultural soils are deficient in zinc, and about half of the world's people (largely in the same geographic areas) eat diets deficient in this micronutrient. Likewise, soils low in sulfur, as occur widely in Africa, Asia, Australia, and parts of North America, produce wheat (or beans, etc.) likely to be low in methionine and cystine, sulfur-containing amino acids essential for the human body to utilize the protein in food. Foods grown in certain areas tend to reflect the low levels of iodine and selenium in local soils, two elements not needed by plants but widely deficient as nutrients for people (causing goiters and Keshan disease, respectively). Other examples abound.

INFECTIOUS DISEASES FROM SOILS

Among the millions of soil-dwelling organisms, a few can bring disease and even death to humans. Two notorious soil pathogenic bacteria are *Clostridium tetani*, which causes tetanus, and *Bacillus anthracis*, which causes anthrax and whose spores may survive in the soil for decades. Such soil-borne infectious bacterial diseases kill millions of people each year, including many babies and mothers who die during childbirth under unsanitary conditions. A less common, but still potentially fatal, soil-borne disease is caused when a soil fungus, *Blastomyces sp.*, infects a cut in the skin or is breathed into the lungs. Blastomycosis is usually associated geographically with localized soil conditions, but it is hard to track down as its pneumonia-like lung symptoms or skin ulcerations may not appear for months after exposure. Cryptococcosis, a fairly rare disease causing brain damage or pneumonia-like lung symptoms, can be contracted by breathing in spores of *Cryptococcus*, another soil fungus. Still other human diseases are caused by microscopic soil animals, such as roundworms, hookworms, and protozoa. An example of the latter is *Cryptosporidium sp.*, which cause widespread outbreaks of cryptosporidiosis, sometimes sickening (but rarely killing) thousands of people in a single city if the protozoa-containing soil or farm manure contaminates drinking water supplies. Another under-recognized health hazard comes from fine dust picked up by desert winds and carried halfway around the world (see Sections 2.2 and 14.11). Airborne dust not only poses a risk of physical irritation of lung tissues that results in cancer but also carries pathogenic soil microorganisms that can remain alive and virulent during the intercontinental journey.

THE CURATIVE POWERS OF SOILS

The aforementioned discussion does not mean that we should never hike in the forest or garden without rubber gloves (though gloves are a good idea if your hand has an open wound). To the contrary—the balance of nature in most soils is overwhelmingly in favor of organisms that provide ecosystem services essential to human welfare. For example, it was recently observed that certain single-celled soil animals called *Paramecium* voraciously eat the spores of the pathogenic fungus *Cryptococcus* just mentioned. In fact, soils play a far greater role in *curing* our diseases than in *causing* them!

Many people are unaware that plants grown in the soil are the source of most of the medicines (both traditional herbals and modern pharmaceuticals) that prevent, alleviate, or cure so many of the diseases and ailments that plagued and often killed our ancestors. The story of Taxol (paclitaxel) illustrates this role quite well. This highly prized anticancer drug was first discovered in the bark of a rare type of yew tree that grows in the Pacific coast soils of Oregon and Washington States. Demand for this drug resulted in the destruction of half a million of these yews before scientists learned to make it from other organisms using molecular culture and gene-transfer techniques.

Soil microorganisms themselves are the source of most of our life-saving antibiotics. Drugs such as penicillin, ciprofloxacin, and neomycin originate from certain soil bacteria (e.g., *Streptomyces*) and fungi (e.g., *Penicillium*) that produce these compounds as part of their defensive strategies against competing soil microbes. See Chapter 11 for more on soil microbes and their antibiotics. Poultices made from soil clays have long been effectively used in traditional medicine to heal skin conditions and fight infections. Some research even suggests that just being in close contact with healthy soils (think avid gardeners) and breathing in certain microorganisms or volatile compounds they produce may give people a sense of well-being through interactions with their brain chemistry (the marked increase in brain cell serotonin in response to the soil bacterium, *Mycobacterium vaccae*, is well documented). Regulation of our immune systems and promotion of our well-being by diverse soil microbes should be considered among the ecosystem services that soils provide.

---

[a]Many scientific papers are available for further reading on soils and human health (see Alloway and Graham (2008); Frager et al. (2010); Griffin (2007); Liu and Khosla (2010); Stokes (2006)). For research that illuminates why soil clays exhibit powers of healing (it's the metals adsorbed to the clays!), see Otto and Haydel (2013). For a review of human immune system regulation by soil (and other environmental) organisms, see Rook (2013).

We have said that billions of organisms made up of thousands of species can coexist in a handful of soil. How is it possible for such a diversity of organisms to live and interact in such a small space? One explanation is the tremendous range of niches and habitats that exist in even a uniform-appearing soil. Some pores of the soil will be filled with water in which organisms such as roundworms, diatoms, rotifers, and bacteria swim. Tiny insects and mites may be crawling about in other larger pores filled with moist air. Micro-zones of good aeration may be only millimeters from areas of **anoxic** conditions. Different areas may be enriched with decaying organic materials; some places may be highly acidic, some more basic. Temperature, too, may vary widely.

Hidden from view in the world's soils are communities of living organisms every bit as complex and intrinsically valuable as their larger counterparts that roam the savannas, forests, and oceans of the Earth. In fact, soils harbor a large part of the Earth's genetic diversity. Soils, like air and water, are important components of larger ecosystems. So it is important to assure that **soil quality** is considered, along with air quality and water quality, in discussions of environmental protection.

1.7 HOW ARE SOILS USED IN BUILDING AND ENGINEERING?

Soil is one of the earliest and the most widely used of building materials. Nearly half the people in the world live in houses constructed from soil. Soil buildings vary from traditional mud huts in Africa to large centuries-old circular apartment houses in China (Figure 1.5) to today's environmentally friendly "rammed-earth" buildings (see http://www.yourhome.gov.au/materials/rammed-earth).

"*Terra firma*, solid ground." We usually think of the soil as being firm and solid, a good base on which to build roads and all kinds of structures. Indeed, most constructed structures do rest on the soil, and many construction projects require excavation into the soil. Unfortunately, as can be seen in Figure 1.6, some soils are not as stable as others. Reliable construction on soils, and with soil materials, requires knowledge of the diversity of soil properties, as discussed in this and later chapters. Designs for roadbeds or building foundations that work well in one location on one type of soil may be inappropriate for another location with different soils.

Figure 1.5 *Soil is among the oldest and most common of building materials, with half the world's people living in homes made of soil. (Left) An elderly African villager weaves a basket outside his house made from red and black clay soil reinforced with small tree branches (a technique termed wattle and daub). (Right) Several round Tulou apartment buildings housing up to 80 families each in Fu-Jian, China. These buildings have 2-m-thick walls made thousands of years ago from compacted yellowish soil mixed with bamboo and stones. These massive "rammed-earth" walls make the buildings warm in winter but cool in summer (see Chapter 7) and resistant to damage from earthquakes.* (Left photo courtesy of Ray R. Weil; right photo courtesy of Lu Zhang, Zhejiang, China)

Figure 1.6 *Better knowledge of the soils on which this road was built may have allowed its engineers to develop a more stable design, thus avoiding this costly and dangerous situation.* (Photo courtesy of Ray R. Weil)

Working with natural soils or excavated soil materials is not like working with concrete or steel. Properties such as bearing strength, compressibility, shear strength, and stability are much more variable and difficult to predict for soils than for manufactured building materials. Chapter 4 provides an introduction to some engineering properties of soils. Many other physical properties discussed will have direct application to engineering uses of soil. For example, Chapter 8 discusses properties of certain types of clay soils that upon wetting expand with sufficient force to crack foundations and buckle pavements. Much of the information on soil properties and soil classification discussed in later chapters will be of great value to people planning land uses that involve construction or excavation.

1.8 THE PEDOSPHERE AND THE CRITICAL ZONE?[3]

The outer layers of our planet that lie between the tops of the tallest trees and the bottom of the groundwater aquifers that feed our rivers comprise what scientists term *the critical zone*. Environmental research is increasingly focused on this zone where active cycles and flows of materials and energy support life on Earth. The soil plays a central role in this critical zone. The importance of the soil derives in large part from its role as an **interface** between the worlds of rock (the **lithosphere**), air (the **atmosphere**), water (the **hydrosphere**), and living things (the **biosphere**). Because all four of these worlds interact in the soil or **pedosphere** (Figure 1.7a), it comprises one of the most complex and productive environments on Earth.

The concept of the soil as interface means different things at different scales. At the scale of kilometers, soils channel water from rain to rivers and transfer mineral elements from bed rocks to the oceans. They also substantially influence the global balance of atmospheric gases. At a scale of a few meters (Figure 1.7b), soil forms the transition zone between hard rock and air, holding both liquid water and oxygen gas for use by plant roots. It transfers mineral elements from the Earth's rock crust to its vegetation. It processes or stores the organic remains of terrestrial plants and animals. At a scale of a few millimeters (Figure 1.7c), soil provides diverse microhabitats for air-breathing and aquatic organisms, channels water and nutrients to plant roots, and provides surfaces and solution vessels for thousands of biochemical reactions. Finally, at the scale of a few micrometers or nanometers (Figure 1.7d), soil provides ordered

---
[3]For a readable introduction to the concept of the Critical Zone, see Fisher (2012).

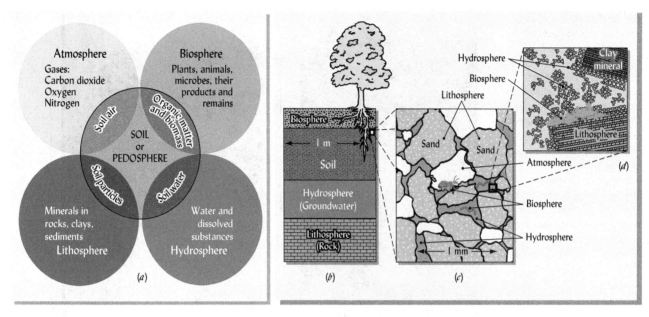

Figure 1.7 *The pedosphere (a), where the worlds of rock (the lithosphere), air (the atmosphere), water (the hydrosphere), and life (the biosphere) all meet. The soil as interface can be understood at many different scales. At the kilometer scale, soil participates in global cycles of rock weathering, atmospheric gas changes, water storage and partitioning, and the life of terrestrial ecosystems. At the meter scale (b), soil forms a transition zone between the hard rock below and the atmosphere above—a zone through which surface water and groundwater flow and in which plants and other living organisms thrive. A thousand times smaller, at the millimeter scale (c), mineral particles form the skeleton of the soil that defines pore spaces, some filled with air and some with water, in which tiny creatures lead their lives. Finally, at the micrometer and nanometer scales (d), soil minerals (lithosphere) provide charges, reactive surfaces that adsorb water and cations dissolved in water (hydrosphere), gases (atmosphere), and bacteria and organic matter (biosphere).* (Diagram courtesy of Ray R. Weil)

and complex surfaces, both mineral and organic, that act as templates for chemical reactions and interact with water and solutes. Its tiniest mineral particles form micro-zones of electromagnetic charge that attract everything from bacterial cell walls to proteins to conglomerates of water molecules. As you read this book, the frequent cross-referencing between chapters will remind you of the importance of such interfacing in the story of soil.

1.9 SOILS AS NATURAL BODIES

You may notice that this book sometimes refers to "soil," sometimes to "the soil," sometimes to "a soil," and sometimes to "soils." These variations of the word "soil" refer to two distinct concepts—*soil* as a material or *soils* as natural bodies. *Soil* is a material composed of minerals, gases, water, organic substances, and microorganisms. Some people (usually *not* soil scientists!) also refer to this material as *dirt*, especially when it is found where it is not welcome (e.g., in your clothes or under your fingernails).

A *soil* is a three-dimensional natural body in the same sense that a mountain, lake, or valley is. *The soil* is a collection of individually different soil bodies, often said to cover the land as the peel covers an orange. However, while the peel is relatively uniform around the orange, the soil is highly variable from place to place on Earth. One of the individual bodies, *a soil*, is to *the soil* as an individual tree is to the Earth's vegetation. Just as one may find sugar maples, oaks, hemlocks, and many other species of trees in a particular forest, so, too, might one find Altamont clays, San Benito clay loams, Diablo silty clays, and other kinds of soils in a particular landscape.

Soils are natural bodies composed of soil (the material just described) *plus* roots, animals, rocks, artifacts, and so forth. By dipping a bucket into a lake you may sample some of its water. In the same way, by making a hole in a soil, you may retrieve some soil. Thus, you can take a sample of soil or water into a laboratory and analyze its contents, but you must go out into the field to study a soil or a lake.

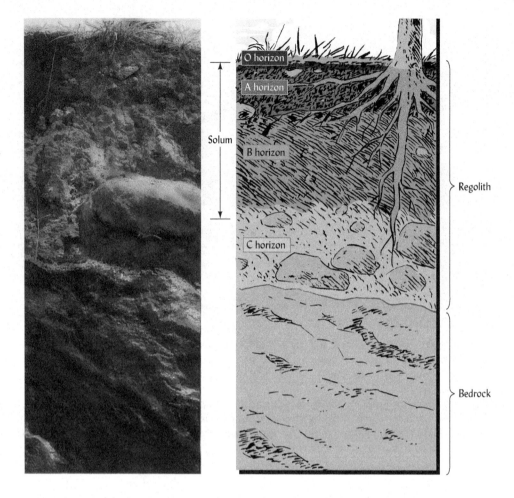

Figure 1.8 *Regolith, soil, and underlying bedrock. The soil is a part of the regolith and the A and B horizons are part of the solum (from the Latin word* solum *meaning soil or land). The C horizon is the part of the regolith that underlies the solum but may be slowly changing into soil in its upper parts. Sometimes the regolith is so thin that it has been changed entirely to soil; in such a case, the solum rests directly on the bedrock.* (Photo courtesy of Ray R. Weil)

In most places, the rock exposed at the Earth's surface has crumbled and decayed to produce a layer of unconsolidated debris overlying the hard, unweathered rock. This unconsolidated layer is called the **regolith** (Figure 1.8) and varies in thickness from virtually nonexistent in some places (i.e., exposed bare rock) to tens of meters in other places. Where the underlying rock has weathered in place to the degree that it is loose enough to be dug with a spade, the term **saprolite** is used. In other cases, regolith materials have been transported many kilometers from the site of its initial formation and then deposited over the bedrock that it now covers. Thus, regolith material may or may not be related to the rock now found below it.

Through their biochemical and physical effects, living organisms such as bacteria, fungi, and plant roots have altered the upper part—and, in many cases, the entire depth—of the regolith. Here, at the interface between the worlds of rock, air, water, and living things, soil is born. The transformation of inorganic rock and debris into a living soil is one of nature's most fascinating displays. Although generally hidden from everyday view, the soil and regolith can often be seen in road cuts and other excavations.

A soil is the product of both destructive and creative (synthetic) processes. Weathering of rock and microbial decay of organic residues are examples of destructive processes, whereas the formation of new minerals and new stable organic-mineral complexes are examples of synthesis. Perhaps the most striking result of synthetic processes is the formation of contrasting layers called **soil horizons**. The development of these horizons in the upper regolith is a unique characteristic of soil that sets it apart from the deeper regolith materials (Figure 1.8).

Soil scientists specializing in **pedology** (*pedologists*) study soils as natural bodies, the properties of soil horizons, and the relationships among soils within a landscape. Other soil scientists, called **edaphologists**, focus on the soil as habitat for living things, especially for plants. For both types of study it is essential to examine soils at all scales and in all three dimensions (especially the vertical dimension).

1.10 THE SOIL PROFILE AND ITS LAYERS (HORIZONS)

Soil scientists often dig a large hole, called a *soil pit* (e.g., Figure 3.37) usually several meters deep and about a meter wide, to expose soil horizons for study. The vertical section exposing a set of horizons in the wall of such a pit is termed a **soil profile**. Road cuts and other ready-made excavations can expose soil profiles and serve as windows to the soil. In an excavation open for some time, horizons may be more clearly seen if a fresh face is exposed by scraping off a layer of material several centimeters thick from the pit wall. Once you have learned to interpret the different horizons (see Chapter 2), soil profiles can tell you much about the environment and history of a site as well as warn you about potential problems in using the land.

Horizons within a soil may vary in thickness and have somewhat irregular boundaries, but generally they parallel the land surface. This alignment is expected since the differentiation of the regolith into distinct horizons is largely the result of influences such as air, water, solar radiation, and plants, originating at the soil–atmosphere interface. Because the weathering of the regolith occurs first at the surface and works its way down, the uppermost layers have been changed the most, while the deepest layers are most similar to the original regolith, which is referred to as the soil's **parent material**. In places where the regolith was originally rather uniform in composition, the material below the soil may have a similar composition to the parent material from which the soil formed. In contrast, where wind, water, or glaciers have transported and deposited the parent material on top of dissimilar material, the regolith found below a soil may be quite different from the upper layer of regolith in which the soil has formed.

In undisturbed ecosystems, especially forests, organic remains of fallen leaves and other plant and animal materials tend to accumulate on the surface. There they undergo varying degrees of physical and biochemical breakdown and transformation so that layers of older, partially decomposed materials may underlie the freshly added debris. Together, these organic layers at the soil surface are designated the **O horizons** (Figure 1.9).

Soil animals and percolating water move some of these organic materials downward to intermingle with the mineral grains of the regolith. These join the decomposing remains of plant roots and microbes to form organic materials that darken the upper mineral layers. Also, because weathering tends to be most intense nearest the soil surface, in many soils the upper layers lose some of their clay or other weathering products by leaching to the horizons below. **A horizons** are the layers nearest the surface that are dominated by mineral particles but have been darkened by the accumulation of organic matter (Figure 1.10).

Figure 1.9 *A piece of the O horizon from a deciduous forest floor in Vermont, USA.* (Photo courtesy of Ray R. Weil)

Figure 1.10 *Decaying plant materials have darkened a thick A horizon that caps this soil along the central California coast.* (Photo courtesy of Ray R. Weil)

The layers underlying the A and O horizons contain comparatively less organic matter than the horizons nearer the surface. Varying amounts of silicate clays, iron and aluminum oxides, gypsum, or calcium carbonate may accumulate in the underlying horizons. The accumulated materials may have been washed down from the horizons above, or they may have been formed in place through the weathering process. These underlying layers are referred to as **B horizons** (Figure 1.11).

Plant roots and microorganisms often extend below the B horizon, especially in humid regions, causing chemical changes in the soil water, some biochemical weathering of the regolith, and the formation of **C horizons**. The C horizons are the least weathered part of the soil profile.

In certain soils, intensely weathered and leached horizons that have not accumulated organic matter occur in the upper part of the profile, usually just below the A horizons. These

Figure 1.11 (Left) *An A Horizon begins to differentiate as materials (such as organic matter) are added to the upper part of the profile. The B horizon forms as other materials (such as salts and clays) are translocated and accumulate in deeper zones. Under certain conditions, usually associated with forest vegetation and high rainfall, a leached E horizon forms between the A and B horizons. If sufficient rainfall occurs, soluble salts will be carried below the soil profile, perhaps all the way to the groundwater (and eventually out to sea). Many soils (e.g., the soil in Figure 1.8) lack one or more of the five horizons shown here.* (Center) *This soil profile was exposed by digging a pit about 2 meters deep in a well-developed soil (called a Hapludalf; see Chapter 3) in southern Michigan. The white string was attached to the profile to demarcate the horizon boundaries. Then a trowel full of soil material was removed from each horizon and placed on a board.* (Right) *The top horizon (A) can be easily distinguished by its darker color. However, some of the horizons in this photo are difficult to discern on the basis of color differences. Yet it is clear from the way the soil either crumbled or held together that soil material from horizons with very similar colors may have very different physical properties.* (Diagram and photo courtesy of Ray R. Weil)

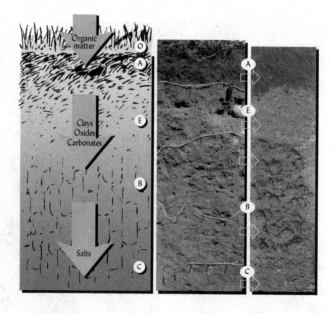

horizons are designated **E horizons** (Figures 1.11) and usually are lighter in color than the adjacent horizons above and below.

In some soil profiles, the component horizons are very distinct in color, with sharp boundaries that can be seen easily by even novice observers. In other soils, the color changes between horizons may be very gradual, and the boundaries more difficult to locate. However, color is only one of many properties by which one horizon may be distinguished from the horizon above or below it (see Figure 1.11, *right*). In addition to seeing the colors in a profile, a soil scientist may feel, smell, and listen[4] to the soil, as well as conduct chemical tests, in order to distinguish the horizons present.

1.11 HOW DOES TOPSOIL DIFFER FROM SUBSOIL?

The organically enriched A horizon at the soil surface is sometimes referred to as **topsoil**. Plowing and cultivating a soil homogenizes and modifies the upper 10 to 25 cm of the profile to form a **plow layer**. The plow layer may remain long after cultivation has ceased. For example, in a forest that has regrown for 100 years on abandoned farmland, you may still see the smooth boundary between the century-old plowed layer and the lighter-colored, undisturbed soil below.

In cultivated soils, the majority of plant roots can be found in the topsoil as that is the zone in which the cultivator can most readily enhance the supply of nutrients, air, and water by mixing in organic and inorganic amendments, loosening the structure, and applying irrigation. Sometimes the plow layer is removed from a soil and sold as topsoil for use at another site. This use of topsoil is especially common to provide a rooting medium suitable for lawns and shrubs around newly constructed buildings, where the original topsoil was removed or buried and the underlying soil layers were exposed during grading operations (Figure 1.12).

The soil layers that underlie the topsoil are referred to as **subsoil**. Although usually hidden from view, the subsoil horizons can greatly influence most land uses. Much of the water needed

Figure 1.12 *The large gray-brown mound of material in the foreground at this construction site consists of topsoil (A horizon material) carefully separated from the lower horizons and pushed aside during initial grading operations. Behind it is a similar sized pile of reddish-brown soil from the lower (subsoil) horizons. In this scene, the stockpiles have just been seeded with grasses to protect from erosion. The subsoil material will be used during construction to fill in low spots and build foundations and roadbeds. After construction activities are complete, the stockpiled topsoil (front pile) will be used in landscaping the grounds around the new buildings.* (Photo courtesy of Ray R. Weil)

---

[4]For example, the grinding sound emitted by wet soil rubbed between one's fingers indicates a sandy nature.

by plants is stored in the subsoil. Many subsoils also supply important quantities of certain plant nutrients. The properties of the topsoil are commonly far more conducive to plant growth than those of the subsoil. In cultivated soils, therefore, productivity is often correlated with the thickness of the topsoil layer. Subsoil layers that are too dense, acidic, or wet can impede root growth. It can be extremely difficult and expensive to physically or chemically modify the subsoil.

Many of the chemical, biological, and physical processes that occur in the upper soil layers also take place to some degree in the C horizons, which may extend deep into the underlying saprolite or other regolith material. Traditionally, the lower boundary of the soil has been considered to occur at the greatest rooting depth of the natural vegetation, but soil scientists are increasingly studying layers below this in order to understand ecological processes of the critical zone, such as groundwater pollution, parent material weathering, and biogeochemical cycles (Box 1.3).

BOX 1.3
USING INFORMATION FROM THE ENTIRE SOIL PROFILE

Soils are three-dimensional bodies that carry out important ecosystem processes at all depths in their profiles. Depending on the particular application, relevant information may come from soil layers as shallow as the upper 1 or 2 cm or as deep as the lowest layers of saprolite (Figure 1.13).

For example, the upper few centimeters of soil often hold the keys to plant growth and biological diversity, as well as to certain hydrologic processes. Here, at the interface between the soil and the atmosphere, living things are most numerous and diverse. Forest trees largely depend for nutrient uptake on a dense mat of fine roots growing in this zone. The physical condition of this thin surface layer may also determine whether rain will soak in or run downhill on the land surface. Certain pollutants, such as lead, are also concentrated in this zone. For many types of soil investigations it will be necessary to sample the upper few centimeters separately so that important conditions are not overlooked.

On the other hand, it is equally important not to confine one's attention to the easily accessible "topsoil," for many soil properties are to be discovered only in the deeper layers. Plant-growth problems are often related to inhospitable conditions in the B or C horizons that restrict the penetration of roots. Similarly, the great volume of these deeper layers may control the amount of plant-available water held by a soil. For the purposes of recognizing or mapping different types of soils, the properties of the B horizons are often paramount. Not only is this the zone of major accumulations of minerals and clays, but the layers nearer the soil surface may be too quickly altered by management and soil erosion to be considered a reliable basis for soil classification.

In deeply weathered regoliths, the lower C horizons and saprolite play important roles. These layers, generally at depths below 1 or 2 meters and often as deep as 5 to 10 meters, greatly affect the suitability of soils for most urban uses that involve construction or excavation. The proper functioning of on-site sewage disposal systems and the stability of building foundations are often determined by regolith properties at these depths. Likewise, processes that control the movement of pollutants to groundwater or the weathering of geologic materials may occur at depths of many meters. These deep layers also have major ecological influences because, although the intensity of biological activity and plant rooting may be quite low, the total impact can be great as a result of the enormous volume of soil that may be involved. This is especially true of forest systems in warm climates.

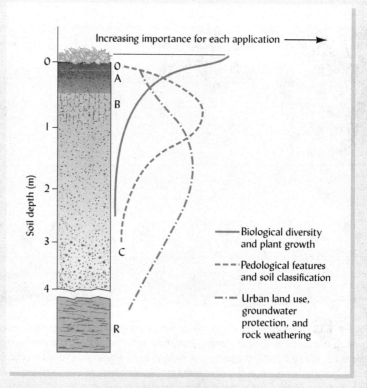

Figure 1.13 *Information important to different soil functions and applications is most likely to be obtained by studying different layers of the soil profile.* (Diagram courtesy of Ray R. Weil)

1.12 SOIL—INTERFACE OF AIR, MINERALS, WATER, AND LIFE

The relative proportions of air, water, mineral matter, and organic matter greatly influence the behavior and productivity of soils. In a soil, these four components are mixed in complex patterns. Figure 1.14 shows the approximate proportions (by volume) of the components found in a **loam** surface soil in good condition for plant growth. Although a handful of soil may at first seem to be a solid thing, it should be noted that only about half the soil volume consists of solid material (mineral and organic); the other half consists of pore spaces filled with air or water. Of the solid material, typically most is mineral matter derived from the rocks of the Earth's crust. Only about 5% of the *volume* in this ideal soil consists of organic matter. Since it is far less dense than mineral matter, the organic matter accounts for only about 2% of the *weight* of this soil. However, the influence of the organic component on soil properties is often far greater than these small proportions would suggest.

The spaces between the particles of solid material are just as important to the nature of a soil as are the particles themselves. It is in these pore spaces that air and water circulate, roots grow, and microscopic creatures live. Plant roots need both air and water. In an optimum condition for most plants, the pore space will be divided roughly equally among the two, with 25% of the soil volume consisting of water and 25% consisting of air. If there is much more water than this, the soil will be waterlogged. If much less water is present, plants will suffer from drought. The relative proportions of water and air fluctuate as water is added or lost. Compared to surface soil layers, subsoils tend to contain less organic matter, less total pore space, and a larger proportion of its pore space is made up of small pores (*micropores*), which tend to be filled with water rather than with air.

1.13 WHAT ARE THE MINERAL (INORGANIC) CONSTITUENTS OF SOILS?

Except in organic soils, most of the soil's solid framework consists of **mineral**[5] particles. Excluding, for the moment, the larger rock fragments such as stones and gravel, soil particles range in size over four orders of magnitude: from 2.0 millimeters (mm) to smaller than

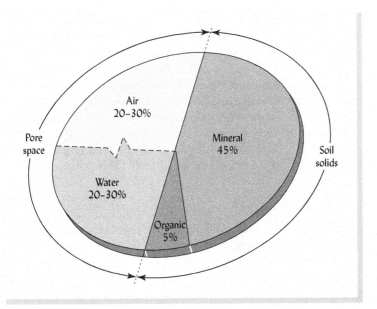

Figure 1.14 *Volume composition of a loam surface soil when conditions are good for plant growth. The broken line between water and air indicates that the proportions of these two components fluctuate as the soil becomes wetter or drier. Nonetheless, a nearly equal proportion of air and water is generally ideal for plant growth.*

---

[5]The word *mineral* is used in soil science in three ways: (1) as a general adjective to describe inorganic materials derived from rocks; (2) as a specific noun to refer to distinct minerals found in nature such as quartz and feldspars (see Chapter 2 for detailed discussions of soil-forming minerals and the rocks in which they are found); and (3) as an adjective to describe chemical elements, such as nitrogen and phosphorus, in their inorganic state in contrast to their occurrence as part of organic compounds.

Table 1.2
Some General Properties of the Three Major Size Classes of Inorganic Soil Particles

Property	Sand	Silt	Clay
1. Range of particle diameters in millimeters	2.0–0.05	0.05–0.002	Smaller than 0.002
2. Means of observation	Naked eye	Microscope	Electron microscope
3. Dominant minerals	Primary	Primary and secondary	Secondary
4. Attraction of particles for each other	Low	Medium	High
5. Attraction of particles for water	Low	Medium	High
6. Ability to hold chemicals and nutrients in plant-available form	Very low	Low	High
7. Consistency when wet	Loose, gritty	Smooth	Sticky, malleable
8. Consistency when dry	Very loose, gritty	Powdery, some clods	Hard clods

0.0002 mm in diameter (Table 1.2). **Sand** particles are large enough (2.0–0.05 mm) to be seen by the naked eye and feel gritty when rubbed between the fingers. Sand particles do not adhere to one another; therefore, sands do not feel sticky. **Silt** particles (0.05–0.002 mm) are too small to be seen without a microscope or to be felt individually, so silt feels smooth but not sticky, even when wet. **Clay** particles are the smallest mineral particles (<0.002 mm) and adhere together to form a sticky mass when wet and hard clods when dry. The smaller particles (<0.001 mm) of clay (and similar-sized organic particles) are termed **colloids** and can be seen only with the aid of an electron microscope. Because of their extremely small size, colloidal particles possess a tremendous amount of surface area per unit of mass. The surfaces of soil colloids (both mineral and organic) exhibit electromagnetic charges that attract positive and negative ions as well as water, making this fraction of the soil the most chemically and physically active (see Chapter 8).

Soil Texture

The proportion of particles in these different size ranges is described by **soil texture**. Terms such as *sandy loam, silty clay,* and *clay loam* are used to identify the soil texture. Texture has a profound influence on many soil properties, and it affects the suitability of a soil for most uses. To understand the degree to which soil properties can be influenced by texture, imagine sunbathing first on a sandy beach (loose sand) and then on a clayey beach (sticky mud). The difference in these two experiences would be due largely to the properties described in Table 1.2.

To anticipate the effect of clay on the way a soil will behave, it is necessary to know the *kinds* of clays as well as the *amount* present. As home builders and highway engineers know all too well, certain clayey soils, such as those high in smectite clays, make very unstable material on which to build because they swell when wet and shrink when dry. This shrink–swell action can easily crack foundations and cause retaining walls to collapse. These clays also become extremely sticky and difficult to work with when they are wet. Other types of clays, formed under different conditions, can be very stable and easy to work with. Learning about the different types of clay minerals will help us understand many of the physical and chemical differences among soils in various parts of the world.

Soil Minerals

Minerals that have persisted with little change in composition since they were extruded in molten lava (e.g., quartz, micas, and feldspars) are known as **primary minerals**. They are prominent in the sand and silt fractions of soils and contain many of the nutrient elements needed by plants. Other minerals, such as silicate clays and iron oxides, were formed by the breakdown and weathering of less resistant minerals as soil formation progressed. These minerals are called **secondary minerals** and tend to dominate the clay and, in some cases, silt

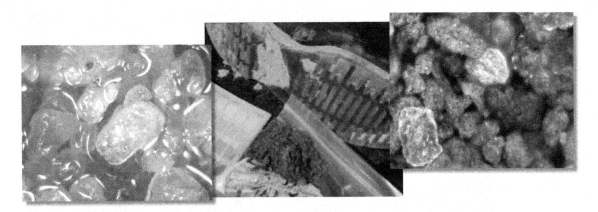

Figure 1.15 *Soil can be useful in solving crime mysteries and is often placed in evidence in court proceedings. The mud clinging to these boots (center) was used to prove the innocence of a man accused of robbery and breaking into a house. In this case the type of sand grains found in the mud from the boots was typical of quartz-rich soils (left), which occurred at the job site of the accused but not anywhere near the crime site, which instead had soils with mica-rich sands (right). The largest sand grains in both photos are about 1 mm in diameter.* (Photo courtesy of Ray R. Weil)

fractions. Soil mineralogical signatures are one of the clues used by forensic soil scientists to locate crime victims or establish guilt by matching soil clinging to shoes, tires, tools, or under fingernails with soil from a crime scene (Figure 1.15).

Soil Structure

Sand, silt, and clay particles can be thought of as the building blocks from which soil is constructed. **Soil structure** describes the way these building blocks are associated together in aggregates of various sizes and shapes (see, e.g., the nature of the soil "clumps" in Figure 1.11, *right*). Soil structure (the way particles are arranged together) is just as important as soil texture (the relative amounts of different sizes of particles) in governing how water and air move in soils. Both structure and texture fundamentally influence many processes in soil, including the growth of plant roots.

1.14 WHAT IS SOIL ORGANIC MATTER LIKE?

Soil organic matter consists of a wide range of organic (carbonaceous) substances, including living organisms (the soil **biomass**), carbonaceous remains of organisms that once occupied the soil, and organic compounds produced by current and past metabolism in the soil. Over time, organic matter is destroyed by microbial respiration and its carbon is lost from the soil as carbon dioxide. Because of such losses, repeated additions of new plant and/or animal residues are necessary to maintain soil organic matter.

Under conditions that favor plant production more than microbial decay, large quantities of atmospheric carbon dioxide used by plants in photosynthesis are sequestered in the abundant plant tissues that eventually become part of the soil organic matter. Since carbon dioxide is a major greenhouse gas whose increase in the atmosphere is warming Earth's climate, the balance between accumulation of soil organic matter and its loss through microbial respiration has global implications. In fact, more carbon is stored in the world's soils than in the world's plant biomass and atmosphere combined.

Even so, organic matter comprises only a small fraction of the mass of a typical soil. By weight, typical well-drained mineral surface soils contain from 1 to 6% organic matter. The organic matter content of subsoils is even smaller. However, the influence of organic matter on soil properties, and consequently on plant growth, is far greater than the low percentage would indicate (see also Chapter 11).

Organic matter binds mineral particles into a granular soil structure that is largely responsible for the loose, easily managed condition of productive soils. Part of the soil organic matter that is especially effective in stabilizing these granules consists of certain glue-like substances produced by various soil organisms, including plant roots (Figure 1.16).

Figure 1.16 *Abundant organic matter, including plant roots, helps create physical conditions favorable for the growth of higher plants as well as microbes (left). In contrast, soils low in organic matter, especially if they are high in silt and clay, are often cloddy (right) and not suitable for optimum plant growth.* (Photo courtesy of Ray R. Weil)

Organic matter also increases the amount of water a soil can hold and the proportion of water available for plant growth (Figure 1.17). In addition, organic matter is a major source of the plant nutrients nitrogen, phosphorus, and sulfur. As soil organic matter decays, these nutrient elements, which are present in organic combinations, are released as soluble ions that can be taken up by plant roots. Finally, organic matter, including plant and animal residues, is the main food that supplies carbon and energy to soil organisms. Without it, biochemical activity so essential for ecosystem functioning would come to a near standstill.

Humus, usually black or brown in color, is a collection of organic compounds that accumulate in soil when partially broken down plant, microbial, and animal residues are protected from complete decay by various factors in the soil environment. Like clay, much of the soil's humus is colloidal in size and exhibits highly charged surfaces. Both humus and clay act as contact bridges between larger soil particles; thus, both play an important role in the formation of soil structure. The surface charges of humus, like those of clay, attract and hold both nutrient ions and water molecules. However, gram for gram, the capacity of humus to hold nutrients and water is far greater than that of clay. Unlike clay, humus may contain components that can make micronutrients more easily used by plants and may even cause hormone-like stimulation of certain plant processes. All in all, small amounts of humus remarkably increase the soil's capacity to promote plant growth.

Figure 1.17 *Soils high in organic matter are darker and can hold more water than soils low in organic matter. The soil in each container has the same mineral composition, but the one on the right has been depleted of much of its organic matter. The same amount of water was applied to each container. The depth of water penetration was less in the high organic matter soil (left) because of its greater water-holding capacity. It required a greater volume of the low organic matter soil to hold the same amount of water.* (Photo courtesy of Ray R. Weil)

1.15 WHY IS SOIL WATER SO DYNAMIC AND COMPLEX?

The soil moisture regime, often reflective of climatic factors, is a major determinant of the productivity of terrestrial ecosystems, including agricultural systems. Movement of water, and substances dissolved in it, through the soil profile impacts both the quality and quantity of local and regional water resources. Water moving through the regolith is also a major driving force in soil formation from parent material (see Box 2.1).

Two main factors help explain why **soil water** is different from, say, drinking water in a glass:

1. Water is held within soil pores where the attraction between water and the surfaces of soil particles greatly restricts the ability of water to flow as it would flow in a drinking glass.
2. Because soil water is never pure water, but contains hundreds of dissolved organic and inorganic substances, it may be more accurately called the **soil solution**. An important function of the soil solution is to serve as a constantly replenished, dilute nutrient solution bringing dissolved nutrient elements (e.g., calcium, potassium, nitrogen, and phosphorus) to plant roots.

When the soil moisture content is optimal for plant growth (Figure 1.14), the water in the large- and intermediate-sized pores can move about in the soil and can easily be used by plants. The plant roots, however, remove water from the largest pores first. Soon the larger pores hold only air, and the remaining water is found only in the intermediate- and smallest-sized pores. The water in the intermediate-sized pores can still move toward plant roots and be taken up by them. However, the water in the smallest pores is so close to solid particles that it may be so strongly held that plant roots cannot pull it away. Consequently, not all soil water is *available* to plants. Depending on the soil, one-sixth to one-half of the water may remain in the soil after plants have wilted or died for lack of moisture.

Soil Solution

The soil solution contains small but significant quantities of soluble organic and inorganic substances, including the plant nutrients listed in Table 1.1. The soil solids, particularly the fine organic and inorganic colloidal particles (clay and humus), release nutrient elements to the soil solution from which they are taken up by plant roots. The soil solution tends to resist changes in its composition even when compounds are added or removed from the soil. This ability to resist change is termed the soil **buffering capacity** and is dependent on many chemical and biological reactions, including the attraction and release of substances by colloidal particles (see Chapter 8).

Many chemical and biological reactions are controlled by the relative amounts of acidity caused by a dominance of hydrogen ions (H^+) and alkalinity caused by a dominance of hydroxyl ions (OH^-) in the soil solution. The **pH** is a logarithmic scale used to express the degree of soil acidity or alkalinity (see Chapter 9, especially Box 9.1 and Figure 9.2). The pH is considered a master variable that influences most chemical processes in the soil and is of great significance to nearly all aspects of soil science.

1.16 SOIL AIR: A CHANGING MIXTURE OF GASES

If we think of the network of soil pores as the ventilation system of the soil connecting airspaces to the atmosphere, we can understand that when pores are filled with water the ventilation system becomes clogged. Think how stuffy the air would become if the ventilation ducts of a windowless classroom became clogged. Because oxygen could neither enter the room nor carbon dioxide leave it, the air in the room would soon become depleted of oxygen and enriched in carbon dioxide and water vapor by the respiration (breathing) of the people in it. In an air-filled soil pore surrounded by water-filled smaller pores, the metabolic activities of plant roots and microorganisms have a similar effect.

Therefore, soil air differs from atmospheric air in several respects. First, the composition of soil air varies greatly from place to place in the soil. In local pockets, some gases are consumed by plant roots and by microbial reactions, and others are released, thereby greatly modifying the composition of the soil air. Second, soil air generally has a higher moisture content than the atmosphere; the relative humidity of soil air approaches 100% unless the soil is very dry. Third, the carbon dioxide (CO_2) content is usually much higher, and that of oxygen (O_2) lower, than found in the atmosphere. Carbon dioxide in soil air is often several hundred times more concentrated than the 0.035% commonly found in the atmosphere. Oxygen decreases accordingly and, in extreme cases, may be 5 to 10%, or even less, compared to about 20% for atmospheric air. In extreme cases, lack of oxygen both in the soil air and dissolved in the soil water may fundamentally alter the chemistry and biology of the soil solution. This is of particular importance to understanding the functions of wetland soils.

The amount and composition of air in a soil are determined to a large degree by the water content of the soil. When water enters the soil, it displaces air from some of the pores; the air content of a soil is therefore inversely related to its water content. As the soil drains from a heavy rain or irrigation, large pores are the first to be filled with air, followed by medium-sized pores, and finally the small pores, as water is removed by evaporation and plant use. This explains the tendency for soils with a high proportion of tiny pores to be poorly aerated.

1.17 HOW DO SOIL COMPONENTS INTERACT TO SUPPLY NUTRIENTS TO PLANTS?

As you read our discussion of each of the four major soil components, you may have noticed that the impact of one component on soil properties is seldom expressed independently from that of the others. Rather, the four components interact with each other to determine the nature of a soil. Thus, soil moisture, which directly meets the needs of plants for water, simultaneously controls much of the air and nutrient supply to the plant roots. The mineral particles, especially the finest ones, attract soil water, thus determining its movement and availability to plants. Likewise, organic matter, because of its physical binding power, influences the arrangement of the mineral particles into clusters and, in so doing, increases the number of large soil pores, thereby influencing the water and air relationships.

Essential Element Availability

Perhaps the most important interactive process involving the four soil components is the provision of essential nutrient elements to plants. Plants absorb essential nutrients, along with water, directly from one of these components: the soil solution. However, the amount of essential nutrients in the soil solution at any one time is sufficient to supply the needs of growing vegetation for only a few hours or days. Consequently, the soil solution nutrient levels must be constantly replenished from inorganic or organic solids in the soil and from fertilizers or amendments added to agricultural soils.

By a series of chemical and biochemical processes, nutrients are released from these solid forms to replenish those in the soil solution. For example, the tiniest colloidal-sized particles—both clay and humus—exhibit negative and positive charges. These charges tend to attract or **adsorb**[6] oppositely charged ions from the soil solution and hold them as **exchangeable ions**. Through ion exchange, elements such as Ca^{2+} and K^+ are released from this state of electrostatic adsorption on colloidal surfaces and escape into the soil solution where they can be readily taken up (absorbed) by plant roots.

Nutrient ions are also released to the soil solution as soil microorganisms decompose organic tissues. Plant roots then absorb these nutrients from the soil solution, provided there is enough O_2 in the soil air to support root metabolism. Only a small fraction of the nutrient content of a soil is present in forms that are readily soluble and available for plant uptake. Table 1.3 provides some idea of the quantities of various essential elements present in different forms in typical soils of humid and arid regions.

---

[6]*Adsorption* refers to the attraction of ions to the surface of particles, in contrast to *absorption*, the process by which ions are taken *into* plant roots. The adsorbed ions are exchangeable with ions in the soil solution.

Table 1.3
QUANTITIES OF SIX ESSENTIAL ELEMENTS FOUND IN UPPER 15 CM OF REPRESENTATIVE SOILS IN TEMPERATE REGIONS

Essential element	Humid Region Soil			Arid Region Soil		
	In solid framework, kg/ha	Exchangeable, kg/ha	In soil solution, kg/ha	In solid framework, kg/ha	Exchangeable, kg/ha	In soil solution, kg/ha
Ca	8,000	2,250	60–120	20,000	5,625	140–280
Mg	6,000	450	10–20	14,000	900	25–40
K	38,000	190	10–30	45,000	250	15–40
P	900	—	0.05–0.15	1,600	—	0.1–0.2
S	700	—	2–10	1,800	—	6–30
N	3,500	—	7–25	2,500	—	5–20

Figure 1.18 illustrates how solid soil components interact with the liquid component (soil solution) to provide essential elements to plants. Plant roots do not ingest soil particles, no matter how fine, but are able to absorb only nutrients that are dissolved in the soil solution. Because elements in larger soil particles are only slowly released into the soil solution over long periods of time, the bulk of most nutrients in a soil is not readily available for plant use. Nutrient elements in the framework of colloid particles are somewhat more readily available to plants, as these particles break down much faster because of their greater surface area. Thus, the structural framework is the major storehouse and, to some extent, a significant source of essential elements in many soils.

The distribution of nutrients among the various components of a fertile soil, as illustrated in Figure 1.18, may be likened to the distribution of financial assets in the portfolio of

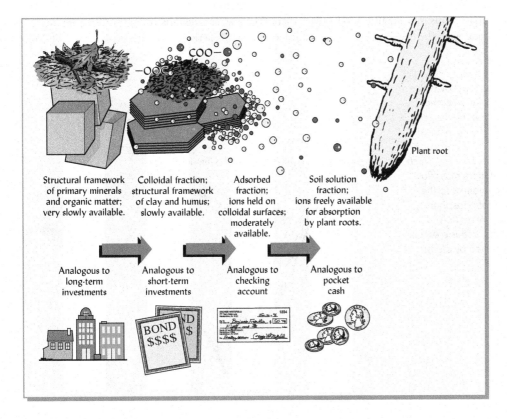

Figure 1.18 Nutrient elements exist in soils in various forms characterized by different accessibility to plant roots. The bulk of the nutrients is locked up in the structural framework of primary minerals, organic matter, clay, and humus. A smaller proportion of each nutrient is adsorbed in a swarm of ions near the surfaces of soil colloids (clay and organic matter). From the swarm of adsorbed ions, a still smaller amount is released into the bulk soil solution, where uptake by plant roots can take place. The lower diagram considers the analogy between financial assets and nutrient assets. (Diagram courtesy of Ray R. Weil)

a wealthy individual. Nutrients readily available for plant use would be analogous to cash in the individual's pocket. A billionaire would likely keep most assets in long-term investments such as real estate or bonds (the coarse fraction solid framework), while investing a smaller amount in short-term stocks and bonds (colloidal framework). For more immediate use, an even smaller amount might be kept in a checking account linked to an automated teller machine (ATM) (exchangeable nutrients), while a tiny fraction of the overall wealth might be carried to spend as currency and coins (nutrients in the soil solution). As the cash is used up, the supply is replenished by making a withdrawal from the ATM. The checking account, in turn, is replenished occasionally by the sale of long-term investments. It is possible for wealthy persons to run short of coins for a vending machine even though they may own a great deal of valuable real estate. In an analogous way, plants may use up the readily available supply of a nutrient even though the total supply of that nutrient in the soil is very large. Luckily, in a fertile soil, the process described in Figure 1.18 can quickly release dissolved essential elements as rapidly as plant roots remove them.

1.18 HOW DO PLANT ROOTS OBTAIN NUTRIENTS?

To be taken up by a plant, the nutrient element must be in a soluble form and must be located *at the root surface*. Often, parts of a root are in such intimate contact with soil particles that a direct exchange may take place between nutrient ions adsorbed on the surface of soil colloids and H^+ ions from the surface of root cell walls. In any case, the supply of nutrients in contact with the root will soon be depleted. So how can a root obtain additional supplies once the nutrient ions at the root surface have all been taken up into the root? There are three basic mechanisms by which the concentration of nutrient ions at the root surface is maintained (Figure 1.19).

First, **root interception** comes into play as roots continually grow into new, undepleted soil. Root exploration in search of nutrients is much enhanced by thin root cell extensions called **root hairs**. In fact, root hair growth was recently discovered to be controlled by a regulatory plant gene that is "turned on" by low nutrient conditions. Even with root hairs extending into tiny water-filled soil pores where nutrients may be dissolved, for the most part, nutrient ions must still travel some distance in the soil solution to reach the root surface. This movement can take place by **mass flow**, as when dissolved nutrients are carried along with the flowing soil water toward a root that is actively drawing water from the soil.

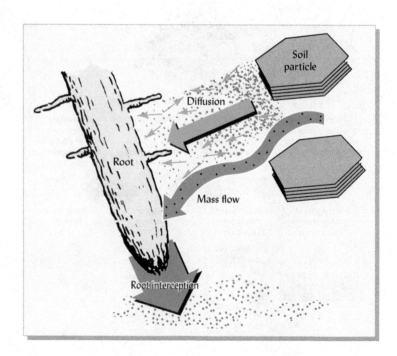

Figure 1.19 Three principal mechanisms by which nutrient ions dissolved in the soil solution come into contact with plant roots. All three mechanisms may operate simultaneously, but one mechanism or another may be most important for a particular nutrient. For example, in the case of calcium, which is generally plentiful in the soil solution, mass flow alone can usually bring sufficient amounts to the root surface. However, in the case of phosphorus, diffusion is needed to supplement mass flow because the soil solution is very low in this element in comparison to the amounts needed by plants. (Diagram courtesy of Ray R. Weil)

In this type of movement, the nutrient ions are somewhat analogous to leaves floating down a stream. On the other hand, plants can continue to take up nutrients even at night, when little, if any, water is absorbed into the roots. Nutrient ions continually move by **diffusion** from areas of greater concentration toward the nutrient-depleted areas of lower concentration around the root surface.

In the diffusion process, the random movement of ions in all directions causes a *net* movement from areas of high concentration to areas of lower concentrations, independent of any mass flow of the water in which the ions are dissolved. Factors such as soil compaction, cold temperatures, and low soil moisture content, which reduce root interception, mass flow, or diffusion, can result in poor nutrient uptake by plants even in soils with adequate supplies of soluble nutrients. Furthermore, the availability of nutrients for uptake can also be negatively or positively influenced by the activities of microorganisms that thrive in the immediate vicinity of roots. Maintaining the supply of available nutrients at the plant root surface is thus a process that involves complex interactions among different soil components.

It should be noted that the plant membrane separating the inside of the root cell from the soil solution is permeable to dissolved ions only under special circumstances. Plants do not merely take up, by mass flow, those nutrients that happen to be in the water that roots are removing from the soil. Nor do dissolved nutrient ions brought to the root's outer surface by mass flow or diffusion cross the root cell membrane and enter the root passively by diffusion. On the contrary, a nutrient is normally taken up into the plant root cell only by reacting with specific chemical binding sites on large protein molecules embedded in the root membrane. These proteins form hydrophilic channels across an otherwise hydrophobic lipid (fatty) membrane. Energy from metabolism in the root cell is used to activate this carrier protein so that it will pass the nutrient ion across the cell membrane and release it into the cell interior. This carrier mechanism allows the plant to accumulate concentrations of a nutrient inside the root cell that far exceed that nutrient's concentration in the soil solution. Because different nutrients are taken up by specific types of protein molecules, the plant is able to exert some control over how much and in what relative proportions essential elements are taken up.

Nutrient uptake being an active metabolic process, conditions that inhibit root metabolism may also inhibit nutrient uptake. Examples of such conditions include excessive soil water content or soil compaction resulting in poor soil aeration, excessively hot or cold soil temperatures, and aboveground conditions that result in low translocation of sugars to plant roots. We can see that plant nutrition involves biological, physical, and chemical processes and interactions among many different components of soils and the environment.

1.19 SOIL HEALTH, DEGRADATION, AND RESILIENCE

Soil is a basic resource underpinning all terrestrial ecosystems. Managed carefully, soils are a *reusable* resource, but in the scale of human lifetimes they cannot be considered a *renewable* resource. As we shall see in the next chapter, most soil profiles are thousands of years in the making. In all regions of the world, human activities are destroying some soils far faster than nature can rebuild them. Growing numbers of people are demanding more and more ecosystem services from the Earth's fixed amount of land. This situation presents soil scientists and humanity with a number of grand challenges that must be met if human civilization and nature are to be sustained side by side (Table 1.4).

In most parts of the world, nearly all of the soils best suited for growing crops are already being farmed. Therefore, as each year brings millions more people to feed, the amount of cropland per person continuously declines. In addition, many of the world's major cities were originally located where excellent soils supported thriving agricultural communities. Without policies to protect farmland, many of the very best soils for farming are lost forever to suburban development as these cities expand.

Finding more land on which to grow food is not easy. Most additional land brought under cultivation comes at the cost of clearing natural forests, savannas, and grasslands. Images of the Earth made from orbiting satellites show the resulting decline in land covered by forests

Table 1.4
GRAND SOIL SCIENCE CHALLENGES FOR 2050

No.	Topic Area	Grand Challenge
1	Food	How can we feed 2 billion *more* people than today without harming our soils or the broader environment?
2	Nutrients	How do we preserve and enhance the fertility of our soils, conserve scarce nutrient resources, and also export nutrients from farms to cities in ever bigger harvests?
3	Fresh water	How can we manage our soils to use dwindling water supplies more efficiently and wisely while managing soils to protect our waters from pollution?
4	Energy	How can we sustainably manage our lands to contribute to energy supplies by integrating biochar use and producing biofuel feedstocks?
5	Climate change	How can we manage soils to mitigate climate change by reducing greenhouse gases while also adapting to climate change by protecting soil productivity and resilience?
6	Biodiversity	How can we better understand and enhance the biotic communities within and on the soil to create more resilient and productive ecosystems and utilize the diverse gene pool?
7	Recycling "wastes"	How can we better use soils as biogeochemical reactors to avoid contamination, detoxify contaminants, and maintain soil productivity?
8	Global perspective	How can we develop a global perspective that still permits us to optimize management of local places, wherever they may be?

Modified from Janzen et al. (2011)

and other natural ecosystems. Thus, as the human population struggles to feed itself, wildlife populations are deprived of vital habitat, and overall biodiversity suffers. Efforts to reduce and even reverse human population growth must be accelerated if our grandchildren are to inherit a livable world. In the meantime, if there is to be space for both people and wildlife, the best of our existing farmland soils will require improved and more intensive management. While soils completely washed away by erosion or excavated and paved over by urban sprawl are permanently lost, many more soils are degraded in quality rather than totally destroyed.

Soil Quality and Health

People who work with the land and depend on soils to perform critical functions want to improve and maintain **soil health**. They recognize that soils are living systems with highly complex and diverse communities of organisms. In their optimal state, these organisms work together to function in a self-regulating and perpetuating manner (Chapter 10). Healthy soils function more efficiently with less need for expensive human interventions and inputs than unhealthy, degraded soils. **Soil quality** is a measure of the ability of a soil to carry out particular ecological functions, such as those described in Sections 1.2–1.7. Soil quality reflects a combination of *chemical, physical*, and *biological* properties. Some of these properties are relatively unchangeable, inherent properties that help define a particular type of soil. Soil texture and mineral makeup (Section 1.13) are examples. Other soil properties, such as structure (Section 1.13) and organic matter content (Section 1.14), can be significantly changed by management. These more changeable soil properties can indicate the status of a soil's quality relative to its potential in much the same way that water turbidity or oxygen content indicates the water-quality status of a river.v

Soil Degradation and Resilience

Mismanagement of forests, farms, and rangeland causes widespread degradation of soil quality by erosion that removes the topsoil, little by little (see Chapter 14). Another widespread cause of soil degradation is the accumulation of salts in improperly irrigated soils in arid regions (see Chapter 9). When people cultivate soils and harvest the crops without returning organic residues and mineral nutrients, the soil's supply of organic matter and nutrients becomes depleted (see Chapters 11 and 12). Contamination of a soil with toxic substances can degrade its capacity to provide habitat for soil organisms, to grow plants that are safe to eat, or to safely recharge ground and surface waters (see Chapter 15). Degradation of soil quality by pollution is usually localized, but the environmental impacts and costs involved are very large.

While protecting soil quality must be the first priority, it is often necessary to restore the quality of soils that have already been degraded. Some soils have sufficient **resilience** to recover from minor degradation if left to regenerate on their own. In other cases, more effort is required to restore degraded soils (see Chapter 15). Organic and inorganic amendments may have to be applied, vegetation may have to be planted, physical alterations by tillage or grading may have to be made, or contaminants may have to be removed. As societies around the world assess the damage already done to their natural and agricultural ecosystems, the science of **restoration ecology** has rapidly evolved to guide managers in restoring plant and animal communities to their former levels of diversity and productivity. The job of **soil restoration**, an essential part of these efforts, requires in-depth knowledge of all aspects of the soil system.

1.20 CONCLUSION

The Earth's soil is comprised of numerous soil individuals, each of which is a three-dimensional natural body in the landscape. Each individual soil is characterized by a unique set of properties and soil horizons as expressed in its profile. The nature of the soil layers seen in a particular profile is closely related to the nature of the environmental conditions at a site.

Soils perform six broad ecological functions: they act as the principal medium for plant growth, regulate water supplies, modify the atmosphere, recycle raw materials and waste products, provide habitat for many kinds of organisms, and serve as a major engineering medium for human-built structures. Soil is thus a major ecosystem in its own right. The soils of the world are extremely diverse, each type of soil being characterized by a unique set of soil horizons. A typical surface soil in good condition for plant growth consists of about half solid material (mostly mineral, but with a crucial organic component, too) and half pore spaces filled with varying proportions of water and air. These components interact to influence a myriad of complex soil functions, a good understanding of which is essential for wise management of our terrestrial resources.

If we take the time to learn the language of the land, the soil will speak to us.

STUDY QUESTIONS

1. As a society, is our reliance on soils likely to increase or decrease in the decades ahead? Explain.
2. Discuss how *a soil*, a natural body, differs from *soil*, a material that is used in building a roadbed.
3. What are the six main roles of soil in an ecosystem? For each of these ecological roles, suggest one way in which interactions occur with another of the six roles.
4. Think back over your activities during the past week. List as many incidents as you can in which you came into direct or indirect contact with soil.
5. Figure 1.14 shows the volume composition of a loam surface soil in ideal condition for plant growth. To help you understand the relationships among the four components, redraw this pie chart to represent what the situation might be after the soil has been compacted by heavy traffic. Then draw another pie chart to show how the four components of the original ideal soil would be related on a mass (weight) basis rather than on a volume basis.
6. Explain in your own words how the soil's nutrient supply is held in different forms, much the way that a person's financial assets might be held in different forms.
7. List the essential nutrient elements that plants derive mainly from the soil.
8. Are all elements contained in plants essential nutrients? Explain.
9. Define these terms: *soil texture*, *soil structure*, *soil pH*, *humus*, *soil profile*, *B horizon*, *soil quality*, *solum*, and *saprolite*.
10. Describe four processes that commonly lead to degradation of soil quality.
11. Compare the pedological and edaphological approaches to the study of soils. Which is more closely aligned with geology and which with ecology?
12. Which of the *grand challenges* listed in Table 1.4 is most exciting and inspiring to you, and why?

REFERENCES

Abrahams, P. W. 2005. "Geophagy and the involuntary ingestion of soil." In O. Selinus (ed.). -*Essentials of Medical Geology*. Elsevier, The Hague, pp. 435–457.

Abrahams, P. W. 2012. "Involuntary soil ingestion and geophagia: A source and sink of mineral nutrients and potentially harmful elements to consumers of earth materials." *Applied Geochemistry* 27: 954–968.

Alloway, B. J., and R. D. Graham. 2008. "Micronutrient deficiencies in crops and their global significance." In B. J. Alloway (ed.). *Micronutrient Deficiencies in Global Crop Production*. Springer, Netherlands, pp. 41–61.

Costanza, R., R. de Groot, P. Sutton, S. van der Ploeg, S. J. Anderson, I. Kubiszewski, S. Farber, and R. K. Turner. 2014. "Changes in the global value of ecosystem services." *Global Environmental Change* 26:152–158.

Epstein, E., and A. J. Bloom. 2005. *Mineral Nutrition of Plants: Principles and Perspectives*. 2nd ed. Sinauer Associates, Sunderland, MA, p. 400.

Fisher, M. 2012. "Investigating the earth's critical zone." *CSA News* 57(1):5–9.

Frager, S. Z., C. J. Chrisman, R. Shakked, and A. Casadevall. 2010. "Paramecium species ingest and kill the cells of the human pathogenic fungus *Cryptococcus neoformans*." *Medical Mycology* 48:775–779.

Griffin, D. W. 2007. "Atmospheric movement of microorganisms in clouds of desert dust and implications for human health." *Clinical Microbiology Reviews* 20:459–477.

Janzen, H. H., P. E. Fixen, A. J. Franzluebbers, J. Hattey, R. C. Izaurralde, Q. M. Ketterings, D. A. Lobb, and W. H. Schlesinger. 2011. "Global prospects rooted in soil science." *Soil Science Society of America Journal* 75:1–8.

Liu, T., and C. Khosla. 2010. "A balancing act for taxol precursor pathways in *E. coli*." *Science* 330:44–45.

McBratney, A. B., C. L. Morgan, and L. E. Jarrett. 2017. "The value of soil's contributions to ecosystem services." In D. J. Field, et al. (eds.). *Global Soil Security*. Springer, Cham, Switzerland, pp. 227–235.

Otto, C. C., and S. E. Haydel. 2013. "Exchangeable ions are responsible for the in vitro antibacterial properties of natural clay mixtures." *PLoS One* 8:e64068.

Rook, G. A. 2013. "Regulation of the immune system by biodiversity from the natural environment: An ecosystem service essential to health." *Proceedings of the National Academy of Sciences* 110:18360–18367.

Stokes, T. 2006. "The earth-eaters." *Nature* 444:543–544.

Young, S. L., P. W. Sherman, J. B. Lucks, and G. H. Pelto. 2011. "Why on earth?: Evaluating hypotheses about the physiological functions of human geophagy." *The Quarterly Review of Biology* 86:97–120.

2
Formation of Soils from Parent Materials

Leaving tracks in the soils of other worlds: Apollo 14 astronaut sampling lunar soil with inset showing tracks of Curiosity Rover in Mars soil. (NASA)

It is a poem of existence . . . not a lyric but a slow epic whose beat has been set by eons of the world's experience. . . .
—JAMES MICHENER, CENTENNIAL

The first astronauts to explore the moon labored in their clumsy pressurized suits to collect samples of rocks and dust from the lunar surface. These they carried back to Earth for analysis. It turned out that moon rocks are similar in composition to those found deep in the Earth—so similar that scientists concluded that the moon itself began when a stupendous collision between a Mars-sized object named Theia and the young Earth spewed molten material into orbit around the planet. The force of gravity eventually pulled this material together to form the moon. On the moon, this rock remained unchanged or crumbled into dust with the impact of meteors. On Earth, the rock at the surface, eventually coming in contact with water, air, and living things, was transformed into something new, into many different kinds of living soils. This chapter reveals the story of how rock and dust become "the ecstatic skin of the Earth."[1]

We will study the processes of soil formation that transform the lifeless regolith into the variegated layers of the soil profile. We will also learn about the environmental factors that influence these processes to produce soils in Belgium so different from those in Brazil, soils on limestone so different from those on sandstone, and soils in the valley bottoms so different from those on the hills.

Every landscape is composed of a suite of different soils, each influencing ecological processes in its own way. Whether we intend to modify, exploit, farm, preserve, or simply understand the landscape, our success will depend on our knowing how soil properties relate to the environment on each site and to the landscape as a whole.

2.1 WEATHERING OF ROCKS AND MINERALS

Weathering breaks up rocks and minerals, modifies or destroys their physical and chemical characteristics, and carries away the finer fragments and soluble products. Nothing escapes it. However, weathering also synthesizes new minerals that influence important properties in soils. The nature of the rocks and minerals being weathered determines the rates and results of the breakdown and synthesis (Figure 2.1).

---

[1]The apt description of soil as "ecstatic skin of the Earth" is from a delightfully readable account of soils by Logan (1995).

Figure 2.1 *Two stone markers, photographed on the same day in the same cemetery, illustrate the effect of rock type on weathering rates. The date and initials carved in the slate marker in 1798 are still sharp and clear, while the date and figure of a lamb carved in the marble marker in 1875 have weathered almost beyond recognition. The slate rock consists largely of resistant silicate clay minerals, whereas the marble consists mainly of calcite, which is much more easily attacked by acids in rainwater.* (Photo courtesy of Ray R. Weil)

Characteristics of Rocks and Minerals

Geologists classify Earth's rocks as igneous, sedimentary, and metamorphic. Igneous rocks are those formed from molten rock (lava or magma) and include such common rocks as granite and diorite (Figure 2.2). Igneous rock is composed of primary minerals that solidify as molten rock cools. Examples of primary minerals include light-colored quartz, muscovite, and feldspars and dark-colored biotite, augite, and hornblende.[2] The mineral grains in igneous rocks interlock and are randomly dispersed, giving a salt-and-pepper appearance if they are coarse enough to see with the unaided eye (Figure 2.3). In general, dark-colored minerals contain iron and magnesium and are more easily weathered. Therefore, dark-colored igneous rocks such as gabbro and basalt are more easily broken down than are granite, syenite, and other lighter-colored igneous rocks.

Sedimentary rocks may contain both primary minerals and secondary minerals, the latter being recrystallized products of the chemical breakdown and/or alteration of primary minerals.

Rock texture	Quartz	Light-colored mineral (e.g., feldspars, muscovite)		Dark-colored minerals (e.g., hornblende, augite, biotite)
Coarse	Granite	Diorite	Gabbro	Peridotite Hornblendite
Intermediate	Rhyolite	Andesite	Basalt	
Fine	Felsite/Obsidian		Basalt glass	

Figure 2.2 *Classification of some igneous rocks in relation to mineralogical composition and the size of mineral grains in the rock (rock texture). Worldwide, light-colored minerals and quartz are generally more prominent than the dark-colored minerals.*

---

[2]*Primary minerals* have not been altered chemically since they formed as molten lava solidified. *Secondary minerals* are recrystallized products of the chemical breakdown and/or alteration of primary minerals.

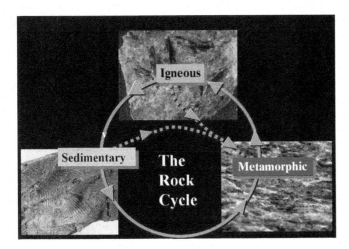 

Figure 2.3 *(Left) The three rock types are interrelated by processes that comprise the rock cycle. (Right) Primary minerals are randomly interlocked in igneous rocks, as in the syenite to the left of the scale. High heat and pressure have deformed and reoriented the crystals and caused lighter minerals to separate from heavier ones, forming the light- and dark-colored bands typical of gneiss, a metamorphic rock on the right of the scale. In this case, the primary mineral content of both rocks is similar, the light-colored minerals being mainly feldspars and the darker ones hornblende. Scale in inches and centimeters.* (Photo courtesy of Ray R. Weil)

Sedimentary rocks form when weathering products released from other, older rocks collect under water as sediment and eventually reconsolidate into new rock. For example, quartz sand weathered from granite and deposited near the shore of a prehistoric sea may become cemented by calcium or iron in the water to become a solid mass called sandstone. Similarly, clays may be compacted into shale. Other important sedimentary rocks are listed in Table 2.1, along with their dominant minerals. The resistance of a given sedimentary rock to weathering is determined by its particular dominant minerals and by the cementing agent. Because most of what is presently dry land was at some time in the past covered by water, sedimentary rocks are the most common type of rock encountered, covering about 75% of the Earth's land surface.

Metamorphic rocks are formed from other rocks by a process of change termed "metamorphism." As Earth's continental plates shift, and sometimes collide, forces are generated that can uplift great mountain ranges or cause huge layers of rock to be pushed deep into the crust. These movements subject igneous and sedimentary rock masses to tremendous heat and pressure. These forces may slowly compress and partially remelt and distort the rocks, as well as break

Table 2.1
SOME OF THE MORE IMPORTANT SEDIMENTARY AND METAMORPHIC ROCKS AND THE MINERALS COMMONLY DOMINANT IN THEM

	Type of Rock	
Dominant mineral	Sedimentary	Metamorphic
Calcite ($CaCO_3$)	Limestone	Marble
Dolomite ($CaCO_3 \cdot MgCO_3$)	Dolomite	Marble
Quartz (SiO_2)	Sandstone	Quartzite
Clays	Shale	Slate
Variable, silicates	Conglomerate[a]	Gneiss[b]
Variable, silicates		Schist[b]

[a]Small stones of various mineralogical makeup are cemented into conglomerate.
[b]Primary minerals present in the igneous rocks from which they metamorphosed commonly dominate these rocks, although some secondary minerals are also present.

the bonds holding the original minerals together. Recrystallization during metamorphism may produce new (usually larger) crystals of the same minerals, or elements from the original minerals may recombine to form new minerals. Igneous rocks like granite may be modified to form gneiss, a metamorphic rock in which light and dark minerals have been reoriented into bands (Figure 2.3, *right*). Sedimentary rocks, such as limestone and shale, may be metamorphosed to marble and slate, respectively (Table 2.1). Slate may be further metamorphosed into phyllite or schist, which typically features mica crystallized during metamorphism.

Metamorphic rocks are usually harder and more strongly crystalline than the sedimentary rocks from which they formed. The particular minerals that dominate a given metamorphic rock influence its resistance to chemical weathering (see Table 2.2 and Figure 2.1).

Weathering: A General Case

Weathering is a biochemical process that involves both destruction and synthesis. Moving from left to right in the weathering diagram (Figure 2.4), the original rocks and minerals are destroyed by both *physical disintegration* and *chemical decomposition*. Without appreciably affecting their composition, physical disintegration breaks down rock into smaller rocks and eventually into sand and silt particles that are commonly made up of individual minerals. Simultaneously, the minerals decompose chemically, releasing soluble materials and synthesizing new minerals, some of which are resistant end products. New minerals form either by minor chemical alterations or by complete chemical breakdown of the original mineral and resynthesis of new minerals. During the chemical changes, particle size continues to decrease, and constituents continue to dissolve in the aqueous weathering solution. The dissolved substances might leave the profile in drainage water, be taken up by plant roots, or recombine into new (secondary) minerals.

Three groups of minerals that remain in well-weathered soils are shown on the right side of Figure 2.4: (1) silicate clays, (2) very resistant end products, including iron and aluminum

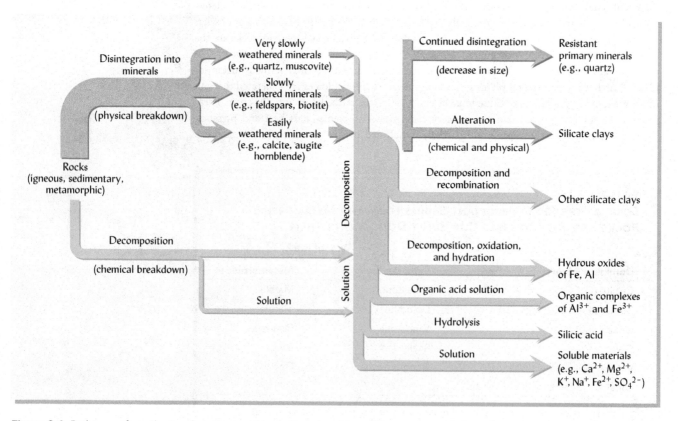

Figure 2.4 *Pathways of weathering that occur under moderately acid conditions common in humid temperate regions. Note that resistant primary minerals, newly synthesized secondary minerals, and soluble materials are products of weathering. In arid regions the physical processes predominate, but in humid tropical areas decomposition and recombination are most prominent.*

Table 2.2
SELECTED PRIMARY AND SECONDARY MINERALS FOUND IN SOILS, LISTED IN ORDER OF DECREASING RESISTANCE TO WEATHERING UNDER CONDITIONS COMMON IN HUMID TEMPERATE REGIONS

Primary minerals		Secondary minerals		
		Goethite	FeOOH	Most resistant
		Hematite	Fe_2O_3	
		Gibbsite	$Al(OH)_3$	
Quartz	SiO_2			
		Clay minerals	Al silicates	
Muscovite	$KAl_3Si_3O_{10}(OH)_2$			
Microcline	$KAlSi_3O_8$			
Orthoclase	$KAlSi_3O_8$			
Biotite	$KAl(Mg,Fe)_3Si_3O_{10}(OH)_2$			
Albite	$NaAlSi_3O_8$			
Hornblende	$Ca_2Al_2Mg_2Fe_3Si_6O_{22}(OH)_2$			
Augite	$Ca_2(Al,Fe)_4(Mg,Fe)_4Si_6O_{24}$			
Anorthite	$CaAl_2Si_2O_8$			
Olivine	$(Mg,Fe)_2SiO_4$			
		Dolomite[a]	$CaMg(CO_3)_2$	
		Calcite[a]	$CaCO_3$	
		Gypsum	$CaSO_4 \cdot 2H_2O$	Least resistant

[a]In semiarid grasslands, dolomite and calcite are more resistant to weathering than suggested because of low rates of acid weathering.

oxide clays, and (3) very resistant primary minerals, such as quartz. In highly weathered soils of humid tropical and subtropical regions, the oxides of iron and aluminum, and certain silicate clays with low Si/Al ratios, predominate because most other constituents have been broken down and removed.

Physical Weathering (Disintegration)

Temperature Rocks exposed to sunlight heat up during the day and cool down at night, causing alternate expansion and contraction of their constituent minerals. As some minerals expand more than others, temperature changes set up differential stresses that eventually cause the rock to crack apart.

Because the outer surface of a rock is often warmer or colder than the more protected inner portions, some rocks may weather by *exfoliation*—the peeling away of outer layers (Figure 2.5). This process may be sharply accelerated if ice forms in the surface cracks. When water freezes, it expands with a force of about 1465 Mg/m^2, disintegrating huge rock masses and dislodging mineral grains from smaller fragments.

Abrasion by Water, Ice, and Wind When loaded with sediment, water has tremendous cutting power, as is amply demonstrated by the gorges, ravines, and valleys around the world. The rounding of riverbed rocks and beach sand grains is further evidence of the abrasion that accompanies water movement.

Windblown dust and sand also can wear down rocks by abrasion, as can be seen in the many picturesque rounded rock formations in certain arid regions. In glacial areas, huge moving ice masses embedded with soil and rock fragments grind down rocks in their path and carry away large volumes of material.

Plants and Animals Plant roots sometimes enter cracks in rocks and pry them apart, resulting in some disintegration. Burrowing animals may also help disintegrate rock somewhat. However, such influences are of little importance in producing parent material when compared to the drastic physical effects of water, ice, wind, and temperature change.

 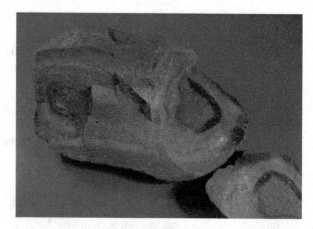

Figure 2.5 Two illustrations of rock weathering. (Left) An illustration of concentric weathering called exfoliation. A combination of physical and chemical processes stimulates the mechanical breakdown, which produces layers that appear much like the leaves of a cabbage. (Right) Concentric bands of light and dark colors indicate that chemical weathering (oxidation and hydration) has occurred from the outside inward, producing iron compounds that differ in color. (Photo courtesy of Ray R. Weil)

Biogeochemical Weathering

While physical weathering is accentuated in very cold or very dry environments, chemical reactions are most intense where the climate is wet and hot. However, both types of weathering occur together, and each tends to accelerate the other. For example, physical abrasion (rubbing together) decreases the size of particles and therefore increases their surface area, making them more susceptible to rapid chemical reactions.

Chemical weathering is enhanced by such *geological* agents as the presence of water and oxygen, as well as by such *biological* agents as the acids produced by microbial and plant-root metabolism. That is why the term **biogeochemical weathering** is often used to describe the process. The various agents act in concert to convert primary minerals (e.g., feldspars and micas) to secondary minerals (e.g., clays and carbonates) and release plant nutrient elements in soluble forms (see Figure 2.4). Had there been no living organisms on Earth, the chemical weathering processes as outlined subsequently would probably have proceeded 1000 times more slowly, with the result that little, if any, soil would have developed on our planet.

 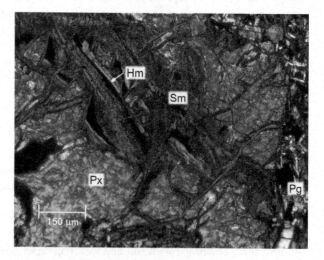

Figure 2.6 Alteration of primary minerals to form clays and other secondary minerals. (Left) Scanning electron micrograph illustrating silicate clay formation from weathering of a granite rock. Potassium feldspar (K-spar) is surrounded by the silicate clays, smectite (Sm), and vermiculite (Vm)(see Chapter 8). (Right) A crossed polarized light microscope image of weathered basalt rock. The image shows replacement of the blue primary mineral pyroxene (Px) by smectite (Sm) and hematite (Hm). Tiny crystals of plagioclase (Pg) also exhibit alteration to smectite (Sm). Note the 30-fold difference in scale between the two images. (Photo Courtesy of Dr. J. Reed Glasmann, Willamette Geological Service, Philomath, OR)

Chemical Weathering Processes One mineral may be changed into another by the binding of intact water molecules by a process called *hydration*. Hydrated oxides of iron and aluminum (e.g., $Al_2O_3 \cdot 3H_2O$) exemplify common products of hydration reactions. In contrast, in *hydrolysis* reactions, water molecules split into their hydrogen and hydroxyl components, and the hydrogen often replaces a cation from the mineral structure. A simple example is the action of water on microcline, a potassium-containing feldspar, releasing soluble potassium and silicic acid (H_4SiO_4). Dissolution reactions occur when water hydrates cations and anions in a mineral until they become dissociated from each other and surrounded by water molecules. For example, gypsum dissolves in water to release calcium (Ca^{2+}) and sulfate (SO_4^{2-}) ions, each in the dissolved state surrounded by water molecules. *Acid reactions* accelerate weathering because acids increase the activity of hydrogen ions in water. For example, when carbon dioxide dissolves in water (a process enhanced by microbial and root respiration) the carbonic acid (H_2CO_3) produced hastens the chemical dissolution of calcite in limestone or marble (Figure 2.7). Soils also contain nitric acid (HNO_3), sulfuric acid (H_2SO_4), and many organic acids and hydrogen ions associated with clays—all of which can react with soil minerals.

Soil biological processes produce organic acids such as oxalic, citric, and tartaric acids, as well as larger organic acid molecules (see Section 11.4). In addition to providing H^+ ions that help solubilize aluminum and silicon, they also promote weathering by *chemical complexation reactions*. The organic molecules form complexes that bind with Al^{3+} ions from within the structure of silicate minerals, causing these minerals to further disintegrate. For example, oxalic acid can form a soluble complex with aluminum from minerals like muscovite, destroying the muscovite structure and releasing dissolved ions of the plant nutrient, potassium.

Minerals that contain iron, manganese, or sulfur are especially susceptible to *oxidation–reduction reactions*. Iron is usually laid down in primary minerals in the divalent Fe(II) (ferrous) form. When rocks containing such minerals are exposed to air and water during soil formation, the iron is easily oxidized (loses an electron) and becomes trivalent Fe(III) (ferric). If iron is oxidized from Fe(II) to Fe(III), the change in valence and ionic radius causes destabilizing adjustments in the crystal structure of the mineral. In other cases, Fe(II) may be released from the mineral and almost simultaneously oxidized to Fe(III). The oxidation and/or removal of iron during weathering is often made visible by changes in the colors of the resulting altered minerals (see Figure 2.5, *right*).

Integrated Weathering Processes The various chemical weathering processes occur simultaneously and are interdependent. For example, hydrolysis of a given primary mineral may

Figure 2.7 *First stages of soil development: Biochemical weathering of a rock under the influence of mosses and lichen (a symbiotic combination of algae and fungi). Lichens are especially effective at pioneering inhospitable sites. The algal component provides energy-rich compounds from photosynthesis, whereas the fungal partner unlocks mineral nutrients from the rock. The fungus produces organic acids that break down the rock by hydrolysis and complexation reactions. Note the pitted surface of the rock where a lichen mat was pulled away to expose the weathering rock surface. The resulting loose mineral material and soluble nutrients, in conjunction with trapped dust and organic debris left by the lichen, will eventually provide a medium for the growth of higher plants. These in turn will further accelerate the process of soil formation. White bar is 1 cm.* (Photo courtesy of Ray R. Weil)

release ferrous iron {Fe(II)} that is quickly oxidized to the ferric {Fe(III)} form, which, in turn, is hydrolyzed to give a hydrous oxide of iron. Hydrolysis or complexation also may release soluble cations, silicic acid, and aluminum or iron compounds. In humid environments, some of the soluble cations and silicic acid are likely to be lost from the weathering mass in drainage waters. The released substances can also be recombined to form silicate clays and other secondary silicate minerals. In this manner, the biochemical processes of weathering transform primary geologic materials into the compounds of which soils are made (Figure 2.4).

2.2 WHAT ENVIRONMENTAL FACTORS INFLUENCE SOIL FORMATION?[3]

We learned in Chapter 1 that *the soil* is a collection of *individual soils*, each with distinctive profile characteristics. This concept of soils as organized natural bodies derived initially from late nineteenth-century field studies by a team of brilliant Russian soil scientists led by V. V. Dokuchaev. They noted similar profile layering in soils hundreds of kilometers apart, provided that the climate and vegetation were similar at the two locations. Such observations and much careful subsequent field and laboratory research led to the recognition by Hans Jenny of five major environmental factors that control the formation of soils. In 1941 he published the now classical state factor model of soil formation:

$$S_i = f(cl,o,r,p,t) \qquad (2.1)$$

which tells us that a particular soil property, S_i, will be determined by a function of five factors that describe the environment in which the soil occurs:

1. **cl:** *climate* (primarily precipitation and temperature)
2. **o:** *organisms* (biota, especially native vegetation, microbes, soil animals, and increasingly human beings)
3. **r:** *relief or topography* (slope, aspect, and landscape position)
4. **p:** *parent materials* (geological or organic precursors to the soil)
5. **t:** *time* (the period of time since the parent materials began to undergo soil formation)

Soils are often defined in terms of these factors as *dynamic natural bodies having properties derived from the combined effects of* **climate** *and* **biotic activities**, *as modified by* **topography**, *acting on* **parent materials** *over periods of* **time**.

We will now examine how each of these five factors affects the outcome of soil formation. However, as we do, we must keep in mind that these factors do not exert their influences independently. Indeed, interdependence is the rule. For example, contrasting climatic regimes are likely to be associated with contrasting types of vegetation, and perhaps differing topography and parent material as well. Nonetheless, in certain situations one of the factors has had the dominant influence in determining differences among a set of soils. Soil scientists refer to such a set of soils as a *lithosequence*, *climosequence*, *biosequence*, *toposequence*, or *chronosequence*.

2.3 PARENT MATERIALS

Geological processes have brought to the Earth's surface numerous parent materials in which soils form (Figure 2.8). The nature of the parent material profoundly influences soil characteristics. For example, a soil might inherit a sandy texture (see Section 4.2) from a coarse-grained, quartz-rich parent material such as granite or sandstone. Soil texture, in turn, helps control the percolation of water through the soil profile, thereby affecting the translocation of fine soil particles and plant nutrients.

The chemical and mineralogical composition of parent material also influences both chemical weathering and the natural vegetation. For example, the presence of limestone in parent material will slow the development of acidity that typically occurs in humid climates.

---

[3]Many of our modern concepts concerning the factors of soil formation are derived from the work of Hans Jenny (1941 and 1980) and E. W. Hilgard (1921), American soil scientists whose books are considered classics in the field.

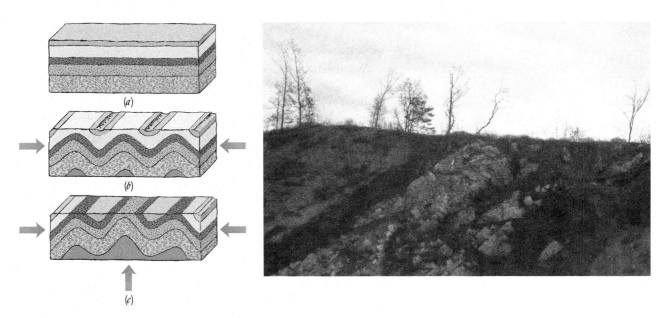

Figure 2.8 *Diagrams showing how geological processes have brought different rock layers to the surface in a given area. (a) Unaltered layers of sedimentary rock with only the uppermost layer exposed. (b) Lateral geological pressures deform the rock layers through a process called* crustal warping. *At the same time, erosion removes much of the top layer, exposing part of the first underlying layer. (c) Localized upward pressures further reform the layers, thereby exposing two more underlying layers. As these four rock layers are weathered, they give rise to the parent materials on which different kinds of soils can form. (Right) Crustal warping that lifted up the Appalachian Mountains tilted these sedimentary rock formations that were originally laid down horizontally. This deep road cut in Virginia illustrates the abrupt change in soil parent material (lithosequence) as one walks along the ground surface at the top of this photograph.* (Courtesy of Ray R. Weil)

The parent material may contain varying amounts and types of clay minerals, perhaps from a previous weathering cycle. The nature of the parent material greatly influences the kinds of clays that can develop as the soil evolves (see Section 8.5). In turn, the nature of the clay minerals present markedly affects the kind of soil that develops.

Classification of Parent Materials

Inorganic parent materials can either be formed in place as residual material weathered from the rock below them, or they can be transported from one location and deposited at another (Figure 2.9). In wet environments (such as swamps and marshes), incomplete decomposition may allow organic parent materials to accumulate from the residues of many generations of vegetation. Although it is their chemical and physical properties that most influence soil development, parent materials are often classified with regard to the mode of placement in their current location, as seen on the right side of Figure 2.9.

Although people sometimes refer to *organic soils, glacial soils, alluvial soils,* and so forth, these terms are quite nonspecific because parent material properties vary widely within each group and because the effect of parent material is modified by the influence of climate, organisms, topography, and time.

Residual Parent Material

Residual parent material develops in place from weathering of the underlying rock. In stable landscapes it may have experienced long and possibly intense weathering. Where the climate is warm and very humid, residual parent materials are typically thoroughly leached and oxidized, and they show the red and yellow colors of various oxidized iron compounds. In cooler and especially drier climates, the color and chemical composition of residual parent material tends to resemble more closely the rock from which it formed (Figure 2.10).

Residual materials are widely distributed on all continents. A great variety of soils occupy the regions covered by residual debris because of the marked differences in the nature

Figure 2.9 *How various kinds of parent material are formed, transported, and deposited.* (Diagram courtesy of N. C. Brady and Ray R. Weil)

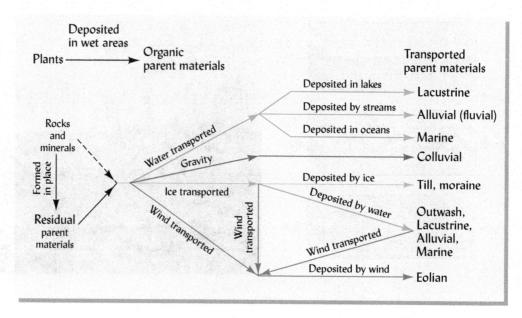

of the rocks from which these materials evolved. The varied soils are also a reflection of wide differences in other soil-forming factors, such as climate and vegetation (Sections 2.4 and 2.5).

Colluvial Debris

Colluvial debris, or **colluvium**, is made up of poorly sorted rock fragments detached from the heights above and carried downslope, mostly by gravity, assisted in some cases by frost action or water. Rock fragment (talus) slopes, cliff rock debris (detritus), and similar heterogeneous materials are examples that can contribute to avalanches on steep slopes.

Figure 2.10 *Rock weathering in humid tropical Liberia (right) and in subhumid, temperate Kansas (left). In the tropical soil, weathering has been deep and the gray rock (not seen) has radically changed in color and composition (due to residual iron oxides). Compare this to the shallower regolith and similar gray-brown colors of the rock and parent material in the subhumid temperate soil. Scale in meters.* (Photo courtesy of Ray R. Weil)

Colluvial parent materials are frequently coarse and stony because physical rather than chemical weathering has been dominant. Stones, gravel, and fine materials are interspersed (not layered), and the coarse fragments are rather angular (Figure 2.11). Packing voids, spaces created when tumbling rocks come to rest against each other (sometimes at precarious angles), help account for the easy drainage of many colluvial deposits and also for their tendency to be unstable and prone to slumping and landslides, especially if disturbed by excavations.

Alluvial Stream Deposits

There are three general classes of alluvial deposits: *floodplains, alluvial fans*, and *deltas*. They will be considered in this order.

Floodplains During flooding, a swollen stream will inundate the adjacent flat land known as the floodplain. The floodwaters deposit sediments on the floodplain, laying down the coarser materials near the river channel where the water flows deeper, faster, and with more turbulence. Finer materials settle out in the calmer floodwaters farther from the channel. Each major flooding episode lays down a distinctive layer of sediment, creating the stratification that characterizes alluvial soils (Figure 2.12).

If, over a period of time, there is a change in grade, a stream may cut down through its already well-formed alluvial deposits. This cutting action leaves **terraces** above the floodplain on one or both sides. Some river valleys feature two or more terraces at different elevations, each reflecting a past period of alluvial deposition and stream cutting.

Major areas of alluvial parent materials are found along the Nile River in Egypt and Sudan; the Euphrates, Ganges, Indus, Brahmaputra, and Hwang Ho river valleys of Asia; and the Amazon River of Brazil. The floodplain along the Mississippi River is the largest in the United States, varying from 30 to 125 km in width. Floodplains of smaller streams also provide parent materials for locally important soil areas.

To some degree, nutrient-rich materials lost by upland soils are deposited on the river floodplain and **delta** (see Delta Deposits, following). Soils derived from alluvial sediments generally have characteristics seen as desirable for human settlement and agriculture. These characteristics include nearly level topography, proximity to water, high fertility, and high productivity. Although many alluvial soils are well drained, others may require artificial drainage if they are to be used for upland crops or for stable building foundations.

While alluvial soils are often uniquely suited to forestry and crop production, their use for home sites and urban development should generally be avoided. Unfortunately, the desirable properties of many alluvial soils have led people to build cities and towns on floodplains.

Figure 2.11 *A productive soil formed in colluvial parent material in the Appalachian Mountains of the eastern United States. The ridge in the background was at one time much taller, but material from the crest tumbled downslope and came to rest in the configuration seen in the soil profile. Note the unstratified mix of particle sizes and the rather angular nature of the coarse fragments.*
(Photo courtesy of Ray R. Weil)

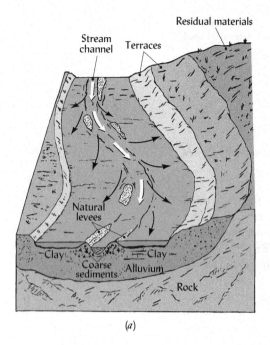

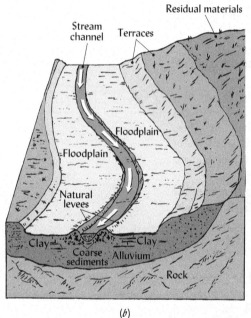

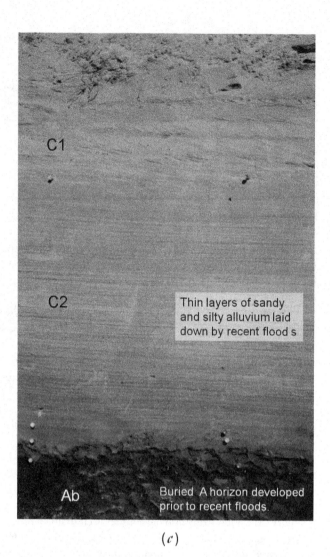

Figure 2.12 *Floodplain development. (a) A stream at flood stage has overflowed its banks and is depositing sediment in the floodplain. The coarser particles are deposited nearest the stream channel where the water is flowing most rapidly, whereas the finer particles settle out where the water is moving more slowly. (b) After the flood the sediments are in place and vegetation is growing. (c) Upper 120 cm of a profile of a soil on the Mississippi River floodplain showing contrasting thin layers of sand and silt sediments in the new C horizon that was laid down over an older A horizon. Each layer resulted from a separate flooding episode.* (Courtesy of Ray R. Weil)

As the many disastrous floods of recent years have illustrated, building on a floodplain, no matter how great the investment in flood-control measures, all too often leads to tragic loss of life and property.

In many areas, installation of systems for drainage and flood protection has proven costly and ineffective. Farmers and the general public pay high costs to keep such areas in agricultural or urban uses. Steps are therefore being taken to reestablish the wetland conditions of certain flood-prone agricultural areas that originally were natural wetlands. These and other alluvial soils can provide natural habitats, such as bottomland forests, which produce vast quantities of timber and support a high diversity of birds and other wildlife.

Alluvial Fans Streams that leave a narrow valley in an upland area and suddenly descend to a much broader valley below deposit sediment in the shape of a fan, as the water spreads out and slows down (see Figure 2.13). The rushing water tends to sort the sediment particles by size, first dropping the gravel and coarse sand, then depositing the finer materials toward the bottom of the alluvial fan.

Alluvial fan debris is found in widely scattered areas in mountainous and hilly regions. The soils derived from this debris often prove very productive, although they may be quite coarse-textured. The Sacramento Valley in California and the Willamette Valley in Oregon are examples of large, agriculturally important areas with alluvial fan materials.

Delta Deposits Much of the finer sediment carried by streams is not deposited in the floodplain but is discharged into the lake, reservoir, or ocean into which the streams flow. Some of the suspended material settles near the mouth of the river, forming a delta. Such delta deposits are by no means universal, being found at the mouths of only a few rivers of the world. A delta often is a continuation of a floodplain (its front, so to speak). It is clayey in nature and is likely to be poorly drained as well.

Delta marshes are among the most extensive and biologically important of wetland habitats. Many of these habitats are today being protected or restored, but civilizations both ancient and modern have also developed important agricultural areas (often specializing in the production of rice) by creating drainage and flood-control systems on the deltas of such rivers as the Amazon, Euphrates, Ganges, Hwang Ho, Mississippi, Nile, Po, and Tigris.

Coastal Sediments

Streams eventually deposit much of their sediment loads in oceans, estuaries, and gulfs. The coarser fragments settle out near the shore and the finer particles at a distance (Figure 2.14). Over long periods of time, these underwater sediments build up, in some cases becoming hundreds of meters thick. Changes in the relative elevations of sea and land may later raise these marine deposits above sea level, creating a coastal plain. The deposits are then subject to a new cycle of weathering and soil formation.

A coastal plain usually has only moderate slopes, being more level in the low-lying parts nearer the coastline and more hilly farther inland, where streams and rivers flowing down the steeper grades have more deeply dissected the landscape. The land surface in the lower coastal portion may be only slightly above the water table during part of the year, so wetland forest and marsh vegetation often dominate areas of such parent materials.

Marine and other coastal deposits are quite variable in texture. Some are sandy, as is the case in much of the Atlantic seaboard Coastal Plain. Others are high in clay, as are deposits found in the Atlantic and Gulf Coastal Flatwoods. Where streams have cut down through layers of marine sediments (as in the detailed block diagram in Figure 2.14), clays, silts, and sand may be encountered side by side. All of these sediments came from the erosion of upland areas, some of which were highly weathered before the transport took place. However, marine sediments generally have been subjected to soil-forming processes for a shorter period of time than

Figure 2.13 Characteristically shaped alluvial fan (central foreground of photo) in a valley in central Nevada. Although alluvial fan areas are usually small and sloping, they can develop into productive, well-drained soils. (Photo courtesy of Ray R. Weil)

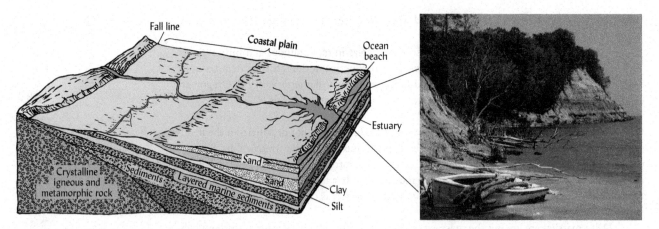

Figure 2.14 *Coastal plain sediments consist of materials laid down in marine waters and washed off interior hills onto coastal areas. The diagram represents the coastal plain of the southeastern United States where such sediments cover older crystalline igneous and metamorphic rocks. Changes in the location of the shoreline and currents over time have resulted in sediment layers consisting alternately of fine clay, silts, coarse sands, and gravels. The photo shows such layering in coastal marine sediments along the Chesapeake Bay in Maryland.* (Courtesy of Ray R. Weil)

their upland counterparts. As a consequence, the properties of the soils that form are heavily influenced by those of the marine parent materials. Because seawater is high in sulfur, many marine sediments are high in sulfur and go through a period of acid-forming sulfur oxidation at some stage of soil formation (see Sections 9.6 and 12.2).

Parent Materials Transported by Glacial Ice and Meltwaters

During the Pleistocene Epoch (about 10^4 to 10^7 years ago), up to 20% of the world's land surface—northern North America, northern and central Europe, and parts of northern Asia—was invaded by a succession of great ice sheets, some more than 1 km thick (Figure 2.15). Present-day glaciers in polar regions and high mountains cover about a third as much area but are not nearly so thick as the glaciers of the Great Pleistocene Ice Age.[4]

In North America, Pleistocene Epoch glaciers covered most of what is now Canada, southern Alaska, and the northern part of the contiguous United States. The southernmost extension went down the Mississippi Valley, where the least resistance was met because of the lower and smoother topography.

As the glacial ice pushed forward, it swept away the existing regolith with its soil mantle, rounded the hills, filled in the valleys, and, in some cases, severely ground and gouged the underlying rocks. Thus, the glacier became filled with rock and all kinds of unconsolidated materials, carrying great masses of these materials as it pushed ahead. Finally, as the ice melted and the glacier retreated, a mantle of glacial debris or drift remained. This provided a new regolith and fresh parent material for soil formation.

Glacial Till and Associated Deposits The name **drift** is applied to all material of glacial origin, whether deposited by the ice or by associated waters. The materials deposited directly by the ice, called **glacial till**, are heterogeneous (unstratified) mixtures of debris, which vary in size from boulders to clay. Till (the adjective "glacial" is optional as there is no non-glacial till) may therefore be somewhat similar in appearance to colluvial materials, except that the coarse fragments are more rounded from their grinding journey in the ice, and the deposits are often much more densely compacted because of the great weight of the overlying ice sheets. Much glacial till is deposited in irregular ridges called **moraines**. Figure 2.15 shows how glacial sheets deposited several types of soil parent materials.

---

[4]Though they are much smaller than paleoglaciers, if all present day glaciers were to melt, the world sea level would rise by about 65 m. Current global warming trends are observably accelerating glacier melting and sea level rise, thus threatening many coastal areas around the world with inundation in coming decades.

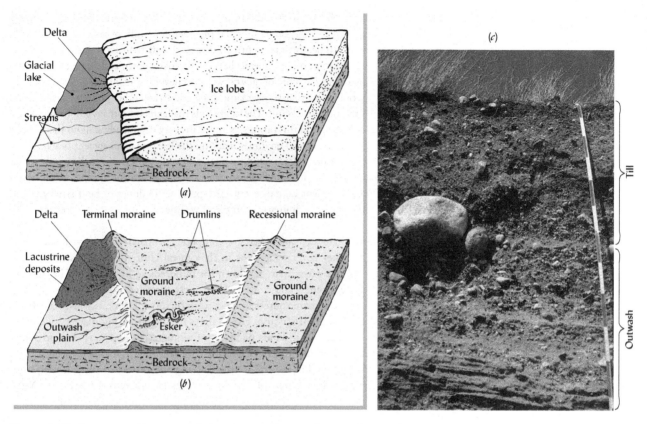

Figure 2.15 *Illustration of how several glacial materials were deposited. (a) A glacier ice lobe moving to the left, feeding water and sediments into a glacial lake and streams and building up glacial till near its front. (b) After the ice retreats, terminal, ground, and recessional moraines are uncovered along with cigar-shaped hills (drumlins), the beds of rivers that flowed under the glacier (eskers), and lacustrine, delta, and outwash deposits. (c) The stratified glacial outwash in the lower part of this soil profile in North Dakota is overlain by a layer of glacial till containing a random assortment of particles, ranging in size from small boulders to clays. Note the rounded edges of the rocks, evidence of the churning action within the glacier. Scale is marked every 10 cm.*
(Courtesy of Ray R. Weil)

Glacial Outwash and Lacustrine Sediments The torrents of water gushing forth from melting glaciers carried vast loads of sediment. In valleys and on plains where the glacial waters were able to flow away freely, the sediment formed an **outwash plain** (Figure 2.15). Such sediments, with sands and gravels sorted by flowing water, are common **valley fills**. Figure 2.15 shows the sorted layering of coarse and fine materials in glacial outwash overlaid by mixed materials of glacial till.

When the ice front came to a standstill, where there was no ready escape for the water, ponding began; ultimately, very large lakes were formed (Figure 2.15). The **lacustrine deposits** formed in these glacial lakes range from coarse delta materials and beach deposits near the shore to larger areas of fine silts and clay deposited from the deeper, more still waters at the center of the lake. Flat areas of inherently fertile (though not always well drained) soils developed from these materials as the lakes dried.

Parent Materials Transported by Wind

Wind is capable of picking up an enormous quantity of material at one site and depositing it at another. Wind can most effectively pick up material from soil or regolith that is loose, dry, and unprotected by vegetation. Dry, barren landscapes have served, and continue to serve, as sources of parent material for soils forming as far away as the opposite side of the globe. The smaller the particles, the higher and farther the wind will carry them. Wind-transported (**eolian**) materials important as parent material for soil formation include, from largest to smallest particle size: **dune sand**, **loess** (pronounced "luss"), and **aerosolic dust**. Windblown **volcanic ash** from erupting volcanoes is a special case that is also worthy of mention.

Dune Sand Along the beaches of the world's oceans and large lakes and over vast barren deserts, strong winds pick up medium and fine sand grains and pile them into hills of sand called *dunes*. The dunes, ranging up to 100 m in height, may continue to slowly shift their locations in response to the prevailing winds. Because most other minerals have been broken down and carried away by the waves, beach sand usually consists mainly of quartz, which is devoid of plant nutrients and highly resistant to weathering action. Nonetheless, over time dune grasses and other pioneering vegetation may take root, and soil formation may begin. The sandy soils that extend for many kilometers east of Lake Michigan provide an example of this process. Some of the very deep sandy soils on the Atlantic coastal plain are thought to have formed on dunes marking the location of an ancient beach.

Desert sands, too, are usually dominated by quartz, but they may also include substantial amounts of other minerals that could contribute more to the establishment of vegetation and the formation of soils, should sufficient rainfall occur. The pure-white dunes of sand-sized gypsum at White Sands, New Mexico, are a dramatic example of weatherable minerals in desert sands.

Loess The windblown materials called *loess* are composed primarily of silt with some very fine sand and coarse clay. They cover wide areas in the central United States, eastern Europe, Argentina, and central China. Loess may be blown for hundreds of kilometers. The deposits farthest from the source are thinnest and consist of the finest particles.

In the United States, the main sources of loess were the great barren expanses of till and outwash left in the Missouri and Mississippi river valleys by the retreating glaciers of the last Ice Age. During the winter months, winds picked up fine materials and moved them southward, covering the existing soils and parent materials with a blanket of loess that accumulated to as much as 8 m thick. Similar dust storms occur today as glaciers recede in Iceland, exposing barren outwash plains from which winds blow dust particles far out into the Atlantic Ocean. This input of mineral dust is thought to be an important source of iron for ocean life.

In central and western China, loess deposits reaching 30–100 m in depth cover some 800,000 km^2. These materials have been windblown from the deserts of central Asia and are generally not associated directly with glaciers. These and other loess deposits tend to form silty soils of rather high fertility and potential productivity but also of high susceptibility to erosion by wind and water.

Aerosolic Dust Very fine particles (about 1–10 μm diameter) are called *aerosolic dust* because they can remain suspended in air for thousands of kilometers before being deposited, usually with rainfall. Although this dust has not blanketed the receiving landscapes as thickly as is typical for loess, it does accumulate at rates that make significant contributions to soil formation. Much of the calcium carbonate in soils of the western United States probably originated as windblown dust. Recent studies have shown that dust, originating in the Sahara Desert of northern Africa and transported over the Atlantic Ocean in the upper atmosphere (Figure 2.16) is the source of much of the calcium and other nutrients found in the highly leached soils of the Amazon basin in South America. Likewise, in the springtime, dust from wind storms in the loess region of China blows across the Pacific Ocean to add soil parent materials (and air pollution) to the western part of North America.

Volcanic Ash During volcanic eruptions, cinders fall in the immediate vicinity of the volcano, whereas fine, often glassy, ash particles may blanket extensive areas downwind. Soils developed from volcanic ash are most prominent within a few hundred kilometers of the volcanoes that ring the Pacific Ocean. Important areas of volcanic ash parent materials occur in Japan, Indonesia, New Zealand, Iceland, western United States (in Hawaii, Montana, Oregon, Washington, and Idaho), Mexico, Central America, and Chile. The soils formed are uniquely light and porous and tend to accumulate organic matter more rapidly than other soils in the area (Section 3.7). The volcanic ash tends to weather rapidly into allophane, a type of clay with unusual properties (see Section 8.5).

Organic Deposits

Organic material accumulates in wet places where plant growth exceeds the rate of residue decomposition. In such areas residues accumulate over the centuries from wetland plants, such as pondweeds, cattails, sedges, reeds, mosses, shrubs, and certain trees. These residues sink into the water, where their decomposition is limited by lack of oxygen. As a result, organic deposits often accumulate up to several meters in depth (Figure 2.17). Collectively, these organic deposits are called **peat**.

Figure 2.16 A great cloud of dust from the Sahara desert in Africa rides the winds out into the Atlantic Ocean and up into Europe. Some of this dust also crosses the southern Atlantic Ocean and contributes to soils in South America. In a similar manner, giant dust clouds generated by wind erosion of loess soils in the Gobi desert region of China take about a week to ride the easterly winds across the Pacific Ocean before contributing to soil parent material (and air pollution) on the west coast of North America. (NASA)

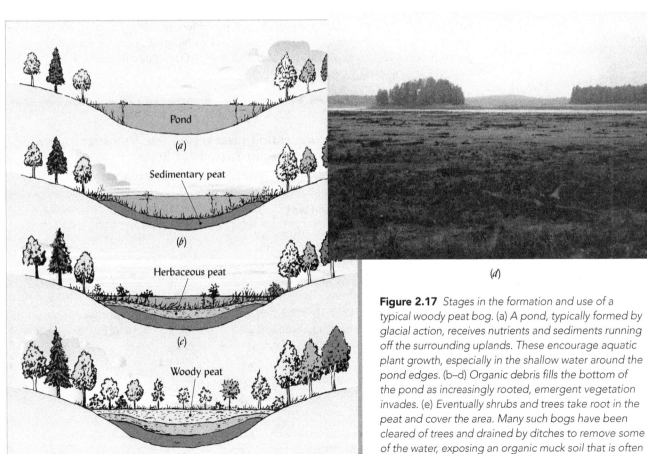

Figure 2.17 Stages in the formation and use of a typical woody peat bog. (a) A pond, typically formed by glacial action, receives nutrients and sediments running off the surrounding uplands. These encourage aquatic plant growth, especially in the shallow water around the pond edges. (b–d) Organic debris fills the bottom of the pond as increasingly rooted, emergent vegetation invades. (e) Eventually shrubs and trees take root in the peat and cover the area. Many such bogs have been cleared of trees and drained by ditches to remove some of the water, exposing an organic muck soil that is often highly productive for vegetable crops. The bog in the photo is in central Michigan. (Courtesy of Ray R. Weil)

Distribution and Accumulation of Peats Peat deposits are found all over the world, but most extensively in the cool climates and in areas that have been glaciated. About 75% of the 340 million hectares of peat lands in the world are found in Canada and northern Russia.

The rate of peat accumulation varies from one area to another, depending on the balance between production of plant material and its loss by decomposition. Cool climates and acidic conditions favor slow decomposition but also slower plant production. Warm climates and alkaline conditions favor rapid losses but also rapid plant production. Enrichment with nutrients may increase the rate of organic production more than it does the rate of decomposition, leading to very high net accumulation rates (Table 2.3). As we shall see in Chapter 11, artificial drainage, used to remove excess water from a peat soil, lets air into the peat and drastically alters the balance between production and decomposition of organic matter, causing a reversal of the accumulation process and a loss or subsidence of the peat soil.

Types of Peat Materials Based on the nature of the parent materials, four kinds of peat are recognized:

1. Moss peat, the remains of mosses such as sphagnum
2. Herbaceous peat, residues of herbaceous plants such as sedges, reeds, and cattails
3. Woody peat, from the remains of woody plants, including trees and shrubs
4. Sedimentary peat, remains of aquatic plants (e.g., algae) and of fecal material of aquatic animals

Organic deposits generally contain two or more of these kinds of peats, either in alternating layers or as mixtures. Because the succession of plants, as the residues accumulate, tends to favor trees (see Figure 2.17), woody peats often dominate the surface layers of organic materials.

In cases where a wetland area has been drained, woody peats tend to make very productive agricultural soils that are especially well suited for vegetable production. While moss peats have high water-holding capacities, they tend to be quite acidic. Sedimentary peat is generally undesirable as an agricultural soil. This material is highly colloidal and compact and

Table 2.3
VARIOUS FRESHWATER WETLANDS, THEIR RATES OF ACCRETION (INCREASING SOIL THICKNESS), AND ACCUMULATION OF CARBON, NITROGEN, AND PHOSPHOROUS IN THEIR PEAT SOILS

Type and location of wetland	Characteristics of wetland types	Accretion rate, mm/yr	Accumulation rate, g/m^2/yr		
			C	N	P
Bogs (Massachusetts)	**Bogs** are peat-accumulating ponds with no outflow and little inflow. They receive little calcium from the surrounding landscape and are therefore quite acidic.	4.3	90	1.2	—
Fens (Michigan)	**Fens** are peat-accumulating ponds with no outflow and little inflow. Calcium-rich mineral matter from the surrounding landscapes makes them relatively alkaline.	0.9	42	3.0	0.11
Pocosin swamps (North Carolina)	**Swamps** are periodically inundated with shallow, slow-moving water and dominated by shrubs and trees.	2.6	127	3.0	0.06
Everglades marsh (Florida), unenriched	**Marshes** are periodically inundated with shallow, slow-moving water and dominated by grasses and herbaceous plants.	1.4	65	4.7	0.06
Everglades marsh (Florida), enriched	Area in Florida with both marshy and swampy components.	6.7	223	16.6	0.46

(Accumulation rates selected from Craft and Richardson [1998])

is rubbery when wet. Upon drying, it resists rewetting and remains in a hard, lumpy condition. Fortunately, it occurs mostly deep in the profile and is unnoticed unless it interferes with drainage of the bog area.

Organic material is called **peat**, or **fibric**, if the residues are sufficiently intact to permit the plant fibers to be identified. If most of the material has decomposed sufficiently so that little fiber remains, the term **muck** or **sapric** is used. In mucky peats (**hemic** materials) only some of the plant fibers can be recognized.

Wetland Preservation Wetland areas are important environmental buffers and natural habitats for wildlife. Drainage of these areas reduces the benefits of wetlands. While organic soils are the foundation for some very productive agricultural systems, environmentalists argue that such use is unsustainable because, once drained, the organic deposits will decompose and disappear after a century or so; therefore, these areas might be better left in (or returned to) their natural state (see Section 7.7).

Recognizing that the effects of **parent materials** on soil properties are modified by the combined influences of **climate**, **biotic activities**, **topography**, and **time**, we will now turn to these other four factors of soil formation, starting with climate.

2.4 HOW DOES CLIMATE AFFECT SOIL FORMATION?

Climate is perhaps the most influential of the four factors acting on parent material because it determines the nature and intensity of the weathering that occurs over large geographic areas. The principal climatic variables influencing soil formation are *effective precipitation* (see Box 2.1) and *temperature*, both of which affect the rates of chemical, physical, and biological processes.

BOX 2.1
EFFECTIVE PRECIPITATION FOR SOIL FORMATION

Water from rain and melting snow is a primary requisite for parent material weathering and soil development. To fully promote soil development, water must not only enter the profile and participate in weathering reactions but also percolate through the profile and translocate soluble weathering products.

Let's consider a site that receives an average of 600 mm of rainfall per year. The amount of water leaching through a soil is determined not only by the total annual precipitation but also by at least four other factors as well (Figure 2.18).

a. **Seasonal distribution** of precipitation. The 600 mm of rainfall distributed evenly throughout the year, with about 50 mm each month, is likely to cause less soil leaching or erosion than the same annual amount of rain falling at the rate of 100 mm per month during a six-month rainy season.

b. **Temperature and evaporation.** In a hot climate, evaporation from soils and vegetation is much higher than in a cool climate. Therefore, in the hot climate, much less of the 600 mm will be available for percolation and leaching. Most or all will evaporate soon after it falls on the land. Thus, 600 mm of rain may cause more leaching and profile development in a cool climate than in a warmer one. Similar reasoning would suggest that rainfall concentrated during a mild winter (as in California) may be more effective in leaching the soil than the same amount of rain concentrated in a hot summer (as in the Great Plains).

c. **Topography.** Water falling on a steep slope will run downhill so rapidly that only a small portion will enter the soil where it falls. Therefore, even though they receive the same rainfall, level or concave sites will experience more percolation and leaching than steeply sloping sites. The effective rainfall can be said to be greater on the level site than on the sloping one. The concave site will receive the greatest effective rainfall because, in addition to direct rainfall, it will collect the runoff from the adjacent sloping site.

d. **Permeability.** Even if the above conditions are the same, more rainwater will infiltrate and leach through a coarse, sandy profile than a tight, clayey one. Therefore the sandy profile can be said to experience a greater effective precipitation, and more rapid soil development may be expected.

BOX 2.1
EFFECTIVE PRECIPITATION FOR SOIL FORMATION (CONTINUED)

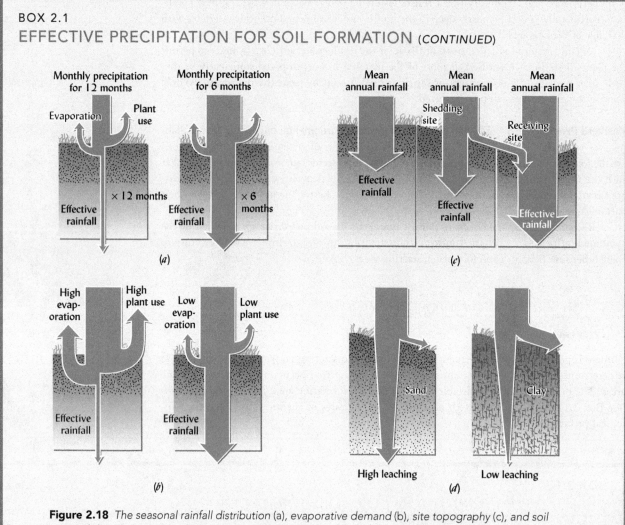

Figure 2.18 *The seasonal rainfall distribution (a), evaporative demand (b), site topography (c), and soil permeability (d) interact to determine how effectively precipitation can influence soil formation.* (Diagram courtesy of Ray R. Weil)

Effective Precipitation

We have already seen that water is essential for all the major chemical weathering reactions. To be effective in soil formation, water must penetrate into the regolith. The greater the depth of water penetration, the greater will be the depth of weathering and soil development. Surplus water percolating through the soil profile transports soluble and suspended materials from the upper to the lower layers. It may also carry away soluble materials in the drainage waters. Thus, percolating water stimulates weathering reactions and helps differentiate soil horizons.

Likewise, a deficiency of water is a major factor in determining the characteristics of soils of dry regions. Soluble salts are not leached from these soils, and in some cases they build up to levels that curtail plant growth. Soil profiles in arid and semiarid regions are also apt to accumulate carbonates and certain types of cracking clays.

Temperature

For every 10°C rise in temperature, the rates of biochemical reactions more than double. Both temperature and moisture influence the organic matter content of soil through their effects on the balance between plant growth and microbial decomposition (see Figure 12.27). If warm

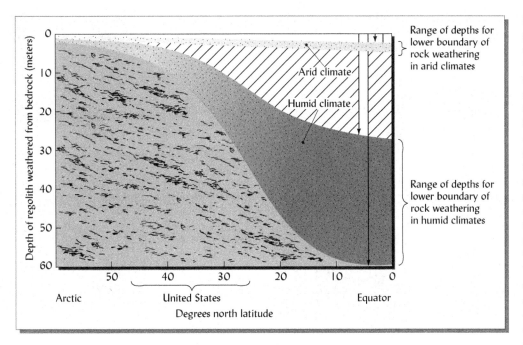

Figure 2.19 *A generalized illustration of how two climatic variables, temperature and precipitation, influence the depth of regolith weathering from bedrock. The stippled areas represent the range of depths to which the regolith typically extends. In cold climates (arctic regions), the regolith is shallow under both humid and arid conditions. In the warmer climates of the lower latitudes, the depth of the residual regolith increases sharply in humid regions but is little affected in arid regions. Under humid tropical conditions, the regolith may be 50 or more meters in depth. The vertical arrows represent depths of weathering near the Equator. Remember that soil depth may not be as great as regolith depth.* (Diagram courtesy of N. C. Brady)

temperatures and abundant water are present in the profile at the same time, the processes of weathering, leaching, and plant growth will be maximized. The very modest profile development characteristic of cold areas contrasts sharply with the deeply weathered profiles of the humid tropics (Figures 2.19 and 2.10).

Considering soils with similar temperature regime, parent material, topography, and age, increasing effective annual precipitation generally leads to increasing clay and organic matter contents, greater acidity, and lower ratio of Si/Al (an indication of more highly weathered minerals). However, many places have experienced climates in past geologic epochs that were very different from the climate evident today. This fact is illustrated in certain old landscapes in arid regions, where highly leached and weathered soils stand as relics of the humid tropical climate that prevailed there many thousands of years ago. Climate also influences the natural vegetation. Humid climates favor the growth of trees. In contrast, grasses are the dominant native vegetation in subhumid and semiarid regions, while shrubs and brush of various kinds dominate in arid areas. Thus, climate exerts its influence partly through the soil-forming factor we will consider next: living organisms.

2.5 HOW DO LIVING ORGANISMS (INCLUDING PEOPLE) AFFECT SOIL FORMATION?

Organic matter accumulation, biochemical weathering, profile mixing, nutrient cycling, and aggregate stability are all enhanced by the activities of organisms in the soil. Vegetative cover reduces the rate of mineral surface soil removal by erosion. Organic acids produced from certain types of plant leaf litter bring iron and aluminum into solution by complexation and accelerate the downward movement of these metals and their accumulation in the B horizon.

Organic Matter Accumulation The effect of vegetation on soil formation can be seen by comparing properties of soils formed under grassland and forest vegetation near the boundary between these two ecosystems (Figure 2.20). In the grassland, much of the organic matter added to the soil is from the deep, fibrous, root systems. By contrast, tree leaves falling on the forest floor are a principal source of soil organic matter in the forest. Another difference is the frequent occurrence in grasslands of fires that destroy large amounts of aboveground biomass and create bits of charcoal that accumulate in the soil. Also, the extreme acidity under certain forests inhibits such soil organisms as earthworms that otherwise would mix much of the

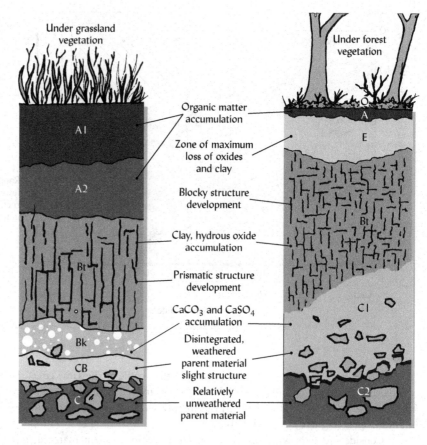

Figure 2.20 *Natural vegetation influences the type of soil eventually formed from a given parent material (calcareous till, in this example). The forested soil exhibits surface layers (O horizons) of leaves and twigs in various stages of decomposition, along with a thin mineral A horizon, into which some of the surface litter has been mixed. In contrast, most organic matter in the grassland is added as fine roots distributed throughout the upper 1 m or so, creating a thick mineral A horizon. Also note that calcium carbonate has been solubilized and has moved down to the lower horizons (Bk) in the grassland soils, while it has been completely removed from the profile in the more acidic, leached forested soil. Under both types of vegetation, clay and iron oxides move downward from the A horizon and accumulate in the B horizon, encouraging the formation of characteristic soil structure. In the forested soil, the zone above the B horizon usually becomes a distinctly bleached E horizon, because organic matter is concentrated in the near-surface layers, and because decomposition of forest litter generates organic acids that remove the brownish iron oxide coatings. Compare these mature profiles to the changes over time discussed in Sections 2.7 and 2.8.* (Diagrams courtesy of Ray R. Weil)

surface litter into the mineral soil. As a result, the soils under grasslands generally develop a thicker A horizon with a deeper distribution of organic matter than in soils under nearby forests, which characteristically store most of their organic matter in the forest floor (O horizons) and a thin A horizon. The microbial community in a typical grassland soil is dominated by bacteria, while that of the forest soil is dominated by fungi (see Chapter 10 for details). Differences in microbial action affect the aggregation of the mineral particles into stable granules and the rate of nutrient cycling. The light-colored, leached E horizon typically found under the O or A horizon of a forested soil results from the action of organic acids generated mainly by fungi in the acidic forest litter. An E horizon is generally not found in a grassland soil.

Role of Natural Vegetation

Cation Cycling by Trees The ability of natural vegetation to accelerate the release of nutrient elements from minerals by biogeochemical weathering, and to take up these elements from the soil, strongly influences the characteristics of the soils that develop. Soil acidity is especially affected. Differences occur not only between grassland and forest vegetation but also between different species of forest trees. Litter falling from coniferous trees (e.g., pines, firs, spruces, and hemlocks) will recycle only a small amount of calcium, magnesium, and potassium compared to that recycled by litter from some deciduous trees (e.g., yellow poplar, beech, oaks, and maples), which take up and store much larger amounts of these cations (Figure 2.21). Conifer tree roots take up less Ca, Mg, and K from minerals weathering deep in the profile and allow more of these nonacidic cations to be lost by leaching. Therefore, soil acidity often develops more strongly in the surface horizons under coniferous vegetation than under most deciduous trees. Furthermore, the acidic, resinous needles from conifer trees resist decomposition and discourage earthworm populations leading to the accumulation of a thick O horizon with distinctly separate layers of fibric (undecomposed) and sapric (highly decomposed) material. The leaves of deciduous trees generally break down more

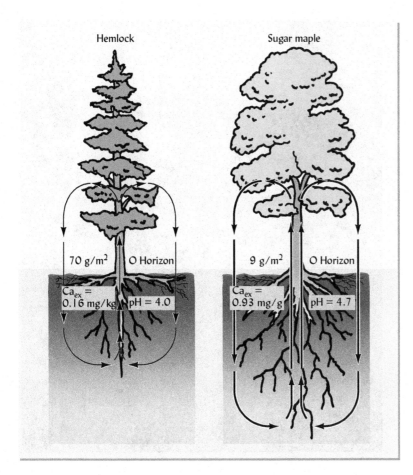

Figure 2.21 Nutrient cycling is an important process by which plants affect the soil in which they grow, altering the course of soil development and the suitability of the soil environment for future generations of vegetation. For example, hemlock (a conifer) and sugar maple (a deciduous hardwood) differ markedly in their ability to accelerate mineral weathering, mobilize nutrient cations, and recycle them to the upper soil horizons. Sugar maple roots are efficient at taking up Ca from soil minerals, and the maple leaves produced contain high concentrations of Ca. When these leaves fall to the ground, they decompose rapidly and release large amounts of Ca^{2+} ions that become adsorbed as exchangeable Ca^{2+} on humus and clay in the O and A horizons. This influx of Ca^{2+} ions may somewhat retard acidification of the surface layers. However, the maple roots' efficient extraction of Ca from minerals in the parent material may accelerate acidification and weathering in deeper soil horizons. In contrast, hemlock needles are Ca-poor, much slower to decompose, and therefore result in a thicker O horizon, greater acidity in the O and upper mineral horizons, but possibly less rapid weathering of minerals in the underlying parent material. [Data for a Connecticut forest reported by Van Breemen and Finzi (1998); for a book with competing theories of how trees affect nutrient cycling and soil formation, see Binkley and Menyailo (2005).]

readily and form a thinner forest floor with less distinction between layers and with more litter mixed into the A horizon. Nutrient cycling by plant roots can even alter the types of clay minerals found in a soil.

Heterogeneity in Rangelands In arid and semiarid rangelands, competition for limited soil water does not permit vegetation to completely cover the soil surface. Scattered shrubs or bunch grasses are interspersed with openings in the plant canopy where the soil is bare or partially covered with plant litter. The widely scattered vegetation alters soil properties in several ways. Plant canopies trap windblown dust that is often relatively rich in silt and clay. Roots scavenge nutrients such as nitrogen, phosphorus, potassium, and sulfur from the interplant areas. These nutrients are then deposited with the leaf litter under the plant canopies. The decaying litter adds organic acids, which lower the soil pH and stimulate mineral weathering. As time goes on, the relatively bare soil areas between plants decline in fertility and may increase in size as they become impoverished and even less inviting for the establishment of plants. Simultaneously, the vegetation creates "islands" of enhanced fertility, thicker A horizons, and often more deeply leached calcium carbonate.

Role of Animals

Animals can have far-reaching and often rapid impacts on the course of soil formation. Large animals such as gophers, moles, and prairie dogs bore into the lower soil horizons, bringing materials to the surface. Their tunnels are often open to the surface, encouraging movement of water and air into the subsurface layers. In localized areas, they enhance mixing of the lower and upper horizons by creating, and later refilling, underground tunnels. For example, dense populations of prairie dogs may completely turn over the upper meter of soil in the course of several thousand years. Old animal burrows in the lower horizons often become filled with soil material from the overlying A horizon, creating profile features known as *crotovinas* (Figure 2.22). In certain situations, animal activity may arrest soil development by accelerating soil erosion.

Figure 2.22 *Abandoned animal burrows in one horizon filled with soil material from another horizon are called* **crotovinas**. *In this Illinois prairie soil, dark, organic-matter-rich material from the A horizon has filled in old prairie dog burrows that extend into the B horizon. The dark circular shapes in the subsoil mark where the pit excavation cuts through these burrows. Scale marked every 10 cm.* (Photo courtesy of Ray R. Weil)

Invertebrate Animals Earthworms, ants, and termites mix the soil as they burrow, significantly affecting soil formation. Earthworms ingest soil particles and organic residues, enhancing the availability of plant nutrients in the material that passes through their bodies. They aerate and stir the soil and enhance soil aggregation. Ants and termites, as they build mounds, also transport soil materials from one horizon to another. In general, the mixing activities of animals, sometimes called **pedoturbation**, tend to undo or counteract the tendency of other soil-forming processes to accentuate the differences among soil horizons. Termites and ants may also retard soil profile development by denuding large areas of soil around their nests, leading to increased loss of soil by erosion. On the other hand, the burrows constructed by ants, termites, and earthworms may significantly increase the amount of water that enters and moves through the soil (see Sections 6.2 and 10.4), thus stimulating biological productivity, weathering, and soil formation.

Human Influences and Urban Soils Human activities widely influence soil formation. For example, it is believed that during the last few thousand years Native Americans regularly set fires to maintain large areas of prairie grasslands in Indiana and Michigan. Similarly, in Africa about 3000 years ago, Bantu-speaking farmers used fire to clear forest land for shifting cultivation and thereby facilitated the abrupt change in the vegetation of central Africa from dense rain forest to a more open tree-scattered grassland (savanna). The sediments that washed off the land into the Atlantic Ocean at that time left evidence of intensified biogeochemical weathering rates resulting from the loss of forest canopy on a wide scale. In more recent times, human destruction of natural vegetation (trees and grass) and subsequent tillage of the soil for crop production have abruptly modified soil formation. Likewise, irrigating a desert soil drastically influences the soil-forming factors, as does adding fertilizer and lime to soils of low fertility. In surface mining and urbanizing areas today, bulldozers may have an effect on soils almost akin to that of the ancient glaciers; they level and mix soil horizons and set the clock of soil formation back to zero.

In other situations, people actually engineer new soils, such as those used in most golf greens and certain athletic fields (see Section 5.6), the cover material used to vegetate and seal completed landfills (see Section 15.9), and the plant media on rooftop gardens. Humans may even reverse the processes of erosion and sedimentation that normally destroy soils and counteracts soil formation (see Section 2.6 and Chapter 14). For example, sediments dredged from the bottom of the bays and rivers can be placed as a thick layer of muddy parent material on land—commonly on barren, highly disturbed land. Within a year, the new parent material will dry out, oxidize, and begin to develop such soil characteristics as granular and prismatic structure (Figure 2.23). It will soon support increasingly lush vegetation, which in turn will facilitate organic matter accumulation and further soil development.

Human impacts on soils are widespread and their potential must be considered in understanding the roles of soils in most landscapes. In urban areas the human influence is pervasive and special consideration will be given to soils in these environments (Section 2.10).

2.6 HOW DOES TOPOGRAPHY AFFECT SOIL FORMATION?

Topography relates to the configuration of the land surface and is described in terms of differences in elevation, slope, and landscape position—in other words, the lay of the land. The topographical setting may either hasten or retard the work of climatic forces (as shown in Box 2.1). Steep slopes generally encourage rapid soil loss by erosion and allow less rainfall to enter the soil before running off. In semiarid regions, the lower effective rainfall on steeper slopes also results in less complete vegetative cover, so there is less plant contribution to soil formation. For all of these reasons, steep slopes prevent the formation of soil from getting very far ahead of soil destruction. Therefore, soils on steep terrain tend to have rather shallow, poorly developed profiles in comparison to soils on nearby, more level sites (Figure 2.24).

In swales and depressions that collect runoff water, the regolith is usually more deeply weathered and soil profile development is more advanced. However, in the lowest landscape positions,

Figure 2.23 *Soil formation taking place in sediments dredged from the bottom of the Potomac River at Washington, DC. The "dredge spoil" was barged several hundred kilometers to a farm in Virginia where the sediments were spread several meters thick on land previously made barren by mining activities. Note the cracks forming during the first year after deposition as the soupy material drains and dries (inset). Within three years, the black iron sulfide coatings begin to oxidize and turn brown in color and cracks begin to define soil structural peds, including large incipient prisms. Within just a few years, good crops could be grown largely because the dredge spoil was initially non-contaminated, non-salty, and low in sulfides. Salty, sulfide-rich marine sediments would have produced a much more problematic soil (see Sections 9.6 and 10.6).* [For more information see *Daniels et al. (2007).*] (Photo courtesy of Ray R. Weil)

Figure 2.24
Topography influences soil properties. (Left) The effect of slope on the profile characteristics and the depth of a soil on which forest trees are the natural vegetation. (Right) The same principle under grassland vegetation. A relatively small change in slope can have a great effect on soil development. See Section 2.9 for explanation of horizon symbols.
(Courtesy of Ray R. Weil)

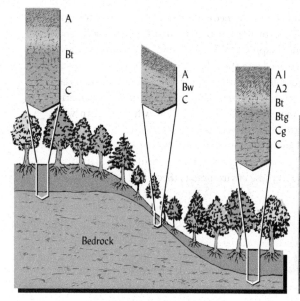

water may saturate the regolith to such a degree that drainage and aeration are restricted. Here the weathering of some minerals and the decomposition of organic matter are retarded, whereas the loss of iron and manganese is accelerated. In such low-lying topography, special profile features characteristic of wetland soils may develop (see Section 7.7 on the soils of wetlands).

Soils that commonly occur together in the landscape in sequence are termed a **catena** (from the Latin meaning *chain*—visualize a length of chain suspended from two adjacent hills with each link in the chain representing a soil). Each member of the catena occupies a characteristic topographic position. Soils in a catena generally exhibit properties that reflect the influence of topography on water movement and drainage (see, e.g., Figure 2.18c). A **toposequence** is a type of catena, in which the differences among the soils result almost entirely from the influence of topography because the soils in the sequence all share the same parent material and have similar conditions regarding climate, vegetation, and time (see, e.g., Figures 2.24 and 2.25).

Figure 2.25 *A soil catena or toposequence in central Zimbabwe. Redder colors indicate better internal drainage. A ditch has been excavated going up the slope, exposing the colors of the B horizons. Inset: B-horizon clods from each soil in the catena.* (Photo courtesy of Ray R. Weil)

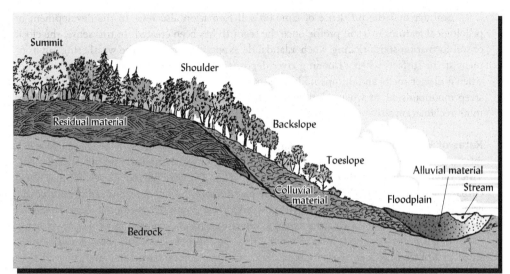

Figure 2.26 *An interaction of topography and parent material as factors of soil formation. The soils on the summit, toeslope, and floodplain in this idealized landscape have formed from residual, colluvial, and alluvial parent materials, respectively.* (Diagram courtesy of Ray R. Weil)

Interaction with Vegetation Topography often interacts with vegetation to influence soil formation. In grassland–forest transition zones, trees are commonly confined to depressions where soil is generally wetter than in upland positions. As would be expected, soil in the depressions is quite different from that in the uplands. The influence of climate in regulating soil development will be much reduced in landscape depressions where water stands for part or all of the year. Some low-lying areas may give rise to peat bogs and, in turn, to organic soils (see Figure 2.17).

Slope Aspect Topography affects the absorbance of solar energy in a given landscape. In the northern hemisphere, south-facing slopes are more perpendicular to the sun's rays and are generally warmer and thereby commonly lower in moisture than their north-facing counterparts. Consequently, soils on the south slopes tend to be lower in organic matter and are not so deeply weathered (see also Figure 7.25). The opposite pertains in the southern hemisphere.

Salt Buildup In arid and semiarid regions, topography influences the buildup of soluble salts. Dissolved salts from surrounding upland soils move on the surface and through the underground water table to the lower-lying areas (see Section 9.11). There they rise to the soil surface as the water evaporates, often accumulating to plant-toxic levels.

Parent Material Interactions Topography can also interact with parent material. For example, in areas of tilted beds of sedimentary rock, the ridges often consist of resistant sandstone, whereas the valleys are underlaid by more weatherable limestone. In many landscapes, topography reflects the distribution of residual, colluvial, and alluvial parent materials, with residual materials on the upper slopes, colluvium covering the lower slopes, and alluvium filling the valley bottom (Figure 2.26).

2.7 HOW DOES TIME AFFECT SOIL FORMATION[5]

Soil-forming processes take time to show their effects. The weathering of rock generally creates only 0.01–0.1 mm of new soil material per year, the rate of soil formation depending on the environmental factors just discussed. However, the rates of rock weathering into soil material tend to be greatest for the thinnest soils on steep slopes and rates as high as 2 mm/year have been measured on steep mountains in high rainfall tropical regions.

---

[5]Work in New Zealand has provided new insights into how extremely fast rock weathering into soil material can keep ahead of the high rate of erosion on mountains in the wet tropics (Larson et al., 2014).

Soil age and the influence of time on soil formation also refer to the development of pedological features in a soil profile once the regolith has been created. In this sense, the clock of soil formation starts ticking when a landslide exposes new rock to the weathering environment at the surface, when a flooding river deposits a new layer of sediment on its floodplain, when a glacier melts and dumps its load of mineral debris, when a landslide tumbles down a steep mountainside, or when a bulldozer cuts and fills a landscape to level a construction or mine-reclamation site.

Rates of Weathering When we speak of a "young" or a "mature" soil, we are not so much referring to the age of the soil in years, as to the degree of weathering and profile development. Time interacts with the other factors of soil formation. For example, on a level site in a warm climate, with much rain falling on permeable parent material rich in reactive minerals, weathering and soil profile differentiation will proceed far more rapidly, yielding a more developed soil profile than on a site with steep slopes and resistant parent material in a cold, dry climate.

In a few instances, soils form so rapidly that the effect of time on the process can be measured in a human life span. For example, dramatic mineralogical, structural, and color changes occur within a few months to a few years when certain sulfide-containing materials are first exposed to air by excavation, wetland drainage, or sediment dredging (see Figure 2.23 and Sections 9.6 and 12.2). Under favorable conditions, organic matter may accumulate to form a darkened A horizon in freshly deposited, fertile alluvium in a mere decade or two. In some cases, incipient B horizons become discernible on humid-region mine spoils in as few as 40 years. Structural alteration and coloring by accumulated iron may form a simple B horizon within a few centuries if the parent material is sandy and the climate is humid. The same degree of B-horizon development would take much longer under conditions less favorable for weathering and leaching. The accumulation of silicate clays and the formation of blocky structure in B horizons usually become noticeable only after several thousand years. Developing in resistant rock, a mature, deeply weathered soil may be hundreds of thousands of years in the making (Figure 2.27).

Example of Soil Genesis over Time Figure 2.27 is worth studying carefully; it illustrates changes that typically take place during soil development on residual rock in a warm, humid

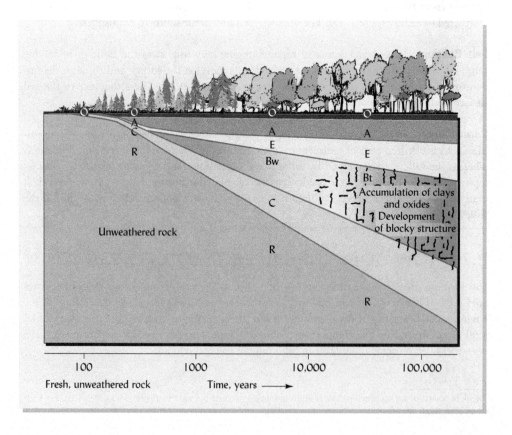

Figure 2.27 *Progressive stages of soil profile development over time for a residual igneous rock in a warm, humid climate that is conducive to forest vegetation. The time scale increases logarithmically from left to right, covering more than 100,000 years. Note that the mature profile (right side of this figure) expresses the full influence of the forest vegetation as illustrated in Figure 2.20. This mature soil might be classified as an Ultisol (see Section 3.14).* (Diagram courtesy of Ray R. Weil)

climate. During the first 100 years, lichens and mosses establish themselves on the bare exposed rock and begin to accelerate its breakdown and the accumulation of dust and organic matter. Within a few hundred years, grasses, shrubs, and stunted trees have taken root in a deepening layer of disintegrated rock and soil, adding greatly to the accumulation of organic materials and to the formation of the A and C horizons. During the next 10,000 years or so, successions of forest trees establish themselves and the activities of a multitude of tiny soil organisms transform the surface plant litter into a distinct O horizon. The A horizon thickens somewhat, becomes darker in color, and develops a stable granular structure. Soon, a bleached zone appears just below the A horizon as soluble weathering products, iron oxides, and clays are moved with water and organic acids percolating down from the litter layer. These transported materials begin to accumulate in a deeper layer, forming a B horizon. The process continues with more silicate clay accumulating and blocky structure forming as the B horizon thickens and becomes more distinct. Eventually, the silicate clays themselves break down, some silica is leached away, and new clays containing less silica form in the B horizon. These clays often become mixed or coated with oxides of iron and aluminum, causing the B horizon to take on a reddish hue. As the entire profile continues to deepen over time, the zone of weathered, unconsolidated rock may become many meters thick.

Chronosequence Most soil-profile features develop so slowly that it is not possible to directly measure time-related changes in their formation. Indirect methods, such as carbon dating or the presence of fossils and human artifacts, must be turned to for evidence about the time required for different aspects of soil development to occur. In a different approach to studying the effects of time on soil development, soil scientists look for a **chronosequence**—a set of soils that share a common climate, parent material, slope, and community of organisms, but differ with regard to the length of time that the materials have been subjected to weathering and soil formation. A chronosequence can sometimes be found among the soils forming on alluvial terraces of differing age. The highest terraces have been exposed for the longest time, those in lower positions have been more recently exposed by the cutting action of the stream, and those on the current floodplain are the youngest, being still subject to periodic additions of new material.

Interaction with Parent Materials Residual parent materials have generally been subject to soil-forming processes for longer periods of time than have materials transported from one site to another. For example, soils forming on glacial materials have generally had far less time to develop than those soils farther south that escaped disturbance by the glaciers. Often the mineralogy and other properties of the "younger" soils in glaciated regions are more similar to those of their parent materials than is the case in the "older" soils of unglaciated regions. Nevertheless, comparisons are complicated because climate, vegetation, and parent material mineralogy also often differ between soils in glaciated and unglaciated areas.

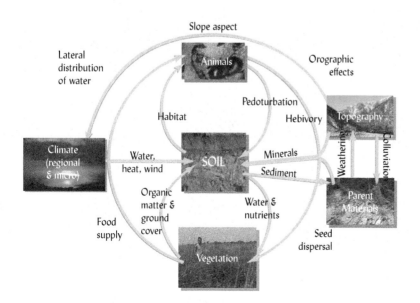

Figure 2.28 Parent material, topography, climate, and organisms (vegetation and animals) do not act independently. Rather, they are linked in many ways and influence the formation of soils in concert. The influence of each factor shown is modified by the length of time *it has been acting, although time as a soil-forming factor is not shown here.* [Adapted from Monger and Bestelmyer (2006)]

The five soil factors of soil formation act simultaneously and interdependently to influence the nature of soils that develop at a site. Figure 2.28 illustrates some of the complex interactions that can help us predict what soil properties are likely to be encountered in a given environment. We will now turn our attention to the *processes* that cause parent materials to change into soils under the influence of these interacting soil-forming factors.

2.8 FOUR BASIC PROCESSES OF SOIL FORMATION[6]

The accumulation of regolith from the breakdown of bedrock or the deposition (by wind, water, ice, etc.) of unconsolidated geologic materials may precede or, more commonly, occur simultaneously with the development of the distinctive horizons of a soil profile. During the formation (**genesis**) of a soil from parent material, the regolith undergoes many profound changes. These changes are brought about by variations of four broad soil-forming processes (Figure 2.29). These four basic processes—often referred to as the soil-forming, or **pedogenic**, processes—help define what distinguishes soils from layers of sediment deposited by geologic processes. They are responsible for soil formation in all kinds of environments—even for **subaqueous soils** that develop in sediments under shallow water (see Box 2.2 and Table 2.4).

Figure 2.29 *A schematic illustration of additions, losses, translocations, and transformations as the fundamental processes driving soil-profile development.* (Diagram courtesy of Ray R. Weil)

---

[6]For the classic presentation of the processes of soil formation, see Simonson (1959). In addition, detailed discussion of these basic processes and their specific manifestations can be found in Birkeland (1999), Fanning and Fanning (1989), Buol et al. (2011), and Schaetzl and Thompson (2015).

> **BOX 2.2**
> ## SUBAQUEOUS SOILS—UNDERWATER PEDOGENESIS
>
> Soil scientists are beginning to study soil profiles that develop under shallow water. In particular, the concept of soils has recently expanded to include the natural bodies that comprise sediment "landscapes" lying under up to 2 m of water in ponds, lakes, and coastal estuaries and bays. These soils differ from wetlands that support emergent hydrophytic plants; subaqueous soils host only submerged aquatic vegetation (SAV—mainly aquatic grasses that are rooted but do not emerge above the water). Working from floating drill rigs, soil scientists have mapped distinct soil bodies in which the four basic soil-forming processes have created distinguishable horizons (Table 2.4).
>
> Note the striking similarities to the soil-forming processes pictured in Figure 2.29. Understanding the different soils in the underwater landscape may help estuary managers inventory and conserve coastal marine resources in much the same way that soil information helps managers of dry land protect and optimize terrestrial resources.
>
> **Table 2.4**
> **SOIL-FORMING PROCESSES OBSERVED TO FORM SUBAQUEOUS SOILS IN SEDIMENTS WITH SUBMERGED VEGETATION UNDER SHALLOW WATER IN A COASTAL ESTUARY**
>
> | Process | Examples observed in subaqueous soils |
> |---|---|
> | Additions | Calcium carbonates added from clam and oyster shells. |
> | | Organic matter added from roots and leaves of aquatic vegetation. |
> | | Mineral sediments washed in by currents. |
> | Losses | Carbon lost by decomposition of organic matter. |
> | | Surface mineral material washed away by storm currents. |
> | Translocations | Oxygen moved into the profile by diffusion from seawater. |
> | | Oxygen and organic material mixed into upper 15 cm by burrowing clams and tubeworms. |
> | Transformations | Formation of humus-like organic matter by microbial processing of plant residues. |
> | | Formation of the mineral pyrite from sulfate in the seawater and iron in the sediments. |
>
> Summarized from Demas and Rabenhorst (2001). For a review of subaqueous soils, see Rabenhorst and Stolt (2012). For updates on subaqueous soil mapping in the northeastern United States, see Turenne (2014)

Transformations occur when soil constituents are chemically or physically modified or destroyed and others are synthesized from the precursor materials. Many transformations involve weathering of primary minerals, disintegrating and altering some to form various kinds of silicate clays. As other primary minerals decompose, the decomposition products recombine into new minerals that include additional types of silicate clays and hydrous oxides of iron and aluminum (see Figure 2.4). Other important transformations involve the decomposition of plant root and shoot litter to form soil organic matter. Still other transformations change the size (e.g., physical weathering to smaller particles) or arrangement (e.g., aggregation) of mineral particles.

Translocations involve the movement of inorganic and organic materials laterally within a horizon or vertically from one horizon up or down to another. Water, either percolating down with gravity or rising up by capillary action (see Section 5.2), is the most common translocation agent. The materials moved within the profile include dispersed fine clay particles, dissolved salts, and dissolved organic substances. Translocations of materials by soil organisms also have a major influence on soil genesis. Important examples include incorporation of surface organic litter into the A and B horizons by certain earthworms, transport of B and C horizon material to the surface by mound-building termites, and the widespread burrowing actions of rodents.

Inputs of materials to the developing soil profile from outside sources are considered **additions**. A very common example is the input of organic matter from fallen plant leaves and sloughed-off roots (the carbon having originated in the atmosphere). Another ubiquitous addition is dust particles falling on the surface of the soil (wind may have blown these particles from a source just a few meters away or across an ocean). Still another example, common in arid regions, is the addition of salts or silica dissolved in the groundwater and deposited near or at the soil surface when the rising water evaporates. Animals and people can also contribute additions, such as manure and fertilizers.

Materials are lost from the soil profile by leaching to groundwater, erosion of surface materials, and volatilization of gases. Evaporation and plant use cause **losses** of water. Leaching and drainage cause the loss of water, dissolved substances such as salts or silica weathered from parent minerals, or organic acids produced by microorganisms or plant roots. Fire and biochemical reactions cause the loss of carbon, nitrogen, and sulfur as gases (Chapters 11 and 12). Erosion, a major loss agent, often removes the finer particles (humus, clay, and silt), leaving the surface horizon relatively sandier and less rich in organic matter than before. Grazing by animals or harvest by people can remove large amounts of both organic matter and nutrient elements.

These processes of soil genesis, operating under the influence of the environmental factors discussed previously, give us a logical framework for understanding the relationships between particular soils and the landscapes and ecosystems in which they function. In analyzing these relationships for a given site, ask yourself: What are the materials being added to this soil? What transformations and translocations are taking place in this profile? What materials are being removed? And how have the climate, organisms, topography, and parent material at this site affected these processes over time?

Soil-Forming Processes in Action: A Simplified Example

Consider the changes that might take place as a soil develops from a thick layer of relatively uniform calcareous loess parent material in a climate conducive to grass vegetation (Figure 2.30). Although some physical weathering and leaching of carbonates and salts may be necessary to allow plants to grow in certain parent materials, soil formation really gets started when plants become established and begin to provide *additions* of litter and root residues on and in the surface layers of the parent material. The plant residues are *transformed* by soil organisms into soil organic matter, including various organic acids. The accumulation of organic matter enhances the capacity of the soil to hold water and nutrients, providing positive feedback for accelerated plant growth and further organic buildup. Earthworms, ants, termites, and a host of smaller animals come to live in the soil and feed on the newly accumulating organic resources. In so doing, they accelerate the organic *transformations*, as well as promote the *translocation* of plant residues, loosening the mineral material as they burrow into the soil.

A-Horizon Development The resulting organic–mineral mixture near the soil surface soon becomes the A horizon. It is darker in color and its chemical and physical properties differ from those of the original parent material. Individual soil particles in this horizon commonly clump together under the influence of organic substances to form granules, differentiating this layer from the deeper layers and from the original parent material. On sloping land, erosion may remove some materials from the newly forming upper horizon, retarding, somewhat, the progress of horizon development.

Formation of B and C Horizons Carbonic and other organic acids are carried by percolating waters into the soil, where they stimulate weathering reactions. The acid-charged percolating water dissolves various minerals (a *transformation*) and leaches the soluble products (a *translocation*) from upper to lower horizons, where they may precipitate. This combination of transformation and translocation creates (eluvial) zones of depletion in the upper layers and (illuvial) zones of accumulation in the lower layers. The dissolved substances include both positively charged ions (cations; e.g., Ca^{2+}) and negatively charged ions (anions; e.g., CO_3^{2-} and SO_4^{2-}) released from the breakdown of minerals and organic matter. In semiarid and arid regions, precipitation of these ions produces horizons enriched in calcite ($CaCO_3$) or gypsum ($CaSO_4 \cdot 2H_2O$), designated as a Bk layer in Figure 2.30.

As the process proceeds, the leached surface layer thickens, and the zone of Ca accumulation is moved downward to the maximum depth of water penetration. Where rainfall is

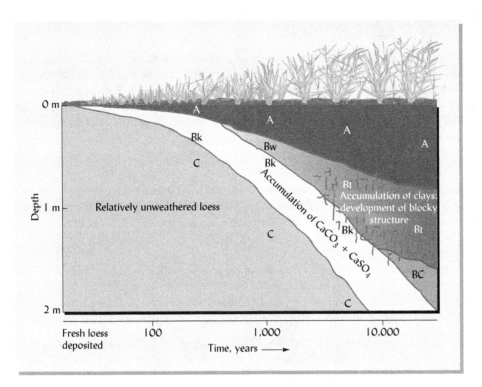

Figure 2.30 *Development of a hypothetical soil in about 2 m of uniform calcareous loess deposits where the warm subhumid climate is conducive to tall grass prairie vegetation. The time scale increases logarithmically from left to right, covering about 20,000 years. In the initial stages, rainwater, charged with organic acids from microbial respiration, dissolves carbonates from the loess and moves them downward to a zone of accumulation (Bk horizon). As this happens, plant roots take hold in the upper layer and add the organic matter necessary to create an A horizon. Ants, beetles, earthworms, and a host of smaller creatures take up residence and actively mix in surface litter and speed the release of nutrients from the minerals and plant residues. Over time the carbonate concentration zone moves deeper, the A horizon thickens, and noncalcareous B horizons develop as changes in color and structure occur in the weathering loess above the zone of carbonate accumulation. Eventually, silicate clay accumulates in the B horizon (giving it the designation Bt), both by stationary weathering of primary minerals and by traveling there with water percolating from the upper horizons. Compare the stages of development and rates of change to those illustrated in Figure 2.27 for a different parent material, vegetation, and climate. Note that the mature profile (right side of this figure) expresses the full influence of the grassland vegetation as illustrated in Figure 2.20. This mature soil would be classified as a Mollisol (see Section 3.12).* (Diagram courtesy of Ray R. Weil)

great enough to cause significant drainage to the groundwater, some of the dissolved materials may be completely removed from the developing soil profile (*losses*), and the zone of accumulation may move below the reach of plant roots or be dissipated altogether. On the other hand, deep-growing plant roots may intercept some of these soluble weathering products and return them, through leaf- and litter-fall, to the soil surface, thus retarding somewhat the processes of acid weathering and horizon differentiation.

The weathering of primary minerals into clay minerals becomes evident only long after the dissolution and movement of Ca is well underway. The newly formed clay minerals may accumulate where they are formed, or they may move downward and accumulate deeper in the profile. As clay is removed from one layer and accumulates in another, adjacent layers become more distinct from each other, and a Bt horizon (one enriched in silicate clay) is formed. When the accumulated clay in the Bt horizon periodically dries out and cracks, block-like or prismatic units of soil structure begin to develop (see Figure 2.23 and Section 4.4). As the soil matures, the various horizons within the profile generally become more numerous and more distinctly different from each other.

2.9 THE SOIL PROFILE

Each location on the land surface has experienced a particular combination of influences from the five soil-forming factors, causing a different set of layers (horizons) to form in each part of the landscape, thus slowly giving rise to the natural bodies we call **soils**. Each soil is characterized by a given sequence of these horizons. A vertical exposure of this sequence is termed a **soil profile**. We will now consider the major horizons making up **soil profiles** and the terminology used to describe them.

The Master Horizons and Layers[7]

Six **master** soil horizons are commonly recognized and are designated using the capital letters O, A, E, B, C, and R (Figure 2.31). Subhorizons may occur within a master horizon and these are designated by lowercase letters (suffix symbols) following the capital master horizon letter (e.g., Bt, Ap, or Oi).

O Horizons   The O horizons are composed of organic layers that generally form above the mineral soil or occur in an organic soil profile. They derive from dead plant and animal residues. Generally absent in grassland regions, O horizons usually occur in forested areas and are commonly referred to as the **forest floor** (see Figure 2.32). Often three subhorizons within the O horizon can be distinguished.

> The **Oi horizon** is an organic horizon of *fibric* materials—recognizable plant and animal parts (leaves, twigs, and needles), only slightly decomposed. (The Oi horizon is referred to as the *litter* or *L layer* by some foresters.)

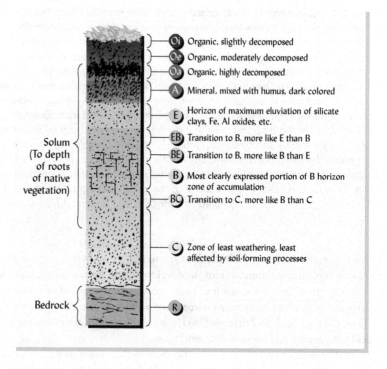

Figure 2.31 *Hypothetical mineral soil profile showing the major horizons that may be present in a well-drained soil in the temperate humid region. Any particular profile may exhibit only some of these horizons, and the relative depths vary. In addition, however, a soil profile may exhibit more detailed subhorizons than indicated here. The solum includes the A, E, and B horizons plus some cemented layers of the C horizon.*

---

[7]In addition to the six master horizons described in this section, L (limnic—Greek *limne̅*, marsh), W (water), V (vesicular), and M (manufactured root-limiting subsoil layers) are also considered to be master horizons or layers. The L horizon occurs only in certain organic soils and includes layers of organic and mineral materials deposited in water or by aquatic organisms (e.g., diatomaceous earth, sedimentary peat, and marl). Layers of water (frozen or liquid) found within (not overlying) certain soil profiles are designated as W master horizons. The V horizons are layers dominated by non-connected, bubble-like pores that occur in or near the surface of desert soils. The M horizons occur mainly in urban soils in which manufactured root-impermeable materials, such as geotextile or asphalt, form a horizontal, near-continuous layer.

The **Oe horizon** consists of **hem**ic materials—finely fragmented residues intermediately decomposed, but still with much fiber evident when rubbed between the fingers. (This layer corresponds to the *fermentation* or *F layer* described by some foresters.)

The **Oa horizon** contains **sa**pric materials—highly decomposed, smooth, amorphous residues that do not retain much fiber or recognizable tissue structures. (This is the *humidified* or *H layer* designated by some foresters.)

A Horizons The topmost mineral horizons, designated A horizons, generally contain enough partially decomposed organic matter to give the soil a color darker than that of the lower horizons (Figure 2.33, *left*). In medium-textured soils, the A horizons are often coarser in texture, having lost some of the finer materials by translocation to lower horizons and by erosion.

E Horizons These are zones of maximum leaching or ***eluviation*** (from Latin *ex* or *e*, out, and *lavere*, to wash) of clay, iron, and aluminum oxides, which leaves a concentration of resistant minerals, such as quartz, in the sand and silt sizes. An E horizon is usually found underneath the A horizon and is generally lighter in color than both the A horizon above it and the horizon below. Such E horizons are quite common in soils developed under forests where they occasionally occur directly under the O horizons (Figure 2.33, *right*). They rarely occur in soils developed under grassland.

B Horizons B horizons form below an O, A, or E horizon and have undergone sufficient changes during soil genesis so that the original parent material structure is no longer discernible. In many B horizons, materials have accumulated, typically by washing *in* from the horizons above, a process termed **illuviation** (from the Latin *il*, in, and *lavere*, to wash). In humid regions, B horizons are the layers of maximum accumulation of materials such as iron and aluminum oxides and silicate clays (Bt horizons), some of which may have illuviated from upper horizons and some of which may have formed in place. In arid and semiarid regions, calcium carbonate or calcium sulfate may accumulate in the B horizon (giving Bk and By horizons, respectively).

Figure 2.32 *The upper horizons of a soil formed under temperate forest vegetation. Note the accumulation of decaying organic material (O horizons) above a thin organic-enriched mineral A horizon, that in turn sit above a redder-colored B horizon. An E horizon, as found in many forested soils, is absent in this profile. Scale in centimeters.* (Courtesy of Ray R. Weil)

Figure 2.33 *Examples of master soil horizons. (Left) A young sandy soil under grassy vegetation exhibits a marked accumulation of organic matter in the A horizon. (Right) Another sandy soil under forest vegetation has an O horizon of organic material directly over a bleached E horizon. Scales in centimeters.* (Photo courtesy of Ray R. Weil)

The B horizons are sometimes referred to loosely as the subsoil, although this is an imprecise term. In soils with shallow A horizons, part of the B horizon may become incorporated into the plow layer and thus become part of the topsoil. In other soils with deep A horizons, the plow layer or topsoil may include only the upper part of the A horizons, and the subsoil would include the lower part of the A horizon along with the B horizon. This emphasizes the need to differentiate between colloquial terms (*topsoil* and *subsoil*) and technical terms used by soil scientists to describe the soil profile.

C Horizon The C horizon is the unconsolidated material underlying the solum (A and B horizons). It may or may not be the same as the parent material from which the solum formed. The C horizon is below the zones of greatest biological activity and has not been sufficiently altered by soil genesis to qualify as a B horizon. In dry regions, carbonates and gypsum may be concentrated in the C horizon. While loose enough to be dug with a shovel, C horizon material that retains some of the structural features of the parent rock or geologic deposits from which it formed is termed **saprolite** (see, e.g., the lower third of the profile shown in Figure 2.34). Its upper layers may in time become a part of the solum as weathering and erosion continue.

R Layers These are consolidated rock, with little evidence of weathering (bottom of Figure 2.34).

Subdivisions Within Master Horizons

Often distinctive layers exist *within* a given master horizon, and these are indicated by a numeral *following* the letter designation. For example, if three different combinations of structure and colors can be seen in the B horizon, then the profile may include a sequence such as B1–B2–B3 (see Figure 2.34).

If two different geologic parent materials (e.g., loess over glacial till) are present within the soil profile, the numeral 2 is placed *in front* of the master horizon symbols for horizons developed in the second layer of parent material. For example, a soil would have a sequence of

Figure 2.34 *A residual soil formed from gneiss parent rock in the Uluguru Mountains of Tanzania. Note the three B horizons, the variable depth of rock weathering, and the rock-like features retained in the Cr-horizon.* (Photo courtesy of Ray R. Weil)

horizons designated O–A–B–2C if the C horizon developed in glacial till whereas the upper horizons developed in loess. Such a change in the type of parent material with depth is referred to as **lithologic discontinuity**.

Where a layer of mineral organic soil material was transported by *humans* (usually using machinery) from a source outside the pedon, the "caret" symbol (^) is inserted before the master horizon designation. For example, suppose a landscaping contractor hauls in and spreads a layer of sandy fill material over an existing soil in order to level a site. The resulting soil might eventually (after enough organic matter had accumulated to form an A horizon) have the following sequence of horizons: ^A-^C-2Ab-2Btb, where the first two horizons formed in the human-transported fill (hence the ^ prefixes), and the last two horizons were part of the underlying, now buried soil (hence the lower case "b" designations).

Transition Horizons

Transitional layers between the master horizons (O, A, E, B, and C) may be dominated by properties of one horizon but also have characteristics of another. The two applicable capital letters are used to designate the transition horizons (e.g., AE, EB, BE, and BC), the dominant horizon being listed before the subordinate one (e.g., Figures 2.31 and 2.34). Letter combinations with a slash such as E/B, are used to designate transition horizons where distinct parts of the horizon have properties of E while other parts have properties of B.

Subhorizon Distinctions

Since the capital letter designates the nature of a master horizon in only a very general way, specific horizon characteristics may be indicated by a lowercase letter following the master horizon designation. For example, three types of O horizons (Oi, Oe, and Oa) are indicated in the profile shown in Figure 2.32, which presents a commonly encountered sequence of horizons. Other subhorizon distinctions include special physical properties and the accumulation of particular materials such as clays and salts. A list of the recognized subhorizon letter designations and their meanings is given in Table 2.5. We suggest that you mark this table for future reference and study it now to get an idea of the distinctive soil properties that can be indicated by horizon designations. As mentioned previously, a Bt horizon is a B horizon characterized by clay accumulation (t from the German *ton*, meaning clay); likewise, in a Bk horizon, carbonates (k) have accumulated. We have already used the suffixes i, a, and e to distinguish different types of O horizons.

Table 2.5
SOME COMMON SUBHORIZON DISTINCTIONS WITHIN MASTER HORIZONS[a]

Lower case symbol	Distinction	Lower case symbol	Distinction
a	Highly decomposed organic matter	n	Accumulation of sodium
b	Buried soil horizon	o	Accumulation of iron and aluminum oxides
c	Concretions or nodules	p	Plowing or other disturbance
co	Coprogenous earth	q	Accumulation of silica
d	Dense unconsolidated materials	r	Weathered or soft bedrock
di	Diatomaceous earth	s	Illuvial organic matter and iron, aluminum oxides
e	Intermediately decomposed organic matter	se	Presence of sulfides
f	Frozen soil	ss	Slickensides (shiny clay wedges)
ff	Dry permafrost	t	Accumulation of silicate clays
g	Strong gleying (mottling)	u	Human-manufactured artifacts
h	Illuvial accumulation of organic matter	v	Plinthite (high iron, red material)
i	Slightly decomposed organic matter	w	Distinctive color or structure without clay accumulation
j	Jarosite (yellow sulfate mineral)	x	Fragipan (high bulk density, brittle)
jj	Cryoturbation (frost churning)	y	Accumulation of gypsum
k	Accumulation of carbonates	yy	Gypsum >50% of soil by mass
kk	Engulfment of carbonates, >50% of soil by mass	z	Accumulation of soluble salts
m	Cementation or induration		

[a]Based on information in Soil Survey Staff (2014)

The letter subhorizon designations are considered before any numerical subdivisions are applied. Thus the subdivision numbers start over with "1" if there is a change in the horizon letter designations; for example, Bt1-Bt2-Btg1-Btg2 is a correctly labeled sequence, but Bt1-Bt2-Btg3-Btg4 is *not* correct. The significance of the subhorizon horizon designations will be discussed further in the next chapter.

Horizons in a Given Profile

It is not likely that the profile of any one soil will show all of the horizons that collectively are shown in Figure 2.31. The ones most commonly found in well-drained soils are Oi and Oe (or Oa) if the land is forested, A or E (or both, depending on circumstances), Bt or Bw, and C. Conditions of soil genesis will determine which horizons are present and their clarity of expression.

When a virgin (never-cultivated) soil is plowed and cultivated for the first time, the upper 15 to 20 cm becomes the plow layer or Ap horizon. Cultivation obliterates the original layered condition of the upper portion of the profile, and the Ap horizon becomes more or less homogeneous. In some soils, the A and E horizons are deeper than the plow layer (Figure 2.33*left*). In other cases where the upper horizons are quite thin, the plow line is just at the top of, or even down in, the B horizon.

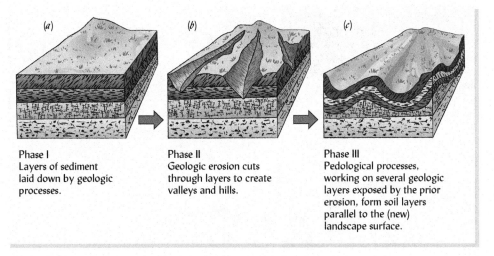

Figure 2.35 *Distinguishing between geologic layers (potential parent materials) and pedogenic layers (soil horizons). (a) Layers of sediment laid down by geologic processes, possibly forming different sedimentary rock layers. (b) The exposed land surface erodes by the action of water (or in dry environments, by wind) forming deep gullies that eventually become swales and valleys. (c) Once the landscape stabilizes, soils begin to form in the layers of geologic materials exposed by erosion. The pedogenic (soil-forming) processes cause the geologic material to differentiate into layers that parallel the new land surfaces.* (Diagram courtesy of Ray R. Weil)

Soil Genesis in Nature

Not every contrasting layer of material found in soil profiles is a **genetic horizon** that developed as a result of the processes of soil genesis such as those just described. The parent materials from which many soils develop contained contrasting layers *before* soil genesis started. For example, such parent materials as glacial outwash, marine deposits, or recent alluvium may consist of various layers of fine and coarse particles laid down by separate episodes of sedimentation. Consequently, in characterizing soils, we must recognize not only the genetic horizons and properties that come into being during soil genesis, but also those layers or properties that may have been inherited from the parent material. Figure 2.35 illustrates how soil horizons that form during soil genesis and follow the surface topography of the land can develop in layers of contrasting materials previously laid down by geologic processes.

2.10 URBAN SOILS[8]

Urban land now accounts for nearly 3% of the Earth's land surface. For the first time in human evolution, more than half of all people live in cities. Therefore, although human influences on soils long have had widespread importance, urban soils are taking on new significance. Soils occurring in urban landscapes will be included in discussions throughout the text, but here we will give special consideration to some of the issues relevant to the formation and morphology of such soils.

Pedological Properties Unique to Urban Soils

Urban soils are often heavily influenced by human bioturbation—that is, the soil materials are reworked by human activities ranging from shoveling to bulldozing as the land surface is excavated and graded. Commonly, materials are trucked in from nearby excavations or from off-site locations and spread as a fill layer (typically an ^Au horizon) on top of the existing soil. This process also creates various buried horizons (typically designated 2Ab, 2Btb, etc.).

---

[8] For a survey of soil properties in urban areas, see Pouyat et al. (2010). For an easy to read introduction to the pedology and processes of urban soils using vacant land in the city of Detroit as an example, see Makarushka (2012). For an accessible review of a broad range of urban soils issues, see Bartens, et al. (2012).

Figure 2.36 *Some features of highly human-modified urban soils. (Left) A soil in a landscaped area showing a thin layer of topsoil (^Ap) spread over a thin layer of sediment from recent erosion (2C) over a thick buried A horizon spread in an earlier grading operation. An "M" master horizon (see footnote 7) is present at 55–75 cm where a thick layer of impermeable asphalt was buried. (Right) An engineered urban soil under a layer of paving stones set in sand. The M horizon in this profile consists of a layer of geotextile that is impermeable to roots and was installed to prevent the sand layer from shifting down into the layer of gravel and artifacts (^2Cu) below it. This is underlain by a thick sand layer (^3C) for drainage and then a layer of compacted clay and rubble fill (^4Cu). The buried remnants of the original soil C horizon can be seen at 150 cm depth.* (Courtesy of Ray R. Weil)

In addition, urban profiles may include layers of any of the following: geotextile fabric, sand or gravel, clay compacted to engineering standards, building rubble and artifacts such as nails, metal and plastic scrapes, broken glass, bricks and concrete blocks, and chunks of gypsum wallboard and paving asphalt (Figure 2.36 *left*). In residential areas, "topsoil" may be brought in from other areas and spread, usually in a very thin layer, on top of unrelated materials comprising the deeper soil layers. For many urban soils, the various human-influenced processes occurred sporadically during different periods of time. This type of history often results in several buried horizons (Figure 2.36 *right*). This profile also exhibits many direct and indirect influences of humans as part of the *organisms* factor of soil formation—influences that are most strongly expressed in urban soils.

In other cases, these processes were carefully orchestrated to "engineer" a new soil for particular purposes. For example, the final cover for solid waste landfills is a special case of an engineered urban soil profile that includes an impermeable layer of compact clay covered by a drainage layer of coarse sand and topped with a layer of loamy material to support protective vegetation (see Section 15.9).

Physical Properties Unique to Urban Soils

Urban soils can become highly compacted because of the weight of structures and pavement and human or vehicular traffic both during and after the placement of soil materials. Another unique feature of urban soils is the abrupt boundaries between materials of highly contrasting texture and porosity, such as layers of gravel buried beneath layers of clay. While these properties may be desirable from the point of view of engineering function, they can create major problems for water percolation and tree root growth.

Chemical Properties Unique to Urban Soils

The chemical properties of soils in urban areas may or may not exhibit much of a relationship to the properties expected from local climatic or parent material conditions. For instance, due to the influence of concrete and cement, some urban soils are surprisingly alkaline in high rainfall areas where acid soils are expected. Soils that occur beneath sealed pavements may receive little percolating water even in high rainfall environments, and therefore may be less thoroughly leached of soluble constituents than might be expected for the local climate.

Figure 2.37 *Two highly human-modified but nonengineered urban soils. (Left) The upper horizons of a soil in a residential area (spatula handle is 10 cm long). A thin layer of "topsoil" was spread on top of very acid, infertile clayey B and C horizon material left after the land surface was graded by heavy machinery. The lawn grasses grew poorly because the soil favorable for rooting is only 2–5 cm thick. (Right) A soil profile formed in construction debris mixed with excavated soil materials used as fill. This is a LaGuardia soil (scale in inches, 1 inch = 2.54 cm) mapped as part of the soil survey of Bronx River watershed, New York City. About 35 to 75% of the volume of the B and C horizons consists of coarse fragments—chunks of concrete, brick, glass, and metal can be seen. The B horizon had developed enough change in soil structure to be designated a Bw horizon.* (Left photo courtesy of Ray R. Weil; right photo courtesy of the Natural Resource Conservation Service)

Urban soils are commonly contaminated with various anthropogenic substances; lead, other toxic metals, organic petroleum-based contaminants, and deicing salts are commonly of concern. In some cases, rusting iron nails or dissolving concrete reacts to reduce the mobility and toxicity of such contaminants as lead. Contamination with nutrients and pathogens from biological wastes is also prevalent in urban soils, either from failed or under-engineered sewage disposal systems, or from stray animals, rats, and from dogs whose owners fail to clean up after them.

Biological Properties Unique to Urban Soils

Urban soils differ from other soils because they are less likely to be covered with a vegetative community reflecting the local climate and parent material. Except in park areas allowed to maintain remnants of natural vegetation communities, most vegetated urban soils support exotic species planted by humans and maintained against natural succession processes. Many urban soils are kept nearly bare by erosion or trampling, whereas others are covered by impermeable pavements. Thus plant inputs of carbon that normally "feed" the soil ecosystem are likely to be highly variable or even absent. The compaction and low organic matter levels may make the soils very inhospitable to both plant roots and microbial communities. In many of the vegetated areas, only a very thin layer of suitable rooting medium may have been spread on top of the chemically and physically inhospitable materials (Figure 2.37). Even the atmosphere in urban areas may impact plants and soil microorganisms in both positive and negative ways. High levels of ozone and other pollutants reduce plant growth and subsequent carbon inputs into the soil, while high levels of carbon dioxide accompanied by warmer temperatures due to "the urban heat island effect" may increase both plant growth and microbial decomposition rates compared to what would occur in nonurban situations.

2.11 CONCLUSION

The parent materials from which soils develop vary widely around the world and from one location to another only a few meters apart. Knowledge of these materials, their sources or origins, mechanisms for their weathering, and means of transport and deposition are essential to understanding soil genesis.

Soil formation is stimulated by *climate* and living *organisms* acting on *parent materials* over periods of *time* and under the modifying influence of *topography*. These five major factors of soil formation determine the kinds of soil that will develop at a given site. When all of these factors are the same at two locations, the kind of soil at these locations should be the same. Although all five factors interact and work together, at times one factor's influence dominates over the others. For example, climate probably dominates in Arctic soils, whereas organisms (namely humans) may dominate in urban soils.

Soil genesis starts when layers or horizons not present in the parent material begin to appear in the soil profile. Organic matter accumulation in the upper horizons, the downward movement of soluble ions, the synthesis and downward movement of clays, and the development of specific soil particle groupings (structure) in both the upper and lower horizons are signs that the process of soil formation is under way. As we have learned, soil bodies are dynamic in nature. Their genetic horizons continue to develop and change. Consequently, in some soils horizon differentiation has only begun, whereas in others it is well advanced.

The four general processes of soil formation (gains, losses, transformations, and translocations) and the five major factors influencing these processes provide us with an invaluable logical framework in site selection and in predicting the nature of soil bodies likely to be found on a particular site. Conversely, analysis of the horizon properties of a soil profile can tell us much about the nature of the climatic, biological, and geological conditions (past and present) at the site.

Characterization of the horizons in the profile leads to the identity of a soil individual, which is then subject to classification—the topic of our next chapter.

STUDY QUESTIONS

1. What is meant by the statement, *weathering combines the processes of destruction and synthesis*? Give an example of these two processes in the weathering of a primary mineral.
2. How is water involved in the main types of chemical weathering reactions?
3. Explain the weathering significance of the ratio of silicon to aluminum in soil minerals.
4. Give an example of how parent material may vary across large geographic regions on one hand but may also vary within a small parcel of land on the other.
5. Name the five factors affecting soil formation. With regard to each of these factors of soil formation, compare a forested Rocky Mountain slope to the semiarid grassland plains far below.
6. How do *colluvium, glacial till,* and *alluvium* differ in appearance and agency of transport?
7. What is *loess*, and what are some of its properties as a parent material?
8. Give two specific examples for each of the four broad processes of soil formation.
9. Assuming a level area of granite rock was the parent material in both cases, describe in general terms how you would expect two soil profiles to differ, one in a warm, semiarid grassland and the other in a cool, humid pine forest.
10. For the two soils described in question 5, make a profile sketch using master horizon symbols and subhorizon suffixes to show the approximate depths, sequence, and nature of the horizons you would expect to find in each soil.
11. Visualize a slope in the landscape near where you live. Discuss how specific soil properties (such as colors, horizon thickness, types of horizons present, etc.) would likely change along the toposequence of soils on this slope.
12. Imagine a soil pit dug in a vacant brown field in an old city industrial area. The horizons you observe include, from the soil surface down: (1) 19 cm of dark-colored organic enriched loamy material with nails and broken glass embedded in it; (2) 32 cm of sandy loam with bits of red brick, some broken glass, some rusting

metal pieces, and rusty red and yellow stains on most of the sand grain surfaces giving the layer a yellow-red color; (3) 15cm of dark brown silt loam material rich in organic matter with an abrupt lower boundary; (4) 50 cm of light brown silty clay loam material with clay coatings in the cracks between blocky chunks of soil material. Draw this profile in a simple diagram and label the four layers with master horizon and subhorizon letters and other appropriate symbols to indicate your best interpretation of soil formation at this site.

REFERENCES

Bartens, J., N. Basta, S. Brown, C. Cogger, B. Dvorak, B. Faucette, P. Groffman, G. Hettairachchi, K. McIvor, R. Pouyat, G. Toor, and J. Urban. 2012. "Soils in the city." *Crop Soils Agronomy News* 57:4–13.

Binkley, D., and O. Menyailo (eds.). 2005. *Tree Species Effects on Soils: Implications for Global Change.* Kluwer Academic Publishers, Dordrecht.

Birkeland, P. W. 1999. *Soils and Geomorphology.* 3rd ed. Oxford University Press, New York.

Buol, S. W., R. J. Southard, R. C. Graham, and P. A. McDaniel. 2011. *Soil Genesis and Classification.* 6th ed. Wiley-Blackwell, New York, p. 560.

Craft, C. B., and C. J. Richardson. 1998. "Recent and long-term soil accretion and nutrient accumulation in the everglades." *Soil Science Society of America Journal* 62:834–843.

Daniels, W. L., G. R. Whittecar, and C. H. Carter, III. 2007. "Conversion of potomac river dredge sediments to productive agricultural soils." Thirty Years of SMCRA and Beyond. American Society of Mining and Reclamation, Gillette, WY. http://www.asmr.us/Publications/Conference%20Proceedings/2007/0183-Daniels-VA.pdf

Demas, G. P., and M. C. Rabenhorst. 2001. "Factors of subaqueous soil formation: A system of quantitative pedology for submerged environments." *Geoderma* 102:189–204.

Fanning, D. S., and C. B. Fanning. 1989. *Soil: Morphology, Genesis, and Classification.* John Wiley and Sons, New York.

Hilgard, E. W. 1921. *Soils: Their Formation, Properties, Composition, and Plant Growth in the Humid and Arid Regions.* Macmillan, London.

Jenny, H. 1941. *Factors of Soil Formation: A System of Quantitative Pedology.* Originally published by McGraw-Hill; Dover, Mineola, NY.

Jenny, H. 1980. *The Soil Resource—Origins and Behavior.* Ecological Studies, Vol. 37. Springer-Verlag, New York.

Larsen, I. J., P. C. Almond, A. Eger, J. O. Stone, D. R. Montgomery, and B. Malcolm. 2014. "Rapid soil production and weathering in the southern alps, New Zealand." *Science* 343:637–640.

Logan, W. B. 1995. *Dirt: The Ecstatic Skin of the Earth.* Riverhead Books, New York.

Makarushka, M. 2012. "Detroit's vacant lots provide 'natural laboratory' for studying soil processes," *Soil Horizons* 53:https://www.soils.org/publications/sh/articles/53/2/6.

Monger, H. C., and B. T. Bestelmeyer. 2006. "The soil-geomorphic template and biotic change in arid and semi-arid ecosystems." *Journal of Arid Environments* 65:207–218.

Pouyat, R. V., K. Szlavecz, I. D. Yesilonis, P. M. Groffman, and K. Schwarz. 2010. "Chemical, physical, and biological characteristics of urban soils." In J. Aitkenhead-Peterson and A. Volder (eds.). *Urban Ecosystem Ecology,* Vol. 55. American Society Agronomy, Crop Science Society of America, Soil Science Society of America Journal, Madison, WI, pp. 119–152.

Rabenhorst, M. C., and M. H. Stolt. 2012. "Subaqueous soils: Pedogenesis, mapping and applications." In H. Lin (ed.). *Hydropedolgy: Synergistic Integration of Soil Science and Hydrology.* Academic Press, Amsterdam, pp. 173–204.

Schaetzl, R. J., and M. L. Thompson. 2015. *Soils: Genesis and Geomorphology.* 2nd ed. Cambridge University Press, Cambridge, p. 795.

Simonson, R. W. 1959. "Outline of a generalized theory of soil genesis." *Soil Science Society of America Journal* 23:152–156.

Soil Survey Staff. 2014. *Keys to Soil Taxonomy.* 12th ed. [Online]. Available by United States Department of Agriculture, Natural Resources Conservation Service. http://www.nrcs.usda.gov/wps/PA_NRCSConsumption/download?cid=stelprdb1252094&ext=pdf.

Turenne, J. 2014. Subaqueous soils page. [Online]. Available by Nesoil.com http://nesoil.com/sas/index.htm (verified 6 June 2014).

van Breemen, N., and A. C. Finzi. 1998. "Plant–soil interactions: Ecological aspects and evolutionary implications." *Biogeochemistry* 42:1–19.

3 Soil Classification

The late Roy Simonson investigates a soil profile. (Photo courtesy of Ray R. Weil)

It is embarrassing not to be able to agree on what soil is. In this the pedologists are not alone. Biologists cannot agree on a definition of life and philosophers on philosophy.
—HANS JENNY, THE SOIL RESOURCE: ORIGIN AND BEHAVIOR

We classify things in order to make sense of our world. We do it whenever we call things by group names, based on their important properties. Imagine a world without classifications. Imagine surviving in the woods knowing only that each plant was a plant, not which are edible by people, which attract wildlife, or which are poisonous. Imagine surviving in a city knowing only that each person was a person, not a child or an adult, a male or a female, a police officer, a hoodlum, a friend, a teacher, a potential date, or any of the other categories into which we classify people. So, too, our understanding and management of soils and terrestrial systems would be hobbled if we knew only that a soil was a soil. How could we organize our information about soils? How could we learn from others' experience or communicate our knowledge to clients, colleagues, or students?

From the time crops were first cultivated humans noticed differences in soils and classified them according to their suitability for different uses. Farmers used descriptive names such as *black cotton soils*, *rice soils*, or *olive soils*. Other soil names still in common use today suggest the parent materials from which the soils are formed: *limestone soils*, *piedmont soils*, and *alluvial soils*. Such terms may convey some valuable meaning to local users, but they are inadequate for helping us to organize our scientific knowledge of soils or for defining the relationships among the soils of the world.

In this chapter we will learn how to classify soils as natural bodies on the basis of their profile characteristics, not merely on the basis of their suitability for a particular use. Such a soil classification system is essential to foster global communications about soils among soil scientists and all people concerned with the management of land and the conservation of the soil resource. Soil classification allows us to take advantage of research and experience at one location to predict the behavior of similarly classified soils at another location. Soil names such as Histosols or Vertisols conjure up similar mental images in the minds of soil scientists everywhere, whether they live in the United States, Europe, Japan, China, or elsewhere. A goal of the classification system is to create a universal language of soils that enhances communication among users of soils around the world.

3.1 CONCEPT OF INDIVIDUAL SOILS

Compared to most sciences, the organized study of soils is rather young, having begun in the 1870s when the Russian scientist V. V. Dokuchaev and his associates first conceived of a system for classifying natural soil bodies. In the United States, it was not until the late 1920s that C. F. Marbut of the U.S. Department of Agriculture (USDA), one of the few scientists who grasped the concept of soils being natural bodies, developed a soil classification scheme based on this concept.

The natural body concept of soils recognizes the existence of individual entities, each of which we call *a soil*. Just as human individuals may be grouped according to characteristics such as gender, height, or hair color, soil individuals having one or more characteristics in common may be grouped together. In turn, we may aggregate these groups into higher-level categories of soils, each having some characteristic that sets them apart from the others. Increasingly broad soil groups are defined as one moves up the classification pyramid from *a soil* to *the soil*.

Pedon and Polypedon[1]

Profile characteristics are unlikely to be exactly the same at any two points within any soil individual you may choose to examine. Consequently, it is necessary to characterize a soil individual in terms of an imaginary three-dimensional unit called a **pedon** (rhymes with "head on," from the Greek *pedon*, ground; see Figure 3.1). It is the smallest sampling unit that displays the full range of properties characteristic of a particular soil. Pedons occupy from about 1 to 10 m^2 of land area. Because it is what is actually examined during field investigation of soils, the pedon serves as the fundamental unit of soil classification.

However, a soil unit in a landscape usually consists of a group of very similar pedons, closely associated together in the field. Such a group of similar pedons, or a **polypedon**, is of sufficient size to be recognized as a landscape component termed a **soil individual**.

Although soil boundaries are drawn on soil maps as sharp lines, we know that soil properties (thickness of horizons, colors, textures, etc.) change gradually across a landscape. In this regard different soils are somewhat like different colors. For example, green and yellow are definitely two different colors, but the boundary between them on the spectrum is rather arbitrarily defined (Figure 3.2).

All the soil individuals in the world that have in common a suite of soil profile properties and horizons that fall within a particular range are said to belong to the same **soil series**. A soil series, then, is a class of soils, not a soil individual, in the same way that *Pinus sylvestris* is a species of tree, not a particular individual tree. Tens of thousands of soil series have been characterized and comprise the basic units used to classify the world's soils. Units delineated on detailed soil maps are not purely one soil, but are usually named for the soil series to which *most* of the pedons within the unit belong.

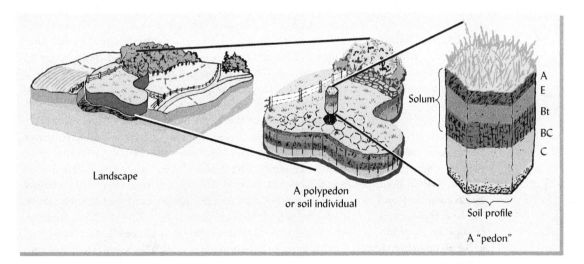

Figure 3.1 *A schematic diagram to illustrate the concept of pedon and of the soil profile that characterizes it. Note that several contiguous pedons with similar characteristics are grouped together in a larger area (outlined by broken lines) called a polypedon or soil individual. Several soil individuals are present in the landscape on the left.* (Diagram courtesy of Ray R. Weil)

---

[1]Although widely used, the polypedon concept is not without its critics. See Ditzler (2005).

Figure 3.2 *Like individual people, each soil is different. However, soil properties (thickness of horizons, colors, textures, etc.) vary more or less gradually across a landscape. Therefore, soils change gradually from one individual into another, despite the fact that soil maps represent soil boundaries as definite lines. Similarly, two different colors such as green and yellow may be rather arbitrarily defined on the light spectrum.* (Group photo of the U.S. national championship soil judging team, courtesy of Karen Vaughan, California Polytechnic State University; other photos courtesy of Ray R. Weil)

Groupings of Soil Individuals

In the concept of soils, the most specific extreme is that of a natural body called *a soil*, characterized by a three-dimensional sampling unit (pedon), related groups of which (polypedons) are included in a soil individual. At the most general extreme is *the soil*, a collection of all these natural bodies that is distinct from deep water, solid rock, and other natural parts of the Earth's crust. Hierarchical soil classification schemes generally group soils into classes at increasing levels of generality between these two extremes.

For thousands of years, most societies were primarily agricultural, and almost everyone worked with soils on a daily basis. Raw survival, on a personal and community level, depended on the food that could be coaxed from the different soils that people found in their environment. By trial and error, people learned which soils were best suited to various crops and which responded best to different kinds of management. As farmers passed their observations and traditions from one generation to the next, they summarized their knowledge about soils by developing unique systems of soil classification. In some regions this local knowledge about soils helped shape agricultural systems that were sustainable for centuries.

Scientific classification of soils began in the late 1800s stemming from the work of Dokuchaev in Russia (see Section 2.2). Australia, Brazil, Canada, China, the United Kingdom, Russia, and South Africa are among the countries that have developed and continue to use their own national soil classification systems.[2] The Food and Agriculture Organization of the United Nations developed World Reference Base (WRB) for Soils to assist global communication about soils and provide a reference by which various national soil classification systems can be compared and correlated. The WRB classifies the world's soils into 32 Soil Reference Groups (see Appendix Table A.1) that are differentiated mainly by the pedogenic process (such as the accumulation of clay in the subsoil) or parent material (such as volcanic ash) that is most responsible for creating the soil properties that typify the particular group.

---

[2]See Appendix A for summaries of the World Reference Base for Soils and the Canadian and Australian Systems of Soil Classification. For more information on other national systems and their interrelationships, see Eswaran et al. (2003), and for scientific as well as indigenous "folk" soil classifications, see chapters in Arnold et al. (2012) and Barrera-Bassols et al. (2006).

Another widely used system of global soil classification is the *Soil Taxonomy* developed by the Soil Survey Staff of the USDA. First published in 1975, *Soil Taxonomy* is used in the United States and approximately 50 other countries. Both the WRB for Soils and *Soil Taxonomy* are widely used in international publications. However, to avoid the confusion of dealing with two quite different systems, only *Soil Taxonomy* will be employed throughout this text.

3.2 *SOIL TAXONOMY*: A COMPREHENSIVE CLASSIFICATION SYSTEM[3]

Soil Taxonomy provides a hierarchical grouping of natural soil bodies. To encourage international acceptance, terminology in this system is not based on any one national language. Instead, the system uses a unique nomenclature that connotes the major characteristics of the soils in question.

Bases of Soil Classification

Although one of the objectives of *Soil Taxonomy* is to group soils that are similar in their genesis, soils are placed in these groups based on properties observable in the soil profile as it is found today. The specific criteria used to group soils within *Soil Taxonomy* involve most of the chemical, physical, and biological properties presented in this text. A few examples are moisture and temperature status throughout the year, as well as color, texture, and structure of the soil. Chemical and mineralogical properties, such as the contents of organic matter, clay, iron (Fe) and aluminum (Al) oxides, silicate clays, salts, the pH, the percentage *base saturation*,[4] and soil depth are other important criteria for classification. While many of the properties used may be observed in the field, others require precise measurements on samples taken to a sophisticated laboratory. This precision makes the system more objective, but in some cases may make the proper classification of a soil quite expensive and time-consuming.

Diagnostic Surface Horizons of Mineral Soils

The presence or absence of certain layers, termed **diagnostic soil horizons**, helps determine where to place a soil in the classification system. Diagnostic horizons that occur at the soil surface are called **epipedons** (from the Greek *epi*, over, and *pedon*, soil). The epipedon includes the upper part of the soil darkened by organic matter, the upper eluvial horizons, or both. It may include part of the B horizon if the latter is significantly darkened by organic matter. Eight are recognized (Table 3.1), but only five occur naturally over wide areas. Two of the others, anthropic and plaggen, result from intensive human use. They are common in parts of Europe and Asia where soils have been modified by people for many centuries.

As can be seen from Table 3.1, each diagnostic horizon is defined by a specific set of properties. For example, the **mollic epipedon** (Latin *mollis*, soft) is a mineral surface horizon noted for its dark color associated with accumulated organic matter (>0.6% organic C throughout), for its thickness, and for its softness even when dry. It has a base saturation greater than 50%. Mollic epipedons are moist at least three months a year when the soil temperature is usually 5°C or higher to a depth of 50 cm. These epipedons are characteristic of soils developed under grassland (Figure 3.3, *left*).

Diagnostic Subsurface Horizons

Many subsurface diagnostic horizons are used to characterize different soils in *Soil Taxonomy* (Figure 3.4). Each diagnostic horizon provides a characteristic that helps place a soil in its proper class in the system. The specific properties of the various subsurface diagnostic horizons are listed in the lower portion of Table 3.1. We will briefly discuss a few of the more commonly encountered subsurface diagnostic horizons.

---

[3]For the latest updates in *Soil Taxonomy*, see Soil Survey Staff (2014). For a complete description of *Soil Taxonomy*, see Soil Survey Staff (1999). The first edition of *Soil Taxonomy* was published as Soil Survey Staff (1975). For an explanation of the earlier U.S. classification system, see USDA (1938).

[4]The percentage base saturation is the percentage of the soil's negatively charged sites (cation exchange capacity) that are satisfied by attracting nonacid (or *base*) cations (such as Ca^{2+}, Mg^{2+}, and K^+) (see Section 9.3).

Table 3.1
MAJOR FEATURES OF DIAGNOSTIC HORIZONS IN MINERAL SOILS USED FOR DIFFERENTIATION AT THE HIGHER LEVELS OF *SOIL TAXONOMY*

The asterisks (*) indicate the five epipedons that are naturally occurring over wide areas.

Diagnostic Horizon (and Typical Genetic Horizon Designation)	Major Features
Surface horizons = epipedons	
Anthropic (A)	Human-transported or modified materials, with artifacts, or high phosphorus or puddled condition (rice paddies)
Folistic (O)	Organic horizon saturated for less than 30 days per normal year
Histic (O)*	A 20- to 60-cm-thick layer of **organic soil materials** (peat or muck) overlying mineral soil; formed in wet areas, black to dark brown color; very low density.
Melanic (A)*	Thick, black, high in organic matter (>6% organic carbon), common in volcanic ash soils
Mollic (A)*	Thick (generally >25 cm), dark-colored, high base saturation, well-developed structure most common in grassland soils
Ochric (A)*	Lighter in color, lower in organic matter or thinner than a mollic; may be hard and massive when dry
Plaggen (A)	Human-made sod-like horizon created by years of manuring, often with artifacts and spade marks
Umbric (A)*	Similar to mollic except low base saturation
Subsurface horizons	
Agric (A or B)	Organic and clay accumulation just below plow layer resulting from cultivation
Albic (E)	Light-colored, clay and iron and aluminum oxides mostly removed
Anhydritic (By)	Accumulation of anhydrite ($CaSO_4$)
Argillic (Bt)	Silicate clay accumulation
Calcic (Bk)	Accumulation of carbonates of calcium and/or magnesium
Cambic (Bw, Bg)	Altered by physical movement, structure development, or by chemical reactions, generally nonilluvial
Duripan (Bqm)	Hard pan, strongly cemented by silica
Fragipan (Bx)	Brittle hard pan, usually loamy textured, dense, and with coarse prisms
Glossic (E)	Whitish eluvial horizon that tongues into a Bt horizon
Gypsic (By)	Accumulation of gypsum ($CaSO_4 \cdot 2H_2O$)
Kandic (Bt)	Accumulation of low-activity clays
Natric (Btn)	Argillic, high in sodium, columnar or prismatic structure
Oxic (Bo)	Highly weathered, primarily mixture of Fe, Al oxides, and nonsticky-type silicate clays
Petrocalcic (Ckm)	Cemented calcic horizon
Petrogypsic (Cym)	Cemented gypsic horizon
Placic (Csm)	Thin pan cemented with iron alone or with manganese and organic matter
Salic (Bz)	Accumulation of salts
Sombric (Bh)	Organic matter accumulation
Spodic (Bh, Bs)	Organic matter, Fe and Al oxide accumulation
Sulfuric (Cj)	Highly acid with Jarosite mottles

Figure 3.3 *Two soil profile illustrating five diagnostic horizons. (Left) The mollic epipedon (a diagnostic horizon) in this soil includes genetic horizons designated Ap, A, and Bt1, all darkened by the accumulation of organic matter. A subsurface diagnostic horizon, the argillic horizon, overlaps the mollic epipedon. The argillic horizon is the zone of illuvial clay accumulation (Bt1 and Bt2 horizons in this profile). Scale marked every 10 cm. (Right) The thin, dark ochric epipedon at the surface of this sandy soil typifies the A horizon under acid forest vegetation. It is underlain by a light-colored eluviated albic horizon, under which a wavy dark layer of illuvial humus and iron has accumulated to form a spodic horizon. The parent material, glacial outwash, can be seen in lower part of the profile. The mollic and spodic horizons are diagnostic for soils in the orders Mollisols and Spodosols, respectively.* (Photo courtesy of Ray R. Weil)

The **argillic horizon** is a subsurface accumulation of silicate clays that have moved downward from the upper horizons or have formed in place. An example is shown in Figure 3.3 (*left*). The clays often are found as coatings on pore walls and surfaces of the structural groupings. The coatings usually appear as shiny surfaces or as clay bridges between sand grains. Termed *argillans* or *clay skins*, they are concentrations of clay translocated from upper horizons (see Figure 3.5).

The **kandic horizon** has an accumulation of iron and aluminum oxides as well as low-activity silicate clays (e.g., kaolinite), but clay skins need not be evident. The clays are low in activity as shown by their low cation-holding capacities (<16 cmol$_c$/kg clay). The epipedon that overlies a kandic horizon has commonly lost much of its clay content.

The **spodic horizon** is an illuvial horizon that is characterized by the accumulation of colloidal organic matter and aluminum oxide (with or without iron oxide). It is commonly found in highly leached forest soils of cool humid climates, typically on sandy-textured parent materials (see Figure 3.3, *right*).

The **albic horizon** is a light-colored eluvial horizon that is low in clay and oxides of iron and aluminum. These materials have largely moved downward from this horizon (see Figure 3.3, *right*).

A number of horizons have accumulations of salt-like chemicals that have leached from upper horizons in the profile. **Calcic horizons** contain an accumulation of carbonates (mostly

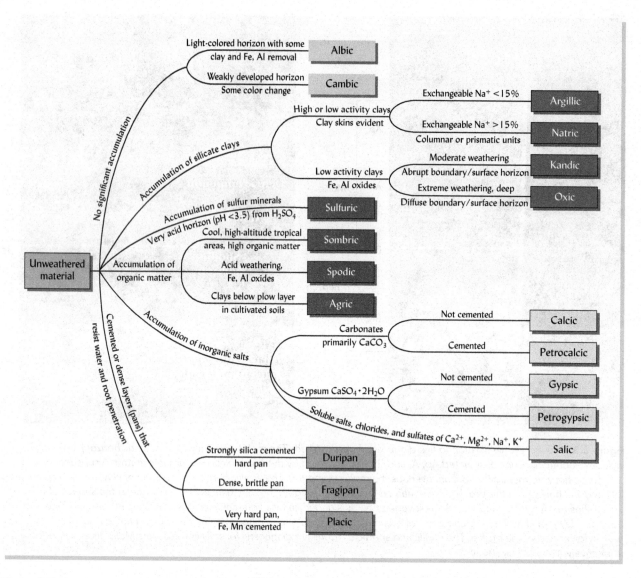

Figure 3.4 *Names and major distinguishing characteristics of subsurface diagnostic horizons. Among the characteristics emphasized is the accumulation of silicate clays, organic matter, iron and aluminum oxides, calcium compounds, and soluble salts, as well as materials that become cemented or highly acidified, thereby constraining root growth. The presence or absence of these horizons plays a major role in determining into which class a soil falls in Soil Taxonomy. See Chapter 8 for a discussion of low- and high-activity clays.*

$CaCO_3$) that often appear as white chalk-like nodules (see the Bk horizon in the lower part of the profile shown in Figure 3.3, *left*). **Gypsic horizons** have an accumulation of gypsum ($CaSO_4 \cdot 2H_2O$), and **salic horizons** have an accumulation of soluble salts. These are found mostly in soils of arid and semiarid regions.

In some subsurface diagnostic horizons, the materials are cemented or densely packed, resulting in relatively impermeable layers called *pans* (**duripan, fragipan,** and **placic horizons**).[5] These can resist water movement and penetration by plant roots. Such pans constrain plant growth and may encourage water runoff and erosion because rainwater cannot move readily downward through the soil. Figure 3.4 explains the genesis of these and the other subsurface diagnostic horizons.

---

[5] Well-developed argillic horizons may present such a great and abrupt increase in clay content that water and root movement are severely restricted; such a horizon is colloquially referred to as a *claypan*.

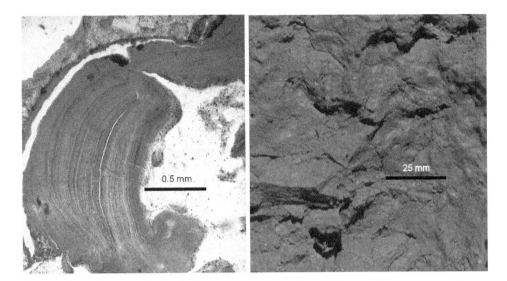

Figure 3.5 *The argillic subsurface diagnostic horizon is characterized by the accumulation of silicate clay, some of which has illuviated into the B horizon from soil layers higher in the profile. The illuvial movement of clay results in the deposition of thin coatings of clay on ped and pore surfaces. These "clay skins" or "argillans" can be seen under a petrographic microscope with plain polarized light (left) and with the naked eye as shiny clay coatings (right).* (Photo courtesy of Carlos Dorransoro Fernandez, University of Granda, Spain (*left*) and Ray R. Weil (*right*))

3.3 CATEGORIES AND NOMENCLATURE OF *SOIL TAXONOMY*

There are six hierarchical categories of classification in *Soil Taxonomy:* (1) *order*, the highest (broadest) category; (2) *suborder*; (3) *great group*; (4) *subgroup*; (5) *family*; and (6) *series* (the most specific category). The lower categories fit within the higher categories (Figure 3.6). Thus, each order has several suborders, each suborder has several great groups, and so forth. This system may be compared with those used for the classification of plants or animals, as shown in Table 3.2. Just as *Trifolium repens* identifies a specific kind of plant, the Miami series identifies a specific kind of soil.

Nomenclature of *Soil Taxonomy*

Although unfamiliar at first sight, the nomenclature system has a logical construction and conveys a great deal of information about the nature of the soils named. The system is easy to learn after a bit of study. The nomenclature is used throughout this book, especially to identify the kinds of soils shown in illustrations. When reading, if you make a conscious effort to identify the parts of each soil class name mentioned and recognize the level of category indicated, the system will become second nature to you.

The names of the classification units are combinations of syllables, most of which are derived from Latin or Greek, and are root words in several modern languages. Since each part of a soil name conveys a concept of soil character or genesis, the name automatically describes the general kind of soil being classified. Names of soil orders are combinations of (1) formative elements, which generally define the characteristics of the soils, and (2) the ending *sols* (Latin *solum*, soil). For example, soils of the order **Aridisols** (from the Latin *aridus*, dry) are characteristically dry soils in arid regions. Those of the order **Inceptisols** (from the Latin *inceptum*, beginning) are soils with only the beginnings or inception of profile development.

The names of suborders automatically identify the order of which they are a part. For example, soils of the suborder **Aquolls** are the wetter soils (from the Latin *aqua*, water) of the Mollisols order. Likewise, the name of the great group identifies the suborder and

Figure 3.6 *The categories of Soil Taxonomy and approximate number of units in each category.*

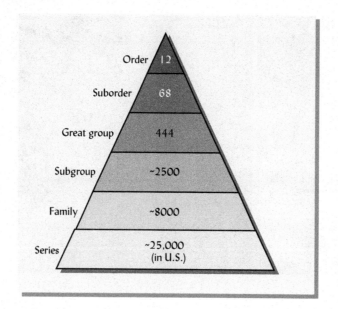

Table 3.2

COMPARISON OF THE CLASSIFICATION OF A COMMON CULTIVATED PLANT, WHITE CLOVER (*TRIFOLIUM REPENS*), AND A SOIL, MIAMI SERIES

Plant Classification			Soil Classification	
Phylum	Pterophyta		Order	Alfisols
Class	Angiospermae		Suborder	Udalfs
Subclass	Dicotyledoneae	Increase specificity →	Great Group	Hapludalfs
Order	Rosales		Subgroup	Oxyaquic Hapludalfs
Family	Leguminosae		Family	Fine loamy, mixed, mesic, active
Genus	*Trifolium*		Series	Miami
Species	*Repens*		Phase[a]	Miami silt loam

[a]Technically not a category in *Soil Taxonomy* but used in field surveying. *Silt loam* here refers to the texture of the A horizon.

order of which it is a part. **Argiaquolls** are Aquolls with clay or argillic (Latin *argilla*, white clay) horizons. If one is given only the subgroup name, the great group, suborder, and order to which the soil belongs are automatically known. In the following illustration, note that the three letters *oll* identify each of the lower categories as being in the M*oll*isols order.

M*oll*isols	Order
Aqu*oll*s	Suborder
Argiaqu*oll*s	Great group
Typic Argiaqu*oll*s	Subgroup

Family names in general identify subsets of the subgroup that are similar in texture, mineral composition, and mean soil temperature at a depth of 50 cm. Thus the name *fine*, *mixed*, *mesic*, and *active Typic Argiaquolls* identifies a family in the Typic Argiaquolls subgroup with a fine texture, mixed clay mineral content, mesic (moderate, 8–15°C) annual mean soil temperature, and clays active in cation exchange.

Soil series are generally named after a geographic feature (town, river, etc.) near where they were first recognized. Thus, names such as *Cecil, Harare, Miami, Norfolk*, and

Ontario identify soil series first described near the town or geographic feature named. Many thousands of soil series have been classified worldwide, with about 25,000 in the United States alone.

In detailed field soil surveying, soil series are sometimes further differentiated on the basis of surface soil texture, degree of erosion, slope, or other characteristics. These practical subunits are called soil **phases**. Names such as *Cecil clay loam* or *Miami loam, eroded phase* are used to identify such phases. Note, however, that soil phases, practical as they may be in local situations, are *not* a category in the *Soil Taxonomy* system.

With this explanation of the nomenclature of *Soil Taxonomy*, we will now briefly consider the general nature of soils in the 12 orders.

3.4 SOIL ORDERS

Each of the world's soils is assigned to one of 12 **orders**, largely on the basis of soil properties that reflect a major course of development, with considerable emphasis placed on the presence or absence of major diagnostic horizons (Table 3.3). As an example, many soils that developed under grassland vegetation have the same general sequence of horizons and are characterized by a mollic epipedon—a thick, dark, surface horizon that is high in nonacid cations. Because of the properties they have in common, these soils are included in the same order: Mollisols. The names and major characteristics of each soil order are shown in Table 3.3. Order names always end in *sols*.

The general conditions that promote the formation of soils in the different orders are shown in Figure 3.7. From soil profile characteristics, soil scientists can ascertain the relative degree of soil development in the different orders, as shown in this figure. Note that soils with essentially no profile layering (Entisols) have the least development, whereas the deeply weathered soils of the humid tropics (Oxisols and Ultisols) show the greatest soil development. The effect

Table 3.3
NAMES OF SOIL ORDERS IN *SOIL TAXONOMY* WITH THEIR DERIVATION AND MAJOR CHARACTERISTICS
The bold letters in the order names indicate the formative element used as the ending for suborders and lower taxa within that order.

Name	Formative Element	Derivation	Pronunciation	Major Characteristics
Alfisols	alf	Nonsense symbol	Ped**alf**er	Argillic, natric, or kandic horizon; high-to-medium base saturation
Andisols	and	Jap. *ando*, black soil	**And**esite	From volcanic ejecta, dominated by allophane or Al-humic complexes
Ari**d**isols	id	L. *aridus*, dry	Ari**d**	Dry soil, ochric epipedon, sometimes argillic or natric horizon
Entisols	ent	Nonsense symbol	Rec**ent**	Little profile development, ochric epipedon common
G**el**isols	el	Gk. *gelid*, very cold	J**el**ly	Permafrost, often with cryoturbation (frost churning)
H**ist**osols	ist	Gk. *histos*, tissue	H**ist**ology	Peat or bog; >20% organic matter
Inc**ept**isols	ept	L. *inceptum*, beginning	Inc**ept**ion	Embryonic soils with few diagnostic features, ochric or umbric epipedon, cambic horizon
M**oll**isols	oll	L. *mollis*, soft	M**oll**ify	Mollic epipedon, high base saturation, dark soils, some with argillic or natric horizons
Oxisols	ox	Fr. *oxide*, oxide	**Ox**ide	Oxic horizon, no argillic horizon, highly weathered
Sp**od**osols	od	Gk. *spodos*, wood ash	P**od**zol; **od**d	Spodic horizon commonly with iron, aluminum oxides and humus accumulation
Ultisols	ult	L. *ultimus*, last	**Ult**imate	Argillic or kandic horizon, low base saturation
V**ert**isols	ert	L. *verto*, turn	Inv**ert**	High in swelling clays; deep cracks when soil is dry

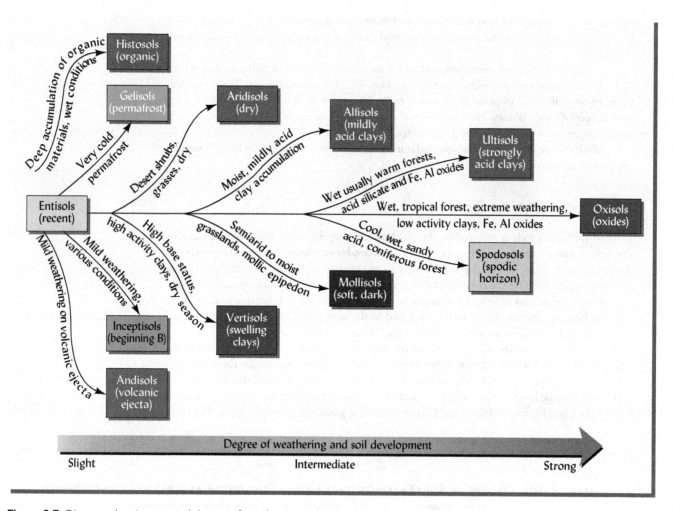

Figure 3.7 *Diagram showing general degree of weathering and soil development in the different soil orders classified in* Soil Taxonomy. *Also shown are the general climatic and vegetative conditions under which soils in each order are formed.* (Diagram courtesy of N. Brady and R. Weil. For more on environmental factors influencing the formation of soil orders, see Lin (2011))

of climate (temperature and moisture) and of vegetation (forests or grasslands) on the kinds of soils that develop is also indicated in Figure 3.7. Study Table 3.3 and Figure 3.7 to better understand the relationship between soil properties and the terminology used in *Soil Taxonomy*.

While only the Gelisols and Aridisols orders are defined directly in relation to climate, Figure 3.8 indicates that orders with the most highly weathered soils tend to be associated with the warmer and wetter climates. A general world map of the 12 soil orders is printed in color on the front endpaper. A more detailed color-coded soil map for the United States can be found on the back endpaper of this book. A typical profile and global distribution map for each soil order are shown at the beginning of Sections 3.5 to 3.16. In studying these maps, try to confirm that the distribution of the soil orders is in accordance with what you know about the climate in various regions of the world.

Although a detailed description of all the lower levels of soil categories is far beyond the scope of this book, a general knowledge of the 12 soil orders is essential for understanding the nature and function of soils in different environments. The simplified key given in Figure 3.9 helps illustrate how *Soil Taxonomy* can be used to key out the order of any soil based on observable and measurable properties of the soil profile. Because certain diagnostic properties take precedence over others, the key must always be used starting at the top, and working down. It will be useful to review this key after reading the following descriptions of the general characteristics, nature, and occurrence of each soil order. We will begin with those orders characterized by little profile development and progress to those with the most highly weathered profiles (as represented from left to right in Figure 3.7).

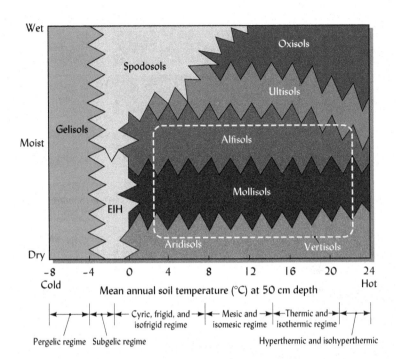

Figure 3.8 General soil moisture and soil temperature regimes that characterize most soils in each of eight soil orders. Soils of the other four orders (Andisols, Entisols, Inceptisols, and Histosols) may be found under any of the soil moisture and temperature conditions (including the area marked EIH). Major areas of Vertisols are found only where clayey materials are in abundance and are most extensive where the soil moisture and temperature conditions approximate those shown inside the box with broken lines. Note that these relationships are only approximate and that less extensive areas of soils in each order may be found outside the indicated ranges. For example, some Ultisols (Ustults) and Oxisols (Ustox) have soil moisture levels that are much lower than this graph would indicate for at least part of the year. (The terms used at the bottom to describe the soil temperature regimes are those used in helping to identify soil families.)

3.5 ENTISOLS (RECENT: LITTLE IF ANY PROFILE DEVELOPMENT)

Weakly developed mineral soils without natural genetic (subsurface) horizons or with only the beginnings of such horizons (Figure 3.10) belong to the Entisols order. Most have an ochric epipedon and a few have human-made anthropic or agric epipedons. Some have albic subsurface horizons. Soil productivity ranges from very high for certain Entisols formed in recent alluvium to very low for those forming in shifting sand or on steep rocky slopes.

This is an extremely diverse group of soils with little in common, other than the lack of evidence for all but the earliest stages of soil formation. Entisols are either young in years or their parent materials have not reacted to soil-forming factors. On such parent materials as fresh lava flows or recent alluvium (Fluvents), there has been too little time for much soil formation. In extremely dry areas, scarcity of water and vegetation may inhibit soil formation. Likewise, frequent saturation or inundation with water (Aquents, Wassents) may delay soil formation. Some Entisols occur on steep slopes, where the rates of erosion may exceed the rates of soil formation, preventing horizon development. Others occur on urbanized areas and construction sites where bulldozers have destroyed or mixed together the original soil horizons, causing the existing soils to become Entisols as the horizon formation process must start anew.

Distribution and Use

Globally, Entisols are found under a wide variety of environmental conditions (Figure 3.10 and endpapers). Entisols developed on alluvial floodplains are among the world's most productive soils. Such soils, with their level topography, proximity to water for irrigation, and periodic nutrient replenishment by floodwater sediments have supported the development of many major civilizations. However, the productivity of most Entisols is restricted by inadequate soil depth, clay content, or water availability. For example, in rocky and mountainous regions, medium-textured Entisols (Orthents, e.g., Figure 3.11) are commonly too shallow to support high plant productivity, especially in dry regions. Sandy Entisols (Psamments; Figure 3.10) are found in parts of the Sahara desert, southern Africa, central Australia, northwest Nebraska, and the southeastern U.S. coastal plain. Psamments in the humid southern United States are successfully used for citrus, vegetable, and peanut production. Poorly drained and seasonally flooded Entisols (Aquents) occur in major river valleys and wetlands, whereas subaqueous Entisols (Wassents) are important for supporting submerged aquatic vegetation in shallow estuaries and bays.

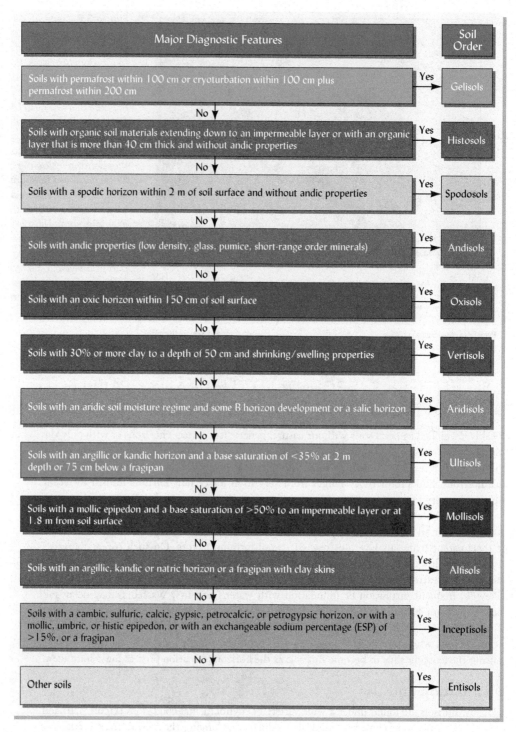

Figure 3.9 A simplified key to the 12 soil orders in Soil Taxonomy. In using the key, always begin at the top. Note how diagnostic horizons and other profile features are used to distinguish each soil order from the remaining orders. Entisols, having no such special diagnostic features, key out last. Also note that the sequence of soil orders in this key bears no relationship to the degree of profile development and adjacent soil orders may not be more similar than nonadjacent ones. See Section 3.2 for explanations of the diagnostic horizons.

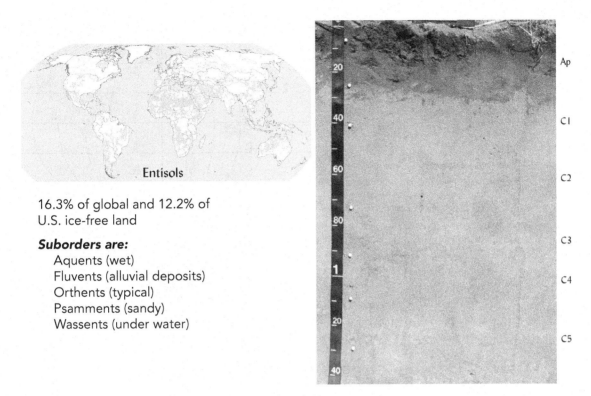

16.3% of global and 12.2% of U.S. ice-free land

Suborders are:
  Aquents (wet)
  Fluvents (alluvial deposits)
  Orthents (typical)
  Psamments (sandy)
  Wassents (under water)

Figure 3.10 *Entisols soil order—global distribution; land area; suborders; and a representative profile, a Typic Udispamment on a flood plain in North Carolina, USA.* (Photo courtesy of Ray R. Weil)

Figure 3.11 *Shallow Entisols (Udorthents) on a steep rocky slope support moderately productive forest in a humid temperate climate.* (Photo courtesy of Ray R. Weil)

3.6 INCEPTISOLS (FEW DIAGNOSTIC FEATURES: INCEPTION OF B HORIZON)

In Inceptisols, the beginning or *inception* of profile development is evident, and some diagnostic features are present (Figure 3.12). However, the well-defined profile characteristics of soils thought to be more mature have not yet developed. For example, a cambic horizon showing some color or structural change is common in Inceptisols, but a more mature illuvial B horizon such as an argillic cannot be present. Other subsurface diagnostic horizons that may be present in Inceptisols include duripans: fragipans; and calcic, gypsic, and sulfuric horizons.

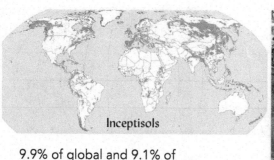

9.9% of global and 9.1% of U.S. ice-free land

Suborders are:
  Aquepts (wet)
  Cryepts (very cold)
  Gelepts (permafrost)
  Udepts (humid climate)
  Ustepts (semiarid)
  Xerepts (dry summers, wet winters)

Figure 3.12 *Inceptisols soil order—global distribution; land area; suborders; and a representative profile, a Typic Eutrodept from Vermont.* (Courtesy of Ray R. Weil)

The epipedon in most Inceptisols is an ochric. Inceptisols show more significant profile development than Entisols but are defined to exclude soils with diagnostic horizons or properties that characterize certain other soil orders. Thus, soils with only slight profile development occurring in arid regions or containing permafrost or andic properties are excluded from the Inceptisols. They fall, instead, in the soil orders Aridisols, Gelisols, or Andisols, as discussed in later sections.

Distribution and Use

As with Entisols, Inceptisols are found in most climatic and physiographic conditions. They are often prominent in mountainous areas. They are also probably the most important soil order in the lowland rice-growing areas of Asia. Inceptisols are found in each of the continents (see front papers). Inceptisols of humid regions, called *Udepts*, often have only thin, surface horizons (ochric epipedons). Wet Inceptisols or Aquepts are found in flood plain areas along major rivers. The natural productivity of Inceptisols is highly variable.

3.7 ANDISOLS (VOLCANIC ASH SOILS)

Andisols are usually formed on ash and cinders deposited by volcanos in recent geological times. They are commonly found near the volcano source or in areas downwind from the volcano, where a sufficiently thick layer of ash has been deposited during eruptions. Andisols have not had time to become highly weathered. The principal soil-forming process has been the rapid weathering (transformation) of volcanic ash to produce amorphous or poorly crystallized silicate minerals such as **allophane** and **imogolite** and the iron oxy-hydroxide, **ferrihydrite**. Some Andisols have a melanic epipedon, a surface diagnostic horizon that has a high organic matter content and dark color (Figure 13.13). The accumulation of organic matter is quite rapid due largely to its protection by complexing with aluminum. Little downward translocation of the colloids, or other profile development, has taken place. Like the Entisols and Inceptisols, Andisols are young soils, usually having developed for only 5,000 to 10,000 years.

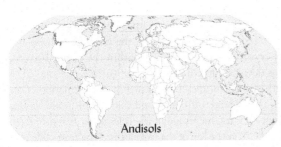

0.7% of global and 1.7% of U.S. ice-free land

Suborders are:
  Aquands (wet)
  Cryands (cold)
  Gelands (very cold)
  Torrands (hot, dry)
  Udands (humid)
  Ustands (moist/dry)
  Vitrands (volcanic glass)
  Xerands (dry summers, moist winters)

Figure 3.13 *Andisols soil order—global distribution; land area; suborders; and a representative profile, a Typic Melanudand from western Tanzania. Scale in decimeters.* (Courtesy of Ray R. Weil)

Unlike the previous two orders of immature soils, Andisols exhibit a unique set of **andic properties** in at least 35 cm of the upper 60 cm of soil due to common types of parent materials. Materials with andic properties are characterized by a high content of volcanic glass and/or a high content of amorphous or poorly crystalline iron and aluminum minerals. The combination of these minerals and the high organic matter results in light, fluffy soils that are easily tilled, yet have a high water-holding capacity and resist erosion by water. They are mostly found in regions where rainfall keeps them from being susceptible to erosion by wind. Andisols are usually of high natural fertility, except that phosphorus (P) availability is severely limited by the extremely high P retention capacity of the andic materials (see Section 12.3). Fortunately, proper management of plant residues and fertilizers can usually overcome this difficulty.

Distribution and Use

Andisols are found in areas where significant depths of volcanic ash and other ejecta have accumulated (Figure 3.13). Globally, they make up less than 1% of the soil area. Andisols having a udic (humid) moisture regime (Udands) and the somewhat drier Ustands are both used intensively for forestry and agriculture. In the Pacific Rim and African Rift areas they are important and productive soils that support intensive agriculture, especially where they occur at high-elevations. Andisols are found to a minor extent in cold climates (Cryands) in Canada and Russia, and in hot, dry climates (Torrands) in Mexico and Syria. In the United States, significant areas of Andisols occur in Washington, Idaho, Montana, and Oregon. Very recent eruptions are giving rise to Vitrands that still have much volcanic glass and lower-water-holding capacities.

3.8 GELISOLS (PERMAFROST AND FROST CHURNING)

Gelisols are young soils with little profile development. Cold temperatures and frozen conditions for much of the year slow the process of soil formation. The principal defining feature of these soils is the presence of a **permafrost** layer (see Figures 3.14–3.15). Permafrost is a layer of material that remains at temperatures below 0°C for more than two consecutive years. It may be a hard, ice-cemented layer of soil material (e.g., designated Cfm in profile descriptions), or, if dry, it may be uncemented (e.g., designated Cff). In Gelisols, the permafrost layer lies within 100 cm of the soil surface, unless **cryoturbation** is evident within the upper 100 cm, in which case the permafrost may begin as deep as 200 cm from the soil surface.

Cryoturbation is the physical disturbance of soil materials caused by the formation of ice wedges and by the expansion and contraction of water as it freezes and thaws. This *frost-churning* action moves the soil material so as to orient rock fragments along the lines of force and to form broken, convoluted horizons (e.g., designated Cjj) at the top of the permafrost. The frost churning also may form patterns on the ground surface, such as hummocks and ice-rich polygons that may be several meters across. In some cases, rocks forced to the surface form rings or netlike patterns.

Gelisols showing evidence of cryoturbation are called *Turbels*. Other Gelisols, often found in wet environments, have developed in accumulations of mainly organic materials, making them **Histels** (Greek *histos*, tissue; Figure 3.15). Most of the soil-forming processes that occur take place above the permafrost in the **active layer** that thaws every year or two. Various types of diagnostic horizons may have developed in different Gelisols, including mollic, histic, umbric, calcic, and, occasionally, argillic horizons.

Distribution and Use

Gelisols are most extensive in Northern Russia, Canada, and Alaska. Blanketed under snow and ice for much of the year, most Gelisols support tundra vegetation of lichens, grasses, and low shrubs that grow during the brief summers. Large areas of Gelisols consist of bogs, some literally floating on layers of frozen or unfrozen water. Millions of caribou, reindeer, and musk ox survive on this vegetation during the summer, then migrate to the boreal forests during the coldest seasons. The many bogs and pools serve as nesting sites for migratory birds, which feed on the thick clouds of biting flies and mosquitoes. Few humans live in these inhospitable environments.

8.6% of global and 7.5% of U.S. ice-free land

Suborders are:
  Histels (organic)
  Orthels (no special features)
  Turbels (cryoturbation)

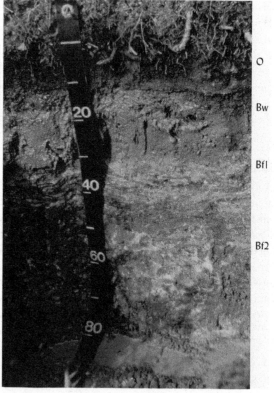

Figure 3.14 *Gelisols soil order—global distribution; land area; suborders; and a representative profile, a Typic Aquaturbel from Alaska. Permafrost is seen below 32 cm in this profile.* (Courtesy of Chien-lu Ping, University of Alaska, Fairbanks)

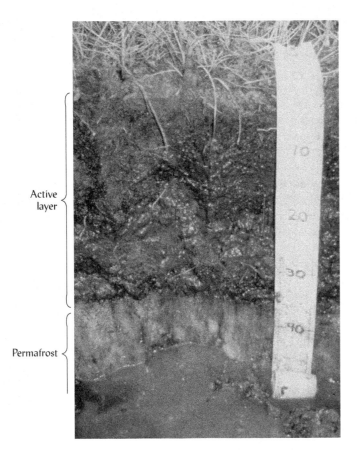

Figure 3.15 *Gelisols in Alaska.* (Left) *The soil is in the suborder Histels and has a histic epipedon and permafrost. This soil was photographed in Alaska in July. Scale in centimeters.* (Right) *Melting of the permafrost under this section of the Alaska Highway caused the soil to lose all bearing strength and collapse. Scale in centimeters.* (Left photo courtesy of Chien-Lu Ping/University of Alaska, Fairbanks; right photo courtesy of John Moore, USDA/NRCS)

Gelisols are rarely used for agriculture. Plant productivity is low because of the extremely short potential growing season in the far northern latitudes, the low levels of solar radiation (except during the fleeting summer), and the waterlogged condition of many Gelisols in which permafrost inhibits internal drainage during the summer thaw (Figure 3.15).

If the vegetation or insulating surface peat layer on Gelisols is disturbed (as by cultivation, forest fires or construction activities), the permafrost layer may melt. The permafrost in the southern part of the Gelisols region may be only 1 or 2°C below freezing, so even small changes can cause melting. Unless the soil is mainly gravel, the melting is likely to cause the soil to completely lose its bearing strength and collapse. This presents many serious engineering difficulties (see Figure 3.15, *right*). Houses built directly on Gelisols may sink into the ground as the heat from inside the building penetrates the soil and melts the permafrost. Oil pipelines must be constructed on stilts, rather than buried as the heat from the flowing oil would melt the permafrost, causing the pipes to sag and rupture.

Permafrost melting in Gelisols is an early symptom of global climate change caused by greenhouse gas emissions (see Section 11.9). Unfortunately, the melting of permafrost and deepening of the active layer in Gelisols appear to be accelerating this trend as the enormous pools of organic carbon and once locked away in the permafrost become exposed to decay, thus releasing yet more greenhouse gases to the atmosphere.

3.9 HISTOSOLS (ORGANIC SOILS WITHOUT PERMAFROST)

Histosols are soils that have undergone little profile development because of the anaerobic environment in which they form. The main process of soil formation evident in Histosols is the accumulation of partially decomposed organic parent material—but without permafrost (which would cause the soil to be classified in the Histels suborder of Gelisols). Histosols consist of one or more thick layers of **organic soil material**. Generally, Histosols have organic soil materials in more than half of the upper 80 cm of soil (Figure 3.16) or in two-thirds of the soil overlying shallow rock.

Organic deposits accumulate in marshes, bogs, and swamps, which are habitats for hydrophilic (water-loving plants) such as sedges, reeds, mosses, shrubs, and even some trees. Generation after generation, the residues of these plants sink into the water, where low oxygen conditions inhibit their decomposition and, consequently, act as a partial preservative (see Figure 2.17).

The organic matter in Histosols ranges from peat to muck. *Peat* is comprised of the brownish, partially decomposed, fibrous remains of plant tissues (see Figures 3.16 and 3.17). Some of these soils are mined and sold as peat, a material widely used in containerized plant production (see Box 11.4). Peat deposits are also used for fuel in some countries, especially in Russia, where several power stations are fueled by this material. *Muck*, on the other hand, is a smooth colloidal black material in which decomposition is much more complete (Figure 3.17). Muck is like black ooze when wet and powdery when dry.

While not all wetlands contain Histosols, all Histosols (except Folists) occur in wetland environments. They can form in almost any moist climate in which plants can grow, from equatorial to arctic regions, but they are most prevalent in cold climates, up to the limit of permafrost. Horizons are differentiated by the type of vegetation contributing the residues, rather than by translocations and accumulations within the profile.

Whether artificially drained for cultivation or left in their natural water-saturated state, Histosols possess unique properties resulting from their high organic matter contents. Histosols are generally black to dark brown in color and extremely lightweight (0.15–0.4 Mg/m^3) when dry, being only about 10 to 20% as dense as most mineral soils. Histosols

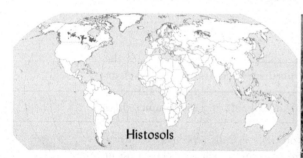

Histosols

2.2% of global and 1.3% of U.S. ice-free land

Suborders are:
Fibrists (fibers of plants obvious)
Folists (leaf mat accumulations)
Hemists (fibers partly decomposed)
Saprists (fibers not recognizable)
Wassists (underwater most of the time)

Figure 3.16 *Histosols soil order—global distribution; land area; suborders; and a representative profile, a Limnic Haplosaprist from southern Michigan, USA. A buried mineral soil can be seen at the bottom of the scale (which is marked in 30 cm increments).* (Courtesy of USDA/NRCS)

Figure 3.17 *A tidal marsh Histosol. The inset shows the fibric (peaty) organic material that contains recognizable roots and rhizomes of marsh grasses that died perhaps centuries ago, the anaerobic conditions having preserved the tissues from extensive decay. The soil core (held horizontally for the photograph) gives some idea of the soil profile, the surface layer being at the right and the deepest layer at the left. The water level is usually at or possibly above the soil surface.* (Photo courtesy of Ray R. Weil)

possess very high capacities to hold both water (up to 400% of soil dry weight) and nutrient cations (see Sections 5.9 and 8.9). The water- and cation-holding capacities are much higher than those of even clay-rich mineral soils on a weight basis, but similar to those of mineral soils rich in 2:1 silicate clays, when considered on a volume basis (water or cations held per liter of soil).

Distribution and Use

Even though they cover only about 1% of the world's land area, Histosols, especially **peat lands**, comprise significant areas in cold, wet regions of Alaska, Canada, Finland, Russia, Iceland, Ireland, and Scotland. Ecologically important Histosols in the United States include glaciated areas in Wisconsin, Minnesota, New York, and Michigan, tule-reed beds of California, the Everglades of Florida, the bayous of Louisiana, and the tidal marshes of the mid-Atlantic states (Figure 3.17). Some Histosols (Wassists) occur in shallow lagoons, bays, and lakes and are completely submerged under water all or most of the time.

Because the ecological roles of natural wetland environments have not always been appreciated (or protected by law), more than 50% of the original wetland area in the lower 48 United States has been drained for agricultural or other uses. Some Histosols make very productive farmlands, but the organic nature of the materials requires liming, fertilization, tillage, and drainage practices quite different from those applied to soils in the other 11 orders.

If other than wetland plants are to be grown, the water table is usually lowered to provide an aerated zone for root growth. This practice, of course, alters the soil environment and causes the organic material to oxidize. In drained Histosols, the land surface is actually lowered by as much as 5 cm of soil per year as a result of both compression and oxidation, a process termed **subsidence** (see Figure 3.18). To slow the loss of valuable soil resources and avoid unnecessarily aggravating the global climate change (see Section 11.9), the water table in forested or agricultural Histosols should be kept no lower than is needed to assure adequate root aeration. A more sustainable approach would be to grow such wetland crops as rice or cranberries or to restore Histosol areas to their native wetland condition (see Section 7.7).

Figure 3.18 *Soil subsidence due to rapid organic matter decomposition after artificial drainage of Histosols in the Florida Everglades. The house was built at ground level, with the septic tank buried about 1 m below the soil surface. Over a period of about 60 years, more than 1.2 m of the organic soil has "disappeared." The loss has been especially rapid because of Florida's warm climate, but artificial drainage that lowers the water table and continually dries out the upper horizons is an unsustainable practice on any Histosol.* (Photo courtesy of George H. Snyder, Everglades Research and Education Center, Belle Glade, FL)

3.10 ARIDISOLS (DRY SOILS)

Aridisols occupy a larger area globally than any other soil order except Entisols. As the name implies, aridity (scarcity of water) is a major characteristic of these soils such that soil moisture sufficient to support plant growth is present for no longer than 90 consecutive days. The natural vegetation consists mainly of scattered desert shrubs and short bunchgrasses. Soil properties, especially in the surface horizons, may differ substantially between interspersed bare and vegetated areas (see Section 2.5).

Aridisols are characterized by an ochric epipedon that is usually light in color and low in organic matter (see Figure 3.19). Although there is generally not enough water to leach soluble materials completely out of the profile, the processes of soil formation have brought about a redistribution of these materials, often causing them to accumulate at a lower level in the profile. These soils may have a horizon of accumulation of calcium carbonate (calcic), gypsum (gypsic), soluble salts (salic), or exchangeable sodium (natric). With time and the addition of

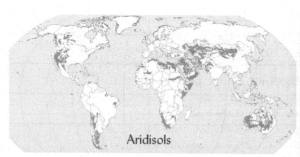

1.7% of global and 8.8% of U.S. ice-free land

Suborders are:
  Argids (clay)
  Calcids (carbonate)
  Cambids (typical)
  Cryids (cold)
  Durids (duripan)
  Gypsids (gypsum)

Figure 3.19 *Aridisols soil order—global distribution; land area; suborders; and a representative profile, a skeletal Ustic Haplocalcid from Nevada, USA. Shovel handle is 60 cm long.* (Courtesy of Ray R. Weil)

Figure 3.20 *Several features characteristic of some Aridisols. (Left) Wind-rounded pebbles have given rise to a desert pavement. The close-up shows a pebble removed to reveal the vesicular pores in the soil underneath. Scale bars = 10 cm. (Right) A petrocalcic horizon cemented with calcium carbonate.* (Photo courtesy of Ray R. Weil)

carbonates from calcareous dust and other sources, a B horizon rich in carbonates may form (*Calcids*). Under certain circumstances, carbonates may cement together the soil particles and coarse fragments in the layer of accumulation, producing hard layers known as **petrocalcic** horizons (Figure 3.20). These hard layers act as impediments to plant root growth and also greatly increase the cost of excavations for building.

Some Aridisols (the *Argids*) have an argillic horizon, most probably formed under a wetter climate that long ago prevailed in many areas that are deserts today. On steeper land surfaces subject to erosion, argillic horizons do not get a chance to form, and the dominant soils are often *Cambids* (Aridisols with only weakly differentiated cambic subsurface B horizons). In stony or gravelly soils, erosion may remove all the fine particles from the surface layers, leaving behind a layer of wind-rounded pebbles that is called *desert pavement* (see Figure 3.20). The surfaces of rocks, including the pebbles in desert pavement, often have a shiny, microbe-derived coating called *desert varnish*.

Except where there is groundwater or irrigation, the soil layers are only moist for short periods during the year. These short, moist periods may be sufficient for drought-adapted desert shrubs and annual plants but not for conventional crop production. If groundwater is present near the soil surface, soluble salts may accumulate in the upper horizons to levels that most crop plants cannot tolerate (see Section 9.15).

Distribution and Use

Vast areas of Aridisols are present in the Sahara desert in Africa, the Gobi and Taklamakan deserts in China, and the Turkestan desert of central Asia. Most of the soils of southern and central Australia are Aridisols, as are those of southern Argentina, southwestern Africa, Pakistan, and much of the Middle East. In North America, Aridisols occur mainly in the western United States and northern Mexico.

Without irrigation, Aridisols are not suitable for growing cultivated crops. Some areas are used for low-intensity grazing, especially with sheep or goats, but the production per unit area is low. Poorly managed grazing of Aridisols leads to increased heterogeneity of both soils and vegetation. The animals graze the relatively even cover of palatable grasses, giving a competitive advantage to various shrubs not eaten by the grazing animals. The scattered shrubs compete against the struggling grasses for water and nutrients. The once-grassy areas become increasingly bare, and the soils between the scattered shrubs succumb to erosion by the desert winds and occasional thunderstorms. The desertification of areas of Africa, Asia, and the western United States is evidence of such degradation.

Some xerophytic plants, such as a jojoba, have been cultivated on Aridisols to produce various industrial feedstocks such as oil and rubber. Where irrigation water and fertilizers are available, some Aridisols can be made highly productive. Irrigated valleys in arid areas are among the most productive in the world. However, they must be carefully managed to prevent the accumulation of soluble salts (see Section 9.11).

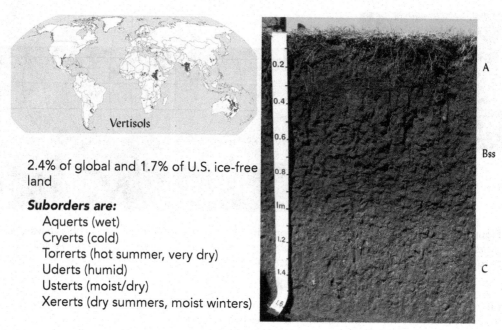

Figure 3.21 *Vertisols soil order—global distribution; land area; suborders; and a representative profile, a Typic Haplustert from Queensland, Australia during the wet season.* (Courtesy of Ray R. Weil)

3.11 VERTISOLS (DARK, SWELLING, AND CRACKING CLAYS)[6]

The main soil-forming process affecting Vertisols is the shrinking and swelling of clay as these soils go through periods of drying and wetting. Vertisols have a high content (>30%) of sticky, swelling, and shrinking-type clays to a depth of 1 m or more. Most Vertisols are dark, even blackish in color, to a similar depth (Figure 3.21). However, unlike for most other soils, the dark color of Vertisols is not necessarily indicative of high organic matter content. The organic matter content of dark Vertisols typically ranges from as much as 5 or 6% to as little as 1%.

Vertisols typically develop from limestone, basalt, or other calcium- and magnesium-rich parent materials. In east Africa, they typically form in landscape depressions that collect the calcium and magnesium leached out of the surrounding upland soils. The presence of these cations encourages the formation of swelling-type clays (see Section 8.3).

Vertisols occur mostly in warm subhumid to semiarid environments, but a few (Cryerts) occur where the average soil temperatures are as low as 0°C (see Figure 3.8). The native vegetation is usually grassland. Vertisols generally occur where the climate features alternating wet and dry periods of several months each. In dry seasons the clay shrinks, causing the soils to develop deep, wide cracks that are diagnostic for this order (Figure 3.22a). The surface soil generally forms granules, of which a significant number may slough off into the cracks, giving rise to a partial inversion of the soil (Figure 3.23a, b). This accounts for the association with the term *invert*, from which this order derives its name.

When the rains come, water entering the large cracks moistens the clay in the subsoils, causing it to swell. The repeated shrinking and swelling of the subsoil clay results in a kind of imperceptively slow "rocking" movement of great masses of soil. As the subsoil swells, blocks of soil shear off from the mass and rub past each other under pressure, giving rise in the subsoil to shiny, grooved, tilted surfaces called **slickensides** (Figure 3.23c). Eventually, this back-and-forth motion may form bowl-shaped depressions with relatively deep profiles surrounded by slightly raised areas in which little soil development has occurred and in which the parent material remains close to the surface (see Figure 3.23b). The resulting pattern of micro-highs and micro-lows on the land surface, called **gilgai**, is usually discernable only where the soil is untilled (Figure 3.22b, c).

---
[6]See Coulombe et al. (1996) for a detailed review of the properties and mode of formation of Vertisols.

Figure 3.22 (a) *Wide cracks formed during the dry season in the surface layers of a Vertisol in Ethiopia. Surface debris can slough off into these cracks and move to subsoil. When the rains come, water can move quickly to the lower horizons, but the cracks are soon sealed, making the soils relatively impervious to the water.* (b) *Once the cracks have sealed, water may collect in the "micro-lows," making the gilgai relief easily visible as in the Texas Vertisol shown here.* (c) *Patterns of different grass species reflect the gilgai features of a Vertisol in the lower part of this central Ethiopia landscape* (Photos (a) and (c) courtesy of Ray R. Weil; (b) courtesy of K. N. Potter, USDA/ARS, Temple, Texas)

Distribution and Use

Globally, Vertisols comprise about 2.5% of the total land area. Large areas of Vertisols are found in India, Ethiopia, the Sudan, and northern and eastern Australia (see front papers). Smaller areas occur in sub-Saharan Africa, east central and southern Texas and California, USA, Mexico, Venezuela, Bolivia, and Paraguay. These latter soils probably are of the Usterts or Xererts suborders, since dry conditions persist long enough for the wide cracks to stay open for periods of three months or longer.

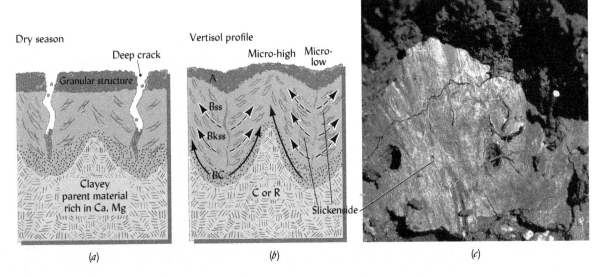

Figure 3.23 *Vertisols are high in swelling-type clay and have developed wedge-like structures in the subsoil horizons. (a) During the dry season, large cracks appear as the clay shrinks upon drying. Some of the surface soil granules fall into cracks under the influence of wind and animals. This action causes a partial mixing, or inversion, of the horizons. (b) During the wet season, rainwater pours down the cracks, wetting the soil near the bottom of the cracks first and then the entire profile. As the clay absorbs water, it swells the cracks shut, entrapping the collected granular soil. The increased soil volume causes lateral and upward movement of the soil mass. The soil is pushed up between the cracked areas. As the subsoil mass shears from the strain, smooth surfaces or slickensides form at oblique angles. These processes result in a Vertisol profile that typically exhibits gilgai, cracks more than 1 m deep and slickensides in a Bss horizon. (c) An example of a slickenside in a Vertisol. Note the grooved, shiny surface. The white spots in the lower right of the photo are calcium carbonate concretions that often accumulate in a Bkss horizon.* (Courtesy of Ray R. Weil)

There are several small but significant areas in the United States (see endpapers) of Vertisols classified in the Uderts suborder because relatively moist conditions prevent cracks from persisting for more than three months of the year. These areas are located in eastern Mississippi and western Alabama (the so-called *black belt*) and along the southeast coast of Texas. An area of Cryerts is located in the Dakotas and Saskatchewan.

The high shrink–swell potential of Vertisols makes them extremely problematic for any kind of roads or buildings (see Figure 4.43). This property also makes agricultural management very difficult. Because they are very sticky and plastic when wet and become very hard when dry, the timing of tillage operations is critical. Some farmers refer to Vertisols as *24-hour soils*, because they are said to be too wet to plow one day and too dry the next.

Even when the soil moisture is near optimal, the energy requirement for tillage is high. In areas such as those in India and the Sudan, where slow-moving animals or human power are commonly used to till the soil, farmers cannot perform tillage operations on time and are limited to the use of very small tillage implements that their animals can pull through the heavy soil.

Research shows that the large areas of Vertisols in the tropics can produce greatly increased yields of food crops with improved soil management practices. Soils in this order are, however, very susceptible to physical degradation and erosion (despite their mainly gentle slopes), and conservation practices or reversion to rangeland are important management options to consider.

3.12 MOLLISOLS (DARK, SOFT SOILS OF GRASSLANDS)

The principal process in the formation of Mollisols is the accumulation of calcium-rich organic matter, largely from the dense root systems of prairie grasses, to form the thick, soft Mollic epipedon that characterizes soils in this order (Figures 3.24). This humus-rich surface horizon is often 60 to 80 cm in depth. Its cation exchange capacity (Section 8.9) is more than 50% saturated with nonacid cations (Ca^{2+}, Mg^{2+}, etc.). Mollisols in humid regions generally have higher organic matter and darker, thicker mollic epipedons than their drier region counterparts (see Section 11.8).

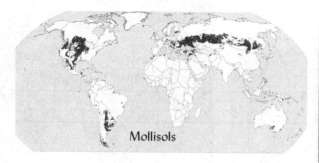

Figure 3.24 *Mollisols soil order—global distribution; land area; suborders; and a representative profile, a Fluventic Calciudoll from Queensland, Australia. The mollic epipedon extends to about 1 m.* (Courtesy of R. Weil)

The surface horizon generally has granular or crumb structures, largely resulting from an abundance of organic matter, fine roots, and swelling-type clays. In many cases, the highly aggregated soil is not hard when dry, hence the name *Mollisol*, implying softness (Table 3.3). In addition to the mollic epipedon, Mollisols may have an argillic (clay), natric, albic, or cambic subsurface horizon but not an oxic or spodic horizon.

Most Mollisols have developed under grass vegetation (Figure 3.25). Grassland soils of the central part of the United States, lying between Aridisols on the west and the Alfisols on the east, typify the central concept of this order. However, a few soils developed under forest vegetation (primarily in depressions) have a mollic epipedon and are included among the Mollisols.

Distribution and Use

The largest area of Mollisols in the world stretches from east to west across the heartlands of Kazakhstan, Ukraine, and Russia. Other sizable areas are found in the United States, Mongolia, and northern China and in northern Argentina, Paraguay, and Uruguay. Mollisols occupy only about 7% of the world's total soil area, but because of their generally high fertility, they account for a much higher percentage of total crop production.

Mollisols are extensive in North America, dominating the Great Plains and as far east as Illinois (see Figure 3.25 and endpapers). Where soil moisture is not limiting, Udolls dominate the uplands and are associated with wet Mollisols, termed Aquolls, in the low-lying places. A region characterized by Ustolls (intermittently dry during the summer) extends from Manitoba and Saskatchewan in Canada to southern Texas. Farther west are found sizable areas of Xerolls (with a Xeric moisture regime, which is very dry in summer but moist in winter). Conservation of soil water is a major consideration in the management of Ustolls, in particular.

In the United States, efforts are underway to preserve the few remnants of the once vast and diverse prairie ecosystem and also to restore native grasses and wildlife habitat. Because the high native fertility of Mollisols makes them among the world's most productive soils, few Mollisols have been left uncultivated in regions with sufficient rainfall for crop production. When they were first cleared and plowed, much of their native organic matter was oxidized,

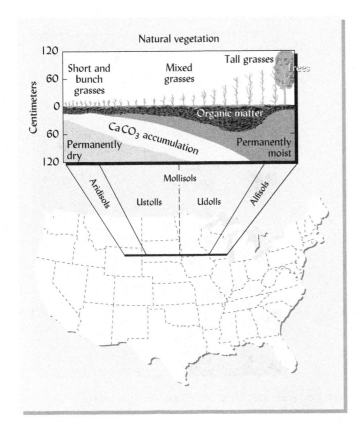

Figure 3.25 (Left) *Correlation between natural grassland vegetation and certain soil orders is graphically shown for a transect across north central United States. The controlling factor, of course, is climate. Note the deeper organic matter and deeper zone of calcium accumulation, sometimes underlain by gypsum, as one proceeds from the drier areas in the west toward the more humid region where prairie soils are found. Alfisols may develop under grassland vegetation, but more commonly occur under forests and have lighter-colored surface horizons.* (Upper, left to right) *Representative profiles of Aridisols, Ustolls, Udolls, Alfisols, and Spodosols.* (Courtesy of Ray R. Weil)

releasing nitrogen and other nutrients in sufficient quantities to produce high crop yields without the use of fertilizers. Even after more than a century of cultivation, these Mollisols are among the most productive soils, although continuous cultivation with row crops continues to lead to serious decline in organic matter, deterioration of soil structure and to soil erosion where the land is sloping.

3.13 ALFISOLS (ARGILLIC OR NATRIC HORIZON, MODERATELY LEACHED)

The Alfisols are more strongly weathered than soils in the orders just discussed, but less so than Spodosols, Ultisols, and Oxisols (see following sections). They are found in cool to hot humid areas (see Figure 3.8) as well as in the semiarid tropics and Mediterranean climates. Most often, Alfisols develop under native deciduous forests, although in some cases, as in California and parts of Africa, savanna (mixed trees and grass) is the native vegetation.

Alfisols are characterized by a subsurface diagnostic horizon in which silicate clay has accumulated by illuviation (see Figure 3.26). Clay skins or other signs of clay movement are present in such a B horizon (see Figure 3.5). In Alfisols, this clay-rich horizon is only moderately leached, and its cation exchange capacity is more than 35% saturated with nonacid cations (Ca^{2+}, Mg^{2+}, etc.). In most Alfisols this horizon is termed *argillic* because of its accumulation of silicate clays. The horizon is termed *natric* if, in addition to having an accumulation of clay, it is more than 15% saturated with sodium and has prismatic or columnar structure (see Figure 4.13). In some Alfisols in subhumid tropical regions, the accumulation is termed a *kandic* horizon (from the mineral kandite) because the clay minerals are more highly weathered and have a low cation exchange capacity.

Alfisols very rarely have a mollic epipedon, for such soils would be classified in the Argiudolls or other suborder of Mollisols with an argillic horizon. Instead, Alfisols typically have a relatively thin, gray to brown ochric epipedon or an umbric epipedon. Those formed under deciduous temperate forests commonly have a light-colored, leached *albic* E horizon immediately under the A horizon (see Figure 1.11).

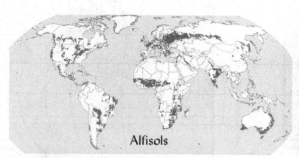

9.6% of global and 14.5% of U.S. ice-free land

Suborders are:
  Aqualfs (wet)
  Cryalfs (cold)
  Udalfs (humid)
  Ustalfs (moist/dry)
  Xeralfs (dry summers, moist winters)

Figure 3.26 *Alfisols soil order—global distribution; land area; suborders; and a representative profile, a Glossic Hapludalf from New York, USA. Scale in centimeters.* (Courtesy of Ray R. Weil)

Figure 3.27 *A landscape in southern Portugal dominated by Xeralfs. Most of the original forest has been cleared and the soils plowed for production of winter cereals and grapes.* (Photo courtesy of Ray R. Weil)

Distribution and Use

Udalfs (humid region Alfisols) dominate large areas in north central United States (see endpapers), as well as in central China, England, France, central Europe, and southeastern Australia. There are sizable areas of Xeralfs (Alfisols in regions of dry summers and moist winters) in central California, southwestern Australia, Italy, and central Spain and Portugal (Figure 3.27). Cryalfs (very cold) can be found in the Rocky Mountains; in south-central Canada; in Minnesota; in northern Europe, extending from the Baltic States through western Russia; and in Siberia. Where summers are hot and dry, including areas in Texas, New Mexico, sub-Saharan Africa, eastern Brazil, eastern India, and southeastern Asia, Ustalfs are prominent. Many Alfisols landscapes include wet depressions characterized by Aqualfs.

In general, Alfisols are productive soils. Good hardwood forest growth and crop yields are favored by their medium to high nonacid cation saturation status, generally favorable texture, and location (except for some Xeralfs) in regions with enough rainfall to support good plant growth for at least part of the year. In the United States these soils rank favorably with the Mollisols and Ultisols in their productive capacity. Many Alfisols, especially the sandier ones, are quite susceptible to erosion by heavy rains if deprived of their natural surface litter. Alfisols in Udic moisture regimes are sufficiently acidic in the A horizon to require amendment with limestone for many kinds of plants (see Chapter 9).

3.14 ULTISOLS (ARGILLIC HORIZON, HIGHLY LEACHED)

The principal processes involved in forming Ultisols are clay mineral weathering, translocation of clays to accumulate in an argillic or kandic horizon, and leaching of nonacid cations from the profile. Most Ultisols have developed under moist conditions in warm to tropical climates. Ultisols are formed on old land surfaces, usually under forest vegetation, although savanna or even swamp vegetation is also common. They often have an ochric or umbric epipedon but are characterized by a relatively acidic B horizon that has less than 35% of the exchange capacity satisfied with nonacid cations. The clay accumulation may be either an argillic horizon or, if the clay is of low activity, a kandic horizon. Ultisols commonly have both an epipedon and a subsoil that is quite acid and low in plant nutrients.

Ultisols form under forest vegetation and are more highly weathered and acidic than Alfisols but less acid than Spodosols and less highly weathered than the Oxisols. Except for the wetter members of the order, their subsurface horizons are commonly red or yellow in color, evidence of accumulations of oxides of iron (see Figure 3.28). Certain Ultisols (as well as certain Alfisols in tropical regions) that formed under fluctuating wetness conditions exhibit horizons

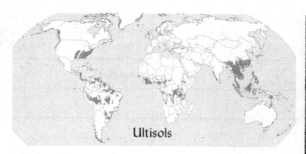
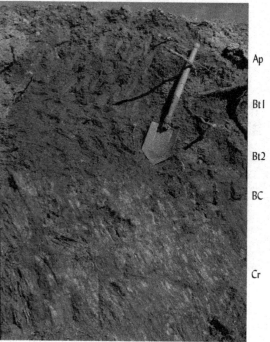

8.5% of global and 9.6% of U.S. ice-free land

Suborders are:
    Aquults (wet)
    Humults (high humus)
    Udults (humid)
    Ustults (moist/dry)
    Xerults (dry summers, moist winters)

Figure 3.28 *Ultisols soil order—global distribution; land area; suborders; and a representative profile, a Typic hapludult from central Virginia, USA, showing rock structure in the saprolite below the 60-cm-long shovel.* (Courtesy of Ray R. Weil)

of iron-rich mottled material called **plinthite** (see Figure 3.29). This material is soft and can be easily dug from the profile so long as it remains moist. When dried in the air, however, plinthite hardens irreversibly into a kind of ironstone that is virtually useless for cultivation (Figure 3.29, *right*) but can be used to make durable bricks for building (Figure 3.29, *left*).

Distribution and Use

Large areas of Udults are also located in southeastern United States, southeastern Asia, and in southern China (see endpapers). Extensive areas of Ultisols are found in the humid tropics in close association with some Oxisols. Important agricultural areas are found in southern Brazil and Paraguay.

Humults (high in organic matter) are found in the United States in Hawaii and in western California, Oregon, and Washington. Humults are also present in the highlands of some tropical countries. Xerults (Ultisols in Mediterranean-type climates) occur locally in southern Oregon and northern and eastern California. Ustults are found in semiarid areas with a marked dry season. Together with the Ustalfs, the Ustults occupy large areas in Africa and India. Ultisols are prominent on the east and northeast coasts of Australia (see front papers).

Figure 3.29 *Plinthite is a feature of some Ultisols (middle) that formed under fluctuating water table conditions such that iron alternately dissolved and precipitated creating red iron-rich and whitish iron-depleted splotches of soil color. The material is soft when in place and moist, but hardens irreversibly when dried in the air. Blocks of the material (also termed laterite) can be cut and dried into durable bricks (left). Plinthite can be a serious limitation for plant growth if the hardened form of the material occurs at a shallow depth. It often occurs as hard cemented layers, a chunk of which is seen on the soil surface in an African field (right).* (Photo courtesy of Ray R. Weil)

Although Ultisols are not naturally as fertile as Alfisols or Mollisols, they respond well to good management. They are located mostly in regions of long growing seasons and of ample moisture for good crop production. The silicate clays of Ultisols are usually of the nonsticky type, which, along with the presence of iron oxides and aluminum, may enhance workability. With adequate management of fertilizers, lime, and organic matter, Ultisols can be quite productive for agriculture and softwood and hardwood forests.

3.15 SPODOSOLS (ACID, SANDY, FOREST SOILS, HIGHLY LEACHED)

Intensive acid leaching of coarse-textured, acid parent materials is the principal process leading to the formation of Spodosols (Figure 3.30). These are mineral soils with a *spodic* horizon, a subsurface accumulation of illuviated organic matter, and an accumulation of aluminum oxides with or without iron oxides. This usually thin, dark, illuvial horizon typically underlies a light, ash-colored, eluvial *albic* horizon. The combination of bright white E and very dark A and B horizons, often occur along wavy wetting fronts, is visually quite striking and makes for some of the most easily recognized, and some would say most beautiful of soil profiles (see Figure 3.30).

Spodosols form under forest vegetation, especially under coniferous species whose needles are low in base-forming cations like calcium and high in acid resins. As this acid litter decomposes, strongly acid organic compounds are released and carried down into the permeable profile by percolating waters. Some of the leaching organic compounds may precipitate and form a black-colored Bh horizon. Leaching organic acids bind with iron and aluminum, removing these metals from the A and E horizons and carrying them downward. This iron and aluminum eventually precipitates in a reddish-brown-colored Bs horizon, usually just below the black-colored Bh horizon. Together, the Bh and Bs horizons constitute the spodic diagnostic horizon that defines the Spodosols. As iron oxides are stripped from the E horizon by the organic leaching process, this horizon may become a nearly white albic diagnostic horizon that consists mainly of clean quartz sand.

2.6% of global and 3.3% of U.S. ice-free land

Suborders are:
  Aquods (wet)
  Cryods (cold)
  Gelods (very cold)
  Humods (humus)
  Orthods (typical)

Figure 3.30 *Spodosols soil order—global distribution; land area; suborders; and a representative profile, a Typic Haplorthod from New Jersey, USA, showing pronounced albic and spodic horizons. Scale marked in 10 cm.*
(Courtesy of Ray R. Weil)

Distribution and Use

Spodosols generally occur in regions with cool, moist climates (see Figures 3.8 and 3.30), but some also can be found in moist tropical and subtropical areas (such as the U.S. Atlantic coastal plain).

Large areas of Spodosols are found in northern United States, northern Europe, and Russia and in central and eastern Canada. Small but ecologically important areas occur in the southern part of South America and in the cool mountainous areas of temperate regions.

Most Spodosols are Orthods, soils that typify the central concept of Spodosols described previously. Some, however, are Aquods because they are seasonally saturated with water and possess characteristics associated with wetness. Important areas of Aquods occur in Florida and other areas with warm climates.

Spodosols are naturally very acid and infertile. When properly fertilized, however, they can become quite productive. For example, most potato-producing soils of northern Europe and Maine are Spodosols, as are some of the vegetable and fruit-producing soils of Florida, Michigan, and Wisconsin. Because of their sandy nature and occurrence in regions of high rainfall, groundwater contamination by leaching of soluble fertilizers and pesticides has proved to be a serious problem where these soils are used in crop production. Many formerly cultivated Spodosols have been abandoned and are now covered mostly with forests, the vegetation under which they originally developed. Most Spodosols should remain as forest habitats. Because they are already quite acid and poorly buffered, many Spodosols and the lakes in watersheds dominated by soils of this order are susceptible to damage from acid rain (see Section 9.6).

3.16 OXISOLS (OXIC HORIZON, HIGHLY WEATHERED)

The Oxisols are the most highly weathered soils in the classification system (see Figure 3.7). They form in hot climates with nearly year-round moist conditions; hence, the native vegetation is generally tropical rain forest. However, some Oxisols (Ustox) are found in areas that are today much drier than was the case when the soils were forming their oxic characteristics. Oxisols' most important diagnostic feature is a deep subsurface oxic horizon (Figure 3.31). Weathering and intense leaching have removed much of the silica from the silicate materials in this horizon. Some quartz and 1:1-type silicate clay minerals remain, but the iron and aluminum hydrous oxide clays are dominant (see Chapter 8 for information on the various clay minerals). The epipedon in most Oxisols is either ochric or umbric. Usually the boundaries between subsurface horizons are indistinct, giving the subsoil a relatively uniform appearance with depth.

The clay content of Oxisols is generally high, but the clays are of the low-activity, non-sticky type. Consequently, when the clay dries out, it is not hard and cloddy but is easily worked. Also, Oxisols are resistant to compaction, so water moves freely through the profile. The depth of weathering in Oxisols is typically much greater than for most of the other soils, 20 m or more having been observed. The low-activity clays have a very limited capacity to hold nutrient cations such as Ca^{2+}, Mg^{2+}, and K^+, so Oxisols are typically of low natural fertility and moderately acid. The high concentration of iron and aluminum oxides also gives these soils a capacity to bind so tightly with what little P is present that P deficiency often limits plant growth once the natural vegetation is disturbed.

Road and building construction is relatively easily accomplished on most Oxisols because these soils are easily excavated, do not shrink and swell, and are physically very stable on slopes (Figure 3.32). The very stable aggregation of the clays, stimulated largely by iron compounds, makes these soils quite resistant to erosion.

Distribution and Use

Oxisols occupy old land surfaces that have not been disturbed by glaciation or erosion. Although nearly all Oxisols occur in the tropics, most tropical soils are *not* Oxisols (see front papers). Some of the areas of the Amazon basin previously mapped as Oxisols are in reality

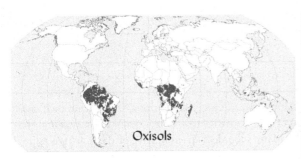

7.6% of global and 0.01% of U.S. ice-free land

Suborders are:
  Aquox (wet)
  Perox (very humid)
  Torrox (hot, dry)
  Udox (humid)
  Ustox (moist/dry)

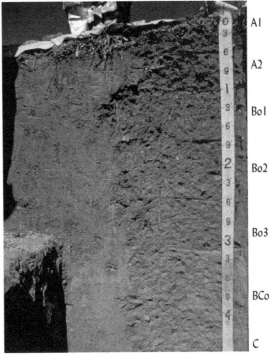

Figure 3.31 *Oxisols soil order—global distribution; land area; suborders; and a representative profile, a Eudeptic Haplorthox from Puerto Rico, USA. Scale marked in feet and inches.* (Courtesy of Soil Science Society of America)

dominated by Ultisols and other soils. Udox (Oxisols having little or no dry season) occur in northern Brazil and neighboring countries as well as in the Caribbean area (Figure 3.32). Important areas of Ustox (hot, dry summers) occur in Brazil to the south of the Udox. Oxisols are prominent in the humid areas of central Africa.

Relatively less is known about Oxisols than about most other soil orders. They occur in large geographic areas, often associated with Ultisols. Millions of people in the tropics depend on them for food and fiber production. However, because of their low natural fertility, most Oxisols have been left under forest vegetation or are farmed by shifting cultivation methods. Nutrient cycling by deep-rooted trees is especially important to the productivity of these soils. Probably the best use of Oxisols, other than supporting rain forests, is the culture of mixed-canopy perennial crops, especially tree crops. Such cultures can restore the nutrient cycling system that characterized the soil–plant relationships before the rain forest was removed.

Figure 3.32 *An Oxisols landscape and profile in a rain-forest area of Puerto Rico illustrating the dusky, red colors, rather homogenous profile and physical stability characteristic of many Oxisols.* (Photo courtesy of Ray R. Weil)

3.17 LOWER-LEVEL CATEGORIES IN *SOIL TAXONOMY*

Suborders

As indicated next to the global distribution maps in previous sections, soils within each order are grouped into suborders on the basis of soil properties that reflect major environmental controls on current soil-forming processes. Many suborders are indicative of the moisture regime or, less frequently, the temperature regime under which the soils are found.

Soil Moisture Regimes (SMRs)

A SMR refers to the presence or absence of either water-saturated conditions (usually groundwater) or plant-available soil water during specified periods in the year in what is termed the *control section* of the soil. The upper boundary of the SMR control section is the depth that 2.5 cm of water will penetrate within 48 hours when added to a dry soil. The lower boundary is the depth that 7.5 cm of water will penetrate. Generally, the control section ranges from 10 to 30 cm for soils high in fine particles (clay) and from 30 to 90 cm for sandy soils. Several moisture regime classes are used to characterize soils.

Aquic. Soil is saturated with water and virtually free of gaseous oxygen for sufficient periods of time for evidence of poor aeration (gleying and mottling) to occur.

Udic. Soil moisture is sufficiently high year-round in most years to meet plant needs. This regime is common for soils in humid climatic regions and characterizes about one-third of the worldwide land area. An extremely wet moisture regime with excess moisture for leaching throughout the year is termed **perudic**.

Ustic. Soil moisture is intermediate between Udic and Aridic regimes—generally winters are relatively dry, but the growing season is characterized by some plant-available moisture between significant periods of drought.

Aridic. The soil is dry for at least half of the growing season and moist for less than 90 consecutive days. This regime is characteristic of arid regions. The term *torric* is used to indicate the same moisture condition in certain soils that are both hot and dry in summer, though they may not be hot in winter.

Xeric. This SMR is found in typical Mediterranean-type climates, with cool, moist winters and warm, dry summers. Like the Ustic regime, it is characterized by having long periods of drought in the summer.

To determine the relationship between suborder names and soil characteristics, refer to Table 3.4. Here the formative elements for suborder names are identified and their connotations given. Thus, the Ustolls are dry Mollisols. Likewise, soils in the Udults suborder are moist Ultisols. Note that other characteristics besides SMR can be used in naming a suborder (e.g., presence of gypsum for Gypsids or extreme sandiness for Psamments).

Great Groups

The great groups are subdivisions of suborders. About 350 great groups are recognized. They are defined largely by the presence or absence of diagnostic horizons and the arrangements of those horizons. These horizon designations are included in the list of formative elements for the names of great groups shown in Table 3.5. Note that these formative elements refer to epipedons such as umbric and ochric (see Table 3.1 and Figure 3.3), to subsurface horizons such as argillic and natric, and to certain diagnostic impervious layers such as duripans and fragipans (see Figure 3.34).

Remember that the great group names are made up of these formative elements attached as prefixes to the names of suborders in which the great groups occur. Thus, Ustolls with a natric horizon (high in sodium) belong to the Natrustolls great group. As can be seen in the example discussed in Box 3.1, soil descriptions at the great group level can provide important information not indicated at the higher, more general levels of classification.

Table 3.4
FORMATIVE ELEMENTS IN NAMES OF SUBORDERS IN *SOIL TAXONOMY*

Formative Element	Derivation	Connotation of Formative Element
alb	L. *albus*, white	Presence of albic horizon (a bleached eluvial horizon)
aqu	L. *aqua*, water	Characteristics associated with wetness
ar	L. *arare*, to plow	Mixed horizons
arg	L. *argilla*, white clay	Presence of argillic horizon (a horizon with illuvial clay)
calc	L. *calcis*, lime	Presence of calcic horizon
camb	L. *cambriare*, to change	Presence of cambic horizon
cry	Gk. *kryos*, icy cold	Cold
dur	L. *durus*, hard	Presence of a duripan
fibr	L. *fibra*, fiber	Least-decomposed stage
fluv	L. *fluvius*, river	Floodplains
fol	L. *folia*, leaf	Mass of leaves
gel	Gk. *gelid*, cold	Cold
gyps	L. *gypsum*, gypsum	Presence of gypsic horizon
hem	Gk. *hemi*, half	Intermediate stage of decomposition
hist	Gk. *histos*, tissue	Presence of histic epipedon
hum	L. *humus*, earth	Presence of organic matter
orth	Gk. *orthos*, true	The common ones
per	L. *per*, throughout time	Of year-round humid climates, perudic moisture regime
psamm	Gk. *psammos*, sand	Sand textures
rend	Modified from Rendzina	Rendzina-like—high in carbonates
sal	L. *sal*, salt	Presence of salic (saline) horizon
sapr	Gk. *sapros*, rotten	Most decomposed stage
torr	L. *torridus*, hot and dry	Usually dry
turb	L. *turbidus*, disturbed	Cryoturbation
ud	L. *udus*, humid	Of humid climates
ust	L. *ustus*, burnt	Of dry climates, usually hot in summer
vitr	L. *vitreus*, glass	Resembling glass
wass	G. *wasser*, water	Positive water potential at the soil surface year round
xer	Gk. *xeros*, dry	Dry summers, moist winters

Not all possible combinations of great group prefixes and suborders are used. In some cases a particular combination does not exist. For example, Aquolls occur in lowland areas but not on very old landscapes. Hence, there are no "Paleaquolls." Also, since *all* Ultisols contain an argillic horizon, the use of terms such as *Argiudults* would be redundant.

Subgroups

Subgroups are subdivisions of the great groups. More than 2,600 subgroups are recognized. The central concept of a great group makes up one subgroup, termed *Typic*. Thus, the Typic Hapludolls subgroup typifies the Hapludolls great group. Other subgroups may have characteristics that *intergrade* between those of the central concept and soils of other orders, suborders, or great groups. A Hapludoll with restricted drainage (but not wet enough to be an Aquoll) would be classified as an Aquic Hapludoll. Some intergrades may have properties in common with other orders or with other great groups. Thus, soils in the Entic Hapludolls subgroup are very weakly developed Mollisols, close to being in the Entisols order.

Table 3.5
FORMATIVE ELEMENTS FOR NAMES OF GREAT GROUPS AND THEIR CONNOTATION
These formative elements combined with the appropriate suborder names give the great group names.

Formative Element	Connotation	Formative Element	Connotation
acr	Extreme weathering	hist	Presence of organic materials
al	High aluminum, low iron	hum	Humus
alb	Albic horizon	hydr	Water
and	Ando-like	kand	Low-activity 1:1 silicate clay
anhy	Anhydrous	kanhapl	Kandic and minimum horizon
aqu	Water saturated	luv, lu	Illuvial
argi	Argillic horizon	melan	Melanic epipedon
calc, calci	Calcic horizon	molli	With a mollic epipedon
camb	Cambic horizon	natr	Presence of a natric horizon
cry	Cold	pale	Old development
dur	Duripan	petr	Cemented horizon
dystr, dys	Low base saturation	plac	Thin pan
endo	Fully water saturated	plagg	Plaggen horizon
epi	Perched water table	plinth	Plinthite
eutr	High base saturation	psamm	Sand texture
ferr	Iron	quartz	High quartz
fibr	Least decomposed	rhod	Dark red colors
fluv	Floodplain	sal	Salic horizon
fol	Mass of leaves	sapr	Most decomposed
fragi	Fragipan	somb	Dark horizon
fragloss	Combination of *fragi* and *gloss*	sphagn	Sphagnum moss
frassi	Inundated but low in salts	sulf	Sulfuric
fulv	Light-colored melanic horizon	torr	Usually dry and hot
gel	Gelic temperature regime	ud	Humid climates
glaci	Glacic layer	umbr	Umbric epipedon
gyps	Gypsic horizon	ust	Dry climate, usually hot in summer
gloss	Tongued	verm	Wormy or mixed by animals
hal	Salty	vitr	Glass
hapl	Minimum horizon	xanthic	Red/yellow colors from iron
hem	Intermediate decomposition	xer	Dry summers, moist winters

Other subgroups are considered to be extragrades rather than intergrades. That is, they exhibit special characterisitics that do not suggest they are close to being in another soil group. For example, a Hapudoll with evidence of intense earthworm activity would fall in the Vermic Hapludolls subgroup.

Subgroups for Human-Influenced Soils Human-influenced (anthropic) soil characteristics are now recognized as subgroup extragrades in Soil Taxonomy. Subgroup adjectives for human-influenced characteristics include Plaggic (having a Plaggen epipedon), Anthropic (having an Anthropic epipedon), Anthraquic (e.g., rice paddy soils), Anthrodensic (human-compacted soils), Anthraltic (formed in human-altered, typically bulldozed, material), and Anthroportic (formed in human-transported material, such as dredge spoil). As an example, see the Anthroportic Udorthent formed in human-transported dredge spoil material in Figure 2.23.

BOX 3.1
GREAT GROUPS, FRAGIPANS, AND ARCHAEOLOGIC DIGS

In this box we will see how misclassification, even at a lower level in *Soil Taxonomy*, such as the great group, can have costly ramifications. In order to preserve our historical and prehistorical heritage, laws require that an archaeological impact statement be prepared prior to starting major construction work. Selected sites are studied by archaeologists, so that at least some of the artifacts can be preserved and interpreted before construction activities obliterate them forever. Only a few relatively small sites can be subjected to actual archaeological digs because of the expensive skilled hand labor involved (Figure 3.33).

Such an archaeological impact study was ordered as a precursor to construction of a new highway in a mid-Atlantic state. In the first phase, a consulting company gathered soils and other information from maps, aerial photographs, and field investigations to determine where Neolithic people may have occupied sites. Then the consultants identified about 12 ha of land where artifacts indicated significant Neolithic activities. The soils in one area were mapped mainly as Typic Dystrudepts. These soils formed in old colluvial and alluvial materials that, many thousands of years ago, had been along a river bank. Several representative soil profiles were examined by digging pits with a backhoe. The different horizons were described, and it was determined in which horizons artifacts were most likely to be found.

What was *not* noted was the presence in these soils of a fragipan, a dense, brittle layer that is extremely difficult to excavate using hand tools. A fragipan is a subsurface diagnostic horizon used to classify soils, usually at the great group or subgroup level (see Figure 3.34). Its presence would distinguish Fragiudepts from Dystrudepts.

When it came time for the actual hand excavation of sites to recover artifacts, a second consulting company was awarded the contract. Unfortunately, their bid on the contract was based on soil descriptions that did not specifically classify the soils as Fragiudepts—soils with very dense, brittle, hard fragipans in the layer that would need to be excavated by hand. So difficult was this layer to excavate and sift through by hand that it nearly doubled the cost of the excavation—an additional expense of about $1 million. Needless to say, there ensued a controversy as to whether this cost would be borne by the consulting firm that bid with faulty soils data, the original consulting firm that failed to adequately describe the presence of the fragipan, or the highway construction company that was paying for the survey.

This episode illustrates the practical importance of soil classification. The formative element *Fragi* in a soil great group name warns of the presence of a dense, impermeable layer that will be very difficult to excavate, will restrict root growth (often causing trees to topple in the wind or become severely stunted), may cause a perched water table (epiaquic conditions), and will interfere with proper percolation in a septic drain field.

Figure 3.33 *An archaeological dig.* (Photo courtesy of Antonio Segovia, University of Maryland)

BOX 3.1
GREAT GROUPS, FRAGIPANS, AND ARCHAEOLOGIC DIGS

Figure 3.34 *A forested fragiudalf in Missouri containing a typical well-developed fragipan with coarse prismatic structure (outlined by gray, iron-depleted coatings). Fragipans (usually Bx or Cx horizons) are extremely dense and brittle. They consist mainly of silt, often with considerable sand, but not very much clay. One sign of encountering a fragipan in the field is the ringing noise that your shovel will make when you attempt to excavate it. Digging through a fragipan is almost like digging concrete. Plant roots cannot penetrate this layer. Yet, once a piece of a fragipan is broken loose, it fairly easily crushes with hand pressure. It does not squash or act in a plastic manner as a claypan would; instead, it bursts in a brittle manner.* (Photo courtesy of Fred Rhoton, Agricultural Research Service, U.S. Department of Agriculture)

Families

Within a subgroup, soils fall into a particular family if, at a specified depth, they have similar physical and chemical properties affecting the growth of plant roots. About 8,000 families have been identified. The criteria used include broad classes of particle size, mineralogy, cation exchange activity of the clay, temperature, and depth of the soil penetrable by roots. Table 3.6 gives examples of the classes used. Terms such as *loamy*, *sandy*, and *clayey* are used to identify the broad particle size classes. Terms used to describe the mineralogical classes include *smectitic*, *kaolinitic*, *siliceous*, *carbonatic*, and *mixed*. The clays are described as *superactive*, *active*, *semiactive*, or *subactive* with regard to their capacity to hold cations. The terms *shallow* and *micro* are sometimes used at the family level to indicate unusual soil depths.

Soil Temperature Regimes

Soil temperature regimes, such as frigid, mesic, and thermic, are used to classify soils at the family level. In addition, the cryic (Greek *kryos*, very cold) temperature regime also distinguishes some higher-level groups. These regimes are based on mean annual soil temperature, mean summer temperature, and the difference between mean summer and winter temperatures, all at 50 cm depth (Table 3.6).

Thus, a Typic Argiudoll from Iowa, loamy in texture, having a mixture of moderately active clay minerals and with annual soil temperatures (at 50 cm depth) between 8 and 15°C, is classed in the *loamy, mixed, active, mesic Typic Argiudolls* family. In contrast, a sandy-textured Typic Haplorthod, high in quartz, and located in a cold area in eastern Canada, is classed in the *sandy, siliceous, frigid Typic Haplorthods* family (note that clay activity classes are not used for soils in sandy textural classes).

Series

The series category is the most specific unit of the classification system. It is a subdivision of the family, and each series is defined by a specific range of soil properties involving primarily the kind, thickness, and arrangement of horizons. Features such as a hard pan within a certain

Table 3.6
PARTICLE-SIZE, MINERALOGY, CATION EXCHANGE ACTIVITY, AND TEMPERATURE CLASSES COMMONLY USED TO DIFFERENTIATE SOIL FAMILIES

The characteristics generally apply to the subsoil or 50 cm depth. Criteria used to differentiate soil families, but not shown here, include calcareous or highly aluminum toxic (allic) properties, extremely shallow depth (shallow or micro), degree of cementation, coatings on sand grains, and the presence of permanent cracks or human artifacts.

Particle-Size Class	Mineralogy Class	Cation Exchange Activity[b]		Soil Temperature Regime Class		
		Term	CEC/% clay	Mean Annual Temperature, °C	>6°C Difference Between Summer and Winter	<6°C Difference Between Summer and Winter
Ashy	Mixed	Superactive	>0.60	<−10	Hypergelic[c]	—
Fragmental	Micaceous	Active	0.4–0.6	−4 to −10	Pergelic[c]	—
Sandy-skeletal[a]	Siliceous	Semiactive	0.24–0.4	+1 to −4	Subgelic[c]	—
Sandy	Kaolinitic	Subactive	<0.24	<+8	Cryic	—
Loamy	Smectitic			<+8	Frigid[d]	Isofrigid
Clayey	Gibbsitic			+8 to +15	Mesic	Isomesic
Fine-silty	Gypsic			+15 to +22	Thermic	Isothermic
Fine-loamy	Carbonatic			>+22	Hyperthermic	Isohyperthermic

[a]Skeletal refers to presence of 35–90% rock fragments by volume.
[b]Cation exchange activity class is not used for taxa already defined by low CEC (e.g., kandic or oxic groups).
[c]Permafrost present.
[d]Frigid is warmer in summer than cryic.

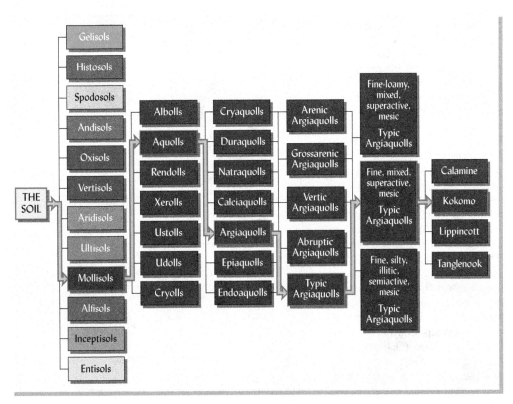

Figure 3.35 *How one soil (Kokomo) keys out in the overall classification scheme. The shaded boxes show that this soil is in the Mollisols order, Aquolls suborder, Argiaquolls great group, and so on. In each category, other classification units are shown in the order in which they key out in Soil Taxonomy. Many more families exist than are shown.* (Diagram courtesy of Ray R. Weil)

distance below the surface, a distinct zone of calcium carbonate accumulation at a certain depth, or striking color characteristics may aid in series identification.

In the United States, and many other countries, each series is given a name, usually from some town, river, or lake such as Fargo, Muscatine, Cecil, Mohave, or Ontario. There are about 25,000 soil series in the United States.

The complete classification of a Mollisol, the Kokomo series, is given in Figure 3.35. This figure illustrates how *Soil Taxonomy* can be used to show the relationship between *the soil*, a comprehensive term covering all soils, and a specific soil series. The figure deserves study because it reveals much about the structure and use of *Soil Taxonomy*. If a soil series name is known, the complete *Soil Taxonomy* classification of the soil may be found on the Internet (http://www.nrcs.usda.gov/wps/portal/nrcs/detail/soils/ref/?cid=nrcs142p2_053587). Box 3.2 illustrates how soil taxonomic information can assist in understanding the nature of a landscape such as shown in Figure 3.36.

BOX 3.2
USING *SOIL TAXONOMY* TO UNDERSTAND A LANDSCAPE

Figure 3.36 depicts a landscape in a humid temperate region (Iowa) where 2 to 7 m of loess overlies leached glacial till and the native vegetation was principally tall grass prairie interspersed with small areas of trees. This landscape demonstrates how diagnostic horizons and other features of soil taxonomy are used to organize soils information and understand landscapes.

Seven soil map units are shown in the block landscape diagram, along with a profile diagram for the dominant soil series in each map unit. The soils include two Alfisols (Fayette and Downs) and five Mollisols (Tama, Wabash, Dinsdale, Muscatine, and Garwin). The particular set of soil horizons present in each profile relates to the (1) parent material, (2) vegetation, and (3) topography and drainage. For example, notice that the Dinsdale soil differs from the Tama because two parent materials (loess and glacial till) contributed to the Dinsdale profile, but the Tama soil is found where the loess layer by itself is thick enough to accommodate the entire profile. Both the Tama and Fayette soils exhibit argillic B horizons, but the Fayette has a thin ochric epipedon and a bleached albic horizon because it formed under forest vegetation, while the Tama has a thick mollic epipedon because it formed under grassland vegetation. The influence of topography can be seen by noting that soils on concave or level positions (the Garwin and Muscatine soils, which have slopes ranging from 0 to 1% and 1 to 3%, respectively) are wetter and less permeable than those on the steeper slopes (Tama and Dinsdale soils). In the less sloping, wetter soils, restricted drainage has retarded the development of an argillic horizon, so that only a gleyed (waterlogged) cambic B horizon (Bg) is present.

The soil taxonomy names reflect these relationships. The formative element *aqu* appears in the taxonomic name at the suborder level (Endoaquolls) for the very wet, poorly drained soils, but only at the subgroup level for the less wet, somewhat poorly drained soil (Aquic Hapludoll). The formative element *argi* is used in the name of two soils (Argiudolls) to indicate that enough clay has accumulated in the B horizon of these Mollisols to develop into an argillic diagnostic horizon. *Argi* does not appear in the names of the two Alfisols, because an accumulation of clay (argillic or similar horizon) is a required feature of *all* Alfisols. The subgroup modifier Cumulic indicates that the Wabash soil has an unusually thick mollic epipedon because soil material washing off the uplands and carried by local streams has accumulated in the low-lying floodplains where this soil is found. The modifier Mollic used for the Downs soil indicates that this soil is transitional between the Alfisols and Mollisols, the A horizon in the Downs soils being slightly too thin to classify as a mollic epipedon.

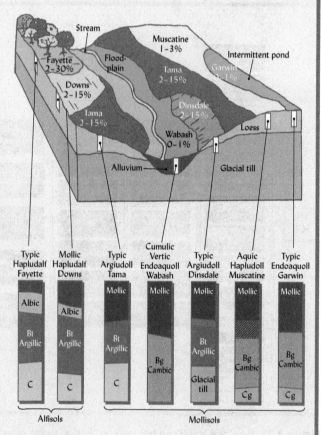

Figure 3.36 *Soil taxonomy reflects soil-landscape relationships.*

3.18 MAPPING THE DIFFERENT SOILS IN A LANDSCAPE[7]

Soil maps are in great demand as tools for practical land planning and management. Many soil scientists therefore specialize in mapping soils. Before beginning the actual mapping process, a soil scientist must learn as much as possible about the soils, geology, landforms, and vegetation in the survey area.

Soil Description[8]

At the heart of soil mapping lies the detailed description and classification of soils as observed in a soil pit (Figure 3.37). A soil pit is a rectangular hole large enough and deep enough to allow one or more people to enter and study a typical pedon (see Section 3.1) as exposed on the pit face. Before describing a profile, it is necessary to clean away the surface material from the pit face to expose fresh soil at all depths. This is done using a sturdy knife or trowel, starting at the ground surface (top of the pit wall) and working down such that the last layer to be cleaned is at the pit bottom. In this way, the soil removed will not fall onto and obscure the previously cleaned layers. In soils with significant structure, individual structural peds should be picked out with a knife point to leave a vertical pit face that follows ped surfaces (e.g., extreme right in Figure 3.37).

One then examines the fresh pit face for colors (using a Munsell color book, Figure 4.1), texture (by feel, Figure 4.9), consistency (Table 4.6), structure (Figure 4.13), and plant rooting patterns. Based on these and other soil features, the next step is to determine which horizons are present and at what depths their boundaries occur.

A soil description is then written in a standard format that facilitates communication with other soil scientists and comparison with other soils. As far as possible at this stage,

Figure 3.37 *A soil pit allows detailed observations to be made of the soil in place. Here, two soil scientists describe a typical pedon for the Pamunkey soil series as revealed on the wall of a soil pit. The scientists are comparing the colors and textures of soil samples removed from several horizons and recording their observations to make a soil profile description. Later, when they are making transects of auger boring to map soils in a landscape (see Figure 3.38), they can identify the soils they encounter by comparing the properties of samples brought up in their augers to the properties listed in detailed written descriptions of the soil pit.* (Photo courtesy of Ray R. Weil)

---

[7]For an internationally oriented guide to all aspects of the process of making soil maps, see Legros (2006). For official procedures for making soil surveys in the United States, see USDA-NRCS (2006).

[8]Several detailed practical guides are available and should be consulted to standardize the process of making soil descriptions globally and in the United States (FAO, 2006; Schoeneberger et al., 2012).

the soil horizons are given master (A, E, B, etc.) and subhorizon (2Bt, Ap, etc.) designations (see Table 2.5). For example, the description of the Ap horizon (Thorndale soil, Typic Fragiaqualfs) might read: "0-20cm; dark grayish brown (2.5Y 4/2) silt loam; weak medium granular structure; friable, slightly sticky, slightly plastic; many fine roots; neutral; clear smooth boundary." Finally, samples of soil material are obtained from each horizon used for chemical, physical, and mineralogical laboratory analyses.

Delineating Soil Boundaries

For obvious reasons, soil scientists cannot dig pits at very many locations on the landscape to determine which soils are present and where their boundaries occur. Instead, they will bring up soil material from numerous small boreholes made with a hand auger or hydraulic probe (Figure 3.38a, b). The texture, color, and other properties of the soil material from various depths can be compared mentally to characteristics of the known soils in the region.

With hundreds of different soils in many regions, this might seem to be a hopeless task. However, the soil scientist is not blindly or randomly boring holes. Rather, he or she is working with a landscape model in mind—an understanding of the soil associations and how the five soil-forming factors determine which soils are likely to be found in which landscape positions. Usually there are only a few soils likely to occupy a particular location, so only a few characteristics must be checked. The soil auger is used primarily to confirm that the type of soil predicted to occur in a particular landscape position is the type actually there. The nature of soil units and the locations of the boundary lines surrounding them are inferred from auger borings at numerous locations across a landscape.

A landscape model that reflects knowledge of the interplay of the soil-forming factors across the land can greatly expedite the soil scientist's work. An efficient soil mapper will select sites to make auger borings based on information obtained from detailed maps or imagery of topography, vegetation, and drainage networks, as well as from observed changes in topography, vegetation, and soil surface colors. A typical approach is to traverse the landscape along selected transects

Figure 3.38 *Soil maps are prepared by soil scientists who examine the soils in the field using such tools as a truck-mounted hydraulic soil probe (a) or a hand-powered soil bucket auger (b). About a 10-cm depth increment of soil is typically retrieved with just two or three revolutions of the typical bucket auger. After pulling the auger from the ground, but before emptying its load of soil, it is advisable to carefully examine the soil exposed at the end of the auger bucket to observe the relatively undisturbed colors and structure of the soil (c, d). Soil scientists may lay each soil increment in order on a sheet of plastic or in a trough to examine later (e). This allows the colors and textures to be efficiently determined while also presenting an overall view of the soil profile.* (Photo (a) courtesy of USDA/NRCS, other photos courtesy of Ray R. Weil)

(straight-line paths), augering only at enough points to confirm expected soil properties and boundaries. In order to more accurately locate soil boundaries, more frequent borings are made near the places where breaks in slope or other landscape clues suggest that soil boundaries will occur.

Online Interactive Soil Survey

Online soils information systems are being developed around the world. The system for the United States is known as Web Soil Survey (WSS). This interactive program is designed to provide soil maps useful in both land planning and modeling. Most areas of the United States can be represented at a mapping scale as detailed as 1:10,000. At this scale, individual trees and houses are discernable on the air photo map base, and areas of soil as small as 1 ha can be delineated (Figure 3.39c). The mapping units are generally not "pure soil series," but mostly consociations—groupings of related soil series and phases named for the dominate soils in them. The maps are intended for natural resource planning and management by landowners, municipalities, and local governments. A map unit may specify the component soils, but not indicate *where* within the map unit area the various components are located. Therefore, these maps should *not* substitute for actual on-site investigation when a specific site is to be chosen for a construction project such as building a house or making a waste lagoon.

Web Soil Survey, like the paper soil survey reports that preceded it, includes more than just the capacity to generate soil maps on satellite imagery backgrounds. Each mapping unit has a wide range of soil properties and interpretations associated with it and this information can be queried and retrieved in tabular form, downloaded as datasets, or displayed as interpretive maps.

How to Use Web Soil Survey

There are four basic steps to producing a desired soil map and interpretative report using the WSS website (http://websoilsurvey.sc.egov.usda.gov). First, the user defines an area of interest (AOI) by zooming in on the map display or by entering either a street address or geographic coordinates and then using the AOI tools to draw either a rectangle or irregular boundaries. The AOI can be as small as 1 ha or as large as 40,000 ha. However, for a very small AOI (generally <10 ha), the WSS will warn that you have zoomed in beyond the scale at which the map was made and you should be aware that mapping is not as precise as it may appear. For a very large AOI (generally >10,000 ha), the soil map created may be too crowded with map units to be legible on the screen or when printed. However, interpretive maps that combine many individual mapping units into a relatively small number of more general groupings can still be quite useful at this scale. The AOI used as an example in Figure 3.39 is 130 ha in size and is located just southeast of the hamlet of Singer, Maryland, USA.

Second, once the AOI has been outlined and enlarged for optimum viewing, the user clicks on the Soil Map tab to generate a detailed soil map of the AOI accompanied by a table listing each of the mapping units present, the map unit symbol, the soil name (usually a series or phase of a series), the land area covered by that soil, and its percentage of the total area in AOI. The soil map displays soil boundaries and labels each delineated unit (as in Figure 3.39). The map unit labels may be a numeral (e.g., map unit 1238 = Urban land-Montebello complex, 3–8% slopes) or an alphanumeric code (e.g., in Figure 3.39a, map unit CcB2 = Chester silt loam, 3–8% slope, moderately eroded). The map can be embellished by adding or removing features such as roads, rivers, town names, etc. The default map background is a natural color air photo, but a topographic map can be added to the background.

Third, one can explore interpretive information using the Soil Data Explorer tab which accesses data and interpretive information associated with the soils in the AOI. The topics for which information is available range from such soil properties as pH, cation exchange capacity, and saturated hydraulic conductivity on the Soil Properties tab to ratings on the Suitability and Limitations for Land Use tab for such land uses as constructing small commercial buildings (see Figure 3.39a), growing forest trees and crops; maneuvering military vehicles, or recycling sewage sludge (see Figure 3.39b), and dozens of other uses.

The ratings given summarize qualitative observations of teams of soil scientists in the field, as well as numerical models, and take into consideration many different variables. As one example, consider the criteria used to rate soils for their limitations as sites for *small commercial buildings*, which the soil survey defines as one or two story structures without basements.

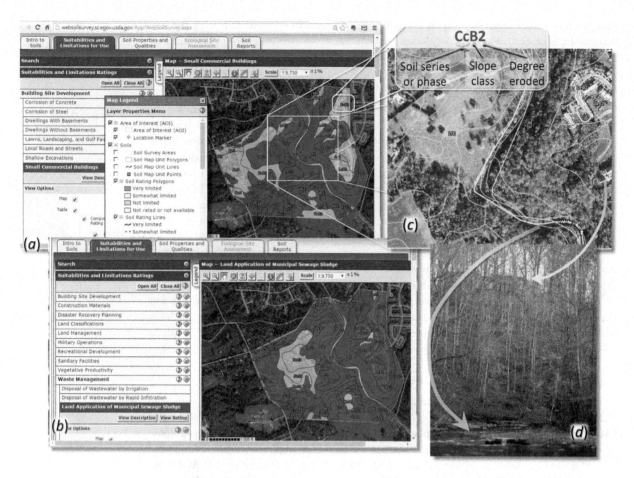

Figure 3.39 Examples of interpretive soil maps generated with Web Soil Survey. (a) A map of soils with unlimited (green), somewhat limited (yellow), and very limited (red) suitability for siting of small commercial buildings. (b) A similar map (with CcB2 map symbol enlarged) showing soils somewhat and severely limited for application of municipal sewage sludge. (c) An enlarged section of the soil map showing an open pasture on the west side of Winters Run road and creek and forestland and a townhouse development on the east side of the creek. (d) A ground view of the creek with map unit Av, steep forested gorge (map unit LfE), and townhouses constructed on the ridgetop (map unit LeB2 which is only somewhat limited for small commercial buildings). Note the left side of each screenshot lists some of the many land uses for which soil limitation ratings are available. [Images generated by Ray R. Weil using WebSoilSurvey (USDA/NRCS, 2017)]

The ratings are based on soil properties that affect the load-bearing capacity of the soil as well as properties that affect excavation and construction costs, including depth to water table, frequency of ponding or flooding, propensity for subsidence, shrink-swell potential (see Section 4.9), compressibility (see unified system of classification of soil, Table 4.6), slope, depth and hardness of bedrock or a cemented pan, and the amount and size of rock fragments.

Soil Explorer also allows classification of the map units according to their *Land Capability Class*, their *Soil Taxonomy family names*, or their *hydrologic group* (information in much demand for modeling urban storm runoff), among other classifications. The maps and tables of interest can be saved by adding a report to the "shopping cart" (even though reports are free of charge). The user can also choose to download the data as files that can be integrated with nonsoils geographic information in complex spatial models.

"There's an App for That"

"SoilWeb" is a program developed jointly by University of California at Davis and USDA/NRCS. It makes data easily usable on mobile devices as well as on any computer (Figure 3.40). SoilWeb uses Google® Inc.'s Google Earth® program to link to the U.S. national SURRGO database, and so can display soil maps similar to those created with Web Soil Survey, but with Google Earth® style and capabilities. SoilWeb can also display average soil profile data for map units in the SURRGO database such as the soil organic matter profile shown in Figure 3.40, *left*.

Figure 3.40 *"SoilWeb" program that couples Google Earth with the USDA soils information database. As a smartphone app, it uses the GPS location information to show the type of soil in the location where the user is standing. The on-the-ground photo (right) shows the view as seen by someone standing at the red spot on the SoilWeb map (left). The smart phone screen (right) shows profile diagrams for the Codorus and Hatboro soils that account for 85% and 15%, respectively, of the Cu map unit on the flood plain of Winters Run where the user is standing. The SoilWeb screen (left) can also display soil properties from the SURRGO database, such as the soil organic matter profile shown.* (Screenshot of WebSoilSurvey, Photo courtesy of Ray R. Weil and USDA/NRCS)

3.19 CONCLUSION

The soil that covers the Earth is actually comprised of many individual soils, each with distinctive properties. Among the most important of these properties are those associated with the layers, or *horizons*, found in a soil profile. These horizons reflect the physical, chemical, and biological processes soils have undergone during their development. Horizon properties greatly influence how soils can and should be used.

Knowledge of the kinds and properties of soils around the world is critical to humanity's struggle for survival and well-being. A soil classification system based on these properties is equally critical if we expect to use knowledge gained at one location to solve problems at other locations where similarly classed soils are found. *Soil Taxonomy*, a classification system based on measurable soil properties, helps fill this need in more than 50 countries. Scientists constantly update the system as they learn more about the nature and properties of the world's soils and the relationships among them. The updates are published online periodically as new editions of *Keys to Soil Taxonomy* (http://www.nrcs.usda.gov/wps/portal/nrcs/main/soils/survey/class/).

Making soil surveys is both a science and an art by which soils classification and information are applied to real world landscapes. Soil maps and descriptive information in the soil survey databases enable planners to make rational decisions about what should go where. In the remaining chapters of this book we will use *Soil Taxonomy* names whenever appropriate to indicate the kinds of soils to which a concept or illustration may apply.

STUDY QUESTIONS

1. Diagnostic horizons are used to classify soils in *Soil Taxonomy*. Explain the difference between a diagnostic horizon (such as an argillic horizon) and a genetic horizon designation (such as a Bt1 horizon). Give a field example of a diagnostic horizon that contains several genetic horizon designations.
2. Explain the relationships among a *soil individual*, a *polypedon*, a *pedon*, and a *landscape*.
3. Rearrange the following soil orders from the *least* to the *most* highly weathered: Oxisols, Alfisols, Mollisols, Entisols, and Inceptisols.
4. What is the principal soil property by which Ultisols differ from Alfisols? Inceptisols from Entisols?
5. Use the key given in Figure 3.9 to determine the soil order of a soil with the following characteristics: a spodic horizon at 30 cm depth; permafrost at 80 cm depth. Explain your choice of soil order.
6. Of the five soil-forming factors discussed in Chapter 2 (parent material, climate, organisms, topography, and time), choose *two* that have had the dominant influence on developing soil properties characterizing each of the following soil orders: Vertisols, Mollisols, Spodosols, and Oxisols.

7. To which soil order does each of the following belong: Psamments, Udolls, Argids, Udepts, Fragiudalfs, Haplustox, and Calciusterts.
8. What's in a name? Write a hypothetical soil profile description and land-use suitability interpretation for a hypothetical soil that is classified in the Aquic Argixerolls subgroup.
9. Explain why *Soil Taxonomy* is said to be a hierarchical classification system.
10. Name the soil taxonomy category and discuss the engineering implications of these soil taxonomy classes: Aquic Paleudults, Fragiudults, Haplusterts, Saprists, and Turbels.
11. What is the main purpose of digging soil pits as part of making a soil survey?
12. Assume you are planning to buy a 4-ha site on which to start a small orchard. Explain, step by step, how you could use the web soil survey to help determine if the prospective site was suitable for your intended use.
13. Try to produce the map shown in Figure 3.39b by visiting websoilsurvey.nrcs.usda.gov/app/. Hints: the location is about six kilometers due south of the town of Bel Air, Maryland (enter these as the city and state under the "Navigate by Address" tab). Be sure to study the instructions available on the Web site.

REFERENCES

Arnold, R., S. Shoba, P. Krasilnikov, and J. J. I. I. Marti. 2012. *A Handbook of Soil Terminology, Correlation and Classification*. Earthscan, London. 448 p.

Barrera-Bassols, N., J. Alfred Zinck, and E. Van Ranst. 2006. "Symbolism, knowledge and management of soil and land resources in indigenous communities: Ethnopedology at global, regional and local scales." *Catena* 65:118–137.

Coulombe, C. E., L. P. Wilding, and J. B. Dixon. 1996. "Overview of vertisols: Characteristics and impacts on society." *Advances in Agronomy* 17:289–375.

Ditzler, C. A. 2005. "Has the polypedon's time come and gone?" *HPSSS Newsletter*, February 2005, pp. 8–11. Commission on History, Philosophy and Sociology of Soil Science, International Union of Soil Sciences. Available at http://www.iuss.org/Newsletter12C4-5.pdf (verified 20 October 2005).

Eswaran, H. 1993. "Assessment of global resources: Current status and future needs." *Pedologie* 43(1):19–39.

Eswaran, H., T. Rice, R. Ahrens, and B. A. Stewart (eds.). 2003. *Soil Classification: A Global Desk Reference*. CRC Press, Boca Raton, FL.

FAO. 2006. Guidelines for soil description, 4th ed. Food and Agriculture Organization of the United Nations, Rome. Available at http://www.fao.org/publications/card/en/c/903943c7-f56a-521a-8d32-459e7e0cdae9/

Gong, Z., X. Zhang, J. Chen, and G. Zhang. 2003. "Origin and development of soil science in ancient China." *Geoderma* 115:3–13.

Legros, J.-P. 2006. *Mapping of the soil*. Science Publishers, Enfield, New Hampshire. 422 p.

Lin, H. 2011. "Three principles of soil change and pedogenesis in time and space." *Soil Science Society of America Journal* 75:2049–2070.

Riecken, F. F., and G. D. Smith. 1949. "Principal upland soils of Iowa, their occurrence and important properties." *Agron 49* (revised). Iowa Agr. Exp. Sta.

Schoeneberger, P. J., D. A. Wysocki, E. C. Benham, and S. S. Staff. 2012. *Field Book for Describing and Sampling Soils*, version 3.0. Natural Resources Conservation Service, National Soil Survey Center, Lincoln, NE. Available at http://www.nrcs.usda.gov/wps/portal/nrcs/detail/soils/research/report/?cid=nrcs142p2_054184.

Soil Survey Staff. 1975. *Soil Taxonomy: A Basic System of Soil Classification for Making and Interpreting Soil Surveys*. Natural Resources Conservation Service, Washington, DC.

Soil Survey Staff. 1999. *Soil Taxonomy: A Basic System of Soil Classification for Making and Interpreting Soil Surveys*. 2nd ed. Natural Resources Conservation Service, Washington, DC.

Soil Survey Staff. 2014. *Keys to Soil Taxonomy*. 12th ed. [Online]. Available from United States Department of Agriculture, Natural Resources Conservation Service. Available at http://www.nrcs.usda.gov/wps/PA_NRCSConsumption/download?cid=stelprdb1252094&ext=pdf.

SSSA. 1984. *Soil Taxonomy, Achievements and Challenges*. SSSA Special Publication 14. Soil Science Society of America, Madison, WI.

U.S. Department of Agriculture. 1938. *Soils and Men*. USDA Yearbook. U.S. Government Printing Office, Washington, DC.

USDA-NRCS. 2006. "National soil survey handbook, title 430-vi." U.S. Department of Agriculture Natural Resources Conservation Service. Available at http://soils.usda.gov/technical/handbook/ (posted 22 March 2006; verified 20 October 2006).

USDA/NRCS. 2015a. "Official soil series descriptions (OSDS)" [Online]. U.S. Department of Agriculture Natural Resources Conservation Service. Available at http://www.nrcs.usda.gov/wps/portal/nrcs/detailfull/soils/home/?cid=nrcs142p2_053587 (verified 1 September 2017).

USDA/NRCS. 2015b. "National soil survey handbook, title 430-vi" [Online]." U.S. Department of Agriculture Natural Resources Conservation Service. Available at http://www.nrcs.usda.gov/wps/portal/nrcs/detail/soils/survey/?cid=nrcs142p2_054242 (verified 1 September 2017).

USDA/NRCS. 2018. "Web soil survey" [Online]. Available by U.S. Department of Agriculture Natural Resources Conservation Service. Available at http://websoilsurvey.sc.egov.usda.gov/App/HomePage.htm (verified 1 April 2018).

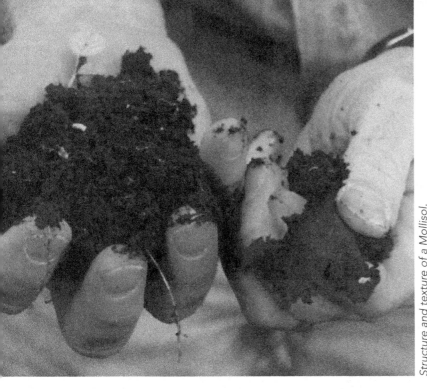

Structure and texture of a Mollisol. (Photo courtesy of Ray R. Weil)

4
Soil Architecture and Physical Properties

And when that crop grew, and was harvested, no man had crumbled a hot clod in his fingers and let the earth sift past his fingertips.
—JOHN STEINBECK, THE GRAPES OF WRATH

Success or failure of agricultural, ecological, and engineering projects often hinges on the physical properties of the soil used. Soil physical properties profoundly influence how soils function in an ecosystem and how they can best be managed. The occurrence and growth of many plant species are closely related to soil physical properties, as is the movement over and through soils of water and its dissolved nutrients and chemical pollutants.

Furthermore, soil scientists use the color, texture, and other physical properties of soil horizons in classifying soil profiles and in making determinations about soil suitability for various types of land uses. Knowledge of basic soil physical properties is not only of great practical value in itself but also helps in understanding many aspects of soils considered in later chapters.

The term *soil architecture* used in the title of this chapter encourages us to think of the soil as an edifice, as if it were a building such as a house. The primary particles in soil are the building blocks from which the house is constructed. **Soil texture** describes the sizes of the soil particles. **Soil structure** describes the manner in which soil particles are aggregated together. This property, therefore, defines the nature of the system of pores and channels in a soil—analogous to the rooms and hallways in a house.

The physical properties considered in this chapter focus on soil solids and on the pore spaces between the solid particles. Together, soil texture and structure help determine the ability of the soil to hold and conduct the water and air necessary for sustaining life. These factors also determine how soils behave when used for highways and building foundations, when manipulated by tillage, or when exposed to erosive wind and rain.

4.1 SOIL COLOR[1]

Color is often the most obvious characteristic of a soil. Although color itself has little effect on the most aspects of soil behavior and use, it does provide visual clues about other soil properties and conditions. To obtain the precise, repeatable description of colors needed for soil classification and interpretation, soil scientists compare a small clump of soil to standard color chips in special Munsell color charts. The Munsell charts use color chips arranged according to the three components of how people see color: the **hue** (in soils, usually

---

[1]For a collection of papers on causes and measurement of soil color, see Bigham and Ciolkosz (1993).

Figure 4.1 A page from the Munsell color book showing the colors of 2.5YR Hue present in soils. The color of the soil clod being held behind the page is closest to the outlined color chip designated 2.5YR4/4. The 2.5YR stands for the hue or pigment (reddish brown), the 4/ indicates the degree of lightness or darkness, going from near black of low values near the bottom of the page to near-white of high values near the top. The /4 designates the chroma column or brightness of the color, going from near grayscale at the spine of the book to very bright intense colors near the outer edge of the page. The color name "reddish brown" is applied to a group of four color chips as shown on the facing page (left). (Photo courtesy of Ray R. Weil)

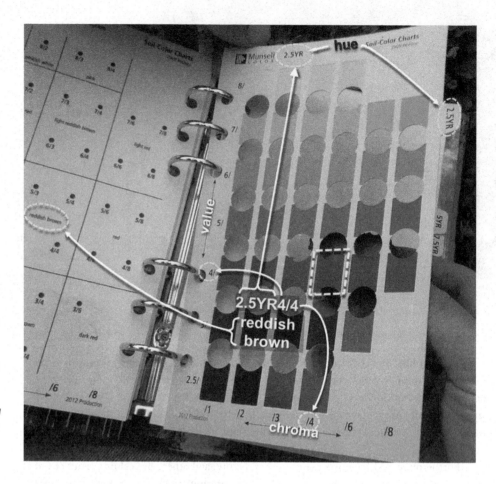

redness or yellowness), the **value** (lightness or darkness, a value of 0 being black), and the **chroma** (intensity or brightness, a chroma of 0 being neutral gray). In a Munsell color book, color chips are arranged on pages with value increasing from the bottom to the top, and the chroma increasing from left to right, while hue changes from one page to another (carefully study Figure 4.1).

Soils display a wide range of reds, browns, yellows, and even greens (Figure 4.2). Some soils are nearly black, others nearly white. Some soil colors are very bright, others are dull grays. Soil colors may vary from place to place in the landscape (see catena in Figure 2.25) as well as with depth through the various layers (horizons) within a soil profile (see Figure 3.30), or even within a single horizon or clod of soil (Figure 4.2). When making field soil descriptions, it is worth noting that horizons in a given profile that differ in chroma and value are often similar in hue.

Causes and Interpretation of Soil Colors

Three major factors influence soil colors: (1) organic matter content, (2) water content, and (3) the presence and oxidation states of iron and manganese oxides in various minerals. Organic matter tends to coat mineral particles, darkening and masking the brighter colors of the minerals themselves (see Figures 3.24 and 4.3). Soils are generally darker (have low color value) when wet than when dry. A more profound indirect long-term effect of water on soil colors is its influence on oxygen levels in the soil, and thereby the rate of organic matter accumulation (which darkens the soil) and the oxidation state of iron (well-oxidized iron compounds impart bright, high chroma reds and browns to the soil). Other minerals that influence soil color include manganese oxide (black) and glauconite (green)—as in Figure 4.3. In dry regions, calcite and soluble salts impart a whitish color to many soils. These colors are in contrast to the gray and bluish colors (low chroma) that reduced iron compounds impart to poorly drained soil profiles (Figure 4.2a,b,d). Under prolonged

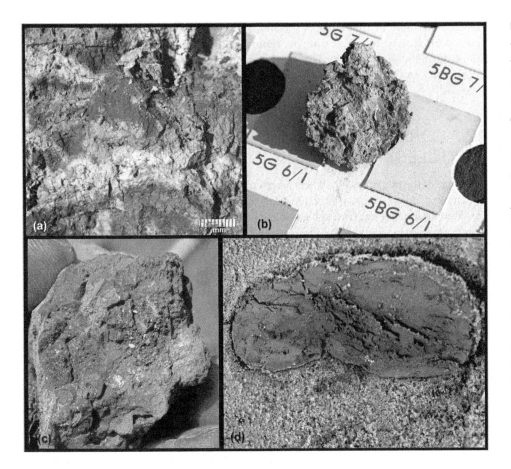

Figure 4.2 Examples of some red, orange, brown, gray, and blue soil colors derived from various iron minerals and influenced by oxidation state. (a) B horizon peds in an Ultisol, where the interped cracks that often hold stagnant water have turned gray, while the ped interiors have remained oxidized and are red and yellow. (b) B horizon ped in an Alfisol showing bluish color where a plant root has used up the oxygen and created localized reducing conditions. (c) A ped from the B horizon of an Alfisol in Chad showing purplish hardened plinthite. (d) A clay inclusion in sandy river bank sediment showing blue and gray colors where iron has become reduced. Scale marked in millimeters. (Photo courtesy of Ray R. Weil)

anaerobic conditions, reduced iron (which is far more soluble than oxidized iron) is removed from particle coatings, often exposing the light gray colors of the underlying silicate minerals. Soil exhibiting gray colors from reduced iron and iron depletion is said to be **gleyed**. If sulfur is present under anaerobic conditions, iron sulfides may color the soil black regardless of the organic matter level.

Color is used as a diagnostic criterion for classifying soils. For example, a mollic epipedon (see Section 3.2) is so dark that both its value and its chroma are three or less. In another example, Rhodic subgroups of certain soil orders have B horizons that are very red, having hues between 2.5YR (the most red of the yellowish red pages) and 10R (the most red in the Munsell color book). The presence in upper horizons of gley

Figure 4.3 Greenish colored clay in a glauconite-rich Marlton soil Bss horizon and blackish coatings from manganese oxides in a Davidson soil B/C horizon. (Photo courtesy of Ray R. Weil)

(low-chroma colors), either alone or mixed in a mottled pattern with brighter colors (see Figure 4.6), is used in delineating **wetlands**, for it is indicative of waterlogged conditions during at least a major part of the plant growing season (see Section 7.7). The depth in the profile at which gley colors are found helps to define the **drainage class** of the soil.

4.2 SOIL TEXTURE (SIZE DISTRIBUTION OF SOIL PARTICLES)

Knowledge of the proportions of different-sized particles in a soil (i.e., the **soil texture**) is critical for understanding soil behavior and management. When investigating soils on a site, the texture of various soil horizons is often the first and most important property to determine, for a soil scientist can draw many conclusions from this information. Furthermore, the texture of a soil in the field is not readily subject to change, so it is considered a basic permanent property of a soil.

Nature of Soil Separates

Diameters of individual soil particles range from boulders (1 m) to submicroscopic clays ($<10^{-6}$ m). Scientists group these particles according to several classification systems, as shown in Figure 4.4. The classification system established by the U.S. Department of Agriculture is used in this text. The size ranges are not purely arbitrary, but reflect major changes in how the particles behave and in the physical properties they impart to soils. Gravels, cobbles, boulders, and other **coarse fragments** greater than 2 mm in diameter may affect the behavior of a soil, but they are not considered to be part of the **fine earth fraction** to which the term *soil texture* properly applies.

Sand. Particles smaller than 2 mm but larger than 0.05 mm are termed *sand*. Sand feels gritty between the fingers. The particles are generally visible to the naked eye and may be rounded or angular (e.g., see Figure 1.15), depending on the degree of weathering and abrasion undergone. Because quartz is the dominant mineral in most sands, few plant nutrients are generally present. The large particle size also greatly slows the release of nutrients into solution from whatever nutrient-containing minerals may be present.

As sand particles are relatively large, so, too, the pores between them are relatively large. The large pores in sandy soils cannot hold much water against the pull of gravity (see Section 5.2) and so drain rapidly and promote entry of air into the soil. The relationship between particle

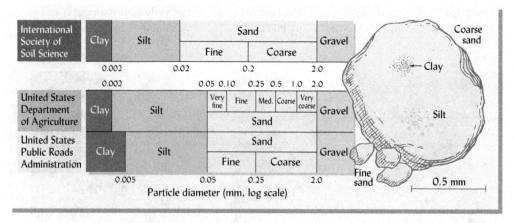

Figure 4.4 *Classification of soil particles according to their size. The scale in the center and the names on the drawings of particles follow the U.S. Department of Agriculture system, which is widely used throughout the world and in this book. The other two systems shown are also widely used by soil scientists and by highway construction engineers. The drawing illustrates the relative sizes (note scale).* (Diagram courtesy of Ray R. Weil)

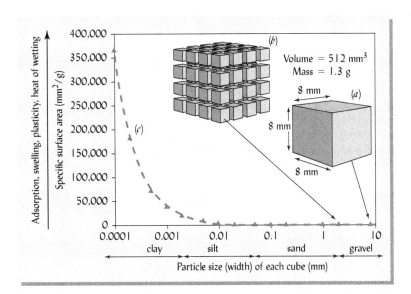

Figure 4.5 *Surface area and particle size. Consider a 1.3 g cube, 8 mm on a side (a). The cube has six faces with a total of 384 mm² surface area (6 faces • 64 mm² per face) or a specific surface of 295 mm²/g (384/1.3). If this cube were cut into smaller cubes of only 2 mm on each side (b), the same mass would now be present as 64 (4 • 4 • 4) cubes, each with 24 mm² of surface area (6 faces • 4 mm² per face) for a total surface area of 1536 mm² (24 mm² per cube • 64 cubes), or a specific surface of 1182 mm²/g (1536/1.3). This is four times as much surface area as the single large cube. The curve (c) explains why nearly all of the adsorbing power, swelling, plasticity, heat of wetting, and other surface area–related properties are associated with the clay fraction in mineral soils.* (Diagram courtesy of Ray R. Weil)

size and **specific surface area** (the surface area for a given mass of particles) is illustrated in Figure 4.5. The large particles of sand have low specific surface areas. Therefore, sand particles possess little capacity to hold water or nutrients and do not stick together into a coherent mass (see Section 4.9). Owing to the just described properties, most sandy soils are well aerated and loose, but also infertile and prone to drought.

Silt. Particles smaller than 0.05 mm but larger than 0.002 mm in diameter are classified as *silt*. Although similar to sand in shape and mineral composition, individual silt particles are so small as to be invisible to the unaided eye. Rather than feeling gritty when rubbed between the fingers, silt feels smooth or silky, like flour. Where silt is composed of weatherable minerals, the relatively small size (and large surface area) of the particles allows weathering rapid enough to release significant amounts of plant nutrients.

The pores between particles in silty material are much smaller (and much more numerous) than those in sand, so silt retains water and lets less drain through. However, even when wet, silt itself does not exhibit much **stickiness** or **plasticity** (malleability). What little plasticity, cohesion, and adsorptive capacity some silt fractions exhibit is largely due to a film of adhering clay. Because of their low stickiness and plasticity, soils high in silt and fine sand can be highly susceptible to erosion by both wind and water. Silty soil is easily washed away by flowing water in a process called **piping** (Box 4.1).

Clay. Clay particles are smaller than 0.002 mm. They therefore have very large specific surface areas, giving them a tremendous capacity to adsorb water and other substances. A spoonful of clay may have a surface area the size of a football field (see Section 8.1). This large adsorptive surface causes clay particles to cohere in a hard mass after drying. When wet, clay is sticky and can be easily molded (exhibits high plasticity).

Fine clay–sized particles are so small that they behave as **colloids**—if suspended in water they do not readily settle out. Unlike most sand and silt particles, clay particles tend to be shaped like tiny flakes or flat platelets. The pores between clay particles are very small and convoluted, so movement of both water and air is very slow. In clayey soil, the pores between particles are tiny in size, but huge in number, allowing the soil to hold a great deal of water; however, much of it may be unavailable to plants (see Sections 5.7 and 5.9). These tiny particles are not just smaller bits, but are comprised of special secondary minerals termed clay minerals (see Chapter 8), each of which imparts different properties to the soils in which it is prominent. Therefore, soil properties such as shrink–swell behavior, plasticity, water-holding capacity, soil strength, and chemical adsorption depend on the *kind* of clay present as well as the *amount*.

BOX 4.1
SILT AND THE FAILURE OF THE TETON DAM[a]

One of the most tragic and costly engineering failures in American history occurred in southern Idaho on 5 June 1976, less than a year after construction was completed on a large earth-filled dam across the Teton River. Eleven people were killed and 25,000 made homeless as the massive wall of water surged through the collapsed dam and the valley below. The dam failed with little warning as small seepage leaks (Figure 4.6, *left*) quickly turned into raging torrents that swept away a team of large bulldozers sent to make repairs.

The Teton dam was built according to a standard, time-tested design for zoned earth-fill embankments. Essentially, after preparing a base in the rhyolitic/basaltic rock below the soil, a core (zone 1) of tightly compacted soil material is constructed and covered with a layer (zone 2) of rocks and coarser alluvial soil material to protect it from water and wind erosion. The core is meant to be the watertight seal that prevents water from seeping through the dam. Normally, clayey material is chosen for the core, since the sticky, plastic qualities of moist clay allow it to be compacted into a malleable, watertight mass that holds together and does not crack so long as it is kept moist. Silt, on the other hand, though it may appear similar to clay in the field, cannot be compacted into a coherent mass and will crack as it settles because it lacks plasticity. If water seeps into these cracks, the silty material will rapidly wash away with the flowing water, enlarging the crack and inviting more water to flow through, which will wash away still more of the silt. This process of rapidly enlarging seepage channels is termed *piping*. Such piping was almost certainly a major cause of the Teton Dam failure, because the engineers built the zone 1 core of the dam using the local windblown silt deposits (termed *loess*—see Section 2.3) rather than clay. This was a tragic but useful lesson about the role of texture in determining soil behavior and the importance of distinguishing clay from silt.

Figure 4.6 *Photographs taken less than two hours apart show how two leaks in the dam rapidly enlarged, leading to the complete collapse and failure of Teton Dam. (Left) Black basaltic rocks form the abutments on either side of the dam. Dark brown muddy water can be seen leaking from the left side of the dam near the black rock abutment. The elongated speck above the leak is a large bulldozer attempting to push material down to plug the hole. This and another dozer were soon swallowed by the rapidly enlarging leak. (Right) Total collapse of the dam allowed a torrent of water to rush out, emptying the entire lake in just a few hours, leaving a path of destruction downstream.* (Photo courtesy of U.S. Bureau of Land Reclamation)

[a]Based, in part, on Arthur (1977) and Sasiharan (2003).

Influence of Surface Area on Other Soil Properties

When particle size decreases, specific surface area and related properties increase greatly, as shown graphically in Figure 4.5. Fine colloidal clay has about 10,000 times as much surface area as the same weight of medium-sized sand. Soil texture influences many other soil properties in far-reaching ways (see Table 4.1) as a result of five fundamental surface phenomena: (1) water is retained in soils as thin films on the surfaces of soil particles. (2) Both gases and dissolved chemicals are attracted to and adsorbed by mineral particle surfaces. (3) Weathering takes place at the surface of mineral particles, releasing constituent elements into the soil solution. (4) Charged particle surfaces and the water films between them tend to attract each other (see Section 4.5) forming aggregates. (5) Microorganisms tend to grow on and colonize particle surfaces.

Table 4.1
GENERALIZED INFLUENCE OF PARTICLE SIZE ON PROPERTIES AND BEHAVIOR OF SOILS[a]

Property/behavior	Rating associated with soil separates		
	Sand	Silt	Clay
Water-holding capacity	Low	Medium to high	High
Aeration	Good	Medium	Poor
Drainage rate	High	Slow to medium	Very slow
Soil organic matter level	Low	Medium to high	High to medium
Decomposition of organic matter	Rapid	Moderate	Slow
Warm-up in spring	Rapid	Moderate	Slow
Compactability	Low	Medium	High
Susceptibility to wind erosion	Moderate (high if fine sand)	High	Low
Susceptibility to water erosion	Low (unless fine sand)	High	Low if aggregated, high if not
Shrink–swell potential	Very low	Low	Moderate to very high
Sealing ponds, dams, and landfills	Poor	Poor	Good
Suitability for tillage after rain	Good	Medium	Poor
Pollutant leaching potential	High	Medium	Low (unless cracked)
Ability to store plant nutrients	Poor	Medium to high	High
Resistance to pH change	Low	Medium	High

[a]Exceptions to these generalizations do occur, especially as a result of soil structure and clay mineralogy.

4.3 SOIL TEXTURAL CLASSES

Within the three broad groups of *sandy soils, clayey soils*, and *loamy soils*, specific **textural class** names convey a more precise idea of the size distribution of particles and the general nature of soil physical properties. The relationship between textural class names and proportions of sand, silt, and clay is commonly shown diagrammatically as a triangular graph (Figure 4.7). It is worthwhile to study Figure 4.7 carefully and learn how to use the **textural triangle** by following the examples given in the caption.

Textural classes form a graduated sequence from the sands, which are coarse in texture to the clays, which are fine. Sands and loamy sands are dominated by the properties of sand, for sand comprises at least 70% of the material by weight and less than 15% of the material is clay (see boundaries on Figure 4.7). Clays, sandy clays, and silty clays are dominated by characteristics of clay. Likewise, silts are dominated by the properties of silt. However, most soils are some type of **loam**.

A **loam** is a mixture of sand, silt, and clay particles that exhibits the *properties* of those separates in about equal proportions. This definition does not mean that the three separates are present in equal *amounts* (that is why the loam class is not exactly in the middle of the triangle in Figure 4.7). This anomaly exists because a relatively small percentage of clay is required to engender clayey properties in a soil, whereas small amounts of sand and silt have a lesser influence on how a soil behaves. A loam in which sand is dominant is classified as a *sandy loam*. In the same way, some soils are classed as *silt loams, silty clay loams, sandy clay loams*, and *clay loams*. Note from Figure 4.7 that a *clay loam* may have as little as 26% clay, but to qualify as *sandy loam* or *silt loam*, a soil must have at least 45% sand or 50% silt, respectively.

If a soil contains a significant proportion of particles larger than sand (termed **coarse fragments**), a qualifying adjective may be used as part of the textural class name. Coarse fragments that range from 2 to 75 mm along their greatest diameter are termed *gravel* or *pebbles*, those ranging from 75 to 250 mm are called *cobbles* (if round) or *channers* (if flat), and those more than 250 mm across are called *stones* or *boulders*. A *gravelly, fine sandy loam* is an example of such a modified textural class.

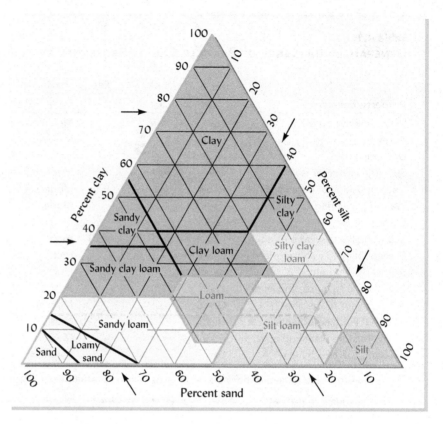

Figure 4.7 Soil textural classes are defined by percentages (g/100 g mineral soil) of sand, silt, and clay according to the heavy boundary lines on the textural triangle. To use the graph, first, find the appropriate clay percentage along the left side of the triangle, then draw a line inward parallel to the base of the triangle. Next, find the sand percentage along the base of the triangle and draw a line inward parallel to the right side of the triangle. The small arrows indicate the direction to draw the lines. The name of the compartment in which these two lines intersect indicates the textural class of the soil sample. If all three percentages are used, the three lines will all intersect at the same point. Because the percentages for sand, silt, and clay add up to 100%, the third percentage can easily be calculated if the other two are known. For example, a soil containing 15% sand and 15% clay must have 70% silt and would be a "Silt loam" (blue dashed arrows). Lines (not shown) for another soil sample with 33% sand, 34% silt, and 33% clay would intersect in the center of the "Clay loam."

Alteration of Soil Textural Class

Over long periods of time, pedologic processes such as illuviation and mineral weathering can alter the textures of certain soil horizons. Likewise, erosion and subsequent deposition downslope can selectively remove or deposit particles of certain sizes. However, management practices generally do not alter the textural class of a soil on a field scale. Changing the texture of a given soil would require mixing it with another soil material of a different textural class. For example, the incorporation of large quantities of sand to change the physical properties of a clayey soil for use in greenhouse pots or for turf grass would be considered to change the soil texture. However, adding peat or compost to a soil while mixing a potting medium does not constitute a change in texture, since the property of texture refers only to the mineral particles. In fact, the term *soil texture* is not relevant to artificial media that contain mainly perlite, peat, styrofoam, or other nonsoil materials.

Great care must be exercised in attempting to ameliorate physical properties of fine-textured soils by adding sand. Where specifications (as for a landscape design) call for soil materials of a certain textural class, it is generally advisable to find a naturally occurring soil that meets the specification, rather than attempt to alter the textural class by mixing in sand or clay. If the sand is not of the proper size and not added in sufficient amounts, it may make matters worse, rather than better. While adjacent coarse sand grains form large pores between them, sand grains embedded in a silty or clayey matrix do not. Mixing in moderate amounts of fine sand or sand ranging widely in size may yield a product more akin to concrete than to a sandy soil. For some applications (such as golf putting greens and athletic fields), the need for rapid drainage and resistance to compaction even when wet may justify the construction of an artificial soil from carefully selected uniform sands. Similarly, where a smooth, hard surface is required, such as for a tennis court, an artificial clay soil may be needed.

Determination of Textural Class by the "Feel" Method

Textural class determination is one of the first field skills a student of soils should develop. Determining the textural class of a soil by its feel is of great practical value in soil survey, land classification, and any investigation in which soil texture may play a role.

Accuracy depends largely on experience, so practice whenever you can, beginning with soils of known texture to "calibrate" your fingers. The textural triangle (see Figure 4.7) should be kept in mind when determining the textural class by the feel method, as explained in Box 4.2 and Figure 4.8.

Note that soils are assigned to textural classes *solely* on the basis of the mineral particles of sand size and smaller; therefore, the percentages of sand, silt, and clay always add up to 100%. The amounts of stone and gravel are rated separately. Organic matter is not considered.

Laboratory Particle-Size Analyses

The first and sometimes most difficult step in laboratory particle-size analysis is the complete dispersion of a soil sample in water, so even the tiniest clumps are broken down into individual, primary particles. Dispersion is usually accomplished using chemical treatments along with a high-speed blender, shaker, or ultrasonic vibrator.

BOX 4.2
A METHOD FOR DETERMINING TEXTURE BY FEEL

The first, and most critical, step in the texture-by-feel method is to knead a walnut-sized sample of moist soil into a uniform puttylike consistency, slowly adding water if necessary. This step may take a few minutes, but a premature determination is likely to be in error as hard clumps of clay and silt may feel like sand grains.

While squeezing and kneading the sample, note its malleability, stickiness, and stiffness, all properties associated with the clay content. A high silt content makes a sample feel smooth and silky, with little stickiness or resistance to deformation. A soil with a significant content of sand feels rough and gritty, and makes a grinding noise when rubbed near one's ear.

Get a feel for the amount of clay by attempting to squeeze a ball of properly moistened soil between your thumb and the side of your forefinger, making a ribbon of soil. Squeeze out the ribbon little by little, making it as long as possible until it breaks from its own weight (see Figure 4.8). Interpret your observations as indicated in Figure 4.9.

A more precise estimate of sand content (and hence more accurate placement in the horizontal dimension of the textural class triangle) can be made by wetting a pea-sized clump of soil in the palm of your hand and smearing it around with your finger until your palm becomes coated with a souplike suspension of soil. The sand grains will stand out visibly and their volume as compared to the original "pea" can be estimated, as can their relative size (fine, medium, coarse, etc.).

Figure 4.8 The "feel" method for determining soil textural class. A moist soil sample is rubbed between the thumb and forefingers and squeezed out to make a "ribbon." (Right) The gritty, noncohesive appearance and short ribbon of a sandy loam containing about 15% clay. (Middle) The smooth, dull appearance and crumbly ribbon characteristic of a silt loam. (Left) The smooth, shiny appearance and long, flexible ribbon of a clay. (Photo courtesy of Ray R. Weil)

(Continued)

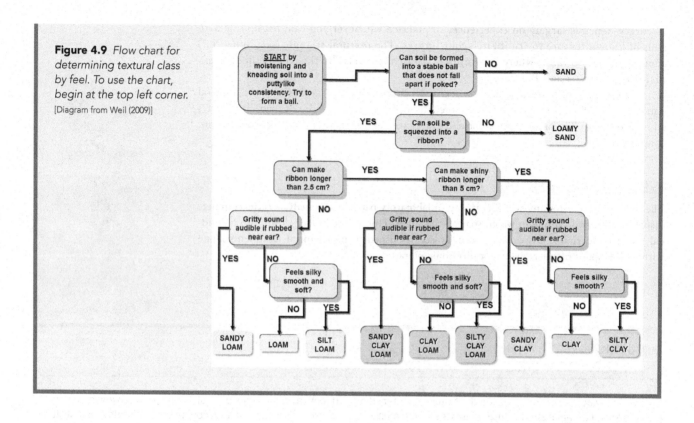

Figure 4.9 *Flow chart for determining textural class by feel. To use the chart, begin at the top left corner.* [Diagram from Weil (2009)]

Once suspended in water, most labs use sedimentation (often in combination with sieving different sand sizes) to determine the distribution of particle sizes in a soil sample (Figure 4.10). The principle involved is simple. Because soil particles are denser than water, they tend to sink, settling at a velocity that is proportional to their size. In other words, "The bigger they are, the faster they fall." The equation that describes this relationship is referred to as *Stokes' law* (Box 4.3).

Figure 4.10 *In performing particle-size analysis, a sample of soil (with organic matter removed) is suspended in water, stirred vigorously, and then allowed to settle. A hydrometer (right) can indicate the mass of particles remaining in suspension after different settling times (it floats higher when more soil is in suspension). Stokes' law is used to calculate the smallest effective diameter of the particles still in suspension at these times. (Left) The layers of sand and silt that have settled out after seven hours.* (Photo courtesy of Ray R. Weil)

> **BOX 4.3**
> **STOKES' LAW AND PARTICLE SIZE DETERMINED BY SEDIMENTATION**
>
> The complete expression of Stokes' law tells us the velocity V of a particle falling through a fluid is directly proportional to the gravitational force g, the difference between the density of the particle and the density of the fluid ($D_s - D_f$), and the square of the effective particle diameter (d^2). The effective diameter is referred to because Stokes' law applies to smooth, round particles. Since most soil particles are neither smooth nor round, sedimentation techniques determine the *effective* diameters, not necessarily the actual diameter of the soil particle. The settling velocity is *inversely* proportional to the viscosity or "thickness" of the fluid. Since velocity equals distance h divided by time t we can write Stokes' law as:
>
> $$V = \frac{h}{t} = \frac{d^2 g (D_s - D_f)}{18n}$$
>
> Where:
>
> g = gravitational force = 9.81 newtons per kilogram (9.81 N/kg)
> n = viscosity of water at 20°C = 1/1000 newton-seconds per m² (10^3 Ns/m²)
> D_s = density of the solid particles, for most soils = 2.65×10^3 kg/m³
> D_f = density of the fluid (i.e., water) = 1.0×10^3 kg/m³
>
> Substituting these values into the equation, we can write:
>
> $$V = \frac{h}{t}$$
>
> $$= \frac{d^2 \times 9.81 \text{ N/kg} \times (2.65 \times 10^3 \text{ kg/m}^3 - 1.0 \times 10^3 \text{ kg/m}^3)}{18 \times 10^{-3} \text{ Ns/m}^2}$$
>
> $$= \frac{9.81 \text{ N/kg} \times 1.65 \times 10^3 \text{ kg/m}^3}{18 \times 10^{-3} \text{ Ns/m}^2} \times d^2$$
>
> $$= \frac{16.19 \times 10^3 \text{ N/m}^3}{0.018 \text{ Ns/m}^2} \times d^2$$
>
> $$= \frac{9 \times 10^5}{\text{sm}} \times d^2 = kd^2, \text{ where } \ldots \quad k = \frac{9 \times 10^5}{\text{sm}}$$
>
> Note that $V = kd^2$ is a highly simplified formula for Stoke's law in which k represents a constant related to the acceleration of gravity and the nature of the liquid.
>
> Let's choose to sample or measure a soil suspension at 0.1 m (10 cm) depth (Figure 4.10). We can calculate the seconds of settling time we must allow if we want the smallest silt particle to have just passed our sampling depth so our sample will contain only clay.
>
> Chosen: $h = 0.1$ m and $d = 2 \cdot 10^{-6}$ m (0.002 mm, smallest silt)
>
> Solving for t we can write:
> Therefore:
>
> $$t = \frac{0.1 \text{ m}}{(2 \times 10^{-6} \text{ m})^2 \times 9 \times 10^5 \text{s}^{-1}/\text{m}^{-1}} = 27{,}777 \text{ sec}$$
> $$= 463 \text{ min} = 7.72 \text{ hours}$$
>
> By comparison, the smallest sand particle ($d = 0.05$ mm) would make the same journey in only 44 seconds.

4.4 STRUCTURE OF MINERAL SOILS[2]

As mentioned in the chapter introduction, a soil can be thought of as a complex building with the texture representing the sizes of building blocks or bricks used in its construction. It is useful to continue this architecture analogy and consider how soil particles (analogous to bricks) are grouped together to create structure (analogous to a house). The arrangement of various sized bricks into a house with associated windows, doors, and hallways represents the formation of soil structure and the associated complex pores and channels (Figure 4.11). The cement that holds the bricks in place represents the microbial glues, roots, and fungal hyphae that stabilize soil structure.

Soil structure describes the *spatial arrangement* of particles to form complex aggregations, pores, and channels. Sand, silt, clay, and organic particles become aggregated

---

[2]For fascinating insights into the relationships between soil structure and soil biology, see Ritz and Young (2011). For a review of the role of fungi, see Singh (2012)

Figure 4.11 *Particles (analogous to bricks) are grouped together to create structure (analogous to a house). The cement and fibers that hold the bricks in place represent the microbial glues, roots, and fungal hyphae that stabilize soil structure and create a complex of windows, hallways, and rooms—"living spaces" for microbes and "storage spaces" for water, gases, nutrients, and other chemicals.* (Photo courtesy of Ray R. Weil)

together due to various forces and at different scales to form distinct structural units called **peds** or **aggregates** (Figure 4.12). When a mass of soil is excavated and gently broken apart, it tends to break into peds along natural zones of weakness. These zones exhibit low tensile strength because particles within a ped or aggregate are more strongly attracted to one another than to the particles of the surrounding soil. Although *aggregate* and *ped* can be used synonymously, the term *ped* is most commonly used to describe the large-scale structure evident when observing soil profiles and involving structural units which range in size from a few mm to about 1 m. At this scale, the attraction of soil particles to one another in patterns that define structural units is influenced mainly by such physical processes as freeze–thaw, wet–dry, shrink–swell, the penetration and swelling of plant roots, the burrowing of soil animals, and the activities of people and machines. Structural peds should not be confused with **clods**—the compressed, cohesive chunks of soil that can form artificially when wet soil is plowed or excavated.

Most large peds are composed of, and can be broken into, smaller peds or aggregates (Figure 4.12). The networks of **pores** within and between the aggregates constitute a key aspect of soil structure (see Section 4.7). The pore network greatly influences the movement of air and water, the growth of plant roots, and the activities of soil organisms, including the accumulation and breakdown of organic matter. We will now examine the nature and types of structure observable in soil profiles.

Types of Soil Structure[3]

Many types or shapes of peds occur in soils, often within different horizons of a particular soil profile. Some soils may exhibit a **single-grained** structural condition in which particles are not aggregated. The loose sand in wind-blown dunes and loose dust accumulations such as freshly deposited loess are examples of this single-grain structural condition. At the opposite extreme, some soils (such as certain clay sediments) occur as large, cohesive masses of material and are described as exhibiting a **massive** structural condition. However, most soils exhibit some type of aggregation and are composed of peds that can be characterized by their shape

---

[3]In the U.S. *Soil Survey Handbook* (USDA-NRCS, 2005), the single-grained and massive conditions are described as *structureless*.

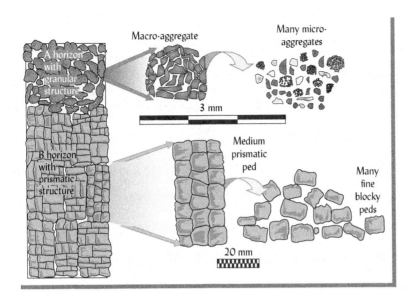

Figure 4.12 *Larger structural units observable in a soil profile each contain many smaller units. The lower example shows how large prismatic peds typical of B horizons break down into smaller peds (and so on). The upper example illustrates how microaggregates smaller than 0.25 mm in diameter are contained within the granular macroaggregates of about 1 mm diameter that typify A horizons. The microaggregates often form around and occlude tiny particles of organic matter originally trapped in the macroaggregate. Note the two different scales for the prismatic and granular structures.*
(Diagram courtesy of Ray R. Weil)

(or *type*), size, and distinctness (or *grade*). The four principal soil ped shapes are *spheroidal, platy, prismlike*, and *blocklike* (see following text and Figure 4.13).

Spheroidal. Granular structure consists of roughly spheroidal **aggregates** that may be separated from each other in a loosely packed arrangement (see Figure 4.13*a*). They typically range from <1 to as large as 10 mm in diameter. In reference to this type of structure, the term *aggregate* is used more commonly than *ped*. Granular structure characterizes many surface soils (usually A horizons), particularly those high in organic matter. Consequently, this is the principal type of soil structure affected by management.

Platelike. Platy structure, characterized by relatively thin horizontal sheetlike peds (plates), may be found in both surface and subsurface horizons. In most instances, the plates have developed as a result of soil-forming processes. However, unlike other structure types, platy structure may also be inherited from soil parent materials, especially those laid down by water or ice. In some cases compaction of clayey soils by heavy machinery can create platy structure (see Figure 4.13*b*).

Blocklike. Blocky peds are irregular, roughly cubelike, and range from about 5 to 50 mm across. The individual blocks are not shaped independently, but are molded by the shapes of the surrounding blocks. When the edges of the blocks are sharp and the rectangular faces distinct, the subtype is designated **angular blocky** (see Figure 4.13*c*). When some rounding has occurred, the peds are referred to as **subangular blocky** (see Figure 4.13*d*). These types are usually found in B horizons, where they promote drainage, aeration, and root penetration.

Prismlike. Columnar and **prismatic** structures are characterized by vertically oriented prisms or pillarlike peds that vary in size among different soils and may have a diameter of 150 mm or more (Figure 4.13). Columnar structure (see Figure 4.13*e*), which has pillars with distinct, rounded tops, is mainly found in subsoils high in sodium (e.g., natric horizons; see Section 3.2). When the tops of the prisms are relatively angular and flat horizontally, the structure is designated as prismatic (see Figure 4.13*f*). Both prismlike structures are often associated with swelling types of clay. Prismatic structure commonly occurs in subsurface horizons in arid and semiarid regions and, when well developed, provides a very striking feature of the profile (see Figure 4.14). In humid regions, prismatic structure sometimes occurs in poorly drained soils, in human-made soils forming from structureless sediments deposited on land, and in fragipans (see Figures 2.23 and 3.34).

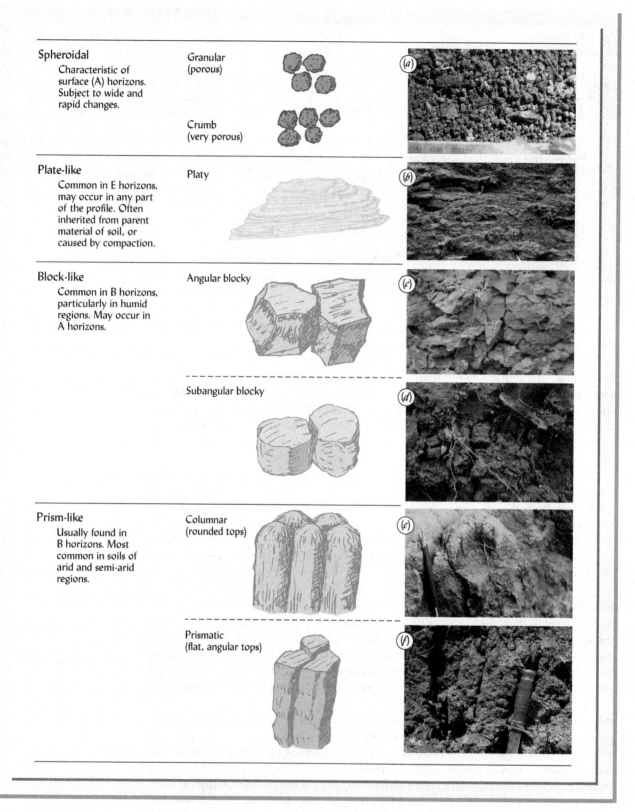

Figure 4.13 *The various structure types (shapes) found in mineral soils. Their typical location is suggested. The drawings illustrate their essential features and the photos indicate how they look in situ. For scale, note the 15-cm-long pencil in (e) and the 3-cm-wide knife blade in (d) and (f).* (Photo (e) courtesy of James L. Arndt/Merjent, Inc.; others courtesy of Ray R. Weil)

Figure 4.14 *An example of a very coarse prismatic ped removed from the Bw horizon of an Ustic Haplocryoll in British Columbia, Canada.* (Photo courtesy of Margaret G. Schmidt, Simon Fraser University)

Description of Soil Structure in the Field

In describing soil structure (see Section 3.18), soil scientists note not only the *type* (shape) of the structural peds present but also the relative *size* (fine, medium, coarse) and degree of development or distinctness of the peds (*grades* such as strong, moderate, or weak). For example, the soil shown in Figure 4.13*d* might be described as having "weak, fine, subangular blocky structure." Generally, the structure of a soil is easier to observe when the soil is relatively dry. When wet, structural peds may swell and press closer together, making the individual peds less well defined. We will now turn our attention to the formation and stabilization of soil structure, particularly the granular aggregates that characterize surface horizons.

4.5 FORMATION AND STABILIZATION OF SOIL AGGREGATES

The granular aggregation of surface soils is a highly dynamic soil property. Some aggregates disintegrate and others form anew as soil conditions change. Generally, smaller aggregates are more stable than larger ones, so maintaining the much-prized larger aggregates requires great care. We will discuss practical means of managing soil structure after we consider the factors responsible for aggregate formation and stabilization.

Hierarchical Organization of Soil Aggregates[4]

Surface horizons are usually characterized by roundish granular structure that exhibits a hierarchy in which relatively large **macroaggregates** (0.25–5 mm in diameter) are comprised of smaller **microaggregates** (2–250 µm). The latter, in turn, are composed of tiny packets of clay and organic matter only a few µm in diameter. You may easily demonstrate the existence of this *hierarchy of aggregation* by selecting a few of the largest aggregates in a soil and gently crushing

---

[4]The hierarchical organization of soil aggregates was first put forward by Tisdall and Oades (1982). For a review of advances in this area, see Six et al. (2004).

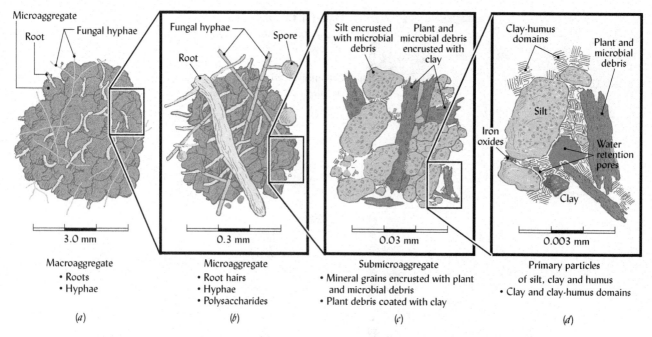

Figure 4.15 Larger aggregates are often composed of an agglomeration of smaller aggregates. This illustration shows four levels in this hierarchy of soil aggregates. The different factors important for aggregation at each level are indicated. (a) A macroaggregate composed of many microaggregates bound together mainly by a kind of sticky network formed from fungal hyphae and fine roots. (b) A microaggregate consisting mainly of fine sand grains and smaller clumps of silt grains, clay, and organic debris bound together by root hairs, fungal hyphae, and microbial gums. (c) A very small submicroaggregate consisting of fine silt particles encrusted with organic debris and tiny bits of plant and microbial debris (called particulate organic matter) encrusted with even smaller packets of clay, humus, and Fe or Al oxides. (d) Clusters of parallel and random clay platelets interacting with Fe or Al oxides and organic polymers at the smallest scale. (Diagram courtesy of Ray R. Weil)

or picking them apart to separate them into many smaller-sized pieces. Then try rubbing the tiniest of these soil crumbs between your thumb and forefinger. For most soils, you will find that even the smallest specks of soil usually break down into a smear of still smaller particles of silt, clay, and humus. At each level in the hierarchy of aggregates, different factors are responsible for binding together the subunits (Figure 4.15).

Factors Influencing Aggregate Formation and Stability in Soils

Both biological and physical–chemical processes are involved in the formation of soil aggregates. Physical–chemical processes tend to be the most important at the smaller end of the scale; biological processes at the larger end. The physical–chemical processes of aggregate formation are associated mainly with clays and, hence, tend to be of greater importance in finer-textured soils. In sandy soils that have little clay, aggregation is almost entirely dependent on biological processes.

Physical–Chemical Processes

Most important among the physical–chemical processes are: (1) flocculation, the mutual attraction among clay and organic molecules; and (2) the swelling and shrinking of clay masses.

Flocculation of Clays and the Role of Adsorbed Cations. Except in very sandy soils that are almost devoid of clay, aggregation begins with the **flocculation** of clay particles into microscopic clumps or *floccules* (Figure 4.16). If two clay platelets come close enough to each other, cations compressed in a layer between them will attract the negative charges on both platelets, thus serving as bridges to hold the platelets together. These processes lead to the

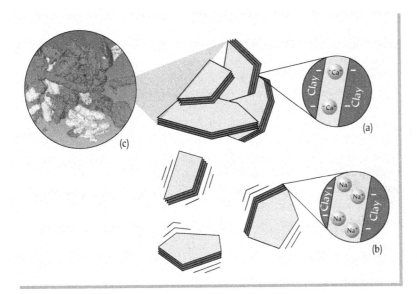

Figure 4.16 *The role of cations and bacteria in clay flocculation. (a) Di- and trivalent cations like Ca^{2+}, Fe^{3+}, and Al^{3+} are tightly adsorbed, form bridges that bring clay particles together, and effectively neutralize negative charges on clay surfaces. (b) Monovalent ions, especially Na^+, can cause clay particles to repel each other and create a dispersed condition because (1) the large hydrated Na^+ ions cannot get close enough to effectively neutralize the negative charges; (2) the single charge on Na^+ is not effective in forming a bridge between clay particles; and (3) compared to di- or trivalent ions, two or three times as many monovalent ions must crowd between clay particles to neutralize the clay surfaces. (c) due to charges on their cells walls and sticky exocellular compounds, bacteria (yellow) further flocculate clay particles (brown).* (Diagrams (a) and (b) courtesy of Ray R. Weil; X-ray tomographic image (c) from Thieme et al. (2003))

formation of a small "stack" of parallel clay platelets, termed a *clay domain*. Other types of clay domains are more random in orientation, resembling a house of cards. These form when the positive charges on the edges of the clay platelets attract the negative charges on the planar surfaces (Figure 4.16). Multivalent cations (especially Ca^{2+}, Fe^{2+}, and Al^{3+}) also complex with hydrophobic organic molecules, allowing them to bind to clay surfaces. Clay/humus domains form bridges that bind to each other and to fine silt particles (mainly quartz), creating the smallest size groupings in the hierarchy of soil aggregates (Figure 4.15d). These domains, aided by the flocculating influence of certain polyvalent cations (again, mainly Ca^{2+}, Fe^{2+}, and Al^{3+}) and humus, provide much of the long-term stability for the smaller (<0.25 mm) microaggregates. In some highly weathered clayey soils (Ultisols and Oxisols) the cementing action of iron oxides and other inorganic compounds produces very stable small aggregates called **pseudosand**.

When certain cations (especially Na^+, but to a lesser degree K^+ and even Mg^{2+}) with less flocculating ability than Ca^{2+} or Al^{3+} are prominent, the attractive forces are not able to overcome the natural repulsion of one negatively charged clay platelet by another (Figure 4.16). The clay platelets cannot approach closely enough to flocculate, so remain dispersed and cause the soil to become gel-like, impervious to water and air, and very undesirable from the standpoint of plant growth. This dispersed condition is most dramatically stimulated by Na^+ ions and is most common in soils of arid and semiarid areas (more details in Section 9.14).

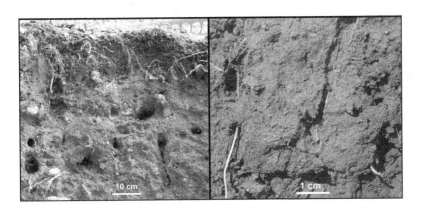

Figure 4.17 *Biopores formed by soil organisms. (Left) Upper horizons in a forested Ultisol in Maryland are shot through with old tree root channels in which the original tree roots have long ago decayed. The channels are lined with dark organic matter and serve as "superhighways" for current root growth. (Right) Earthworm burrows in the Bt horizon (60 cm depth) of an Inceptisol in Pennsylvania managed with cover crops and no-till techniques. Nutrient-rich organic materials line the burrows, which provide easy access to subsoil water for these sweet pepper roots.* (Photo courtesy of Ray R. Weil)

Volume Changes in Clayey Materials. As a soil dries out and water is withdrawn, the platelets in clay domains move closer together, causing the domains and, hence, the soil mass to shrink in volume. As a soil mass shrinks, cracks will open up along zones of weakness. Over the course of many cycles (as occur between rain or irrigation events in the field) the network of cracks becomes better defined. In one of many ways in which physical and biological soil processes interact, plant water uptake dries the root zone and accentuates the physical aggregation processes associated with wetting and drying.

Freezing and thawing cycles have a similar effect, since the formation of ice crystals is a drying process that also draws water out of clay domains. The swelling and shrinking actions that accompany freeze–thaw and wet–dry cycles in soils create fissures and pressures that alternately break apart large soil masses and compress soil particles into defined structural peds. The aggregating effects of these water and temperature cycles are most pronounced in soils with a high content of swelling-type clays (see Chapter 8), especially Vertisols, Mollisols, and some Alfisols (see Chapter 3).

Biological Processes

Activities of Soil Organisms. Among the biological processes of aggregation, the most prominent are: (1) the burrowing and molding activities of soil animals; (2) the enmeshment of particles by sticky networks of roots and fungal hyphae; and (3) the production of organic glues by microorganisms, especially bacteria and fungi. Earthworms (and termites) move soil particles about, often ingesting them and forming them into pellets or casts (see Chapter 10). In some forested soils, the surface horizon consists primarily of aggregates formed as earthworm castings (see, e.g., Figure 4.13*a*). Plant roots also move particles about as they push their way through the soil. This movement forces soil particles to come into close contact with each other, encouraging aggregation. At the same time, the channels created by plant roots and soil animals serve as large conduits for new root growth. The channels also break up clods and help to define larger soil structural units (see Figure 4.17).

Plant roots (particularly root hairs) and fungal hyphae exude sugarlike polysaccharides and other organic compounds, forming sticky networks that bind together individual soil particles and tiny microaggregates into larger macroaggregates (see Figures 4.15*a* and 4.18). The threadlike fungi that associate with plant roots (called *mycorrhizae*; see Section 10.9) produce a sticky sugar–protein called **glomalin**, which is thought to be an effective cementing agent.

Figure 4.18 *The influence of organic matter (OM) on the stability of soil aggregates against slaking (falling apart) when wetted. The two soil samples were collected from adjacent plots on a Beltsville silt loam (Fragiudults). The lower OM soil grew conventionally tilled corn grain crops every year for 20 years. The higher OM soil grew bluegrass sod during the same period, resulting in about 9 g/1000 g (0.9%) more OM in the soil. Although both soils appeared well aggregated when dry (left), when the same amount of water was added to each the aggregates in the low OM soil rapidly fell apart while those in the higher OM soil remained intact. For data associated with these plots, see Weil and Magdoff (2004).* (Photo courtesy of Ray R. Weil)

As they decompose plant residues, bacteria also produce organic glues such as the polysaccharides shown in Figure 4.19 intermixed at a very small scale with clay. Many of these root and microbial organic glues resist dissolution by water and so not only enhance the formation of soil aggregates but also help ensure their stability over a period of months to a few years. These processes are most notable in surface soils, where root and animal activities and organic matter accumulation are greatest.

Influence of Organic Matter. In most temperate zone soils, the formation and stabilization of granular aggregates is primarily influenced by soil organic matter (see Figure 4.18). Organic matter provides the energy substrate that makes possible the previously mentioned biological activities. During the aggregation process, soil mineral particles (silts and fine sands) become coated and encrusted with bits of decomposed plant residue and other organic materials. Organic polymers resulting from decay chemically interact with particles of silicate clays and iron and aluminum oxides. These compounds help orient the clays into packets (domains), which form bridges between individual soil particles, thereby binding them together in water-stable aggregates (see Figure 4.15d). In Figure 4.19 we can directly observe bacterial polymers and organomineral domains that bind soil particles.

Influence of Tillage. Tillage can both promote and destroy aggregation. If the soil is not too wet or too dry, tillage can break large clods into natural aggregates, creating a temporarily loose, porous condition conducive to the easy growth of young roots and the emergence of tender seedlings (see Box 4.4). Tillage can also incorporate organic amendments into the soil and kill weeds.

Over longer periods, however, tillage greatly hastens the oxidative loss of soil organic matter, thus weakening soil aggregates. Tillage operations, especially if carried out when the soil is wet, also tend to crush or smear soil aggregates, resulting in loss of macroporosity and the creation of a *puddled* condition.

Influence of Iron/Aluminum Oxides. Many of the highly weathered soils of the tropics (especially Oxisols) have large amounts of iron and aluminum sesquioxides (in largely amorphous forms) that coat soil particles and cement soil aggregates, thereby preventing their ready breakdown when the soil is tilled or wetted. Compared to soils of more temperate regions with similar amounts of organic matter, such tropical soils tend to have much greater aggregate stability, and their aggregation is less dependent on soil organic matter (Figure 4.22).

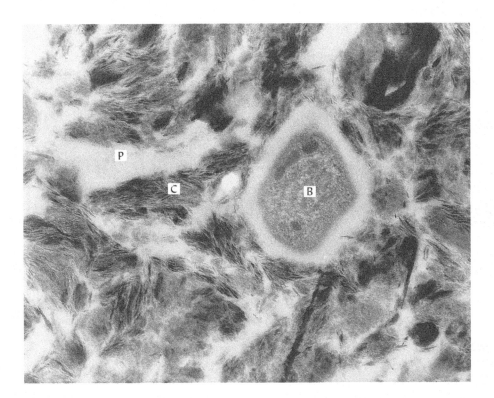

Figure 4.19 *A transmission electron micrograph illustrating the interaction among organic materials and silicate clays in a water-stable aggregate. The dark-colored materials (C) are groups of clay particles that are interacting with organic polysaccharides (P). A bacterial cell (B) is also surrounded by polysaccharides. Note the generally parallel orientation of the clay particles, an orientation characteristic of clay domains.* (From Emerson et al. (1986); Photograph provided by R. C. Foster, CSIRO, Glen Osmond, Australia; Soil Science Society of America)

BOX 4.4
PREPARING A GOOD SEEDBED

Early in the growing season, one of the main activities of a farmer or gardener is the preparation of a good seedbed to ensure that the sowing operation goes smoothly, that the seeds germinate quickly, and that the plants come up well spaced.

A good seedbed should be packed firmly enough to ensure the seed can easily imbibe water to begin the germination process. The seedbed soil should also be loose enough to allow easy root elongation and seedling emergence. The seedbed should also be relatively free of large clods between which small seeds could fall and become lodged without sufficient soil contact for germination and too deeply for proper emergence.

Tillage is commonly used to loosen compacted soil, help control weeds, and, in cool climates, help the soil dry out so it will warm more rapidly. On the other hand, the objectives of seedbed preparation may be achieved with little or no tillage if soil and climatic conditions are favorable and a mulch, herbicide, or cover crop is used to control early weeds (Figure 4.20).

Mechanical planters can assist in maintaining a good seedbed, even if no tillage is used. No-till planters are equipped with coulters (sharp steel disks) designed to cut a path through plant residues on the soil surface. No-till planters usually follow the coulter with a pair of sharp cutting wheels called a *double disk opener* that opens a groove in the soil into which the seeds can be dropped (Figure 4.21). Most planters also have a press wheel that follows behind the seed dropper and packs the loosened soil just enough to ensure that the groove is closed and the seed is pressed into contact with moist soil.

Ideally, only a narrow strip in the seed row is packed down to create a seed germination zone, while the soil between the crop rows is left as loose as possible to provide a good rooting zone. The surface of the interrow rooting zone may be left in a rough condition to encourage water infiltration and discourage erosion. The above principles also apply to the home gardener who may be sowing seeds by hand. Instead of a press wheel, the gardener can ensure good seed–soil contact by pressing a board over the seed row or even walking carefully over the seeds, while avoiding compaction in the interrow areas.

Figure 4.20 *Cotton seedlings emerging in three types of seedbed in a Texas Mollisol (Argiustolls). (Left) A conventionally tilled seedbed prepared by stalk shredding, disking, bedding, cultivation to incorporate herbicide, and rod weeding before planting (and further weed control cultivation later on). (Middle) A no-till seedbed in which seeds were mechanically inserted into soil beneath dead residues of the previous cotton crop (note white cotton tufts). Herbicides were sprayed to control weeds without soil disturbance. (Right) A no-till seedbed protected by a cover crop, which was killed by herbicides shortly before the new crop was planted.* (Photo courtesy of Paul DeLaune, Texas A and M University)

Figure 4.21 *A large no-till planter (left) sowing corn seed from yellow hoppers into a field covered by cover crop residues. Very little soil is exposed by the process. (Right) A close-up of another no-till planter showing how it works to sow seeds with little disturbance of the "O horizon" covering of residues from the previous crop. The coulters cut a narrow "V"-shaped slot into the soil and the press wheels closed the slot after the seed is inserted. The complex planter rig also can apply fertilizer and chemicals at the same time that it sows seeds. The fertilizer delivery tube is visible (blue arrow), but the seed is dropped into the soil slot from a tube (yellow arrow, hidden behind the frame) that extends between a pair of coulters called a "double disk opener."* (Photos courtesy of Lynn Betts/Publication Services, LLC (left) and Ray R. Weil (right))

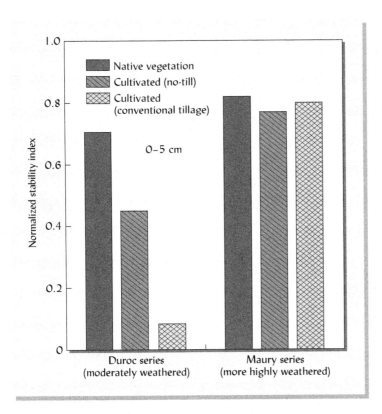

Figure 4.22 *The interaction of soil organic matter with the clay fraction accounts for most of the aggregate stability of such moderately weathered soils as the Duroc series. Consequently, the stability of soil aggregates declines when cultivation, especially with conventional tillage, decreases soil organic matter levels. In more highly weathered soils such as the Maury series, aggregate stability is less dependent on organic matter levels than on the interaction of iron oxide compounds with silicate clays such as kaolinite. The tillage system used therefore has less effect on aggregate stability in these more highly weathered soils. Cultivated soils of the highly weathered tropics may have greater aggregate stability than their counterparts in temperate zones.* (Redrawn from Six et al. (2000))

4.6 TILLAGE AND STRUCTURAL MANAGEMENT OF SOILS

When protected under dense vegetation and undisturbed by trampling or tillage, most soils (except perhaps some sparsely vegetated soils in arid regions) possess a surface structure sufficiently stable to allow rapid infiltration of water and to prevent crusting. However, for the manager of cultivated soils, the development and maintenance of stable surface soil structure is a major challenge. Many studies have shown that aggregation and associated desirable soil properties such as water infiltration rate decline under long periods of tilled row-crop cultivation.

Tillage and Soil Tilth

Tilth refers to the physical condition of the soil in relation to plant growth. Tilth is a highly dynamic soil property that depends not only on aggregate formation and stability but also on such factors as bulk density (see Section 4.7), soil moisture content, degree of aeration, rate of water infiltration, drainage, and capillary water capacity.

A major aspect of tilth is soil **friability** (see also Section 4.9). Soils are said to be **friable** if their clods are not sticky or hard, but rather crumble easily, revealing their constituent aggregates. Generally, **soil friability** is enhanced when the **tensile strength** of individual aggregates (i.e., the force required to pull them apart) is relatively high compared to the tensile strength of the clods. This condition allows tillage or excavation forces to easily break down the large clods, while the resulting aggregates remain stable. As might be expected, friability changes markedly with changes in soil water content, especially for fine-textured soils. Each soil typically has an optimal water content for greatest friability (Figure 4.23).

Clayey soils are especially prone to puddling and compaction because of their high plasticity and cohesion. When puddled clayey soils dry, they usually become dense and hard. Proper timing of trafficking is more difficult for clayey than for sandy soils, because the former take much longer to dry to a suitable moisture content and may also become too dry to work easily. Increased soil organic matter content usually enhances soil friability and can partially alleviate the susceptibility of a clay soil to structural damage during tillage and traffic.

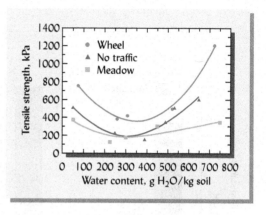

Figure 4.23 Influence of tillage and wheel traffic on soil strength at various water contents. Tensile strength is the force (measured in kilopascals, kPa) required to break apart soil clods. The lower this force, the easier it is to crumble the soil and the greater the soil friability and workability. The upper two curves represent soil from the wheel tracks (●) and the nontrafficked portion (▲) of a cultivated calcareous clay soil in England. The lowest curve represents the same soil in an adjacent long-term uncultivated meadow (■). There appears to be an optimum water content at which a soil has the lowest tensile strength and is therefore most friable (near 300 g/kg for this soil). Also, the less traffic and tillage, the more friable the soil (the lower the tensile strength), especially when very dry or very wet. (Drawn from data in Watts and Dexter (1997))

Some clayey soils of humid tropical regions are much more easily managed than those just described. The clay fraction of these soils is dominated by hydrous oxides of iron and aluminum, which are not as sticky, plastic, and difficult to work. These soils may have very favorable physical properties, since they hold large amounts of water but have such stable aggregates that they respond to tillage after rainfall much like sandy soils.

Farmers in temperate regions typically find their soils too wet for tillage just prior to planting time (early spring), while farmers in tropical regions may face the opposite problem of soils too dry for easy tillage just prior to planting (end of dry season). In tropical and subtropical regions with a long dry season, farmers often till soil in a very dry state in order to prepare the land for planting with the onset of the first rains. Tillage under such dry conditions can be very energy intensive and can result in hard clods if the soils contain much clay.

In contrast to the situations just discussed, rice farmers often purposely till extensively when their soils are saturated with water (Figure 4.24). They do this to destroy aggregation and greatly reduce water permeability, thus making their soils more suited for holding water in paddies where rice is grown under flooded conditions.

Conventional Tillage and Crop Production

Since the middle ages, the moldboard plow has been the primary tillage implement most used in the western world.[5] Its purpose is to lift, twist, and invert the soil while incorporating crop residues and animal wastes into a 10- to 20-cm-thick *plow layer* of soil (Figure 4.25, *left*). The moldboard plow is often supplemented by the disk harrow, which has rolling steel disks that cut up and partially incorporate residues while partially inverting a thinner (5–10 cm thick) layer of soil. Other implements use steel tines to fracture and stir the soil to various depths,

Figure 4.24 A farmer purposely puddles the soil to prepare his rice paddy for flooding and planting in northern Tanzania. The tillage will destroy the soil structure and reduce soil permeability—results that would normally be avoided on upland agricultural fields, but that are desirable where rice is to be grown on inundated paddies. (Photo courtesy of Ray R. Weil)

---

[5]For an early but still valuable critique of the moldboard plow, see Faulkner (1943).

but with less complete burial of plant residues. In conventional practice, such primary tillage is followed by a number of secondary tillage operations, such as harrowing to kill weeds and to break up clods, thereby preparing a fine seedbed. After the crop is planted, the soil may receive further secondary tillage to control in-season weeds. While these tillage operations temporarily loosen the soil, break up clods, and suppress weeds, they also have serious detrimental effects: (1) tillage speeds the loss of soil organic matter and thereby the weakening of soil structure, and (2) most kinds of tillage leave the soil naked without a natural blanket of plant litter to protect the soil surface from sun, rain, and wind.

Conservation Tillage and Soil Tilth

During the past half century, agricultural land-management systems have been developed that minimize the need for soil tillage and leave the soil surface largely covered by plant residues, thereby maintaining soil biological habitat, stabilizing soil structure, conserving soil organic matter, and physically protecting the soil from drying sun, scouring wind, and beating rain (see Section 14.6 for a detailed discussion). For these reasons, the tillage practices followed in these systems are called *conservation tillage*. The U.S. Department of Agriculture defines *conservation tillage* as that which leaves at least 30% of the soil surface covered by residues after planting. Figure 4.25 (*right*) illustrates a no-till operation, where one crop is planted into the residue of another, with virtually no tillage. Other minimum-tillage systems such as chisel plowing permit some stirring of the soil, but still leave a high proportion of the plant residues as a protective cover on the soil surface.

Soil Crusting

Falling drops of water during heavy rain or sprinkler irrigation can beat apart aggregates exposed at the soil surface. In some soils the dilution of salts by this water stimulates the dispersion of clays (Section 9.14). Once the aggregates become dispersed, small particles and dispersed clay tend to wash into and clog the soil pores. The remaining coarse particles at the soil surface become densely packed with little pore space under the influence of beating raindrops. Soon the soil surface is covered with a thin, partially cemented, low-permeability layer material called a **surface seal**. The surface seal inhibits water infiltration and increases erosion losses.

As the surface seal dries, it forms a hard **crust** (Figure 4.26). Seedlings, if they emerge at all, can do so only through cracks in the crust. Formation of a crust soon after a crop is sown may allow so few seeds to emerge that the crop has to be replanted. In addition, soil sealing and crusting can have disastrous consequences because high runoff losses leave little water available to support plant growth.

Figure 4.25 *Contrasting tillage systems. (Left) A moldboard plow slices, twists, and partially inverts the plow layer of soil, burying all plant residues and leaving the soil "naked." (Right) An extreme conservation tillage system in which tomato seedlings are being transplanted into a mulch of killed cover crop residue with almost no soil disturbance at all. This no-tillage system keeps a residue cover on the soil at all times, protects the soil from erosion, reduces fuel costs, saves time, and allows operations to proceed under conditions that would be prohibitive if tillage were required. In this case the no-till transplanter is operating in the rain with no problems of compaction and mud. The black lines are irrigation tubing being lain down.* (Left photo courtesy of Ray R. Weil. Right photo courtesy of Steve Groff)

Crusting can be minimized by keeping some vegetative or mulch cover on the land to reduce the impact of raindrops. Once a crust has formed, it may be necessary to rescue a newly planted crop by breaking up the crust with light tillage (as with a rotary hoe), preferably while the soil is still moist.

Soil Conditioners

Improved management of soil organic matter and use of certain soil amendments can "condition" the soil and help prevent clay dispersion and crust formation (see also Section 9.18).

Gypsum. Gypsum (calcium sulfate) is widely available in its relatively pure mined form, or as a major component of various industrial byproducts. Gypsum has been shown to be effective in improving the physical condition of many types of soils, from some highly weathered acid soils to some low-salinity, high-sodium soils of semiarid regions (see Chapter 9). The more soluble gypsum products provide enough electrolytes (dissolved cations and anions) to promote flocculation and inhibit the dispersion of aggregates, thus preventing surface crusting. Field trials have shown that gypsum-treated soils permit greater water infiltration and are less subject to erosion than untreated soils. Similarly, gypsum can reduce the strength of hard subsurface layers, thereby allowing greater root penetration and subsequent plant uptake of water from the subsoil.

Organic Polymers. Certain synthetic organic polymers can stabilize soil structure in much the same way as do natural organic polymers such as polysaccharides. For example, polyacrylamide (PAM) is effective in stabilizing surface aggregates when applied at rates as low as 1 to 15 mg/L of irrigation water or sprayed on at rates as low as 1 to 4 kg/ha. Combining the use of PAM and gypsum products can nearly eliminate irrigation-induced erosion.

Other Soil Conditioners. Several species of algae and various humate products (most derived from soft coal) are marketed as soil conditions for application at low rates (10–100 kg/ha). Some, but not all, humate products have been shown to significantly improve aggregate stability after 3 to 5 years of application.

General Guidelines for Managing Soil Tilth

Although each soil presents unique problems and opportunities, the following suggestions are generally relevant to managing soil tilth:

1. Minimizing tillage, especially moldboard plowing, disk harrowing, or rototilling, reduces the loss of aggregate-stabilizing organic matter.
2. Timing traffic activities to occur when the soil is as dry as possible and restricting tillage to periods of optimum soil moisture conditions will minimize destruction of soil structure.

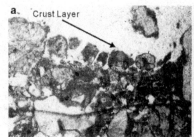 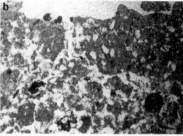 

Figure 4.26 *Physical soil crusts. Vertical section of the surface 3 mm of an Alfisol (yellow-brown soil in Chinese classification) with sandy loam texture from Hubei Province, China, exhibiting a typical crusted surface condition (a) and a noncrusted surface condition (b). The crust in (a) formed after the soil was exposed to 15 minutes of heavy rain. Note that the aggregates in the immediate surface have been destroyed, the fine particles have been washed downward, and the coarse particles near the surface have become tightly packed with few pores (white areas). A bean seedling (c) struggles to break a soil crust as it emerges from a seedbed.* (Photos (a) and (b) from Hu Xia; Shunjiangli; photo (c) courtesy of Ray R. Weil.) See also Hu et al. (2012)

3. Mulching the soil surface with crop residues or plant litter adds organic matter, encourages earthworm activity, and protects aggregates from beating rain and direct solar radiation.
4. Adding crop residues, composts, and animal manures to the soil is effective in stimulating microbial supply of the decomposition products that help stabilize soil aggregates.
5. Including perennial sod crops in the rotation favors stable aggregation by helping to maintain soil organic matter, providing maximal aggregating influence of fine plant roots, and ensuring a period without tillage.
6. Using cover crops and green manure crops, where practical, provides another good source of root action and decomposable organic matter for structural management.
7. Applying gypsum by itself or in combination with synthetic polymers can be very useful in stabilizing surface aggregates, especially in irrigated soils.

A high degree of aggregation helps the soil to perform critical ecosystem functions because most of these functions are influenced by soil porosity and density, properties which we shall now consider.

4.7 SOIL DENSITY

Particle Density

Soil **particle density** D_p is defined as the mass per unit volume of soil *solids* (in contrast to the volume of the *soil*, which would also include spaces between particles). Thus, if 1 cubic meter (m^3) of soil solids weighs 2.6 megagrams (Mg), the particle density is 2.6 Mg/m^3 (which can also be expressed as 2.6 grams per cubic centimeter).[6]

Particle density is essentially the same as the **specific gravity** of a solid substance. The chemical composition and crystal structure of a mineral determines its particle density. Particle density is *not* affected by pore space, and therefore is not related to particle size or to the arrangement of particles (soil structure).

Particle densities for most mineral soils vary between the narrow limits of 2.60 and 2.75 Mg/m^3 because quartz, feldspar, micas, and the colloidal silicates that usually make up the major portion of mineral soils all have densities within this range. For general calculations concerning arable mineral surface soils (1–5% organic matter), a particle density of about 2.65 Mg/m^3 may be assumed if the actual particle density is not known. This number would be adjusted upward to 3.0 Mg/m^3 or higher when large amounts of high-density minerals such as magnetite or hornblende are present. Likewise, it would be reduced for soils known to be high in organic matter, which has a particle density of only 0.9 to 1.4 Mg/m^3.

Bulk Density

A second important mass measurement of soils is **bulk density** D_b, which is defined as the mass of a unit volume of dry soil. This volume includes both solids and pores. A careful study of Figure 4.27 should make clear the distinction between particle and bulk density. Both expressions of density use only the mass of the solids in a soil; therefore, any water present is excluded from consideration.

There are several methods of determining soil bulk density by obtaining a known volume of soil, drying it to remove the water, and weighing the dry mass. A special coring instrument (Figure 4.28) can obtain a sample of known volume without disturbing the natural soil structure. For surface soils, perhaps the simplest method is to dig a small hole, dry and weigh all the excavated soil, and then determine the soil volume by lining the hole with plastic film and filling it completely with a measured volume of water. This method is well adapted to stony soils in which it is difficult to use a core sampler.

Factors Affecting Bulk Density

Soils with a high proportion of pore space to solids have lower bulk densities than those that are more compact and have less pore space. Consequently, any factor that influences soil pore space will affect bulk density. Typical ranges of bulk density for various soil materials and

---

[6]Since 1 Mg = 1 million grams and 1 m^3 = 1 million cubic centimeters, 1 Mg/m^3 = 1 g/cm^3.

Figure 4.27 Bulk density D_b and particle density D_p of soil. Particle density describes the weight of solid particles in a given volume of those solid particles, for example, the four blocks (particles) of solid rock stacked tightly with no pore space (a). Bulk density is the weight of the solid particles in a given volume of dry soil (which includes both solids and pore space occupied by air). Consider the diagram of the same four blocks of rock, now loosely arranged inside a cylinder to form "soil" that includes both the solid blocks and the spaces between them (b). For a second example presented, follow the calculations carefully and the terminology should be clear. In this particular case (c,d) the bulk density is one-half the particle density, and the porosity is 50%. (Diagrams courtesy of Ray R. Weil)

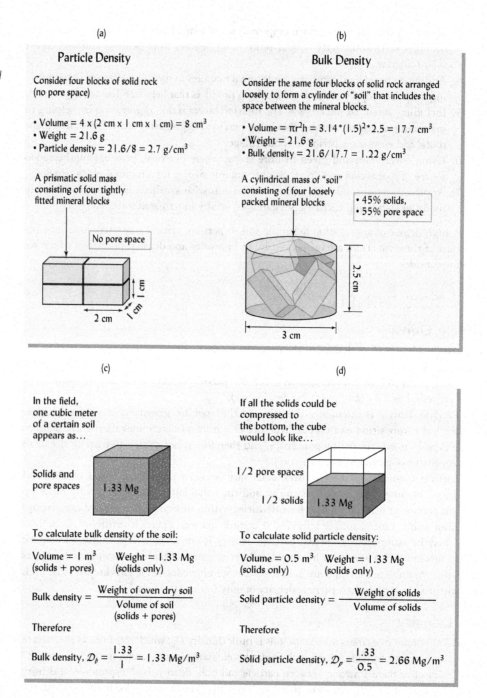

conditions are illustrated in Figure 4.29. It would be worthwhile to study this figure until you have a good feel for bulk density values.

Effect of Soil Texture. Although people commonly refer to "light sandy soils" and "heavy clay soils" soils, Figure 4.29 suggests that fine-textured soils such as silt loams, clays, and clay loams generally have lower bulk densities than do sandy soils. This fact may seem counterintuitive at first because. The terms *heavy* and *light* in this context refer not to the mass per unit volume of the soils, but to the amount of effort that must be exerted to manipulate these soils with tillage implements—the sticky clays being much more difficult to till.

The solid particles of the fine-textured soils tend to be organized in porous granules, especially if adequate organic matter is present. In these aggregated soils, pores exist both between *and* within the granules. This condition ensures high total pore space and a low bulk density. In

Figure 4.28 *A special sampler designed to remove a cylindrical core of soil without causing disturbance or compaction (a). The sampler head contains an inner cylinder and is driven into the soil with blows from a drop hammer. The inner cylinder (b) containing an undisturbed soil core is then removed and trimmed on each end with a knife to yield a core whose volume can easily be calculated from its length and diameter. The weight of this soil core is then determined after drying in an oven.*
(Photo courtesy of Ray R. Weil)

sandy soil, however, organic matter contents generally are low, the solid particles are less likely to be aggregated, and the bulk densities are commonly higher than in the finer-textured soils. Similar amounts of large pores are present in both sandy and well-aggregated fine-textured soils, but sandy soils have few of the fine, within-ped pores, and so have less total porosity (Figure 4.30).

While sandy soils generally have high bulk densities, the packing arrangement of the sand grains also affects their bulk density (see Figure 4.31). Loosely packed grains may fill as little as 52% of the bulk volume, while tightly packed grains may fill as much as 75% of the volume. If we assume that grains consist of quartz with a particle density of 2.65 Mg/m^3, then the corresponding range of bulk densities for loose to tightly packed sand would be 1.38 to 1.99 Mg/m^3, not too different from the range actually encountered in very sandy soils. The bulk density is generally lower if the sand particles are mostly of one size class (i.e., *well-sorted* sand), while a mixture of different sizes (i.e., *well-graded* sand) is likely to have an especially high bulk density. In the latter case, the smaller particles partially fill in the spaces between the larger particles. The most dense materials are those characterized by both a mixture of sand sizes and a tight packing arrangement.

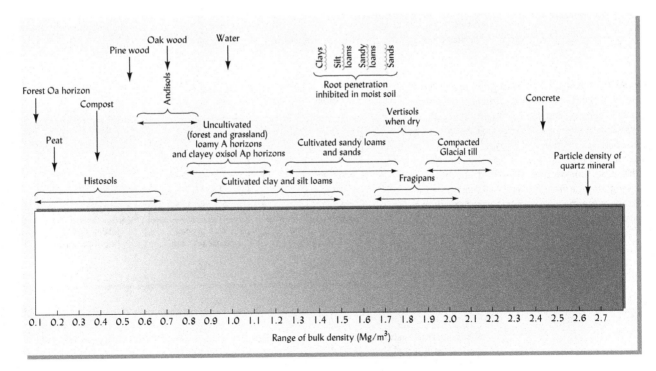

Figure 4.29 *Bulk densities typical of a variety of soils and soil materials.*

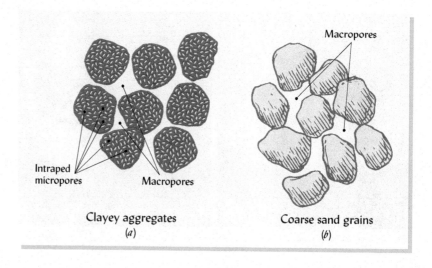

Figure 4.30 *A schematic comparison of sandy and clayey soils showing the relative amounts of large (macro-) pores and small (micro-) pores in each. There is less total pore space in the sandy soils than in the clayey one because the clayey soil contains a large number of fine pores within each aggregate (a), but the sand particles (b), while similar in size to the clayey aggregates, are solid and contain no pore spaces within them. This is the reason why, among surface soils, those with coarse texture are usually denser than those with finer textures.* (Diagram courtesy of Ray R. Weil)

Depth in Soil Profile. Deeper in the soil profile, bulk densities are generally higher, usually as a result of compaction caused by the weight of the overlying layers, lower organic matter contents, less aggregation, and fewer biopores. Very compact subsoils may have bulk densities of 2.0 Mg/m³ or even greater. Many soils formed from glacial till (see Section 2.3) have extremely dense subsoils as a result of past compaction by the enormous masses of ice overlaying them.

Useful Density Figures

For engineers involved with moving soil during construction, or for landscapers bringing in topsoil by the truckload, knowledge of the bulk density of various soils is useful in estimating the weight of (dry!) soil to be moved. A typical medium-textured mineral soil might have a bulk density of 1.25 Mg/m³, or 1,250 kilograms in a cubic meter.[7] People are often surprised by how heavy soil is. Imagine driving your pickup truck to a nursery where natural topsoil is sold in bulk and filling your truck bed (1.8 m × 2.4 m) with a nice, rounded load of about 4.3 m³. Of course, you would not really want to do this, as you certainly would not be able to drive away with the load. A typical "half-ton" (1,000 lb or 454 kg) pickup truck load capacity would be equivalent to 0.36 m³ of this soil (less if the soil were moist), even though the truck bed has room for about 12 times this volume of material.

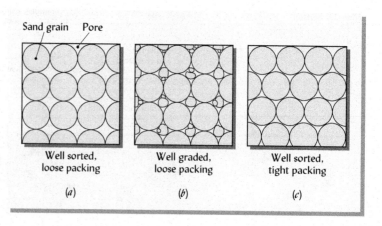

Figure 4.31 *The uniformity of grain size and the type of packing arrangement significantly affect the bulk density of sandy materials. Materials consisting of all similar-sized grains are termed* well sorted *(or* poorly graded*). Those with a variety of grain sizes are* well graded *(or* poorly sorted*). In either case, compaction of the particles into a tight packing arrangement markedly increases the bulk density of the material and decreases its porosity.* (Diagram courtesy of R. Weil)

---

[7] Most commercial landscapers and engineers in the United States still use English units. To convert values of bulk density given in units of Mg/m³ into values of lb/yd³, multiply by 1686. Therefore, 1 yd³ of a typical medium-textured mineral soil with a bulk density of 1.25 Mg/m³ would weigh over a ton (1686 × 1.25 = 2108 lb/yd³). Any water in the soil would be additional.

The mass of soil in 1 ha to a depth of normal plowing (15 cm) can be calculated from soil bulk density. If we assume a bulk density of 1.3 Mg/m^3 for a typical arable surface soil, such a hectare-furrow slice 15 cm deep weighs about 2 million kg.[8] This estimate of the mass of surface soil in a hectare of land is very useful in calculating lime and fertilizer application rates and stocks of carbon, nitrogen, or other soil components (see Boxes 8.4, 9.3 and 12.1 for detailed examples). However, this estimated mass should be adjusted if the bulk density is other than 1.3 Mg/m^3 or the depth of the layer under consideration is more or less than 15 cm.

Management Practices Affecting Bulk Density

Changes in bulk density for a given soil are easily measured and can alert soil managers to changes in soil quality and ecosystem function. Increases in bulk density usually indicate a poorer environment for root growth, reduced aeration, and undesirable changes in hydrologic function, such as reduced water infiltration.

Forest Lands. The surface horizons of most forested soils have rather low bulk densities (see Figure 4.29). Tree growth and forest ecosystem function are particularly sensitive to increases in bulk density. Conventional timber harvest generally disturbs and compacts 20 to 40% of the forest floor (Figure 4.32) and is especially damaging along the skid trails where logs are dragged and at the landing decks—areas where logs are piled and loaded onto trucks. An expensive, but effective, means of moving logs while minimizing compaction of forest lands is the use of cables strung between towers or hung from large balloons.

Intensive recreational and transport use of soils in forests and other areas with natural vegetation can also lead to increased bulk densities. Such effects can be seen where access roads, trails, and campsites are found (Figure 4.33). An important consequence of increased bulk density is a diminished capacity of the soil to take in water, hence increased losses by surface runoff. Damage from hikers can be minimized by restricting foot traffic to well-designed, established trails that may include a thick layer of wood chips, or even a raised boardwalk in the case of heavily traveled paths over very fragile soils, such as in wetlands.

Figure 4.32 *Timber harvest can compact forest soils. (Left) Heavy equipment (a skidder and logging truck) at a landing deck where the soil is severely impacted under wet conditions in a Southern Appalachian hardwood forest. (Right) A similar conventional rubber-tired skidder in a boreal forest in western Alberta, Canada. Such practices cause significant soil compaction that can impair soil ecosystem functions for many years. Timber harvest practices that can reduce such damage to forest soils include selective cutting, use of flexible-track vehicles or overhead cable transport of logs, and abstaining from harvest during wet conditions.* (Photo courtesy of Ray R. Weil (left) and Andrei Startsev, Alberta Environmental Center (right))

---

[8] 10,000 m^2/ha × 1.3 Mg/m^3 × 0.15 m = 1950 Mg/ha, or about 2 million kg per ha to a depth of 15 cm. A comparable figure in the English system is 2 million lb per acre—furrow slice 6 to 7 in. deep.

Figure 4.33 *Impact of campers on the bulk density of forest soils, and the consequent effects on rainwater infiltration rates and runoff losses (see white arrows). At most campsites the high impact area extends for about 10 m from the fire circle or tent pad. Managers of recreational land must carefully consider how to protect sensitive soils from compaction that may lead to death of vegetation and increased erosion.* (Data from Vimmerstadt et al. (1982))

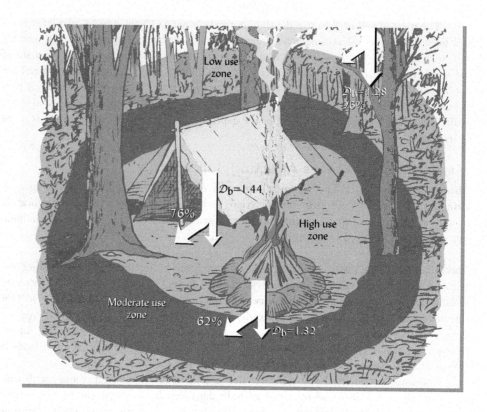

Urban Soils. In urban areas, trees planted for landscaping purposes must often contend with severely compacted soils. While it is usually not practical to modify the entire root zone of a tree, several practices can help (see also Section 7.6). First, making the planting hole as large as possible will provide a zone of loose soil for early root growth. Second, a thick layer of mulch spread out to the drip line (but not too near the trunk) will enhance root growth, at least near the surface. Third, a rigid but porous cover, such as an iron grill, can be installed around the tree to protect against compaction while letting in water and air. Fourth, the tree roots may be given paths for expansion by digging a series of narrow trenches radiating out from the planting hole and backfilled with loose, enriched soil.

In some urban settings, it may be desirable to create an "artificial soil" that includes a skeleton of coarse angular gravel to provide strength and stability, and a mixture of loam-textured topsoil and organic matter to provide nutrient- and water-holding capacities. Also, large quantities of sand and organic materials are sometimes mixed into the upper few centimeters of a fine-textured soil in which turf grass is to be grown for golf putting greens.

Green Roofs. Soil bulk density is critical in the design of rooftop gardens. The mass of soil involved must be minimized in order to design a cost-effective structure of sufficient strength to carry the soil load. One might choose to grow only such shallow-rooted plants as sedums or turf grasses so that a relatively thin layer of soil (say, 15 cm) could be used, keeping the total mass of soil from being too great. It may also be possible to reduce the cost of construction by selecting a natural soil having a relatively low bulk density, such as some well-aggregated loams or peat soils. Often an artificial growing medium is created from such lightweight materials as perlite and peat. However, such very low-density materials may require a surface netting system to prevent wind from blowing them off the roof, and this type of media will not perform the anchorage function of soils as a plant medium (see Section 1.2) for tall plants such as trees.

Agricultural Land. Although tillage may temporarily loosen the surface soil, in the long-term intense tillage increases soil bulk density by depleting soil organic matter and weakening soil structure (see Section 4.6). The data in Table 4.2 illustrate this trend. These data are from long-term studies in different locations where relatively undisturbed soils were compared to adjacent areas that had been cultivated for 12 to 80+ years. In all cases, cropping

Table 4.2
BULK DENSITY AND PORE SPACE OF SURFACE SOILS FROM CULTIVATED AND NEARBY UNCULTIVATED AREAS

With cultivation, the bulk density was increased, and the pore space proportionately decreased, in every case.

Soil	Texture	Years cropped	Bulk density, Mg/m³ Cultivated soil	Bulk density, Mg/m³ Uncultivated soil	Pore space, % Cultivated soil	Pore space, % Uncultivated soil
Mean of 6 Ustolls (South Dakota)	Silt loam	80+	1.30	1.10	50.9	58.5
Mean of 2 Udults (Maryland)	Sandy loam	50+	1.59	0.84	40.0	66.4
Mean of 2 Udults (Maryland)	Silt loam	50+	1.18	0.78	55.5	68.8
Mean of 3 Ustalfs (Zimbabwe)	Clay	20–50	1.44	1.20	54.1	62.6
Mean of 3 Ustalfs (Zimbabwe)	Sandy loam	20–50	1.54	1.43	42.9	47.2

Data for South Dakota soils from Eynard et al. (2004), for Maryland from Lucas and Weil (2012), and for Zimbabwe from Weil (unpublished).

increased the bulk density of the topsoils. The effect of cultivation can be minimized by adding crop residues or farm manure in large amounts and rotating cultivated crops with a grass sod.

In mechanized agriculture, the wheels of heavy machines used to pull implements, apply amendments, or harvest crops can create yield-limiting soil compaction. Certain tillage implements, such as the moldboard plow and the disk harrow, compact the soil below their working depth even as they lift and loosen the soil above. Use of these implements or repeated trips over the field by heavy machinery can form **plow pans** or **traffic pans**, dense zones immediately below the plowed layer (Figure 4.34). Other tillage implements, such as the chisel plow and the spring-tooth harrow, do not press down upon the soil beneath them, and so are useful in breaking up plow pans and stirring the soil with a minimum of compaction. Large chisel-type plows (Figure 4.35a) can be used in **subsoiling** to fracture soils to considerable depth and break up dense subsoil layers, thereby enhancing deep root penetration (Figure 4.35b). These implements should loosen the soil without producing horizontal compacted layers, and cut through crop residues while causing little disturbance of the soil surface. However, in most soils the effects of subsoiling are quite temporary and any tillage tends to reduce soil strength,

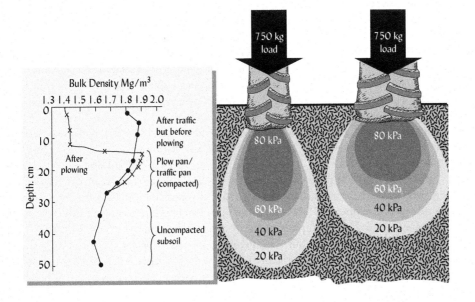

Figure 4.34 *Vehicle tires compact soil to considerable depths. (Left) Representative bulk densities associated with traffic compaction on a sandy loam soil. Plowing can temporarily loosen the compacted surface soil (plow layer), but usually increases compaction just below the plow layer. (Right) Vehicle tires (750 kg load per tire) compact soil to about 50 cm. The more narrow the tire, the deeper it sinks and the deeper its compactive effect. The tire diagram shows the compactive pressure in kPa. For tire designs that reduce compaction, see Tijink and Van der Linden (2000).* (Diagram courtesy of Ray R. Weil)

Figure 4.35 Two approaches to the alleviation of subsoil compaction. (a) A heavy chisel plow, also known as a subsoiler or ripper, is used when the soil is dry to cause compacted soil to shatter leaving a network of cracks that enhance water, air, and root movement (b). An alternative approach involves growing taprooted plants (such as the forage radish shown) in the fall and spring when the subsoil is relatively wet and easily penetrated by roots (c). The taproots then decay, leaving semipermanent channels in which roots of subsequent crops can grow to pass through the compacted subsoil zone, even during the summer when the soil is relatively dry and hard. (d) Corn planted after forage radish had twice as many roots reaching the subsoil (below 30 cm) as corn planted after the fibrous rooted rye, and nearly ten times as many as corn planted in soil that had no cover crop during the winter. (Photos (a) and (c) courtesy of Ray R. Weil; photo (b) courtesy of USDA National Tillage Laboratory; data based on study by Chen and Weil (2011))

thus making the soil less resistant to subsequent compaction. In moist fine-textured soils, subsoiling can cause planes of compaction as the implements slide through the soil.

In cold climates, repeated cycles of freezing and thawing during winter can break up compacted soil near the surface. However, even where it is cold enough to freeze the soil down to 50 or 100 cm, the repeated freeze–thaw cycles occur only on the upper 10 to 20 cm. Thus, if heavy equipment driven on wet soils has caused deep compaction, one cannot rely on cold winters to alleviate the problem.

Certain plants have been shown to alleviate deep subsoil compaction. A recently developed alternative to deep tillage is the use of cover crop plants that can penetrate compacted layers during moist periods and leave deep root channels for subsequent crops' roots to follow (Figure 4.35c,d). Species with thick taproots are best suited to this use.

In many parts of the world, farmers use hand hoes or animal-drawn implements to stir the soil. Although humans and draft animals are not as heavy as tractors, their weight is applied to the soil in a relatively small area (foot- or hoof-print), and so can also cause considerable compaction.

Traffic is particularly damaging on wet soil. Generally, with heavier loads and on wetter soils, compactive effects are more pronounced and penetrate more deeply into the profile. To prevent compaction, which can result in yield reductions and loss of profitability, the number of tillage operations and heavy equipment trips over the field should be minimized and timed to avoid periods when the soil is wet. Unfortunately, traffic on wet agricultural soils is sometimes unavoidable in humid temperate regions in spring and fall.

Another approach for minimizing compaction is to carefully restrict all wheel traffic to specific lanes, leaving the rest of the field (usually 90% or more of the area) free from compaction. Such **controlled traffic** systems are widely used in Europe, especially on clayey soils. Gardeners can practice controlled traffic by establishing permanent footpaths between planting beds. The paths may be enhanced by covering with a thick mulch, planting to sod grass, or paving with flat stones.

Some managers attempt to reduce compaction using an opposite strategy in which special wide tires are fitted to heavy equipment so as to spread the weight over more soil surface, thus reducing the force applied per unit area (Figure 4.36, *right*). Wider tires do lessen the compactive effect, but they also increase the percentage of the soil surface that is impacted. In a practice analogous to using wide wheels, home gardeners can avoid concentrating their body

Figure 4.36 One approach to reducing soil compaction is to spread the applied weight over a larger area of the soil surface. Examples are extra-wide wheels on heavy vehicles used to apply soil amendments (left) and standing on a wooden board while preparing a garden seedbed in early spring (right). (Photo courtesy of Ray R. Weil)

weight on just the few square centimeters of their footprints by standing on wooden boards when preparing seedbeds in relatively wet soil (Figure 4.36, *left*).

Influence of Bulk Density on Soil Strength and Root Growth

High bulk density may occur as a natural soil profile feature (e.g., a fragipan), or it may be an indication of human-induced soil compaction. In any case, root growth is inhibited by excessively dense soils for a number of reasons, including the soil's resistance to penetration, poor aeration, slow movement of nutrients and water, and the buildup of toxic gases and root exudates.

Roots penetrate the soil by pushing their way into pores. If a pore is too small to accommodate the root cap, the root must push the soil particles aside and enlarge the pore. To some degree, the density *per se* restricts root growth, as the roots encounter fewer and smaller pores. However, root penetration is also limited by **soil strength**, the property of the soil that causes it to resist deformation. One way to quantify soil strength is to measure the force needed to push a standard pointed rod (a **penetrometer**) into the soil (see also Section 4.9). Compaction generally increases both bulk density and soil strength.

Soil water content and bulk density both affect *soil strength* (see Section 4.9 and Figure 4.37). Soil strength is increased when a soil is compacted to a higher bulk density, and also when finer-textured soils dry out and harden. Root can penetrate a greater bulk density when the soil is more moist. For example, a traffic pan having a bulk density of 1.6 Mg/m^3 may completely prevent the penetration of roots when the soil is rather dry, yet roots may readily penetrate this same layer when it is in a moist condition.

The more clay present in a soil, the smaller the average pore size, and the greater the resistance to penetration at a given bulk density. Therefore, if the bulk density is the same, roots more easily penetrate a moist sandy soil than a moist clayey one. The growth of roots

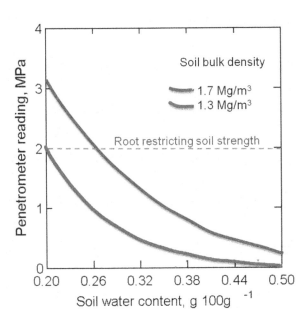

Figure 4.37 Both water content and bulk density affect soil strength as measured by penetrometer resistance. The data are for the clay textured Bt horizon of a Tatum soil in Virginia (Hapludults), which was either severely compacted (bulk density 1.7 Mg/m^3) or not compacted (bulk density 1.3 Mg/m^3). Note that soil strength decreases as water content increases and is very low regardless of bulk density when the soil is nearly saturated with water. (Graph based on study by Gilker et al. (2002))

into moist soil is generally limited by bulk densities ranging from 1.45 Mg/m³ in clays to 1.85 Mg/m³ in loamy sands (see Figure 4.29).

As might be expected, such land uses as row-crop agriculture, pasture, rangeland, forestry, or off-road trafficking often markedly and simultaneously affect soil bulk density and strength in ways that restrict or enhance root growth and water movement. It is not always appreciated that tillage and traffic can cause compaction quite deep into the subsoil. It may take many years of restorative management for the subsoil to recover its natural degree of porosity and friability. Figure 4.38 illustrates such a case.

4.8 PORE SPACE OF MINERAL SOILS[9]

For soils with the same particle density, the lower the bulk density, the higher the percent pore space (**total porosity**). See Box 4.5 for derivation of the formula expressing this relationship.

Factors Influencing Total Pore Space

In Chapter 1 (Figure 1.14), we noted that for an "ideal" medium-textured, well-granulated surface soil in good condition for plant growth, approximately 50% of the soil volume would consist of pore space, and that the pore space would be about half-filled with air and half-filled with water. Actually, total porosity varies widely among soils for the same reasons that bulk density varies. Values range from as low as 25% in compacted subsoils to more than 65% in well-aggregated, high-organic-matter surface soils. As is the case for bulk density, management can exert a decided influence on the pore space of soils (see Table 4.2). Data from a wide range of soils show that cultivation tends to lower the total pore space compared to that of uncultivated soils. This reduction usually is associated with a decrease in organic matter content, and consequently, of **aggregation**.

Size of Pores

Bulk density values help us predict only *total* porosity. However, soil pores occur in a wide variety of sizes and shapes that largely determine what role the pore can play in the soil. Pores can be grouped by size into macropores, mesopores, micropores, and so on. We will simplify our discussion at this point by referring only to **macropores** (larger than about 0.08 mm) and **micropores** (smaller than about 0.08 mm).

Macropores. The macropores characteristically allow the ready movement of air and the drainage of water. They also are large enough to accommodate plant roots and the wide range

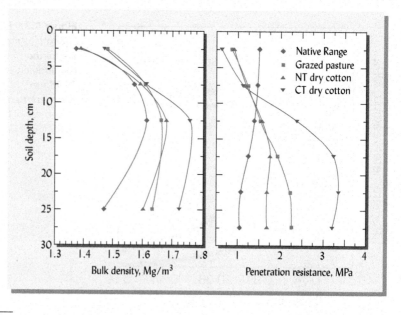

Figure 4.38 Bulk density and penetration resistance after about 30 years of no-till (NT) and conventional tillage (CT) dryland cotton, or 15 years of grazed pasture compared to native rangeland. Cotton with CT resulted in the greatest bulk density and penetration resistance, especially below the depth of plowing (15–20 cm), suggesting the formation of a plow pan in the CT cotton soil. After 15 to 30 years without tillage, the pasture and NT cotton system had partially recovered the favorable physical properties of the native rangeland. The penetration resistance levels for all systems were quite high (2 MPa can restrict root growth) because water contents of these semiarid region soils were quite low at the time of measurement. However, soil water was relatively uniform among the systems, at least in the lower layers where the major differences in penetration resistance occurred. Amarillo loamy fine sand (Aridic Paleustalfs) in the southern high plains of Texas. (Drawn from data in Halfmann (2005))

---

[9]For a review of how soil pores interact with hydrology, see Lin (2012).

of tiny animals that inhabit the soil (see Chapter 10). Several types of macropores are illustrated in Figure 4.39. Macropores can occur as the spaces between individual sand grains in coarse-textured soils. Thus, even though a sandy soil has relatively low total porosity, the movement of air and water through such a soil is surprisingly rapid because of the dominance of the macropores. In well-structured soils, the macropores are generally found between peds. These **interped pores** may occur as spaces between loosely packed granules or as the planar cracks between tight-fitting blocky and prismatic peds (see Figure 4.39d). Macropores created by roots, earthworms, and other organisms constitute a very important type of pores termed **biopores**. These are usually tubular in shape and may be continuous for lengths of a meter or more (see Table 4.3 and Figure 4.17). In some clayey soils, biopores are the principal form of macropores, greatly facilitating the growth of plant roots (Figure 4.35). Perennial plants, such as forest trees and certain cover crops, are particularly effective at creating channels that serve as conduits for roots, long after the death and decay of the roots that originally created them.

Micropores. In contrast to macropores, micropores are usually filled with water in field soils. Even when not water-filled, they are too small to permit much air movement. Water movement in micropores is slow, and much of the water retained in these pores is not available to plants (see Chapter 5). Fine-textured soils, especially those without a stable granular structure, may have

---

BOX 4.5
CALCULATION OF PERCENT PORE SPACE IN SOILS

The bulk density of a soil can be easily measured and particle density can usually be assumed to be 2.65 Mg/m³ for most silicate-dominated mineral soils. Direct measurement of the pore space in soil requires the use of much more tedious and expensive techniques. Therefore, it is often desirable to calculate soil porosity from data on bulk and particle densities.

The derivation of the formula used to calculate the percentage of total pore space in soil follows:

Let
D_b = bulk density, Mg/m³
D_p = particle density, Mg/m³
W_s = Weight of soil (solids), Mg
V_s = volume of solids, m³
V_p = volume of pores, m³
$V_s + V_p$ = total soil volume V_t

By definition,

$$\frac{W_s}{V_s} = D_p \text{ and } \frac{W_s}{V_s + V_p} = D_b$$

Solving for W_s gives

$$W_s = D_p \times V_s \text{ and } W_s = D_b(V_s + V_p)$$

Therefore

$$D_p \times V_s = D_b(V_s + V_p) \text{ and } \frac{V_s}{V_s + V_p} = \frac{D_b}{D_p}$$

Since

$$\frac{V_s}{V_s + V_p} \times 100 = \% \text{ solid space} \quad \text{then}$$

$$\% \text{ solid space} = \frac{D_b}{D_p} \times 100$$

Since % pore space + % solid space = 100%, and, rearranging, % pore space = 100% − % solid space, then

$$\% \text{ pore ace} = 100\% - \left(\frac{D_b}{D_p} \times 100\right)$$

EXAMPLE

Consider a cultivated clay soil with a bulk density determined to be 1.28 Mg/m³. In the absence of direct measurement, we assume that the particle density is approximately that of the common silicate minerals (i.e., 2.65 Mg/m³). We calculate the percent pore space using the formula derived above:

$$\% \text{ pore space} = 100\% - \left(\frac{1.28 \text{ Mg/m}^3}{2.65 \text{ Mg/m}^3} \times 100\right)$$

$$= 100\% - 48.3 = 51.7$$

This value of pore space, 51.7%, is quite close to the typical percentage of air and water space described in Figure 1.14 for a well-granulated, medium- to fine-textured soil in good condition for plant growth. This simple calculation tells us nothing about the relative amounts of large and small pores, however, and so must be interpreted with caution.

As an example of a soil rich in high-density minerals, let us consider the uncultivated clay soils from Zimbabwe (Ustalfs) described in Table 4.2. These are red-colored clays high in iron oxides. The particle density for these soils was determined to be 3.21 Mg/m³ (not shown in Table 4.2). Using this value and the bulk density value from Table 4.3, we calculate the pore space as follows:

$$\% \text{ pore space} = 100\% - \left(\frac{1.20}{3.21} \times 100\%\right)$$

$$= 100 - 37.4 = 62.6$$

Such a high percentage pore space is an indication that this soil is in an uncompacted, very well-granulated condition typical for soils found under undisturbed natural vegetation.

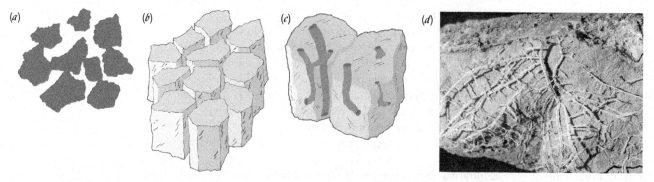

Figure 4.39 *Various types of soil pores. (a) Many soil pores occur as packing pores, spaces left between primary soil particles. The size and shape of these spaces is largely dependent on the size and shape of the primary sand, silt, and clay particles and their packing arrangement. (b) In soils with structural peds, the spaces between the peds form interped pores. These may be rather planar in shape, as with the cracks between prismatic peds, or they may be more irregular, like those between loosely packed granular aggregates. (c) Biopores are formed by organisms such as earthworms, insects, and plant roots. Most of these are long, sometimes branched channels, but some are round cavities left by insect nests and the like. (d) Plant roots proliferating along the face of a large prismatic ped in a fragipan. Note that the roots have become squeezed flat as the crack between adjacent prisms swelled closed during wet periods. No roots had penetrated the dense interior of the prism.* (Photo courtesy of Ray R. Weil)

a preponderance of micropores, thus restricting gas and water movement, despite the relatively large total pore space. While the larger micropores accommodate plant root hairs and microorganisms, the smaller micropores (sometimes termed *ultramicropores* and *cryptopores*) are too small to permit the entrance of even the smallest bacteria or even some enzymes molecules produced by the bacteria. Such pores can act as hiding places for some adsorbed organic compounds (both naturally occurring and pollutants), thereby protecting them from breakdown for long periods of time.

Cultivation and Pore Size

Continuous cropping, particularly of soils originally high in organic matter, often results in a reduction of macropore spaces (see Table 4.4). When native prairie lands are plowed and planted to row crops such as corn or soybeans, soil organic matter contents and total pore space are reduced. But most striking is the effect of cropping on the amount of macropore space that is so critical for ready air movement.

In recent years, conservation tillage practices, which minimize plowing and associated soil manipulations, have been widely adopted in North and South America (see Sections 4.6

Table 4.3
DISTRIBUTION OF DIFFERENT-SIZED LOBLOLLY PINE ROOTS IN THE SOIL MATRIX AND IN OLD ROOT CHANNELS IN THE UPPERMOST METER OF AN ULTISOL IN SOUTH CAROLINA

The root channels were generally from 1 to 5 cm in diameter and filled with loose surface soil and decaying organic matter. Such channels are easy for roots to penetrate and have better fertility and aeration than the surrounding soil matrix.

	Numbers of roots counted per 1 m² of the soil profile		
Root size, diameter	Soil matrix	Old root channels	Comparative increase in root density in the old channels, %
Fine roots, <4 mm	211	3617	94
Medium roots, 4–20 mm	20	361	95
Coarse roots, >20 mm	3	155	98

Calculated from Parker and Van Lear (1996).

and 14..17). Because of increased accumulation of organic matter near the soil surface and the development of a long-lived network of macropores (especially biopores), some conservation tillage systems lead to greater macroporosity of the surface layers. These benefits are particularly likely to accrue in soils with extensive production of earthworm burrows, which may remain undisturbed in the absence of tillage. Such improvements in porosity are most likely to occur in soils managed with both cover crops and no-tillage techniques.

4.9 SOIL PROPERTIES RELEVANT TO ENGINEERING USES

Soil properties are obviously important to engineering. In the words of Richard Hand:

Virtually every (built) structure is supported by soil or rock. Those that aren't either fly, float, or fall over.

Field Rating of Soil Consistence and Consistency

Consistence. Soil **consistence** describes the ease with which a soil can be reshaped or ruptured. Observations are made on the amount of force needed to crush a clod of soil and on the manner in which the soil responds to the force. The degree of cementation of the soil by such materials as silica, calcite, or iron is also considered in identifying soil consistence. Moisture content greatly influences how a soil responds to stress; hence, moist and dry soils are given separate consistence ratings (Table 4.5). A dry, clayey soil that cannot be crushed between the thumb and forefinger but can be crushed easily underfoot would be designated as *very hard*. The same soil, when wet, would exhibit much less resistance to deformation, and would be termed *plastic*. The degrees of *stickiness* and *plasticity* (malleability) of soil in the wet condition are often included in describing soil consistence (although not shown in Table 4.5). As described in Section 4.6, a moist clod that crumbles with only light pressure is said to be friable. Friable soils are easily excavated or tilled.

Consistency. Engineers use the term **consistency** to describe how a soil resists *penetration* by an object, while the soil scientist's consistence describes resistance to *rupture*. Instead of crushing a clod of soil, the engineer attempts to penetrate it with either the blunt end of a pencil or thumbnail (Table 4.5). Consistency, then, is a kind of simple field estimation of soil strength or penetration resistance (see Section 4.7).

Field observations of both consistence and consistency provide valuable information to guide decisions about loading and manipulating soils. For construction purposes, however, more precise measurements are needed to predict how a soil will respond to applied stress.

Table 4.4
CONTINUOUS CROPPING AFFECTS MACROPORE AND MICROPORE SPACES

Compared to undisturbed prairie soil, the cultivated soil has far less macropore space, but has gained some micropore space as aggregates were destroyed, changing large interped pores into much smaller micropores. Loss of organic matter made the aggregates more susceptible to damage by tillage. Houston clay (Hapludert).

Soil history	Soil depth, cm	Organic matter, %	Total pore space, %	Macropore space, %	Micropore space, %	Bulk density, Mg/m^3
Prairie	0–15	5.6	58.3	32.7	25.6	1.11
Tilled 50 years	0–15	2.9	50.2	16.0	34.2	1.33
Prairie	15–30	4.2	56.1	27.0	29.1	1.16
Tilled 50 years	15–30	2.8	50.7	14.7	36.0	1.31

Data from Laws and Evans (1949).

Table 4.5
SOME FIELD TESTS AND TERMS USED TO DESCRIBE THE CONSISTENCE AND CONSISTENCY OF SOILS
The consistency of cohesive materials is closely related to, but not exactly the same as, their consistence. Conditions of least coherence are represented by terms at the top of each column, those of greater coherence near the bottom.

	Soil consistence[a]			Soil consistency[b]	
Dry soil	Moist to wet soil	Soil dried then submerged in water	Field rupture (crushing) test	Soil at in situ moisture	Field penetration test
Loose	Loose	Not applicable	Specimen not obtainable	Soft	Blunt end of pencil penetrates deeply with ease
Soft	Very friable	Noncemented	Crumbles under very slight force between thumb and forefinger	Medium firm	Blunt end of pencil can penetrate about 1.25 cm with moderate effort
Slightly hard	Friable	Extremely weakly cemented	Crumbles under slight force between thumb and forefinger	Firm	Blunt end of pencil can penetrate about 0.5 cm
Hard	Firm	Weakly cemented	Crushes with difficulty between thumb and forefinger	Very firm	Blunt end of pencil makes slight indentation; thumbnail easily penetrates
Very hard	Extremely firm	Moderately cemented	Cannot be crushed between thumb and forefinger, but can be crushed slowly underfoot	Hard	Blunt end of pencil makes no indentation; thumbnail barely penetrates

[a]Abstracted from USDA-NRCS (2005).
[b]Modified from McCarthy (1993).

Soil Strength and Sudden Failure

Engineers define soil **bearing strength** as the capacity of a soil mass to withstand stresses without rupturing or becoming deformed. Failure of a soil to withstand stress can result in a building toppling over as its weight exceeds the soil's bearing strength. Similarly, an earthen dam or levee might give way under the pressure of impounded water, or pavements and structures might slide down unstable hillsides (see Figure 1.6).

Cohesive Soils. Two components of strength apply to **cohesive soils** (essentially soils with a clay content of more than about 15%): (1) the inherent electrostatic attractive forces between clay platelets and between clay surfaces and the water in very fine pores (see *clay flocculation* in Section 4.5), and (2) the frictional resistance to movement between soil particles of all sizes. While many different laboratory tests are used to estimate soil strength, perhaps the simplest to understand is the direct **unconfined compression test** using the apparatus illustrated in Figure 4.40a. A cylindrical specimen of cohesive soil is placed vertically between two flat, porous stones (which allow water to escape from the compressed soil pores) and a slowly increasing downward force is applied. The soil column will first bulge out a bit and then give way suddenly and collapse—when the force exceeds the soil strength.

The strength of cohesive soils declines dramatically if the material is very wet and the pores are filled with water. Then the particles are forced apart so that neither the cohesive nor the frictional component is very strong, making the soil prone to failure, often with dramatic results (such as mudslides and soil creep, Figure 4.41, or levee failures, Box 4.6). On the other hand, if cohesive soils become more compacted or dry down, their strength increases as particles are forced into closer contact with one another—a result that has implications for plant root growth as well as for engineering (see Section 4.7).

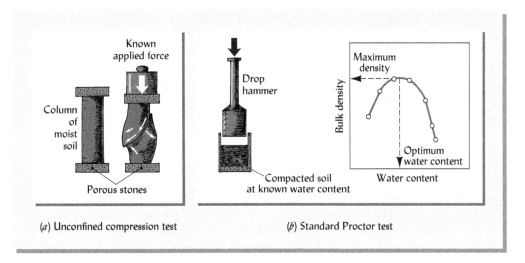

Figure 4.40 *Two important tests to determine engineering properties of soil materials. (a) An unconfined compression test for soil strength. (b) The Proctor test for maximum density and optimum water content for compaction control.*

Noncohesive Soils. The strength of dry, noncohesive soil materials such as loose sand depends entirely on frictional forces, including the interlocking of rough particle surfaces. One reflection of such interparticle friction is the **angle of repose**, the steepest angle to which a material can be piled without slumping. Smooth, rounded sand grains cannot be piled as steeply as can rough, interlocking sands. If a small amount of water bridges the gaps between particles, electrostatic attraction of the water for the mineral surfaces will increase the soil strength. Interparticle water bridges explain why cars can drive along the edge of the beach where the sand is moist, but their tires sink in and lose traction on loose, dry sand or in saturated quicksand.

Collapsible Soils. Certain soils that exhibit considerable strength at low *in situ* water contents lose their strength suddenly if they become wet. Such soils may collapse without warning under a roadway or building foundation. A special case of soil collapse is **thixotropy**, the sudden liquefaction of a wet soil mass when subjected to vibrations, such as those accompanying earthquakes and blasting.

Figure 4.41 *As soils become wetter, both electrostatic and frictional components of soil strength may be reduced to the point that mass movement may occur either suddenly as in a mudslide (left) or very gradually as in soil creep (right). Mudslides occur when the soil becomes so wet as to behave almost as a liquid, flowing downslope, even over retaining walls. Soil creep is too slow to observe directly, but one can see the bowed tree trunks that result as the tree growth attempts to compensate for the downslope movement of soil.* (Photo courtesy of Ray R. Weil)

Most **collapsible soils** are noncohesive materials in which loosely packed sand grains are cemented at their contact points by small amounts of gypsum, clay, or water under tension. These soils usually occur in arid and semiarid regions, where such cementing agents are relatively stable. Many collapsible soils have derived their open particle arrangement from the process of sedimentation beneath past or present bodies of water. When these soils are wetted, excess water may dissolve cements such as gypsum or disperse clays that form bridges between particles, causing a sudden loss of strength.

Settlement—Gradual Compression

Compaction Control. Most foundation problems result from slow, often uneven, vertical subsidence or **settlement** of the soil. Soils to be used for a foundation or roadbed are compacted on purpose using heavy rollers (Figure 4.42) or vibrators. Compaction occurring after construction would result in uneven settlement and cracked pavements or foundations.

Some soil particles, such as certain silicate clays and micas, can be compressed when a load is placed upon them. If that load is removed, these particles tend to regain their original shape, in effect reversing their compression. As a result, soils rich in these particles are not easily compacted into a stable base for roads and foundations.

The **Proctor test** is used to guide efforts at compacting soil materials before construction. A specimen of soil is mixed to a given water content and placed in a holder, where it is compacted by a drop hammer. The bulk density (usually referred to as the *dry density* by engineers) is then measured. The process is repeated with increasing water contents until a *Proctor curve* (Figure 4.40b) can be drawn from the data. The curve indicates the maximum bulk density achievable and the soil water content that maximizes compactability. On construction sites, tanker trucks may spray water to bring the soil water content to the determined optimum level before heavy equipment (such as that shown in Figure 4.42) compacts the soil to the desired density.

Compressibility. A consolidation test may be conducted on a soil specimen to determine its **compressibility**—how much its volume will be reduced by a given applied force. Because of the relatively low porosity and equidimensional shape of the individual mineral grains, very sandy soils resist compression once the particles have settled into a tight packing arrangement. They make excellent soils for foundations. The high porosity of clay floccules and the flakelike shape of clay particles give clayey soils much greater compressibility. Soils consisting mainly of organic matter (peats) have the highest compressibilities and generally are unsuitable for foundations.

In the field, compression of wet, clayey soils may occur very slowly after a load (e.g., a building) is applied because compression can occur only as fast as water can escape from the soil pores—which for the fine pores in clayey materials is not very fast. Perhaps the most famous example of uneven settlement due to slow compression is the Leaning Tower of Pisa in Italy. Unfortunately, most cases of uneven settlement result in headaches, not tourist attractions.

Figure 4.42 Compaction of soils used as foundations and roadbeds is accomplished by heavy equipment such as this sheepsfoot roller. The knobs ("sheepsfeet") concentrate the mass on a small impact area, punching and kneading the loose, freshly graded soil to optimum density. (Photo courtesy of Ray R. Weil)

Expansive Soils

Some clays, particularly the smectites (see Section 8.3), swell when wet and shrink when dry. Expansive soils are rich in these types of clay. The electrostatic charges on clay surfaces attract water molecules from larger pores into the micropores within clay domains. Also, the adsorbed cations associated with the clay surfaces tend to hydrate, drawing in additional water. The water pushes apart the layers of clay, causing the mass of soil to swell in volume. The reverse of these processes occurs when the soil dries and water is withdrawn from packets of clay platelets, causing shrinkage and cracking. After a prolonged dry spell, soils high in smectites can be recognized in the field by the crisscrossed pattern of wide, deep cracks. The swelling and shrinkage cause sufficient movement of the soil to crack building foundations, burst pipelines, and buckle pavements.

Construction Hazards of Expansive Soils.[10]   If preventative design measures (e.g., Figure 4.43) are not taken during the construction of houses on smectite clays, the building foundation is likely to move with the swelling and shrinking of the soil, misaligning doors and windows and eventually cracking foundations, walls, and pipes. During extended dry periods soils under and around building foundations become unusually dry resulting in major shrinkage even in moderately expansive soils. The cost of building homes on smectitic soils may be double that of building on soils dominated by nonswelling clays, for which conventional foundation designs can be safely used.

Damage caused by expansive soils only rarely makes news headlines, although in many industrialized countries the total cost annually exceeds that caused by tornados, floods, or earthquakes. For example, expansive clays occur on about 20% of the land area in the United States and cause upwards of $10 billion in damages annually to pavements, foundations, and utility lines. The damages can be severe in certain sites in all parts of the country, but are most extensive in regions that have long dry periods alternating with periods of rain (see distribution of Vertisols, endpapers).

Atterberg Limits

Soil that presents as a hard, rigid solid in the dry state becomes a crumbly (friable) semisolid when a certain moisture content (termed the **shrinkage limit**) is reached. If it contains expansive clays, the soil also begins to swell in volume as this moisture content is exceeded.

Figure 4.43 *The different swelling tendencies of two types of clay are illustrated at lower left. All four cylinders initially contained dry, sieved clayey B horizon soil, the two on the left are a kaolinitic soil, the two on the right are smectitic. An equal amount of water was added to the two center cylinders. The kaolinitic soil settled a bit and was not able to absorb much of the water. The smectitic soil swelled about 25% in volume and absorbed nearly all the added water. The scenes to the right and above show a practical application of knowledge about these clay properties. The California Vertisols shown make very poor building sites. The normal-appearing homes (upper) are actually built on deep, reinforced-concrete pilings (lower right) that rest on nonexpansive substrata. The 15 to 25 such pilings needed for each home more than doubles the cost of construction.* (Photo courtesy of Ray R. Weil)

---

[10] For news coverage of homeowner problems with soil shrinkage during severe drought, see Salter (2012).

Conversely, as a wet soil containing expansive clay dries out, it will shrink until it's water content reaches this shrinkage limit. Increasing the water content beyond the **plastic limit** will transform the soil into a malleable, plastic mass and cause additional swelling. The soil will remain in this plastic state until its **liquid limit** is exceeded, causing it to transform into a viscous liquid that will flow when jarred. These critical water contents (measured in units of percent) are termed the **Atterberg limits**.

Plasticity Index (PI).
The PI is the difference between the plastic limit (PL) and liquid limit (LL) and indicates the water-content range over which the soil has plastic properties:

$$PI = LL - PL$$

Soils with a plasticity index greater than about 25 are usually expansive clays that make unstable roadbeds or foundations. Figure 4.44 shows the relationship among the Atterberg limits and the changes in soil volume associated with increasing water contents for a hypothetical soil.

Coefficient of Linear Extensibility (COLE).
The expansiveness of a soil (and therefore the hazard of its destroying foundations and pavements) can be quantified as the COLE. Figure 4.44 indicates how the volume change used to calculate the COLE relates to the Atterberg limits. Suppose a sample of soil is moistened to its plastic limit and molded into the shape of a bar with length L_M. If the bar of soil is allowed to air dry, it will shrink to length L_D. The COLE is the percent reduction in length of the soil bar upon shrinking:

$$COLE = \frac{L_M - L_D}{L_M} \times 100$$

The COLE and the PI are two measures of soil expansiveness (see Section 8.3). Both are much higher for smectitic soils than kaolinitic ones. Relatively pure, mined smectite clay (often sold as "bentonite") can have far greater potential for swelling and plasticity than the impure clays in soil. Smectite clays (see Section 8.3) generally have high liquid limits and plasticity indices, especially if saturated with sodium ions. Kaolinite and other nonexpansive clays have low liquid limit values. The tendency of expansive clay soils to literally flow down steep slopes when the liquid limit is exceeded, producing mass wasting and landslides, is illustrated in Figure 4.41).

Unified Classification System for Soil Materials

The U.S. Army Corps of Engineers and the U.S. Bureau of Reclamation have established a widely used system of classifying soil materials to predict their behavior in engineering applications (Table 4.6). The system first groups soils into coarse- and fine-grained soils. The coarse materials are further divided on the basis of grain size (gravels and sands), amount of fines present, and uniformity of grain size (well or poorly graded). The fine-grained soils are divided

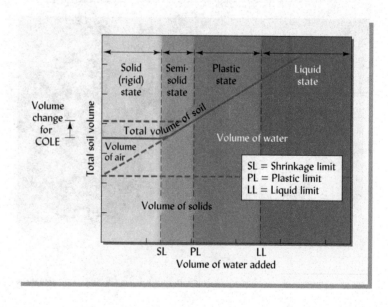

Figure 4.44 A common depiction of the Atterberg limits, which mark major shifts in the behavior of a cohesive soil as its water content changes (from left to right). As water is added to a certain volume of dry solids, first air is displaced; then, as more water is added, the total volume of the soil increases (if the soil has some expansive properties). When the shrinkage limit (SL) is reached, the once rigid, hard solid becomes a crumbly semisolid. With more water, the plastic limit (PL) is reached, after which the soil becomes plastic and can be molded. It remains in a plastic stage over a range of water contents until the liquid limit (LL) is exceeded, at which point the soil begins to behave as a viscous liquid that will flow when jarred. The volume change for calculating the coefficient of linear extensibility (COLE) is shown at the left. (Diagram courtesy of Ray R. Weil)

Table 4.6
THE UNIFIED SYSTEM OF CLASSIFICATION USED TO CLASSIFY SOIL MATERIALS (NOT NATURAL SOIL BODIES) FOR ENGINEERING USES[a]

Coarse-grained soils 50% or more retained on No. 200 (0.075 mm) sieve							
Gravels more than 50% of coarse fraction retained on No. 4 (2 mm) sieve				Sands more than 50% of coarse fraction passes No. 4 (2 mm) sieve			
Clean gravels		Gravels with fines		Clean sands		Sands with fines	
GW	GP	GM	GC	SW	SP	SM	SC
Well-graded gravels and gravel-sand mixtures, little to no fines	Poorly graded gravels and gravel-sand mixtures, little to no fines	Silty gravels and gravel-sand mixtures	Clayey gravels and gravel-sand mixtures	Well-graded sands and gravelly sands, little to no fines	Poorly graded sands and gravelly sands, little to no fines	Silty sands, sand-silt mixtures	Clayey sands, sand-clay mixtures

Fine-grained soils 50% or more passes No. 200 (0.075 mm) sieve						
Silts and clays Liquid limit 50% or less			Silts and clays Liquid limit greater than 50%			Highly organic soils
ML	CL	OL	MH	CH	OH	Pt
Inorganic silts, very fine sands, rock flour, silty or clayey fine sands	Inorganic clays of low to medium plasticity, gravelly, sandy or, silty clays, lean clays	Organic silts and organic silty clays of low plasticity	Inorganic silts, micaceous or diatomaceous fine sand or silts, elastic silts	Inorganic clays of high plasticity, fat clays	Organic clays of medium to high plasticity	Peat, muck, other highly organic soils

Coarse-grained soils 50% or more retained on No. 200 (0.075 mm) sieve							
Gravels more than 50% of coarse fraction retained on No. 4 (2 mm) sieve				Sands more than 50% of coarse fraction passes No. 4 (2 mm) sieve			
Clean gravels		Gravels with fines		Clean sands		Sands with fines	
GW	GP	GM	GC	SW	SP	SM	SC
sands, rock flour, silty or clayey fine sands	medium plasticity, gravelly, sandy or, silty clays, lean clays	silty clays of low plasticity	micaceous or diatomaceous fine sand or silts, elastic silts	high plasticity, fat clays	medium to high plasticity	soils	

[a] The two-letter designations (SW, MH, etc.) help engineers predict the behavior of the soil material when used for construction purposes. The first letter is one of the following: G = gravel, S = sand, M = silts, C = clays, O = organic-rich materials. The second letter indicates whether the sand or gravels are well graded (W) or poorly graded (P), and whether the silts, clays, and organic-rich materials have a high plasticity index (H) or a low plasticity index (L). Among the fine-grained materials, those closer to the right side of the table are the most troublesome materials for foundations and roadbeds.

into silts, clays, and organic materials. These classes are further subdivided on the basis of their liquid limit (above or below 50) and their plasticity index. Each type of soil is then given a two-letter designation based primarily on its particle-size distribution (texture), Atterberg limits, and organic matter content (e.g., GW for well-graded gravel, SP for poorly graded sands, CL for clay of low plasticity, and OH for organic-rich clays of high plasticity). This classification of soil materials helps engineers predict the soil strength, expansiveness, compressibility, and other properties so that appropriate engineering designs can be made for the soil at hand (Box 4.6).

BOX 4.6
TRAGEDY IN THE BIG EASY—A LEVEE DOOMED TO FAIL[a]

After Hurricane Katrina hit New Orleans in 2005, a 4-m-high storm surge breached the city's levees in several places, causing one of the worst disasters in American history with some 100,000 homes flooded and over 1,000 people killed. Some of the greatest flooding occurred when the 17th Street levee failed (Figure 4.45). Investigations later showed that the 17th Street levee was doomed to fail by a faulty levee design that did not properly deal with underlying layers of organic soils and sands. Poor design, combined with poor levee maintenance, allowed water to seep under the levee and weaken the soil at its base.

The levee, essentially a gently sloping mound of compacted clay soils, was covered on the landward side with a thin veneer of topsoil to support a protective grass mantle. Hard armor of rock or concrete (see Section 14.10) was not used in most places. To hold back floodwater and storm surges, engineers had constructed a concrete seawall along the crest of the levee. The seawall was attached to long steel pilings driven deep into the soil of the levee. The pilings were meant to both anchor the seawall and to prevent water from seeping through or under the levee. Out of sight, under the layers of clayey and loamy materials from which the levee was constructed, several layers of peat (buried Histosols) and sand provided for the weak link in the levee design.

Organic soils are highly compressible and have very low bearing and shear strengths. Both peats and sands are also highly permeable and conduct water readily. To perform their intended functions, the steel pilings attached to the seawall had to be long enough to penetrate through the peat/sand layers and into the more cohesive, higher-strength soil below. Unfortunately, the pilings in the 17th Street levee were too short (less than 6 m) and failed to penetrate through the peat layer (Figure 4.46). Whenever storm surges—or even high flows—raised the water level in the canal, seepage would occur through the peat/sand layer under the levee. The seepage water would then rise to saturate the soil at the foot of levee. When saturated, the soil would lose most of its shear strength and resistance to compression. Apparently, the engineers designing the levee had data on the peat layers, but based their design on the *average* soil properties, rather than on the weakest soils present.

On August 21, 2005, with seepage water saturating the soil at the levee base and turning the peat layer into little more than "soup," Katrina's storm surge "snapped the chain" at the weakest link, toppling a 140 m long section of the seawall, pushing both it and the levee some 14 m inland. The storm-churned waters poured through the breach, inundating the city of New Orleans.

Figure 4.45 A large helicopter attempts emergency repairs to the 17th Street levee breach several days after Hurricane Katrina hit New Orleans and toppled this section of the levee and seawall. Large chunks of the levee can be seen some 14 m inland. (Photo courtesy of U.S. Army Corps of Engineers)

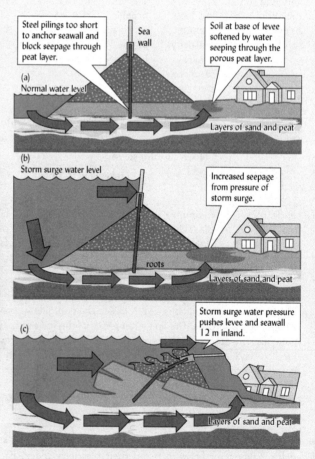

Figure 4.46 Illustration (not to scale) of how the buried layers of low-strength organic soil (a) allowed seepage of storm surge water (b) dooming the levee and its seawall to failure (c). (Diagram courtesy of Ray R. Weil)

---
[a]Based on forensic engineering investigation information in Seed et al. (2005) and reporting by Marshall (2005) and Vartabedian and Braun (2006).

4.10 CONCLUSION

Soils present an incredibly complex physical network of solid surfaces, pores, and interfaces that provides the setting for a myriad of chemical, biological, and physical processes. These in turn influence plant growth, hydrology, environmental management, and engineering uses of soil. The nature and properties of the individual particles, their size distribution, and their arrangement in soils determine the total volume of nonsolid pore space, as well as the pore sizes, thereby impacting on water and air relationships.

The properties of individual particles and their proportionate distribution (soil texture) are subject to little human control in field soils. However, it is possible to exert some control over the arrangement of these particles into aggregates (soil structure) and on the stability of these aggregates. Tillage and traffic must be carefully controlled to avoid undue damage to soil tilth, especially when soils are rather wet. Generally, nature takes good care of soil structure, and humans can learn much about soil management by studying natural systems. Vigorous and diverse plant growth, generous return of organic residues, and minimal physical disturbance are attributes of natural systems worthy of emulation. Proper plant species selection; crop rotation; and management of chemical, physical, and biological factors can help ensure maintenance of soil physical quality. In recent years, these management goals have been made more practical by the advent of conservation tillage systems that minimize soil manipulations while decreasing soil erosion and water runoff.

Particle size, moisture content, and plasticity of the colloidal fraction all help determine the stability of soil in response to loading forces from traffic, tillage, or building foundations. The physical properties presented in this chapter greatly influence nearly all other soil properties and uses, as discussed throughout this book.

STUDY QUESTIONS

1. If you were investigating a site for a proposed housing development, how could you use soil colors to help predict where problems might be encountered?

2. You are considering the purchase of some farmland in a region with variable soil textures. The soils on one farm are mostly sandy loams and loamy sands, while those on a second farm are mostly clay loams and clays. List the potential advantages and disadvantages of each farm as suggested by the texture of its soils.

3. Revisit your answer to question 2. Explain how soil structure in both the surface and subsurface horizons might modify your opinion of the merits of each farm.

4. Two different timber-harvest methods are being tested on adjacent forest plots with clay loam surface soils. Initially, the bulk density of the surface soil in both plots was 1.1 Mg/m^3. One year after the harvest operations, plot A soil had a bulk density of 1.48 Mg/m^3, while that in plot B was 1.29 Mg/m^3. Interpret these values with regard to the relative merits of systems A and B, and the likely effects on the soil's function in the forest ecosystem.

5. What are the textural classes of two soils, the first with 15% clay and 45% silt, and the second with 80% sand and 10% clay? (Hint: Use Figure 4.7.)

6. For the forest plot B in question 4, what was the change in percent pore space of the surface soil caused by timber harvest? Would you expect that most of this change was in the micropores or in the macropores? Explain.

7. Discuss the positive and negative impacts of tillage on soil structure. What is another physical consideration that you would have to take into account in deciding whether or not to change from a conventional to a conservation tillage system?

8. What would you, as a home gardener, consider to be the three best and three worst things that you could do with regard to managing the soil structure in your home garden?

9. What does the Proctor test tell an engineer about a soil, and why would this information be important?

10. In a humid region characterized by expansive soils, a homeowner experienced burst water pipes, doors that no longer closed properly, and large vertical cracks in the brick walls. The house had had no problems for over 20 years, and a consulting soil scientist blamed the problems on a large tree that was planted near the house some ten years before the problems began to occur. Explain.

REFERENCES

Arthur, H. G. 1977. *Teton dam failure*. The Evaluation of Dam Safety (Engineering Foundation Conference Proceedings. Asilomar). American Society of Civil Engineers, New York, pp. 61–71. http://www.geol.ucsb.edu/faculty/sylvester/Teton_Dam/narrative.html.

Bigham, J. M., and E. J. Ciolkosz (eds.). 1993. *Soil color*. SSSA Special Publication 31. Soil Science Society of America, Madison, WI.

Chen, G., and R. R. Weil. 2011. "Root growth and yield of maize as affected by soil compaction and cover crops." *Soil and Tillage Research* 117:17–27.

Emerson, W. W., R. C. Foster, and J. M. Oades. 1986. "Organomineral complexes in relation to soil aggregation and structure." p 521–548 in P. M. Huang and M. Schnitzer (eds.). *Interaction of soil minerals with natural organics and microbes*. SSSA Special Publication 17. Soil Science Society of America, Madison, WI.

Eynard, A., T. E. Schumacher, M. J. Lindstrom, and D. D. Malo. 2004. "Porosity and pore-size distribution in cultivated Ustolls and Usterts." *Soil Science Society of America Journal* 68:1927–1934.

Faulkner, E. H. 1943. *Plowman's Folly*. University of Oklahoma Press, Norman, OK.

Gilker, R. E., R. R. Weil, D. T. Krizek, and B. Momen. 2002. "Eastern gamagrass root penetration in adverse subsoil conditions." *Soil Science Society America Journal* 66:931–938.

Halfmann, D. 2005. "Management system effects on water infiltration and soil physical properties." Master's Thesis, Texas Tech University, Lubbock, TX.

Hu, X., L.-Y. Liu, S.-J. Li, Q.-G. Cai, Y.-L. LÃœ, and J.-R. Guo. 2012. "Development of soil crusts under simulated rainfall and crust formation on a loess soil as influenced by polyacrylamide." *Pedosphere* 22:415–424.

Laws, W. D., and D. D. Evans. 1949. "The effects of long-time cultivation on some physical and chemical properties of two rendzina soils." *Soil Science Society of America Proceedings* 14:15–19.

Lin, H. 2012. "Understanding soil architecture and its functional manifestations across scales." In H. Lin (ed.). *Hydropedolgy: Synergistic Integration of Soil Science and Hydrology*. Academic Press, Amsterdam, pp. 41–74.

Lucas, S. T., and R. R. Weil. 2012. "Can a labile carbon test be used to predict crop responses to improve soil organic matter management?" *Agronomy Journal* 104:1160–1170.

Marshall, B. 2005. "17th Street Canal levee was doomed—report blames Corps: Soil could never hold." *The Times-Picayune*, Wednesday, November 30, New Orleans.

McCarthy, D. F. 1993. *Essentials of Soil Mechanics and Foundations*. 4th ed. Prentice Hall, Englewood Cliffs, NJ.

Parker, M. M., and D. H. Van Lear. 1996. "Soil heterogeneity and root distribution of mature loblolly pine stands in Piedmont soils." *Soil Science Society of America Journal* 60:1920–1925.

Ritz, K., and I. Young (eds.). 2011. *Architecture and Biology of Soils: Life in Inner Space*. CABI Cambridge, Massachusetts, pp. 1–416.

Salter, J. 2012. "U.S. drought damage: Homes see cracking due to parched soil." Huffington Post, 08/31/12. http://www.huffingtonpost.com/2012/08/31/us-drought-damage-homes_n_1846712.html.

Sasiharan, N. 2003. "The failure of Teton dam – a new theory based on soil mechanics." Washington State University, Pullman, WA.XX.

Seed, R. B., et al. 2005. "Preliminary report on the performance of the New Orleans levee systems in hurricane Katrina on August 29, 2005—Preliminary findings from field investigations and associated studies shortly after the hurricane." Report UCB/CITRIS – 05/01. University of California at Berkeley and the American Society of Civil Engineers, Berkeley, CA.

Singh, P. K. 2012. "Role of glomalin related soil protein produced by arbuscular mycorrhizal fungi: A review." *Agricultural Science Research Journal* 2:119–125.

Six, J., H. Bossuyt, S. Degryze, and K. Denef. 2004. "A history of research on the link between (micro) aggregates, soil biota, and soil organic matter dynamics." *Soil Tillage Research* 79:7–31.

Six, J., E. T. Elliott, and K. Paustian. 2000. "Soil structure and soil organic matter: II. A normalized stability index and the effect of mineralogy." *Soil Science Society of America Journal* 64:1042–1049.

Thieme, J., G. Schneider, and C. Knochel. 2003. "X-ray tomography of a microhabitat of bacteria and other soil colloids with sub-100 nm resolution." *Micron* 34:339–344.

Tijink, F. G. J., and J. P. Van Der Linden. 2000. "Engineering approaches to prevent compaction in cropping systems with sugar beet." In R. Horn et al. (eds.). *Subsoil Compaction: Distribution, Processes and Consequences*. Catena Verlag, Reiskirchen, Germany, pp. 442–452.

Tisdall, J. M., and J. M. Oades. 1982. "Organic matter and water-stable aggregates in soils." *Soil Science Society of America Journal* 33:141–163.

USDA-NRCS. 2005. *National soil survey handbook*, title 430-vi. U.S. Department of Agriculture, Natural Resources Conservation Service. http://soils.usda.gov/technical/handbook/pos.html ted September 2005; verified 23, 2005).

Vartabedian, R., and S. Braun. 2006. "Fatal flaws: Why the walls tumbled in New Orleans." *Los Angeles Times*, Los Angeles, CA.

Vimmerstadt, J., F. Scoles, J. Brown, and M. Schmittgen. 1982. "Effects of use pattern, cover, soil drainage class, and overwinter changes on rain infiltration on campsites." *Journal of Environmental Quality* 11:25–28.

Watts, C. W., and A. R. Dexter. 1997. "The influence of organic matter in reducing the destabilization of soil by simulated tillage." *Soil Tillage Research* 42:253–275.

Weil, R. R. 2009. *Laboratory Manual for Introductory Soils*. 8th ed. Kendall/Hunt, Dubuque, IO. p. 228.

Weil, R. R., and F. Magdoff. 2004. "Significance of soil organic matter to soil quality and health." In F. Magdoff and R. R. Weil (eds.). *Soil Organic Matter in Sustainable Agriculture*. CRC Press, Boca Raton, FL, pp. 1–43.

5
Soil Water: Characteristics and Behavior

Soil conveys and receives water. (Photo courtesy of Ray R. Weil)

When the earth will ... drink up the rain as fast as it falls.
—H. D. Thoreau, The Journal

One of nature's simplest chemical compounds, water is a vital component of every living cell. Its unique properties promote a wide variety of physical, chemical, and biological processes. These processes greatly influence almost every aspect of soil development and behavior, from the weathering of minerals to the decomposition of organic matter, from the growth of plants to the pollution of groundwater.

We are all familiar with water. We drink it, wash with it, and swim in it. But water in the soil is something quite different from water in a drinking glass. In the soil, the intimate association between water and soil particles changes the behavior of both. Water causes soil particles to swell and shrink, to adhere to each other, and to form structural aggregates. Water participates in innumerable chemical reactions that release or tie up nutrients, create acidity, and wear down minerals so that their constituent elements eventually contribute to the saltiness of the oceans.

Certain soil water phenomena seem to contradict our intuition about how water ought to behave. Attraction to solid surfaces restricts the free movement of water molecules, making water less liquid and more solid-like in its behavior. In the soil, water can flow up as well as down. Plants may wilt and die in a soil whose profile contains a million kilograms of water in a hectare. A layer of sand or gravel in a soil profile may actually inhibit drainage, rather than enhance it.

Soil–water interactions determine the rates of water loss by leaching, surface runoff, and evapotranspiration, the balance between air and water in soil pores, the rate of change in soil temperature, the rate and kind of metabolism of soil organisms, and the capacity of soil to store and provide water for plant growth.

The characteristics and behavior of water in the soil comprise a common thread that interrelates nearly every chapter in this book. The principles contained in this chapter will help us understand why mudslides occur in water-saturated soils (Chapter 4), why earthworms may improve soil quality (Chapter 10), why wetlands contribute to global ozone depletion (Chapter 12), and why famine stalks humanity in certain regions of the world. Mastery of the principles presented in this chapter is fundamental to your working knowledge of the soil system.

5.1 STRUCTURE AND RELATED PROPERTIES OF WATER[1]

The ability of water to influence so many soil processes is determined primarily by the structure of the water molecule. Water is a simple compound, its individual molecules containing one oxygen atom and two much smaller hydrogen atoms. The elements are bonded together covalently, each hydrogen atom sharing its single electron with the oxygen. Instead of lining up symmetrically on either side of the oxygen atom (H—O—H), the hydrogen atoms are attached to the oxygen in a V-shaped arrangement at an angle of 105° (Figure 5.1, *left*). Consequently, water molecules exhibit **polarity**; that is, the charges are not evenly distributed. Rather, the side on which the hydrogen atoms are located tends to be electropositive and the opposite side electronegative.

Through a phenomenon called **hydrogen bonding**, a hydrogen atom of one water molecule is attracted to the oxygen end of a neighboring water molecule, thereby forming a low-energy bond between the two molecules. This type of bonding accounts for the polymerization of water. Hydrogen bonding also accounts for the relatively high boiling point, specific heat, and viscosity of water compared to the same properties of other hydrogen-containing compounds, such as H_2S, which has a higher molecular weight but no hydrogen bonding.

Polarity also explains why water molecules are attracted to electrostatically charged ions and to colloidal surfaces. Cations such as H^+, Na^+, K^+, and Ca^{2+} become hydrated through their attraction to the oxygen (negative) end of water molecules. Likewise, negatively charged clay surfaces attract water, this time through the hydrogen (positive) end of the molecule. Polarity of water molecules also encourages the dissolution of salts in water since the ionic components have greater attraction for water molecules than for each other.

When water molecules become attracted to electrostatically charged ions or clay surfaces, they are more closely packed than in pure water. In this packed state, their freedom of movement is restricted and their energy status is lower than in pure water. Thus, when ions or clay particles become hydrated, energy is released. That released energy is evidenced as **heat of solution** when ions hydrate or as **heat of wetting** when clay particles become wet.

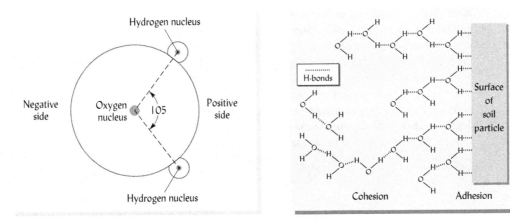

Figure 5.1 (Left) Two-dimensional representation of a water molecule showing a large oxygen atom and two much smaller hydrogen atoms. The H—O—H angle of 105° results in an asymmetrical arrangement. The side with the two hydrogens is electropositive; the opposite side is electronegative. This accounts for the polarity of water and its cohesive and adhesive forces. (Right) The forces of cohesion (between water molecules) and adhesion (between water and solid surface) in a soil–water system are largely a result of H-bonding, shown as broken lines. The adhesive or adsorptive force diminishes rapidly with distance from the solid surface. The cohesion of one water molecule to another results in water molecules forming temporary clusters that are constantly changing in size and shape as individual water molecules break free or join up with others. The cohesion between water molecules also allows the solid to indirectly restrict the freedom of water for some distance beyond the solid–liquid interface.

---

[1] For more in-depth discussions of water–soil interactions, see pp. 91–167 in Hillel (2004) or Warrick (2001).

Cohesion Versus Adhesion

Hydrogen bonding accounts for two basic forces responsible for water retention and movement in soils: the attraction of water molecules for each other (**cohesion**) and the attraction of water molecules for solid surfaces (**adhesion**). By adhesion (also called *adsorption*), some water molecules are held rigidly at the surfaces of soil solids. In turn, these tightly bound water molecules hold, by cohesion, other water molecules farther removed from the solid surfaces (Figure 5.1, *right*). Together, the forces of adhesion and cohesion allow soil solids to retain water and control its movement and use. Adhesion and cohesion also make possible the property of plasticity possessed by clays (see Section 4.9).

Surface Tension

Another important property of water that markedly influences its behavior in soils is that of **surface tension**. This property is commonly evidenced at liquid–air interfaces and results from the greater attraction of water molecules for each other (cohesion) than for the air above. The net effect is an inward force at the surface that causes water to behave as if its surface were covered with a stretched elastic membrane allowing certain insects to "walk on water." Because of the relatively high attraction of water molecules for each other, water has a high surface tension which, in turn, contributes to the phenomenon of **capillarity**.

5.2 CAPILLARY FUNDAMENTALS AND SOIL WATER

The movement of water up a wick typifies the phenomenon of capillarity. Two forces cause capillarity: (1) the attraction of water for the solid (adhesion or adsorption), and (2) the surface tension of water, which is due largely to the attraction of water molecules for each other (cohesion).

Capillary Mechanism

Capillarity can be demonstrated by placing one end of a fine (< 1 mm diameter), clean glass tube in water. The water rises in the tube; the smaller the tube bore, the higher the water rises. The water molecules are attracted to the sides of the tube (adhesion) and start to spread out along the glass in response to this attraction. At the same time, the cohesive forces hold the water molecules together and create surface tension, causing a curved surface (called a *meniscus*) to form at the interface between water and air in the tube. Lower pressure under the meniscus in the glass tube (P2) allows the higher pressure (P1) on the free water to push water up the tube. The process continues until the water in the tube has risen high enough that its weight just balances the pressure differential across the meniscus.

The height of rise in a capillary tube is inversely proportional to the tube radius r. If we limit our consideration to water at a given temperature (e.g., 20°C), then other factors (liquid density, surface tension, adhesive force) can be combined into a single constant, and we can use a simple capillary equation to calculate the height of rise h:

$$h(\text{cm}) = \frac{0.15(\text{cm})^2}{r(\text{cm})} \tag{5.1}$$

where both h and r are expressed in centimeters. This equation tells us that the smaller the tube bore, the greater the capillary force and the higher the water rise in the tube.

As one would expect, capillary rise will only occur if the tube is made of hydrophilic material. If a *hydrophobic* tube (such as one with a waxed surface) is dipped into a pool of water, the meniscus will be convex rather than concave to the air, so that the situation is reversed and capillary *depression* rather than capillary *rise* will occur. This is the case in certain water-repellent soil layers (see Figure 7.23).

Height of Rise in Soils

Capillary forces are at work in all moist soils. However, the rate of movement and the rise in height are less than one would expect on the basis of soil pore size alone. One reason is that soil pores are not straight, uniform openings like glass tubes. Furthermore, some soil pores are

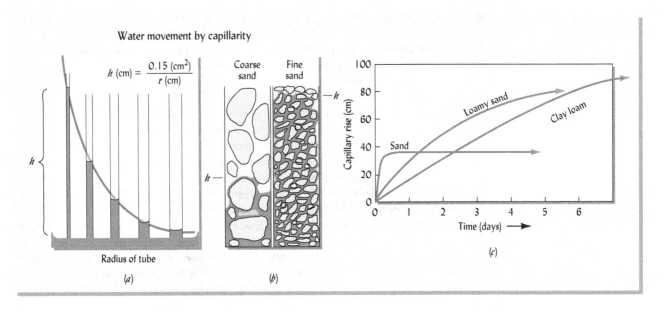

Figure 5.2 Upward capillary movement of water through tubes of different bore and soils with different pore sizes. (a) The capillary equation can be graphed to show that the height of rise h doubles when the tube inside radius is halved. The same relationship can be demonstrated using glass tubes of different bore size. (b) The same principle also relates pore sizes in a soil and height of capillary rise, but the rise of water in a soil is rather jerky and irregular because of the tortuous shape and variability in size of the soil pores (as well as because of pockets of trapped air). (c) The finer the soil texture, the greater the proportion of small-sized pores and, hence, the higher the ultimate rise of water above a free-water table. However, because of the much greater frictional forces in the smaller pores, the capillary rise is much slower in the finer-textured soil than in the sand. (Diagrams courtesy of Ray R. Weil)

filled with air, which may be entrapped, slowing down or preventing the movement of water by capillarity (see Figure 5.2b).

Since capillary movement is determined by pore size, it is the pore-size distribution discussed in Chapter 4 that largely determines the amount and rate of movement of capillary water in the soil. The abundance of medium- to large-sized capillary pores in sandy soils permits rapid initial capillary rise, but limits the ultimate height of rise (Figure 5.2c). Clays have a high proportion of very fine capillary pores, but frictional forces slow down the rate at which water moves through them. Consequently, in clays the capillary rise is slow initially, but in time it generally exceeds that of sands. Loams exhibit capillary properties between those of sands and clays.

Capillarity is traditionally illustrated as an upward adjustment. But movement in any direction takes place, since the attractions between soil pores and water are as effective in forming a water meniscus in horizontal pores as in vertical ones (Figure 5.3). The significance of capillarity in controlling water movement in small pores will become evident as we turn to soil water energy concepts.

Figure 5.3 Capillary water movement in the field can be both vertical and horizontal. (Left) Capillary flow has caused water in a small surface runoff collection basin to move both up and horizontally away from the pool of collected water. (Right) Capillary rise above the water level in a stream bank. (Photo courtesy of Ray R. Weil)

5.3 SOIL WATER ENERGY CONCEPTS

The retention and movement of water in soils, its uptake and translocation in plants, and its loss to the atmosphere are all energy-related phenomena. *Kinetic energy* is certainly an important factor in the rapid, turbulent flow of water in a river, but the movement of water in soil is so slow that the kinetic energy component is usually negligible. *Potential energy* is most important in determining the status and movement of soil water. For the sake of simplicity, in our discussion of soil water, we will use the term *energy* to refer to potential energy.

Every day we can see that things tend toward a lower energy state (and that it takes an input of energy—work—to prevent them from doing so). Use your cell phone and its battery runs down from a fully charged high potential energy state to a discharged, low energy state. If you should drop your phone, it would fall from its state of relatively high potential energy in your hand to a lower potential energy state on the floor (where it is closer to the source of gravitational pull). The difference in energy levels (that is how high off the floor you are holding the phone) determines how forcefully transition will occur. In this respect, soil water is no different—it tends to move from a higher to a lower energy state. Therefore, if we know the pertinent energy levels at various points in a soil, we can predict the direction of water movement. It is the *differences* in energy levels from one contiguous site to another that influence this water movement.

Forces Affecting Potential Energy

The discussion of the structure and properties of water in the previous section suggests three important forces affecting the energy level of soil water. First, adhesion, or the attraction of water to the soil solids (matrix), provides a **matric** force (responsible for adsorption and capillarity) that markedly reduces the energy state of water near particle surfaces. Second, the attraction of water to ions and other solutes, resulting in **osmotic** forces, tends to reduce the energy state of water in the soil solution. Osmotic movement of pure water across a semipermeable membrane into a solution (osmosis) is evidence of the lower energy state of water in the solution. The third major force acting on soil water is **gravity**, which always pulls the water downward. The energy level of soil water at a given elevation in the profile is thus higher than that of water at some lower elevation. This difference in energy level causes water to flow downward.

Soil Water Potential

The *difference* in energy level of water from one site or one condition to another (e.g., between wet soil and dry soil) determines the direction and rate of water movement in soils and in plants. In a wet soil, most of the water is retained in thick water films around particles or in large pores. Therefore, most of the water molecules in a wet soil are not very close to a particle surface and so are not held very tightly by the soil solids (the matrix). In this condition, the water molecules have considerable freedom of movement, so their energy level is near that of water molecules in a pool of pure water outside the soil. In a drier soil, however, the water that remains is located in thin water films, in small pores, and pore corners (Figure 5.4). Thus, the water molecules in a drier soil are close to solid surfaces where they are held tightly and have little freedom of movement. Their energy level is much lower than that of the average water molecules in wet soil. If wet and dry soil samples are brought in touch with each other, water will move from the wet soil (higher energy state) to the drier soil (lower energy).

Relative values of soil water energy can be used to predict how water will move in soils and in the environment. Usually the energy status of soil water in a particular location in the profile is compared to that of pure water at standard pressure and temperature, unaffected by the soil and located at some reference elevation. The *difference* in energy levels between this pure water in the reference state and that of the soil water is termed soil **water potential**, the term *potential*, like the term *pressure*, implying a difference in energy status.

If all water potential values under consideration have a common reference point (the energy state of pure water), differences in the water potential of two soil samples in fact reflect differences in their absolute energy levels. This means that water will move from a soil zone having a high soil water potential to one having a lower soil water potential. This fact should always be kept in mind when thinking about the behavior of water in soils.

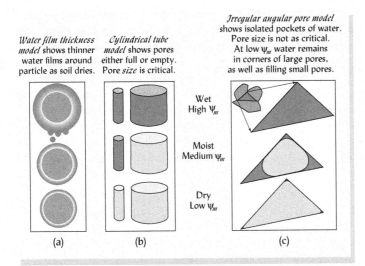

Figure 5.4 *Three models of water distribution in wet, moist, or nearly dry soils. (a) Water films around soil particles are thicker when the soil is wetter. (b) Water is held in soil pores of varying size; the wetter the soil, the larger the pores that are filled with water. As the soil dries, the larger pores are emptied first, leaving water in smaller pores where it is held more tightly. (c) Many soil pores are angular with corners in which capillary forces can hold water, even when most of the pore is emptied of water. As the soil dries, the water in the middle of large angular pores is removed first as it is least tightly held by mineral surfaces. Decreasing amounts of water are held in the corners as drying continues. In most soils all three models operate simultaneously and explain why matric potential changes as soil water content changes.* (Diagrams courtesy of Ray R. Weil)

Several forces are implicated in soil water potential, each of which is a component of the **total soil water potential** ψ_t. These components are due to differences in energy levels resulting from gravitational, matric, submerged hydrostatic, and osmotic forces and are termed **gravitational potential** ψ_g, **matric potential** ψ_m, **hydrostatic potential** ψ_h, and **osmotic potential** ψ_o, respectively. All of these components act simultaneously to influence water behavior in soils. The general relationship of soil water potential to potential energy levels can be expressed as:

$$\psi_t = \psi_g + \psi_m + \psi_o + \psi_h + \cdots \quad (5.2)$$

where the ellipsis (...) indicates the possible contribution of additional potentials not yet mentioned.

Gravitational Potential

The force of gravity acts on soil water the same as it does on any other body, the attraction being toward the Earth's center. The gravitational potential ψ_g of soil water may be expressed mathematically as:

$$\psi_g = gh \quad (5.3)$$

where g is the acceleration due to gravity and h is the height of the soil water above a reference elevation. The reference elevation is usually chosen within the soil profile or at its lower boundary to ensure that the gravitational potential of soil water above the reference point will always be positive.

Following heavy precipitation or irrigation, gravity plays an important role in removing excess water from the upper horizons and in recharging groundwater below the soil profile. It will be given further attention when the movement of soil water is discussed (see Section 5.5).

Pressure Potential (Including Hydrostatic and Matric Potentials)

This component accounts for effects on soil water potential other than gravity and solute levels. Pressure potential most commonly includes: (1) the positive hydrostatic pressure due to the weight of overlying water in saturated soils and aquifers and (2) the negative pressure due to the attractive forces between the water and the soil solids or the soil matrix.[2]

The hydrostatic pressures give rise to what is often termed the **hydrostatic potential** ψ_h, a component that is operational only for water in saturated zones below the water table. Anyone who has dived to the bottom of a swimming pool has felt hydrostatic pressure on their eardrums.

---

[2]In addition to matric and hydrostatic forces, in some situations the weight of the overburdened soil and the pressure of air in the soil also make a contribution to the total soil water potential.

The attraction of water to solid surfaces gives rise to the **matric potential** ψ_m, which is always negative because the water attracted by the soil matrix has an energy state lower than that of pure water. (These negative pressures are sometimes referred to as *suction* or *tension*. If these terms are used, their values are positive.) The matric potential is operational in unsaturated soil above the water table (Figure 5.5).

Matric potential ψ_m, which results from adhesive forces and capillarity, influences both the retention and movement of soil water. Differences between the ψ_m of two adjoining soil zones encourage the movement of water from wetter (high-energy state) areas to drier (low-energy state) areas or from large pores to small pores. Although this movement may be slow, it is extremely important in supplying water to plant roots and in engineering applications.

Osmotic Potential[3]

The osmotic potential ψ_o is attributable to the presence of both inorganic and organic substances dissolved in water. As water molecules cluster around solute ions or molecules, the freedom of movement (and therefore the potential energy) of the water is reduced. The greater the concentration of solutes, the more osmotic potential is lowered. As always, water will tend to move to where its energy level will be lower, in this case to the zone of higher solute concentration. However, liquid water will move in response to differences in osmotic potential (the process termed **osmosis**) only if a *semipermeable membrane* exists between the zones of high and low osmotic potential, allowing water through but *preventing the movement of the solute*. If no membrane is present, the solute, rather than the water, generally moves to equalize concentrations.

Because soil zones are *not* generally separated by membranes, the osmotic potential ψ_o has little effect on the mass movement of water in soils. Its major effect is on the uptake of water by plant root cells that *are* isolated from the soil solution by their semipermeable cell membranes. In soils high in soluble salts, ψ_o may be lower (have a greater negative value) in the soil solution than in plant root cells. This leads to constraints in the uptake of water by plants. In very salty soil, the soil water osmotic potential may be lower than the plant cytoplasm osmotic potential, thus causing cells in the plant roots to collapse (plasmolyze) as water moves out of the cells to the lower osmotic potential zone in the soil. The drought tolerance of a plant is largely related to its ability to maintain high salt levels in its cells, and thus maintain turgor in leaves even in the face of such low soil water potentials.

The random movement of water molecules causes a few of them to escape a body of liquid water, enter the atmosphere, and become water vapor. Since the presence of solutes restricts the movement of water molecules, fewer water molecules escape into the air as the solute concentration of liquid water is increased. Therefore, water vapor pressure is lower in the air over salty water than in the air over pure water. By affecting water vapor pressure, ψ_o affects the movement of water vapor in soils (see Section 5.5).

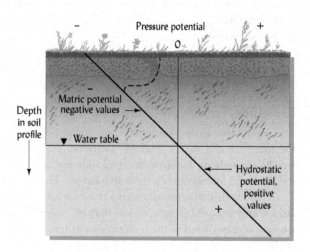

Figure 5.5 *The matric potential and hydrostatic potential are both pressure potentials that may contribute to total water potential. The matric potential is always negative and the hydrostatic potential is positive. The top of the saturated zone is termed the water table. Above the water table, the soil is unsaturated and its water subject to the influence of matric potentials. Water below the water table in saturated soil is subject to hydrostatic potentials. In the example shown, the matric potential decreases linearly with elevation above the water table, signifying that water rising by capillary attraction up from the water table is the only source of water in this profile. Rainfall or irrigation (see dotted line) would alter or curve the straight line, but would not change the fundamental relationship described.* (Diagram courtesy of Ray R. Weil)

---

[3]For a discussion of how osmotic potential govern drought tolerance of plants in global ecosystems, see Bartlett et al. (2012).

Table 5.1
EXPRESSIONS OF SOIL WATER POTENTIAL AND EQUIVALENT DIAMETER OF PORES EMPTIED

Height of unit column of water, cm	Soil water potential, bars	Soil water potential,[a] kPa	Equivalent diameter of pores emptied,[b] μm
0	0	0	—
10.2	−0.01	−1	300
102	−0.1	−10	30
306	−0.3	−30	10
1,020	−1.0	−100	3
15,300	−15	−1,500	0.2
31,700	−31	−3,100	0.097
102,000	−100	−10,000	0.03

[a]The SI unit kilopascal (kPa) is equivalent to 0.01 bars.
[b]Smallest pore that can be emptied by equivalent tension as calculated by rearranging Eq. (5.1).

Units Used to Quantify Water Potentials

Several units can be used to express differences in energy levels of soil water. One is the *height of a water column* (usually in centimeters) whose weight just equals the potential under consideration. We have already encountered this means of expression since the h in the capillary equation (Section 5.2) tells us the matric potential of the water in a capillary pore. A second unit is the standard *atmosphere* pressure at sea level, which is 760 mm Hg or 1,020 cm of water. Another unit termed *bar* approximates the pressure of a standard atmosphere. Energy may be expressed per unit of mass (**joules/kg**) or per unit of volume (**newtons/m^2**). In the International System of Units (SI), 1 pascal (Pa) equals 1 newton (N) acting over an area of 1 m^2. In this text, we will use Pa or kilopascals (kPa) to express soil water potential. Since other publications may use other units, Table 5.1 is provided to show the equivalency among common means of expressing soil water potential.

Combined Potentials

If the soil water contains salts and other solutes, the mutual attraction between water molecules and these chemicals reduces the *osmotic* potential of the water. Similarly, the mutual attraction between soil solids (soil matrix) and soil water molecules also reduces the water's *matric* potential. Since both of these interactions reduce the water's potential energy level compared to that of pure water, the changes in energy level (osmotic potential and matric potential) are both considered to be negative. In contrast, differences in energy due to gravity (*gravitational potential*) are always positive because the reference elevation of the pure water is purposely designated at a site in the soil profile below that of the soil water. A plant root attempting to remove water from a moist soil would have to overcome all three forces simultaneously.

5.4 SOIL WATER CONTENT AND SOIL WATER POTENTIAL

The previous discussions suggest an inverse relationship between the water content of soils and the tenacity with which the water is held in soils. Many factors affect the relationship between water potential ψ and water content θ in soil. A few examples will illustrate this point.

Soil Water Versus Energy Curves

The relationship between soil water potential ψ and water content θ of four different soils is shown in Figure 5.6. Such curves are sometimes termed *water release characteristic curves*, or simply *water characteristic curves*. The absence of sharp breaks in the curves indicates a continuous range of pore sizes and therefore a gradual change in the water potential with increased soil water content.

The clay soil holds much more water at a given potential than does the loam or sand. Likewise, at a given moisture content, the water is held much more tenaciously in the clay than in the other two soils (note that soil water potential is plotted on a log scale). The amount of clay in a soil largely determines the proportion of very small micropores in that soil. As we shall see, about half of the water held by clay soils is held so tightly in these micropores that it cannot be removed by growing plants. Soil texture clearly exerts a major influence on soil moisture retention.

Soil structure also influences soil water content–energy relationships, as does large amounts of organic matter. A well-granulated soil has more total pore space and greater overall water-holding capacity than one with poor granulation or one that has been compacted. Soil aggregation especially increases the relatively large interaggregate pores (Section 4.5) in which water is held with little tenacity. In contrast, a compacted soil will hold less total water but is likely to have a higher proportion of small- and medium-sized pores that hold water with greater tenacity than do larger pores. Therefore, soil structure predominantly influences the shape of the water characteristic curve in the portion where the potentials are between 0 and about 100 kPa. The shape of the remainder of the curve generally reflects the influence of soil texture.

The soil water characteristic curves in Figure 5.6 have great practical significance for various field measurements and processes. It will be useful to refer back to these curves as we consider the applied aspects of soil water behavior in the following sections.

Measurement of Soil Water Status

The soil water characteristic curves just discussed highlight the importance of making two general kinds of soil water measurements: the *amount* of water present (water content) and the *energy status* of the water (soil water potential). In order to understand or manage water supply and movement in soils, it is essential to have information (directly measured or inferred) on *both* types of measurements. For example, a soil water potential measurement might tell us whether water will move toward the groundwater, but without a corresponding measurement of the soil water content, we would not know the possible significance of the contribution to groundwater.

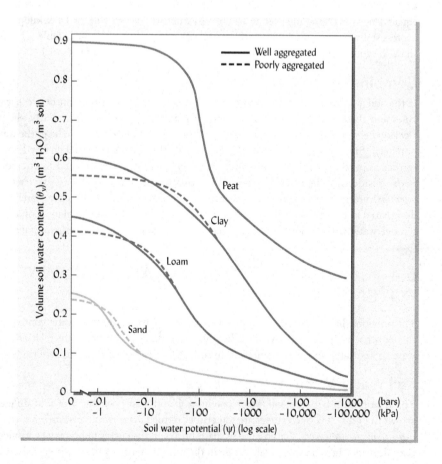

Figure 5.6 *Soil water potential curves for three representative mineral soils and an organic peat. The curves show the relationship obtained by slowly drying completely saturated soils. The dashed lines show the effect of compaction or poor aggregation in mineral soils. Note that the soil water potential ψ (which is negative) is plotted on a log scale in units of bars (upper scale) and kilopascals (kPa) (lower scale). Soil water content is plotted on the vertical scale on a volumetric basis (left axis). The curves do not account for possible soil volume changes (shrinkage) and are generally representative of data in Rawls et al. (1982, 2004) and Schwärzel et al. (2002).*

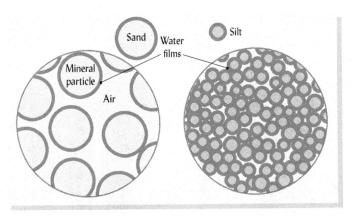

Figure 5.7 *Fine- and coarse-textured soils at the same water potential but very different water contents. (Left) Sandy soil at −10 kPa potential in which particles are surrounded by water films of a uniform thickness. Most of the pore space is filled with air. (Right) Silty soil at −10 kPa potential with much smaller particles surrounded by uniform films of water of the same thickness as those in the sandy soil. However, the particles and pores are more numerous and smaller than in the sandy soil, so water films of the same thickness cause most of the porespace in the silty soil to be filled with water, not air. Because both soils have the same water potential they would feel similarly moist and could easily supply water to plants, but it is visually obvious that the silty soil has a much greater water content than the sandy soil.* (Diagram courtesy of Ray R. Weil)

Generally, the behavior of soil water is most closely related to the energy status of the water, not to the amount of water in a soil. Thus, a silt loam and a loamy sand will both feel moist and will easily supply water to plants when the ψ_m is, say, −10 kPa. However, the amount of water held by the silt loam, and thus the length of time it could supply water to plants, would be far greater at this potential than would be the case for the loamy sand (Figure 5.7).

We will briefly consider several methods for making each of these two types of soil water measurements. Researchers, land managers, and engineers may use a combination of several of these methods to study the storage and movement of water in soil, manage irrigation systems, and predict the physical behavior of soils.

Volumetric Water Content

The **volumetric water content** θ_v is defined as the volume of water associated with a given volume (usually 1 m^3) of dry soil (see Figure 5.6). A comparable expression is the **mass water content** θ_m, or the mass of water associated with a given mass (usually 1 kg) of dry soil. Both of these expressions have advantages for different uses. In most cases we shall use the volumetric water content θ_v in this text.

As compaction reduces total porosity, it also increases θ_v (assuming a given θ_m), therefore often leaving too little air-filled pore space for optimal root activity. However, if a soil is initially very loose and highly aggregated (such as the forested A horizons described in Figure 5.8), moderate compaction may actually benefit plant growth by increasing the volume of pores that hold water between 10 and 1,500 kPa of tension (water that plants can utilize). Figure 5.8 is worth careful study to understand the relationships among mass and volume water contents, bulk density, and porosity.

We think of plant root systems as exploring a certain depth of soil. We measure precipitation (and sometimes irrigation) as a depth of water (e.g., mm of rain). Therefore, it is often necessary to express the volumetric water content as a depth of water per unit depth of soil. Conveniently, the numerical values for these two expressions are the same. For example, for a soil containing 0.1 m^3 of water per m^3 of soil (10% by volume) the depth ratio of water is 0.1 m of water per meter of soil depth (see also Section 5.8).[4]

---

[4] When measuring amounts of water added to soil by irrigation, it is customary to use units of volume such as m^3 and hectare-meter (the volume that would cover a hectare of land to a depth of 1 m). Many people in the United States use the English units ft^3 or acre-foot (the volume needed to cover an acre of land to a depth of 1 ft).

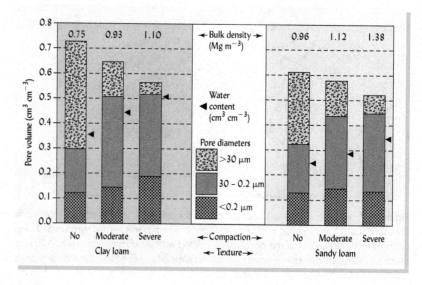

Figure 5.8 Compaction increases volumetric soil water content. Compaction of both soils decreased total porosity, mainly by converting the largest (usually air-filled) pores into smaller pores that hold water more tightly. These forested A horizon soils were initially so loose that moderate compaction benefited plants by increasing the volume of water-holding 0.2 to 30 μm pores. On the other hand, the water originally in the uncompacted soil takes up a greater proportion of the pore volume (indicated as ◀ cm water/cm soil) when the soil is compacted, possibly leading to nearly water saturated conditions. For example, the clay loam with severe compaction contains 0.52 cm³ water, but only 0.04 cm³ air per cm³ soil, less than the ~ 10% air porosity (see Section 7.2) needed for good plant growth. (Modified from Shestak and Busse (2005) with permission of The Soil Science Society of America)

Measuring Soil Water Status

Gravimetric Method. The gravimetric method is a direct measurement of soil water content and is therefore the standard method by which all indirect methods are calibrated. The water associated with a given mass (and, if the bulk density of the soil is known, a given volume) of dry soil solids is determined. A sample of moist soil is weighed and then thoroughly dried and finally

---

BOX 5.1
GRAVIMETRIC DETERMINATION OF SOIL WATER CONTENT

The gravimetric procedures for determining mass soil water content θ_m are relatively simple. Assume that you want to determine the water content of a 100 g sample of moist soil. You dry the sample in a convection oven at 80°C for 24 hours followed by one hour at 105°C. At this point the soil is completely "oven dry" and it is weighed again. Assume that the dried soil then weighs 70 g, which indicates that 30 g of water has been removed from the moist soil. Expressed in kilograms, this is 30 kg water associated with 70 kg dry soil.

Since the mass soil water content θ_m is commonly expressed in terms of kg water associated with 1 kg *dry* soil (*not* 1 kg of *wet* soil!), it can be calculated as follows:

$$\frac{30 \text{ kg water}}{70 \text{ kg dry soil}} = \frac{X \text{ kg water}}{1 \text{ kg dry soil}}$$

$$X = \frac{30}{70} = 0.428 \text{ kg water/kg dry soil} = \theta_m$$

To calculate the volumetric soil water content θ_v, we need to know the bulk density of the dried soil, which in this case we shall assume to be 1.3 Mg/m³. In other words, a cubic meter of this soil (*when dry*) has a mass of 1,300 kg. From the above calculations we know that the mass of water associated with this 1,300 kg of dry soil is 0.428 × 1300 or 556 kg.

Since 1 m³ of water has a mass of 1,000 kg, the 556 kg of water will occupy 556/1000 or 0.556 m³.

Thus, the volumetric water content is 0.556 m³/m³ of dry soil:

$$\frac{1300 \text{ kg soil}}{\text{m}^3 \text{ soil}} \times \frac{\text{m}^3 \text{ water}}{1000 \text{ kg water}} \times \frac{0.428 \text{ kg water}}{\text{kg soil}}$$

$$= \frac{0.556 \text{ m}^3 \text{ water}}{\text{m}^3 \text{ soil}}$$

Assuming a soil that does not swell when wet, the relationship between the mass and volumetric water contents can be summarized as:

$$\theta_v = D_b \times \theta_m \qquad (5.4)$$

Table 5.2
SOME METHODS OF MEASURING SOIL WATER
More than one method may be needed to cover the entire range of soil moisture conditions.

Method	Measures soil water — Content	Measures soil water — Potential	Useful range, kPa	Used mainly in — Field	Used mainly in — Lab	Comments
1. Gravimetric	×		0 to < −10,000		×	Destructive sampling; slow (1–2 days) unless microwave used. The standard for calibration.
2. Neutron scattering	×		0 to < −1,500	×		Radiation permit needed; expensive equipment; not good in high-organic-matter soils; requires access tube.
3. Time-domain reflectometry (TDR)	×		0 to < −10,000	×	×	Can be automated; accurate to ±1 to 2% volumetric water content; very sandy, clayey, or salty soils need separate calibration; requires wave guides; expensive instrument.
4. Capacitance	×		0 to < −1,500	×		Can be automated; accurate to ±2 to 4% volumetric water content; sands or salty soils need special calibration; simple, inexpensive sensors and recording instruments.
5. Resistance blocks		×	−90 to < −1,500	×		Can be automated; not sensitive near optimum plant water contents; may need calibration.
6. Tensiometer		×	0 to −85	×		Can be automated; accurate to ±0.1 to 1 kPa; limited range; inexpensive; needs periodic servicing to add water to tensiometer.
7. Thermocouple psychrometer		×	50 to < −10,000	×	×	Moderately expensive; wide range; accurate only to ±50 kPa.
8. Pressure membrane apparatus		×	50 to < −10,000		×	Used with gravimetric method to construct drier part of water characteristic curve.
9. Tension table		×	0 to −50		×	Used with gravimetric method to construct wetter part of water characteristic curve.

weighed again. The weight loss represents the soil water. Box 5.1 provides examples of how θ_v and θ_m can be calculated. The gravimetric method is a *destructive* method (a soil sample must be collected for each measurement) and cannot be easily automated, thereby making it poorly suited to monitoring temporal changes in soil moisture. Several indirect methods of measuring soil water content are nondestructive, easily automated, and very useful in the field (see Table 5.2).

Instrumental Methods for Water Content. A variety of instruments (Figure 5.9) are now available to measure soil water content using sensors that determine the water content of soil by measuring the manner in which electrical impulses or even sub-atomic particles (neutrons) move through the soil. These include **capacitance** sensors, **time-domain reflectometry** (TDR) wave guides, and **neutron scattering** probes, each of which can be wired to a specialized meter that converts the signal into water content values for monitoring and datalogging (Table 5.2).

Instrumental Methods for Water Potential The tenacity with which water is attracted to soil particles is an expression of matric potential ψ_m. Field **tensiometers** (Figure 5.10) measure this attraction or *tension*. The tensiometer is basically a water-filled tube closed at the bottom

Figure 5.9 *A time-domain reflectometry (TDR) meter, and sensor (left) and a capacitance sensor and meter (right) are shown. The TDR meter calculates volumetric water content of soil by analyzing the velocity at which electromagnetic waves move through the soil and the reflected wave patterns generated. Electrical capacitance probes determine volumetric water content by measuring the amount of charge required to raise the voltage between two conductors separated by the soil.* (Photo courtesy of Ray R. Weil)

with a porous ceramic cup and at the top with an airtight seal. Once placed in the soil, water in the tensiometer moves through the porous cup into the adjacent soil until the water potential in the tensiometer is the same as the matric potential in the soil. As the water is drawn out, a vacuum develops under the top seal, which can be measured by a vacuum gauge or an electronic transducer. If rain or irrigation rewets the soil, water will enter the tensiometer through the ceramic tip, reducing the vacuum or tension recorded by the gauge.

Electrical resistance blocks are made of porous gypsum ($CaSO_4 \cdot 2H_2O$), suitably embedded with electrodes. When placed in moist soil, the fine pores in the block absorb water in proportion to the soil water potential. The more tightly the water is being held in the soil, the less water the block will be able to absorb. The resistance to flow of electricity between

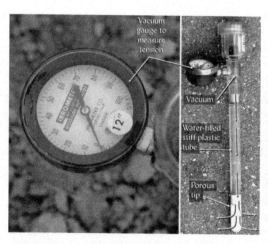

Figure 5.10 *Instruments that measure soil water potential in the field include the tensiometer (center) and electrical resistance blocks (right). The tensiometer tube is filled with water through the screw-off top. Once the instrument is tightly sealed, the white porous tip and the lower part of the plastic tube is inserted into a snug-fitting hole in the soil. The vacuum gauge (close up, left) will directly indicate the "tension" or negative potential generated as the soil draws the water out (curved arrows) through the porous tip. Note the scale goes up to only 100 centibars (= 100 kPa) tension at the driest. (Right) A cutaway view of soil with a gypsum electrical resistance block installed about 45 cm below the surface. Insulated wires lead from the block to the surface, where they can be connected to a special resistance meter. Another gypsum block has been broken open (inset) to reveal two concentric metal screen cylinders that serve as the electrodes between which moistened gypsum conducts a small electric current. The resistance to current flow is inversely proportional to the wetness of the gypsum block.* (Photo courtesy of Ray R. Weil)

the electrodes embedded in the block decreases in proportion to how much water has been absorbed in the block (Figure 5.10). These devices must be calibrated for each soil, and the accuracy and range of soil moisture contents measured are limited (Table 5.2). However, they are very inexpensive and can be used to measure approximate changes in soil moisture during one or more growing seasons. It is possible to connect them to data loggers or electronic switches so that irrigation systems can be turned on and off automatically at set soil moisture levels.

5.5 HOW DOES WATER MOVE IN SOIL?

Water movement within the soil occurs as: (1) saturated flow, (2) unsaturated flow, and (3) vapor movement.[5] In all cases, water flows in response to energy gradients with water moving from a zone of higher to one of lower water potential. *Saturated flow* takes place when the soil pores are completely filled (or saturated) with water. *Unsaturated flow* occurs when the larger pores in the soil are filled with air, leaving only the smaller pores to hold and transmit water. *Vapor movement* occurs as vapor pressure differences develop in relatively dry soils.

Saturated Flow Through Soils

Under some conditions, at least part of a soil profile may be completely saturated; that is, all pores, large and small, are filled with water. The lower horizons of poorly drained soils are often saturated, as are portions of well-drained soils above water-restricting layers of clay. During and immediately following heavy rain or irrigation, pores in the upper soil zones are often filled entirely with water.

The quantity of water per unit of time Q/t that flows through a column of saturated soil (Figure 5.11) can be expressed by Darcy's law, as follows:

$$\frac{Q}{t} = A K_{sat} \frac{\Delta \psi}{L} \tag{5.5}$$

where A is the cross-sectional area of the column through which the water flows, K_{sat} is the **saturated hydraulic conductivity**, $\Delta \psi$ is the change in water potential between the ends of the column (e.g., $\psi_1 - \psi_2$), and L is the length of the column. For a given column, the rate of flow is determined by the ease with which the soil transmits water (K_{sat}) and the amount of force driving the water, namely, the **water potential gradient** $\Delta \psi / L$. For saturated flow, this force may also be called the **hydraulic gradient**. By analogy, think of pumping water through a

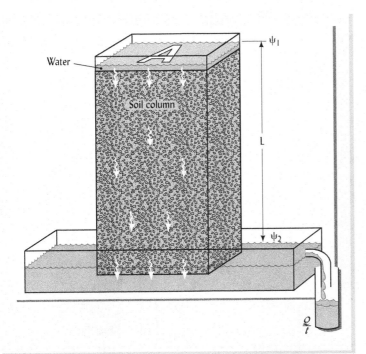

Figure 5.11 *Saturated flow (percolation) in a column of soil with cross-sectional area A cm². All soil pores are filled with water. At lower right, water is shown running off into a container to indicate that water is actually moving down the column. The force driving the water through the soil is the water potential gradient, $\psi_1 - \psi_2 / L$, where both water potentials and length are expressed in cm (see Table 5.1). If we measure the quantity of water flowing out Q/t as cm³/s we can rearrange Darcy's law (from Eq. 5.5) to calculate the saturated hydraulic conductivity of the soil K_{sat} in cm/s as:*

$$K_{sat} = \frac{Q}{A \cdot t} \frac{L}{\psi_1 - \psi_2} \tag{5.6}$$

Remember that the same principles apply where the water potential gradient moves the water in a horizontal direction.

---

[5] A fourth type of flow, **hydraulic redistribution** by plants can move significant amounts of water in the soil profile but occurs within the plant roots rather than in the soil, per se, will be considered in Section 6.3.

Table 5.3
SOME APPROXIMATE VALUES OF SATURATED HYDRAULIC CONDUCTIVITY (IN VARIOUS UNITS) AND INTERPRETATIONS FOR SOIL USES

K_{sat}, cm/s	K_{sat}, cm/h	K_{sat}, in./h	Comments
1×10^{-2}	36	14	Typical of beach sand.
5×10^{-3}	18	7	Typical of very sandy soil, too rapid to effectively filter pollutants in wastewater.
5×10^{-4}	1.8	0.7	Typical of moderately permeable soils, K_{sat} between 1.0 and 15 cm/h considered suitable for most agricultural, recreational, and urban uses calling for good drainage.
5×10^{-5}	0.18	0.07	Typical of fine-textured, compacted, or poorly structured soils. Too slow for proper operation of septic tank drain fields, most types of irrigation, and many recreational uses such as playgrounds.
$< 1 \times 10^{-8}$	$< 3.6 \times 10^{-5}$	$< 1.4 \times 10^{-5}$	Extremely slow; typical of compacted clay. K_{sat} of 10^{-5}–10^{-8} cm/h may be required where nearly impermeable material is needed, as for lagoon lining or landfill cover material.

garden hose, with K_{sat} representing the diameter of the hose (water flows more readily through a larger hose) and $\Delta\psi/L$ representing the size of the pump that drives the water through the hose.

The units in which K_{sat} is measured are length/time, typically cm/s or cm/h. The K_{sat} is an important property that helps determine how well a soil or soil material will perform in such uses as irrigated cropland, sanitary landfill cover material, wastewater storage lagoon lining, and septic tank drain fields (Table 5.3).

It should not be inferred from Figure 5.11 that saturated flow occurs only down the profile. The hydraulic force can also cause horizontal and even upward flow, as occurs when groundwater wells up under a stream (see Section 6.6). The rate of such flow is usually not quite as rapid, however, since the force of gravity does not assist horizontal flow and hinders upward flow. Downward and horizontal flows are illustrated in Figure 5.12, which illustrates characteristic flow of water

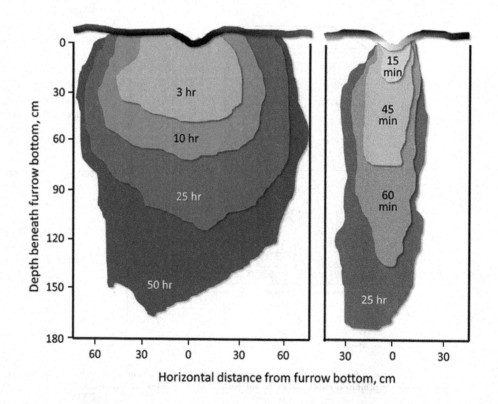

Figure 5.12 *Movement of irrigation water into typical clay loam (left) and sandy loam (right) soils. Note the much more rapid and downward focused movement in the sandy loam in comparison to the slower, but more horizontal movement into the clay loam.* [Redrawn from Cooney and Peterson (1955)]

from irrigation furrows into two soils, a sandy loam and a clay loam. The water moved down much more rapidly in the sandy loam than in the clay loam. On the other hand, horizontal movement (which would have been largely by unsaturated flow) was much more evident in the clay loam.

Factors Influencing the Hydraulic Conductivity of Saturated Soils

Macropores. Any factor affecting the size and configuration of soil pores will influence hydraulic conductivity. The total flow rate in soil pores is proportional to the fourth power of the radius. Thus, flow through a pore 1 mm in radius is equivalent to that in 10,000 pores with a radius of 0.1 mm (even though it takes only 100 pores of radius 0.1 mm to give the same cross-sectional area as a 1 mm radius pore). As a result, macropores (radius >0.08 mm) account for nearly all water movement in saturated soils. Because they usually have more macropore space, sandy soils generally have higher saturated conductivities than finer-textured soils. Likewise, soils with stable structure conduct water much more rapidly through interped cracks and pores than do those with unstable structural units, which break down upon being wetted. However, air trapped in rapidly wetted soils can block pores and thereby reduce hydraulic conductivity. Similarly, the *interconnectedness* of pores is important, as noninterconnected pores are like "dead-end streets" to flowing water. Vesicular pores in certain desert soils are examples (Figure 3.20).

Biopores, such as root channels and earthworm burrows (typically >1 mm in radius), are a special type of macropore and their presence has a marked influence on the saturated hydraulic conductivity of different soil horizons (Table 5.4 and Figure 4.39). Perennial vegetation creates a network of stable biopores, while tillage for production of annual plants destroys these pores and cuts them off from the soil surface. Therefore, saturated conductivity of soils under perennial grassland or forest vegetation is commonly much greater than where annual crop plants are cultivated, and among cultivated soils, saturated hydraulic conductivity is usually greater under no-tillage management than under conventional tillage.

Preferential Flow. Scientists have been surprised to find more extensive pollution of groundwater from pesticides and other toxicants than would be predicted from traditional hydraulic conductivity measurements that assume uniform soil porosity. Apparently solutes (dissolved substances) are carried downward rapidly by water that moves through large macropores such as cracks and biopores, often before the bulk of the soil is thoroughly wetted. Mounting evidence suggests that this type of nonuniform water movement, referred to as **preferential flow**, greatly increases the chances of groundwater pollution (see Figure 5.13).

Preferential flow in fine-textured soils is enhanced by clay shrinkage, which leaves open cracks and fissures that can extend down into the lower subsoil horizons. In some clay soils, water from the first rain storm after a dry spell moves rapidly down the profile, carrying with it soluble pesticides or nutrients that may be on the soil surface (Table 5.5). Recognition that heterogeneity of field soils leads to preferential flow is stimulating more aggressive approaches

Table 5.4
THE SATURATED HYDRAULIC CONDUCTIVITY K_{sat} AND RELATED PROPERTIES OF VARIOUS HORIZONS IN A TYPIC HAPLUDULT PROFILE

The upper horizons had many biopores (mainly earthworm burrows), which resulted in high values of K_{sat} as well as extreme variability from sample to sample. The presence of a clay-enriched argillic horizon resulted in reduced K_{sat} values. Apparently most large biopores in this soil did not extend below 30 cm.

Horizon	Depth, cm	Clay, %	Bulk density, Mg/m³	Mean K_{sat}, cm/h	Range of K_{sat} values,[a] cm/h
Ap	0–15	12.6	1.42	22.4	0.80–70
E	15–30	11.1	1.44	7.9	0.50–24
E/B	30–45	14.5	1.47	0.93	0.53–1.33
Bt	45–60	22.2	1.40	0.49	0.19–0.79
Bt	60–75	27.2	1.38	0.17	0.07–0.27
Bt	75–90	24.1	1.28	0.04	0.01–0.07

[a]For each soil layer K_{sat} was determined on five soil cores of 7.5 cm diameter.
(Data from Waddell and Weil (1996))

Figure 5.13 *Preferential flow of contaminated water downward to the water table. Assume a contaminant (pink color) was spilled on the clayey soil. Uniform movement of water through the soil (a) could be slow enough that the chemical would degrade before it reached the water table. However, with preferential flow (b) contaminated water could quickly reach the water table, in this example via macropores made by the cracking of swelling clay as deep-rooted vegetation dried the soil. Because of these cracks, the first heavy rain after the spill would carry the chemical into the groundwater by preferential flow before the soil could swell and shut the cracks.* (Diagram courtesy of Ray R. Weil)

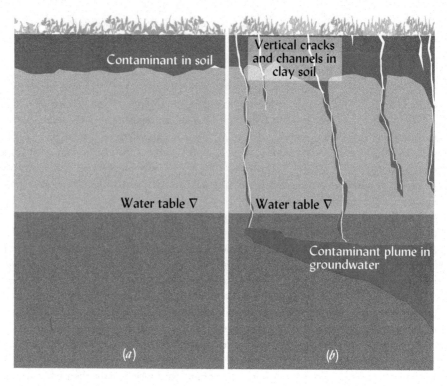

to controlling groundwater contamination. When chemicals and animal wastes are applied to the soil surface, transport of chemicals and fecal bacteria by preferential flow can threaten human health, as well as environmental quality.

Macropores with continuity from the soil surface down through the profile encourage preferential flow. Burrowing animals (e.g., worms, rodents, and insects) as well as decayed plant roots leave tubular channels through which water can flow rapidly. In very sandy soils, hydrophobic organic coatings on sand grains repel water, preventing it from soaking in uniformly. Where these coatings are absent or wear off, water rapidly enters and produces "fingers" of rapid wetting. This "finger flow" probably is responsible for the finger-like shapes of the spodic horizon in some Spodosol profiles (e.g., Figure 3.30). Fingering also occurs in stratified sandy layers that underlie finer-textured materials (see Section 5.6).

Unsaturated Flow in Soils

Most of the time, water movement takes place when upland soils are *unsaturated*. Such movement occurs in a more complicated environment than that which characterizes saturated water flow. In saturated soils, essentially all the pores are filled with water, although the

Table 5.5
LEACHING OF PESTICIDES BY PREFERENTIAL FLOW IN A SLOWLY PERMEABLE ALFISOL
Most leaching took place following the first major storm of the year.

	Leaching of pesticide applied, % (3-year average)		
Chemical	First storm	Spring season	First storm/spring season total, %
Carbofuran	0.22	0.25	88
Atrazine	0.037	0.053	68
Cyanazine	0.02	0.02	100

(Calculated from Kladivko et al. (1999))

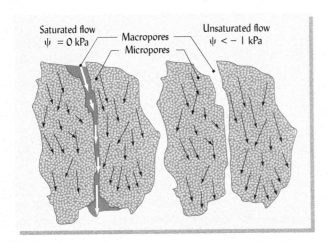

Figure 5.14 *Schematic comparison of saturated and unsaturated water flow in soils. Note that when the water potential is even slightly negative, large macropores will have drained empty of water and therefore cannot contribute to water flow.* (Diagram courtesy of Ray R. Weil)

most rapid water movement is through the large and continuous pores. But in unsaturated soils, these macropores are filled with air, leaving only the finer pores to accommodate water movement (Figure 5.14). Also, in unsaturated soils the water content and, in turn, the tightness with which water is held (water potential) can be highly variable. This influences the rate and direction of water movement and also makes it more difficult to measure the flow of soil water.

As was the case for saturated water movement, the driving force for unsaturated water flow is differences in water potential. This time, however, the difference in the matric potential, not gravity, is the primary driving force. This **matric potential gradient** is the difference in the matric potential of the moist soil areas and nearby drier areas into which the water is moving. Movement will be from a zone of thick moisture films (high matric potential, e.g., −1 kPa) to one of thin films (lower matric potential, e.g., −100 kPa). Generally, if a dry soil is wetted (e.g., by a slow leak in a buried water pipe), water will flow in all directions from the wet soil into the adjacent dry soil. This flow will be rapid at first, but then will slow down as the initially large matric potential gradient declines. The gradient declines mainly because the zone of wettest soil and the as-yet unwetted dry soil zone become farther apart as the water advances—that is, the distance L in Eq. (5.5) becomes greater, thus reducing the potential gradient $\Delta \psi / L$.

Influence of Texture. Near zero potential (which characterizes the saturated flow region), the hydraulic conductivity is thousands of times greater than at potentials that characterize typical unsaturated flow (−10 kPa and below). At high potential levels (high-moisture contents), hydraulic conductivity is higher in a sand than in a clay. The opposite is true at low potential values (low-moisture contents). This relationship is to be expected because sandy soil contains many large pores that are water-filled when the soil water potential is high (and the soil is quite wet), but most of these have been emptied by the time the soil water potential becomes lower than about −10 kPa. Clay soil has many more micropores that are still water-filled at lower soil water potentials (drier soil conditions) and can participate in unsaturated flow.

5.6 INFILTRATION AND PERCOLATION

A special case of water movement is the entry of free water into the soil at the soil–atmosphere interface. As we shall explain in Chapter 6, this is a pivotal process in landscape hydrology that greatly influences the moisture regime for plants and the potential for soil degradation, chemical runoff, and down-valley flooding. The free water at the soil surface may be from rainfall, snowmelt, irrigation, or a liquid chemical spill.

Infiltration

The process by which water enters the soil pore spaces and becomes soil water is termed *infiltration*, and the rate at which water can enter the soil is termed the *infiltrability i*:

$$i = \frac{Q}{A \times t} \quad (5.7)$$

where Q is the volume quantity of water (m^3) infiltrating, A is the area of the soil surface (m^2) exposed to infiltration, and t is time (s). Since m^3 appears in the numerator and m^2 in the denominator, the units of infiltration can be simplified to m/s or, with appropriate conversions, to cm/h. The infiltration rate is not constant over time, but generally decreases during an irrigation or rainfall episode. If the soil is quite dry when infiltration begins, all the macropores open to the surface will be available to conduct water into the soil. In soils with expanding types of clays, the initial infiltration rate may be particularly high as water pours into the network of shrinkage cracks. However, as infiltration proceeds, many macropores fill with water, and shrinkage cracks close up. The infiltration rate declines sharply at first and then tends to level off, remaining fairly constant thereafter (Figure 5.15).

Measurement. The infiltration capacity of a soil may be easily measured using a simple device known as an **infiltrometer**. One common type is the falling head **double-ring infiltrometer** which is comprised of two heavy metal cylinders, one smaller in diameter than the other, that are pressed partially into the soil so that the smaller cylinder is inside the larger (see Figure 5.15). A layer of plastic film is placed inside the center ring to protect the soil surface from disturbance, and water is poured into both cylinders. The plastic film is then quickly removed and the depth of water in the central cylinder recorded periodically as the water infiltrates the soil. The water infiltrating in the outer cylinder is not measured, but it ensures that the surrounding soil will be equally moist and that the movement of water from the central cylinder will be principally downward, not horizontal.

Percolation

Infiltration is a transitional phenomenon that takes place at the soil surface. Once the water has infiltrated the soil, the water moves downward into the profile by the process termed **percolation**. Both saturated and unsaturated flow are involved in percolation of water down the profile, and rate of percolation is related to the soil's hydraulic conductivity. In the case of water that

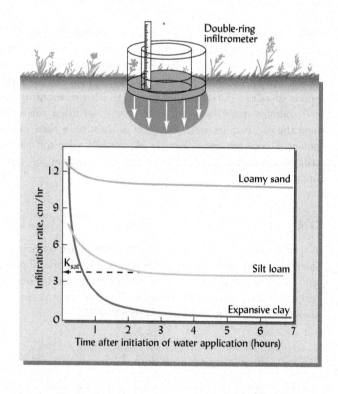

Figure 5.15 The potential rate of water entry into the soil, or infiltration capacity, can be measured by recording the drop in water level in a double-ring infiltrometer (top). Changes in the infiltration rate of several soils during a period of water application by rainfall or irrigation are shown (bottom). Generally, water enters a dry soil rapidly at first, but its infiltration rate slows as the soil becomes saturated. The decline is least for very sandy soils with macropores that do not depend on stable structure or clay shrinkage. In contrast, a soil high in expansive clays may have a very high initial infiltration rate when large cracks are open, but a very low infiltration rate once the clays swell with water and close the cracks. Most soils fall between these extremes, exhibiting a pattern similar to that shown for the silt loam soil. The dashed arrow indicates the level of K_{sat} for the silt loam illustrated.
(Diagram courtesy of Ray R. Weil)

has infiltrated a relatively dry soil, the progress of water movement can be observed by the darkened color of the soil as it becomes wet. There usually appears to be a sharp boundary, termed a **wetting front**, between the dry underlying soil and the soil already wetted. During an intense rain or heavy irrigation, water movement near the soil surface occurs mainly by saturated flow in response to gravity. At the wetting front, however, water is moving into the underlying drier soil in response to matric potential gradients as well as gravity. During a light rain, both infiltration and percolation may occur mainly by unsaturated flow as water is drawn by matric forces into the fine pores without accumulating at the soil surface or in the macropores.

Water Movement in Stratified Soils

The fact that water moves by unsaturated flow at the wetting front has important ramifications for how percolating water behaves when it encounters an abrupt change in pore sizes, due to such layers as fragipans or claypans or sand and gravel lenses. In some cases, such pore-size stratification may be created by soil managers, as when coarse plant residues are plowed under in a layer or a layer of gravel is placed under finer soil in a planting container. In all cases, the effect on water percolation is similar—that is, the downward movement is impeded—even though the causal mechanism may vary. The contrasting layer acts as a barrier to water flow and results in much higher field-moisture levels above the barrier than what would normally be encountered in freely drained soils. It is not surprising that percolating water should slow down markedly when it reaches a layer with finer pores, which therefore has a lower hydraulic conductivity. However, the fact that a layer of *coarser* pores will temporarily stop the movement of water may not be obvious.

In Figure 5.16, a layer of coarse sand underlies a fine-textured soil. Intuitively, one might expect the sand layer to speed the downward percolation of water. However, the coarser sand layer has just the opposite effect; it actually impedes the flow of water. The macropores of the sand offer less attraction for the water than do the finer pores of the overlying material. Since water always moves from higher to lower potential (to where it will be held more tightly), the wetting front cannot move readily into the sand. If it cannot move laterally, as it would in the case of a sloping coarse layer, the downward-moving water will eventually accumulate above the sand layer and saturate the pores at the soil–sand interface. The matric potential of the water at the wetting front will then fall to nearly zero or even become positive. Once this occurs, the water will be so loosely held by the fine-textured soil that gravity or hydrostatic pressure will force the water into the sand layer.

Interestingly, a coarse sand layer in an otherwise fine-textured soil profile would also inhibit the *rise* of water from moist subsoil layers up to the surface soil, a situation that could be illustrated by turning Figure 5.16*b* upside down. The large pores in the coarse layer will not be able to support capillary movement up from the smaller pores in a finer layer. Consequently, water rises by capillarity up to the coarse-textured layer but cannot cross it to supply moisture to overlying layers. Thus, plants growing on some soils with buried gravel lenses are subject

Figure 5.16 *Downward water movement in soils with a stratified layer of coarse material. (Left) Water is applied to the surface of a medium-textured topsoil. Note that after 40 min, downward movement is no greater than movement to the sides, indicating that in this case the gravitational force is insignificant compared to the matric potential gradient between dry and wet soil. (Center) The downward movement stops when a coarse-textured layer is encountered. After 110 min, no movement into the sandy layer has occurred. The macropores of the sand provide less attraction for water than the finer-textured soil above, so the water accumulates and moves laterally above the sand layer. (Right) Only when the water content (and in turn the matric potential gradient) is raised sufficiently will the water move into the sand. After 400 min, the water content of the overlying layer becomes sufficiently high to give a water potential close to 0 kPa or slightly positive, and the water breaks through into the coarse material abruptly and unevenly.* (Photos Courtesy of W. H. Gardner, Washington State University)

to drought since they are unable to exploit water in the lower soil layers. This principle also allows a layer of gravel to act as a capillary barrier under a concrete slab foundation to prevent water from soaking up from the soil and through the concrete floor of a home basement.

The fact that coarse-textured layers (e.g., gravel, sand, coarse organic materials, or geotextile fabrics) can hinder both downward and upward unsaturated flow of water must be considered—and may be used to advantage—when dealing with such materials in planting containers, landscape drainage schemes (see Section 6.7), or engineering works (see Box 5.2).

BOX 5.2
PRACTICAL APPLICATIONS OF UNSATURATED WATER FLOW IN CONTRASTING LAYERS

Unsaturated flow is interrupted where soil texture abruptly changes from relatively fine to coarse. Capillary water held tightly by matric attraction in the smaller pores of the finer-textured layer cannot move into the larger pores in the underlying coarser-textured layer unless under positive pressure. That is, a larger pore cannot "pull" water from a smaller pore. In fact, unsaturated water flow always occurs in the reverse direction, from larger to smaller pores. A wetting front moving down a soil profile along the matric potential gradient therefore ceases downward movement when it encounters pores much larger than those through which it has been traversing. Instead of continuing downward, water will move laterally in the finer layer. If water is entering the system more rapidly than lateral capillarity can carry it away, a perched water table may develop above the interface between the two layers (Figure 5.17).

This phenomenon is applied in the design of soil profiles for golf course putting greens. The soil specified for the rooting zone consists almost entirely of sand in order to promote rapid infiltration of water and to resist compaction by foot traffic. However, water normally drains so fast through sand that too little is held to meet the needs of growing grass. This situation is remedied to some extent by constructing the putting green with a layer of gravel underneath the sand rooting zone. The large pores in the gravel temporarily stop the downward movement of water. The resulting perched water table (Figure 5.18) causes the sand layer to retain more water than it would otherwise, but still allows for rapid drainage of excess water—when a buildup of positive pressure allows gravitational forces to override matric forces.

Figure 5.17 A water-saturated layer resulting from the interruption of water flow where fine soil is underlain by coarser material. Water logging (indicated by the gleyed, gray-colored soil layer) often occurs during wet weather where the silt loam upper layers meet the coarse loamy sand lower layers in this soil profile. Note the reddish colors indicating good oxidation in both the silt loam soil layers above and in the sandy layer below the perched water table. (Photo courtesy of C. White, The Pennsylvania State University)

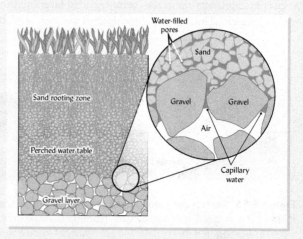

Figure 5.18 A gravel layer under a putting green is used to increase the water available to turfgrass roots in the sand rooting zone, while allowing rapid drainage if saturation occurs. (Diagram courtesy of Ray R. Weil)

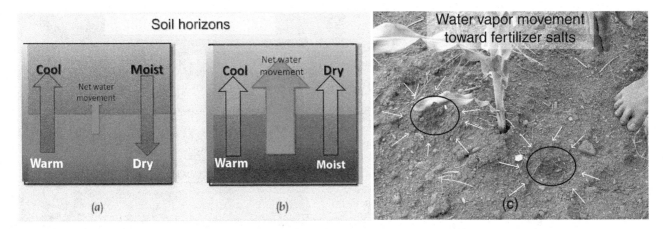

Figure 5.19 *Vapor movement caused by vapor pressure gradients between soil zones differing in temperature, moisture, or salinity. In (a) the gradients nearly negate each other, resulting in only a small net vapor pressure gradient and net water vapor movement (small vertical arrow). In (b) the moisture and temperature gradients are coordinated, and considerable vapor transfer is expected if the liquid water in the soil capillaries does not interfere. In (c) two small spots of soils have been darkened by moisture. Below each of these spots, the African farmer had buried a spoonful of soluble fertilizer in the relatively dry soil. The fertilizer began to dissolve, creating a salty solution with very low water vapor pressure around it, thus setting up a vapor pressure gradient that caused water vapor to move from the nonsalty surrounding soil toward the salty soil where the fertilizer had been applied.* (Courtesy of Ray R. Weil)

Water Movement in Stratified Soils

Two types of water vapor movement are associated with soils, *internal* and *external*. Internal movement takes place within the soil, that is, in the soil pores. External movement occurs at the land surface, and water vapor is lost by surface evaporation (see Section 6.4).

Water vapor moves from one point to another within the soil in response to differences in vapor pressure. Thus, water vapor will move from a moist soil where the soil air is nearly 100% saturated with water vapor (high vapor pressure) to a drier soil where the vapor pressure is somewhat lower. Also, water vapor will move from a zone of low salt content to one with a higher salt content (e.g., around a fertilizer granule). The salt lowers the vapor pressure of the water and encourages water movement from the surrounding soil.

If the temperature of one part of a uniformly moist soil is lowered, the vapor pressure will decrease and water vapor will tend to move toward this cooler part. Heating will have the opposite effect in that heating will increase the vapor pressure, and the water vapor will move away from the heated area. Figure 5.19 illustrates these relationships.

Even though the amount of water vapor is small, its movement in soils can be of some practical significance. For example, seeds of some plants can absorb sufficient water vapor from the soil to stimulate germination. Likewise, in dry soils, water vapor movement may be of considerable significance to drought-resistant desert plants (*xerophytes*), many of which can exist at extremely low soil water contents. For instance, at night the surface horizon of a desert soil may cool sufficiently to cause vapor movement up from deeper layers. If cooled enough, the vapor may then condense as dewdrops in the soil pores, supplying certain shallow-rooted xerophytes with water for survival.

5.7 QUALITATIVE DESCRIPTION OF SOIL WETNESS

As an initially water-saturated soil dries down, both the soil as a whole and the soil water it contains undergo a series of gradual changes in physical behavior and in their relationships with plants. These changes are due mainly to the fact that the water remaining in the drying soil is found in smaller pores, in thinner films or in smaller pore corners (review Figure 5.4) where the water potential is lowered principally by the action of matric forces. Matric potential therefore accounts for an increasing proportion of the total soil water potential, while the proportion attributable to gravitational potential decreases.

Figure 5.20 Volumes of water and air associated with 100 g of soil solids in a representative well-granulated silt loam. The top bar shows the situation when the soil is completely saturated with water. This situation will usually occur for short periods of time when water is being added. Water will soon drain out of the larger pores (macropores). The soil is then said to be at field capacity. Plants will remove water from the soil quite rapidly until they begin to wilt. When permanent wilting of the plants occurs, the soil water content is said to be at the wilting coefficient. There is still considerable water in the soil, but it is held too tightly to permit its absorption by plant roots. The water lost between field capacity and wilting coefficient is considered to be the soil's plant available water-holding capacity (AWHC). A further reduction in water content to the hygroscopic coefficient is illustrated in the bottom bar. At this point the water is held very tightly, mostly by the soil colloids. (Diagram courtesy of Ray R. Weil and N. Brady)

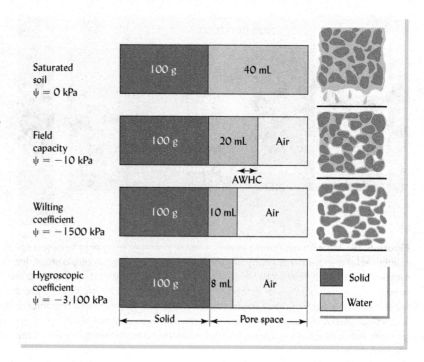

To study these changes and introduce the terms commonly used to describe varying degrees of soil wetness, we shall follow the moisture and energy status of soil during and after wetting. The terms to be introduced describe various stages along a continuum of soil wetness and should not be interpreted to imply that soil water exists in different "forms." Because these terms are essentially qualitative and lack a precise scientific basis, some soil physicists object to their use. However, it would be a serious disservice to the reader to omit them from this text as they are widely used in practical soil management and help communicate important facts about soil water behavior.

Maximum Retentive Capacity

When all soil pores are filled with water, the soil is said to be *water-saturated* (Figure 5.20) and at its **maximum retentive capacity**. The matric potential is close to zero, nearly the same as that of free water. The volumetric water content is essentially the same as the total porosity. The soil will remain at maximum retentive capacity only so long as water continues to infiltrate, for the water in the largest pores (sometimes termed **gravitational water**) will percolate downward, mainly under the influence of **gravitational forces**. Data on maximum retentive capacities and the average depth of soils in a watershed are useful in predicting how much rainwater can be stored in the soil temporarily, thus possibly avoiding downstream floods.

Field Capacity

Once the rain or irrigation has ceased, water in the largest soil pores will drain downward quite rapidly in response to the hydraulic gradient (mostly gravity). After one to three days, this rapid downward movement will become negligible as matric forces play a greater role in the movement of the remaining water. The soil then is said to be at its **field capacity**. In this condition, water has moved out of the macropores and air has moved in to take its place. The micropores or capillary pores are still filled with water and can supply plants with needed water. The matric potential will vary slightly from soil to soil but is generally in the range of −10 to −30 kPa, assuming drainage into a less-moist zone of similar porosity.[6] Water

---

[6]Note that because of the relationships pertaining to water movement in stratified soils (see Section 5.6), soil in a flower pot will cease drainage while much wetter than field capacity.

movement will continue to take place by unsaturated flow, but the rate of movement is very slow since it now is due primarily to capillary forces, which are effective only in micropores (Figures 5.14). The water found in pores small enough to retain it against rapid gravitational drainage, but large enough to allow capillary flow in response to matric potential gradients, is sometimes termed **capillary water**.

While all soil water is affected by gravity, the term *gravitational water* refers to the portion of soil water that readily drains away between the states of maximum retentive capacity and field capacity. Most soil leaching occurs as gravitational water that drains from the larger pores before field capacity is reached. Gravitational water therefore includes much of the water that transports chemicals such as nutrient ions, pesticides, and organic contaminants into the groundwater and, ultimately, into streams and rivers.

Field capacity is a very useful term because it refers to an approximate degree of soil wetness at which several important soil properties are in transition:

1. At field capacity, a soil is holding the maximal amount of water useful to plants. Additional water, while held with low energy of retention, would be of limited use to plants because it would remain in the soil for only a short time before draining, and, while in the soil, it would occupy the larger pores, thereby reducing soil aeration. Drainage of gravitational water from the soil is generally a requisite for optimum plant growth (hydrophilic plants, such as rice or cattails, excepted).
2. At field capacity, the soil is near its lower plastic limit—that is, the soil behaves as a crumbly semisolid at water contents below field capacity, and as a plastic puttylike material that easily turns to mud at water contents above field capacity (see Section 4.9). Therefore, field capacity approximates the upper limit of soil wetness for easy tillage or excavation.
3. At field capacity, sufficient pore space is filled with air to allow optimal aeration for most aerobic microbial activity and for the growth of most plants (see Section 7.5).

Permanent Wilting Percentage or Wilting Coefficient

Once an unvegetated soil has drained to its field capacity, further drying is quite slow, especially if the soil surface is covered to reduce evaporation. However, if plants are growing in the soil, they will remove water from their rooting zone, and the soil will continue to dry. The roots will remove water first from the largest water-filled pores where the water potential is relatively high. As these pores are emptied, roots will draw their water from the progressively smaller pores and thinner water films in which the matric water potential is lower and the forces attracting water to the solid surfaces are greater. Hence, it will become progressively more difficult for plants to remove water from the soil at a rate sufficient to meet their needs.

As the soil dries, the rate of plant water removal may fail to keep up with water losses. Plants will then close the openings (**stomata**) in their leaves by which they transpire water and take in carbon dioxide (CO_2) for photosynthesis. Plant growth will therefore cease (due to starvation for CO_2) and plant leaves may wilt during the daytime as their cells lose turgor. At first the plants will regain their turgor at night when water is not being lost through the leaves, and the roots can catch up with the plants' demand. Ultimately, however, herbaceous plants will remain wilted night and day when the roots cannot generate water potentials low enough to coax the remaining water from the soil. Woody plants may not wilt *per se*, but their leaves will go flaccid and the water columns in their xylem tissues will be broken and water uptake will be disrupted or cease. Although not yet dead, the plants will soon die if water is not provided. For most annual herbaceous plants this condition develops when the soil water potential ψ has a value of about $-1{,}500$ kPa (-15 bars). Many trees and a few herbaceous plants, especially xerophytes (desert-type plants), can continue to remove water at even $-2{,}000$ to $-3{,}000$ kPa, but the amount of water available between $-1{,}500$ kPa and $-3{,}000$ kPa is very small (Figure 5.21).

The water content of the soil at this stage is called the **wilting coefficient** or **permanent wilting percentage** and by convention is taken to be that amount of water retained

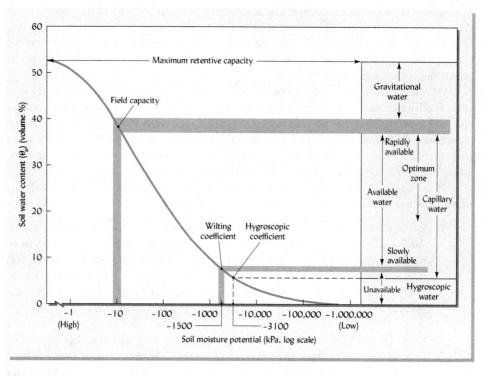

Figure 5.21 Water content–matric potential curve of a loam soil as related to different terms used to describe water in soils. The shaded bars in the diagram to the right suggest that concepts such as field capacity are only approximations. The gradual change in potential with soil moisture change discourages the concept of different "forms" of water in soils. At the same time, such terms as gravitational and available assist in the qualitative description of moisture utilization in soils.

by the soil when the water potential is −1,500 kPa. The soil will appear to be dusty dry, although some water remains in the smallest of the micropores and in very thin films (perhaps only 10 molecules thick) around individual soil particles (see Figure 5.20). As illustrated in Figure 5.21, plant **available water** is considered to be that water retained in soils between the states of field capacity and wilting coefficient (conventionally taken to be between −10 kPa and −1,500 kPa). The amount of capillary water remaining in the soil that is unavailable to most plants can be substantial in fine-textured soils and those high in organic matter.

Hygroscopic Coefficient

Although plant roots do not generally dry the soil beyond the permanent wilting percentage, if the soil is exposed to the air, water will continue to be lost by evaporation. When soil moisture is lowered below the wilting point, the water molecules that remain are very tightly held, mostly being adsorbed by colloidal soil surfaces. This state is approximated when the atmosphere above a soil sample is nearly saturated with water vapor (98% relative humidity) and equilibrium is established at a water potential of −3,100 kPa. The water is thought to be in films only four or five molecules thick and is held so rigidly that much of it is considered nonliquid and can move only in the vapor phase. The moisture content of the soil at this point is termed the **hygroscopic coefficient**. Soils high in colloidal materials (clay and humus) will hold more water under these conditions than will sandy soils that are low in clay and humus (Table 5.6). Soil water considered to be *unavailable* to plants includes the hygroscopic water as well as that portion of capillary water retained at potentials below −1,500 kPa (see Figure 5.21).

Table 5.6
VOLUMETRIC WATER CONTENT θ_v AT FIELD CAPACITY, WILTING COEFFICIENT, AND HYGROSCOPIC COEFFICIENT FOR THREE REPRESENTATIVE SOILS AND THE CALCULATED AVAILABLE AND CAPILLARY WATER

Note that the clay soil retains most water at the field capacity, but much of that water is held tightly in the soil at −3,100 kPa potential by soil colloids (hygroscopic coefficient).

	Volume % (θ_v)				
Soil	Field capacity −10 to −30 kPa	Wilting coefficient, −1,500 kPa	Hygroscopic coefficient, −3,100 kPa	Plant-available water, col. 1 − col.2	Capillary water, col. 1 − col. 3
Sandy loam	12	4	3	8	9
Silt loam	30	15	10	15	20
Clay	35	22	18	13	17

5.8 FACTORS AFFECTING AMOUNT OF PLANT-AVAILABLE SOIL WATER

The amount of soil water available for plant uptake is determined by a number of factors, including the relationship between water content and water potential for each soil horizon, soil strength and density effects on root growth, soil depth, rooting depth, and soil stratification or layering. Each will be discussed briefly.

Water Content–Potential Relationship

As illustrated in Figure 5.21, a given soil will exhibit a specific relationship between the water potential and the amount of water held in the soil. In particular, the plant-available water-holding capacity is the difference in the amount of water held when the soil water potential is indicative of field capacity and permanent wilting percentage. This energy–controlling concept should be kept in mind as we consider the various soil properties that affect the amount of water a soil can hold for plant use.

The general influence of texture on field capacity, wilting coefficient, and **available water-holding capacity** is shown in Figure 5.22. Note that as fineness of texture increases, there is a general increase in available moisture storage from sands to loams and silt loams. Plants growing on sandy soils are more apt to suffer from drought than are those growing on a silt loam in the same area. However, clay soils frequently provide less available water than do well-granulated silt loams, since the clays tend to have a high wilting coefficient.

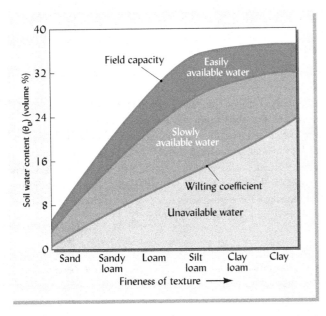

Figure 5.22 *General relationship between soil water characteristics and soil texture. Note that the wilting coefficient increases as the texture becomes finer. The field capacity increases until we reach the silt loams, then levels off. Remember these are representative curves; individual soils would probably have values different from those shown.* (Diagram courtesy of Ray R. Weil and N. Brady)

Soil organic matter exerts both direct and indirect influences that contribute increased soil water availability. The direct effects are due to the very high volumetric water-holding capacity of organic matter, which, when the soil is at the field capacity, is much higher than that of an equal volume of mineral matter (see curve for peat soil in Figure 5.6). Even though the water held by organic matter at the wilting point is also somewhat higher than that held by mineral matter, the amount of water available for plant uptake is still greater from the organic fraction. Organic matter indirectly affects the amount of water available to plants mainly by enhancing soil structure (see Section 4.5), which in turn increases both water infiltration and water-holding capacity. Recognition of the beneficial effects of organic matter on plant-available water is essential for wise soil management.

Compaction Effects on Matric Potential, Aeration, and Root Growth

Soil compaction generally reduces the amount of water that plants can take up. Four factors account for this negative effect. First, compaction crushes many of the macropores and large micropores into smaller pores, and the bulk density increases. As the clay particles are forced closer together, soil strength may increase beyond about 2,000 kPa, the level considered to limit root penetration. Second, compaction decreases the total pore space, which generally means that less water is retained at field capacity. Third, reduction in macropore size and numbers generally means less air-filled pore space when the soil is near field capacity. Fourth, the creation of more very fine micropores will increase the permanent wilting coefficient and so decrease the available water content.

Least Limiting Water Range. We have already defined **plant-available water** as that held with matric potentials between field capacity (−10 to −30 kPa) and the permanent wilting point (−1,500 kPa). Thus, plant-available water is that which is not held too tightly for roots to take up and yet is not held so loosely that it freely drains away by gravity. The **least limiting water range** is that range of water contents for which soil conditions do not severely restrict root growth. According to the least limiting water range concept, soils are too *wet* for normal root growth when so much of the soil pore space is filled with water that less than about 10% remains filled with air. At this water content, lack of oxygen for respiration limits root growth. In loose, well-aggregated soils, this water content corresponds quite closely to field capacity. However, in a compacted soil with very few large pores, oxygen supply may become limiting at lower water contents (and potentials) because some of the smaller pores will be needed for air.

The least limiting water range concept tells us that soils are too *dry* for normal root growth when the soil strength (measured as the pressure required to push a pointed rod through the soil) exceeds about 2,000 kPa. This level of soil strength occurs at water contents near the wilting point in loose, well-aggregated soils, but may occur at considerably higher water contents if the soil is compacted (see Figure 5.23). To summarize, the least limiting water range concept suggests that root growth is limited by lack of oxygen at the wet end of the range and by the inability of roots to physically push through the soil at the dry end. Thus, compaction effects on root growth are greatest in dry soils (Figure 5.23).

Osmotic Potential

The presence of soluble salts, either from applied fertilizers or as naturally occurring compounds, can influence plant uptake of soil water. For soils high in salts, the osmotic potential ψ_o of the soil solution will contribute to the total moisture stress along with the matric potential. The osmotic potential tends to reduce available moisture in such soils because more water is retained in the soil at the permanent wilting coefficient than would be the case due to matric potential alone. In most humid-region soils these osmotic potential effects are insignificant, but they become of considerable importance for certain soils in dry regions that may accumulate soluble salts through irrigation or natural processes (see Section 9.11).

Soil Depth and Layering

Our discussion thus far has referred to available water-holding capacity as the percentage of the soil volume consisting of pores that can retain water at potentials between field capacity

Figure 5.23 *Compaction (dashed curve) reduces the range of soil water contents suitable for root growth, thus narrowing the least limiting water range.* (Diagram courtesy of Ray R. Weil)

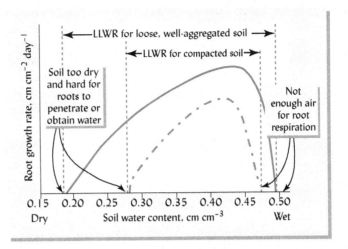

and wilting percentage. The total volume of available water will depend on the total volume of soil explored by plant roots. This volume may be governed by the total depth of soil above root-restricting layers, by the greatest rooting depth characteristic of a particular plant species, or even by the size of a pot chosen for containerized plants. The depth of soil available for root exploration is of particular significance for deep-rooted plants (see Figure 1.2), especially in subhumid to arid regions where perennial vegetation depends on water stored in the soils for survival during long periods without precipitation.

Soil stratification or layering can markedly influence the available water and its movement in the soil. Impervious layers drastically slow down the rate of water movement and also restrict the penetration of plant roots, thereby reducing the soil depth from which moisture is drawn. Sandy layers also act as barriers to soil moisture movement from the finer-textured layers above, as explained in Section 5.6.

The capacity of soils to store available water determines to a great extent their usefulness for plant growth (see Figures 5.24). This capacity provides a buffer between an

Figure 5.24 *Soil depth and competition for soil water between trees and turfgrass in lawn landscapes during an extended period of hot, dry weather. (Left) The brown, drought-stressed grass near the line of trees shows where tree roots have used most of the available water in the topsoil some 10–15 m beyond the tree canopy itself. The trees have an advantage in competing for water partly because water uptake by their roots is driven by the large canopy leaf area. (Right) In this scene, with the greenest grass nearest the trees, the pattern of drought stress in the grass appears to be opposite from that in the scene on the left. What accounts for the difference? The rectangular area of brown grass is actually a green roof atop a below-ground building. The soil on this vegetated roof is less than 20 cm thick and so has been depleted of it's total profile store of available water. The trees are planted beyond the edge of the green roof in native soil where both the grass and the tree roots can exploit the much larger total available water supply held by a deeper soil profile.* (Photo courtesy of Ray R. Weil)

adverse climate and plant production. The productivity of forest sites is often related to soil water-holding capacity. In irrigated soils, it helps determine the frequency with which water must be applied. Soil water-holding capacity is becoming increasingly important for global ecosystem management as the use of water for all purposes—industrial and domestic, as well as irrigation—begins to tax the supply of this critical natural resource. To estimate the water-holding capacity of a soil, each soil horizon to which roots have access may be considered separately and then summed to give the total water-holding capacity for the profile (see Box 5.3).

BOX 5.3
TOTAL AVAILABLE WATER-HOLDING CAPACITY OF A SOIL PROFILE

The total amount of water available to a plant growing in a field soil can be estimated from the rooting depth of the plant and the amount of water held between field capacity and wilting percentage in each of the soil horizons explored by the roots. For each soil horizon, the mass available water-holding capacity (AWHC) is estimated as the difference between the mass water content at field capacity θ_{mFC} (g water per 100 g soil at field capacity) and that at permanent wilting percentage, θ_{mWP}. This value can be converted into a volumetric water content θ_v by multiplying by the ratio of the bulk density D_b of the soil to the density of water D_w. Finally, this volume ratio is multiplied by the thickness of the horizon to give the total centimeters of available water capacity AWHC in that horizon.

$$AWHC = (\theta_{mFC} - \theta_{mWP}) \times D_b/D_w \times L \quad (5.8)$$

For the first horizon described in Table 5.7 we can substitute values (with units) into Eq. (5.8):

$$AWHC = \left(\frac{22\,g}{100\,g} - \frac{8\,g}{100\,g}\right) \times \frac{1.2\,Mg}{m^3} \times \frac{1\,m^3}{1\,Mg} \times 20\,cm$$
$$= 3.36\,cm$$

Note that all units cancel out except cm, resulting in the depth of available water (cm) held by the horizon. In Table 5.7 the AWHC of all horizons within the rooting zone are summed to give a total AWHC for the soil–plant system. Since no roots penetrated to the last horizon (1.0 to –1.25 m), this horizon was not included in the calculation. We can conclude that for the soil–plant system illustrated, 14.13 cm of water could be stored for plant use. At a typical warm weather water-use rate of 0.5 cm of water per day, this soil could hold about a four-week supply.

Table 5.7
CALCULATION OF ESTIMATED SOIL PROFILE AVAILABLE WATER-HOLDING CAPACITY (AWHC)

Soil depth, cm	Relative root length	Soil depth increment, cm	Soil bulk density, Mg/m³	Field capacity (FC), g/100 g	Wilting percentage (WP), g/100 g	AWHC, cm
0–20	xxxxxxxxx	20	1.2	22	8	$20 \times 1.2\left(\frac{22}{100} - \frac{8}{100}\right) = 3.36$ cm
20–40	xxxx	20	1.4	16	7	$20 \times 1.4\left(\frac{16}{100} - \frac{7}{100}\right) = 2.52$ cm
40–75	xx	35	1.5	20	10	$35 \times 1.5\left(\frac{20}{100} - \frac{10}{100}\right) = 5.25$ cm
75–100	xx	25	1.5	18	10	$25 \times 1.5\left(\frac{18}{100} - \frac{10}{100}\right) = 3.00$ cm
100–125	—	25	1.6	15	11	No roots
Total						$3.36 + 2.52 + 5.25 + 3.00 = 14.13$ cm

5.9 MECHANISMS BY WHICH PLANTS ARE SUPPLIED WITH WATER

At any one time, only a small proportion of the soil water is adjacent to absorptive plant–root surfaces. How then do the roots get access to the immense amount of water (see Section 6.3) needed by vigorously growing plants? Two phenomena seem to account for this access: the capillary movement of the soil water to plant roots and the growth of the roots into moist soil.

Rate of Capillary Movement

When plant rootlets absorb water, they reduce the water content in the soil immediately surrounding them. Thus, roots set up a water potential gradient that causes water to move from the more moist, higher potential bulk soil toward the drier, lower potential soil adjacent to the roots. The rate of movement depends on the magnitude of the potential gradients developed and the conductivity of the soil pores. With some sandy soils, the adjustment may be comparatively rapid and the flow appreciable if the soil is near field capacity. In fine-textured and poorly granulated clays, the movement will be sluggish and only a meager amount of water will be delivered. However, in drier conditions with water held at lower potentials, a clay soil will eventually be able to deliver more water by capillarity than a sandy soil because the latter will have very few capillary pores still filled with water.

The total distance that water flows by capillarity on a day-to-day basis may be only a few centimeters. This might suggest that capillary movement is not a significant means of enhancing moisture uptake by plants. However, if the roots have penetrated much of the soil volume so that individual roots are rarely more than a few centimeters apart, movement over greater distances may not be necessary. Even during periods of hot, dry weather when evaporative demand is high, capillary movement can be an important means of providing water to plants. It is of special significance during periods of low moisture content when plant root extension is low.

Rate of Root Extension

Capillary movement of water is complemented by rapid rates of root extension, which ensure that new root–soil contacts are constantly being established. Such root penetration may be rapid enough to take care of most of the water needs of a plant growing in a soil at optimum moisture. The mats of roots, rootlets, and root hairs in a forest floor or a meadow sod exemplify successful adaptations of terrestrial plants for exploitation of soil water storage.

The primary limitation of root extension is the small proportion of the soil with which roots are in contact at any one time. Even though the root surface is considerable, root–soil contacts commonly account for less than 1% of the total soil surface area. This suggests that most of the water must move from the soil to the root even though the distance of movement may be no more than a few millimeters. It also suggests the complementarity of capillarity and root extension as means of providing soil water for plants.

Root Distribution

The distribution of roots in the soil profile determines to a considerable degree the plant's ability to absorb soil water. Most plants, both annuals and perennials, have the bulk of their roots in the upper 25–30 cm of the profile. For most plants in limited rainfall regions, roots explore relatively deep soil layers, but usually 95% of the entire root system is contained in the upper 2 m of soils. As Figure 5.25 shows, in more humid regions rooting tends to be somewhat shallower. Perennial plants, both woody and herbaceous, produce some roots that grow very deeply (>3 m) and are able to absorb a considerable proportion of their moisture from deep subsoil layers. Even in these cases, however, it is likely that much of the root absorption is from the upper layers of the soil, provided these layers are well supplied with water. On the other hand, if the upper soil layers are moisture-deficient, even annual row crops such as cotton, sunflower, corn, and soybeans will absorb much of their water from the lower horizons (Figure 5.26), provided that adverse physical or chemical conditions do not inhibit their exploration of these horizons.

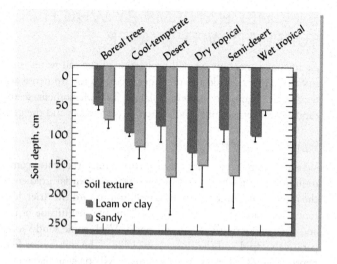

Figure 5.25 *The soil depth above which 95% of all roots are located under different vegetation and soil types. The analysis used 475 root profiles reported from 209 geographic locations. The deepest rooting depths were found mainly in water-limited ecosystems. Within all but the wettest ecosystem, rooting was deeper in the sandier soils. Globally, 9 out of 10 profiles had at least 50% of all roots in the upper 30 cm of the soil profile and 95% of all roots in the upper 2 m (including any O horizons present).* (Redrawn from Schenk and Jackson (2002))

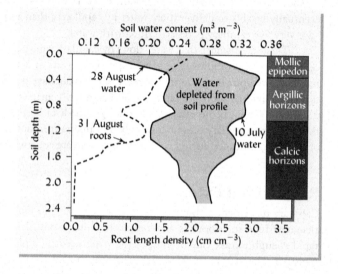

Figure 5.26 *Distribution of roots and depletion of water in the profile of a soil growing sunflowers under dryland conditions in Texas. Notice the relationship between root density (cm of root length per cm³ of soil) and amount of water used during the hot, dry weather between 10 July and 28 August. The water content on 28 August is near the wilting point in the upper 1 m of this Pullman clay loam (Torrertic Paleustolls).* (Drawn from data in Moroke et al. (2005))

Root–Soil Contact

As roots grow into the soil, they move into pores of sufficient size to accommodate them. Contact between the outer cells of the root and the soil permits ready movement of water from the soil into the plant in response to differences in energy levels. When the plant is under moisture stress, however, the roots tend to shrink in size as their cortical cells lose water in response to this stress. Such conditions exist during hot, dry weather and are most severe during the daytime, when water loss through plant leaves is at a maximum. The diameter of roots under these conditions may shrink by 30%. This reduces the direct root–soil contact considerably, as well as the movement of liquid water and nutrients into the plants. While water vapor can still be absorbed by the plant, its rate of absorption is too low to keep any but the most drought-tolerant plants alive.

5.10 CONCLUSION

Water impacts all life. The interactions and movement of this simple compound in soils help determine whether these impacts are positive or negative. An understanding of the principles that govern the attraction of water for soil solids and for dissolved ions can help land managers maximize the positive impacts while minimizing the less desirable ones.

The water molecule has a polar structure that results in electrostatic attraction of water to both soluble cations and soil solids. These attractive forces tend to reduce the potential energy level of soil water below that of pure water. The extent of this reduction, called soil water potential ψ, has a profound influence on a number of soil properties, but especially on the movement of soil water and its uptake by plants.

The water potential due to the attraction between soil solids and water (the matric potential ψ_m) combines with the force of gravity ψ_g to largely control water movement. This movement is relatively rapid in soils high in moisture and with an abundance of macropores. In drier soils, however, the adsorption of water on the soil solids is so strong that its movement in the soil and its uptake by plants are greatly reduced. As a consequence, plants can die for lack of water—even though there are still significant quantities of water in the soil—because that water is unavailable to plants.

Water is supplied to plants by capillary movement toward the root surfaces and by growth of the roots into moist soil areas. In addition, vapor movement may be of significance in supplying water for drought-resistant desert species (xerophytes). The osmotic potential *o* becomes important in soils with high soluble salt levels that can impede plant uptake of water from the soil. Such conditions occur most often in soils with restricted drainage in areas of low rainfall, in irrigated soils, and in potted indoor plants.

The characteristics and behavior of soil water are quite complex. As we have gained more knowledge, however, it has become apparent that soil water is governed by relatively simple, basic physical principles. Furthermore, researchers are discovering the similarity between these principles and those governing the movement of groundwater and the uptake and use of soil moisture by plants—the subject of the next chapter.

STUDY QUESTIONS

1. What is the role of the *reference state of water* in defining soil water potential? Describe the properties of this reference state of water.
2. Imagine a root of a cotton plant growing in the upper horizon of an irrigated soil in California's Imperial Valley. As the root attempts to draw water molecules from this soil, what forces (potentials) must it overcome? If this soil were compacted by a heavy vehicle, which of these forces would be most affected? Explain.
3. Using the terms *adhesion, cohesion, meniscus, surface tension, atmospheric pressure*, and *hydrophilic surface*, write a brief essay to explain why water rises up from the water table in a mineral soil.
4. Suppose you were hired to design an automatic irrigating system for a wealthy homeowner's garden. You determine that the flower beds should be kept at a water potential above −60 kPa, but not wetter than −10 kPa, as the annual flowers here are sensitive to both drought and lack of good aeration. The rough turf areas, however, can do well if the soil dries to as low as −300 kPa. Your budget allows either tensiometers or electrical resistance blocks to be hooked up to electronic switching valves. Which instruments would you use and where? Explain.
5. Suppose the homeowner referred to in question 4 increased your budget and asked to use the capacitance method to measure soil water contents. What additional information about the soils, not necessary for using the tensiometer, would you have to obtain to use the capacitance instrument to meet the criteria stated in question 4? Explain.
6. A greenhouse operator was growing ornamental woody plants in 15-cm-tall plastic containers filled with a loamy sand. He watered the containers daily with a sprinkler system. His first batch of 1,000 plants yellowed and died from too much water and not enough air. As an employee of the greenhouse, you suggest that he use 30-cm-tall pots for the next batch of plants. Explain your reasoning.
7. Suppose you measured the following data for a soil:

Horizon	Bulk density, Mg/m³	θ_m at different water tensions, kg water/kg dry soil		
		−10 kPa	−100 kPa	−1,500 kPa
A (0–30 cm)	1.28	0.28	0.20	0.08
Bt (30–70 cm)	1.40	0.30	0.25	0.15
Bx (70–120 cm)	1.95	0.20	0.15	0.05

Estimate the total available water-holding capacity (AWHC) of this soil in centimeters of water.

8. A forester obtained a cylindrical ($L = 15$ cm, $r = 3.25$ cm) core of soil from a field site. She placed all the soil in a metal can with a tight-fitting lid. The empty metal can weighed 300 g and when filled with the field-moist soil weighed 972 g. Back in the lab, she placed the can of soil, with lid removed, in an oven for several days until it ceased to lose weight. The weight of the dried can with soil (including the lid) was 870 g. Calculate both θ_m and θ_v.

9. Give four reasons why compacting a soil is likely to reduce the amount of water available to growing plants.
10. Since even rapidly growing, finely branched root systems rarely contact more than 1 or 2% of the soil particle surfaces, how is it that the roots can utilize much more than 1 or 2% of the water held on these surfaces?
11. For two soils subjected to "no," "moderate," or "severe" compaction, Figure 5.8 shows the volume fraction (cm^3/cm^3) of pores in three size classes. The symbol ◄ indicates the volume fraction of water, θ_v (cm^3/cm^3), in each soil. The figure indicates that $\theta_v \approx 0.35$ for the uncompacted clay loam (bulk density = 0.75 g/cm^3). Show a complete calculation (with all units) to demonstrate that ◄ in the figure correctly indicates that $\theta_v \approx 0.36$ for the severely compacted sandy loam (bulk density = 1.10 g/cm^3).
12. Saturated flow is one of three types of water movement in soils. Infiltration of water into an irrigated furrow is one example of this type of water flow. Name the other two types of water flow in soils and for each type, briefly describe three real-world examples.
13. Fill in the blank cells of this table to show the cm of available water capacity in the entire 90 cm profile. Show complete calculations for the first (*upper left*) and last (*lower right*) of the blank cells.

Soil depth, cm	Bulk density, D_b, (Mg/m^3)	Field capacity θ_m [a]%	Field capacity θ_d, [b]cm	Wilting point θ_m, %	Wilting point θ_d, cm	Available water θ_m, %	Available water θ_d, cm
0–30	1.48	27.1	___	17.9	___	9.2	___
30–60	1.51	27.5	___	18.1	___	9.4	___
60–90	1.55	27.1	___	20.0	___	7.1	___
0–90	—	—	___	—	___	—	___

[a] θ_m is water content by mass and is reported here as percent (%), which is equivalent to g water/100 g dry soil.
[b] θ_d is water content by depth and is reported a mm of water held in the soil layer indicated.

REFERENCES

Bartlett, M. K., C. Scoffoni, and L. Sack. 2012. "The determinants of leaf turgor loss point and prediction of drought tolerance of species and biomes: A global meta-analysis." *Ecology Letters* 15:393–405.

Cooney, J. J., and J. E. Peterson. 1955. *Avocado Irrigation*. Leaflet 50. California Agricultural Extension Service, Davis, CA.

Hillel, D. 2004. *Introduction to environmental soil physics*. Elsevier Science, New York. p. 494.

Kladivko, E. J., et al. 1999. "Pesticide and nitrate transport into subsurface tile drains of different spacing." *Journal of Environmental Quality* 28:997–1004.

Moroke, T. S., R. C. Schwartz, K. W. Brown, and A. S. R. Juo. 2005. "Soil water depletion and root distribution of three dryland crops." *Soil Science Society of America Journal* 69:197–205.

Rawls, W. J., D. L. Brakensiek, and K. E. Saxton. 1982. "Estimation of soil water properties." *Transactions of the ASAE* 25:1316–1320, 1328.

Rawls, W. J., A. Nemes, Y. Pachepsky. 2004. "Effect of soil organic carbon on soil hydraulic properties." Vol. 30. *Developments in soil science*. Elsevier, New York, pp. 95–114.

Schenk, H., and R. Jackson. 2002. "The global biogeography of roots." *Ecological Monographs* 72:311–328.

Schwärzel, K., M. Renger, R. Sauerbrey, and G. Wessolek. 2002. "Soil physical characteristics of peat soils." *Journal of Plant Nutrition and Soil Science* 165:479–486.

Shestak, C. J., and M. D. Busse. 2005. "Compaction alters physical but not biological indices of soil health." *Soil Science Society of America Journal* 69:236–246.

Waddell, J. T., and R. R. Weil. 1996. "Water distribution in soil under ridge-till and no-till corn." *Soil Science Society of America Journal* 60:230–237.

Warrick, A. W. 2001. *Soil Physics Companion*. CRC Press, Boca Raton, FL. p. 400.

6
Soil and the Hydrologic Cycle

Water cycles from soil to clouds and back in the Teton Mountains. (Photo courtesy of Ray R. Weil)

Both soil and water belong to the biosphere, to the order of nature, and—as one species among many, as one generation among many yet to come— we have no right to destroy them.
—DANIEL HILLEL, OUT OF EARTH

One of the most striking—and troubling—features of human society is the yawning gap in wealth between the world's rich and poor. The life experience of the one group is quite incomprehensible to the other. So it is with the distribution of the world's water resources. The rain forests of the Amazon and Congo basins are drenched by more than 2,000 mm of rain each year, while the deserts of North Africa and central Asia get by with less than 200. South America and the Caribbean receive nearly one-third of the annual global precipitation, while Australia receives only 1%. Nor is the supply of water distributed evenly throughout the year. Rather, periods of high rainfall and flooding alternate with dry spells or periods of drought.

Yet, one could say that everywhere the supply of water is adequate to meet the needs of the plants and animals native to the natural communities of the area. Of course, this is so only because the plants and animals have adapted to the local availability of water. Early human populations, too, adapted to the local water supplies by settling where water was plentiful from rain or rivers, by learning which gourds and plant stems could quench one's thirst, by developing techniques to harvest water for agriculture and store it in underground cisterns, and by adopting nomadic lifestyles that allowed them and their herds to follow the rains and the grass supply.

But "civilized" humans have not been willing to adapt their needs and cultures to their environment. Rather, they have joined in organized efforts to adapt their environment to their desires. Hence, the ancients tamed the flows of the Tigris and Euphrates. We moderns dig wells in the Sahel, bottle up the mighty Nile at Aswan, channel the waters of the Colorado to the chaparral region of southern California, pump out the aquifers under farms and suburbs, and create sprawling cities (with swimming pools and bluegrass lawns!) in the deserts of the American Southwest or on the sands of Arabia. Truly, cities like Las Vegas are gambling in more ways than one.

There is plenty of room for improvement in managing water resources, and many of the improvements are likely to come as a result of better management of soils. The soil plays a central role in the cycling and use of water. For instance, by serving as a massive reservoir, soil helps moderate the adverse effects of excesses and deficiencies of water. It takes in water during times of surplus and then releases it in due time, either to satisfy the transpiration (T) requirements of plants or to replenish groundwater. Stable structure at the surface of the soil ensures that a large fraction of the precipitation received will move slowly into the groundwater and from there to nearby streams or to deeper reservoirs under the earth. The soil can help us purify and reuse wastewaters from animal, domestic, and industrial sources. The flow of water through the soil in these and other circumstances connects the chemical pollution of soils to the possible contamination of groundwater.

In Chapter 5, we considered the nature and movement of water in soils. In this chapter, we will see how those characteristics apply to the management of water as it cycles between the soil, atmosphere, and vegetation. We will then examine the unique role of the soil in water resource management.

6.1 THE GLOBAL HYDROLOGIC CYCLE

Global Stocks of Water

There are nearly 1,400 million km^3 of water on Earth—enough (if it were all aboveground at a uniform depth) to cover the Earth's surface to a depth of some 3 km. Most of this water, however, is saline and/or inaccessible and is not active in the annual cycling of water that supplies rivers, lakes, and living things. Almost 97% is found in oceans where the water is not only salty, but has an average **residence time** of several thousand years. Only the near-surface ocean layers take part in annual water cycling. An additional 1.7% of the water is in glaciers and ice caps of mountains with similarly long residence times (about 10,000 years). Some 1.7% is found in groundwater, most of which is more than 750 m underground and about half of which is saline. Except where it is pumped by humans, it, too, has a long average residence time measured in centuries.

The water with shorter residence times (that which cycles more actively) is in the surface layer of the oceans, in shallow groundwater, in lakes and rivers, in the atmosphere, and in the soil. Although the combined volume is a tiny fraction of the water on Earth, these pools of water are accessible for movement in and out of the atmosphere and from one place on Earth's surface to another. The average residence time for water in the atmosphere is about 10 days, that for the longest rivers is 20 days or less, and that for soil moisture is about one month.

The Hydrologic Cycle[1]

Solar energy drives the cycling of water from Earth's surface to the atmosphere and back again in what is termed the *hydrologic cycle* (Figure 6.1). About one-third of the solar energy that reaches Earth is absorbed by water, stimulating *evaporation* (E)—the conversion of liquid water into water vapor. The water vapor moves up into the atmosphere, eventually forming clouds that can move from one region of the globe to another. Within an average of about 10 days,

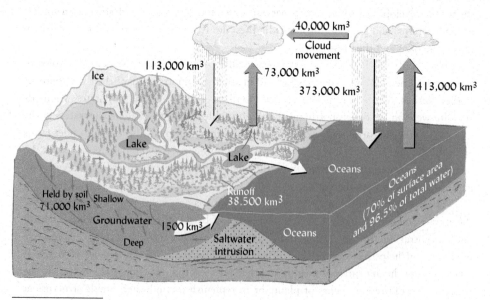

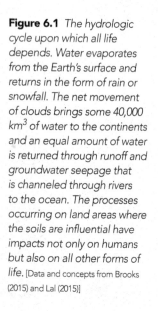

Figure 6.1 *The hydrologic cycle upon which all life depends. Water evaporates from the Earth's surface and returns in the form of rain or snowfall. The net movement of clouds brings some 40,000 km^3 of water to the continents and an equal amount of water is returned through runoff and groundwater seepage that is channeled through rivers to the ocean. The processes occurring on land areas where the soils are influential have impacts not only on humans but also on all other forms of life.* [Data and concepts from Brooks (2015) and Lal (2015)]

---

[1]For an in-depth review of the global water cycle, see Oki (2011). For assessments of environmental pressures related to global water use, see Jeswani and Azapagic (2011) and Pfister et al. (2011).

pressure and temperature differences in the atmosphere cause the water vapor to condense into liquid droplets or solid particles, which return to the Earth as rain or other precipitation.

About 483,000 km^3 of water are evaporated from the Earth's surfaces and vegetation each year, some 113,000 km^3 of which falls as rain or snowfall on the continents. Some of the water falling on land runs off the surface of the soil, and some infiltrates into the soil. About 65% of the water falling on land is bound as soil moisture, most of which is eventually transpired by plants or drains into the groundwater. Both the surface runoff and groundwater seepage enter streams and rivers that, in turn, flow into the oceans. The volume of water returned in this way is about 40,000 km^3, which balances the same quantity of water that is transferred annually in clouds from the oceans to the continents.

Water Balance Equation

It is often useful to consider the components of the hydrologic cycle as they apply to a given **watershed**, an area of land drained by a single system of streams and bounded by ridges that separate it from adjacent watersheds (Figure 6.2). All the precipitation falling on a watershed is either stored in the soil, returned to the atmosphere (see Section 6.2), or discharged from the watershed as surface or subsurface runoff. Water is returned to the atmosphere either by **evaporation** from the land surface (vaporization of water from the surfaces of soil, rock, or leaves) or, after plant uptake and use, by vaporization through the pores (called stomata) in plant leaves (a process termed **transpiration**). Together, these two pathways of evaporative loss to the atmosphere are called **evapotranspiration** (ET).

Water Balance. The disposition of water in a watershed is often expressed by the water-balance equation, which in its simplest form is:

$$P = ET + SS + D \qquad (6.1)$$

where P = precipitation, ET = evapotranspiration, SS = soil storage, and D = discharge (Figure 6.2).

Rearranging Eq. (6.1) (D = P − ET − SS) shows that discharge can be increased only if ET and/or SS are decreased, changes that may or may not be desirable. For a forest watershed

Figure 6.2 A watershed with a visualization of the water balance equation showing how discharge D (by surface and below ground runoff to streams) is controlled by the proportion of incoming precipitation P that is evapotranspired (ET) from leaf and land surfaces or stored by soils (SS). The yellow arrows represent the general direction of water flow and the dotted yellow line represents the boundary of one particular watershed in a landscape near San Francisco, California. A dam has been built across the outlet of the watershed to impound runoff from snowmelt and rainfall for use by the city's residents. After being used for various purposes, the runoff will eventually find its way to the Pacific Ocean in the background. (Courtesy of Ray R. Weil)

(sometimes termed a *catchment*) managers may aim to maximize D so as to provide more water to downstream users. Clear-cutting the trees will almost certainly decrease ET and therefore increase discharge. In the case of an irrigated field, water applied in irrigation would be included on the left side of Eq. (6.1). Irrigation managers may want to save water applied by minimizing unnecessary losses in D and allowing negative values for SS (withdrawals of soil storage water) during parts of the year (see Section 6.9).

6.2 FATE OF INCOMING WATER

In most climates, a significant portion of incoming precipitation is intercepted by plant foliage and returned to the atmosphere by evaporation without ever reaching the soil. Generally, the proportion of precipitation so intercepted increases with higher temperature and decreases with greater humidity and intensity of rainfall. In some forested areas, **interception** may prevent 30–50% of the precipitation from reaching the soil. Interception and subsequent sublimation (vaporization directly from the solid state) of snow are especially important in coniferous forests.

When water from rain or irrigation reaches the soil surface, it encounters a very important "fork in the road" along the hydrologic cycle. On the one hand, water may follow the path of **infiltration** and enter the soil itself, becoming soil water in the process. Water that reaches the ground is more likely to take this path and infiltrate if the soil surface structure is loose and open. On the other hand, water may take the path of overland flow across the soil surface and become **runoff**, promoting stream flow, but not plant growth and ecosystem productivity.

If the rate of rainfall, irrigation, or snowmelt exceeds the soil's infiltration capacity, some ponding may result and considerable runoff and erosion may take place, thereby reducing the proportion of water that moves into the soil. In some cases, more than 50% of the precipitation may be lost as **surface runoff**, which usually carries with it dissolved chemicals, nutrients, and detached soil particles (**sediment**; see Chapter 14).

Once water penetrates into the soil, some of it is subject to downward percolation and eventual loss from the root zone by **drainage**. In humid areas and on irrigated land, up to 50% of the water input may be lost as drainage below the root zone. However, during subsequent periods of low rainfall, some of this water may move back up into the plant-root zone by the rise of **capillary water**. Such movement is important to plants growing in deep soils, especially in dry climates.

The water retained by the soil is referred to as **soil storage** water, some of which eventually moves upward by capillarity and is lost by evaporation from the soil surface. Much of the remainder is absorbed by plants and then moves through the roots and stems to the leaves, where it is lost by transpiration. The water thus lost to the atmosphere by evapotranspiration may later return to the soil as precipitation or irrigation water, and the cycle starts again.

Timing of precipitation greatly influences the amount of water moving through each of the pathways just discussed. Heavy rainfall, even if of short duration, can supply water faster than most soils can absorb it. This accounts for the fact that in some arid regions with only 200 mm of annual rainfall, a cloudburst that brings 20–50 mm of water in a few minutes can result in flash flooding and gully erosion. A larger amount of precipitation spread over several days of gentle rain could move slowly into the soil, thereby increasing the stored water available for plant absorption, as well as replenishing the underlying groundwater. As illustrated in Figure 6.3, the timing of snowfall in early winter can affect the partitioning of spring snowmelt water between surface runoff and infiltration. A blanket of snow is good insulation and may keep the soil from freezing.

Effects of Vegetation and Soils on Infiltration

Type of Vegetation. Plants help determine the proportion of water that runs off versus infiltrates the soil. The vegetation and surface residues of perennial grasslands and dense forests protect the porous soil structure from the beating action of raindrops. Therefore, they encourage water infiltration and reduce the likelihood that soil will be carried off by any runoff that does occur. In general, very little runoff occurs from land under undisturbed forests or well-managed turfgrass. However, differences in plant species, even among grasses, can influence runoff.

Stem Flow. Many plant canopies direct rainfall toward the plant stem, thus altering the spatial distribution of rain reaching the soil. Under a forest canopy, more than half of the rainfall

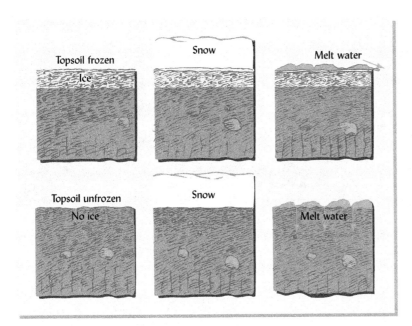

Figure 6.3 The relative timing of freezing temperatures and snowfall in the fall in some temperate regions drastically influences water runoff and infiltration into soils in the spring. The upper three diagrams illustrate what happens when the surface soil freezes before the first heavy snowfall. The snow insulates the soil so that it is still frozen and impermeable as the snow melts in the spring. The lower sequence of diagrams illustrates the situation when the soil is unfrozen in the fall when it is covered by the first deep snowfall.

may trickle down the leaves, twigs, and branches to the tree trunk, there to progress downward as **stem flow**. Likewise, certain crop canopies, such as that of corn, funnel a large proportion of the rainfall to the soil in the crop row (Figure 6.4). The concentration of water in limited zones around plant stems increases the opportunity for saturated flow to occur. Stem flow must be considered in studying the hydrology and nutrient cycling in many plant ecosystems.

Soil Management. Encouraging infiltration rather than runoff is usually a major objective of soil and water management. One approach is to allow more time for infiltration to take place by enhancing soil surface storage (Figure 6.5, *left*). A second approach is to maintain dense vegetation during periods of high rainfall. For example, **cover crops**, plants established between the principal crop-growing seasons, can greatly enhance water infiltration by creating open root channels,

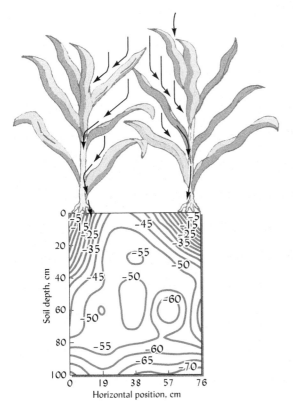

Figure 6.4 Vertical and horizontal distribution of soil water resulting from stem flow. The contours indicate the soil water potential in kPa between two corn rows in a sandy loam soil. During the previous two days, 26 mm of rain fell on this field. Many plant canopies, including that of the corn crop shown, direct a large proportion of rainfall toward the plant stem. Stem flow results in uneven spatial distribution of water in the soil. In cropland, this may have ramifications for the leaching of chemicals such as fertilizers, depending on whether they are applied in or away from the zone of highest wetting near the plant stems. The concentration of water by stem flow may also increase the likelihood of macropore flow in soils after a moderate rainfall. [Data from Waddell and Weil (1996); used by permission of the Soil Science Society of America]

Figure 6.5 Managing soils to increase infiltration of rainwater. (Left) Small furrow dikes on the left side of this field in Texas retain rainwater long enough for it to infiltrate rather than run off. (Right) This saturated soil under a winter cover crop of hairy vetch is riddled with earthworm burrows that greatly increased the infiltration of water from a recent heavy rain. Scale in centimeters. (Photo (left) courtesy of O. R. Jones, USDA Agricultural Research Service, Bushland, Texas; photo (right) courtesy of Ray R. Weil)

encouraging earthworm activity, and protecting soil surface structure (Figure 6.5, *right*). However, remember that cover crops also transpire water. If the following crop will be dependent on soil storage water, care may be needed to kill the cover crop before it can dry out the soil profile.

A third approach is to maintain soil structure by minimizing compaction, whether by people in parks, heavy equipment on farm fields, or cattle on rangeland (see Section 14.9). For example, soil compaction by heavy forestry equipment used to skid logs or clear forestland for agricultural use can seriously impair soil infiltration capacity and watershed hydrology. While the ill effects of compaction can sometimes be partially overcome by deep tillage (see Section 4.7), environmental stewardship requires that caution be exercised in the use of heavy equipment on any landscape and that disturbance of the natural soil and vegetation be completely avoided on as much of the land as possible.

Urban Watersheds. Built-up landscapes are prime examples where soil compaction affects watershed hydrology. The use of heavy equipment to prepare land for urban development can severely curtail soil infiltration capacity and saturated hydraulic conductivity (Figure 6.6). Soil compaction by construction activity therefore results in drastically

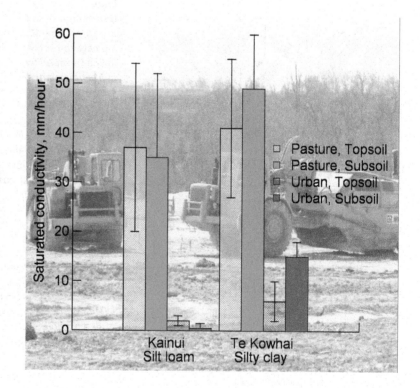

Figure 6.6 Grading and excavating land in preparation for urban/suburban development can greatly reduce soil permeability and hydraulic conductivity. The impairment results from compaction, damage to soil structure, and profile truncation (removal of A horizon). Both topsoil and subsoil horizons in these two New Zealand soils suffered three- to tenfold reductions in saturated hydraulic conductivity when the land was converted from a pasture to a housing development. The vertical lines indicate the variation among repeated measurements in the same soil, this variation being especially large in the pasture soils because of the random presence of earthworm channels. [Data from Zanders (2001), used with permission from Landcare Research NZ Ltd.; photo courtesy of Ray R. Weil]

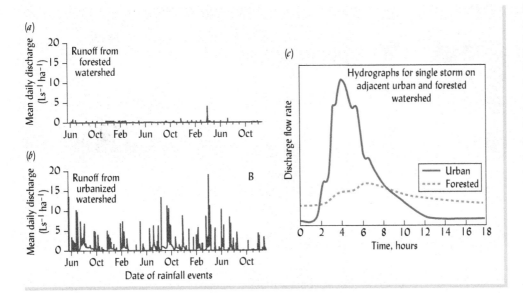

Figure 6.7 Hydrographs showing much greater, flashier, and more frequent runoff events from an urbanized watershed (b) compared to an adjacent forested watershed (a) subject to the same rainfall events near Atlanta, Georgia. (c) A representative single storm hydrograph showing the difference in intensity of discharge and level of base flow from adjacent urbanized and forested watersheds. [Graphs (a) and (b) Based on Nagy et al. (2011), graph (c) courtesy of Ray R. Weil]

increased surface runoff during storms (Figure 6.7). The runoff burden carried by streams in urbanized watersheds is further increased because a large proportion of the land is covered with completely impermeable surfaces (rooftops, paved streets, and parking lots). Erosion of stream banks, toppling of trees, scouring of streambeds, and exposure of once-buried pipelines (Figure 6.8) are typical signs of the environmental degradation that ensues as streams become overwhelmed. Damage to streams in urban watersheds is accentuated because storm sewers and street gutters rush excess water off the land, requiring the stream to carry a huge volume of runoff concentrated in a short period of time. In recognition of the severe environmental problems caused by such excessive and concentrated runoff, urban planners and engineers are now working with soil scientists to reduce disruptions of the hydrologic cycle by urban development. Features such as permeable pavers that allow some infiltration even in parking lots (Figure 6.9, *left*) and rain gardens that catch runoff and release it slowly by infiltration and percolation (Figure 6.9, *right*) are part of what is termed *low impact* urban design.

Soil Properties. Inherent soil properties also affect the fate of precipitation. If the soil is loose and open (e.g., sands and well-granulated soils), a high proportion of the incoming water will infiltrate the soil, and relatively little will run off. In contrast, heavy clay soils with

Figure 6.8 Signs of excessive runoff from urbanized land. (Left) Exposure of a once-buried storm sewer manhole along a highly degraded stream that collects runoff from part of Baltimore, Maryland. The drastically increased storm runoff volume from the largely impermeable urban watershed has caused the stream to erode its banks and deeply cut into the hillside. The round manhole cover indicates the former land surface. (Right) The toppled tree and scoured soil are signs that this placid-appearing small stream is overwhelmed by raging torrents of water during heavy rainfalls because the urbanized watershed conducts the water more directly and rapidly to the stream than did the natural watershed that formed the stream. (Photo courtesy of Ray R. Weil)

Figure 6.9 Two methods of increasing infiltration and slowing runoff in urbanized watersheds. (Left) Permeable pavers allow grass to grow and water to infiltrate a parking lot while cars can still park without compacting soil or forming mud. (Right) Water entering one of several small rain gardens that captures runoff from a suburban parking lot. The water is directed to a small depression. The small pond that forms is ephemeral, holding the runoff water only temporarily. Permeable soil placed under the depression along with buried drain pipes are designed to allow the water to infiltrate and seep away over a period of a few hours. The depression is planted with a variety of native plants that provide both beauty and wildlife habitat. (Photo courtesy of Ray R. Weil)

unstable structures or soils subjected to compaction resist infiltration and encourage runoff (see Figure 4.33. Other factors that influence the balance between infiltration and runoff include the slope of the land (steep slopes favoring runoff over infiltration; see Box 2.1), the type and completeness of vegetative cover, and the presence of impermeable layers within the soil profile. Such impermeable layers as fragipans (see Figure 3.34) and clay pans can restrict infiltration and increase surface runoff once the upper horizons become saturated, even if the surface soil has intact structure and high porosity (Figure 6.10).

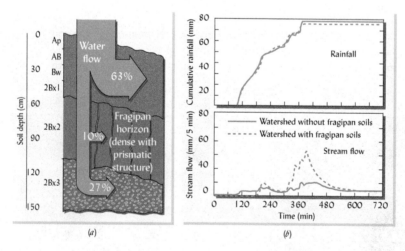

Figure 6.10 Soil profile characteristics largely determine the vertical and lateral movement of water. In this example, a fragipan (left, also see Figure 3.34) provides a barrier to both root growth and downward percolation of water. The zone above the fragipan may become saturated during wet weather, giving rise to a perched water table. As more rain falls, little can enter the already saturated surface horizons. Most of the water flows laterally, either through the soil above the fragipan or as surface runoff. The graphs (right) show the cumulative rainfall (upper) and volume of stream flow (lower) during and after a storm in which nearly 80 mm of rain fell in a 300-minute period. Of the two small (13–20 ha) watersheds represented, one has fragipan-containing soils (Fragiochrepts) formed in colluvium near the stream channel; the other watershed has no such fragipans. Although the rainfall was almost identical for the two nearby watersheds, stream flow (both the maximum and the overall volume) was much greater for the watershed with fragipans. [Soil profile based on data in Day et al. (1998); graphs from Gburek et al. (2006), with permission of Elsevier Science, Oxford, United Kingdom]

The burrowing and nesting activities of certain soil animals that produce a network of macropores connected to the soil surface can also enhance infiltration, reduce runoff losses, and thereby improve water availability for plants. Such enhancement of soil hydrologic function by earthworms is a well-known phenomenon in humid temperate regions. Ants and termites can similarly increase soil productivity in dry tropical regions. More information on the activities of soil animals can be found in Sections 11.4 and 11.5.

All the soil and plant factors just discussed can result in some parts of a landscape contributing more runoff than others. Such spatial heterogeneity of infiltration and runoff is particularly important to ecosystem function in dry regions (see Box 10.1).

6.3 THE SOIL–PLANT–ATMOSPHERE CONTINUUM (SPAC)[2]

The flow of water through the SPAC is a major component of the overall hydrologic cycle. Figure 6.12 ties together many of the processes we have just discussed: *interception, surface runoff, percolation, drainage, evaporation, plant water uptake, ascent of water to plant leaves,* and *transpiration* of water from the leaves back into the atmosphere.

Water Potentials. In studying the SPAC, scientists have discovered that the same basic principles govern the retention and movement of water, whether it is in soil, in plants, or in the atmosphere. In Chapter 5, we learned that water in soil moves to where its potential energy level will be lower. This principle applies to water movement between the soil and the plant root and between the plant and the atmosphere (see Figure 6.11). If a plant is to absorb water from the soil, the water potential must be lower (greater negative value) in the plant root than in the soil adjacent to the root. Likewise, movement up the root and stem to the leaf cells is in response to differences in water potential, as is the movement from leaf surfaces to the atmosphere. To illustrate the movement of water to sites of lower and lower water potential, Figure 6.11 shows that water potentials might drop from −50 kPa in the soil, to −70 kPa in the root, to −500 kPa at the leaf surfaces, and, finally, to −20,000 kPa in the atmosphere.

Evapotranspiration

While it is relatively easy to measure the total change in soil water content due to vapor losses, it is quite difficult to determine just how much of that loss occurred directly from the soil (by E) and how much occurred from the leaf surfaces after plant uptake (by T). Therefore, information is most commonly available on ET, the combination of these two processes. The evaporation component of ET may be viewed as a "waste" of water from the standpoint of plant productivity. On the other hand, at least some of the transpiration component is essential for plant growth, providing the water that plants need for cooling, nutrient transport, photosynthesis, and turgor maintenance.

The **potential evapotranspiration** (PET) rate tells us how fast water vapor *would* be lost from a densely vegetated plant–soil system *if* soil water content were continuously maintained at an optimal level. The PET is largely determined by the *vapor pressure gradient* between a wet soil, leaf, or body of water and the atmosphere. This gradient is in turn influenced by solar radiation and such related climatic variables as temperature, relative humidity, cloud cover, and wind speed.

A number of mathematical models have been devised to estimate PET from climatological data, but in practice, PET can be most easily estimated by applying a correction factor to the amount of water evaporated from an open pan of water of standard design (a class A evaporation pan). Because of the resistances to water flow just mentioned, water loss by transpiration from well-watered, dense vegetation is typically only about 65% as rapid as loss by evaporation from an open pan; hence, the correction factor for dense vegetation such as a lawn is typically 0.65 (it is lower for less dense vegetation):

$$\text{PET} = 0.65 \times \text{pan evaporation} \qquad (6.2)$$

---

[2] The physics of water movement in plants has been quite controversial at times. For strong evidence that confirms that water moves up tall trees by the same capillary forces that control its movement in soil, see Tyree (2003).

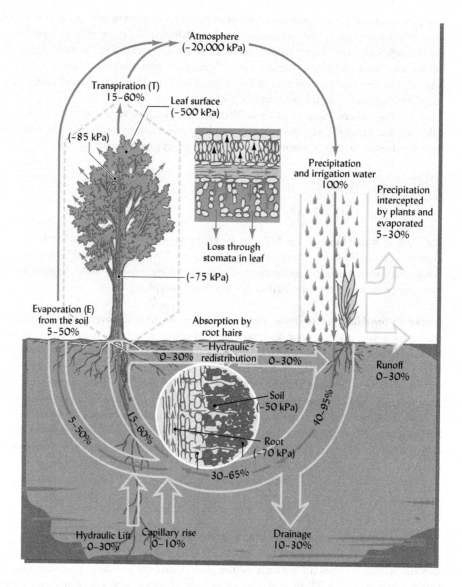

Figure 6.11 *Soil–plant–atmosphere continuum (SPAC) showing water movement from soil to plants to the atmosphere and back to the soil. As described for soil water movement in Chapter 5, water in the entire SPAC moves from a higher to a lower moisture potential. Note that the moisture potential in the soil is −50 kPa, dropping to −70 kPa in the root, declining still further as it moves upward in the stem and into the leaf, and is very low (−500 kPa) at the leaf–atmosphere interface, from whence it moves into the atmosphere, where the moisture potential is −20,000 kPa. Note the suggested ranges for partitioning of the precipitation and irrigation water as it moves through the continuum. Most of the water absorbed by plant roots is eventually transpired as water vapor, but a significant proportion may be exuded by roots into soil zones that are drier than the zones from which the water was initially taken up—a process termed hydraulic redistribution.* (Diagram courtesy of R. Weil and N. Brady)

Values of PET range from more than 1500 mm per year in hot, arid areas to less than 40 mm in very cold regions. During the winter in temperate regions, PET may be less than 1 mm per day. By contrast, hot, dry wind will continually sweep away water vapor from a wet surface, creating a particularly steep vapor pressure gradient and PET levels as high as 10–12 mm per day.

Effect of Soil Moisture. Evaporation from the soil surface at a given temperature is determined largely by soil surface wetness and by the ability of the soil to replenish this surface water as it evaporates. In most cases, the upper 15–25 cm of soil provides most of the water for surface evaporation. Unless a shallow water table exists, the upward capillary movement of water is very limited and the surface soil soon dries out, greatly reducing further evaporation loss.

However, because plant roots penetrate deep into the profile, a significant portion of the water lost by evapotranspiration can come from the subsoil layers. As discussed in Chapter 5 (see Figure 5.25), native vegetation rooting depths suggest that water stored deep in the profile is especially important in regions having alternating moist and dry seasons (such as Ustic or Xeric moisture regimes). Water stored in the subsoil during rainy periods is available for evapotranspiration during dry periods. Vegetation may go dormant—or die—because of the inability of a shallow soil to hold sufficient water during prolonged dry periods (see Figure 5.24 illustrating the death of a rooftop lawn).

Plant Water Stress. For dense vegetation growing in soil well supplied with water, ET will nearly equal PET. When soil water content is less than optimal, the plant will not be able to withdraw water from the soil fast enough to satisfy PET. If water evaporates from the leaves faster than it enters the roots, the leaves will lose turgor pressure and wilt. Under these conditions, actual evapotranspiration is less than potential evapotranspiration and the plant experiences *water stress*. The difference between PET and actual ET is termed the *water deficit*. A large deficit is indicative of high water stress and aridity.

Under water-stress conditions, plants first close the *stomata* (openings) on their leaf surfaces to reduce the vapor loss of water and prevent wilting or possible cell collapse. However, the closure of the stomata has two detrimental side effects: (1) plant growth is arrested because CO_2 for photosynthesis cannot pass through the closed stomata, and (2) the reduction in evaporative cooling results in detrimental *heating* of leaves as they continue to absorb solar radiation. The latter effect allows infrared-detecting instruments to estimate water stress in plants by sensing the increase in leaf temperature over air temperature. Stomatal closure in response to low leaf water potentials may also be "too little, too late" for plant survival.

Plant Characteristics. In vegetated areas, as the leaf area per unit land area (a ratio termed the **leaf area index** (LAI)) increases, more radiation will be absorbed by the foliage to stimulate transpiration and less will reach the soil to promote evaporation. For a monoculture of annual plants, the LAI value typically varies from 0 at planting to a peak of perhaps 3–5 at flowering, then declines as the plant senesces, and finally drops back to 0 when the plant is removed at harvest (assuming no weeds are allowed to grow). In contrast, perennial vegetation, such as in pastures and forests, has very high leaf area indices both early and late in the growing season. Where leaf litter has accumulated on the forest floor, very little direct sunlight strikes the soil, and evaporation is very low throughout the year. Other plant characteristics, including rooting depth, length of life cycle, and leaf morphology, can influence the amount of water lost by evapotranspiration over a growing season. Figure 6.12 illustrates the comparative competitiveness of mature trees and annual crops for soil water supplies.

Water-Use Efficiency. Water-use efficiency may be expressed in terms of plant dry matter produced per unit of water transpired (*T efficiency*) or per unit of water lost by evapotranspiration (*ET efficiency*). Agriculturalists commonly express water-use efficiency as kg of grain produced per m^3 of water used in evapotranspiration. For irrigation systems analysis, water-use efficiency may consider additional aspects such as losses of water in storage reservoirs or leakage from canals (see Section 6.10).

Whatever the units of expression, data from studies around the world show that huge amounts of water are needed to produce the human food supply and that water-use use efficiency is largely driven by climatic factors. In arid regions, crops may use 1,000 to more than 5,000 kg of water to produce a single kg of grain. Ironically, in more humid regions where water and

Figure 6.12 *Water-stressed soybeans show the extent of tree root competition for water in the surface soil.* (Photo courtesy of Ray R. Weil)

rain are plentiful, much less water is needed for each kg of grain produced because the evaporative demand is far lower. When we take into account the water used to grow grains, fruits, vegetables, and feed for cattle, almost 7,000 L (1,700 gal) of water are required to grow a *single day's* food supply for one adult in the United States!

ET Efficiency. Since evapotranspiration includes both transpiration from plants and evaporation from the soil surface, evapotranspiration efficiency is more subject to management than transpiration efficiency. Highest ET efficiency is attained where plant density and other growth factors minimize the proportion of ET attributable to evaporation from the soil. A positive relationship is generally found between crop yields and water-use efficiency. Higher yielding plants generally send their roots deeper into the soil profile and produce a denser canopy (higher LAI) that allows less solar radiation to pass through and cause evaporation from the soil surface. We can conclude that as long as the supply of water is not too limited, maintaining optimum conditions for plant growth (by closer plant spacing, fertilization, or selection of more vigorous varieties) increases the efficiency of water use by plants. If irrigation water is not available and the period of rainfall is very short, one should apply this principle with caution. The increase in evapotranspiration by more vigorously growing plants may deplete stored soil water, resulting in serious water stress or even plant death before any harvestable yield has been produced!

Hydraulic Redistribution.[3] Plant roots not only remove water from the soil, but they also can release water back into the soil. During the day when leaf stomata are open and atmospheric conditions favor evaporation, transpiration water losses through the leaves generate low soil water potentials (large negative values) inside the plant stem and root cells. As the water potential in the root becomes lower than that in the adjacent soil, water moves into the roots and is taken up by the plant. In turn, as water is removed from the soil, the potential of the remaining soil water decreases (exhibits larger negative values). During the night when atmospheric humidity is high and the leaf stomata are closed, the process is reversed. Without the "pull" of transpiration, the water potential in the roots is increased, increasing turgor pressure inside the plant cells. As the water potential inside the roots become higher than that in the surrounding soil, water begins to move from the root out into the soil where its potential will be lower (Figure 6.13).

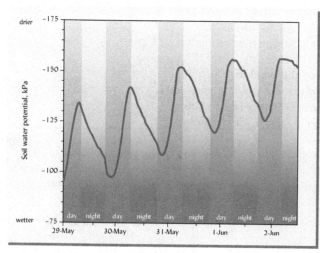

Figure 6.13 (Left) The daily fluctuations of soil water tension measured at 50-cm depth in the root zone of a corn crop reflect the cycle of water uptake from the soil during the day and exudation into the soil during the night. Note that the overall trend during the 5-day rainless period shown is one of soil drying (increasing tension = decreasing potential). (Right) Water is being exuded by a cut plant stem in another soil. The plant is growing in soil that has high water content at depth, but has dried out in the surface horizons. The plant stem was cut on a cool, cloudy day with low evaporative demand such that a high (positive) water potential had built up in the cells. The plant's shallow roots can be expected to also exude water into the surface soil, though more slowly than is the case for the cut stem. [Graphed from data of Chen and Weil (2011). Photo courtesy of Ray R. Weil]

[3]For a review of ecosystem implications of hydraulic redistribution of soil water by plants, see Prieto et al. (2012).

Recent research by soil and plant scientists has clearly shown that the mechanisms just described allow plant roots to serve as conduits for moving significant amounts of water from moist soil zones to dry soil zones, a process termed **hydraulic redistribution**. The importance of vertical hydraulic redistribution, or **hydraulic lift**, has been especially well documented for woody plants. In contrast to the competitive situation shown in Figure 6.12, deep-rooted trees can transfer water from moist soil layers many meters deep to surface soil layers. During long dry periods, this added water ensures the survival and growth of shallower rooted understory grass and herbaceous vegetation as well as the functioning of soil microbial systems that improve plant growth, including that of the trees. For example, sage brush, a deep-rooted woody plant found in the deserts of south western North America, conducts enough hydraulic redistribution to enhance the microbial mineralization of nitrogen from organic matter in surface soil and supply one-third to one-half of the water used by relatively shallow-rooted crested wheatgrass growing nearby.

Research in Japan and elsewhere suggests that the process of hydraulic lift may be manipulated to create "self-irrigating" agroecosystems in which rows of deep-rooted "water donor" plants alternate with water-receiving areas planted to shallow rooted but high-value vegetable crops.

In summary, water losses from the soil surface and from transpiration are determined by: (1) climatic conditions, (2) plant cover in relation to soil surface (LAI), (3) efficiency of water use by different plants and management regimes, and (4) length and season of the plant-growing period. In addition, the ability of plant roots to redistribute water from moist soil to drier soil can alter natural vegetation communities and provide enough water to shallow-rooted species to ensure their survival and possibly improve crop production.

6.4 CONTROL OF ET

Because transpiration is closely related to the total leaf area exposed to solar radiation and evaporation is related to the amount of soil surface so exposed, measures taken to reduce these exposures can bring ET into closer balance with the available water supply and may, to some degree, lessen water stress on desirable vegetation.

Control of Transpiration

In various agricultural situations it may be necessary to limit transpiration by desirable plants (e.g., crops) and/or unwanted (weedy) vegetation. Where rapid crop growth might prematurely deplete the water available, it may be wise to limit plant growth factors such as nutrient supply to only moderate levels, thus keeping LAI in check. The LAI of desired plants can also be limited by spacing plants farther apart, as nature does in arid environments. It should be noted, however, that plants growing farther apart tend to individually produce a greater leaf area, compensating somewhat for the lower density. Also, wider spacing allows greater evaporation losses from the soil.

Unwanted Vegetation. It is largely through their heavy use of soil water that weeds interfere with the establishment and growth of desirable forest, range, and crop plants. Weeds in cropland have traditionally been controlled by cultivation (light soil tillage) designed to uproot young weeds and leave them to dry out and die. Disadvantages of cultivation include likely damage to roots of nearby desired plants and the exposure of bare soils, which in turn increases evaporative water loss and eventually encourages runoff and erosion.

Herbicides. Use of weed-killing chemicals called herbicides (see Section 15.2) provides several advantages over cultivation, among which are that it requires less labor and energy, and it allows the soil to be left undisturbed, with plant residues covering the soil surface (see below). Chemical weed control also has serious disadvantages in some situations, including high material costs, eventual evolution of weed resistance to specific compounds, damage to desired plants (including subsequently planted crops or cover crops), and risks of environmental toxicity. Herbicides may be toxic to soil organisms, fish, or land animals, including carcinogenicity to humans. In most cases, herbicides have accumulated to undesirable levels in downstream waterways and underground drinking water sources (see Section 15.3).

Alternative Weed Controls. Consequently, alternative weed control methods are being sought. For example, some weeds can be held in check by biological controls, that is, encouraging specific insects or diseases that attack only the weeds. Undesirable vegetation in rangeland and some forests can be suppressed by prescribed fires if the timing and temperature levels are carefully controlled to minimize damage to desirable vegetation. Well-timed mowing or grazing can also reduce weed problems in pastures and lawns.

Although cover crops also transpire water they can be used to reduce weed pressure in subsequent crops through competitive and chemical effects on weed seed germination. In most cases, the cover crops are terminated and weeds controlled with the help of herbicides. Certain cover crops can be rolled down and killed mechanically to produce a thick weed-smothering mat of residues. These cover crops can be combined with weed control using new implements designed to cultivate through the high levels of plant residue by undercutting the roots of emerging weeds without disturbing the soil surface. Such techniques hold promise for controlling weeds in humid regions without herbicides while still maintaining the desirable surface residue cover (Figure 6.14).

Fallow in Dryland Cropping. Farming systems that alternate bare **fallow** (unvegetated period) one year with traditional cropping the next have been used to conserve soil moisture in some low-rainfall environments. Transpiration water losses during the fallow year are minimized by controlling weeds with light tillage and/or with herbicides. Some of the water saved during the fallow year remains in the profile the following year when a crop is planted, resulting in higher yields of that second-year crop than if the soil had been cropped every year. This cropping practice is responsible for the checkerboard of dark fallow soils and golden ripening wheat that can be seen when flying over such semiarid "breadbaskets" as the Great Plains region of the United States and Canada.

Figure 6.14 Cover crops and new mechanical weed control implements can help minimize vapor losses of water from row-crop soil while planting the crop without tillage and without herbicides. (a) Mechanical termination of a rye cover crop using a roller-crimper mounted on the front of the tractor to avoid leaving rye unkilled in the tire tracks. A no-till planter mounted on the rear of the tractor is simultaneously sowing soybean seed into the soil under freshly rolled and crimped cover crop residue mulch. (b) The resulting thick mulch of killed rye residue all but eliminates evaporation losses of water from the soil between soybean rows and suppresses most weed growth early in the season. The weeds that eventually did emerge through the mulch can be killed using a high residue cultivator that slices and lifts the soil under the mulch while leaving the surface mulch largely undisturbed. (c) The high residue cultivator works by cutting a slit through the mulch with sharp rolling coulters followed by a V-shaped sweep that slices and lifts the soil under the mulch to destroy the roots of young weeds growing between the soybean rows. This technology is of particular relevance to organic farming in which use of synthetic weed-killing chemicals is not allowed. (Photo © Rodale Institute)

The main disadvantage of this system is the soil degradation that occurs, largely because of a negative soil organic matter balance (see Section 11.7) and wind erosion (see Section 14.11) during the fallow years. The growing use of conservation tillage practices (see below and Section 14.6) in semiarid areas has led to sufficient savings in soil moisture to minimize the need for fallow cropping. Conservation tillage practices (including "no-till") leave most of the crop residues on the soil surface where they reduce evaporation and protect the soil. From research at many sites, we can now conclude that long-term productivity, profitability, and soil quality are likely to be optimized with cropping systems that use conservation tillage and keep the soil continuously vegetated with a diversity of crops.

Control of Surface Evaporation

More than half the precipitation in semiarid and subhumid areas is usually returned to the atmosphere by E directly from the soil surface. In natural rangeland systems, E is a large part of ET because plant communities tend to self-regulate toward configurations that minimize the deficit between PET and ET—generally low plant densities with large unvegetated areas between scattered patches of shrubs or bunchgrass. In addition, plant residues on the soil surface are sparse. Evaporation losses are also high in arid-region irrigated soil, especially if inefficient practices are used (see Section 6.9). Even in humid-region rainfed areas, E is significant during hot, rainless periods. Such moisture losses rob the plant community of much of its growth potential and reduce the water available for discharge to streams. Careful study of Figure 6.15 (*left*) will clarify these relationships and the principles just discussed. Note that the gap between PET and ET represents the *soil water deficit*. The global map in Figure 6.15 (*right*) suggests where and to what degree the soil water supply is limiting for plant productivity.

For arable soils, the most effective practices aimed at controlling E are those that cover the soil. This cover can best be provided by mulches and by selected conservation tillage practices that leave plant residues on the soil surface, mimicking the soil cover of natural ecosystems.

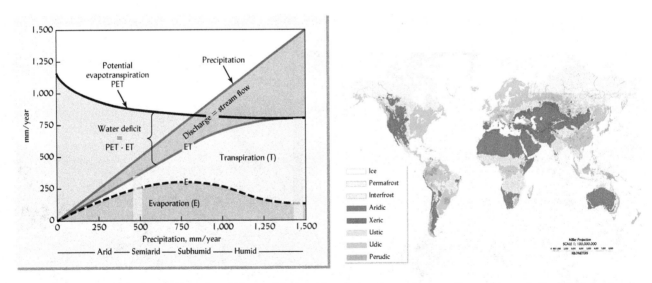

Figure 6.15 (Left) *Partitioning of liquid water losses (discharge) and vapor losses (evaporation and transpiration) in regions varying from low (arid) to high (humid) levels of annual precipitation. The example shown assumes that temperatures are constant across the regions of differing rainfall. Evaporation (E) represents a much greater proportion of total vapor losses (ET) in the drier regions due to sparse plant cover caused by interplant competition for water. The greater the gap between PET and ET, the greater the deficit and the more serious the water stress to which plants are subjected. (Right) The purple, yellow, and especially red colors on the global map of soil moisture regimes indicate the regions of the world where water stress is most limiting to plant productivity on an annual basis. Only the blue and green areas in the map receive enough precipitation so that there is little or no gap between what the soil can offer and what the atmosphere demands in a year. See Section 3.2 for more information on soil moisture regimes.* (Diagram courtesy of Ray R. Weil, Map courtesy of USDA/NRCS)

Vegetative Mulches. A *mulch* is a material used to cover the soil surface. Examples of *organic* mulches include straw, leaves, and crop residues. Mulches can be highly effective in checking evaporation, but they may be expensive and labor-intensive to produce or purchase, transport to the field, and apply to the soil. Mulching is therefore most practical for small areas (gardens and landscaping beds) and for high-value crops such as cut flowers, berries, fruit trees, and certain vegetables. It is much less labor-intensive to produce a mulch "in place" by growing a cover crop or cash crop and letting the residues remain on the soil surface (see Crop Residue and Conservation Tillage section, following). Natural vegetation often produces such a mulch, which ecologists refer to as *litter* (O horizon).

In addition to reducing evaporation, vegetative mulches may provide these benefits: (1) reduce soil-borne diseases spread by splashing water; (2) provide a clean path for foot traffic; (3) reduce weed growth (if applied thickly); (4) moderate soil temperatures, especially preventing overheating in summer months (see Section 7.11); (5) increase water infiltration; (6) provide organic matter and, possibly, plant nutrients to the soil; (7) encourage earthworm populations; and (8) reduce soil erosion. Most of these ancillary benefits do *not* accrue from the use of plastic mulches, discussed next.

Plastic Mulches. Plastic films (and specially prepared paper) are also used as mulch to control evaporative water losses and warm the soil to encourage rapid early growth. If the plastic is opaque, the mulch can also help control weeds, eliminating the need for cultivation. Plastic mulch can be cost-effective with high-value crops. The mulch is often applied by machine over raised soil beds, and plants grow through holes made in the film (Figure 6.16). These mulches are widely used for vegetable and small fruit crops and in landscaping beds, where they are often covered with a layer of tree-bark mulch or gravel for longer life and a more pleasing appearance. As long as the ground is covered, evaporation is checked, and in some cases remarkable increases in plant growth have been reported. Rainwater can usually infiltrate somewhat through the plant holes and in the uncovered interrow spaces, but irrigation tubes are often installed under the plastic. On the negative side, plastic mulches add no organic matter to soils and often encourage runoff and soil erosion between the mulched beds. It can be difficult to completely remove the plastic at the end of the growing season, so scraps of plastic accumulate in the soil, causing an unsightly mess and interfering with water movement and cultivation. Properly disposing of or recycling dirty shredded plastic is also problematic. Some manufacturers now offer biodegradable and light-degradable plastic

Figure 6.16 *For crops with high cash value, plastic mulches are commonly used. Red and black colored mulches are compared in an experimental field (left). Both colors control weeds, but the red plastic reflects red light up into the crop canopy encouraging rapid growth and flowering. A field of strawberries in California (right) is grown with black plastic which encourages not only rapid crop growth, but also rapid chemical runoff and erosion.* (Photo courtesy of Ray R. Weil)

Figure 6.17 *Conservation tillage leaves plant residues on the soil surface, reducing both evaporation losses and erosion. (Left) In a semiarid region (South Dakota), the straw from the previous year's wheat crop was only partially buried to anchor it against the wind while still allowing it to cover much of the soil surface. In the next year, the left half of the field, now shown growing wheat, will be stubble mulched and the right half sown to wheat. (Right) No-till planted corn in a more humid region grows up through the residues left on the surface by previous corn, soybean and wheat crops. Note that with no-till, virtually none of the soil surface is directly exposed to solar radiation, rain, or wind. Weeds are normally controlled with herbicide sprays.* (Photo courtesy of Ray R. Weil)

mulches (made from plant products) that remain intact for a month or so and then break down almost completely.

Crop Residue and Conservation Tillage. We have seen that plant residues on the soil surface conserve soil moisture by reducing evaporation and increasing infiltration. *Conservation tillage* practices leave a high percentage of the residues from the previous crop on or near the surface. A conservation tillage practice widely used in semiarid regions is ***stubble mulch*** tillage (Figure 6.17, *left*). With this method, special tillage implements permit much of the plant residue to remain on or near the surface. Conservation tillage planters are capable of planting through the stubble and allow much of it to remain on the surface during the establishment of the next crop. In semiarid regions, summer fallow is commonly combined with stubble mulching to conserve soil moisture.

Other conservation tillage systems that leave residues on the soil surface include no-tillage (see Figure 6.17, *right*), where the new crop is planted directly into the sod or residues of the previous crop, with almost no soil disturbance (see also Figure 4.25). The long-term soil water conserving benefits of such tillage systems are shown in Box 6.1. Conservation tillage systems will receive further attention in Section 14.6.

6.5 LIQUID LOSSES OF WATER FROM THE SOIL

In our discussion of the hydrologic cycle we noted two types of liquid losses of water from soils: (1) water lost by percolation or drainage, and (2) water lost by surface runoff (see Figure 6.1). *Percolation water* recharges the groundwater and moves dissolved chemicals out of the soil. Surface *runoff water* often carries appreciable amounts of soil particles (erosion) as well as dissolved chemicals.

Percolation and Leaching

When the amount of rainfall entering a soil exceeds the water-holding capacity of the soil, losses by percolation will occur. Percolation losses are influenced by: (1) the amount of rainfall and its distribution, (2) runoff from the soil, (3) evaporation, (4) the character of the soil, and (5) the nature of the vegetation.

BOX 6.1
WATER CONSERVATION PAYS OFF

Productivity of grain crops per unit land area has increased dramatically around the world during the past few decades. Three major beneficial factors have stimulated these increases: (1) improved crop genetic potential (from plant breeding), (2) increased nutrient availability (mainly from fertilizers), and (3) increased amounts of available water (mostly from increased irrigation).

Figure 6.18 illustrates the contribution that conserved soil moisture has made to these yield increases, especially in semiarid areas of the United States. The sorghum grain yields from numerous plots at a USDA research facility in Lubbock, Texas, increased some 325% over a 60-year period.

Further research suggests that about one-third of this yield increase could be attributed to improved varieties, the remaining two-thirds to other factors. Increased nutrient availability was ruled out as a yield-inducing factor since no fertilizers were applied to these plots.

Ruling out nutrient limitations left increased water availability as the prime factor likely contributing to the higher yields. Analyses of rain and snowfall records showed no increase in precipitation that might have accounted for the increased yields. However, soil moisture measurements made at planting time (Figure 6.18, *right*) were found to be much higher during the period from 1972 to 1997 than in the earlier period (1956–1972).

Records showed that tillage and residue management were the primary factors that accounted for this soil moisture difference. During the earlier period (1956–1972) the plots were tilled between crops to control weeds, and little residue remained on the soil. During the later period (1972–1997), there was a major shift to no-tillage crop production, which maintained crop residues on the soil surface. Together with improved weed control using herbicides, the no-till systems reduced evaporative losses, leaving extra moisture in the soil to accommodate the increased sorghum yields.

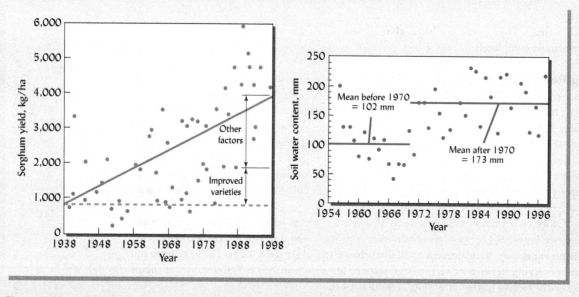

Figure 6.18 (Left) *Sorghum yield increases 1938–1998.* (Right) *Changes in soil water storage 1954–1997.* From Unger and Baumhardt (1999).

Percolation–Evaporation Balance

Figure 6.19 illustrates the relationships among precipitation, runoff, soil storage, soil water depletion, and percolation for representative humid and semiarid temperate regions and for an irrigated arid region. In the humid temperate region, the rate of water infiltration into the soil (precipitation minus runoff) is greater, at least during certain seasons, than the rate of evapotranspiration. As soon as the soil field capacity is reached, percolation into the substrata occurs.

In the example shown in Figure 6.19a, maximum percolation occurs during the winter and early spring, when evaporation is lowest. During the summer, little percolation occurs. In

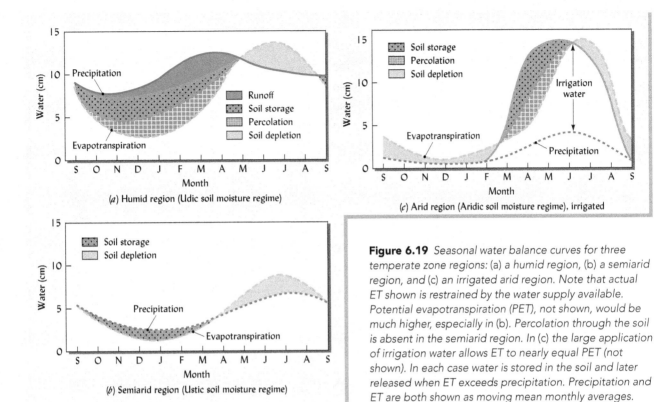

Figure 6.19 Seasonal water balance curves for three temperate zone regions: (a) a humid region, (b) a semiarid region, and (c) an irrigated arid region. Note that actual ET shown is restrained by the water supply available. Potential evapotranspiration (PET), not shown, would be much higher, especially in (b). Percolation through the soil is absent in the semiarid region. In (c) the large application of irrigation water allows ET to nearly equal PET (not shown). In each case water is stored in the soil and later released when ET exceeds precipitation. Precipitation and ET are both shown as moving mean monthly averages. (Diagrams courtesy of N. Brady and Ray R. Weil)

fact, evapotranspiration exceeds precipitation, resulting in a depletion of soil water. Normal plant growth is possible only because of water stored in the soil from the previous winter and early spring.

In the semiarid region, as in the humid region, water is stored in the soil during the winter months and is used to meet the moisture deficit in the summer. But because of the low rainfall, little runoff and essentially no percolation out of the profile occur. Water may move to the lower horizons, but it is absorbed by plant roots and ultimately is lost by transpiration.

The irrigated soil in the arid region (Figure 6.19c) shows a unique pattern. Irrigation in the early spring, along with a little rainfall, provides more water than is being lost by evapotranspiration. The soil profile is charged with water, and some percolation may occur. As we shall see in Section 9.16, irrigation systems must provide enough water for some percolation, in order to remove excess soluble salts. During the summer, fall, and winter months, the stored water is depleted because the amount added is less than that removed by the very high evapotranspiration that takes place in response to large vapor pressure gradients.

The situations depicted in Figure 6.19 are typical of temperate zones where PET varies seasonally with temperature. In the tropics, where temperatures are somewhat higher and less variable, PET is somewhat more uniform throughout the year, although it does vary with seasonal changes in humidity. In very high rainfall tropical areas (perudic moisture regimes), much more runoff and somewhat more percolation occur than shown in Figure 6.19. In irrigated areas of the arid tropics, the relationships are similar to those shown for arid temperate regions.

The comparative losses of water by evapotranspiration and percolation through soils found in different climatic regions are shown in Figure 6.20. These differences should be kept in mind while reading the following section on percolation and groundwater.

Figure 6.20 *Percentage of the water entering the soil that is lost by downward percolation and by evapotranspiration. Representative figures are shown for different climatic regions. The examples shown are for a nearly level terrain for which surface runoff can be assumed to be negligible.*

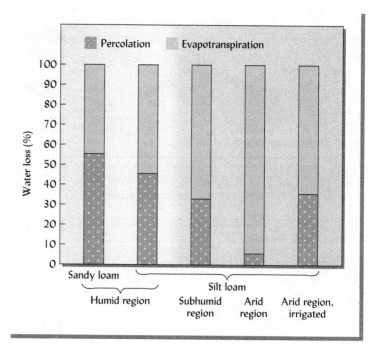

6.6 PERCOLATION AND GROUNDWATER

When drainage water moves downward through the soil and regolith, it eventually encounters a zone in which the pores are all saturated with water. Often, this saturated zone lies above an impervious soil horizon (Figure 6.21) or a layer of impermeable rock or clay. The upper surface of this zone of saturation is known as the **water table**, and the water within the saturated zone is termed **groundwater**. The water table (Figure 6.22) is commonly only 1–10 m below the soil surface in humid regions but may be several hundred or even thousands of meters deep in arid regions. In swamps, it is essentially at the land surface.

The unsaturated zone above the water table is termed the **vadose zone** (see Figures 6.21 and 6.22). The vadose zone may include unsaturated materials underlying the soil profile, and so may be considerably deeper than the soil itself. In some cases, however, the saturated zone may be sufficiently near the surface to include the lower soil horizons, with the vadose zone confined to the upper soil horizons.

Shallow groundwater receives downward-percolating drainage water. Most of the groundwater, in turn, seeps laterally through porous geological materials (termed **aquifers**) until it is discharged into springs and streams. Groundwater may also be removed by pumping for domestic and irrigation uses. The water table will move up or down in response to the balance between the amount of drainage water coming in through the soil and the amount lost through pumped wells and natural seepage to springs and streams. In humid temperate regions, the water table is usually highest in early spring following winter rains and snowmelt, but before evapotranspiration begins to withdraw the water stored in the soil above.

Groundwater Resources

Groundwater is a significant source of water for domestic, industrial, and agricultural use. For example, some 20% of all water used in the United States comes from groundwater sources, and about 50% of the people use groundwater to meet at least some of their needs. A shallow water-bearing layer that is not separated from the soil surface by any overlying impermeable layer is called an **unconfined aquifer**. Unconfined aquifers, which are replenished annually, commonly provide water for farm and rural dwellings (see Figure 6.21). The larger groundwater stores in deeper aquifers, which may take decades or centuries to recharge, are commonly pumped to meet municipal, industrial, and irrigation needs. In most cases, deep wells are used to tap water from aquifers

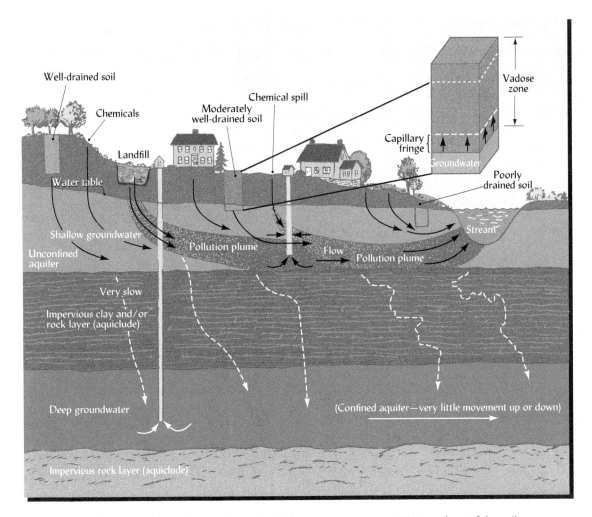

Figure 6.21 *The water table and groundwater in relation to water movement into and out of the soil. Precipitation and irrigation water percolate down the soil profile, ultimately reaching the water table and underlying shallow groundwater. The unsaturated zone above the water table is known as the* vadose zone *(upper right). Groundwater moves up from the water table by capillarity into the capillary fringe. Groundwater also moves horizontally down the slope toward a stream, carrying with it chemicals that have leached through the soil, including plant nutrients (N, P, Ca, etc.) as well as pesticides and other pollutants. A shallow well pumps groundwater from the unconfined aquifer near the surface. A deeper well extracts groundwater from a deep, confined aquifer. Two plumes of pollution are shown, one originating from landfill leachate, the other from a chemical spill. The former appears to be contaminating the shallow well.* (Diagram courtesy of Ray R. Weil)

confined between impermeable strata (termed **aquicludes**). Water is replenished by slow, mainly horizontal seepage from **recharge areas** where the aquifer is exposed to the soil.

Water draining from soils is the main source of replenishment for most underground water resources. So long as water is not withdrawn faster than percolation allows it to be replenished, groundwater may be considered to be a renewable resource. However, in many areas of the world, people are pumping water out faster than it can be replenished, an activity that is lowering water tables and depleting the resource in a manner akin to mining an ore. When overpumping occurs in coastal areas, sea water may push its way into the aquifer, a process termed **saltwater intrusion** (as illustrated in Figure 6.1). Soon, deep municipal wells begin to pull in salty water instead of freshwater.

Shallow Groundwater

Groundwater that is near the surface can serve as a reciprocal water reservoir for the soil. Water removed from the soil by plants may be replaced by upward capillary movement from a shallow water table. The zone of wetting by capillary movement is known as the **capillary fringe**

Figure 6.22 *The water table, capillary fringe, zone of unsaturated material above the water table (vadose zone), and groundwater are illustrated in this photograph. The groundwater can provide significant quantities of water for plant uptake.* (Photo courtesy of Ray R. Weil)

(see Figure 6.22). Such movement can provide a steady and significant supply of water that enables plants to survive during periods of low rainfall. However, capillary rise may also bring a steady supply of dissolved salts to the surface, a process that often leads to serious soil degradation in dry regions (see Section 9.11).

Movement of Chemicals in the Drainage Water

Percolation of water through the soil to the water table not only replenishes the groundwater, but also dissolves and carries downward—or *leaches*—a variety of inorganic and organic chemicals found in the soil or on the land surface. Chemicals leached from the soil to the groundwater (and eventually to streams) in this manner include elements weathered from minerals, natural organic compounds resulting from decay or metabolism in the soil, plant nutrients derived from natural and human sources, and various synthetic chemicals applied intentionally or inadvertently to soils.

In the case of plant nutrients, especially nitrogen, downward movement through the soil and into underlying groundwater has at least three serious implications. First, the leaching of these chemicals represents a depletion of plant nutrients from the root zone (see Section 13.2). Second, accumulation of these chemical nutrients in ponds, lakes, reservoirs, and groundwater downstream may stimulate a process called *eutrophication*, which ultimately depletes the oxygen content of the water, with disastrous effects on fish and other aquatic life (see also Section 12.3). Third, in some areas underground sources of drinking water may become contaminated with excess nitrates to levels unsafe for human consumption (see Section 12.1).

Of even more concern is the contamination of groundwater with human pathogens and various toxic substances, such as pesticides, hormones, drugs, or chemicals leached out of waste disposal sites (see Chapter 15 for a detailed discussion of these pollution hazards). Figure 6.21 illustrates how groundwater can be charged with these contaminants, and how a plume of contamination can spread to downstream wells and bodies of water.

Chemical Movement Through Macropores

The pore configuration in most soils is nonuniform. Old root channels, earthworm burrows, and clay shrinkage cracks commonly contribute large macropores that may provide channels for surprisingly rapid water flow from the soil surface to depths of 1 m or more. Once chemicals are

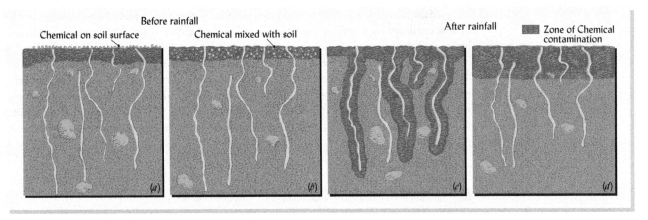

Figure 6.23 Preferential or bypass flow in macropores transports soluble chemicals downward through a soil profile. Where the chemical is on the soil surface (left) and can dissolve in surface-ponded water when it rains, it may be transported rapidly down cracks, earthworm channels, and other macropores. Where the chemical is dispersed within the soil matrix in the upper horizon (right), most of the water moving down through the macropores will bypass the chemical, and thus little of the chemical will be carried downward. Note that channels not open all the way to the surface do not conduct water or contaminants by preferential flow.

carried below the zone of greatest root and microbial activity, they are less likely to be removed or degraded before being carried further down to the groundwater. This means that chemicals that are normally broken down in the soil by microorganisms within a few weeks may move down through large macropores to the groundwater before their degradation can occur.

Preferential or Bypass Flow. In some cases, leaching of chemicals is most serious if the chemicals are merely applied on the soil surface. Chemicals may be washed from the soil surface into large pores, through which they can quickly move downward (Figure 6.23a,c). Research suggests that most of the water flowing through large macropores does not come into contact with the bulk of the soil. Such flow is sometimes termed **preferential** or **bypass flow**, as it tends to move rapidly around, rather than through, the soil matrix. As a result, if chemicals have been incorporated into the upper few centimeters of soil, their movement into the larger pores is reduced, and downward leaching is greatly curtailed (Figure 6.23b,d).

Intensity of Rain or Irrigation. During a high-intensity rainfall event, water and its dissolved chemicals move downward rapidly through channels and cracks (macropores), bypassing the bulk of the soil. In contrast, a gentle rain that in time may provide as much water as the more intense event would likely wet the upper soil aggregates thoroughly, thereby minimizing the rapid downward percolation of both water and the chemicals and pathogens it carries. Table 6.1 illustrates the effects of rainfall intensity on the leaching of pesticides applied to turfgrass.

Table 6.1
INFLUENCE OF WATER APPLICATION INTENSITY ON THE LEACHING OF PESTICIDES THROUGH THE UPPER 50 CM OF A GRASS SOD–COVERED MOLLISOL

Soil columns with the natural structure undisturbed received heavy rain (four doses of 2.54 cm each) or light rain (16 doses of 0.64 cm each). Metalaxyl is far more water soluble than Isazofos, but in each case the heavy rain stimulated much more pesticide leaching through the macropores.

	Pesticide leached as percentage of that applied to surface	
Pesticide	Heavy rains	Light rains
Isazofos	8.8	3.4
Metalaxyl	23.8	13.9

Data from Starrett et al. (1996).

Manipulations of soil macropores and irrigation intensity provide us with two examples of how environmental quality can be impacted by human activities that influence the hydrologic cycle. Likewise, we have considered steps that can be taken to influence the infiltration of water into soils and to maximize plant biomass production from the water stored in soils. The picture would not be complete if we did not consider briefly three other anthropogenic modifications to the hydrologic cycle: (1) the use of artificial drainage to enhance the downward percolation of excess water from some soils; conversely, (2) the application of wastewater to soils as a means of disposal, and (3) the use of irrigation to supplement water available for plant growth.

6.7 ENHANCING SOIL DRAINAGE[4]

Certain soils tend to be water-saturated in the upper part of their profile for extended periods during part or all of the year. In some soils, the prolonged saturation may be related to a low-lying landscape position, such that the regional water table is at or near the soil surface for extended periods. In other soils, water may accumulate above an impermeable layer in the soil profile, creating a *perched water table* (see Figure 6.24). Soils with either type of saturation may be components of wetlands, transitional ecosystems between land and water that are characterized by anaerobic (no oxygen) conditions (see Section 7.7).

Reasons for Enhancing Soil Drainage

Water-saturated, poorly aerated soil conditions are essential to the normal functioning of wetland ecosystems; however, for most other land uses, these conditions are a distinct detriment.

Engineering Problems. During construction, the muddy, low-bearing-strength conditions of saturated soils make it very difficult to operate machinery (see Section 4.9). By the same token, soils used for recreation can withstand trampling much better if they are well drained. Houses built on poorly drained soils may suffer from uneven settling and flooded basements during wet periods. Similarly, a high water table will result in capillary rise of water into roadbeds and around foundations, lowering the soil strength and leading to

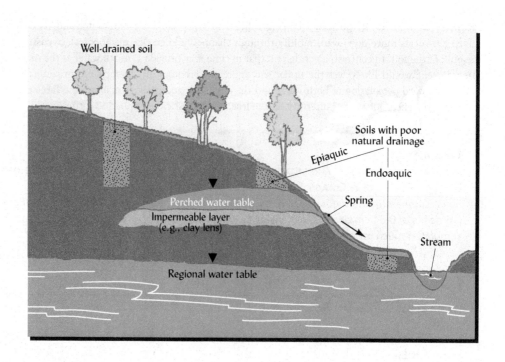

Figure 6.24 *Cross section of a landscape showing the regional and perched water tables in relation to three soils, one well-drained and two with poor internal drainage. By convention, a triangle (▼) is used to identify the level of the water table. The soil containing the perched water table is wet in the upper part, but unsaturated below the impermeable layer, and therefore is said to be epiaquic (Greek epi, upper), while the soil saturated by the regional water table is said to be endoaquic (Greek endo, within). Artificial drainage can help to lower both types of water tables.* (Diagram courtesy of Ray R. Weil)

---

[4]For an update on controlled drainage for agriculture, see Skaggs et al. (2012).

damage from frost-heaving (see Section 7.8) if the water freezes in winter. Heavy trucks traveling over a paved road underlain by a high water table create potholes and eventually destroy the pavement.

Plant Production. Water-saturated soils make the production of most upland crops and forest species difficult, if not impossible. In wet soil, farm equipment used for planting, tillage, or harvest operations may bog down. Except for a few specially adapted plant species (bald cypress trees, rice, cattails, etc.), most crop and forest species grow best in well-drained soils since their roots require adequate oxygen for respiration (see Section 7.1). Furthermore, a high water table early in the growing season will confine the plant roots to a shallow layer of partially aerated soil; the resulting restricted root system can lead to water stress later in the year, when the weather turns dry and the water table drops rapidly (Figure 6.25).

For these and other reasons, artificial drainage systems have been widely used to remove excess (gravitational) water and lower the water table in poorly drained soils. Land drainage is practiced in select areas in almost every climatic region, but is most widely used to enhance the agricultural productivity of clayey alluvial and lacustrine soils. Drainage systems are also a vital, if sometimes neglected, component of arid region irrigation systems, where they are needed to remove excess salts and prevent waterlogging (see Section 9.17). About 33% of the world's and 25% of North America's cropland require drainage. Once the soils are drained, these lands are typically among the most productive of croplands.

Artificial drainage is a major alteration of the soil system, and the following potential beneficial and detrimental effects should be carefully considered. In many instances, laws designed to protect wetlands require that a special permit be obtained for the installation of a new artificial drainage system.

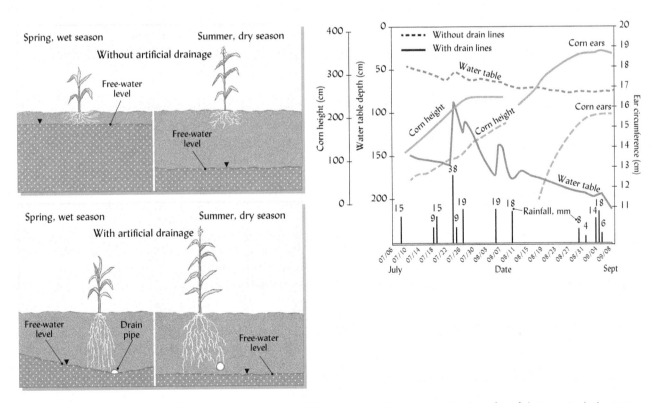

Figure 6.25 (Left) *Effects of artificial drainage on seasonal development of plant roots. The benefits of drainage include more vigorous early growth, as well as a more extensive root system capable of exploiting a larger volume of soil when the profile dries out in summer.* (Right) *Actual measurements of water table depth and corn growth (height and ear circumference) in two adjacent no-till fields in Wisconsin, one with drain lines installed and one without drain lines. In the field without drain lines, the water table dropped very slowly, remaining within 50–75 cm of the soil surface throughout the season. In the drained field, the water table dropped below 150 cm except during a rainy period in late July.* [Graph at right based on data in Kemper et al. (2012)]

Benefits of Artificial Drainage

1. Increased bearing strength and improved soil workability, which allow timelier field operations and greater access to vehicular or foot traffic.
2. Less frost-heaving of foundations, pavements, and plants (e.g., see Figure 7.22).
3. Enhanced rooting depth, growth, and productivity of most upland plants due to improved oxygen supply and, in acid soils, lessened toxicity of manganese and iron (see Sections 7.3 and 7.5).
4. Reduced levels of fungal disease infestation in seeds and on young plants.
5. More rapid soil warming, resulting in earlier maturing crops (see Section 7.11).
6. Less production of methane and nitrogen gases that cause global environmental damages (see Sections 11.9 and 12.1).
7. Removal of excess salts from irrigated soils and prevention of salt accumulation by capillary rise in areas of salty groundwater (see Section 9.16).

Detrimental Effects of Artificial Drainage

1. Loss of wildlife habitat, especially waterfowl breeding and overwintering sites.
2. Reduction in nutrient assimilation and other biochemical functions of wetlands (see Section 7.7).
3. Increased loss of nutrients (especially N and P) and contaminants to ground and surface waters (see Sections 12.1 and 12.2).
4. Accelerated loss of soil organic matter, leading to subsidence of certain soils (see Sections 3.9 and 11.8).
5. Increased frequency and severity of flooding due to loss of runoff water retention capacity.
6. Greater cost of damages when flooding occurs on alluvial lands developed after drainage.

Artificial drainage systems are designed to promote two general types of drainage: (1) *surface drainage* and (2) internal or *subsurface drainage*. Each will be discussed briefly.

Surface Drainage Systems

Surface drainage is extensively used, especially where the landscape is nearly level and soils are fine-textured with slow internal drainage (percolation). Its purpose is to remove water from the land before it infiltrates the soil.

Surface Drainage Swales. Most surface drainage systems involve the construction of shallow ditches or swales with gentle side slopes that hasten runoff of surface water but do not interfere with equipment traffic. If there is some slope on the land, the shallow ditches are usually oriented across the slope and across the direction of planting and cultivating, thereby permitting the interception of water as it runs off down the slope. These ditches can be made at low cost with simple equipment. For removing surface water from landscaped lawns, this system of drainage can be modified by constructing gently sloping mounds and swales rather than ditches.

Land Smoothing. Often, surface drainage ditches are combined with *land smoothing* to eliminate the ponding of water and facilitate its removal from the land. High spots are cut down and depressions are filled in, often using precision laser–guided field-leveling equipment. The resulting land configuration permits excess water to move at a controlled rate over the soil surface to an outlet ditch and then on to a natural drainage channel. Land smoothing is also commonly used to prepare a field for flood irrigation (see Section 6.9).

Subsurface (Internal) Drainage

The purpose of subsurface drainage systems is to remove the groundwater from within the soil and to subsequently lower the water table (Figure 6.26). They require channels such as deep ditches, underground pipes, or "mole" tunnels into which excess water can flow.

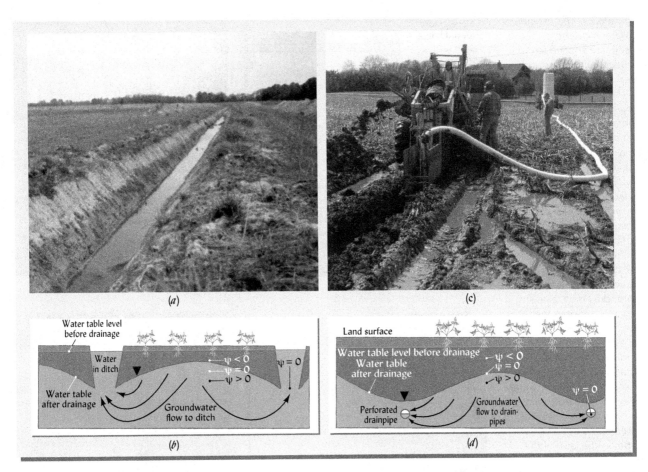

Figure 6.26 *Two types of subsurface drainage system. (a,b) Open ditches are used to lower the water table in a poorly drained soil. The wet season levels of the water table before and after ditch installation are shown. The water table is deepest next to the ditch, and the drainage effect diminishes with distance from the ditch. (c,d) Buried "tile lines" made of perforated plastic pipe act very much as the ditches in (a), but have two advantages: they are not visible after installation and they do not present any obstacle for surface equipment. Note the flow lines indicating the paths taken by water moving to the drainage ditches or pipes in response to the water potential gradients between the submerged water ($\omega > 0$) and free water in the drainage ditch and pipes ($\omega = 0$).* (Courtesy of Ray R. Weil)

Internal drainage occurs only when the pathway for drainage is located below the level of the water table. The flowpath of water from a saturated soil into a drainage outlet may first go down and then up toward the outlet pipe in response to potential gradients. In the same way, most groundwater enters a tile drain or a stream from the bottom (see Figure 6.21). Box 6.2 provides an example of how knowledge of these basic principles of soil water movement and interaction with soil can be applied in designing a system to alleviate drainage problems in an ornamental garden.

Depending on the nature of the area to be drained, a network of drainage channels may be laid out in several types of patterns, ranging from a highly geometric series of parallel drain lines to a more random placement of few lines designed to drain specific low spots or intercept hillside seepage. Where the landscape is too level or at too low an elevation to provide for sufficient fall for gravity removal, expensive pumping operations are sometimes used to remove the drainage water (e.g., in parts of Florida and the Netherlands).

Deep Open-Ditch Drainage. If a ditch is excavated to a depth below the water table (Figure 6.26a, b), water will seep from the saturated soil, where it is under a positive pressure, into the ditch, where its potential will be essentially zero. Once in the ditch, the water can flow rapidly off the field, as it no longer must overcome the frictional forces that would delay its twisting journey through tiny soil pores. However, the ditches, being 1 m or more

Figure 6.27 *The principles of water movement in soils are applied to protect foundations and keep basements dry. (a) The water table around a building foundation with poor drainage. (b) Measures that guide surface water away from the house foundation, including grading to make the soil surface slope away from the house and a black plastic drainpipe that extends the roof downspout away from the foundation. (c) Proper foundation drainage includes installation of a footer drainpipe, a layer of gravel between the concrete slab floor and the soil, and correct surface grading. (d) Photo showing installation of a footer drainpipe that carries water away from the bottom of the foundation, effectively lowering the water table to below the level of the basement floor. Soil will be back-filled to cover the foundation and black flexible drainpipe.* (Courtesy of Ray R. Weil)

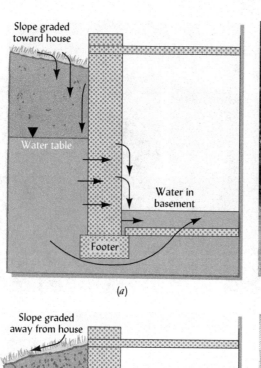
(a)

(b)

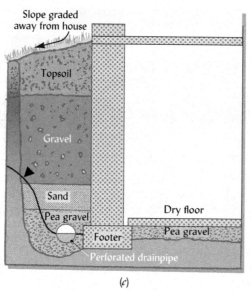
(c)

(d)

BOX 6.2
TOPIARY GARDEN SUCCESS HINGES ON DRAINAGE DESIGN

Dozens of hemlock trees, all carefully pruned into ornamental shapes (topiary) as part of an intricate landscape design, were dying again because of poor drainage. A very expensive effort to improve the drainage under the hemlocks had proved to be a failure, and the replanted hemlocks were again causing an unsightly blemish in an otherwise picture-perfect garden.

Finally, the landscape architect for the world-famous ornamental garden called in a soil scientist to assist her in finding a solution to the dying hemlock problem. Records showed that in the previous failed attempt to correct the drainage problem, contractors had removed all the hemlock trees in the hedge and had dug a trench under the hedge some 3 m deep (a). They had then backfilled the trench with gravel up to about 1 m from the soil surface, completing the backfill with a high-organic-matter, silt loam topsoil. It was into this silt loam that the new hemlock trees had been planted. Finally, a bark mulch layer had been applied to the surface.

The landscaper who had designed and installed the drainage system apparently had little understanding of the various soil horizons and their relation to the local hydrology. When the soil scientist examined the problem site, he found an impermeable claypan that was causing water from upslope areas to move laterally into the hemlock root zone (Figure 6.28a).

(continued)

BOX 6.2
TOPIARY GARDEN SUCCESS HINGES ON DRAINAGE DESIGN (CONTINUED)

Basic principles of soil water movement could have predicted that water would not drain from the fine pores in the silt loam topsoil into the large pores of the gravel (compare the situation to that in Figures 5.16 and 5.18). In fact, the water moving laterally over the impermeable layer created a perched water table that poured water into the gravel-filled trench, soon saturating both the gravel and the silt loam topsoil.

To cure the problem, the previous "solution" had to be undone (b). The dead hemlocks were removed, the ditches were re-excavated, and the gravel was removed from the trenches. Then the trench, except for the upper ½ m, was filled with a sandy loam subsoil to provide a suitable rooting medium for the replacement evergreen trees (a different species was chosen for reasons unrelated to drainage). The upper ½ m of the trench was filled with a sandy loam topsoil, which was also acidic but contained a higher level of organic matter. The interface between the subsoil and surface soil was mixed so that there would be no abrupt change in pore configuration. This would allow an unsaturated wetting front to move down from the upper to the lower layers, drawing down any excess water.

About 1 m uphill from the tree planting a small trench was excavated through the impermeable clay layer that had been guiding water to the area. A perforated drainage pipe surrounded by a layer of gravel was laid in the bottom of this trench with about a 1% slope to allow water to flow away from the area to a suitable outlet. This *interceptor*

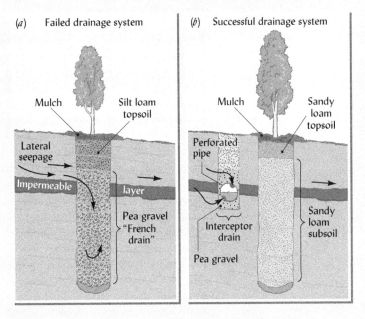

Figure 6.28 *Failed and successful drainage designs.* (Diagrams courtesy of Ray R. Weil)

drain prevented the water moving laterally over the impermeable soil layer from reaching the evergreen hedge root zone (Figure 6.28b).

Even though the replanting of the hedge was followed by an exceptionally rainy year, the new drainage system kept the soil well aerated, and the trees thrived. The principles of water movement explained in Sections 5.6 and 6.7 of this text were applied successfully in the field.

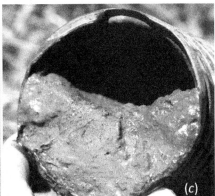

Figure 6.29 *Subsurface "tile" drainage systems. (a) Two outlet pipes empty drainage water from a tile drain system into an open ditch. Such an outlet is essential for a gravity drainage system to function. (b) Flexible plastic drainpipe perforated with holes on one side. These holes must be on the bottom of the drainpipe when it is installed, to avoid soil entering and clogging the pipe as shown in (c), where the pipe was installed with the holes on the top.* (Photo courtesy of Ray R. Weil)

deep, present barriers to equipment. Therefore, deep-ditch drainage is generally practical only for sandy soils in which the high-saturated hydraulic conductivity ensures that the water table will be lowered for a considerable distance from each ditch and therefore the ditches may be placed far apart.[5] Open ditches need regular maintenance to control vegetation and sediment buildup.

Buried Perforated Pipes (Drain Tiles).[6] A network of perforated plastic pipes can be laid underground using specialized equipment (see Figure 6.26c). Water moves into the pipe through the perforations. The pipe should be laid with the slotted or perforated side *down*. This allows water to flow up into the pipe but protects against soil falling into and clogging the pipe (see Figure 6.29c). For pipe with perforations all around, a filter "sock" made of geotextile may be needed to encase the pipe and prevent sediment from entering. Sediment buildup will also be avoided if the pipe has the proper slope (usually a 0.5–1% drop) so water flows rapidly to the outlet ditch or stream. A gate or wire mesh should cover the pipe outlet to prevent the entrance of rodents in dry weather but still allow the free flow of water.

If a control structure is installed at the outlet of a buried pipe (tile) drain network, the elevation of the outlet can be adjusted by raising or lowering an overflow barrier. This adjustment allows the manager to control the flow of water out of the system and the water table depth under the drained field. During the winter when root zone saturation is not a problem, the manager can restrict drainage flows and thus greatly reduce the loss of nutrients such as nitrates to receiving waters. Lowering the barrier to increase drainage will enhance crop growth early in the season. Later in the season, raising the barrier again can maintain a more shallow water table and conserve soil profile water for crop use during the relatively hot dry summer. Such managed drainage practices should be more widely adopted as they offer clear advantages for both productivity and environmental quality.

Building Foundation Drains. Surplus water around building foundations can cause serious damage. The removal of this excess water is commonly accomplished using buried perforated pipe, placed alongside and slightly below the foundation or underneath the floor (Figure 6.27b,d). The perforated pipe must be sloped to allow water to move rapidly to an outlet ditch or sewer. If the drain successfully prevents the water table from rising above the floor level, water will not seep into the basement for the same reason that it will not seep into a drainpipe placed above the water table.

Mole Drainage. A **mole drain** system can be created by pulling a pointed shank followed by an attached bullet-shaped steel plug about 7–10 cm in diameter through the soil at the desired depth. The compressed wall channel thus formed provides a pathway for the removal of excess water, similar to a buried pipe. Mole drainage is quite inexpensive to install, but is efficient only in fine-textured soils in which the channel is likely to remain open for a number of years.

6.8 SEPTIC TANK DRAIN FIELDS

Many ordinary suburbanites get their first exposure to the importance of water movement through soils when they apply for a permit to build their new dream house. The local authorities will usually not allow a home to be built until arrangements are made for wastewater treatment. Typically, a soil scientist will inspect the soils at the homesite and judge their suitability for use as a septic tank **drain field**. If the soils are found unsuitable, the landowner may be denied the permit to build.

Thus, in many regions, it is through their wastewater treatment function that soil properties influence the value of land and the spread of residential development. For consulting soil

---

[5]The required spacing of ditches and buried drainpipes ranges from about 10 m apart for very low-permeability soils to more than 100 m apart for high-permeability soils such as peats and sands.
[6]Before the adoption of plastic perforated pipe in the 1960s, short sections of ceramic pipe known as *tiles* were laid down end-to-end with small gaps between them through which water could enter. The terms *drain tile* or *tile line* are still often used in reference to the newer plastic pipe.

scientists in industrialized nations, determination of soil suitability for septic tank drain fields is probably the single most commonly rendered service. Furthermore, in rapidly suburbanizing areas, improperly sited septic tank drain fields may contribute significantly to pollution of groundwater and streams. For all these reasons, our discussion of practical water management in soil systems would be incomplete without consideration of the soils in on-site wastewater treatment in areas not served by a centralized sewage treatment system.

Operation of a Septic System

The most common type of on-site wastewater treatment for homes not connected to municipal sewage systems is the *septic tank* and associated *drain field* (sometimes called *filter field* or *absorption field*). This method of sewage and wastewater treatment depends on several soil processes, of which the most fundamental is the movement of water through the soil.

In essence, a septic drain field operates like artificial soil drainage in reverse (Figure 6.30). A network of perforated underground pipes is laid in trenches, very much like the network of drainage pipes used to lower the water table in a poorly drained soil. But instead of draining water away from the soil, the pipes in a septic drain field carry wastewater *to* the soil, the water entering the soil via slits or perforations in the pipes. In a properly functioning septic drain field, the wastewater will enter the soil and percolate downward, undergoing several purifying processes before it reaches the groundwater. One of the advantages of this method of sewage treatment is that it has the potential to replenish local groundwater supplies for other uses.

The Septic Tank. Water carrying wastes from toilets, sinks, and bathtubs flows by gravity through sealed pipes to a large underground concrete box, called the *septic tank* (Figure 6.30b). Baffles in the tank cause the inflowing wastewater to slow down and drop most (70%) of its load of suspended solid materials, which subsequently settle to the bottom of the septic tank.

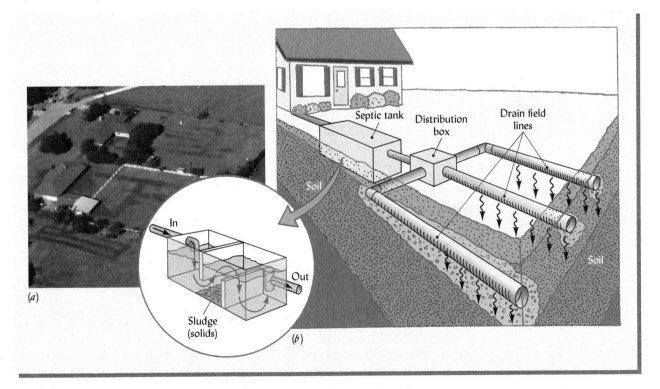

Figure 6.30 *On-site wastewater treatment with a soil-based septic drain field system. (a) During dry weather, the lawn grasses around houses in Texas exhibit darker green bands where the grass is responding to water and nutrients released by underground septic drain field lines. (b) A diagrammatic representation of a septic system showing the main parts, including three drain field lines made from perforated pipe embedded in gravel at the bottom of trenches. The wastewater trickles from these pipes into the native soil for purification. The inset diagram shows a cut-away view of the septic tank which traps most of the solids in the waste stream.* (Courtesy of Ray R. Weil)

As the organic solids partially decompose by microbial action in the septic tank, their volume is reduced so that many years may pass before the septic tank becomes too full and the accumulated **sludge** (also called *septage*) has to be pumped out. Because these systems depend on active microorganisms and have limited physical storage capacities, one should avoid introducing harmful chemicals or nonbiodegradable materials.

The Drain Field. The water exiting the septic tank via a pipe near the top is termed the septic tank **effluent**. Although its load of suspended solids has been much reduced, it still carries organic particles, dissolved chemicals (including nitrogen), and microorganisms (including pathogens). The flow is directed to one or more buried pipes that constitute the **drain field**. Blanketed in gravel and buried in trenches about 0.6–2 m under the soil surface, these pipes are perforated on the bottom to allow the wastewater to seep out and enter the soil. It is at this point that soil properties play a crucial role. Septic systems depend on the soil in the drain field to: (1) keep the effluent out of sight and out of contact with people, (2) treat or purify the effluent, and (3) conduct the purified effluent to the groundwater.

As the wastewater percolates, soil microbial action removes the organic materials and pathogens, but releases nitrates that are subject to leaching into the groundwater. Some 4–5 kg of nitrogen per year per person using the system may be released in this manner. This suggests that septic tanks from a group of family homes in near proximity to each other may release to the groundwater sufficient nitrates to be of environmental concern.

Soil Properties Influencing Suitability for a Septic Drain Field

The soil should have a *saturated hydraulic conductivity* (see Section 5.5) that will allow the wastewater to enter and pass through the soil profile rapidly enough to avoid backups that might saturate the surface soil with effluent, but slowly enough to allow the soil to purify the effluent before it reaches the groundwater. The soil should be sufficiently *well aerated* to encourage *microbial breakdown* of the wastes and *destruction of pathogens*. The soil should have some fine pores and clay or organic matter to adsorb and filter contaminants from the wastewater.

Soil properties that may disqualify a site for use as a septic drain field include impermeable layers such as a fragipan or a heavy claypan, gleying in the upper horizons indicating extended saturated conditions, too steep a slope, or excessively drained sand and gravel.

Septic tank drain fields installed where soil properties are not appropriate may result in extensive pollution of groundwater and in health hazards caused by seepage of untreated wastewater. In Figure 6.30a the dark green lines in the lawn show where the grass has responded to the water and nitrogen in the waste stream. Such signs indicate that the soil has too slow a percolation rate or too high a water table and that the wastewater has moved upward rather than downward.

Suitability Rating. The suitability of a site for septic drain field installation depends largely on soil properties that affect water movement and the ease of installation (see Table 6.2). For example, too steep a slope may interfere both with the ease of installation and with the operation of a septic drain field. A septic drain field laid out on a slope greater than 15% may allow considerable lateral movement of the percolating water such that at some point downslope, the wastewater will seep to the surface and present a potential health hazard.

Perc Test. This is a test that determines the *percolation rate* (which is related to the saturated hydraulic conductivity described in Section 5.5) expressed in millimeters (or other unit of depth) of water entering the soil per hour. A hole is augered to the depth of the proposed drain field, the bottom is lined with gravel, a known amount of water is added, and then the depth of the water is monitored over time. The percolation rate indicates whether or not the soil can accept wastewater rapidly enough to provide a practical disposal medium (Table 6.2). The test is simple to conduct and should be carried out during the wettest season of the year.

To some degree, a low percolation rate can be compensated for by increasing the total length of drain field pipes, and hence increasing the area of land devoted to the drain

Table 6.2
SOIL PROPERTIES INFLUENCING SUITABILITY FOR A SEPTIC TANK DRAIN FIELD
Note that most of these soil properties pertain to the movement of water through the soil profile.

Soil property[a]	Limitations		
	Slight	Moderate	Severe
Flooding	—	—	Floods frequent to occasional
Depth to bedrock or impermeable pan, cm	>183	102–183	<102
Ponding of water	No	No	Yes
Depth to seasonal high water table, cm	>183	122–183	<122
Permeability (perc test) at 60–152 cm soil depth, mm/h	50–150	15–50	<15 or >150[b]
Slope of land, %	<8	8–15	>15
Stones >7.6 cm, % of dry soil by weight	<25	25–50	>50

[a]Assumes soil does not contain permafrost and has not subsided more than 60 cm.
[b]Soil permeability (as determined by a perc test) greater than 150 mm/h is considered too fast to allow for sufficient filtering and treatment of wastes.
Based on Soil Survey Staff (1993), Table 620–17.

field. The size of the septic drain field is also influenced by the amount of wastewater that is likely to be generated (which may be estimated by the number of bedrooms in the house being served).

Alternative Systems. Environmentally friendly alternatives include self-contained composting toilets that convert human wastes directly into a humus-like soil amendment rather than using water to flush wastes away into the soil. Although alternatives to septic drain fields are not currently allowed in some localities, compost toilets and other systems that provide for actually recycling the nutrients in human wastes (especially the phosphorus) will be required for our long-term sustainability on Earth.

6.9 IRRIGATION PRINCIPLES AND PRACTICES[7]

While an adult human requires about 2–4 liters of water for drinking each day, it requires some 3,000–7,000 liters of water to grow the food eaten by that person in a day. This water requirement includes the use of both "green" water and "blue" water resources (Table 6.3). In Sections 6.2–6.6 we focused primarily on effective management of green water resources—natural precipitation water stored in the soil profile for plant uptake. In irrigated agriculture, this green water is supplemented by the application of blue water resource—water withdrawn from lakes, rivers, or underground aquifers.

In most regions of the world, insufficient water is the prime limitation to agricultural productivity. In semiarid and arid regions, intensive crop production is all but impossible without supplementing the meager rainfall provided by nature. However, if given supplemental water through irrigation, the sunny skies and fertile soils of some arid regions stimulate extremely high crop yields. The history of irrigation is nearly as old as the history of

---

[7]For a fascinating account of water resources and irrigation management in the Middle East, see Hillel (1995). For a practical manual on small-scale irrigation with simple but efficient microtechnology, see Hillel (1997). For extensive information on methods of irrigating crops, see Lascano and Sojka (2007). For perspectives on optimal irrigation and crop management when water supply is either scarce enough to be fixed or plentiful enough to be nonlimiting, see Basso and Ritchie (2012).

Table 6.3
SOME TERMS USED TO DESCRIBE VARIOUS POOLS AND USES OF THE WATER RESOURCE

Water term	Explanation
Blue water	Freshwater available for removal from surface water bodies (rivers, lakes) and aquifers.
Green water	Water stored in the soil as soil moisture and used by plants in evapotranspiration.
Gray water	Mildly contaminated wastewater from industrial or municipal uses that can safely be reused for nonpotable purposes such as irrigation or flushing toilets.
Black water	Wastewater that has been contaminated with human feces and is unsafe to reuse or discharge into streams without extensive treatment.
Irrigation water	Blue water consumed by the production of plants (crops, lawns, etc.).
Water withdrawals	Water removal for use from surface water bodies or subsurface aquifers.
Nonconsumptive water use	Freshwater used and then returned to any freshwater source where it becomes available for further use.
Consumptive water use	Freshwater withdrawals which are not immediately available for further use because they have been evaporated, embodied in products and waste, or discharged after use into different watersheds or the sea.
Water degradation	Water which is discharged in the same watershed from which it was withdrawn, but after the quality of water has been compromised.

agriculture itself. Rice producers in Asia, wheat and barley producers in the Middle East, and corn producers in Central and South America were irrigating their crops well over 2,000 years ago. The ancient Mesopotamian civilizations of the Tigris and Euphrates river valleys created complex networks of canals and ditches to divert water from the great rivers to extensive areas of cultivated land. In other places, farmers filled simple buckets with water from a stream or open well and carried it to the thirsty plants in their gardens. Today, these methods of water conveyance can still be seen, although in many agricultural and landscaping systems they have been replaced by modern pumps and pipes.

Importance of Irrigation Today

Food Production. During the twentieth century, the area of irrigated cropland expanded greatly in many parts of the world. However, in recent decades, the total area under irrigation seems to be leveling off at about 275 million hectares worldwide. Application of irrigation water typically increases the productivity of the land by 100–400%, accounting for the fact that the 20% of cropland that is irrigated produces about 40% of the world's crop yield.

Expanded and improved irrigation, especially in Asia, has been a major factor in helping global food supplies keep up with, and even surpass, the global growth in population. As a result, agricultural irrigation remains the largest *consumptive* use of water resources, accounting for about 80% of all water consumed worldwide, in both developing and developed countries. Water withdrawn for irrigation and lost to the atmosphere by evapotranspiration is a consumptive use, but water withdrawn to cool an electric generation turbine and then returned to the river from which it came is a nonconsumptive use (Table 6.3). On average, about 30–35% of the water withdrawals for irrigation are returned to water sources for reuse and are not part of the consumptive use. By contrast, 90–95% of the withdrawals for industrial use are returned to the same watershed and so are not considered consumptive use.

Landscaping. Irrigation is an integral part of such landscaping installations as golf courses, flower beds, municipal parks, and home lawns. Irrigation serves to keep the grass green during summer dry spells in humid regions, but is necessary almost year-round to keep certain species thriving in drier regions. In arid regions, irrigation can alter a glaring, brown landscape into one of cooling shade trees and colorful flowers. In most cases, the use of irrigation for landscaping is predicated on the desire to maintain vegetation that conforms to an ideal notion of perpetual green lushness. In contrast, lawns, landscapes, and golf courses in many parts of the world utilize only native or adapted vegetation that is capable of surviving, without irrigation, the periods of dry weather and other adverse conditions characteristic of the local climate. Increasing environmental awareness has engendered a growing trend toward more xerophytic landscaping (utilizing desert plants and rocks) in arid regions and generally toward more use of locally native vegetation that requires little or no irrigation.

Future Prospects. Water resources for irrigation are most rapidly becoming scarce and/or expensive in arid regions where they are most needed. Reducing waste, re-using water and achieving greater efficiency of water use in irrigation are increasingly important aspects that will be emphasized in this section. One of the other major problems associated with irrigation, the salinization of soils and drainage waters, is considered in Chapter 9.

Water-Use Efficiency

Various measures of water-use efficiency are used to compare the relative benefits of different irrigation practices and systems. The most meaningful overall measure of efficiency would compare the output of a system (crop biomass or value of marketable product) to the amount of water allocated as an input into the system. There are many factors to consider (such as type of plants grown and reuse of "wasted" water by others downstream), so such comparisons must be made with caution.

Application Efficiency. A simpler measure of water-use efficiency, sometimes termed the **water application efficiency**, compares the amount of water available or allocated to irrigate a field to the amount of water actually used in transpiration by the irrigated plants. In this regard, most irrigation systems are quite inefficient, with only about 10–30% of the water that is taken from the source transpired by the desired plants. Table 6.4 gives estimates of the various water losses that typically occur between allocation and transpiration. The table compares average losses in semiarid regions under both rainfed and irrigated agriculture.

Much of the water loss occurs by evaporation and leakage in the reservoirs, canals, and ditches used to store and deliver water to irrigated fields. To reduce these losses during water distribution, ditches can be lined with concrete or plastic (see Figure 6.31). Evaporative losses can be all but eliminated during distribution by the use of pipelines instead of open canals, but this is very much more expensive.

Figure 6.31 *Unlined ditches* (left) *lose much water to adjacent soil areas or to the groundwater. Note the evidence of capillary movement above the water level in the unlined ditch. Concrete-lined irrigation ditches to carry water long distances* (right) *or within a farm* (center) *can increase the efficiency of water delivery to the field. Standard-sized siphon pipes efficiently transfer water from a distribution ditch to the cropped field itself* (center). (Photo courtesy of Ray R. Weil)

Figure 6.32 *Estimates of average water losses in traditional irrigated and rainfed agriculture in semiarid areas of the world. Available water for irrigated agriculture = water stored in reservoirs or pumped from groundwater, available water for rainfed agriculture = rainfall. Evaporation includes vaporization from both open water and soil surfaces. Some of the water lost by drainage and runoff may be reused by downstream irrigators. Note that an average of 30% of the allocated water is lost in storage or conveyance before it reaches the field.*
[Recalculated and graphed from data in Wallace (2000)]

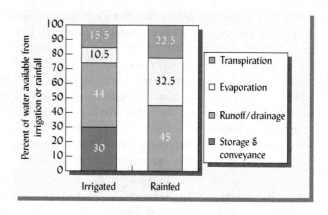

Field Water Efficiency.

Water-use efficiency *in the field* may be expressed as

$$\text{Field water efficiency, \%} = \frac{\text{Water tranpired by the crop}}{\text{Water applied to the field}} \times 100 \qquad (6.3)$$

Values are usually less than 50%. As shown in Table 6.4, field water efficiency for traditional irrigation systems used in semiarid regions is often as low as 20–25%. For example, if transpiration uses 18% out of the total allocated water of the 70% that reaches the field, the field water efficiency = 100(18/70) = 25.7%. The water delivered to the field that is not transpired by the crop is lost as surface runoff, deep percolation below the root zone, and/or evaporation from the soil surface. Achieving a high level of field water efficiency (or a low level of water wastage) is highly dependent on the skill of the irrigation manager and on the methods of irrigation used. As shown in Table 6.4, different irrigation systems vary greatly in their field water efficiencies. We will now turn our attention to a brief description of the principal methods of irrigation in use today.

Surface Irrigation

In these systems, water is applied to the upper end of a field and allowed to distribute itself by gravity flow. Usually, the land must be leveled and shaped so that the water will flow uniformly across the field. The water may be distributed in **furrows** graded to a slight slope

Table 6.4
SOME CHARACTERISTICS OF THE THREE PRINCIPAL METHODS OF IRRIGATION

Methods and specific examples	Direct costs of installation,[a] 2018 dollars/ha	Labor requirements	Field water efficiency,[b] %	Suitable soils
Surface: basin, flood, furrow	750–1,100	High to low, depending on system	20–50	Nearly level land; not too sandy or rocky
Sprinkler: center pivot, movable pipe, solid set	1,000–4,000	Medium to low	60–70	Level to moderately sloping; not too clayey
Microirrigation: drip, porous pipe, spitter, bubbler	1,300–5,000	Low	80–90	Steep to level slopes; any texture, including rocky or gravelly soils

[a]Average ranges from many sources. Costs of required drainage systems not included.
[b]Field water efficiency = 100 × (water transpired by crop/water applied to field).

so that water applied to the upper end of the field will flow down the furrows at a controlled rate (Figure 6.33). In **border irrigation** systems, the land is shaped into broad strips 10–30 m wide, bordered by low dikes.

Water Control. Water is usually brought to surface-irrigated fields in supply ditches or gated pipes (such as are shown in Figure 6.31). The amount of water that enters the soil is determined by the permeability of the soil and by the length of time a given spot in the field is inundated with water. Achieving a uniform infiltration of exactly the required amount of water is very difficult and depends on controlling the slope and length of the irrigation runs across the field. If the soil is highly permeable (e.g., some loamy sands), too much water may infiltrate near the upper end of the field and too little may reach the lower end (Figure 6.33). On the other hand, for a fine-textured soil, infiltration may be so slow that water flows across the field and ponds up or runs off the lower end without a sufficient amount soaking into the soil. Therefore, in the very sandy soils, leaching loss of water and chemicals is a problem at the upper end of the field. In clayey soils, erosion, runoff, and waterlogging may be problems at the lower end.

The **level basin** technique of surface irrigation, as is used for paddy rice and certain tree crops, alleviates these problems because each basin has no slope and is completely surrounded by dikes that allow water to stand on the area until infiltration is complete. This method is not practical for highly permeable soils into which water would infiltrate so fast that the basin would never fill. On sloping land, terraces can be built in a modification of the level-basin method (Figure 6.34).

Variants of the surface systems have been in use for 5,000 years. They require little equipment and are relatively inexpensive to operate. The principal capital cost is usually the initial land shaping, which may be quite expensive if the land is not nearly level to begin with. However, control over leaching and runoff losses is difficult, and the entire soil surface is wetted so that much water is lost by evaporation from the soil and by weed transpiration.

Sprinkler Systems

In sprinkler irrigation, water is sprayed through the air onto a field, simulating rainfall. Thus, the entire soil surface, as well as plant foliage (if present), is wetted. This leads to evaporative losses similar to those described for surface systems. Furthermore, an additional 5–20% of the applied water may be lost by evaporation or windblown mist as the drops fly through the air. One advantage is that plants often respond positively to the cooler, better-aerated sprinkler water. A disadvantage is that wet leaves may increase the incidence of fungal diseases in some plants, such as grapes, fruit trees, and roses, so sprinkler systems are not often used for these plants.

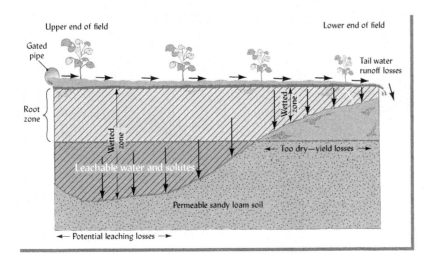

Figure 6.33 *Penetration of water into a coarse-textured soil under surface irrigation. The high infiltration rate causes most of the water to soak in near the gated pipe at the upper end of the field. The uneven penetration of water results in the potential for leaching losses of water and dissolved chemicals at the upper end of the field, while plants at the lower end may receive insufficient water to moisten the entire potential root zone. On a less permeable soil, or on a field with a steep slope, the tail water runoff losses would likely be greater at the lower end and leaching potential less at the upper end.* (Diagram courtesy of Ray R. Weil)

Figure 6.34 *The level-basin type of surface irrigation. (Upper left) Nut trees are flood irrigated on level land in Arizona. (Lower left) Level basin irrigation is modified for sloping land in South Asia by construction of terraces on which paddy rice is grown. (Right) Whole landscapes can be transformed with level terraces supporting flood irrigation. This type of irrigation is practical only on soils of low permeability and where water supplies are inexpensive.* (Photo courtesy of Ray R. Weil)

Water Control. A sprinkler system should be designed to deliver water at a rate that is less than the infiltration capacity of the soil, so that runoff or excessive percolation will not occur. In practice, runoff and erosion may be problems if the soil infiltration capacity is low, the land is relatively steep, or too much water is applied in one place. Water is sometimes lost by deep percolation because more water falls near the sprinkler than farther from it. Overlapping of spray circles can help achieve a more even distribution of water. Because of better control over application rates, the field water-use efficiency is generally higher for sprinkler systems than for surface systems, especially on coarse-textured soils.

Suitable Soils. Sprinkler irrigation is practical on a wider range of soil conditions than is the case for the surface systems. Various types of sprinkler systems are adapted to moderately sloping as well as level land (Figure 6.35c). They can be used on soils with a wide range of textures, even those too sandy for surface irrigation systems. Advanced center pivot systems use computer-controlled nozzles to vary the amount of water applied according to the soil properties encountered as the rig rotates around a field (Figure 6.35). For instance, rocky or very sandy spots with low potential for

Figure 6.35 *Center pivot irrigation systems. (a) A center pivot irrigation system with a heavy duty motor used to pump the water up from the groundwater. Also shown are a large tank of liquid fertilizer and the computer controls that allow nutrients to be injected into the water at precise rates during irrigation events, a process termed fertigation. (b) This system is seen from just outside the irrigated circle where the system is making a low-energy, precision application of water to a soybean crop and is rotating slowly toward the left. (c) Many such irrigation circles as seen from the air have transformed the hydrology of this arid Texas landscape.* (Photo courtesy of Ray R. Weil)

Figure 6.36 *Four examples of microirrigation that ensure the root zone is kept almost continuously at optimal moisture content by application of small quantities of water at high frequency (such as once or twice a day). (a) Drip or trickle irrigation with a single emitter for each seedling in a cabbage field in Malawi. (b) Irrigation applied drop by drop to an individual tree in a young apple orchard in Mexico. (c) Microirrigation is often used in conjunction with plastic mulch to control the loss of water by evaporation and weed transpiration. In this tomato field, a main water supply line runs perpendicular to the crop rows, while smaller drip lines run from the supply line under the plastic with emitters near each plant. (d) Microirrigation lines in a no-till pumpkin field mulched by thick cover crop residues. The white object is an in-line filter that removes sediment particles before they can clog the fine emitter holes. In each case, irrigation wets only the small portion of the soil in the immediate root zone of individual plants, thus limiting the amount of water required and reducing weed growth and evaporation in the nonmoistened area.* (Photo courtesy of Ray R. Weil)

crop growth and low water holding capacity may receive little or no water, while areas of better soils with high production potential may receive larger amounts of water. In addition, soil water sensors (such as the capacitance sensor shown in Figure 5.9) can be embedded in the soil at various locations around a field and wirelessly connected to a computer system that can then control the amount of water applied to maintain optimal soil water content at each location.

Equipment. The equipment costs for sprinkler systems are higher than those for surface-flow systems. Large pressure pumps and specialized pipes and nozzles are required. Some types of sprinkler systems are set in place, others are moved by hand, and still others are self-propelled, either moving in large circles around a central pivot or rolling slowly across a rectangular field. Most systems can be automated and adapted to deliver doses of pesticides or soluble fertilizers to plants (Figure 6.35a).

Microirrigation

The most efficient irrigation systems in use today are those using microirrigation, whereby only a small portion of the soil is wetted (Figure 6.36a,c) in contrast to the complete wetting accomplished by most surface and sprinkler systems. The "micro" in microirrigation may also refer to the tiny amounts of water applied at any one time and the miniature size of the equipment involved.

Perhaps the best-established microirrigation system is *drip* (or *trickle*) *irrigation*, in which tiny emitters attached to plastic tubing apply water to the soil surface alongside individual plants. In some cases, the tubing and emitters are buried 20–50 cm deep so the water soaks directly into the root zone. In either case, water is applied at a low rate (sometimes drop by drop) but at a high frequency, with the objective of maintaining optimal soil water availability in the immediate root zone while leaving most of the soil volume dry (Figure 6.36). Table 6.4 illustrates the high field-water efficiency of drip irrigation.

Other forms of microirrigation that are especially well adapted for irrigating individual trees include *spitters* (microsprayers) and *bubblers* (small vertical standpipes). The bubblers (and usually the spitters) require that a small level basin be formed in the soil under each tree.

Water Control. Water is normally carried to the field in pipes, run through special filters (Figure 6.36d) to remove any grit or chemicals that might clog the tiny holes in the emitters (filtering is not necessary for bubblers), and then distributed throughout the field by means of a network of plastic tubes. Soluble fertilizers may be added to the water as needed.

If properly maintained and managed, microirrigation allows much more control over water application rates and spatial distribution than do either surface or sprinkler systems. Losses by supply-ditch seepage, sprinkler-drop evaporation, runoff, drainage (in excess of that needed to remove salts), soil evaporation, and weed transpiration can be greatly reduced or eliminated. Once in place, the labor required for operation is modest and the system can be largely automated.

Microirrigation often produces healthier plants and higher crop yields because the plant is never stressed by low water potentials or low aeration conditions that are associated with the feast-or-famine regime of infrequent, heavy water applications made by all surface irrigation systems and most sprinkler systems. A disadvantage or risk is that there is very little water stored in the soil at any time, so even a brief breakdown of the system could be disastrous in hot, dry weather.

Equipment. The capital costs per hectare for microirrigation tend to be higher than for other systems (see Table 6.4), but the differences are not so great if the cost of drainage systems for control of salinity and waterlogging in surface systems, the costs of high-pressure pumping in sprinkler systems, and the real value of wasted water are taken into account. Because of its high water-use efficiency, microirrigation is most profitable where water supplies are scarce and expensive and where high-valued plants such as fruit trees or vegetables are being grown.

6.10 CONCLUSION

The hydrologic cycle encompasses all movements of water on or near the Earth's surface. It is driven by solar energy, which evaporates water from the ocean, soil, and vegetation. The water cycles into the atmosphere, returning elsewhere to the soil and the oceans in rain and snow.

The soil is an essential component of the hydrologic cycle. It receives precipitation from the atmosphere, rejecting some of it, which is then forced to run off into streams and rivers, and absorbing the remainder, which then moves downward to be either transmitted to the groundwater, taken up and later transpired by plants, or evaporated directly from soil surfaces and returned to the atmosphere.

Water moves within the soil as a liquid by the processes of gravitational drainage, capillary action, and redistribution by plant roots. The behavior of water is governed by the same set of principles in both soils and plants: water moves in response to differences in energy levels, moving from higher to lower water potential. These principles can be used to manage water more effectively and to increase the efficiency of its use.

Management practices should encourage movement of water into well-drained soils while minimizing evaporative losses from the soil surface. These two objectives will provide as much water as possible for plant uptake and groundwater recharge. "Green water" as water from the soil is called, must satisfy the T requirements of healthy leaf surfaces; otherwise, plant growth will be limited by water stress. Practices that leave plant residues on the soil surface and that maximize plant shading of this surface will help achieve high efficiency of water use by minimizing evaporation losses.

Extreme soil wetness, characterized by surface ponding and saturated conditions, is a natural and necessary condition for wetland ecosystems. However, extreme wetness is detrimental for most other land uses. Drainage systems have therefore been developed to hasten the removal of excess water from soil and lower the water table so that upland plants can grow without aeration stress, and so the soil can better bear the weight of vehicular and foot traffic.

A septic tank drain field operates as a drainage system in reverse. Septic wastewaters can be disposed of and treated by soils if the soils are freely draining. Soils with low permeability or high water tables may indicate good conditions for wetland creation or appropriate sites for installation of artificial drainage for agricultural use, but they are not generally suited for septic tank drain fields.

Blue water, the term for water withdrawn from streams or wells, can be applied to the soil as irrigation, greatly enhancing plant growth, especially in regions with scarce precipitation. With increasing competition for limited global and local water resources, it is essential that irrigators manage water with maximal efficiency so that the greatest production can be achieved with the least waste of water resources. Such efficiency is encouraged by practices that favor transpiration over evaporation, such as mulching and the use of microirrigation.

The operation of the hydrologic cycle causes changes in soil water content, which in turn influence other soil properties, most notably soil aeration and temperature. These soil properties will be the subjects of the next chapter.

STUDY QUESTIONS

1. You know that the forest vegetation that covers a 120 km^2 wildland watershed uses an average of 4 mm of water per day during the summer. You also know that the soil averages 150 cm in depth and at field capacity can store 0.2 mm of water per mm of soil depth. However, at the beginning of the season the soil was quite dry, holding an average of only 0.1 mm/mm. As the watershed manager, you are asked to predict how much water will be carried by the streams draining the watershed during the 90-day summer period when 450 mm of precipitation falls on the area. Use the water balance equation to make a rough prediction of the stream discharge as a percentage of the precipitation and in cubic meters of water.

2. Draw a simple diagram of the hydrologic cycle using a separate arrow to represent these processes: *evaporation, transpiration, infiltration, interception, percolation, surface runoff,* and *soil storage*.

3. Describe and give an example of the *indirect* effects of plants on the hydrologic balance through their effects on the soil.

4. State the basic principle that governs how water moves through the SPAC. Give two examples, one at the soil–root interface and one at the leaf–atmosphere interface.

5. Define *potential evapotranspiration* and explain its significance to water management.

6. What is the role of evaporation (E) from the soil in determining water-use efficiency, and how does it affect ET? List three practices that can be used to control losses by E.

7. Weed control should reduce water losses by what process?

8. Hydraulic lift is a special case of hydraulic redistribution. Describe this phenomenon and the conditions that favor its operation.

9. Comment on the relative advantages and disadvantages of organic versus plastic mulches.

10. What does conservation tillage conserve? How does it do it?

11. The small irrigation project you manage collects 2,000,000 m^3 of water annually in a reservoir. Of this, 20% evaporates from the reservoir surface during the year. Of the remaining water, 25% is lost by evaporation and percolation into the soil during distribution via unlined canals before the water reaches the fields. The water is then applied by furrow irrigation, with averages 20% of the water applied percolating below the crop root zone, 20% running off into collection canals at the low end of the field, and 30% evaporating from the soil surface. By the time of crop harvest, ET has dried the soil to about the same water content it had before irrigation began. Average ET is 7 mm/day for a 180-day irrigation season and crop water-use efficiency averages 1.1 kg of dry matter/m^3 water transpired. Show calculations to estimate:
    (a) the overall water-use efficiency of the project (kg output/m^3 water allocated),
    (b) the application water efficiency for the project,
    (c) the field water efficiency for the project,
    (d) the number of hectares that can be irrigated in this project.

12. Explain under what circumstances earthworm channels might increase downward saturated water flow, but not have much effect on the leaching of soluble chemicals applied to the soil.

13. What will be the effect of placing a perforated drainage pipe in the capillary fringe zone just above the water table in a wet soil? Explain in terms of water potentials.

14. What soil features may limit the use of a site for a septic tank drain field?

15. Which irrigation systems are likely to be used where: (a) water is expensive and the market value of crops produced per hectare is high, and (b) the cost of irrigation water is subsidized and the value of crop products that can be produced per hectare is low? Explain.

REFERENCES

Basso, B., and J. T. Ritchie. 2012. "Assessing the impact of management strategies on water use efficiency using soil–plant–atmosphere models." *Vadose Zone Journal* 11:doi:10.2136/vzj2011.0173.

Brooks, J. R. 2015. "Water, bound and mobile." *Science* 349:138–139.

Chen, G., and R. R. Weil. 2011. "Root growth and yield of maize as affected by soil compaction and cover crops." *Soil and Tillage Research* 117:17–27.

Day, R. L. et al. 1998. "Water balance and flow patterns in a fragipan using in situ soil block." *Soil Science* 163:517–528.

Gburek, W. J., B. A. Needelman, and M. S. Srinivasan. 2006. "Fragipan controls on runoff generation: Hydropedological implications at landscape and watershed scales." *Geoderma* 131:330–344.

Hillel, D. 1995. *The Rivers of Eden*. Oxford University Press, New York.

Hillel, D. 1997. *Small-Scale Irrigation for Arid Zones*. FAO Development Series 2. U.N. Food and Agriculture Organization, Rome.

Jeswani, H. K., and A. Azapagic. 2011. "Water footprint: Methodologies and a case study for assessing the impacts of water use." *Journal of Cleaner Production* 19:1288–1299.

Kemper, W. D., C. E. Bongert, and D. M. Marohn. 2012. "Corn response to tillage and water table depth." *Journal of Soil and Water Conservation* 67:31A–36A.

Lal, R. 2015. "World water resources and achieving water security." *Agronomy Journal* 107:1526–1532.

Lascano, R. J., and R. E. Sojka (eds.). 2007. *Irrigation of Agricultural Crops*. Agronomy Monograph 30. American Society of Agronomy, Madsion, WI, pp. 1–664.

Nagy, R. C., B. G. Lockaby, L. Helms, L. Kalin, and D. Stoeckel. 2011. "Water resources and land use and cover in a humid region: The southeastern United States." *Journal of Environmental Quality* 40:867–878.

Oki, T. 2011. "Global hydrology." In W. Peter (ed.). *Treatise on Water Science*. Vol. 2. Elsevier, Oxford, pp. 3–25.

Pfister, S., P. Bayer, A. Koehler, and S. Hellweg. 2011. "Environmental impacts of water use in global crop production: Hotspots and trade-offs with land use." *Environmental Science & Technology* 45:5761–5768.

Prieto, I., C. Armas, and F. I. Pugnaire. 2012. "Water release through plant roots: New insights into its consequences at the plant and ecosystem level." *New Phytologist* 193:830–841.

Skaggs, R. W., N. R. Fausey, and R. O. Evans. 2012. "Drainage water management." *Journal of Soil and Water Conservation* 67:167A–172A.

Soil Survey Staff. 1993. *National Soil Survey Handbook*. Title 430-VI. USDA Natural Resources Conservation Service, Washington, DC.

Starrett, S. K., N. E. Christians, and T. Al Austin. 1996. "Movement of pesticides under two irrigation regimes applied to turfgrass." *Journal of Environmental Quality* 25:566–571.

Tyree, M. T. 2003. "The ascent of water." *Nature* 423:923.

Unger, P. W., and R. L. Baumhardt. 1999. "Factors related to dryland grain sorghum yield increases: 1939 through 1997." *Agronomy Journal* 91:870–875.

Waddell, J., and R. Weil. 1996. "Water distribution in soil under ridge-till and no-till corn." *Soil Science Society of America Journal* 60:230–237.

Wallace, J. S. 2000. "Increasing agricultural water use efficiency to meet future food production." *Agriculture, Ecosystems and Environment* 82:105–119.

Zanders, J. 2001. "Urban development and soils." *Soil Horizons—A Newsletter of Landcare Research New Zealand, Ltd.* 6 (Nov.):6.

Adventitious tree roots in a wetland soil. (Photo courtesy of Ray R. Weil)

7 Soil Aeration and Temperature

The naked earth is warm with Spring.
—JULIAN GRENFELL, INTO BATTLE

It is a central maxim of ecology that "everything is connected to everything else." This interconnectedness is one reason why soils are such fascinating (and challenging) objects of study. In this chapter we shall explore two aspects of the soil environment, aeration and temperature; they are not only closely connected to each other but are also intimately influenced by many of the soil properties discussed in other chapters.

Since air and water share the pore space of soils, it is not surprising that much of what we learned about the texture, structure, and porosity of soils (Chapter 4) and the retention and movement of water in soils (Chapters 5 and 6) will have direct bearing on soil aeration. In addition, chemical and biological processes also affect, and are affected by, soil aeration.

For the growth of plants and the activity of microorganisms, soil aeration status can be just as important as soil moisture status and can sometimes be even more difficult to manage. In most forest, range, agricultural, and ornamental applications, a major management objective is to maintain a high level of oxygen in the soil for root respiration. Yet it is also vital that we understand the chemical and biological changes that take place when the oxygen supply in the soil is depleted.

Soil temperatures affect plant and microorganism growth and also influence soil drying by evaporation. The movement and retention of heat energy in soils are often ignored, but they hold the key to understanding many important soil phenomena, from frost-damaged pipelines and pavements to the spring awakening of biological activity in soils. We will see that warm soil temperatures influence soil aeration largely through their stimulating effects on the growth of plants and soil organisms and on the rates of biochemical reactions. Nowhere are these interrelationships more critical than in the water-saturated, oxygen depleted soils of wetlands, ecosystems that will therefore receive special attention in this chapter.

7.1 SOIL AERATION—THE PROCESS

Soil ventilation allows the exchange of gases between the soil and the atmosphere to supply enough oxygen (O_2) for respiration while preventing the potentially toxic accumulation of gases such as carbon dioxide (CO_2), methane (CH_4), and ethylene (C_2H_4). Soil aeration status involves the rate of such ventilation, as well as the proportion of pore spaces filled with air, the composition of that soil air, and the resulting chemical oxidation or reduction potential in the soil environment.

Soil Aeration in the Field

Oxygen availability in field soils is regulated by three principal factors: (1) *soil macroporosity* (as affected by texture and structure), (2) *soil water content* (as it affects the proportion of porosity that is filled with air), and (3) O_2 *consumption* by respiring organisms (including plant roots and microorganisms). The term *poor soil aeration* refers to a condition in which the availability of O_2 in the root zone is insufficient to support optimal growth of upland plants and aerobic microorganisms. Typically, poor aeration becomes a serious impediment to plant growth when O_2 concentration drops below 0.1 L/L. This often occurs when more than 80 to 90% of the soil pore space is filled with water. The high soil water content not only leaves less than 10 to 20% of the pore space for air storage, but, more important, the water blocks the pathways by which gases could exchange with the atmosphere. Compaction can also cut off gas exchange by decreasing pore size and total pore space, even if the soil is not very wet and has a relatively large percentage of air-filled pores.

Excess Moisture

When all (or nearly all) of the soil pores are filled with water, the soil is said to be ***water saturated*** or ***waterlogged***. Waterlogged soil conditions are typical of wetlands and may also occur for short periods of time on upland sites. In well-drained soils, saturated conditions may occur temporarily during a heavy rain or irrigation water, or if wet soil has been compacted by plowing or by heavy machinery.

Plants adapted to life in waterlogged soils are termed ***hydrophytes***. Examples are sedges and certain grass species, including rice, that transport oxygen for respiration down to their roots via hollow structures in their stems and roots known as ***aerenchyma*** tissues. Mangrove and other hydrophytic trees produce aerial roots and other structures that allow their roots to obtain O_2 while growing in water-saturated soils.

Most plants, however, are dependent on a supply of oxygen from the soil to their roots and suffer dramatically if good soil aeration is not maintained by drainage or other means (Figure 7.1). Some plants succumb to O_2 deficiency within hours after the soil becomes saturated. Other upland plants are capable of sensing low O_2 levels and adapting their physiology or morphology (for instance, by developing aerenchyma or slowing their oxygen use) within a few days or weeks to alleviate the aeration stress.

Gaseous Interchange

The more rapidly roots and microbes use up oxygen and release carbon dioxide, the greater is the need for ventilation—the exchange of gases between the soil and the atmosphere. This exchange is facilitated by two mechanisms, ***mass flow*** and ***diffusion***. Mass flow of air is enhanced by fluctuations in soil moisture content that force air in or out of the soil and by wind

Figure 7.1 Most plants depend on the soil to supply oxygen for root respiration and therefore are disastrously affected by even relatively brief periods of soil saturation during which oxygen becomes depleted. (Left) Sugar beets on a clay loam soil dying where the soil has become water saturated in a compacted area. (Right) Conifer trees dying in a sandy soil area that has become saturated as a result of flooding by beavers. (Photos courtesy of Ray R. Weil)

and changes in barometric pressure. However, the great bulk of gaseous interchange in soils occurs by *diffusion*.

Through diffusion, each gas moves in a direction determined by its own partial pressure. The *partial pressure* of a gas in a mixture is simply the pressure this gas would exert if it alone were present in the volume occupied by the mixture. Thus, if the pressure of air is 100 kPa (~1 atmosphere), then the partial pressure of oxygen, which makes up about 21% (0.21 L/L) of the air by volume, is approximately 21 kPa.

Diffusion allows extensive gas movement from one area to another even though there is no overall pressure gradient for the total mixture of gases. There is, however, a concentration gradient for each individual gas, which may be expressed as a *partial pressure gradient*. As a consequence, the higher concentration of oxygen in the atmosphere will result in a net movement of this particular gas into the soil. Carbon dioxide and water vapor normally move in opposite directions, since the partial pressures of these two gases are generally higher in the soil air than in the atmosphere.

7.2 MEANS OF CHARACTERIZING SOIL AERATION

The aeration status of a soil can be characterized in several ways, including (1) the content of oxygen and other gases in the soil atmosphere, (2) the air-filled soil porosity, and (3) the chemical oxidation–reduction (redox) potential.

Gaseous Composition of the Soil Air

Oxygen. The atmosphere above the soil contains nearly 21% O_2, 0.04% CO_2, and more than 78% N_2. In comparison, soil air has about the same level of N_2 but is consistently lower in O_2 and higher in CO_2. The O_2 content may be only slightly below 20% in the upper layers of a soil with an abundance of macropores. It may drop to less than 5% or even to near zero in the lower horizons of a poorly drained soil with few macropores. Once the supply of O_2 is exhausted, the soil environment is said to be **anaerobic**.

Low O_2 contents are typical of wet soils. Even in well-drained soils, marked reductions in the O_2 content of soil air may follow a heavy rain, especially if oxygen is being rapidly consumed by actively growing plant roots or by microbes decomposing readily available supplies of organic materials (Figure 7.2). Oxygen depletion in this manner occurs most rapidly when the soil is warm.

Carbon Dioxide. Since the N_2 content of soil air is relatively constant, there is a general inverse relationship between the contents of the other two major components of soil air— O_2 and CO_2—with O_2 decreasing as CO_2 increases. Although the actual differences in CO_2 amounts may not be impressive, they are significant, comparatively speaking. Thus, when the soil air contains only 0.4% CO_2, this gas is about ten times as concentrated as it is in the atmosphere. In cases where the CO_2 content becomes as high as 10%, it may be toxic to some plant processes.

Other Gases. Soil air usually is much higher in water vapor than atmospheric air, being essentially saturated except at or very near the surface of the soil (see Section 5.7). Also, under waterlogged conditions, the concentrations of gases such as methane (CH_4) and hydrogen sulfide (H_2S), which are formed as organic matter decomposes, are notably higher in soil air. Another gas produced by anaerobic microbial metabolism is ethylene (C_2H_4). This is very bioactive and is involved in how some plants sense the depletion of O_2 in their root zones. When gas exchange rates between the atmosphere and the soil are so slow that ethylene accumulates, even in concentrations as low as 1 μL/L (0.0001%), root growth of a number of plants has been shown to be inhibited.

Figure 7.2 *Oxygen diffusion through the soil (squiggly arrows) is inhibited by water-filled pores because the diffusion coefficient (D) for O_2 is about 10,000 times smaller through water than through air. In a moist soil the micropores inside aggregates are mostly filled with water, making it difficult for O_2 to reach the center of such aggregates.* (Diagram courtesy of Ray R. Weil)

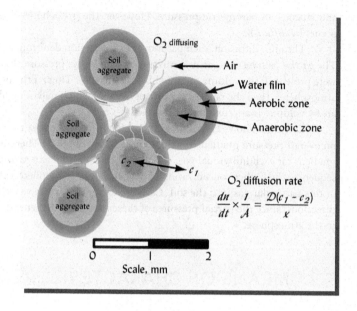

Air-Filled Porosity

In Chapter 1 we noted that the ideal soil composition for plant growth would include close to a 50:50 mix of air and water in the soil pore space, or about 25% air in the soil, by volume (assuming a total porosity of 50%) (Figure 1.14). Many researchers believe that microbiological activity and plant growth become severely inhibited in most soils when air-filled porosity falls below 20% of the pore space or 10% of the total soil volume (with correspondingly high water contents).

One of the principal reasons that high water contents cause oxygen deficiencies for roots is that water-filled pores block the diffusion of oxygen into the soil to replace that used by respiration (Figure 7.2). In fact, oxygen diffuses 10,000 times faster through a pore filled with air than through a similar pore filled with water.

7.3 OXIDATION–REDUCTION (REDOX) POTENTIAL[1]

Important chemical characteristics related to soil aeration are the reduction and oxidation states of the chemical elements.

Redox Reactions

The reaction that takes place when the reduced state of an element is changed to the oxidized state may be illustrated by the oxidation of two-valent iron [Fe^{2+} or Fe(II)] in FeO to the trivalent form [Fe^{3+} or Fe(III)] in FeOOH:

$$\underset{\text{Fe(II)}}{\overset{(2+)}{2FeO}} + 2H_2O \rightleftharpoons \underset{\text{Fe(III)}}{\overset{(3+)}{2FeOOH}} + 2H^+ + 2e^- \tag{7.1}$$

As reaction (7.1) proceeds to the *right*, each Fe(II) loses an electron (e^-) to become Fe(III) and forms H^+ ions by hydrolyzing H_2O. These H^+ ions lower the pH. When the reaction proceeds to the *left*, FeOOH acts as an **electron acceptor** and the pH rises as H^+ ions are consumed. The tendency or potential for electrons to be transferred from one substance to another in such reactions is termed the **redox potential** (E_h), which can be measured using a platinum electrode attached to a millivolt meter.

The redox potential is usually expressed in volts or millivolts. As is the case for water potential (Section 5.3), redox potential is related to a reference state; in this case, the redox potential of the

---

[1] For reviews of redox reactions in soils, see Bartlett and James (1993) and Bartlett and Ross (2005).

hydrogen couple ½H$_2$ ↔ 2H$^+$ + e$^-$ is arbitrarily taken as zero. If a substance will accept electrons easily, it is known as an *oxidizing agent*; if a substance supplies electrons easily, it is a *reducing agent*.

Role of Oxygen Gas

Oxygen gas (O$_2$) is an important example of a strong oxidizing agent, since it rapidly accepts electrons from many other elements. All aerobic respiration requires O$_2$ to accept the electrons released as living organisms oxidize organic carbon to provide energy for life.

Oxygen can oxidize both organic and inorganic substances. Keep in mind, however, that as it oxidizes another substance, O$_2$ is in turn reduced. This reduction process can be seen in the following reaction:

$$\overset{(0)}{½O_2} + 2H^+ + 2e^- \rightleftharpoons \overset{(2-)}{H_2O} \qquad (7.2)$$

Note that the oxygen atom which has zero charge in elemental O$_2$ accepts two electrons, taking on a charge of −2 when it becomes part of the water molecule. These electrons could have been donated by two molecules of FeO undergoing oxidation, as shown in reaction (7.1). If we combine Equations (7.1) and (7.2), we can see the overall effect of oxidation and reduction:

$$\begin{array}{r}
2FeO + 2H_2O \rightleftharpoons 2FeOOH + 2H^+ + 2e^- \\
½O_2 + 2H^+ + 2e^- \rightleftharpoons H_2O \\
\hline
2FeO + ½O_2 + H_2O \rightleftharpoons 2FeOOH
\end{array} \qquad (7.3)$$

The donation and acceptance of electrons (e$^-$) and H$^+$ ions on each side of the equations have balanced each other and therefore do not appear in the combined reaction. However, for the specific reduction and oxidation reactions, they are both very important.

In a well-aerated soil with plenty of gaseous O$_2$, the E_h is in the range of 0.4 to 0.7 volt (V). As aeration is reduced and gaseous O$_2$ is depleted, the E_h declines to about 0.3 to 0.35 V. If organic-matter-rich soils are flooded under warm conditions, E_h values as low as −0.3 V can be found.

Within a day or two after a warm soil becomes water saturated, aerobic and facultative microorganisms oxidizing organic carbon in the soil (see Section 10.2) respire most of the O$_2$ initially present, thereby lowering both the O$_2$ content and the E_h of the soil solution.

Other Electron Acceptors

As O$_2$ becomes depleted and E_h drops, reducing conditions become established. With no O$_2$ available, only **anaerobic** microorganisms can survive. They must use substances other than O$_2$ as the terminal electron acceptors for their metabolism. For example, they may use iron in soil minerals. As they reduce iron, the E_h drops still further because electrons are consumed. The pH simultaneously rises because the reaction consumes H$^+$ ions:

$$\underset{Fe(III)}{Fe(OH)_3} + e^- + 3H^+ \longrightarrow \underset{Fe(II)}{Fe^{2+}} + 3H_2O \qquad (7.4)$$

When these reduction reactions proceed, soil colors change from the reds of iron oxides to the grays of reduced iron and minerals exposed by the removal of iron oxide coatings (see Figure 7.3 and Sections 4.1 and 7.7). Similar reactions involve the reduction or oxidation of C, N, Mn, S, and other elements from the solution, organic matter, and mineral components of the soil.

The E_h value at which a particular oxidation–reduction (**redox**) reaction can occur depends on the chemical to be oxidized or reduced and the pH at which the reaction takes

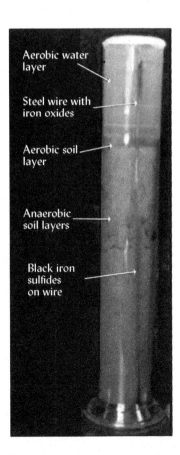

Figure 7.3 Oxidized (orange iron oxide) and reduced (black iron sulfide) zones along a steel wire embedded in a waterlogged soil column. Soil was mixed with organic materials and calcium sulfate (to supply sulfur), blended with water to make a thick slurry and poured into a large glass cylinder such that several centimeters of water covered the soil which settled to the bottom. A steel wire was then inserted near the glass to act as a simple redox indicator. The cylinder was then covered with parafilm to prevent evaporation and allowed to stand in sunlight. After several weeks the action of soil microorganisms resulted in a redox gradient and the chemical changes shown. (Photo courtesy of Ray R. Weil)

Figure 7.4 *The effect of pH on the redox potential E_h at which several important reduction–oxidation reactions theoretically take place. In each reaction the oxidized form of the element is on the left side and the reduced form is on the right side with the arrows indicating that the reactions can go either way. If a soil becomes saturated, the reactions will occur in order from the top down, as microorganisms first use the substances that most readily accept electrons. For example, if the pH is 6.5 (indicated by the vertical dotted line), any gaseous O_2 will be reduced first, followed by nitrogen (NO_3^-), then manganese (MnO_2), then iron in $Fe(OH)_3$, iron in FeOOH, and finally sulfur (in SO_4^{2-}) and carbon (in CO_2).* [Modified from McBride (1994); used with permission of Oxford University Press]

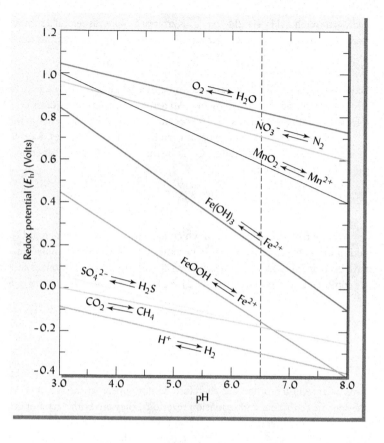

place, as illustrated by the sloping lines in Figure 7.4. At E_h values above and below a given line, the corresponding chemical element would be found in oxidized and reduced forms, respectively. The fact that these lines are sloping downward suggests that the E_h for a reaction is lower when the pH of the soil solution is higher. Since both pH and E_h are easily measured, it is not too difficult to ascertain whether a certain reaction could occur in a given soil under specific conditions.

The order of E_h lines from the top to the bottom of Figure 7.4 illustrates the *sequence* in which the various reduction reactions are likely to occur when a well-aerated soil becomes saturated with water. It is important to note, however, that E_h lines are theoretical and based on the assumptions that all the *reactions are at equilibrium* and that the electrons and protons from one reaction are perfectly free to participate in any of the other reactions. In real soils, these assumptions are hardly ever met.

Figure 7.5 provides further insights into the sequence of reduction reactions that typically occur after a soil is inundated. Once the soil becomes essentially devoid of O_2, the soil redox potential falls below levels of about 0.38 to 0.32 V (at pH 6.5). After O_2, the next most easily reduced substance present is usually the N^{5+} in nitrate (NO_3^-). If the soil contains much NO_3^-, the E_h will remain near 0.28 to 0.22 V as the nitrate is reduced:

$$\underset{N(V)}{\overset{(5+)}{NO_3^-}} + 2e^- + 2H^+ \longleftrightarrow \underset{N(III)}{\overset{(3+)}{NO_2^-}} + H_2O \tag{7.5}$$

Once all the N^{5+} in nitrate has been transformed into NO_2^-, N_2, and other N species, Mn, Fe, and S (in SO_4^{2-}) and C (in CO_2) accept electrons and become reduced, predominantly in the order shown in Figure 7.5.

In other words, transformation of different elements requires different degrees of reducing conditions. The soil E_h must be lowered to -0.2 V before methane is produced, but reduction of NO_3^- to N_2 gas takes place when the E_h is as high as $+0.28$ V. We can conclude that soil aeration helps determine the specific chemical species present in soils and, in turn, the availability, mobility, and possible toxicity of many chemical elements.

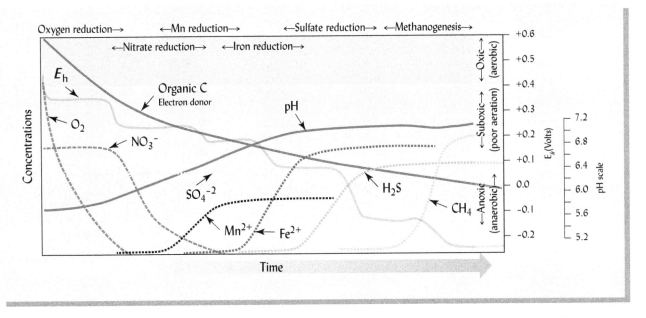

Figure 7.5 *Generalized changes in soil chemistry following water saturation of a soil with plentiful organic matter. The time period depicted could range from a few days (if warm and high in sugars) to a few weeks (if cool and low in easily oxidized C). In the first period, aerobic and facultative microorganisms digest (oxidize) the organic carbon via respiration, consuming (reducing) most of the dissolved O_2, thereby lowering the E_h of the soil solution (solid blue line). As O_2 is depleted and E_h drops, anaerobic microorganisms use substances in the soil solution such as nitrate (NO_3^-) as the terminal acceptor for electrons released by their metabolism. This process reduces the nitrate and lowers the E_h further. As E_h drops, microbes then reduce manganese [Mn(III,IV)] and later iron [Fe(III)] oxides in soil minerals, causing the reduced metal ions to appear in solution. Such reactions change the color of the soil by dissolving certain iron-containing minerals. As depicted by the solid red line in the graph, the pH tends to rise as H^+ ions are consumed in reactions that reduce Fe^{3+} (see reactions (7.1) and (7.4)). Once most of the iron has been reduced and the E_h has dropped well below 0 V, the S(+VI) in sulfate (SO_4^{-2}) is reduced to sulfide (S^{-2}) producing hydrogen sulfide gas (H_2S). Finally, carbon compounds are reduced by methanogens microbes to produce methane gas (CH_4).* (Diagram courtesy of Ray R. Weil)

7.4 FACTORS AFFECTING SOIL AERATION AND E_H

Rates of Respiration in the Soil

The concentrations of both O_2 and CO_2 are largely dependent on microbial activity, which in turn depends on the availability of organic carbon compounds as food. Incorporation of large quantities of animal manure, crop residues, compost, or sewage sludge may alter the soil air composition appreciably. Likewise, the cycling of plant residues by leaf fall, root mass decay, and root excretion provides the main substrates for microbial activity in natural ecosystems. Respiration by plant roots and enhanced respiration by soil organisms near the roots are also significant processes. All these processes are very much enhanced as soil temperature increases (see Section 7.8).

Depth in the Soil Profile

Subsoils are usually lower in oxygen than are topsoils. Not only is the water content usually higher (in humid climates), but the total pore space, as well as the macropore space, is generally much lower in the deeper horizons. In addition, gases in deep horizons have a longer pathway for diffusion into and out of the soil than gases near the soil surface. However, the subsoil may still be aerobic because O_2 can diffuse fast enough to replace that used by respiration if organic substrates are in low supply. For this reason, certain recently flooded soils may become anaerobic in the upper 50 to 100 cm, but remain aerobic deeper in the soil profile. Understanding the diffusion of gases up and down a soil profile is also important for modeling such environmentally influential processes as the emission of gases from landfills (Figure 7.6).

Figure 7.6 Simulated changes in the composition of the soil air as landfill gases (methane CH_4 and carbon dioxide CO_2) diffuse up and oxygen O_2 and nitrogen N_2 diffuse down through the soil covering a landfill (see also Section 15.9). In the surface horizon (upper 20 cm) the O_2 and N_2 concentrations are similar to those in the atmosphere, but with the concentration of CO_2 rising with depth at the expense of O_2. Deeper in the soil (closer to the buried garbage) O_2 is virtually absent as it has all been respired by soil microbes, while CH_4 and CO_2 produced by microbial processes in the anaerobic decay of the garbage rise up through the soil by diffusion. In the deeper layers the N_2 concentration is reduced because of dilution by the other gases. [Redrawn from Molins et al. (2008)]

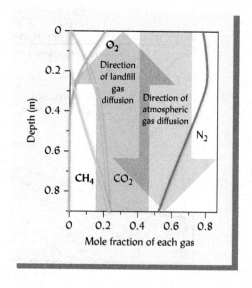

Drainage of Excess Water

Drainage of gravitational water out of the profile and concomitant diffusion of air into the soil takes place most readily in macropores. Soil texture, bulk density, aggregate stability, organic matter content, and biopore formation are among the soil properties that help determine macropore content and, in turn, soil aeration (see Section 4.8). Artificial drainage influences aeration by essentially creating gigantic macropores in the soil system that can rapidly carry off excess water and lower the water table (see Section 6.7). On the other hand, water saturation and eventually anoxic conditions often occur just above abrupt profile changes in pore size, such as caused by layers of gravel or clay (see Box 5.2).

Small-Scale Soil Heterogeneity

Tillage. One cause of soil heterogeneity is tillage, which has both short-term and long-term effects on soil aeration. In the short term, stirring the soil often allows it to dry out faster and also mixes in large quantities of air. These effects are especially evident on somewhat compacted, fine-textured soils, on which plant growth often responds immediately after a cultivation to control weeds or "knife in" fertilizer. In the long term, however, tillage may reduce macroporosity (see Section 4.6).

Pore Sizes. Very small pores in a clayey or compacted soil may slow the diffusion of O_2 and thereby lower redox potentials within that layer, whether or not the soil is very wet. Similarly, O_2 will diffuse much more slowly through the small, largely water-filled pores within a soil aggregate than through the largely air-filled pores between aggregates. Therefore, anaerobic conditions may occur in the center of an aggregate, only a few millimeters from well-aerated conditions near the surface of the same aggregate (Figures 7.2 and 7.7).

In some upland soils, large subsoil pores such as cracks between peds and old root channels may periodically fill with water, causing localized zones of poor aeration. This condition is expressed by gray, reduced surfaces on peds with reddish, oxidized interiors (see, e.g., Figure 4.39d). In a normally saturated soil, such large pores may cause the opposite effect (peds with oxidized faces but reduced interiors), as they facilitate O_2 diffusion into the soil during dry periods.

Plant Roots. Respiration by the roots of upland plants usually depletes the O_2 in the soil just outside the root. The opposite can occur in wet soils growing hydrophytic plants. Aerenchyma tissues in these plants may transport surplus O_2 into the roots, allowing some to diffuse into the soil and produce an oxidized zone in an otherwise anaerobic soil (see Section 7.7).

For all these reasons, aerobic and anaerobic processes may proceed simultaneously and in very close proximity to each other in the same soil. This heterogeneity of soil aeration should be kept in mind when attempting to understand the role that soil plays in elemental cycling and ecosystem function.

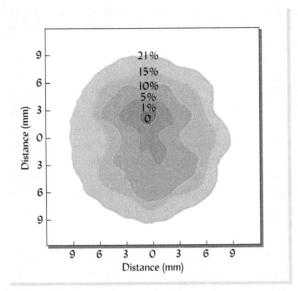

Figure 7.7 The oxygen content of soil air in a wet aggregate in an Aquic Hapludoll (Muscatine silty clay loam) from Iowa. The measurements were made with a unique microelectrode. Note that the oxygen content near the aggregate center was zero, while that near the edge of the aggregate was 21%. Thus, pockets of oxygen deficiency (anaerobic zones) can be found in a soil whose overall oxygen content may not be low. Refer also to Figure 7.4. [Diagram from Sexstone et al. (1985)]

Seasonal Differences

In many soils, aeration status varies markedly with the seasons. In temperate humid regions, soils are often wet early in spring and opportunities for ready gas exchange are poor. But due to low soil temperatures, the utilization of O_2 and release of CO_2 by plant roots and microorganisms is restricted. In summer, these soils are commonly lower in moisture content, and the opportunity for gas exchange is increased. However, plant roots and microorganisms, stimulated by the warmer temperatures, quickly deplete O_2 and release copious quantities of CO_2.

Effects of Vegetation

In addition to the root respiration effects mentioned previously, vegetation may affect soil aeration by removing large quantities of water via transpiration, enough to lower the water table in some poorly drained soils.

7.5 ECOLOGICAL EFFECTS OF SOIL AERATION[2]

Effects on Organic Residue Degradation

Soil aeration influences many soil reactions and, in turn, many soil properties. The most obvious of these reactions are associated with microbial activity, especially the breakdown of organic residues and other microbial reactions. Poor aeration slows down the rate of decay, as evidenced by the relatively high levels of organic matter that accumulate in poorly drained soils.

The nature as well as the rate of microbial activity is determined by the O_2 content of the soil. Where O_2 is present, aerobic organisms are active, see Section 11.2. In the absence of gaseous oxygen, anaerobic organisms take over. Much slower breakdown occurs through reactions such as the following:

$$\underset{\text{Sugar}}{C_6H_{12}O_6} \rightarrow 2CO_2 + \underset{\text{Ethanol}}{2CH_3CH_2OH} \qquad (7.6)$$

Poorly aerated soils therefore tend to contain a wide variety of only partially oxidized products such as ethylene gas (C_2H_4), alcohols, and organic acids, many of which can be toxic

---

[2] For interdisciplinary reviews of the influence of soil aeration and redox reactions on plant and microbial ecology, see Husson (2012) and of how plants sense and adapt to low O_2, see Bailey-Serres et al. (2012).

to higher plants and to many decomposer organisms. The latter effect helps account for the formation of Histosols in wet areas where inhibition of decomposition allows thick layers of organic matter to accumulate. In summary, the presence or absence of oxygen gas completely modifies the nature of the decay process and its effect on plant growth.

Oxidation–Reduction of Elements

Nutrients. The level of soil oxygen and the redox potential largely determine the forms of several inorganic elements, as shown in Table 7.1. The oxidized states of the nitrogen and sulfur are readily utilizable by higher plants. In general, oxidized conditions are desirable for iron and manganese nutrition of most plants in acid soils of humid regions because, in these soils, the reduced forms of these elements are so soluble that toxicities may occur. In drier areas with neutral-to-alkaline soils, oxidized forms of iron and manganese are tied up in highly insoluble compounds, resulting in plant deficiencies of these elements. In these situations, restricted aeration resulting in reduced forms of elements such as iron and manganese can improve soil fertility. Some reduction of iron may also be beneficial in acid soils as it will release phosphorus from insoluble iron-phosphate compounds. Such phosphorus release has implications for eutrophication when it occurs in saturated soils or in underwater sediments. These examples illustrate the interaction of soil E_h and pH in supplying available nutrients to plants (see also Chapter 12).

Toxic Elements. Redox potential determines the species of such potentially toxic elements as chromium, arsenic, and selenium, markedly affecting their impact on the environment and food chain (see Section 18.8). Reduced forms of arsenic are most mobile and toxic, giving rise to toxic levels of this element in drinking water, a serious human health problem in many places around the world, but especially in Bangladesh. In contrast, it is the oxidized hexavalent form of chromium (Cr^{6+}) that is mobile and very toxic to humans. In neutral to acid soils, easily decomposed organic materials can be used to reduce the chromium to the less dangerous Cr^{3+} form, which is not subject to ready reoxidation (Figure 7.8).

Soil Colors. As was discussed in Section 4.1, soil color is influenced markedly by the oxidation status of iron and manganese. Colors such as red, yellow, and reddish brown are characteristic of well-oxidized conditions. More subdued shades such as grays and blues predominate if under reduced conditions. Soil color is used in the field to classify wetland soils. Soil zones of alternating oxidized and reduced conditions are characterized by contrasting small areas of oxidized and reduced materials often associated with ped interiors and ped faces. Such a "mottled" appearance indicates conditions not conducive to the optimum growth of upland plants.

Greenhouse Gases. The production of nitrous oxide (N_2O) and methane (CH_4) in wet soils is of global significance. These two gases, along with carbon dioxide (CO_2), are responsible for about 80% of the anthropogenic global warming caused by the "greenhouse effect" (see Section 11.9). The atmospheric concentrations of these gases have been increasing at alarming rates each year for the past half century or more.

Table 7.1
OXIDIZED AND REDUCED FORMS OF SEVERAL IMPORTANT ELEMENTS

Element	Normal form in well-oxidized soils	Reduced form found in waterlogged soils
Carbon	CO_2, $C_6H_{12}O_6$	CH_4, C_2H_4, CH_3CH_2OH
Nitrogen	NO_3^-	N_2, NH_4^+
Sulfur	SO_4^{2-}	H_2S, S^{2-}
Iron	Fe^{3+} [Fe(III) oxides]	Fe^{2+} [Fe(II) oxides]
Manganese	Mn^{4+} [Mn(IV) oxides]	Mn^{2+} [Mn(II) oxides]

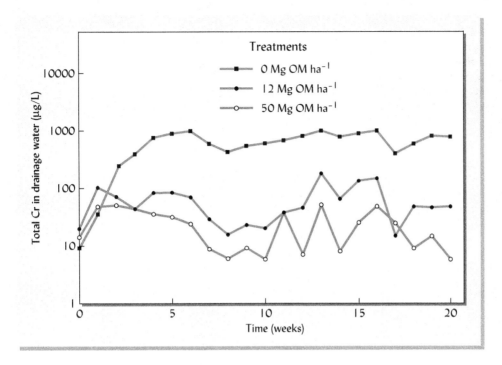

Figure 7.8 Effect of adding decomposable organic matter (dried cattle manure) on the concentration of chromium in water draining from a chromium-contaminated soil. As the manure oxidized, it caused the reduction of the toxic, mobile Cr^{6+} to the relatively immobile, nontoxic Cr^{3+}. Note the log scale for the Cr in the water, indicating that the addition of 50 Mg manure ha^{-1} caused Cr to be lowered approximately 100-fold. The coarse-textured soil was a Typic Torripsamment in California. [Data from Losi et al. (1994)]

Methane gas is produced by the reduction of CO_2. Its formation occurs when the E_h is −0.2 V, a condition common in natural wetlands and in rice paddies. It is estimated that wetlands in the United States emit about 100 million metric tons of methane annually. Nitrous oxide is also produced in large quantities by wetland soils, as well as sporadically by upland soils. Because of the biological productivity and diversity of wetlands (Section 7.7), soil scientists are seeking means of managing greenhouse gases from these wetland soils without resorting to draining (i.e., destroying) them. Fortunately, it may be possible to minimize production of all three major greenhouse gases by maintaining soil E_h in a moderately low range that is feasible for many rice paddies and natural wetlands (Figure 7.9).

Effects on Activities of Higher Plants

It is usually the lack of oxygen in the root environment rather than the excessive wetness itself that impairs plant growth in flooded or overly wet soils. This fact explains why flooding a soil with stagnant water is generally much more damaging to plants—even for some hydrophytes—than flooding with flowing water. The O_2 content of the soil is continually replenished by flowing water, but roots and microbes will completely deplete the O_2 supply under stagnant water.

Plant Growth. The lack of oxygen in the soil alters root function and changes the metabolism of the entire plant. Plant species vary in their ability to tolerate poor aeration (Table 7.2). Among crop plants, sugar beet is an example of a species that is very sensitive to poor soil aeration (see Figure 7.1, *left*). At the opposite extreme, paddy rice (Figure 12.15) is an example of a species that can grow with its roots completely submerged in water. Furthermore, for a given species of plant, the young seedlings may be more tolerant of low soil aeration porosity than are older plants. A case in point is the tolerance of red pine to restricted drainage during its early development and its poor growth or even death on the same site at later stages (see Figure 7.2, *right*). The occurrence of native plants specially adapted to anaerobic conditions is useful in identifying wetland sites (see Section 7.7). Knowledge of plant tolerance to poor aeration is useful in choosing appropriate species to revegetate wet sites.

Nutrients and Water Uptake. Low O_2 levels constrain root respiration and impair root function. The root cell membrane may become less permeable to water, so that plants may actually have difficulty taking up water and some species will wilt and desiccate in a waterlogged

Figure 7.9 *Soil redox potentials (E_h) and emissions of three "greenhouse" gases from soils. Because the three gases differ widely in their potential for global warming per mole of gas, their emissions are expressed here as CO_2 equivalents. Note the low global warming potential from all three gases when E_h is between −0.15 and +0.18. Although this study used small containers of flooded soil in the lab, it may be desirable to manage the aeration status of flooded soils in rice paddies and wetlands with these results in mind. Manipulating organic matter additions, water table levels, water flow rates, and the duration of flooding might allow managers to maintain the soil within the "window" where E_h is too high to stimulate microbial methanogenesis (CH_4 production) but also too low to stimulate production of much N_2O or CO_2. Such management, if feasible, could potentially reduce the contribution of wet soils to global warming.* [From Yu and Patrick (2004) with permission of The Soil Science Society of America]

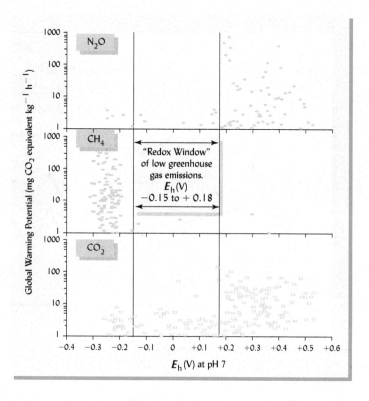

soil. Likewise, plants may exhibit nutrient deficiency symptoms on poorly drained soils even though the nutrients may be in good supply. Furthermore, toxic substances (e.g., hydrogen sulfide or ethylene gas) produced by anaerobic microorganisms may harm plant roots and adversely affect plant growth.

Soil Compaction. Soil compaction does decrease the exchange of gases; however, the negative effects of soil compaction are not all owing to poor aeration. Soil density and strength itself can impede the root growth even if adequate oxygen is available (see Sections 4.7 and 5.9).

Table 7.2
EXAMPLES OF PLANTS WITH VARYING DEGREES OF TOLERANCE TO A SHALLOW WATER TABLE AND ACCOMPANYING RESTRICTED AERATION
Certain species (left most) thrive in wetlands, while others (rightmost) are very sensitive to poor aeration.

Plants adapted to grow well with a water table at the stated depth				
<10 cm	15 to 30 cm	40 to 60 cm	75 to 90 cm	>100 cm
Bald cypress	Alsike clover	Birdsfoot trefoil	Beech	Arborvitae
Black spruce	Bermuda grass	Black locust	Birch	Beans
Common cattail	Black willow	Bluegrass	Cabbage	Cherry
Cranberries	Cottonwood	Mulberry	Corn	Hemlock
Phragmites grass	Creeping bentgrass	Pin oak	Hairy vetch	Oats
Mangrove	Eastern gamagrass	Red maple	Millet	Peach
Reed canary grass	Ladino clover	Sorghum	Peas	Sugar beets
Rice	Loblolly pine	Sycamore	Pecan	Walnut
Skunk cabbage	Orchard grass	Weeping love grass	Red oak	Wheat
Spartina grass	Tall fescue	Willow oak	Soybean	White pine

7.6 SOIL AERATION IN URBAN LANDSCAPES

For crops in the field, aeration may be enhanced by implementing the principles outlined in Sections 4.5–4.7 regarding the maintenance of soil aggregation and tilth. Equally important are systems to increase both surface and subsurface drainage (see Section 6.7) and encourage the production of vertically oriented biopores (e.g., earthworm and root channels) that are open to the surface (Sections 4.5 and 6.7). Here we will briefly consider steps that can be taken to avoid aeration problems for container-grown plants, landscape trees, and lawns.

Container-Grown Plants

Plants grown in containers require especially careful aeration management. The root zone of containerized plants has a limited volume to store water, is exposed to fluctuations in air temperature, and lacks pore continuity to a subsoil layer. These characteristics lead potted plants to become subject to extremes of both drought and waterlogging more frequently than their in-ground counterparts (Figure 7.10). Potting soils (or *growing media*) are engineered to meet the requirements of containerized plants and minimize waterlogging. Mineral soil generally makes up no more than one-third of the volume of most potting mixes, the remainder being composed of inert, lightweight, and coarse-grained materials such as perlite (expanded volcanic glass), vermiculite (expanded mica—not soil vermiculite), or pumice (porous volcanic rock). In order to achieve maximum aeration and minimum weight, some potting mixes contain no mineral soil at all. Most modern potting soils also contain some peat, shredded bark, wood chips, compost, or other stable organic material that adds macroporosity and holds water. Some media are made up entirely of such organic materials.

Despite the fact that most planting containers have holes to allow for drainage of excess water, the bottom of the container still creates a perched water table. As is the case with stratified soils in the field (Section 5.6), water drains out of the holes at the bottom of the pot *only* once the soil at the bottom is saturated with water and the water potential is positive. Most soil pores near the bottom of the container remain filled with water, leaving no room for air, and anaerobic conditions soon prevail. The situation is aggravated if the potting medium contains much mineral soil. In any case, the use of as tall a pot as possible will allow for better aeration in the upper part of the medium. Small, frequent applications of water should be made once the soil *near the bottom of the container* has begun to dry. Thus, feeling the soil at the surface is a poor guide as to when water should be added. It is better to use a sensor probe or soil corer to determine the moisture status in the bottom one-third of the container.

Figure 7.10 *Plants grown in containers are susceptible to extremes of drought and waterlogging. Containerized soils require special management to control aeration and water, but it is easier to apply such management to soil in containers than to field soils.* (Photo courtesy of Ray R. Weil)

Figure 7.11 *Providing a good supply of air to tree roots can be problematic, especially when trees are planted in fine-textured, compacted soils of urban areas. (Lower left) A machine-dug hole with smooth sides will act as a "tea cup" and fill with water, suffocating tree roots. This was the fate of the tree in the upper left photo with only a dead stump remaining. (Right) Breather tubes, a larger rough-surfaced hole, and a layer of surface mulch in which some fine tree roots can grow are all measures that can improve the aeration status of the root zone.* (Courtesy of Ray R. Weil)

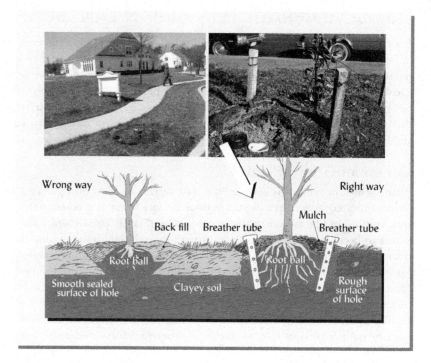

Tree and Lawn Management

Care must be taken when transplanting trees or shrubs into compacted soils that are especially conducive to creating poorly aerated, waterlogging conditions immediately around the young roots. Figure 7.11 illustrates the right and the wrong way to transplant trees into compacted soils.

The aeration of well-established, mature trees must also be safeguarded. If operators push surplus excavated soil around the base of a tree during landscape grading, serious consequences are soon noticed (Figure 7.12). The tree's feeder roots near the original soil surface become deficient in oxygen even if the overburden is no more than 5 to 10 cm in depth. One should build a protective wall (a *dry well*) or install a fence around the base of a valuable tree before grading

Figure 7.12 *Protection of valuable trees during landscape grading operations. Even a thin layer of soil spread over a large tree's root system can suffocate the roots and kill the tree. (Left) Therefore in the situation shown, the effort to be careful grading to preserve valuable trees is doomed to failure. (Right) To ensure that tree feeder roots can obtain sufficient oxygen to survive, the area immediately around the trunk should be protected by a temporary fence or a dry well may be constructed. These measures avoid compaction and preserve the original ground surface near a tree so feeder roots are not buried and suffocated. The dry well may be incorporated into the final landscape design or filled in at a rate of a few centimeters per year.* (Photos courtesy of Ray R. Weil)

Figure 7.13 One way of increasing the aeration of compacted soil is by core cultivation. The machine removes small cores of soil, leaving holes about 2 cm in diameter and 5 to 8 cm deep. This method is commonly used on high-traffic turf areas. Note that the machine removes the cores and does not simply punch holes in the soil, a process that would increase compaction around the hole and impede air diffusion into the soil. (Photos courtesy of Ray R. Weil)

operations begin to preserve the original soil surface and avoid compaction in a zone around the trunk, so the size of the protected zone varying with the size of the tree. This measure will allow the tree's roots to continue to access the O_2 they need. Failure to observe these precautions can easily kill a large, valuable tree, although it may take a year or two to do so.

Management systems for heavily trafficked lawns commonly include the installation of perforated drainage pipes (tiles) as well as other practices that enhance soil aeration (Section 6.7). For example, *core cultivation* can improve aeration in compacted lawn areas. This procedure removes thousands of small cores of soil from the surface horizon, thereby permitting gas exchange to take place more easily (Figure 7.13). Spikes that merely punch holes in the soil are much less effective than corers, since compaction is increased in the soil surrounding a spike.

7.7 WETLANDS AND THEIR POORLY AERATED SOILS[3]

Poorly aerated areas called **wetlands** cover approximately 14% of the world's ice-free land, with the greatest areas occurring in the cold regions of Canada, Alaska, and Russia (Table 7.3). Since environmental consciousness has become a force in modern societies, wetland preservation has become a major issue.

Defining a Wetland[4]

Wetland is a scientific term for ecosystems that are transitional between land and water (Figure 7.14). These systems are neither strictly terrestrial (land-based) nor aquatic (water-based). While there are many different types of wetlands, they all share a key feature, namely, *soils that are water saturated near the surface for prolonged periods when soil temperatures and other conditions are such that plants and microbes can grow and remove the soil oxygen, thereby assuring anaerobic conditions.* It is largely the prevalence of anaerobic conditions that determines the kinds of plants, animals, and soils found in these areas. There is widespread agreement that the wetter end of a wetland occurs where the water is too deep for rooted,

---

[3]Two well-illustrated, nontechnical, yet informative publications on wetlands are Welsh et al. (1995) and CAST (1994). For a compilation of technical papers on hydric soils and wetlands, see Vepraskas and Craft (2016).
[4]In 1987, the U.S. Army Corps of Engineers and the U.S. Environmental Protection Agency agreed on the following definition to be used in enforcing the Clean Water Act: "The term wetlands means those areas that are inundated or saturated by surface or ground water at a frequency and duration sufficient to support, and that under normal circumstances do support, a prevalence of vegetation typically adapted for life in saturated soil conditions."

Table 7.3
MAJOR TYPES OF WETLANDS AND THEIR GLOBAL AREAS
Altogether wetlands constitute perhaps 14% of the world's land area.

Wetland areas	Global area, 1000s km^2	Percent of ice-free land area	Percent of all wetland type
Inland (swamps, bogs, etc.)	5415	3.9	28.8
Riparian or ephemeral	3102	2.3	16.5
Organic (Histosols)	1366	1.0	7.3
Salt-affected, including coastal	2230	1.6	11.9
Permafrost-affected (Histels)	6697	4.9	35.6

Data from Eswaren et al. (1996).

emergent vegetation to take hold. The difficulty is in precisely defining the so-called *drier end* of the wetland, the boundary beyond which exist upland systems in which the plant–soil–animal community is no longer predominantly influenced by the presence of anaerobic soil conditions.

The controversy is probably more political than scientific. Since uses and management of wetlands are regulated by governments in many countries, billions of dollars are at stake in determining what is and what is not protected as a wetland. (Consider, e.g., a developer who wants to buy a 100-ha tract of land on which to build a shopping center. If 20 of those hectares are declared off-limits to development, how will that affect the developer's willingness to pay?)

Because so much money is at stake, thousands of environmental professionals are employed in the process of **wetland delineation**—finding the exact drier-end boundaries of wetlands on the ground. Wetland delineation is *not* done in front of a computer screen, but is a sweaty, muddy, tick- and mosquito-ridden business that those trained in soil science are uniquely qualified to carry out.

What are the characteristics these scientists look for to indicate the existence of a wetland system? Most authorities agree that to be considered as a wetland, an area should exhibit three characteristics: (1) a wetland hydrology or water regime, (2) hydric soils, and (3) hydrophytic plants. We shall briefly examine each.

Wetland Hydrology

Water Balance. Water flows into wetlands from surface runoff (e.g., bogs and marshes), groundwater seepage, and direct precipitation. It flows out by surface and subsurface flows, as well as by evaporation and transpiration (see Section 6.3). The balance between inflows and outflows, as well as the water storage capacity of the wetland itself, determines the elevation of the water table within or above the soil.

Figure 7.14 *The golden brown color of hydrophytic vegetation in winter dormancy characterizes this wetland near New York City. The function of wetlands as a connection between land and water is illustrated by this landscape.* (Photo courtesy of Ray R. Weil)

Hydroperiod. The temporal pattern of water table changes is termed the **hydroperiod**. For a coastal marsh, the hydroperiod may be daily, as the tides rise and fall. For inland swamps, bogs, or marshes, the hydroperiod is more likely to be seasonal. Some wetlands may be flooded for only a month or so each year, while some may never be flooded, although they are saturated within the upper soil horizons.

Also, if the period of saturation occurs when the soil is too cold for microbial or plant–root activity to take place, oxygen may be dissolved in the water or entrapped in aggregates within the soil. Consequently, true anaerobic conditions may not develop, even in flooded soils. Remember, it is the anaerobic condition, not just the saturation with water, that makes a wetland a wetland.

Residence Time. The more slowly water moves through a wetland, the longer the *residence time* and the more likely that wetland functions and reactions will be carried out. For this reason, actions that speed water flow, such as creating ditches or straightening stream meanders, generally degrade a wetland and should be avoided.

Indicators. All wetlands are water saturated some of the time, but many are not saturated all of the time. Systematic field observations, assisted by instruments to monitor the changing level of the water table, may be required to ascertain the frequency and duration of flooding or saturated conditions.

In the field, even during dry periods, there are many signs one can look for to indicate where saturated conditions frequently occur (see Chapter 7 opening photo). Trees with extensive root masses above ground indicate an adaptation to saturated conditions. Past periods of flooding will leave water stains on trees and rocks and a coating of sediment on the plant leaves and litter. Drift lines of once-floating branches, twigs, and other debris also suggest previous flooding. But perhaps the best indicator of saturated conditions is the presence of **hydric soils**.

Hydric Soils[5]

In order to assist in delineating wetlands, soil scientists developed the concept of **hydric soils**. In *Soil Taxonomy* (see Chapter 3), these soils are mostly (but not exclusively) classified in the order Histosols, in Aquic suborders such as Aquents, Aquepts, and Aqualfs, or in Aquic subgroups. These soils generally have an Aquic or Peraquic moisture regime (see Section 3.2).

Defined. The U.S. Department of Agriculture Natural Resources Conservation Service defines a hydric soil as one "that formed under conditions of saturation, flooding, or ponding long enough during the growing season to develop anaerobic conditions in the upper part." Three properties help define hydric soils. First, they are subject to *periods of saturation* that inhibit the diffusion of O_2 into the soil. Second, for substantial periods of time they undergo *reduced conditions* (see Section 7.3); that is, electron acceptors other than O_2 are reduced. Third, they exhibit certain features termed *hydric soil indicators*. Such indicators result from the reduced conditions and are discussed in Box 7.1.

Hydrophytic Vegetation

Although the vast majority of plant species cannot thrive in conditions characteristic of wetlands, there are varied and diverse communities of plants that have evolved special mechanisms to adapt to life in saturated, anaerobic soils. These plants comprise the **hydrophytic vegetation** that distinguishes wetlands from other systems.

Typical adaptive features include hollow aerenchyma tissues that allow plants like spartina grass to transport O_2 down to their roots. Certain trees (such as bald cypress and mangrove) produce adventitious roots, buttress roots, "knees," or **pneumatophores** (see Chapter 7 opener photo). Other species spread their roots in a shallow mass on or just under the soil

---

[5]For an illustrated field guide to features that indicate hydric soils, see USDA/NRCS (2017). A searchable database of soil series in the United States classified as hydric soils can be found at www.nrcs.usda.gov/wps/portal/nrcs/main/soils/use/hydric/.

BOX 7.1
HYDRIC SOIL INDICATORS

Hydric soil indicators are features associated (sometimes only in specific geographic regions) with the occurrence of saturation and reduction. Most of the indicators can be observed in the field by using a tiling spade to dig a slice of soil to a depth of 30 cm. They principally involve the loss or accumulation of various forms of Fe, Mn, S, or C. The carbon (organic matter) accumulations are most evident in Histosols, but thick, dark surface layers in other soils can also be indicators of hydric conditions in which organic matter decomposition has been inhibited (Figure 7.15).

Iron, when reduced to Fe(II), becomes sufficiently soluble that it migrates away from reduced zones and may precipitate as Fe(III) compounds in more aerobic zones. Zones where reduction has removed the iron coatings from mineral grains are termed *redox depletions*. They commonly exhibit the gray, low-chroma colors of the bare, underlying minerals (see Section 4.1 for an explanation of chroma). Also, iron itself turns to gray or blue-green when reduced. The contrasting colors of redox depletions or reduced iron and zones of reddish oxidized iron result in unique mottled redoximorphic features (see, e.g., Figures 4.2 and 4.6). Other redoximorphic features involve reduced Mn. These include the presence of hard black *nodules* that sometimes resemble shotgun pellets. Under severely reduced conditions the entire soil matrix may exhibit low-chroma colors, termed gley. Colors with a chroma of 1 or less quite reliably indicate reduced conditions (Figure 7.16).

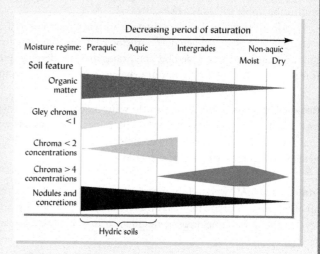

Figure 7.16 *The relationship between the occurrence of some soil features and the annual duration of water-saturated conditions. The absence of iron concentrations (mottles) with colors of chroma >4 and the presence of strong expressions of the other features are indications that a soil may be hydric. Peraquic refers to a moisture regime in which soils are saturated with water throughout the year. For other moisture regimes, see Section 3.2.* [Data from Veneman et al. (1999)]

Figure 7.15 *Wetlands characteristically have hydric soils which can be recognized as such by the presence of certain features considered to be hydric soil indicators. Three of the most common of these indicators are shown here: a gleyed (iron-depleted) soil matrix with low chroma gray colors; a thick, black accumulation of organic matter; and reddish oxidized coatings on roots and root channels made by the influence of oxygen transported into the soil by aerenchyma on hydrophytic plants.* (Photo courtesy of Ray R. Weil)

Always keep in mind that redoximorphic features are indicative of hydric soils only when these features occur in the upper horizons. Many soils of upland areas exhibit redoximorphic features only in their deeper horizons, due to the presence of a fluctuating water table at depth. Upland soils that are saturated or even flooded for short periods, especially if during cold weather, are not wetland (hydric) soils.

A unique redoximorphic feature associated with certain wetland plants is the presence, in an otherwise gray matrix, of reddish oxidized iron around root channels where O_2 diffused out from the aerenchyma-fed roots of a hydrophyte (see Figure 7.15). These *oxidized root zones* exemplify the close relationship between hydric soils and hydrophytic vegetation.

Although identifying hydric soils can be difficult and require a good deal of field experience, the following set of simple questions can be useful in many situations. First, look around to assess if the soil is located in or near an inundated swamp or marsh, or in a flood plain or depression. Examine a spadeful of soil about 18 cm deep (not including any O horizons) consisting of identifiable plant litter. If near a coastal marsh, can you detect the rotten egg odor of hydrogen sulfide? Using a Munsell color book, can you see features with chroma <2? Does the soil exhibit easily visible (distinct or prominent) redox concentrations as concretions, pore linings, root traces, or ped faces? Does the soil exhibit layers of smooth black (mucky) organic material? If none of the just mentioned features exist, the soil is probably not hydric.

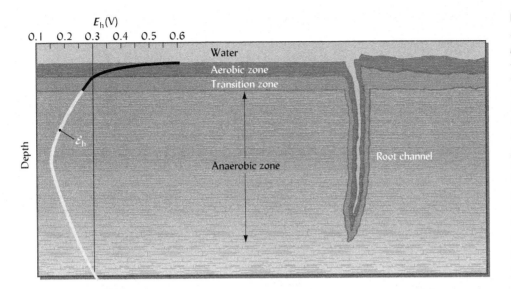

Figure 7.17 *Representative redox potentials within the profile of an inundated hydric soil. Many of the biological and chemical functions of wetlands depend on the close proximity of reduced and oxidized zones in the soil. The changes in redox potential at the lower depths depend largely on the vertical distribution of organic matter. In some cases, low subsoil organic matter results in a second oxidized zone beneath the reduced zone.* (Diagram courtesy of Ray R. Weil)

surface, where some O_2 can diffuse even under a layer of ponded water. The leftmost column in Table 7.3 lists a few common hydrophytes. Not all the plants in a wetland are likely to be hydrophytes, but most usually are.

Wetland Chemistry

The central characteristic of wetland chemistry is the low redox potential (see Section 7.3) that pertains. Furthermore, many wetland functions depend on *variations* in the redox potential; that is, in certain zones or for certain periods of time oxidizing conditions alternate with reducing ones.

Low Oxygen. For example, even in a flooded wetland, O_2 will be able to diffuse from the atmosphere or from oxygenated water into the upper 1 or 2 cm of soil, creating a thin *oxidizing zone* (see Figure 7.17). The diffusion of O_2 within the saturated soil is extremely limited, so that a few centimeters deeper into the profile, O_2 is eliminated and the redox potential becomes low enough for reactions such as nitrate reduction to take place. The close proximity of oxidized and anaerobic zones allows water passing through wetlands to be stripped of N as the ammonium N is oxidized to nitrate N, and the nitrate is then reduced to various nitrogen gases that escape into the atmosphere (see Section 12.1).

Redox. To be considered a wetland, redox potentials should become low enough for iron reduction to produce redoximorphic features. When E_h is even lower, sulfate reduction will produce rotten-egg-smelling hydrogen sulfide (H_2S) gas. The anaerobic zone may extend downward or in some cases may be limited to the upper horizons where microbial activity is high. The anaerobic carbon reactions discussed in Section 7.5 are characteristic of this zone, including methane (swamp gas) production. These and other chemical reactions involving the cycling of C, N, and S are explained in Sections 11.2, 12.1, and 12.2. Toxic elements such as chromium and selenium undergo redox reactions that may help remove them from the water before it leaves the wetland. Acids from industry or mine drainage may also be neutralized by reactions in hydric soils. This array of unique chemical reactions contributes greatly to the benefits that society and the environment gain from wetlands (see Box 7.2).

Constructed Wetlands

Realizing all the beneficial functions of wetlands, scientists and engineers have begun not only to find ways to preserve natural wetlands but also to construct artificial ones for specific purposes, such as wastewater treatment. Another reason for attempting to construct wetlands is the provision in several regulations that allows for the destruction of certain natural wetland areas, provided that an equal or larger area of new wetlands is constructed or that previously degraded wetlands are restored. In some cases, wetland banks are established (Figure 7.18) where a developer can invest in the construction of new wetlands to mitigate against the loss

> **BOX 7.2**
> **WETLAND FUNCTIONS OF VALUE TO ECOSYSTEMS AND SOCIETY**
>
> Once considered to be nothing but disease-breeding wastelands, wetlands are now widely recognized as performing many extremely valuable functions. In fact, a group of scientists and economists (Costanza et al., 1997) has estimated that on average, one hectare of wetlands provides nonmarket services worth about 15 times more than those provided by one hectare of forestland. Globally, wetlands annually perform needed ecosystem services that would cost some $14 trillion (in 2018 dollars) to replace. We can summarize at least six types of benefits derived from wetlands, the recognition of which has motivated the intensification of wetland study and protective regulation.
>
> 1. *Species habitat.* Wetlands provide special environmental conditions required by a wide array of wild species. Plants like bald cypress trees and pitcher plants, and animals like salamanders and muskrats, use wetlands as their primary habitat (i.e., they live in wetlands). Furthermore, about 40% of all endangered and threatened species depend on wetlands in some way (e.g., for food or shelter). About one-third of all bird species also depend on wetlands.
> 2. *Water filtration.* Water is filtered and purified as it passes through wetlands on its way from the land to rivers, bays, lakes, estuaries, and groundwater aquifers. Wetlands physically filter out most sediment, remove a high proportion of plant nutrients (especially N and P) that could otherwise cause eutrophication of aquatic systems (see Section 12.2), and break down many organic substances that could cause toxic effects or deplete O_2 supplies in aquatic systems and drinking water.
> 3. *Flooding reduction.* Wetlands act as giant reservoirs to hold back large volumes of stormwater runoff from the land. They then release the water slowly, either to surface flow or to groundwater, thereby avoiding high peak flows that lead to floods and damage to homes and developed land along rivers. Studies have shown that floods are far less severe and less frequent where river systems have been allowed to retain their undisturbed wetlands.
> 4. *Shoreline protection.* Coastal wetlands serve as a buffer between high-energy ocean waves and the shore, preventing the rapid shoreline erosion that is observed when the wetlands are drained or filled in.
> 5. *Commercial and recreational activities.* In many parts of the world, people spend a great deal of money every year to catch, hunt, view, or photograph birds, animals, and fish that either live in wetlands or depend on wetlands for their food supply or for nesting sites. As a result, some of the first and most ardent proponents of wetland conservation have been organizations devoted to duck hunting or fishing.
> 6. *Natural products.* Valuable products such as timber, blueberries, cranberries, fish, and wild rice can be sustainably harvested from certain wetlands under careful management. Scenic beauty could also be considered a natural product of wetlands that people can enjoy without damaging the system.

of other wetlands destroyed or degraded by drainage or urban development. As would be expected, this process, termed ***wetland mitigation***, has been only partially successful in terms of function, as scientists cannot be expected to create what they do not fully understand.

We have seen how greatly soil aeration is influenced by soil water. We will now turn our attention to soil temperature, another physical soil property that is closely related to both soil water and aeration.

Figure 7.18 *A wetland bank in Virginia where developers can pay to have new areas of wetlands created in order to mitigate for the destruction or degradation of wetlands by a development project. Usually, the wetlands created must be of the same type (freshwater or marsh) and in the same watershed as the destroyed wetlands.* (Photo courtesy of Ray R. Weil)

7.8 PROCESSES AFFECTED BY SOIL TEMPERATURE

The temperature of a soil greatly affects the physical, biological, and chemical processes occurring in that soil and in the plants growing on it (Figure 7.19).

Plant Processes

Most plants are actually more sensitive to *soil* temperature than to aboveground *air* temperature, but this is not often appreciated since air temperature is more commonly measured. Research also shows that, contrary to what one might expect, adverse soil temperature generally

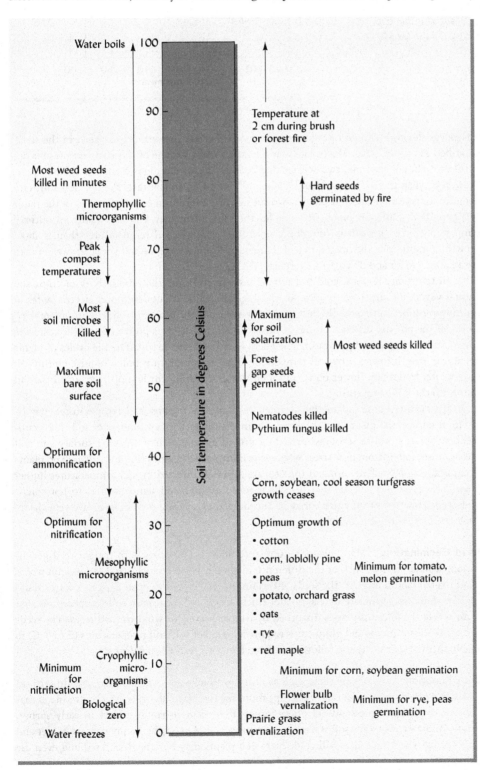

Figure 7.19 *Soil temperature ranges associated with various soil processes.* (Diagram courtesy of Ray R. Weil)

Figure 7.20 Effects of air and soil temperatures on turfgrass quality. Heat stress is a major problem for bent grass (Agrostis spp.) grown on golf greens in warmer climates. Researchers in this study grew bent grass for 60 days while separately controlling soil and air temperatures. Turfgrass quality (color, vigor, etc.) was rated from 0 (dead) to 10 (best quality). Compare the small decline in turf quality caused by increasing air temperature from 20 to 35°C to the much greater decline caused by the same increase in soil temperature. High temperature in both air and soil caused the worst effects. A fine spray of water combined with a large fan can cool both air and soil on a golf green. [Data from Xu and Huang (2000)]

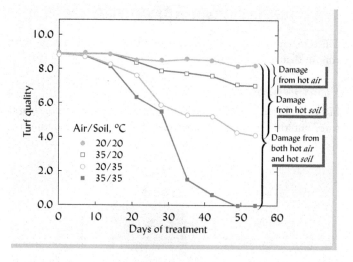

influences shoot growth and photosynthesis more than root growth (as was seen in the study described in Figure 7.20). Most plants have a rather narrow range of soil temperatures for optimal growth. For example, two species that evolved in warm regions, corn and loblolly pine, grow best when the soil temperature is about 25 to 30°C. In contrast, the optimal soil temperature for cereal rye and red maple, two species that evolved in cool regions, is in the range of 12 to 18°C. Different plant processes for the same plant may also have different optimal temperatures; root growth is commonly optimized at lower soil temperatures than is shoot growth. For example, the optimum soil temperature ranges for root and shoot growth of corn are about 23 to 25 and 25 to 30°C, respectively.

In temperate regions, cold soil temperature often limits the productivity of crops and natural vegetation. In these regions, artificial soil warming by underground electric wires or warm water pipes can markedly increase crop yields, but the economic and environmental viability of the practice depends on the value of the additional crops produced and the costs and environmental impacts of generating the energy used to heat the soil. The life cycles of plants are also greatly influenced by soil temperature. For example, tulip bulbs require chilling in early winter to develop flower buds, although flower development is suppressed until the soil warms up the following spring.

In warm regions, and in the summer in temperate regions, soil temperatures may be too high for optimal plant growth, especially in the upper few centimeters of soil. For example, bent grass, a cool-season grass prized for providing an excellent "playing surface" on golf greens, often suffers from heat stress when grown in warmer regions (Figure 7.20). Even plants of tropical origin, such as corn and tomato, are adversely affected by soil temperatures higher than 35°C. Seed germination may also be reduced by high soil temperatures. In hot conditions, root growth near the surface may be encouraged by shading the soil with live vegetation, plant residues, or organic mulch.

Seed Germination. Many plants require specific soil temperatures to trigger seed germination, accounting for much of the difference in species between early- and late-season weeds in cultivated land. Likewise, the seeds of certain plants adapted to open gaps in a forest stand are stimulated to germinate by the greater fluctuations and maximum soil temperatures that occur where the forest canopy is disturbed by timber harvest or wind-thrown trees. The seeds of certain prairie grasses and grain crops require a period of cold soil temperatures (2 to 4°C) to enable them to germinate the following spring, a process termed *vernalization*.

Root Functions. Root functions such as nutrient uptake and water uptake are sluggish in cool soils with temperatures below the optimum for the particular species. One result is that nutrient deficiencies, especially of phosphorus, often occur in young plants in early spring, only to disappear when the soil warms later in the season. On bright, sunny days in winter and early spring when the soil is still cold, evergreen plants may become desiccated and even die because the slow water uptake by roots in the cold soils cannot keep up with the high evaporative demand of bright sun on the foliage. This "winter burn" problem can be prevented by covering the shrubs with a shade cloth.

Microbial Processes

Microbial processes are influenced markedly by soil temperature changes (Figure 7.21). Although it is commonly assumed that microbial activity ceases at temperatures that freeze water (<0°C), low rates of soil microbial activity and organic matter decomposition have been measured in the permafrost layers of Gelisols at temperatures as low as −20°C. In fact, given that some 80% of the Earth's biosphere is colder than 5°C, it should not be surprising that microbes have widely adapted to life at cold temperatures.

Nonetheless, microbial activity is far greater at warm temperatures; the rates of microbial processes such as respiration typically more than double for every 10°C rise in temperature. The optimum temperature for microbial decomposition processes may be 35 to 40°C, considerably higher than the optimum for plant growth. The dependence of microbial respiration on warm soil temperatures has important implications for soil aeration (see Section 7.7) and for the decomposition of plant residues and, hence, the cycling of the nutrients they contain. The productivity of northern (boreal) forests is probably limited by the inhibiting effect of low soil temperatures on microbial recycling and release of nitrogen from the tree litter and soil organic matter.

The microbial oxidation of ammonium ions to nitrate ions, which occurs most readily at temperatures near 30°C, is also negligible when the soil temperature is low, below about 8 to 10°C. Farmers in cool regions can take advantage of this fact by injecting ammonia fertilizers into cold soils in the early spring, expecting that ammonium ions will not be readily oxidized to leachable nitrate ions until the soil temperature rises. Unfortunately, in some years a warm spell allows the production of nitrates earlier than expected, with the result that much nitrate is lost by leaching, to the detriment of both the farmer and the downstream water quality.

In environments with hot, sunny summers (maximum daily air temperatures >35°C), a heating process called *soil solarization* can be used to control pests and diseases in some high-value crops. In this process, the ground is covered with a clear plastic film that traps enough heat to raise the temperature of the upper few centimeters of soil to as high as 50 to 60°C. Such high temperatures can effectively suppress certain fungal pathogens, weed seeds, and insect pests.

As we shall see in Chapter 15, warm soil temperatures are also critical for new pollution remediation technologies that utilize specialized microorganisms to degrade petroleum products, pesticides, and other organic contaminants in soils.

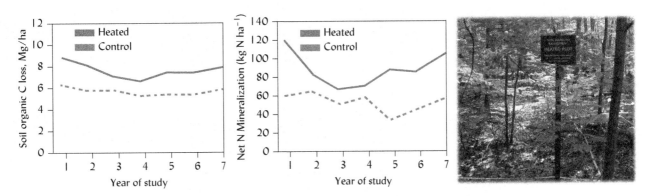

Figure 7.21 *Elevated soil temperature accelerates biological processes, such as (right graph) nitrogen release from organic matter by mineralization and (left graph) carbon dioxide release by soil microorganisms respiring organic matter. In this mid-latitude hardwood forest, electric heating cables were buried in certain plots to maintain soil temperature at 5°C above normal throughout the year. Cables were also installed in the control plots (photo) to assure an equal degree of physical disturbance, but no electricity was applied. The biological processes represented are discussed in detail in Chapters 12 and 13.* [Graphs from data in Melillo et al. (2011); photo courtesy of Ray R. Weil]

Freezing and Thawing

When soil temperatures fluctuate above and below 0°C, the water in the soil undergoes cycles of freezing and thawing. Alternate freezing and thawing subject the soil aggregates to pressures as zones of pure ice, called *ice lenses*, form within the soil and as ice crystals form and expand. These pressures alter the physical structure of the soil. In a saturated soil with a puddled structure, the frost action breaks up the large masses and greatly improves granulation. In contrast, for soils with good aggregation to begin with, freeze–thaw action when the soil is very wet can lead to structural deterioration. When the surface of a frozen soil thaws, the upper few millimeters may be completely saturated with water, making it easy for spring rains to detach soil particles that are then easily removed by erosion. In some soils of temperate regions more than 50% of the annual erosion losses take place from thawing soils.

Alternate freezing and thawing can force objects upward in the soil, a process termed *frost heaving*. Objects subject to heaving include stones, fence posts, and perennial tap-rooted plants (Figure 7.22). This action is most severe where the soil is silty in texture, wet, and lacking a covering of snow, vegetative mulch or dense vegetation. Freezing can drastically reduce stands of perennial tap-rooted legumes, and heave shallow foundations, roads, and runways that have fine material as a base. Gravels and pure sands are normally resistant to frost damage, but silts and sandy soils with modest amounts of finer particles are particularly susceptible. To avoid damage by freezing soil temperatures, foundation footings (as well as water pipelines) should be set into the soil below the maximum depth to which the soil freezes—a depth that ranges from less than 10 cm in subtropical zones, such as south Texas and Florida, to more than 200 cm in very cold climates.

Permafrost[6]

Perhaps the most significant global event involving soil temperatures is the thawing in recent years of some of the permafrost (permanently frozen ground) in northern areas of Canada, Russia, Iceland, China, Mongolia, and Alaska. Nearly 25% of the land areas of the Earth are

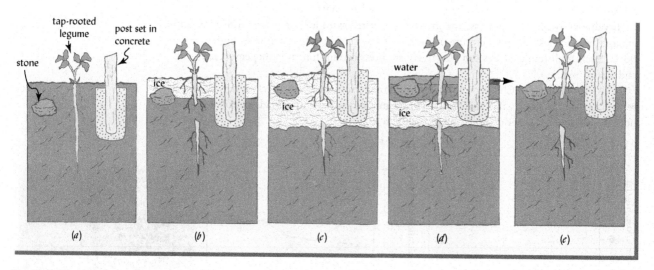

Figure 7.22 *How frost heaving moves objects upward. (a) Position of the object (stone, plant, or fence post) before the soil freezes. (b) As lenses of pure ice form in the freezing soil by attraction of water from the unfrozen soil below, the frozen soil tightens around the upper part of the object, lifting it somewhat—enough to break the root in the case of the plant. (c) The objects are lifted upward as ice-lens formation continues with deeper penetration of the freezing front. (d) As for freezing, thawing commences from the surface downward. Water from thawing ice lenses escapes to the surface because it cannot drain downward through the frozen soil. The soil surface subsides while the heaved objects are held in the "jacked-up" position by the still-frozen soil around their lower parts. (e) After complete thaw, the stone is closer to the surface than previously (although rarely at the surface unless erosion of the thawed soil has occurred), and the upper part of the broken plant's root is exposed, so that the plant is likely to die.* (Diagram courtesy of Ray R. Weil)

---

[6]For a discussion of permafrost degradation in the Artic, see Liljedahl et al. (2016).

underlain by permafrost. Rising temperatures since the late 1980s have caused some of the upper layers of permafrost to thaw. In parts of Alaska, for example, temperatures in top layers of permafrost have risen about 3.5°C since the late 1980s, resulting in melting rates of about a meter in a decade. Such melting has serious implications since it can drastically affect the flow of rivers and physical stability of buildings and roads, as well as the stability of root zones of forests and other vegetation in the region. Trees fall and buildings collapse as the frozen layers melt. The thawing of arctic permafrost is expected to further accelerate global warming, as decomposition of organic materials long trapped in the frozen layers of Histels releases vast quantities of carbon dioxide into the atmosphere (see Figure 3.15).

Soil with ice lenses may contain much more water than would be needed to saturate the soil in the unfrozen state. When the ice lenses thaw, the soil becomes supersaturated because the excess water cannot drain away through the underlying still-frozen soil. Soil in this condition readily turns into noncohesive mud that is very susceptible to erosion and movement by mudslides.

Soil Heating by Fire

Fire is one of the most far-reaching ecosystem disturbances in nature. In addition to the obvious aboveground effects of forest, range, or crop-residue fires, the brief but sometimes dramatic changes in soil temperature also may have lasting impacts below ground. Unless the fire is artificially stoked with added fuel, the temperature rise from wildfires is usually very brief and is limited to the upper few centimeters of soil. But the temperatures resulting from "slash and burn" practices in the tropics may be sufficiently high in the upper few millimeters of soil to cause the loss of nearly all organic natter and even the breakdown of minerals such as gibbsite and kaolinite.

The heat from wildfires may also cause the breakdown and movement of organic compounds, especially in sandy soils (Figure 7.23). The high temperatures (>125°C being common) essentially distill certain fractions of the organic matter, with some of the volatilized hydrocarbon compounds moving quickly through the soil pores to deeper, cooler areas. As these compounds reach cooler soil particles deeper in the soil, they condense (solidify) on the surface of the soil particles and fill some of the surrounding pore spaces. Some of the condensed compounds are water-repellent (hydrophobic) hydrocarbons. Consequently, when rain comes, water infiltration in such burned over sandy soil may be greatly reduced in comparison to unburned areas. This effect of soil temperature is quite common on pine forest and chaparral lands in semiarid regions and may be responsible for the disastrous mudslides that occur when the layer of soil above the hydrophobic zone becomes saturated with rainwater (see Figure 7.23*d*).

Fires also affect the germination of certain seeds, which have hard coatings that prevent them from germinating until they are heated above 70 to 80°C. The burning of straw in wheat fields generates similar soil temperatures, but with the effect of killing most of the weed seeds near the surface and thus greatly reducing subsequent weed infestation. The heat and ash from low-intensity fires may also hasten the cycling of plant nutrients. In fact, grassland fires often stimulate plant regrowth so much that soil organic matter increases with fire as the enhanced plant biomass more than compensates for the loss of soil organic matter during the fire. On the other hand, fires set to clear land of timber slash may burn long and hot enough to seriously deplete soil organic matter and kill so many soil organisms that forest regrowth is inhibited.

Contaminant Removal

The removal of certain organic pollutants from contaminated soils can be accomplished by raising the soil temperature, causing the offending compounds to volatilize and diffuse out of the soil in the gaseous state. The process may be prohibitively expensive if the soil has to be excavated and hauled to and from a heated extraction facility. Techniques are under development to warm the soil in place in the field using electromagnetic radiation. The resulting temperatures are sufficiently high to vaporize some contaminants, such as toxic components in diesel fuel, which can then be flushed from the soil by a stream of injected air.

Figure 7.23 (a) *Wildfires heat up the surface layers of a sandy soil sufficiently to volatilize organic compounds from the soil organic matter and surface litter. (b) Some of the volatilized organics then diffuse away from the heat down into the soil and condense (solidify) on the surface of cooler soil particles. (c) These condensed compounds are waxlike hydrocarbons that are water repellent and drastically reduce the infiltration of water into the soil for a period of years. (d) An ecologist investigates the soil under chaparral vegetation in southern California where dry season fires have created a water repellant layer a few centimeters below the surface. The inset photos show water beading up on this soil layer exposed by scrapping aside the upper 8 cm of loose, nonrepellent soil.* [Concepts from DeBano (2000), bar graph from data in Dryness (1976); photos courtesy of Ray R. Weil]

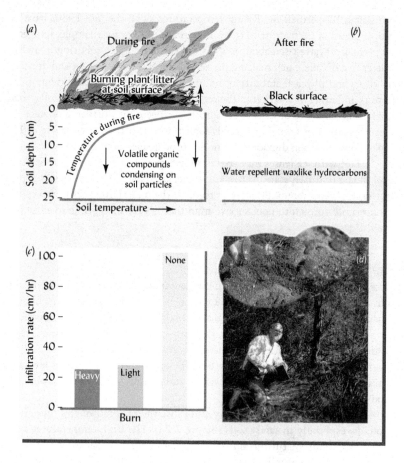

7.9 ABSORPTION AND LOSS OF SOLAR ENERGY[7]

The temperature of soils in the field is directly or indirectly dependent on at least three factors: (1) the net amount of heat energy the soil absorbs, (2) the heat energy required to bring about a given change in the temperature of a soil, and (3) the energy required for processes such as evaporation, which are constantly occurring at or near the surface of soils.

Solar radiation is the primary source of energy to heat soils. But clouds and dust particles intercept the sun's rays and absorb, scatter, or reflect most of the energy (Figure 7.24). Only about 35 to 40% of the solar radiation actually reaches the Earth in cloudy humid regions and 75% in cloud-free arid areas. The global average is about 50%.

Little of the solar energy reaching the Earth actually results in soil warming. The energy is used primarily to evaporate water from the soil or leaf surfaces or is radiated or reflected back to the sky. Only about 10% is absorbed by the soil to warm it. Even so, this warming is critical to soil processes and to plants growing in soils.

Albedo The fraction of incident radiation that is reflected by the land surface is termed the **albedo** and ranges from as low as 0.1 to 0.2 for dark-colored, rough soil surfaces to as high as 0.5 or more for smooth, light-colored surfaces. Vegetation may affect the surface albedo either way, depending on whether it is dark green and growing or yellow and dormant.

The fact that dark-colored soils absorb more energy than lighter-colored ones does not necessarily imply, however, that dark soils are always warmer. In fact, the opposite is often true. In most landscapes, the darkest soils are those found in the low spots where excessive wetness has caused organic matter to accumulate. Therefore, the darkest soils are also usually the wettest. The water in these soils will absorb much energy with only little change in temperature (see Section 7.10), and it also cools the soil when it evaporates.

---

[7]For application of these principles to the role of soil moisture in models of global warming, see Lin et al. (2003).

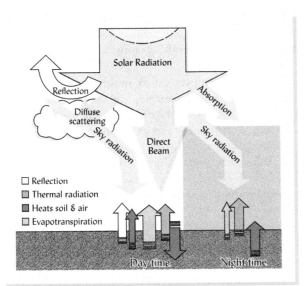

Figure 7.24 *Schematic representation of the radiation balance in daytime and nighttime in the spring or early summer in a temperate region. About half the solar radiation reaches the Earth, either directly or indirectly, from sky radiation. Most radiation energy that strikes the Earth in the daytime is used to drive evapotranspiration or is radiated back to the atmosphere. Only a small portion, perhaps 10%, actually heats the soil. At night the soil loses some heat, and some evaporation and thermal radiation occur.* (Diagram courtesy of Ray R. Weil)

Slope Angle and Aspect

The angle at which the sun's rays strike the soil also influences soil temperature. If the incoming path of the rays of solar energy is perpendicular to the soil surface, energy absorption (and soil temperature increase) is greatest. The effect of the direction of slope, or **aspect**, on forest species is illustrated in the photo in Figure 7.25.

Rain Mention should also be made of the effect of rain or irrigation water on soil temperature. In the summer, rainfall cools the soil, since it is often cooler than the soil it penetrates. On the

Figure 7.25 *(Inset diagram) Effect of slope aspect on solar radiation received per unit land area. Slope (a) is north facing and receives solar radiation at an angle of 45° to the ground surface so only 5 units of solar radiation (arrows) hit the unit of land area. The same land area on the south-facing slope (b) receives 7 units of radiation at a 90° angle to the ground. In other words, if a given amount of radiation from the sun strikes the soil at right angles, the radiation is concentrated in a relatively small area, and the soil warms quite rapidly. This is one of the reasons why north slopes tend to have cooler soils than south slopes. It also accounts for the colder soils in winter than in summer. (Photo) A view looking eastward toward a forested mountain in Virginia illustrates the temperature effect. The main ridge (left to right) runs north–south and the smaller side ridges run east–west (up and down). The dark patches are pine trees in this predominantly hardwood deciduous forest. The pines dominate the southern slopes on each east–west ridge. The soils on the southern slopes are warmer and therefore drier, less deeply weathered, and lower in organic matter.* (Photo and diagram courtesy of Ray R. Weil)

other hand, in temperate zones, spring rain definitely warms the surface soil in the short term as the relatively warm water moves into cold soil. However, spring rain, by increasing the amount of solar energy used in evaporating water from the soil, can also accentuate low temperatures.

Soil Cover Whether the soil is bare or is covered with vegetation, mulch, or snow is another factor markedly influencing the amount of solar radiation reaching the soil. Bare soils warm up more quickly and cool off more rapidly than those covered with vegetation, snow, or plastic mulches. Frost penetration during the winter is considerably greater in bare, noninsulated land.

Even low-growing vegetation such as turfgrass has a very noticeable influence on soil temperature and on the temperature of the surroundings. Much of the cooling effect is due to heat dissipated by transpiration of water. To experience this effect, on a blistering hot day, try having a picnic on an asphalt parking lot instead of on a growing green lawn! The asphalt surface may reach 70°C while the turfgrass is only 30°C.

The effect of a dense forest is universally recognized. Timber-harvest practices that leave sufficient canopy to provide about 50% shade will likely prevent undue soil warming that could hasten the loss of soil organic matter or the onset of anaerobic conditions in wet soils. The effect of timber harvest on soil temperature as deep as 50 cm is seen in the case presented in Table 7.4, where tree removal warmed the soil in spring, even though it also raised the soil water content, since trees were no longer taking up water. However, as we shall see in the next section, a higher water content normally slows the warming of soils in spring.

7.10 THERMAL PROPERTIES OF SOILS

Specific Heat of Soils

A dry soil is more easily heated than a wet one. This is because the amount of energy required to raise the temperature of water by 1°C (its heat capacity) is much higher than that required to warm soil solids by 1°C. When heat capacity is expressed per unit mass—for example, in calories per gram (cal/g)—it is called **specific heat**, or heat capacity, c. The specific heat of pure water is defined as 1.00 cal/g (or 4.18 joules per gram, J/g); that of dry soil is about 0.2 cal/g (0.8 J/g).

The specific heat largely controls the degree to which soils warm up in the spring, wetter soils warming more slowly than drier ones (see Box 7.3). Furthermore, if the water does not drain freely from the wet soil, it will be evaporated, a process that is very energy consuming, as the next section will show.

Table 7.4

TREE REMOVAL EFFECT ON SOIL AERATION AND TEMPERATURE IN A SUBTROPICAL PINE FOREST

Removal of 55-year-old loblolly pine trees reduced transpiration and shading, resulting in a higher water table, warmer spring temperatures, and lower redox potentials in this Vertic Ochraqualf. Because warmer temperatures stimulated microbial activity, E_h was lower in spring, even though the soil was wetter in winter.

Site treatment	Time soil is saturated, %	Soil temperature, °C		Soil redox potentials E_h, V	
		Winter	Spring	Winter	Spring
Measured at 50-cm depth					
Undisturbed pine stand	39	11.8	18.3	0.83	0.65
Trees cut, soil not compacted	71	11.7	20.5	0.51	0.11
Measured at 100-cm depth					
Undisturbed pine stand	57	13.3	17.3	0.83	0.49
Trees cut, soil not compacted	69	13.2	18.7	0.54	0.22

Data from Tiarks et al. (1996).

> **BOX 7.3**
> **CALCULATING THE SPECIFIC HEAT (HEAT CAPACITY) OF MOIST SOILS**
>
> Soil water content markedly impacts soil temperature changes through its effect on the specific heat, or heat capacity, c of a soil. For example, consider two soils with comparable characteristics, soil A, a relatively dry soil with 10 g water/100 g soil solids, and soil B, a wetter soil with 30 g water/100 g soil solids.
>
> We can assume the following values for specific heat:
>
> Water = 1.0 cal/g and dry mineral soil = 0.2 cal/g
>
> For soil A with 10 g water/100 g dry soil, or 0.1 g water/g dry soil, the number of calories required to raise the temperature of 0.1 g of water by 1°C is as follows: 0.1 g × 1.0 cal/g = 0.1 cal. The corresponding figure for the 1.0 g of soil solids is as follows:
>
> $$1\ g \times 0.2\ cal/g = 0.2\ cal.$$
>
> Thus, a total of 0.3 cal (0.1 + 0.2) is required to raise the temperature of 1.1 g (1.0 + 0.1) of the moist soil by 1°C. Since the specific heat is the number of calories required to raise the temperature of 1 g of moist soil by 1°C, we can calculate the specific heat of soil A as follows:
>
> $$c_{\text{soil A}} = \frac{0.3}{1.1} = 0.273\ cal/g$$
>
> These calculations can be expressed as a simple equation to calculate the weighted average specific heat of a mixture of substances:
>
> $$c_{\text{moist soil}} = \frac{c_1 m_1 + c_2 m_2}{m_1 + m_2} \quad (7.7)$$
>
> where c_1 and m_1 are the specific heat and mass of substance 1 (the dry mineral soil, in this case), and c_2 and m_2 are the specific heat of substance 2 (the water, in this case).
>
> Applying this equation to soil A, we again calculate that $c_{\text{soil A}}$ is 0.273 cal/g, as follows:
>
> $$c_{\text{soil A}} = \frac{0.2\ cal/g \times 1.0\ g + 1.0\ cal/g \times 0.10\ g}{1.0\ g + 0.10\ g}$$
> $$= \frac{0.30\ cal}{1.1\ g} = 0.273\ cal/g$$
>
> In the same manner, we calculate the specific heat of the wetter soil B:
>
> $$c_{\text{soil B}} = \frac{0.2\ cal/g \times 1.0\ g + 1.0\ cal/g \times 0.30\ g}{1.0\ g + 0.30\ g}$$
> $$= \frac{0.50\ cal}{1.3\ g} = 0.385\ cal/g$$
>
> The wetter soil B has a specific heat c_B of 0.385 cal/g, whereas the drier soil A has a specific heat c_A of 0.273 cal/g. Because it must absorb an additional 0.112 cal (0.385 − 0.27) of solar radiation for every degree of temperature rise, the wetter soil will warm up much more slowly than the drier soil.

The high specific heat of soils is also used in the design of energy-efficient geothermic temperature control systems that both warm and cool buildings. To maximize heat-exchange contact with the soil, a network of pipes is laid underground near the building to be heated and cooled. Advantage is taken of the fact that subsoils are generally warmer than the atmosphere in the winter and cooler than the atmosphere in the summer. Fluid circulating through the network of pipes absorbs heat from the soil during the winter and releases it to the soil in the summer. The high specific heat of soils combined with their enormous mass permits a large exchange of energy to take place without greatly modifying the soil temperature.

Heat of Vaporization

The evaporation of water from soil surfaces requires a large amount of energy, 540 kilocalories (kcal) or 2.257 megajoules (mJ) for every kilogram of water vaporized. This use of energy for evaporation cools the soil, much the way it chills a person who comes out from the water after swimming on a windy day.

For example, if the amount of water associated with 100 g of dry soil was reduced by evaporation from 25 to 24 g (only about a 1% decrease) and if all the thermal energy needed to evaporate the water came from the moist soil, the soil would be cooled by about 12°C. Such a figure is hypothetical because only a part of the heat of vaporization comes from the soil itself. Nevertheless, it indicates the tremendous cooling influence of evaporation.

The low temperature of a wet soil may be due partially to evaporation and partially to high specific heat. The temperature of the upper few centimeters of wet soil is commonly 3 to 6°C lower than that of a drier soil. This is a significant factor in the spring in a temperate zone, when a few degrees will make the difference between the germination or lack of germination of seeds, or the microbial release or lack of release of nutrients from organic matter.

Thermal Conductivity of Soils

As shown in Section 7.9, some of the solar radiation that reaches the Earth slowly penetrates the profile largely by conduction; this is the same process by which heat moves to the handle of a cast-iron frying pan. The movement of heat in soil is analogous to the movement of water (see Section 5.5), the rate of flow being determined by a driving force and by the ease with which heat flows through the soil. This can be expressed as Fourier's law:

$$Q_h = \lambda \times \frac{\Delta T}{x} \qquad (7.8)$$

where Q_h is the *thermal flux*, the quantity of heat transferred across a unit cross-sectional area in a unit time; λ is the **thermal conductivity** of the soil; and $\Delta T/x$ is the temperature gradient over distance x that serves as the driving force for the conduction of heat.

The thermal conductivity, λ, of soil is influenced by a number of factors, the most important being the moisture content of the soil and the degree of compaction (see Figure 7.26). Heat passes through water many times faster than through air. As the water content increases in a soil, the air content decreases, and the transfer resistance is decidedly lowered. When sufficient water is present to form a bridge between most of the soil particles, further additions will have little effect on heat conduction. Heat moves through mineral particles even faster than through water, so when particle-to-particle contact is increased by soil compaction, heat-transfer rates are also increased. Therefore, a wet, compacted soil would be the poorest insulator or the best conductor of heat. Here again the interconnectedness of soil properties is demonstrated.

Relatively dry soil makes a good insulating material. Buildings built mostly underground can take advantage of both the low thermal conductivity and relatively high heat capacity of large volumes of soil.

Because of relatively low thermal conductivities, changes in subsoil temperature lag behind those of the surface layers. Moreover, temperature changes are of a smaller magnitude in the subsoil. In temperate regions, surface soils in general are expected to be warmer in summer

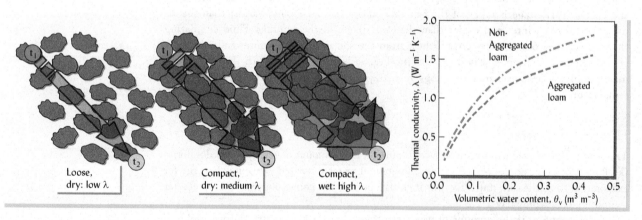

Figure 7.26 *Bulk density and water content affect the rate at which heat is transferred through soils from a warm zone (X_1) to a cooler zone (X_2). The arrows represent heat transfer through the soil, the rate of transfer being in proportion to the soil thermal conductivity, λ, represented by the arrow thickness. Higher bulk density from soil compaction increases particle-to-particle contact, which in turn hastens heat transfer because the thermal conductivity of mineral particles is much higher than that of air. If the remaining gaps between particles become filled with water instead of air, the soil thermal conductivity increases still more because water also conducts heat better than air. Therefore wet, compacted soils transfer heat most rapidly. The graph at right shows measurements of thermal conductivity as affected by soil aggregation and increasing water content for a loam soil. Lack of aggregation has a similar effect to compaction.* [Diagram courtesy of Ray R. Weil, graph modified from Ju et al. (2011)]

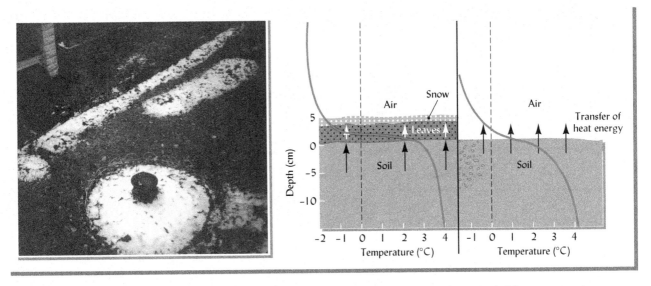

Figure 7.27 Transfer of heat energy from soil to air. The scene, looking down on a garden after an early fall snowstorm, shows snow on the leaf-mulched flower beds, but not on areas where the soil is bare or covered with thin turf. The reason for this uneven accumulation of snow can be seen in the temperature profiles. Having stored heat from the sun, the soil layers are often warmer than the air as temperatures drop in fall (this is also true at night during other seasons). On bare soil, heat energy is transferred rapidly from the deeper layers to the surface, the rate of transfer being enhanced by high moisture content or compaction, which increases the thermal conductivity of the soil. As a result, the soil surface and the air above it are warmed to above freezing, so snow melts and does not accumulate. The leaf mulch, which has a low thermal conductivity, acts as an insulating blanket that slows the transfer of stored heat energy from the soil to the air. The upper surface of the mulch is therefore hardly warmed by the soil, and the snow remains frozen and accumulates. A heavy covering of snow can itself act as an insulating blanket. (Courtesy of Ray R. Weil)

and cooler in winter than the subsoil, especially the lower horizons of the subsoil. Soil thermal conductivity can also affect air temperature above the soil, as shown in Figure 7.27.

Variation with Time and Depth

Soil temperature at any time and place depends on the ratio of the energy absorbed to that being lost. The constant change in this relationship is reflected in seasonal, monthly, and daily temperature fluctuations. Figure 7.28 illustrates the considerable seasonal variations of soil temperature that occur in northern hemisphere temperate regions. In the subsoil, the seasonal temperature increases and decreases lag behind changes registered in the surface soil and in the air. Accordingly, Figure 7.28 shows surface soil temperatures in March

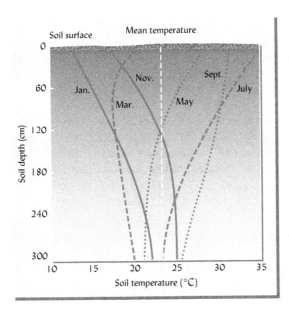

Figure 7.28 Average monthly soil temperatures for 6 of the 12 months of the year at different soil depths at College Station, TX (1951–1955). Note the lag in soil temperature change at the lower depths. [Graph based on data selected from Fluker (1958)]

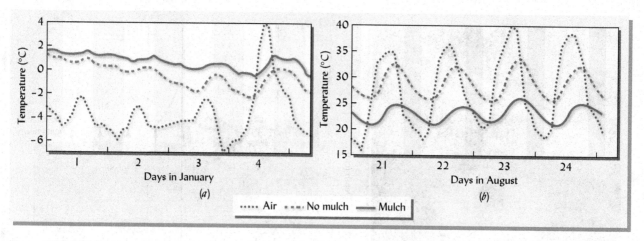

Figure 7.29 (Right) Influence of straw mulch (8 tons/ha) on air temperature at a depth of 10 cm during an August hot spell in Bushland, TX. Note that the soil temperatures in the mulched area are consistently lower than where no mulch was applied. (Left) During a cold period in January, the soil temperature was higher in the mulched than in the unmulched area. The shaded bars represent nighttime. Note that air temperatures fluctuate much more widely than do soil temperatures and that peak soil temperature is reached several hours after the air temperature peaks. [Redrawn from Unger (1978); used with permission of American Society of Agronomy]

responding to the warming of spring, while temperatures of the subsoil still reflect the cold of winter. Subsoil temperatures are less variable than air and surface soil temperatures, although there is some temperature fluctuation even at 300 cm deep.

Compared to the surface soil and the air, deep soil layers are generally warmer in the late fall and winter and cooler in the spring and summer. On a daily basis (see Figure 7.29), soil reaches its maximum temperature later in the afternoon or evening than does the air, the lag time being greater and the fluctuation less pronounced for greater depths. Deeper than 4 to 5 m, the temperature changes little and approximates the mean annual air temperature (a fact experienced by people visiting deep caverns).

7.11 SOIL TEMPERATURE CONTROL

The temperature of field soils is usually not subject to radical human regulation. However, two kinds of management practice have significant effects on soil temperature: those that affect the vegetative or mulch covering the soil surface and those that affect soil moisture. These effects have meaningful biological implications.

Organic Mulches and Plant-Residue Management

Figure 7.29 shows that mulches effectively buffer extremes in soil temperatures. In periods of hot weather, they keep the surface soil cooler than where no cover is present; in contrast, during cold weather they keep the soil warmer than it would be if bare.

The forest floor is a prime example of a natural temperature-modifying mulch. It is not surprising, therefore, that timber harvest practices can markedly affect forest soil temperature regimes (Figure 7.30 and Table 7.4). Disturbance of the leaf mulch, changes in water content due to reduced evapotranspiration, and compaction by machinery are all factors that influence soil temperatures through thermal conductivity. Reduced shading after tree removal also lets in more solar radiation.

Mulch from Conservation Tillage. Until fairly recently, the labor and expense of carrying and spreading mulch materials limited their use in modifying soil temperature extremes mostly to home gardens and flower beds. Although these uses are still important, the use of mulches has been extended to field-crop culture in areas that have adopted conservation tillage practices. Conservation tillage leaves most or all of the residues from crops and cover crops at or near the soil surface, thereby allowing the farmer to grow mulch in place, rather than transporting it to the field.

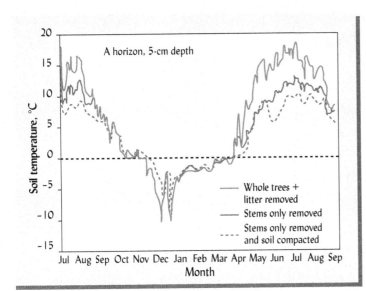

Figure 7.30 *Soil temperature in an aspen-spruce boreal forest after two levels of harvest and soil compaction. One harvest procedure removed only the stems (tree trunks) with branches and foliage left on the soil, while a second procedure removed whole trees and stripped woody materials and litter to expose the mineral soil (this was done to simulate the kind of damage often inflicted by poorly managed harvest equipment). The soils were either left undisturbed during harvest, or they were severely compacted. The compacted treatment is shown only for the whole-tree removal procedure as compaction did not affect soil temperature where harvest removed only tree trunks. Exposure of the mineral soil A horizon resulted in much warmer temperatures in summer and somewhat colder soil in winter. Compaction of this soil mainly slowed warming in summer, partly because of a higher water content (and therefore a higher heat capacity). The Aquepts (Luvic Gleysols in the Canadian soil classification) at this site in British Columbia included about 20 to 30 cm of silt loam material over a clay loam.* [Data from Tan et al. (2005)]

Concerns in Cool Climates. In no-tillage systems all residues are left on the surface as a mulch. While mulch provides great control over erosion (see Section 17.6), it also causes the soil to warm more slowly during spring than where the soil is left bare. In cold climates the soil-temperature-depressing effects of mulch may require delayed planting and can have serious negative impacts on crop productivity. The lower soil temperatures in spring resulting from mulch practices inhibit seed germination, seedling performance, and, often, the yields of such warm season crops as corn or soybean. The effect of the residue mulch is most pronounced in lowering the midday maximum temperature and has much less effect on the minimum temperature reached at night (Figure 7.29). The addition of row cleaners to the no-tillage planter can alleviate this problem by pushing aside the residues in just a 10 to 15 cm-wide band from over the seed row. Another approach to solving this problem is to ridge the soil, permit water to drain out of the ridge, and then plant on the drier, warmer ridgetop (or on the south side of the ridge—see Section 7.9).

Advantages in Warm Climates. In warm regions, delayed planting is not a problem. In fact, the cooler near-surface soil temperatures under a mulch may reduce heat stress on roots during summer. Plant-residue mulches also conserve soil moisture by decreasing evaporation. The resulting cooler, moist surface layer of soil is an important part of no-tillage systems because it allows roots to proliferate in this zone, where nutrient and aeration conditions are optimal.

Plastic Mulches

One of the reasons for the popularity of plastic mulches for gardens and high-value specialty crops is their effect on soil temperature (see Section 6.4 for their effect on soil moisture). In contrast to the cooling effects of organic mulches just discussed, plastic mulches generally increase soil temperature. Clear plastic has a greater heating effect than black plastic. In temperate regions, this effect can be used to extend the growing season or to hasten production to take advantage of the higher prices offered by early-season markets. Figure 7.31 shows the use of clear plastic mulch for winter-grown strawberries in southern California.

Major disadvantages of both clear and black plastic mulch are the nonrenewable fossil fuels used in their manufacture, the difficulty of removing the material from the field at the end of the season, and the problem of properly disposing of all that soil-encrusted, shredded

Figure 7.31 *These winter-grown southern California strawberries will come to market when prices are still high because of the effect of the clear plastic mulch on soil temperature.* (Photo courtesy of Ray R. Weil)

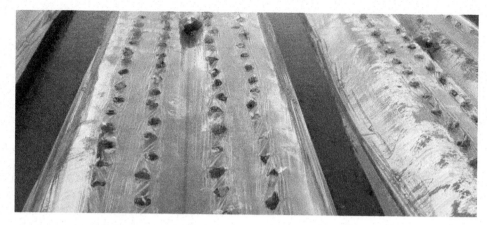

plastic waste. One solution may be found in newer biodegradable plastic films that have been manufactured from such natural renewable raw materials as corn starch.

In warmer climates, and during the summer months, the soil-heating effect of plastic mulches may be quite detrimental, inhibiting root growth in the upper soil layers and sometimes seriously decreasing crop yields. Table 7.5 provides an example in which the benefits of weed control and moisture conservation by plastic mulch were outweighed by the detrimental effects of excessive soil heating.

Moisture Control

Another means of exercising some control over soil temperature is by controlling soil moisture. Poorly drained soils in temperate regions that are wet in the spring have temperatures 3 to 6°C lower than comparable well-drained soils. By removing excess water, warmer soils can be achieved. Water can be removed by installing drainage systems using ditches and underground pipes (see Section 6.7). An alternative to artificial drainage is the construction of raised beds or the ridging systems of tillage just referred.

As is the case with soil air, the controlling influence of soil water on soil temperature is apparent everywhere. Whether a problem concerns capture of solar energy, loss of energy to the atmosphere, or the movement of heat within the soil, the amount of water present is always important. Water regulation seems to be a key to what little practical temperature control is possible for soils in the field.

Table 7.5
SOIL TEMPERATURE AND TOMATO YIELD WITH STRAW OR BLACK PLASTIC MULCH[a]

Means for two years of tomato production on a sandy Ultisol in Georgia, USA. The straw kept the surface soil below detrimental temperatures, while increasing rainwater infiltration and reducing compaction. Daily drip irrigation could not overcome the temperature effects of black plastic mulch.

	Not irrigated		Irrigated daily	
	Straw mulch	Plastic mulch	Straw mulch	Plastic mulch
Average soil temperature, °C	24	37	24	35
Tomato yield, Mg/ha	68	30	70	24

[a]Soil temperature measured at 5 cm below the soil surface, average of weeks 2–10 of the growing season. Data calculated from Tindall et al. (1991).

7.12 CONCLUSION

Soil aeration and soil temperature critically affect the quality of soils as habitats for plants and other organisms. Most plants have definite requirements for soil oxygen along with limited tolerance for carbon dioxide, methane, and other such gases found in poorly aerated soils. Some microbes, such as the nitrifiers and general-purpose decay organisms, are also constrained by low levels of soil oxygen. Through its effect on soil redox potential (E_h) and acidity (pH), soil aeration status helps determine the forms present, availability, mobility, and possible toxicity of such elements as nitrogen, sulfur, carbon, iron, manganese, chromium, and many others.

Soils with extremely wet moisture regimes are unique with respect to their morphology and chemistry and support unique plant communities. Such hydric soils are characteristic of wetlands and help these ecosystems perform a myriad of valuable ecosystem functions.

Plants as well as microbes are also quite sensitive to differences in soil temperature, particularly in temperate climates where low temperatures can limit essential biological processes. Soil temperature also impacts the use of soils for engineering purposes, again primarily in the cooler climates. Frost action, which can move perennial plants such as alfalfa out of the ground, can also cause damage to building foundations, fence posts, sidewalks, and highways.

Soil water exerts a major influence over both soil aeration and soil temperature. It competes with soil air for the occupancy of soil pores and interferes with the diffusion of gases into and out of the soil. Soil water also resists changes in soil temperature by virtue of its high specific heat and its high energy requirement for evaporation. Fire, mulching, and drainage are other factors that markedly influence soil temperatures.

STUDY QUESTIONS

1. What are the two principal gases involved with soil aeration, and how do their relative amounts change as one samples deeper into a soil profile?
2. What is aerenchyma tissue, and how does it affect plant–soil relationships?
3. If the redox potential for a soil at pH 6 is near zero, write two reactions that you would expect to take place. How would the presence of a great deal of nitrate compounds affect the occurrence of these reactions?
4. It is sometimes said that organisms in anaerobic environments will use the combined oxygen in nitrate or sulfate instead of the oxygen in O_2. Why is this statement incorrect? What actually happens when organisms reduce sulfate or nitrate?
5. If an alluvial forest soil were flooded for 10 days, and you sampled the gases evolving from the wet soil, what gases would you expect to find (other than oxygen and carbon dioxide)? In what order of appearance? Explain.
6. Explain why warm weather during periods of saturation is required in order to form a hydric soil.
7. If you were in the field trying to delineate the so-called drier end of a wetland area, what are three soil properties and three other indicators that you might look for?
8. For each of these gases, write a sentence to explain its relationship to wetland conditions: *ethylene, methane, nitrous oxide, oxygen,* and *hydrogen sulfide.*
9. What are the three major components that define a wetland?
10. Discuss four plant processes that are influenced by soil temperature.
11. Explain how a brush fire might lead to subsequent mudslides, as often occurs in California.
12. If you were to build a house below ground in order to save heating and cooling costs, would you firmly compact the soil around the house? Explain your answer.
13. If you measured a daily maximum air temperature of 28°C at 1 P.M., what might you expect the daily maximum temperature to be at a 15-cm depth in the soil? At about what time of day would the maximum temperature occur at this depth? Explain.
14. In relation to soil temperature, explain why conservation tillage has been more popular in Missouri than in Minnesota.

REFERENCES

Bailey-Serres, J., T. Fukao, D. J. Gibbs, M. J. Holdsworth, S. C. Lee, F. Licausi, P. Perata, L. A. C. J. Voesenek, and J. T. van Dongen. 2012. "Making sense of low oxygen sensing." *Trends in Plant Science* 17:129–138.

Bartlett, R. J., and B. R. James. 1993. "Redox chemistry of soils." *Advances in Agronomy* 50:151–208.

Bartlett, R. J., and D. S. Ross. 2005. "Chemistry of redox process in soils." In A. Tabatabai and D. Sparks (eds.).

Chemical Processes in Soils. SSSA Book Series No. 8. Soil Science Society of America, Madison, WI, pp. 461–487.

CAST. 1994. *Wetland Policy Issues*. Publication No. CC1994-1. Council for Agricultural Science and Technology, Ames, IA.

Costanza, R., R. D'arge, R. De Groot, S. Farber, M. Grasso, B. Hannon, K. Limburg, S. Naeem, R. V. O'Neill, J. Paruelo, R. G. Raskin, P. Sutton, and M. Van Den Belt. 1997. "The value of the world's ecosystem services and natural capital." *Nature* 387:253–260.

DeBano, L. F. 2000. "The role of fire and soil heating on water repellency in wildland environments: A review." *Journal of Hydrology* 231/232:195–206.

Dryness, C. T. 1976. "Effects of wildfire on soil wettability in the high cascades of Oregon." USDA Forest Service Research Paper PNW-202. USDA, Washington, DC.

Eswaren, H., P. Reich, P. Zdruli, and T. Levermann. 1996. "Global distribution of wetlands." *American Society of Agronomy Abstracts* 328.

Fluker, B. J. 1958. "Soil temperature." *Soil Science* 86:35–46.

Husson, O. 2012. "Redox potential (E_h) and pH as drivers of soil/plant/microorganism systems: A transdisciplinary overview pointing to integrative opportunities for agronomy." *Plant and Soil* XX:1–29.

Ju, Z., T. Ren, and C. Hu. 2011. "Soil thermal conductivity as influenced by aggregation at intermediate water contents." *Soil Science Society of America Journal* 75:26–29.

Liljedahl, A. K., J. Boike, R. P. Daanen, A. N. Fedorov, G. V. Frost, G. Grosse et al. 2016. "Pan-Arctic ice-wedge degradation in warming permafrost and its influence on tundra hydrology." *Nature Geosci* 9:312–318. doi:10.1038/ngeo2674

Lin, X., J. E. Smerdon, A. W. England, and H. N. Pollack. 2003. "A model study of the effects of climatic precipitation changes on ground temperatures." *Journal of Geophysics Research* 108(D7):4230. doi:10.1029/2002JD002878.

Losi, M. E., C. Amrhein, and W. T. Frankenberger, Jr. 1994. "Bioremediation of chromatic contaminated groundwater by reduction and precipitation in surface soils." *Journal of Environmental Quality* 23:1141–1150.

McBride, M. B. 1994. *Environmental Chemistry of Soils*. Oxford University Press, New York.

Melillo, J. M., S. Butler, J. Johnson, J. Mohan, P. Steudler, H. Lux, E. Burrows, F. Bowles, R. Smith, L. Scott, C. Vario, T. Hill, A. Burton, Y.-M. Zhou, and J. Tang. 2011. "Soil warming, carbon-nitrogen interactions, and forest carbon budgets." *Proceedings of the National Academy of Sciences* 108:9508–9512.

Molins, S., K. U. Mayer, C. Scheutz, and P. Kjeldsen. 2008. "Transport and reaction processes affecting the attenuation of landfill gas in cover soils." *Journal of Environmental Quality* 37:459–468.

Sexstone, A. J., N. P. Revsbech, T. B. Parkin, and J. M. Tiedje. 1985. "Direct measurement of oxygen profiles and denitrification rates in soil aggregates." *Soil Science Society of America Journal* 49:645–651.

Tan, X., S. X. Chang, and R. Kabzems. 2005. "Effects of soil compaction and forest floor removal on soil microbial properties and N transformations in a boreal forest long-term soil productivity study." *Forest Ecology and Management* 217:158–170.

Tiarks, A. E., W. H. Hudnall, J. F. Ragus, and W. B. Patterson. 1996. "Effect of pine plantation harvesting and soil compaction on soil water and temperature regimes in a semi-tropical environment." In A. Schulte and D. Ruhiyat (eds.). *Proceedings of International Congress on Soils of Tropical Forest Ecosystems 3rd Conference on Forest Soils: Vol. 3, Soil and Water Relationships*. Mulawarmon University Press, Samarinda, Indonesia.

Tindall, J. A., R. B. Beverly, and D. E. Radcliff. 1991. "Mulch effect on soil properties and tomato growth using micro-irrigation." *Agronomy Journal* 83:1028–1034.

Unger, P. W. 1978. "Straw mulch effects on soil temperatures and sorghum germination and growth." *Agronomy Journal* 70:858–864.

United States Department of Agriculture, N.R. and C. Service. 2017. Field Indicators of Hydric Soils in the United States USDA, NRCS in cooperation with the National Technical Committee for Hydric Soils: p. 45. https://www.nrcs.usda.gov/Internet/FSE_DOCUMENTS/nrcs142p2_053171.pdf.

Veneman, P. L. M., D. L. Lindbo, and L. A. Spokas. 1999. "Soil moisture and redoximorphic features: A historical perspective." In M. J. Rabenhorst, J. C. Bell, and P. A. McDaniel (eds.). *Quantifying Soil Hydromorphology*. Special Publication No. 54. Soil Science Society of America, Madison, WI.

Vepraskas, M. J., and C. B. Craft. 2016. *Wetland Soils: Genesis, Hydrology, Landscapes, and Classification*. 2nd ed. CRC Press, Boca Raton, FL, p. 508.

Welsh, D., D. Smart, J. Boyer, P. Minkin, H. Smith, and T. McCandless (eds.). 1995. *Forested Wetlands: Functions, Benefits, and Use of Best Management Practices*. USDA Forest Service, Radnor, PA.

Xu, Q., and B. Huang. 2000. "Growth and physiological responses of creeping bentgrass to changes in air and soil temperatures." *Crop Science* 40:1363–1368.

Yu, K., and W. H. Patrick, Jr. 2004. "Redox window with minimum global warming potential contribution from rice soils." *Soil Science Society of America Journal* 68:2086–2091.

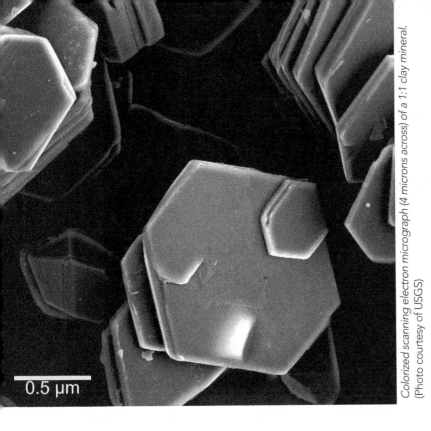

Colorized scanning electron micrograph (4 microns across) of a 1:1 clay mineral. (Photo courtesy of USGS)

8
The Colloidal Fraction: Seat of Soil Chemical and Physical Activity

Not even the ... clay can keep a secret if memory is stronger than time.
—RICK BRAGG IN SOMEBODY TOLD ME

Why is it more difficult to restore productivity after logging a tropical rain forest on Oxisols than a temperate forest on Alfisols? Why do utility poles lean every which way across landscapes of black clay soils? How can using sewage effluent for irrigation contribute to the safe recharge of groundwater aquifers? Why would a nuclear power plant accident seriously contaminate food grown on some downwind soils, but not on others? The answers to these and other environmental mysteries lie in the nature of the smallest of soil particles, the clay and humus **colloids**. These particles are not just extra-small fragments of rock and organic matter. They are highly reactive materials with electrically charged surfaces. Because of their size and shape, they give the soil an enormous amount of reactive **surface area**. It is the colloids, then, that allow the soil to serve as nature's great electrostatic chemical reactor.

Each tiny colloid particle carries a swarm of positively and negatively charged ions (cations and anions) that are attracted to electrostatic charges on its surface. The ions are held tightly enough by the soil colloids to greatly reduce their loss in drainage waters, but loosely enough to allow plant roots access to the nutrients among them. Other modes of adsorption bind ions more tightly so that they are no longer available for plant uptake, reaction with the soil solution, or leaching loss to the environment. In addition to plant nutrient ions, soil colloids also bind with water molecules, biomolecules (e.g., DNA, hormones, or antibiotics), viruses, toxic metals, pesticides, and a host of other mineral and organic substances in the environment. Hence, soil colloids greatly impact nearly all ecosystem functions.

We shall see that different soils are endowed with different types of clays that, along with humus, elicit very different types of physical and chemical behaviors. Certain clay minerals are much more reactive than others. Some are more dramatically influenced than others by the acidity of the soil and other environmental factors. Studying the soil colloids in some detail will deepen your understanding of soil architecture (Chapter 4) and soil water (Chapters 5 and 6). Knowledge of the structure, origin, and behavior of the different types of soil colloids will also help you understand soil chemical and biological processes (Chapters 9–13) so you can make better decisions regarding the use of soil resources.

8.1 GENERAL PROPERTIES AND TYPES OF SOIL COLLOIDS

Size

The clay and humus particles in soils are referred to collectively as the **colloidal fraction** because of their extremely small size and colloid-like behavior. So small, they are visible only with an electron microscope. Particles behave as colloids if they are less than about 1 μm (0.000001 m) in diameter (although 2 μm defines the boundary between silt and clay particle size fractions).

Surface Area

As discussed in Section 4.2, the smaller the size of the particles in a given mass of soil, the greater the surface area exposed for adsorption, catalysis, precipitation, microbial colonization, and other surface phenomena. Soil colloids expose a large **external surface** area per unit mass, more than 1,000 times the surface area of the same mass of sand particles. Certain silicate clays also possess extensive **internal surface** area between the layers of their platelike crystal units. To grasp the relative magnitude of the internal surface area, remember that these clays are structured much like this book. If you were to paint the external surfaces of this book (the covers and edges), a single brush of paint would do. However, to cover the internal surfaces (both sides of each page in the book) you would need a very large amount of paint.

The total surface area of soil colloids ranges from 10 m^2/g for clays with only external surfaces to more than 800 m^2/g for clays with extensive internal surfaces. To put this in perspective, we can calculate that the surface area exposed within 1 ha (about the size of a football field) of a 1.5-m-deep fine-textured soil (45% clay) might be as great as 8,700,000 km^2 (the land area of the entire United States).

Surface Charges

The internal and external surfaces of soil colloids carry positive and/or negative electrostatic charges. For most soil colloids, electronegative charges predominate, although some mineral colloids in very acid soils have a net electropositive charge. As we shall see in Sections 8.3–8.7, the amount and origin of surface charge differ greatly among the different types of soil colloids and, in some cases, is influenced by changes in chemical conditions, such as soil pH. The charges on the colloid surfaces attract or repulse substances in the soil solution as well as neighboring colloid particles. These reactions, in turn, greatly influence soil chemical and physical behavior.

Adsorption of Cations and Anions

Of particular significance is the attraction of positively charged ions (**cations**) to the surfaces of negatively charged soil colloids. Each colloid particle attracts thousands of Al^{3+}, Ca^{2+}, Mg^{2+}, K^+, H^+, and Na^+ ions and lesser numbers of other cations. In moist soils, the cations exist in the hydrated state (surrounded by a shell of water molecules), but for simplicity in this text, we will show just the cations (e.g., Ca^{2+} or H^+) rather than the hydrated forms (e.g., $Ca(H_2O)_6^{2+}$ or the hydronium ion, H_3O^+). These hydrated cations constantly vibrate about in a swarm near the colloid surface, held there by electrostatic attraction to the colloid's negative charges. Frequently, an individual cation will break away from the swarm and move out to the soil solution. When this happens, another cation of equal charge will simultaneously move in from the soil solution and take its place. This process, termed **cation exchange**, will be discussed in detail (Section 8.8) because of its fundamental importance in nutrient cycling and other environmental processes. The cations swarming about near the colloidal surface are said to be **adsorbed** (loosely held) on the colloid surface. Because these cations can *exchange places* with those moving freely about in the soil solution, the term **exchangeable ions** is also used to refer to the ions in this adsorbed state.

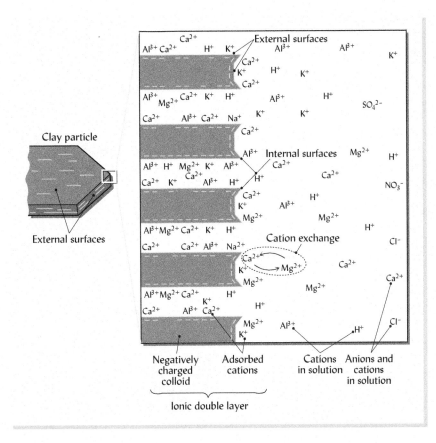

Figure 8.1 *Simplified representation of a silicate clay crystal, its complement of adsorbed cations, and ions in the surrounding soil solution. The enlarged view (right) shows that the clay comprises sheetlike layers with both external and internal negatively charged surfaces. The negatively charged particle acts as a huge anion and a swarm of positively charged cations is adsorbed to it because of attraction between charges of opposite sign. Cation concentration decreases with distance from the clay. Anions (such as Cl^-, NO_3^-, and SO_4^{2-}), which are repulsed by the negative charges, can be found in the bulk soil solution farthest from the clay (far right). Some clays (not shown) also exhibit positive charges that can attract anions.*

The colloid with its adsorbed cations is sometimes described as an **ionic double layer** in which the negatively charged colloid acts as a huge anion constituting the inner ionic layer, and the swarm of adsorbed cations constitutes the outer ionic layer (Figure 8.1). Because cations from the soil solution are constantly trading places with those that are adsorbed to the colloid, the ionic composition of the soil solution reflects that of the adsorbed swarm. For example, if Ca^{2+} and Mg^{2+} dominate the exchangeable ions, they will also dominate the soil solution. Under natural conditions, the proportions of specific cations present are largely influenced by the soil parent material and the degree to which the climate has promoted the loss of cations by leaching (see Section 9.1).

Anions such as Cl^-, NO_3^-, and SO_4^{2-} (also surrounded by water molecules, though, again, we do not show these water shells) may also be attracted to certain soil colloids that have *positive* charges on their surfaces. While adsorption of **exchangeable anions** is not as extensive as that for exchangeable cations, we shall see (Section 8.11) that it is an important mechanism for holding negatively charged constituents, especially in acidic subsoils.

Adsorption of Water

In addition to adsorbing cations and anions, soil colloids attract and hold a large number of water molecules. Generally, the greater the external surface area of the soil colloids, the greater the amount of hydroscopic water held when the soil is air-dry. While this water may not be available for plant uptake (see Section 5.8), it does play a role in the survival of soil microorganisms. The charges on the colloid surfaces attract the oppositely charged end of the polar water molecule. Some water molecules are attracted to the exchangeable cations, each of which is hydrated with a shell of water molecules. Water adsorbed between the clay layers can cause the layers to move apart, making the clay more plastic and swelling its volume (see Sections 4.9 and 8.14). As a soil colloid dries, water adsorbed on the external and internal surfaces is removed, so the particles and their constituent layers are brought closer together, shrinking the soil volume.

Types of Soil Colloids[1]

The colloids most important in soils can be grouped in four major types, each with its particular composition, structure, and properties (Table 8.1).

Crystalline Silicate Clays. These clays are the dominant type in most soils (except in Andisols, Oxisols, and Histosols—see Chapter 3). Their crystalline structure is layered much like pages in a book (clearly visible in Figure 8.2a). Each layer (page) consists of two to four sheets of closely packed and tightly bonded oxygen, silicon, and aluminum atoms. Although all are predominately negatively charged, silicate clay minerals differ widely with regard to their particle shapes (**kaolinite**, a **fine-grained mica**, and a **smectite** are shown in Figure 8.2a–c), intensity of charge, stickiness, plasticity, and swelling behavior.

Noncrystalline Silicate Clays. These clays also consist mainly of tightly bonded silicon, aluminum, and oxygen atoms, but they do not exhibit ordered, crystalline sheets. The two principal clays of this type, **allophane** and **imogolite**, usually form from volcanic ash and are characteristic of Andisols (Section 3.7). They have high amounts of both positive and negative charge, and high water-holding capacities. Although malleable (plastic) when wet, they exhibit a very low degree of stickiness. Allophane and imogolite are also known for their extremely high capacities to strongly adsorb phosphate and other anions, especially under acid conditions.

Iron and Aluminum Oxides. These are found in many soils, but are especially important in the more highly weathered soils of warm, humid regions (e.g., Alfisols, Ultisols, and Oxisols). They consist mainly of either iron or aluminum atoms coordinated with oxygen atoms (the latter are often associated with hydrogen ions to make hydroxyl (OH) groups). Some, like **gibbsite** (an Al oxide) and **goethite** (an Fe oxide) consist of crystalline sheets. Other oxide minerals are

Table 8.1
MAJOR PROPERTIES OF SELECTED SOIL COLLOIDS

Colloid	Type	Size, μm	Shape	Surface area, m^2/g External	Surface area, m^2/g Internal	Interlayer Spacing,[a] nm	Net charge,[b] $cmol_c/kg$
Smectite	2:1 silicate	0.01–1.0	Flakes	80–150	550–650	1.0–2.0	−80 to −150
Vermiculite	2:1 silicate	0.1–0.5	Plates, flakes	70–120	600–700	1.0–1.5	−100 to −200
Fine mica	2:1 silicate	0.2–2.0	Flakes	70–175	—	1.0	−10 to −40
Chlorite	2:1 silicate	0.1–2.0	Variable	70–100	—	1.41	−10 to −40
Kaolinite	1:1 silicate	0.1–5.0	Hexagonal crystals	5–30	—	0.72	−1 to −15
Gibbsite	Al oxide	<0.1	Hexagonal crystals	80–200	—	0.48	+10 to −5
Goethite	Fe oxide	<0.1	Variable	100–300	—	0.42	+20 to −5
Allophane and Imogolite	Noncrystalline silicates	<0.1	Hollow spheres or tubes	100–1000	—	—	+20 to −150
Humus	Organic	0.1–1.0	Amorphous	Variable[c]	—	—	−100 to −500

[a] From the top of one layer to the next similar layer, 1 nm=10^{-9} m=10 Å.
[b] Centimoles of unbalanced or net charge per kilogram of colloid ($cmol_c/kg$), a measure of ion exchange capacity (see Section 8.9).
[c] It is very difficult to determine the surface area of organic matter. Different procedures give values ranging from 20 to 800 m^2/g.

---

[1] For a review of the structures and properties of the clays, see Meunier (2005), and for properties of clay and humus colloids in soils, see Dixon and Weed (1989).

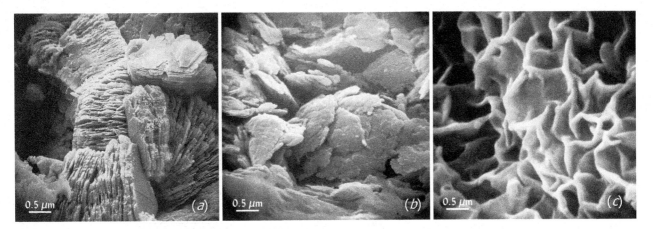

Figure 8.2 Crystals of three silicate clay minerals. (a) Kaolinite (note hexagonal crystal at upper right). (b) A fine-grained mica. (c) Montmorillonite, a smectite group mineral. (Bohor, B. F., and R. E. Hughes. 1971. "Scanning electron microscopy of clays and clay minerals." *Clays and Clay Minerals* 19:49–54)

noncrystalline, often occurring as **amorphous** coatings on soil particles. The oxide colloids are relatively low in plasticity and stickiness. Their net charge ranges from slightly negative to moderately positive. Although for simplicity we will use the term *Fe, Al oxides* for this group, many are actually hydroxides or oxyhydroxides because of the presence of hydrogen ions.

Organic (Humus). Organic colloids are important in nearly all soils, especially in the upper parts of the soil profile. Humus colloids are not minerals, nor are they crystalline. Instead, they consist of a wide variety of partially decomposed cell walls and biomolecules derived mainly from microorganisms and bits of partially decomposed tissues derived mainly from plants (see Section 11.4). Much of this material has complex chains and rings of carbon atoms bonded to hydrogen, oxygen, and nitrogen. Humus particles are often among the smallest of soil colloids and exhibit very high capacities to adsorb water, but almost no plasticity or stickiness. Because humus is noncohesive, soils composed mainly of humus (Histosols) have very little bearing strength and are unsuitable for making building or road foundations. Humus has high amounts of both negative and positive charge per unit mass, but the net charge is always negative and varies with soil pH. The negative charge on humus is extremely high in neutral to alkaline soils.

8.2 FUNDAMENTALS OF LAYER SILICATE CLAY STRUCTURE[2]

To understand why soils rich in, say, kaolinite clay, behave so very differently from soils dominated by, say, smectite clay, it is necessary to know something about the silicate clay minerals. We will begin by examining the main building blocks from which the layer silicates are constructed, then consider the particular arrangements that give rise to the critically important surface charges.

Tetrahedral and Octahedral Sheets

The most important silicate clays are known as **phyllosilicates** (Greek *phyllon*, leaf) because of their leaflike or planar structure. As shown in Figure 8.3, they are composed of two kinds of horizontal **sheets**.

Tetrahedral Sheets. This kind of sheet consists of two **planes** of oxygens with mainly silicon in the spaces between the oxygens. The basic building block for the tetrahedral sheet is a unit composed of one silicon atom surrounded by four oxygen atoms. It is called a **tetrahedron**

---

[2]The authors are indebted to Dr. Darrel G. Schultze of Purdue University for kindly providing structural models for the silicate clay minerals.

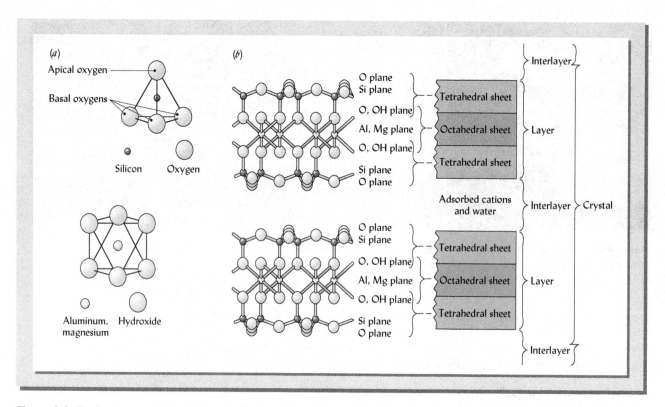

Figure 8.3 The basic structural components of silicate clays. (a) A single tetrahedron, a four-sided building block composed of a silicon ion surrounded by four oxygen atoms; and a single eight-sided octahedron, in which an aluminum (or magnesium) ion is surrounded by six oxygen atoms or hydroxy groups. (b) In clay crystals, thousands of these tetrahedral and octahedral building blocks are connected to give planes of silicon and aluminum (or magnesium) ions. These planes alternate with planes of oxygen atoms and hydroxy groups. Note that apical oxygen atoms are common to adjoining tetrahedral and octahedral sheets. The silicon plane and associated oxygen–hydroxy planes make up a tetrahedral sheet. Similarly, the aluminum–magnesium plane and associated oxygen–hydroxy planes constitute the octahedral sheet. Different combinations of tetrahedral and octahedral sheets are termed layers. In some silicate clays these layers are separated by interlayers in which water and adsorbed cations are found. Many layers are found in each crystal.

because (as shown in Figure 8.3, *top left*) the oxygens define the apices of a *four*-sided geometric solid that resembles a pyramid (having three "sides" and a bottom). An interlocking array of such tetrahedra, each sharing its basal oxygens with its neighbor, makes a **tetrahedral sheet**.

Octahedral Sheets. Six oxygen atoms coordinating with a central aluminum or magnesium atom form the shape of an *eight*-sided geometric solid, or **octahedron**. Numerous octahedra linked together horizontally constitute the **octahedral sheet**.

The tetrahedral and octahedral sheets are the fundamental structural units of silicate clays. Two to four of these sheets may be stacked together in sandwich-like arrangements, with adjacent sheets strongly bound together by sharing some of the same oxygen atoms (see Figure 8.3). The specific nature and combination of sheets in these layers vary from one type of clay to another and largely control the physical and chemical properties exhibited. The relationship between *planes*, *sheets*, and *layers* shown in Figure 8.3 should be carefully studied.

The structural arrangements just described suggest a very simple relationship among the elements making up silicate clays. In nature, however, clays have formulas that are more complex. During the weathering of rocks and minerals, many different elements are present in the weathering solution. As clay minerals or their precursors crystallize, cations of similar size may substitute for silicon, aluminum, and magnesium ions in the respective tetrahedral and octahedral sheets.

An aluminum atom is only slightly larger than silicon. Consequently, aluminum can fit into the center of the tetrahedron in the place of the silicon without much change in the basic structure of the crystal. This process by which one element fills a position usually filled by another of similar size is called **isomorphous substitution**. Isomorphous substitution can also occur in the octahedral sheets.

Source of Charges

Isomorphous substitution is of vital importance because it is the primary source of both negative and positive charges of silicate clays. For example, the Mg^{2+} ion is only slightly larger than the Al^{3+} ion, but it has one less positive charge. If a Mg^{2+} ion substitutes for an Al^{3+} ion in an octahedral sheet, there will be insufficient positive charges to balance the negative charges from the oxygens; hence, the lattice is left with a -1 net charge. Similarly, every Al^{3+} that substitutes for a Si^{4+} in a tetrahedral sheet creates a net negative charge at that site because the negative charges from the four oxygens will be only partially balanced.

The net charge associated with a clay crystal is the sum of the positive and negative charges. In most silicate clays, the negative charges predominate (as will be discussed in Section 8.8). As we shall see (Sections 8.3 and 8.6), additional, more temporary charges can also develop on the edges of the tetrahedral and octahedral surfaces.

8.3 MINERALOGICAL ORGANIZATION OF SILICATE CLAYS

Based on the number and arrangement of tetrahedral (Si) and octahedral (Al, Mg, Fe) sheets contained in the crystal units or layers, crystalline clays may be classed into two main groups: **1:1 silicate clays**, in which each layer contains *one* tetrahedral and *one* octahedral sheet, and **2:1 silicate clays**, in which each layer has *one* octahedral sheet sandwiched between *two* tetrahedral sheets.

1:1-Type Silicate Clays

Kaolinite is by far the most common 1:1 silicate clay in soils (others include *halloysite, nacrite*, and *dickite*). As implied by the term *1:1 silicate clay*, each kaolinite layer consists of one silicon tetrahedral sheet and one aluminum octahedral sheet. The two types of sheets are tightly held together because the apical oxygen atom (the oxygen atom that forms the apex or tip of the "pyramid") in each tetrahedron also forms a bottom corner of one or more of the octahedra in the adjoining sheet (Figure 8.4). Note that because a kaolinite crystal layer consists of these two sheets, it exposes a plane of oxygen atoms on the bottom surface, but a plane of hydroxyls on the upper surface.

This arrangement has two very important consequences. First, as will be discussed in Section 8.6, where the hydroxyl plane is exposed on the clay particle surface, changes in the pH of the soil cause the removal or addition of hydrogen ions, producing either positive or negative charges. The exposed **hydroxylated surface** can also react with and strongly bind specific anions. Second, when the layers consisting of alternating tetrahedral and octahedral sheets are stacked on top of one another, the hydroxyls of the octahedral sheet in one layer are adjacent to

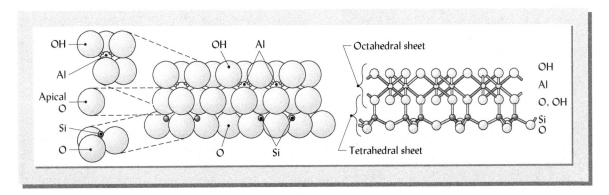

Figure 8.4 *Models of the 1:1-type clay kaolinite. The primary elements of the octahedral (upper far left) and tetrahedral (lower far left) sheets are depicted as they might appear separately. Note a kaolinite layer consists of adjacent octahedral and tetrahedral sheets—hence, the designation 1:1. The octahedral and tetrahedral sheets are bound together (center) by mutually shared (apical) oxygen atoms. The result is a layer with hydroxyls on one surface and oxygens on the other. To permit us to view the front silicon atoms, we have not shown some basal oxygen atoms that are normally present. The diagram at right shows the bonds between atoms. The kaolinite mineral is comprised of a stacked series of these flat layers tightly held together with no interlayer spaces.*

the basal oxygens of the tetrahedral sheet of the next layer. Therefore, adjacent layers are bound together by **hydrogen bonding** (see Section 5.1).

Because of the interlayer hydrogen bonding, the structure of kaolinite is fixed, and no expansion can occur between the layers when the clay is wetted. Cations and water generally do not enter between the structural layers of a 1:1 mineral particle. The effective surface of kaolinite is thus restricted to its outer faces or external surface area. This fact and the lack of significant isomorphous substitution in this mineral account for the relatively small capacity of kaolinite to adsorb exchangeable cations (see Table 8.1).

Kaolinite crystals are usually hexagonal in shape (see chapter opener image and Figure 8.2a) and larger than most other clays (Table 8.1). Compared to some 2:1 silicate clays, 1:1 clays like kaolinite exhibit less plasticity, stickiness, cohesion, shrinkage, and swelling and can also hold less water. Because of these properties, soils dominated by 1:1 clays are relatively easy to cultivate for agriculture and, with proper nutrient management, can be quite productive. Kaolinite-containing soils are well suited for use in roadbeds and building foundations. The nonexpanding 1:1 structure gives kaolinite clays properties good for making bricks and ceramics and for many other uses (see Box 8.1).

BOX 8.1
KAOLINITE CLAY—THE STORY OF WHITE GOLD[a]

Kaolinite, the most common of the 1:1 clay minerals, has been used for thousands of years to make pottery, roofing tiles, and bricks. The basic processes have changed little to this day. The clayey material is saturated with water, kneaded, and molded or thrown on a potter's wheel to obtain the desired shape, and then hardened by drying or firing (Figure 8.5). The mass of cohering clay platelets hardens irreversibly when fired and the nonexpanding nature of kaolinite allows it to be fired without cracking from shrinkage. The heat also changes the typical gray color of the soil material to "brick red" because of the irreversible oxidation and crystallization of the iron–oxyhydroxides that often coat soil kaolinite particles. In contrast, kaolinite mined from pure deposits fires to a light, creamy color. Kaolinite is not as plastic (moldable) as some other clays, however, and so is usually mixed with more plastic types of clays for making pottery.

It was in seventh-century China that pure kaolinite deposits were first used to make objects of a translucent, lightweight, and strong ceramic called porcelain. The name *kaolinite* derives from the Chinese words *kao* and *ling*, meaning "high ridge," as the material was first mined from a hillside in Kiangsi Province. The Chinese held a monopoly on porcelain-making technology (hence the term *china* for porcelain dishes) until the early 1700s. English colonists, in what is now Georgia in the United States, noted outcrops of white kaolinite clay in areas of rather unproductive soil. The colonialists soon were exporting this kaolinite as the main ingredient for making porcelain in England, where the now-famous pottery was first manufactured from the Georgia kaolinite clay. Sometimes called "white Georgia dirt," pure white kaolinite is still sold locally in rural Georgia (and now online, for instance, at www.whitedirt.com) for purposes ranging from geophagy (see Box 1.1) to shampooing hair.

The market for pure, white kaolinite clay greatly expanded when paper manufacturers started using kaolinite clay to make sizing, the coating that makes high-quality papers smoother, whiter, and more printable. Other industrial uses now include paint pigments, fillers in plastic manufacture, and ceramic materials used for electrical insulation and heat shielding (as on the belly of the space shuttle). The kaolinite in kaopectin-type medications lines the stomach walls and inactivates diarrhea-causing bacteria by adsorbing them on the clay particle surfaces.

A development that is likely to increase the demand for kaolinite is its use as a spray-on coating that provides a nontoxic (approved for organic farming) but effective physical barrier to protect foliage and fruits from insect pests and fungal diseases.

Figure 8.5 *Kaolinite clay in African pottery and early nineteenth-century English china (inset).* (Photo courtesy of Ray R. Weil)

[a]For more on social and commercial aspects of kaolinite, see Seabrook (1995) and Windham (2007). For information on kaolin as pest barrier, see Reddy (2013).

Expanding 2:1-Type Silicate Clays

The four general groups of 2:1 silicate clays are characterized by *one* octahedral sheet sandwiched between *two* tetrahedral sheets. Two of these groups, **smectite** and **vermiculite**, include expanding-type minerals; the other two, **fine-grained micas (illite)** and **chlorite**, are relatively nonexpanding.

Smectite Group. The flake-like crystals of smectites (see Figure 8.2c) have a high amount of mostly negative charge resulting from isomorphous substitution. Most of the charge derives from Mg^{2+} ions substituted in the Al^{3+} positions of the octahedral sheet, but some also derives from substitution of Al^{3+} ions for Si^{4+} in the tetrahedral sheets (Figure 8.6). Because of these substitutions, the capacity to adsorb cations is very high—about 20–40 times that of kaolinite.

In contrast to kaolinite, smectites have a 2:1 structure that exposes a layer of oxygen atoms at both the top and bottom planes. Therefore, adjacent layers are only loosely bound to each other by very weak oxygen-to-oxygen and cation-to-oxygen linkages and the space between is variable (Figure 8.6). The internal surface area exposed between the layers by far exceeds the external surface area of these minerals and contributes to the very high total **specific surface area** (Table 8.1). Exchangeable cations and associated water molecules are attracted to the spaces between the interlayer spaces.

Flake-like smectite crystals tend to pile upon one another, forming wavy stacks that contain many extremely small *ultramicropores*. When soils high in smectite are wetted, adsorption of water in these ultramicropores leads to significant swelling; when they re-dry, the soils shrink in volume (see Section 4.9). The expansion upon wetting contributes to the high degree of plasticity, stickiness, and cohesion that make smectitic soils very difficult to cultivate or excavate. Wide cracks commonly appear during the drying of smectite-dominated soils (such as Vertisols, Figure 3.22). The shrink/swell behavior makes smectitic soils quite undesirable for most construction activities, but they are well suited for a number of applications that require a high adsorptive capacity and the ability to form seals of very low permeability (see Box 8.2). **Montmorillonite** is the most prominent of the smectites in soils.

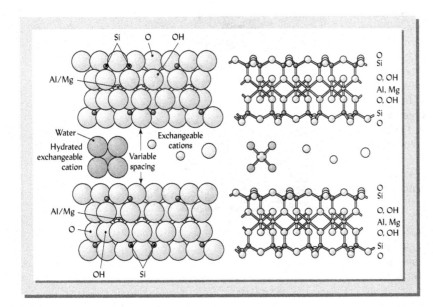

Figure 8.6 *Model of two crystal layers and an interlayer characteristic of montmorillonite, a smectite expanding lattice 2:1-type clay mineral. Each layer is made up of an octahedral sheet sandwiched between two tetrahedral sheets with shared apical oxygen atoms. There is little attraction between oxygen atoms in the bottom tetrahedral sheet of one unit and those in the top tetrahedral sheet of another. This permits a variable space between layers, which is occupied by water and exchangeable cations. The internal surface area thus exposed far exceeds the surface around the outside of the crystal. Note that magnesium has replaced aluminum in some sites of the octahedral sheet. Likewise, some silicon atoms in the tetrahedral sheet may be replaced by aluminum (not shown). These substitutions give rise to a negative charge, which accounts for the high cation exchange capacity of this clay mineral. A ball-and-stick model of the atoms and chemical bonds is at the right.*

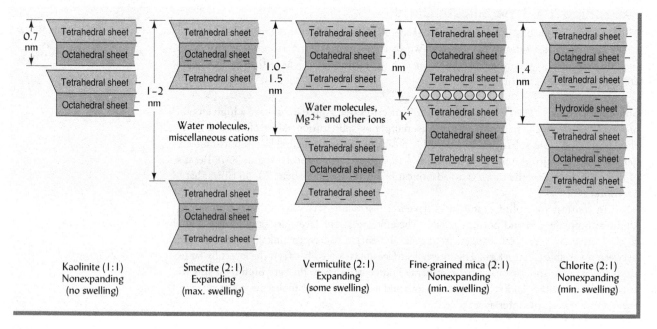

Figure 8.7 *Schematic drawing illustrating the organization of tetrahedral and octahedral sheets in one 1:1-type mineral (kaolinite) and four 2:1-type minerals. Note that kaolinite is nonexpanding, the layers being held together by hydrogen bonds. Maximum interlayer expansion is found in smectite, with somewhat less expansion in vermiculite because of the moderate binding power of numerous Mg^{2+} ions. Fine-grained mica and chlorite do not expand because K^+ ions (fine-grained mica) or an octahedral-like sheet of hydroxides of Al, Mg, Fe, and so forth (chlorite) tightly bind the 2:1 layers together. The interlayer spacings are shown in nanometers (1 nm = 10^{-9} m).*

BOX 8.2
ENVIRONMENTAL USES OF SMECTITE-TYPE EXPANDING CLAYS

The chemical and physical properties of smectite-type clays make them extremely useful in certain environmental engineering applications. A common use of swelling clays—especially a mined mixture of smectite clays called bentonite—is as a sealant layer placed on the bottom and sides of ponds, waste lagoons, and landfill cells (Figure 8.8). The clay material expands when wetted and forms a highly impermeable barrier to the movement of water as well as organic and inorganic contaminants contained in the water. The contaminants are thus held in the containment structure and prevented from polluting the groundwater.

A more exotic use for swelling clays is proposed in Sweden for the final repository of that country's highly radioactive and toxic nuclear power plant wastes. The plan is to place the wastes in large (about 5 m × 1 m) copper canisters and bury them deep underground in chambers carved from solid rock. As a final defense against leakage of the highly toxic material to the groundwater, the canisters will be surrounded by a thick buffer layer of bentonite clay. The clay is packed dry around the canisters and is expected to absorb water to saturation during the first century of storage, thus gradually swelling into a sticky, malleable mass that will fill any cavities or cracks in the rock. The clay buffer will serve three protective functions: (1) cushion

Figure 8.8 *A large waste lagoon under construction is being lined with a layer of dark-colored smectitic clay. The very low hydraulic conductivity of the expansive clay will seal in the stored liquid and retard percolation of polluted water. The high CEC of the clay will retard leaching of pollutants by retaining dissolved pollutants, even when some water movement takes place. Straw-colored erosion blankets are also being used to establish erosion controlling grasses on the berms.* (Photo courtesy of Ray R. Weil)

> **BOX 8.2**
> **ENVIRONMENTAL USES OF SMECTITE-TYPE EXPANDING CLAYS** (CONTINUED)
>
> the canister against small (10-cm) movements in the rock formation, (2) form a seal of extremely low permeability to keep corrosive substances in the groundwater away from the canister, and (3) act as a highly efficient electrostatic filter (see Section 8.8) to adsorb and trap cationic radionuclides that might leak from the canister in some far future time.
>
> Figure 8.9 shows how bentonite is used as a plug or sealant to prevent leakage around an environmental groundwater monitoring well. For most of the well depth, the gap between the bore hole wall and the well tube is back-filled with sand to support the tube and allow vertical movement of the groundwater to be sampled. About 30 cm below the soil surface, the space around the well casing is filled instead with dry granulated bentonite (white substance being poured from bucket in the photograph). As the bentonite absorbs water, it swells markedly, taking on an almost rubbery consistency and forming an impermeable seal that fits tightly against both the well casing and the soil bore hole wall. This seal prevents contaminants in or on the surface soil from leaking down the outside of the well casing. In the case of groundwater contaminated with volatile organics like gasoline, the bentonite also prevents vapors from escaping without being properly sampled.
>
> Increasingly, environmental scientists are using swelling-type clays for the removal of organic chemicals from water by partitioning (see Section 8.12). For example, where there has been a spill of toxic organic chemicals, a deep trench may be dug across the slope and back-filled with a slurry of smectite clay and water to intercept a plume of polluted water. The swelling nature of the smectites prevents the rapid escape of the contaminated water while the highly reactive colloid surfaces chemically sorb the contaminants, purifying the groundwater as it slowly passes by.
>
>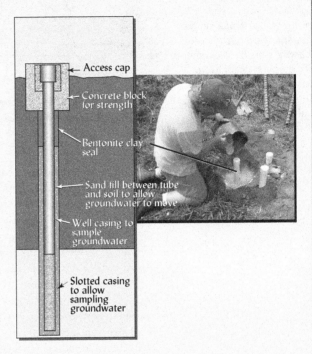
>
> **Figure 8.9** Use of swelling clay as seal for environmental monitoring well. (Courtesy of Ray R. Weil)

Vermiculite Group. The most common **vermiculites** are 2:1-type minerals in which the octahedral sheet is aluminum dominated. The tetrahedral sheets of most vermiculites have considerable substitution of aluminum in the silicon positions, giving rise to a cation exchange capacity (CEC) that usually exceeds that of all other silicate clays, including smectites (Table 8.1).

The interlayer spaces of vermiculites usually contain strongly adsorbed water molecules, Al-hydroxy ions, and cations such as magnesium (Figure 8.7). However, these interlayer constituents act primarily as bridges to hold the units together, rather than wedges driving them apart. The degree of swelling and shrinkage is, therefore, considerably less for vermiculites than for smectites. For this reason, vermiculites are considered limited-expansion clays, expanding more than kaolinite, but much less than the smectites.

Nonexpanding 2:1 Silicate Minerals

The main nonexpanding 2:1 minerals are the **fine-grained micas** and the **chlorites**. Biotite and muscovite are examples of unweathered micas typically found in the sand and silt fractions. The more weathered **fine-grained micas**, such as **illite** and **glauconite**, are found in the clay fraction of soils. Their 2:1-type structures are quite similar to those of their unweathered cousins. Unlike in smectites, the main source of charge in fine-grained micas is the substitution of Al^{3+} in about 20% of the Si^{4+} sites in the tetrahedral sheets. This results in a high net

Figure 8.10 *Model of a 2:1-type nonexpanding lattice mineral of the fine-grained mica group. The general constitution of the layers is similar to that in the smectites, one octahedral sheet between two tetrahedral sheets. However, potassium ions are tightly held between layers, giving the mineral a more or less rigid type of structure that prevents the movement of water and cations into the space between layers. The internal surface and cation exchange capacity of fine-grained micas thus are far below those of the smectites.*

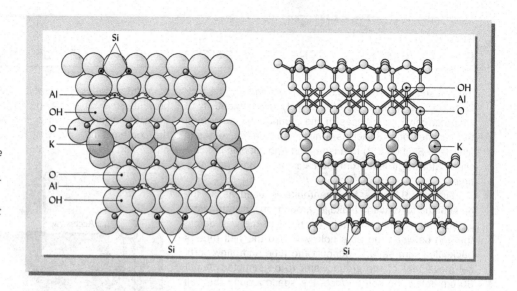

negative charge in the tetrahedral sheet, even higher than that found in vermiculites. The negative charge attracts cations, among which potassium (K^+) is just the right size to fit snugly into certain hexagonal "holes" between the tetrahedral oxygen groups (Figures 8.7 and 8.10) and thereby get very close to the negatively charged sites. By their mutual attraction for the K^+ ions in between, adjacent layers in fine-grained micas are strongly bound together. Hence, the fine-grained micas are quite **nonexpansive** and more like kaolinite than smectites with regard to their capacity to adsorb water and their degree of plasticity and stickiness.

In most soil **chlorites**, iron or magnesium, rather than aluminum, occupy many of the octahedral sites. Commonly, a magnesium-dominated trioctahedral hydroxide sheet is sandwiched in between adjacent 2:1 layers (Figure 8.7). Thus, chlorite is sometimes said to have a 2:1:1 structure. Chlorites are nonexpansive because the hydroxylated surfaces of an intervening Mg-octahedral sheet are hydrogen-bonded to the oxygen atoms of the two adjacent tetrahedral sheets, binding the layers tightly together. The colloidal properties of the chlorites are therefore quite similar to those of the fine-grained micas (Table 8.1).

8.4 STRUCTURAL CHARACTERISTICS OF NONSILICATE COLLOIDS

Iron and Aluminum Oxides

These clays consist of modified octahedral sheets with either iron (Fe^{3+}) or aluminum (Al^{3+}) in the cation positions. They have neither tetrahedral sheets nor silicon in their structures. Isomorphous substitution by ions of varying charge rarely occurs, so these clays do not have a large negative charge. The small amount of net charge these clays possess (positive and negative) is caused by the removal or addition of hydrogen ions at the surface oxy-hydroxyl groups. The presence of these bound oxygen and hydroxyl groups enables the surfaces of these clays to strongly adsorb and combine with anions such as phosphate or arsenate. The oxide clays are nonexpansive and generally exhibit relatively little stickiness, plasticity, and cation adsorption. They make quite stable materials for construction purposes (see Figure 3.32).

Gibbsite [$Al(OH)_3$], the most common soil aluminum oxide, is a prominent constituent of highly weathered soils (e.g., Oxisols and Ultisols). Figure 8.11 shows gibbsite to consist of a series of aluminum octahedral sheets linked to one another by hydrogen-bonding between their hydroxyls. Note that a plane of hydroxyls is exposed at the upper and lower surfaces of gibbsite crystals. These hydroxylated surfaces can strongly adsorb certain anions.

Other oxide-type clays have iron instead of aluminum in the central cation positions, and their octahedral structures are somewhat distorted and less regular than that of gibbsite.

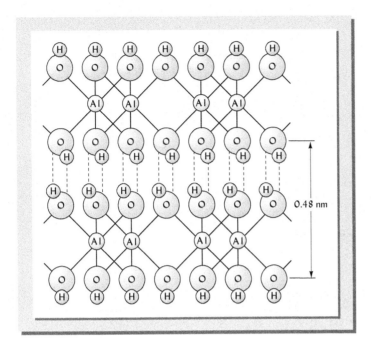

Figure 8.11 A simplified diagram showing the structure of gibbsite, an aluminum oxide clay common in highly weathered soils. This clay consists of dioctahedral sheets (two are shown) that are hydrogen-bonded together. Other oxide-type clays have iron instead of aluminum in the octahedral positions, and their structures are somewhat less regular and crystalline than that shown for gibbsite. The surface plane of covalently bonded hydroxyls gives this, and similar clays, the capacity to strongly adsorb certain anions (see Section 8.8).

Goethite (FeOOH) and **ferrihydrite** ($Fe_2O_3 \cdot nH_2O$) are common iron oxide clays in temperate regions, accounting for the yellow-brown colors of many soils. **Hematite** (Fe_2O_3) is common in drier environments and gives redder colors to well-drained soils, especially in hot, dry climates.

In many soils, iron and aluminum oxide minerals are mixed with silicate clays. The oxides may form coatings on the external surfaces of the silicate clays, or they may occur as "islands" in the interlayer spaces of such 2:1 clays as vermiculites and smectites. In either case, the presence of iron and aluminum oxides can substantially alter the colloidal behavior of the associated silicate clays by masking charge sites, interfering with shrinkage and swelling, and providing anion-retentive surfaces.

Humus

As mentioned in Section 8.1, humus is comprised of noncrystalline heterogeneous organic substances whose chemical composition varies considerably, but generally contains 40–60% C, 30–50% O, 3–7% H, and 1–5% N. A small part of the humus in soils may consist of very large polymerized molecules of so-called humic acids with molecular weights in the 1000s of g/mol. The actual structures of these humic acids have not been identified and many soil chemists have concluded that most of the colloids in humus are comprised of tiny bits of partially decomposed or charred plant tissue, microbial cell walls and films of biomolecules closely associated with mineral colloids (Section 11.4). Many of the partially decomposed plant tissues and biomolecules contain many ring structures and carbon chains with chemically active groups of atoms that expose hydroxyl groups to the soil solution. Figure 8.12 provides a simplified diagram to illustrate the three main types of —OH groups thought to be responsible for the high amount of charge associated with these colloids. Negative or positive charges on the humus colloid develop as H^+ ions are either lost or gained by these groups. Both cations and anions are therefore attracted to and adsorbed by the humus colloid. The negative sites always outnumber the positive ones, and a very large *net* negative charge is associated with humus (Table 8.1). Because of its great surface area and many hydrophilic (water-loving) groups, humus can adsorb very large amounts of water per unit mass. However, humus also contains many hydrophobic sites and therefore can strongly adsorb a wide range of hydrophobic, nonpolar organic compounds (see Section 8.12). Because of its extraordinary influence on soil properties and behavior, we will delve much more deeply into the nature and function of soil organic matter in Chapter 11.

Figure 8.12 *A simplified diagram showing the principal chemical groups responsible for the high amount of negative charge on humus colloids. The three groups highlighted all include —OH that can lose its hydrogen ion by dissociation and thus become negatively charged. Note that the carboxylic, phenolic, and alcoholic groups on the right side of the diagram are shown in their disassociated state, while those on the left side still have their associated hydrogen ions. Note also that association with a second hydrogen ion causes a site to exhibit a net positive charge.* (Diagram courtesy of Ray R. Weil)

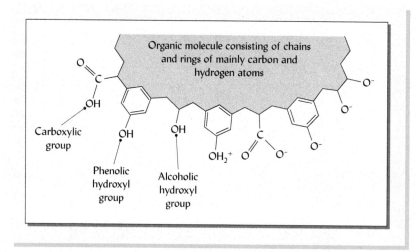

8.5 GENESIS AND GEOGRAPHIC DISTRIBUTION OF SOIL COLLOIDS

Genesis of Colloids

The silicate clays develop from the weathering of a wide variety of minerals by at least two distinct processes: (1) a slight physical and chemical **alteration** of certain primary minerals and (2) a **decomposition** of primary minerals with the subsequent **recrystallization** of certain of their products into the silicate clays.

Relative Stages of Weathering. Specific conditions conducive to the formation of important clay types are shown in Figure 8.13. Note that fine-grained micas and magnesium-rich chlorites represent earlier weathering stages of the silicates, and kaolinite and (ultimately) iron and aluminum oxides the most advanced stages. The smectites (e.g., montmorillonite) represent intermediate stages. Different weathering stages may occur across climatic zones or landscapes. Several members of this weathering series may even be present in a single soil profile, with the less weathered clay in the C horizon and the more

Figure 8.13 *Formation of layer silicate clays and oxides of iron and aluminum. Fine-grained micas, chlorite, and vermiculite are formed through rather mild weathering of primary aluminosilicate minerals, whereas kaolinite and oxides of iron and aluminum are products of much more intense weathering. Intermediate weathering intensity favors smectite. In each case, silicate clay genesis is accompanied by the removal in solution of such elements as Si, K, Na, Ca, and Mg.*

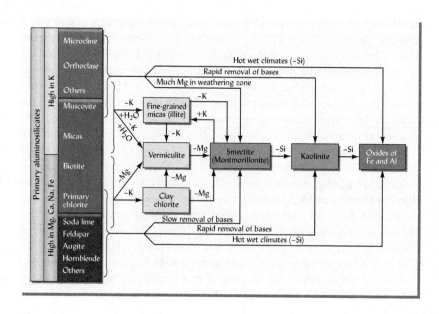

weathered clay minerals in the B or A horizons. As noted in Section 2.1, silicon tends to be lost as weathering progresses, leaving a lower Si:Al ratio in more highly weathered soil horizons.

Mixed and Interstratified Layers. In a given soil, it is common to find several silicate clay minerals in an intimate mixture. In fact, the properties and compositions of some mineral colloids are intermediate between those of the well-defined minerals described in Section 8.3. For example, a **mixed layer** or **interstratified** clay mineral in which some layers are more like mica and some more like vermiculite might be called *fine-grained mica-vermiculite.*

Iron and Aluminum Oxides. Iron oxides often are produced by the weathering of iron-containing primary silicate minerals or by precipitation of iron from soil solutions. Under aerated weathering conditions, divalent iron (Fe^{2+}) oxidizes rapidly to trivalent iron (Fe^{3+}), either while still within the structure of primary minerals or after its release into the soil solution. By reacting with oxygen atoms and water, the Fe^{3+} forms stable oxides and hydroxides (e.g., **goethite**, FeOOH, and **hematite**, Fe_2O_3). Hematite is most characteristic of red soils formed under relatively dry, warm, well-oxidized conditions, but it can also be inherited from such soil parent materials as red shales. Other iron oxides precipitate under wet, poorly oxygenated conditions.

Aluminum oxides, mainly **gibbsite** [$Al(OH)_3$], are produced by strong weathering environments in which acid leaching rapidly removes the Si released from the breakdown of primary and secondary silicate minerals. During weathering, hydrogen ions replace the K^+, Mg^{2+}, and other such ions in the crystal, causing the framework to break down, releasing the silicon and aluminum. Gibbsite is extremely stable in soils and typically represents the most advanced stage of weathering in soils.

Allophane and Imogolite. Relatively little is known of factors influencing the formation of allophane and imogolite. While they are commonly associated with materials of volcanic origin, they are also formed from igneous rocks and are found in some Spodosols. Apparently, volcanic ashes release significant quantities of $Si(OH)x$ and $Al(OH)x$ materials that precipitate as gels in a relatively short period of time. These minerals are generally poorly crystalline in nature, imogolite being the product of a more advanced state of weathering than that which produces allophane. Both types of minerals have a pronounced capacity to strongly retain anions as well as to bind with humus, protecting it from decomposition.

Humus. The production of plant tissues (especially roots) and their subsequent breakdown by microorganisms and the concurrent stabilization of plant and microbial materials by protective interaction with other soil components and microenvironments result in the formation of the dark-colored, largely colloidal organic material called *humus* (see Section 11.4 for details), especially in soils high in adsorptive clay surfaces (such as Mollisols or Andisols) or low oxygen environments (such a Histosols). Various functional groups on the stabilized organic compounds provide charged sites for the attraction of both cations and anions.

Distribution of Clays by Geography and Soil Order

The kinds of clay that develop in a soil depends not only on climatic influences and profile conditions but also on the nature of the parent material. Nevertheless, some broad generalizations are possible. Table 8.2 shows the dominant clay minerals in different soil orders, descriptions of which were given in Chapter 3. The well-drained and highly weathered Oxisols and Ultisols of warm humid and subhumid tropics tend to be dominated by kaolinite, along with oxides of iron and aluminum. The smectite, vermiculite, and fine-grained mica groups are more prominent in Vertisols, Mollisols, and temperate region Alfisols where weathering is less intense. If the parent material is high in micas, fine-grained micas such as illite are apt to form. Parent materials high in metallic cations (particularly magnesium) or subject to restricted drainage, encourage smectite formation.

Table 8.2
PROMINENCE OF CLAY MINERALS IN DIFFERENT SOIL ORDERS AND TYPICAL ENVIRONMENTS FOR THESE SOILS

Soil order[a]	General weathering intensity	Typical location/ environment	Fe, Al oxides	Kaolinite	Smectite	Fine-grained mica	Vermiculite	Chlorite	Intergrades
Aridisols	Low	Dry areas			XX[b]	XX		X	X
Vertisols	↑	Dry & wet seasons, warm, high Ca, Mg			XXX				X
Mollisols		Prairie, steppe		X	XX	X	X	X	X
Alfisols		Cool or semiarid forests		X	X	X	X	X	X
Spodosols		Boreal forests	X	X					
Ultisols	↓	Warm, wet forests	XX	XXX			X	X	X
Oxisols	High	Hot, wet tropics	XX	XXX					

[a]See Chapter 3 for soil descriptions.
[b]X = moderate presence; XX = major presence; XXX = dominant presence.

8.6 SOURCES OF CHARGES ON SOIL COLLOIDS

There are two major sources of charges on soil colloids: (1) the *permanent* or *constant* charge imbalance brought about in some clay crystal structures by the isomorphous substitution of one cation by another of similar size but different charge, and (2) hydroxyls and other functional groups on the surfaces of the colloidal particles that may release or accept H^+ ions thus providing either negative or positive charges, *depending* on the solution pH.

Constant Charges on Silicate Clays

Negative Charges. An example of net negative charge is found in minerals where there has been an isomorphous substitution of a lower-charged ion (e.g., Mg^{2+}) for a higher-charged ion (e.g., Al^{3+}). Such substitution commonly occurs in aluminum-dominated octahedral sheets of smectite, vermiculite, and chlorite clays and leaves an unsatisfied negative charge. A second example is the substitution of an Al^{3+} for an Si^{4+} in the tetrahedral sheet, which also leaves one unsatisfied negative charge from the tetrahedral oxygen atoms. Such a substitution is common in the fine-grained micas, vermiculites, and even some smectites.

Positive Charges. Isomorphous substitution can also be a source of positive charges if the substituting cation has a higher charge than the ion for which it substitutes. In magnesium dominated octahedral sheet, there are three magnesium ions surrounded by oxygen and hydroxy groups, and the sheet has no charge. However, if an Al^{3+} ion substitutes for one of the Mg^{2+} ions, a positive charge results.

The net charge in these clays is the balance between the negative and positive charges. In all 2:1-type silicate clays, however, the **net charge** is negative since those substitutions leading to negative charges far outweigh those producing positive charges (see Figure 8.14).

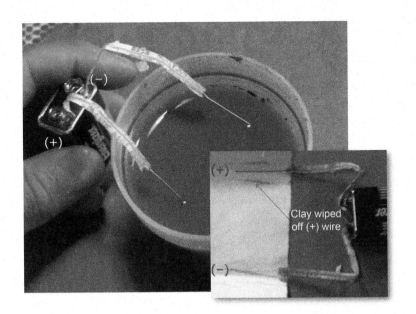

Figure 8.14 Simple demonstration of the negatively charged nature of clay. Wires connected to the (−) and (+) terminals of a 9-volt battery are dipped for a few minutes in a suspension of clayey soil in water. The wires are then wiped on a piece of paper (inset), showing that the wire on the (+) terminal has attracted the clay while the (−) wire has not. [Adapted from Weil (2015)]

pH-Dependent Charges

Most of the charges associated with humus, 1:1-type clays, the oxides of iron and aluminum, and allophane are dependent on the soil pH and consequently is termed **variable** or **pH-dependent**. Both negative and positive charges come from this source.

Negative Charges. The pH-dependent charges are associated primarily with OH groups on the surfaces of the inorganic and organic colloids. Broken edges of mineral colloids also generate pH-dependent charges (see Figure 8.15). The OH groups or oxygen atoms are attached to iron and/or aluminum in the inorganic colloids (e.g., $\text{Al}-\text{OH}$) and to the carbon in humus (e.g., $-\text{C}-\text{OH}$). Under moderately acid conditions, there is little or no charge on these particles, but as the pH increases, the hydrogen dissociates from the colloid OH group, and negative charges result.

$$\text{Al}-\text{OH} + \text{OH}^- \rightleftharpoons \text{Al}-\text{O}^- + \text{H}_2\text{O} \tag{8.1}$$

No charge (soil solids)   (soil solution)   Negative charge (soil solids)   (soil solution)

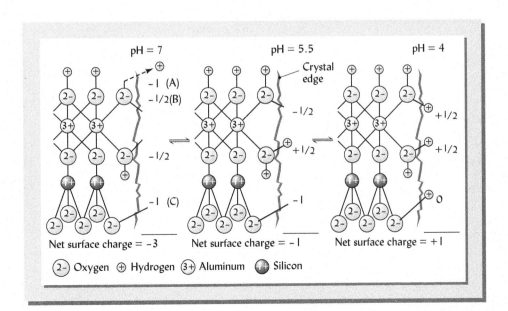

Figure 8.15 How pH-dependent charges develop at the broken edge of a kaolinite crystal. Three sources of net negative surface charge at a high pH are illustrated (left): (A) One (−1) charge from octahedral oxygen that has lost its H^+ ion by dissociation (the H broke away from the surface hydroxyl group and escaped into the soil solution). Note that such dissociation can generate negative charges all along the surface hydroxyl plane, not just at a broken edge. (B) One half (−1/2) charge from each octahedral oxygen that would normally be sharing its electrons with a second aluminum. (C) One (−1) charge from a tetrahedral oxygen atom that would normally be balanced by bonding to another silicon if it were not at the broken edge. The middle and right diagrams show the effect of acidification (lowering the pH), which increases the activity of H^+ ions in the soil solution. At the lowest pH shown (right), all of the edge oxygens have an associated H^+ ion, giving rise to a net positive charge on the crystal. These mechanisms of charge generation are similar to those illustrated for humus in Figure 8.12. (Diagram courtesy of Ray R. Weil)

As indicated by the ⇌ arrows, such reactions are reversible. If the pH increases, more OH^- ions are available to force the reactions to the right, and the negative charge on the particle surfaces increases. If the pH is lowered, OH^- ion concentrations are reduced, the reaction goes back to the left, and the negative charge is reduced.

Another source of increased negative charges as the pH is increased is the removal of positively charged complex aluminum hydroxy ions [e.g., $Al(OH)_2^+$]. At low pH levels, these ions block negative sites on the silicate clays (e.g., vermiculite) and make them unavailable for cation exchange. As the pH is raised, $Al(OH)_2^+$ ions react with OH^- ions in the soil solution to form insoluble $Al(OH)_3$, thereby freeing negatively charged sites.

$$\underset{\substack{\text{Negative charged}\\\text{site is blocked}}}{\diagdown\!\!\!\text{Al}\!\!-\!\!(OH)_2^-\,Al(OH)_2^+} + OH^- \longrightarrow \underset{\substack{\text{Negative charge}\\\text{site is freed}}}{\diagdown\!\!\!\text{Al}\!\!-\!\!(OH)_2^-} + \underset{\text{No charge}}{Al(OH)_3} \qquad (8.2)$$

Positive Charges. Under moderate to extreme acid soil conditions, humus, Fe, Al oxides, and some silicate clays may develop positive charges by **protonation**—the attachment of H^+ ions to the surface OH groups (Figure 8.15, *right*).

Since a mixture of humus and several inorganic colloids is usually found in soil, it is not surprising that positive and negative charges may be exhibited at the same time. In most soils of temperate regions, the negative charges far exceed the positive ones (Table 8.2). However, in some acid soils high in Fe, Al oxides, or allophane, the overall net charge may be positive. The effect of soil pH on positive and negative charges on such soils is illustrated in Figure 8.16.

The charge characteristics of selected soil colloids are shown in Table 8.3. Note the high percentage of constant negative charges in some 2:1-type clays (e.g., smectite and vermiculites). Humus, kaolinite, allophane, and Fe, Al oxides have mostly variable (pH-dependent) negative charges and exhibit modest positive charges at low pH values. We will now turn our attention to how the charges on soil colloids facilitate the adsorption of oppositely charged ions from the soil solution.

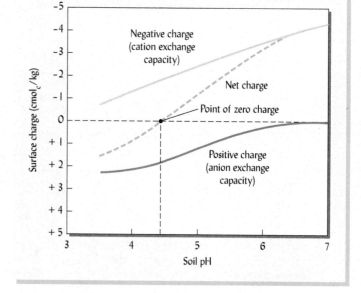

Figure 8.16 *Relationship between soil pH and positive and negative charges on an Oxisol surface horizon in Malaysia. The negative charges (cation exchange capacity) increase and the positive charges (anion exchange capacity) decrease with increasing soil pH. In the case illustrated, the positive and negative charges are in balance (zero net charge) at about pH 4.4.* [Redrawn from Shamshuddin and Ismail (1995)]

Table 8.3
COMPARATIVE LEVELS OF PERMANENT (CONSTANT) AND pH-DEPENDENT NEGATIVE CHARGES AS WELL AS pH-DEPENDENT POSITIVE CHARGES IN REPRESENTATIVE COLLOIDS

Colloid type	Negative charge			Positive charge, $cmol_c/kg$
	Total at pH 7, $cmol_c/kg$	Constant, %	pH dependent, %	
Organic	200	10	90	0
Smectite	100	95	5	0
Vermiculite	150	95	5	0
Fine-grained micas	30	80	20	0
Chlorite	30	80	20	0
Kaolinite	8	5	95	2
Gibbsite (Al)	4	0	100	5
Goethite (Fe)	4	0	100	5
Allophane	30	10	90	15

8.7 ADSORPTION OF CATIONS AND ANIONS

In soil, the negative and positive surface charges on the colloids attract and hold a complex swarm of cations and anions. In Figure 8.1, ion adsorption was illustrated in a simplified manner, showing positive cations held on the negatively charged surfaces of a soil colloid. Actually, both cations and anions are usually attracted to the same colloid. In temperate-region soils, anions are commonly adsorbed in much smaller quantities than cations because these soils generally contain mainly 2:1-type silicate clays on which negative charges predominate. In the tropics, where soils are more weathered, acid, and rich in 1:1 clays and Fe, Al oxides, the amount of negative charge on the colloids is not so high, and positive charges are more abundant. Therefore, the adsorption of anions is more prominent in these soils.

Figure 8.17 shows how both cations and anions may be attracted to the same colloid if it has both positively and negatively charged sites. This figure also illustrates that adsorption of ions on colloidal surfaces occurs by the formation of two quite different general types of colloid–ion complexes referred to as *outer-sphere* and *inner-sphere* complexes.

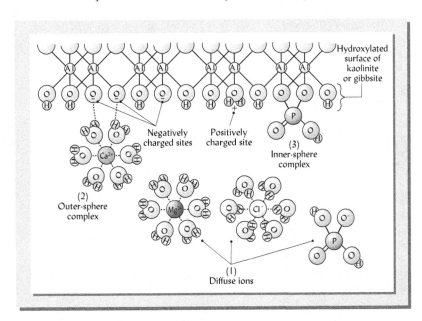

Figure 8.17 Adsorption of ions as outer-sphere and inner-sphere complexes. (1) Water molecules surround diffuse cations and anions (such as the Mg^{2+}, and Cl^- shown) in the soil solution. (2) In an outer-sphere, complex water molecules form a bridge between the adsorbed cation (Ca^{2+} shown) and the charged colloid surface. (3) In the case of an inner-sphere complex, no water molecules intervene, and the cation or anion (such as $H_2PO_4^-$ anion shown) binds directly with the metal atom (aluminum in this case) in the colloid structure. Outer-sphere complexes typify easily exchangeable ions that satisfy, in a general way, the net charge on the colloid surface. Inner-sphere complexes are not easily replaced from the colloid surface, as they represent strong bonding of specific ions to specific sites on the colloid. Adsorption on a hydroxylated surface octahedral sheet, such as in kaolinite or gibbsite, is shown. (Diagram courtesy of Ray R. Weil)

Outer- and Inner-Sphere Complexes

Remembering that water molecules surround (hydrate) cations and anions in the soil solution, we can visualize that in an **outer-sphere complex** water molecules form a bridge between the adsorbed ion and the charged colloid surface. Sometimes several layers of water molecules are involved. Thus, the ion itself never comes close enough to the colloid surface to form a bond with a specific charged site. Instead, the ion is only weakly held by electrostatic attraction, the charge on the oscillating hydrated ion balancing, in a general way, an excess charge of opposite sign on the colloid surface. Ions in an outer-sphere complex are therefore easily replaced by other similarly charged ions.

In contrast, adsorption via formation of an **inner-sphere complex** does *not* involve any intervening water molecules. Therefore, one or more direct bonds are formed between the adsorbed ion and the atoms in the colloid surface. One example already discussed is the case of the K^+ ions that fit so snugly into the spaces between silicon tetrahedra in a mica crystal (see Figure 8.10). Since there are no intervening water molecules, the K^+ ions are directly bonded by sharing electrons with the negatively charged tetrahedral oxygen atoms. Similarly, strong inner-sphere complexes may be formed by reactions of Cu^{2+} or Ni^{2+} with the oxygen atoms in silica tetrahedra.

Another important example, this time involving an anion, occurs when a $H_2PO_4^-$ ion is directly bonded by shared electrons with the octahedral aluminum in the colloid structure (Figure 8.17). Other ions cannot easily replace an ion held in an inner-sphere complex because this type of adsorption involves relatively strong bonds that are dependent on the compatible nature of specific ions and specific sites on the colloid.

Table 8.4 lists cations and anions commonly adsorbed by inner and outer sphere complexation. The adsorption of these ions by soil colloids greatly affects their mobility and biological availability, thereby influencing both soil fertility and environmental quality. Note

Table 8.4
CATIONS AND ANIONS COMMONLY ADSORBED TO SOIL COLLOIDS AND IMPORTANT IN PLANT NUTRITION AND ENVIRONMENTAL QUALITY

The listed ions form inner- and/or outer-sphere complexes with soil colloids. Ions marked by an asterisk (*) are among those that predominate in most soil solutions. Many other ions may be important in certain situations.

Cation	Formula	Comments	Anion	Formula	Comments
Ammonium	NH_4^+	Plant nutrient	Arsenate	AsO_4^{3-}	Toxic to animals
Aluminum	Al^{3+}, $AlOH^{2+}$, $Al(OH)_2^+$	Toxic to many plants	Borate	$B(OH)_4^-$	Plant nutrient, can be toxic
Calcium*	Ca^{2+}	Plant nutrient	Bicarbonate	HCO_3^-	Toxic in high-pH soils
Cadmium	Cd^{2+}	Toxic pollutant	Carbonate*	CO_3^{2-}	Forms weak acid
Cesium	Cs^+	Radioactive contaminant	Chromate	CrO_4^{2-}	Toxic pollutant
Copper	Cu^{2+}	Plant nutrient, toxic pollutant	Chloride*	Cl^-	Plant nutrient, toxic in large amounts
Hydrogen*	H^+	Causes acidity	Fluoride	Fl^-	Toxic, natural, and pollutant
Iron	Fe^{2+}	Plant nutrient	Hydroxyl*	OH^-	Alkalinity factor
Lead	Pb^{2+}	Toxic to animals, plants	Nitrate*	NO_3^-	Plant nutrient, pollutant in water
Magnesium*	Mg^{2+}	Plant nutrient	Molybdate	MoO_4^{2-}	Plant nutrient, can be toxic
Manganese	Mn^{2+}	Plant nutrient	Phosphate	HPO_4^{2-}	Plant nutrient, water pollutant
Nickel	Ni^{2+}	Plant nutrient, toxic pollutant	Selenate	SeO_4^{2-}	Animal nutrient and toxic pollutant
Potassium*	K^+	Plant nutrient	Selenite	SeO_3^{2-}	Animal nutrient and toxic pollutant
Sodium*	Na^+	Used by animals, some plants, can damage soil	Silicate*	SiO_4^{4-}	Mineral weathering product, used by plants
Strontium	Sr^{2+}	Radioactive contaminant	Sulfate*	SO_4^{2-}	Plant nutrient
Zinc	Zn^{2+}	Plant nutrient, toxic pollutant	Sulfide	S^{2-}	In anaerobic soils, forms acid on oxidation

that the soil solution and colloidal surfaces in most soils are dominated mainly by just a few of the cations and anions, the others being found in much smaller amounts or only in special situations such as contaminated soils.

8.8 CATION EXCHANGE REACTIONS

The phenomenon of cation exchange in soils was stumbled upon in 1840s England by two farmers and then further investigated by J. Thomas Way, a consulting agricultural chemist (Way, 1850). They poured smelly brown cow urine and other solutions containing *ammonium* onto columns of soil. To their surprise, the liquid "went in manure and came out water." The water draining out from the bottom of the column was not smelly or colored and did not contain ammonium, but instead was clear and contained mainly *calcium*. Apparently, the ammonium (NH_4^+) had exchanged places with the calcium (Ca^{2+}).

To understand what was happening, let us consider the case of an outer-sphere complex between a negatively charged colloid surface and a hydrated cation (such as the Ca^{2+}, Mg^{2+}, and K^+ ions shown in Figures 8.17 and 8.18). Such an outer-sphere complex is only loosely held together by electrostatic attraction, and the adsorbed ions remain in constant motion near the colloid surface. For microseconds, an adsorbed cation may vibrate out a bit farther than usual from the colloid surface. Such a moment provides an opportunity for another hydrated cation from the soil solution (say a copper Cu^{2+} ion as shown in Figure 8.18) to diffuse into a position a bit closer to the negative site on the colloid. The instant this should occur, the second ion would replace the formerly adsorbed ion—freeing the latter to diffuse out into the soil solution. In this manner, an *exchange of cations* between the adsorbed and diffuse state takes place. If water is percolating through the soil profile, cations applied to the surface of a soil (such as the Cu^{2+} from copper sulfate in Figure 8.18) may be adsorbed and retained in the upper soil horizons by exchanging with some of the cations originally adsorbed on the colloids there (such as the Ca^{2+} ions shown in Figure 8.18). As a result, water draining from the profile contains ions desorbed from the soil instead of the ions applied to the surface. The degree to

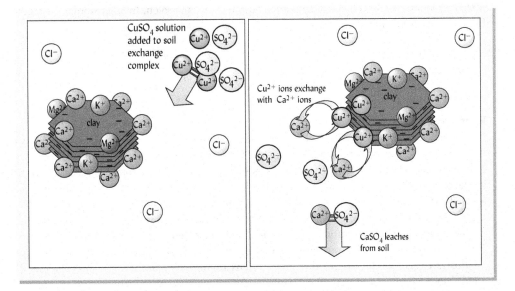

Figure 8.18 *A highly simplified visualization of cation exchange in soils. (Left) A chemical compound is added to the soil; in this example, copper sulfate from a fungicide spray. The Cu^{2+} cations migrate toward the negatively charged clay particle. As the exchangeable ions vibrate within their outer-sphere complex with the clay, the Cu^{2+} ions in the solution are also in constant motion nearby. (Right) Occasionally, an exchangeable Ca^{2+} ion vibrates to the outer edge of the clay's attractive charge field at the same instant that a Cu^{2+} ion gyrates into the attractive field such that the Cu^{2+} becomes captured by the negative charge while the Ca^{2+} is freed to diffuse away from the clay in the soil solution. Thus the two ions have exchanged places. Meanwhile, anions such as the sulfate SO_4^{2-} ions are repelled by the negatively charged clay. The Ca^{2+} freed into the solution by exchange with the Cu^{2+} soon pairs up with a SO_4^{2-} anion and may leach down through the soil profile with percolating water as copper sulfate (lower right).* (Diagram courtesy of Ray R. Weil)

which such processes occur can have important implications for the movement of contaminants from land to water and of nutrients from soils to plant roots.

The process just described is referred to as **cation exchange** (see also Section 8.1). If a hydrated *anion* similarly replaces another hydrated anion at a *positively* charged colloid site, the process is termed **anion exchange**. The ions held in outer-sphere complexes from which they can be replaced by such exchange reactions are said to be **exchangeable** cations or anions. As a group, all the colloids in a soil, inorganic and organic, capable of holding exchangeable cations or anions are termed the cation or anion **exchangeable complex**.

Principles Governing Cation Exchange Reactions

Figure 8.18 is a very simplified visualization of cation exchange. In fact, if we are to have a more sophisticated understanding of exchange reactions and the roles they play, we will need to become aware of a number of principles that govern how these reactions take place.

Reversibility. We can illustrate the process of cation exchange using a simple reaction in which a hydrogen ion displaces a sodium ion from its adsorbed state on a colloid surface:

$$\boxed{\text{Colloid}}\,Na^+ + \underset{\text{(soil solution)}}{H^+} \rightleftharpoons \boxed{\text{Colloid}}\,H^+ + \underset{\text{(soil solution)}}{Na^+} \quad (8.3)$$

The reaction takes place rapidly and, as shown by the double arrows, it is reversible. It will go to the left if sodium is added to the system. This reversibility is a fundamental principle of cation exchange.

Charge Equivalence. In cation exchange reactions, the exchange takes place on a *charge-for-charge* basis. Therefore, although one H^+ ion is exchanged with *one* Na^+ ion in the reaction just shown, it would require *two* singly charged H^+ ions to exchange with *one* divalent Ca^{2+} ion. If the reaction is reversed, one Ca^{2+} ion will displace two H^+ ions. Extending this concept, it would require three Na^+ ions to replace a single Al^{3+} ion, and so on. In other words, x charges from one cation species replace x charges from the other:

$$\boxed{\text{Colloid}}\,Ca^{2+} + \underset{\text{(soil solution)}}{2\,H^+} \rightleftharpoons \boxed{\text{Colloid}}\begin{matrix}H^+\\H^+\end{matrix} + \underset{\text{(soil solution)}}{Ca^{2+}} \quad (8.4)$$

Ratio Law. Consider an exchange reaction between two similar cations, say Ca^{2+} and Mg^{2+}. If there are many Ca^{2+} ions adsorbed on a colloid and some Mg^{2+} is added to the soil solution, the added Mg^{2+} ions will begin displacing the Ca^{2+} from the colloid. This will bring more Ca^{2+} into the soil solution and these Ca^{2+} ions will, in turn, displace some of the Mg^{2+} back off the colloid. Theoretically, these exchanges will continue back and forth until equilibrium is reached. At this point, there will be no further *net* change in the number of adsorbed Ca^{2+} and Mg^{2+} ions (although the exchanges will continue each balancing the other). The **ratio law** tells us that, at equilibrium, the ratio of Ca^{2+} to Mg^{2+} on the colloid will be the same as the ratio of Ca^{2+} to Mg^{2+} in the solution and both will be the same as the ratio in the overall system. To illustrate this concept, assume that 20 Ca^{2+} ions are initially adsorbed on a soil colloid and 5 Mg^{2+} ions are added to the system:

$$\boxed{\text{Colloid}}\,20\,Ca^{2+} + \underset{\text{(soil solution)}}{5\,Mg^{2+}} \rightleftharpoons \boxed{\text{Colloid}}\begin{matrix}16\,Ca^{2+}\\4\,Mg^{2+}\end{matrix} + \underset{\text{(soil solution)}}{1\,Mg^{2+} + 4\,Ca^{2+}} \qquad \text{Ratio: 4 Ca:1 Mg} \quad (8.5)$$

If the two exchanging ions are not of the same charge (e.g., K^+ exchanging with Mg^{2+}), the reaction becomes somewhat more complicated and a modified version of the ratio law would apply.

Up to this point, our discussion of exchange reactions has assumed that both ionic species (elements) exchanging places take part in the exchange reaction in exactly the same way. This assumption must be modified to take into account three additional factors if we are to understand how exchange reactions actually proceed in nature.

Anion Effects on Mass Action. In the reactions just discussed, we have not mentioned the anions that always accompany cations in solution. We also showed the exchange reactions as being completely reversible, with an equal chance of proceeding to the right or the left. In reality, the laws of **mass action** tell us that an exchange reaction will be more likely to proceed to the right if the released ion is prevented from reacting in the reverse direction. This may be accomplished if the released cation on the right side of the reaction either *precipitates, volatilizes*, or *strongly associates* with an anion. In each case, most of the displaced cations will be removed from solution and so will not be able to reverse the exchange. To illustrate this concept, consider the displacement of H^+ ions on an acid colloid by Ca^{2+} ions added to the soil solution, either as calcium carbonate:

$$\boxed{Colloid}\begin{matrix}H^+\\H^+\end{matrix} + CaCO_3 \rightleftharpoons \boxed{Colloid}Ca^{2+} + H_2O + CO_2\uparrow \qquad (8.6)$$

(added)   Water   (gas)

Where $CaCO_3$ is added, when a hydrogen ion is displaced off the colloid, it combines with an oxygen atom from the $CaCO_3$ to form water. Furthermore, the CO_2 produced is a gas, which can volatilize out of the solution and leave the system. The removal of these products pulls the reaction to the right. Therefore, much more Ca will be adsorbed on the colloid and much more H displaced if $CaCO_3$, rather than an equivalent amount of $CaCl_2$, is added to the hydrogen-dominated soil. This principle helps explain why $CaCO_3$ (i.e., limestone) is effective in neutralizing an acid soil, while calcium chloride is not (see Section 9.8).

Cation Selectivity. Up to now, we have assumed that both cation species taking part in the exchange reaction are held with equal tenacity by the colloid and therefore have an equal chance of displacing each other. In reality, some cations are held much more tightly than others and so are less likely to be displaced from the colloid. In general, the higher the charge and the smaller the hydrated radius of the cation, the more strongly it will adsorb to the colloid. The order of strength of adsorption for selected cations is:

$$Al^{3+} > Sr^{2+} > Ca^{2+} > Mg^{2+} > Cs^+ > K^+ = NH_4^+ > Na^+ > Li^+$$

The less tightly held cations oscillate farther from the colloid surface and therefore are the most likely to be displaced into the soil solution and carried away by leaching. This series therefore explains why the soil colloids are dominated by Al^{3+} (and other aluminum ions) and Ca^{2+} in humid regions and by Ca^{2+} in drier regions, even though the weathering of minerals in many parent materials provides relatively larger amounts of K^+, Mg^{2+}, and Na^+ (see Section 8.10). The strength of adsorption of the H^+ ion is difficult to determine because hydrogen-dominated mineral colloids break down to form aluminum-saturated colloids.

The relative strengths of adsorption order may be altered on certain colloids whose properties favor adsorption of particular cations. An important example of such colloidal "preference" for specific cations is the very high affinity for K^+ ions (and the similarly sized NH_4^+ and Cs^+ ions) exhibited by vermiculite and fine-grained micas, which attract these ions to intertetrahedral spaces exposed at weathered crystal edges (Section 8.3). The influence of different colloids on the adsorption of specific cations impacts the availability of cations for leaching or plant uptake (see Section 12.4). Certain metals such as copper, mercury, and lead have very high affinities for sites on humus and iron oxide colloids, making most soils quite efficient at removing these potential pollutants from leaching water.

Table 8.5
SOME EXCHANGEABLE IONS, THEIR HYDRATED SIZE, AND THEIR EXPECTED REPLACEMENT BY NH_4^+ IONS
Among ions of a given charge, the larger the hydrated radius, the more easily it is replaced.

Element	Ion	Hydrated ionic radius,[a] nm	Likely replacement of ion initially saturating a kaolinite clay if $cmol_c$ NH_4^+ added = CEC of the soil,[b] %
Lithium	Li^+	1.00	80
Sodium	Na^+	0.79	67
Ammonium	NH_4^+	0.54	50
Potassium	K^+	0.53	49
Rubidium	Rb^+	0.51	48
Cesium	Cs^+	0.50	47
Magnesium	Mg^{2+}	1.08	31
Calcium	Ca^{2+}	0.96	29
Strontium	Sr^{2+}	0.96	29
Barium	Ba^{2+}	0.88	26

[a] Not to be confused with nonhydrated radii, hydrated radii are from Evangelou and Phillips (2005).
[b] Based on empirical data from various sources and assumes no special affinity by kaolinite for any of the listed ions.

Complementary Cations. Soil colloids are always surrounded by many different adsorbed cation species. The likelihood that a given adsorbed cation will be displaced from a colloid is influenced by how strongly its neighboring cations (sometimes called **complementary ions**) are adsorbed to the colloid surface. For example, consider an adsorbed Mg^{2+} ion. An ion diffusing in from the soil solution is more likely to displace one of the neighboring ions rather than the Mg^{2+} ion, if the neighboring adsorbed ions are loosely held. If they are tightly held, then the chances are greater that the Mg^{2+} ion will be displaced. The influence of complementary ions on the availability of nutrient cations for plant uptake will be discussed in Section 8.10.

8.9 CATION EXCHANGE CAPACITY (CEC)

We now turn to a consideration of the quantitative CEC. This property is defined simply as the sum total of the exchangeable cationic charges that a soil can absorb. The CEC is expressed as the number of moles of positive charge adsorbed per unit mass. In order to be able to deal with whole numbers of a convenient size, many publications, including this textbook, report CEC values in centimoles of charge per kilogram ($cmol_c$/kg). Some publications still use the older unit, milliequivalents per 100 g (me/100 g), which gives the same value as $cmol_c$/kg (1 me/100 g = 1 $cmol_c$/kg). A particular soil may have a CEC of 15 $cmol_c$/kg, indicating that 1 kg of the soil can hold 15 $cmol_c$ of H^+ ions, for example, and can exchange this number of charges from H^+ ions for the same number of charges from any other cation. This means of expression emphasizes that exchange reactions take place on a charge-for-charge (not an ion-for-ion) basis. The concept of a mole of charges and its use in CEC calculations are reviewed in Box 8.3.

Methods of Determining CEC

The CEC is an important soil chemical property that is used for classifying soils in *Soil Taxonomy* (e.g., in defining an Oxic, Mollic, or Kandic diagnostic horizon, Section 3.2) and for assessing their fertility and environmental behavior. Several different standard methods can be used to determine the CEC of a soil. In general, a concentrated solution of a particular exchanger cation (e.g., Ba^{2+}, NH_4^+, or Sr^{2+}) is used to leach the soil sample. This provides an overwhelming number of the exchanger cations that can completely replace all the exchangeable cations initially in the soil. Then, the CEC can be determined by measuring either the number of

> **BOX 8.3**
> **CHEMICAL EXPRESSION OF CATION EXCHANGE**
>
> One mole of any atom, molecule, or charge is defined as 6.02×10^{23} (Avogadro's number) of atoms, molecules, or charges. Thus, 6.02×10^{23} negative charges associated with the soil colloidal complex would attract 1 mole of positive charge from adsorbed cations such as Ca^{2+}, Mg^{2+}, and H^+. The number of moles of the positive charge provided by the adsorbed cations in any soil gives us a measure of the CEC of that soil.
>
> The CEC of soils commonly varies from 0.03 to 0.5 mole of positive charge per kilogram (mol_c/kg). Expressing this same range of CEC values in centimoles (1/100s of a mole) gives convenient whole numbers: 3–50 $cmol_c$/kg.
>
> ## CALCULATING MASS FROM MOLES
>
> Using the mole concept, it is easy to relate the moles of charge to the mass of ions or compounds involved in cation or anion exchange. Consider, for example, the exchange that takes place when adsorbed sodium ions in an alkaline arid-region soil are replaced by hydrogen ions:
>
> $$\boxed{Colloid}\ Na^+ + H^+ \rightleftharpoons \boxed{Colloid}\ H^+ + Na^+$$
>
> If 1 $cmol_c$ of adsorbed Na^+ ions per kilogram of soil were replaced by H^+ ions in this reaction, how many grams of Na^+ ions would be replaced?
>
> Since the Na^+ ion is singly charged, the mass of Na^+ needed to provide 1 mole of charge (1 mol_c) is the gram atomic weight of sodium, or 23 g (see periodic table in Appendix B). The mass providing 1 *centimole* of charge ($cmol_c$) is 1/100 of this amount; thus, the mass of the 1 $cmol_c$ Na^+ replaced is 0.23 g Na^+/kg soil. The 0.23 g Na^+ would be replaced by only 0.01 g H, which is the mass of 1 $cmol_c$ of this much lighter element.
>
> Another example is the replacement of H^+ ions when hydrated lime [$Ca(OH)_2$] is added to an acid soil. This time, assume that 2 $cmol_c$ H^+/kg soil is replaced by the $Ca(OH)_2$, which reacts with the acid soil as follows:
>
> $$\boxed{Colloid}\ \begin{matrix}H^+\\H^+\end{matrix} + Ca(OH)_2 \rightleftharpoons \boxed{Colloid}\ Ca^{2+} + 2H_2O$$
>
> Since the Ca^{2+} ion in each molecule of $Ca(OH)_2$ has two positive charges, the mass of $Ca(OH)_2$ needed to replace 1 mole of charge from the H^+ ions is only one-half of the gram molecular weight of this compound, or 74/2 = 37 g. A comparable figure for 1 centimole is 37/100, or 0.37 grams. The mass of $Ca(OH)_2$ needed to replace 2 $cmol_c$ H^+/kg soil is:
>
> $$2\ cmol_c\ Ca(OH)_2/kg \times 0.37\ g\ Ca(OH)_2/cmol_c$$
> $$= 0.74\ g\ Ca(OH)_2/kg\ soil$$
>
> The 0.74 g $Ca(OH)_2$/kg soil can be converted to the amount of $Ca(OH)_2$ needed to replace 2 $cmol_c$ H^+/kg from the surface 15 cm of soil in a 1 ha field, remembering from Chapter 4 (Section 4.7, footnote 8) that this depth of soil typically weighs 2 million kg/ha
>
> $$0.74\ g/kg \times 2 \times 10^6\ kg = 1.48 \times 10^6\ g$$
> $$= 1.48 \times 10^3\ kg = 1.48\ Mg$$
>
> ## CHARGE AND CHEMICAL EQUIVALENCY
>
> In each preceding example, the number of charges provided by the replacing ion is equivalent to the number associated with the ion being replaced. Thus, 1 mole of negative charges attracts 1 mole of positive charges, whether the charges come from H^+, K^+, Na^+, NH_4^+, Ca^{2+}, Mg^{2+}, Al^{3+}, or any other cation. Keep in mind, however, that only one-half the atomic weights of divalent cations, such as Ca^{2+} or Mg^{2+}, and only one-third the atomic weight of trivalent Al^{3+} are needed to provide 1 mole of charge. This *chemical equivalency* principle applies to both cation and anion exchange.

exchanger cations adsorbed or the amounts of each of the displaced elements originally held on the exchange complex (usually Ca^{2+}, Al^{3+}, Mg^{2+}, K^+, and Na^+).

Some CEC procedures use a solution buffered to maintain a certain pH (usually either pH 7.0 using ammonium as the exchanger cation or pH 8.2 using barium as the exchanger cation). If the native soil pH is less than the pH of the buffered solution, then these methods measure not only the cation exchange sites active at the pH of the particular soil, but also any pH-dependent exchange sites (see Section 8.6) that *would* become negatively charged *if* the soil pH were raised to pH 7.0 or 8.2. Thus these methods exaggerate the true CEC of acid soils.

An alternative CEC procedure uses unbuffered solutions to allow the exchange to take place at the actual pH of the soil. The buffered methods (NH_4^+ at pH 7.0 or Ba^{2+} at pH 8.2) measure the *potential* or *maximum* cation exchange capacity of a soil. The unbuffered method measures only the **effective cation exchange capacity** (ECEC), which can hold exchangeable cations at the pH of the soil as sampled. As the different methods may yield significantly different values of CEC, it is important that the method used be known when comparing soils based on their CEC. This is especially significant if the soil pH is much below the buffer pH used.

Cation Exchange Capacities of Soils

The CEC of a given soil sample is determined by the relative amounts of different colloids in that soil and by the CEC of each of these colloids. Figure 8.19 illustrates the common range of CEC for different soils and other organic and inorganic exchange materials. Note that sandy soils, which are generally low in all colloidal material, have low CECs compared to those exhibited by silt loams and clay loams. Also note the very high CECs associated with humus compared to those exhibited by the inorganic clays, especially kaolinite and Fe, Al oxides. The CEC coming from humus generally plays a very prominent role, and sometimes a dominant one, in cation exchange reactions in A horizons. For example, in a clayey Ultisol (pH = 5.5) containing 2.5% humus and 30% kaolinite, about 75% of the CEC is associated with humus. The contribution of organic matter to the soil CEC increases at higher soil pH levels. Using the CEC ranges from Figure 8.19 or Table 8.6, it is possible to estimate the CEC of a soil if the quantities of the different soil colloids in the soil are known (see Box 8.4).

Figure 8.19 Ranges in the cation exchange capacities (at pH 7) that are typical of a variety of soils and soil materials. The high CEC of humus shows why this colloid plays such a prominent role in most soils and especially those high in kaolinite and Fe, Al oxides, and clays that have low CECs. (Diagram courtesy of Ray R. Weil)

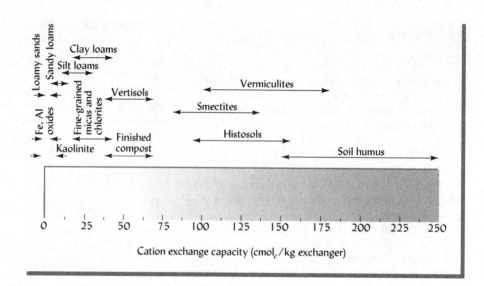

Table 8.6
COMMON RANGES OF POTENTIAL CATION EXCHANGE CAPACITIES IN THE SURFACE LAYERS (A AND B HORIZONS) OF SOILS IN DIFFERENT SOIL ORDERS

Organic colloids give Histosols a very high CEC. Compare to data in Tables 8.2 and 8.3 to see the relationship between the average CEC and the main types of colloids in the other soil orders. The wide range of CEC within each soil order is largely a result of variations in the amount of clay and organic matter in the individual soils.

Soil order	Common range of CEC,[a] $cmol_c$/kg	Soil order	Common range of CEC, $cmol_c$/kg
Histosols	110–170	Inceptisols & Entisols	5–37
Vertisols	33–67	Aridisols	7–29
Andisols	13–49	Alfisols	4–26
Spodosols	2–57	Ultisols	3–15
Mollisols	12–36	Oxisols	2–13

[a]The CEC was measured on a total of approximately 1,000 soils using methods buffered at pH 7.0 or 8.2 and therefore may overestimate CEC for more acid soils and underestimate CEC for more alkaline soils. The range given can be expected to include roughly 70% of the soils in the indicated soil order. Gelisols were not listed, but were included in either the Histosols or Inceptisols.
From Essington (2004).

> **BOX 8.4**
> ## ESTIMATING CEC AND CLAY MINERALOGY
>
> Data on clay mineralogy and cation exchange capacity are rather time-consuming to obtain and not always available. Fortunately, one can often estimate one of these types of data from the other, assuming that easily obtained data on soil pH, clay content, and organic matter level are available.
>
> 1. *Example of Estimating CEC from Data on Mineralogy*
>
> Assume you know that a cultivated Mollisol in Iowa contains 20% clay and 4% organic matter (OM) and its pH = 7.0. The dominant clays in Mollisols are likely 2:1 types such as vermiculite and smectite (see Table 8.3). We estimate the average CEC of the clays of these types to be about 100 $cmol_c$/kg clay (Tables 8.1 and 8.3). At pH 7.0, the CEC of OM is about 200 $cmol_c$/kg (Table 8.3). Since 1 kg of this soil has 0.20 kg (20%) of clay and 0.04 kg (4%) of OM, we can calculate the CEC associated with each of these sources.
>
> From the clays in this Mollisol: 0.2 kg × 100 $cmol_c$/kg = 20 $cmol_c$
>
> From the OM in this Mollisol: 0.04 kg × 200 $cmol_c$/kg = 8 $cmol_c$
>
> The total CEC of this Mollisol: 20 + 8 = 28 $cmol_c$/kg soil
>
> 2. *Example of Estimating the Clay Mineralogy from Information on CEC*
>
> Assume you know that a soil contains 60% clay and 4% organic matter and the pH = 4.2. You also know the CEC is 5.8 $cmol_c$/kg. You want to estimate the types of clays present. At pH 4.2 the CEC of the organic matter would be comparatively low, about 100 $cmol_c$/kg (Figure 8.21). Therefore, we estimate:
>
> CEC from OM in 1 kg soil = 0.04 kg OM × 100 $cmol_c$/kg OM = 4.0 $cmol_c$
>
> The remaining portion of the CEC contributed by the clay can be estimated as:
>
> CEC from the clay in 1 kg soil = 5.8 $cmol_c$ − 4.0 $cmol_c$ = 1.8 $cmol_c$
>
> Since this 1.8 $cmol_c$/kg soil is provided by 0.60 kg of clay (60% of 1 kg soil), we can estimate:
>
> CEC of the pure clay = 1.8 $cmol_c$/kg soil × 1 kg soil/0.60 kg clay = 3 $cmol_c$/kg clay
>
> From Tables 8.1 and 8.4, we see that the Fe and Al oxides at pH 7 have a CEC of about 4 $cmol_c$/kg clay. We know (from Section 8.6) their CEC would be lower at pH 4. Likewise, kaolinite has a CEC of about 8 $cmol_c$/kg clay at pH 7 and perhaps only about 4 $cmol_c$/kg clay at pH 4. The CEC values for the other types of clay listed in Table 8.4 are far higher. *Therefore, it is reasonable to conclude that the clays in this soil consist mainly of Fe and Al oxides and kaolinite.*

Data in Table 8.6 show the range of CEC values characteristic for soils in different soil orders. Note the very high CEC for the Histosols, reflecting the high CEC of the organic colloids. The Vertisols, which are very high in swelling-type clays (mostly smectite), have the highest average CEC of the mineral soils. Andisols and Spodosols are commonly high in organic matter and, in the case of Andisols, allophane and smectitic-type clays. Ultisols and Oxisols, whose clays are dominantly kaolinite and hydrous oxides of iron and aluminum, have relatively low CEC values. Despite large variations in soil organic matter and texture, these data appear to reflect the quantities and kinds of colloids found in the soils.

pH and Cation Exchange Capacity

In previous sections, it was pointed out that the cation exchange capacity of most soils increases with pH. At very low pH values, the cation exchange capacity is also generally low (Figure 8.20). Under these conditions, only the permanent charges of the 2:1-type clays (see Section 8.6) and a small portion of the pH-dependent charges of organic colloids, allophane, and some 1:1-type clays hold exchangeable ions. As the pH is raised, the negative charges on some 1:1-type silicate clays, allophane, humus, and even Fe, Al oxides increases, thereby increasing the cation exchange capacity. As noted earlier, to obtain a measure of this maximum retentive capacity, the CEC is commonly determined at a pH of 7.0 or 8.2. At neutral or slightly alkaline pH, the CEC reflects most of those pH-dependent charges as well as the permanent ones.

Figure 8.20 Influence of pH on representative cation exchange capacities of two clay minerals and humus. Below pH 6.0, the charge for smectite has a fairly constant charge that is due mainly to isomorphic substitution and is considered permanent. Above pH 6.0, the charge on smectite increases somewhat with pH (shaded area) because of the ionization of hydrogen from the exposed hydroxyl groups at crystal edges. In contrast, the charges on kaolinite and humus are all variable, increasing with increasing pH. Humus carries a far greater number of charges than kaolinite. At low pH, kaolinite carries a net negative CEC because its positive charges outnumber the negative ones. (Diagram courtesy of N. Brady and Ray R. Weil)

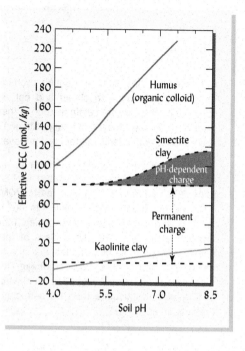

8.10 EXCHANGEABLE CATIONS IN FIELD SOILS

Climatic region influences the exchangeable cations associated with soil colloids; Ca^{2+}, Al^{3+}, complex aluminum hydroxy ions, and H^+ are most prominent in humid regions, while Ca^{2+}, Mg^{2+}, and Na^+ dominate soil in low-rainfall areas (Table 8.7). The cations that dominate the exchange complex have a marked influence on soil properties.

Table 8.7
CATION EXCHANGE PROPERTIES TYPICAL FOR UNAMENDED CLAY LOAM SURFACE SOILS IN DIFFERENT CLIMATIC REGIONS

Note that soils with coarser textures would have less clay and organic matter and therefore lower amounts of exchangeable cations and lower CEC values.

Property	Warm, humid region (Ultisols)[a]	Cool, humid region (Alfisols)	Semiarid region (Ustolls)	Arid region (Natrargids)[b]
Exchangeable H^+ and Al^{3+}, cmolc/kg (% of CEC)	7.5 (75%)	5 (28%)	0 (0%)	0 (0%)
Exchangeable Ca^{2+}, $cmol_c$/kg (% of CEC)	2.0 (20%)	9 (50%)	17 (65%)	13 (50%)
Exchangeable Mg^{2+}, $cmol_c$/kg (% of CEC)	0.4 (4%)	3 (17%)	6 (23%)	5 (19%)
Exchangeable K^+, $cmol_c$/kg (% of CEC)	0.1 (1%)	1 (5%)	2 (8%)	3 (12%)
Exchangeable Na^+, $cmol_c$/kg (% of CEC)	Tr	0.02 (0.1%)	1 (4%)	5 (19%)
CEC,[c] $cmol_c$/kg	10	18	26	26
Probable pH	4.5–5.0	5.0–5.5	7.0–8.0	8–10
Nonacid cations (% of CEC)[d]	25%	68%	100%	100%

[a]See Chapter 3 for explanation of soil group names.
[b]Natrargids are Aridisols high in exchangeable sodium, as explained in Section 9.12.
[c]The sum of all the exchangeable cations measured at the pH of the soil (see Section 8.9).
[d]Traditionally referred to as "base" saturation.

In a given soil, the proportion of the cation exchange capacity satisfied by a particular cation is termed the *saturation percentage* for that cation. Thus, if 50% of the CEC is satisfied by Ca^{2+} ions, the exchange complex is said to have a *calcium saturation percentage* of 50. This terminology is especially useful in identifying the relative proportions of sources of acidity and alkalinity in the soil solution. Thus, the percentage saturation with Al^{3+} and H^+ ions gives an indication of the acid conditions, while increases in the percentage **nonacid cation saturation** (sometimes referred to as the *base saturation percentage*[3]) indicate the tendency toward neutrality and alkalinity. These relationships will be discussed further in Section 9.3.

Cation Saturation and Nutrient Availability

Exchangeable cations generally are available for uptake by both higher plants and microorganisms. By cation exchange, hydrogen ions from root hairs and microorganisms replace nutrient cations from the exchange complex. The nutrient cations are forced into the soil solution, where they can be assimilated by the adsorptive surfaces of roots and soil organisms, or they may be removed by drainage water. Cation exchange reactions affecting the mobility of organic and inorganic pollutants in soils will be discussed in Section 8.12. Here we focus on the plant nutrition aspects.

The percentage saturation of essential nutrient cations such as Ca^{2+} and K^+ greatly influences the uptake of these elements by growing plants. For example, if the percentage calcium saturation of a soil is high, the displacement of this cation is comparatively easy and rapid. Thus, 6 cmol/kg of exchangeable Ca^{2+} in a soil whose exchange capacity is 8 cmol/kg (75% Ca^{2+} saturation) probably would mean ready availability, but 6 cmol/kg when the total exchange capacity of a soil is 30 cmol/kg (20% Ca^{2+} saturation) would produce lower availability. This is one reason that, calcium-loving plants such as alfalfa grow best when the Ca^{2+} saturation of at least part of the soil approaches 75–85%.

Influence of Complementary Cations

A second factor influencing plant uptake of a given cation is the effect of the complementary ions held on the colloids. As was discussed in Section 8.8, the strength of adsorption of common cations on most colloids is in the order: $Al^{3+} > Ca^{2+} > Mg^{2+} > K^+ = NH_4^+ > Na^+$.

For example, if the complementary ions surrounding a K^+ ion are held tightly (i.e., they oscillate very close to the colloid surface), then a H^+ ion from a root is less likely to "find" a complementary ion than "bump into" and replace a K^+ ion (see Figure 8.21). If, on the other hand, the complementary ions are loosely held (oscillating quite far from the colloid surface), then the H^+ is more likely to "bump into" and replace one of the

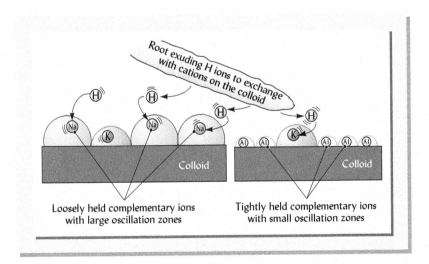

Figure 8.21 Effect of complementary ions. The half spheres show that the more loosely held ions move within larger zones of oscillation. For simplicity, the water molecules that hydrate each ion are not shown. (Left) H^+ ions from the root are more likely to encounter and exchange with loosely held Na^+ ions rather than the more tightly held K^+ ion. (Right) The likelihood that H^+ ions from the root will encounter and exchange with a K^+ ion is increased by the inaccessibility of the neighboring tightly held Al^{3+} ions. The K^+ ion on the right colloid is comparatively more vulnerable to being replaced and sent into the soil solution and is therefore more available for plant uptake or leaching than the K^+ ion on the left colloid. (Diagram courtesy of Ray R. Weil)

---

[3]Nonacid cations such as Ca^{2+}, Mg^{2+}, K^+, and Na^+ are not true bases. When adsorbed by soil colloids in the place of Al^{3+} and H^+ ions, however, they reduce acidity and increase the soil pH. For that reason, they are traditionally referred to as *bases* and the portion of the CEC that they satisfy is often termed *base saturation percentage*.

complementary ions and less likely to "find" the K^+. Consequently, K^+ is more likely to be replaced off the colloid if the complementary ions are mainly tightly held Al^{3+} or H^+ (as in acid soils) than if they are mainly Mg^{2+} and Na^+ (as in neutral to alkaline soils). There are also some nutrient antagonisms in certain soils related to competition in the plant uptake process than to soil exchange properties. For example, potassium uptake by plants is limited by high levels of calcium. Likewise, high potassium levels can limit the uptake of magnesium.

Effect of Type of Colloid

Differences exist in the tenacity with which several types of colloids hold specific cations and in the ease with which they exchange cations. At a given percentage base saturation, smectites—which have a high charge density per unit of colloid surface—hold calcium much more strongly than does kaolinite (low charge density). As a result, smectite clays must be raised to about 70% base saturation before calcium will exchange easily and rapidly enough to satisfy many plants. In contrast, a kaolinite clay exchanges calcium much more readily, serving as a satisfactory source of this constituent at a much lower percentage base saturation. The need to add limestone to the two soils will be somewhat different, partly because of this factor. We will now turn our attention from the exchange of cations to the exchange of anions.

8.11 ANION EXCHANGE

Anions are held (**sorbed**) by soil colloids in two major ways. First, they are held by anion adsorption mechanisms similar to those responsible for cation exchange. Second, they may actually react with surface oxides or hydroxides, forming more definitive **inner-sphere complexes**. We shall consider anion adsorption first.

The principles of **anion exchange** are similar to those of cation exchange, except the charges on the colloids are positive and the exchange is among negatively charged anions. The positive charges associated with the surfaces of kaolinite, iron and aluminum oxides, and allophane attract anions such as Cl^-, SO_4^{-2} and NO_3^- which can then undo anion exchange:

$$\boxed{\text{Colloid}}\ NO_3^- + Cl^- \rightleftharpoons \boxed{\text{Colloid}}\ Cl^- + NO_3^- \tag{8.7}$$

(positively charged soil solid)   (soil solution)   (positively charged soil solid)   (soil solution)

Just as in cation exchange, *equivalent* quantities of NO_3^- and Cl^- are exchanged; the reaction can be reversed, and nutrient anions so released can be absorbed by plants.

In contrast to cation exchange capacities, anion exchange capacities (AEC) of soils generally increase with *decreasing* pH. In some very acid tropical soils that are high in kaolinite and iron and aluminum oxides, the anion exchange capacity may actually exceed the cation exchange capacity.

Anion exchange is very important in making anions available for plant growth while at the same time retarding the leaching of such anions from the soil. For example, anion exchange restricts the loss of sulfates from subsoils in the southern United States (Section 12.2). Even the leaching of nitrate may be retarded by anion exchange in the subsoil of certain highly weathered soils of the humid tropics. Similarly, the downward movement into groundwater of some charged organic pollutants found in organic wastes can be retarded by such anion and/or cation exchange reactions.

Inner-Sphere Complexes

Some anions, such as phosphates, arsenates, molybdates, and sulfates, can react with particle surfaces, forming **inner-sphere complexes** (see Figure 8.17). For example, the $H_2PO_4^-$ ion may react with the protonated hydroxyl group rather than remain as an easily exchanged anion

$$\text{Al}-OH_2^+ + H_2PO_4^- \rightarrow \text{Al}-H_2PO_4 + H_2O \tag{8.8}$$

(soil solid)   (soil solution)   (soil solid)   (soil solution)

This reaction actually reduces the net positive charge on the soil colloid. Also, the $H_2PO_4^-$ is held very tightly by the soil solids and is not readily available for plant uptake.

Anion adsorption and exchange reactions regulate the mobility and availability of many important ions. Together with cation exchange they largely determine the ability of soils to hold nutrients in forms that are accessible to plants and to retard movement of pollutants in the environment.

Weathering and CEC/AEC Levels

Clays developed under mild weathering conditions (e.g., smectites and vermiculites) have much higher CEC levels than those developed under more extreme weathering pressures (Figure 8.22). The AEC levels, in turn, tend to be much higher in clays developed under strong weathering conditions (e.g., kaolinite) than in those found under milder weathering. In addition to the climate, the nature of the parent material and the time allowed for the weathering to occur also influence the clay types present and the CEC/AEC relations. The generalized relationships in Figure 8.22 are helpful in obtaining a first approximation of CEC and AEC levels in soils of different soil orders.

These relationships are validated by many studies using real soils to adsorb specific ions, such as the one featured in Figure 8.23 showing adsorption of toxic cadmium cations (Cd^{2+}) on

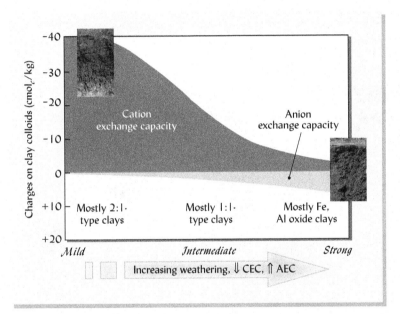

Figure 8.22 *The effect of weathering intensity on the charges on clay minerals and, in turn, on their cation and anion exchange capacities (CECs and AECs). Note the high CEC and very low AEC associated with mild weathering, which has encouraged the formation of 2:1-type clays such as fine-grained micas, vermiculites, and smectites. More intense weathering destroys the 2:1-type clays and leads to the formation of first kaolinite and then oxides of Fe and Al.* (Diagram courtesy of Ray R. Weil)

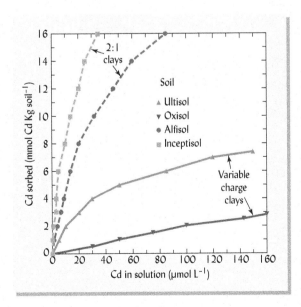

Figure 8.23 *Sorption of cadmium (Cd, a toxic heavy metal) on four Chinese soils varying in their degree of weathering, and therefore type of clay mineralogy. The more Cd^{2+} is sorbed to (held by) the soil colloids, the less it is likely to be leached to groundwater or taken up by vegetation. The amount of Cd^{2+} sorbed onto the soils increased as each soil was equilibrated with increasingly concentrated solutions of $Cd(NO_3)_2$, with the amount leveling off as the maximum sorptive capacity of each soil was approached. Such curves are termed sorption isotherms and illustrate adsorption by both outer- and inner-sphere complexation.* [Redrawn from Hu et al. (2007)]

four soils of different clay mineralogy in China. The two soils (Alfisols and Inceptisols) highest in 2:1 clays exhibited much higher CEC and greater capacities to adsorb cations from the soil solution. For all four soils, the amount of Cd^{2+} adsorbed increased as each soil was equilibrated with increasingly concentrated solutions of Cd^{2+}, with the amount adsorbed leveling off as the maximum sorptive capacity of each soil was approached. Such curves are termed **sorption isotherms**.

8.12 SORPTION OF ORGANIC MOLECULES TO SOIL COLLOIDS[4]

Soil colloids can sorb charged organic molecules as well as the inorganic cations emphasized in this chapter so far. Thus, the retentive properties of soil colloids help control the biological activity of such organic compounds as pesticides and retard their movement into groundwater (Figure 8.24). The retention of these chemicals by soil colloids can allow time for soil microbes to break the chemicals down into less toxic by-products. Retention by soil colloids may also inactivate organic compounds (such as soil-applied herbicides), especially those that have positively charged sites that can interact with the soil CEC.

By accepting or releasing protons (H^+ ions), groups such as —OH, —NH_2, and —COOH in the chemical structure of many organic compounds provide positive or negative charges that participate in anion or cation exchange reactions. Other organic compounds

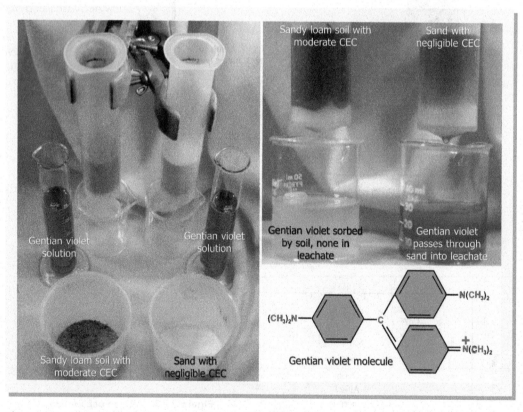

Figure 8.24 *Retention of a soluble positively charged organic compound (the purple dye, gentian violet) by the process of cation exchange in soil. (Left) The experimental setup is shown before adding the gentian violet solution to the soil in the columns. The left-hand beaker and column contain a sandy loam soil with a moderate CEC (10 cmolc/kg), while the right-hand beaker and column contain a sand with negligible CEC (<1 cmolc/kg). (Right) After the gentian violet solutions were poured into the columns and leached through the soils, the water draining from the soil is clear with no purple color, indicating that the CEC of the soil has removed all the gentian violet from the solution. In contrast, the water draining from the sand column is still purple; only a small portion of the compound was retained by the very small CEC of the sand. It requires little imagination to see how soils can thereby protect groundwater from pesticides and other organic contaminants.* (Photo courtesy of Ray R. Weil)

---

[4]For a review of the occurrence and fate of veterinary antibiotics in soil, see Kim et al. (2011).

participate in inner-sphere complexation and adsorption reactions just as do the inorganic ions we have discussed. However, it is common for organic compounds to be **absorbed** within the soil organic colloids by a process termed **partitioning**. The soil organic colloids tend to act as a solvent for the applied chemicals, thereby partitioning their concentrations between those molecules held on/in the soil colloids and those left in the soil solution.

Since we seldom know for certain the exact involvement of the adsorption, complexation, or partitioning processes, we use the general term **sorption** to describe the retention by soils of these organic compounds. Nonionic organic compounds are **hydrophobic**, being repelled by water. As a result, moist clays contribute little to their partitioning since their adsorbed water molecules prevent the movement of the nonionic organic chemicals into or around the clay particles. The hydrated metal cations (e.g., Ca^{2+}) that are adsorbed on the surface of smectites can be replaced with large organic cations, giving rise to what are termed **organoclays**. Such clay surfaces are more friendly toward applied organic compounds, making it possible for the clay to participate in partitioning. Environmental scientists use this phenomenon by making smectite organoclays that can remove organic contaminants from wastewaters and from contaminated groundwaters (see Section 15.5).

Distribution Coefficients

The tendency of a pesticide or other organic compound to leach into the groundwater is determined by the solubility of the compound and by the ratio of the amount of chemical sorbed by the soil to that remaining in solution. This ratio is known as the **soil distribution coefficient** K_d

$$K_d = \frac{\text{mg chemical sorbed/kg soil}}{\text{mg chemical/L solution}} \tag{8.9}$$

The K_d therefore is typically expressed in units of L/kg. Researchers have found that the K_d for a given compound may vary widely depending on the nature of the soil in which the compound is distributed. The variation is related mainly to the amount of organic matter (organic carbon) in the soils. Therefore, most scientists prefer to use a similar ratio that focuses on sorption by organic matter. This ratio is termed the *organic carbon distribution coefficient* K_{oc}:

$$K_{oc} = \frac{\text{mg chemical sorbed/kg organic carbon}}{\text{mg chemical/L solution}} = \frac{K_d}{\text{g org. C/g soil}} \tag{8.10}$$

The K_{oc} can be calculated by dividing the K_d by the fraction of organic C (g/g) in the soil. This relationship can be seen in Table 8.8, which shows both K_d and K_{oc} for several commonly used herbicides and metabolites. Higher K_d or K_{oc} values indicate the chemical is more

---

Table 8.8
PARTITIONING COEFFICIENTS FOR SOIL (K_d) AND FOR ORGANIC CARBON (K_{OC}) FOR SEVERAL WIDELY USED HERBICIDES

Three of the listed compounds are metabolites that form when microorganisms decompose Atrazine. Higher K_d or K_{oc} values indicate stronger attraction to the soil solids and lower susceptibility to leaching loss. The values were measured for a particular soil (an Ultisol in Virginia). Using the relationship between K_d and K_{oc}, it can be ascertained that this soil contained 0.013 g C/g soil (1.3%).

Herbicide	K_d	K_{oc}
Atrazine	1.82	140
Diethyl atrazine	0.99	80
Diisopropyl atrazine	1.66	128
Hydroxy atrazine	7.92	609
Metolachlor	2.47	190

Data from Seybold and Mersie (1996).

tightly sorbed by the soil and therefore less susceptible to leaching and movement to the ground-water. On the other hand, if the management objective is to wash the chemical out of a soil, this will be more easily accomplished for chemicals with lower coefficients. Equations (8.9) and (8.10) emphasize the importance of the sorbing power of the soil colloidal complex, and especially of humus, in the management of organic compounds added to soils.

Binding of Biomolecules to Clay and Humus

The enormous surface area and number of charged sites on the clay and humus in soils attract and bind many types of organic molecules. These molecules include such biologically active substances as DNA (genetic code material), enzymes, antibiotics, hormones, toxins, and even viruses. The initial attraction may be between charged colloidal surfaces and positively or negatively charged functional groups on the biomolecule, similar to the ion exchange reactions just discussed. Adsorption of biomolecules can take place rapidly (in less than a minute) and the amount adsorbed is related to the type of clay mineral. The bond between the biomolecule and the colloid is often quite strong so that the biomolecule cannot be easily removed by washing or by exchange reactions. In most cases, biomolecules bound to clays do not enter the interlayer spaces, but are attached to the outer planar surfaces and edges of the clay crystals.

The nonexchangeable binding of biomolecules to soil colloids in this manner has important environmental implications for two reasons. First, such binding may protect biomolecules from enzymatic attack, meaning that the molecules will persist in the soil much longer than studies of unbound biomolecules might suggest. Apparently, interaction with the charged colloid surface changes the three-dimensional shape of the biomolecule and its electron distribution so that microbial enzymes, which would normally cleave (cut) the biomolecule, are unable to recognize and react with their target sites. Second, it has been shown that many biomolecules retain their biological activity in the bound state. Toxins remain toxic to susceptible organisms, enzymes continue to catalyze reactions, viruses can lyse (break open) cells or transfer genetic information to host cells, and DNA strands retain the ability to transform the genetic code of living cells, even while bound to colloidal surfaces and protected from decay.

When genetically modified organisms (GMOs) are introduced to the soil environment, the cryptic (hidden) genes may represent an undetected potential for transfer of genetic information to organisms for which it was not intended. A similar concern exists regarding plants genetically modified to produce such compounds as human drugs ("pharma crops") or insecticidal toxins. For example, millions of hectares are planted each year with corn and cotton plants genetically engineered to have a bacterial gene that codes for production of the insecticidal toxin Bt. The Bt toxin is released into the soil by root excretion and decomposition of plant residues containing the toxin. We know little about what effect the toxins may have on soil ecology if it accumulates in soils in the colloid-bound, but still active state (see Sections 10.2–10.5).

Antibiotics comprise another class of important organic compounds that sorb to soil colloids. These unique natural chemicals are irreplaceable life-saving compounds. Animal manures produced on most industrial-style farms have been found to be laden with antibiotics that have passed through the animals' digestive systems. When these manures are applied to farmland, the antibiotics become sorbed on the soil colloids and may accumulate with repeated manure applications. Apparently, the sorption is very strong. For example, K_d values as high as 2,300 L/kg have been reported for the antibiotic tetracycline in some soils (compare this K_d value to those of the herbicides listed in Table 8.8). Many antibiotics appear to interact with soil colloids via the process of cation exchange (Figure 8.25), the antibiotic compounds developing positively charged sites, especially in acid soils. Increasingly, research shows that even though sorption to soil colloids may reduce their efficacy somewhat, the soil-bound antibiotics still work against bacteria. In addition, there is evidence that at least some antibiotics can be taken up from soil by food crops and so enter the human food supply. These revelations raise concerns that the huge amounts of antibiotics exposed in the environment will select for resistant strains of "super bacteria" (including human pathogens), which would then no longer be controllable by these (once) life-saving drugs. It is clear that soil colloids and soil science have important roles to play with respect to environmental and human health.

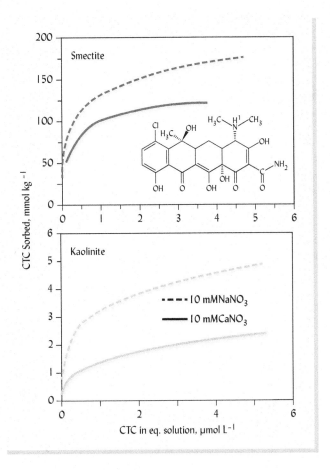

Figure 8.25 Adsorption isotherms illustrating the retention of chlortetracycline (CTC) by (a) montmorillonite (pH 4.3–4.7) and (b) kaolinite (pH 5.1–5.6) as a function of ionic background cation. Binding to colloids slowed antibiotic release to microorganisms. Decreasing CTC retention with the inclusion of Ca(NO$_3$)$_2$ as the background electrolyte (relative to NaNO$_3$) may also be interpreted to indicate ion exchange because Ca^{2+} is more competitive with CTC than Na$^+$ for exchange sites. Greater sorption of CTC with Na background than with Ca background suggests that sorption took place by cation exchange mechanisms since Ca is more tightly held by cation exchange and would be more difficult for CTC to displace. [Redrawn from data in Essington et al. (2010)]

8.13 CONCLUSION

The complex structures, enormous surface area (both internal and external), and tremendous numbers of charges associated with soil colloids combine to make these tiniest of soil particles the seats of chemical and physical activity in soils. The physical activity of the colloids, their adsorption of water, swelling, shrinking, and cohesion are discussed in detail in Chapters 4–6. Here we focused on the chemical activity of the colloids, activity that results largely from charged sites on or near colloid surfaces. These charged sites attract oppositely charged ions and molecules from the soil solution. The negative sites attract positive ions (cations) such as Ca^{2+}, Cu^{2+}, K$^+$, or Al^{3+} and the positive sites attract negative ions (anions) such as Cl$^-$, SO$_4^{-2}$, HPO$_4^{2-}$, or NO$_3^-$. In most soils, the negative charges far outnumber the positive.

Most elements weathered from rocks or added as soil amendments will eventually end up in the oceans, accounting for the increasing saltiness of seawater. Fortunately, colloidal attraction greatly slows that journey and allows soils to accumulate the stocks of nutrients necessary to support forests, grasslands, crops, and, ultimately, civilizations. Colloidal attraction also enables soils to act as effective filters, sinks, and exchangers, protecting groundwater and food chains from excessive exposure to many pollutants.

When ions are attracted to a colloid, they may enter into two general types of relationships with the colloid surface. If the ion bonds directly to atoms of colloidal structure with no water molecules intervening, the relationship is termed an *inner-sphere complex*, and is not easily reversed. In contrast, ions that keep their hydration shell of water molecules around them are generally attracted to colloidal surfaces with excess opposite charge, called *outer-sphere complexation*. The adsorbed ion and its shell of water molecules oscillate or move about within a zone of attraction in a less specific and easily reversed interaction. Ions in such a state of dynamic adsorption are termed *exchangeable ions* because they break away from the colloid whenever another ion from the solution moves in closer and takes over, neutralizing the colloid's charges.

The replacement of one ion for another in the outer-sphere complex is termed *ion exchange*. Cation and anion exchange reactions are reversible and balanced charge-for-charge (rather than ion-for-ion). The extent of the reaction is influenced by mass action, the relative charge and size of the hydrated ions, the nature of the colloid, and the nature of the other (complementary) ions already adsorbed on the colloid.

The colloids in soils are both organic (humus) and mineral (clays) in nature. Organic matter contributes most of the charges in surface soils, while clay provides the majority of charges in subsoil. The total number of negative colloid charges per unit mass is termed the *cation exchange capacity* (CEC), and is dependent on the amount of humus in the soil, and the amount and type of clays present. Kaolinite and other 1:1-type silicate clays and aluminum oxides are low-activity clays, while smectite and vermiculite and other 2:1 clays are high-activity. The CEC of different colloids varies from about 1 to over 200 $cmol_c/kg$ and that of whole mineral soils commonly varies from about 1 to 50 $cmol_c/kg$.

The differing ability of soil colloids to adsorb ions and molecules is key to managing soils, both for plant production and to take advantage of how CEC may regulate movement and bio-availability of both nutrients and toxins in the environment. Among the important properties influenced by colloids is the acidity or alkalinity of the soil, the topic of the next chapter.

STUDY QUESTIONS

1. Describe the *soil colloidal complex*, indicate its various components, and explain how it tends to serve as a "bank" for plant nutrients.
2. How do you account for the difference in surface area associated with a grain of kaolinite clay compared to that of montmorillonite, a smectite?
3. Contrast the difference in crystalline structure among *kaolinite, smectites, fine-grained micas, vermiculites,* and *chlorites.*
4. There are two basic processes by which silicate clays are formed by weathering of primary minerals. Which of these would likely be responsible for the formation of (1) fine-grained mica, and (2) kaolinite from muscovite mica? Explain.
5. If you wanted to find a soil high in kaolinite, where would you go? The same for (1) smectite and (2) vermiculite?
6. Which of the silicate clay minerals would be *most* and *least* desired if one were interested in (1) a good foundation for a building, (2) a high cation exchange capacity, (3) an adequate source of potassium, and (4) a soil on which hard clods form after plowing?
7. Which of the following would you expect to be *most* and *least* sticky and plastic when wet: (1) a soil with significant sodium saturation in a semiarid area, (2) a soil high in exchangeable calcium in a subhumid temperate area, or (3) a well-weathered acid soil in the tropics? Explain your answer.
8. A soil at pH 7.0 contains 4% humus, 10% montmorillonite, 10% vermiculite, and 10% Fe, Al oxides. What is its approximate cation exchange capacity?
9. Calculate the number of grams of Al^{3+} ions needed to replace 10 $cmol_c$ of Ca^{2+} ion from the exchange complex of 1 kg of soil.
10. A 1,000 g sample of soil has been determined to contain exchangeable cations in these amounts: $Ca^{2+} = 9$ $cmol_c$, $Mg^{2+} = 3$ $cmol_c$, $K^+ = 1$ $cmol_c$, $Al^{3+} = 3$ $cmol_c$. (a) What is the CEC of this soil? (b) What is the aluminum saturation of this soil?
11. A 100 g sample of a soil has been determined to contain the exchangeable cations in these amounts: $Ca^{2+} = 90$ mg, $Mg^{2+} = 35$ mg, $K^+ = 28$ mg, $Al^{3+} = 60$ mg. (a) What is the CEC of this soil? (b) What is the aluminum saturation of this soil?
12. A 100 g sample of a soil was shaken with a strong solution of $BaCl_2$ buffered at pH 8.2. The soil suspension was then filtered, the filtrate was discarded, and the soil was thoroughly leached with distilled water to remove any nonexchangeable Ba^{2+}. Then the sample was shaken with a strong solution of $MgCl_2$ and again filtered. The last filtrate was found to contain 10,520 mg of Mg^{2+} and 258 mg of Ba^{2+}. What is the CEC of the soil?
13. Explain the importance of K_d and K_{oc} in assessing the potential pollution of drainage water. Which of these expressions is likely to be most consistently characteristic of the organic compounds in question regardless of the type of soil involved? Explain.
14. An accident at a nuclear power plant has contaminated soil with strontium-90 (Sr^{2+}), a dangerous radionuclide. Health officials order forages growing in the area to be cut, baled, and destroyed. However, there is concern that as the forage plants regrow, they will take up the strontium from the soil and cows eating this contaminated forage will excrete the strontium into their milk. You are the only soil scientist assigned to a risk assessment team consisting mainly of distinguished physicians and statisticians. Write a brief memo to your colleagues explaining how the properties of the soil in the area, especially those related to cation exchange, could affect the risk of contaminating the milk supply.
15. Explain why there is environmental concern about the adsorption by soil colloids of such normally beneficial substances as antibiotic drugs and natural insecticides.

REFERENCES

Dixon, J. B., and S. B. Weed. 1989. *Minerals in Soil Environments.* 2nd ed. Soil Science Society of America, Madison, WI.

Essington, M. E., J. Lee, and Y. Seo. 2010. "Adsorption of antibiotics by montmorillonite and kaolinite." *Soil Science Society of America Journal* 74:1577–1588.

Evangelou, V. P., and R. E. Phillips. 2005. "Cation exchange in soils." In A. Tabatabai and D. Sparks (eds.). *Chemical Processes in Soils.* SSSA Book Series No. 8. Soil Science Society of America, Madison, WI, pp. 343–410.

Kim, K. R., G. Owens, S. I. Kwon, K. H. So, D. B. Lee, and Y. Ok. 2011. "Occurrence and environmental fate of veterinary antibiotics in the terrestrial environment." *Water, Air, & Soil Pollution* 214:163–174.

Meunier, A. 2005. *Clays.* Springer-Verlag, Berlin, p. 470.

Reddy, P. P. 2013. "Disguising the leaf surface." no edtor- *Recent Advances in Crop Protection.* Springer India, New York, pp. 91–102.

Seabrook, C. 1995. *Red Clay, Pink Cadillacs and White Gold: The Kaolin Chalk War.* Longstreet Press, Atlanta.

Shamsuddin, J., and H. Ismail. 1995. "Reactions of ground magnesium limestone and gypsum in soils with variable-charged minerals." *Soil Science Society of America Journal* 59:106–112.

Way, J. T. 1850. "On the power of soils to absorb manure." *Journal of the Royal Agricultural Society of England* 11:313–379.

Weil, R. R. 2015. *Laboratory Manual for Introductory Soils.* 9th ed. Kendall/Hunt, Dubuque, IA.

Windham, B. 2007. "'White dirt' is part of culture, commerce." TuscaloosaNews.com, Tuscaloosa, AL. http://www.tuscaloosanews.com/article/20071028/news/710280337?p=1&tc=pg

9
Soil Acidity, Alkalinity, Salinity, and Sodicity

What have they done to the rain?
—SONG BY MALVINA REYNOLDS

*An acid-soil community.
(Photo courtesy of Ray R. Weil)*

The degree of soil acidity or alkalinity, expressed as pH, act as a *master variable* that affects a wide range of soil chemical and biological properties. This chemical variable greatly influences the likelihood that plant roots will take up both nutrient and toxic elements. The pH also has pronounced impacts on communities of soil microorganisms and their activities. It is also a major determinant of which plant species will grow well or even grow at all at a given site.

Soil pH affects the *mobility* of many pollutants in soil by influencing the rate of their biochemical breakdown, their solubility, and their adsorption to colloids. Thus, soil pH is a critical factor in predicting the likelihood that a given pollutant will contaminate groundwater, surface water, and food chains. Furthermore, there are certain situations in which so much acidity is generated that the acid itself becomes a significant environmental pollutant.

The balance between acid and nonacid cations on colloid surfaces and the balance between H^+ and OH^- ions in the soil solution are the main determinant of soil pH. These balances, in turn, are largely controlled by the nature of the soil colloids and cation exchange (see Chapter 8). Acidification reaches its greatest expression in humid regions where rainfall is sufficient to encourage the production of H^+ ions and the thorough leaching of nonacid cations from the soil profile. A major problem in strongly acidic soil is the toxicity of the nonnutrient element, aluminum.

By contrast, leaching is much less extensive in drier regions, allowing soils to retain enough nonacid Ca^{2+}, Mg^{2+}, K^+, and Na^+ to prevent a buildup of acid cations. Soils in semiarid and arid regions, therefore, tend to have **alkaline** pH levels (i.e., pH > 7), sometimes accompanied by high levels of soluble salts (**salinity**) and exchangeable sodium (**sodicity**). Conditions associated with alkalinity, salinity, and sodicity can lead to serious problems in the physical condition and fertility of soils in these regions.

In this chapter, we will first learn why soils become acidic in humid regions and how to manage this acidity. Then we will turn our attention to the causes and management of salinity and sodicity that plague soils of drier regions where for centuries people in many lands have acted as their own worst enemies by unwittingly adding salt to their soils.

9.1 WHAT PROCESSES CAUSE SOIL ACIDITY AND ALKALINITY?

Acidity and alkalinity are all about the balance between hydrogen ions (H^+) and hydroxyl ions (OH^-) which is usually quantified using the pH scale (Box 9.1). It will be useful to develop a feel for the range of pH values found in various soils and to compare these to the pH values of other common substances (Figure 9.2). The degree of acidification that actually occurs in a given soil is determined by the balance between those processes that produce H^+ ions and other processes that *consume* them (Table 9.1).

BOX 9.1
SOIL pH, SOIL ACIDITY, AND ALKALINITY

Whether a soil is acid, neutral, or alkaline is determined by the concentrations (or activities) of H^+ and OH^- ions. Pure water provides these ions in equal concentrations:

$$H_2O \rightleftharpoons H^+ + OH^-$$

The equilibrium for this reaction is far to the left; only about 1 out of every 10 million water molecules is dissociated into H^+ and OH^- ions. The ion product of the concentrations of the H^+ and OH^- ions is a constant (K_w), which at 25°C is known to be 1×10^{-14}:

$$[H^+] \times [OH^-] = K_w = 10^{-14}$$

Since in pure water the concentration H ions [H^+] must be equal to that of OH^- ions [OH^-], this equation shows that the concentration of each is 10^{-7} ($10^{-7} \times 10^{-7} = 10^{-14}$). It also shows the inverse relationship between the concentrations of these two ions (Figure 9.1). As one increases, the other must decrease proportionately.

Scientists have simplified the means of expressing the very small concentrations of H^+ and OH^- ions by using the *negative logarithm of the H^+ ion concentration*, termed the *pH*. Thus, if the H^+ concentration in an acid medium is 10^{-5}, the pH is 5; if it is 10^{-9} in an alkaline medium, the pH is 9. Every one unit step down the pH scale represents a tenfold change in acidity. Thus a solution at pH 5.0 has 10 times as many H^+ ions as one at pH 6.0 and 100 times as many as one at pH 7.0.

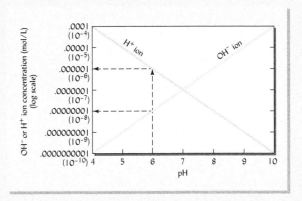

Figure 9.1 *Relationship between pH, pOH, and concentrations of hydrogen and hydroxyl ions in water solution.*

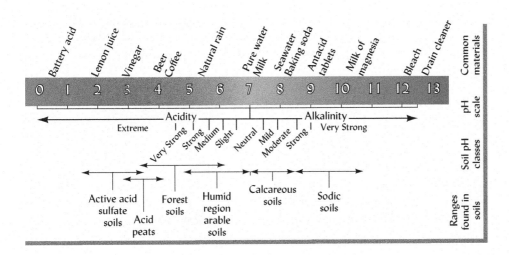

Figure 9.2 *Some pH values for familiar substances (top) compared to terms used to describe soil acidity or alkalinity and ranges of pH typical for various types of soils (bottom). (Diagram courtesy of courtesy of R. Weil)*

Table 9.1
THE MAIN PROCESSES THAT PRODUCE OR CONSUME HYDROGEN IONS (H^+) IN SOIL SYSTEMS
Production of H^+ ions increases soil acidity, while consumption of H^+ ions delays acidification and leads to alkalinity. The pH level of a soil reflects the long-term balance between these two types of processes.

Acidifying (H^+ ion–producing) processes	Alkalinizing (H^+ ion–consuming) processes
Formation of carbonic acid from CO_2	Input of bicarbonates or carbonates
Acid dissociation:	Anion protonation:
$RCOOH \rightarrow RCOO^- + H^+$	$RCOO^- + H^+ \rightarrow RCOOH$
Oxidation of N, S, and Fe compounds	Reduction of N, S, and Fe compounds
Atmospheric H_2SO_4 and HNO_3 deposition	Atmospheric Ca, Mg deposition
Cation uptake by plants	Anion uptake by plants
Accumulation of acidic organic matter	Specific (inner sphere) adsorption of anions (especially SO_4^{2-})
Cation precipitation:	Cation weathering from minerals:
$Al^{3+} + 3H_2O \rightarrow 3H^+ + Al(OH)_3^0$	$3H^+ + Al(OH)_3^0 \rightarrow Al^{3+} + 3H_2O$
$SiO_2 + 2Al(OH)_3 + Ca^{2+} \rightarrow CaAl_2SiO_6 + 2H_2O + 2H^+$	$CaAl_2SiO_6 + 2H_2O + 2H^+ \rightarrow SiO_2 + 2Al(OH)_3 + Ca^{2+}$
Deprotonation of pH-dependent charges	Protonation of pH-dependent charges

Acidifying Processes That Produce Hydrogen Ions

Carbonic and Other Organic Acids. Many organic acids, some weak and others quite strong, are produced by biological activities. The most ubiquitous of these is **carbonic acid** which is formed when carbon dioxide gas dissolves in water. Root and microbial respiration produce high levels of CO_2 in soil air, pushing the following reaction to the right, releasing H^+ ions:

$$CO_2 + H_2O \longrightarrow H_2CO_3 \rightleftharpoons HCO_3^- + H^+ \qquad pK_a = 6.35 \qquad (9.1)$$

Accumulation of Organic Matter. Organic matter forms soluble complexes with nonacid nutrient cations such as Ca^{2+} and Mg^{2+}, thus facilitating the loss of these cations by leaching. Organic matter also is a source of H^+ ions because it contains numerous acid functional groups from which these ions can dissociate (see Section 8.4).

Oxidation of Nitrogen (Nitrification). Ammonium ions (NH_4^+) from organic matter or from most fertilizers are subject to microbial oxidation that converts the N to nitrate ions (NO_3^-). The reaction with oxygen, termed *nitrification*, releases two H^+ ions for each NH_4^+ ion oxidized (see also Section 12.1). The NO_3^- produced is the anion of a **strong acid** (nitric acid, HNO_3):

$$NH_4^+ + 2O_2 \rightleftharpoons H_2O + H^+ + \underbrace{H^+ + NO_3^-}_{\text{Dissociated nitric acid}} \qquad (9.2)$$

Oxidation of Sulfur. The decomposition of plant residues commonly involves the oxidation of organic –SH groups to yield sulfuric acid (H_2SO_4). Another important source of this strong acid is the oxidation of reduced sulfur in minerals such as pyrite. This and related reactions are responsible for producing large amounts of acidity in certain soils when oxygen levels are increased by drainage or excavation (see Section 9.6)

$$FeS_2 + 3\tfrac{1}{2}O_2 + H_2O \rightleftharpoons \underset{\text{Ferrous sulfate}}{FeSO_4} + \underbrace{2H^+ + SO_4^{2-}}_{\text{Dissociated sulfuric acid}} \quad (9.3)$$
$\underset{\text{Pyrite}}{}$

Acids in Precipitation. Rain, snow, fog, and dust contain a variety of acids that contribute H^+ ions to the soil receiving the precipitation. As raindrops fall through unpolluted air, they dissolve CO_2 and form enough carbonic acid to lower the pH of the water from 7.0 (the pH of pure water) to about 5.6. Varying amounts of sulfuric and nitric acids form in precipitation from certain sulfur and nitrogen gases produced by lightning, volcanic eruptions, forest fires, and the combustion of fossil fuels.

Plant Uptake of Cations. For every positive charge taken in as a cation, a root can maintain charge balance either by taking up a negative charge as an anion or by exuding a positive charge as a different cation. When they take up far more of certain cations (e.g., K^+, NH_4^+, and Ca^{2+}) than they do of anions (e.g., NO_3^-, SO_4^{2-}), plants usually exude H^+ ions into the soil solution to maintain charge balance. In the first two of the following examples, plant nutrient uptake results in the addition of H^+ ions to the soil solution:

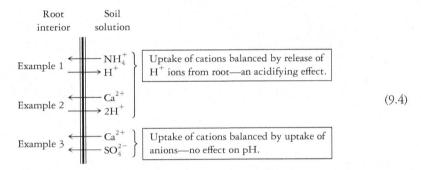

(9.4)

Acid and Nonacid Cations. While H^+ ions are the direct cause of acidity, aluminum ions (e.g., Al^{3+}) also cause acidity when they hydrolyze water molecules (see Section 9.2). Both are considered **acid cations**. Most other cations common on the exchange complex are nonhydrolyzing and so are termed **nonacid cations** (mainly Ca^{2+}, Mg^{2+}, K^+, NH_4^+, and Na^+). Rather, their effect in water is neutral,[1] and soils dominated by them have a pH about 7 unless certain *anions* are present in the soil solution.

Leaching Removal of Nonacid Cations. Acid cations added to the soil solution may replace **nonacid cations** on the cation exchange sites of humus and clay. The soil slowly becomes more acid if the displaced nonacid cations leach out of the profile faster than mineral weathering can release a new supply of these cations (Figure 9.3). Thus, the formation of acid soils is favored by high rainfall, parent materials low in Ca, Mg, K, and Na, and a high degree of biological activity (favoring formation of H_2CO_3). The global distribution of acid and alkaline soils demonstrates the positive relationship between annual rainfall and soil acidity (Figure 9.4).

Alkalizing Processes That Consume Hydrogen Ions or Produce Hydroxyl Ions

Alkaline soils are simply those with a pH above 7.0. *Alkalinity* refers to the concentration of OH^- ions, much as *acidity* refers to that of H^+ ions. In dry regions where water is scarce and organic biomass production is low, soils become alkaline (i.e., their pH rises above 7) because

---

[1] The cations Ca^{2+}, Mg^{2+}, K^+, Na^+, and NH_4^+ have been traditionally called *base* or *base-forming* cations as a convenient way of distinguishing them from the acid cations, H^+ and Al^{3+}. However, since Ca^{2+} and associated cations generally do not produce OH^- ions in solution and do not accept protons, they are technically not bases. Consequently, it is less misleading to refer simply to acid cations (H^+, Al^{3+}) and nonacid cations (most other cations). Likewise, the term *nonacid saturation* should be used rather than *base saturation* to refer to the percentage of the exchange capacity satisfied by nonacid cations (usually Ca^{2+}, Mg^{2+}, K^+, and Na^+) on the exchange complex (see also Section 9.3).

Figure 9.3 Soils become acid when H⁺ ions added to the soil solution exchange with nonacid Ca²⁺, Mg²⁺, K⁺, and Na⁺ ions held on humus and clay colloids. The nonacid cations can then be exported in leaching water along with accompanying anions. As a result, the exchange complex (and therefore also the soil solution) becomes increasingly dominated by acid cations (H⁺ and Al³⁺). Because of this sequence of events, H⁺ ion–producing processes acidify soils in humid regions where leaching is extensive, but cause little long-term soil acidification in arid regions. In the latter case, the Ca²⁺, Mg²⁺, K⁺, and Na⁺ are mostly not removed by leaching but remain in the soil and reexchange with the acid cations, preventing a drop in pH level. (Diagram courtesy of Ray R. Weil)

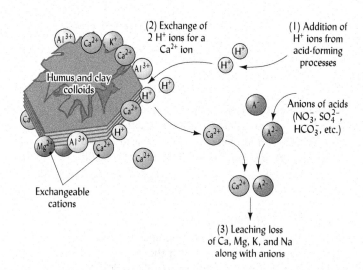

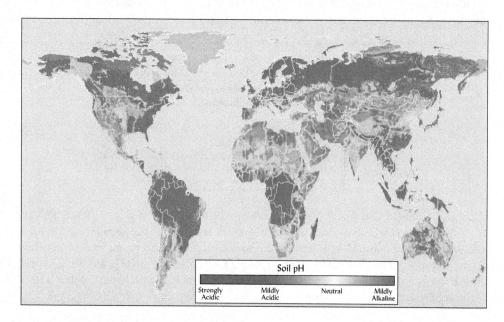

Figure 9.4 The global distribution of soil acidity and alkalinity is largely related to annual rainfall and its excess evapotranspiration. Except for the frozen artic regions, the distribution of acid soils (red areas on the map) corresponds quite closely to the distribution of forest vegetation. Compare to Figure 6.15, right. (IGBP-DIS. 2000. Soildata: A Program for Creating Global Soil-Property Databases. Global Soil Data Task Force, International Geosphere-Biosphere Programme, Data and Information System, Potsdam, Germany. Available from: http://www.daac.ornl.gov)

more H⁺ ions are consumed than generated (see right side of Table 9.1) and there is not enough rain to wash away the nonacid cations weathered from minerals.

Accumulation of Nonacid Cations. The weathering of nonacid cations from minerals (Section 2.1) is a slow but very important H⁺ ion–consuming process that counteracts acidification. An example is the weathering of calcium from a silicate mineral:

$$\text{Ca-silicate} + 2\text{H}^+ \longrightarrow \text{H}_4\text{SiO}_4 + \text{Ca}^{2+} \tag{9.5}$$

Role of Carbonate and Bicarbonate Anions. The basic hydroxyl (OH⁻)-generating anions are principally **carbonate** (CO_3^{2-}) and **bicarbonate** (HCO_3^-). These anions originate from the dissolution of such minerals as calcite ($CaCO_3$) or from the dissociation of carbonic acid (H_2CO_3). In this series of linked equilibrium reactions, carbonate and bicarbonate act as bases because they react with water to form hydroxyl ions and thus raise the pH. The importance of these reactions in **soil buffering** (resistance to pH change) is discussed in Section 9.4.

$$CaCO_3 \rightleftharpoons \underset{\text{(dissolved in water)}}{Ca^{2+}} + \underset{\text{(dissolved in water)}}{CO_3^{2-}} \quad (9.6)$$
Calcite (solid)

$$CO_3^{2-} + H_2O \rightleftharpoons HCO_3^- + OH^- \quad (9.7)$$

$$HCO_3^- + H_2O \rightleftharpoons H_2CO_3 + OH^- \quad (9.8)$$

$$\underset{\text{Carbonic acid}}{H_2CO_3} \rightleftharpoons H_2O + \underset{\text{(gas)}}{CO_2 \uparrow} \quad (9.9)$$

Carbon Dioxide and Carbonates. The direction of the overall reaction [Reactions (9.6)–(9.9)] determines whether OH^- ions are consumed (proceeding to the left) or produced (proceeding to the right). The reaction is controlled mainly by the precipitation or dissolution of calcite on the one end and by the production (by respiration) or loss (by volatilization to the atmosphere) of carbon dioxide at the other end. Biological activity in soil tends to lower the pH by increasing concentration of CO_2 and driving the reaction series to the left (see Section 7.2). The other process that limits the rise in pH is the precipitation of $CaCO_3$ that occurs when the soil solution becomes saturated with respect to Ca^{2+} ions.

Such precipitation removes Ca from the solution, again driving the reaction series to the left (lowering pH). Because of the limited solubility of $CaCO_3$, the pH of the solution cannot rise above 8.4 when the CO_2 in solution is in equilibrium with that in the atmosphere. The pH at which $CaCO_3$ precipitates in soil is typically only about 7.0–8.0, depending on how much the CO_2 concentration is enhanced by biological activity. This is an important point to remember because it suggests that if other carbonate minerals more soluble than $CaCO_3$ (e.g., Na_2CO_3) were present, the pH would rise considerably higher (see Section 9.13). In fact where sodium is abundant, the pH can rise to 10 or 11. It is fortunate for plants that Ca^{2+}, not Na^+, ions are dominant in most soils.

Excess Anion Uptake by Plants. Consider, for example, the case in which plant uptake of an anion such as NO_3^- exceeds the uptake of associated cations. In this case, the root exudes the anion, bicarbonate (HCO_3^-), to maintain charge balance. The increase in the concentration of bicarbonate ions tends to reverse the dissociation of carbonic acid (see Eq. [9.1]), and thereby *consume* H^+ ions and raise the pH of the soil solution:

$$\begin{array}{c} \text{Root} \quad \text{Soil} \\ \text{interior} \quad \text{solution} \end{array} \quad \begin{array}{c} \leftarrow NO_3^- \\ \rightarrow HCO_3^- \end{array} \Big\} \boxed{\text{Uptake of anion balanced by release of bicarbonate ion—alkalizing effect.}} \quad (9.10)$$

Another H^+ ion–consuming process involving nitrogen is the reduction of nitrate to nitrogen gases under anaerobic conditions (see denitrification, Sections 7.3 and 12.1).

9.2 WHAT ROLE DOES ALUMINUM PLAY IN SOIL ACIDITY?

Aluminum is a major constituent of most soil minerals, including clays. When H^+ ions are adsorbed on a clay surface, they attack the structure of the minerals, releasing Al^{3+} ions in the process. Exchangeable Al^{3+} ions, in turn, are in equilibrium with dissolved Al^{3+} in the soil solution. The exchangeable and soluble Al^{3+} ions play two critical roles in the soil acidity story. First, aluminum is *highly toxic* to most organisms and is responsible for much of the deleterious impact of soil acidity on plants and aquatic animals. We will discuss this role in Section 9.7. Second, Al^{3+} ions have a strong tendency to hydrolyze, splitting water molecules into H^+ and OH^- ions. The aluminum combines with the OH^- ions, leaving the H^+ to lower the pH of the soil solution. For this reason, Al^{3+} and H^+ together are considered **acid cations**. A single Al^{3+} ion can thus release up to three H^+ ions as the following reversible reaction series proceeds to the right in stepwise fashion:

$$Al^{3+} \underset{H_2O \quad H^+}{\overset{H_2O \quad H^+}{\rightleftharpoons}} AlOH^{2+} \underset{H_2O \quad H^+}{\overset{H_2O \quad H^+}{\rightleftharpoons}} Al(OH)_2^+ \underset{H_2O \quad H^+}{\overset{H_2O \quad H^+}{\rightleftharpoons}} Al(OH)_3^0 \text{ Gibbsite or amorphous (solid)} \quad (9.11)$$

$$pK_a = 5.0 \qquad pK_a = 5.1 \qquad pK_a = 6.7$$

Most of the hydroxy-aluminum ions $[Al(OH)_x^{y+}]$ formed as the pH increases are strongly adsorbed to clay surfaces or complexed with organic matter and so mask much of the colloid's potential cation exchange capacity. As the pH is raised and more of the hydroxyl-aluminum ions precipitate as uncharged $Al(OH)_3^0$, the negative sites on the colloids become available for cation exchange. This is one reason for the increase in soil CEC as the pH is raised from 4 to 7 (at which pH virtually all the aluminum cations have precipitated as $Al(OH)_3^0$).

9.3 POOLS OF SOIL ACIDITY

Principal Pools of Soil Acidity

Three major pools of acidity are common in soils: (1) a small amount of **active acidity** due to the H^+ ions in the soil solution; (2) a larger amount of **salt-replaceable (exchangeable) acidity**, involving the aluminum and hydrogen that are *easily exchangeable* by other cations in a simple unbuffered salt solution, such as KCl; and (3) an often very large pool of **residual acidity**, which can be neutralized by limestone or other alkaline materials but cannot be detected by salt replacement. These types of acidity all add up to the **total acidity** of a soil (see Figure 9.5). In addition, a less common, but sometimes very important fourth pool, namely, **potential acidity**, can arise upon the oxidation of sulfur compounds in certain acid sulfate soils (see Section 9.6).

The chemical equivalent of salt-replaceable acidity in strongly acid soils is commonly 1,000 times that of active acidity in the soil solution. Even in moderately acid soils, the limestone needed to neutralize this type of acidity is commonly more than 100 times that needed to neutralize the soil solution (active acidity). At a given pH value, exchangeable acidity is generally highest in smectite clays, intermediate for vermiculites, and lowest for kaolinite.

The **residual acidity** is generally associated with hydrogen and aluminum ions (including the aluminum hydroxy ions) that are bound in nonexchangeable forms by organic matter and clays (see Figure 9.5). As the pH increases, the bound hydrogen dissociates and the bound aluminum ions are released and precipitate as amorphous $Al(OH)_3^0$. These changes free up negatively charged cation exchange sites and increase the CEC.

The residual acidity is commonly far greater than either the active or salt-replaceable acidity. It may be 1,000 times greater than the soil solution or active acidity in a sandy soil and 50,000 or even 100,000 times greater in a clayey soil high in organic matter.

For most soils (not potential acid sulfate soils) the total acidity that must be overcome to raise the soil pH to a desired value can be defined as:

$$\text{Total acidity} = \text{active acidity} + \text{salt-replaceable acidity} + \text{residual acidity} \quad (9.12)$$

We can conclude that the pH of the soil solution is only the tip of the iceberg in determining how much lime may be needed to overcome effects of soil acidity.

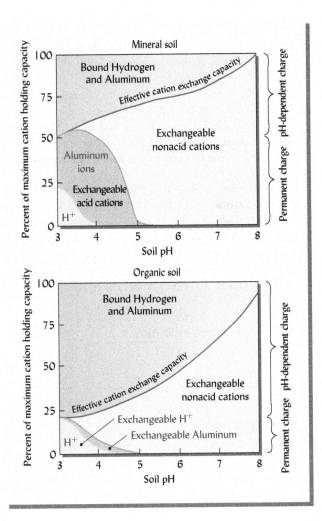

Figure 9.5 General relationships between soil pH and bound cations in representative soils. (Left) A mineral soil with mixed mineralogy and a moderate organic matter level exhibits a moderate decrease in effective CEC as pH is lowered, suggesting that pH-dependent charges and permanent charges (see Section 8.6) each account for about half of the maximum CEC. Above pH 5.5, concentrations of exchangeable aluminum and H^+ ions are negligible, and the effective CEC is essentially 100% saturated with exchangeable nonacid cations (Ca^{2+}, Mg^{2+}, K^+, and Na^+, the so-called base cations). As pH drops from 7.0 to about 5.5, the effective CEC is reduced because hydrogen and aluminum ions are tightly bound to some of the pH-dependent charge sites. As pH is further reduced from 5.5 to 4.0, Al^{3+} ions, along with some H^+ ions, occupy an increasing portion of the remaining exchange sites. (Lower) The CEC of an organic soil is dominated by pH-dependent (variable) charges with only a small amount of permanent charge. Therefore, as pH is lowered, the effective CEC of the organic soil declines more dramatically than the effective CEC of the mineral soil. At low pH levels, exchangeable H^+ ions are more prominent and Al^{3+} less prominent on the organic soil than on the mineral soil. (Diagram courtesy of Ray R. Weil)

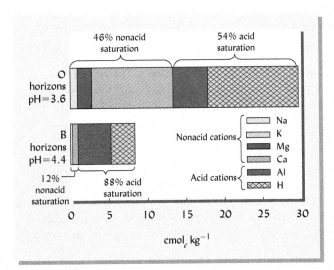

Figure 9.6 Saturation of the exchange capacity with acid and nonacid cations for O horizons and B horizons of soils in the Adirondack Mountains of New York. The effective cation exchange capacity (ECEC), the sum of all the exchangeable cations, was almost 30 $cmol_c/kg$ in the O horizons compared to only about 8 $cmol_c/kg$ in the B horizons. As is typical of temperate forested soils, the O horizons (which were about 90% organic) exhibited an extremely acid pH but a relatively low acid saturation, and the acid cations were mainly H^+. By contrast, the B horizons (which were about 90% mineral) had a more moderate pH but were 88% acid-saturated, and most of the acid cations were aluminum. Averages from 150 pedons in 144 watersheds. [Modified from Sullivan et al. (2006)]

Effective CEC and pH. Note that in both soils illustrated in Figure 9.5, the effective CEC increases as the pH level rises. This change in effective CEC results mainly from two factors: (1) the binding and release of H^+ ions on pH-dependent charge sites (as explained in Section 8.6), and (2) the hydrolysis reactions of aluminum species (as explained in Section 9.2). The change in effective CEC will be most dramatic for organic soils (Figure 9.5, *lower*) and highly weathered mineral soils dominated by iron and aluminum oxide clays.

Cation Saturation Percentages

The proportion of the CEC occupied by a given ion is termed its **saturation percentage**. Consider a soil with a CEC of 20 cmol$_c$/kg holding these amounts of exchangeable cations (in cmol$_c$/kg): 10 of Ca^{2+}, 3 of Mg^{2+}, 1 of K$^+$, 1 of Na$^+$, 1 of H$^+$, and 4 of Al^{3+}. This soil with 10 cmol$_c$ Ca^{2+}/kg and a CEC of 20 cmol$_c$/kg is said to be 50% calcium saturated. Likewise, the aluminum saturation of this soil is 20% (4/20 = 0.20 or 20%). Together, the 4 cmol$_c$/kg of exchangeable Al^{3+} and 1 cmol$_c$/kg of exchangeable H$^+$ ions give this soil an **acid saturation** of 25% [(4 + 1)/20 = 0.25]. Similarly, the term **nonacid saturation** can be used to refer to the proportion of Ca^{2+}, Mg^{2+}, K$^+$, Na$^+$, etc., on the CEC. Thus, the soil in our example has a nonacid saturation of 75% [(10 + 3 + 1 + 1)/20 = 0.75].

Traditionally, the nonacid cations have been referred to as "base" cations and their proportion on the CEC the **percent "base" saturation**. Cations such as Ca^{2+}, Mg^{2+}, K$^+$, and Na$^+$ do not hydrolyze as Al^{3+} and Fe^{3+} do, and therefore are not acid-forming cations. However, they are also *not* bases and do not necessarily form bases in the chemical sense of the word.[2] Because of this ambiguity, it is more straightforward to refer to *acid saturation* when describing the degree of acidity on the soil cation exchange complex. Figure 9.6 uses these concepts to characterize acidified soil in the Adirondack Mountains of New York. The relationships among these terms can be summarized as follows:

$$\text{Percent acid saturation} = \frac{\text{cmol}_c \text{ of exchangeable Al}^{3+} + \text{H}^+}{\text{cmol}_c \text{ of CEC}} \quad (9.13)$$

$$\text{Percent nonacid saturation} = \text{Percent "base" saturation} = \frac{\text{cmol}_c \text{ of exchangeable Ca}^{2+} + \text{Mg}^{2+} + \text{K}^+ + \text{Na}^+}{\text{cmol}_c \text{ of CEC}}$$

$$= 100 - \frac{\text{percent acid}}{\text{saturation}} \quad (9.14)$$

Acid (or Nonacid) Cation Saturation and pH

The percentage saturation of a particular cation (e.g., Al^{3+} or Ca^{2+}) or class of cations (e.g., nonacid cations or acid cations) is often more closely related to the nature of the soil solution than is the absolute amount of these cations present. Generally, when the acid cation percentage increases, the pH of the soil solution decreases. However, a number of factors can modify this relationship.

Effect of Types of Colloid and Adsorbed Nonacid Cations. The type of clay minerals or organic matter present influences the pH of different soils at the same percent acid saturation. This is due to differences in the ability of various colloids to furnish H$^+$ ions to the soil solution. For example, the dissociation of adsorbed H$^+$ ions from smectites is much higher than that from Fe and Al oxide clays. Consequently, the pH of soils dominated by smectites is appreciably lower than that of the oxides at the same percent acid saturation. The dissociation of adsorbed hydrogen from 1:1-type silicate clays and organic matter is intermediate between that from smectites and from the hydrous oxides. The relative amount of each of the nonacid cations present on the colloidal complex is another factor influencing soil pH. For example, soils with high sodium saturation (usually highly alkaline soils of arid and semiarid regions) have much higher pH values than those dominated by calcium and magnesium.

Effect of Method of Measuring CEC. An unfortunate ambiguity in the cation saturation percentage concept is that the actual percentage calculated depends on whether the effective CEC (which itself changes with pH) or the maximum potential CEC (which is a constant for a given soil) is used in the denominator. The different methods of measuring CEC are explained in Section 8.9.

---

[2] A base is a substance that combines with H$^+$ ions, while an acid is a substance that releases H$^+$ ions. The anions OH$^-$ and HCO$_3^-$ are strong bases because they react with H$^+$ to form the weak acids, H$_2$O and H$_2$CO$_3$, respectively.

When the concept of cation saturation was first developed, the percent nonacid saturation (then termed "base saturation") was calculated by dividing the level of these exchangeable cations by the *potential* cation exchange capacity that is measured at high pH values (7.0 or 8.2). Thus, if a representative mineral soil such as shown in Figure 9.5 has a potential CEC of 20 $cmol_c/kg$, and at pH 6 has a nonacid exchangeable cation level of 15 $cmol_c/kg$, the percent nonacid cation saturation would be calculated as 15 $cmol_c$/20 $cmol_c \times 100 = 75\%$. The percent "base" saturation determined by this method is still used to classify soils in *Soil Taxonomy*.

A second method relates the exchangeable cation levels to the *effective* CEC at the pH of the soil. As Figure 9.5 shows, the effective CEC of the representative soil at pH 6 would be only about 15 $cmol_c/kg$. At this pH level, essentially all the exchangeable sites are occupied by nonacid cations (15 $cmol_c/kg$). Using the effective CEC as our base, we find that the nonacid cation saturation is 15 $cmol_c$/15 $cmol_c \times 100 = 100\%$. Thus, this soil at pH 6 is either 75% or 100% saturated with nonacid cations, depending on whether we use the potential CEC or the effective CEC in our calculations.

Uses of Cation Saturation Percentages. Which nonacid saturation percentage just described is the correct one? It depends on the purpose at hand. The first percentage (75% of the potential CEC) indicates that significant acidification has occurred and is used in soil classification (e.g., by definition Ultisols must have a nonacid or "base" saturation of less than 35%). The second percentage (100% of the effective CEC) is more relevant to soil fertility and the availability of nutrients. It indicates what proportion of the total exchangeable cations at a given soil pH is accounted for by nonacid cations. For example, when the effective CEC of a mineral soil is less than 80% saturated with nonacid cations (i.e., more than 20% acid saturated), aluminum toxicity is likely to be a problem.

9.4 BUFFERING OF pH IN SOILS[3]

The soil solution pH resists change when either acid or base is added. This resistance to change is called **buffering** and can be demonstrated by comparing the *titration curves* for pure water with those for various soils (Figure 9.7).

The titration curves shown in Figure 9.7 suggest that the soils are most highly buffered when aluminum compounds (at low pH) and carbonates (at high pH) are controlling the buffer reactions. The soil is least well buffered at intermediate pH levels where H^+ ion dissociation

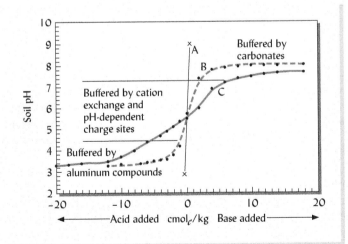

Figure 9.7 Buffering of soils against changes in pH when acid or base is added. A well-buffered soil (C) and a moderately buffered soil (B) are compared to unbuffered water (A). Most soils are strongly buffered at low pH by the hydrolysis and precipitation of aluminum compounds and at high pH by the precipitation and dissolution of calcium carbonate. Most of the buffering at intermediate pH levels (pH 4.5–7.5) is provided by cation exchange and protonation or deprotonation (gain or loss of H^+ ions) of pH-dependent exchange sites on clay and humus colloids. The well-buffered soil (C) would likely have a higher amount of organic matter and/or more highly charged clay than the moderately buffered soil (B). [Based on data in Magdoff and Bartlett (1985) and Lumbanraja and Evangelou (1991)]

---

[3]For a detailed discussion of the chemical principles behind this and related topics, see Bloom et al. (2005).

Figure 9.8 *Equilibrium relationships among residual, salt-replaceable (exchangeable), and soil solution (active) acidity in a soil with organic and mineral colloids. Adsorbed (exchangeable) and residual (bound) ions are much more numerous than those in the solution. Bound aluminum is held tightly on the surfaces of the clay or humus. Remember that aluminum ions, by hydrolysis, also supply H^+ ions in the soil solution. Neutralizing only the hydrogen and aluminum ions in the soil solution will be of little consequence as they will be quickly replaced by ions associated with the colloid.* (Diagram courtesy of Ray R. Weil)

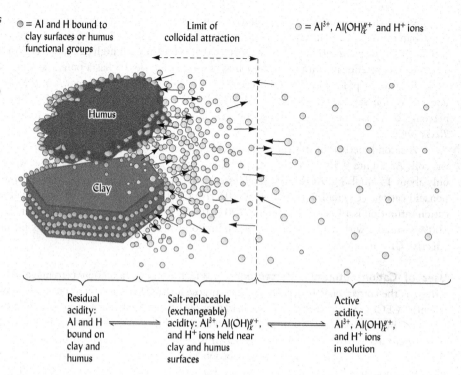

and cation exchange are the primary buffer mechanisms. Different soils exhibit different titration curves reflecting differences among soils with regard to the amounts and types of dominant colloids, carbonates, and bound Al–hydroxy complexes.

For soils with intermediate pH levels (5–7), buffering can be explained in terms of the equilibrium that exists among the three principal pools of soil acidity: active, salt-replaceable, and residual (Figure 9.8). If just enough base (lime, for example) is applied to neutralize the H^+ ions in the soil solution, they are largely replenished as the reactions move to the right, thereby minimizing the change in soil solution pH (Figure 9.8). Likewise, if the H^+ ion concentration of the soil solution is increased (e.g., by organic decay or fertilizer applications) the reactions in Figure 9.8 are forced to the left, consuming most of the added H^+ and again minimizing changes in soil solution pH. Because of the involvement of residual and exchangeable acidity, we can see that soils with higher clay and organic matter contents are likely to be better buffered in this pH range.

Why Is Soil pH Buffering Important?

Soil buffering is important for two primary reasons. First, buffering tends to ensure some stability in the soil pH, preventing drastic fluctuations that might be detrimental to plants, soil microorganisms, and aquatic ecosystems. For example, well-buffered soils resist the acidifying effect of acid rain, preventing the acidification of both the soil and the drainage water. Second, buffering influences the amount of amendments, such as lime or sulfur, required to bring about a desired change in soil pH.

Soils vary greatly in their buffering capacity. Other things being equal, the higher the CEC of a soil, the greater its buffering capacity. This relationship exists because in a soil with a high CEC, more reserve and exchangeable acidity must be neutralized or increased to affect a given change in soil pH. Thus, a clay loam soil containing 6% organic matter and 20% of a 2:1-type clay would be more highly buffered than a sandy loam with 2% organic matter and 10% kaolinite (Figure 9.9).

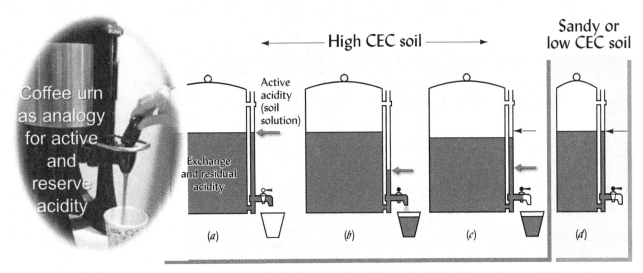

Figure 9.9 The buffering capacity of soils can be described by using the analogy of a coffee dispensing urn. (a) The active acidity, which is represented by the coffee in the indicator tube on the outside of the urn, is small in quantity. (b) When H^+ ions are removed, this active acidity falls rapidly. (c) The active acidity is quickly restored to near the original level by movement from the exchange and residual acidity. By this process, the active acidity resists change. (d) A second soil with the same active acidity (pH) level but much less exchange and residual acidity would have a lower buffering capacity. Much less coffee would have to be added to raise the indicator level in the last dispenser. So too, much less liming material must be added to a soil with a small buffering capacity in order to achieve a given increase in the soil pH.

9.5 HOW CAN WE MEASURE SOIL pH?

More may be inferred regarding the chemical and biological conditions in a soil from the pH value than from any other single measurement. Soil pH can be easily and rapidly measured in the field by colorimetric or potentiometric (glass electrode) methods.

Potentiometric Methods

The most accurate method of determining soil pH is with a pH-sensitive *glass* electrode and a standard reference electrode (or a probe that combines both electrodes in one). The pH electrode is inserted into a soil–water suspension (usually at a ratio of 1:1 or 1:2.5). The difference between the H^+ ion activities in the soil suspension and in the glass electrode gives rise to an electrometric potential that is related to the soil solution pH. A special millivolt meter (called a *pH meter*) measures the electrometric potentials and converts them into pH readings (Figure 9.10). Be advised that certain metallic soil probes on the market that do *not* contain a glass electrode cannot measure pH as claimed and may give highly misleading readings.

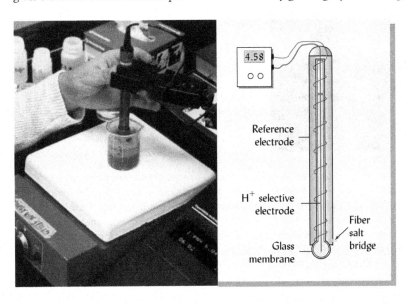

Figure 9.10 Soil pH is measured accurately and inexpensively in the laboratory using a pH meter (a sensitive millivolt meter that converts readings to pH) connected to a combination pH electrode that consists of a reference electrode constructed around a H^+-sensing glass membrane electrode (diagram). A mixture of soil and water (in beaker) is stirred, allowed to settle, and then the electrode is immersed in the liquid above the sediment. The H^+ ions create an electrical potential once the salt bridge completes the circuit. One technician can make hundreds of determinations in a day. (Courtesy of Ray R. Weil)

Most soil-testing laboratories in North America measure the pH of a suspension of soil in water. This is designated the pH$_{water}$. Other labs (mainly in Europe and Asia) suspend the soil sample in a salt solution of either 0.02 M CaCl$_2$(pH$_{CaCl}$) or 1.0 M KCl (pH$_{KCl}$). For normal low-salt soils, the pH$_{CaCl}$ and pH$_{KCl}$ readings are typically about 0.5 and 1.0 units lower than for pH$_{water}$. Therefore, if subsamples of a soil were sent to three labs, the labs might report that the soil pH was 6.5, 6.0, or 5.5 if the labs used methods for pH$_{water}$, pH$_{CaCl}$, and pH$_{KCl}$, respectively. All three pH values indicate the same level of acidity—and a suitable pH for most crops. Therefore, to interpret soil pH readings or compare reports from different laboratories, it is essential to know the method used. In this textbook (as in most North American publications), we will report values for pH$_{water}$, unless specified otherwise.

Variability in the Field

Spatial Variation. Soil pH may vary dramatically over very small distances (millimeter or smaller). For example, plant roots may raise or lower the pH in their immediate vicinity, making the soil pH around the roots quite different from that in the bulk soil just a few mm away (Figure 9.11). Thus, the root may experience a very different chemical environment from that indicated by lab measurements of bulk soil samples. Such mm-scale variability may account in part for the great diversity in microbial species present in normal soils (see Chapter 10).

Applications of soil amendments or ashes from forest fires may cause sizeable pH variations within the space of a few centimeters to a few meters. Other factors, such as erosion or topography-related drainage, may cause pH to vary considerably over larger distances (hundreds of m), possibly ranging over two pH units within a few hectares. Any sampling scheme for measuring soil pH should take into account such variations from place to place, as well as those with depth and time.

With Soil Depth. Different horizons, or even parts of horizons, within the same soil may exhibit substantial differences in pH. Acidifying processes usually proceed initially near the soil surface and slowly work their way down the profile. Examples include the acid input from rainfall, the oxidation of nitrogen applied as fertilizer to the soil surface, and the decomposition of plant litter falling on the soil surface. Reinforcing this vertical pH trend, many natural alkalizing processes such as mineral weathering are typically most active in the lower soil horizons where weatherable minerals from the parent material are most plentiful.

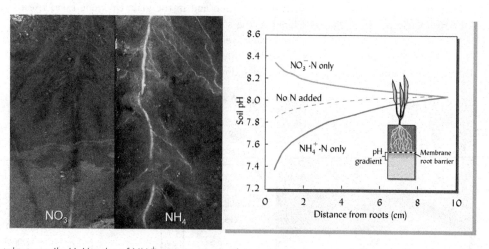

Figure 9.11 Effect of root N uptake on soil pH. Uptake of NH$_4^+$ cations causes the roots to release equivalent positive charges (H$^+$ cations), which lower the pH (see Reaction 9.4) causing the pH-sensitive dye to turn yellow (middle). When a NO$_3^-$ anion is taken up, the roots release a bicarbonate anion (HCO$_3^-$), which raises the pH (see Reaction 9.10) causing a purple color in the pH dye (far left). In another experiment (right) using a calcareous sandy clay loam with pH = 8.1, a barrier membrane allowed soil solution to pass through, but prevented root growth into the lower soil where pH was measured. In this experiment, soil pH was measured using a tiny pH sensor. Plants were watered from the bottom by capillary rise. [Diagram drawn from data in Zhang et al. (2004) with permission of the Soil Science Society of America. Left photos courtesy of Joseph Heckman, The Rutgers University]

Human application of liming materials is one of the more obvious exceptions to the already described trends since this practice raises the pH mainly in the upper horizons into which the lime is incorporated. The incorporation of liming materials into the Ap horizon by tillage usually results in relatively uniform pH readings within the top 15–20 cm of cultivated soil. However, severe acidity may occur in the subsoil beyond the depth of lime incorporation but within the reach of most plant roots. Untilled soils—including croplands managed with no-till practices, unplowed grasslands, lawns, and forest land—often show marked vertical variation in soil pH, with most of the changes in pH occurring in the upper few cm. For all these reasons, it is often advisable to obtain soil samples from various depth increments within the root zone and determine the pH level for each. Otherwise, serious pH issues may be overlooked.

Effects of Season and Time. A buildup of salts near the surface during dry periods versus the leaching of these salts during wet periods often produce seasonal variations in soil pH_{water}. Other causes of seasonal variation include periods of intense organic decay with the onset of warm temperatures or first rains. Because of such variations, successive soil samples should be collected at the same time of year if changes in pH are to be monitored over a number of years. Most acidification processes are quite slow and must overcome soil buffering, so field soil pH generally changes slowly over a period of years or decades. However, if finely ground, limestone or sulfur are applied, changes of 1 or more pH units may occur within two to six months (see Section 9.8).

9.6 HOW DO HUMANS ACIDIFY SOILS?[4]

The natural processes of soil acidification are greatly (and usually inadvertently) accelerated by human activities. We will consider three major types of human-influenced soil acidification: (1) nitrogen amendments, (2) acid precipitation, and (3) acid sulfate soils.

Nitrogen Fertilization

During the past 100 years, agricultural activities have greatly increased the amount of nitrogen cycling through the world's soils (Section 12.1). This intensification of nitrogen cycling, largely by the use of chemical fertilizers, has helped global food supplies stay ahead of population growth, but has also accelerated acidification of the world's cropland soils. Widely used ammonium-based fertilizers, such as ammonium sulfate [$(NH_4)_2SO_4$] and urea [$CO(NH_2)_2$], are oxidized in soil by microbes to produce strong inorganic acids by reactions such as the following:

$$(NH_4)_2SO_4 + 4O_2 \rightleftharpoons 2HNO_3 + H_2SO_4 + 2H_2O \qquad (9.15)$$
$$\text{Nitric acid} \quad \text{Sulfuric acid}$$

These strong acids provide H^+ ions that lower pH; however, since H^+ ions are consumed by the bicarbonate released when plants take up nitrate anions (Reaction [9.10]), net soil acidification results largely from that portion of applied nitrogen that is *not* actually used. Thus soil acidification is often tied to the *excessive* use of nitrogen fertilizer, as popularized in many countries during the past half century (Sections 12.1 and 13.1). For example, as farmers in China increased N fertilizer use to excessive levels during the past three decades, the acidification from ammonium oxidation and subsequent nitrate leaching was reinforced by increased crop yields and removal of nonacid cations (Ca, Mg, and K) (Figure 9.12). The use of nitrogen-fixing legumes in crop rotations and as cover crops can also acidify soils via the production of ammonium nitrogen and a similar imbalance between anion and cation uptake.

Acid-Forming Organic Materials. Application of organic materials such as animal manures, composts, sewage sludge, and some types of plant litter can decrease soil pH, both

---

[4]Rice and Herman (2012) review human-influenced acidification of Earth's air, waters, and soils. Ritchey et al. (2015) provide a clear example of acidification by a legume cover crop.

Figure 9.12 Net H^+ ion production indicates acidification of wheat-maize cropland in China. The main drivers of acidification were nitrogen cycling (mainly ammonium oxidation and leaching of nitrate) and crop uptake and harvest of nonacid cations (mainly Ca, Mg, and K). These acidifying processes were only slightly counterbalanced by H^+ ion consumption with the crop uptake of S and P as anions. [Drawn from data in Guo et al. (2010)]

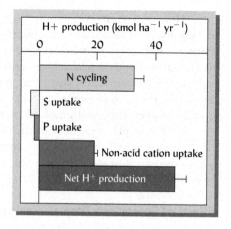

by oxidation of the ammonium nitrogen released and by organic and inorganic acids formed during decomposition. Some types of compost or plant material contain enough nonacid cations to partially or completely replace those lost by acid leaching. Such organic materials may not lower, or may even increase, pH when applied to soils.

Acid Deposition from the Atmosphere

Origins of Acid Rain. Combustion of fossil fuels and the smelting of sulfur-containing metal ores emit enormous quantities of nitrogen and sulfur-containing gases into the atmosphere (Figure 9.18). Other sources of these gases include forest fires and the burning of crop residues. The gases (mainly sulfur dioxide and oxides of nitrogen) react with water and other substances in the atmosphere to form HNO_3 and H_2SO_4. These strong acids are then returned to the Earth in *acid rain*, a popular term which we will use here to include all forms of acidified precipitation: rain, snow, fog, and dry deposition. Normal rainwater that is in equilibrium with atmospheric carbon dioxide has a pH of about 5.5. The pH of acid rain commonly reaches 4.0 or even lower. The greatest amount of acidity currently falls on regions downwind from major coal-fired power plants.

Soil Acidification. The incoming strong acids mobilize aluminum in the soil minerals, and the Al^{3+} displaces Ca^{2+} and other nonacid cations from the exchange complex. The presence of the strong acid anions (SO_4^{2-} and NO_3^-) facilitates the leaching of the displaced Ca^{2+} ions (as explained in Figure 9.3). Soon Al^{3+} and H^+ ions, rather than Ca^{2+} ions, become dominant on the exchange complex, as well as in the soil solution and drainage waters. It is not easy to sort out how much acidification is due to natural processes internal to the soil ecosystem (see left side of Table 9.1) and how much is due to acid rain. Various studies estimate that acid deposition has caused 30–80% of the recently observed acidification in humid regions receiving highly acid rain. The acidity falling on a given hectare in a year is usually not enough to significantly change soil pH in the short term (see Box 9.2); however, the cumulative acid deposition negatively impacts soils, the plants growing in them, and the aquatic ecosystems receiving their drainage waters (Figure 9.13).

Effects on Forests. Some scientists are concerned that trees, which have a high requirement for calcium to synthesize wood, may suffer from insufficient supplies of this and other nutrient cations in acidified soils. Research has shown reductions in Ca^{2+}, Mg^{2+}, and K^+ and declines in pH levels over several decades in forested watersheds subject to acid rain deposits. The leaching of calcium and the mobilization of aluminum may result in Ca/Al ratios (mol_c/mol_c) of less than 1.0 in both the soil solution and on the exchange complex- a ratio widely considered a threshold for aluminum toxicity, reduced calcium uptake, and reduced survival for forest vegetation. The calcium supply in most forested soils in humid temperate regions is being depleted as the rate of calcium loss by leaching, tree uptake, and harvest exceeds the rate of

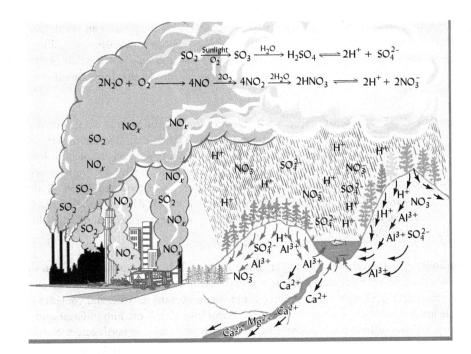

Figure 9.13 *Simplified diagram showing the formation of acid rain in urban areas and its impact on distant watersheds. Combustion of fossil fuels emits sulfur gases and nitrogen gases that are oxidized in the clouds to form sulfuric and nitric acid. These acids then return to Earth in precipitation and in dry deposition. The H^+ cations and NO_3^- and SO_4^{2-} anions accelerate soil acidification, soil aluminum mobilization, and losses of calcium and magnesium. The mobilized aluminum percolates through the soil mantle, eventually reaching lakes and streams. The principal ecological effects in sensitive watersheds are (1) decline in forest health and (2) decline or death of aquatic ecosystems.* (Diagram courtesy of Ray R. Weil)

BOX 9.2
HOW MUCH ACIDITY FALLS IN ACID RAIN?

To use an example typical of the area most impacted by acid rain in North America, consider 1 m² of land in a humid region with 1,000 mm of annual rainfall at pH 4 (i.e., H^+ concentration of 0.0001 mol/L). This area of land would annually receive 1,000 L of rain (1 m³ of water) carrying 0.0001 mol of H^+ ion/L for a total of $1000 \times 0.0001 = 0.1$ mol of H^+ ions per m² of land. Assume that this rain reacts mainly with the top 20 cm of soil (0.2 m³), consisting of 10 cm of organic O_i, O_e horizons underlain by 10 cm of sandy E horizon. In that 1 m², the O horizons might weigh about 50 kg, assuming a bulk density = 0.5 Mg/m³, and the E horizon might weigh about 140 kg assuming a bulk density = 1.4 Mg/m³. The soil, a Spodosol typical of many areas sensitive to acid rain, would be already quite acid, say, pH = 4.0. It might have an effective CEC in the O horizons of about 50 cmol$_c$/kg and in the E horizon of about 4 cmol$_c$/kg. Therefore, in this case, the acidity in the annual rainfall (10 cmol H^+/m²) would be the equivalent of about 0.4% of the O horizons' CEC or about 1.8% of the E horizon's CEC:

Acid input as percentage of O horizon CEC

Acid input: 0.1 mol H^+/50 kg soil
$= 2 \times 10^{-3}$ mol H^+/kg $= 2$ mmol H^+/kg $= 0.2$ cmol$_c H^+$/kg

Percentage of CEC:
$(0.2 \text{ cmol}_c H^+/kg)/(50 \text{ cmol}_c CEC/kg) \times 100 = 0.4\%$

Acid input as percentage of E horizon CEC

Acid input:
0.1 mol H^+/140 kg soil $= 7 \times 10^{-4}$ mol H^+/kg
$= 0.7$ mmol H^+/kg
$= 0.07$ cmol$_c H^+$/kg

Percentage of CEC:
$(0.07 \text{ cmol}_c H^+/kg)/(4 \text{ cmol}_c CEC/kg) \times 100 = 1.8\%$

The degree to which the incoming H^+ ions actually would replace exchangeable nonacid cations would depend on the conditions of cation exchange (see Section 8.8) and on such factors as how much of this acidity could be consumed by weathering Ca and other cations from the soil minerals, etc. Remember that acidification will occur only to the degree that H^+ ion generation exceeds H^+ ion consumption (by the processes listed in Table 9.1).

In our example, a H^+ ion input of the magnitude just considered would cause only a very small change in residual acidity and almost no measurable change in pH from year to year. However, in a more poorly buffered soil (e.g., if some disturbance had caused our Spodosol to lose its O horizons), residual acidity would increase and the pH of the soil solution and drainage water would soon decline significantly.

calcium deposition. In some cases forest ecosystems have responded to applications of calcium compounds (carbonates or silicates) by increased tree nitrogen capture and utilization and improved stream water quality. However, it seems that even in very acid soils low in exchangeable Ca^{2+}, the weathering of soil minerals often releases sufficient calcium for good tree growth—at least in the short term.

Effects on Aquatic Ecosystems. The water draining from acidified soils often contains elevated levels of aluminum, as well as sulfate and nitrate. When this aluminum-charged soil drainage reaches streams and lakes, these bodies of water become lower in calcium, less well-buffered, more acid, and higher in aluminum. The aluminum is directly toxic to fish, and as water pH drops to about 6.0, acid-sensitive organisms in the aquatic food web die off. With a further drop in water pH to about 5.0, virtually all fish are killed. Although the acidified water may look crystal clear (in part due to the flocculating influence of aluminum), the lake is essentially "dead."

Sensitive Soils. Ecological damage from acid rain is most likely to occur where the rain is most acid and the soils are most susceptible to acidification. Soil susceptibility to acidification increases with a soil's acid saturation percentage and decreases with its CEC and content of weatherable minerals. Evidence is mounting for the importance of soil calcium in forest acidification and nitrogen utilization. Sensitive ecosystems with sandy soils in northeastern North America and Northern Europe are recovering slowly from past acid precipitation. Eastern China and Brazil are among those most likely to be damaged in the future if effective steps to control acid rain are not undertaken.

Exposure of Potential Acid Sulfate Materials

Potential Acidity from Reduced Sulfur. A large pool of potential soil acidity may occur in certain soils or sediments that contain reduced sulfur. If drainage, excavation, or other disturbance introduces oxygen into these normally anaerobic soils, oxidation of the sulfur may produce large amounts of acidity. The adjective **sulfidic** is used to describe such materials with enough reduced sulfur to markedly lower the pH within two months of becoming aerated. The term **potential acidity** refers to the acidity that could be produced by such reactions.

Drainage of Certain Coastal Wetlands. Due to the microbial reduction of sulfates originally in seawater, many coastal sediments contain significant quantities of pyrite (FeS_2), iron monosulfides (FeS), and elemental sulfur (S). Coastal wetland areas in the southeastern United States, Southeast Asia, coastal Australia, northern Europe, and West Africa commonly contain soils formed in such sediments. So long as waterlogged conditions prevail, the *potential* acid sulfate soils retain the sulfur and iron in their reduced forms. However, if these soils are drained for agriculture, forestry, or other development, air enters the soil pores and both the sulfur (S^0, S^-, or S^{2-}) and the iron (Fe^{2+}) are oxidized, changing the potential acid sulfate soils into *active* acid sulfate soils. Ultimately, such soils earn their name by producing prodigious quantities of sulfuric acid, resulting in soil pH values below 3.5 and in some cases as low as 2.0. The principal reactions involved are:

$$\underset{\text{Pyrite}}{Fe^{II}S^{-I}_2} + 3\tfrac{1}{2}O_2 + H_2O \rightleftharpoons \underset{\substack{\text{Ferrous}\\\text{sulfate}}}{Fe^{II}S^{VI}O_4} + \underset{\substack{\text{Sulfuric}\\\text{acid}}}{H_2S^{VI}O_4}$$

$$\underset{\substack{\text{Ferrous}\\\text{sulfate}}}{Fe^{II}SO_4} + \tfrac{1}{4}O_2 + 1\tfrac{1}{2}H_2O \rightleftharpoons \underset{\substack{\text{Iron}\\\text{oxyhydroxides}}}{Fe^{III}OOH} + \underset{\substack{\text{Sulfuric}\\\text{acid}}}{H_2SO_4} \quad (9.16)$$

$$\underset{\text{Elemental S}}{S^0} + 1\tfrac{1}{2}O_2 + H_2O \rightarrow \underset{\substack{\text{Sulfuric}\\\text{acid}}}{H_2SO_4} \quad (9.17)$$

Note that both oxidation of the S in pyrite and oxidation and hydrolysis of the Fe^{II} in FeSO$_4$ produce acidity. The oxidation reactions can occur by purely chemical means, but generally they are facilitated by microorganisms that make them proceed thousands of times faster, especially when warm, moist conditions favor microbial activity.

Iron sulfide in potential acid sulfate soils often give these soils a black color (Figure 9.14a,d). The black color has sometimes misled, with disastrous results, those seeking black, organic-matter-rich "topsoil" material for landscaping installations. The pH of the potential acid sulfate soils may be near 7.0 while under reduced conditions, but drops precipitously within days or weeks of the soil being exposed to air. When in doubt, the pH of the soil should be monitored for several weeks while a sample is incubated in a moist, well-aerated, warm condition.

Excavation of Pyrite-Containing Materials. In addition to coastal marsh soils, sediments dredged to deepen coastal harbors may also contain high concentrations of reduced sulfur compounds (Figure 9.14a). Furthermore, many sulfide-rich types of geological sediment long ago became sedimentary rock or parent materials for upland soils. Since coal seams usually occur between layers of sedimentary rock, it is not surprising that coal-mining operations often uncover large quantities of pyrite-containing shale and similar rocks. When these once deeply buried materials are exposed to air and water, the reduced sulfur and iron compounds oxidize to form sulfuric acid in large quantities (Figure 9.14d).

As water percolates through such oxidizing materials, it becomes an extremely acid and toxic brew known as **acid mine drainage**. Typical acid mine drainage has a pH in the range of 0.5–2.0. When this drainage water reaches a stream (as in Figure 9.14c), iron sulfates dissolved in the drainage water continue to produce acid by oxidation and hydrolysis. The aquatic community can be devastated by the pH shock and the iron and aluminum that are mobilized. Similar problems occur when road cuts or building excavations expose buried sulfide-containing layers (Figure 9.14b).

Figure 9.14 Active acid sulfate soils can cause serious ecosystem disturbances when they release some of their potential acidity into the environment. (a) Sulfidic sediments dredged from the bottom of Baltimore harbor used as fill material on land are beginning to dry out and oxidize. The black sulfidic materials turn light brown as they oxidize and release orange-colored acidic drainage water. (b) Similar orange-colored acid is draining across a sidewalk in a new housing development where road grading exposed sulfidic layer of marine sediment in the substratum beneath these coastal plain soils. The area to the right where the drainage originates has failed to grow any vegetation as the soil pH was 3.4. (c) Orange-colored acid water draining from acid sulfate soils exposed by coal mining has killed all life in this mountain stream. (d) A close-up view of the active acid sulfate soil that was the source of the polluting drainage water. Again, the blue-black material is rich in iron sulfide which forms sulfuric acid as it oxidizes to the light brown color in the foreground. (Photo courtesy of Ray R. Weil)

Avoidance as the Best Solution. The amount of sulfur present in such soil material is considered an indication of its **potential acidity**. If calcite is either naturally occurring in the soil or is added as a neutralizing agent, sulfuric acid from S oxidation may react with it to form gypsum:

$$H_2SO_4 + CaCO_3 + H_2O \longrightarrow CaSO_4 \cdot 2H_2O + CO_2 \quad (9.18)$$

Sulfuric acid   Calcite   Gypsum   Carbon dioxide

Equation (9.18) indicates that 1 mole of liming material ($CaCO_3$) would be required to eventually neutralize the sulfuric acid produced by the oxidation of 1 mole of reduced S. Since 1 mole of $CaCO_3$ equals 100 g and 1 mole of S equals 32 g, about 3 kg of limestone would be needed to neutralize the acidity from the oxidation of 1 kg of S. The enormous amounts of $CaCO_3$ required often make neutralization impractical in the field. Usually the best approach to solving this environmental challenge is to *prevent* the S oxidation in the first place. This means that sulfide-bearing wetland soils are best maintained under saturate conditions. In the case of mining or other excavation, any sulfide-bearing materials exposed must be identified and eventually deeply reburied to prevent their oxidation.

9.7 HOW DOES SOIL pH AFFECT LIVING THINGS?

The pH of the soil solution exerts a critical influence on all the plants, animals, and microbes that live in the soil. To nonadapted plants, strongly acid soil presents a host of problems. These include toxicities of aluminum, manganese, and hydrogen, as well as deficiencies of calcium, magnesium, molybdenum, and phosphorus. As it is difficult to separate one problem from another, the situation is sometimes simply referred to as the acid soil "headache."

Aluminum Toxicity[5]

Aluminum toxicity stands out as the most common and severe problem associated with acid soils. Not only plants are affected; many bacteria, such as those that carry out transformations in the nitrogen cycle, are also adversely impacted by the high levels of Al^{3+} and $AlOH^{2+}$ that come into solution at low soil pH. As can be deduced from Figure 9.15, aluminum toxicity is rarely a problem when the soil pH is above about 5.2 (above pH_{CaCl} 4.8) because little aluminum exists in the solution or exchangeable pools above this pH level. Figure 9.15 (*left*) shows an exponential increase in solution Al^{3+} concentration typical of mineral soils as pH level drops from 5 to 4. At comparable pH levels in most organic soil horizons (O horizons) and organic soils (Histosols), much less Al^{3+} comes into solution because there is far less total aluminum in these soils.

Effects on Plants. When aluminum, which is not a plant nutrient, is taken into the root, most remains there and little is translocated to the shoot (except in aluminum accumulator plants such as tea, which may contain as much as 5,000 ppm in the dry leaves). Therefore, analysis of leaf tissue is rarely a good diagnostic technique for aluminum toxicity. In the root, aluminum damages membrane sites where calcium is normally taken in and restricts cell wall expansion so roots cannot grow properly (Figure 9.16). Aluminum also interferes with the metabolism of phosphorus-containing compounds essential for energy transfers (ATP), genetic coding (DNA), and seed germination (Figure 9.15, *right*).

Symptoms. The most common symptom of aluminum toxicity is a stunted root system with short, thick, stubby roots that show little branching or growth of laterals (Figure 9.16). The root tips and lateral roots often turn brown. In some plants, the leaves may show chlorotic (yellowish) spots. Because of the restricted root system, plants suffering from aluminum toxicity often show symptoms of drought stress and phosphorus deficiency (stunted growth, dark green foliage, and purplish stems).

---

[5]Several good reviews explain the soil chemistry of aluminum toxicity and the mechanisms of plant tolerance to aluminum (Hoekenga and Magalhaes, 2011; Ryan and Delhaize, 2012).

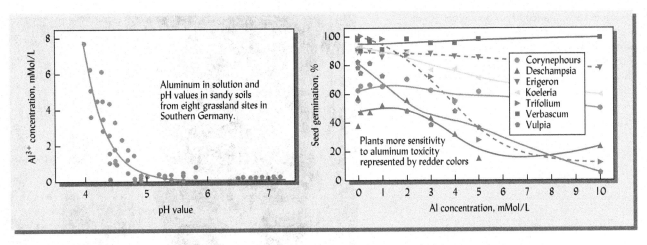

Figure 9.15 Low soil pH levels bring toxic aluminum into solution and influence plant ecology. In a study of eight grassland sites with sandy soils in southern Germany, the soil pH level was closely related to the amount of Al^{3+} in the soil solution (left). Depending on the plant's genetic adaptation, Al^{3+} concentration typical in mineral soils with H levels below 5.0 can cause dramatically stunted, dysfunctional root systems and other toxic effects, including inhibition of seed germination (right). Some plant species (green lines) were little affected by even high levels of Al^{3+} and therefore tended to dominate the most acid soils. Seed germination for the most aluminum-sensitive plants (red lines) declined dramatically when Al^{3+} concentrations exceeded 1 mM. [Graphed from selected data in Abedi et al. (2013)]

Tolerance. Among and within plant species there exists a great deal of genetic variability in sensitivity to aluminum toxicity. Major differences among plants with regard to the sensitivity to aluminum toxicity of such functions as root growth and seed germination (Figure 9.15, *right*) can result in distinctly different natural plant communities flourishing on sites that are nearby but whose soils exhibit different soil pH levels. Agricultural crops that originated in areas dominated by acid soils (such as most humid regions) tend to be less sensitive than species originating in areas of neutral to alkaline soils (such as the Mediterranean region). Fortunately, plant breeders have been able to find genes that confer tolerance to aluminum even in species that are typically sensitive to this toxicity (Figure 9.16). Most Al-tolerant plants avoid the problem by excluding aluminum from their roots. To do this, some species raise the pH of the soil just outside the root, causing the aluminum to precipitate. Others excrete organic mucilage that complexes with the aluminum, preventing its uptake into the root. Still others produce certain organic acids that combine with the aluminum to form nontoxic compounds (see Section 9.9).

Manganese, Hydrogen, and Iron Toxicity to Plants

Manganese Toxicity. Although not as widespread as aluminum toxicity, **manganese toxicity** is a serious problem for plants in acid soils derived from manganese-rich parent minerals. Unlike Al, Mn is an essential plant nutrient (see Section 12.8) that is toxic only when taken up in excessive quantities. Manganese becomes increasingly soluble as pH drops, with toxicity possible below pH_{water} 5.6 (about 0.5 units higher than for aluminum). The leaf tissue content of Mn usually correlates with toxicity beginning at levels that range from 200 mg/kg in sensitive plants to over 5,000 mg/kg in tolerant plants (Figure 9.17).

Since the reduced form [Mn(II)] is far more soluble than the oxidized form [Mn(IV)], toxicity is greatly increased by low oxygen conditions associated with a combination of oxygen-demanding, decomposable organic matter and wet conditions. Manganese toxicity is also common in certain high-organic-matter surface horizons of volcanic soils (e.g., Melanudands). Unlike for Al, the solubility and toxicity of Mn are commonly accentuated, rather than restricted, by higher soil organic matter (Table 9.2).

Figure 9.16 *Influence of pH on the growth of shoots (a) and roots (b) from two wheat varieties, one sensitive and one tolerant to aluminum. Note the stunted shoot growth and extremely stunted, stubby root system of the sensitive variety in the low-pH treatment.* (Photos courtesy of C. D. Foy, USDA/ARS, Beltsville, MD)

Figure 9.17 *Plant responses to toxicity of manganese at low soil pH. Plant shoot growth (the average of bean and cabbage) declined and Mn content of foliage increased at low pH levels in manganese-rich soils from East Africa (average data for an Andisol and an Alfisol).* [Redrawn from data in Weil (2000)]

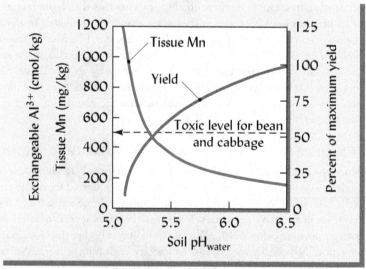

Nutrient Availability to Plants

Figure 9.18 shows in general terms the relationship between the pH of mineral soils and the availability of plant nutrients and aluminum. Note that in strongly acid soils, the availability of the macronutrients (Ca, Mg, K, P, N, and S) as well as the two micronutrients, Mo and B, is curtailed. In contrast, availability of the micronutrient cations (Fe, Mn, Zn, Cu, and Co) is increased by low soil pH, even to the extent of toxicity.

Table 9.2
RELATIONSHIP BETWEEN SOIL PROPERTIES ASSOCIATED WITH ACIDITY AND SURVIVAL OF SUGAR MAPLE SEEDLINGS

Data are averages for 18 forested sites dominated by overstory sugar maples. Al was most abundant in the low-organic-matter B horizons and Mn most abundant in the high-organic-matter O horizons. The critical ratio associated with tree mortality of Ca/Al <1.0 occurred only in the B horizons, while the critical ratio of Ca/Mn <30 occurred in all horizons and was especially low in the highly organic O horizons.

Seedlings present?	Exchangeable cations, mg/kg			Ratio, molc/molc		Soil pH$_{water}$
	Mn	Ca	Al	Ca/Al	Ca/Mn	
O Horizons						
No	188	2738	53	23.1	20	4.02
Yes	89	6371	38	74.1	98	4.45
A Horizons						
No	59	1252	143	3.9	28	4.34
Yes	33	2755	142	8.5	114	4.58
B Horizons						
No	15	305	279	0.5	28	4.62
Yes	8	1061	202	2.3	180	4.90

[Data from Demchik et al. (1999b)] Ca/Al and Ca/Mn ratios calculated here to give units shown.

In slightly to moderately alkaline soils, molybdenum and all of the macronutrients (except phosphorus) are amply available, but Fe, Mn, Zn, Cu, and Co are so unavailable that plant growth is constrained. Because phosphorus ends to be tied up with calcium, its availability is likewise reduced in alkaline soil. Boron deficiency is common at high pH levels in both sandy soils (because of low boron content) and clayey soils (because the boron is tightly held by the clay). On the other hand, since alkaline arid region soils do not lose boron by leaching as humid region acid soils do, boron toxicity is also a problem in alkaline soils, especially where irrigation water containing boron is used.

It appears from Figure 9.18 that the pH range of 5.5–7.0 may provide the most satisfactory plant nutrient levels overall. However, this generalization may not be valid for all soil and plant combinations. For example, certain micronutrient deficiencies are common in some plants when sandy mineral soils are limed to pH values of only 6.5–7.0. Also, in organic-rich soils such as peats, forest O horizons, and many potting mixes, the pH for optimal nutrient availability is about 1 full unit lower than shown in Figure 9.18 and aluminum toxicity is not a problem.

Microbial Effects

Fungi are particularly versatile, flourishing satisfactorily over a wide pH range. Fungal activity tends to predominate in low pH soils because bacteria are strong competitors and tend to dominate the microbial activity at intermediate and higher pH. Individual microbial species exhibit pH optima that may differ from this generality, but bacterial communities tend to be less active and less diverse in strongly acid soils (Figure 9.19). Manipulation of soil pH can alter microbial communities and help control certain soil-borne plant diseases (see Section 10.13).

Optimal pH Conditions for Plant Growth

Plants vary considerably in their tolerance to acid and/or alkaline conditions (Figure 9.20). For example, certain legume crops such as alfalfa and sweet clover grow best in near-neutral or alkaline soils, and most humid-region mineral soils must be limed to grow these crops

Figure 9.18 *Relationships existing in mineral soils between pH and the availability of plant nutrients. The relationship with activity of certain microorganisms is also indicated. The width of the bands indicates the relative microbial activity or nutrient availability. The jagged lines between the P band and the bands for Ca, Al, and Fe represent the effect of these metals in restraining the availability of P. When the correlations are considered as a whole, a pH range of about 5.5 to perhaps 7.0 seems to be best to promote the availability of plant nutrients. In short, if the soil pH is suitably adjusted for phosphorus, the other plant nutrients, if present in adequate amounts, will be satisfactorily available in most mineral soils. In organic-rich soils such as peats, forest O horizons and many potting mixes, the pH for optimal nutrient availability is about 1 full unit lower and Al is not a problem.* (Diagram courtesy of N. C. Brady and Ray R. Weil)

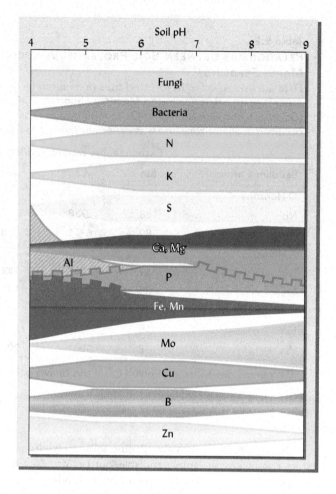

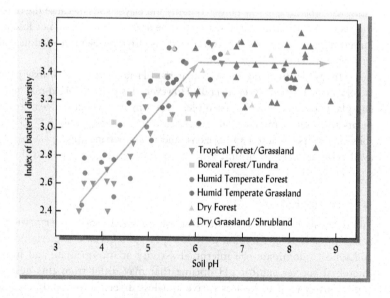

Figure 9.19 *Soil pH greatly influences the diversity of bacteria. In 98 different ecosystems in North and South America, ribosomal DNA fingerprinting was used to estimate the diversity of the bacterial communities. The index of diversity was high in soils with pH_{water} above 6, but was much reduced by more acid soil conditions. Fungal diversity was not studied here, but could be expected to show the opposite trend.* [Modified from Fierer and Jackson (2006)]

satisfactorily. Asparagus and cantaloupe are two food crops that have a high calcium requirement and grow best at pH levels near or above 7.0.

Because forests exist mainly in humid regions where acid soils predominate, many forest plants are at the opposite end of the scale. Forest species such as rhododendrons, azaleas, blueberries, larch, some oaks, and most pines are inefficient in taking up the iron they need. Since high soil pH and high calcium saturation reduce the availability of iron, these plants will show

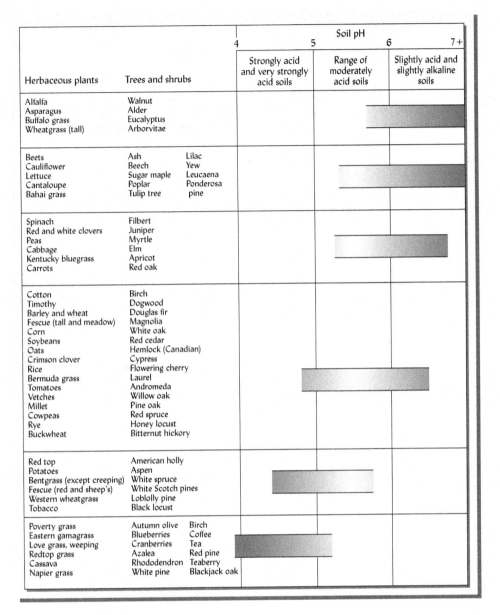

Figure 9.20 Ranges of pH in mineral soils that present appropriate conditions for optimal growth of various plants. Note that the pH ranges are quite broad, but that plant requirement for calcium and sensitivity to aluminum toxicity generally decrease from the top group to the bottom group. Other factors, such as fertility level, will influence the actual relationships between plant growth and soil pH in specific cases. However, this chart may help in choosing which species to plant or when deciding whether to change the pH with soil amendments.

iron deficiency **chlorosis** (yellowing of the interveinal part of the newest leaves) under these soil conditions (see Section 12.8). Even forest trees differ in their tolerance of soil acidity and high aluminum (Table 9.3). For example, elm, poplar, honey locust, and the tropical legume tree *Leucaena* are known to be less tolerant to acid soils than are many other forest species.

Most cultivated crop plants (except those such as sweet potato, cassava, and others that originated in the humid tropics) grow well on soils that are moderately acid to near neutral. Therefore, for most grain and vegetable crops, a soil pH in the range of perhaps 5.5–7.0 is most suitable.

Soil pH and Organic Molecules

Soil pH influences environmental quality in many ways, but we will discuss only one example here—the influence of pH on the mobility of ionic organic molecules in soils. The molecular structure of certain ionic herbicides includes such chemical groups as $-NH_2$ and $-COO^-$. If soil pH is low, the excess H^+ ions (protons) present in solution are attracted to and bond with these chemical groups, neutralizing negative charges and creating positively charged sites on the molecule. This process is called *protonation*.

Atrazine, a chemical that is widely used to control weeds in corn fields, is an example of a chemical whose mobility is greatly influenced by soil pH. In a low pH environment, the atrazine molecule protonates, developing a positive charge. The positively charged

Table 9.3
FOREST SPECIES RESPONSE TO CALCIUM OXIDE ON AN EXTREMELY ACID SOIL

Adding 1 g of CaO to 180 g of this soil (a Dystrudept in Pennsylvania) raised the pH from 3.8 to 6.8 and the Ca/Al ratio from 0.24 to 15.5. The treated soil was overlimed with respect to such acid-loving species as Teaberry, White pine, and Chestnut oak which were among trees showing reduced growth at the higher pH.

Positive response to CaO (Al-sensitive)		Negative or no response to CaO	
Plant species	Root response to CaO addition, %	Plant species	Root response to CaO addition, %
Sugar maple	+150	Black locust	0
Pin oak	+104	Mountain laurel	−15
Quaking aspen	+83	Blueberry	−21
Bitternut hickory	+52	Norway spruce	−31
Red cedar	+46	Black birch	−47
Grey dogwood	+41	Chestnut oak	−52
White spruce	+40	White pine	−53
Honey locust	+17	Teaberry	−68

[Data selected from Demchik et al. (1999a)]

molecule is then adsorbed on the negatively charged soil colloids, where it is held until it can be decomposed by soil organisms (see also Figures 8.24–8.25). The adsorbed pesticide is less likely to move downward and into the groundwater. At pH values above 5.7, however, the adsorption is greatly reduced, and the tendency for the herbicide to move downward in the soil is increased. Of course, the adsorption in acidic soils also reduces the availability of atrazine to weed roots, thus reducing its effectiveness as a weed killer. Farmers using such chemicals may notice excessive weediness in their fields as the first sign that their soils need liming.

9.8 RAISING SOIL pH BY LIMING
Agricultural Liming Materials

To decrease soil acidity (raise the pH), the soil is usually amended with alkaline materials that provide conjugate bases of weak acids. Examples of such conjugate bases include carbonate (CO_3^{2-}), hydroxide (OH^-), and silicate (SiO_3^{2-}). These conjugate bases are anions that are capable of consuming (reacting with) H^+ ions to form weak acids (such as water). For example:

$$CO_3^{2-} + 2H^+ \longrightarrow CO_2 + H_2O \tag{9.19}$$

Most commonly, these bases are supplied in their calcium or magnesium forms ($CaCO_3$, etc.) and are referred to as **agricultural limes**. Some liming materials contain oxides or hydroxides of alkaline earth metals (e.g., CaO or MgO), which form hydroxide ions in water.

$$CaO + H_2O \longrightarrow Ca(OH)_2 \longrightarrow Ca^{2+} + 2OH^- \tag{9.20}$$

Unlike fertilizers, which are used to supply plant nutrients in relatively small amounts for plant nutrition, *liming materials are used to change the chemical makeup of a substantial part of the root zone.* Therefore, lime must be added in large enough quantities to chemically react with a large volume of soil. This requirement dictates that inexpensive, plentiful materials are normally used for liming soils—most commonly finely ground limestone or materials derived from it (see Table 9.4). Wood ashes (which contain oxides of Ca, Mg, and K) are also effective. As a practical matter, the choice of which liming material to use is often based mainly on the relative costs of transporting large amounts of these materials from their source to the site to be limed.

Table 9.4
COMMON LIMING MATERIALS: THEIR COMPOSITION AND USE

The two limestones are by far the most commonly used. Use of the other materials is largely dependent on the need for fast reaction, cost, and local availability. The relative amounts needed of the different materials can be judged by comparing the respective CaCO₃ equivalent values.

Common name of liming material	Chemical formula (of pure materials)	% CaCO₃ equivalent	Comments on manufacture and use
Calcitic limestone	$CaCO_3$	100	Natural rock ground to a fine powder. Low solubility; may be stored outdoors uncovered. Noncaustic, slow to react.
Dolomitic limestone	$CaMg(CO_3)_2$	95–108	Natural rock ground to a fine powder; somewhat slower reacting than calcitic limestone. Supplies Mg to plants.
Burned lime (oxide of lime)	CaO (+ MgO if made from dolomitic limestone)	178	Caustic, difficult to handle, fast-acting, can burn foliage, expensive. Made by heating limestone. Protect from moisture.
Hydrated lime (hydroxide of lime)	$Ca(OH)_2$ (+ $Mg(OH)_2$ if made from dolomitic limestone)	134	Caustic and difficult to handle. Fast-acting, can burn foliage, expensive. Made by slaking hot CaO with water. Protect from moisture.
Basic slag	$CaSiO_3$	70–90	By-product of pig-iron industry. Must be finely ground. Also contains 1–7% P.
Wood ashes	CaO, MgO, K_2O, $K(OH)$, etc.	40–80	Caustic, fast-acting, water-soluble so must be kept dry during storage.

Dolomitic limestone products should be used if soil magnesium levels are low. In some highly weathered soils, small amounts of lime may improve plant growth, more because of the enhanced calcium or magnesium nutrition than from a change in pH.

How Do Liming Materials React to Raise Soil pH?

Most liming materials—whether they be oxide, hydroxide, or carbonate—react with carbon dioxide and water to yield bicarbonate when applied to an acid soil. The carbon dioxide partial pressure in the soil, usually several hundred times greater than that in atmospheric air, is generally high enough to drive such reactions to the right. For example:

$$CaMg(CO_3)_2 + 2H_2O + 2CO_2 \rightleftharpoons Ca + 2HCO_3^- + Mg + 2HCO_3^- \quad (9.21)$$

Dolomitic limestone; Bicarbonate; Bicarbonate

The Ca and Mg bicarbonates are much more soluble than are the carbonates, so the bicarbonate formed is quite reactive with the exchangeable and residual soil acidity. The Ca^{2+} and Mg^{2+} replace H^+ and Al^{3+} on the colloidal complex:

$$\left[\text{Clay or humus}\right]\begin{matrix}H^+\\Al^{3+}\end{matrix} + 2Ca^{2+} + 4HCO_3^- \rightleftharpoons \left[\text{Clay or humus}\right]\begin{matrix}Ca^{2+}\\Ca^{2+}\end{matrix} + Al(OH)_3 + H_2O + 4CO_2\uparrow \quad (9.22)$$

(Bicarbonate; solid)

$$\left[\text{Clay or humus}\right]\begin{matrix}H^+\\Al^{3+}\end{matrix} + 2CaCO_3 + H_2O \rightleftharpoons \left[\text{Clay or humus}\right]\begin{matrix}Ca^{2+}\\Ca^{2+}\end{matrix} + Al(OH)_3 + 2CO_2\uparrow \quad (9.23)$$

(Limestone (calcite); solid)

The insolubility of Al(OH)$_3$, the weak dissociation of water, and the release of CO$_2$ gas to the atmosphere all pull these reactions to the right. In addition, the adsorption of the calcium and magnesium ions lowers the percentage acid saturation of the colloidal complex, and the pH of the soil solution increases correspondingly.

Greenhouse Gas Emissions. The reactions just discussed also point out the important fact that much of the carbon in limestone applied to acid soils will eventually end up in the atmosphere as CO$_2$. As a result, agricultural liming contributes significantly to the carbon dioxide emissions responsible for anthropogenic climate change. For example, the 30 Tg (teragrams, 10^{12} g) of agricultural limestone used annually in the United States is thought to release between 6 and 12 Tg of CO$_2$.

Silicates can be used as liming materials that do not contain carbon and therefore do not release CO$_2$ into the atmosphere when they react with acid soils. The most commonly used silicates are calcium silicate from blast furnace slag (a by-product of steel making) and naturally occurring minerals such as Wollastonite (see Section 12.7). The calcium silicates increase soil pH by consuming H$^+$ ions and replacing acid cations with Ca^{2+} ions on the exchange complex:

$$CaSiO_3 + H_2O + Colloid\text{-}H^+ \rightarrow Colloid\text{-}Ca^{2+} + H_4SiO_4 \tag{9.24}$$

Lime Requirement: How Much Lime Is Needed to Do the Job?

The amount of liming material required to ameliorate acid soil conditions is determined by several factors, including: (1) the change required in the pH or exchangeable Al saturation, (2) the buffer capacity of the soil, (3) the amount or depth of soil to ameliorate, (4) the chemical composition of the liming materials to be used, and (5) the fineness of the liming material. Because of greater buffering capacity, the lime requirement for a clay loam is much higher than that of a sandy loam with the same pH value (see Figure 9.21 and Section 9.4).

Within the pH range of 4.5–7.0, the pH change induced by adding lime to an acid soil is determined by the buffering capacity of the particular soil (Figure 9.7). In turn, the capacity to buffer pH is reflected in the CEC and closely related to organic matter and clay contents. Therefore, a titration curve (such as the two examples shown in Figure 9.7) make it possible to estimate approximate amount of liming material needed to raise the soil pH to a desired (see Box 9.3).

Exchangeable Aluminum. Liming to eliminate exchangeable aluminum, rather than to achieve a certain soil pH, has been found appropriate for highly weathered soils such as Ultisols and Oxisols. By this approach, the required amount of lime can be calculated using values for the initial CEC and the percent Al saturation. For example, if a soil has a

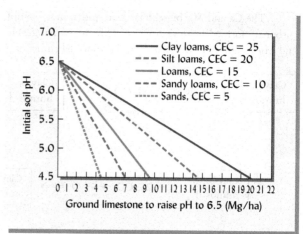

Figure 9.21 Effect of soil textural class on the amount of limestone required to raise the pH of soils from their initial level to pH 6.5. Note the very high amounts of lime needed for fine-textured soils that are strongly buffered by their high levels of clay, organic matter, and CEC. The chart is most applicable to soils in cool, humid regions where 2:1 clays predominate. In warmer regions where organic matter levels are lower and clays provide less CEC, the target pH would likely be closer to 5.8 and the amounts of lime required would be one-half to one-third of those indicated here. In any case, it is unwise to apply more than 7–9 Mg/ha (3–4 tons/acre) of liming materials in a single application. If more is needed, subsequent applications can be made at 2- to 3-year intervals until the desired pH is achieved.

BOX 9.3
CALCULATING LIME NEEDS BASED ON pH BUFFERING

Your client wants to grow a high-value crop of asparagus in a 2-ha field. The soil is a sandy loam with a current pH of 5.0. Asparagus is a calcium-loving crop that requires a high pH for best production (see Figure 9.26), so you recommend that the soil pH be raised to 6.8. Since the soil texture is a sandy loam, we will assume that its buffer curve is similar to that of the moderately buffered soil B in Figure 9.7. In actual practice, a soil test laboratory using this method to calculate the lime requirement should have buffer curves for the major types of soils in its service area.

1. Extrapolating from curve B in Figure 9.7, we estimate that it will require about 2.5 $cmol_c$ of lime/kg of soil to change the soil pH from 5.0 to 6.8. (Draw a horizontal line from 5.0 on the y-axis of Figure 9.7 to curve B, then draw a vertical line from this intersection down to the x-axis. Repeat this procedure beginning at 6.8 on the y-axis. Then use the x-axis scale to compare the distance between where your two vertical lines intersect the x-axis.)

2. Each molecule of $CaCO_3$ neutralizes 2 H^+ ions:

$$CaCO_3 + 2H^+ \rightarrow \rightarrow Ca^{2+} + CO_2 + H_2O$$

3. The mass of 2.5 $cmol_c$ of pure $CaCO_3$ can be calculated using the molecular weight of $CaCO_3 = 100$ g/mol:

$(2.5\ cmol_c CaCO_3/\text{kg soil}) \times (100\ \text{g/mol}\ CaCO_3)$
$\times (1\ \text{mol}\ CaCO_3/2\ mol_c) \times (0.01\ mol_c/cmol_c)$
$= 1.25\ \text{g}\ CaCO_3/\text{kg soil}$

4. Using the conversion factor of 2×10^6 kg/ha of surface soil (see footnote 8 in Chapter 4), we calculate the amount of pure $CaCO_3$ needed per hectare:

$(1.25\ \text{g}\ CaCO_3/\text{kg soil}) \times (2 \times 10^6\ \text{kg soil/ha})$
$= 2,500,000\ \text{g}\ CaCO_3/\text{ha}$

$(2,500,000\ \text{g}\ CaCO_3/\text{ha}) \times (1\ \text{kg}\ CaCO_3/1000\ \text{g}\ CaCO_3)$
$= 2,500\ \text{kg}\ CaCO_3/\text{ha}$

$2,500\ \text{kg}\ CaCO_3/\text{ha}$
$= 2.5\ \text{Mg/ha}$ (or about 1.1 tons/acre)

5. Since our limestone has a $CaCO_3$ equivalence of 90%, 100 kg of our limestone would be the equivalent of 90 kg of pure $CaCO_3$. Consequently, we must adjust the amount of our limestone needed by a factor of 100/90:

$2.5\ \text{Mg pure}\ CaCO_3 \times 100/90 = 2.8\ \text{Mg limestone/ha}$

6. Finally, because not all the $CaCO_3$ in the limestone will completely react with the soil, the amount calculated from the laboratory buffer curve is usually increased by a factor of 2:

$(2.8\ \text{Mg limestone/ha}) \times 2 = 5.6\ \text{Mg limestone/ha}$

(using Appendix B, this value can be converted to about 2.5 tons/acre)

Note that this result is very similar to the amount of lime indicated by the chart in Figure 9.21 for this degree of pH change in a sandy loam:

$5\ cmol_c/\text{kg} \times (100\ \text{g/mol}\ CaCO_3)$
$\times (1\ \text{mol}\ CaCO_3/2\ mol_c) \times (0.01\ mol_c/cmol_c)$
$= 2.5\ \text{g}\ CaCO_3/\text{kg soil}.$

This amount is equivalent to 5,000 kg/ha (2.5 g/kg × 2×10^6 kg/ha). Experience suggests that to assure a complete reaction in the field, the amount of limestone so calculated must be multiplied by a factor of 1.5 or 2.0 to give the actual amount of lime to apply.

CEC of 10 $cmol_c$/kg and is 50% Al-saturated, then 5 $cmol_c$/kg of Al^{3+} ions must be displaced (and their acidity from Al hydrolysis neutralized). This would require 5 $cmol_c$/kg of $CaCO_3$:

Influence of Lime Composition, Fineness, and Depth of Incorporation. Liming materials are usually applied at rates based on pure CaO or $CaCO_3$. The general values in Table 9.4 suggest, for example, that a recommendation for 1.0 Mg of $CaCO_3$ equivalent could be met by using either 0.56 Mg of burned lime (1.0/1.78 = 0.56) or 1.43 Mg of basic slag (1.0/0.7 = 1.43). In addition, for limestone, materials should be ground finely so at least 50% of the particles can pass through a 60-mesh screen (smaller than 0.25 mm in diameter). Coarsely ground limestone reacts so slowly with the soil as to have little ameliorating effect on soil acidity. Finally, the recommendation must be adjusted if the depth of incorporation is other than that assumed or specified. In the example just given, if the recommendation was for incorporation to 18 cm, but you plan to incorporate to only 9 cm, then half as much liming material should be used.

How Lime Is Applied

Frequency. Liming materials slowly react with soil acidity, gradually raising the pH to the desired level over a period ranging from a few weeks in the case of hydrated lime to a year or so with finely ground limestone. In humid regions acidification proceeds relentlessly so application of lime to arable soils is not a one-time proposition, but must be repeated every 3–5 years.

Depth of Incorporation. Liming will be most beneficial to acid-sensitive plants if as much as possible of the root environment is altered. The Ca^{2+} and Mg^{2+} ions provided by limestone replace acid cations on the exchange complex and do not move readily down the profile. Therefore, for soil with a high CEC, the short-term effects of limestone are mainly limited to the soil layer into which the material was incorporated, usually the upper 15–20 cm.

Since subsoils are low in organic matter, subsoil acidity is often accompanied by Al toxicity causing plant roots to be restricted to the upper, limed surface soil layer. The effects of subsoil acidity will be especially detrimental during dry periods when the plants would most benefit if their roots had access to the huge volume of water stored in deeper soil horizons.

Overliming. Another practical consideration is the danger of **overliming**; it is easy to add a little more lime later, but quite difficult to counteract the results of applying too much. The detrimental results of excess lime include deficiencies of iron, manganese, copper, and zinc; reduced availability of phosphate; and constraints on the plant absorption of boron from the soil solution. Overliming is not very common on fine-textured soils with high buffer capacities, but it can occur easily on low organic matter coarse-textured soils. Therefore, liming materials should be added conservatively to poorly buffered soils. For some Ultisols and Oxisols, overliming may occur if the pH is raised even to 6.0.

Special Liming Situations

Liming Forests. It is rarely practical to spreading limestone on forested watersheds (see Section 9.6) as these areas are not accessible to ground-based spreaders, and manual application is too tedious and expensive. In some cases, landowners use helicopters for aerial spreading of small quantities of liming material. For very acid sandy soils, such small applications can ameliorate soil acidity and provide sufficient calcium for the trees.

Untilled Soils. Some soil–plant systems, such as no-till farming, make it difficult to incorporate and mix limestone with the soil. Fortunately, the undisturbed residue mulches in these systems tend to encourage earthworm activity, which, as we have just mentioned, can help distribute lime down the profile. Although apparently not critical on low CEC soils (Figure 9.22), it is considered advisable to incorporate a generous application of limestone as deeply as possible to correct any root-limiting acidity before commencing a no-till system on an acid high-CEC soil. Subsequent surface applications of limestone should adequately combat the acidity, which forms near the surface due to the decomposition of residues and the application of nitrogen fertilizers.

In lawns, golf greens, and other turfgrass areas, it is also not possible to till the needed limestone into the soil. Again, the time to correct any problems with soil acidity is during soil preparation for the initial establishment of the grass. By proper timing of future liming applications with annual aeration tillage operations that leave openings down into the soil (see Figure 7.13), some downward movement of the lime can be achieved. In untilled systems, which require surface application of lime, frequent application of small quantities is most effective. In orchards, incorporation of lime into the planting hole, followed by periodic surface applications, is recommended for acid soils.

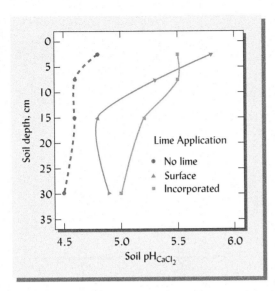

Figure 9.22 Soil pH 5 years after lime was applied to the soil surface or incorporated in preparation for converting grass pasture to no-till cropland. Dolomitic limestone was applied at 4.5 Mg/ha, enough to eliminate exchangeable Al^{3+} in the upper 20 cm of this clayey Oxisol in Brazil. Four months before the first crop was sown, three treatments were established: (1) no lime, (2) lime applied on the soil surface, and (3) lime incorporated into the soil, half plowed down to 20 cm and the remainder harrowed into the upper 10 cm. No other tillage was used during the 5 years of cropping. The incorporated lime raised the pH significantly down to 40 cm, but most markedly in the 20 cm zone of incorporation. The surface-applied lime also raised the pH to a considerable depth, thanks to the earthworm activity and biopore leaching characteristic of pasture soils. Exchangeable Al was negligible at $pH_{CaCl_2} > 4.8$. The cumulative crop yield over 5 years was higher where lime was applied, but was not affected by the method of application. Profitability was greatest for the surface-applied lime because the cost of tillage to incorporate lime was nearly equal to the cost of the lime itself. [Graphed from data in Caires et al. (2006)]

9.9 AMELIORATING ACIDITY WITHOUT LIME

Where subsoil acidity is a problem or where liming materials are not affordable, several approaches to combating the negative effects of soil acidity may be appropriate for use with or without traditional liming. Particularly deserving of attention are the use of gypsum and organic materials to reduce aluminum toxicity and the use of plant species or genotypes that tolerate acid conditions. Also keep in mind that it may be easier solve soil acidity problems by choosing a more acid tolerant plant rather than by trying to change the soil chemistry.

Using Gypsum

Gypsum ($CaSO_4 \cdot 2H_2O$) is a widely available material found in natural deposits or as an industrial by-product from the manufacture of high-analysis fertilizers, from flue-gas desulfurization by air-pollution "scrubbers" and in dry-wallboard wastes from construction or demolition of buildings. Gypsum can ameliorate aluminum toxicity despite the fact that it does not increase soil pH. In fact, gypsum has been found more effective than lime in reducing exchangeable aluminum in subsoils and thereby in improving root growth and crop yields (see Figure 9.23).

Using Organic Matter

Practices such as the application of organic wastes, production of cover crops (see Section 13.2), addition of organic mulch, return of crop residues, and protection of the surface soil from erosion help maintain high levels of organic matter in soil. In so doing they can ameliorate the effects of soil acidity in at least three ways:

1. High-molecular-weight organic matter can bind tightly with aluminum ions and prevent them from reaching toxic concentrations in the soil solution.
2. Low-molecular-weight organic acids produced by microbial decomposition or root exudation can form soluble complexes with aluminum ions that are nontoxic to plants and microbes. Table 9.5 provides an example of the ameliorating effect of adding decomposable organic material.
3. Many organic amendments contain substantial amounts of calcium held in organic complexes that leach quite readily down the soil profile. Therefore, if such amendments as legume residues, animal manure, or sewage sludge are high in Ca, they can effectively combat aluminum toxicity and raise Ca and pH levels, not only in the surface soil where they are incorporated, but also quite deep into the subsoil (see Figure 9.23).

Figure 9.23 *Aluminum saturation percentage in the subsoil of a fine-textured Hawaiian Ultisol after treatment of the surface soil with chicken manure, limestone, or gypsum. The soil was slowly leached with the 380 mm of water. The lime raised the pH and thereby effectively reduced Al saturation, though only in the upper 10–20 cm. The effect of gypsum extended somewhat more deeply. Gypsum does not raise the pH or increase the CEC, but is more soluble than lime and provides the SO_4^{2-} anions to accompany Ca^{2+} cations as they leach downward in solution. The greatest and deepest reduction in Al saturation resulted from the chicken manure. Another Ca-rich organic amendment, sewage sludge, gave similar results (not shown). The manure probably formed soluble organic complexes with Ca^{2+} ions, which then moved down the profile where Ca exchanged with Al to form nontoxic organic Al complexes.* [Redrawn from Hue and Licudine (1999); used with permission of the American Society of Agronomy]

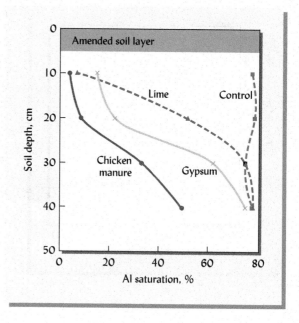

Table 9.5
EFFECTS OF ORGANIC RESIDUES AND LIME ON SOIL ACIDITY, SOIL ALUMINUM, SOLUBLE CARBON, AND THE GROWTH OF A LEGUME FORAGE PLANT, *DESMODIUM INTORTUM* IN ACID TROPICAL SOILS

Note that the organic material (cowpea leaves) raised the pH only 0.5 unit, but caused a dramatic reduction in the various forms of aluminum in the soil and soil solution, comparable to that caused by the slaked lime. The ash from cowpea leaves was less effective in reducing the aluminum. Only the cowpea leaves raised the amount of oxidizable carbon in solution, which in turn complexed the Al leaving much less in solution than otherwise expected at pH 4.9. The ash and lime each doubled the Desmodium yield, but the cowpea leaves tripled it. The data are means of two soils from Hawaii, an Andisol and an Ultisol.

Soil amendment type	Soil amendment amount[a]	Soil pH	Exchangeable Al	Σ Inorganic Al ions in solution	Al^{3+} in solution	Oxidizable carbon in solution	Desmodium shoot dry weight
			$cmol_c$/kg soil	— μM —		mM	g/pot
Control	None	4.55	2.49	18.25	4.18	0.12	2.86
Calcium hydroxide	4 $cmol_c$ as $Ca(OH)_2$/kg	5.35	0.14	2.51	0.02	0.13	6.05
Cowpea leaves	10 g ground dry cowpea leaves/kg[a]	4.90	0.73	2.42	0.09	2.67	8.82
Cowpea ash	Ash from 10 g dry cowpea leaves/kg	4.95	1.55	8.85	0.62	0.17	5.97

Compiled from data in Hue (2011).
[a] The 10 g dry cowpea leaves contained 0.91 $cmol_c$ as $Ca(OH)_2$. All pots were fertilized with all nutrients except Ca.

Figure 9.24 *These corn plants from two areas in the same well-fertilized and irrigated field illustrate the dramatic effect that aluminum toxicity can have on plant growth. The plant on the left produced no grain and was less than one-third the size of the plant on the right, as indicated by the brackets showing the stem diameter. Its root system is stunted, with little root branching and no visible thin, white, healthy roots. The stunted plant is suffering from aluminum toxicity because of excessively acid soil with reduced organic matter. In contrast to the noneroded parts of the field where soil pH was 5.8, soil in the eroded spots in the field exhibited pH 4.7, well below the critical level for aluminum toxicity. The eroded soil also had less organic matter content to detoxify the aluminum, as evidenced from the redder, lighter soil color than the less eroded soil on the right.* (Photo courtesy of Ray R. Weil)

Enhancing these organic matter reactions may be more practical than standard liming practices for resource-poor farmers or those in areas far from limestone deposits (Figure 9.24). Green manure crops (vegetation grown specifically for the purpose of adding organic matter to the soil) and mulches can provide the organic matter needed to stimulate such interactions and thereby reduce the level of Al^{3+} ions in the soil solution, even if they do not raise the soil pH. Aluminum-sensitive crops can then be grown following the green manure crop. One caution regarding the use of organic amendments to ameliorate aluminum toxicity is that the amounts of these materials needed for this purpose may provide nutrients in excess of what is suggested by nutrient management guidelines, potentially polluting water from excessive leaching and runoff losses of nitrogen and phosphorus (see Chapter 13).

9.10 LOWERING SOIL pH

It is often desirable to reduce the pH of highly alkaline soils. Furthermore, some acid-loving plants, for example, blueberries, rhododendrons, and azaleas, grow best on soils having pH values of 5.0 and below. To accommodate such plants, it is sometimes desirable to amend soils with acid-forming organic and inorganic materials.

Acid Organic Matter

As organic residues decompose, organic and inorganic acids are formed. These can reduce the soil pH if the organic material is low in calcium and other nonacid cations. Leaf mold from coniferous trees, pine needles, pine sawdust, and acid peat are quite satisfactory organic materials to add around ornamental plants (but see Section 11.3 for nitrogen considerations with these materials). However, farm manures (particularly poultry manures), lime-stabilized sewage sludge, and leaf mold of such high-base trees as beech and maple may be alkaline and may increase the soil pH, as discussed in Section 9.9.

Inorganic Chemicals

If nitrogen is needed for fertility, its supply as ammonium sulfate $((NH_4)_2SO_4)$ will promote rapid acidification. When the addition of large amounts of nitrogen or acid organic matter is not desirable or feasible, inorganic chemicals such as aluminum sulfate $(Al_2(SO_4)_3)$ or ferrous sulfate $(Fe^{II}SO_4)$

Figure 9.25 *Aluminum sulfate was applied to the soil on the left side of this hydrangea bush and lime was applied to the soil on the right side. When the soil pH is below 5.0–5.5, sufficient aluminum is absorbed and combined with the plants anthocyanin pigment to form a bright blue color in the blossoms (especially the large showy sepals surrounding each tiny flower). When the soil pH is 6.0 or above, almost no aluminum is in solution and the plant uses iron in the pigment to produce a pink-colored blossom. White hydrangeas contain no pigment in their sepals, so color changes in their tiny true flowers is not very noticeable.* (Photo courtesy of Ray R. Weil)

may be used (Figure 9.25). The latter chemical provides available iron (Fe^{2+} ions) for the plant and, upon hydrolysis, enhances acidity by reactions such as the following:

$$FeSO_4 + \tfrac{1}{4}O_2 + 1\tfrac{1}{2}H_2O \longrightarrow FeOOH + \underset{\text{Sulfuric acid}}{H_2SO_4} \qquad (9.25)$$

Another material often used to increase soil acidity is elemental sulfur (Box 9.4). As the sulfur undergoes microbial oxidation in the soil (see Section 13.20), 2 moles of acidity (as sulfuric acid) are produced for every mole of S oxidized:

$$2S + 3O_2 + 2H_2O \longrightarrow 2H_2SO_4 \qquad (9.26)$$

Under favorable conditions, sulfur is 4–5 times more effective, kilogram for kilogram, in developing acidity than is ferrous sulfate. Although ferrous sulfate brings about

---

BOX 9.4
COSTLY AND EMBARRASSING SOIL pH MYSTERY

Work was nearing completion on an elaborate, 1-ha public garden in the heart of Washington, D.C. A ribbon-cutting ceremony planned for the following week would include the wealthy donors and powerful politicians who had supported the creation of the multimillion-dollar garden. The horticulturalist in charge paced nervously in the light June rain as half a dozen soil scientists worked feverishly to collect soil and plant samples. Something had gone terribly wrong. Large ugly areas of brown, dying turfgrass were scattered throughout the garden.

Now, after so many rare and exotic trees, shrubs, and flowers had been planted, the horticulturalist feared the worst—the soil might be toxic, need to be removed, and the whole garden started over. The garden was built partly over an underground museum. Beneath the pleasing, undulating surface topography, the "topsoil" lay over a meter thick in places, covering a tangle of pipes, conduits, and wires on the museum roof. If the soil needed to be removed, it would have to be done the slow and expensive way—by hand. The horticulturalist suspected some toxic factor in the soil was killing the grass and would soon start damaging other plants as well. He knew that the growing media installed was not exactly "topsoil" in the usual sense of the word. The landscape design specifications had called for "a natural friable soil ... with 2% organic content ... USDA textural class of loam and ... pH 5.5 to 7.0." The lowest bidder had offered to make a "topsoil" using sediments dredged from a nearby tidal river, modified with enough lime and sand to meet the pH and texture requirements. The consulting engineers had run their lab tests and determined that the material met the specifications.

(continued)

> BOX 9.4
> ## COSTLY AND EMBARRASSING SOIL pH MYSTERY (CONTINUED)
>
> In late April, as the grand opening date approached, the grass began to turn brown and die in small patches. Although the shrubs, trees, and flowers in the cultivated flower beds were still looking good, the dead patches of turfgrass grew larger with every passing day. Turf specialists were called in, but could find no diseases or pests to account for the dead patches. Now, in desperation, the horticulturalist had hired several soil scientists to check the soil.
>
> Some of the soil scientists knew where the "topsoil" had come from. Those trained as pedologists augured deep, looking in vain for telltale signs of acid sulfate weathering (Section 9.6) they suspected might solubilize toxic aluminum and heavy metals from the river bottom. Others, noting that it was the shallow-rooted turfgrass that seemed to suffer first, went shallow instead of deep (see Box 1.3), obtaining samples from several depths, including a separate set of samples from the topmost 3 cm. They collected pairs of soil samples from several low spots of dead turf and from adjacent high spots where the turf was relatively healthy. Back at the lab, they stirred each soil sample in water and measured the pH. The results were completely normal until they got to the samples of that thin surface layer. Then they couldn't believe their eyes—all the samples of the upper 3 cm from the dead turf areas gave pH readings below pH 3, one as low as pH 1.9 (see bar graph in Figure 9.26). Looking closely at the 0–3 cm samples, they noticed small yellow flecks that smelled like sulfur. The pieces to the puzzle began to fall into place.
>
> What was going on? The previous summer, shortly after the sod had been installed, the horticulturalist had pulled "normal" 20 cm deep soil cores for soil testing (see Chapter 13). The results indicated the soil pH was 7.2, a bit higher than the initial tests and considerably above the pH 6.0–6.5 recommended for the fine fescue turfgrass. Therefore, the horticulturalist had applied about 1,000 kg/ha of elemental sulfur powder, as recommended to lower the pH by about 1 unit. He pulled another set of 20 cm deep soil samples about two months later and found the pH was still about 7.0. Therefore, he repeated the sulfur application. The lawn looked healthy during the cool, rainy winter, while the landscapers installed the valuable trees and shrubs. What he had failed to consider was that sufficient time and warm weather would be needed for the soil microorganisms to oxidize the sulfur and produce sulfuric acid to acidify the soil. So the second sulfur application had been an overresponse to the normal delay and had doubled the amount of S available to oxidize. Sulfur powder is quite water repellent and buoyant, so rain easily washed much of it off the high areas into the low spots, thus doubling or tripling the already doubled application—giving five or six times the recommended S concentration in those areas. When warm, wet weather the following spring stimulated the S-oxidizing bacteria to go into high gear, extreme acidity was produced in the thin surface layer of soil where the S was located and most of the turfgrass roots proliferated. Thankfully, the remedy would be simple and inexpensive: remove the sod along with about 5 cm of soil and lay down new sod. This they did and everyone at the opening ceremony was impressed by the beautiful lawn and garden.
>
> The lessons learned? (1) Many soil processes are biological in nature—be patient and realize they will respond to environmental conditions with time. (2) Taking deep soil samples may "dilute out" evidence of extreme conditions near the soil surface. Therefore, be sure to sample the upper few centimeters separately for untilled soils, especially if amendments have been applied to the surface.
>
>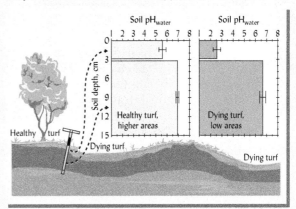
>
> **Figure 9.26** *Dying turf and soil acidity in a rooftop garden.* (Diagram courtesy of Ray R. Weil)

more rapid plant response, sulfur is less expensive and much less is needed. The quantities of ferrous sulfate or sulfur that should be applied will depend upon the buffering capacity of the soil and its original pH level. Figure 9.7 suggests that for each unit drop in pH desired, a well-buffered soil (e.g., a silty clay loam with 4% organic matter) will require about 4 $cmol_c$ of S per kilogram of soil. This is about 1,200 kg S/ha (since 2 mol of H^+ ion are produced by each mole of S oxidized, 1 $cmol_c$ of S = 0.32/2 = 0.16 g).

9.11 DEVELOPMENT OF SALT-AFFECTED SOILS[6]

Salt-affected soils are widely distributed throughout the world (Figure 9.27), typically in areas with precipitation-to-evaporation ratios of 0.75 or less and in low, flat areas with high water tables that may be subject to seepage from higher elevations. Salt-affected soils cover approximately 7% of the Earth's total land area, about 23% of its cultivated agricultural land, and almost 50% of its irrigated land. In some regions, the area of land so affected is growing by about 10% annually.

Many soils become salt-affected because changes in the local water balance, usually brought about by human activities, increase the input of salt-bearing water more than they increase the output of drainage water. Increased evaporation, waterlogging, and rising water tables usually result. It is worth remembering the irony that *salts usually become a problem when too much water is supplied, not too little*.

Accumulation of Salts in Nonirrigated Soils

Large areas of soils in arid and semiarid regions are affected by some degree of salinity. The salts are primarily chlorides and sulfates of calcium, magnesium, sodium, and potassium. These salts accumulate naturally in some surface soils because there is insufficient rainfall to flush them from the upper soil layers. In coastal areas, sea spray and inundation with seawater can be locally important sources of salt in soils.

Other localized but important sources are fossil deposits of salts laid down during geological time in the bottom of now-extinct lakes or oceans or in underground saline water pools. These fossil salts can be dissolved in groundwater that moves horizontally over underlying impervious geological layers and ultimately rises to the surface of the soil in the low-lying parts of the landscape. The salts then concentrate near or on the surface of the soil in these low-lying areas, creating a saline soil. The low-lying areas where the saline groundwater emerges are termed **saline seeps**.

Saline seeps occur naturally in some locations, but their formation may increase dramatically when the water balance in a semiarid landscape is disturbed by bringing land under cultivation (Figure 9.28). Replacement of native, deep-rooted perennial vegetation with annual crop species reduces the annual evapotranspiration, especially if the cropping system includes periods of fallow during which the soil is bare of vegetation. The decreased evapotranspiration allows more rainwater to percolate through the soil, thus raising the water table and increasing the flow of groundwater to lower elevations. Eventually, capillary rise begins to contribute a continuous stream of salt-laden water to replace the water lost at the surface by evaporation. The evaporating water leaves behind the salts, which soon accumulate to levels that inhibit plant growth. Year by year the evaporation zone creeps up the slope, and the barren area becomes larger and more saline. Millions of hectares of land in North and South America, Australia, and other semiarid regions have been degraded in this fashion.

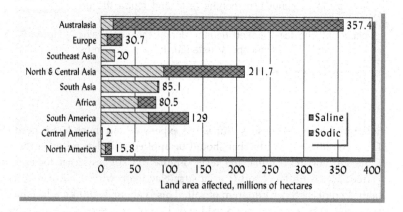

Figure 9.27 Area extent of salt-affected soils (saline or sodic) in different regions of the world. In total, some 933 million hectares are affected by salt globally, of which 352 million are saline and 581 million sodic. The numbers represent the sum of saline and sodic lands. [Graphed from tabular data in Szabolcs (1989) as cited in Rengasamy (2006)]

---

[6]For a discussion of these soils, which are also referred to as *halomorphic soils*, see Rengasamy (2006) and Wichelns and Qadir (2015).

Figure 9.28 Saline seep formation in a semiarid area where the salt-rich substrata are underlain by an impermeable layer. (a) Under deep-rooted perennial vegetation, transpiration is high and the water table is kept low. (b) After conversion to agriculture, shallow-rooted annual crops take up much less water, especially if fallow is practiced, allowing more water to percolate through the salt-bearing substrata. Consequently, in lower-elevation landscape positions, the wet-season water table rises close to the soil surface. This allows the salt-laden groundwater to rise by capillary flow to the surface, from which it evaporates, leaving behind an increasing accumulation of salts. Note that the diagrams greatly exaggerate vertical distances. (c) A spreading saline seep area in eastern Montana where wheat fallow cropping has replaced natural deep-rooted prairie vegetation. (d) Close-up of salt crust over moist soil. (Courtesy of Ray R. Weil)

Irrigation-Induced Salinity and Alkalinity

The amount of salt brought in with irrigation water may seem negligible, but the amounts of water applied over the course of time are huge. Again, pure water is lost by evaporation, but the salt stays behind and accumulates (Figure 9.29). The effect is accentuated in arid regions for two reasons: (1) the water available from rivers or from groundwater is relatively high in salts because it has flowed through dry-region soils that typically contain large amounts of easily weatherable minerals, and (2) the dry climate creates a relatively high evaporative demand, so large amounts of water are needed for irrigation. An arid-region farmer may need to apply 90 cm of water to grow an annual crop. Even if this is good-quality

Figure 9.29 Salinization, the accumulation of soluble salts in soils, can be observed in the potting medium of houseplants (left) and in irrigated fields (middle and right). The salt accumulates because of evaporative water loss from soil that is repeatedly supplied with water that contains dissolved salts, even if in low concentrations. Only pure water evaporates; the salts dissolved in the water do not. Note that the salt tends to concentrate at the highest points of the soil surface, from which evaporation loss is greatest and to which the soil solution flows by capillary action. (Photo courtesy of Ray R. Weil)

water relatively low in salts, it will likely dump more than 6 Mg/ha (3 tons/acre) of salt on the land every year (see Section 9.16). These principles also apply to the production of plants indoors where rainfall is essentially zero and plants are irrigated with salt-containing water (such as urban tap water).

As the human population quadrupled during the past half century, people greatly expanded the area of land growing food under irrigation. Initially, the expanded irrigation stimulated phenomenal increases in food-crop production. Unfortunately, many irrigation projects failed to provide for adequate drainage. As a result, the process of *salinization* accelerated, and in some areas, sodic soils have been created.

Throughout human history, salts eventually accumulated on irrigated land to such a degree that crop production declined and land had to be abandoned. Today, societies around the world are repeating the mistakes of the past. Some observers believe that each year, the area of previously irrigated land degraded by severe salinization is greater than the area of land newly brought under irrigation. Truly, the world needs to give serious attention to the large-scale problems associated with salt-affected soils.

9.12 MEASURING SALINITY AND SODICITY

Salinity

Total Dissolved Solids (TDS). In concept, the simplest way to determine the total amount of dissolved salt in a sample of water is to heat it in a container until all of the water has evaporated and then weigh the dry residue that remains. The TDS can then be expressed as milligrams of solid residue per liter of water evaporated (mg/L).[7]

Electrical Conductivity. As a practical matter, TDS is rarely measured by actual evaporation; rather the **electrical conductivity** (EC) of water or a soil–water mixture is measured and the salinity so measured is expressed either directly in units of conductivity or sometimes used to estimate mg/L TDS. Pure water is a poor conductor of electricity, but conductivity increases as more and more salt is dissolved in the water. Thus, the EC of the soil solution gives us an indirect measurement of the salt content. The EC can be measured both on samples of soil mixed with water or on the bulk soil in situ (Table 9.6). It is expressed in terms of deciSiemens per meter (dS/m).[8]

The **saturation paste extract** method is the most commonly used lab procedure and the standard to which the others are usually compared. A soil sample is saturated with distilled water and mixed to the consistency of a paste, allowed to stand overnight, extracted by suction filtration and then the electrical conductivity (EC_e) of the extracted solution is measured. A quicker variant of this method involves measuring the EC of the solution extracted from a 1:2 soil–water mixture after 0.5 h of shaking (EC_w). Values for EC_w can be converted to total dissolved solids if the type of salt is known (see footnote, Table 9.6).

Two more **rapid methods** that are well correlated to the salinity of the soil solution are: (1) the conductivity of the saturated soil paste itself (EC_p) and (2) the conductivity of a 1:2 soil–water mixture (EC_s). Because the tedious extraction and filtration steps are eliminated, EC_p and EC_s can easily be determined on samples in the field.

Mapping EC in the Field. Advances in instrumentation now allow rapid, continuous field measurement of bulk soil conductivity, which, in turn, is directly related to soil salinity (see Table 9.6) as well as several other soil properties affecting conductivity, such as texture, density, and water content. The method involves inserting four or more carefully spaced electrodes into moist soil to make direct measurements of **apparent electrical conductivity** in the field (EC_a). The depth to which the electrodes sense the EC_a is related to the spacing between

---

[7] In this case, the unit mg/L has the same meaning as parts per million (ppm), since there are a million mg in a kg of water (which has the volume of 1 L). Although ppm is often used in commercial literature, it is an ambiguous expression and should be avoided.

[8] Formerly expressed as millimhos per centimeter (mmho/cm). Since 1 S = 1 mho, 1 ds/m = 1 mmho/cm.

Table 9.6
DIFFERENT MEASUREMENTS FOR ESTIMATING SOIL SALINITY

The methods are well correlated, so each can be converted to any other. The EC_e is the standard method.

Measured on a soil sample	
EC_e	Conductivity of the solution extracted from a water-saturated soil paste
EC_p	Conductivity of the water-saturated soil paste itself
EC_w	Conductivity of the solution extracted from a soil–water mixture (usually either 1:2 or 1:5)
EC_s	Conductivity of a 1:1, 1:2, or 1:5 soil–water mixture itself
TDS	Total dissolved solids in water or the solution extracted from a water-saturated soil paste[a]
Measured on bulk soil in place	
EC_a	Apparent conductivity of bulk soil sensed by metal electrodes in soil
EC_a^*	Electromagnetic induction of an electric current using surface transmitter and receiving coils

[a]Note that TDS (mg/L) can be converted to EC_w using these relationships between 0 and 5 dS/m: for Na salts, TDS = 640 × EC_w based on a 1:2 soil–water mixture; for Ca salts, TDS = 800 × EC_w. The dilution effect of varying soil:water ratios must be taken into account when comparing EC_s and EC_w data.

the electrodes. Such an on-the-go apparatus for measuring EC_a can be constructed with rolling colter blades that serve as electrodes that move through the soil as a tractor pulls the apparatus across a field. This technique is rapid, simple, and practical, and gives values that can be correlated with EC_e as well as with other soil properties. If a geopositioning system (GPS) receiver is integrated with the apparatus, the resulting data can be transformed into a map showing the spatial variation of soil salinity (and other soil properties) across a parcel of land (Figure 9.30). Similar results can be obtained with a second indirect method that **employs electromagnetic (EM)** induction of electrical current in the body of the soil, the level of which is related to electrical conductivity and, in turn, to soil salinity.

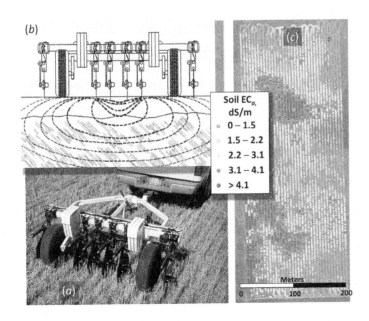

Figure 9.30 *Mapping soil electrical conductivity in the field. (a) A mobile apparatus with six rolling colter electrodes and an integrated geopositioning system (GPS) generates a continuous readout of apparent soil conductivity (EC_a) as it is pulled across a field. (b) A diagram illustrating how the rolling colters serve as electrodes to send electrical currents through the soil and to detect the flow of these currents as affected by the soil EC. (c) Soil electrical conductivity data from a series of transects of the apparatus up and down an agricultural field in North Dakota are presented as a map showing distinct zones of high EC_a (redder colors) where salinity will likely be a problem for crop growth. In this case, the EC_a readings by the mobile apparatus were highly correlated with lab analyses of saturated paste extract and could be converted into EC_e by multiplying by 2. [Images (a) and (b) courtesy of Veris Technologies Inc., Salina, KS; image (c) modified from Hopkins et al. (2012) courtesy of David Hopkins, North Dakota State University]*

Such measurements reflect the electrical conductivity of all the soil components. Therefore, the EC_a measured is not only related to the salinity of the soil solution, but also to how wet the soil is, the texture of the soil, and a number of other soil properties that influence electrical conductivity. While this fact complicates the data interpretation, when EC_a levels are below those of concern for salinity, it allows EC_a maps to be used for a variety of soil management purposes. The information from these maps can be used as part of **precision agriculture** techniques (see Section 13.11), which are capable of applying corrective measures tailored to match the degree of salinity in each small part of a large field.

The values of EC_e, EC_w, EC_s, EC_p, EC_a, or EC_a^* obtained by these procedures will not be identical; however, these values are all well correlated with each other such that if soil clay and water contents are known, the results from any of the methods can be reasonably well converted to the standard EC_e for interpretation.

Sodium Status (Sodicity)

The tendency of certain cations, especially sodium, to cause deterioration of soil physical properties traditionally is termed **sodicity** and is characterized primarily by several equations based on the relative amounts and flocculating ability of the various cations present.

Two expressions are commonly used to characterize the sodium status of soils. The **exchangeable sodium percentage (ESP)** identifies the degree to which the exchange complex is saturated with sodium:

$$\text{ESP} = \frac{\text{Exchangeable sodium, cmol}_c/\text{kg}}{\text{Cation exchange capacity, cmol}_c/\text{kg}} \times 100 \tag{9.27}$$

ESP levels greater than 15 are associated with severely deteriorated soil physical properties and pH values of 8.5 and above. Figure 9.31 provides two example calculations of ESP.

The **sodium adsorption ratio (SAR)** is a second, more easily measured property that is becoming even more widely used than ESP. The SAR gives information on the comparative concentrations of Na^+, Ca^{2+}, and Mg^{2+} in soil solutions. It is calculated as follows:

$$\text{SAR} = \frac{[Na^+]}{(0.5[Ca^{2+}] + 0.5[Mg^{2+}])^{1/2}} \tag{9.28}$$

where $[Na^+]$, $[Ca^{2+}]$, and $[Mg^{2+}]$ are the concentrations (in mmol of charge per liter) of the sodium, calcium, and magnesium ions in the soil solution. An SAR value of 13 for the solution extracted from a saturated soil paste is approximately equivalent to an ESP value of 15. The SAR of a soil extract takes into consideration that the adverse effect of sodium is moderated by the presence of calcium and magnesium ions. The SAR also is used to characterize irrigation water applied to soils (see Section 9.16).

9.13 CLASSES OF SALT-AFFECTED SOILS

Using EC, ESP (or SAR), and soil pH, salt-affected soils are classified as **saline**, **saline–sodic**, and **sodic** (Figure 9.32). Soils that are not greatly salt-affected are classed as **normal**.

Saline Soils

The processes that result in the accumulation of neutral soluble salts are referred to as **salinization**. The salts are mainly chlorides and sulfates of calcium, magnesium, potassium, and sodium. The concentration of these salts sufficient to interfere with plant growth (see Section 9.15) is generally defined as that which produces an electrical conductivity in the saturation extract (EC_e) greater than 4 dS/m. However, some sensitive plants are adversely affected when the EC_e is only about 2 dS/m.

Saline soils are those soils that contain sufficient salinity to give EC_e values greater than 4 dS/m, but have an ESP less than 15 (or an SAR less than 13) in the saturation extract. Thus, the exchange complex of saline soils is dominated by calcium and magnesium, not sodium. The pH of saline soils is usually below 8.5. Because soluble salts help prevent dispersion of soil colloids, plant growth on saline soils is not generally constrained by poor infiltration,

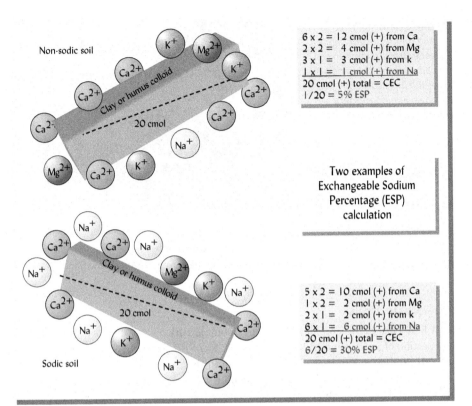

Figure 9.31 *An illustration of exchangeable sodium percentage calculation for two soils, one sodic and the other not.* (Diagram courtesy of Ray R. Weil)

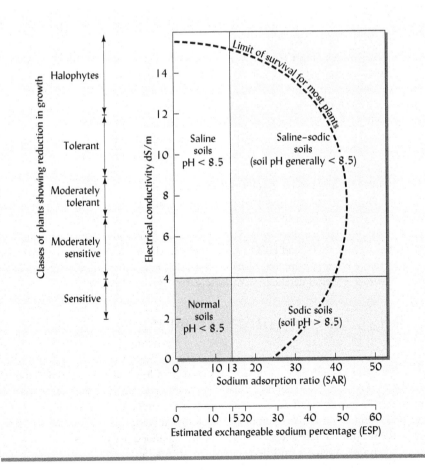

Figure 9.32 *Diagram illustrating the classification of normal, saline, saline–sodic, and sodic soils in relation to soil pH, electrical conductivity, sodium adsorption ratio (SAR), and exchangeable sodium percentage (ESP). Also shown are the ranges for different degrees of sensitivity of plants to salinity.*

Figure 9.33 *The upper profile of a sodic soil (a Natrustalf) in a semiarid region of western Canada. Note the thin A horizon (knife handle is about 12 cm long) underlain by columnar structure in the natric (Btn) horizon. The white, rounded "caps" of the columns are comprised of soil dispersed because of the high sodium saturation. The dispersed clays give the soil an almost rubbery consistency when wet.* [Photo courtesy of Agriculture Canada, Canadian Soils Information System (CANSIS)]

aggregate stability, or aeration. Evaporation of water commonly creates a white salt crust on the soil surface (see Figure 9.29), which accounts for the name *white alkali* that was previously used to designate saline soils.

Saline–Sodic Soils

Soils that have both detrimental levels of neutral soluble salts (EC_e greater than 4 dS/m) *and* a high proportion of sodium ions (ESP greater than 15 or SAR greater than 13) are classified as **saline–sodic soils** (see Figure 9.32). Plant growth in these soils can be adversely affected by both excess salts and excess sodium levels. The high concentration of neutral salts moderates the dispersing influence of the sodium by helping to keep the colloidal particles associated with each other in floccules. This situation is subject to rather rapid change if the soluble salts are leached from the soil, especially if the SAR of the leaching waters is high. In such a case, salinity will drop, but the exchangeable sodium percentage will increase, and the saline–sodic soil will become a sodic soil.

Sodic Soils

Sodic soils are, perhaps, the most troublesome of the salt-affected soils. While their levels of neutral soluble salts are low (EC_e less than 4.0 dS/m), they have relatively high levels of sodium on the exchange complex (ESP and SAR values are above 15 and 13, respectively). Some sodic soils in the order Alfisols (Natrustalfs) have a very thin A horizon overlying a clayey layer with columnar structure, a profile feature closely associated with high sodium levels (Figure 9.33). The pH values of sodic soils generally exceed 8.5 because sodium carbonate is much more soluble than calcium or magnesium carbonate and so maintains high concentrations of CO_3^{2-} and HCO_3^- in the soil solution.

The extremely high pH levels in sodic soils may cause the soil organic matter to disperse and/or dissolve. The dispersed and dissolved humus moves upward in the capillary water flow and, when the water evaporates, can give the soil surface a black color. The name *black alkali* was previously used to describe these soils. Plant growth on sodic soils is often constrained by specific toxicities of Na^+, OH^-, and HCO_3^- ions. However, the main reason for the poor plant growth—often to the point of complete barrenness—is that few plants can tolerate the chemically induced extreme physical degradation of sodic soils.

9.14 PHYSICAL DEGRADATION IN SOIL–SODIC SOILS

The high sodium and low salt levels in sodic soils (and, to a lesser degree, in some "normal" soils) can cause serious clay dispersion, degradation of aggregate structure, and loss of macroporosity such that the movement of water and air into and through the soil is severely restricted. This structural degradation is most commonly measured in terms of the readiness of water movement—the infiltration rate and saturated hydraulic conductivity of the soil (Section 5.5) may be reduced almost to zero. A **puddled** condition results as water fails to soak into the soil.

Slaking, Swelling, and Dispersion

The low permeability related to sodic conditions has three common underlying causes. First, exchangeable sodium increases the tendency of aggregates to break up or *slake* upon becoming wet. The clay and silt particles released by slaking aggregates clog soil pores as they are washed down the profile. Second, when expanding-type clays (e.g., montmorillonite) become highly Na^+-saturated, their degree of swelling is increased. As these clays expand, the larger pores responsible for water drainage in the soil are squeezed shut. Third, and perhaps most important, the combination of high sodium and low ionic strength leads to soil dispersion. Dispersion is the opposite of flocculation. In normal soils, clay particles flocculate together, giving rise to tiny clumps (floccules) that create pores between them and promote the formation of larger aggregates (see Section 4.5). In dispersed soils, the clay particles separate from one another, creating an almost gel-like condition. While these phenomena are most pronounced in sodic soils, they can occur to some degree in other dry-region soils subjected to low electrolyte water (e.g., rain!) and/or moderately high levels of monovalent cations.

Two Causes of Soil Dispersion

Two chemical conditions promote dispersion. One is a high proportion of Na^+ ions on the exchange complex. The second is a low concentration of electrolytes (salt ions) in the soil water.

High Sodium. Exchangeable Na^+ ions promote dispersion because their single charge and large hydrated size cause them to be only weakly attracted to soil colloids; so they spread out to form a relatively broad swarm of ions held in very loose outer-sphere complexes around the colloids (see Section 8.7). It is worth noting that, to a lesser degree, other weakly adsorbed ion species such as K^+ or Mg^{2+} can also favor dispersion as compared to the strongest flocculating cations such as Ca^{2+} and Al^{3+}.

Secondly, compared to a swarm of divalent cations (which have two positive charges each), twice as many monovalent ions (with only one charge each) are needed to provide enough positive charges to counter the negative charges on a clay surface. As illustrated in Figure 9.34,

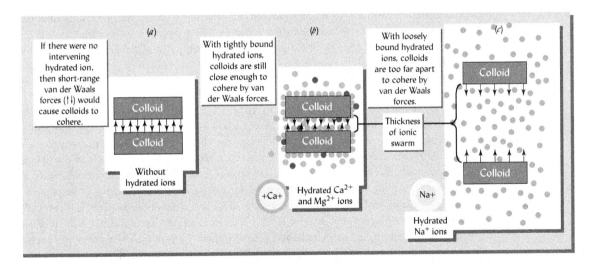

Figure 9.34 Conceptual diagrams showing how the type of cations present on the exchange complex influences clay dispersion. If colloids could approach closely (a), they would be held together (cohere) by short-range van der Waals forces. In soil, the colloids are surrounded by a swarm of hydrated exchangeable ions, which prevent the colloids from approaching so closely. If these are strongly attracted calcium and magnesium ions (b), they do not keep the colloids very far apart, so cohesive forces still have some effect. However, if they are sodium ions (c), the more spread-out ionic swarm keeps the colloids too far apart for cohesive forces to come into play. Sodium ions cause a spread-out ion swarm for two reasons: (1) their large hydration shell of water allows them to be only loosely attracted to the colloids, and (2) twice as many monovalent Na^+ ions as divalent (Ca^{2+} or Mg^{2+}) ions are attracted to a given colloid charge. When the colloid particles are separated from each other, the soil is in a dispersed condition. (Diagram courtesy of Ray R. Weil)

the layer of exchangeable monovalent Na^+ ions is therefore much thicker than that which would form with the more strongly attracted divalent ions such as Ca^{2+}. The highly sodium-saturated colloids are kept so far apart that the forces of cohesion cannot come into play to attract one colloid surface to another. Instead, the poorly balanced electronegativity of each colloidal surface repels other electronegative colloids, and the soil becomes dispersed. Figure 9.34 therefore illustrates the relationships expressed mathematically by the SAR (Eq. [9.30].

Low Salt Concentration. A low ionic concentration in the bulk soil solution simultaneously increases the gradient causing exchangeable cations to diffuse away from the clay surface while it decreases the gradient causing anions to diffuse toward the clay. The result is a thick ionic layer or swarm of absorbed cations. Adding *any* soluble salt would increase the ionic concentration of the soil solution and encourage the opposite effects—resulting in a compressed ionic layer that allows the clay particles to come close enough together to form floccules.

It is worth remembering that *low salt (ion) concentrations and weakly attracted ions (e.g., sodium) encourage soil dispersion and puddling, while high salt concentrations and strongly attracted ions (e.g., calcium) promote clay flocculation and soil permeability.*

9.15 BIOLOGICAL IMPACTS OF SALT-AFFECTED SOILS[9]

How Salts Affect Plants

In addition to the nutrient-deficiency problems associated with high pH (see Section 10.1), high levels of soluble salts affect plants by two primary mechanisms: **osmotic effects** and **specific ion effects**.

Osmotic Effects. Soluble salts lower the osmotic potential of the soil water (see Section 5.3), making it more difficult for roots to remove water from the soil. For established plants (and soil bacteria), this condition requires the expenditure of more energy on osmotic adjustments—accumulating organic and inorganic solutes to lower the osmotic potential *inside* the cells to counteract the low osmotic potential of the soil solution outside. The lost energy results in reduced growth. For annual crops, the lowered osmotic potential can result in more frequent wilting and reduced water uptake from the soil profile. Plants are most susceptible to salt damage in the early stages of growth. Salinity may delay, or even prevent, the germination of seeds (Figure 9.35). In response to excessive soil salinity, many plants become severely stunted and exhibit small dark-bluish green leaves with dull surfaces.

Specific Ion Effects. The *kind* of salt can make a big difference in how plants respond to salinity. Certain ions, including Na^+, Cl^-, $H_3BO_4^-$, and HCO_3^- are quite toxic to many plants. However, plant species, and even strains within a species, differ widely in their sensitivity to these ions. In addition to specific toxic effects, high levels of Na^+ can cause imbalances in the uptake and utilization of other cations. For example, Na^+ competes with the essential nutrient ion K^+ in the process of transport across the cell membrane during uptake, making it difficult for plants to obtain the K^+ they need from saline–sodic or sodic soils. The presence of adequate Ca^{2+} helps the plant to discriminate against Na^+ and for K^+. High levels of sodium or chloride typically produce scorching or necrosis of the leaf margins and tips (see Figure 9.35, *inset*). These symptoms appear first and most severely on the oldest leaves as these have been transpiring water and accumulating salts for the longest period. Salt-stressed plants may also lose their leaves prematurely.

Physical Effects of Sodicity. Deterioration of physical properties is also a major factor in determining which plants can grow in sodic soils. The colloidal dispersion caused by sodicity may harm plants in at least two ways: (1) oxygen becomes deficient due to the breakdown of soil structure and the very limited air movement that results, and (2) water relations are poor due largely to the very slow infiltration and percolation rates.

---

[9]Effects of salinity and water stress are reviewed by Munns (2002) for plants and by Yan et al. (2016) for microorganisms.

Figure 9.35 Foliar symptoms on oldest leaves (inset), reduced germination, and stunted growth of soybean plants with increasing levels of soil salinity due to additions of NaCl to a sandy soil. The large numbers written on the pots indicate the electrical conductivity (EC_e) of the soil. Note that serious growth reductions occurred for this sensitive cultivar even at EC_e levels considered "normal." The soybean cultivar used (Jackson) is more sensitive to salinity than most other cultivars of soybean. (Photo courtesy of Ray R. Weil)

Selective Tolerance of Higher Plants to Saline and Sodic Soils

Satisfactory plant growth on salty soils depends on a number of interrelated factors, including the physiological constitution of the plant, its stage of growth, and its rooting habits. For example, older, deeper-rooted plants are generally more tolerant to salt-affected soils than younger, more shallow-rooted ones.

Plant Sensitivity. Table 9.7 classifies selected plants according to how severely soil salinity, as measured by electrical conductivity (EC_e), affects their relative productivity. Approximately 1% of plant species are *halophytes*, plants that can tolerate high levels of salinity, some as high as sea water. They are quite a diverse group and plant scientists speculate that many could be domesticated into crops for salt-tolerant agriculture systems that can produce food, feed, fiber, and biofuels on the world's increasing area of salinized land. Halophytes are also valuable for restoring disturbed or degraded land under saline conditions.

Salt Problems Not Related to Arid Climates

Deicing Salts. In areas where deicing salts are used to keep roads and sidewalks free of snow and ice during winter months, these salts may impact roadside soils and plants. Repeated application of deicing salts can result in salinity levels sufficiently high as to adversely affect plants (especially trees) and soil organisms living alongside highways or sidewalks. In humid regions, such salt contamination is usually temporary, as the abundant rainfall leaches out the salts in a matter of weeks or months. To reduce the specific chemical and physical problems associated with sodium salts, many municipalities have switched from NaCl to KCl or $MgCl_2$ for deicing purposes. Sand can be used to improve traction, thereby reducing the need for deicing salt.

Containerized Plants. Salinity can also be a serious problem for potted plants, particularly perennials that remain in the same pot for long periods (Figure 9.35, *left*). Greenhouse operators producing containerized plants must carefully monitor the quality of the water used for irrigation. Salts in the water, as well as those applied in fertilizers, can build up if care is not taken to flush them out occasionally with excess water. Chlorinated urban tap water used for indoor plants should be left overnight in an open container to allow some of the dissolved chlorine to escape as chlorine gas, thus reducing the load of Cl^- ions added to the potting soil.

Table 9.7
RELATIVE SALT TOLERANCE OF SELECTED PLANTS

Approximate EC_e resulting in a 10% reduction in plant growth for the most sensitive species in each column.

Tolerant, 12 dS/m	Moderately tolerant, 8 dS/m	Moderately sensitive, 4 dS/m	Sensitive, 2 dS/m
Alkali grass, Nutall	Ash (white)	Alfalfa	Alders
Alkali sacaton	Asparagus	Arborvitae	Almond
Barley (grain)	Aspen	Broad bean	Apple
Bent grass	Beet (garden)	Cabbage	Azalea
Bermuda grass	Black cherry	Celery	Beech
Bougainvillea	Broccoli	Clover (ladino, red)	Bean
Boxwood, Japanese	Cowpea	Corn	Birch
Canola (rapeseed)	Elm	Cucumber	Blackberry
Cotton	Fescue (tall)	Grape	Carrot
Date	Hydrangea	Lettuce	Dogwood
Jojoba	Juniper	Locust (black)	Hibiscus
Oak (red and white)	Kale	Maple (red)	Linden
Olive	Locust (honey)	Pea	Maple (sugar and red)
Redwort	Oak (red and white)	Peanut	Onion
Rescue grass	Oats	Radish	Orange
Rosemary	Orchard grass	Rice (paddy)	Peach
Rugosa Rye (grain)	Pomegranate	Soybean (sens. var.)	Pear
Salt grass (desert)	Ryegrass (perennial)	Squash	Pine (red and white)
Sugar beet	Safflower	Sugar cane	Pineapple
Tamarix	Sorghum	Sweet potato	Potato
Wheat grass (crested)	Soybean (tol. var.)	Timothy	Raspberry
Wheat grass (tall)	Squash (zucchini)	Tomato	Rose
Wild rye (Russian)	Sudan grass	Turnip	Strawberry
Willow	Wheat	Vetch	Tomato

9.16 WATER-QUALITY CONSIDERATIONS FOR IRRIGATION[10]

Whether in a single field or in a large regional watershed, understanding the **salt balance** is a basic prerequisite for wise management of salt-affected soils and waters. To achieve salt balance and therefore prevent deterioration of soil quality, the amount of salt coming in must be matched by the amount being removed. Meeting this condition is a fundamental challenge to the long-term sustainability of irrigated agriculture. In irrigated areas, this principally means managing the quality and amount of the irrigation water brought in and the quality and amount of soil drainage water removed.

Irrigation Water Quality. Table 9.8 provides some guidelines on water quality for irrigation. Monitoring the chemical quality of water added to salt-affected soils is a prime management strategy. For example, if the salt content of irrigation water is high, salt balance will be difficult to achieve. However, even very salty water can be used successfully if soils are sufficiently well

---
[10]For contrasting overviews of water-quality sand conservation problems facing advanced irrigated agriculture in places such as California, see Letey (2000) and Gleick et al. (2011).

Table 9.8
WATER-QUALITY GUIDELINES FOR IRRIGATION

Note that higher total salinity (EC_w) in irrigation water compensates, somewhat, for increasing sodium hazard (SAR). Also note that water low in salts (low EC_w) avoids problems of restricted water availability to plants, but it may worsen soil physical properties, especially if the SAR is high.

Water property	Units	Degree of restriction on use		
		None	Slight to moderate	Severe
Salinity (affects crop water availability)				
EC_w	dS/m	<0.7	0.7–3.0	>3.0
TDS	mg/L	<450	450–2,000	>2,000
Physical structure and water infiltration (evaluate using EC_w and SAR together)				
SAR = 0–3 and EC_w =	dS/m	>0.7	0.7–0.2	<0.2
SAR = 3–6 and EC_w =	dS/m	>1.2	1.2–0.3	<0.3
SAR = 6–12 and EC_w =	dS/m	>1.9	1.9–0.5	<0.5
SAR = 12–20 and EC_w =	dS/m	>2.9	2.9–1.3	<1.3
SAR = 20–40 and EC_w =	dS/m	>5.0	5.0–2.9	<2.9
Sodium (Na) specific ion toxicity (affects sensitive crops)				
Surface irrigation	mmol/L	<3	3–9	>9
Sprinkler irrigation	mmol/L	<3	>3	
Chloride (Cl) specific ion toxicity (affects sensitive crops)				
Surface irrigation	mmol/L	<4	4–10	>10
Sprinkler irrigation	mmol/L	<3	>3	
Boron (B) specific ion toxicity (affects sensitive crops)	mg/L	<0.7	0.7–3.0	>3.0

Modified from Abrol et al. (1988) with permission of the Food and Agriculture Organization of the United Nations.

drained to allow careful management of salt inputs and outputs. Where irrigation water is low in salts but has a high SAR, the formation of sodic soils is likely to accelerate. In addition, irrigation water high in carbonates or bicarbonates can reduce Ca^{2+} and Mg^{2+} concentrations in the soil solution by precipitating these ions as insoluble carbonates. This leaves a higher proportion of Na^+ in the soil solution and can increase its SAR, moving the soil toward the sodic class.

Drainage Water Salinity. Since some portion of added water must be drained away to combat salt buildup, the quality and disposition of *irrigation drainage waters* must also be carefully monitored and controlled to minimize potential harm to downstream users and habitats. In any irrigation system, the drainage water leaving a field will be considerably more concentrated in salts than the irrigation water applied to the same field (Figure 9.36 explains why). What to do with the increasingly saline drainage water presents a major challenge to the sustainability of irrigated agriculture.

Perhaps the most efficient approach is to collect the drainage water, keep it isolated from the relatively high-quality canal water, and reuse it to irrigate a more salt-tolerant crop in a lower field. This approach provides both high-quality canal water and lower-quality drainage water for use on appropriate crops. Often, some fresh water must be mixed with the recycled drainage water to bring its salinity level down to what can be tolerated by even the salt-tolerant crop. After several cycles of reuse, the drainage water must be disposed of, as it will have become too saline for irrigating even the most salt-tolerant species. In many cases, once it is useable for irrigation, wastewater is eventually channeled into shallow ponds that allow the water to evaporate and the salts to collect.

Figure 9.36 *Evapotranspiration and salt balance together ensure that the drainage water from irrigated fields is much saltier than the irrigation water applied. In this example, the irrigation water contains 250 mg salts per liter. Some 75% of the applied water is lost to the atmosphere by evapotranspiration. About 25% of the water applied is used for drainage, which is necessary to maintain the salt balance (prevent the buildup of salts) in the field. The added salts are leached away with the drainage water, which then contains the same amount of salt as was added, but in only 25% of the added amount of water. The concentration of salt in the drainage water is thereby four times as great (1,000 mg/L) as in the irrigation water. Disposal and/or reuse of the highly saline drainage water present challenges for any irrigation project.* (Diagram courtesy of Ray R. Weil)

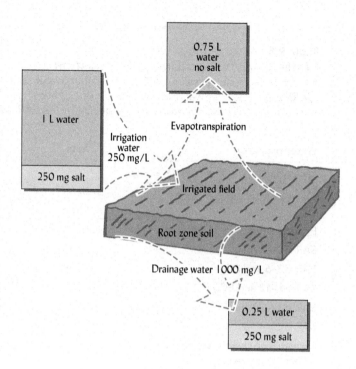

Toxic Elements in Drainage Water. If either the irrigation water or the soil of the irrigated fields contains significant quantities of certain toxic trace elements, these too will become increasingly concentrated in the drainage water. The elements of concern include molybdenum (Mo), arsenic (As), boron (B), and selenium (Se). Molybdenum and selenium are necessary nutrients for animals and humans in trace amounts, but all four elements can be toxic to cattle, wildlife, or people if they become concentrated in water or food.

9.17 RECLAMATION OF SALINE SOILS

The restoration of soil chemical and physical properties conducive to high productivity is referred to as soil *reclamation*. Reclamation of saline soils is largely dependent on the provision of effective drainage and the availability of good-quality irrigation water (see Table 9.8) so that salts can be leached from the soil. In areas where irrigation water is not available, such as in saline seeps in nonirrigated dry regions, the leaching of salts may not be practical. In these areas, deep-rooted vegetation may be used to lower the water table and reduce the upward movement of salts.

If the natural soil drainage is inadequate to accommodate the leaching water, an artificial drainage network must be installed. Intermittent applications of excess irrigation water may be required to effectively reduce the salt content to a desired level. The process can be monitored by measuring the soil's EC, using either the saturation extract procedure or one of the field instruments described in Section 9.12.

Leaching Requirement (LR)

The amount of water needed to remove the excess salts from saline soils, called the **LR**, is determined by the characteristics of the crop to be grown, the irrigation water, and the soil. As demonstrated in Box 9.5, an approximation of the *LR* is given for relatively uniform salinity conditions by the ratio of the salinity of the irrigation water (expressed as its EC_{iw}) to the maximum acceptable salinity of the soil solution for the crop to be grown (expressed as EC_{dw}, the EC of the drainage water).

$$LR = \frac{EC_{iw}}{EC_{dw}} \qquad (9.29)$$

> **BOX 9.5**
> **LEACHING REQUIREMENT FOR SALINE SOILS**
>
> The salt balance in a field can be described by equating the salt inputs and outputs:
>
> $$\underbrace{S_{iw} + S_p + S_f + S_m}_{\text{Salt inputs}} = \underbrace{S_{dw} + S_{dw} + S_c + S_{ppt}}_{\text{Salt outputs}} \quad (9.30)$$
>
> The salt inputs include those from irrigation water (S_{iw}), atmospheric deposition (S_p), fertilizers (S_f), and the weathering or dissolution of existing soil minerals (S_m). The salt outputs are due to drainage water (S_{dw}), crop removal (S_c), and chemical precipitation (S_{ppt}) of carbonates and sulfates. Usually, S_{iw} and S_{dw} are far larger than the other terms in Eq. (9.30), so the main concern is to balance the salt coming in with the irrigation water and that leaving with the drainage water:
>
> $$\underbrace{S_{iw}}_{\text{Salt in with irrigation water}} = \underbrace{S_{dw}}_{\text{Salt out with drainage water}} \quad (9.31)$$
>
> We can estimate the quantity of salt S carried in drainage or irrigation as the product of the volume of the water (expressed as cm depth applied to an area of land) and the concentration of salt in that water (as approximated by its EC). Therefore, we can rewrite Eq. (9.31) as follows (using the same subscripts as before):
>
> $$D_{iw} \times EC_{dw} = D_{dw} \times EC_{dw} \quad (9.32)$$
>
> Rearranging the terms, we obtain the following expression that defines the LR or the ratio of drainage water depth to irrigation water depth (D_{dw}/D_{iw}) needed to maintain salt balance:
>
> $$\frac{EC_{iw}}{EC_{dw}} = \frac{D_{dw}}{D_{iw}} = LR \quad (9.33)$$
>
> The LR tells farmers how much irrigation water (in excess of that required to wet the soil) they should apply for sufficient leaching. The goal is usually to assure that the upper two-thirds of the root zone does not accumulate salts beyond the level acceptable for a particular crop.
>
> From Eq. (9.33), we see that LR also equals the ratio of the irrigation water EC to the drainage water EC:
>
> $$LR = \frac{EC_{iw}}{EC_{dw}} \quad (9.34)$$
>
> where EC_{dw} is an acceptable level for the crop being grown. What is considered an acceptable EC_{dw} is open to interpretation. An acceptable EC_{dw} might be interpreted to mean the EC_e that allows 90% of maximum crop yield. A more conservative interpretation of acceptable EC_{dw} is the threshold EC_e at which the growth of the particular crop just begins to decline (usually 1–2 dS/m lower than that which gives the 90% yield level).
>
> As an example, consider the situation where the irrigation water has an EC_{iw} of 2.5 dS/m and a moderately tolerant crop (e.g., broccoli) is to be grown. If information that is more specific is unavailable, the acceptable EC_{dw} for the crop can be (roughly) estimated from the column heading in Table 9.7. For a moderately tolerant crop, we can use 8 dS/m as the acceptable EC_{dw} to produce 90% of the maximum yield. Then,
>
> $$LR = \frac{2.5 \text{ dS/m}}{8 \text{ dS/m}} = 0.31 \quad (9.35)$$
>
> If this LR (0.31) is multiplied by the amount of water needed to wet the root zone—let us suppose it is 12 cm of water—the amount of water to be leached would be 3.7 cm (12 cm × 0.31). This is the minimum amount of water that must be leached through a water-saturated soil to maintain the root zone salinity at the acceptable level. A more sensitive crop could be grown in this soil, but it would have a lower acceptable EC_{dw} and therefore would require the application of a greater amount of water for leaching.

The *LR* indicates the water that should be added in excess of that needed to thoroughly wet the soil and meet the crop's evapotranspiration needs. Note that if EC_{iw} is high and a salt-sensitive crop is chosen (dictating a low EC_{dw}), a very large leaching requirement *LR* will result. As mentioned in Section 9.16, disposal of the drainage water that has leached through the soil can present a major problem. Therefore, it is generally desirable to use management techniques that minimize the *LR* and the amount of drainage water that requires disposal.

Management of Soil Salinity

Management of irrigated soils should aim to simultaneously minimize drainage water and protect the root zone (usually the upper meter of soil) from damaging levels of salt accumulation. These two goals are obviously in conflict. The irrigator can attempt to find the best compromise between the two and can use certain management techniques that allow plants to tolerate the presence of higher salt levels in the soil profile. One option is to plant salt-tolerant species or choose the most salt-tolerant varieties within a crop species (as discussed in Section 9.15).

Irrigation Timing. The timing of irrigation is extremely important on saline soils, particularly early in the growing season. Germinating seeds and young seedlings are especially sensitive to salts. Therefore, irrigation should precede or immediately follow planting to move the salts downward and away from the seedling roots. The irrigator can use high-quality water to keep root-zone salinity low during the sensitive early growth stages and then switch to lower-quality water as the maturing plants become more salt-tolerant.

Location of Salts in the Root Zone. Tillage or surface-residue management practices (such as mulches or conservation tillage) that reduce evaporation from the soil surface should also reduce the upward transport of soluble salts. Likewise, specific techniques for applying irrigation water that direct salt concentrations away from young plant roots can allow higher levels of salt to accumulate without damage to the crop. Applying water in every other furrow and asymmetrically planting only on the wet side of the furrows can provide significant protection to young plants (Figure 9.37).

Frequent application of water, as with sprinkler- or drip-irrigation systems (see Section 6.9), can help move salts away from plant roots. Proper placement of water emitters in drip-irrigation systems is crucial in establishing a low-salt zone around sensitive young plants. Buried drip-emitter lines, in particular, can cause problems by moving salts to the soil surface where seeds are germinating (Figure 9.37e). The water emitted below the surface moves toward the soil surface by capillary rise, carrying dissolved salts with it. When the water evaporates or is taken up by plant roots, the salts are left to concentrate at the soil surface. The size and shape of the low-salinity zone created by drip irrigation is also dependent on the rate of water application and the texture of the soil (Figure 9.38).

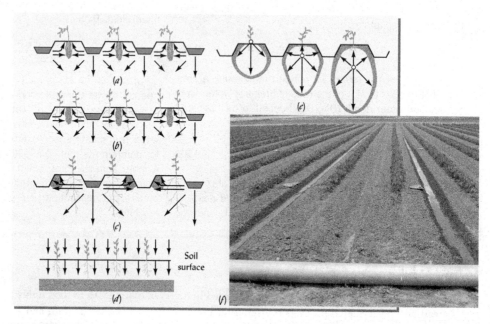

Figure 9.37 Effect of irrigation techniques on salt movement and plant growth in saline soils. (a) With irrigation water applied to furrows on both sides of the row, salts move to the center of the ridge and damage young plants. (b) Plants on the edges of the bed avoid the most concentrated salts. (c) Application of water to every other furrow and placement of plants near the water help plants avoid highest salt concentrations. (d) Sprinkler irrigation or uniform flooding temporarily moves salts downward out of the root zone, but the salts will return as the soil surface dries out and water moves up by capillary flow. (e) Drip irrigation at low rates provides a nearly continuous flow of water, creating a low-salt soil zone with the salts concentrated at the wetting front. The placement of the drip emitters largely determines whether the salts are moved toward or away from the plant roots. (f) An example of alternate furrow irrigation in California, but with tomatoes planted in the center, rather than along the irrigated furrow side of the beds where salts are lowest. (Diagram courtesy of Wesley M. Jarrell; photo courtesy of Guihua Chen)

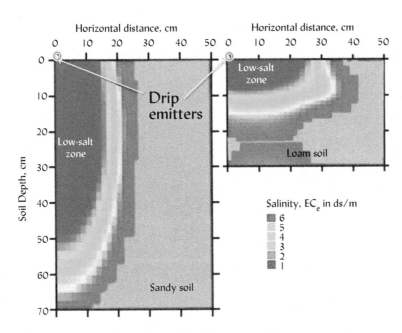

Figure 9.38 Effect of soil texture on distribution of salts after irrigation from surface drip emitters. The salts were originally distributed evenly within each soil, but the slowly applied irrigation water (4 L/h for 3 hours) dissolved the salts and moved them away from the emitters toward the border between the wet and dry soil (the wetting front). Note that the low-salt zone in the sandy soil is much deeper, but more narrow, than in the loam. Repeated irrigation at short intervals may keep the salts away from the root zone. However, if the soil is allowed to dry, the dissolved salts will begin to return to the root zone as the matric potential gradient draws water toward the drier soil. [Redrawn from data in Bresler (1975)]

Spatial Variability. The actual leaching fraction applied to a field may exceed the theoretical amount calculated from the LR if soil salinity levels vary from spot to spot in a field. In order to avoid yield losses from salinity, a farmer may irrigate a field with the amount of water determined by the LR of the *most saline* parts of the field. This amount will assure that the most saline parts of the field are properly leached, but it will waste valuable water and create more than the necessary amount of drainage water in the less saline parts of the field.

It may be feasible to avoid such inefficiencies by using *precision agriculture* technology. For example, irrigation systems (such as sprinkler or drip) could be designed with variable rate controls and used in conjunction with soil-salinity sensing devices (described in Section 9.12) so that they apply just the right amount of leaching water for each part of a field.

Some Limitations of the Leaching Requirement Approach[11]

The leaching requirement approach to managing irrigated soils is only an approximation and has several inherent weaknesses. First, additional leaching may be needed, in some cases, to reduce the excess concentration of specific elements, such as boron. Second, the LR by itself does not take into account the rise in the water table that is likely to result from increased leaching, and so it may lead to waterlogging and, eventually, increased salinization. Third, using a simple LR approach to manage irrigation usually over-applies water because an entire field is treated to avoid salt damage in its most saline spots. Fourth, the LR method does not consider salts that may be picked up from fossil salt deposits already in the soil and substrata. Fifth, it assumes that the EC of the drainage water is known, but in fact this may be largely unknown, since it may take years or even decades for the water applied in irrigation to reach the main drains where it can be easily sampled. In other words, the drainage water sampled today may represent the leaching conditions of several months or years ago.

An alternative approach would be to closely monitor the salinity in the soil profile by taking repeated measurements across the field, using the EM sensor or four-electrode methods discussed in Section 9.12. The sufficiency of leaching and the dominant direction of water movement could then be judged from the type of salinity profile observed (Figure 9.39). This more complex approach, combined with site-specific management techniques, seems to hold promise for more efficient management and reclamation of salt-affected soils under irrigation.

---

[11] For a discussion of how detailed spatial assessment of salinity can be used as an alternative to the LR approach, see Rhoades et al. (1999) and Corwin and Lesch (2013).

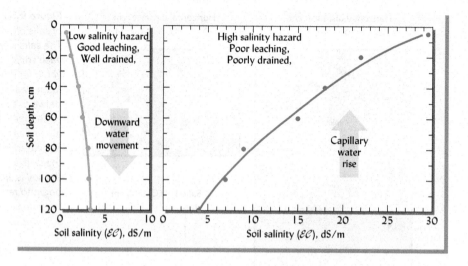

Figure 9.39 *Soil salinity profile curves show the levels of salts throughout the root zone. The shapes of the curves also indicate whether leaching has been sufficient and whether saline water is rising from a shallow water table. The arrows indicate the direction of water flow for the low- and high-salinity soils. The curve on the right shows increasing salinity near the soil surface, a pattern that is typical of saline seeps and waterlogged irrigated fields. The leftmost curve is similar in shape to the salinity profile of a dry, well-drained arid region soil in which there is no influence of groundwater* (Diagram courtesy of Ray R. Weil)

9.18 RECLAMATION OF SALINE–SODIC AND SODIC SOILS

Saline–sodic soils have some of the adverse properties of both saline and sodic soils. If attempts are made to leach out the soluble salts in saline–sodic soils, as was discussed for saline soils, the exchangeable Na^+ level as well as the pH would likely increase, and the soil would take on adverse characteristics of sodic soils. Consequently, for both saline–sodic and sodic soils, attention must first be given to reducing the level of exchangeable Na^+ ions and then to the problem of excess soluble salts.

Gypsum

Removing Na^+ ions from the exchange complex is most effectively accomplished by replacing them with either Ca^{2+} or H^+ ions. Providing Ca^{2+} in the form of gypsum ($CaSO_4 \cdot 2H_2O$) is the most practical way to bring about this exchange. The relatively high solubility of gypsum will also increase the ionic strength, thus further helping to flocculate the soil. When gypsum is added, reactions such as the following take place:

$$2NaHCO_3 + CaSO_4 \rightarrow CaCO_3 + \underset{\text{(leachable)}}{Na_2SO_4} + CO_2\uparrow + H_2O \qquad (9.36)$$

$$\underset{Na^+}{\overset{Na^+}{\boxed{\text{Colloid}}}} + CaSO_4 \rightleftharpoons Ca^{2+}\boxed{\text{Colloid}} + Na_2SO_4 \qquad (9.37)$$

Note that in each case the soluble salt Na_2SO_4 is formed, which can be easily leached from the soil as was done in the case of the saline soils. Several tons of gypsum per hectare are usually necessary to achieve reclamation. The soil must be kept moist to hasten the reaction, and the gypsum should be thoroughly mixed into the surface by cultivation—not simply plowed under. The treatment must be supplemented later by a thorough leaching of the soil with irrigation water to leach out most of the sodium sulfate.

Sulfur and Sulfuric Acid

Elemental sulfur and sulfuric acid can be used to advantage on sodic soils, especially where sodium bicarbonate abounds. The sulfur, upon biological oxidation (see Sections 9.6 and 12.2), yields sulfuric acid, which not only changes the sodium bicarbonate to the less harmful and more leachable sodium sulfate but also decreases the pH. The reactions of sulfuric acid with the compounds containing sodium may be shown as follows:

$$2NaHCO_3 + H_2SO_4 \rightarrow 2CO_2\uparrow + 2H_2O + \underset{\text{(leachable)}}{Na_2SO_4} \qquad (9.38)$$

$$\underset{Na^+}{\overset{Na^+}{\boxed{\text{Colloid}}}} + H_2SO_4 \rightleftharpoons \underset{H^+}{\overset{H^+}{\boxed{\text{Colloid}}}} + \underset{\text{(leachable)}}{Na_2SO_4} \qquad (9.39)$$

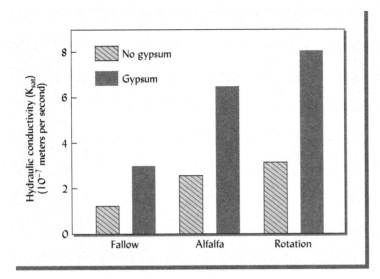

Figure 9.40 *The influence of gypsum and growing crops on the hydraulic conductivity (K_{sat}) of the upper 20 cm of a saline–sodic soil (Natrustalf) in Pakistan. The use of gypsum to increase water conductivity in all plots was more effective when deep-rooted alfalfa and rotation of sesbania-wheat were grown.* [Drawn from selected data from Ilwas et al. (1993); used with permission of the Soil Science Society of America]

Not only is the sodium bicarbonate changed to sodium sulfate, a mild neutral salt, but the carbonate anion is also removed from the system. When gypsum is used, however, a portion of the carbonate may remain as a calcium compound ($CaCO_3$). In research trials, sulfur and even sulfuric acid have proven to be very effective in the reclamation of sodic soils, especially if large amounts of $CaCO_3$ are present. In practice, however, gypsum is much more widely used than these acid-forming materials.

Physical Condition

The effects of gypsum and sulfur on the physical condition of sodic soils is perhaps more spectacular than are the chemical effects. Sodic soils are almost impermeable to water, since the soil colloids are largely dispersed and the soil is essentially void of stable aggregates. When the exchangeable Na^+ ions are replaced by Ca^{2+} or H^+, soil aggregation and improved water infiltration results (Figure 9.40). The neutral sodium salts (e.g., Na_2SO_4) formed when the exchange takes place can then be leached from the soil, thereby reducing both salinity and sodicity. Some research suggests that aggregate-stabilizing synthetic polymers may be helpful in at least temporarily increasing the water infiltration capacity of gypsum-treated sodic soils. Data in Table 9.9 from one experiment show the possible potential of these soil conditioners, especially when used in combination with gypsum.

Deep-Rooted Vegetation. The reclamation effects of gypsum or sulfur are greatly accelerated by plants growing on the soil. Crops that have some degree of tolerance to saline and sodic soils, such as sugar beets, cotton, barley, sorghum, berseem clover, or rye, can be grown initially. Their roots help provide channels through which gypsum can move downward into the soil. Deep-rooted crops, such as alfalfa, are especially effective in improving the water conductivity of gypsum-treated sodic soils. Figure 9.40 illustrates the ameliorating effects of the combination of gypsum and deep-rooted crops.

Air Injection. In addition to reduced diffusion of air because of soil dispersion, irrigated soils may provide less than optimal oxygen availability to roots because irrigation water—especially applied by drip systems—contains much less dissolved oxygen than rainwater. One approach to improving the aeration of the root zone in heavy-textured, irrigated soils is to mechanically add air. This can be easily accomplished by injecting air into drip irrigation lines and may be practical for high-value crops and ornamental landscape containers.

Management of Reclaimed Soils

Once salt-affected soils have been reclaimed, prudent management steps must be taken to be certain that the soils remain productive. For example, surveillance of the EC and SAR and trace element composition of the irrigation water is essential. Management adjustment is needed to

Table 9.9
USE OF GYPSUM AND SYNTHETIC POLYMERS ON RECLAMATION OF A SODIC SOIL

Adding gypsum to samples of a fine-textured (clay) saline–sodic soil, a Mollisol from California, increased both the hydraulic conductivity and the salts leached in the experiment while decreasing the exchangeable sodium percentage (ESP). Adding two experimental synthetic polymers (T4141 and 21J) with the gypsum gave even greater increases in hydraulic conductivity and leached salts.

	Characteristic measured					
	Hydraulic conductivity, mm/h		Total salts leached, mg/kg		ESP, %	
Gypsum added? →	No	Yes	No	Yes	No	Yes
Polymer treatment						
No polymer	0.0	0.06	0.0	4.7	22.9	9.6
T4141	0.0	0.28	0.0	10.1	25.4	9.6
21J	0.0	0.28	0.0	9.7	25.5	9.6

Data from Zahow and Amrhein (1992).

accommodate any change in water quality that could affect the soil. The number and timing of irrigation episodes helps determine the balance of salts entering and leaving the soil. Likewise, the maintenance of good internal drainage is essential for the removal of excess salts.

Steps should also be taken to monitor appropriate chemical characteristics of the soils, such as pH, EC, and SAR, as well as specific levels of such elements as boron, chlorine, molybdenum, and selenium that could lead to chemical toxicities. These measurements will help determine the need for subsequent remedial practices and/or chemicals.

Crop and soil fertility management for satisfactory yield levels is essential to maintain the overall quality of salt-affected soils. The crop residues (roots and aboveground stalks) will help maintain organic matter levels and good physical condition of the soil. To maintain high yields, micronutrient and phosphorus deficiencies characteristic of other high-pH soils will need to be overcome by adding appropriate organic and inorganic sources.

9.19 CONCLUSION

There are few reactions involving the soil or its biological inhabitants that are not influenced by soil pH. Acidification is a natural process in soil formation that is accentuated in humid regions where processes that produce H^+ ions outpace those that consume them. Natural acidification is largely driven by the production of organic acids (including carbonic acid) and the leaching away of the nonacid cations (Ca^{2+}, Mg^{2+}, K^+, and Na^+) that the H^+ ions from the acids displace from the exchange complex. Soil erosion, emissions from power plants and vehicles, as well as inputs of nitrogen into agricultural systems are the principal means by which human activities accelerate acidification. Accelerated acidification is the second only to soil erosion as a globally widespread form of soil degradation.

Aluminum is the other principal acid cation besides hydrogen. Its hydrolysis reactions produce H^+ ions and its toxicity comprises one of the main detrimental effects of soil acidity. The total acidity in soils is the sum of the active, exchangeable (salt replaceable), and residual pools of soil acidity. Changes in the soil solution pH (the active acidity) are buffered by the presence of the other two pools. In certain anaerobic soils and sediments, the presence of reduced sulfur provides the potential for enormous acid production if the material is exposed to air by drainage or excavation.

Soil pH is largely controlled by the humus and clay fractions and their associated exchangeable cations. The maintenance of satisfactory soil fertility levels in humid regions depends considerably on the judicious use of lime to balance the losses of calcium and magnesium from the soil. Liming not only maintains the levels of exchangeable calcium and magnesium but in so doing also provides a chemical and physical environment that encourages the growth of most common plants. Gypsum and organic matter (either applied or grown) represent other tools that can be used to ameliorate soil acidity instead of, or in addition to, liming. On the other hand, it is sometimes most judicious to use acid-tolerant plants, rather than attempt to change the chemistry of the soil.

Knowing how pH is controlled, how it influences the supply and availability of essential plant nutrients as well as toxic elements, how it affects higher plants and human beings, and how it can be ameliorated, is essential for the conservation and sustainable management of soils throughout the world. Understanding soil processes that occur at high pH levels is especially important in arid regions.

Arid and semiarid regions predominantly feature alkaline and salt-affected soils. These soils typically exhibit above-neutral pH values throughout their profiles and often have calcic (calcium carbonate-rich) or gypsic (gypsum-rich) horizons at some depth. Such soils cover vast areas and support important—though water-limited—desert and grassland ecosystems. Irrigation of alkaline soils of arid regions almost inevitably leads to the accumulation of salts, which must be carefully managed.

Since their pH levels are typically quite high, management practices used on these soils are quite different from those applied to acid-soil regions. The high-pH, calcium-rich conditions of alkaline soils commonly lead to deficiencies of certain essential micronutrients (especially iron and zinc) and macronutrients (especially phosphorus). In localized areas, boron, molybdenum, and selenium may be so readily available as to accumulate in plants to levels that can harm grazing animals.

Based on their total salt content (indicated by EC) and the proportion of sodium among the cations (indicated by either the SAR or the ESP), scientists have grouped salt-affected soils into three classes: *Saline soils*, *Saline–sodic*, and *Sodic soils*. The physical conditions of saline and saline–sodic soils are satisfactory for plant growth, but the colloids in sodic soils are largely dispersed, the soil is puddled and poorly aerated, and the water infiltration rate is extremely slow.

Agricultural systems tend to be inherently disharmonious with natural conditions regarding acidity and salinity. The humid-region farmer growing crops that require soil pH near 6 must be ever on guard to counteract acidification. So, too, the arid-region irrigator must be ever vigilant to combat salinization. Whereas the humid-region farmer periodically uses lime to restore a favorable balance between H^+ ion consumption and production, the irrigator uses periodic leaching to restore the balance between the import and export of salts. If irrigators in arid regions apply only enough water to meet plant evapotranspiration needs, salts will build up relentlessly in the soil until the land becomes too salinized to grow crops. Furthermore, there can be no effective leaching if drainage is insufficient to carry away the leaching water. Thus, leaching and drainage are both essential components of any successful irrigation scheme. However, the leaching of salts is not a simple matter, and it inevitably leads to additional problems both in the field being leached and in sites further downstream. At best, salinity management in irrigation agriculture provides a compromise between unavoidable evils.

The reclamation of saline–sodic and sodic soils requires an additional process before leaching of excess salinity can be achieved. In order to make these soils permeable enough for leaching to take place, the excess exchangeable sodium ions must first be removed from the exchange complex. This is accomplished by replacing the Na^+ ions with either Ca^{2+} or H^+ ions. The Ca^{2+} or H^+ ions then stimulate flocculation and increased permeability to the point that the replaced sodium and other salts can be leached downward and out of the profile. Gypsum ($CaSO_4 \cdot 2H_2O$) and elemental sulfur (S) are two amendments that can supply the Ca^{2+} and H^+ ions needed. Monitoring the chemical content of both the irrigation water and the soil is essential to achieve this goal of removing the Na^+ ions from the exchange complex and, ultimately, the soil.

About 20% of the world's farmland is irrigated, up from only 10% in 1960. Impressively, this land produces some 45% of our food supply. The world increasingly depends on irrigated agriculture in dry regions for the production of food for its growing population. Unfortunately, irrigated agriculture is in inherent disharmony with the nature of arid-region ecology and is therefore fraught with difficulties that require constant and careful management. Acidity and salinity also constitute major influences on the habitat for micro- and not so micro- organisms that live in the soil, which we will learn about in the next chapter.

STUDY QUESTIONS

1. Soil pH gives a measure of the concentration of H^+ ions in the soil solution. What, if anything, does it tell you about the concentration of OH^- ions? Explain.
2. Describe the role of aluminum and its associated ions in soil acidity. Identify the ionic species involved and the effect of these species on the CEC of soils.
3. If you could somehow extract the soil solution from the upper 16 cm of 1 ha of moist acid soil (pH = 5), how many kg of pure $CaCO_3$ would be needed to neutralize the soil solution (bring its pH to 7.0)? Under field conditions, up to 6 Mg of limestone may be required to bring the pH of this soil layer to a pH of 6.5. How do you explain the difference in the amounts of $CaCO_3$ involved?
4. What is meant by *buffering*? Why is it so important in soils, and what are the mechanisms by which it occurs?
5. What is acid rain, and why does it seem to have greater impact on forests than on commercial agriculture?
6. Calculate the amount of pure $CaCO_3$ that could theoretically neutralize the H^+ ions in a year's worth of acid rain if a 1-ha site received 500 mm of rain per year and the average pH of the rain was 4.0.
7. Discuss the significance of soil pH in determining specific nutrient availabilities and toxicities, as well as species composition of natural vegetation in an area.
8. How much limestone with a $CaCO_3$ equivalent of 90% would you need to apply to eliminate exchangeable aluminum in an Ultisol with CEC = 8 $cmol_c$/kg and an aluminum saturation of 60%?
9. Based on the buffer pH of your soil sample, a lab recommends that you apply 2 Mg of $CaCO_3$ equivalent to your field and plow it in 18 cm deep to achieve your target pH of 6.5. You actually plan to use the lime to prepare a large lawn and till it in only 12 cm deep. The lime you purchase has a carbonate equivalent = 85%. How much of this lime do you need for 2.5 ha?
10. A landscape contractor purchased 10 dump-truck loads of "topsoil" excavated from a black, rich-appearing soil in a coastal wetland. Samples of the soil were immediately sent to a lab to be sure they met the specified properties (silt loam texture, pH 6–6.5) for the topsoil to be used in a landscaping job. The lab reported that the texture and pH were in the specified range, so the topsoil was installed and an expensive landscape of beautiful plants established. Unfortunately, within a few months all the plants began to die. Replacement plants also died. The topsoil was again tested. It was still a silt loam, but now its pH was 3.5. Explain why this pH change likely occurred and suggest appropriate management solutions.
11. A neighbor complained when his azaleas were adversely affected by a generous application of limestone to the lawn immediately surrounding the azaleas. To what do you ascribe this difficulty? How would you remedy it?
12. The ill effects of acidity in subsoils can be ameliorated by adding gypsum ($CaSO_4 \cdot 2H_2O$) to the soil surface. What are the mechanisms responsible for this effect of the gypsum?
13. What are the primary sources of alkalinity in soils? Explain.
14. Compare the availability of the following essential elements in alkaline soils with that in acid soils: (1) iron, (2) nitrogen, (3) molybdenum, and (4) phosphorus.
15. The iron analysis of an arid-region soil showed an abundance of this element, yet a peach crop growing on the soil showed serious iron deficiency symptoms. What is a likely explanation?
16. A soil with an abundance of $CaCO_3$ may have a pH no higher than about 8.3, while a nearby soil with high Na_2CO_3 content has a pH of 10.5. What is the primary reason for this difference?
17. An arid-region soil, when it was first cleared for cropping, had a pH of about 8.0. After several years of irrigation, the crop yield began to decline, the soil aggregation tended to break down, and the pH had risen to 10. What is the likely explanation for this situation?
18. What physical and chemical treatments would you suggest to bring the soil described in question 5 back to its original state of productivity?
19. You receive a lab analysis (as shown in the accompanying table) of the exchangeable cations in alkaline soils from four sites you are managing. Calculate the CEC of the soil and the exchangeable sodium percentage (ESP) for each soil:

Site I.D.	Exch. cation concentration, $cmol_c$ kg^{-1}					
	Na	K	Ca	Mg	CEC	ESP
A	1	1.2	3	3		
B	1	1.2	17.4	6		
C	4.3	1.2	3	3		

20. In your position as consultant you receive the data in the accompanying table for the soil water extracted from three soils under your management. Calculate the SAR for each soil.

Soil I.D.	Concentrations of cations, mg L^{-1}				
	Na	K	Ca	Mg	SAR
A	230	46	40	24	
B	230	46	230	48	
C	990	46	40	24	

21. What are some of the adverse consequences of using wetlands as recipients of irrigation wastewater?
22. Calculate the leaching requirement to prevent the buildup of salts in the upper 45 cm of a soil if the EC_{dw} of the drainage water is 6 dS/m and the EC_{iw} of the irrigation water is 1.2 dS/m.
23. Explain how the same soil amendment (gypsum) can be used to ameliorate adverse conditions in both highly weathered acid soils and certain relatively unweathered alkaline arid region soils.
24. Show the chemical reactions that take place when using gypsum ($CaSO_4 \cdot 2H_2O$) in the reclamation of a sodic soil.

REFERENCES

Abedi, M., M. Bartelheimer, and P. Poschlod. 2013. "Aluminium toxic effects on seedling root survival affect plant composition along soil reaction gradients – A case study in dry sandy grasslands." *Journal of Vegetation Science* 24:1074–1085.

Abrol, I. P., J. S. P. Yadov, and F. I. Massoud. 1988. "Salt-affected soils and their management." *FAO Soils Bulletin 39*. Food and Agriculture Organization of the United Nations, Rome. Complete text available at www.fao.org/docrep/x5871e/x5871e00.htm.

Bloom, P. R., U. L. Skyllberg, and M. E. Sumner. 2005. "Soil acidity." In A. Tabatabai and D. Sparks (eds.). *Chemical Processes in Soils*. SSSA Book Series No. 8. Soil Science Society of America, Madison, WI, pp. 411–459.

Bresler, E. 1975. "Two-dimensional transport of solutes during non-steady infiltration from a trickle source." *Soil Science Society of America Proceedings* 39:604–613.

Caires, E. F., G. Barth, and F. J. Garbuio. 2006. "Lime application in the establishment of a no-till system for grain crop production in southern Brazil." *Soil Tillage Research* 89:3–12.

Corwin, D. L., and S. M. Lesch. 2013. "Protocols and guidelines for field-scale measurement of soil salinity distribution with ECa-directed soil sampling." *Journal of Environmental & Engineering Geophysics* 18:1–25.

Demchik, M. C., W. E. Sharpe, T. Yangkey, B. R. Swistock, and S. Bubalo. 1999a. "The effect of calcium/aluminum ratio on root elongation of twenty-six Pennsylvania plants." In W. E. Sharpe and J. R. Drohan (eds.). *The Effects of Acidic Deposition on Pennsylvania's Forests*. Proceedings of the Sept. 14–16, 1998 PA Acidic Deposition Conference. Environmental Resources Research Institute, Pennsylvania State University, University Park, PA, pp. 211–217.

Demchik, M. C., W. E. Sharpe, T. Yangkey, B. R. Swistock, and S. Bubalo. 1999b. "The relationship of soil Ca/Al ratio to seedling sugar maple population, root characteristics, mycorrhizal infection rate, and growth and survival." In W. E. Sharpe and J. R. Drohan (eds.). *The Effects of Acidic Deposition on Pennsylvania's Forests*. Proceedings of the Sept. 14–16, 1998 PA Acidic Deposition Conference. Environmental Resources Research Institute, Pennsylvania State University, University Park, PA, pp. 201–210.

Fierer, N., and R. B. Jackson. 2006. "The diversity and biogeography of soil bacterial communities." *Proceedings of the National Academy of Science* 103:626–631.

Gleick, P. H., J. Christian-Smith, and H. Cooley. 2011. "Water-use efficiency and productivity: Rethinking the basin approach." *Water International* 36:784–798.

Guo, J. H., X. J. Liu, Y. Zhang, J. L. Shen, W. X. Han, W. F. Zhang, P. Christie, K. W. T. Goulding, P. M. Vitousek, and F. S. Zhang. 2010. "Significant acidification in major Chinese croplands." *Science* 327:1008–1010.

Hoekenga, O. A., and J. V. Magalhaes. 2011. "Mechanisms of aluminum tolerance." In A. Costa de Oliveira and R. K. Varshney (eds.). *Root Genomics*. Springer-Verlag, Berlin, pp. 133–153.

Hopkins, D., K. Chambers, A. Fraase, Y. He, K. Larson, L. Malum, L. Sande, J. Schulte, E. Sebesta, D. Strong, E. Viall, and R. Utter. 2012. "Evaluating salinity and sodium levels on soils before drain tile installation: A case study." *Soil Horizons* 53:24–29.

Hue, N.V. 2011. "Alleviating soil acidity with crop residues." *Soil Science* 176:543–549.

Hue, N. V., and D. L. Licudine. 1999. "Amelioration of subsoil acidity through surface applications of organic manures." *Journal of Environmental Quality* 28:623–632.

IGBP-DIS. 2000. *Soildata: A Program for Creating Global Soil-Property Databases*. Global Soil Data Task Force, International Geosphere-Biosphere Programme, Data and Information System, Potsdam, Germany. Available from: http://www.daac.ornl.gov

Ilyas, M., R. W. Miller, and R. H. Qureski. 1993. "Hydraulic conductivity of saline-sodic soil after gypsum application." *Soil Science Society of America Journal* 57:1580–1585.

Letey, J. 2000. "Soil salinity poses challenges for sustainable agriculture and wildlife." *California Agriculture* 54(2):43–48.

Lumbanraja, J., and V. P. Evangelou. 1991. "Acidification and liming influence on surface charge behavior of Kentucky subsoils." *Soil Science Society of America Journal* 54:26–34.

Magdoff, F. R., and R. J. Barlett. 1985. "Soil pH buffering revisited." *Soil Science Society of America Journal* 49:145–148.

Munns, R. 2002. "Comparative physiology of salt and water stress." *Plant, Cell and Environment* 25:239–250.

Rengasamy, P. 2006. "World salinization with emphasis on Australia." *Journal of Experimental Botany* 57:1017–1023.

Rhoades, J. D., F. Chanduvi, and S. Lesch. 1999. *Soil Salinity Assessment Methods and Interpretation of Electrical Conductivity Measurements*. FAO Irrigation and Drainage Paper No. 57. Food and Agriculture Organization of the United Nations, Rome.

Rice, K. C., and J. S. Herman. 2012. "Acidification of earth: An assessment across mechanisms and scales." *Applied Geochemistry* 27:1–14.

Ritchey, E. L., D. D. Tyler, M. E. Essington, M. D. Mullen, and A. M. Saxton. 2015. "Nitrogen rate, cover crop, and tillage practice alter soil chemical properties." *Agronomy Journal* 107:1259–1268.

Ryan, P. R., and E. Delhaize. 2012. "Adaptations to aluminum toxicity." In S. Shabala (ed.). *Plant Stress Physiology*. CAB International, London, pp. 171–195.

Sullivan, T. J., I. J. Fernandez, A. T. Herlihy, C. T. Driscoll, T. C. McDonnell, N. A. Nowicki, K. U. Snyder, and J. W. Sutherland. 2006. "Acid-base characteristics of soils in the Adirondack mountains, New York." *Soil Science Society of America Journal* 70:141–152.

Szabolcs, I. 1989. *Salt-Affected Soils*. CRC Press, Boca Raton, FL, p. 274.

Weil, R. R. 2000. "Soil and plant influences on crop response to two African phosphate rocks." *Agronomy Journal* 92:1167–1175.

Wichelns, D., and M. Qadir. 2015. "Achieving sustainable irrigation requires effective management of salts, soil salinity, and shallow groundwater." *Agricultural Water Management* 157:31–38.

Yan, N., P. Marschner, W. Cao, C. Zuo, and W. Qin. 2016. "Influence of salinity and water content on soil microorganisms." *International Soil and Water Conservation Research* 4: doi:10.1016/j.iswcr.2015.11.003

Zahow, M. F., and C. Amrhein. 1992. "Reclamation of a saline sodic soil using synthetic polymers and gypsum." *Soil Science Society of America Journal* 56:1257–1260.

Zhang, F., S. Kang, J. Zhang, R. Zhang, and F. Li. 2004. "Nitrogen fertilization on uptake of soil inorganic phosphorus fractions in the wheat root zone." *Soil Science Society of America Journal* 68:1890–1895.

10
Organisms and Ecology of the Soil[1]

Soil mite *Ametroproctus oresbios*, center, lends physical jaw-power to the litter decomposition efforts of the other invertebrates and fungi. (Scanning Electron Micrograph by Byron Lee, AAFC Lethbridge Research Centre, Lethbridge, AB, Canada, other images courtesy of Ray R. Weil)

Under the silent, relentless chemical jaws of the fungi, the debris of the forest floor quickly disappears...
—A. FORSYTH AND K. MIYATA,
TROPICAL NATURE

The terms *ecosystem* and *ecology* usually call to mind scenes of lions stalking vast herds of wildebeest on the grassy savannas of East Africa or the interplay of phytoplankton, fish, and fishermen in some great estuary. Like a savanna or an estuary, a soil is an ecosystem in which thousands of different kinds of creatures interact and contribute to the global cycles that make all life possible. This chapter will introduce some of the actors in the living drama staged largely unseen in the soil beneath our feet.

If our bodies were small enough to enter the tiny passages in the soil, we would discover a world populated by a wild array of creatures all fiercely competing for every leaf, root, fecal pellet, and dead body that reaches the soil. We would also find predators of all kinds lurking in dark passages, some with fearsome jaws to snatch unwary victims, others whose jellylike bodies simply engulf and digest their hapless prey. But in that world without light, our main sensory apparatus—eyesight—would be useless; we would have to feel our way along the sometimes rough, sometimes slimy surfaces, listen for vibrations, sniff the humid air, and taste the water for chemical signs that might mean food—or danger ahead. Soil organisms find their food and communicate with one another by sensing vibrations, surfaces, chemical gradients, and even electrical fields.

The vast majority of belowground individuals are single microscopic cells whose "jaws" are chemical enzymes that eat away at organic substances left in the soil by their coinhabitants. Some of these enzymes are located inside microbial cells, others may be found in the guts of tiny animals, but many—perhaps most—are found adsorbed to soil particles and in the solution surrounding them, having been excreted by microorganisms into their immediate environment.

The diversity of substrates and environmental conditions found in every handful of soil spawns a diversity of adapted organisms that staggers the imagination. The collective vitality, diversity, and balance among these organisms make possible the functions of a healthy soil. We will learn how these organisms interact with one another, what they eat, how they affect the soil, and how soil conditions affect them. We will come to appreciate how vascular plants live with "one foot in each world"—connecting the above and belowground realms, and how plant

---

[1] For stories about how life underground affects everything on Earth, see Baskin (2005); for soil ecology with emphasis on the meso- and microfauna, see Coleman et al. (2018); for reviews of soil microbiology, see Sylvia et al. (2005).

roots are soil organisms that play many critical roles in the life of the soil community. The central theme will be how this community of organisms assimilates plant and animal materials, creating soil organic matter, recycling carbon and mineral nutrients, and supporting plant growth. A subtheme will be how people can manage soils to encourage a healthy, diverse soil community that efficiently makes nutrients available to plants, protects plant roots from pests and disease, and helps protect the global environment from some of the excesses of the human species.

10.1 THE DIVERSITY OF ORGANISMS IN THE SOIL[2]

Soil organisms are creatures that spend all or part of their lives in the soil environment. Every handful of soil is likely to contain billions of organisms, with representatives of nearly every phylum of living things. In this book we will emphasize organisms' activities rather than the scientific classification; consequently, we will consider only very broad, simple taxonomic categories (Table 10.1).

Table 10.1
SELECTED SOIL ORGANISMS GROUPED BY SIZE AND FOOD SOURCE

Generalized grouping by body width and food source	Major taxonomic groups	Examples
Macro-organisms (>2 mm)		
All heterotrophs: herbivores, detritivores, fungivores, bacterivores, and predators	Vertebrates	Gophers, moles, snakes, salamanders
	Arthropods	Ants, beetles and their larvae, centipedes, grubs, maggots, millipedes, spiders, termites, large collembola
All heterotrophs: herbivores, detritivores, fungivores, bacterivores	Annelids	Earthworms
	Mollusks	Snails, slugs
Largely autotrophs	Vascular plants	Feeder roots
	Bryophytes	Mosses
Meso-organisms (0.1–2 mm)		
All heterotrophs: detritivores, fungivores, bacterivores, and predators	Arthropods	Mites, collembola (springtails), pseudoscorpions
All heterotrophs: detritivores, fungivores, bacterivores	Annelids	Enchytraeid (pot) worms
Microorganisms (<0.1 mm)		
All heterotrophs: detritivores, predators, fungivores, bacterivores	Nematoda	Nematodes
	Rotifera	Rotifers
	Tardigrades	Water bears, *Macrobiotus sp.*
	Protozoa	Amoebae, ciliates, flagellates
Largely autotrophs	Vascular plants	Root hairs
	Algae	Greens, yellow-greens, diatoms
Largely heterotrophs	Fungi	Yeasts, mildews, molds, rusts, mushrooms
Heterotrophs and autotrophs	Bacteria	Acidobacteria, proteobacteria
	Cyanobacteria	Blue-green algae
	Actinobacteria	Streptomyces
	Archaea	Methanotrophs, *Thermoplasma sp.*, halophiles

[2]For beautifully illustrated, up-close and personal introduction to the actors in the soil drama, see Nardi (2007).

Based on similarities in genetic material, biologists classify all living organisms into three primary domains: **Eukarya** (which includes all plants, animals, and fungi), **Bacteria**, and **Archaea**. Organisms can also be grouped by what they "eat." Some organisms subsist on living plants (**herbivores**), others on dead plant debris (**detritivores**). Some consume animals (**predators**), some devour fungi (**fungivores**) or bacteria (**bacterivores**), and some live off of, but do not consume, other organisms (**parasites**). **Heterotrophs** rely on organic compounds (originally synthesized by other life forms) for their carbon and energy needs while **autotrophs** obtain their carbon mainly from carbon dioxide and their energy from photosynthesis or oxidation of various elements.

Sizes of Organisms. Soil organisms range in size from the tiniest virus to vertebrates some 10,000,000 times larger (Figure 10.1). The animals (*fauna*) of the soil range from **macrofauna** (>2 mm, moles, gophers, earthworms, and millipedes) through **mesofauna** (0.1–2 mm, springtails mites and other arthropods) to **microfauna** (<0.1 mm, nematodes, rotifers and single-celled protozoans). Plants (*flora*) include the roots of higher plants, as well as single-cell algae and diatoms. **Microorganisms** (too small to be seen without the aid of a microscope) belonging to the fungi, bacteria, and archaea tend to predominate in terms of numbers, mass, and metabolic capacity.

A typical, healthy soil might contain several species of vertebrate animals (mice, gophers, snakes, etc.), a half dozen species of earthworms, 50 species of arthropods (mites, collembola, beetles, ants, etc.), 100 species of nematodes, hundreds of species of fungi, and perhaps thousands of species of bacteria and archaea. As in any ecosystem, organism abundance (number of individuals) is generally inversely related to size.

Geographic Diversity. Microbial communities almost everywhere contain many of the same species and certain common microbial species can be found in almost any soil sample (see Section 10.10). Studies of sites from around the world using molecular techniques to identify species (or genetically distinct organisms) present and have found that geographic and soil-related variations in organism diversity are more apparent for soil animals than for microorganisms (Figure 10.2).

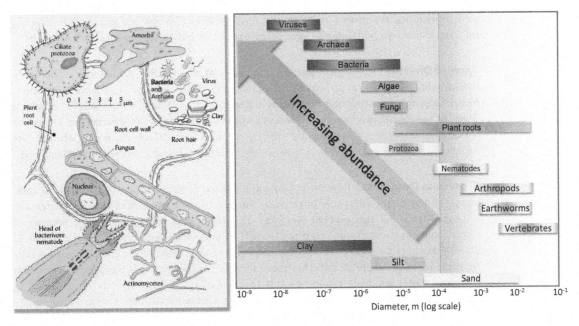

Figure 10.1 (Left) *Representative groups of soil microorganisms, showing their approximate relative sizes. The large white-outlined structure in the center background is a plant root cell.* (Right) *The size ranges represented by various groups of soil organisms on a logarithmic scale, with size ranges for sand, silt, and clay included for reference. The blue background indicates the microscopic range. The large arrow indicates that smaller organisms are more numerous.* (Diagrams courtesy of Ray R. Weil)

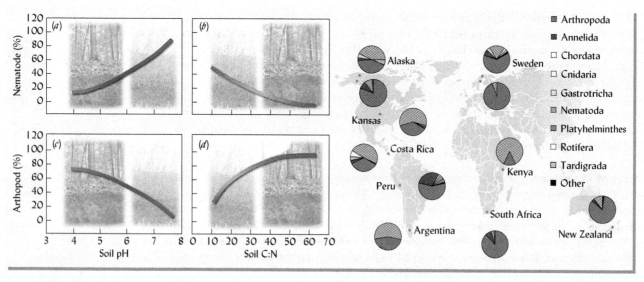

Figure 10.2 *Global distribution of soil animal species is influenced by soil pH and the ratio of C to N in the soil organic matter. In a study of 42 plots at 11 locations around the world (map) some 2,259 species of soil animals from ten different major groups (phyla) were identified by molecular techniques. At all the locations, the great majority of species belonged to just two phyla, arthropods (mites and collembola) and/or nematodes (red and green in the small pie charts). In forested soils with low pH levels and high C/N ratios a high percentage of the species present were arthropods, while in grassland soils with higher pH and low C/N ratios more of the species were nematodes.* [Modified from Wu et al. (2011)]

Diversity and Isolation. Tremendous diversity is possible because of the nearly limitless variety of foods and the wide range of habitat conditions found in soils. Within a handful of soil there may be areas of good and poor aeration, high and low acidity, cool and warm temperatures, moist and dry conditions, large and small hiding places, and localized concentrations of dissolved nutrients, organic substrates, and competing organisms. The populations of soil organisms tend to be concentrated in zones of favorable conditions, rather than evenly distributed throughout the soil. The soil aggregate (Section 4.5) can be considered a fundamental unit of habitat for meso- and microorganisms, providing a complex range of hiding places, food sources, environmental gradients, and genetic isolation on a microscale. Due to physical and genetic isolation, microorganisms living in different aggregates may go down their separate evolutionary paths.

Types of Diversity. Biological diversity is an indicator of soil health. A high **species diversity** indicates that the organisms present are fairly evenly distributed among a large number of species. Most ecologists believe that such complexity and species diversity are usually paralleled by a high degree of **functional diversity**—the capacity to utilize a wide variety of substrates and carry out a wide array of processes. Ecologists often turn to the soil as a model ecosystem with which to test such fundamental tenets of ecology such as how the level of biodiversity affects ecological functions.

Ecosystem Dynamics. In most healthy soil ecosystems there are several—and in some cases many—different species capable of carrying out each of the thousands of different enzymatic or physical processes that proceed every day. This **functional redundancy**—the presence of several organisms to carry out each task—leads to both ecosystem **stability** and **resilience**. *Stability* describes the ability of soils, even in the face of wide variations in environmental conditions and inputs, to continue to perform such functions as the cycling of nutrients, assimilation of organic wastes, and maintenance of soil structure. *Resilience* describes the ability of the soil to "bounce back" to functional health after a severe disturbance has disrupted normal processes.

Given a high degree of diversity, no single organism is likely to become completely dominant. By the same token, the loss of any one species is unlikely to cripple the entire system. Nonetheless, for certain soil processes, such as ammonium oxidation (see Section 12.1), methane oxidation (see Section 11.10), or the creation of aeration macropores (see Section 4.8), primary responsibility may fall to only one or a few species. The activity and abundance of a few **keystone species** (e.g., certain nitrifying bacteria or burrowing earthworms) may indicate the health of the entire soil ecosystem.

Genetic Resources. The diversity of organisms in a soil is important for reasons in addition to the safeguarding of ecological functions—soils also make an enormous contribution to **global biodiversity**. Many scientists believe that Earth's land areas host more species belowground than aboveground and that most of those species have yet to be discovered and described. The soil is therefore a major storehouse of the genetic innovations that nature has written into the DNA code over hundreds of millions of years. Humans have always found ways to make use of some of the genetic material in soil organisms (making beer and yogurt are examples). In modern times antibiotic medicines derived from soil microbes have played a major role in saving human lives. Advances in bioengineering are opening up the possibility that genes from soil organisms may be used to produce useful compounds at industrial scale and create plants and animals of superior utility to the human community.

10.2 ORGANISMS IN ACTION[3]

Trophic Levels and the Soil Food Web

As one organism "eats" or metabolizes its food, the nutrients and carbon contained therein are said to be passed from one **trophic level** to a higher one. The first trophic level is that of the **primary producers** such as green plants that *produce* energy-rich organic substances from abiotic components of the environment (sunlight, CO_2, etc.). The second trophic level consists of the **primary consumers** that eat those producers or their dead products. The third trophic level would be those **predators** and **microbial feeders** that eat the primary consumers; the fourth level would be predators that eat predators, and so on. The complex web of arrows in Figure 10.3 illustrate why ecologists describe these relationships with the term **food** *web* rather than the commonly used term food *chain*.

Sources of Energy and Carbon

Soil organisms may be classified as either **autotrophic** or **heterotrophic** based on where they obtain the *carbon* needed to build their cell constituents (Table 10.2). Heterotrophic soil organisms obtain their carbon from the breakdown of organic materials previously produced

Table 10.2
METABOLIC GROUPING OF SOIL ORGANISMS ACCORDING TO THEIR SOURCE OF METABOLIC ENERGY AND THEIR SOURCE OF CARBON FOR BIOCHEMICAL SYNTHESIS

	Source of energy	
Source of carbon	Biochemical oxidation	Solar radiation
Combined organic carbon	Chemoheterotrophs: All animals, plant roots, fungi, actinomycetes, and most bacteria Examples: Earthworms *Aspergillus sp.* *Azotobacter sp.* *Pseudomonas sp.*	Photoheterotrophs: A few algae
Carbon dioxide or carbonate	**Chemoautotrophs**: Some bacteria, many archaea Examples: Ammonia oxidizers—*Nitrosomonas sp.* Sulfur oxidizers—*Thiobacillus denitrificans*	**Photoautotrophs**: Plant shoots, algae, and cyanobacteria Examples: *Chorella sp.* *Nostoc sp.*

---

[3]For a succinct and well-illustrated summary of soil communities and ecological dynamics, see USDA/NRCS (2000). For a discussion of some basic principles of soil ecology and function, see Fierer et al. (2009). For a fascinating video of micro- and mesofauna in action, see https://www.youtube.com/watch?v=VuHznslr8aI.

by other organisms. Nearly all heterotrophs also obtain their *energy* from the oxidation of the carbon in organic compounds. They are responsible for organic decay and most soil food web processes. In most soils, these organisms, which include all soil animals, the fungi, actinomycetes, and most other bacteria, are far more numerous than the autotrophs.

The autotrophs obtain their carbon from simple carbon dioxide gas (CO_2) or carbonate minerals, rather than from carbon already fixed in organic materials. Autotrophs can be

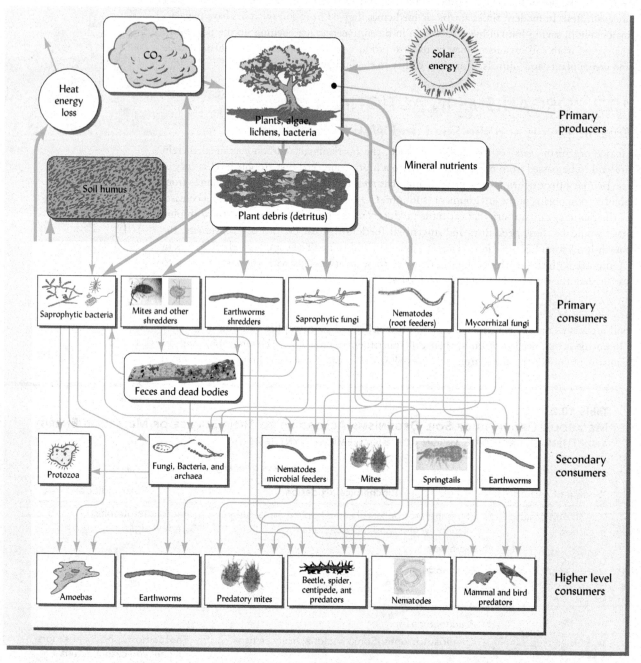

Figure 10.3 *Generalized diagram of the soil food web involved in the breakdown of plant tissue, the formation of humus, and the cycling of carbon and nutrients. The large shaded compartment represents the community of soil organisms. The rectangular boxes represent various groups of organisms; the arrows represent the transfer of carbon from one group to the next as predator eats prey. The thick arrows entering the top of the shaded compartment represent primary consumption of carbon originating from the tissue of the producers—green plants, algae, and cyanobacteria. The rounded boxes represent other inputs that support, and outputs that result from, the soil food web. Although all groups shown play important roles in the process, some 80 to 90% of the total metabolic activity in the food web can be ascribed to the microbes: fungi, bacteria, and archaea. As a result of this metabolism, soil is created and stabilized, and carbon dioxide, heat energy, and mineral nutrients are released into the soil environment.* (Diagram courtesy of Ray R. Weil)

further classified based on how they obtain energy. Some use solar energy (photoautotrophs), while others use energy released by the oxidation of inorganic elements such as nitrogen, sulfur, and iron (chemoautotrophs). While the autotrophs are in the distinct minority, their carbon fixation and inorganic oxidation reactions allow them to play crucial roles in the soil system. The autotrophs are mainly algae, cyanobacteria, and certain other bacteria and archaea. The aboveground shoots of plants are autotrophic, while the belowground roots behave somewhat as heterotrophs, metabolizing organic compounds synthesized by the shoot and transported to the root.

Primary Producers

As in most aboveground ecosystems, vascular plants play the principal role as primary producers. By combining carbon from atmospheric carbon dioxide with water, using energy from the sun, the producer organisms make organic molecules and living tissues. In this, they are commonly joined by mosses, algae, lichens, and photosynthesizing bacteria. The producers form the food base for the entire soil food web.

Primary Consumers

As soon as a leaf, a stalk, or a piece of bark drops to the ground, it is subject to coordinated attack by microbes and by macro- and mesofauna (see Figure 10.3). Animals, including mites (see chapter opener image), collembola, woodlice, and earthworms, chew or shred the tissue, opening it up to more rapid attack by the microbes. The animals and microbes that use the energy stored in the plant residues are termed *primary consumers*.

Herbivores. Soil organisms that eat live plants are called **herbivores**. Examples are parasitic nematodes and insect larvae that attack plant roots, as well as termites, ants, beetle larvae, woodchucks, and mice that devour above- or belowground plant parts. Because they attack living plants that may be of value to humans, many of these soil herbivores are considered pests.

Detritivores. For the vast majority of soil organisms the principal source of food is the debris of dead tissues left by plants on the soil surface and within the soil pores. This debris is called **detritus**, and the animals that directly feed on it are called **detritivores**. Both herbivores and detritivores that eat the tissues of primary producers are considered primary consumers. However, many of the animals that chew up plant detritus actually get most of their nutrition from the microorganisms that live in the detritus, not from the dead plant tissues themselves. Such animals are not really primary consumers (second trophic level), but belong to higher trophic levels as they metabolize either primary consumers microbes or the detritus from other animals (fecal pellets, dead bodies) (see "Secondary Consumers," following).

Saprotrophic Microorganisms. On balance, most actual decomposition of dead plant and animal debris is carried out by **saprotrophic** (feeding on dead tissues) microorganisms: fungi, bacteria, and archaea. With the assistance of the animals that physically shred and chew the plant debris (see following), the saprotrophs break down all kinds of plant and animal compounds, from simple sugars to woody materials (Figure 10.3). Saprotrophs feed on detritus, dead animals (corpses), and animal feces. These microorganisms commonly grow to form large colonies of microbial cells on the decaying material. These microbial colonies, in turn, soon provide nutrition to a myriad of other soil organisms—the secondary consumers.

Secondary Consumers

Predators and Microbial Feeders. The bodies (cells) of primary consumers become food sources for an array of predators and parasites in the soil. These *secondary consumers* include microbes, such as bacteria and fungi, as well as **carnivores**, animals which consume other animals. Examples of carnivores include centipedes (Figure 1.4) and mites that attack small insects or nematodes, spiders, predatory nematodes, and snails.

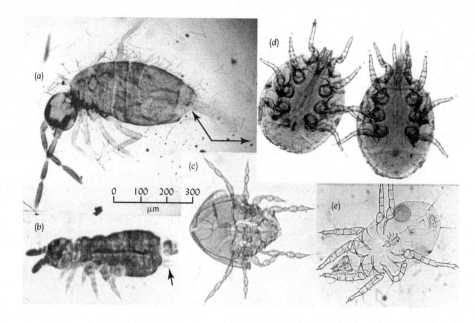

Figure 10.4 Collembola (a,b) and mites (c,d,e) are prominent among the mesofauna that play important roles in soil food webs. Collembola are also called springtails because of their springlike furcula or "tail" (arrows). The larger springtail (a) lives in and on the O horizons where there is light to see (hence the operational eyes) and room to hop (hence the well-developed furcula). The smaller springtail (b) lives in the mineral soil horizons where there is no light and no room to hop (hence it has only tiny vestigial "eyes" and furcula). Both springtails shown probably eat detritus and the fungi growing in it. The Orabatid mite (c) also feeds on detritus. The Mesostigmatic mites (d,e) are predators of smaller arthropods, nematodes, and other denizens of the soil. Note that mites (arachnids) have eight legs, while springtails (insects) have six. (Photo courtesy of Ray R. Weil)

Microbivorous feeders, organisms that use microbes as their source of food, include certain collembola (springtails), mites (Figure 10.4), termites, certain nematodes (see chapter opening photo), and protozoa. These microphytic feeders exert considerable influence over the activity and growth of fungal and bacterial populations. The grazing by these meso- and microfauna on microbial colonies may stimulate faster growth and activity among the microbes, in much the same way that grazing animals can stimulate the growth of pasture grasses. In other cases, the attack of the microphytic feeders may kill off or inhibit the microorganisms.

Stimulation of Decomposition. The mesofauna and macrofauna enhance the activity of the microbes in several ways. First, their chewing action fragments the litter, cutting through the resistant waxy coatings on many leaves to expose the more easily decomposed cell contents for microbial digestion. Second, the chewed plant tissues are thoroughly mixed with microorganisms in the animal gut, where conditions are ideal for microbial activity. Third, the mobile animals carry microorganisms with them and help the latter to disperse and find new food sources to decompose.

Tertiary Consumers

Predators. In the next level of the food web, the secondary consumers are prey for *tertiary consumers*. For example, ants consume centipedes, spiders, mites, and scorpions—all of which can themselves prey on primary or secondary consumers. Many species of birds specialize in eating soil animals such as beetles and earthworms. Robins are widely known to pull earthworms from their burrows, while other birds such as gulls and blackbirds commonly follow behind a farmer's plow to feed on the exposed earthworms and other soil macrofauna. Such mammals as moles can be effective predators of macrofauna. Whether the prey is a nematode or an earthworm, predation serves important ecological functions, including the release of nutrients tied up in the living cells.

Microbial Decomposers. Microorganisms are intimately involved in every level of decomposition. Some live freely on soil particles and plant root surfaces; others exist within larger organisms in the soil. Thus, some microbes directly attack plant tissues (as primary consumers), while other microorganisms active within the digestive tracts of soil animals help these animals digest more resistant organic materials. Microbes also attack the finely shredded organic material in animal feces, as well as decompose the bodies of dead animals. For these reasons, while soil animals may chew on organic materials, the microbes are referred to as the ultimate decomposers.

Diversification Within Broad Groups. As indicated in Figure 10.3, broad groups of organisms, such as fungi or mites, include some species that are primary consumers, some that are secondary consumers, and others that are tertiary consumers. For example, some mites attack detritus directly, while others eat mainly fungi or bacteria that grow on the detritus. Still others attack and devour the mites that eat fungi. Two-way interactions exist between many groups of organisms. For example, some nematodes eat fungi, and some fungi attack nematodes.

Ecosystem Engineers[4]

Certain organisms make major alterations to their physical environment that influence the habitats of many other organisms in the ecosystem. These organisms are *ecosystem engineers*. For example, some of these "engineer" species are microorganisms that create an impermeable surface crust that spatially concentrates scarce water supplies in certain desert soils.

Burrowing Animals. Others ecological engineers are burrowing animals that create opportunities and challenges for other organisms by digging channels that greatly alter air and water movement in soils. Termites (see Section 10.5) and ants, for example, may literally invert the soil profile in local areas (bringing subsoil material to the surface) and denude the soil surface in wide areas. Earthworms create extensive channels through the soil and incorporate plant residues into the mineral soil by passing them through their bodies along with mineral soil particles (see Section 10.4). Larger animals, such as gophers (see Figure 2.22), moles, prairie dogs, and rats, also burrow into the soil and bring about considerable soil mixing and granulation. Not only do such burrows encourage more water and air to enter the soil, they also provide passages that plant roots can easily follow to penetrate dense subsurface soil layers. Furthermore, the underground chambers and burrows made by the engineers provide new kinds of habitat in which other organisms (from frogs to fungi) soon take up residence.

Dung Beetles. Certain beetles of the *Scarabaeidae* family greatly enhance nutrient cycling by burying animal dung in the upper soil horizons. Many of these **dung beetles** cut round balls from large mammal feces, enabling them to roll the dung balls to a new location (Figure 10.5). The female dung beetle then lays her eggs in the ball of dung and buries it in the soil. Dispersal and burial of the dung not only provides a food source for the beetle larvae, but also protects the nutrients in the manure from easy loss by runoff or volatilization—fates to which the nutrients would most likely succumb if left on the soil surface. Dung beetles therefore play important roles in nutrient cycling and conservation in many grazed ecosystems.

Several thousand different dung beetle species are known worldwide, many having evolved to specialize in the burial of dung from particular mammal species (elephant, cattle,

Figure 10.5 A dung beetle (Scarabaeidae) rolls a ball that it has fashioned out of cow dung on a pasture soil in Tanzania. The female will lay her eggs in the ball of dung and bury it in the soil. Burying the dung is very important for making the nutrients therein available to the soil food web. Dung burial also prevents the reproduction of carnivorous flies and other pests of dung-producing mammals. Different dung beetles have evolved to specialize in the burial of dung from particular species of animals. (Photo courtesy of Ray R. Weil)

---

[4]For a discussion of the ecosystem engineer concept in soils, see Jouquet et al. (2006) and Lavelle et al. (1997). For applications of the concept to coral reef and desert soils, see Alper (1998).

buffalo, etc.). Rapid burial of dung by beetles also prevents the reproduction of carnivorous flies and other pests of large dung-producing mammals. Where native dung-burying "engineer" species are lacking in a grazed ecosystem (such as a savanna, a prairie, or a pasture), scientists have found that the introduction of appropriate species of dung beetles or earthworms can greatly increase the amount of vegetation produced and the number of grazing animals supported.

10.3 ABUNDANCE, BIOMASS, AND METABOLIC ACTIVITY

Soil organism numbers are influenced primarily by the amount and quality (especially the carbon and nitrogen content) of food available. Physical factors (e.g., moisture and temperature), biotic factors (e.g., predation and competition), and chemical characteristics of the soil (e.g., pH, dissolved nutrients, and salinity) also affect their numbers. Cultivated fields are generally lower than undisturbed native lands in numbers and biomass of soil organisms, especially the fauna, partly because tillage destroys much of their soil habitat.

Total **soil biomass**, the living fraction of the soil, is generally related to the amount of organic matter present. On a dry-weight basis, the living portion is usually between 1 and 5% of the total soil organic matter (see Section 11.6). To put this amount of biomass in perspective, one hectare of surface soil containing 3% soil organic matter might contain 600–3,000 kg (1–5 pickup truck loads) of living creatures—not counting the plant roots!

Comparative Organism Activity

The importance of specific groups of soil organisms is commonly identified by: (1) the numbers of individuals in the soil, (2) their weight (biomass) per unit volume or area of soil, and (3) their metabolic activity (often measured as the amount of carbon dioxide given off in respiration). The numbers and biomass of groups of organisms that commonly occur in soils are shown in Table 10.3. Although the relative metabolic activities are not shown, they are generally related

Table 10.3
NUMBERS AND BIOMASS OF ORGANISMS COMMONLY FOUND IN SURFACE SOIL HORIZONS
Microorganisms and earthworms dominate the biomass of most humid region soils.

Organisms	Biodiversity Taxa[c]	Number[a] Per m²	Per g	Biomass[b] kg/ha	g/m²
Microorganisms					
Bacteria and Archaea[d]	1–9,000/g	10^{14}–10^{15}	10^9–10^{10}	400–5,000	40–500
Actinomycetes		10^{12}–10^{13}	10^7–10^8	400–5,000	40–500
Fungi	1–300/g	10^6–10^8 m	10–10^3 m	1,000–15,000	100–1,500
Algae	—	10^9–10^{10}	10^4–10^5	10–500	1–50
Fauna					
Protista (protozoa)	1–5,000/g	10^7–10^{11}	10^2–10^6	20–300	2–30
Nematodes	10–1,000/m²	10^5–10^7	1–10^2	10–300	1–30
Mites	100–500/m²	10^3–10^6	1–10	2–500	0.2–5
Collembola	10–100/m²	10^3–10^6	1–10	2–500	0.2–5
Earthworms	2–10/m²	10–10^3		100–4,000	10–400
Other fauna	30–3,000/m²	10^2–10^4		10–100	1–10

[a] A fungus individual is hard to define, so abundance is given as meters hyphal length.
[b] Biomass values are on a live weight basis. Dry weights are about 20 to 25% of these values.
[c] Species or genome equivalents per unit of soil (g or m² of A horizon).
[d] Estimated numbers of bacteria and archaea from Torsvik et al. (2002); biodiversity estimates based on Bardgett and van der Putten (2014); other estimates from many sources.

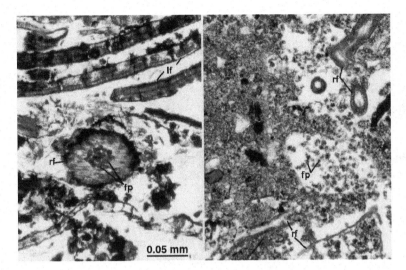

Figure 10.6 *Influence of mesofauna on soil structure in an O horizon (left) and an A horizon (right) of a forested Ultisol in Tennessee. The labeled features include leaf fragments (lf), root fragments (rf), and fecal pellets (fp). In the O horizon, the fecal pellets were left by mites that were eating out the inside of a pine tree needle (seen in cross section). In the A horizon, fecal pellets are seen in an oval biopore.* (Photo courtesy of Debra Phillips/Queen's University Belfast)

to the biomass of the organisms. Concentrations of microbial activity (hotspots) occur in the immediate vicinity of living plant roots and decaying bits of detritus, in the organic material lining earthworm burrows, in fecal pellets of soil fauna (Figure 10.6), and in other favored soil environments. Microorganisms dominate the biological activity in most soils. Despite their relatively small total biomass, such microfauna as nematodes and protozoa play important roles in nutrient cycling by preying on bacteria and fungi.

In this chapter we will focus largely on both microorganisms and certain keystone or ecological engineering macrofauna. We will begin with the earthworms.

10.4 EARTHWORMS[5]

Earthworms (along with their much smaller cousins, the **enchytraeid worms**) are egg-laying *hermaphrodites* (organisms without separate male and female genders) that eat detritus, soil organic matter, and, most importantly, microorganisms found on these materials (Figure 10.7). They do not eat living plants or their roots, and so do not act as pests to crops. Earthworms are often keystone species in soil ecosystems and can have major impacts on soil function.

Epigeic, Endogeic, and Anecic Earthworms. The 7,000 or so species of earthworms reported worldwide can be grouped according to their burrowing habits and habitat. The relatively small **epigeic** earthworms live in the litter layer or in the organic-rich soil very near the surface. Epigeic earthworms, which include the common compost worm, *Eisenia foetida*, hasten decomposition of litter but do not mix it into the mineral soil. **Endogeic** earthworms, such as the pale, pink *Allolobophora caliginosa* (known as "red worm"), live mainly in the upper 10 to 30 cm of mineral soil where they make shallow, largely horizontal burrows. Finally, the large **anecic** earthworms make vertical, relatively permanent burrows as much as several meters deep. They emerge in wet weather or at night to forage on the surface for pieces of litter that they drag back into their burrows, often covering the entrance to their burrow with a **midden** of leaves (Figure 10.7d). The well-known "night crawler" (*Lumbricus terrestris*) was accidentally introduced to North America from Europe (perhaps in the root balls of fruit trees) and is now the most common anecic earthworm on both continents.

Burrows. Earthworms burrow through the soil by first elongating and narrowing their flexible bodies, "worming" into an existing crack or plane of weakness, and then exerting

---

[5]For extensive information on earthworm ecology, biology, and distribution, see Edwards (2004). For a review of earthworm impacts on soil ecosystem services, see Blouin et al. (2013). Edwards and Arancon (2004) provide a good discussion of earthworms in relation to agriculture, soil organic matter, and soil microbiology.

Figure 10.7 Anecic earthworm species such as these Lumbricus terrestris (a) come to the surface to feed on litter (c), excrete soil casts (a, arrows), and reproduce (b). They incorporate large amounts of plant litter into the soil and gather plant debris into piles called middens to cover their burrow entrances (d, 10 cm scale markings). Earthworms are perhaps the most significant macroorganism in soils of humid temperate regions, particularly in relation to their effects on the physical conditions of soils. [Photos (a) and (d) courtesy of Ray R. Weil; photos (b) and (c) Steve Groff/Cedar Meadow Farm, LLC.]

considerable lateral pressure to thicken their bodies and so widen the passage as they move ahead. In so doing, they create extensive systems of channels that may be empty or filled with earthworm casts. These channels offer important pathways by which plant roots can penetrate dense soil layers, especially in compacted soils (Figure 10.8b). Their extensive physical activity, which is particularly important in untilled soils (including grasslands and no-till croplands) has earned earthworms the title of "nature's tillers" (see Figure 10.8). The middens of plant litter and the circular burrow entrances (often hidden by middens) are important signs used to assess earthworm activity (Figure 10.7d).

Influence on Soil Fertility, Productivity, and Environmental Quality

Casts. In addition to plant litter, earthworms also ingest a great deal of soil—perhaps to 2 to 30 times their own weight of soil in a single day. In a single year, the earthworms in grassland and forest ecosystems typically ingest and process between 20 and 1,000 Mg of soil, the higher figure occurring on fertile soils in moist, tropical climates. After passing through the earthworm gut, ingested soil is expelled as globules called **casts** (Figure 10.7a). During the passage through the earthworm's gut, organic materials are thoroughly shredded and soil microaggregates are broken down, with new aggregates formed from the mineral and organic materials. Probably because of enhanced bacterial activity, earthworm casts are usually high in polysaccharides, which are credited with stabilizing the casts into granular structure (see Section 4.5). The casting behavior of earthworms generally enhances nutrient availability to plants. The casts deposited within the soil profile or on the soil surface, depending on the species of earthworm, are another sign used to assess earthworm activity in soil.

Nutrients. Earthworms hasten the cycling and increase the availability of mineral nutrients to plants in three ways. First, as soil and organic materials pass through an earthworm, they

Figure 10.8 Burrowing activities of earthworms and their effects on plant roots and the distribution of nitrogen from plant litter. (a) A glass cylinder was packed with layers of A and B horizon soil and wrapped to exclude light. The soil was moistened, a few grass seeds planted, a layer of leaves placed on the soil surface, and two anecic earthworms added. The photo, taken one week later, shows the soil riddled with burrows (some occupied by grass roots). (b) The profile of an Inceptisol after15 years of no-till crop production using cover crops. The close-up shows tomato roots following organic matter–lined earthworm channels through the subsoil at about 60 to 70 cm deep. (c) Data from soil columns (called mesocosms) in which corn litter labeled with ^{15}N (isotope of nitrogen) was incubated for three weeks with or without an earthworm (Lumbricus terrestris). After the incubation, the earthworms and remaining litter were removed, and corn plants were grown for 30 days. Without earthworms, most of the N remained in the litter. In contrast, the earthworms made nearly all the N available for biological processes, including plant uptake (which nearly doubled) and gaseous losses (which increased fivefold) (see Sections 13.6 and 13.9). Such studies demonstrate that earthworms dramatically hasten the cycling of nutrients in the soil–plant system. [Photo courtesy of Ray R. Weil; charts based on data in Amador and Gorres (2005)]

are ground up physically as well as attacked chemically by the digestive enzymes of the earthworm and its gut microorganisms. Compared to the bulk soil, the casts are significantly higher in bacteria, organic matter, and available plant nutrients. Roots growing down earthworm burrows also find rich sources of nutrients in the casts and burrow lining material.

Second, although earthworms may feed on detritus and soil organic matter of relatively low nitrogen, phosphorus, and sulfur concentrations, their own body tissues have high concentrations of these nutrients. When the earthworms die and decay, the nutrients in their bodies are readily released into plant-available form. Studies have shown that where earthworm populations are large, a major proportion of the N taken up by plants (50–90 kg N/ha) can be made available by this mechanism.

Third, physical incorporation of animal droppings and plant litter into the soil reduces the loss of nutrients, especially of nitrogen, by erosion and ammonia gas volatilization. However, the same action may increase losses by other pathways (Figure 10.8, *pie charts*).

Beneficial Physical Effects. Earthworms are important in other ways. The holes left in the soil (see, e.g., Figure 6.5) serve to increase aeration and drainage, an important consideration in plant productivity and soil development. In turfgrass, the mixing activity of earthworms can alleviate or reduce compaction problems and nearly eliminate the formation of an undesirable thatch layer. Under conditions of heavy rainfall, earthworm burrows may greatly increase the infiltration of water into the soils and reduce erosion of agricultural soils, but these benefits are dependent on farming practices that return large amounts of crop residues (Figure 10.9).

Deleterious Effects of Earthworms

Exposure of Soil on the Surface. Not all effects of earthworms are beneficial. For example, in the process of building its middens, *Lumbricus terrestris* has been observed to leave some 60% of the soil surface bare of residues (see Figure 10.7d). To some extent this action leaves

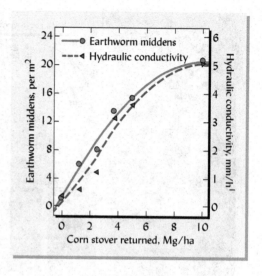

Figure 10.9 *Influence of crop residue on earthworm activity and soil hydraulic conductivity. The lower rates of residue return represent the situations when some or all of the corn stover is collected and removed for the production of ethanol biofuel or corn silage. The graph shows the effect of just 1 year of differing stover returns on numbers of earthworm middens present and the soil hydraulic conductivity (rate of water infiltration). The data suggest two conclusions: (1) there exists a close positive relationship between earthworm activity and soil infiltration rate, and (2) if more than about one-third of the corn stover is removed, significant damage may be done to soil quality needed to sustain future production of crops and biofuels. Unfortunately, many models of potential biofuel production fail to take this relationship into account. The data are averages for three medium- to fine-textured soils under long-term no-till management in Ohio.* [Graphed from data in Blanco-Canqui et al. (2007)]

the exposed soil susceptible to the impact of raindrops, crust formation, and increased erosion. In other situations, the middens themselves may be considered a nuisance—for example, on closely cropped golf greens. Even the burrowing and litter incorporation activities may not always be welcome—as in forests ecosystems with thick litter layers (see Box 10.1).

Influence on Chemical Leaching. Another aspect of concern is that water rapidly percolating down vertical earthworm burrows may carry potential pollutants toward the groundwater (see Figure 6.34). In artificially drained fields, earthworm channels may provide direct paths from the soil surface to the drain tiles, leading to increased leaching losses of soluble agrichemicals and nutrients. However, the organic matter–enriched material lining earthworm burrows has two to five times as great a capacity to adsorb certain herbicides as the bulk soil. Therefore, transport of such pollutants through earthworm burrows may be much less than is suggested by the mass flow of water through these large biopores.

Increased Greenhouse Gas Emissions.[6] Earthworms present something of a soil conundrum because the same activities that promote organic matter incorporation and nutrient cycling and aid plant growth also promote increased emissions of carbon dioxide and nitrous oxide, two major gases that drive global warming (see Sections 11.10 and 12.1). This mixed blessing of earthworm activity is illustrated by the data in Figure 10.8c where earthworm incorporation of plant litter substantially increased plant uptake of nitrogen, but also stimulated the loss of nitrogen as nitrous oxide gas.

On balance, earthworms are usually very beneficial, and soil managers would do well to encourage their activity in most, but not all, situations.

Factors Affecting Earthworm Activity

Earthworms prefer cool, moist, but well-aerated soils well supplied with decomposable organic materials, preferably supplied as surface mulch. In temperate regions they are most active in the spring and fall, often curling into a tight ball (aestivating) to ride out hot, dry periods in summer. They do not live under anaerobic conditions. Most earthworms thrive best where the soil is not too acid (pH 5.5–8.5) and has an abundant supply of calcium (which is an important component of their mucilage excretions). Enchytraeid worms are much more tolerant of acid conditions and are more active than earthworms in some forested Spodosols. Most earthworms are quite sensitive to excess salinity.

---

[6] A review of 237 observations from 57 published studies (Lubbers et al., 2013) showed that the presence of earthworms usually increased both CO_2 and N_2O gas emissions while having little effect on soil C accumulation.

BOX 10.1
GARDENERS' FRIEND NOT ALWAYS SO FRIENDLY[a]

Earthworms are legendary for the many ways they enhance the productivity of gardens, pastures, and croplands. However, in other ecological settings, "nature's tillers" can be quite damaging. In particular, these active immigrants are starting to wreak havoc in some of North America's forest ecosystems. Since the Pleistocene glaciers wiped out the native earthworms some 10,000 years ago, the plant–soil communities in these boreal forests have evolved without substantial earthworm populations. Without the soil-mixing action of earthworms, these forests have developed a thick, stratified forest floor, usually consisting of several distinct O horizons. This loose, thick litter layer is essential habitat for certain native plants (mayflowers, wood anemone, and trillium) and animals (millipedes, ground-nesting birds, and salamanders).

Recently, scientists have observed that the forest floor has all but disappeared from certain boreal forest stands (Figure 10.10). They think they know the culprit—invading populations of *Lumbricus terrestris* and similar *anecic*-type earthworms. This type of earthworm is particularly damaging to the forest ecosystem for precisely the same actions that make it so beneficial in gardens and pastures—the incorporation of surface litter into the soil and production of deep vertical burrows. These actions rapidly destroy the forest floor O horizons and greatly accelerate the normally conservative cycling of nutrients.

Some of the worst impacts are being felt in Minnesota where European earthworms (including *Aporrectodea sp.*, *Lumbricus rubellus*, and *L. terrestris*) have invaded the lake-dotted boreal forests. One approach to slowing the invasion of these uninvited soil engineers might be to surround the forest with buffer zones of habitat unsuitable for earthworms. Researchers believe the advance of these earthworms has been greatly abetted by their use as fishing bait. The take-home message: bring those bait worms home and dump them in your compost pile, not on the shore of your favorite fishing lake!

Figure 10.10 *Upper soil horizons and forest floor before (left) and after (middle) invasion of Minnesota forest by exotic anecic earthworms. Note the thick O horizon of decomposed litter before, and its absence after the earthworms did their work. The rapid decomposition caused the soil level to subside and tree roots to become exposed (right)* (Photos courtesy of University of Minnesota, David L. Hansen)

---
[a]For more on invasive earthworms, see Hendrix et al. (2008) and Hopfensperger et al. (2011).

Other factors that depress earthworm populations include predators (moles, mice, and certain mites and millipedes), very sandy soils (partly because of the abrasive effect of sharp sand grains), direct contact with ammonia fertilizer, application of certain insecticides (especially carbamates), and tillage. The last factor is often the overriding deterrent to earthworm populations in agricultural soils. Minimum tillage, with year-round vegetative cover and plenty of plant litter left as a mulch on the soil surface, is ideal for encouraging earthworms.

10.5 ANTS AND TERMITES[7]

We will now turn focus on two groups of macrofauna insects, ants and termites, which are considered to be important ecosystem engineers.

Ants

Nearly 9,000 species of soil-inhabiting ants have been identified. They are perhaps most functionally prominent in temperate semiarid grasslands. Ants also play important roles in forests from the tropics to the taiga. Some ant species act as detritivores, others as herbivores, and still others as predators.

Some ants feed on the sugary fluid (*honeydew*) produced by aphids as these sucking insects attack vascular plants. The ants may "raise" the aphids in their nest or "herd" them while they feed on plant sap. Ants belonging to the genus *Formica* have been shown to offer forest trees significant relief from canopy-feeding aphids, a single nest of these ants collecting and devouring some 10 kg of aphids in a single summer. Ant nest-building activity can improve soil aeration, increase water infiltration, and modify soil pH. The nests, and the microbial populations they encourage, also stimulate the cycling of soil nitrogen. Although ants occur widely and play vital and varied ecological roles, their cousins, the termites, have been much more intensively studied and will be considered next.

Termites

Termites live in complex social colonies found in about two-thirds of the world's land area (Figure 10.11). They are most prominent in the grasslands (savannas) and forests of tropical and subtropical areas (both humid and semiarid). Globally their activity is on a scale comparable to that of earthworms. In the drier tropics (less than 800 mm annual rainfall) termites surpass earthworms as the dominant soil fauna.

Diversity and Diet. Termites are sometimes called *white ants*, although they are quite distinct from ants (one can distinguish an ant by its narrow "waist" between the abdomen and thorax; compare in Figures 10.11). There are thought to be about 3,500 species of termites, most of which use cellulose as their primary food. Yet most termites cannot themselves digest cellulose. Instead, many termites depend on a symbiosis with certain protozoa and bacteria living in their gut. These termite gut microorganisms produce the enzymes that degrade

Figure 10.11 Termites and ants in the humid tropics. (Left) *Predatory army ants (black) attack and overwhelm two soldier termites (brown) guarding a nest.* (Right) *Soldier termites take up defensive positions inside their colony's mound in Liberia. Note the chambers in which fungi are cultivated and the soil particles glued together with termite saliva.* (Photos courtesy of Ray R. Weil)

---

[7] Ants are receiving increasing attention in the soil ecology literature (Anderson and Majer, 2004; Goheen and Palmer, 2010; Sanders and van Veen, 2011). Many descriptions of termite biology and ecological impacts are available (Ackerman et al., 2007; Eggleton, 2011; Sileshi et al., 2010).

cellulose and allow the termite to derive energy from it. Termites (and their gut microorganisms) are major contributors to the breakdown of organic materials near or on the soil surface. The metabolic action by certain Archaea that takes place under anaerobic conditions in termite guts also accounts for a substantial fraction of the global production of methane (CH_4), an important greenhouse (global warming) gas (see Section 11.10).

Most termites chew up dead plant material (decaying logs, fallen leaves, etc.), but some attack the sound wood in standing trees. The latter groups have become quite infamous because of their habit of invading (and subsequently destroying) the houses people build of wood. Most tree- and house-invading termites build protective tubes made of compacted soil and termite feces, which enable them to return to the soil for their daily water supply. Termite inspectors look for these earthen tunnels (about 1 cm in diameter) running up foundation walls under a house to indicate an infestation.

Mound-Building Activities. Termites are social animals that live in very complex labyrinths of nests, passages, and chambers that they build both below and above the soil surface. Termite mounds built from soil particles and feces cemented with saliva are characteristic features of many landscapes in Africa, Latin America, Australia, and Asia (Figure 10.12, *left*). Several species, such as *Macrotermes spp.* in Africa, use their mounds to "cultivate" fungi that they feed with cellulose-rich plant materials. The fungi have the enzymes necessary to digest the cellulose and the termites then harvest and eat the fungi from these "gardens."

In building their mounds, termites transport soil from lower layers to the surface, thereby extensively mixing the soil and incorporating into it the plant residues they use as food. Scavenging a large area around each mound, these insects remove up to 4,000 kg/ha of leaf and woody material annually, a substantial portion of the plant litter produced in many tropical ecosystems (Figure 10.13, *right*). They can also annually move more than 1,000 kg/ha of soil in their mound-building activities. These activities have significant impacts on soil formation, as well as on current soil fertility and productivity.

Each mound can provide a home for 1 million termites. Certain species grow wings and migrate by air to form new colonies (Figure 10.12, *right*). The mounds may be abandoned after 10 to 20 years and can then be broken down by farmers to level the land for crop production. Attempts to level an occupied mound are usually frustrating, as the termites rebuild very rapidly unless the queen (egg-laying) termite is destroyed.

Plant growth on active mounds may be poor or absent (see Figure 10.12, *left*), however, once the termite mound is abandoned, the resulting "island" of soil high in clay, nutrients, and sometimes organic matter is often vegetated by a more diverse and vigorously growing plant community than exists on the nontermite affected part of the landscape.

Figure 10.12 *Termites build nests above- and belowground and live in complex social colonies. (Left) A termite mound in a semiarid tropical savanna. The mound is constructed of subsoil material cemented hard with termite saliva. (Right) At the beginning of the rainy season, dozens of worker termites build funnel-shaped structures though which their colony mates—having grown wings—will exit their underground metropolis for migration.* (Photos courtesy of Ray R. Weil)

Figure 10.13 (Left) *Termites in semiarid Texas prevent the accumulation of a plant litter mulch by rapidly covering plant debris with a coating of cemented soil and then incorporating the material into their nests.* (Right) *Individual worker termites (striped) drag cut pieces of leaves into their underground nest as soldier termites (large heads and mandibles) stand guard.* (Photos courtesy of Ray R. Weil)

10.6 SOIL MICROANIMALS[8]

From the viewpoint of microscopic animals (microfauna), soils present many habitats that are essentially aquatic, at least intermittently so. For this reason, the soil microfauna are closely related to the microfauna found in lakes and streams. The two groups exerting the greatest influence on soil processes are the nematodes and protozoa.

Nematodes

Nematodes live in almost all soils, often in surprisingly great numbers (see Table 10.3) and diversity. Some 20,000 species have been identified of the 100,000 nematode species thought to exist. These unsegmented roundworms are highly mobile creatures about 4–100 μm in cross section and 40–1,000 μm in length. They wriggle their way through the labyrinth of soil pores, sometimes swimming in water-filled pores (like their aquatic cousins), but more often pushing off the moist particle surfaces of partially air-filled pores. The latter mode of locomotion helps explain why some nematodes are active and reproduce when the soil water potential is so low (see Section 5.4) that only pores smaller than 1 μm (far too small for nematodes to enter) are filled with water. Nematodes are very sensitive to soil water content and porosity. Moist, well-aggregated, or sandy soils typically have especially high nematode populations, as these soils contain abundant pores large enough to accommodate their movements. When the soil becomes too dry, nematodes survive by coiling up into a **cryptobiotic** or resting state, in which they seem to be nearly impervious to environmental conditions and use no detectable oxygen for respiration.

Feeding Habits. Most nematodes feed on fungi, bacteria, and algae or prey on other soil animals. Perhaps 10 to 20% of the nematodes in most soils are primary consumers that feed on living plants. The different trophic groups of nematodes can often be distinguished by the type of mouth parts present (Figures 10.14 and 10.15). Grazing by nematodes can have a marked effect on the growth and activities of fungal and bacterial populations; much like a fruit tree is affected by pruning, the microbial activity can be stimulated by light grazing or reduced by heavy grazing. Since bacterial cells contain more nitrogen than the nematodes can use, microbial-feeding nematodes excrete considerable soluble nitrogen. This nematode activity accounts for 30 to 40% of the plant-available nitrogen released in some ecosystems.

---

[8]For a detailed discussion of soil nematode ecology, see Neher (2010).

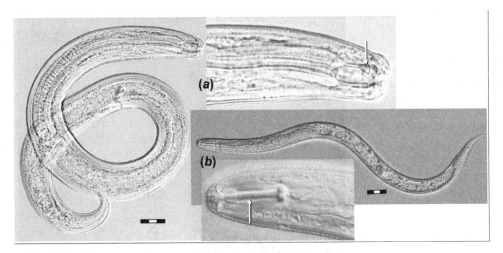

Figure 10.14 *A predator nematode (a, Mononchidae) and a plant root parasite (b, soybean cyst nematode). The head and mouth parts of a nematode often reflect its trophic role, as the inset enlargements of these nematodes illustrate. Predators usually have hard teeth (a, arrow) and a large mouth for capturing and swallowing prey. Nematodes that feed on plant roots or fungal hyphae have a retractable spearlike mouth part (b, arrow) that pierces the targeted cell to feed on the liquid contents. Scale bars marked in 10 µm units.* (Photos by Lisa Stocking Gruver, University of Maryland, courtesy of Joel Gruver, Western Illinois University)

Predatory nematodes, like wolves or lions aboveground, can have a major "top-down" influence on the whole food web community. Unlike many aboveground predators, most predatory nematodes are not larger than their prey—probably because the predators must be able to squeeze into the same small soil pores in which the prey is hiding. The prey can include other nematodes, protozoa, or insect larvae. In fact, nematodes that kill the larvae of insect pests in the soil are sold for use as biological control agents, an environmentally beneficial alternative to the use of toxic pesticides. These nematodes that attack insect larvae carry symbiotic (to the nematode) bacteria to kill the much larger insect by disease—hence these nematodes are referred to as entomopathogenic (insect disease causing). Once established, they can provide effective, long-term control of such soilborne insect pests such as the corn rootworm or the grubs (Japanese beetle larvae) that destroy homeowners' lawns (Figure 10.16).

Plant Parasites. Some nematodes, especially those of the genus *Heterodera*, can infest the roots of practically all plant species by piercing the plant cells with a sharp, spearlike mouth part. These wounds often allow infection by secondary pathogens and may cause the formation of knotlike growths on the roots. Minor nematode infestations are nearly ubiquitous and often have little deleterious effect on the host plant. In fact, low levels of root feeding may even stimulate greater root

Figure 10.15 *Head of bacteria-feeding nematode (Acrobeles ciliatus). The ornate structures swirl around during feeding to sweep bacteria into the mouth.* (Photo courtesy of Sven Boström, Swedish Museum of Natural History)

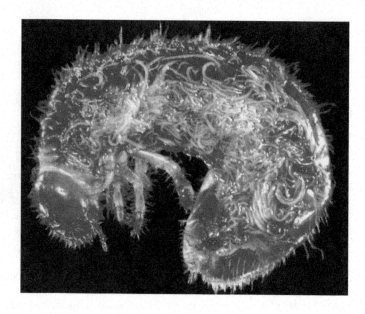

Figure 10.16 *Nematodes feeding inside an insect larva. These nematodes are entomopathogenic—a few nematodes invade the insect larvae, carrying bacteria that cause the larvae to quickly die of disease while the nematodes reproduce, multiply, and eat both the bacteria and the insect tissue. After a week or so, the insect larva has been consumed and the horde of new nematodes and their bacterial allies return to the soil from the insect and search for a new victim.* (Photo courtesy of Yi Wang & Randy Gaugler)

growth and plant health. However, infestations beyond a certain threshold level result in serious stunting of the plant. Cyst- (egg sac) forming nematodes are major pests of soybeans, while root-knot-forming nematodes cause widespread damage to fruit trees and solanaceous crops.

Until recently, the principal methods of controlling crop damaging nematodes were long rotations with nonhost crops (often a period of 5 years is required for the parasitic nematode populations to sufficiently dwindle), use of genetically resistant crop varieties, and soil fumigation with highly toxic chemicals (**nematicides**). The use of soil nematicides, such as methyl bromide, has been sharply restricted because of undesirable environmental effects.

New, less dangerous approaches to nematode control include the use of hardwood bark for containerized plants and interplanting or rotating susceptible crops with plants such as marigolds that produce root exudates with nematicidal properties (Figure 10.17). Certain plants in the Brassicaceae family (such as mustards, radishes, and rapeseed) contain sulfurous compounds called glucosinolates. When the residues of these plants break down in the soil, their glucosinolates produce volatile biofumigants that can kill certain nematodes. The use of these plants as green manures has successfully substituted for the use of more broadly toxic synthetic chemicals to control nematodes in potato, strawberry, and other nematode-sensitive crops. Progress has also been made in the development of nematode-resistant varieties of such plants as soybeans. Individually, each of these measures may provide only partial crop protection, but when applied together, such protective measures can be highly effective.

Protozoa

Protozoa are mobile, single-celled creatures that capture and engulf their food. With some 50,000 species in existence, they are the most varied and numerous microfauna in most soils (Table 10.3). Most are considerably larger than bacteria (see Figure 10.1), having a diameter range of 4 to 250 μm. Their cells do not have true cell walls and have a distinctly more complex organization than bacterial cells. Soil protozoa include amoebas (which move by extending and contracting pseudopodia), ciliates (which move by waving hairlike structures; Figure 10.18), and flagellates (which wave whiplike appendages called *flagella*). They swim about in the water-filled pores and water films in the soil and can form resistant resting stages (called *cysts*) when the soil dries out or food becomes scarce.

Sometimes as many as 40 or 50 different protozoa species may occur in a single sample of soil. The liveweight of protozoa in surface soil ranges from 20 to 300 kg/ha (see Table 10.3). A considerable number of serious animal and human diseases are attributed to infection by protozoa, but mainly by those that are waterborne, rather than soilborne. Most soil-inhabiting protozoa prey upon soil bacteria, exerting a significant influence on bacterial populations and decomposition pathways in soils.

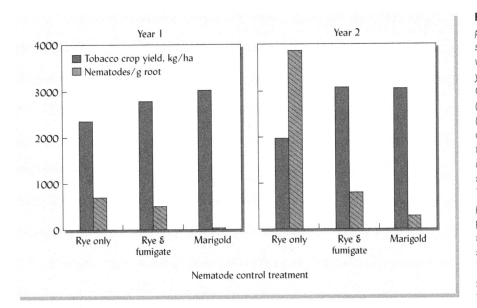

Figure 10.17 Marigolds control plant-parasitic nematodes. A susceptible host plant (tobacco) was grown in the summers of years 1 and 2 on sandy soil in Ontario, Canada. Untreated plots (rye only) and fumigated plots (rye and fumigate) were cover cropped with rye each winter. The fumigated plots were injected in years 1 and 2 with a chemical fumigant (1,3-dichloropropene). The remaining plots were planted to marigolds (Tagetes patula cv. "Creole") only in the summer of the year before the susceptible crop was first planted. The nematode infestation grew worse in the second year that a susceptible crop was grown (especially in rye-only plots) and without nematode control the crop yield declined substantially. Growing marigolds in 1 year controlled nematodes in susceptible crops for the following 2 years. [Based on data from Reynolds et al. (2000); used with permission of the American Society of Agronomy. For a review of nematode control using marigolds, see Hooks et al. (2010)]

Protozoa generally thrive best in moist but well-aerated soils and are most numerous in surface horizons. Protozoa are especially active in the area immediately around plant roots. Their main influence on organic matter decay and nutrient release is through their effects on bacterial populations. In pursuit of their bacterial prey, some soil-dwelling protozoa are adapted to squeezing into soil pores with openings as small as 10 μm. Still, soil aggregates often provide even smaller pores where bacterial cells can hide to escape predation. This protection from protozoa predation is a characteristic of the soil environment that helps explain the greater diversity of bacteria in soils than in aquatic habitats where such hiding places do not exist.

Other Fascinating Soil Microcreatures

Space limitations do not allow us to describe in detail the many other invertebrate animals of the soil. However, one soil life-form is simply too fascinating to omit: the ***Dictyostelium*** slime mold. These amoeba-like, eukaryotic single-celled microorganisms live in the litter layers and the upper few centimeters of soil, feeding on bacteria, much as protozoa do. However, when the local food supply runs out, a truly amazing phenomenon occurs that is illustrative of the degree to which soil organisms often act together. The individual ***Dictyostelium*** amoeboid cells begin to congregate. In response to a complex pattern of chemical signals, some 50,000 of these individual organisms stream together to form a mound about 0.1 mm in size. The cells

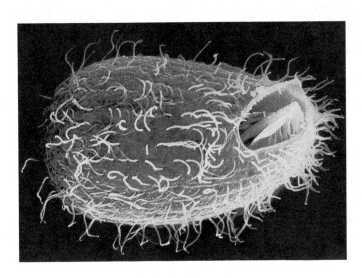

Figure 10.18 A ciliated protozoan (Glaucoma scintillans), a microanimal found in soils. Predation by protozoa exerts a major influence on the abundance, diversity, and activity of soil bacteria. (Photo courtesy of Dr. John O. Corliss)

in this mound then differentiate themselves into two types, most conglomerating together to form a "foot," while the others form a kind of "head"—the 50,000 cells now behaving as a single, multicellular organism! This tiny "creature" crawls, slug-like, up to the soil surface, where the cells undergo yet another rearrangement. Cells of the slug's "foot" transform into the stalk of a fruiting body, atop which the slug's "head" transforms into spores—which can now disperse through the air to new hunting grounds.

10.7 PLANT ROOTS[9]

Green plants store the sun's energy and are the primary producers of organic matter (see Figure 10.3). Their roots live and die in the soil and are therefore considered as soil organisms. They typically occupy about 1% of the volume of surface soil horizons and may be responsible for a quarter to a third of the respiration occurring in a soil. Roots usually compete for oxygen, but they also supply much of the carbon and energy needed by the soil community of fauna and microbes. The activities of plant roots greatly influence soil chemical and physical properties, the specific effects depending on the type of soil and plant in question (see, e.g., Section 7.7). As we shall see, plant roots interact with other soil organisms in varied and complex ways.

Root Morphology

Fine feeder roots range in diameter from 100 to 400 μm, while root hairs are only 10 to 50 μm in diameter—similar in size to the strands of microscopic fungi (see Figure 10.1). Certain cells in the root's outer skin, the epidermis, can elongate to form long, thin protuberances termed root hairs (Figure 10.19). One function of root hairs is to anchor the root as it pushes its way through the soil. They also maintain direct contact with soil particle and water films and increase the amount of root surface area available to absorb water and nutrients from the soil solution.

Roots grow by forming and expanding new cells at the growing point (meristem), which is located just behind the root tip. This growing point can exert surprisingly high pressures to enable it to push ahead through the soil; generally thicker roots can exert greater pressures than thin ones. To lubricate the root's push through the soil, certain cells near the tip exude a nonwater soluble gel-like substance termed **mucigel**. The root tip itself is shielded by a protective cap of expendable cells that slough off as the root pushes through the soil, leaving a trail of broken cells imbedded in the mucigel coating, the combination of which both attracts and provides food sources to throngs of mainly beneficial microorganisms (Figure 10.19).

Root morphology is affected by both, type of plant and soil conditions. For example, fine roots may proliferate in localized areas of high nutrient concentrations. Root-hair formation is stimulated by contact with

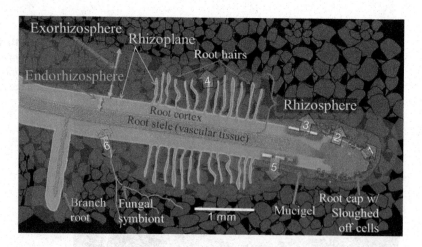

Figure 10.19 *Diagram of a root showing the rhizosphere (light brown shaded area) and its three zones: the endorhizosphere consisting of the root cortex where the solution is in contact with the soil solution, the exorhizosphere consisting of the soil in the immediate vicinity of the root and profoundly under its influence, and the rhizoplane consisting of the outer surface of the root including its root hairs. Six ways that roots add organic compounds to the rhizosphere are also shown (broad numbered arrows): (1) sloughing off of root cap cells; (2) excretion of mucigel; (3) spilling of cell contents when epidermal cells are lysed (broken); (4) exudation of specific compounds produced by root hairs; (5) exudation of various compounds by cortical cells; (6) export of organic compounds to symbiotic fungi (see Section 10.9).* (Diagram courtesy of Ray R. Weil)

[9]For a complete scientific review of all aspects of roots and their soil interactions, see Eshel and Beeckman (2013).

soil particles and by low nutrient supply. When soil water (or nutrient supply) is scarce, plants typically put more energy into root growth than into shoot growth, decreasing the shoot-to-root ratio—thus increasing the uptake of water and minimizing its loss by transpiration. Roots may become thick and stubby in response to high soil bulk density or high aluminum concentrations in soil solution (see Sections 4.7 and 9.7).

How Roots Alter Soil Conditions

Physically, roots alter the arrangement of particles, widening some pore spaces and creating new channels as they push through the soil. Roots follow paths of least resistance, growing between soil peds and into existing cracks and channels. Once extended into a pore, the root matures and expands, exerting lateral forces that enlarge that pore while compressing the soil around it. Roots also encourage soil shrinkage and cracking by removing moisture from the soil. This action, along with the chemical and biological effects discussed next, increases stable soil aggregation.

Roots chemically alter the soil around them by what they take out of it (depletion of nutrients and water) and what they put into it (root exudates and tissues). Root exudates help to further stabilize soil aggregates, both directly and indirectly by supporting a myriad of fungi and bacteria. In addition, when roots die and decompose, they provide input to soil organic matter, not only in the A horizon, but also throughout the soil profile. Generally, rooting is deepest in hot, dry climates and most shallow in boreal or wet tropical environments (see Figure 5.26).

Root contributions to maintain soil organic matter levels are generally more important than the more visible (to humans) aboveground plant residues. In grasslands, about 50 to 60% of the net primary production (total plant biomass) is commonly in the roots. In addition, grass fires may remove most of the aboveground biomass, so that the deep, dense root systems are the main source of organic matter added to these soils. In forests, 40 to 70% of the total biomass production may be in the form of tree roots. In arable soils, the mass of roots remaining in the soil after crop harvest is commonly 15 to 40% that of the aboveground crop. The mass of organic compounds contributed by roots is seen to be even greater when the rhizosphere effects discussed in the following subsections are considered.

Rhizosphere[10]

The zone of soil significantly influenced by living roots is termed the **rhizosphere** and extends out to about 2 to 3 mm from the root surface (Figure 10.19). The entire rhizosphere is sometimes considered to include three zones: (1) the **endorhizosphere** (inside the root) consisting of the root cortex where the cells are loosely packed and the intercellular spaces are filled with an extension of the soil solution and easily colonized by soil bacteria, (2) the **rhizoplane** consisting of the outer surface of the root including its root hairs adhering soil, and (3) the **exorhizosphere** consisting of the soil close enough to the root to be profoundly under its influence. The chemical and biological characteristics of the exorhizosphere can be very different from those of the bulk soil. Soil acidity is commonly 10 times (1 pH unit) higher or lower in the rhizosphere than in the bulk soil (e.g., see Figure 9.11). Roots greatly affect the nutrient supply in this zone. On one hand roots withdraw dissolved nutrients and so deplete this zone. On the other hand, roots exude compounds that hasten the release of nutrients from soil minerals and organic matter. By these and other means, roots affect the mineral nutrition of soil microbes, just as the microbes affect the nutrients available to the plant roots.

Rhizodeposition

Roots add organic compounds to the rhizosphere in at least six ways (see broad numbered arrows in Figure 10.19): (1) cells from the root cap and epidermis continually slough off as the root grows and enrich the rhizosphere with a wide variety of cell contents; (2) cells near the root tip excrete large quantities of a nonwater soluble mucilage gel (mucigel); (3) epidermal cells are lysed (broken) by friction or microbial attack and spill their cell contents into the adjacent soil;

---

[10]For a well-illustrated basic introduction to the rhizosphere with insights only an expert could provide, see McNear Jr. (2013).

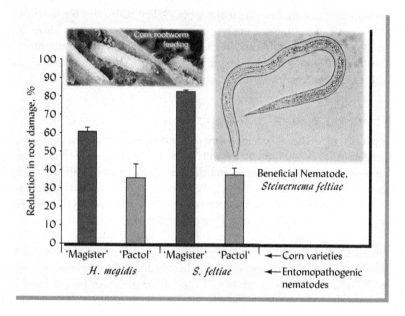

Figure 10.20 Example of signaling compound rhizodeposition. The Western corn rootworm (Diabrotica virgifera) feeds on corn roots and severely damages this major food crop in many parts of the world. As a defense, roots of some corn plants can release a volatile chemical ((E)-b-caryophyllene) when they first suffer damage from this insect larvae pest. This signaling compound diffuses into the soil and specifically "recruits" certain entomopathogenic nematodes that can attack and kill corn rootworms. However, the ability to release this signaling compound has been inadvertently lost from many high-yielding corn hybrids. Such nonsignaling hybrids get less help from nematodes and are more reliant on pesticides to minimize corn root worm damage. Under field conditions the beneficial nematodes H. megidis and S. feltiae reduced root worm damage much more effectively on a corn variety ("Megister") that produces the belowground signal than on a variety ("Pactol") that does not. [Bar graph based on data from Hiltpold et al. (2010)]

(4) epidermal cells, including root hairs, actively produce and exude specific compounds; (5) cortical cells passively leak—or in some cases actively exude—a wide variety of plant metabolites; and (6) certain cortical cells export organic compounds directly to symbiotic fungi. Many of the exuded compounds enable roots to communicate with other soil organisms (Figure 10.20).

The high-molecular-weight mucilages secreted by root-cap cells and epidermal cells near apical zones form a substance called **mucigel** when mixed with microbial cells and clay particles. Mucigel appears to have several beneficial functions: It lubricates the root's movement through the soil; it improves root–soil contact, especially in dry soils when roots may shrink in size and lose direct contact with the soil; its high polysaccharide content helps stabilize soil aggregates thus improving the physical conditions for both root and microbial growth; it may protect the root from certain toxic chemicals in the soils; and its agar-like properties provide an ideal environment for the growth of the rhizosphere microorganisms.

Taken together with the constant death and decay of older roots, these types of chemical additions to the soil are termed **rhizodeposition** and typically account for 2 to 30% of total dry-matter production in young plants. Sometimes when a plant is carefully uprooted from a

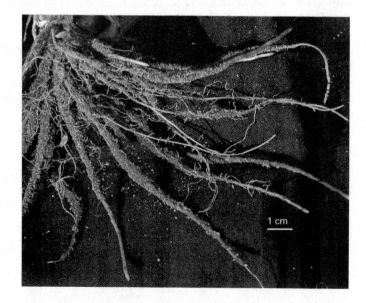

Figure 10.21 The rhizosphere is greatly enriched in organic compounds excreted by the roots. These exudates and the microorganisms they support as well as root hairs have caused the rhizosphere soil to adhere to these grass roots as a sheath, except near the root tips where the bare white root is visible. The soil is a sandy loam. (Photo courtesy of Ray R. Weil)

loose soil, these root exudates cause a thin layer of soil, approximating the rhizosphere, to form a sheath around the actively growing root tips (Figure 10.21). Because older roots are constantly dying and new roots constantly growing, the amount of organic material deposited in the rhizosphere during the growing season of annual plants may be more than twice the amount remaining in the root system at the end of the growing season. Rhizodeposition decreases with plant age but increases with soil stresses, such as compaction and low nutrient supply.

Because of the rhizodeposition of carbon substrates and specific growth factors (such as vitamins and amino acids), microbial numbers in the rhizosphere are typically 2 to 10 times as great as in the bulk soil (sometimes expressed as a *R/S ratio*). This effect and processes just described explain why plant roots are among the most important organisms in the soil ecosystem.

10.8 SOIL ALGAE

Like vascular plants, algae consist of eukaryotic cells, those with nuclei organized inside a nuclear membrane. (Organisms formerly called *blue-green algae* are prokaryotes and therefore will be considered with the bacteria.) Also like vascular plants, algae are equipped with chlorophyll, enabling them to carry out photosynthesis. As photoautotrophs, algae need light and are therefore mostly found very near the surface of the soil. Some species can also function as heterotrophs in the dark. A few species are photoheterotrophs that use sunlight for energy but cannot synthesize all of the organic molecules they require (see Table 10.3). In addition to producing a substantial amount of organic matter in some fertile soils, certain algae excrete polysaccharides that have strong favorable effects on soil aggregation (see Section 4.5).

Most soil algae range in size from 2 to 20 μm; many are motile and swim about in soil pore water, some by means of flagella (whiplike "tails"). Most prefer moist to wet conditions, but some are also very important in hot or cold desert environments. Certain algae (as well as cyanobacteria) form *lichens*, symbiotic associations with fungi. These are important in colonizing bare rock and other low organic matter environments (see Figure 2.7). In unvegetated patches in deserts, algae commonly contribute to the formation of **microbiotic crusts** (see Section 10.14).

10.9 SOIL FUNGI[11]

Fungi are eukaryotes (meaning their cell have a nuclear membrane and cell walls). Biochemically and genetically they comprise several groups of rather unrelated organisms. As heterotrophs, they depend on living or dead organic materials for both their carbon and their energy. Fungi typically either absorb small soluble organic molecules such as simple sugars or amino acid into their cells, or they first exude extracellular enzymes that break a complex substrate into such simple molecules prior to their absorption. Fungi are generally aerobic organisms, although some can tolerate the rather low oxygen concentrations and high levels of carbon dioxide found in wet or compacted soils. Strictly speaking, fungi are not entirely microscopic, since some of these organisms, such as mushrooms, form macroscopic structures that can easily be seen without magnification.

For convenience of discussion about their gross morphology and behavior in soils (but not their evolutionary relationships or current taxonomy), fungi have been traditionally divided into three groups: (1) yeasts, (2) molds, and (3) mushroom fungi. *Yeasts*, which are single-celled organisms, live principally in waterlogged, anaerobic soils. *Molds* and *mushroom fungi* are both considered to be filamentous fungi, because they are characterized by long, threadlike, branching chains of cells. Individual fungal filaments, called **hyphae** (Figure 10.22), are often twisted together to form **mycelia** that appear somewhat like woven ropes. Fungal mycelia are often visible as thin, white or colored strands running through decaying plant litter (Figure 10.23). Filamentous fungi reproduce by means of spores, often formed on fruiting bodies, which may be microscopic (e.g., molds, Figure 10.22) or macroscopic (such as that shown in Figures 10.24 and 10.25).

---

[11]For an in-depth but accessible and well-illustrated book on everything fungi—from ecological roles to how to grow them as culinary delights, see Stamets (2005).

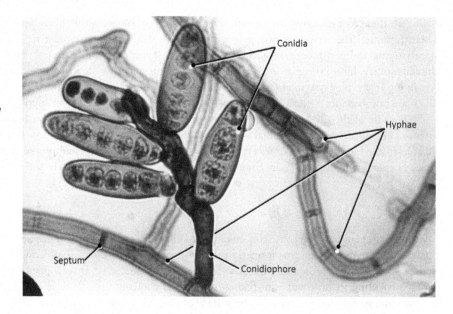

Figure 10.22 *A highly magnified soil fungus in the genus Helminthosporium, showing the long chains of cells called hyphae as well as a reproductive structure termed a conidiophore. The septa that separate the cells of the hyphae are clearly visible, as are the five multicellular fruiting bodies or conidia attached to the conidiophore.* (Photo by Hardin, courtesy of U.S. Department of Health and Human Services, Centers for Disease Control and Prevention)

Figure 10.23 *Fungal mycelia, consisting of bundles of microscopic hyphae, grow up from the soil into the leaves and woody debris of the forest floor. The ability of fungi to "reach out" from the soil in this manner helps explain why they dominate the decay of surface litter and mulch, while bacteria are more prominent in decaying organic material incorporated into the soil. Scale marked in cm, and see quarter dollar coin in upper left.* (Photo courtesy of Ray R. Weil)

Molds

The molds are distinctly filamentous, microscopic, or semimacroscopic fungi that play a much more important role in soil organic matter breakdown than the mushroom fungi. Molds develop vigorously in acid, neutral, or alkaline soils. Some are favored, rather than harmed, by lowered pH. Consequently, they may dominate the microbial community in acid surface soils, where bacteria and actinomycetes offer only mild competition.

Many genera of molds are found in soils. Four of the most common are *Penicillium, Mucor, Fusarium,* and *Aspergillus*. The complexity of the organic compounds available as food seems to determine which particular mold (or molds) prevail. Their biomass fluctuates greatly with soil conditions. In some cases much of the hyphal length measured may no longer be living.

Figure 10.24 White rot fungi include many species of Basidiomycota and are known for their ability to produce enzymes that breakdown lignin, cellulose, and hemicellulose. These fungi are able to metabolize materials that most microorganism cannot, including woody debris in forests. Fortunately, their extracellular enzymes are not specific for lignin, but can also be put to work cleaning up soils contaminated by other complex aromatic compounds such as synthetic dyes and pesticides. (Left) Fruiting bodies of white rot fungi decomposing a fallen tree trunk. (Right) Fruiting bodies of a Bird's Nest fungus (Cyathus sp.) contain disc-shaped spores ("eggs") that are scattered forcefully when the cone is hit by a raindrop. Like other white rot fungi, this fungus produces enzymes that easily break down such woody materials as forest litter or corncobs left on agricultural fields after harvest. Individual "nests" are about 1 cm in diameter. (Photos courtesy of Ray R. Weil)

Mushroom Fungi

These fungi are associated with forest and grassland vegetation where moisture and organic residues are ample. Although the mushrooms of many species are extremely poisonous to humans, some are edible—and a few have been domesticated and are grown as popular food items.

The aboveground fruiting body of most mushrooms is only a small part of the total organism. An extensive network of hyphae permeates the underlying soil or organic residue. While mushrooms are not as widely distributed as the molds, these fungi are very important, especially in the breakdown of woody tissue, and because some species form a symbiotic relationship with plant roots (see "Mycorrhizae," following).

Activities of Fungi

As saprotrophs decomposing organic materials in soil, fungi are the most versatile and persistent of any group. Cellulose, starch, gums, and lignin, as well as the more easily metabolized proteins and sugars, succumb to their attack. Otherwise resistant materials such a wood (lignin) and cellulose and even complex synthetic ring-structured compounds are rapidly broken down by a group of fungi termed **white rot fungi** (Figure 10.24) because of the whitish residue and mycelial growth that mark their decay. Fungi play major roles in the formation and stabilization of soil organic matter (see Section 11.4) and soil aggregates (see Section 4.5). The ability of fungi to tolerate low pH is especially important in decomposing organic residues in acid forest soils. Their ability to grow mycelia out from the soil (Figure 10.23) helps them decompose surface litter both in forests and in no-till agricultural fields.

Fungi are quite efficient in using the organic materials they metabolize. Up to 50% of the substances decomposed by fungi may become fungal tissue, compared to about 20% for bacteria. Soil *fertility* depends in no small degree on nutrient cycling by fungi, since they continue to decompose complex organic materials after most bacteria and actinomycetes have essentially ceased to function. Soil *tilth* also benefits from fungi as their hyphae stabilize soil structure (see Figure 4.15). Some of the nutrient cycling and ecological activities of soil fungi are made easily visible in the case of "fairy rings," commonly seen on lawns and pastures in early spring (Figure 10.25).

In addition to the processing of organic residues and cycling of nutrients, numerous other fungal activities have significant impact on soil ecology. A few species even trap and consume nematodes (see Figure 10.26). Soil fungi can synthesize a wide range of complex organic compounds in addition to those associated with soil humus. Certain fungi produce compounds that kill other fungi or bacteria and provide a competitive edge over rival microorganisms in the soil. These fungi have proved to be highly beneficial to human society (see Sections 10.10 and 10.14).

Unfortunately, not all the compounds produced by soil fungi benefit humans or higher plants. A few fungi produce chemicals (**mycotoxins**) that are highly toxic to plants or animals (including humans). An important example of the latter is the production of highly carcinogenic aflatoxin by the fungus *Aspergillus flavus* growing on seed crops such as corn or peanuts,

Figure 10.25 A "fairy ring" of fungal growth (a) and the fungi's fruiting bodies (mushrooms) (b,c). As the fungi (most commonly *Marasmius* spp.) metabolize grass thatch and residues, they release excess nitrogen that stimulates the lush green growth of the grass, which is especially visible in early spring. Later, bacteria decompose the aging and dead fungi, producing a second release of nitrogen. The fungus produces a chemical that is toxic to itself. Therefore, each generation must grow into uncolonized soil, producing an ever-expanding circle of fungi and decay marked by an ever-larger ring of dark-green grass. The grass in the center of the ring is often water-stressed because the fungi render the upper soil layers somewhat hydrophobic. (Photos courtesy of Ray R. Weil)

especially when the seeds are exposed to soil and moisture. Other fungi produce compounds that allow them to invade the tissues of higher plants (see Section 10.13), causing such serious plant diseases as wilts (e.g., *Verticillium*) and root rots (e.g., *Rhizoctonia*).

On the other hand, efforts are now underway to develop the potential of certain fungi (such as *Beauveria*) as biological control agents against some insects and mites that damage higher plants. These examples merely hint at the impact of the complex array of fungal activities in the soil.

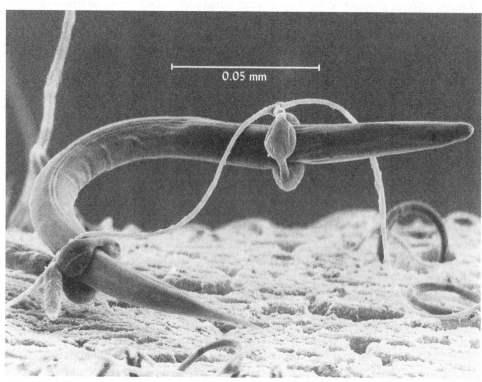

Figure 10.26 Several species of fungi prey on soil nematodes—often on those nematodes that parasitize higher plants. Some species of nematode-killing fungi attach themselves to, and slowly digest, the nematodes. Others, like this *Arthrobotrys anchonia*, make loops with their hyphae and wait for a nematode to swim through these lasso-like structures. The loop is then constricted, and the nematode is trapped. The nematode shown here is being crushed by two such fungal loops. Additional loops can be seen in their nonconstricted configuration. (Photo courtesy of George L. Barron, University of Guelph)

Mycorrhizae[12]

One of the most ecologically and economically important aspects of soil fungi is the mutually beneficial association (**symbiosis**) between certain fungi and the roots of higher plants. This association is called **mycorrhizae**, a term meaning "fungus root." Mycorrhizae are the rule, not the exception, for most plant species, including the majority of economically important plants. In natural ecosystems most plants are quite dependent on mycorrhizal relationships and cannot survive without them.

Mycorrhizal fungi derive an enormous survival advantage from teaming up with plants. Instead of having to compete with all the other soil heterotrophs for decaying organic matter, the mycorrhizal fungi obtain sugars directly from the plant's root cells. This represents an energy cost to the plant, which may devote as much as 5 to 30% of its total photosynthate production to its mycorrhizal fungal symbiont.

In return, plants receive some extremely valuable benefits from the fungi. The fungal hyphae grow out into the soil many centimeters from the infected root, reaching farther and into smaller pores than could the plant's own root hairs. This extension of the plant root system increases its efficiency, providing perhaps 10 times as much absorptive surface as the root system of an uninfected plant.

The most frequently reported benefit to plants from the mycorrhizal association is the greatly enhanced ability of plants to take up phosphorus from low-phosphorus soils. The fungi also assist with plant acquisition of other nutrients that are relatively immobile and present in low concentrations in the soil solution. Water uptake may also be improved by mycorrhizae, making plants more resistant to drought and salinity stress (Table 10.4). In extremely acid soils, this fungal relationship protects plants from aluminum toxicity (see Section 9.7), while in soils contaminated with high levels of metals, mycorrhizae protect the plants from excessive uptake of these potential toxins (see Section 15.5). There is evidence that mycorrhizae also protect plants from certain soilborne diseases and parasitic nematodes.

Ectomycorrhiza. Two types of mycorrhizal associations are of considerable practical importance: **ectomycorrhiza** and **endomycorrhiza**. The ectomycorrhiza group includes hundreds of different fungal species associated primarily with temperate- or semiarid-region trees and shrubs, such as pine, birch, hemlock, beech, oak, spruce, and fir. These fungi, stimulated by root exudates, cover the surface of feeder roots with a fungal mantle. Their hyphae penetrate

Table 10.4
EFFECT OF ARBUSCULAR MYCORRHIZAE (AM) ON TOMATO ROOT COLONIZATION, FRUIT YIELD, AND NUTRIENT UPTAKE WITH NONSALINE OR SALINE WATER

A bit of inoculum was added to potting mix in seedling trays. Mycorrhizal inoculation increased all parameters, but the greatest benefits accrued under saline conditions. AM-inoculated plants under saline conditions yielded 5.3 kg fruit/m², not statistically different from the 5.8 kg fruit/m² yield of noninoculated plants under nonsaline conditions.

Irrigation water treatment	AM root colonization	Fruit yield	Nutrient content of plant shoot					
			Percent increase resulting from inoculation with AM fungi					
			P	K	Na	Cu	Fe	Zn
Nonsaline (EC$_w$ = 0.5 dS/m)	166	29	44	33	21	93	33	51
Saline (EC$_w$ = 2.4 dS/m)	293	60	192	138	7	193	165	120

Data selected from Al-Karaki (2006).

---

[12] For a forward looking review of potential agricultural manipulation of mycorrhizae, see French (2017). For meta-analysis of research on the efficacy of inoculation with mycorrhizal fungi in restoration ecology, see Hoeksema et al. (2010). For brief overviews of current thinking about mycorrhizal associations and their ecological implications, see Forsberg et al. (2012) and van der Heijden et al. (2015).

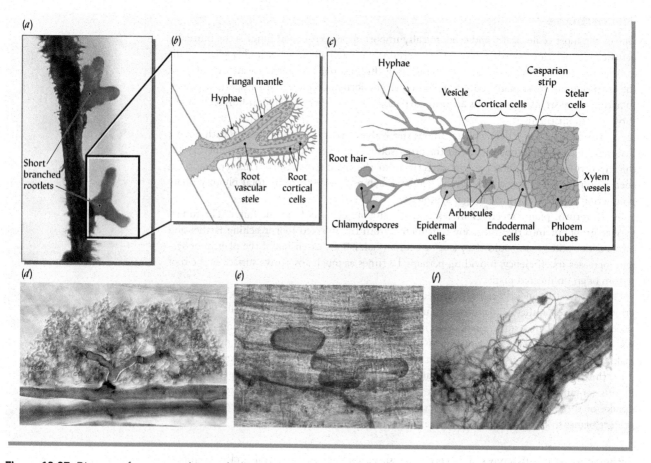

Figure 10.27 Diagram of ectomycorrhiza and arbuscular mycorrhiza (AM) associations with plant roots. (a) The ectomycorrhiza association produces short branched rootlets that are covered with a fungal mantle. (b) Fungal hyphae grow between the root cells, forming the Hartig net, but do not penetrate the cells. The hyphae also extend far out into the soil. (c) In contrast, the AM fungi penetrate not only between cells but into certain root cortical cells as well. Within these cells, the fungi form arbuscules (d), structures that transfer nutrients between the plant and the fungus. Many AM fungi also form vesicles (e), structures in which the fungi stores its cache of sugars from the plant. The symbiotic fungi form extensive networks of hyphae that act to greatly extend the reach of the plant root system (f). In both types of association, the host plant provides sugars and other food for the fungi and receives in return essential mineral nutrients that the fungi absorb from the soil. [Diagrams courtesy of Ray R. Weil; photos courtesy of (a) Ray R. Weil, (d) Mark Brundrett, (e) Charles White (f) R. P. Schreiner, USDA-ARS]

the roots and develop in the free space around the cells of the cortex but do not penetrate the cortex cell walls (hence the term *ecto*, meaning outside). Ectomycorrhizae cause the infected root system to consist primarily of stubby, white structures with a characteristic Y shape (Figure 10.27a,b). These Y-shaped structures provide evidence of mycorrhizal infection that is visible to the naked eye if one carefully excavates the roots.

Endomycorrhiza. The most important members of the endomycorrhiza group are called **arbuscular mycorrhizae (AM)**. When forming AM, fungal hyphae actually penetrate the cortical root cell walls and, once inside the plant cell, form small, highly branched structures known as **arbuscules** (Figure 10.27d). These structures serve to transfer mineral nutrients from the fungus to the host plants and sugars from the plant to the fungus. Other structures, called **vesicles**, are usually also formed and serve as storage organs for the mycorrhizae (Figure 10.27e).

Nearly 100 identified species of fungi form these endomycorrhizal associations in soils from the tropics to the arctic. Most native plants and agricultural crops can form AM associations and do not grow well on unfertilized soil in their absence. Two economically important groups of plants that do *not* form mycorrhizae are the *Cruciferae* (mustards, cabbage, radish, rapeseed) and the *Chenopodiaceae* (beet and spinach). Mycorrhizal fungi are agriculturally most important where soils are low in nutrients, especially phosphorus (Figure 10.28). Mycorrhizal

Figure 10.28 *The efficacy of inoculating with mycorrhizal fungi (+ AM fungi) largely depends on the soil phosphorus status before mycorrhiza are added (NM Control). (Right) Grape vines grown in Jory soil (low P availability Ultisol) show a dramatic growth response to addition of mycorrhizal inoculum. (Left) The same variety of grape vine is unaffected when grown in Chehalis soil, which is a high P fertility Mollisol.* (Photo courtesy of R. P. Schreiner, USDA-ARS)

hyphal interconnections among nearby plants have been observed in forest, grassland, and pasture ecosystems. These hyphae can transfer nutrients from one plant to another, sometimes resulting in a complex, four-way symbiotic relationship (Figure 10.29). The ecological significance of these AM-mediated nutrient transfers is not yet well understood.

Managing Mycorrhizae. Because of the near ubiquitous distribution of native mycorrhizal fungi, adding mycorrhizal inoculum rarely makes a difference in normal, biologically active soils. However, there are steps that can ensure good mycorrhizal colonization. Soil tillage destroys hyphal networks; therefore physical soil disruption is likely to decrease the effectiveness of native mycorrhizae. It is also best to avoid too-frequent use of nonhost species, long periods without vegetation, or heavy fertilization with phosphorus. In addition, the buildup of effective mycorrhizae in soils is favored by growing a diversity of host plant species as continuously as possible. For example, after forest clear cutting, newly planted Douglas firs survive and grow much better if some hardwood saplings are allowed to remain to provide host continuity and serve as an inoculum source for AM fungi.

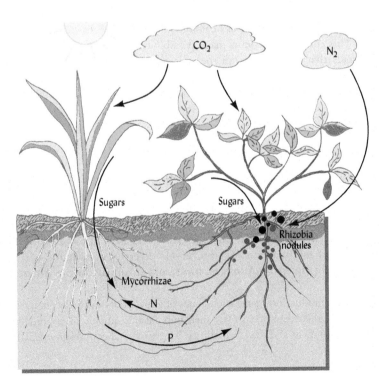

Figure 10.29 *Mycorrhizal fungi, rhizobia bacteria, legumes, and nonlegume plants can all interact in a four-way, mutually beneficial relationship. Both the fungi and the bacteria obtain their energy from sugars supplied through photosynthesis by the plants. The rhizobia form nodules on the legume roots and enzymatically capture atmospheric nitrogen, providing the legume with nitrogen to make amino acids and proteins. The mycorrhizal fungi infect both types of plants and form hyphal interconnections between them. The mycorrhizae then not only assist in the uptake of phosphorus from the soil, but can also directly transfer nutrients from one plant to the other. Nitrogen is transferred from the nitrogen-fixing legume to the nonlegume (e.g., grass) plant, and phosphorus is mostly transferred to the legume from the nonlegume. Research indicates that some direct transfer of nutrients via mycorrhizal connections occurs in many mixed plant communities, such as in forest understories, grass–legume pastures, and mixed cropping systems.* (Diagram courtesy of Ray R. Weil)

Applying Mycorrhizal Inoculum. There may be a need to inoculate soils with mycorrhizal fungi where native populations are very low or conditions for infection are unusually adverse. Examples include soils that have been subjected to broad spectrum fumigation; extreme soil heating, drying, or salinization; drastic disturbance such that subsoil layers are brought to the surface; or long periods without vegetative cover (such as surface soil stockpiled during mining or construction activities as shown in Figure 1.12). Successful restoration of healthy vegetation to such denuded soils often requires inoculation with effective mycorrhizal fungi.

10.10 SOIL PROKARYOTES: BACTERIA AND ARCHAEA

The organisms described in the previous sections—from mammals to molds—all belong to the Eukarya domain. The organisms in the other two domains of life, the Bacteria and the Archaea, are prokaryotes—their cells lack a nucleus surrounded by a membrane. However, despite their similar appearance under the microscope, archaea are evolutionarily quite distinct from bacteria. Genetic analysis suggests that Archaea may be as closely related to plants or people as they are to bacteria! Until recently, the archaea were thought of as rare and primitive creatures that live in only the most extreme and unusual environments on Earth—deeply frozen ice, boiling hot water, anaerobic sediments, and the like. However, molecular techniques now suggest that archaea are also common in more "normal" environments and probably represent about 10% of the microbial biomass in typical upland soils. Another recent insight provided by molecular identification techniques is that although we previously had no idea how little we knew, now we do (see Box 10.2). We will consider the archaea together with the bacteria in this section, calling them prokaryotes when the discussion applies to members of both domains.

Characteristics

Prokaryotes range in size from 0.5 to 5 µm, considerably smaller in diameter than most fungal hyphae (Figure 10.30). The smaller ones approach the size of the average clay particle (see Figure 10.1). Prokaryotes are found in various shapes: nearly round (coccus), rodlike (bacillus), or spiral (spirillum). Many prokaryotes are motile, swimming about in the soil water films by means of hairlike cilia or whiplike flagella. Others heavily colonize the nutrient-rich surface of plant roots (Figure 10.31). In many situations large numbers of individual prokaryotes cells act together in a coordinated manner to form what some have called a "superorganism," somewhat analogous to the ant colonies or slime mold slugs discussed in Sections 10.5 and 10.6. This collective action is made possible by communication via both exocellular chemical signals that diffuse through the soil solution and apparently also by proteins shared among cells via intercellular connections. As one result, bacterial cells often form biofilms—dense colonies with cells embedded in layers of excreted polysaccharides. Such biofilms may alter the soil environment by clogging pores, and covering mineral and root surfaces.

Prokaryote Diversity in Soils

The numbers of prokaryotes in soils are enormous but extremely variable, ranging from a few billion to more than a trillion in each gram of soil. A biomass of 400–5,000 kg/ha live weight is commonly found in the upper 15 cm of fertile soils (see Table 10.3). Their small size and ability to form extremely resistant resting stages that survive dispersal by winds, sediments, ocean currents, and animal digestive tracts have allowed prokaryotes to spread to almost all soil environments. The prokaryote diversity in a handful of soil is said to be comparable to the diversity of insects, birds, and mammals in the Amazon Basin!

Their extremely rapid reproduction (generation times of a few hours in the lab to a few days in favorable soil) enables prokaryotes to increase their populations quickly in response to favorable changes in soil environment and food availability. The microbial community usually reflects the soil environment rather than geography. For example, the prokaryote community in a well-drained soil under grassland vegetation with a neutral pH is likely to be quite similar to a well-drained neutral soil under grassland on another continent, but quite different from the community in a geographically nearby soil that is very acid or poorly drained. By the same token, if a new substrate is added—even an industrial waste—populations of prokaryotes capable of feeding on it will likely soon emerge.

BOX 10.2
SOIL MICROBIOLOGY IN THE MOLECULAR AGE

GUEST COMMENTARY BY STEPHANIE YARWOOD[a]

The study of soil microorganisms has dramatically changed in recent decades. From the 1850s to the early 1990s, scientists studied soil microbes by isolating them from soil, growing them on various media in the lab, and studying them in pure culture. This approach yielded many important insights. For example, in 1887 Sergei Winogradsky described the metabolic strategies of chemolithotrophy (autotrophic feeding on minerals) and in 1900 Martinus Beijerinck isolated nitrogen-fixing bacteria from legume root nodules. Early on, soil microbiologists noted that when bacteria from a soil sample were enumerated by direct count under a microscope, the numbers were 100s to 1,000s of times higher compared to when bacteria from the same sample were enumerated by growing on media in a petri dish.

Resolving this "great plate count anomaly" required a new approach that did not rely on microscopes or petri dishes. Technologies were developed focusing on molecular components of the cell: DNA, RNA, lipid, and protein. The most used today is DNA. Vigdis Torsvik was the first to successfully extract DNA directly from soil. By the early 1990s soil microbiologists were using extracted DNA in a variety of molecular-based approaches. Using DNA as the marker for soil microorganisms revealed that each gram of soil may contain 10^9 individual bacterial cells and, although the number is still debated, that same gram of soil may contain about 4,000 species. These methods also led to the discovery of entire groups of microbes never before described, including a new phylum called Acidobacteria that may represent 25% of all the bacteria in some soil types. In fact, scientists now believe that the nine genera of bacteria (*Agrobacterium, Bacillus, Pseudomonas, Streptomyces,* etc.) considered in the late 1970s to be of significance in soils, actually account for only 5% of soil bacteria.

Studies of soil DNA were initially focused on genes like the bacterial 16S rRNA that contains the genetic information needed to make cell ribosomes. The 16S is a good target to examine bacterial diversity because every living cell needs ribosomes in order to make protein. The code for this gene (the sequence of the bases A, C, T, G) is different in each species, allowing microbiologists to identify which microbes are present. Unfortunately, because every microbe has the 16S rRNA gene, this method does not reveal the *function* of the organisms within the soil profile. The next step, therefore, has been to target functional genes. The gene coding for nitrous oxide reductase, for example, can be used to characterize denitrifying bacteria (see Section 12.1). Functional genes can be quantified to determine the population size, and DNA sequencing allows identification of the organisms involved.

Today, with the decreased cost of DNA sequencing and the increased power of computers, soil microbiologists can identify or sequence the entire complement of genes—the genome—in a given organism. In fact, they are now sequencing *all* the sets of genes for *all* the organisms in a whole microbial community—the metagenome. Metagenomics refers to the collection and analysis of all the genetic information in a community's metagenome (such as all the genes in a complex soil sample). The goals of this approach are to determine what individual species are capable of doing and understand what conditions favor their growth.

These new molecular methods also help microbiologists create improved approaches to culturing microbes in the lab. Although less than 1 to 2% of soil bacterial species are in culture today, soil microbiologists have had recent success in culturing difficult-to-grow microbes, including *Acidobacteria*. Culturing remains a valuable approach, because it helps confirm functions and life strategies predicted by the DNA sequencing data. The combination of culturing and molecular methods is revealing an increasingly detailed view of the most diverse microbial community on Earth. With the advent and rapid expansion of molecular methods, soil microbiologists are working toward a day when knowledge of which soil microbes are present in a particular soil will help us predict how that soil will function.

---

[a]Dr. Yarwood is a soil microbial ecologist at the University of Maryland.
For a review of advances in molecular soil microbiology, see Nesme et al. (2016). The first isolation of bacterial DNA from soil was reported in the classic paper by Torsvik (1980).

Figure 10.30 *Fungal hyphae associated with much smaller rod-shaped bacteria.* [Soil and Water Conservation Society (SWCS). 2000. *Soil Biology Primer.* Rev. ed. Ankeny, IA: Soil and Water Conservation Society]

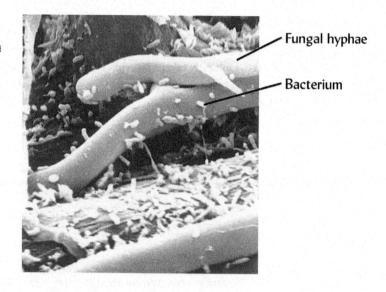

Figure 10.31 *Bacteria colonies on the surface of a wheat root (left) and a corn root (right). The bacteria on the wheat root are embedded in their own secreted mucilage. The rhizosphere soil and the root surface itself are usually crowded with bacteria that are integral to the plant–soil–microbe system.* (Cryo scanning electron micrographs courtesy of Margaret McCully, CSIRO Plant Industry, Canberra, Australia)

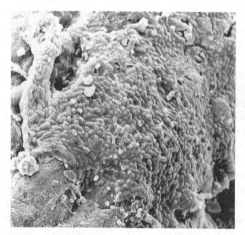 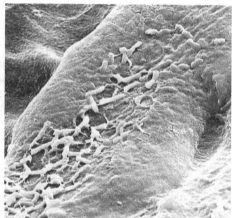

Sources of Energy

Soil prokaryotes are either autotrophic or heterotrophic (see Section 10.2). The autotrophs obtain their energy from sunlight (photoautotrophs) or from the oxidation of inorganic constituents such as ammonium, sulfur, and iron (chemoautotrophs) and obtain their carbon from carbon dioxide or dissolved carbonates. Most soil bacteria are heterotrophic—both their energy and their carbon come from organic matter. Heterotrophic bacteria, along with fungi, account for the general breakdown of organic matter in soil. The bacteria often predominate on easily decomposed substrates, such as animal wastes, starches, and proteins. Where oxygen supplies are depleted, as in wetlands, nearly all decomposition is mediated by prokaryotes and Archaea play a larger role. Certain gaseous products of anaerobic metabolism, such as methane and nitrous oxide, greatly shape the global environment (see Sections 10.9 and 12.9).

Importance of Prokaryotes

Prokaryotes hold near monopolies in the oxidation or reduction of certain chemical elements in soils (see Sections 7.3 and 7.4). Some autotrophic prokaryotes obtain their energy from such inorganic oxidations, while anaerobic and facultative bacteria reduce a number of substances other than oxygen gas. Many of these biochemical oxidation and reduction reactions have significant implications for environmental quality as well as for plant nutrition. For example, through nitrogen oxidation (nitrification), selected bacteria oxidize relatively stable ammonium nitrogen to the much more mobile nitrate form of nitrogen. The Archaea are the most important group in the breakdown of hydrocarbon compounds, such as petroleum products. Likewise, certain archaeans oxidize sulfur, yielding plant-available sulfate ions, but also potentially damaging sulfuric acid (see Sections 9.6 and 12.2).

Prokaryote oxidation and reduction of inorganic ions such as iron and manganese not only influence the availability of these elements to other organisms (see Section 12.8), but also help determine soil colors (see Section 4.1). A critical process in which bacteria are prominent is nitrogen fixation—the biochemical combining of atmospheric nitrogen with hydrogen to form organic nitrogen compounds usable by plants (see Section 12.1).

Previously classified as blue-green algae, **cyanobacteria** contain chlorophyll, which allows them to photosynthesize like plants. Cyanobacteria are especially numerous in rice paddies and other wetland soils and fix appreciable amounts of atmospheric nitrogen when such lands are flooded (see Section 12.1). These organisms also exhibit considerable tolerance to saline and arid environments and are important in forming microbiotic crusts on desert soils (Section 10.14).

Soil Actinomycetes

Actinomycetes is the term traditionally used for bacteria in the order *Actinomycetales* within the phylum **Actinobacteria**. Filamentous and often profusely branched (see Figure 10.32), they may appear somewhat like tiny fungi, but their genetic makeup and cellular properties clearly place them in the bacteria domain—they have no nuclear membrane, are about the same diameter as other bacteria, and often break up into spores that closely resemble cocci bacterial cells. Where the acidity is not too great, actinomycetes are especially numerous in soils high in organic matter, such as old meadows or pastures.

The earthy aroma of organic-rich soils and freshly plowed land is credited mainly due to actinomycete-produced *geosmins*, volatile derivatives of terpene. Actinomycetes develop best in moist, warm, well-aerated soil. However, they tolerate low osmotic potential and are active in arid-region, salt-affected soils and during periods of drought. They are generally rather sensitive to acid soil conditions, with optimum development occurring at pH values between 6.0 and 7.5. In forest ecosystems, much of the nitrogen supply depends on actinomycetes that fix atmospheric nitrogen gas into ammonium nitrogen that is then available to plants (see Section 12.1). Many actinomycete species, especially in the genus *Streptomyces*, produce compounds that kill other microorganisms, and these "antibiotics" have become extremely important in human medicine (see Box 10.3).

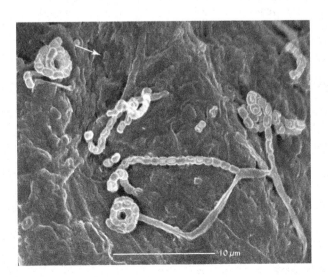

Figure 10.32 Strands of an actinomycete, a type of filamentous bacteria, growing on the surface of a soil biopore (an old root channel). The filaments, some breaking into the beadlike spores by which this organism reproduces, are about 0.8 μm in diameter. Some of the actinomycete filaments are embedded in mucilage of the soil pore (e.g., at arrow). The image is from 1.5 m deep in a clayey soil (poorly structured Alfisols) of a wheat field in eastern New South Wales, Australia. The biopore surfaces are generally smooth, coated with illuvial clay and residues of old alfalfa roots. Although fungi commonly occupy old root channels, few were found in this soil, perhaps because of the antibiotic and chitinase secretions of the actinomycetes. (Cryo-SEM image courtesy of Margaret McCully, CSIRO Plant Industry, Canberra, Australia)

BOX 10.3
A POST-ANTIBIOTIC AGE ON THE HORIZON?[a]

Antimicrobial compounds—commonly called antibiotics—inhibit or kill specific microorganisms. Most antibiotics kill by interfering with a few physiological reactions specific to bacteria, for instance, the generation of bacterial cell walls; therefore they are not toxic to unrelated organisms, such as humans and other animals. Certain soil bacteria and fungi have evolved the capability of producing these compounds as part of their struggle for survival in the soil. Being bathed in their own chemical warfare agents, most of these spore-forming microbes have also evolved immunities to many types of antibiotics. With the mid-twentieth-century advent of antibiotic "miracle drugs," these same compounds enabled humans to all but conquer infectious bacterial diseases, which up until that time were the most common cause of human deaths. Penicillin, the first antibiotic drug, was derived from a soil fungus (*Penicillium spp.*) that contaminated some laboratory petri dishes in 1928. In 1943, streptomycin was discovered, leading to the first of many antibiotic drugs synthesized by soil bacteria belonging to the genus *Streptomyces*. Likely you are alive today because an antibiotic produced by a soil organism (e.g., erythromycin, tetracycline) was available to save your life when you came down with a bacterial infection (e.g., pneumonia or a dirty wound).

Unfortunately, resistant strains of pathogenic bacteria are eroding the efficacy of these drugs. Nearly seven times as many people die from infection acquired in hospitals today as compared to 20 years ago and most of the bacteria causing these illnesses show resistance to at least one antibiotic. The potentially fatal human pathogens Enterococci and Staphylococci have now developed resistance to virtually every antibiotic drug available in the medical arsenal. What is causing the alarming rise in resistance that threatens to return humankind to the pre-antibiotic era?

Overexposure to antibiotic drugs has exerted tremendous selection pressure for resistance in the pathogen populations. Antibiotics now permeate the environment—some 18 million kgs are used annually in the United States alone. Part of the problem stems from overuse and misuse of human drugs by doctors, patients (did you stop taking *all* the pills prescribed when you felt better, letting the most resistant bacteria live on?), hospitals, and consumers (does soap really need an antibiotic when the soap itself kills bacteria by lysing their cell walls?). Worryingly, the resistant bacteria may spread in the environment; antibiotic-resistant bacteria have been found in rivers near sewage outfalls.

But the principal source of antibiotics in the environment is the enormous amount used for nonmedical

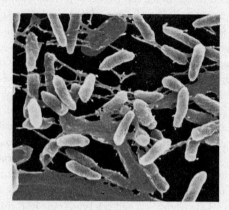

Figure 10.33 *Scanning electron micrograph of a common soil bacteria,* Pseudomonas, *harboring genes for antibiotic resistance and transfer them to human pathogen bacteria.* (Photo courtesy of Gautam Dantas and Kevin Forsberg)

purposes. In fact, in the United States, some 87% of the antimicrobial drugs produced is devoted to nonhuman uses, most of this as a growth-promoting feed additive for healthy (not sick) poultry, hogs, and cattle in industrial-style farms. Such subtherapeutic feed additive use of antibiotics is several times greater in China, the world's biggest antibiotics producer. Much of the antibiotics ingested by the livestock passes through the digestive tract unchanged and accumulates in the manure that is eventually spread on farm fields. Once in the soil, the compounds are known to retain their antibiotic activity even if adsorbed for long periods to clay. Crops (corn, onion, and cabbage) growing on soils fertilized with such manure can take up small amounts of the antibiotic, presenting the possibility that the antibiotics added to livestock feed may end up causing human allergic reactions and selecting for resistance in the human digestive tract. Scientists using molecular techniques have shown that soils provide a large reservoir and breeding ground for antibiotic resistance. Common soil decomposer bacteria (Figure 10.33) are increasingly developing genes for resistance that they readily pass on to other completely unrelated bacteria—such as human pathogens.

Antibiotics in industrial animal facilities and in manured soils almost certainly hastens the evolution of antibiotic resistance in bacteria, including in human pathogens. While the problem has been known by scientists since the 1980s, policymakers in industrialized countries have been slow to realize the need to eliminate such careless use of these life-saving substances. Discovery in a soil from Maine of a new class of antibiotics that seems less likely to engender resistance should not lull society into inaction!

[a]For more on this topic, see D'Costa et al. (2006), Chander et al. (2007), and Zhu et al. (2013); on transfer of antibiotic resistance from soil microbes to human pathogens, see Pehrsson et al. (2013); on discovery of a new soil produced antibiotic named Teixobactin, which inhibits bacterial cell wall synthesis, see Ling et al. (2015).

10.11 CONDITIONS AFFECTING THE GROWTH AND ACTIVITY OF SOIL MICROORGANISMS

Organic Resources

The addition of almost any energy-rich organic substance, including the compounds excreted by plant roots, stimulates microbial growth and activity. If organic materials are left on the soil surface (as in forest litter), fungi dominate the microbial activity (Figure 10.23). Bacteria commonly play a larger role if the substrates are mixed into the soil, as by earthworms, root distribution, or tillage. In addition, many bacteria adapted to life in the plant rhizosphere are sensitive to growth factors and signaling compounds released by plant roots. Many bacteria also communicate with organic compounds they produce to enable coordinated group action with other individuals of the same species or different species.

Oxygen Requirements

While most soil microorganisms are *aerobic* and use O_2 as the electron acceptor in their metabolism, some prokaryotes are *anaerobic* and use substances other than O_2 (e.g., NO_3^-, SO_4^-, or other electron acceptors). *Facultative* bacteria can use either aerobic or anaerobic forms of metabolism. All three of these types of metabolism are usually carried out simultaneously in different habitats within a soil. The zone of greatest microbial activity usually occurs just a few cm below the soil surface where oxygen is high and the soil is not too dry (Figure 10.34).

Moisture and Temperature

Optimum moisture potential for higher plants (−10 to −70 kPa) is also usually best for aerobic microbes. Too high a water content will limit the oxygen supply. Microbial activity is generally greatest when temperatures are 20 to 40°C. The warmer end of this range tends to favor actinomycetes. Ordinary soil temperature extremes seldom kill bacteria and commonly only temporarily suppress their activity. However, except for certain **psychrophilic** species, most microorganisms cease metabolic activity below 3 to 5°C, a temperature sometimes referred to as *biological zero* (see Section 7.8). Nonetheless, the concept of biological zero must be used with caution; bacterial and archeal activities have been documented at subzero temperatures in and under polar ice.

Exchangeable Calcium and pH

Levels of exchangeable calcium and pH help determine which specific organisms thrive in a particular soil. Although some prokaryote species will thrive in any chemical condition found in soils, high calcium and near-neutral pH generally result in the largest, most diverse bacterial populations.

Figure 10.34 A wood fence post pulled up after several years in the soil. Lines on the post indicate the soil surface when it was set in place. Decay of wood was greatest a few cm below the soil surface, where oxygen and moisture are in plentiful supply and create a zone of maximum biological activity. Keeping this phenomenon in mind (commonly referred to as the "fence post principle"), farmers often avoid deep incorporation of organic amendments. (Photo courtesy of Ray R. Weil)

Regardless of other soil properties, bacterial diversity has been found to increase dramatically with soil pH, from very acid soils to slightly alkaline ones (see Figure 9.19). If pH is acidic, fungi become dominant. The effect of pH and calcium helps explain why fungi tend to dominate in forested soils, while bacteria dominate in most subhumid to semiarid prairie and rangeland soils.

10.12 BENEFICIAL EFFECTS OF SOIL ORGANISMS ON PLANT COMMUNITIES

Soil Organic Matter Formation and Nutrient Cycling

Perhaps the most significant biological function of soils is the decomposition of dead plant tissues. Soil organisms also assimilate wastes from animals (including human sewage) and other organic materials added to soils. As a by-product of their metabolism, microbes synthesize new compounds, some of which help to stabilize soil structure. When microbes die, their cell walls often bind to mineral particles forming an important part of soil humus. The bacteria, archaeans, and fungi assimilate some of the N, P, and S in the organic materials they digest. Excess amounts of these nutrients may be excreted into the soil solution in inorganic form either by the microbes themselves or by the nematodes and protozoa that feed on them. In this manner, the soil food web converts organically bound forms of N, P, and S into mineral forms that are taken up once again by higher plants.

Breakdown of Toxic Compounds

Many organic compounds toxic to plants or animals find their way into the soil. Some of these toxins are produced by soil organisms as metabolic byproducts, some are applied purposefully by humans as agrichemicals to kill pests, and some are deposited in the soil because of unintentional environmental contamination. If these compounds accumulated unchanged, they would do enormous ecological damage. Fortunately, most biologically produced toxins do not remain long in the soil since soil ecosystems include organisms that not only are unharmed by these compounds but can use them as food.

Some toxins are **xenobiotic** (literally *strangers to life*) compounds foreign to biological systems, and these may resist attack by commonly occurring microbial enzymes. Soil prokaryotes and fungi are especially important in helping maintain a nontoxic soil environment by breaking down toxic compounds (see Section 15.6). Chemical weed killers or herbicides are examples of xenobiotic compounds that can have undesirable residual effects on future crops if they do not break down rapidly in the soil. Microbial breakdown of xenobiotics is usually fastest in the surface layers of soil, where microbial numbers are concentrated in response to the greater availability of organic matter and oxygen.

Inorganic Transformations

Nitrates, sulfates, and, to a lesser degree, phosphate ions are present in soils primarily due to inorganic transformations, such as the oxidation of sulfide to sulfate or ammonium to nitrate stimulated by microorganisms. Likewise, the availabilities of other essential elements, such as iron and manganese, are determined largely by microbial action. In well-drained soils, these elements are oxidized by autotrophic organisms to their higher valence states, in which forms they are quite insoluble. This keeps iron and manganese mostly in low solubility and nontoxic forms, even under fairly acid conditions (Section 12.8). If such oxidation did not occur, plant growth would be jeopardized because of toxic quantities of these elements in solution. Microbial oxidation also controls the potential for toxicity in soil contaminated with selenium or chromium (see Section 12.8).

Nitrogen Fixation

Elemental nitrogen gas cannot be used directly by higher plants. The fixation of elemental nitrogen gas into compounds usable by plants is one of the most important microbial processes in soils (see Section 12.1). Actinomycetes in the genus *Frankia* fix major amounts of nitrogen

Table 10.5
RICE PLANTS RESPOND TO INOCULATION WITH GROWTH-PROMOTING RHIZOBACTERIA

Rice plants were grown in pots with clay soil that was puddled and flooded with water. All pots were fertilized with adequate N fertilizer, but only some were inoculated with various strains of rhizobia or bradyrhizobia bacteria. The bacteria colonized the rice rhizosphere (hence the term rhizobacteria) and changed the physiology of the rice plant, partly by producing the plant growth hormone IAA. Inoculated roots were more efficient at nutrient uptake. Analysis of N isotope tracers showed that the rhizobacteria did not cause significant amounts of N fixation.

	Grain yield, g/pot	Uptake of nutrients by rice plants, mg/pot				IAA[a] in the treatment rhizosphere, mg/L
		N	P	K	Fe	
Control—no inoculation	36.7	488	111	902	18.9	1.0
Inoculated with rhizobacteria	44.3	612	134	1020	23.6	2.1
Percent change	+21	+25	+21	+13	+25	+110

[a]Indol-3-acetic acid, a plant growth hormone.
Data calculated from Biswas et al. (2000).

in forest ecosystems; cyanobacteria are important in flooded rice paddies, wetlands, and deserts; and rhizobia bacteria are the most important group for the capture of gaseous nitrogen in grassland and agricultural soils. By far the greatest amount of nitrogen fixation by these organisms occurs in root nodules or in other associations with plants.

Rhizobacteria[13]

The zone immediately around plant roots supports a dense population of microorganisms (Section 10.7 and in Figure 10.19). Bacteria especially adapted to living in this zone are termed **rhizobacteria**, many of which are beneficial to higher plants (the so-called **plant growth–promoting rhizobacteria**). In nature, root surfaces are almost completely encrusted with bacterial cells, so little interaction between the soil and root can take place without some intervening microbial influence. In addition to those that ward off plant diseases (see Section 10.13), certain rhizobacteria promote plant growth in other ways, such as enhanced nutrient uptake or hormonal stimulation (for an example, see Table 10.5). The world of rhizobacteria is still largely uncharted, but advances in microbiological and chemical analysis tools are beginning to reveal an incredible level of complexity and interdependence.

Plant Protection

Certain soil organisms attack higher plants, but others act to protect plant roots from invasion by soil parasites and pathogens. Plant diseases and the protective action of the soil microbes will be discussed in the next section.

10.13 SOIL ORGANISMS AND PLANT DAMAGE

Although most of the activities of soil organisms are vital to a healthy soil ecosystem and economic plant production, some soil organisms affect plants in detrimental ways that cannot be overlooked. For example, soil organisms successfully compete with plants for soluble nutrients (especially for nitrogen), as well as for oxygen in poorly aerated soils. Here we focus on the soil organisms that act as herbivores, parasites, or pathogens.

---

[13]For an example of the complex and intimate relationship between one plant species and one of its rhizosphere bacteria, see Zamioudis et al. (2013).

Plant Pests and Parasites

Soil Fauna. Many herbivorous soil fauna are injurious to the plants they feed on. Some rodents may severely damage young trees and farm crops. Snails and slugs in some climates are dreaded pests, especially of young leafy vegetables. Some ants are herbivorous (e.g., leaf-cutting ants), while others transfer aphids onto plants and so contribute to plant damage. Undoubtedly, the greatest damage to plants by soil fauna is caused by the feeding of nematodes and insect larvae. In nature, plants commonly sustain low levels of nematode and insect infestation, but under certain circumstances, infestation may be so great that the plant is killed or severely stunted. In agriculture, such infestations are often associated with lack of diversity that attends monocropping and insufficient organic matter addition. To prevent or diminish such infestations, large amounts of nematicide and insecticide chemicals are used in agriculture, often with unintended ecological results (see Section 10.14).

Microbes and Plant Disease. Disease infestations occur in great variety and are induced by many different microorganisms. Although bacterial blights and wilts are common, the fungi are responsible for the majority of soilborne plant diseases. Fungi of the genera *Pythium, Fusarium, Phytophthora,* and *Rhizoctonia* are especially prominent as soilborne agents of plant diseases described by such symptoms as *damping-off, root rots, leaf blights,* and *wilts*. Soils are easily infested with disease organisms, which are transferred from soil to soil by many means, including tillage or planting implements, transplant material, manure from animals that were fed infected plants, soil erosion, and wind-borne fungal spores and bacteria. Strict quarantine systems can restrict the transfer of soilborne pathogens from one area to another. Once a soil is infested, it is apt to remain so for a long time.

Deleterious Rhizobacteria. Some bacteria that live in the rhizosphere or on the rhizoplane inhibit root growth and function by various noninvasive chemical interactions. These nonparasitic **deleterious rhizobacteria** can cause stunting, wilting, foliar discoloration, nutrient deficiency, and even death of affected plants, but often the effects are subtle and difficult to detect. Their buildup may contribute to yield declines during long-term monoculture and aggravate problems in situations such as planting new trees in old orchards. On the other hand, by management that favors the deleterious rhizobacteria associated with certain weeds, scientists hope to be able to reduce weed seed germination and seedling growth and thereby reduce the use of herbicide (weed-killer) sprays on cropland and rangeland.

Plant Disease Control by Soil Management[14]

A detailed consideration of soil-related plant diseases is beyond the scope of this book text, but consideration of a few principles and examples of how soils may be managed to shift soil microbial community composition and plant nutrition in favor of plant health. Taking pro-active steps to build a diverse, healthy soil–plant community is generally preferable to reactively combating diseases once they become evident or employing preventative spraying. An integrated systems approach will ensure continuous vegetation by diverse plant species, as well as physical and chemical soil conditions that favor healthy plants and plant-beneficial microbial communities. Crop rotation can be very important in controlling a disease by growing nonsusceptible plants for several years between susceptible crops. A combination of organic matter (especially manure) addition, soil pH control (with sulfur) and rotation with nonsusceptible grasses has proved effective in preventing and suppressing the fungus *Phymatotrichopsis ominvorum*, which causes devastating root rots on some 2,000 species of economically important plants (including trees, shrubs, forages, grains, and vegetables).

Tillage may help by burying plant residues on which fungal spores might overwinter. However, disease problems are often lessened in no-tillage systems in which the soil surface

---

[14]For a general overview of disease control by management of the soil ecosystem, see Stone et al. (2004); for a review of the scientific literature on disease and pest management using organic amendments, see Litterick and Harrier (2004). For an in-depth analysis of the influence on plant disease of soil factors affecting Mn and nitrate availability, see Huber and Haneklaus (2007).

remains mulched with plant residues that maintain favorable soil temperature and moisture and a diverse soil community. Maintaining a surface mulch will also prevent the splashing of soil and fungal spores onto foliage by rain or irrigation water. Soil splash on an unmulched soil surface is a major cause of plant disease initiation and spread. Residues from certain green manure crops have been shown to chemically inhibit specific plant diseases. Direct management of soil physical and chemical properties can also be useful in disease control.

Soil pH. Regulation of soil pH is effective in controlling some diseases. For example, keeping the pH low (<5.2) can control both the actinomycete-caused *potato scab* and the fungal disease of turfgrass known as *spring dead spot*. Raising soil pH to about 7.0 can control *clubroot* disease in the cabbage family, because the spores of the fungal pathogen germinate poorly, if at all, under neutral to alkaline conditions.

Soil Nutrients. Healthy, vigorous plants usually can resist or outgrow diseases better than weaker plants, so provision of an optimal level of balanced nutrition is an important step in disease management. High levels of nitrogen fertilization tend to increase plants' susceptibility to fungal diseases; high levels of ammonium (as compared to nitrate) nitrogen especially increase wilt diseases caused by *Fusarium* fungi. On the other hand, high levels of nitrate (as compared to ammonium) tend to aggravate such diseases as potato scab, rice blast, wheat take-all, and corn stalk rot. Potassium fertilizers often reduce fungal disease severity, as do relatively high levels of calcium and manganese. Nutrient imbalances and micronutrient deficiencies can make plants especially susceptible to attack.

Organic Toxins. As an alternative to the use of synthetic, broad-spectrum fungicides or fumigants, certain natural organic antifungals can be introduced to the soil via microbial breakdown of organic amendments. An example is the rotation of cauliflower with broccoli such that the broccoli leaf residues left after harvest are tilled into the soil where they break down and release volatile compounds specifically toxic to the *Verticillium dahliae* fungus. This "biofumigation" can provide a level of disease control in the following cauliflower crop equal to that achieved by synthetic fumigants.

Soil Physical Properties. Soil compaction often aggravates fungal root diseases by slowing root growth, inducing more root excretions that attract the pathogens, and by promoting wet, poorly aerated conditions. Wet, cold soils favor some seed rots and seedling diseases such as *damping-off*. Good drainage and planting on ridges can help control these diseases. High soil temperature can be used to control a number of pathogens. **Solarization**, the use of sunlight to heat soil under clear plastic sheeting, is a practical way to partially sterilize the upper few cm of soil in some field situations. Steam or chemical sterilization is a practical method of treating greenhouse potting media. It should be remembered, however, that sterilization kills beneficial microorganisms, such as mycorrhizal fungi, as well as pathogens, and so may do more harm than good!

Disease-Suppressive Soils[15]

Research on plant diseases ranging from *Phytophthora root rot* in eucalyptus forests to *Fusarium wilts* in banana plantations has documented the existence of **disease-suppressive soils**, an exceptional condition in which a disease fails to develop *even though both the virulent pathogen and a susceptible host are present* (Figure 10.35). The reason that certain soils become disease-suppressive is not entirely understood, but much evidence suggests that the pathogenic organisms are inhibited by **antagonism** from beneficial bacteria and fungi. Two broad types of disease suppression are recognized: general and specific.

General Suppression. General disease suppression is caused by high levels of overall microbial activity in a soil, especially at times critical in the development of a disease, such as when the pathogenic fungus is generating propagules or preparing to penetrate the plant cells. The

---

[15] For an example of a molecular investigation of the mechanisms and organisms behind disease suppressiveness in soil, see Mendes et al. (2011).

Figure 10.35 *Disease-suppressive soil. Sugar beets were grown in pots filled with sandy soil from a field observed to be suppressive of damping off, a fungal disease that can rapidly kill small seedlings. All the pots were inoculated with the disease-causing fungus,* Rhizoctonia solani. *The graph shows the percent of sugar beet seedlings that showed symptoms of the disease during the following 20 days. In the suppressive soil, less than 5% of the seedlings exhibited disease symptoms, while over 60% of the seedlings in the nonsuppressive (conducive) soil succumbed to the disease. Where the suppressive soil was partially heat-sterilized, the disease incidence was low for a few days, but within two weeks the disease was as bad as or worse in those pots than in the conducive soil. As is true in the field, sterilization kills most of the beneficial as well as pathogenic organisms—but the pathogen population tends to grow back much faster. The biological nature of the suppressiveness is also supported by the fact that a mixture of 10% suppressive soil and 90% disease-conducive soil reduced the disease incidence by nearly half.* [Graphed from data in supplementary materials for Mendes et al. (2011)]

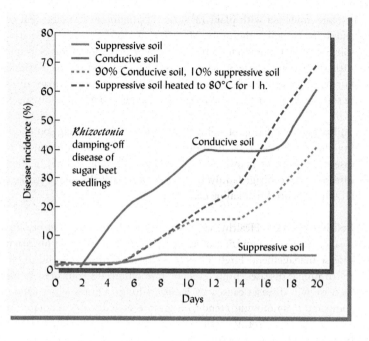

presence of particular organisms is less important than the total level of activity. The mechanisms responsible for general suppression are thought to include: (1) competition by beneficial microorganisms in the rhizosphere for carbon (energy) sources, (2) competition for mineral nutrients (such as nitrogen and iron), (3) colonization and decomposition of pathogen propagules (e.g., spores), (4) antibiotic production by varied actinomycete and fungal populations (see Section 10.10), and (5) lack of suitable root infection sites due to surface colonization by beneficial bacteria or previous infection by beneficial mycorrhizal fungi.

Specific Suppression. Specific suppression is attributable to the actions of a single species or a narrow group of microorganisms that inhibit or kill a particular pathogen. The effective presence of the specific suppressing organism may result from the same types of organic matter management just described or from introduction of an inoculum containing high numbers of the desired organism. Specific organisms known to be antagonistic to pathogens include *Trichoderma viride* fungi and certain fluorescent *Pseudomonas* bacteria, which produce antibiotics specific against pathogens or produce compounds that bind so tightly with iron that the pathogen spores cannot get enough of this nutrient to germinate. Treating plant seeds with specific microbes that fend off seed rot pathogens may be one of the most practical applications of the inoculation approach (Box 10.4). Successful suppression is usually limited not by the absence of a particular organism, but by appropriate conditions that will stimulate the extremely high population densities of the organism needed for effective suppression.

Use of Composts. Horticulturalists have been able to control *Fusarium* diseases in containerized plants by replacing traditional potting mixes with growing media made mainly from certain well-aged **composts**. Apparently, large numbers of beneficial antagonistic organisms colonize the organic material during the final stages of composting (see Section 11.10), and the stabilized organic substrate stimulates the activity of indigenous beneficial organisms without stimulating the pathogens. Similar success in practical disease suppression has been experienced with the use of composted materials on turfgrass, especially for replacing peat (which is relatively inert and does not stimulate disease suppression) in topdressing golf course greens (see Figure 10.37).

Induced Systemic Resistance.[16] The role of soil ecology in protecting plants from disease is not limited to belowground infections. Beneficial rhizobacteria have an intriguing mode of

---

[16] For molecular and ecological insights into induced systemic resistance, see Bakker et al. (2013).

BOX 10.4
CHOOSING SIDES IN THE MICROBE WARS

Recent advances in our understanding of microbial ecology, combined with new techniques for enumerating and isolating microbial species, have opened many interesting and promising opportunities to harness the antagonisms long observed among microorganisms.

Seed treatments are now commercially available that use antagonistic soil bacteria instead of chemical fungicide to protect seed from rot-causing soil pathogens. Strains of *Pseudomonas fluorescens* have been specially selected and enhanced to respond to root signals, colonize the rhizosphere soil, and protect plant roots from soilborne diseases. When a preparation of these bacteria is applied as seeds are sown, the seeds germinate safely, even in soil known to be infested with seed-rotting fungi. Seeds sown without the treatment decay before they can germinate. Similar antagonistic organisms have been used for rot diseases in crops as varied as beans and tea bushes (Figure 10.36). Another approach is to select for plants that are most effective in sending out chemical signals that recruit protective bacteria to their rhizoplane to fend of attacking pathogens. Widespread use of such biologically based disease protection in the place of conventional fungicidal chemicals could reduce the loading of toxic compounds into the environment and avoid "killing the good with the bad" denizens of the soil.

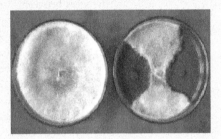

Figure 10.36 Pseudomonas aeruginosa *isolated from tea plantation soil in India was added to the agar plate on the right and showed biocontrol against the tea root pathogen* Fomes lamoensis. *The plate on the left shows the pathogen's growth without the P. aerufinosa.* [Data from Morang et al. (2012)]

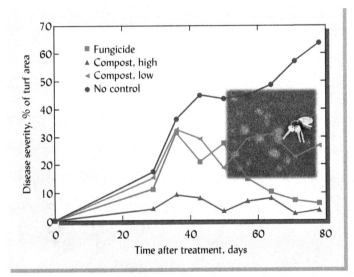

Figure 10.37 *Topdressing with compost as a nontoxic means of suppressing dollar spot disease on bentgrass putting greens (inset photo). The disease, caused by the fungus* Sclerotinia homoeocarpa, *was controlled as well or better by the high rate of compost (0.5 kg/m² topdressed every three weeks) as by the synthetic fungicide (Chlorthalonil, sprayed on every two weeks). Even the lower rate (0.1 kg/m²) of compost provided some control of the disease. All the turfgrass plots were inoculated with the disease organism. The researchers suggest that topdressing with compost provided a general type of disease suppression since they found little difference among composts made from many different materials. The means from 2 years' data are shown.* [Graphed from data in Boulter et al. (2002). Photo courtesy of John Kaminski]

action called **induced systemic resistance**, which helps plants ward off infection by diseases or insect pests both above and below ground. The process begins when a plant root system is colonized by beneficial rhizobacteria that cause the accumulation of a signaling chemical. The chemical signal is translocated up to the shoot, where it induces leaf cells to mount a chemical defense against a specific pathogen, even before the pathogen has arrived on the scene. When the pathogen (perhaps a fungal spore) does arrive on the leaf, its infection process is aborted almost before it can begin.

These examples merely hint at the complexity of the plant–soil communities in nature and the potential for controlling plant diseases and pests through ecological management rather than applications of toxic chemicals.

10.14 ECOLOGICAL RELATIONSHIPS AMONG SOIL ORGANISMS

Mutualistic Associations

We have already mentioned a number of mutually beneficial associations between plant roots and other soil organisms (e.g., mycorrhizae and nitrogen-fixing nodules) and between several microorganisms (e.g., lichens). Other examples of such associations abound in soils. For example, photosynthetic algae reside within the cells of certain protozoans. We will now focus on several types of cyanobacterial–algal–fungal associations that are important components of desert ecosystems.

Biocrusts[17]

In relatively undisturbed arid- and semiarid-region ecosystems, where the vegetation cover is quite patchy, it is common to find an irregular, usually dark-colored crust covering the soil in the areas between clumps of grasses and shrubs. In many cases this forms a sort of miniature landscape of tiny, jagged pinnacles only a few centimeters tall (see Figure 10.38). This crust is not at all like the physical–chemical crusts associated with degraded soils that have a hard, smooth surface seal (see Section 4.6). Rather, the **biocrusts** of arid lands are intricate living systems that greatly benefit the associated natural vegetation communities. They consist of mutualistic associations that usually include algae or cyanobacteria along with fungi, mosses, bacteria, and/or liverworts. An intact microbiotic crust is considered a sign of a healthy ecosystem.

Such crusts provide considerable protection against erosion by wind and water by binding soil particles together, protecting the soil from raindrop impact, and increasing surface roughness, which reduces wind velocity. They also improve arid-region ecosystem productivity by: (1) helping to conserve and cycle nutrients, (2) increasing nitrogen supplies via the nitrogen-fixing activities of the cyanobacteria, (3) enhancing water supplies in some cases by increasing infiltration and reducing evaporation, and (4) contributing to net organic matter production by crust photosynthesis, which may continue during environmental conditions

Figure 10.38 Tiny pinnacles of a microbiotic crust seem to reflect the larger pinnacles of an arid landscape. These crusts consist of algae, cyanobacteria, fungi, and other organisms living mutualistically. (Right) A scanning electron micrograph of cyanobacteria filaments that make up the crust's backbone. (Left) A crust on salty soils in semiarid Australia. Biocrusts cover the soil in unvegetated patches between clumps of desert shrubs and grasses, providing considerable protection against erosion by wind and water. They improve desert productivity by adding nitrogen and conserving water and nutrients. The fragile biocrusts can be easily destroyed by wheels, feet, and hooves. (Photo courtesy of Jayne Belnap/USGS)

---

[17]Other names used to refer to these **biocrusts** include *microbiotic crusts, biological crusts, cryptograms, cryptobiotic crusts, microfloral crusts,* and *microphytic crusts*. All refer to the same thing. See Belnap (2003) for a brief overview and further reference citations on the ecology of microbial crusts. For an overview of technical review of scientific and practical knowledge about biocrusts, see Belnap (2013).

that inhibit photosynthesis by higher plants in the ecosystem. The filamentous cyanobacteria make a particularly important contribution to these functions, as they not only photosynthesize but also fix from 2 to 40 kg/ha of nitrogen annually and form sticky polysaccharide coatings or sheaths that catch nutrient-rich dust and bind soil particles. Unfortunately, the biocrusts can be easily destroyed by trampling or burial under windblown soil and are very slow to reestablish. Off-road vehicles, off-trail hiking, and overgrazing are especially damaging to these important ecosystem components.

Effects of Agricultural Practices on Soil Organisms

Changes in environment affect both the number and kinds of soil organisms. Clearing forests or grasslands for cultivation drastically changes the soil environment. Monocultures or even common two-crop rotations greatly reduce the number of plant species and so provide a much narrower range of plant materials and rhizosphere environments than nature provides in less managed ecosystems. Changing the crops grown, or even the choice of winter cover crop can have far-reaching and long-lasting impacts in the belowground community.

While agricultural practices have different effects on different organisms, a few generalizations can be made (Table 10.6). For example, some agricultural practices (e.g., extensive tillage and monoculture) generally reduce the diversity of soil organisms as well as the abundance (number) of individuals. However, monoculture may *increase* the population of a few specially adapted species.

Adding lime and fertilizers (either organic or inorganic) to an infertile acid soil generally will increase microbial and fauna activity, largely due to the increase in the plant biomass that is likely to be returned to the soil as roots, root exudates, and shoot residues. Tillage, on the other hand, is a drastic disturbance of the soil ecosystem, disrupting fungal hyphae networks and earthworm burrows, as well as speeding the loss of organic matter. Reduced tillage therefore tends to increase the role of fungi at the expense of the bacteria and usually increases overall organism numbers as organic matter accumulates in the upper horizons (Table 10.7). Addition of animal manure or compost stimulates even higher microbial and faunal (especially earthworm) activity.

Pesticides are highly variable in their effects on soil ecology (see Section 15.4). Soil fumigants and nematicides can sharply reduce organism numbers, especially for fauna, at least on

Table 10.6
SOIL-MANAGEMENT PRACTICES AND THE DIVERSITY AND ABUNDANCE OF SOIL ORGANISMS

Practices that tend to enhance biological diversity and activity in soils are also those associated with efforts to make agricultural systems more sustainable.

Decreases biodiversity and populations	Increases biodiversity and populations
Fumigants	Proper fertilizer use
Nematicides	Lime on acid soils
Some insecticides and herbicides	Proper irrigation
Compaction	Improved drainage and aeration
Soil erosion	Animal manures and composts
Industrial wastes and heavy metals	Domestic (clean) sewage sludge
Intensive tillage	Reduced or zero tillage
Monocropping	Complex crop rotations
Row crops	Grass–legume pastures
Bare fallows	Cover crops or mulch fallows
Residue burning or removal	Residue return to soil surface
Plastic mulches	Organic mulches

Table 10.7
EFFECT OF TILLAGE SYSTEMS ON BIOMASS CARBON OF MICROBIAL AND FAUNAL GROUPS IN SOIL

Researchers in Georgia grew grain sorghum in summer and a rye cover crop each winter on a sandy loam (Kanhapludults). With plow tillage, plant residues were mixed into the soil, but with no-till system residues were left as surface mulch. Decomposer, microphytic feeder, and detritivore functions were dominated by larger organisms (fungi, microarthropods, and earthworms) in the no-till system, while smaller organisms (bacteria, protozoa, and nematodes) were more prominent in the plowed system.

| | | | | Carbon, kg/ha[a] | | | | |
| | | | | Nematodes | | | | |
Tillage	Fungi	Bacteria	Protozoa	Fungivores	Bacterivores	Microarthropods	Enchytraeids	Earthworms
No-till	360	260	24	0.14	0.82	1.31	5.55	60
Plowed	240	270	39	0.47	1.27	0.49	4.79	21

[a]Depth of sampling was 0 to 5 cm for microbes and arthropods, 0 to 1 cm for nematodes, 0 to 15 cm for worms. Calculated from data of various sources presented by Beare (1997).

a temporary basis. Use of herbicides often reduces biological diversity, mainly because they are so effective in eliminating weeds that otherwise would add more and diverse plant inputs. On the other hand, application of a particular pesticide often stimulates the population of specific microorganisms, either because the organisms can use the pesticide as food or, more likely, because the predators of that organism have been killed. Figure 10.39 illustrates how a management practice (insecticide application in this case) that affects one group of organisms will likely affect other groups as well and will eventually impact the productivity and functioning of the whole soil ecosystem. Such indirect and often surprising effects occur in all ecosystems. It is wise to remember that the interrelationships among soil organisms are intricate, and the effects of any perturbation of the system are difficult to predict.

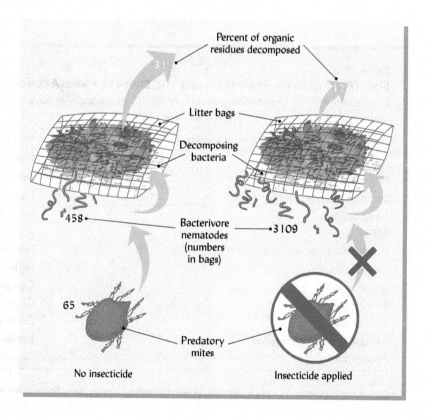

Figure 10.39 The indirect effects of insecticide treatment on the decomposition of creosote bush litter in desert ecosystems. Litter bags filled with creosote bush leaves and twigs were buried in desert soils in Arizona, Nevada, and California, either with or without an insecticide (chlordane) treatment. The insecticide killed virtually all the insects and mites. Without predatory mites to hold them in check, bacterivore nematodes multiplied rapidly and devoured a large portion of the bacterial colonies responsible for litter decomposition and nutrient cycling. Thus, the insecticide reduced the rate of litter decomposition nearly in half, not by any direct effect on the bacteria, but by the indirect effect of killing the predators of their predators. [Diagram courtesy of Ray R. Weil, based on data calculated from Whitford et al. (1982)]

10.15 CONCLUSION

The soil is a complex ecosystem with a highly diverse community of organisms that are vital to the cycle of life on Earth. As plant and animal tissues and compounds are added to the soil, the belowground community processes them, returning carbon dioxide to the atmosphere, where it can be recycled through plant photosynthesis. Simultaneously, the soil organisms use portions of these materials to create soil organic matter. During digestion of organic substrates, soil microbes and their predators release essential plant nutrients in inorganic forms that can be readily absorbed by plant roots or may be leached from the soil. The microbes also mediate the redox reactions that influence soil colors, nutrient cycling, and the production of gases that contribute to global warming.

Animals, particularly earthworms, ants, and termites, mechanically incorporate residues into the soil and leave open channels through which water and air can flow. As such, they are examples of soil ecosystem engineers that change the soil environment for all its inhabitants and create niches in which other organisms can live.

Plants form a bridge connecting the above- and belowground worlds. Plant–soil systems are highly integrated and neither plants nor soil organisms would exist as we know them without the other. Just as humans carry with them and rely upon complex communities of microorganism in their gut and on their skin, plants stimulate and rely upon the complex community of microbes that have adapted to life in the rhizosphere. This zone of soil adjacent to plant roots is one of the most energy-rich and biologically diverse habitats on earth.

Certain rhizosphere microorganisms, such as mycorrhizal fungi and Rhizobia bacteria, form well-known symbiotic associations with certain plants, playing special roles in plant nutrition and nutrient cycling. However, scientific understanding is just beginning to scratch the surface of the complex root–microbes communities beneath our feet. Apparently the rhizosphere is filled with hundreds of specific chemical signaling compounds by which plant and microbes communicate with each other.

Plants use such signals to enlist the aid of soil fauna or microbes in fending off potential pests and pathogens. Numerous specific fungi and bacteria produce antibiotic compounds that help them compete, can rescue plants from disease-causing organisms, and form the basis for life-saving human drugs.

Microorganisms such as fungi, archaea, and bacteria are responsible for most organic decay, although their activity is greatly influenced by the soil fauna that physically shred the plant remains and provide transport and a suitable environment that enhances the function of many of the microbes. To obtain energy and nutrients, soil organisms break down plant litter. As they do so, they stabilize a portion as soil organic matter, and leave behind compounds that are useful to plants. Soil organic matter, its decomposition and influence on soil function are topics of the next chapter.

STUDY QUESTIONS

1. What is *functional redundancy*, and how does it help soil ecosystems continue to function in the face of environmental shocks such as fire, clear-cutting, or tillage?
2. In the example illustrated in Table 10.7, identify the organisms, if any, that play the roles of *primary producers*, *primary consumers*, *secondary consumers*, and *tertiary consumers*.
3. Describe some of the ways in which mesofauna play significant roles in soil metabolism even though their biomass and respiratory activity is only a small fraction of the total in the soil.
4. What are the four main types of metabolism carried out by soil organisms relative to their sources of energy and carbon?
5. What role does O_2 play in aerobic metabolism? What elements take its place under anaerobic conditions?
6. A *mycorrhiza* is said to be a symbiotic association. What are the two parties in this symbiosis, and what are the benefits derived by each party?
7. In what ways is soil improved as a result of earthworm activity? Are there possible detrimental effects as well?
8. What is the *rhizosphere*, and in what ways does the soil in the rhizosphere differ from the rest of the soil?
9. Explain and compare the effects of tillage and manure application on the abundance and diversity of soil organisms.
10. What is *induced systemic resistance*, and how does it work?
11. What is a disease-suppressive soil? Explain the difference between general and specific forms of suppression.
12. Discuss the value and limitations of using specific inoculants for (a) mycorrhizae and (b) disease suppression.

For each type of inoculation, describe a situation for which the chances would be very good for improving plant growth.
13. What are the main food web roles played by nematodes and how can we visually (with a microscope) distinguish among nematodes that play these roles?
14. In what ways are actinomycetes like other groups of bacteria, and in what ways are they special?
15. Through appropriate extractions and counting you determine that there are 58 nematodes in a one-gram sample of soil. How many nematodes would occur in 1.0 m² area of this soil? In one hectare? Assume the samples came from the upper 10 cm of soil with a bulk density = 1.3 Mg/m³.
16. Explain the concept of an *ecosystem engineer* with two soil examples.

REFERENCES

Alper, J. 1998. "Ecosystem 'engineers' shape habitats for other species." *Science* 280:1195–1196.

Amador, J. A., and J. H. Gorres. 2005. "Role of the anecic earthworm *Lumbricus terrestris* L. in the distribution of plant residue nitrogen in a corn (*zea mays*)–soil system." *Applied Soil Ecology* 30:203–214.

Anderson, A. N., and J. D. Majer. 2004. "Ants show the way down under: Invertebrates as bioindicators in land management." *Frontiers in Ecology and the Environment* 2:291–298.

Bakker, P. A. H. M., R. F. Doornbos, C. Zamioudis, R. L. Berendsen, and C. M. J. Pieterse. 2013. "Induced systemic resistance and the rhizosphere microbiome." *Plant Pathology Journal* 29:136–143.

Bardgett, R. D. 2005. *The Biology of Soil: A Community and Ecosystem Approach (Biology of Habitats)*. Oxford University Press, Oxford, p. 254.

Bardgett, R. D., and W. H. van der Putten. 2014. "Belowground biodiversity and ecosystem functioning." *Nature* 515:505–511.

Baskin, Y. 2005. *Under Ground: How Creatures of Mud and Dirt Shape Our World*. Island Press, Washington, D.C., p. 237.

Beare, M. H. 1997. "Fungal and bacterial pathways of organic matter decomposition and nitrogen mineralization in arable soils." In L. Brussaard and R. Ferrera-Cerrato (eds.). *Soil Ecology in Sustainable Agricultural Systems*. Lewis Publishers, Boca Raton, FL.

Belnap, J. 2003. "The world at your feet: Desert biological soil crusts." *Frontiers in Ecology and the Environment* 1:181–189.

Belnap, J. 2013. "Some like it hot, some not." *Science* 340:1533–1534.

Biswas, J. C., J. K. Ladha, and F. B. Dazzo. 2000. "Rhizobia inoculation improves nutrient uptake and growth of lowland rice." *Soil Science Society of America Journal* 64:1644–1650.

Blanco-Canqui, H., R. Lal, W. M. Post, R. C. Izaurralde, and M. J. Shipitalo. 2007. "Soil hydraulic properties influenced by corn stover removal from no-till corn in Ohio." *Soil and Tillage Research* 92:144–155.

Blouin, M., M. E. Hodson, E. A. Delgado, G. Baker, L. Brussaard, K. R. Butt, J. Dai, L. Dendooven, G. Peres, J. E. Tondoh, D. Cluzeau, and J. J. Brun. 2013. "A review of earthworm impact on soil function and ecosystem services." *European Journal of Soil Science* 64:161–182.

Boulter, J. I., G. J. Boland, and J. T. Trevors. 2002. "Evaluation of composts for suppression of dollar spot (*Sclerotinia homoeocarpa*) of turfgrass." *Plant Disease* 86:405–410.

Chander, Y., S. C. Gupta, S. M. Goyal, and K. Kumar. 2007. "Antibiotics: Has the magic gone?" *Journal of the Science of Food and Agriculture* 87:739–742.

Coleman, D. C., Mac Callaham, and J. D. Crossley. 2004. *Fundamentals of Soil Ecology*. 3rd ed., Elsevier Academic Press, London, p. 376.

D'Costa, V. M., K. M. McGrann, D. W. Hughes, and G. D. Wright. 2006. "Sampling the antibiotic resistome." *Science* 311:374–377.

Edwards, C. A. (ed.). 2004. *Earthworm Ecology*. CRC Press, Boca Raton, FL, p. 448.

Edwards, C. A., and N. Q. Arancon. 2004. "Interactions among organic matter, earthworms and microorganisms in promoting plant growth." In F. Magdoff and R. R. Weil (eds.). *Soil Organic Matter in Sustainable Agriculture*. CRC Press, Boca Raton, FL.

Eggleton, P. 2011. "An introduction to termites: Biology, taxonomy and functional morphology." In D. E. Bignell et al. (eds.). *Biology of Termites: A Modern Synthesis*. Springer Netherlands, Amsterdam, pp. 1–26.

Eshel, A., and T. Beeckman (eds.). 2013. *Plant Roots: The Hidden Half*. CRC Press, Boca Raton, FL, pp. 1–848.

Fierer, N., A. S. Grandy, J. Six, and E. A. Paul. 2009. "Searching for unifying principles in soil ecology." *Soil Biology and Biochemistry* 41:2249–2256.

Forsberg, K. J., A. Reyes, B. Wang, E. M. Selleck, M. O. A. Sommer, and G. Dantas. 2012. "The shared antibiotic resistome of soil bacteria and human pathogens." *Science* 337:1107–1111.

French, K. E. 2017. "Engineering mycorrhizal symbioses to alter plant metabolism and improve crop health." *Frontiers in Microbiology* 8:1403.

Goheen, J., and T. Palmer. 2010. "Defensive plant-ants stabilize megaherbivore-driven landscape change in an African savanna." *Current Biology* 20:1–5.

Hendrix, P. F., J. Mac, A. Callaham, J. M. Drake, C.-Y. Huang, S. W. James, B. A. Snyder, and A. W. Zhang. 2008. "Pandora's box contained bait: The global problem of introduced earthworms." *Annual Review of Ecology, Evolution, and Systematics* 39:593–613.

Hiltpold, I., S. Toepfer, U. Kuhlmann, and T. Turlings. 2010. "How maize root volatiles affect the efficacy of entomopathogenic nematodes in controlling the western corn rootworm?" *Chemoecology* 20:155–162.

Hoeksema, J. D., V. B. Chaudhary, C. A. Gehring, N. C. Johnson, J. Karst, R. T. Koide, A. Pringle, C. Zabinski, J. D. Bever, J. C. Moore, G. W. T. Wilson, J. N. Klironomos, and J. Umbanhowar. 2010. "A meta-analysis of context-dependency in plant response to inoculation with mycorrhizal fungi." *Ecology Letters* 13:394–407.

Hooks, C. R. R., K.-H. Wang, A. Ploeg, and R. McSorley. 2010. "Using marigold (*Tagetes* spp.) as a cover crop to protect crops from plant-parasitic nematodes." *Applied Soil Ecology* 46:307–320.

Huber, D. M., and S. Haneklaus. 2007. "Managing nutrition to control plant disease." *Landbauforschung Völkenrode* 57:313–322.

Jouquet, P., J. Dauber, J. Lagerlöfe, P. Lavelle, and M. Lepage. 2006. "Soil invertebrates as ecosystem engineers: Intended and accidental effects on soil and feedback loops." *Applied Soil Ecology* 32:153–164.

Lavelle, P., D. Bignell, M. Lepage, V. Wolters, P. Roger, P. Ineson, O. W. Heal, and S. Dhillion. 1997. "Soil function in a changing world: The role of invertebrate ecosystem engineers." *European Journal of Soil Biology* 33:159–193.

Ling, L. L., T. Schneider, A. J. Peoples, A. L. Spoering, I. Engels, B. P. Conlon, A. Mueller, T. F. Schaberle, D. E. Hughes, S. Epstein, M. Jones, L. Lazarides, V. A. Steadman, D. R. Cohen, C. R. Felix, K. A. Fetterman, W. P. Millett, A. G. Nitti, A. M. Zullo, C. Chen, and K. Lewis. 2015. "A new antibiotic kills pathogens without detectable resistance." *Nature* 517:455–459.

Litterick, A. M., and L. Harrier. 2004. "The role of uncomposted materials, composts, manures, and compost extracts in reducing pest and disease incidence and severity in sustainable temperate agricultural and horticultural crop production: A review." *Critical Reviews in Plant Sciences* 23:453–479.

Lubbers, I. M., K. J. V. Groenigen, S. J. Fonte, J. Six, L. Brussaard, and J. W. V. Groenigen. 2013. "Greenhouse-gas emissions from soils increased by earthworms." *Nature Climate Change* 3:187–194.

McNear Jr., D. H. 2013. "The rhizosphere - roots, soil and everything in between." *Nature Education Knowledge* 4(3):1–7.

Mendes, R., M. Kruijt, I. de Bruijn, E. Dekkers, M. van der Voort, J. H. M. Schneider, Y. M. Piceno, T. Z. DeSantis, G. L. Andersen, P. A. H. M. Bakker, and J. M. Raaijmakers. 2011. "Deciphering the rhizosphere microbiome for disease-suppressive bacteria." *Science* 332:1097–1100.

Nardi, J. B. 2007. *Life in the Soil: A Guide for Naturalists and Gardeners*. University of Chicago Press, Chicago, p. 336.

Neher, D. A. 2010. "Ecology of plant and free-living nematodes in natural and agricultural soil." *Annual Review of Phytopathology* 48:371–394.

Nesme, J., W. Achouak, S.N. Agathos, M. Bailey, P. Baldrian, D. Brunel et al. 2016. "Back to the future of soil metagenomics." *Frontiers in Microbiology* 7:73. doi:10.3389/fmicb.2016.00073.

Paul, E. A. 2014. *Soil Microbiology, Ecology and Biochemistry*. 4th ed. Academic Press, San Diego, p. 600.

Pehrsson, E. C., K. J. Forsberg, M. K. Gibson, S. Ahmadi, and G. Dantas. 2013. "Novel resistance functions uncovered using functional metagenomic investigations of resistance reservoirs." *Frontiers in Microbiology* 4:145.

Reynolds, L. B., J. W. Potter, and B. R. Ball-Coelho. 2000. "Crop rotation with *Tagetes* sp. is an alternative to chemical fumigation for control of root-lesion nematodes." *Agronomy Journal* 92:957–966.

Sanders, D., and F. J. F. van Veen. 2011. "Ecosystem engineering and predation: The multi-trophic impact of two ant species." *Journal of Animal Ecology* 80:569–576.

Sileshi, G. W., M. A. Arshad, S. Konaté, and P. O. Y. Nkunika. 2010. "Termite-induced heterogeneity in African Savanna vegetation: Mechanisms and patterns." *Journal of Vegetation Science* 21:923–937.

Stamets, P. 2005. *Mycelium Running: How Mushrooms Can Help Save the World*. Ten Speed Press, Berkeley, CA, p. 339.

Stone, A. G., S. J. Scheuerell, and H. M. Darby. 2004. "Suppression of soil-borne fungal diseases in field agricultural systems: Organic matter management, cover cropping, and cultural practices." In F. Magdoff and R. R. Weil (eds.). *Soil Organic Matter in Sustainable Agriculture*. CRC Press, Boca Raton, FL.

Sylvia, D. M., J. J. Fuhrmann, P. G. Hartel, and D. A. Zuberer. 2005. *Principles and Applications of Soil Microbiology*. Prentice Hall, Upper Saddle River, NJ, p. 640.

Torsvik, V. L. 1980. "Isolation of bacterial DNA from soil." *Soil Biology and Biochemistry* 12:15–21.

Torsvik, V., L. Ovreas, and T. F. Thingstad. 2002. "Prokaryotic diversity—Magnitude, dynamics, and controlling factors." *Science* 296:1064–1066.

USDA/NRCS. 2000. "Soil biology primer." Rev. Ed. [Online]. *Soil and Water Conservation Society*. Available at soils.usda.gov/sqi/concepts/soil_biology/biology.html (posted 18 June 2007; verified 1 July 2013).

van der Heijden, M.G.A., F.M. Martin, M.-A. Selosse, and I.R. Sanders. 2015. "Mycorrhizal ecology and evolution: The past, the present, and the future." *New Phytologist* 205:1406–1423.

Whitford, W. G., D. W. Freckman, P. F. Santos, N. Z. Elkins, and L. W. Parker. 1982. "The role of nematodes in decomposition in desert ecosystems." In D. Freckman (ed.). *Nematodes in Soil Ecosystems*. University of Texas Press, Austin, TX, pp. 98–116.

Wu, T., E. Ayres, R. D. Bardgett, D. H. Wall, and J. R. Garey. 2011. "Molecular study of worldwide distribution and diversity of soil animals." *Proceedings of the National Academy of Sciences* 108:17720–17725.

Zamioudis, C., P. Mastranesti, P. Dhonukshe, I. Blilou, and C. M. J. Pieterse. 2013. "Unraveling root developmental programs initiated by beneficial Pseudomonas spp. bacteria." *Plant Physiology* 162:304–318.

Zhu, Y.-G., T. A. Johnson, J.-Q. Su, M. Qiao, G.-X. Guo, R. D. Stedtfeld, S. A. Hashsham, and J. M. Tiedje. 2013. "Diverse and abundant antibiotic resistance genes in Chinese swine farms." *Proceedings of the National Academy of Sciences* 110:3435–3440.

11
Soil Organic Matter

Carbon accumulates in a forest soil. (Photo courtesy of Ray R. Weil)

Their host consumed, themselves in death
Their substance too return to earth.
The forest thus sustains its wealth
And turns decay to surging life
—CAROLINE PRESTON (NATURAL RESOURCES CANADA)

In most soils, the percentage of soil organic matter[1] (SOM) is small, but its effects on soil function are profound. This ever-changing soil component exerts a dominant influence on many soil physical, chemical, and biological properties and ecosystem functions of soils. Soil organic matter provides much of the soil's cation exchange capacity (CEC) (Chapter 8) and water-holding capacity (Chapter 5). Certain components of soil organic matter are largely responsible for the formation and stabilization of soil aggregates (Chapter 4). Soil organic matter also contains large quantities of plant nutrients and acts as a slow-release nutrient storehouse, especially for nitrogen (N) (Chapter 12). Furthermore, organic matter supplies energy and cellular constituents for most of the organisms whose nature and activities were discussed in Chapter 10. In addition to enhancing plant growth through the just-mentioned effects, certain organic compounds found in soils have direct growth-stimulating effects on plants. For all these reasons, the quantity and quality of soil organic matter are key to building **soil health**.

All organic substances, by definition, contain the element **carbon**, and, on average, carbon comprises about half of the mass of soil organic matter. Organic matter in the world's soil profiles contains four to six times as much carbon as is found in all the world's vegetation. Soil organic matter, therefore, plays a critical role in the global carbon balance—a balance that largely controls **global climate change**.

We will first examine the role of soil organic matter in the **global carbon cycle** and the process of **decomposition** of organic materials in soils. Next, we will focus on inputs and losses with regard to soil carbon in specific ecosystems. Finally, we will study the processes and consequences involved in soil organic matter management.

11.1 THE GLOBAL CARBON CYCLE

The element *carbon* is the foundation of all life. From cellulose to chlorophyll, the compounds that comprise living tissues are made of carbon atoms arranged in chains or rings and associated with many other elements. The cycle of carbon on Earth is the story of life on this planet. The carbon cycle is all-inclusive because it involves the soil, microbes, plants of

---

[1] For a broad review of the nature, function, and management of organic matter in agricultural soils, see Magdoff and Weil (2004). For information on functions of organic matter in soils, sediments, and other environmental systems, as well on the analysis of organic matter by modern instrumental techniques, see Senesi et al. (2009).

every description, and all animal life, including humans. Disruption of the carbon cycle would mean disaster for all living organisms (Box 11.1).

Basic Processes

The basic processes involved in the global carbon cycle are shown in Figure 11.3. Plants take in carbon dioxide from the atmosphere. Then, through the process of photosynthesis, the energy of sunlight is trapped in the carbon-to-carbon bonds of organic molecules. Some of these organic molecules are used as a source of energy (via respiration) by the plants themselves

---

BOX 11.1
CARBON CYCLING—UP CLOSE AND PERSONAL

Imagine that you were one of the eight biospherians living a scientific game of survival in Biosphere 2, a giant 1.3-ha sealed glass building in the Arizona desert (Figure 11.1). Biosphere 2 contained a miniature ocean, coral reef, marsh, forest, and farm in a self-contained, self-supporting ecosystem in which the biospherians could live as part of the ecosystem they were studying. Instruments throughout the structure constantly monitored environmental parameters. What a great physical model (rather than a computer model) to study how a balanced ecosystem *really* works!

But it didn't take long for trouble to develop. First, the biospherians found it was no easy task to grow all the food they needed (Figure 11.2). Dependent on their meager harvests, they began to lose weight. Then, as if slowly starving were not bad enough, they soon began feeling short of breath. Instruments showed the oxygen level of the air was falling from its normal 21% to levels typical of high mountaintops (it eventually fell to as low as 14.2%). But unlike in "thin" air at high elevations, the carbon dioxide content of the air was rising. This wasn't supposed to be happening. Weren't all the green plants supposed to *use up* the carbon dioxide and *replenish* the oxygen supply? Unusual changes began talking place in the biospherians' blood chemistry and metabolism—changes eerily like those of a bear in hibernation (during which both oxygen and food are limited). Eventually, the atmosphere became so low in oxygen and high in carbon dioxide that engineers had to give up on the "fully self-contained" aspect of the project and pump in oxygen and remove carbon dioxide from the air.

What had they overlooked? It turned out that the ecosystem was thrown out of kilter by the organic-matter-rich soil hauled in for the Biosphere farm. The soil, made from a mixture of pond sediment (1.8% C), compost (22% C), and peat moss (40% C), was installed uniformly about 1 m deep. This artificial soil contained about 2.5% organic C at all depths—far more than the 0.5% C or less expected in a typical desert soil. Had the designers read this book, they would have understood that peat might be stable in a boreal wetland (cool and anaerobic), but that aerobic soil microorganisms would rapidly use up oxygen and give off carbon dioxide as they metabolized organic matter in warm, moist garden soil aerated by tillage (see Section 11.2). This tale reminds us of the importance of soils in cycling C within the real biosphere of Earth! (For more on Biosphere 2, see: http://www.biospheres.com/, Torbert and Johnson (2001), and Walford (2002).)

Figure 11.1 *The Biosphere 2 structure, a huge, sealed, ecological laboratory.* (Photo by C. Allen Morgan. © 1995 by Decisions Investments Corp. Reprinted with permission)

Figure 11.2 *Biospherians at work growing their own food supply in the intensive agriculture biome with compost-amended soils rich in organic matter.* (Photo by Pascale Maslin. © 1995 by Decisions Investments Corp. Reprinted with permission)

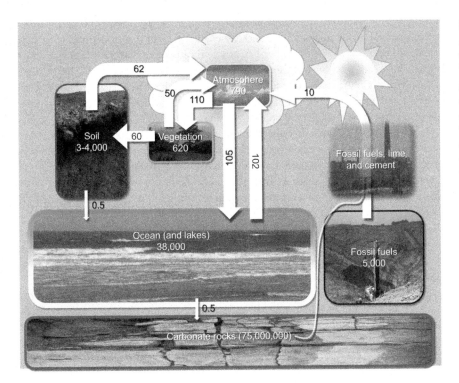

Figure 11.3 *The global carbon cycle emphasizing those pools of C which interact with the atmosphere. The numbers in boxes indicate petagrams (Pg = 10^{15} g) of C stored in the major pools. The numbers by arrows show amounts of C annually flowing (Pg/yr) by various processes between the pools. Note that the soil contains almost twice as much C as the vegetation and the atmosphere combined. Imbalances caused by human activities can be seen in the flow of C to the atmosphere from fossil fuel burning (10) and in the fact that more C is leaving (62 + 0.5) than entering (60) the soil. These imbalances are only partially offset by increased absorption of C by the oceans. The end result is that some 225 Pg/yr enters the atmosphere while only 215 Pg/yr of C is removed. It is easy to see why carbon dioxide levels in the atmosphere are rising.* [Data from IPCC (2007) and Pan et al. (2011); soil C estimate from Batjes (1996), Haddix et al. (2011), Lal (2018)]

(especially by the plant roots), with the carbon being returned to the atmosphere as carbon dioxide. The remaining organic materials are stored temporarily as constituents of the standing vegetation, most of which is eventually added to the soil as plant litter (including crop residues) or root deposition (see Section 10.7). Some plant material may be eaten by animals (including humans), in which case about half of the consumed carbon is exhaled into the atmosphere as carbon dioxide. The carbon not returned to the atmosphere is eventually returned to the soil as bodily wastes or body tissues. Once deposited on or in the soil, these plant or animal tissues are metabolized (digested) by soil organisms that gradually return this C to the atmosphere as carbon dioxide.

Carbon dioxide also reacts in the soil to produce carbonic acid (H_2CO_3) and the carbonates and bicarbonates of calcium, potassium, magnesium, and sodium. The bicarbonates are readily soluble and may be removed in drainage. The carbonates, such as calcite ($CaCO_3$), are much less soluble and tend to accumulate in soils under alkaline conditions. Although this chapter focuses on the organic C in soils, the inorganic C content of soils (mainly as carbonates) may be substantial, especially in arid regions (Table 11.1). Eventually, as with the C in soil organic matter, most of the bicarbonate C and some of the carbonate C in soils is returned to the atmosphere as CO_2.

Some of the partially decomposed plant tissues and microbial cell debris is adsorbed onto soil colloid surfaces or occluded inside soil aggregates where it is protected from further microbial metabolism for decades or even centuries before the C in them is returned to the atmosphere as carbon dioxide. Such protection from decay allows organic matter to accumulate in soils.

Carbon Sources

The original source of SOM is plant tissue, and the amount of organic matter accumulated in soils is partly a function of the net plant productivity that provides this material. Animals are secondary sources of organic matter. As they eat the original plant tissues, they contribute waste products, and they leave their own bodies when they die (review Figure 10.3). Certain forms of animal life, especially earthworms, termites, ants, and dung beetles, also play an important role in the incorporation and translocation of organic residues.

Globally, at any one time, approximately 3,000 petagrams (Pg or 10^{15} g) of C are stored in soil profiles as SOM (excluding surface litter), about one-third of that at depths below 1 m. An additional 940 Pg are stored in the upper 1 m of soil as carbonates that can release CO_2 upon weathering (as with organic C, there are significant additional stores of carbonate C below 1 m). Altogether, more than twice as much C is stored in the soil as in the world's vegetation and atmosphere *combined* (see Figure 11.1). Of course, this C is not equally distributed among all types of soils (Table 11.1). About half of the total organic carbon is contained in soils of just three orders, Histosols, Inceptisols, and Gelisols. Because many Gelisols consist of organic matter accumulations several meters thick, evidence now suggests that there may be much more C in this order than previously thought—perhaps as much as all other soil orders combined! Inceptisols (and nonhistic Gelisols) contain only moderate concentrations of C, but cover vast areas of the globe. The reasons for the varying amounts of organic C in different soils will be detailed in Section 11.8.

In a mature natural ecosystem or a stable agroecosystem, the release of C as carbon dioxide by oxidation of SOM (mostly by microbial respiration) is balanced by the input of C into the soil as plant residues (and, to a far smaller degree, animal residues). However, as discussed in Section 11.8, certain perturbations of the system, such as deforestation, some types of fires, tillage, and artificial drainage, result in a net loss of C from the soil system.

Figure 11.3 shows that, globally, the release of C from soils into the atmosphere is about 62 Pg/yr, while only about 60 Pg/yr enter the soils from the atmosphere via plant residues.

Table 11.1
MASS OF ORGANIC AND INORGANIC CARBON IN THE UPPER 1 M OF THE WORLD'S SOILS

Inorganic C is present mainly as calcium carbonates in soils of dry regions. Wetland soils as a group contain about 500 Pg of organic C, some 30.3% of the total organic C in global soils. Values for the upper 1 m represent 60 to 90% of the C in most soil profiles. Considerable additional C is stored in soil layers deeper than 1 m, especially in Gelisols and Histosols.

| Soil order | Global area, 10^3 km² | Global C[a] in upper 100 cm | | | |
| | | Organic | Inorganic | Total | Total % of global soil C |
		Pg			
Entisols	21,137	90	263	353	14.2
Inceptisols	12,863	190	34	224	9.0
Histosols	1,526	179	0	180	7.2
Andisols	912	20	0	20	0.8
Gelisols	11,260	316[b]	7	323	12.9
Vertisols	3,160	42	21	64	2.6
Aridisols	15,699	59	456	515	20.6
Mollisols	9,005	121	116	237	9.5
Spodosols	3,353	64	0	64	2.6
Alfisols	12,620	158	43	201	8.0
Ultisols	11,052	137	0	137	5.5
Oxisols	9,810	126	0	126	5.1
Misc. land	18,398	24	0	24	1.0
Total	130,795	1,526	940	2,468	100.0

[a] Soil organic matter values may be estimated as 2.0 times soil organic C, although the multiplier traditionally used is 1.72. Organic N may also be estimated from organic C values by dividing by 12 for most soils, but see Section 11.3. Pg = Petagram = 10^{15} g.
[b] The value for Gelisols is likely an underestimate. Evidence-based estimates in Harden et al. (2012) range up to 1,800 Pg organic C in Gelisols to a 3-m depth.
Data selected from Eswaran et al. (2000).

This imbalance of about 2 Pg/yr, along with about 10 Pg/yr of C released by the burning of fossil fuels (for energy) and limestone rock (to make cement), is only partially offset by increased absorption of atmospheric carbon dioxide by the ocean. Fossil fuel burning and degrading land-use practices have increased the concentration of carbon dioxide in the atmosphere at an accelerating rate since the beginning of the industrial revolution four centuries ago. The levels have increased from 290 to 400 ppm during the past century alone. The implications of net emissions of carbon dioxide and other gases on climate change (the greenhouse effect) will be discussed in Section 11.10, after we consider the processes by which carbon cycles in the plant–soil–atmosphere system.

11.2 ORGANIC DECOMPOSITION IN SOILS

Composition of Plant Residues

Plant material is the principal source of SOM. Fresh plant tissues contain from 60 to 90% water by weight. If plant tissues are dried to remove all water, the dry matter remaining consists at least 90 to 95% of carbon, oxygen, and hydrogen, with carbon accounting for approximately 40% of the dry matter.

During photosynthesis, plants obtain these elements from carbon dioxide and water. If plant dry matter is burned (oxidized), these elements become carbon dioxide and water once more. Some 5 to 10% of the dry matter burned will become ash and smoke. The ash and smoke contain many nutrient elements originally taken up from the soil. Even though these elements are present in relatively small quantities, they play a vital role in plant, animal, and microbial nutrition. The essential nutrient elements found in the ash, such as nitrogen, sulfur, phosphorus, potassium, and micronutrients, will be given detailed consideration in Chapter 12.

Organic Compounds in Plant Residues. Tissues from different plant species, as well as from different parts (leaves, roots, stems, etc.) of a given plant, differ considerably with regard to broad classes of organic compounds. As a group, carbohydrates are usually the most plentiful of plant organic compounds. Carbohydrates range from highly bioavailable simple sugars and starches to less easily broken down cellulose. Certain plant parts, especially seeds and leaf coatings, contain significant amounts of fats, oils, and waxes, some of which decompose rather slowly because of their highly water-repellent nature. Root tissues are especially significant as sources of SOM and tend to decompose more slowly than shoot tissues.

Lignins, complex compounds with multiple ring-type or *phenol* structures, are components of plant cell walls. The content of lignin increases as plants mature and is especially high in woody tissues. Other **polyphenols**, such as tannins, may comprise as much as 6 or 7% of the leaves, root, and bark of certain plants (e.g., the red-brown colors of both steeped tea and leaf prints left by wet oak leaves on concrete sidewalks is due to tannins). In certain tree leaves, much of the cellulose is encased in lignin. Lignin is often observed to decompose much more slowly than other litter components, as relatively few microbes can produce the enzymes necessary to break it down.

Proteins contain about 16% N and smaller amounts of other essential elements, such as sulfur, manganese, copper, and iron. Simple proteins decompose and release their N easily, while complex crude proteins are more resistant to breakdown.

Rate of Decomposition. Organic compounds may be listed in terms of ease of decomposition as follows:

1. Sugars, starches, and simple proteins   Rapid decomposition
2. Crude proteins
3. Hemicellulose
4. Cellulose
5. Fats and waxes
6. Lignins and phenolic compounds   Very slow decomposition

Decomposition of Organic Compounds in Aerobic Soils

Decomposition involves the breakdown of large organic molecules into smaller, simpler components. When organic tissue is added to an aerobic soil, four general processes take place: oxidation, release, synthesis, and protection. Although in most cases mechanical shredding by soil fauna or physical processes must occur before these processes can efficiently take place, the reactions themselves result mainly from microbial activity.

1. Plant carbon compounds are enzymatically *oxidized* to produce energy for the decomposer as well as carbon dioxide and water.
2. Essential nutrient elements such as nitrogen, phosphorus, and sulfur are *released* and/or immobilized by a series of specific reactions that are relatively unique for each element.
3. New compounds are *synthesized* by microbes as cellular constituents or as breakdown products or secondary metabolites.
4. Some of the original plant compounds, their breakdown products and microbial compounds become physically or chemical *protected* from further microbial decay via interactions with the soil environment.

Decomposition: An Oxidation Process. In a well-aerated soil, all of the organic compounds found in plant residues are subject to oxidation. Since the organic fraction of plant materials is composed largely of carbon and hydrogen, the oxidation of the organic compounds in soil can be represented as:

$$\underset{\substack{\text{Carbon- and}\\\text{hydrogen-containing}\\\text{compounds}}}{R\text{—}(C, 4H)} + 2O_2 \xrightarrow{\text{Enzymatic oxidation}} CO_2\uparrow + 2H_2O + \text{energy (478 kJ mol}^{-1}\text{ C)} \tag{11.1}$$

This basic reaction accounts for most of the organic matter decomposition in the soil, as well as for the oxygen consumption and CO_2 release. Not shown are the many intermediate steps involved in this overall reaction, as well as accompanying side reactions that involve elements other than carbon and hydrogen.

Breakdown of Cellulose and Starch. Cellulose and starch are polysaccharides—long chains (polymers) of sugar molecules. Enzymatic degradation proceeds in steps: first the long chains are broken down by rather specialized organisms into short chains, then into individual sugar (glucose) molecules, which many different organisms can metabolize as in Eq. (11.1). Cellulose is the most abundant polysaccharide on the Earth. Because the C—O—C chemical bonds linking the sugar molecules of cellulose together are much more difficult to break than those in starch, the initial chain-breaking step in cellulose decomposition requires the activity of specialized organisms that produce the enzyme *cellulase*.

Breakdown of Proteins. Plant proteins also succumb to microbial decay, yielding not only carbon dioxide and water, but amino acids such as glycine (CH_2NH_2COOH) and cysteine ($CH_2HSCHNH_2COOH$). In turn, these nitrogen and sulfur compounds are further broken down, eventually yielding such simple inorganic ions as ammonium (NH_4^+), nitrate (NO_3^-), and sulfate (SO_4^{2-}), forms readily available for plant nutrition.

Breakdown of Lignin. Lignin molecules are very large and complex, consisting of hundreds of interlinked phenolic ring subunits. Only a few microorganisms (mainly *white rot fungi*, see Figure 10.24) can break them down. Lignin is traditionally thought of as highly resistant to decomposition by most microbes, especially in the initial stages of decay. However research suggests that lignin will actually degrade in the early stages of decomposition as long as the supply of readily bioavailable C is great enough to fuel the production of lignin-degrading enzymes. In the field, lignin decomposition is also assisted by the physical activities of soil fauna. Once the lignin subunits are separated, many types of microorganisms participate in their breakdown. Still, significant amounts of lignin and lignin-like compounds remain in soils as organic matter for long periods.

Example of Organic Decay

The process of organic decay in time sequence is illustrated in Figure 11.4. Assume the soil has not been disturbed or amended with plant residues for some time. Initially, little or no readily decomposable materials are present. Competition for food is severe and microbial activity is relatively low, as reflected in the low **soil respiration** rate or level of CO_2 emission from the soil. The supply of soil C is steadily being depleted. Small populations of microorganisms survive by slowly digesting the resistant, stable SOM. Soil ecologists consider these microorganisms to exhibit a K-strategy for survival, so named because they have developed enzymes with high affinity constants (K) for specific types of resistant substrates. The **K-strategists** have a competitive advantage when the soil is poor in easily digested organic materials. These organisms maintain a low but fairly constant population by carrying out specialized reactions.

Now suppose that deciduous trees in a forest begin to lose their leaves in fall or a farmer terminates a cover crop. The appearance of easily decomposable and often water-soluble compounds, such as sugars, starches, and amino acids, stimulates an almost immediate increase in metabolic activity among the soil microbes. Soon the slower-acting K-strategists are overtaken by rapidly multiplying populations of *opportunist* or *colonizing* organisms that have been awakened from their dormant state by the presence of new food supplies. These organisms are known as **r-strategists**, so named for their rapid rate (r) of growth and reproduction that allows them to take advantage of a sudden influx of food.

Microbial numbers and carbon dioxide evolution from microbial respiration both increase exponentially in response to the new food resource (upper panel in Figure 11.4). Soon microbial activity is at peak intensity, energy is being rapidly liberated, and carbon dioxide is being formed in large quantities. As organisms multiply, they increase the **microbial biomass**. The microbial biomass at this point may account for as much as one-sixth of the organic matter in the soil. The intense microbial activity may even stimulate the breakdown of some of the more protected SOM, a phenomenon known as the **priming effect**.

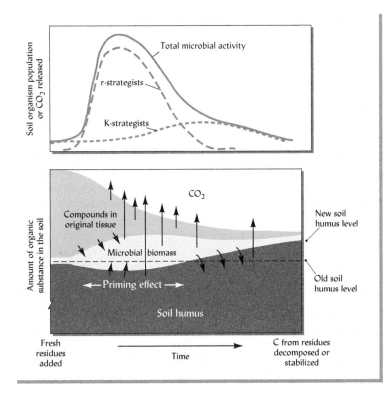

Figure 11.4 Schematic of the general changes occurring when fresh plant residues are added to a soil. The arrows indicate transfers of C among compartments. The upper panel shows the relative growth or activity of r-strategist (opportunist), K-strategist (more specialized) microorganisms, and the sum of these two groups. The time required for the process will depend on the nature of the residues and the soil. Most of the C released during the initial rapid breakdown of the residues is converted to carbon dioxide. Smaller amounts of C are converted into microbial biomass and, eventually, dead microbial cell walls and other forms of nonliving organic materials known as humus. The peak level of microbial activity appears to accelerate the decay of the original humus, a phenomenon known as the priming effect. However, the humus level is increased by the end of the process. Where vegetation, environment, and management remain stable for a long time, the soil humus content will reach an equilibrium level at which the C added to the humus pool through the decomposition of plant residues is balanced by C lost through the decomposition of existing soil humus. (Diagram courtesy of Ray R. Weil)

With all this frenetic microbial activity, the supply of easily accessible and decomposable compounds becomes depleted. In addition, microbial-feeding protozoa and nematodes move in to gorge on the masses of bacteria and fungi. While the specialized K-strategists continue their slow work, degrading cellulose and lignin, r-strategists begin to die of starvation and predation. As microbial populations plummet, the dead cells' contents provide a readily digestible food source for the survivors, which continue to evolve carbon dioxide and water. The decomposition of the dead microbial cells and nematode excretions are also associated with the **mineralization** or release of simple inorganic products, such as nitrates and sulfates. As food supplies are further reduced, microbial activity continues to decline, and the general-purpose r-strategists again sink back into comparative quiescence and most of the microbial feeders die off (and decompose) or move on. A small amount of the original residue material persists, mainly as tiny particles that have become **physically protected** from decay by lodging inside soil pores too tight to allow access by most organisms. Some of the remaining C compounds, including microbial cell wall materials and plant biomolecules have been **chemically protected** from rapid decay by binding tightly with the surfaces of mineral particles. This remaining organic matter, referred to as humus, is largely a dark-colored, heterogeneous, mostly colloidal-sized mixture of bits of dead plant tissues and microbial cells, modified lignin and other plant and microbial compounds. It is considered stabilized organic matter because it is much less water soluble than the sugars and starches that were the first to be digested by the microbial community and because its association with the mineral fraction of the soil makes it less susceptible to enzymatic attack. Thus, a small percentage of the C originally present in the added residues has been retained, increasing slightly the pool of relatively stable soil organic matter (humus). If undisturbed for many years, an ecosystem will come into equilibrium such that the small increase in retained C will likely be offset during each annual cycle by slow, steady K-strategy decomposition, resulting in little net change in the level of SOM from year to year.

Production of Simple Inorganic Products

As proteins are attacked by microbes, the long chains of amino acids are broken, and individual amino acids appear in the soil solution along with dissolved CO_2. The amide (—R—NH_2) and sulfide (—R—S) groups of the amino acids, in turn, are broken off to produce, first, ammonium (NH_4^+) and sulfide (S^{2-}) compounds and, finally, nitrates (NO_3^-) and sulfates (SO_4^{2-}). Similar decomposition of other organic compounds releases these and other inorganic (mineral) nutrient ions. The process that releases elements from organic compounds to produce inorganic (mineral) forms is known as **mineralization**; it is of major importance in supplying nitrogen, sulfur, phosphorus, and other essential elements for plant growth.

Decomposition in Anaerobic Soils

Oxygen supplies may become depleted when soil pores filled with water prevent the diffusion of O_2 into the soil from the atmosphere. Without sufficient oxygen present, aerobic organisms cannot function, so anaerobic or facultative organisms become dominant. Wet, anaerobic soils tend to accumulate large amounts of partially decomposed organic matter for two reasons: First, under low-oxygen or anaerobic conditions, decomposition takes place much more slowly than when oxygen is plentiful. Second, certain products of anaerobic metabolism are toxic to many microbes, acting as a preservative for organic matter.

The products of anaerobic decomposition include a wide variety of partially oxidized organic compounds, such as organic acids, alcohols, and methane gas. Anaerobic decomposition releases relatively little energy for the organisms involved; therefore, the end products still contain much energy. (For this reason, alcohol and methane, which are produced by anaerobic decomposition, can serve as fuel.) Some of the products of anaerobic decomposition are of concern because they produce foul odors or inhibit plant growth. The methane gas produced in wet soils is greenhouse gas and a major contributor to climate change (Section 11.10). The following reactions are typical of those carried out in wet soils by various **methanogenic archaea**:

$$4C_2H_5COOH + 2H_2O \xrightarrow{Archaea} 4CH_3COOH + CO_2 \uparrow + 3CH_4 \uparrow \quad (11.2)$$
$$\text{Propionate} \qquad\qquad\qquad \text{Acetate} \qquad \text{Carbon} \qquad \text{Methane}$$

$$CH_3COOH \xrightarrow{Archaea} CO_2 \uparrow + CH_4 \uparrow \quad (11.3)$$

$$CO_2 + 4H_2 \xrightarrow{Archaea} 2H_2O + CH_4 \uparrow \quad (11.4)$$

11.3 FACTORS CONTROLLING RATES OF RESIDUE DECOMPOSITION AND MINERALIZATION[2]

It may take from days to years for the just described processes of decomposition and mineralization to run their course, depending mainly on two broad factors: (1) the environmental conditions in the soil, and (2) the quality of the added residues as a food source for soil organisms.

The environmental conditions conducive to rapid decomposition and mineralization (see also Sections 10.11 and 11.8) include sufficient soil moisture but with good aeration (about 60% of the soil pore space filled with water), warm temperatures (up to about 35°C), and a near-neutral pH. Ironically, periodic stresses such as episodes of severe drying actually accelerate overall mineralization due to the dramatic burst of microbial activity that occurs each time the soil re-wets (e.g., Figure 12.6). These conditions were discussed in Section 10.11 in relation to microbial activity and will be considered again in Section 11.8 as they affect the levels of organic matter accumulating in soils. Here we will focus on factors that determine the quality of the residues as a food resource for microbes, including the physical condition of the residues, their C/N ratio, and their content of lignins and polyphenols.

Physical Factors Influencing Residue Quality

The location of residues in or on the soil is a physical factor that has a critical impact on decomposition rates. Surface placement of plant residues, as in forest litter or conservation tillage mulch, usually results in slower, more variable rates of decomposition than where similar residues are incorporated into the soil by root deposition, faunal action, or tillage. Surface residues are subject to rapid drying, as well as extremes of temperature. Nutrient elements mineralized from surface-applied residues are also more susceptible to loss in runoff or by volatilization than are those from incorporated residues. Surface residues are physically out of reach for most soil organisms, so fungal mycelia (see Figure 10.24) and larger fauna such as earthworms have a special role to play. Compared to surface residue, incorporated residues experience much more constant moisture and temperature and are in intimate contact with soil moisture and soil organisms. The incorporated residues therefore, decompose more quickly and uniformly, and may lose nutrients more easily by leaching.

Residue particle size is another important physical factor—the smaller the particles, the more rapid the decomposition. Small particle size may result from the nature of the residues (e.g., twigs versus branches), from mechanical treatment (grinding, chopping, tillage, etc.), or from the chewing action of soil fauna. Shredding residues into smaller particles physically exposes more surface area to decomposition and also breaks up lignacious cell walls and waxy outer coatings on leaves exposing more readily digested tissues and cell contents. In addition, some organic materials, including tiny bits of plant tissue lodged in aggregates, exhibit hydrophobicity (water repellency), making them slow to wet and resistant to attack by water-soluble microbial enzymes.

Plant root exudates and dead roots, which can account for 20 to 50% of total plant residues (see Section 10.7), are obviously located in the soil, often very deep in the profile. Root residues generally decompose more slowly than aboveground residues that have been incorporated into the soil.

---

[2]For an excellent collection of papers dealing with litter decomposition in soils, see Cadisch and Giller (1997). Experimental values and mechanistic modeling of the influence of soil moisture, and residue physical placement and quality can be found in Coppens et al. (2007).

Carbon/Nitrogen Ratio of Organic Materials and Soils

The C content of typical plant dry matter is about 40 to 45%. The N content of plant residues is much lower and more variable (ranging from <1 to >6%). The ratio of carbon to nitrogen (C/N) in organic residues applied to soils is important for two reasons: (1) intense competition among microorganisms for available soil N occurs when residues having a high C/N ratio are added to soils, and (2) the residue C/N ratio helps determine the rate of decay and the rate at which N is made available to plants.

C/N Ratio in Plants and Microbes. The C/N ratio in plant residues ranges from between 8:1 to 30:1 in legumes and young green leaves to higher than 500:1 in some kinds of sawdust and charred materials (Table 11.2). Generally, as plants mature, the proportion of protein in their tissues declines, while the proportion of lignin and cellulose, and the C/N ratio increase. Different plant residues have differing proportions of their C distributed between two generalized pools: (1) a metabolic pool with high rates of initial decomposition representing mainly cytoplast contents such as starch, sugars, and proteins; and (2) a structural pool with slower initial decomposition that represents mainly cell wall constituents such as lignin and cellulose. The C/N ratio of the metabolic pool is generally < 25:1 while that of the structural pool is usually > 100:1. In the bodies and cells of microorganisms, the C/N ratio is not only less variable than in plant tissues, but also much lower, ordinarily falling between 5:1 and 10:1. Among microorganisms, bacteria are generally richer in protein than fungi and, consequently, have a lower C/N ratio.

C/N Ratio in Soils. The C/N ratio in the organic matter of arable (cultivated) surface (Ap) horizons commonly ranges from 8:1 to 15:1, the median being near 12:1. The ratio is generally lower for subsoils than for surface layers in a soil profile. In a given climatic region, little variation occurs in the C/N ratio for similarly managed soils. For instance, in calcium-rich soils of semiarid grasslands (e.g., Mollisols and tropical Alfisols), the C/N ratio is relatively narrow. In

Table 11.2
TYPICAL CARBON AND NITROGEN CONTENTS AND C/N RATIOS OF SOME ORGANIC MATERIALS COMMONLY ASSOCIATED WITH SOILS

Organic material	% C	% N	C/N
Hardwood sawdust	46	0.1	400
Newspaper	39	0.3	120
Wheat straw	38	0.5	80
Corn stover	45	0.75	60
Maple leaf litter	48	1.4	34
Rye cover crop, vegetative stage	40	1.5	26
Mature alfalfa hay	40	1.8	25
Rotted barnyard manure	41	2.1	20
Bluegrass from fertilized lawn	42	2.2	20
Broccoli residues	35	1.9	18
Young alfalfa hay	40	3.0	13
Hairy vetch cover crop	40	3.5	11
Class A municipal biosolids (sewage sludge)	35	6.1	6
Soil microorganisms			
Bacteria	50	10.0	5
Fungi	50	5.0	10
Soil organic matter			
Average forest O horizons	50	1.3	45
Average forest A horizons	50	2.8	20
Mollisol Ap horizon	56	4.9	11
Average B horizon	46	5.1	9

more severely leached and acidic A horizons in humid regions, the C/N ratio is relatively wide; C/N ratios greater 20:1 are not uncommon. Forest O horizons commonly have C/N ratios of 30 to 40. When such soils are brought under cultivation and limed to increase their pH and calcium content, the enhanced decomposition tends to lower the C/N ratio to near 12:1.

Influence of Carbon/Nitrogen Ratio on Residue Decomposition

Soil microbes, like other organisms, require a balance of nutrients from which to build their cells and extract energy. The majority of soil organisms metabolize carbonaceous materials to obtain C for building essential organic compounds and to obtain energy for life processes. However, no creature can multiply and grow on C alone. Organisms must also obtain sufficient N to synthesize N-containing cellular components, such as amino acids, enzymes, and DNA.

On the average, soil microbes must incorporate into their cells about eight parts of C for every one part of N (i.e., assuming the microbes have an average C/N ratio of 8:1). Because only about one-third of the C metabolized by microbes is incorporated into their cells (the remainder is respired and lost as CO_2), the microbes need to find about 1 g of N for every 24 g of C in their "food."

This requirement results in two extremely important practical consequences. First, if the C/N ratio of organic material added to soil exceeds about 25:1, the soil microbes will have to scavenge the soil solution to obtain enough N. Thus, the incorporation of high C/N residues will deplete the soil's supply of soluble N, causing plants to suffer from N deficiency. Second, the decay of organic materials can be delayed if sufficient N to support microbial growth is neither present in the material undergoing decomposition nor available in the soil solution. These concepts are illustrated by the example in Figure 11.5.

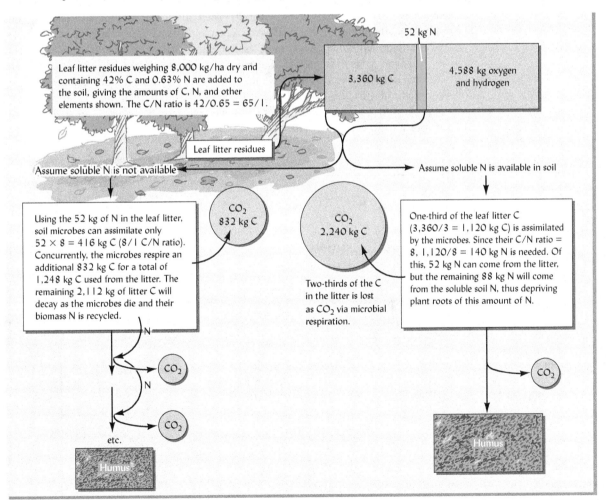

Figure 11.5 *A simplified, quantitative example of plant residue decay illustrating the fates of C and N and the consequences for decomposition and soil N availability. Note that if the supply of N is low, the decomposition process is slowed, and eventually less C remains as stabilized humus.* (Diagram courtesy of Ray R. Weil)

Examples of Inorganic Nitrogen Release During Decay

The practical significance of the C/N ratio becomes apparent if we compare the changes that take place in the soil when residues of either high or low C/N ratio are added (Figure 11.6). Consider a soil with a moderate level of soluble N (mostly nitrates). General-purpose decay organisms are at a low level of activity in this soil, as evidenced by low carbon dioxide production. If no N were lost or taken up by plants, the level of nitrates would very slowly increase as the native SOM decays.

Low Nitrogen Material. Now consider what happens when a large quantity of readily decomposable organic material is added to this soil. If this material has a C/N ratio greater than 25, changes will occur according to the pattern shown in Figure 11.6 (*top*). In the example shown, the initial C/N ratio of the residues is about 55, typical for many kinds of leaf litter and also for cornstalks. As soon as the residues contact the soil, the microbial community responds to the new food supply (see Section 11.2). Heterotrophic r-strategist microorganisms become active, multiply rapidly, and yield carbon dioxide in large quantities. Because of the microbial demand for N, little or no mineral N (NH_4^+ or NO_3^-) is available to higher plants during this period.

Nitrate Depression. This condition, often called the **nitrate depression period**, persists until the activities of the decay organisms gradually subside due to lack of easily oxidizable carbon. As their numbers decrease, carbon dioxide formation drops off, and N demand by microbes becomes less acute. As decay proceeds, the C/N ratio of the remaining plant material decreases because C is being lost (by respiration) and N is being conserved (by incorporation into microbial cells). Generally, one can expect mineral N to begin to be released when the C/N ratio of the remaining material drops below about 20. Then, nitrates appear again in quantity, and the original conditions prevail, except that the soil is somewhat richer in both N and humus.

The nitrate depression period may last for a few days, a few weeks, or even several months. A longer, more severe period of nitrate depression is typical when added residues are easily decomposed and have a higher C/N ratio and when a larger quantity of residues is added. To avoid producing that are stunted, nitrogen-starved seedlings, planting should be delayed until after the nitrate depression period or additional sources of N should be applied to satisfy the nutritional requirements of both the microbes and the plants.

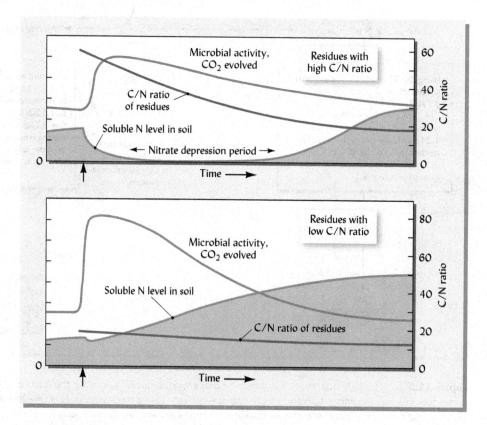

Figure 11.6 *Changes in microbial activity, in soluble N level, and in residual C/N ratio following the addition of either high (top) or low (bottom) C/N ratio organic materials. Where the C/N ratio of added residues is above 25, microbes digesting the residues must supplement the N contained in the residues with soluble N from the soil. During the resulting nitrate depression period, competition between higher plants and microbes would be severe enough to cause N deficiency in the plants. Note that in both cases soluble N in the soil ultimately increases from its original level once the decomposition process has run its course. The soluble N trends shown are for soils without growing plants, which, if present, would continually remove a portion of the N as soon as it is released.* (Diagram courtesy of R. Weil)

High Nitrogen Material. The effects on soil nitrate level will be quite different if residues with a C/N ratio lower than 20 are added (represented by Figure 11.6, *bottom*). With organic materials of low C/N ratio, more than enough N is present to meet the needs of the decomposing organisms. Therefore, soon after decomposition begins, some of the N from organic compounds is released into the soil solution, augmenting the level of soluble N available for plant uptake. Generally, nitrogen-rich materials decompose quite rapidly, resulting in a period of intense microbial growth and activity, but no nitrate depression period.

Influence of Soil Ecology

In nature, the process of N mineralization involves the entire food web (see Section 10.2), not just the saprotrophic bacteria and fungi. For example, when organic residues are added to soil, bacteria and fungi grow rapidly on this food source, producing a large biomass of bacterial and fungal cells that contain much of the N originally in the residues. Until the microbial biomass begins to die off, this N is immobilized and not available to plants. However, a healthy soil ecosystem is likely to contain certain nematodes, protozoa, and earthworms that feed on bacteria and fungi. As these animals feed, they respire most of the C in the microbial cells, using only a small fraction to grow on (or produce eggs). Since the C/N ratio of these animals is not too different from that of their microbial food, and since most of the C is converted to CO_2 by respiration, the animals soon ingest more N than they can use. They then excrete the excess N, mainly as NH_4^+, into the soil solution as plant-available mineral N. The microbial feeding activity of soil animals may increase the rate of N mineralization dramatically as shown in Figure 11.7 for bacterial-feeding nematodes and fungal-feeding collembola. Soil management that favors a complex food web (Section 10.14) with many trophic levels can be expected to enhance the cycling and efficient use of N (and of other nutrients).

Influence of Lignin and Polyphenol Content

The lignin contents of plant litter range from less than 2% to more than 50%. Those materials with high lignin content usually decompose very slowly. Polyphenol compounds found in plant litter may also inhibit decomposition. These phenolics are often water-soluble and may

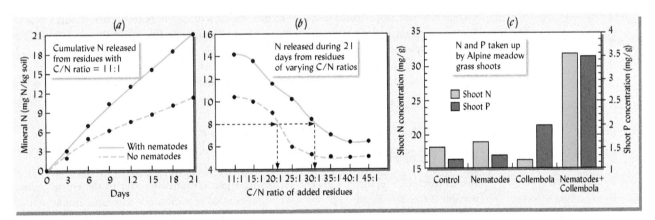

Figure 11.7 *Microbial-feeding fauna enhance the release of plant-available mineral nutrients from organic residues. (a and b) Columns of sandy soil with well-established bacterial communities were amended with ground alfalfa tissue (low C/N) and cellulose (high C/N) in varying proportions to give the indicated C/N ratios. Some of the soil columns were inoculated with bacteria-feeding nematodes (Cephalobus persegnis) and some were free of nematodes. (a) By feeding on the bacteria, the nematodes nearly doubled the amount of mineral N released from the added residues over a 21-day period. Without nematodes, much of the N was tied up in the bacterial biomass, but when the nematodes ate the bacteria, they excreted the excess N into the soil solution as NH_4^+. (b) Nematodes influenced the release of mineral N from residues with different C/N ratios. Enhanced mineralization in the presence of bacteria-feeding nematodes suggests that a level of mineral N satisfactory for plant growth (say 8 mg/kg) could be maintained with residues of relatively high C/N ratio (about 32:1), but that without these nematodes, a more N-rich type of residue (C/N ratio about 22:1) would be required. In an alpine meadow soil (c) bacteria-feeding nematodes worked synergistically with mainly fungal-feeding collembola to increase the amount of both N and P released and subsequently taken up by grass plants.* [Graphs (a) and (b) drawn from data in Ferris et al. (1998); graph (c) drawn from data in Bardgett and Chan (1999)]

Figure 11.8 *Temporal patterns of N release from organic residues differing in quality based on their C/N ratios and contents of lignin and polyphenols. Lignin contents greater than 20%, polyphenol contents greater than 3%, and C/N ratios greater than 30 would all be considered high in the context of this diagram, the combination of these properties characterizing litter of poor quality—that is, litter that has a limited potential for rapid microbial decomposition and mineralization of plant nutrients.* (Diagram courtesy of Ray R. Weil)

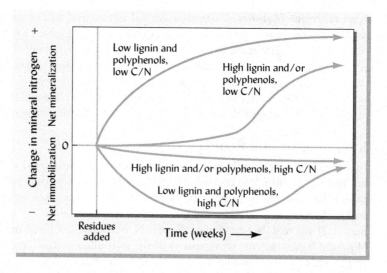

be present in concentrations as high as 5 to 10% of the dry weight. By forming highly resistant complexes with proteins during residue decomposition, these phenolics can dramatically slow the rates of both N mineralization and C oxidation.

Litter Quality. Because they support relatively low levels of microbial activity and biomass, residues high in phenols and/or lignin are considered to be *poor quality resources* for the soil organisms that cycle C and nutrients. The production of such slow-to-decompose residues by certain forest plants may help explain the accumulation of extremely high levels of stable organic N and C in the soils of mature boreal forests.

The lignin and phenol contents also influence the decomposition and release of N from **green manures**—plant residues used to enrich agricultural soils. For example, in the leaves of certain legume trees, the C/N ratio is quite low, but the phenol content is quite high, so that when these leaves are added to soil, N is released only slowly—often too slowly to keep up with the needs of a growing crop. Similarly, residues with a lignin content of more than 20 to 25% will decompose too slowly to be effective as green manure for rapidly growing annual crops. However, for perennial crops, or forests, the slow release of N from such residues may be advantageous in the long run, as the N may be less subject to losses. By the same token, the slow decomposition of phenol- or lignin-rich materials means that even if their C/N ratio is very high, the nitrate depression will not be pronounced. Figure 11.8 illustrates the combined effects of C/N ratio and lignin or phenol content on the balance between immobilization and mineralization of N during plant residue decomposition.

11.4 GENESIS AND NATURE OF SOIL ORGANIC MATTER AND HUMUS[3]

In this textbook, we use the general term **soil organic matter** to refer to the entire organic portion of the soil (Figure 11.9), recognizing that SOM is not a single substance but a complex mixture of substances that exist in association with other soil components. Although surface residues (litter) are not universally considered to be part of the SOM, we include them (mainly in the detritus fraction) in this textbook because they comprise a major part of the O horizons in many soil profiles.

Since the element carbon (C) plays a prominent role in the chemical structure of all organic substances, it is not surprising that the term **soil organic carbon** (SOC) is also often used to refer to the C component of SOM. This term is particularly appropriate for quantitative discussions of SOM because most methods of determining SOM actually measure the C in the

---

[3]Our understanding of the nature and genesis of soil humus has advanced greatly since the turn of the century, requiring that some long-accepted concepts be revised or abandoned, as explained in detail by Schmidt et al. (2011) and Lehmann and Kleber (2015).

material and then use a conversion factor to estimate the organic matter. Since SOM commonly contains about half carbon by weight (50% C), it is usually appropriate to estimate SOM as two times the organic C (SOM = 2 × SOC).[4] However, the C content of SOM does vary, so caution must be used when comparing values reported as SOM and SOC!

As already mentioned, the term soil organic matter encompasses all the organic components of a soil: (1) living **biomass** (intact plant and animal tissues and microorganisms); (2) plant litter or residues—bits of dead roots and other plant residues in various stages of decay and of various sizes (although in practice, particles that do not pass 2-mm sieve openings are often excluded from consideration); (3) dissolved organic biomolecules ranging widely from plant amino acids to microbial enzymes; and (4) a complex mixture of biomolecule agglomerations on particle surfaces, tiny bits of no longer identifiable tissue occluded inside microaggregates, and bits of plant material that have been blackened (charred) by fire. As in Figure 11.9, we will refer to the various components in this fourth broad category of organic material as **soil humus** (pronounced *hew-muss*), and those in the first three categories as the **labile carbon** pool (or group). The word *labile* (*liable to change; easily altered*) implies that the C-containing materials in this pool are subject to rapid oxidation by soil organisms over periods of days to years. In contrast, the carbon in the humus pool appears to be stabilized by various mechanisms that enable it to remain in the soil for relatively long periods (centuries or even millennia). Within each of the two pools, Figure 11.9 defines several organic matter fractions based on chemical and physical criteria. Organic matter is separated into labile and humus pools largely because of the degree to which their component compounds are physically protected from decay by the soil environment, especially by association with soil mineral particles and aggregates. Thus certain chemically similar organic matter fractions may appear to be labile when in a "free"

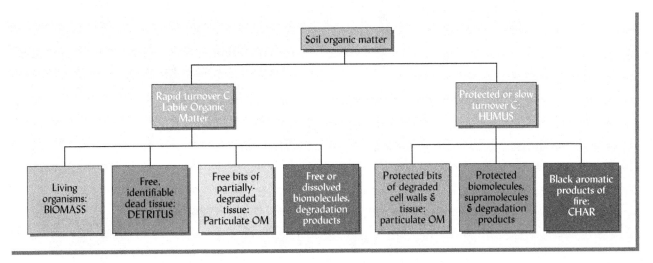

Figure 11.9 *Substances that comprise SOM are classified on the basis of their availability to microbial turnover into two pools of carbon: (1) a rapidly oxidizing labile pool (taking weeks to years to disappear) and (2) a more stable, relatively slow-to-decompose pool (lasting for centuries or millennia) collectively referred to as humus. Within each of the two pools, several organic matter fractions are defined by chemical and physical criteria. Organic matter behaves either as labile C or as humus partly because of differences in molecular structures, but more importantly because of the degree to which their component compounds are physically protected from decay by the soil environment, especially by association with soil mineral particles and aggregates. Thus, certain chemically similar components appear in both the labile (free or unprotected) and humus (protected) pools; examples include particulate organic matter (POM) and certain biomolecules (e.g., lipids, cellulose, glycoproteins, and lignin). Char, a type of black carbon material produced by fire, is mostly (but not in all cases) very slow to decompose and so is included with humus (far-right compartment). This diagram is worth studying as reference will be made to these SOM fractions throughout this and other chapters.* (Diagram courtesy of Ray R. Weil)

---

[4]Traditionally SOM has been estimated as 1.72 × SOC, a conversion that assumes 58% C in the soil organic matter. Although certain materials, such as humic acids, found in soil extracts do contain that much C, and there is little basis for the general use of this value; most organic matter in soils contains about 50% C, so a conversion factor of 2 is recommended.

or unprotected condition and appear to be part of the more stable humus pool when in a protected condition. We can summarize the concept of two organic matter pools with a business analogy: *the labile C largely serves as ready cash flow to pay the workers (feed the soil food web) while humus can be seen as the (carbon) capital of the system (accumulated organic matter), which builds important chemical and physical aspects of the soil.*

Microbial Transformations

Microbial decomposition of plant residue leads to carbon cycling, mineral nutrient release, and the formation and accumulation of various forms of SOM. During this process, microbes carry out two primary types of reactions: the oxidation of carbon and the breakdown of complex molecules into simpler compounds.

In the presence of O_2 oxidation reactions provide energy for microorganisms and create functional chemical groups like —COOH. These functional groups become charged when they dissociate (lose a H^+ ion), increasing the polarity of the molecule, which in turn increases its solubility and reactivity with mineral surfaces. The polarity of functional groups resulting from oxidation reactions therefore tends to transform many hydrophobic, nonpolar biomolecules such as lipids, lignin, and some proteins, into molecules that are more hydrophilic, more soluble, and more readily sorbed onto mineral surfaces.

The breakdown of large biopolymers into their component subunits—for example, cellulose into sugars or large lignin molecules into phenolic subunits—allows the soil microbes to then metabolize and oxidize the resulting simpler compounds. Using some of the carbon not lost as carbon dioxide in respiration, along with most of the nitrogen, sulfur, and phosphorus from these compounds, the microorganisms synthesize new cellular components and biomolecules. Some of these new microbial compounds and decomposition products, along with some modified plant compounds, will continue to be subject to microbial oxidation until completely transformed into CO_2, and mineral nutrients. Other compounds will interact with the soil environment in ways that slow or prevent their further decay.

The accumulation of organic matter in stabilized forms is controlled both by the particular soil environment and by the activity of the particular microbial community (Figure 11.10). The microbial biomass produces exocellular enzymes that break down both plant residues and microbial compounds from previous decomposition cycles, transforming components into dissolved organic carbon (DOC) compounds that the microbes can then utilize for energy. As these compounds are metabolized, most of the C is released as CO_2 while the other elements they contain (N, P, S, etc.) may be released into the soil solution.

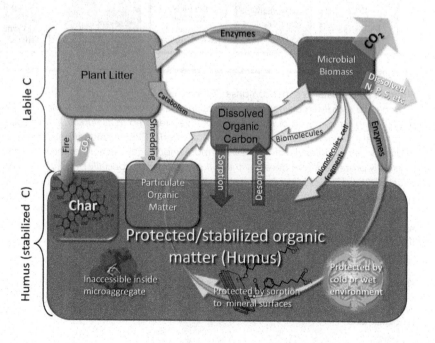

Figure 11.10 *Carbon cycling, mineral nutrient release and formation of SOM. The microbial biomass (upper right) produces (mainly exocellular) enzymes that break down plant residues and microbial compounds from previous decomposition cycles, transforming components into dissolved organic C compounds. When these compounds are metabolized by microbes to obtain energy, most of their C is released as CO_2 while their other elements (N, P, S, etc.) may be released into the soil solution. The labile C (upper half of diagram) is subject to rapid metabolism, some components become stabilized (lower half of diagram). The soil environment and microbial activity control this stabilization and accumulation of organic matter. Bits of plant tissue and microbial cell walls (particulate organic matter) may become inaccessible to microbial attack as microaggregates form around them. Microbial oxidation creates zones of polarity in formerly hydrophobic biomolecules, allowing their C to become protected by bonding to mineral surfaces. Low temperatures or oxygen levels preserve plant and microbial compounds. Fires produce black carbon materials, which may resist enzymatic oxidation. Over time an individual C atom may move among any number of these pools.* (Diagram courtesy of Ray R. Weil)

Some plant and microbial components become protected in various ways by the soil environment. For example, some bits of plant tissue and microbial cell walls—known as particulate organic matter (POM) become inaccessible to microbial attack when soil microaggregates form around them. Microbial oxidation creates some zones of polarity in formerly hydrophobic plant and microbial biomolecules allowing them to become protected from further decay by bonding tightly in layers onto mineral surfaces. In some soils, anaerobic (wet) or freezing cold conditions prevent complete oxidation and preserve plant and microbial compounds. In many soils, the charring process that occurs during fires produces black carbon materials that are quite immune to enzymatic attack and oxidation (see Box 11.2). By a combination of some or all of these mechanisms, a portion of the C entering the soil is preserved, allowing organic matter to accumulate over years and centuries.

BOX 11.2
CHAR: IS BLACK THE NEW GOLD?[a]

During wildfires 1 to 3% of the potential fuel is charred (blackened) rather than burned. The char is left in and on the soil (Figure 11.11). This black carbon (think of charcoal) is a product of **pyrolysis**—the smoldering or charring that takes place when organic material is heated to 280 to 500°C under low-oxygen conditions. This contrasts with complete combustion that produces CO_2 and whitish mineral ash under high-oxygen conditions. Since fires occur in most ecosystems, char is present in most soils. In forests 5 to 10% of the SOC may be char. However, in soils formed under grasslands some 40 to 50% of the SOC is commonly present in this form. Scientists think that such high amounts of char occur in grassland soils because of the high frequency of fires and also because mammalian burrowing activity—bioturbation—rapidly

Figure 11.11 Black carbon or char. (a) The flames of a tropical grassland fire leave the soil covered in black charred residues. (b) A handful of prairie surface soil with visible charred carbon. (c) A scanning electron micrograph (SEM) of centuries-old char from a boreal forest fire, with fungal hyphae growing in it. [Photos courtesy of Ray R. Weil; SEM from Marie-Charlotte Nilsson Hegethorn]

[a]A number of papers review char as a natural component of SOM (Hart and Luckai, 2013; Schmidt and Noack, 2000; Singh, et al.,2012).

(continued)

> **BOX 11.2**
> **CHAR: IS BLACK THE NEW GOLD?** *(CONTINUED)*
>
> incorporates the char left on the soil surface protecting it from complete combustion in subsequent fires.
>
> Despite its wide occurrence, scientists have largely ignored black carbon (char and the like) in their efforts to model carbon cycling in soils and analyze SOM and its functions. Historically, this oversight may stem from the fact there were no simple methods to specifically measure char and the most widely used method to measure SOC (oxidation with potassium dichromate in heated strong acid) is not sensitive to char. The recognition of the importance of char really did not take off until the discovery of small patches of Terra Preta (*dark earth*) soils in the central Amazon rain forest. These soils occurred where humans had added char materials more than 800 years ago. Although the sites have not seen human settlement or char inputs for many centuries, to this day the Terra Preta soils remain much darker in color and much more fertile than the surrounding Oxisols that had not been amended with char.
>
> The carbon content of char (70–85% C) is much higher than most soil organic matter (~50% C) because much of the O and H in plant residues is lost during pyrolysis. The chemical structure of char is generally made up of a complex assortment of aromatic (double bonded hexagonal) rings. Partial oxidation of surface aromatic rings results in many negative-charged carboxylic (COO^-) functional groups that endow the char with high cation and anion exchange capacities. The macrostructure reflects the original plant cell walls and gives the material an extremely high porosity (Figure 11.11c). This structure also endows char with specific surface area several times greater than that of even expansive clays—up to 2,500 m^2/g (compare to values in Table 8.1). The aromatic ring structures are also very stable and resist breakdown. Some char components are soluble and move with leaching water to the oceans. About 5 to 10% of char decomposes at moderate rates. However, most of char is thought to be among the most long-lived soil organic components, especially when it is protected by association with mineral surfaces and inside microaggregates. The longevity of char in soils is indicated by its chemistry (resistance to decay), ^{14}C isotopic dating studies, and archaeological evidence (e.g., remains from ancient cooking fires).
>
> The just-described characteristics of char seem much like those traditionally associated with humic substances; in fact it appears that char may be the source of some of the material found when soils are extracted for humic substances (see Box 11.3). While there is much still to learn about black carbon (char and related fire products), it is likely that in fire-prone ecosystems this material makes a major contribution to the dark soil color, high water holding capacity, reduced bulk density, and high CEC associated with soil organic matter. Where wildfires have been suppressed by humans, prescribed burns may help to maintain the char content and productivity of soils.

One year after plant residues are added to the soil, most of the carbon in them has returned to the atmosphere as CO_2, but approximately 2 to 5% is likely to remain in the soil as live biomass and another 3–10% as nonliving labile C compounds. Perhaps 10 to 30% will have become stabilized humus, the upper value being more representative for root residues and the lower value for shoot residues.

Examples of Biomolecules in Soil Organic Matter

Most of the SOM is comprised of specific biomolecules, some of plant origin (albeit usually modified by microbial enzymes), while others are compounds synthesized by soil microbes as by-products or as part of their cells. Polysaccharides and glomalin, microbial compounds produced by bacteria and fungi, respectively, are important in enhancing soil aggregate stability (see Section 4.5). Soil organic matter was once thought to be comprised mainly of **humic substances**, which were visualized as very large and complex polymers. However, evidence now suggests that, despite their name, they are not the major components of soil humus that they have long been thought to be (see Box 11.3).

Colloid Characteristics of Humus

The nature and functional group chemistry of colloidal humus was described in Section 8.4. Because of the many types of chemical groups that, among them, absorb nearly all wavelengths of visible light, humus imparts a black color to soil as is most characteristic of O, A, and Bh horizons (see Figures 3.3, 3.13, 3.21). Humus colloids include tiny particles of char as well as multiple layers of organic compounds with varying degrees of polarity sorbed to clay surfaces.

BOX 11.3
WHERE HAVE ALL THE HUMICS GONE?[a]

Anyone who has used drain-clearing chemicals to remove hair or grease from a clogged drain knows that strong alkali solutions (like NaOH) are very good at dissolving organic compounds that may not be soluble in water. For more than 100 years, soil scientists have used strong alkali (pH ~13) to extract organic matter from soils so they could quantify it and determine its nature in the lab (Figure 11.12). The dark-colored solutions obtained are commonly acidified and the portion of organic matter precipitated in the acid solution (pH ~1.0) is considered to be humic acid while the material soluble in both alkali and acid is termed fulvic acid. The material not soluble in either acid or alkali is termed **"humin,"** and thought to be highest in molecular weight, darkest in color, and most resistant to microbial attack. Despite a century of research on these materials collectively called humic substances, their exact molecular structures remained elusive. However, chemists determined that humic and fulvic acids are comprised of super-sized polymers (macromolecules) with many aromatic and phenol ring structures, and with properties quite unlike any biomolecules known in plant or microbial cells. It was hypothesized that these huge, complex, and somewhat mysterious molecules were synthesized by soil microbes from plant compounds like lignin and that their sheer complexity made them "recalcitrant" or resistant to decomposition. These complex humic substances in the chemists' extractions have been credited with controlling many soil properties, including the long persistence of organic matter in soils. Analyses made using the alkali extractions indicate that 60 to 80% of SOM existed in the form of humic substances. The term *humic substances* used to describe such molecules should not be confused with the term *humus*, which is used to describe all of the organic molecules and the very small particles of organic materials in soils that are no longer alive and no longer recognizable as tissues.

During the past two decades, new isotopic, spectroscopic, and molecular-marker analytical techniques have allowed soil chemists to study the nature of organic matter *in situ*—that is, while it is still in the soil, without having to extract it first. To their surprise, the new direct measurements found very little in the way of humic macromolecules in mineral soils. Instead, evidence suggests that the alkali extraction process itself actually creates giant polymers from smaller biomolecules. It is now thought that humic substances in soil extracts do *not* represent the nature of most of the organic material as it exists in soils. The complex humic substances that formed in the extracts do, however, have rather unique properties and may be quite bio-active. Although the effects have been quite variable, in some cases humic substances extracted from soil or, more commonly, those from mined partially fossilized organic deposits (termed **humates**) may considerably enhance certain aspects of plant growth when sprayed on the plants or used as soil amendments (see Section 11.5).

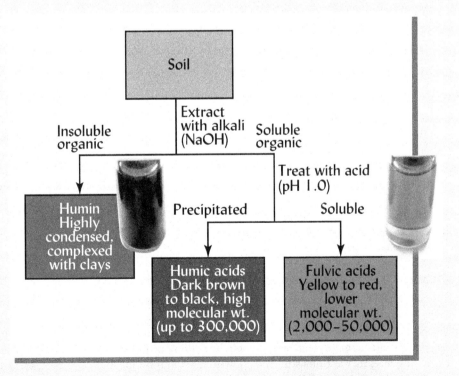

Figure 11.12 *The traditional scheme for classifying so-called humic substances found to occur in highly alkaline extractions of soil.*
(Diagram courtesy of Ray R. Weil)

[a] For a detailed historical and chemical explanation of this view of humic substances, see Kleber and Johnson (2010).

These materials exhibit very high levels of surface area and negative charge—similar per unit volume to those of high-activity clay, but much greater than clay when compared per unit mass. Depending on the pH, the cation exchange capacity of humus may range from about 150 to as high as 500 cmol$_c$/kg (about 40–120 cmol$_c$/L). The water-holding capacity of humus on a mass basis (but not on a volume basis) is four to five times that of the silicate clays. Humus promotes aggregate formation and stability, and, in turn, the aggregate structures help protect the humus from decay.

Stability of Humus

Protection of at least some C from microbial respiration enables soils to accumulate increasing amounts of organic matter over years and centuries (Section 11.8). Studies using the radioactive carbon isotope ^{14}C have shown that some organic C incorporated into plant tissue by photosynthesis thousands of years ago is still present in certain soils. Thus at least some C atoms in soils have escaped microbial respiration to CO^2, although they may have undergone many transformations. In fact, the date given by ^{14}C analysis tells us when the C was first fixed by photosynthesis, nothing more. For example, a ^{14}C date of 800 years may mean the C became part of a tree 800 years ago, then was in a twig in the litter layer for 20 years, then in a fungal hyphae for 2 years, then in an aggregate as a bit of fungal cell wall for 500 years, then something else, etc. It would be a misinterpretation to think the date means the specific carbon compound has remained in the soil unchanged for 800 years.

The protection of biomolecules from microbial degradation is important in maintaining SOM levels and in protecting associated N and other essential nutrients against rapid mineralization and loss from the soil. For example, root hairs that grow into, and eventually die inside, microaggregates may be protected from decay both by physical inaccessibility to microbes and by the localized anaerobic conditions that may exist in the microaggregate interior (see Section 7.4). The stability of humus in many soils is also related to the aromatic structure of char (Box 11.2). Humus that is entrapped in the ultra-micropores (<1 μm) formed by clay particles is physically inaccessible to decomposing organisms. Yet, despite such protection, the compounds that comprise humus are subject to some continual slow decomposition. Without annual additions of sufficient plant residues, slow microbial oxidation of humus results in declining levels of SOM.

11.5 INFLUENCES OF ORGANIC MATTER ON PLANT GROWTH AND SOIL FUNCTION

Long ago, the observation that plants generally grow better on organic-matter-rich soils led people to think that plants derive much of their nutrition by absorbing humus (organic carbon) from the soil. We now know that vascular plants derive their carbon from carbon dioxide and that most of their nutrients come from inorganic ions dissolved in the soil solution. In fact, plants can complete their life cycles growing totally without humus, or even without soil (as in soilless or **hydroponic** production systems using only aerated nutrient solutions). This is not to say that SOM is less important to plants than was once supposed, but rather that most of the benefits accrue to plants indirectly through the many influences of organic matter on soil properties. These will be discussed later in this section, after we consider two types of direct organic matter effects on plants.

Direct Influence of Humus on Plant Growth

It is well established that certain organic compounds are absorbed by higher plants. For example, plants can absorb a varying proportion of their nitrogen and phosphorus needs as soluble organic compounds. In addition various growth-promoting compounds such as vitamins, amino acids, and hormones (e.g., auxins and gibberellins) are formed as organic matter decays. These substances may at times stimulate growth in both higher plants and microorganisms.

Small quantities of both fulvic and humic acids (see Figure 11.12) added to soils are known to enhance certain aspects of plant growth. Some evidence suggests that the application of extracts from mined lignite (a coal-like carbonaceous deposit) can stimulate plant growth by improving the availability of micronutrients, especially iron and zinc. Small biomolecules, such as citric acid, dissolved in the soil solution can have similar effect on micronutrient availability. Some scientists suggested that the humic substances may act as hormone-like regulators of specific plant-growth functions such as cell elongation or lateral root initiation. Several plant hormones, including indoleacetic acid (IAA) and isopentenyladenosine (a cytokinin), have been isolated from alkaline extracts of mined lignite and earthworm casts. It is possible that small amounts of somewhat similar humic substances are dissolved in the soil solution in some soils. Such humic substances may have some practical potential as soil amendments or plant growth stimulants (see Figure 11.13).

Commercial humate products (solid salts of humic acid) are marketed with claims that small amounts enhance plant growth, but scientific tests of many of these products have given very mixed results, often showing no benefit from their use on soils well endowed with organic matter. The lack of amendment effects may be due to the natural presence in most soils of humic substances or their active components at levels sufficient to carry out the desired functions.

Allelochemical Effects[5]

Allelopathy is the process by which one plant infuses the soil with a chemical that affects the growth of other plants. The plant may do this by directly exuding **allelochemicals**, or the compounds may be leached out of the plant foliage by throughfall rainwater. In other cases, microbial metabolism of dead plant tissues (residues) forms the allelochemicals. The term *allelochemical* is also applied to plant chemicals that inhibit microorganisms. In principle, the interactions are much like the antagonistic relationships among certain microorganisms discussed in Section 10.14. In fact, allelochemicals often serve several different functions in the rhizosphere and the concept overlaps with that of "signaling" chemicals, which plants use to communicate with other plants and certain soil organisms (Section 10.7).

Allelochemicals present in the soil may be responsible for some of the effects observed when various plants grow in association with one another. Because they produce such chemicals, certain weeds (e.g., Johnsongrass and giant foxtail) damage crops far out of proportion to the size and number of weeds present. Crop residues left on the soil surface may inhibit the germination and growth of the next crop planted (e.g., wheat residues often inhibit sorghum plants). Farmers may select such cover crops as rye or rapeseed, in part, because of the inhibiting effect the allelochemicals from their residues may have on weed seed germination.

Other allelopathic interactions influence competition and the succession of species in natural ecosystems. Allelopathy may be partially responsible for the invasiveness of certain exotic plant species that rapidly dominate a new ecosystem to which they have been recently introduced. The invaders' allelochemicals may be more effective in the new ecosystem than they were in their territory of origin where neighboring plant species had time to evolve tolerance.

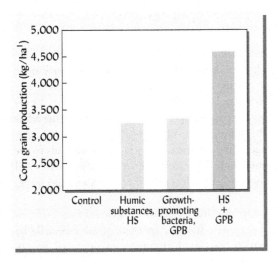

Figure 11.13 *Synergistic effect of spraying a field corn crop with humic substances (HS) from vermicompost extracts and growth promoting bacteria (GPB) under low nitrogen fertility (all plots received just 50 kg N/ha as fertilizer). Either the HS or the GPB added alone increased yield by ~500 kg/ha, but the two components added together increased yield by almost 1,800 kg/ha compared to the control. The HS were from extracts of vermicomposted (earthworm worked) manure applied at about 6 g carbon/ha. The GPB were* Herbaspirillum seropedicae, *a nitrogen-fixing and growth-stimulating species that colonizes grass plant roots. It is thought that the HS altered the corn root growth in a way that made colonization easier for the bacteria. The soil was a low fertility Ultisol in Brazil.* [Graphed from data in Canellas et al. (2013)]

---

[5]For a review of allelopathy as it relates to plants and soil microbes, see Cipollini et al. (2012). For roles of allelopathy in plant species invasiveness and community ecology, see Inderjit et al. (2011).

Figure 11.14 *Variable and genotype specific nature of allelopathy. (a) Positive and negative allelopathic effects of winged bean on grain amaranth plants. The average dry weight of the amaranth plants for each treatment is shown. All pots were watered with a complete nutrient solution so any nutrients from the winged bean should not have had an effect. In the pot on the left (T4) amaranth is growing in fresh soil (no association with winged bean). In the center pot (T20) amaranth is growing in soil previously used to grow winged bean (positive effect). In the pot on the right (T28) the amaranth is growing in fresh soil, but the plant was watered three times with a water extract of winged bean tissue (negative effect). (b) Corn seeds germinated with a water extract of leaves from a Thai variety of winged bean. (c) Corn seeds germinated with a water extract of leaves from a Sri Lankan variety of winged bean.* (Photo courtesy of Ray R. Weil)

Allelopathic interactions are usually very specific, involving only certain species, or even varieties, on both the producing and receiving ends (Figure 11.14). The effects of allelopathic chemicals are many and varied. Although the term *allelopathy* most commonly refers to negative effects, allelochemical effects can also be positive (as in certain **companion plantings**). Because most of the active compounds can be rapidly destroyed by soil microorganisms or easily leached out of the root zone, effects are usually relatively short-lived once the source is removed. While they vary in composition, most allelochemicals are relatively simple phenolic or organic acid compounds that could be included among the biomolecules found in the labile carbon pool of SOM.

Influence of Organic Matter on Soil Properties and Indirectly on Plants[6]

Soil organic matter affects so many soil properties and processes that a complete discussion of the topic is beyond the scope of this chapter. Indeed, in almost every chapter in this book there is mention of the roles of soil organic matter. Figure 11.15 summarizes some of the more important effects of organic matter on soil properties and on soil–environment interactions. Often one effect leads to another, so that a complex chain of multiple benefits results from the addition of organic matter to soils. For example (beginning at the upper left in Figure 11.15), adding organic mulch to the soil surface encourages earthworm activity, which in turn leads to the production of burrows and other biopores, which in turn increases the infiltration of water and decreases its loss as runoff, a result that finally may lead to less pollution of streams and lakes.

Influence on Soil Physical Properties. Organic matter tends to give surface horizons dark brown to black colors. Granulation and aggregate stability are encouraged, especially by the bacteria polysaccharides and fungal glomalin-associated glycoproteins produced during

---

[6]For a readable introduction to increased soil organic matter improves soils physically, chemically, and biologically, see Blanco-Canqui et al. (2013).

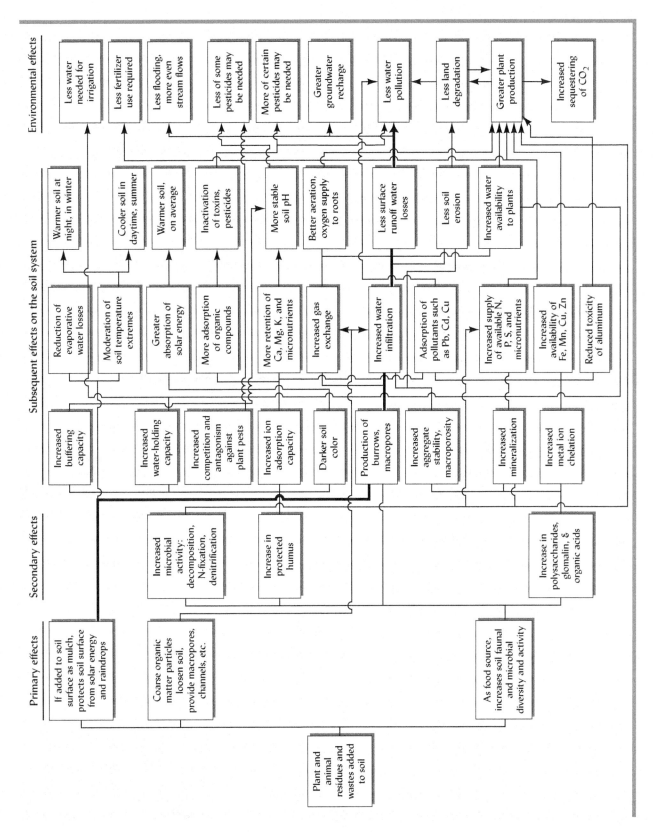

Figure 11.15 Some of the ways in which SOM influences soil properties, plant productivity, and environmental quality. Many of the effects are indirect, the arrows indicating the cause-and-effect relationships. It can readily be seen that the influences of SOM are far out of proportion to the relatively small amounts present in most soils. Many of these influences are discussed in this and other chapters in this book. The thicker line shows the sequence of effects referred to in the text in this section. It will be worthwhile to follow the paths of other chains of cause and effect shown, stopping to ask yourself whether the connections make sense with respect to what you have learned in other parts of this book. (Diagram courtesy of Ray R. Weil)

decomposition (see Section 4.5). The protected particulate organic matter, char, and sorbed biomolecules in the humus help reduce the plasticity, cohesion, and stickiness of clayey soils, making these soils easier to manipulate.

A major reason why soils higher in organic matter tend to be more productive is their enhanced ability to supply plants with water because organic matter increases both infiltration rate and water-holding capacity. Organic matter has an especially pronounced effect on the water-holding capacity of very sandy soils, which can often be improved by adding stable organic amendments. Improved water retention is a major reason why sand-based golf greens are commonly amended and top-dressed with peat. However, the use of large amounts of peat mined from sphagnum wetlands cannot be considered environmentally sustainable (see Box 11.4), so compost made from various organic wastes (Sections 11.11 and 10.13) would be a more environmentally responsible amendment for this use.

Influence on Soil Chemical Properties. Humus generally accounts for 50 to 90% of the cation-adsorbing power of mineral surface soils. Like clays, humus colloids and high surface area char hold nutrient cations (potassium, calcium, magnesium, etc.) in easily exchangeable form, wherein they can be used by plants but are not too readily leached out of the profile by percolating waters. Through its cation exchange capacity and acid and base functional groups, organic matter also provides much of the pH buffering capacity in soils (see Section 9.4). In addition, nitrogen, phosphorus, sulfur, and micronutrients are stored as constituents of soil organic matter, from which they are slowly released by mineralization.

Organic acids also attack soil minerals and accelerate their decomposition, thereby releasing essential nutrients as exchangeable cations. Small molecular weight organic acids, as well as polysaccharides and certain polar biomolecules, are especially effective in attracting such cations as Fe^{3+}, Cu^{2+}, Zn^{2+}, and Mn^{2+} from the edges of mineral structures and **chelating** or binding them in stable organomineral complexes. Some of these metals are made more available to plants as micronutrients because they are kept in soluble, chelated form (see Chapter 12). In very acid soils, organic matter alleviates aluminum toxicity by binding the aluminum ions in nontoxic complexes (see Sections 9.2 and 9.9).

Biological Effects. Soil organic matter—especially the detritus fraction—provides most of the food for the community of heterotrophic soil organisms described in Chapter 11. In Section 11.3 it was shown that the quality of plant litter and SOM markedly affects initial decomposition rates. The type and diversity of organic residues added to a soil can influence the type and diversity of organisms that make up the soil community (as discussed in Section 10.14).

11.6 AMOUNTS AND QUALITY OF ORGANIC MATTER IN SOILS[7]

Labile Organic Matter

The organic matter of the labile carbon pool (described in Section 11.4) provides most of the readily accessible food for soil organisms and most of the readily mineralizable N. It is responsible for most of the beneficial effects on structural stability that lead to enhanced infiltration of water, resistance to erosion, and ease of tillage. The labile organic matter can be readily increased by the addition of fresh plant and animal residues, but it is also very readily lost when such additions are reduced or tillage is intensified. Labile organic matter rarely comprises more than 10 to 20% of the total SOM.

Protected or Stable Organic Matter (Humus)

The more stable and well-protected portion of SOM was referred to as humus in Section 11.4. Its carbon is not readily accessible to microbes and remains in the soil for hundreds or even

---

[7]For a review of simulation models that predict changes in soil organic matter, see Campbell and Paustian (2015).

thousands of years. This pool includes most of the physically protected organic matter in clay–humus complexes, POM protected inside microaggregates, and chemically stable bits of char. Humus accounts for 60 to 90% of the organic matter in most soils, and its quantity is increased or diminished rather slowly. It is most closely associated with the colloidal properties of soil humus and is responsible for most of the cation- and water-holding capacities contributed to the soil by organic matter.

It should be remembered that these pools are determined more by the environmental conditions and microbial community in a soil, than by the chemistry of its constituents. For example, experiments have shown that under warm conditions a greater proportion of a soil's carbon will turn over at rates rapid enough to be characterized as labile than would be the case when the same soil is kept cold.

Changes in Labile and Humus Pools with Soil Management

Soil scientists have consistently observed that productive soils managed with conservation-oriented practices contain relatively high amounts of organic matter fractions associated with the labile pool, including microbial biomass, free particulate organic matter, and easily oxidizable organic matter. Despite the analytical difficulties and conditional definitions, computer models that assume the existence of labile (fast) and humus (slow) pools have proven quite useful in explaining and predicting real changes in SOM levels and in attendant soil properties.

Studies on the dynamics of SOM have established that the different pools of SOM play quite different roles in the soil system and in the carbon cycle. The presence of a resistant (structural) pool of carbon in plant residues, as well as an easily decomposed (metabolic) pool, explains the initially rapid but decelerating rate of decay that occurs when plant tissues are added to a soil (see Figure 11.4). Similarly, the existence of a pool of chemically or physically protected SOM (slow or humus pool), as well as a pool of easily metabolized SOM (labile pool), explains why conversion of native forests or grassland into cultivated cropland results in a very rapid decline in SOM during the first few years, followed by a much slower decline thereafter (see, e.g., Figure 11.16). However, it is widely acknowledged that such models will need to

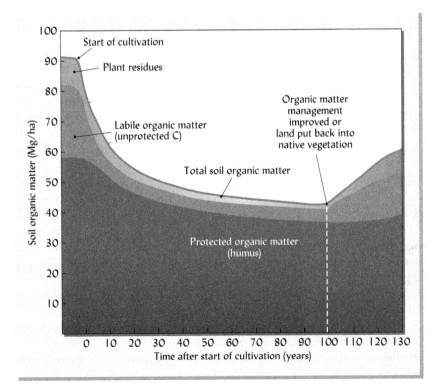

Figure 11.16 *Changes in SOM in the upper 30 cm of a representative soil after bringing virgin land under cultivation. Initially, the rapidly decomposing labile pool accounted for about 23 Mg, or about 25% of the total SOM. After about 50 years of cultivation, the slow humus pool had declined by about 30% from 91 to about 40 Mg/ha, while the labile pool had lost 90% of its mass, declining to only 2 Mg/ha. Thus most of the organic matter loss came from the labile pool. This was also the pool that most quickly increased when improved management was adopted after the hundredth year. This behavior explains why even relatively small changes in total SOM can produce dramatic changes in soil properties, such as aggregate stability and N mineralization.* (Diagram courtesy of Ray R. Weil)

be improved with better consideration of the roles of black carbon (char), root deposition (as opposed to surface litter), and changes in soil conditions such as aggregation and temperature that can shift the "boundaries" of the pools themselves.

Soil management practices that cause only small changes in total SOM often cause rather pronounced alterations in aggregate stability, N mineralization rate, or other soil properties attributed to organic matter. This occurs because the relatively small pool of labile organic matter may undergo a large percentage increase or decrease without having a major effect on the much larger pool of total organic matter. Figure 11.16 shows how the pools of SOM are affected by changing management (in this case, cultivating a previously undisturbed soil). The figure also indicates how each pool contributes to the total SOM level.

Accumulated plant residues and labile organic matter are the first to be affected by changes in land management, accounting for most of the early losses in SOM when cultivation of virgin soil begins. In contrast, losses from the protected humus pool are more gradual. As a result, the SOM remaining after some years is far less effective in promoting structural stability and nutrient cycling than the original organic matter in the virgin soil. If a favorable change in environmental conditions or management regime occurs, the plant litter and labile pools of SOM are also the first to positively respond (see right-hand portion of Figure 11.16).

11.7 CARBON BALANCE IN THE SOIL-PLANT-ATMOSPHERE SYSTEM

Whether the goal is to mitigate climate change by sequestering carbon and reducing greenhouse gas emissions, or to enhance soil quality and plant production, proper management of SOM requires an understanding of the factors and processes influencing the cycling and balance of carbon in an ecosystem. Although each type of ecosystem—forest, prairie, wetland, or wheat field—will emphasize particular compartments and pathways in the carbon cycle, consideration of a specific example, such as that described in Box 11.4, can help us develop a conceptual model that can be applied to many different situations.

The rate at which SOM either increases or decreases largely depends on the balance between *gains* and *losses* of carbon. In this regard, the level of SOM is analogous to the level of water in a tank, which is determined by the rates of inflow through the faucet and outflow through the drain. If the drain is fully opened, the water level will be low—even if the faucet is turned on fully. In soils, the organic matter gains come primarily from plant residues and exudates grown in place and from applied organic materials (in managed soils). The losses are due mainly to respiration (CO_2 emissions), erosion, leaching of dissolved C, and (in managed soils) plant removals (Figure 11.17). Table 11.3 provides some practical strategies for maximizing these gains and minimizing these losses so as to achieve increased SOM in managed terrestrial ecosystems such as urban landscapes, croplands, forest plantations, and pastures or rangelands.

Agroecosystems

Conservation of Soil Carbon. In order to halt or reverse the net carbon loss shown in Figure 11.18, management practices would have to be implemented that would either *increase the additions* of carbon to the soil or *decrease the losses* of carbon from the soil. Since all crop residues and animal manures in the example are already being returned to the soil, additional carbon inputs could most practically be achieved by growing more plant material (i.e., increasing crop yields or growing cover crops during the winter).

Specific practices to reduce carbon losses would include better control of soil erosion and the use of conservation tillage. Using a no-till production system would leave crop residues as mulch on the soil surface where they would decompose much more slowly. Refraining from tillage might also reduce the annual respiration losses from the original 2.5% to perhaps 1.5%. A combination of these changes in management would convert the system in our example from one in which SOM is degrading (declining) to one in which it is accumulating (increasing).

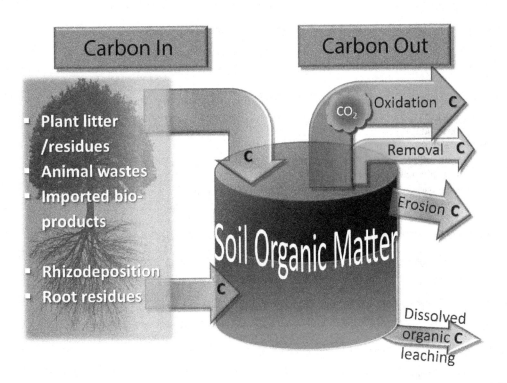

Figure 11.17 The level of organic matter present in a soil is largely determined by the balance between carbon in and carbon out. Soil management therefore needs to focus on strategies to maximize the inputs and minimize outputs, with the exception of the necessary economic harvest removals. Some practices appropriate for these strategies are listed in Table 11.3. (Diagram courtesy of Ray R. Weil)

Table 11.3
HOW TO MAINTAIN A POSITIVE BALANCE BETWEEN GAINS AND LOSSES OF ORGANIC MATTER
Most of these practices will improve soil quality as well as increase carbon stored in soils.

Inputs	Specific practices to maximize	Losses	Specific practices to minimize
Above-ground plant residues	• Return all residues • Fertilize for optimal production • Optimize plant spacing, varieties, timing • Add cover crops in off-season • Grow perennials rather than annuals • Use complex crop rotations • Controlled grazing/mowing • Manage fire to maximize char	Oxidation	• Reduce or eliminate tillage • Moderate soil temperature (mulch, irrigate) • Grow plants with recalcitrant litter • Grow high root/shoot ratio plants • Mulch rather than incorporate organic amendments • Avoid excessive N fertilizer • Pyrolyze organic wastes before applying to soil • Manage fire to minimize losses
Animal inputs	• Recycle manure and bedding • Urine and manure from grazing	Removal	• Remove only economic plant parts • Mow turf to leave clippings
Recycled offsite bio-materials	• Apply organic "waste" materials • Sewage sludge (biosolids) application • Use composted municipal solid wastes • Apply organic mulch • Biochar application	Erosion	• Use no-till • Use surface mulches • Grow cover crops between cash crops seasons • Grow perennial cover crops between tree rows • Use terraces, grassed water ways
Root residues and rhizo-deposition	• Plant high root biomass crops • Grow perennials rather than annuals • Controlled grazing/mowing • Grow cover crops • Manage for enhanced rhizodeposition • Manage soil microbial community	Leaching	• Maximize evapotranspiration during wet periods • Grow cover crops • Manage soil water • Manage plants for root exudates • Manage for mycorrhiza

BOX 11.4
CARBON BALANCE—AN AGROECOSYSTEM EXAMPLE

The principal carbon pools and annual flows in a terrestrial ecosystem are illustrated in Figure 11.18 using a hypothetical cornfield in a warm temperate region. During a growing season the corn plants produce (by photosynthesis) 17,500 kg/ha of dry matter containing 7,500 kg/ha of carbon (C). This C is equally distributed (2,500 kg/ha each) among the roots, grain, and unharvested aboveground residues. In this example, the harvested grain is fed to animals, which oxidize and release as CO_2 about 50% of this C (1,250 kg/ha), assimilate a small portion as weight gain, and void the remainder (1,100 kg/ha) as manure. The corn stover and roots are left in the field and, along with the manure from the animals, are incorporated into the soil by tillage or by earthworms.

The soil microbes decompose the crop residues (including the roots) and manure, releasing as CO_2 some 75% of the manure C, 67% of the root C, and 85% of the C in the surface residues. The remaining C in these pools is assimilated into the SOM. Thus, during the course of one year, some 1,475 kg/ha of C enters the SOM pool (825 kg from roots, plus 375 from stover, plus 275 from manure). These values can vary widely among different soil conditions and ecosystems.

At the beginning of the year, the upper 30 cm of soil in our example contained 65,000 kg/ha C in soil organic matter. Such a soil cultivated for row crops in a temperate region would typically lose about 2.5% of its organic C by soil respiration each year. In our example this loss amounts to some 1,625 kg/ha of C. Smaller losses of soil organic C occur by soil erosion (160 kg/ha), leaching (10 kg/ha), and formation of carbonates and bicarbonates (10 kg/ha).

Comparing total losses (1,805 kg/ha) with the total gains (1,475 kg/ha), we see that the soil in our example suffered a net annual loss of 330 kg/ha of C, or 0.5% of the total C stored in the organic matter soil. If this rate of loss were to continue, degradation of soil quality and productivity would surely result.

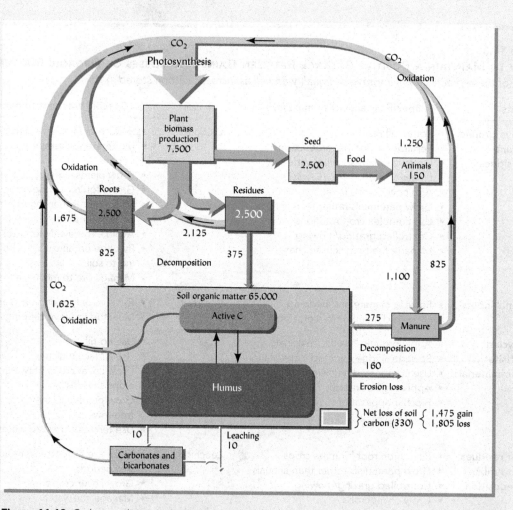

Figure 11.18 *Carbon cycling in an agroecosystem.* (Diagram courtesy of Ray R. Weil)

Natural Ecosystems

Upland Forests. Those interested in natural ecosystems may want to compare the carbon cycle of a natural forest with that of the cornfield in Figure 11.18. If the forest soil fertility were not too low, the total annual biomass production would probably be similar to that of the cornfield. The standing biomass, on the other hand, would be much greater in the forest since the tree crop is not removed each year. While some litter would fall to the soil surface, much of the annual biomass production would remain stored in the woody tissue of the trees.

The rate of organic matter oxidation in the undisturbed forest would be considerably lower than in the tilled field because the litter would not be incorporated into the soil through tillage, and the absence of physical disturbance would result in slower soil respiration. The litter from certain tree species may also be rich in phenolics and lignin, factors that greatly slow initial decomposition and C losses (see Figure 11.8). In forest soils, decomposition of leaf litter produces copious quantities of DOC compounds such as organic acids and tannins, so 5 to 40% of the total C losses may occur by leaching—a much greater proportion than from all but the most heavily manured cropland soils. However, losses of organic matter through soil erosion would be much smaller on the forested site. Taken together, these factors allow annual net gains in SOM in a young forest and maintenance of high SOM levels in mature forests.

Grasslands. Similar trends occur in natural grasslands, although the total biomass production is likely to be considerably less, depending mainly on the annual rainfall. Among the principles illustrated in Box 11.4, and applicable to most ecosystems, is the dominant role that plant root biomass plays in maintaining SOM levels. A relatively large proportion of the net annual productivity from perennial grass occurs below ground as root growth, regeneration, and exudation (see Section 10.7). This belowground plant biomass tends to decay more slowly than above ground residues. Therefore, a greater proportion of the total biomass produced tends to accumulate as SOM, and this soil organic C is distributed more uniformly with depth under grassland than under forest vegetation.

Another factor influencing accumulation of C in grassland soils and not considered in Box 11.4, is the frequent occurrence of fire. Fires that burn dead aboveground plant material are usually thought to reduce SOM inputs to the soil, but in perennial grasslands studies have shown that increased root growth stimulated by the fires may contribute at least as much carbon to the soil as that lost in the blaze itself. Also, while fire instantly converts even lignacious organic C into CO_2, it also produces varying amounts of slow-to-decompose char (see Box 11.2).

Wetlands. Wetlands, whether dominated by woody or herbaceous vegetation, exhibit among the highest levels of primary productivity of any ecosystems. However, microbial decomposition is severely retarded by lack of oxygen in the strongly anaerobic wetland soils. In addition, certain products of anaerobic decay (alcohols, organic acids, etc.) actually act as preservatives and inhibit even anaerobic organisms. As a result, organic carbon accumulates rapidly (300–3,000 kg/ha annually) and may continue to do so for thousands of years in some cases (see Sections 7.7 and 11.9). This prodigious level of carbon sequestration can be reduced or even reversed by practices, such as ditching, drainage, or peat mining, that increase the soil's oxygen content. Converting wetlands to agricultural use, except perhaps for some types of flooded rice or cranberry production, dramatically increases losses and decreases gains of carbon. The effect of prescribed burning on marsh soil carbon is still under study, but may be similar to that just described for grasslands.

11.8 ENVIRONMENTAL FACTORS INFLUENCING SOIL ORGANIC CARBON LEVELS

As just indicated in Section 11.7, the extent to which organic matter accumulates in soils is determined by the balance of gains and losses of organic carbon. Three environmental factors stand out in broadly determining the *concentration* of organic carbon in surface soils; these are *temperature*, *moisture*, and *soil texture*. When soil organic carbon *stocks* the amount stored per unit area of land surface are considered, then soil depth distribution is a fourth major determining factor. In this section we will examine how these factors and related biological and management factors influence the patterns and extent of organic carbon accumulation in soils.

Differences Among Soil Orders

Even though SOC contents vary as much as tenfold within a single soil order, a few generalizations are possible. In considering these, it will be instructive to compare the global soil carbon map (Figure 11.19) with the global soil orders map (front end papers) as well as to review the soil order descriptions in Chapter 3. Aridisols (dry soils) are generally the lowest in organic matter, and Histosols (organic soils) are definitely the highest (compare Figures 3.16 and 3.19). Contrary to popular myth, forested soils in humid tropical regions (e.g., Oxisols and some Ultisols) contain similar concentrations organic carbon to those in humid temperate regions (e.g., Alfisols and Spodosols). Andisols (volcanic ash soils) generally have some of the highest organic carbon contents of any mineral soils, probably because association with allophane clay in these soils protects the organic carbon from oxidation. Among cultivated soils in humid and subhumid regions, Mollisols (prairie soils) are known for their dark, organic, carbon-rich surface layers. The amount of SOC in mineral surface soils thus varies from a mere trace (sandy, desert soils) to as high as 10 or 20% (some forested or poorly drained A horizons) and up to 50% in some Histosols.

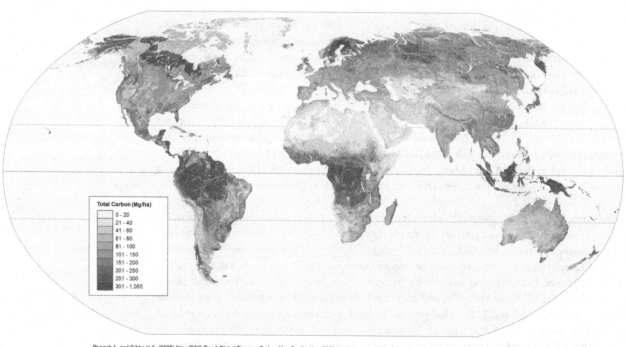

Figure 11.19 *Global distribution of organic carbon in the upper 1 m of the Earth's soils. The influences of temperature and moisture (as a function of annual precipitation/potential evapotranspiration and topographic wetness) and are well illustrated. The dark brown areas in the artic show the importance of Histosols and Gelisols as major repositories of organic C and illustrate the influence of cold temperatures, often combine with waterlogged soils. The dark brown areas in the tropics (mainly the Amazon, Congo, and Indonesian rain forests) illustrate the impact of high rainfall and resulting vegetative productivity. In contrast the light-colored areas in North Africa and Central Asia illustrate how hot and dry conditions produce desert soils very low in organic carbon.* [Adapted from Scharlemann et al. (2010)]

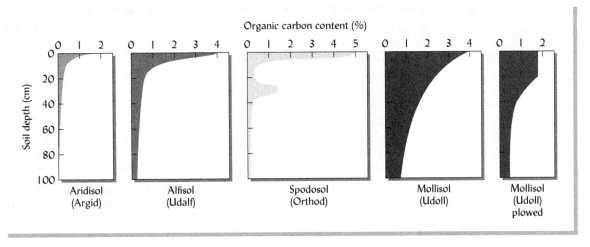

Figure 11.20 Vertical distribution of organic carbon in well-drained soils of four soil orders. Note the higher content and deeper distribution of organic carbon in the soils formed under grassland (Mollisols) compared to the Alfisol and Spodosol, which formed under forests. Also note the bulge of organic carbon in the Spodosol subsoil due to illuvial humus in the spodic horizon (see Figure 3.3). The Aridisol has very little organic carbon in the profile, as is typical of dry-region soils. (Diagram courtesy of N. C. Brady and Ray R. Weil)

Because of the distribution of plant roots and litter, organic matter tends to accumulate in the upper layers with much lower organic carbon contents in the subsurface horizons (Figure 11.20). The organic carbon content decreases less abruptly with depth in grassland soils than in forested ones, because much of the annual carbon addition in grasslands comes in the form of fibrous roots extending deep into the profile (see Section 2.5). In forest ecosystems, deep soil profiles may also hold a substantial amount of carbon below the A horizons, as can be calculated by simply multiplying the rather low carbon concentrations in the subsoil by the very large volume of these deep soil layers.

Influence of Climate

Temperature. Like any biological process, the respiration loss of soil carbon takes place at increasing rates as soil temperatures increase. The temperature response is thought to be greater for respiration of labile soil carbon than for carbon in the more protected or resistant humus fractions (Figure 11.21). The data also show that temperature response interacts with soil moisture, being small when water is limiting, and also declining when soils are so wet that oxygen becomes limiting to the process.

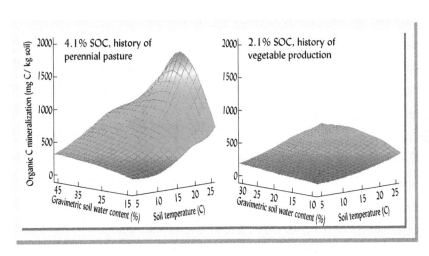

Figure 11.21 Soil temperature and water content (horizontal axes) interact to influence organic carbon mineralization to CO_2 (vertical axis, higher rates are redder). The temperature effect was greater where labile C was high due to many years under grass pasture (Left). The decline in mineralization rate at high water contents in the warm pasture soil indicate that oxygen became limiting. (Right) The temperature effect was much smaller where labile C had been depleted by many years of high-tillage vegetable cultivation. Compared to the vegetable soil, mineralization rates in the pasture soil were almost four times as great although total SOC was only twice as great. [Graphed from data in Curtin et al. (2012)]

The processes of organic matter production (plant growth) and destruction (microbial decomposition) respond differently to increases in temperature. At low temperatures, plant growth outstrips decomposition and organic matter accumulates. However, the opposite is true where mean annual temperature exceeds approximately 25 to 35°C. At high temperatures, decomposition surpasses plant growth, so nutrient release is rapid, but organic matter accumulation is lower than in cooler soils. Some of the most rapid rates of organic matter decomposition occur in irrigated soils of hot desert regions.

Within zones of comparable moisture and vegetation, the average SOC and N contents increase from two to three times for each 10°C decline in mean annual temperature. This temperature effect can be readily observed by noting the greater organic carbon content of well-drained surface soils as one travels from south (east Texas) to north (Minnesota) in the humid grasslands of the North American Great Plains region. Similar changes in SOC are evident as one climbs from warm lowlands to cooler highlands in mountainous regions.

Moisture. Under comparable conditions, SOC and N increase as the effective moisture becomes greater. The C/N ratio also tends to be higher in the more thoroughly leached soils of the higher rainfall areas. These relationships are evidenced by the darker and thicker A horizons encountered as one travels across the North American Great Plains region from the drier zones in the West (Colorado) to the more humid East (Illinois). The explanation lies mainly in the sparser vegetation of the drier regions. The lowest levels of SOM and the greatest difficulty in maintaining those levels are found where annual mean temperature is high and rainfall is low. These relationships are extremely important to the challenge of sustainable natural resource management. It should be remembered that SOM levels are not only influenced by climatic temperature and precipitation, but by a number of environmental factors, some of which in turn affect soil temperature or moisture in the soil.

Influence of Natural Vegetation

Climate and vegetation usually act together to influence the soil content of organic carbon. The greater plant productivity engendered by a well-watered environment leads to greater additions to the pool of SOM. Grasslands generally dominate the subhumid and semiarid areas, while trees are dominant in humid regions. In climatic zones where the natural vegetation includes both forests and grasslands, the total organic matter is higher in soils developed under grasslands than under forests (see Figure 11.22), as explained in Section 11.7.

Effects of Texture and Drainage

While climate and natural vegetation affect SOM over broad geographic areas, soil texture and drainage are often responsible for marked differences in SOM within a local landscape. Under aerobic conditions, soils high in clay and silt are generally richer in organic matter than are nearby sandy soils. The finer-textured soils accumulate more organic matter for several reasons: (1) they produce more plant biomass, (2) they lose less organic matter because they are less well aerated, and (3) more of the organic material is protected from decomposition by being bound

Figure 11.22

Distribution of organic carbon in four soil profiles, two well drained and two poorly drained. Poor drainage results in higher organic carbon content, particularly in the surface horizon. (Diagram courtesy of N. C. Brady and Ray R. Weil)

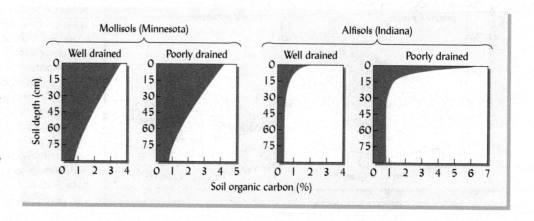

to clay surfaces (see Section 11.4) or sequestered inside soil aggregates. A given amount and type of clay can be expected to have a finite capacity to stabilize organic matter in organomineral complexes. Once this capacity is saturated, further additions of organic matter are likely to add little to humus accumulation, as they will remain readily accessible to microbial decomposition. Thus, soil texture is thought to limit the level of organic C that can accumulate in the soil for given ecosystem.

Drainage-Aeration Effects. In poorly drained soils, the high moisture supply promotes plant dry-matter production and relatively low oxygen or even anaerobic conditions inhibit organic-matter decomposition. Poorly drained soils therefore generally accumulate much higher levels of organic matter and nitrogen throughout their profiles than do similar but better-aerated soils (Figure 11.22).

11.9 SOIL ORGANIC MATTER MANAGEMENT

Influence of Agricultural Management and Tillage

Except where barren deserts are brought under irrigation, it is safe to generalize that cultivated land contains much lower levels of organic matter than do comparable areas under natural vegetation. This is not surprising; under natural conditions all the organic matter produced by the vegetation is returned to the soil, the vegetation is perennial with living roots present all year long, and the soil is not disturbed by tillage. By contrast, in cultivated areas annual plants must start each year from seed, so much of the year may be unproductive. Much of the plant material produced is removed in harvest and relatively little finds its way back to the land. Also, tillage aerates the soil and breaks up organic residues and aggregates, making the carbon more accessible to microbial decomposition.

Conversion to Cropland. A very rapid decline in SOM occurs when a virgin soil is brought under cultivation. Eventually, the gains and losses of organic carbon reach a new equilibrium, and the SOM content stabilizes at a much lower value (Figure 11.16). Similar declines in SOM are seen when tropical rain forests are cleared; however, the losses may be even more rapid because of the higher soil temperatures involved. Declining SOM is a major factor driving a downward spiral of soil degradation and poverty that plagues many of the world's 1 billion subsistence farmers. There is some hope that efforts to introduce conservation tillage, increase plant nutrient inputs, and better manage crop residue will enable farmers to reverse this trend (see following and Section 14.6). Organic matter losses are not so dramatic if forests or prairies are converted to well-managed pasture or hay production because the latter systems still feature perennial vegetation with high root productivity and no tillage. Conversely, in many expanding urban areas, SOM is increased during the years after turfgrass lawns are established on land that formerly had been depleted by crop cultivation.

Conservation Tillage. In terms of soil disturbance and the proportion of residues coverage left on the soil surface, there is a continuum of practices that range from conventional inversion tillage, which causes maximal disturbance and leaves the soil almost bare of residue cover, to no-till practices, which disturb only a tiny fraction of the soil during seed placement and leave the surface nearly completely covered by residues. Practices that fall in between include stubble mulching, some forms of chisel tillage, vertical tillage, and strip tillage. These practices along with no-till are referred to as *conservation tillage* (see Section 14.6) because they conserve soil by protecting against erosion and conserve organic carbon by discouraging the rapid decomposition of crop residues that conventional inversion plow tillage stimulates in the surface soil. These conservation practices can therefore help maintain or restore high surface SOC levels and associated soil quality properties such as aggregate stability (Figure 11.23). As in many similar studies that compare various tillage practices to untilled or no-till systems, the effects tend to be most pronounced very near the surface.

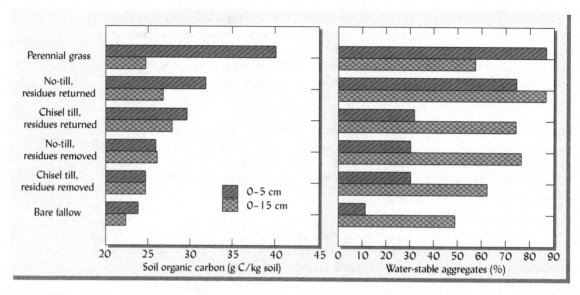

Figure 11.23 Effects of agricultural disturbance intensity and plant residue removal on organic carbon and aggregate stability at two depths of Waukegan silt loam soil (Typic Hapludoll) from east-central Minnesota. The study was designed to understand how the harvest of corn stover for cellulosic biofuel production might affect soil quality, depending on how such land was managed. Experimental treatments applied for 19 years prior to the data collection ranged from maximum positive C balance with untilled perennial grass to maximum negative balance with soil kept continuously bare. There were also several corn production systems with intermediate C balances that depended on various combinations of tillage (chisel or no-till) and residue removal (all returned or 90% removed as for biofuel feedstock). [Graphed from data in Laird and Chang (2013)]

Influence of Rotations, Residues, and Plant Nutrients

Figure 11.24 illustrates changes in SOC during 140 years after the native prairie was first plowed and several different cropping systems imposed. The soils are Mollisols, and the plots can still be seen on the University of Illinois campus. These, and similar long-term experimental plots elsewhere, support the following conclusions. A complex rotation (corn, oats, and clovers) achieved higher SOC levels than monocropping (continuous corn), regardless of fertility inputs, probably because the rotation used tillage less frequently and produced more root residues. Systems that maintain soil fertility with manure, lime, and phosphorus stimulate much higher organic carbon levels, especially where a complex rotation was also followed. This result was likely related to the greater additions of organic matter in the manure and in the residues from higher-yielding crops. Application of lime and fertilizers (N, P, and K) to previously unfertilized and unmanured plots (dashed lines starting in 1955) noticeably increased SOM levels, probably due to the production and return of larger amounts of crop residues and the addition of sufficient N to compliment the carbon in humus formation.

Another famous long-term experiment was begun even earlier (in 1855) at Rothamsted, England, on land that had been farmed to wheat for centuries (Figure 11.24b). Because the previous history of land use differed from that of the Morrow plots, the Rothamsted experiment provides lessons that are applicable to changes in old long-established agricultural systems. Because this soil was already in equilibrium with the carbon and nitrogen gains and losses characteristic of unfertilized, small-grain cropping, continued production (and harvest) of barley and wheat resulted in little change, or only a slow decline, in the already low level of SOC. Annual applications of animal manure at rates sufficient to supply all needed N resulted in a dramatic initial rise in SOC, until a new equilibrium state was approached at a much higher level. In the plots where manure was applied for only the first 20 years of the experiment, the SOC began to decline as soon as the manure applications were suspended, but the positive effect of the manure was still evident some 120 years later!

Results from such long-term experimental plots demonstrate that soils kept highly productive by supplemental applications of nutrients, lime, and manure and by the choice of high-yielding, diverse cropping systems are likely to have more organic matter than

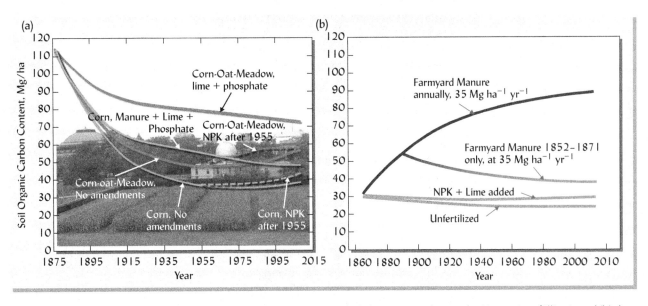

Figure 11.24 Soil organic carbon contents of selected treatments of (a) the Morrow plots at the University of Illinois and (b) the classical Hoosfield experiment at Rothamsted Experiment Station in England. The Morrow plots were begun on virgin grassland soil in 1876 and so suffered rapid loss of organic carbon in the early years of the experiment. The Rothamsted plots were established on soils with a long history of previous cultivation. As a result, the soil at Rothamsted had reached an equilibrium level of organic carbon characteristic of the unfertilized small-grains (barley and wheat) cropping system traditionally practiced in the area. Note the large impact of farmyard manure application and the residual effect of manure some 135 years after the last application was made. [Graphed using data recalculated from Nafziger and Dunker (2011), Jenkinson and Johnson (1977), and Johnston et al. (2009); used with permission of the Rothamsted Experiment Station, Harpenden, England. Photo courtesy of Ray R. Weil]

comparable, less productive soils. High productivity is sustainable if it involves not only greater economic harvests, but also larger amounts of roots and shoots returned to the soil.

The Conundrum of Soil Organic Matter Management

Land managers are faced with an inherent conflict between the need to *use* SOM and the need to *conserve* it. While the total carbon stabilized in the soil is important in relation to the global climate change (see Section 11.10), many beneficial effects on soil productivity and ecosystem function are realized only when the organic matter is destroyed by microbial metabolism.

In this regard, SOM is a lot like money—while it may be nice to accumulate large amounts, the benefits of having money are realized by spending it, not by keeping it "under the mattress." Most desirable would be an income large enough to allow both spending and saving. Similarly, soil management needs to simultaneously save carbon losses (by reducing losses from tillage, erosion, and aggregate disruption) while also providing a large income of preferably diverse, high-quality organic materials. Although contradictory, the goals of using and conserving must be pursued simultaneously. The *decomposition* of SOM is necessary for its use as a source of nutrients for plant and microbial growth and organic compounds that promote biological diversity, disease suppression, aggregate stability, and metal chelation. In contrast, the *accumulation* of SOM is necessary for these functions in the long term, as well as for the sequestering of C, the enhancement of soil water-holding capacity, the adsorption of exchangeable cations, the immobilization of pesticides, and the detoxification of metals.

General Guidelines for Managing Soil Organic Matter

A continuous supply of plant residues (roots and tops), animal manures, composts, and other materials must be added to the soil to maintain an appropriate level of SOM, especially in the active pool. It is almost always preferable for the soil to be vegetated than for it to be bare for any length of time. Even if some plant parts are removed in harvest, vigorously growing plants provide below- and aboveground residues as major sources of organic matter for the soil. Moderate applications of lime and nutrients may be needed to help free plant growth from the constraints imposed by chemical toxicities and nutrient deficiencies. Where climate permits, cover crops often present a tremendous opportunity to provide protective cover and additional organic material for the soil.

There is no "ideal" amount of soil organic matter. It is generally not practical to try to maintain higher SOM levels than the soil–plant–climate control mechanisms dictate. For example, 1.5% organic matter might be an excellent level for a sandy soil in a warm climate, but would be indicative of a very poor condition for a finer-textured soil in a cool climate.

Adequate N is requisite for adequate organic matter because of the relationship between nitrogen and carbon in stabilized organic matter and because of the positive effect of N on plant productivity. Accordingly, the inclusion of leguminous plants, animal manures, and the judicious use of nitrogen-containing fertilizers to enhance high plant productivity are desirable practices. Excessive N applied from mineral sources sometimes accelerates the loss of carbon in high C/N soils. Steps must also be taken to minimize the loss of N by leaching, erosion, or volatilization (see Sections 12.1 and 13.2).

Tillage should be eliminated or limited to that needed to control weeds and to maintain adequate soil aeration. The more tillage that is performed, the faster organic matter is lost from the surface horizons and the more difficult it become to maintain high soil quality.

Perennial vegetation, especially natural ecosystems, should be encouraged and maintained wherever feasible. Improved agricultural production on existing farmlands should be pursued to allow land currently supporting natural ecosystems to be left relatively undisturbed. In addition, there should be no hesitation about taking land out of cultivation and encouraging its return to natural vegetation where such a move is appropriate.

11.10 SOILS AND CLIMATE CHANGE[8]

Soil is a major component of the Earth's system of self-regulation that has created (and, we hope, will continue to maintain) the environmental conditions necessary for life as we know it on this planet. Biological processes occurring in soils have major long-term effects on the composition of the Earth's atmosphere, which in turn influences all living things, including those in the soil.

Global Climate Change

Of particular concern today are increases in the levels of certain gases in the Earth's atmosphere, known as **greenhouse gases**. Like the glass panes of a greenhouse, these gases allow short-wavelength solar radiation in but trap much of the outgoing long-wavelength radiation. This heat-trapping **greenhouse effect** of the atmosphere is a major determinant of global temperature and, hence, global climate change. Gases produced by biological processes, such as those occurring in the soil, account for approximately half of the rising greenhouse effect. Of the four primary greenhouse gases, only the fluorine gases (e.g., chlorofluorocarbons or CFCs) are exclusively of industrial origin.

While it is certain that the concentrations of most greenhouse gases are increasing, there is less certainty about how rapidly global temperatures are actually rising and about how these increases are likely to affect climate in different regions of the world. Predicting changes in global temperature is complicated by numerous factors, such as cloud cover and volcanic dust, which can counteract the heat-trapping effects of the greenhouse gases. Most scientists believe that the average global temperature has increased by approximately 1.5°C during the past century, and predict that it is likely to increase by another 3 to 4°C in this century. If this increase in fact takes place, major changes in the Earth's climate are sure to result, including changes in rainfall distribution and growing season length, increases in sea level, and greater frequency and severity of storms and droughts. The rise in sea level alone, as predicted by some climate models, would threaten the homes of hundreds of millions of people living in coastal areas, mainly in Asia and North America. Through national programs and international agreements much effort is currently being directed at reducing the anthropogenic (human-caused)

---

[8]For a review of the potential for soil management to mitigate climate changes, see Paustian and Babcock (2004), Mackey et al. (2013), and Lal (2018).

contributions to climate change. Soil science has the potential to contribute greatly to our ability to deal with climate change and mitigate the levels of greenhouse gases.

Carbon Dioxide

In 2017, the atmosphere contained about 420 ppm CO_2, as compared to about 280 ppm before the Industrial Revolution. Levels are increasing at about 0.5% per year. Although the burning of fossil fuels is a major contributor, much of the increase in atmospheric CO_2 levels has come from a net loss of organic matter from the world's soils. Through aerobic decomposition, the carbon in plant biomass and soil organic matter—carbon that originated from CO_2 in the atmosphere—is eventually converted back into CO_2 and returned to the atmosphere. Box 11.1 illustrated the importance of SOM in regulating atmospheric CO_2 levels. Research (e.g., Figure 11.25) indicates that the feedback between the soil and atmosphere works both ways—changes in the levels of gases beneficial or harmful to plants influence the rate at which carbon accumulates in SOM.

Gains in SOM occur first in the active fractions, but eventually some of the carbon moves into the stable passive fraction, where it may be *sequestered* for hundreds or thousands of years. The opportunities for sequestering carbon are greatest for degraded soils that currently contain only a small portion of the organic matter levels they contained originally under natural conditions. Reforestation of denuded areas is one such opportunity. Conversion of cultivated land to perennial vegetation may sequester carbon at twice these rates.

Hundreds of studies have shown that reducing or eliminating tillage generally enhances soil quality and surface soil carbon levels in agricultural soils. On average, switching

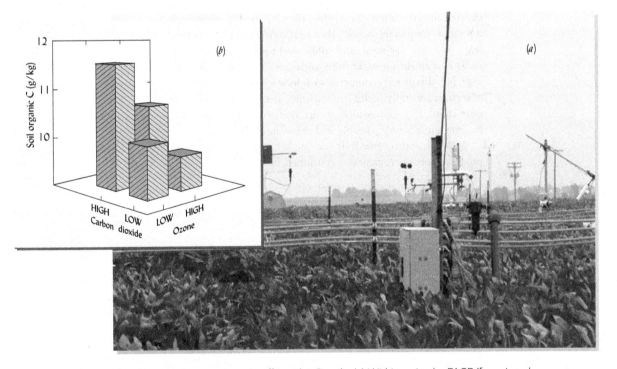

Figure 11.25 *Changing atmospheric composition affects the C cycle. (a) Within a circular FACE (free air carbon enrichment) experimental plot on Mollisols at the University of Illinois, soybeans experience air altered to simulate the atmospheric composition expected in 50 years. Horizontal tubing and vertical poles supply computer-controlled concentrations of carbon dioxide and/or ozone, which are continuously monitored by various sensors. Increasing atmospheric CO_2 from low (350 mg/L, the ambient level) to high (500 mg/L, the level expected by the year 2050) enhanced photosynthesis and plant growth, thus increasing the amount of fixed carbon available for translocation to the roots and eventually to the soil. (b) Data from a different carbon–ozone enrichment experiment on Ultisols in Maryland show measurably increased SOC after 5 years of soybean and wheat crops grown in high CO_2 air. Increased root:shoot ratio, rhizodeposition of carbon compounds, and overall greater litter inputs all played a role. Ozone, a pollutant at ground level, injures plants, reduces photosynthesis, and therefore impacts the soil in a manner opposite to that of CO_2. The data suggest the full effect of CO_2 is seen only when ozone remains at low levels.* [Data from Weil et al. (2000); photo courtesy of Ray R. Weil]

the management of cropland from conventional plow tillage to no-tillage appears to sequester 0.2 to 0.5 Mg/ha of carbon in the surface soil annually during the first 10 to 20 years until a new, higher equilibrium carbon level is reached. Highest rates of C sequestration are likely where an integrated conservation farming system is used that combines minimization of tillage disturbance with diverse crop rotations and cover crops, integrated fertility management, and retention of plant residues on the soil surface. However, most of the data we have on the effect of tillage practices are based on sampling just the upper 5 to 20 cm of soil. The few studies that have analyzed soil down to 1 m tend to question whether, by itself, no-till management of annual row crops significantly increases carbon stored in soils per unit of land area when the entire soil profile depth is considered (Figure 11.26).

By slowly increasing SOM to near precultivation levels, improved soil management could significantly enhance society's efforts to stem the rise in atmospheric CO_2 and at the same time improve soil quality and plant productivity. Some estimates suggest that during a 50-year period, improved management of agricultural lands could provide about 15 to 25% of the CO_2 emission reductions that need to be made. It should be noted, however, that in accordance with the factors discussed in Sections 11.8–9, soils have only a finite capacity to assimilate carbon into stable SOM. Therefore, carbon sequestration in soils can only buy time while other kinds of actions (shifts to renewable energy sources, increased fuel efficiency, etc.) are fully implemented to reduce carbon emissions to levels that will not threaten climate stability.

Biofuels Production. Soils are deeply involved with *biofuels*, both because good soils will be needed to grow the plant biomass feedstocks and also because removal of this plant material can greatly alter the C balance in soils. Biofuels from plant material make attractive substitutes for gasoline and diesel because, in theory, they are both renewable and carbon neutral (the CO_2 emitted into the atmosphere when they are used is taken back out of the atmosphere to produce more biofuel feedstock). Biodiesel consists of fuel-grade oil extracted from the same crops (e.g., soybeans, rapeseed, and sunflower) used to provide cooking oil. Ethanol has long been made by fermenting sugar from sugarcane or starch from corn grain (much like making whiskey), but this process competes with food and feed uses of the sugar and starch and, in the case of corn, is generally inefficient. Instead, attention is turning toward the production of ethanol from cellulosic residues such as corn stover (the stalks and leaves) and switchgrass, which can be grown much more cheaply and abundantly than grain.

Much controversy still surrounds the question as to whether such fuels will actually produce more energy than is consumed in the fossil fuel used to grow and process the crop.

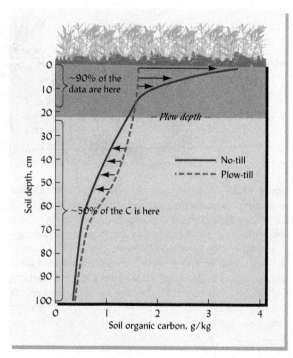

Figure 11.26 *Representative trends showing that conversion of plow-tilled cropland to no-till management encourages the accumulation of high levels of SOC in the surface few cm of soil, stimulating aggregation, high infiltration, and resistance to erosion. Most research on the impact of tillage on SOC has studied only the upper 10 to 25 cm of soil. However, some of the small number of studies that sampled much greater depths suggest that, in the subsoil, plowed soils may accumulate more SOC than under no-till, possibly due to differences in crop rooting depth between tillage systems. Therefore it is uncertain what the effect of no-till may be in the overall carbon storage in the soil.* (Based on many sources, diagram courtesy of Ray R. Weil)

Table 11.4
DETERIORATION OF SOIL ORGANIC CARBON AND STRUCTURAL PROPERTIES IN THE UPPER 5 TO 10 CM OF THREE OHIO SOILS AFTER 1 YEAR OF CORN STOVER REMOVAL FOR BIOFUEL PRODUCTION[a]

Soil texture, great group	Cropping history	Stover returned, Mg/ha[b]	Soil organic C, g/kg	Soil bulk density, Mg/m^3	Tensile strength of soil aggregates, kPa	Water stable aggregates <4.75 mm diameter, %
silt loam (Hapludults)	33-yr continuous no-till corn	5	29	1.46	140	18
		0	19	1.50	50	13
clay loam (Epiaqualfs)	8-yr minimum-till corn/soybean	5	26	1.31	380	20
		0	22	1.49	120	8
silt loam (Hapludalfs)	15-yr no-till corn/soybean	5	28	1.25	225	36
		0	21	1.42	80	13

[a]Data selected from Blanco-Canqui et al. (2006).
[b]5 Mg/ha corn stover represented the normal practice of returning all aboveground residues after corn grain harvest. Return of 0 Mg/ha represented removal of all stover for biofuel production.

However, one aspect that should *not* be controversial is the critical importance of leaving enough crop residues on the soil to maintain the levels of SOM consistent with high quality and continued productivity of the soil resource. Research suggests that SOM levels, aggregate stability, and other soil properties critical to sustaining productivity may rapidly suffer if all, or even half, of the corn stover is removed to make biofuels (Table 11.4). Unfortunately, the need to "share" the plant residues with the soil has not always been recognized by energy engineers and planners advising policy makers on biofuel production strategies. Some have advised not only that all aboveground residues be harvested to make the biofuels, but that the crops be genetically altered to reduce the ratio of roots to shoots—just the opposite of what sustainable soil management requires!

Wetland Soils. Discussion up to this point has focused on organic matter in upland mineral soils. Under waterlogged conditions in bogs and marshes, the oxidation of plant residues is so retarded that the soils in these environments consist mainly of organic matter and many are classified as Histosols in *Soil Taxonomy* (see Chapter 3). Although they cover only about 2% of the world's land area, Histosols (and Histels—permafrost soils with organic surface horizons) are important in the global carbon cycle because they hold a large proportion of global soil carbon. Drainage of these soils for high-value horticultural production speeds the oxidation of organic matter, which over time destroys the soil itself. Draining Histosols, or mining them for peat (Box 11.5), can make major contributions to the rise in atmospheric CO_2. On the other hand, rising sea levels associated with global warming may lead to more Histosol formation, which in turn may buffer climate change by sequestering more CO_2. Although storms increasingly destroy coastal marshes, the rising sea level will flood more coastal land area, creating conditions conducive to the formation of more Histosols in tidal marshes and mangrove forests along tropical coasts (Figure 11.27). The organic matter accumulation in these newly formed Histosols would represent a significant sequestering of CO_2 that might help to buffer the global warming trend (as would the increased absorption of CO_2 by the greater volume of ocean waters). However, such predictions are complicated by the involvement of another greenhouse gas, methane, which is emitted by some Histosols wetland soils.

Figure 11.27 *A tropical mangrove ecosystem, like most forest ecosystems, stored most of its C in the soil rather than in the vegetation. This, combined with wetland condition conductive to high levels of soil C storage, explains why tropical mangrove forests contain such large soil C stocks per unit area.* [Photo courtesy of Ray R. Weil; graphed from supplementary data in Donato et al. (2011)]

Methane. Methane (CH_4) occurs in the atmosphere in far smaller amounts than CO_2. However, methane's contribution to the greenhouse effect is nearly half as great as that from CO_2 because each molecule of CH_4 is about 25 to 30 times as effective as CO_2 in trapping outgoing radiation. The level of CH_4 is rising at about 0.6% per year. In 2015, there was about 1.85 ppm CH_4 in the atmosphere, more than double the preindustrial level. Soils serve as both a source and a sink for CH_4—that is, they both add CH_4 and remove it from the atmosphere.

Biological soil processes account for much of the methane emitted into the atmosphere. When soils are strongly anaerobic, as in wetlands and rice paddies, prokaryote metabolism can produce CH_4, rather than CO_2, as it decomposes organic matter (see Sections 7.3 and 11.2). Among the factors influencing the amount of CH_4 released to the atmosphere from wet soils are: (1) the maintenance of a redox potential (Eh) near 0 mV; (2) the availability of easily oxidizable carbon, either in the SOM or in plant residues returned to the soil; and (3) the nature and management of the plants growing on these soils (70–80% of the CH_4 released from flooded soils escapes to the atmosphere through the hollow stems of wetland plants). Figure 11.28 shows how these factors can influence CH_4 emissions from flooded rice paddies. Although not part of the study illustrated here, it has been demonstrated that periodically draining rice paddies prevents the development of extremely anaerobic conditions and therefore can substantially decrease CH_4 emissions. Such management practices should be given serious consideration, as rice paddies are thought to be responsible for up to 25% of global CH_4 production.

Wetland soils are not the only ones that contribute to atmospheric CH_4. Significant quantities of methane are also produced by the anaerobic decomposition of cellulose in the guts

Figure 11.28 *Factors affecting the emission of methane, a very active greenhouse gas, from a flooded rice paddy soil in California. The methane is generated by microbial metabolism in the soil and transported to the atmosphere through the rice plant. Emissions were greatest during the period of most active rice growth in mid-season. The sharp increase in methane near the end of the season was due to a rapid release of accumulated methane as the soil dried down and cracks opened up in the swelling clay. Very little methane was released if no rice was planted. Moderate amounts were released if rice was planted but rice residues from the previous crop had been burned off. The highest amounts of methane were released where rice was planted and the straw from the previous crop had been incorporated into the soil. Once the soil was drained at the end of the season, no more methane was produced.* [Redrawn from Redeker et al. (2000), with permission of the American Association for the Advancement of Science]

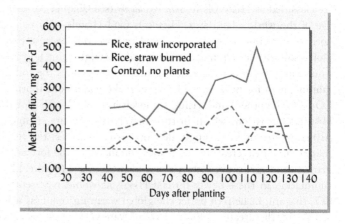

BOX 11.5
PEAT FOR POTS AND POWER: *UNSUSTAINABLE SOIL MINING*[a]

The peat you may have purchased in large bags at the garden center was probably produced by literally mining wetland soils. More than half of the world's wetlands are considered to be *peatlands*, deep Histosols that occur mainly in cool climates (especially in Canada) and contain layers of *woody*, *sedimentary*, and *fibrous peat*. The fibrous peat is most desirable for horticultural uses such as mulching flower beds and topdressing or amending turfgrass, especially in sand-based golf greens (see Section 11.7). Most containerized plants are grown in soilless media consisting largely of peat. Peat and similar organic materials are lightweight, have great water-holding capacity (10–20 times their dry weight), contribute considerable cation exchange capacity for holding available plant nutrients, and provide large pores to promote drainage and aeration.

With the phenomenal growth in demand by horticultural industries, mining of Canadian Histosols for peat is increasing. This activity increases greenhouse gas emissions for two reasons. First, mining destroys the wetlands and eliminates their C sequestration function. Second, peat used for horticultural applications eventually decomposes, and its carbon is returned to the atmosphere. Even if peatland restoration efforts are successful and the cutover peatlands are restored to their role as net carbon sinks, research estimates it would take 2,000 years to sequester the C lost during the past 50 years.

An even more direct conversion of sequestered C in peat to CO_2 is the extensive use of peat for fuel. In Scotland and Ireland, rural people still cut peat blocks from bog soils and dry them for use in home heating (Figure 11.29). In Russia, peat is mined on a much larger scale to provide fuel for electricity generation. While such peat utilization may be profitable in the short term, it is an inherently destructive and unsustainable use of the Histosol resource. It is also incompatible with global efforts to slow climate change by reducing CO_2 emissions.

Alternatives do exist. Organic wastes stabilized by composting can be used in place of peat for potting mixes and in turfgrass amendments. Composted materials provide most of the same benefits as peat, but, in addition, they may also suppress plant diseases (Section 10.13) and release significant levels of nutrients as they slowly decompose. Use of compost made from sewage sludge, farm manure, or municipal garbage not only provides a low-cost, renewable, greenhouse-neutral organic material for horticulture, it also solves some vexing waste disposal problems. Biochar made from crop residues or wood wastes may be another potential substitute for peat. With the mining of large quantities of peat continuing, the shift to such alternative materials needs to be encouraged.

Figure 11.29 *Peat blocks cut from the Histosol profile (see trench) are stacked to dry so they can be burned to heat rural homes in western Scotland.* (Photo courtesy of Josh Weil)

[a]Carlile et al. (2015) discusses practical and sustainable alternatives to peat for organic growing media.

---

of termites living in well-aerated soils (see Section 10.5) and of garbage buried deep in landfills (see Section 15.9).

In well-aerated soils, certain **methanotrophic bacteria** produce the enzyme *methane monooxygenase*, which allows them to oxidize methane as an energy source:

$$\underset{\text{Methane}}{CH_4} + \tfrac{1}{2}O_2 \rightarrow \underset{\text{Methanol}}{CH_3OH} \tag{11.5}$$

This reaction, which is largely carried out in soils, reduces the global greenhouse gas burden by about 1 billion Mg of methane annually. Unfortunately, the long-term use of inorganic (especially ammonium) nitrogen fertilizer on cropland, pastures, and forests has been shown to reduce the capacity of the soil to oxidize methane (see also Section 12.1). The evidence suggests that the rapid availability of ammonium from fertilizer stimulates ammonium-oxidizing bacteria at the expense of the methane-oxidizing bacteria. Long-term experiments in Germany, Belgium, and England indicate that supplying N in organic form (as compost or manure) actually enhances the soil's capacity for methane oxidation (Figure 11.30).

Figure 11.30 Effect of agricultural soil management on methane oxidation rates in a sandy loam soil in Belgium. The data are from samples taken in June of the third year of an experiment comparing monocropped corn grown with and without synthetic agrochemicals. Using herbicides (light green) as compared to manual weeding (dark green) had no effect. On the other hand, as in other experiments around the world, the methane oxidation capacity of the soil was dramatically reduced by the use of inorganic N fertilizer as compared to an organic source of N (compost in this case). [Data from Seghers et al. (2005)]

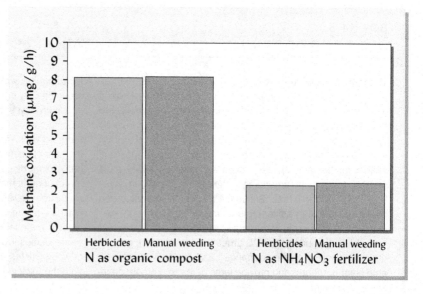

Nitrous Oxide. Nitrous oxide (N_2O) is another greenhouse gas—one with even more global warming power than methane. The impact of this gas is some 300 times greater than that of an equal mass of carbon dioxide. About 75% of nitrous oxide emissions in the United States originate in soils (including manure destined for soil application). It is produced by microorganisms under low oxygen conditions—such as in poorly aerated soils or in the interior of organic nitrogen-rich soil aggregates (Section 7.5). Its production by soil prokaryotes requires a ready supply of energy, usually from easily decomposed carbon. Emissions of this greenhouse gas are often associated with a combination in time and space of highly fertile soil conditions, with warm summer temperatures and excessive wetness. Since nitrous oxide is a component of the N cycle, it will be further discussed in the next chapter (see Section 12.1).

Because the soil can act as a major source or sink for carbon dioxide, methane, and nitrous oxide, it is clear that, together with steps to modify industrial and transportation section outputs of these gases, soil management has a major role to play in controlling the atmospheric levels of greenhouse gases. Figure 11.31 puts into perspective the relative potential

Figure 11.31 Relative global warming mitigation potential of the main low-carbon agricultural strategies to mitigate climate change. To account for the much higher heat absorption capacity of methane and nitrous oxide compared to carbon dioxide, the mitigation potential is shown as Mg of CO_2 equivalent reduced per year. It is clear that among the many strategies, improved management of soils under croplands, grazing lands, and wetlands are the three most important. [Graph adapted from Norse (2012)]

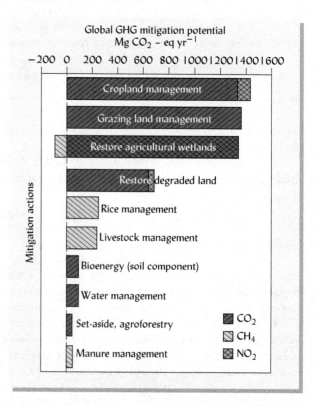

of the main agricultural soil-related strategies to mitigate climate change. The mitigation potential is shown as CO_2 equivalent climate forcing so that effects on emissions of the three greenhouse gases can be fairly compared and summed together. It is clear that among the many strategies, sequestering more carbon into soils under croplands and grazing lands and restoring formerly drained wetlands are the most important. Fortunately, these measures are quite feasible as their implementation would not only reduce greenhouse gases, but also improve soil quality and provide the benefits of enhanced soil function and productivity to land owners.

11.11 COMPOSTS AND COMPOSTING

Composting is the practice of creating partially stabilized useful organic decay products outside of the soil by mixing, piling, or otherwise storing organic materials under conditions conducive to aerobic decomposition and nutrient conservation (Figure 11.32). The decomposition processes and organisms involved are only somewhat similar to those already described for the formation of humus in soils. Important differences include the fact that with composting, decay occurs outside of the soil and commonly undergoes periods of high temperatures not found in soils. Also, in compost there is little opportunity for sorption onto clay or occlusion inside aggregates. The finished product, **compost**, is popular as a mulching material, as an ingredient for potting mixes, as an organic soil conditioner, and as a slow-release fertilizer.

High-quality compost can be made at ambient temperatures by a slow-decomposition process, or more quickly by a process called **vermicomposting**, in which certain litter-dwelling (epigeic) earthworms are added to help transform the material. Vermicompost essentially consists of the casts made by earthworms eating the raw organic materials in moist, aerated piles. The piles are kept shallow to avoid heat buildup that could kill the worms.

We will focus, however, on the most commonly practiced type of composting, in which intense decomposition activity occurs within large, well-aerated piles. This approach is called **thermophilic composting** because the large mass of rapidly decomposing material, combined with the insulating properties of the pile, results in a considerable buildup of heat.

Composting Process. Thermophilic compost typically undergoes a three-stage process of decomposition: (1) During a brief, initial **mesophilic** stage, sugars and readily available microbial

Figure 11.32 *A special machine turns large-scale compost windrows (direction of travel is away from the photographer) to mix the material and maintain well-aerated conditions at a facility in North Carolina where university dining hall food scraps are processed into compost for campus landscaping.* (Photo courtesy of Ray R. Weil)

food sources are rapidly metabolized, causing the temperature in the compost pile to gradually rise from ambient levels to over 40°C. (2) A *thermophilic* stage occurs during the next few weeks or months, during which temperatures rise to 50 to 75°C, while oxygen-using thermophilic organisms decompose cellulose and other more resistant materials. Frequent mixing during this stage is essential to maintain oxygen supplies and ensure even heating of all the material. The easily decomposed compounds are used up, and humuslike compounds are formed during this stage. (3) A second mesophilic or *curing stage* follows for several weeks to months, during which the temperature falls back to near ambient, and the material is recolonized by mesophilic organisms, including certain beneficial microorganisms that produce plant-growth-stimulating compounds or are antagonistic to plant pathogenic fungi (see Section 10.14).

Nature of the Compost Produced. As raw organic materials are decomposed in a compost pile, the content of simple plant molecules like sugars and cellulose declines, and the content of lower energy, complex molecules increases markedly. During the composting process, the C/N ratio of organic materials in the pile decreases until a fairly stable ratio, in the range of 10:1 to 20:1, is achieved. The CEC of the organic matter may increase to about 50 to 70 cmol$c$/kg of compost.

Although 50 to 75% of the carbon in the initial material is typically lost during composting, mineral nutrients are mostly conserved. The proportion that is mineral content (referred to as *ash* content) therefore increases over time, making finished compost more concentrated in nutrients than the initial combination of raw materials used. Properly finished compost should be free of viable weed seeds and pathogenic organisms, as these are generally destroyed during the thermophilic phase. However, inorganic contaminants such as heavy metals are *not* destroyed by composting. Proper management of compost is essential if the finished product is to be desirable for use as a potting media or soil amendment.

Benefits of Composting. Although making compost may involve more work and expense than applying uncomposted organic materials directly to the soil, the process offers several distinct advantages: (1) Composting provides a means of safely storing organic materials with a minimum of odor release until it is convenient to apply them to soils. (2) Compost is easier to handle than the raw materials as a result of the 30 to 60% smaller volume and greater uniformity of the resulting material. (3) For residues with a high initial C/N ratio, proper composting ensures that any nitrate depression period (see Figure 11.6) will occur in the compost pile, not in the soil, thereby avoiding induced plant N deficiency. (4) When applied to the soil, composted materials generally decompose and mineralize much more slowly than uncomposted organic materials. (5) **Co-composting** low-C/N-ratio materials (such as livestock manure and sewage sludge) with high-C/N-ratio materials (such as sawdust, wood chips, senescent tree leaves, or municipal solid waste) provides sufficient carbon for microbes to immobilize the excess N and minimize any nitrate leaching hazard (see Section 12.8) from the low-C/N materials. It also provides sufficient N to speed the decomposition of the high-C/N materials. (6) High temperatures during the thermophilic stage in well-managed compost piles kill most weed seeds and pathogenic organisms in a matter of a few days. Under less ideal conditions, temperatures in parts of the pile may not exceed 40 to 50°C, so weeks or months may be required to achieve the same results. (7) Most toxic compounds that may be in organic wastes (pesticides, natural phytotoxic chemicals, etc.) are destroyed by the time the compost is mature and ready to use. Composting is even used as a method of biological treatment of polluted soils and wastes (see Section 15.5). (8) Some composts contain microorganisms that can effectively suppress soilborne plant diseases (see Section 10.13). Most success in disease suppression has occurred when well-cured compost is used as a main component of potting mixes for greenhouse-grown plants. (9) Because compost is made from organic waste materials that recently used up CO_2 in the process of their production (plant photosynthesis), compost is considered carbon-neutral, making it a much more environmentally sustainable choice than peat.

Disadvantages of Compost. Compost often has low nutrient contents and very slow availability of the nutrients present. It also usually has a relatively high P to N ratio in comparison to plant needs. Because of this ratio, attempts to use compost as the principal source of plant nutrients can easily result in the application of potentially polluting levels of phosphorus (see Section 13.2). In addition, because the more labile organic substances are broken down during the composting process, compost usually provides less benefit to soil aggregation than would the fresh residues from which the compost was made.

11.12 CONCLUSION

Organic matter is a complex and dynamic soil component that exerts a major influence on soil behavior, properties, and functions in the ecosystem. Because of the enormous amount of carbon stored in SOM and the dynamic nature of this soil component, soil management may be an important tool for moderating human-driven climate change.

Organic residue decay, nutrient release, and humus formation are controlled by environmental factors, soil conditions, microbial activities, and by the quality of the organic materials. High contents of lignin and polyphenols, along with high C/N ratios, markedly slow the initial decomposition process, causing organic residues to remain longer while reducing the availability of nutrients.

Soil organic matter can be thought of as comprising two major pools of organic compounds. The *active pool* consists of microbial biomass, soluble carbon compounds, unprotected bits of plant and microbial cells (free POM), and relatively easily decomposed compounds, such as polysaccharides and other high-energy content biomolecules. Although only a small percentage of the total carbon, the active pool plays a major role in nutrient cycling, micronutrient chelation, maintenance of structural stability and soil tilth, and as a food source underpinning biological diversity and activity in soils.

Most of the organic matter is in the *passive pool*, which mainly contains materials that are protected in such a way as to be inaccessible to microbial attack and may persist in the soil for centuries. In many soils, bits of fire-blackened char are also important components of the passive pool. Most char is quite stable and resistant to decomposition, as well as highly porous. Taken together, the passive pool provides cation exchange and water-holding capacities, but is relatively inert biologically. Its relative inertness can change if soil conditions change. When soil is cleared of natural vegetation and brought under cultivation, the initial decline in SOM is principally at the expense of the active pool. The passive pool is depleted more slowly and over very long periods of time.

The carbon-to-nitrogen (C/N) ratio of most soils is relatively constant, generally near 12:1. This means that the level of organic matter will be partially determined by the level of N available for assimilation into humus. Soil management for enhancing long-term organic matter levels must, therefore, include some means of supplying N, for example, by the inclusion of leguminous plants.

The level of SOM is influenced by climate (being higher in cool, moist regions), drainage (being higher in poorly drained soils), and by vegetation type (being generally higher where root biomass is greatest, as under perennial grasses).

The maintenance of SOM, especially the active pools, in mineral soils is one of the great challenges in natural resource management around the world. By encouraging vigorous growth of crops or other vegetation, abundant residues (which contain both C and N) can be returned to the soil directly or through feed-consuming animals. Also, the rate of destruction of SOM can be minimized by restricting soil tillage, controlling erosion, and keeping most of the plant residues at or near the soil surface.

For some purposes it is advantageous to manage the decomposition of organic matter outside of the soil in a process known as *composting*. Composting transforms various organic waste materials into a dark colored, pleasant smelling product that can be used as a soil amendment or a component of potting mixes. The aerobic decomposition in a compost pile can conserve nutrients while avoiding certain problems, such as noxious odors and the presence of either excessive or deficient quantities of soluble N, which can occur if fresh organic wastes are applied directly to soils.

The decay and mineralization of SOM is one of the main processes governing the economy of nitrogen and sulfur in soils—the subjects of the next chapter.

STUDY QUESTIONS

1. Compare the amounts of carbon in Earth's standing vegetation, soils, and atmosphere.
2. If you wanted to apply an organic material that would make a long-lasting mulch on the soil surface, you would choose an organic material with what chemical and physical characteristics?
3. Describe how the addition of certain types of organic materials to soil can cause a nitrate depression period. What are the ramifications of this phenomenon for plant growth?
4. What types or categories of organic materials are found in soils?
5. Some scientists include plant litter (surface residues) in their definition of SOM, while others do not. Write two brief paragraphs, one justifying the inclusion of litter as SOM and one justifying its exclusion.
6. What soil properties are mainly influenced by the active and passive pools, respectively, of organic matter?
7. In this book and elsewhere, the terms *SOC* and *soil organic matter* are used to mean almost the same thing. How are these terms related, conceptually and quantitatively? Why is the term *organic carbon* generally more appropriate for quantitative scientific discussions?
8. Explain, in terms of the balance between gains and losses, why cultivated soils generally contain much lower levels of organic carbon than similar soils under natural vegetation.
9. What is char, how is it produced, and what roles does it play as a component of SOM?
10. In what ways are soils involved in the greenhouse effect that is warming up the Earth? What are some common soil-management practices that could be changed to reduce the negative effects and increase the beneficial effects of soils on the greenhouse effect?
11. Explain why compost is more environmentally sustainable than peat for use in potting media and as an amendment for golf course greens.

REFERENCES

Bardgett, R. D., and K. F. Chan. 1999. "Experimental evidence that soil fauna enhance nutrient mineralization and plant nutrient uptake in Montane grassland ecosystems." *Soil Biology and Biochemistry* 31:1007–1014.

Batjes, N. H. 1996. "Total carbon and nitrogen in the soils of the world." *European Journal of Soil Science* 47:151–163.

Blanco-Canqui, H., R. Lal, W. M. Post, R. C. Izaurralde, and L. B. Owens. 2006. "Rapid changes in soil carbon and structural properties due to stover removal from no-till corn plots." *Soil Science* 171:468–482.

Blanco-Canqui, H., C. A. Shapiro, C. S. Wortmann, R. A. Drijber, M. Mamo, T. M. Shaver, and R. B. Ferguson. 2013. "Soil organic carbon: The value to soil properties." *Journal of Soil and Water Conservation* 68:129A–134A.

Cadisch, G., and K. E. Giller (eds.). 1997. *Driven by Nature—Plant Litter Quality and Decomposition.* CAB International, Wallingford, UK.

Campbell, E. E., and K. Paustian. 2015. "Current developments in soil organic matter modeling and the expansion of model applications: A review." *Environmental Research Letters* 10:123004.

Canellas, L., D. Balmori, L. Médici, N. Aguiar, E. Campostrini, R. Rosa, A. Façanha, and F. Olivares. 2013. "A combination of humic substances and *Herbaspirillum seropedicae* inoculation enhances the growth of maize (*Zea mays* L.)." *Plant and Soil* 366:119–132.

Carlile, W. R., C. Cattivello, and P. Zaccheo. 2015. "Organic growing media: Constituents and properties." *Vadose Zone Journal* 14.

Cipollini, D., C. Rigsby, and E. K. Barto. 2012. "Microbes as targets and mediators of allelopathy in plants." *Journal of Chemical Ecology* 38:714–727.

Coppens, F., P. Garnier, A. Findeling, R. Merckx, and S. Recous. 2007. "Decomposition of mulched versus incorporated crop residues: Modelling with PASTIS clarifies interactions between residue quality and location." *Soil Biology and Biochemistry* 39:2339–2350.

Curtin, D., M. H. Beare, and G. Hernandez-Ramirez. 2012. "Temperature and moisture effects on microbial biomass and soil organic matter mineralization." *Soil Science Society of America Journal* 76:2055–2067.

Donato, D. C., J. B. Kauffman, D. Murdiyarso, S. Kurnianto, M. Stidham, and M. Kanninen. 2011. "Mangroves among the most carbon-rich forests in the tropics." *Nature Geoscience* 4:293–297.

Eswaran, H., P. F. Reich, J. Kimble, F. H. Beinroth, E. Padmanabhan, and P. Moncharoen. 2000. "Global carbon stocks." In R. Lal et al. (eds.). *Global Climate Change and Pedogenic Carbonates.* Lewis Publishers, Boca Raton, FL, pp. 15–26.

Ferris, H., R. C. Venette, H. R. van der Meulen, and S. S. Lau. 1998. "Nitrogen mineralization by bacterial-feeding nematodes: Verification and measurement." *Plant and Soil* 203:159–171.

Haddix, M. L., A. F. Plante, R. T. Conant, J. Six, J. M. Steinweg, K. Magrini-Bair, R. A. Drijber, S. J. Morris, and E. A. Paul. 2011. "The role of soil characteristics on temperature sensitivity of soil organic matter." *Soil Science Society of America Journal* 75:56–68.

Harden, J. W., C. D. Koven, C.-L. Ping, G. Hugelius, A. D. McGuire, P. Camill, T. Jorgenson, P. Kuhry, G. J. Michaelson, J. A. O'Donnell, E. A. G. Schuur, C. Tarnocai, K. Johnson, and G. Grosse. 2012. "Field information links permafrost carbon to physical vulnerabilities of thawing." *Geophysical Research Letters* 39:L15704.

Hart, S., and N. Luckai. 2013. "Charcoal function and management in boreal ecosystems." *Journal of Applied Ecology* 50:1197–1206.

Inderjit, D., A. Wardle, R. Karban, and R. M. Callaway. 2011. "The ecosystem and evolutionary contexts of allelopathy." *Trends in Ecology & Evolution* 26:655–662.

IPCC. 2007. "Climate change 2007: The physical science basis. Summary for policymakers." [Online]. Intergovernmental Panel on Climate Change, United Nations. Available at http://www.ipcc.ch/report/ar4/wg1/.

Jenkinson, D. S., and A. E. Johnson. 1977. "Soil organic matter in the Hoosfield barley experiment." *Report of Rothamsted Experimental Station 1976* 2:87–102.

Johnston, A. E., P. R. Poulton, K. Coleman, and L. S. Donald. 2009. "Soil organic matter: Its importance in sustainable agriculture and carbon dioxide fluxes." *Advances in Agronomy* 101. Academic Press, London, pp. 1–57.

Kleber, M., and M. G. Johnson. 2010. "Advances in understanding the molecular structure of soil organic matter: Implications for interactions in the environment." *Advances in Agronomy* 106:77–142.

Laird, D. A., and C.-W. Chang. 2013. "Long-term impacts of residue harvesting on soil quality." *Soil and Tillage Research* 134:33–40.

Lal, R. 2018. "Digging deeper: A holistic perspective of factors affecting soil organic carbon sequestration in agroecosystems." *Global Change Biology* (doi:10.1111/gcb.14054).

Lehmann, J., and M. Kleber. 2015. "The contentious nature of soil organic matter." *Nature* 528:60–68.

Mackey, B., I. C. Prentice, W. Steffen, J. I. House, D. Lindenmayer, H. Keith, and S. Berry. 2013. "Untangling the confusion around land carbon science and climate change mitigation policy." *Nature Climate Change* 3:552–557.

Magdoff, F., and R. R. Weil (eds.). 2004. *Soil Organic Matter in Sustainable Agriculture*. CRC Press, Boca Raton, FL, p. 398.

Nafziger, E. D., and R. E. Dunker. 2011. "Soil organic carbon trends over 100 years in the morrow plots." *Agronomy Journal* 103:261–267.

Norse, D. 2012. "Low carbon agriculture: Objectives and policy pathways." *Environmental Development* 1:25–39.

Pan, Y., R. A. Birdsey, J. Fang, R. Houghton, P. E. Kauppi, W. A. Kurz, O. L. Phillips, A. Shvidenko, S. L. Lewis, J. G. Canadell, P. Ciais, R. B. Jackson, S. W. Pacala, A. D. McGuire, S. Piao, A. Rautiainen, S. Sitch, and D. Hayes. 2011. "A large and persistent carbon sink in the world's forests." *Science* 333:988–993.

Paustian, K., and B. Babcock. 2004. "Climate change and greenhouse gas mitigation: Challenges and opportunities." Task Force Report 141. Council on Agricultural Science and Technology, Ames, IA, p. 120.

Redeker, K., N. Wang, J. Low, A. McMillan, S. Tyler, and R. Cicerone. 2000. "Emissions of methyl halides and methane from rice paddies." *Science* 290:966–969.

Scharlemann, J. P. W., R. Hiederer, V. Kapos, and C. Ravilious. 2010. "Updated global carbon map." United Nations Environment Programme World Conservation Monitoring Centre, Cambridge, UK. http://esdac.jrc.ec.europa.eu/ESDB_Archive/octop/Resources/Global_OC_Poster.pdf.

Schmidt, M. W. I., and A. G. Noack. 2000. "Black carbon in soils and sediments: Analysis, distribution, implications, and current challenges." *Global Biogeochemical Cycles* 14:777–793.

Schmidt, M. W. I., M. S. Torn, S. Abiven, T. Dittmar, G. Guggenberger, I. A. Janssens, M. Kleber, I. Kogel-Knabner, J. Lehmann, D. A. C. Manning, P. Nannipieri, D. P. Rasse, S. Weiner, and S. E. Trumbore. 2011. "Persistence of soil organic matter as an ecosystem property." *Nature* 478:49–56.

Seghers, D., S. D. Siciliano, E. M. Top, and W. Verstraete. 2005. "Combined effect of fertilizer and herbicide applications on the abundance, community structure and performance of the soil methanotrophic community." *Soil Biology and Biochemistry* 37:187–193.

Senesi, N., B. Xing, and P.M. Huang. 2009. *Biophysico-Chemical Processes Involving Natural Nonliving Organic Matter in Environmental Systems*. John Wiley & Sons, Hoboken, NJ, p. 876.

Singh, N., S. Abiven, M. S. Torn, and M. W. I. Schmidt. 2012. "Fire-derived organic carbon in soil turns over on a centennial scale." *Biogeosciences* 9:2847–2857.

Torbert, H., and H. Johnson. 2001. "Soil of the intensive agriculture biome of Biosphere 2." *Journal of Soil Water Conservation* 56:4–11.

Walford, R. L. 2002. "Biosphere 2 as voyage of discovery: The serendipity from inside." *BioScience* 52:259–263.

Weil, R. R., K. R. Islam, and C. L. Mulchi. 2000. "Impact of elevated CO_2 and ozone on C cycling processes in soil." *Agronomy Abstracts*. American Society of Agronomy, Madison, WI, p. 47.

12
Nutrient Cycles and Soil Fertility

A rich black mold that gave promise of wonderful fertility.
—EDVART ROLVAAG, FROM
GIANTS IN THE EARTH

Rhizobium-Crimson Clover nodules cycle nitrogen in soil. (Photo courtesy of Ray R. Weil)

Soils are at the very hub of the biogeochemical cycles that transform, transport, and renew the supplies of the mineral nutrients so essential for the growth of terrestrial plants. As each nutrient cycles through the soil, a given atom may appear in many different chemical forms, each with its own properties, behaviors, and consequences for soil development and for the ecosystem. For some elements, such as nitrogen (N) and sulfur (S), the cycles are exceedingly complex, involving many different biologically mediated transformations and movement into and out of the soil as solid particles, in solution, and as gases. The cycling of phosphorus (P) also involves a fascinating set of complex interactions among chemical and biological processes. For calcium, magnesium, and potassium, the weathering of minerals and cation exchange reactions dominate the cycles. For the micronutrients, mobility and bioavailability are controlled mainly by soil pH, redox potential, and reactions with soluble organic compounds produced by plants and microbes.

These cycles impact not only soil fertility but also the health of aquatic ecosystems and the health and survival of humans on Earth. This cycling of nutrients explains why vegetation (and, indirectly, animals) can continue to remove nutrients from a soil for millennia without depleting the soil of its supply of essential elements. The biosphere does not quickly run out of such nutrients as N or magnesium because it uses the same supply over and over again. When human activities shortcircuit or break these cycles, soils do become impoverished, as do the people who depend upon them. It is also worth noting that the impacts of mismanaging nutrient cycles in soils are not confined to terrestrial systems. In fact, undue leakage of N and P from the soil phase of their cycles is responsible for some of the most devastating water pollution problems on the planet.

Scarcity of N is the most widely occurring nutritional limit on the productivity of terrestrial ecosystems. Nitrogen is also the element most widely overapplied to agroecosystems and the most widely responsible for deterioration of water quality. Phosphorus is the second most widely limiting nutrient and is generally even more scarce than N. It, too, has a history of overapplication in modern agriculture and therefore has become responsible for widespread pollution of aquatic systems. As crop production expands to keep pace with the rapidly growing global demand for food, the increased removal of S and micronutrients is leading to more widespread deficiencies of these nutrients, as well. In this chapter we discuss the soil processes and principles governing the nutrient cycles. We will also develop the understandings needed to recognize and deal effectively with the varied plant deficiencies and toxicities encountered in a wide range of soil–plant systems. Then in Chapter 13, we will apply this knowledge to the practical management of soil fertility and environmental quality and in Chapter 15 we will consider soil contamination by toxic levels of trace elements, including certain of the micronutrients.

12.1 NITROGEN IN THE SOIL SYSTEM

Nitrogen and Plant Growth and Development

Healthy plant foliage generally contains 2.0 to 4.0% N, mainly as an integral component of essential plant compounds. It is a major part of all proteins—including the enzymes that control virtually all biological processes. Other critical nitrogenous plant components include the nucleic acids and chlorophyll.

Plants deficient in N tend to exhibit chlorosis (yellowish or pale green leaf colors), a stunted appearance, and thin, spindly stems (Figure 12.1). The older leaves of N-starved plants are the first

Figure 12.1 *Symptoms of too little and too much nitrogen. (a) The N-starved bean plant (right) shows the typical chlorosis of lower leaves and markedly stunted growth compared to the normal plant on the left. (b) The oldest leaves at right near the base of this potted cucurbit vine are chlorotic because N has been transferred to the newest leaves on the left which are dark green. (c) The oldest bottom corn leaves yellowed, beginning at the tip and continuing down the midrib—a pattern typical of N deficiency in corn. (d) Asian rice heavily fertilized with N. The traditional tall cultivar in the left field has lodged (fallen over), but the modern rice cultivar at right yields well with high N inputs because of its short stature and stiff straw.* (Photo courtesy of Ray R. Weil)

to turn yellowish, typically becoming prematurely senescent and dropping off (Figures 12.1a,b). Nitrogen-deficient plants often show low shoot/root ratio and early maturity.

When too much N is available, excessive vegetative growth occurs, top-heavy plants are prone to falling over (lodging, Figure 12.1d), plant maturity may be delayed, and the plants may become more susceptible to disease (particularly fungal disease) and to insect pests. These problems are especially noticeable if other nutrients, such as potassium, are in relatively low supply.

Forms of Nitrogen Taken Up by Plants. Plant roots take up N from the soil principally as dissolved nitrate (NO_3^-) and ammonium (NH_4^+) ions. As explained in Figure 8.11, uptake of ammonium markedly lowers the pH of the rhizosphere soil, while uptake of nitrate tends to have the opposite effect. These pH changes, in turn, influence the uptake of other ions such as phosphates and micronutrients. Numerous plants have been shown to also take up dissolved organic N compounds, a phenomenon of particular significance in natural grasslands and forests.

Distribution and Cycling of Nitrogen[1]

The financial and environmental stakes for the management of soil N are very high, indeed. The quest by billions of humans for protein—which contains about one-sixth N—is fueling demand for N fertilizer and transforming both global agriculture and the global N cycle. Nitrogen excesses are now as common as deficiencies and can adversely affect both human and ecosystem health. Nitrogen leaking from overly enriched soils can lead to nitrate pollution of groundwater and eutrophication of aquatic systems. In the latter, N stimulates algae blooms, declining levels of dissolved oxygen, and subsequent death of fish and other aquatic species. Yet another way in which N links soils to the wider environment is the ozone-destroying and climate-forcing action of nitrous oxide gas emitted from soils.

The atmosphere, which is 78% gaseous nitrogen (N_2), appears to be a virtually limitless reservoir of this element. However, the extremely stable triple bond between two nitrogen atoms ($N \equiv N$) makes this gas extremely inert and not directly usable by plants or animals. Little N would be found in soils and little vegetation would grow in terrestrial ecosystems around the world were it not for certain natural processes (principally N fixation by microbes and lightning) that can break this triple bond and form **reactive nitrogen**, which includes any N compound that readily participates biological and chemical reactions.

The nitrogen cycle has long been the subject of intense scientific investigation, because understanding the translocations and transformations of this element is fundamental to solving many environmental, agricultural, and natural resource-related problems. The **biosphere** does not run out of N (or other nutrient elements) because it uses the same atoms over and over again. This cycling of N is highly visible in Figure 10.25, which shows a dark green circle of grass plants that have responded to the fungal release of N from the residues of previous grass growth.

The principal pools and forms of N, and the processes by which they interact in the cycle, are illustrated in Figure 12.2. This figure deserves careful study; we will refer to it frequently as we discuss each of the major processes of the N cycle. Ammonium (NH_4^+) and nitrate (NO_3^-) are two critical forms of inorganic N which are subject to various fates in the N cycle (see the arrows *leaving* the nitrate and ammonium boxes in Figure 12.2).

Most of the N in terrestrial systems is found in the soil, with A horizons normally ranging from 0.02 to 0.5% N by weight. Nitrogen is present in soils mainly as part of organic molecules, and therefore, the distribution of soil N closely parallels that of soil organic matter (SOM), which typically contains about 5% N (see Section 11.3). Except where large amounts of chemical fertilizers or livestock manure have been applied, inorganic (i.e., mineral) N seldom accounts for more than 1 to 2% of the total N in the soil. Unlike the bulk of organic N, most mineral forms of N are quite soluble in water and can be lost from soils through leaching and volatilization.

In addition to its possible loss by erosion and runoff (as well as some leaching), the N in NH_4^+ can be subject to at least six fates: (1) *immobilization* by microorganisms; (2) removal by *plant uptake*; (3) *anammox*—the anaerobic oxidation of NH_4^+ in conjunction with nitrite

---

[1]To read about the evolution of Earth's nitrogen cycle, its disruption during the past century by industrial agriculture, and its inevitable microbial rebalancing, with or without human adaptation, see Canfield et al. (2010).

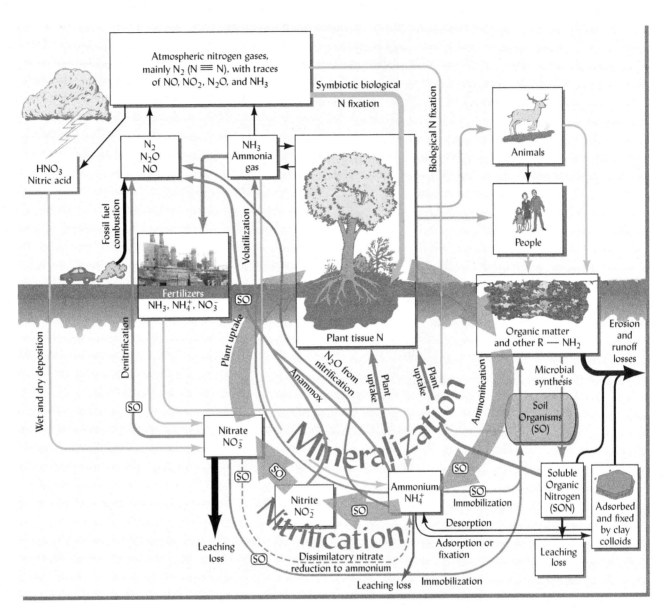

Figure 12.2 The N cycle encompasses many translocations and transformations. Here it is shown emphasizing the primary cycle in aerobic soils (thick olive green arrows) in which microbes mineralize organic N from plant residue, plants take up the mineral N, and eventually return organic N to the soil as fresh plant residues. In this regard N is like any of the other plant nutrients that cycle through soils. However, because N exists in many valence states and in all three phases of matter (gas, liquid, and solid), its interactions and cycling pathways are much more complex than just this basic cycle. The boxes represent various forms of N. The arrows represent processes by which one form is transformed into another. The name of a process (e.g., Mineralization or Anammox) is given alongside the arrow. Note the processes by which N is lost from the soil (thick blue and black arrows) and by which it is replenished (bright green arrows). The blue arrows represent anaerobic processes. Soil organisms, whose enzymes drive most of the reactions in the cycle, are represented as rounded boxes labeled "SO." (Diagram courtesy of Ray R. Weil)

(NO_2^-) to produce N_2O gas; (4) *volatilization* after being transformed into ammonia gas; (5) *nitrification*—the microbial oxidation of ammonium to nitrite and subsequently to nitrate; and (6) *fixation* or strong sorption in the interlayers of certain 2:1 clay minerals.

Similarly, in addition to loss by erosion and runoff, the N in NO_3^- is subject to six possible fates (note that some of these fates are shared with the N in NH_4^+): (1) *immobilization* by microorganisms; (2) removal by *plant uptake*; (3) microbial reduction to NO_2^- followed by conversion, along with ammonium, by *anammox* to N_2O gas; (4) *denitrification*—microbial reduction forming N_2 and other N-containing gases which are lost to the atmosphere; (5) *dissimilatory reduction* by microbes to ammonium; and (6) loss to groundwater by *leaching* in drainage water.

Nitrogen Immobilization and Mineralization

Most N containing organic compounds are protected from loss but also unavailable for plant uptake. Microbial decomposition breaks large, insoluble N-containing organic molecules into smaller and smaller units with the eventual production of simple amine groups ($R-NH_2$). Then the amine groups are hydrolyzed, and the N is released as ammonium ions (NH_4^+) which can be oxidized to the nitrate form. The enzymes that bring about this process are produced mainly by microorganisms. Most often they are excreted by the microbes and work extracellularly in the soil solution or while adsorbed to colloidal surfaces. This enzymatic process termed **mineralization** (Figure 12.2) may be indicated as follows, using an amino compound ($R-NH_2$) as an example of the organic N source:

$$\xrightarrow{\quad\quad\quad\quad\text{Mineralization}\quad\quad\quad\quad}$$

$$R-NH_2 \underset{-2H_2O}{\overset{+2H_2O}{\rightleftharpoons}} OH^- + R-OH + NH_4^+ \underset{-O_2}{\overset{+O_2}{\rightleftharpoons}} 4H^+ + \text{energy} + NO_2^- \underset{-\frac{1}{2}O_2}{\overset{+\frac{1}{2}O_2}{\rightleftharpoons}} \text{energy} + NO_3^- \quad\quad (12.1)$$

$$\xleftarrow{\quad\quad\quad\quad\text{Immobilization}\quad\quad\quad\quad}$$

Typically, about 1.5 to 3.5% of the organic N in A horizon soil mineralizes annually (Box 12.1). In most soils, this rate of mineralization provides sufficient mineral N for normal growth of natural vegetation; in soils with relative high levels of organic matter, mineralization

---

BOX 12.1
CALCULATION OF NITROGEN MINERALIZATION

If the organic matter content of a soil, management practices, climate, and soil texture are known, it is possible to make a rough estimate of the amount of N likely to be mineralized each year. To do so, we make the following assumptions:

- The soil organic matter (SOM) concentration in the upper 15 cm of soil (in kg SOM per 100 kg soil or % SOM) may range from close to zero to over 75% (in a Histosol) (see Section 3.9). Values between 0.5 and 5% are most common. ☞ Use a value of 2.5% (2.5 kg SOM/100 kg soil) in our example.
- Most N used by plants is likely to come from the upper horizon. If this is 15 cm deep, 2×10^6 kg/ha is a reasonable estimate of its mass. See Section 4.7 to calculate the weight of such a soil layer if bulk density of a soil is known. In some soils, it may be important to also consider deeper layers. ☞ Use 2×10^6 kg soil/ha 15 cm deep for our example.
- Assume a typical concentration of N in the SOM (see Section 12.3) of about 5 kg N/100 kg SOM.
- The amount of SOM likely to be mineralized in one year for a given soil depends upon the soil texture, climate, and management practices. Values of around 2% are typical for a fine-textured soil, while values of around 3.5% are typical for coarse-textured soils. Slightly higher values are typical in warm climates; slightly lower values are typical in cool climates. ☞ Assume a value of 2.5 kg SOM mineralized/100 kg SOM for our example.

The amount of N likely to be released by mineralization during a typical growing season may be calculated by substituting the given example values:

$$\frac{\text{kg N mineralized}}{\text{ha 15 cm deep}} = \left(\frac{2.5 \text{ kg SOM}}{100 \text{ kg soil}}\right)\left(\frac{2 \times 10^6 \text{ kg soil}}{\text{ha 15 cm deep}}\right)$$
$$\left(\frac{5 \text{ kg N}}{100 \text{ kg SOM}}\right)\left(\frac{2.5 \text{ Kg SOM mineralized}}{100 \text{ Kg SOM}}\right)$$

$$\frac{\text{kg N mineralized}}{\text{ha}} = \left(\frac{2.5}{100}\right)\left(\frac{2 \times 10^6}{1}\right)\left(\frac{5}{100}\right) \quad (12.2)$$
$$\left(\frac{2.5}{100}\right) = 62.5 \text{ kg N/ha}$$

Most N mineralization occurs during the growing season when the soil is relatively moist and warm. Contributions from the deeper layers of this soil might be expected to bring total N mineralized in the root zone of this soil during a growing season to over 120 kg N/ha.

These calculations estimate the N mineralized annually from a soil that has not had large amounts of organic residues added to it. Animal manures, legume residues, or other N-rich organic soil amendments would mineralize much more rapidly than the native SOM and thus would substantially increase the amount of N available in the soil.

may provide enough mineral N for good crop yields, as well. Furthermore, *isotope tracer studies* of farm soils that have been amended with synthetic N fertilizers show that mineralized soil N constitutes a major part of the N taken up by fertilized crops.

The opposite of mineralization is **immobilization**, the conversion of inorganic N ions (NO_3^- and NH_4^+) into organic forms (see Eq. (12.1) and Figure 12.2). Most N immobilization occurs biologically when microorganisms decomposing organic residues require more N than they can obtain from the residues they are metabolizing. The microorganisms then scavenge NO_3^- and NH_4^+ ions from the soil solution to incorporate into their cellular component, leaving the soil solution essentially devoid of mineral N. When the organisms die, some of the dead cells and the organic N compounds they contain may be become sorbed to clay surfaces and stabilized as part of the SOM. However, some organic N may be released as NO_3^- and NH_4^+ ions. Mineralization and immobilization occur simultaneously in the soil; whether the *net* effect is an increase or a decrease in the mineral N supply depends primarily on the C/N ratio in the organic residues undergoing decomposition (see Section 11.3).

Ammonium Fixation by Clay Minerals

Like other positively charged ions, ammonium ions are attracted to the negatively charged surfaces of clay and humus, where they are held in exchangeable form, available for plant uptake, but partially protected from leaching (see Section 8.7). However, because of its particular size, the ammonium ion (as well as the similarly sized potassium ion) can become entrapped or *fixed* within cavities in the crystal structure of certain 2:1-type clay minerals. Vermiculite has the greatest capacity for such ammonium (and potassium) fixation, followed by fine-grained micas and some smectites (see Figures 8.16 and 12.43 and 12.44). Ammonium and potassium ions "fixed" in the rigid part of a crystal structure are held in a nonexchangeable form, from which they are released only slowly.

Ammonium fixation by clay minerals is generally greater in subsoil than in topsoil, due to the higher clay content of subsoils. In soils with considerable 2:1 clay content, interlayer-fixed NH_4^+ typically accounts for 5 to 10% of the total N in the surface soil and up to 20 to 40% of the N in the subsoil. While ammonium fixation may be considered an advantage because it provides a means of conserving N, the rate of release of the fixed ammonium is often too slow to be of much practical value in fulfilling the needs of fast-growing annual plants.

Dissolved Organic Nitrogen[2]

Plant uptake and leaching losses in both natural and agroecosystems may involve N-containing organic compounds that either are dissolved in the soil solution (dissolved organic nitrogen, DON) or can be easily extracted out of soil samples using water or simple salt solutions (soluble organic nitrogen, SON). The DON generally accounts for between 0.1 and 3% of the total N in soils; the SON pool is usually two to three times larger.

Plant Uptake of DON In nitrogen-limited ecosystems, such as those in strongly acidic and infertile soils (including some organic soils), DON may be the primary source of absorbed N. This helps explain the fact that plant growth, particularly of some forest species, is considerably greater than one would expect based on the limited supply of inorganic N at any one time. Low molecular weight compounds in the DON, such as amino acids, may be taken up directly by plant roots, or they may be assimilated through mycorrhizal associations. Microbial uptake of both DON and mineral N may occur simultaneously in a soil and may cause direct competition between plants and microbes for both forms of N.

Leaching Losses of DON Dissolved organic N is also a significant component of the N lost by leaching. For example, DON may comprise nearly all the N leached from some low-nitrogen grasslands, forests and urban turfgrass systems (Figure 12.3), and typically 30 to 60% of that leached

---

[2]For reviews of the nature, significance, and analyses of soluble organic nitrogen, see Ros et al. (2009) for a focus on natural ecosystems and van Kessel et al. (2009) for agricultural soils.

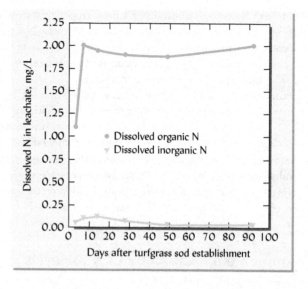

Figure 12.3 *Concentrations of dissolved organic and inorganic N in leachate from a sandy soil supporting young St. Augustine grass sod. Values represent means of three fertilizer treatments as there was no significant difference between the unfertilized turf and turf fertilized with urea or ammonium sulfate. Rain plus irrigation provided at least 2.5 cm water per week to this low nitrate system.* [Graphed from data in Lusk et al. (2018)].

from cropland on livestock farms. It constitutes some 25% of the N carried by the Mississippi River to the Gulf of Mexico. Thus, DON likely contributes to the environmental problems downstream and should be studied along with nitrate N to understand and solve N-pollution problems.

Ammonia Volatilization

Ammonia gas (NH_3) can be produced from the breakdown of plant residues, animal excrement, and from such fertilizers as anhydrous ammonia and urea. This gas may diffuse into the atmosphere, resulting in a loss of valuable N from the soil and environmentally detrimental increases in N deposition from the atmosphere. The ammonia gas is in equilibrium with ammonium ions according to the following reversible reaction:

$$\underset{\text{Dissolved ions}}{NH_4^+ + OH^-} \rightleftharpoons \underset{\text{Gas}}{H_2O + NH_3\uparrow} \qquad (12.3)$$

From Reaction (12.3) we deduce that 1) ammonia volatilization will be more pronounced at high pH levels (i.e., OH^- ions drive the reaction to the right) and 2) ammonia gas-producing amendments or the addition of water will drive the reaction to the left, increasing the concentration of OH^- ions and thus raising the pH of the solution in which they are dissolved.

Soil colloids, both clay and humus, adsorb ammonia gas, so ammonia losses are greatest where low quantities of these colloids are present or where the ammonia is not in close contact with the soil. For these reasons, ammonia losses can be quite large from sandy soils and from alkaline or calcareous soils, especially when the ammonia-producing materials are left at or near the soil surface and when the soil is drying out. High temperatures, as often occur on the surface of the soil, also favor the volatilization of ammonia.

Incorporation of manure and fertilizers into the top few centimeters of soil can reduce ammonia losses by 25 to 75% from those that occur when the materials are left on the soil surface. In natural grasslands and pastures, incorporation of animal wastes by earthworms and dung beetles is critical in maintaining a favorable N balance and a high animal-carrying capacity in these ecosystems (see Section 10.2). A well-timed application of irrigation water can greatly reduce the loss of NH_3 gas from surface-applied fertilizers.

Wetlands, including rice paddies, can lose much ammonia gas, especially on warm days when photosynthesizing algae use up all the dissolved CO_2 from the inundation water, thus removing carbonic acid and increasing the pH, commonly to above pH 9.0. As with upland soils, this ammonia loss can be reduced in rice paddies if the fertilizer is placed below the soil surface.

By the reverse of the ammonium loss mechanism just described, both soils and plants can absorb ammonia from the atmosphere. Thus the soil–plant system can help cleanse ammonia from the air, while deriving usable N for plants and soil microbes.

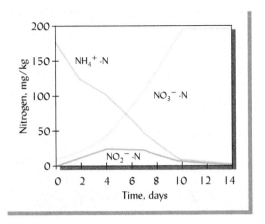

Figure 12.4 *Transformation of ammonium into nitrite and nitrate by nitrification. On day zero, the silt loam soil was amended with enough $(NH_4)_2SO_4$ to supply 170 mg of N/kg soil. It then underwent a warm, well-aerated incubation for 14 days. Every second day, soil samples were extracted and analyzed for various forms of N. Note that the increase in nitrate-N (NO_3^--N) almost mirrored the decline in ammonium-N (NH_4^+-N), except for the small amount of nitrite-N (NO_2^--N) that accumulated temporarily between days 2 and 10. This pattern is consistent with the two-step process depicted by Eqs. (12.4) and (12.5). No plants were grown during the study.* [Data selected from Khalil et al. (2004)]

Nitrification

As ammonium ions appear in the soil, they are generally oxidized quite rapidly by certain soil bacteria and archaea, yielding first nitrites and then nitrates. The prokaryotes that carry out this enzymatic oxidation are classed as **autotrophs** because they obtain their energy from oxidizing the ammonium ions rather than organic matter. The process termed **nitrification** (see Figure 12.2) consists of two main sequential steps. The first step results in the oxidation of ammonium to nitrite. Molecular research has shown that certain archaea (Crenarchaeota) may be active ammonia-oxidizing organisms in soils. In other soils, a specific group of autotrophic bacteria of the genus *Nitrosomonas* is thought to carry out most of the ammonia oxidation. In any case, the nitrite so formed is then immediately acted upon by a second group of autotrophs (the best known being the bacteria of the genus *Nitrobacter*). Therefore, when NH_4^+ is released into the soil it is usually converted rapidly into NO_3^- (see Figure 12.4). The enzymatic oxidation releases energy and may be represented very simply as follows:

Step 1

$$NH_4^+ + 1\tfrac{1}{2}O_2 \xrightarrow[\text{bacteria}]{\textit{Nitrosomonas}} NO_2^- + 2H^+ + H_2O + 275 \text{ kJ energy} \quad (12.4)$$
$$\text{Ammonium} \qquad\qquad\qquad\qquad\qquad \text{Nitrite}$$

Step 2

$$NO_2^- + \tfrac{1}{2}O_2 \xrightarrow[\text{bacteria}]{\textit{Nitrobacter}} NO_3^- + 76 \text{ kJ energy} \quad (12.5)$$
$$\text{Nitrite} \qquad\qquad\qquad \text{Nitrate}$$

So long as conditions are favorable for both reactions, the second transformation is thought to follow the first closely enough to prevent accumulation of much nitrite. This is fortunate, because even at concentrations of just a few mg/kg, nitrite is quite toxic to most plants. When oxygen supplies are marginal, the nitrifying bacteria may also produce some NO and N_2O, which are potent greenhouse gases. The chemical reaction shown in Eq. (12.6), which combines steps 1 and 2, illustrates the production of hydroxylamine and possible production of N_2O gas between the initial oxidation of NH_4^+ and the production of nitrite and nitrate during nitrification:

$$NH_4^+ \xrightarrow{O_2} \underset{\text{hydroxylamine}}{NH_2OH} \xrightarrow[\text{Ammonia oxidation}]{\nearrow N_2O \quad \tfrac{1}{2}O_2} NO_2^- \xrightarrow[\text{Nitrite oxidation}]{\nearrow N_2O \quad \tfrac{1}{2}O_2} NO_3^- \quad (12.6)$$
$$H_2O + 2H^+$$

Figure 12.5 (Left) Rates of nitrification, ammonification, and denitrification are closely related to the availability of oxygen and water as depicted by percentage of water-filled pore space. Both nitrification and ammonification proceed at their maximal rates near 55 to 60% water-filled pore space; however, ammonification proceeds in soils too waterlogged for active nitrification. Only a small overlap exists in the conditions suitable for nitrification and denitrification. (Right) Nitrous oxide (N_2O) is mainly produced by denitrification, but it is also a minor by-product of nitrification, with abrupt shift from one process to the other at water-filled porosities between 60 and 70%. (Bar graph from Bateman and Baggs (2005))

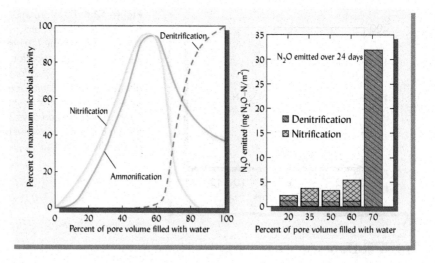

Regardless of the source of ammonium (i.e., ammonia-forming fertilizer, sewage sludge, animal excreta, or any other organic N source), nitrification will significantly increase soil acidity by producing H^+ ions, as shown in Reaction (12.6). See also Section 9.6.

Soil Conditions Affecting Nitrification

The nitrifying bacteria are much more sensitive to environmental conditions than the broad groups of heterotrophic organisms responsible for the release of ammonium from organic N compounds (**ammonification**). Nitrification requires a supply of ammonium ions, but excess NH_4^+ can be toxic to *Nitrobacter*. The nitrifying organisms, being aerobic, require oxygen to make NO_2^- and NO_3^- ions and are therefore favored in well-drained soils. The optimum moisture for these organisms is about the same as that for most plants (about 60% of the pore space filled with water, Figure 12.5). Since they are autotrophs, their carbon sources are bicarbonates and CO_2. They perform best when temperatures are between 20 and 30°C and perform very slowly if the soil is cold (below 5°C). A number of compounds, including several commercial agrochemicals, can inhibit or slow down the nitrification process, thereby reducing the potential for N loss by nitrate leaching and denitrification (see Figure 12.5).

Provided that all the just described conditions are favorable, nitrification proceed rapidly such that nitrate is generally the predominant mineral form of N in upland soils. Irrigation of an initially dry arid-region soil, the first rains after a long dry season, the thawing and rapid warming of frozen soils in spring, and sudden aeration by tillage are examples of environmental fluctuations that typically cause a flush of soil nitrate production (Figure 12.6). The growth patterns of natural vegetation and the optimum planting dates for crops are greatly influenced by such seasonal changes in nitrate levels.

Figure 12.6 Seasonal patterns of nitrate-N concentration in the A horizons of representative soils with and without growing plants in tropical region with four rainy months followed by eight months of dry, hot weather. A large flush of NO_3^--N appears when the rains first moisten the dry soil. This nitrate flush is caused by the activation of microbial enzyme production as well as the rapid mineralization of the dead cells of microorganisms previously killed by the dry, hot conditions. Note that soil nitrate is lower when plants are growing, because much of the nitrate formed is removed by plant uptake. (Diagram courtesy of Ray R. Weil)

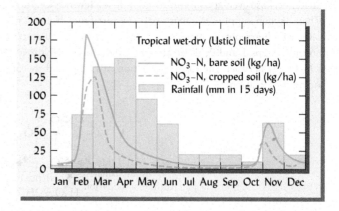

Gaseous Losses by Denitrification[3]

Nitrogen is commonly lost to the atmosphere when nitrate ions are converted to gaseous forms of N by a series of widely occurring biochemical reduction reactions termed **denitrification**.[4] The organisms that carry out this process are mostly facultative anaerobic bacteria and are generally present in large numbers. Most are *heterotrophs*, which obtain their energy and carbon from the oxidation of organic compounds, but some denitrifying bacteria are *autotrophs*, such as *Thiobacillus denitrificans*, which obtain their energy from the oxidation of sulfide. In the reaction, NO_3^- [N(V)] is reduced in a series of steps to NO_2^- [N(III)], and then to nitrogen gases that include NO [N(II)], N_2O [N(I)], and eventually N_2 [N(0)]:

$$2NO_3^- \xrightarrow{-2O} 2NO_2^- \xrightarrow{-2O} 2NO\uparrow \xrightarrow{-O} N_2O\uparrow \xrightarrow{-O} N_2\uparrow \quad (12.7)$$

Nitrate ions (+5) — Nitrite ions (+3) — Nitric oxide gas (+2) — Nitrous oxide gas (+1) — Dinitrogen gas (0) ← Valence state of nitrogen

Although not shown in the simplified reaction given here, the oxygen released at each step would be used to form CO_2 from organic carbon (or SO_4^{2-} from sulfides if *Thiobacillus* is the nitrifying organism).

These reactions require sources of energy (such as C in plant residues, or S in sulfide minerals), low to no oxygen conditions and warm temperatures (2 and 50°C, with optimum 25 to 35°C). Generally, when oxygen levels are very low, the end product released from the overall denitrification process is dinitrogen gas (N_2). However, NO and N_2O are commonly also released during denitrification (Figure 12.6), especially under the fluctuating aeration conditions that often occur in the field. Very strong acidity (pH < 5.0) inhibits rapid denitrification and favors the formation of N_2O.

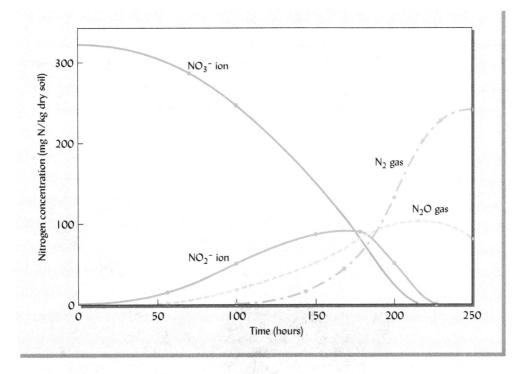

Figure 12.7 Changes in various forms of N during the process of denitrification in a moist soil incubated in the absence of atmospheric oxygen. [Data from Leffelaar and Wessel (1988)]

---

[3] For a review of N cycle pathways of importance in ecosystems, see Thamdrup (2012).
[4] Dissimilatory nitrate reduction to ammonium is another anaerobic bacterial process that reduces NO_3^- to NO_2^- and then to NH_4^+. Nitrate can also be reduced to nitrite and to nitrous oxide gas by nonbiological chemical reaction and under some circumstances by nitrifier bacteria, but these reactions are relatively minor in comparison with biological denitrification.

Anammox[5]

Another, less well-known, bacterial process, the **anaerobic oxidation of ammonium** (*anammox*), converts ammonium to N_2 gas using nitrite as an electron acceptor. The nitrite required for this reaction may come from nitrifying bacteria or archaea that carry out aerobic ammonium oxidation as described:

$$NH_4^+ + 1\tfrac{1}{2}O_2 \rightarrow NO_2^- + 2H^+ + H_2O \quad \text{(aerobic)} \tag{12.4}$$

$$NH_4^+ + NO_2^- \rightarrow N_2 + 2H_2O \quad \text{(anaerobic)} \tag{12.8}$$

$$2NH_4^+ + 1\tfrac{1}{2}O_2 \rightarrow N_2 + 2H^+ + 3H_2O \quad \text{(combined net anammox reaction)} \tag{12.9}$$

Anammox bacteria use CO_2 as their carbon source for growth and therefore a supply of organic carbon is not required for anammox to occur. As suggested by the anammox equations, the process requires the presence of linked oxidizing and reducing environments and typically occurs in redox transition zones such as the upper layers of saturated soils, the capillary fringe above the water table, the rhizosphere of plant roots, or the interior of macroaggregates in moderately wet soils (see Figure 7.10).

Atmospheric Pollution and Greenhouse Gas Emissions

Denitrification and anammox processes are ecologically important both because of the loss of N and because of the nitrogen gases emitted into the atmosphere. Nitrogen losses from soil during even a brief period of denitrification can bring about serious N deficiency in plants (Figure 12.8), while losses from N-laden groundwater or surface water can markedly improve the water quality. Dinitrogen (N_2) gas is quite inert and environmentally harmless, but the oxides of nitrogen are very reactive gases and have the potential to do serious environmental damage in at least four ways. First, when N_2O rises into the upper atmosphere, it contributes to climate change by absorbing infrared radiation (as much as 300 times the effect of an equal amount of CO_2) that would otherwise escape into space (see Section 11.9). Second, the nitrogen oxide gases can react with volatile organic pollutants to form ground-level ozone, a major air pollutant in the photochemical smog that plagues many urban areas. Third, NO and N_2O released into the atmosphere by denitrification can contribute to the formation of nitric acid, one of the principal components of acid rain. Fourth, N_2O from denitrification in soils makes a major contribution to the destruction of the Earth's ozone layer that protects us from cancer-causing ultraviolet radiation.

Figure 12.8 *Foliar symptoms indicate severe N deficiency in the corn growing on these well-fertilized, high organic matter Udolls in central Illinois. Heavy rains during warm weather ponded water on the low-lying parts of the field resulting in perfect conditions for denitrification to cause large losses of N. Some leaching of nitrate and dissolved organic N undoubtedly also occurred.*
(Photo courtesy of Ray R. Weil)

---

[5]For a review of the unique characteristics of anammox reactions and bacteria, see van Niftrik and Jetten (2012).

Quantity of Nitrogen Lost Through Denitrification In forest ecosystems during periods of adequate soil moisture, denitrification results in a slow but relatively steady loss of N. In agricultural soils the losses are highly variable in both time and space. The greater part of the annual N loss often occurs during just a few days in summer, when rain or irrigation water temporarily waterlog the warm, N-fertilized, carbon-rich soils (Figure 12.9).

Low-lying, organic-rich areas and other hot spots may lose N 10 times as fast as the average rate for a typical field. Although as much as 10 kg/ha of N may be lost in a single day from the sudden saturation of a well-drained, humid-region soil, such soils rarely lose more than 5 to 15 kg N/ha annually by denitrification. But where aeration is restricted and large amounts of N fertilizer are applied, losses of 30 to 60 kg N/ha/yr of N have been observed (Figure 12.8).

Denitrification in Flooded Soils In soils subject to alternate periods of wetting and drying, nitrates that are produced by nitrification during the dry periods are often subject to denitrification when the soils are submerged. Even when submerged, the soil permits both reactions to take place at once—nitrification occurs at the soil–water interface where some oxygen derived from the water is present and denitrification occurs at lower soil depths (see Figure 12.10). The resulting rapid loss of N is considered to be a beneficial function of wetlands, in that the process protects estuaries and lakes from the eutrophying effects of too much N. On the other hand, rice paddies are managed to minimize N losses by keeping the soil flooded and by deep placement of the fertilizer into the reduced zone of the soil. In this zone, because there is insufficient oxygen to allow nitrification to proceed, N remains in the ammonium form and is not susceptible to loss by denitrification.

Denitrification in Groundwater In many cases studied, contaminated groundwater loses most of its nitrate N load as it flows through the **riparian** zone on its way to the stream. Most of nitrate is believed to be lost by denitrification, stimulated by soluble C leached from the decomposing forest litter and by the anaerobic conditions that prevail in the wet riparian

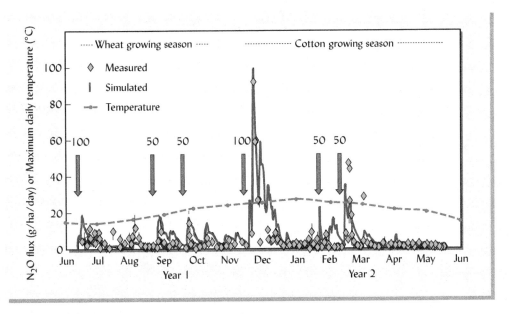

Figure 12.9 The episodic nature of denitrification in an agricultural soil. Most of the N_2O gas was emitted during just one month (December) when the soil was warm, a large amount of N had just been applied, the just-planted crop was not yet using much water, and irrigation made the soil quite wet. The arrows indicate the timing and amount (kg N/ha) of N fertilizer applications to a two-year wheat–cotton crop sequence on a sprinkler-irrigated heavy clay Vertisol in Queensland, Australia. The data points indicate the actual measured fluxes of N_2O, a powerful greenhouse climate-forcing gas produced by denitrification. The solid green line is the N_2O flux calculated by DayCent, a computer model that simulates flows of carbon, nutrients, and trace gases between the atmosphere, soil, and plants. The molecular nitrogen (N_2) gas losses are not shown, but were 20 to 100 times as great as the N_2O losses. The orange data points and line indicate the monthly maximum daily temperature (°C), with Australia's hot season occurring December–March. [Graph redrawn from data in Scheer et al. (2014)]

Figure 12.10 *Nitrification–denitrification reactions and dynamics of the related processes controlling N loss from the aerobic–anaerobic layers of a flooded soil system. Nitrates, which form in the thin aerobic soil layer just below the soil–water interface, diffuse into the anaerobic (reduced) soil layer below and are denitrified to the N_2 and N_2O gaseous forms, which are lost to the atmosphere. In the case of fertilized rice paddies, placing the urea or ammonium-containing fertilizers deep in the anaerobic layer prevents N oxidation of ammonium ions to nitrates, thereby greatly reducing N loss.* [Modified from Patrick (1982)]

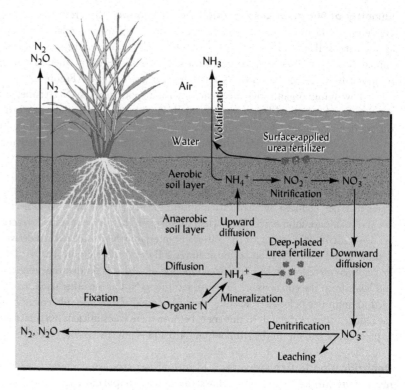

zone soils. The apparent removal of nitrate may be quite dramatic, whether the nitrate source is suburban septic drainfields, livestock feeding operations, or fertilized cropland. Constructed wetlands or biofilters specially engineered with lots of decomposable organic material and controlled inundation can remove 50 to 95% of the nitrates from surface water or groundwater before the nitrate can enter a stream channel and cause eutrophication.

We have just discussed a number of processes (e.g., nitrate leaching, ammonia volatilization, N_2 production by denitrification) that lead to loss of N from the soil system. Were such losses not matched by N inputs, ecosystems would have long ago run out of reactive N and life in them would have ground to halt. To understand why such a dire situation has not come to pass, we will now consider how soil N is replenished.

Biological Nitrogen Fixation in Soils Biological nitrogen fixation converts the inert dinitrogen gas of the atmosphere (N_2) to reactive nitrogen that becomes available to all forms of life through the N cycle. The process is carried out by a limited number of bacteria, including several species of *Rhizobium*, actinomycetes, and cyanobacteria (formerly termed blue-green algae). Globally, terrestrial ecosystems fix an estimated 139 million Mg of N annually. However, the amount that is fixed in the manufacture of fertilizers is now even greater.

Regardless of the organisms involved, the key to biological N fixation is the enzyme *nitrogenase*, which catalyzes the reduction of dinitrogen gas to ammonia:

$$N_2 + 8H^+ + 6e^- \xrightarrow[\text{(Fe,Mo)}]{\text{(Nitrogenase)}} 2NH_3 + H_2 \qquad (12.10)$$

The ammonia, in turn, is combined with organic acids to form amino acids and, ultimately, proteins. The site of N_2 reduction is the enzyme **nitrogenase**. Nitrogen-fixing organisms have a relatively high requirement for molybdenum, iron, phosphorus, and S, because these nutrients are part of the nitrogenase molecule and are also needed for its synthesis and use.

Breaking the strong N≡N triple bond in N_2 gas requires a great deal of energy. Therefore, this microbial process is greatly enhanced when it is carried out in association with plants, which can supply energy from photosynthesis. Nitrogenase is destroyed by free O_2, so organisms that fix N must protect the enzyme from exposure to oxygen.

Table 12.1
INFORMATION ON DIFFERENT SYSTEMS OF BIOLOGICAL NITROGEN FIXATION

N-fixing systems	Organisms involved	Plants involved	Site of fixation
Symbiotic			
Obligatory			
Legumes	Rhizobial bacteria *Rhizobia, Bradyrhizobia*, others	Legumes (Fabaceae)	Root nodules, stem nodules
Nonlegumes (angiosperms)	Actinomycetes bacteria (*Frankia*)	Nonlegume angiosperms (Betulaceae, Casuarinaceae and Myricaceae, Rosaceae, Eleagnaceae, Rhamnaceae, Datiscaceae and Coriariaceae)	Root nodules
Associative			
Symbiotic	Cyanobacteria	Various species of Angiosperms (flowering plants), Gymnosperms (conifers), Pteridophytes (ferns, *Azolla*), and Bryophytes (mosses, liverworts)	Stem glands, leaf and root nodules, special leaf cavities, gametophyte cavities, or root environment
	Plant growth-promoting rhizobacteria (PGPR)	Many families of plants, including Poaceae (grasses, rice, corn).	Root environment
Nonsymbiotic		Not involved with plants	Soil, water independent of plants

Leghemoglobin, a compound that gives actively fixing nodules a red interior color, binds oxygen in such a way as to protect the nitrogenase while making oxygen available for respiration in other parts of the nodule tissue. Leghemoglobin is virtually the same molecule as the hemoglobin that gives human blood its red color while carrying oxygen to our body cells. The reduction reaction is end product inhibited—for example, an accumulation of ammonia will inhibit N fixation. Also, too much nitrate in the soil will inhibit the formation of nodules (see following).

Biological N fixation occurs through a number of microbial systems that may or may not be directly or indirectly associated with plants (Table 12.1). Although the legume–bacteria symbiotic systems have received the most attention, the other systems involve many more families of plants worldwide and may supply large amounts of biologically fixed N to the soil. Each major system will be discussed briefly.

Symbiotic Nitrogen Fixation with Legumes

Plants of the legume family (Fabaceae) are famous for providing biologically fixed N in natural and agroecosystems. They do so in association with several genera of bacteria (in the subclass Alpha-Proteobacteria) collectively termed **rhizobial bacteria**. The legume plants and rhizobial bacteria form a **symbiosis** (a mutually beneficial relationship) in which the host plant supplies the bacteria with carbohydrates for energy, and the bacteria reciprocate by supplying the plant with reactive N with which to make essential plant compounds such as proteins and chlorophyll. In a complex biochemical "conversation" involving many specific signaling compounds, the rhizobial bacteria find and infect the legume plant root hairs and colonize the cortical cells. Here the rhizobial bacteria ultimately induce the formation of **root nodules** that serve as the site of N fixation (Figure 12.11).

Figure 12.11 Nodules on legume plants. (a) Sunn hemp (Crotalaria juncea) used as a cover crop to add N to soils. (b) Close-up of a few nodules on soybean roots with one nodule sliced open to expose the red color of its oxygenated leghemoglobin. (c) A scanning electron micrograph showing a single plant cell within a soybean nodule stuffed with the Bradyrhizobium japonicum bacteria specific to symbiosis with soybean. (Photos courtesy of Ray R. Weil; SEM courtesy of W. J. Brill, University of Wisconsin)

Organisms Involved A given rhizobial bacteria species will infect some legumes but not others. For example, *Rhizobium trifolii* inoculates *Trifolium* species (most clovers), but not sweet clover, which is in the genus *Melilotus*. Likewise, *Rhizobium phaseoli* inoculates *Phaseolus vulgaris* (beans), but not soybeans, which are in the genus *Glycine*. This specificity of interaction is one basis for classifying rhizobia (see Table 12.2). Legumes that can be inoculated by a given *Rhizobium* species are included in the same **cross-inoculation group**.

In soils which have grown a given legume for several years, the appropriate species of rhizobial bacteria is probably present. However, if the natural rhizobial population in the soil is too low or the strain of the rhizobial bacteria species present is not effective, special mixtures of the appropriate rhizobial bacteria may be applied as an **inoculant**. Rhizobial inoculants are available commercially and can be used by coating the legume seeds or by applying directly to the soil. You may want to refer to Table 12.2 when planting legume plants or purchasing commercial inoculant.

Quantity of Nitrogen Fixed Many natural plant communities and certain agricultural systems (generally involving legumes) derive the bulk of their N needs from biological fixation. Those systems involving nodules, which supply energy from photosynthates and protect the nitrogenase enzyme system, may fix 30 to 300 kg N ha^{-1} y^{-1}. Nonnodulating or nonsymbiotic systems generally fix relatively small amounts of N (5 to 30 kg N ha^{-1} y^{-1}). High levels of available N, whether from the soil or added in fertilizers, tend to depress biological N fixation (Figure 12.12). Apparently, plants make the heavy energy investment required for symbiotic N fixation only when supplies of mineral N from the soil are quite limited. However, some legume species fix N so inefficiently (e.g., *Phaseolus* bean) that supplementation with N fertilizer is recommended for high yields.

Effect on Soil Nitrogen Level Over time, the presence of N-fixing species can significantly increase the N content of the soil and benefit nonfixing species grown in association with fixing species (see Figure 12.13). Although some direct transfer may take place via mycorrhizal hyphae connecting two plants, most of the transfer results from mineralization of N-rich compounds in root exudates and in sloughed-off root and nodule tissues. Ammonium and nitrate thus released into the soil are available to any plant growing in association with the legume.

In the case of legume crops harvested for seed or hay, most of the N fixed is removed from the field with the harvest. Such crops should be considered as nitrogen *savers* for the soil rather than nitrogen builders. On the other hand, considerable buildup of soil N can be achieved by perennial legumes (such as alfalfa or kudzu) and by annual legumes (such as hairy vetch) if the entire growth is returned to the soil as **green manure**. If managed to maximize

Table 12.2
CLASSIFICATION OF RHIZOBIA BACTERIA AND ASSOCIATED LEGUME CROSS-INOCULATION GROUPS

The genera Rhizobium and Ensifer contain fast-growing, acid-producing bacteria, while those of Bradyrhizobium are slow growers that do not produce acid.

Bacteria		
Genus	**Species/subgroup**	**Host legume**
Ensifer	E. meliloti	Melilotus (sweet clovers), Medicago (alfalfa), Trigonella spp. (fenugreek)
Rhizobium	R. leguminosarum	
	bv. viceae	Vicia spp. (vetches), Pisum (peas), Lens (lentils), Lathyrus (sweet pea), Vicia faba (faba bean)
	bv. trifolii	Trifolium spp. (most clovers)
	bv. phaseoli	Phaseolus spp. (dry bean, string bean, etc.)
	R. Fredii	Glycine spp. (e.g., soybean)
	R. spp.	Securigera varia (crown vetch)
	R. spp.	Trees in Leucaena group: Leucaena spp. Sesbania grandiflora; Calliandra calothyrsus; Gliricidia sepium; Prosopis spp.
	R. lupini	Lupinus spp. (lupins)
Bradyrhizobium	B. japonicum	Glycine spp. (e.g., soybean)
	B. spp.	Vigna (cowpeas, mung bean), Arachis (peanut), Cajanus (pigeon pea), Pueraria (kudzu), Crotalaria (crotalaria), and many other tropical legumes; Phaseolus lunatus (lima bean) Acacia spp. (acacia trees), Desmodium spp., Stylosanthes spp., Centrosema sp., Lablab purpureus (Lablab bean), Pueraria phaseoloides
Mesorhizobium	M. loti	Lotus (trefoils), Lupinus (lupins), Cicer (chickpea), Anthyllis, Leucaena, and many other tropical trees
Azorhizobium	A. spp.	Produces stem nodules on Sesbania rostrata

the return of their high N biomass, legume green manures can be used to replace most or even all of the N fertilizer typically used in certain crop rotations. In any case, the N contributions from legumes should be taken into account when estimating N fertilizer needs for optimum plant production with minimal environmental pollution (Section 12.15).

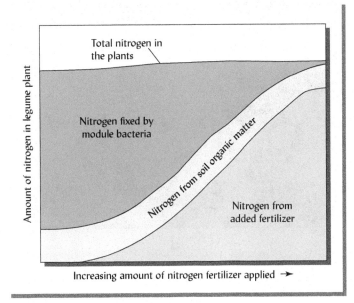

Figure 12.12 *Influence of adding inorganic N on the N found in a representative legume plant. As more N fertilizer is added, the plant obtains less of its N by biological fixation. The inorganic N merely replaces biologically fixed N, saving the plant some energy that would have been allocated to the nodule bacteria. Plant growth (not shown) is also little affected by the fertilizer application. The uptake of inorganic N released by mineralization (middle pool in diagram) may or may not be much affected by the fertilizer.* (Diagram courtesy of Ray R. Weil)

Figure 12.13 *Nitrogen concentrations of five field cuttings of ryegrass grown alone or with ladino clover. For the first two harvests, N fixed by the clover was not available to the ryegrass, and the N concentration of the ryegrass forage was low. In subsequent harvests, the fixed N apparently was available and was taken up by the ryegrass. This was probably due to the mineralization of dead ladino clover root tissue.* [Data from Broadbent et al. (1982)]

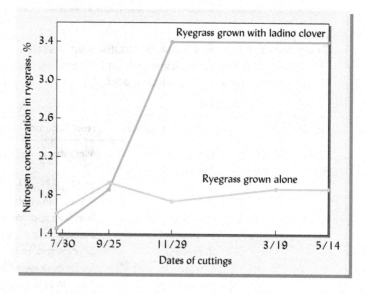

N Fixation in Nodule-Forming Nonlegumes[6]

Some 220 species from eight plant families are known to develop nodules and to accommodate symbiotic N fixation when their root hairs are invaded by soil actinomycetes of the genus *Frankia*. Most of these *actinorhizal* plants are woody shrubs and trees that form distinctive nodules (Figure 12.14, inset). Table 12.3 lists several of the more important actinorhizal associations in certain forests and wetlands.

The rates of N fixation per hectare compare favorably with those of the legume–*Rhizobium* associations. On a worldwide basis, the total N fixed in this way may even exceed that fixed by agricultural legumes. Because of their N-fixing ability, certain of the tree–actinomycete

Figure 12.14 *Soil actinomycetes of the genus Frankia can nodulate the roots of certain woody plant species and form a N-fixing symbiosis that rivals the legume–rhizobia partnership in efficiency. The actinomycete-filled root nodule (a) is the site of N fixation. The red alder tree (b) is among the first pioneer tree species to revegetate disturbed or badly eroded sites in high-rainfall areas of the Pacific Northwest in North America. This young alder is thriving despite the N-poor, eroded condition of the soil because it is not dependent on soil N for its needs.* (Photo courtesy of Ray R. Weil)

---

[6]For a scientific review of microbial N fixation in association with nonleguminous plants, see Santi et al. (2013).

Table 12.3
MAJOR ACTINORHIZAL PLANTS: ACTINOMYCETE-NODULATED NONLEGUME ANGIOSPERMS
About 220 actinorhizal species are known as compared to some 13,000 legume species.

Genus	Family	Geographic distribution
Alnus	Betulaceae	Cool regions of the northern hemisphere
Ceanothus	Rhamnaceae	North America
Myrica	Myricaceae	Many tropical, subtropical, and temperate regions
Casuarina	Casuarinaceae	Tropics and subtropics
Elaeagnus	Elaeagnaceae	Asia, Europe, North America
Coriaria	Coriariaceae	Many subtropical, and temperate regions

associations are able to colonize infertile soils and newly forming soils on disturbed lands, which may have extremely low fertility as well as other conditions that limit plant growth (Figure 12.14). Once N-fixing plants become established and begin to build up the soil N supply through leaf litter and root exudation, the land becomes more hospitable for colonization by other species. *Frankia* thus play a very important role in the N economy of areas undergoing succession, as well as in established wetland forests.

Symbiotic Nitrogen Fixation Without Nodules

Among the most significant nonnodule N-fixing systems are those involving cyanobacteria. One system of considerable practical importance is the *Azolla–Anabaena* complex, which flourishes in certain rice paddies of tropical and semitropical areas (Figure 12.15). The *Anabaena* cyanobacteria inhabit cavities in the leaves of the floating fern *Azolla* and fix quantities of N comparable to those of the more efficient *Rhizobium*–legume complexes (see Table 12.3).

A more widespread but less intense N-fixing phenomenon is that which occurs in the *rhizosphere* of certain nonlegume plants, especially tropical grasses. **Rhizobacteria** (bacteria

Figure 12.15 *Transplanting rice seedlings into paddies containing the floating water fern,* Anabaena *(see insets), and its N-fixing cyanobacteria symbiont* Nostoc.
(Photo courtesy of Ray R. Weil)

adapted to life in the rhizosphere) that benefit plant growth and development are referred to as plant growth–promoting rhizobacteria (PGPR). Certain PGPR use energy from the exudates of plant roots to power N fixation. In some cases, inoculation of soil with certain GPB has stimulated sufficient N fixation under field conditions to significant increase productivity of food crops such as corn, rice, and wheat.

Nonsymbiotic Fixation by Heterotrophs[7]

Certain free-living microorganisms present in soils and water are able to fix N. Because these organisms are not directly associated with higher plants, the transformation is referred to as *nonsymbiotic* or *free-living*.

Several different groups of bacteria and cyanobacteria are able to fix N nonsymbiotically. In upland mineral soils, the major fixation is brought about by species of several genera of heterotrophic aerobic bacteria, *Azotobacter* and *Azospirillum* (in temperate zones) and *Beijerinckia* (in tropical soils). Certain anaerobic bacteria of the genus *Clostridium* are also active in fixing N. Because pockets of low oxygen supply exist within aggregates even in well-drained soils (see Section 7.4), aerobic and anaerobic bacteria probably work side by side in many well-drained soils. These organisms obtain their carbon either from root exudates in the rhizosphere or by saprophytic decomposition of SOM, and they operate best where soil N is limited.

The amount of N fixed by these heterotrophs varies greatly with the pH, soil N level, and sources of organic matter available. In some natural ecosystems these organisms undoubtedly make an important contribution to the N needs of the plant community. Because of limited carbon supplies, in conventional agricultural systems they probably fix only 5 to 20 kg N/ha/yr; however, with proper organic matter management, it is thought that the rates may be considerably higher. If agriculturalists are able to take advantage of these organisms, the benefits would go beyond increased crop yields to include reduced need to manufacture N fertilizer (and therefore lowering the amount of reactive N circulating in the environment) and reductions in N_2O emissions.

Fixation by Autotrophs

In the presence of light, certain photosynthetic bacteria and cyanobacteria are able to fix carbon dioxide and nitrogen simultaneously. The contribution of the photosynthetic bacteria is uncertain, but that of cyanobacteria is thought to be of some significance, especially in wetlands (including in rice paddies). In some cases, cyanobacteria contribute a major part of the N needs of rice, but nonsymbiotic species rarely fix more than 20 to 30 kg N/ha/yr. Nitrogen fixation by cyanobacteria in upland soils also occurs (including in the desert microbiotic crustsdiscussed in Section 11.14), but at much lower levels than found under wetland conditions.

Nitrogen Deposition from the Atmosphere

Reactive N in the atmosphere consists of ammonia and nitrogen oxide gases that originated as emissions from soils, oceans, vegetation, and fossil fuel combustion (especially in vehicle engines). These N gases are generally transformed into NH_4^+ or NO_3^- by reactions in clouds, and additional nitrates are formed in the atmosphere by reaction of N_2 and O_2 during lightning flashes. The term *nitrogen deposition* refers to the addition of these atmosphere-borne reactive N compounds to soils through rain, snow, dust, and gaseous absorption.

Global N deposition totals some 105 Tg N yr^{-1}. Deposition is greatest in high-rainfall areas downwind from cities (nitrate from nitrogen oxides in car exhaust and coal burning power plants), concentrated animal-feeding operations (CAFOs) (ammonium volatilized from manure), and fertilized wetland rice production. Although the deposited N may stimulate greater plant growth in agricultural systems, the effects on forests, grasslands, and aquatic ecosystems are quite damaging. Nitrates in particular are associated with acidification of rain(as discussed in Section 9.6), but since the ammonium soon nitrifies, both forms lead to soil acidification.

Nitrogen deposition as ammonium and nitrate from the atmosphere (or by fertilization) also impacts another soil process with important global change implications, namely,

---

[7]For insights into the exploitation of nonsymbiotic N fixation for agriculture, see Kennedy et al. (2004).

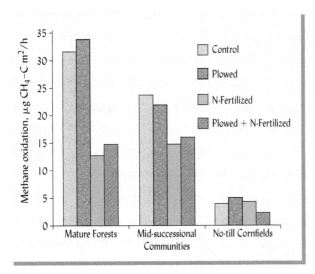

Figure 12.16 Mineral N reduced the capacity of soils to oxidize methane and thereby remove this potent greenhouse gas from the atmosphere. Forested soils exhibited the highest rates of methane oxidization and the greatest impairment due to the addition of N. A one-time physical disturbance (plowing) had little impact. Nitrogen was applied as a solution of ammonium nitrate (100 kg N/ha) that simulated high N deposition. The study was on sandy loam soils (Typic Hapludalfs) in southern Michigan. [Modified from Suwanwaree and Robertson (2005)]

methane oxidation (Figure 12.16). Methane is an important greenhouse gas affecting climate change, and its removal from the atmosphere by soil oxidation helps maintain its global balance. Forested soils have particularly high rates of methane oxidation, but also may be hardest hit by increased nitrogen.

Effects on Forest and Other Ecosystems Nitrogen in precipitation might be considered beneficial fertilizer when it falls on farmland, but it can be a serious pollutant when chronically added to some forested soils. Most forest soils are N limited—that is, they contain a surplus of carbon so that any N added is quickly tied up by microbial and chemical immobilization and very little nitrate is lost by leaching. However, a condition known as *nitrogen saturation* has been found to result from high levels of N deposition on certain mature forests in northern Europe and to a lesser extent in North America. Nitrogen saturation refers to the inability of the forest system to retain the N received by deposition, leading to the leaching of nitrates and the associated soil acidification and loss of calcium and magnesium (as described in Sections 9.6 and 12.20). Nitrogen deposition greater than about 8 kg N/ha/yr can be expected to eventually cause damage to sensitive forests (Figure 12.17), grasslands, and aquatic ecosystems.

Figure 12.17 Drastic effects of chronic high N additions on a pine forest ecosystem in Massachusetts. The experimental addition of 150 kg/ha/yr (=15 g m^2/yr) of N as NH$_4$NO$_3$ since 1988 has decimated the tree canopy (a) and understory (b) compared to the unamended plot (c,d). By the fourteenth year of the study, 56% of the trees had died in the high N plot compared to 12% in the control plot. The high N treatment decreased soil microbial biomass by 40% and soil respiration by 35%. The fungal/bacterial ratio and microbial diversity also decreased. Much more N (inorganic and organic) was lost in leaching water from the high N plots. Foliage analyses showed significantly lower leaf calcium in the high N plots, suggesting that soil acidification and loss of calcium may have played a role in the trees' demise. The Montauk stony sandy loam soils (Typic Dystrochrepts) formed from glacial till. [Photos courtesy of Ray R. Weil; data from Magill et al. (2004)]

The Nitrate Leaching Problem Subsurface flow of dissolved N is commonly the pathway accounting for the greatest losses of N from upland ecosystems, especially agroecosystems (Figure 12.18). Although in some cases subsurface flow carries substantial amounts of DON, the main form of N subject to leaching loss is nitrate. In contrast to positively charged ammonium ions, negatively charged nitrate ions are not adsorbed by the negatively charged colloids that dominate most soils. Therefore, nitrate ions move downward freely with drainage water and are readily leached from the soil. The loss of N in this manner is of concern for three basic reasons: (1) the loss of this valuable nutrient is a waste that impoverishes the ecosystem, (2) leaching of nitrate anions stimulates the acidification of the soils and the co-leaching of such cations as Ca^{2+}, Mg^{2+}, and K^+ (as described in Section 9.6), and (3) the movement of nitrate to groundwater causes several serious water-quality problems downstream.

A major water quality problem attributed to N is the leaching of nitrate to groundwater. Nitrate may contaminate drinking water, causing health hazards for people as well as livestock. The key factor for health hazards is *concentration* of nitrate in the drinking water and the level of exposure (amount of water ingested, especially over long periods). The nitrates may also eventually flow underground to surface waters, such as streams, lakes, and estuaries. The damages to surface waters are even more widespread, impairing water quality and the health of aquatic ecosystems, especially those with salty or brackish water. The key factor for this kind of damage is often the *total load* (mass flux) of N delivered to the sensitive ecosystem where it triggers the process of eutrophication.

Management to Reduce N Losses Even in regions of high leaching potential, careful soil management can prevent excessive nitrate losses. Proper types of fertilizer and manure (Right source) should be applied only in modest amounts to which the crop will actually respond (Right rate), placed in the soil where runoff and gaseous N losses can be minimized and crop roots have optimal access (Right place), and timed to provide soluble N when actively growing plants are able to absorb it (Right time). These principles are sometime referred to as "four Rs" of fertilizer management. Among the modern tools available regarding the "right source" are special fertilizer additives that can slow the loss of N by interrupting particular processes in the N cycle (Figure 12.19) including 1) slowing the dissolution of fertilizer granules, 2) inhibiting the urease enzyme that releases NH_4^+ from the urea molecule, and 3) inhibiting the nitrification process so that the formation of nitrate is delayed.

Nitrogen can be further conserved by planting N-scavenging winter cover crops (Section 13.2) immediately following summer annual crop harvest (or even inter-seeding them into the summer crop *before* harvest) to take up the unused nitrates before they can leach away (see Figures 13.11 and 13.48). Crop rotations and inter-planting with deep rooted trees can also capture N on its way down the soil profile and bring it back to the surface soil where

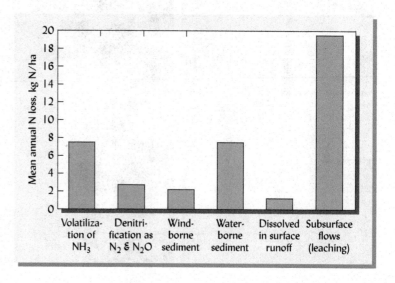

Figure 12.18 *Average annual N loss by each of six major loss pathways in the Upper Mississippi River basin. Leaching through the soil profile to groundwater is the dominant path for N loss.*
[Data from USDA/NRCS (2012)]

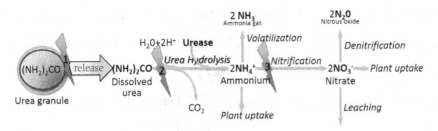

Figure 12.19 *Three types of inhibitors (jagged blades) designed to make fertilizer N less prone to loss: (1) coatings that slow the release of urea from fertilizer granules; (2) chemicals that inhibit the urease enzyme thus slowing hydrolysis of urea to ammonium thus limiting formation of ammonia gas and nitrates; and (3) nitrification inhibitors that slow nitrate formation from ammonium, thus reducing losses by leaching and denitrification.* [Diagram courtesy of Ray R. Weil. For a meta-analysis of research on nitrification inhibitors and their potential benefits, see Qiao et al. (2015)]

it can be used by subsequent crops. If practices such as just described are implemented, N leaching may be kept to a minimum (<5 to 10% of that applied).

The N cycle processes just discussed suggest several basic strategies for achieving a rational reduction of excessive N inputs while maintaining or improving production levels and profitability in agricultural enterprises. These approaches include (1) taking into account the N contribution from *all* sources and reducing the amount of fertilizer applied accordingly; (2) improving the efficiency with which fertilizer and organic amendments (e.g., manure) are used; (3) avoiding overly optimistic yield goals that lead to fertilizer application rates designed to meet crop needs that are much higher than actually occur in most years; and (4) improving crop response knowledge, which identifies the lowest N application that is likely to produce optimum profit. These and other strategies of nutrient management will be discussed further in Chapter 13.

12.2 SULFUR IN THE SOIL SYSTEM[8]

Sulfur (S) is a macronutrient element essential for life. Plants use S in amounts similar to those of phosphorus. Yet S often is forgotten as discussions of soil fertility management tend to focus on N, P, and K. In addition, S is also associated with several environmental problems, including acid precipitation, certain types of forest decline, acid mine drainage, acid sulfate soils, and even some toxic effects in drinking water.

Roles of Sulfur in Plants and Animals

Sulfur is a constituent of the vitamins biotin, thiamine, and B1 and the amino acids methionine, cysteine, and cystine, deficiencies of which result in serious human malnutrition. Sulfur is in many plant enzymes that regulate such activities as photosynthesis and N fixation. Sulfur-to-sulfur bonds link certain sites on long chains of amino acids, causing proteins to assume the specific three-dimensional shapes key to their catalytic action. Sulfur works with nitrogen in the processes of protein and enzyme synthesis. Sulfur is also an essential ingredient of the aromatic oils that give the cabbage and onion families of plants their characteristic odors and flavors. It is not surprising that the legume (beans), brassica (rapeseed, cabbage), and lily (onion) families require especially large amounts of S.

Agronomic Deficiencies of Sulfur Healthy plant foliage generally contains 0.15 to 0.45% S, or approximately one-tenth as much S as N. Plants deficient in S tend to become spindly, slow growing and exhibit chlorotic (light green or yellow) foliage (Figure 12.20). However, unlike N, S is relatively immobile in the plant, so the chlorosis develops first on the youngest leaves as S supplies are depleted (in N-deficient plants, chlorosis develops first on the older leaves). Sulfur-deficient leaves on some plants show interveinal chlorosis or faint striping that

---

[8]Several reviews provide more detail on the subject of sulfur in soils (Eriksen, 2009; Jez, 2008; Kovar and Grant, 2011; Scherer, 2009).

Figure 12.20 Deficiency of S in (a) peanut, (b and c) corn, (d) young wheat, and (e) tea plants. Deficiency of S is often mistaken for that of N as symptoms for both include spindly plants with yellowish (chlorotic) leaves. However, unlike for N, S deficiency causes chlorosis either uniformly throughout the plant or more typically the younger, uppermost leaves to become most chlorotic. Photo (b) illustrates plant response to S applied to soils as gypsum ($CaSO_4 \cdot 7H_2O$). Sulfur deficiency was first reported in Malawi as "tea yellows" (d). (Photo courtesy of Ray R. Weil)

distinguishes them from N-deficient leaves. Also, unlike N-deficient plants, S-deficient plants tend to have low sugar but high nitrate contents in their sap.

Sulfur deficiencies in agricultural plants have become increasingly common as a result of three independent trends: cleaning up of sulfur dioxide air pollution, elimination of most S "impurities" from N–P–K fertilizers, and greatly increased S removals by higher-yielding crops. Sulfur deficiencies are most prevalent where soil parent materials are inherently low in S (e.g. sandy soils), where extreme weathering and leaching have removed this element, and where there is little replenishment of S from air pollution. In many tropical countries, one or more of these conditions prevail and S-deficient areas are common. Soils of dry savannas are particularly deficient in sulfur as a result of the annual burning that converts most of the S in the plant residues to sulfur dioxide.

Natural Sources of Sulfur in Soils

The three major natural sources of S that can become available for plant uptake are (1) *organic matter*, (2) *soil minerals*, and (3) *sulfur gases in the atmosphere*. In natural ecosystems where most of the S taken up by plants is eventually returned to the same soil, these three sources combined are usually sufficient to supply the 5 to 25 kg S per ha needed by growing plants.

Sulfur in Soil Organic Matter[9] In humid region surface soils, 90 to 98% of the S is usually present in organic forms (see Figure 12.21). Three principal groups of S compounds in SOM are characterized by S in +6, 0, and −2 oxidation states. The most reduced S is bonded to carbon in sulfides, disulfides, thiols, and thiophenes, including proteins. Compounds with intermediate redox states include sulfoxides and sulfonates in which the S is bonded to carbon but also to oxygen (C—S—O). Highly oxidized forms of S occur in ester sulfates (C—O—S). Over time, soil microorganisms break down these organic S compounds into soluble forms analogous to the release of ammonium and nitrate from organic matter, discussed in Section 12.3. As with N, most soil S is organic with only a very small percentage in the mineralized (sulfate) form, even in the sandy Spodosols.

In dry regions, less organic matter is present in the surface soils. Therefore, the proportion of organic S is not likely to be as high in arid and semiarid region soils as it is in humid region soils. This is especially true in the subsoils, where organic S may constitute only a small fraction of the S present and where inorganic S in the form of gypsum ($CaSO_4 \cdot 2H_2O$) is often present.

Sulfur in Soil Minerals Inorganic S compounds can supply soluble S for the nutrition of plants and soil microbes. Most inorganic S is present as either sulfates (oxidized with S^{6+}) or sulfides (reduced with S^{2-}). The sulfate minerals are most easily solubilized, and the sulfate ion (SO_4^{2-}) is easily assimilated by plants. Sulfate minerals are most common in regions of low rainfall, where they accumulate in the lower horizons of some Mollisols and Aridisols (see Figure 12.21, *left*). The most common sulfate mineral, gypsum, may also accumulate when sulfate-laden soil water evaporates near the soil surface. Sulfate salts may contribute to salinity in soils of arid and semiarid regions.

Sulfides are found in some humid-region wetland soils, especially those formed from marine sediments. Sulfides must be oxidized to the sulfate form before the S can be assimilated by plants. When waterlogged sulfide-rich soil layers are drained or exposed by excavation,

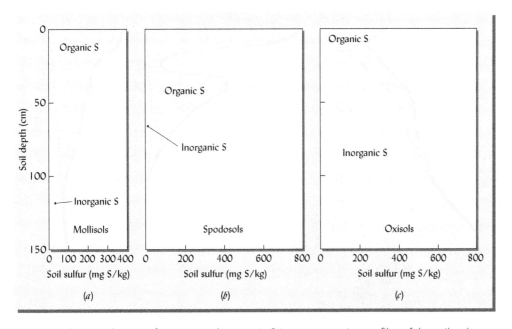

Figure 12.21 *Distribution of organic and inorganic S in representative profiles of the soil orders (a) Mollisols, (b) Spodosols, and (c) Oxisols. In each, organic forms dominate the surface horizon. Considerable inorganic S, as adsorbed sulfate and calcium sulfate minerals, exists in the lower horizons of Mollisols. Relatively little inorganic S exists in Spodosols. However, the bulk of the profile S in the humid tropics (Oxisols) is present as sulfate adsorbed to colloidal surfaces in the subsoil.* (Diagram courtesy of Ray R. Weil)

---

[9]For a study of organic sulfur forms in a range of soils, see Zhao et al. (2006).

oxidation will occur, and ample available sulfate-S will be released. In fact, so much S may be oxidized that problems of extreme acidity may result (see following).

Another mineral source of S is the clay fraction of some soils high in Fe, Al oxides, and kaolinite. These clays are able to strongly adsorb sulfate from soil solution, especially at low pH, and subsequently release it slowly by anion exchange. Oxisols and other highly weathered soils of the humid tropics and subtropics may contain large stores of sulfate sorbed in their subsoil horizons (see Figure 12.21, *right*). Considerable sulfate may also be bound by the metal oxides in the spodic horizons under certain temperate and boreal forests.

Atmospheric Sulfur[10] The atmosphere contains varying quantities of carbonyl sulfide (COS), hydrogen sulfide (H_2S), sulfur dioxide (SO_2), and other sulfur gases, as well as S-containing dust particles. These atmospheric forms of S arise from volcanic eruptions, volatilization from soils, ocean spray, biomass fires, and industrial plants (such as metal smelters and electric-generation stations fired by high-S coal). During the past century, the contribution from industrial sources has dominated S deposition in certain locations. In the atmosphere, most of the S materials are eventually oxidized to sulfates, forming H_2SO_4 (sulfuric acid). About half of the S is returned to the Earth as *dry deposition* (dry particles and gases) and half as *wet deposition* (in snow and rain). Atmospheric S becomes part of the soil–plant system mainly as SO_4^{2-} ions dissolve in soil solution, sorb to soil surfaces, or are directly absorbed by plant foliage (along with some SO_2 gas).

The acid precipitation caused partly by atmospheric S (and N) is a serious threat to the health of lakes, forests, and agroecosystems. In areas immediately downwind from industrial plants, S deposition may be great enough to cause direct toxicity to trees and crops (as well as respiratory problems in people). As far as 1,000 km downwind, the deposited sulfate may mobilize toxic soil aluminum, acidify lakes, and deplete soils of needed calcium. After watching forests and lakes become seriously damaged by acid rain in the 1970s and 1980s, governments in North America and Europe established regulatory programs to reduce S emissions. As a result, S emissions in these regions declined by more than half (although nitrogen oxide emissions have not been equally addressed). By 2008 in the eastern United States, annual S deposition was commonly less than 6 to 12 kg S/ha (Figure 12.22). In contrast, S emissions have been on the rise in China, India, and other industrializing regions, where burning of coal and oil cause as much as 50 to 75 kg S/ha to come down in a year.

In areas little affected by industrial emissions (e.g., western United States, central Australia, and most of Africa), deposition is generally only 1 to 5 kg S/ha/yr. While plants can obtain a significant part of their S directly from the atmosphere, recent reductions in S emissions have resulted in S deficiencies becoming increasingly common in more regions, especially

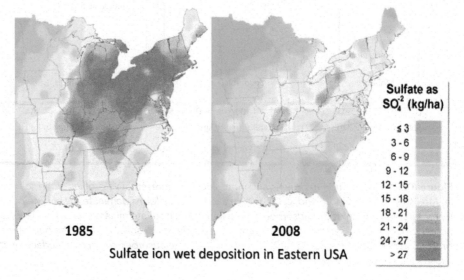

Figure 12.22 *The dramatic reductions in S deposition achieved through environmental policies in the United States during a 23-year period.* (Modified from USEPA)

---

[10]For a discussion of global sulfur deposition trends, see Klimont et al. (2013).

for high yield potential agricultural crops. In these areas, improved S recycling through cover crops and agroforestry may help provide enough of this element for high levels of crop production. Alternatively, farmers must spend more money on S-containing fertilizers and other amendments.

Cycling of Sulfur in Soils

The major transformations that S undergoes in soils are shown in Figure 12.23. The inner circle shows the relationships among the four major forms of this element: (1) *sulfides*, (2) *sulfates*, (3) *organic S*, and (4) *elemental S*. The outer portions show the most important sources of S and how this element is lost from the system.

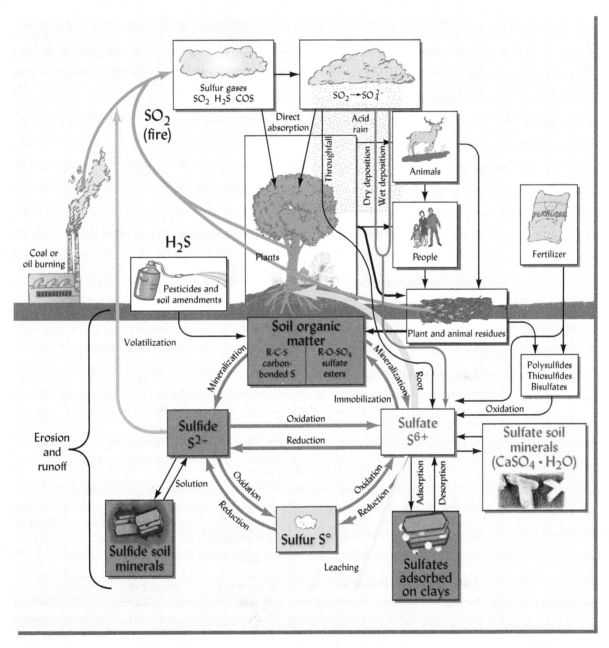

Figure 12.23 *The S cycle, showing some of the transformations that occur as this element is cycled through the soil–plant–animal–atmosphere system. In the surface horizons of all but a few types of arid-region soils, the great bulk of S is in organic forms. However, in deeper horizons or in excavated soil materials, various inorganic forms may dominate. The oxidation and reduction reactions that transform S from one form to another are mainly mediated by soil microorganisms. Sulfur can escape into the atmosphere by microbial reduction to H_2S and by oxidation, by fire to SO_2.*
(Diagram courtesy of Ray R. Weil)

Considerable similarity to the nitrogen cycle is evident (compare Figures 12.2 and 12.23). In each case, the atmosphere is an important source of the element in question. Both elements are held largely in SOM, both are subject to microbial oxidation and reduction, both can enter and leave the soil in gaseous forms, and both are subject to some degree of leaching in the anionic form. Microbial activities are responsible for many of the transformations that determine the fates of both N and S.

Sulfur Mineralization Sulfur behaves much like N as it is absorbed by plants and microorganisms and moves through the S cycle. The organic forms of S are generally mineralized by soil organisms before the S is used by plants. The rate at which this occurs depends on the same environmental factors that affect N mineralization, including moisture, aeration, temperature, and pH. When conditions are favorable for general microbial activity, the amount of inorganic S made available in a soil is directly related to the amount of organic S present. Some of the more easily decomposed organic compounds in the soil are sulfate esters, from which microorganisms release sulfate ions directly. However, in much of the SOM, S in the reduced state is bonded to carbon atoms in protein and amino acid compounds. In the latter case the mineralization reaction might be expressed as follows:

$$\underset{\substack{\text{Proteins and} \\ \text{other organic} \\ \text{combinations}}}{\text{Organic sulfur}} \rightarrow \underset{\substack{\text{H}_2\text{S and other} \\ \text{sulfides are} \\ \text{simple examples}}}{\text{decay products}} \xrightarrow{O_2} \underset{\text{Sulfates}}{SO_4^{2-} + 2H^+} \qquad (12.11)$$

Because this release of available sulfate is mainly dependent on microbial processes, the supply of available sulfate in soils fluctuates with seasonal, and sometimes daily, changes in environmental conditions. These fluctuations lead to the same difficulties in predicting and measuring the amount of S available to plants as were discussed in the case of N. It should be noted that for some plants, such as those in the Brassica family, a considerable amount of S exists in actively growing plant tissue as sulfate or esters of sulfate that hydrolyze to chemically release soluble sulfate almost immediately when the plants are killed and added to soil as a green manure

Sulfur Immobilization Immobilization of inorganic forms of S occurs when organic materials relatively high in C but low in S are added to soils. As for N, the energy-rich material stimulates microbial growth, and the inorganic nutrient (sulfate in this case) is assimilated into microbial tissue. The critical C/S ratio of organic substrates above which immobilization of S is likely to occur seems to be between 300/1 and 400/1 (Figure 12.24). The pattern of S immobilization in soils suggests that S in SOM may be associated with soil organic C and N in a relatively constant C/N/S ratio of about 85:7:1.

During the microbial breakdown of organic materials, hydrogen sulfide (H_2S), carbon disulfide (CS_2), carbonyl sulfide (COS), and methyl mercaptan (CH_3SH) gases are formed. Hydrogen sulfide is commonly produced in waterlogged soils from reduction of sulfates by anaerobic bacteria. All are more prominent in anaerobic soils and give wetland soils their characteristic rotten-egg odor.

Sulfur Oxidation Processes During the microbial decomposition of organic carbon-bonded S compounds, sulfides are formed along with other incompletely oxidized substances. These reduced substances are subject to oxidation (similar to the ammonium compounds formed when nitrogenous materials are decomposed). The oxidation reactions may be illustrated as follows, with hydrogen sulfide and elemental sulfur:

$$H_2S + 2O_2 \rightarrow H_2SO_4 \rightarrow 2H^+ + SO_4^{2-} \qquad (12.12)$$

$$2S + 3O_2 + 2H_2O \rightarrow 2H_2SO_4 \rightarrow 4H^+ + SO_4^{2-} \qquad (12.13)$$

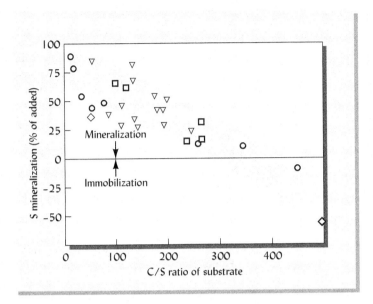

Figure 12.24 The amount of S mineralized or immobilized when organic material are added to soil is largely dependent on the C/S ratio of the added materials. The data suggest that if the C/S ratio is much above 300 or 400, net immobilization is likely. However, the release of sulfate can also be influenced by forms of S (oxidized ester that easily release sulfate or more reduced C-bonded sulfur compounds) in the organic substrate. [Data are for sewage sludge, ○, farmyard manure, □, and various green cover crops, ▽ (Eriksen, 2009); cabbage leaves and wheat straw, ◇ (Nziguheba et al., 2006); and rape, oat, pea wheat residues, ○ (Churka Blum et al., 2013)]

Most sulfur oxidation in soils is *biochemical* in nature, carried out over a wide range of soil conditions by a certain autotrophic bacteria, including species of the genus *Thiobacillus*. For example, sulfur oxidation may occur at pH values ranging from <2 to >9. This flexibility contrasts with the comparable nitrogen oxidation process, nitrification, which requires a rather narrow pH range closer to neutral.

The oxidation and reduction reactions of inorganic S compounds play an important role in determining the quantity of sulfate (the plant-available nutrient form of S) present in soils at any one time. Also, the state of S oxidation is an important factor in the acidity of soil and water draining from soils.

Reactions 12.12 to 12.14 show that, like N oxidation, S oxidation is an acidifying process. These reactions explain why elemental S and iron sulfide can be applied to lower soil pH if it is higher than desired (see Section 9.20). Along with N, the S in the atmosphere forms strong acids that acidify rainwater to a pH of 4 or even lower from the normal pH of 5.6 or higher. Section 9.6 explains how "acid rain" forms and how it damages soils, forests, and lakes in many regions. Part of the damage to these ecosystems stems from the leaching of sulfate anions, which can promote serious losses of calcium and magnesium.

Sulfur Reduction Processes Like nitrate ions, sulfate ions tend to be unstable in anaerobic environments. They are reduced to sulfide ions by a number of bacteria of two genera, *Desulfovibrio* and *Desulfotomaculum*. A representative reaction showing the reduction of S coupled with organic matter oxidation is as follows:

$$2R\text{—}CH_2OH + SO_4^{2-} \rightarrow 2R\text{—}COOH + 2H_2O + S^{2-} \quad (12.14)$$
$$\text{Organic alcohol} \quad \text{Sulfate} \quad \text{Organic acid} \quad \text{Sulfide}$$

Under anaerobic conditions, sulfide ions rapidly react with iron or manganese, which are typically present in their reduced forms. By tying up the soluble reduced iron, the formation of iron sulfides helps prevent iron toxicity in rice paddies and marshes. This reaction may be expressed as follows:

$$Fe^{2+} + S^{2-} \rightarrow FeS \quad (12.15)$$
$$\text{Dissolved} \quad \text{Sulfide} \quad \text{Iron sulfide}$$
$$\text{ferrous iron} \quad \quad \text{(solid)}$$

Sulfide ions will also undergo hydrolysis to form gaseous hydrogen sulfide. Sulfur reduction may take place with S-containing ions other than sulfates. For example, sulfites (SO_3^{2-}), thiosulfates ($S_2O_3^{2-}$), and elemental sulfur (S^0) are readily reduced to the sulfide form by bacteria and other organisms.

Sulfur Retention and Exchange

The sulfate anion (SO_4^{2-}) is the form in which plants absorb most of their S from soils. Since many sulfate compounds are quite soluble, the sulfate would be readily leached from the soil, especially in humid regions, were it not for its adsorption by the soil colloids. As was pointed out in Chapter 8, most soils have some anion exchange capacity that is associated with iron and aluminum oxide coatings on soil particles and, to a limited extent, with 1:1-type silicate clays. Sulfate anions are attracted by the positive charges that characterize acid soils containing these clays. They also react directly with hydroxy groups exposed on the surfaces of these clays. Figure 12.25 illustrates sulfate adsorption mechanisms on the surface of some Fe, Al oxides, and 1:1-type clays. Note that adsorption increases at lower pH values as positive charges that become more prominent on the particle surfaces attract the sulfate ions. Some sulfate reacts with the clay particles, becoming tightly bound, and is only slowly available for plant uptake and leaching.

In warm, humid regions, surface soils are typically quite low in S. However, because of the anion sorption mechanisms just discussed, much sulfate may be held in the subsoil horizons of Ultisols and Oxisols of these regions (see Figure 12.21). For example, symptoms of S deficiency commonly occur early in the growing season on Ultisols with sandy, low-organic-matter surface horizons. However, the symptoms may disappear as the crop matures and its roots reach the deeper horizons where sulfate is retained. In other cases the zone of sorbed sulfate occurs too deep to be accessed by the roots of most annual crops, and deep-rooted perennial vegetation is needed to "pump" the S to the surface soil for crop access.

Extreme Soil Acidity The acidifying effect of S oxidation can bring about extremely acid soil conditions that cause serious soil management problems and broader environmental pollution. Certain soils and sedimentary geologic materials are termed *sulfidic* because they contain high levels of reduced sulfur (sulfides), usually inherited from their present or past association with seawater (which is high in S). A common form of reduced S is the mineral pyrite (iron disulfide, FeS_2). The sulfides in these *potential acid-sulfate* materials are stable so long as oxygen is not present, but if submerged or buried materials are drained or excavated, the sulfides and/or elemental S quickly oxidize and form sulfuric acid, causing pH levels as low as 1.5 (see Section 9.6, Reactions 9.18 and 9.19). Plants cannot grow under these conditions

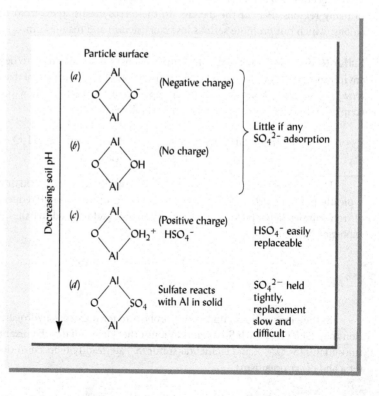

Figure 12.25 Effect of decreasing soil pH on the adsorption of sulfates by 1:1-type silicate clays and oxides of Fe and Al (reaction with a surface-layer Al is illustrated). At high pH levels (a), the particles are negatively charged, the cation exchange capacity is high, and cations are adsorbed. Sulfates are repelled by the negative charges. As pH drops (b), the H^+ ions are attracted to the particle surface and the negative charge is satisfied, but the SO_4^{2-} ions are still not attracted. At still lower pH values (c), more H^+ ions are attracted to the particle surface, resulting in a positive charge that attracts the SO_4^{2-} ion. This is easily exchanged with other anions. At still lower pH levels, the SO_4^{2-} reacts directly with Al and becomes a part of the crystal structure. Such sulfate is tightly bound, and it is removed very slowly, if at all. (Diagram courtesy of Ray R. Weil)

(Figure 12.26). The quantity of limestone needed to neutralize the acidity is so high that it is impractical to remediate these soils by liming(for reactions, solutions, and other details see Section 9.8). If allowed to proceed unchecked, the acids may wash into nearby streams. Thousands of kilometers of streams have been seriously polluted in this manner, the water and rocks in such streams often exhibiting orange colors from the iron compounds in the acid drainage (see Figure 9.21).

Sulfur Fertility Maintenance

The problem of maintaining adequate quantities of S for mineral nutrition of plants is becoming increasingly important. Cleaner air as a result of environmental controls along with the increasing crop removal of S makes it essential that farmers be attentive to prevent deficiencies of this element. In some parts of the world (especially in certain semiarid grasslands), S is already the next most limiting nutrient after N, deficiencies of S being even more common than those of P and K. However, the S deficiencies often go unrecognized and untreated as the plant symptoms can be mistaken for N deficiency and traditional soil testing often does not include tests for S.

Crop residues and farmyard manures can help replenish the S removed in crops, but these sources generally can help recycle only those S supplies that already exist within a farm. In regions with low S soils, greater dependence must be placed on regular applications of S-containing materials such as ammonium sulfate, gypsum, or Epsom salt ($MgSO_4$). Fortunately, optimal plant growth on low S soils can often be obtained with rather small application of about 5 to 15 kg S per hectare. The necessity for sulfur management will certainly increase in the future.

Sulfate Adsorption and Leaching of Nonacid Cations When the sulfate ion leaches from the soil, it is usually accompanied by equivalent quantities of cations, including Ca and Mg and other nonacid cations. In soils with high sulfate adsorption capacities, sulfate leaching is low and the loss of companion cations is also low. In contrast, sulfate-leaching losses from low-sulfate-adsorbing soils are commonly high and take with them considerable quantities of nonacid cations. Sulfur is thus seen as an indirect conserver of these cations in the soil solution. This is of considerable importance in soils of forested areas that receive acid rain.

Figure 12.26 *Construction of this highway cut through several layers of sedimentary rock. One of these layers contained reduced sulfide materials. Now exposed to the air and water, this layer is producing copious quantities of sulfuric acid as the sulfide materials are oxidized. Note the failure of vegetation to grow below the zone from which the acid is draining.*
(Photo courtesy of Ray R. Weil)

12.3 PHOSPHORUS[11] IN PLANT NUTRITION AND SOIL FERTILITY

Among the nutrient elements, phosphorus (P) is second only to N in its impact on the productivity and health of terrestrial and aquatic ecosystems. The total quantity of phosphorus in most native soils is low, with most of what is present occurring in forms quite unavailable to plants. Phosphorus is so scarce in natural ecosystems that archaeologists often test soils for the presence of high P concentrations as an indication of prehistoric human habitations.

Phosphorus and Plant Growth

Neither plants nor animals can live without phosphorus. It is an essential component of the organic compound **adenosine triphosphate** (ATP), which is the *energy currency* that drives most biochemical processes. Phosphorus, like nitrogen, is literally "in our DNA," being an essential component of **deoxyribonucleic acid** (DNA), the seat of genetic inheritance, and of **ribonucleic acid** (RNA), which directs protein synthesis in both plants and animals. Phospholipids, which play critical roles in cellular membranes, are another class of universally important P-containing compounds. Bones and teeth are made of the calcium-phosphate compound apatite. In fact, ground bone (called bone meal) has been used for centuries as a phosphorus fertilizer. In healthy plants, leaf tissue P content is usually about 0.2 to 0.4% of the dry matter, about 1/10th the comparable levels for N and similar to the levels for S.

Symptoms of Phosphorus Deficiency in Plants A P-deficient plant is usually stunted, thin-stemmed, and spindly, but its foliage, rather than being pale, is often dark, almost bluish-green. Thus, unless much larger, healthy plants are present to make a comparison, phosphorus-deficient plants often seem quite normal in appearance (Figure 12.27a–c). Phosphorus-deficient plants are also characterized by delayed maturity, sparse flowering, and poor seed quality. In severe cases, P deficiency can cause yellowing and senescence of leaves. Many plants develop purple colors in their leaves and stems (Figure 12.27d–g) as a result of P deficiency. Older leaves show deficiency symptoms first.

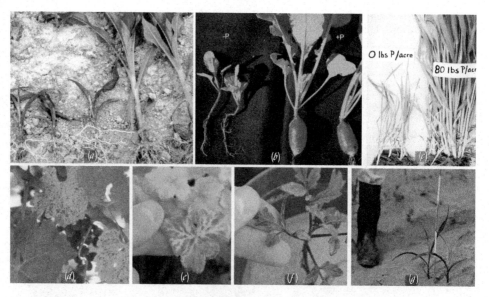

Figure 12.27 *Plant symptoms indicative of phosphorus deficiency. In (a) to (c), P-deficient corn, radish, and cereal rye plants are shown next to plants fertilized with adequate P. Severe stunting (g), sometimes with no special foliar symptoms (b), is typical. Reduction in root carbohydrate storage (b) is also common. In (d) to (g), grape, geranium, tomato, and corn, respectively, exhibit red or purple colors produced by P-deficiency, especially on leaf sheathes and the underside of older leaves.* (Photo courtesy of Ray R. Weil)

---

[11] For a fascinating historical account about all aspects of this element, see Emsley (2002). For a readable reviews of environmental, biogeochemical, agricultural, and social aspects of P, see Butusov and Jernelöv (2013). For a review of the plant availability of this element, see Sharpley (2000).

The Phosphorus Problem in Soil Fertility and Environmental Quality[12]

Phosphorus presents a soil fertility problem in three ways. *First*, the total P content of soils is relatively low, ranging from 500 to 10,000 kg P in the upper 50 cm of 1 ha of soil. *Second*, the P compounds commonly found in soils are mostly unavailable for plant uptake, often because they are highly insoluble. *Third*, when soluble sources of P, such as those in fertilizers and manures, are added to soils, they may become fixed (changed to unavailable forms) and in time form highly insoluble compounds.

Fixation[13] reactions in low-P soils may allow only a small fraction (10 to 15%) of the P applied in fertilizers and manures to be taken up by plants in the year of application. Early research showed that fixation reactions with soil minerals allowed only 10 to 15% of the P applied in fertilizers and manures to be taken up by plants in the year of application. Consequently, farmers who could afford to do so typically applied two to four times as much phosphorus as was removed in the crop harvest. Concentrated livestock production requiring the importation of P-containing deed also contributes the problem of soils becoming over-enriched in P. Repeated over many years, such practices can saturate the P-fixation capacity and build up the level of available soil P to the point where it can cause eutrophication of lakes and streams.

Where poor farmers cannot afford to buy fertilizer, underuse rather than overuse of fertilizer P is the rule. In most of sub-Saharan Africa, soils are depleted of P year after year, such that in some areas, the decline in per-capita food production will not likely be reversed until the critical phosphorus deficiency problems are solved. The productivity of forests and grasslands can also be limited by low P soils. In fact, P deficiency stunts crops and native vegetation on 1 to 2 billion ha of the world's land. The growth, nodulation, and N_2-fixation by legume plants, including forest trees, may be dramatically inhibited by very low levels of soil P (Figure 12.28). Thus two major environmental problems related to soil phosphorus are **land degradation** caused by too little available phosphorus and **accelerated eutrophication** of lakes and streams caused by too much (see Chapter 13).

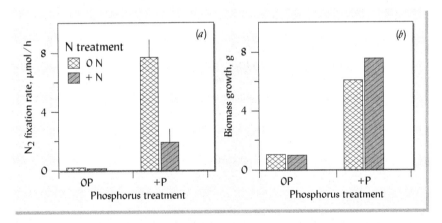

Figure 12.28 *Phosphorus controls symbiotic N fixation and growth by tropical rainforest tree Inga punctata. Biomass was very small and N fixation almost nonexistent when the trees were grown in P-poor soil (Typic Eutrudept) from a Panamanian rainforest (0P, left side of each graph). Even though the tree roots exuded phosphatases and were colonized by mycorrhizal fungi (Section 12.9), they were not able to extract sufficient P from the unfertilized soil. When fertilizer P was added, N fixation and growth were very rapid. Nitrogen fertilizer added with P increased tree biomass, but inhibited symbiotic N fixation, as explained in Section 12.1.* [Modified from Batterman et al. (2012)]

---

[12]For an excellent set of research and review papers on practical and innovative measures to control losses of P from agricultural lands, see *Journal of Environmental Quality* 29:1–181, 2000. For a forward-thinking approaches to problems of soil P in Africa, see Buresh et al. (1997) and Sanchez (2002).

[13]Note that the term *fixation* as applied to P has the same general meaning as the chemical fixation of potassium or ammonium ions; that is, the chemical being fixed is bound, entrapped, or otherwise held tightly by soil solids in a form that is relatively unavailable to plants. In contrast, the fixation of gaseous nitrogen refers to the *biological* conversion of N_2 gas to reactive forms that plants can use.

Table 12.4
INFLUENCE OF WHEAT PRODUCTION AND TILLAGE ON ANNUAL LOSSES OF PHOSPHORUS (P) IN RUNOFF WATER AND ERODED SEDIMENTS COMING FROM SOILS IN THE SOUTHERN GREAT PLAINS

The total P lost includes the P dissolved in runoff water and P adsorbed to eroded particles. Although cattle grazing probably increased losses of P from the natural grasslands, the losses from the agricultural watersheds were about ten times as great. The no-till wheat fields lost much less particulate P, but more dissolved P, than did the conventionally tilled wheat fields. Wheat was fertilized with up to 23 kg/ha of P each fall.

Location and soil	Management	kg P/ha/yr		
		Dissolved P	Particulate P	Total P
El Reno, Oklahoma Paleustolls, 3% slope	Wheat with conventional plow and disk	0.21	3.51	3.72
	Wheat with no-till	1.04	0.43	1.42
	Native grass, heavily grazed	0.14	0.10	0.24
Woodward, Oklahoma Ustochrepts, 8% slope	Wheat with conventional sweep plow and disk	0.23	5.44	5.67
	Wheat with no-till	0.49	0.70	1.19
	Native grass, moderately grazed	0.02	0.07	0.09

Data from Smith et al. (1991).

Phosphorus Loss in Runoff Disturbing the soil surface, such as by overgrazing or tillage, generally increases the amount of P carried away on eroded sediment (i.e., particulate P). On the other hand, increased losses of P dissolved in the runoff water (i.e., **dissolved P**) are stimulated by timber harvest or wildfires in forests and fertilizer or manure applications left unincorporated on the surface of cropland or pastures. The decision to use tillage to incorporate P-bearing soil amendments involves a trade-off between the advantages of incorporating P into the soil (less P dissolves in the runoff water and more is available for plant uptake) and the disadvantages associated with disturbing the surface soil (increased loss of soil particles by erosion and usually increased amounts of runoff water; see Table 12.4). In no-tillage systems, surface application of manure without incorporation may result in lower total loss of P, because this type of management achieves substantial reductions in soil erosion and total runoff. The effect of these reductions may or may not outweigh the effect of the increased P concentration in the relatively small volume of water that does run off. If the equipment is available to do so, the best option might be to handle the high-P amendment as a liquid and inject it into the soil with a minimum of disturbance to the soil surface(Section 13.4).

The Phosphorus Cycle

If we are to successfully manage P for economic plant production, ecosystem services, and environmental protection, we need to understand the nature of the different forms of P found in soils and the manner in which these forms of P interact with the soil and the larger environment. The cycling of P within the soil, from the soil to plants and back to the soil, is illustrated in Figure 12.29.

Phosphorus in Soil Solution Compared to other macronutrients, the soil solution concentration of P is very low, generally ranging from 0.001 mg/L in very infertile soils to about 1 mg/L in rich, heavily fertilized soils. Plant roots absorb P dissolved in the soil solution as phosphate ions (HPO_4^{2-} and $H_2PO_4^-$), but some soluble organic P compounds may also be taken up. Not only are phosphate ions in low supply in the soil solution, they are also quite immobile because they react strongly with the surfaces of soil particles. Therefore, phosphate ions are very slow to move downward through the soil matrix with percolating water and leaching losses are generally quite small.

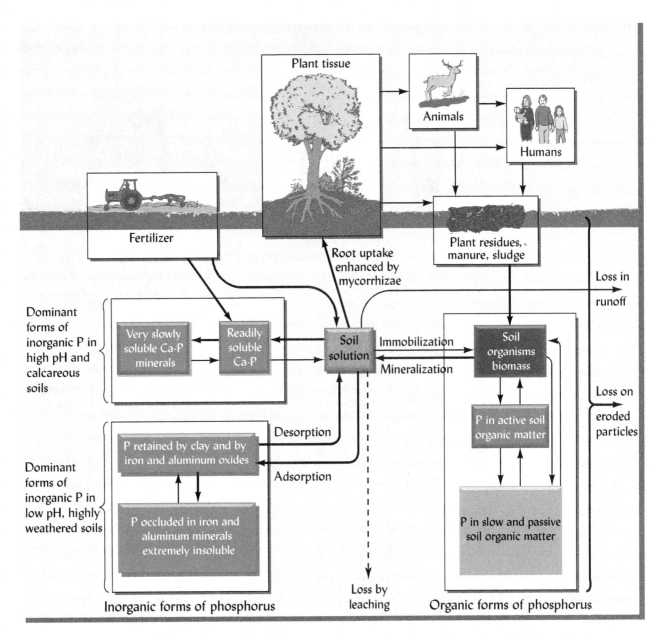

Figure 12.29 *The phosphorus cycle in soils. The boxes represent pools of the various forms of P in the cycle, while the arrows represent translocations and transformations among these pools. The three largest white boxes indicate the principal groups of P-containing compounds found in soils. Within each of these groups, the less soluble, less available forms tend to dominate. Thick arrows represent the principal pathways.* (Diagram courtesy of Ray R. Weil)

Uptake by Roots and Mycorrhizae[14]   Slow P diffusion to root surfaces curtails root uptake and a **depletion zone** with greatly reduced P concentration adjacent to roots and root hairs (Figure 12.30, *left*). Plant roots therefore generally find themselves growing in a soil solution depleted in P compared to the bulk soil. To compensate, roots must continually extend into and proliferate in new soil zones not yet depleted.

Fortunately, for most plants, the roots are not alone in their struggle to obtain sufficient P from soils—they are aided by a symbiotic partnership with mycorrhizal fungi (see also Section 10.9, Figures 10.33 and 10.34). The microscopic, threadlike mycorrhizal hyphae colonize root cortical cells and extend out into the soil several centimeters from the root surface, exploiting phosphorous from a much larger volume of soil than could the roots by itself (Figure 12.30, *right*).

---

[14]For a review of advances in our knowledge of the complex plant–fungal interaction affecting P nutrition, see Smith et al. (2011).

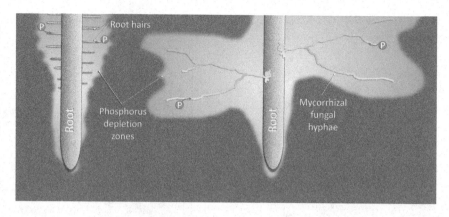

Figure 12.30 *Mycorrhizal root colonization enlarges the volume of soil (the depletion zone) where phosphate ions are taken up faster than they can diffuse in from nearby zones of higher concentration. Mycorrhizal roots may obtain P by both root hair uptake and hyphal uptake, but in some cases the fungi inactivate root P uptake mechanisms.* (Diagram courtesy of Ray R. Weil)

This mycorrhizal association is generally best developed where host plants are growing in undisturbed soils with low phosphorus availability. However, the fungi have been observed to benefit early-season growth of annual plants even in soils testing high in available P if the soil is kept vegetated with plants that can serve as suitable hosts (e.g., Table 12.5). Colonization by AM fungi usually, but not always, enhances total P uptake and plant growth. The plant "pays" the fungi for this service by sharing 5 to 20% of its photosynthate production.

Decomposition of Plant Residues   Once in the plant root, a portion of the P is translocated to sites where it is incorporated into plant tissues. As the plants shed leaves and their roots die, or when they are eaten by animals (including people), P returns to the soil in the form of plant residues, leaf litter, and animal wastes. As previously explained for N and S, microbial mineralization then releases inorganic P that can once again be taken up by plants.

Chemical Forms in Soils   In most soils, the amount of P available to plants from the soil solution at any one time is very low, seldom exceeding about 0.01% of the total P in the soil. The bulk of the soil P exists in three general groups of compounds—namely, *organic phosphorus, calcium-bound inorganic phosphorus*, and *iron- or aluminum-bound inorganic phosphorus* (see Figure 12.29). The organic phosphorus is distributed among the biomass, *labile* and *protected* or *passive* fractions of SOM (see Section 11.6). Of the inorganic phosphorus, the calcium compounds predominate in most alkaline soils, while the iron and aluminum forms are most important in

Table 12.5
PREVIOUS LAND USE (FALLOW VS. CROP) AFFECTS EARLY SEASON MYCORRHIZAL COLONIZATION, AND CORN CHARACTERISTICS ON A SOIL TESTING HIGH IN AVAILABLE PHOSPHORUS

Lack of a continuous host for mycorrhizal fungi due to previous fallow resulted in reduced corn root colonization at the three-leaf stage. This depressed growth and P uptake by the six-leaf stage and led to lower grain yields. Practices that encourage mycorrhizae may help plants take off quickly without starter fertilizer.

	Previous land use	Mycorrhizal root colonization on three-leaf stage corn (%)	Shoot dry wt. at six-leaf stage (kg/ha)	P concentration at six-leaf stage (%)	P uptake at six-leaf stage (g/ha)	Grain yield (kg/ha)
Year 1	Host crop	20.2	193	0.284	563	2,903
	Fallow	11.0	142	0.228	337	2,378
Year 2	Host crop	46.9	103	0.262	273	7,176
	Fallow	12.8	81	0.178	148	6,677
Year 3	Host crop	17.2	261	0.336	882	5,495
	Fallow	8.0	158	0.293	469	4,980

Data selected from Bittman et al. (2006).

acidic soils. All three groups of compounds slowly contribute P to the soil solution, but most of the P in each group is of very low solubility and not readily available for plant uptake.

Unlike N and S, P is not generally lost from the soil in gaseous form. Because soluble inorganic forms of P are strongly adsorbed by mineral surfaces, leaching losses of inorganic phosphorus are generally very low, but they still may be sufficient under certain circumstances to stimulate eutrophication in downstream waters.

Gains and Losses The principal pathways by which P is lost from the soil system are plant removal (5 to 50 kg ha^{-1} yr^{-1}), erosion of P-carrying soil particles (0.1 to 10 kg ha^{-1} yr^{-1} on organic and mineral particles), P dissolved in surface runoff water (0.01 to 3.0 kg ha^{-1} yr^{-1}), and leaching to groundwater (0.0001 to 0.5 kg ha^{-1} yr^{-1}). For each pathway, the higher figures cited for annual P loss would most likely apply to cultivated soils. The amount of P that enters the soil from the atmosphere (sorbed on dust particles) is quite small (0.05 to 0.5 kg ha^{-1} yr^{-1}), but may nearly balance the losses from the soil under natural vegetation.

As already discussed, in agroecosystems on low P soils, optimal crop production may initially require the input from fertilizer to exceed the removal in crop harvest. However, this level of P input can be justified only until enough P accumulates to reduce the P-fixing capacity of the soil. The level of soil fertility and severity of environmental P pollution generated by farmland are largely determined by the balance—or lack of balance—between P imports to a farm in fertilizer and feed and exports in plant and animal products. Urban ecosystems are characterized by complex P imports (mainly food and chemicals) and exports (mainly wastewater and runoff), and also commonly get out of balance such that phosphorus resources are wasted and streams and lakes become polluted.

Organic Phosphorus in Soils[15]

Plants use P from both inorganic and organic soil components. The organic fraction generally constitutes 25 to 75% of the total P in surface soil horizons. The deeper horizons may hold large proportions of inorganic P, especially in soils from arid and semiarid regions (Figure 12.31). For many decades, scientists focused most of their attention on the inorganic rather than on the organic P in soils, and our knowledge of the specific nature of much of the organic-bound P in soils is quite limited. Modern spectrographic methods show the bulk of soil organic P that occurs as mono- and di-esters, in which P atoms are bonded to carbons indirectly via oxygens (Figure 12.31). Three broad groups of organic P compounds are quite common in soils: (1) inositol phosphates or phosphate mono-esters of a sugarlike compound, inositol [$C_6H_6(OH)_6$]; (2) phosphate di-esters such as nucleic acids from DNA and RNA; and (3) phospholipids, partially derived from microbial and plant cell membranes.

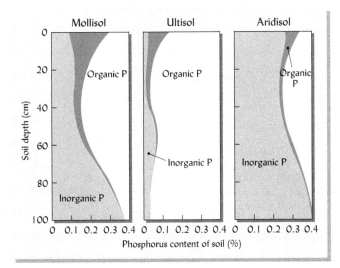

Figure 12.31 Phosphorus contents of representative soil profiles from three soil orders. All three soils contain a relatively high proportion of organic phosphorus in their surface horizons. The Aridisol has a high inorganic phosphorus content throughout the profile because rainfall during soil formation was insufficient to leach much of the inorganic phosphorus compounds from the soil. The increased phosphorus in the subsoil of the Ultisol is due to adsorption of inorganic phosphorus by iron and aluminum oxides in the B horizon. In both the Mollisol and Aridisol, most of the subsoil phosphorus is in the form of inorganic calcium-phosphate compounds. (Diagram courtesy of Ray R. Weil and N. Brady)

---

[15] For a review of organic P and its management in agricultural soil systems, see Dodd and Sharpley (2015).

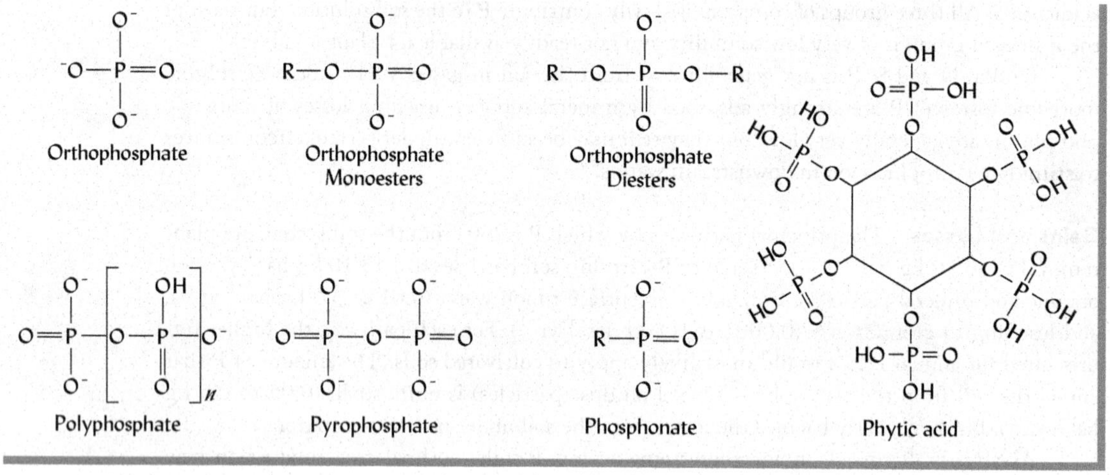

[Chemical structure diagram showing: Orthophosphate, Orthophosphate Monoesters, Orthophosphate Diesters, Polyphosphate, Pyrophosphate, Phosphonate, Phytic acid]

Figure 12.32 *Types of organic phosphorus compounds commonly found in soils, where "R" stands for the rest of a larger organic molecule. Most organic soil phosphorus occurs as some form of phosphate ester. The inorganic orthophosphate ion is also represented (far left, upper). Phytic acid (far right), a plant storage molecule and example of a phosphate mono-ester, is one of the most plentiful organic phosphorus compounds in soils. The nucleic acids that comprise the genetic coding molecule, DNA, are examples of a phosphate di-esters common in small amounts in soils.* (Diagram courtesy of Ray R. Weil)

Inositol phosphates are the most abundant of the known organic P compounds, making up 10 to 60% of the total organic P. One of the most common inositol phosphates in soils is *phytic acid*, a compound in which plants store P in their seeds (including grains like corn). The fact that nonruminant animals cannot digest phytic acid means that grain-fed swine and poultry require supplemental P in their feed and that their manure becomes artificially high in P.

Traditional colorimetric methods (e.g., molybdenum blue) measure only inorganic orthophosphate forms of dissolved P. However, recent research suggests that 50% or more of the water-soluble P in the soil solution and in leachates or runoff water from soils is present as **organic phosphorus** compounds (Figure 12.33). This is especially true where livestock have

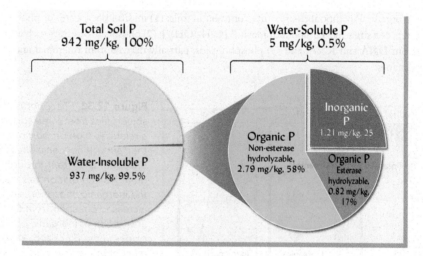

Figure 12.33 *Forms of water-soluble phosphorus in slightly acid sandy loam and silt loam surface soils under unfertilized pastures. (Left) Of a total concentration of 942 mg P per kg soil (dry mass basis), only 0.5% (5 mg/kg) was water soluble (extractable by shaking dried soils in distilled water). (Right) Of the water-soluble P, only 17% (dark blue slice) was inorganic phosphate measurable by traditional colorimetric chemistry. The remainder was organic, about ¼ of which was hydrolysable by phosphate esterase enzymes and therefore could be used by algae in waterways. To accurately predict how soluble P may influence eutrophication and plant nutrition both organic and inorganic P must be measured.* [Diagram courtesy of Ray R. Weil using means for 14 alluvial Vermont Inceptisols in Young et al. (2013)]

been grazed or manure has been applied. Dissolved organic phosphorus (DOP) is generally more mobile than soluble inorganic phosphates, probably because it is not so readily adsorbed by organic matter, clays, Fe/Al oxide, and calcium carbonate in the soil. As a consequence, in soils with high P levels and shallow water tables, the organic P can move with the groundwater to nearby drainage ditches, streams, and lakes where it may contribute significantly to eutrophication.

Mineralization of Organic P Phosphorus held in organic forms can be mineralized and immobilized by the same general processes that release N and S from SOM (see Section 12.1). Net immobilization of soluble P is most likely to occur if residues added to the soil have a C/P ratio greater than 300:1, while net mineralization is likely if the ratio is below 200:1. Mineralization of organic P in soils is subject to many of the same influences that control the general decomposition of SOM—such as temperature, moisture, and tillage (see Section 12.3). In temperate regions, mineralization of organic P in soils typically releases 5 to 20 kg P/ha annually, most of which is readily absorbed by growing plants. These values can be compared to the annual uptake of P by most crops, trees, and grasses, which generally ranges from 5 to 30 kg P/ha.

When forested soils are first brought under cultivation in tropical climates, the amount of P annually released by mineralization may exceed 50 kg/ha, but unless P is added from outside sources these high rates of mineralization will soon decline due to the depletion of readily decomposable SOM. In Florida, rapid mineralization of organic matter in Histosols (Saprists) drained for agricultural use is estimated to release about 80 kg P/ha/yr. Unlike most mineral soils, these organic soils possess little capacity to retain dissolved P, so water draining from them is quite concentrated in P (0.5 to 1.5 mg P/L) and is thought to be contributing to the degradation of the Everglades wetland system.

Contribution of Organic P to Plant Needs The readily decomposable or easily soluble fractions of soil *organic phosphorus* are often the most important factor in supplying P to plants in *highly weathered soils* (e.g., Ultisols and Oxisols), even though the total organic matter content of these soils may not be especially high. The inorganic P in the highly weathered soils is far too insoluble to contribute much to plant nutrition. Apparently plant roots and mycorrhizal hyphae are able to obtain phosphate released from organic molecules before it forms inorganic compounds that quickly become insoluble. In contrast, it appears that the more soluble *inorganic forms* of P play the biggest role in the P fertility of *less weathered soils* (e.g., Mollisols and Vertisols), even though these soils generally contain relatively high amounts of SOM. Cover crop residues left on the soil surface as a mulch can increase P availability in both groups of soils. In temperate regions, freezing and thawing can lyse cells in growing plants and fresh residues, rapidly releasing soluble P from the cell cytoplasm. Under certain soil and weather conditions, some of this soluble organic P may be lost in winter runoff water before the spring flush of plant growth can take it up.

Inorganic Phosphorus in Soils[16]

Of all the macronutrients found in soils, P has by far the smallest quantities in solution or in readily soluble forms in mineral soils. In addition, inorganic P in mineral soils is notoriously immobile. Two phenomena tend to control the concentration of P in the soil solution and the movement of P in soils: (1) the solubility of P-containing minerals, and (2) the adsorption or fixation of phosphate ions on the surface of soil particles. In practice, it is difficult to separate the influence of these two types of reactions or even determine the exact nature of inorganic P compounds present in a particular soil.

Fixation and Retention The tendency for soils to fix P in relatively insoluble, unavailable forms has far-reaching consequences for phosphorus management. For example, P fixation may be viewed as troublesome if it prevents plants from using all but a small fraction of fertilizer P applied. On the other hand, P fixation can be viewed as a benefit if it causes most of the

---

[16]See Devau et al. (2009) for a discussion of the effect of pH on P reactions in soil.

BOX 12.2
PHOSPHORUS REMOVAL FROM WASTEWATER

Environmental soil scientists and engineers remove P from municipal wastewater by taking advantage of some of the same reactions that bind P in soils. After primary and secondary sewage treatment that removes solids and oxidizes most of the organic matter, tertiary treatment in huge, specially designed tanks (Figure 12.34) causes P to precipitate through reactions with iron or aluminum compounds, such as the following:

$$\underset{\text{Alum}}{Al(SO_4)_3 \cdot 14H_2O} + 2PO_4^{3-} \rightarrow 2AlPO_4 + 3SO_4^{2-} + 14H_2O^- \quad (12.17)$$
$$\underset{\text{Soluble phosphate}}{} \underset{\text{Insoluble AlP}}{}$$

$$\underset{\substack{\text{Ferric}\\\text{chloride}}}{FeCl_3} + \underset{\substack{\text{Soluble}\\\text{phosphate}}}{PO_4^{3-}} \rightarrow \underset{\text{Insoluble Fep}}{FePO_4} + 3Cl^- \quad (12.18)$$

The insoluble aluminum and iron phosphates settle out of solution and are later mixed with other solids from the wastewater to form sewage sludge. The low-P water, after additional processing, is returned to the river.

Other less-expensive tertiary treatment approaches involve the spraying of the wastewater on vegetated soils. Natural soil and plant processes clean the P and other constituents out of the wastewater. In some *infiltration* systems, the water percolates through relatively permeable soils.

Figure 12.34 *Sewage plants remove P from wastewater with chemical reactions similar to those affecting P availability in soils.* (Photo courtesy of Ray R. Weil)

dissolved P to be removed from P-rich wastewater applied to a soil (Box 12.2). The fixation reactions responsible in both situations will be discussed as they apply to the solubility of P under acidic and alkaline soil conditions. We will begin by describing the various inorganic compounds and their solubility.

Inorganic Phosphorus Compounds As indicated by Figure 12.29, most inorganic P compounds in soils fall into one of two groups: (1) those containing calcium, and (2) those containing iron and aluminum (and, less frequently, manganese).

As a group, the calcium phosphate compounds become more soluble as soil pH decreases; hence, they tend to dissolve and disappear from acid soils. On the other hand, the calcium phosphates are quite stable and very insoluble at higher pH and so become the dominant forms of inorganic P present in neutral to alkaline soils. Of the common calcium compounds containing P, the **apatite** minerals are the least soluble and are therefore the least plant-available source of P. Some apatite minerals (e.g., fluorapatite) are so insoluble that they persist even in weathered (acid) soils. The simpler mono- and dicalcium phosphates are readily available for

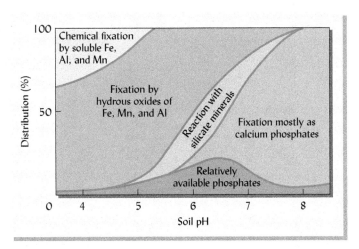

Figure 12.35 *Inorganic fixation of added phosphates at various soil pH values. Average conditions are represented and any particular soil would have a somewhat different distribution. The actual proportion of the P remaining in an available form will depend upon contact with the soil, time for reaction, and other factors. It should be kept in mind that some of the added P may be changed to organic forms in which it would be temporarily unavailable but subject to mineralization.*
(Diagram courtesy of N. Brady and Ray R. Weil)

plant uptake. Except on recently fertilized soils, however, these compounds are present in only extremely small quantities because they easily revert to the more insoluble forms.

In contrast to calcium phosphates, the iron and aluminum hydroxy phosphate minerals, **strengite** ($FePO_4 \cdot 2H_2O$) and **variscite** ($AlPO_4 \cdot 2H_2O$), have very low solubilities in strongly acid soils and become more soluble as soil pH rises. These minerals would therefore be quite unstable in alkaline soils, but are prominent in acid soils, in which they are quite insoluble and stable.

Effect of Aging on Inorganic Phosphate Availability In both acid and alkaline soils, P tends to undergo sequential reactions that produce P-containing compounds of progressively lower solubility. Therefore, the longer that P remains in the soil, the less soluble—and, therefore, less plant-available—it tends to become. Usually, when soluble P is added to a soil, a rapid reaction removes the P from solution (*fixes* the P) in the first few hours. Slower reactions then continue to gradually reduce P solubility for months or years as the phosphate compounds age. The freshly fixed P may be slightly soluble and of some value to plants. With time, the solubility of the fixed P tends to decrease to extremely low levels.

Solubility of Inorganic Soil Phosphorus The particular types of reactions that fix P in relatively unavailable forms differ from soil to soil and are closely related to soil pH (Figure 12.35). In acid soils these reactions involve mostly Al, Fe, or Mn, either as dissolved ions, oxides, or hydrous oxides. Many soils contain such hydrous oxides as coatings on soil particles and as interlayer precipitates in silicate clays. In alkaline and calcareous soils, the reactions primarily involve precipitation as various calcium phosphate minerals or adsorption to the iron impurities on the surfaces of carbonates and clays. At moderate pH values, adsorption on the edges of kaolinite or on the iron oxide coating on kaolinite clays plays an important role.

Precipitation by Iron, Aluminum, and Manganese Ions Probably the easiest type of P-fixation reaction to visualize is the simple reaction of $H_2PO_4^-$ ions with dissolved Fe^{3+}, Al^{3+}, and Mn^{3+} ions to form insoluble hydroxy phosphate precipitates (Figure 12.36a). In strongly acid soils, enough soluble Al, Fe, or Mn is usually present to cause the chemical precipitation of nearly all dissolved $H_2PO_4^-$ ions by reactions such as the following (using the aluminum cation as an example):

$$\underset{\text{(soluble)}}{Al^{3+} + H_2PO_4^- + 2H_2O} \rightleftharpoons 2H^+ + \underset{\text{(insoluble)}}{Al(OH)_2H_2PO_4} \qquad (12.16)$$

Freshly precipitated hydroxy phosphates are slightly soluble because they have a great deal of surface area exposed to the soil solution. Therefore, the P contained in them is, initially at least, somewhat available to plants. Over time, however, as the precipitated hydroxy phosphates age, they become less soluble and the P in them becomes almost completely unavailable to most plants.

Reaction with Hydrous Oxides and Silicate Clays Most of the P fixation in acid soils probably occurs when $H_2PO_4^-$ ions react with, or become adsorbed to, the surfaces of

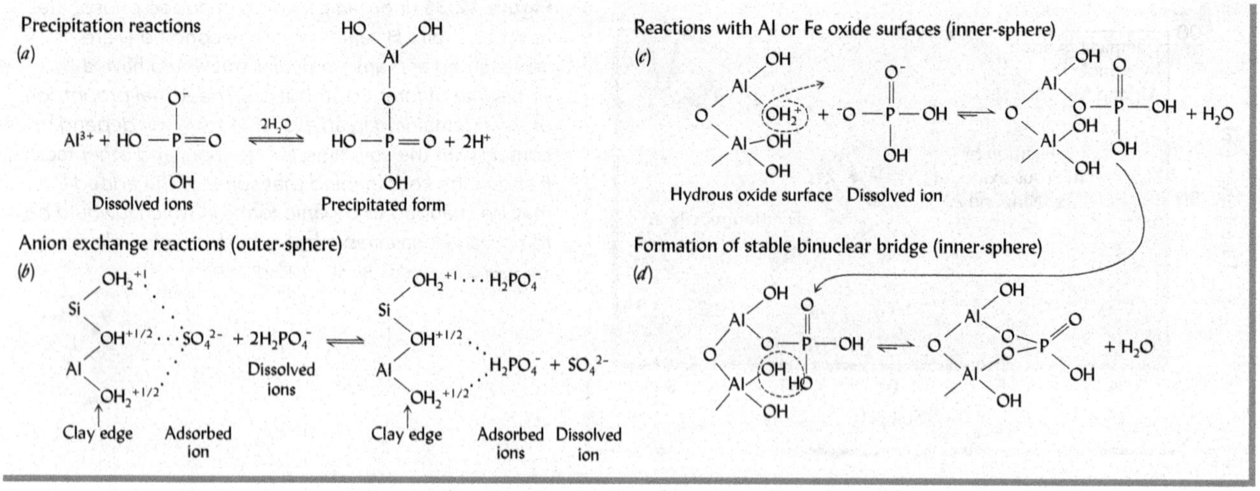

Figure 12.36 *Several of the reactions by which phosphate ions are removed from soil solution and fixed by the iron and aluminum in various hydrous oxides. Freshly precipitated aluminum, iron, and manganese phosphates (a) are relatively available, though over time they become increasingly unavailable. In (b) the phosphate is reversibly adsorbed by anion exchange. In reactions of the type shown in (c), a phosphate ion replaces an —OH_2 or an —OH group in the surface structure of Al or Fe hydrous oxide minerals. In (d) the phosphate further penetrates the mineral surface by forming a stable binuclear bridge. The adsorption reactions (b, c, d) are shown in order from those that bind phosphate with the least tenacity to the most tenacity (from the most to the least reversible and plant-available). Phosphate ions added to a soil may undergo this entire sequence of these reactions, becoming increasingly unavailable. Note that (b) illustrates an outer-sphere complex, while (c) and (d) are examples of inner-sphere complexes (see Figure 8.25).* (Diagram courtesy of Ray R. Weil)

insoluble oxides of iron, aluminum, and manganese, such as gibbsite ($Al_2O_3 \cdot 3H_2O$) and goethite ($Fe_2O_3 \cdot 3H_2O$; see Figure 12.23) and with 1:1 type silicate clays. These hydrous oxides occur as crystalline and noncrystalline particles and as coatings on the interlayer and external surfaces of clay particles. Fixation of P by clays probably takes place over a relatively wide pH range (see Figure 12.35). The large quantities of Fe, Al oxides and 1:1 clays present in many soils make possible the fixation of extremely large amounts of P by these reactions.

The $H_2PO_4^-$ anion may be attracted to positive charges that develop under acid conditions on the surfaces of iron and aluminum oxides and the broken edges of kaolinite clays (see Figure 12.36b). The adsorbed $H_2PO_4^-$ anions form outer-sphere complexes and are subject to anion exchange with certain other anions, such as OH^-, SO_4^{2-}, MoO_4^{2-}, or organic acids (R—COO^-; see Section 8.7). Since this type of adsorption of $H_2PO_4^-$ ions is reversible, the P may slowly become available to plants. Availability of such adsorbed $H_2PO_4^-$ may be increased by (1) liming the soil to increase the hydroxyl ions, or (2) adding organic matter to increase organic acids (anions) capable of replacing $H_2PO_4^-$.

The phosphate ion may also replace a structural hydroxyl to form an inner-sphere complex with the oxide (or clay) surface (see Figure 12.36c). This reaction, while reversible, binds the phosphate too tightly to allow its ready replacement by other anions. The availability of phosphate bound in this manner is very low. Over time, a second oxygen of the phosphate ion may replace a second hydroxyl, so that the phosphate becomes chemically bound to two adjacent aluminum (or iron) atoms in the hydrous oxide surface (see Figure 12.36d). With this step, the phosphate becomes an integral part of the oxide mineral, and the likelihood of its release back to the soil solution is extremely small.

Finally, as more time passes, the precipitation of additional iron or aluminum hydrous oxide may bury the phosphate deep inside the oxide particle. Such phosphate is termed *occluded* and is the least available form of P in most acid soils.

Precipitation reactions similar to those just described are responsible for the rapid reduction in availability of P added to soil as soluble $Ca(H_2PO_4)_2 \cdot H_2O$ in fertilizers. As already mentioned, (see Box 12.2) this type of reaction can also be used to control the solubility of P in wastewater.

Effect of Iron Reduction Under Wet Conditions Phosphorus bound to iron oxides by the mechanisms just discussed is very insoluble under well-aerated conditions. However, prolonged anaerobic conditions can reduce the iron in these complexes from Fe^{3+} to Fe^{2+}, making the iron–phosphate complex much more soluble and causing it to release P into solution. The release of P from iron phosphates by means of the reduction and subsequent solubilization of iron improves the P availability in soils used for paddy rice.

These reactions are also of special relevance to water quality. Phosphorus bound to soil particles may accumulate in underwater sediments, along with organic matter and other debris. As the sediments become anoxic, the reducing environment may cause the gradual release of P held by hydrous iron oxides. Thus, the P eroded from soils today may aggravate the problem of eutrophication for years to come, even after the erosion and loss of P from the land has been brought under control.

Reaction with Ca and Mg in Alkaline Soils In alkaline soils, soluble $H_2PO_4^-$ quickly reacts with calcium to form a sequence of products of decreasing solubility. For instance, highly soluble monocalcium phosphate [$Ca(H_2PO_4)_2 \cdot H_2O$] added as concentrated superphosphate fertilizer rapidly reacts with calcium carbonate in the soil to form first moderately soluble dicalcium phosphate ($CaHPO_4 \cdot 2H_2O$) and then low solubility tricalcium phosphate [$Ca_3(PO_4)_2$]. Although the latter compound is quite insoluble, it may form even more insoluble compounds, such as hydroxy-, oxy-, carbonate-, and fluor-apatites. These compounds are thousands of times less soluble than freshly formed tricalcium phosphates. These solubility relationships mean that powdered phosphate rock (which consists mainly of apatite minerals) may be an effective source of P for plants in strongly acid soils, while the same material would likely be of little utility if used on neutral or alkaline soils.

Because of the various reactions with $CaCO_3$, phosphorus availability tends to be nearly as low in the Aridisols, Inceptisols, and Mollisols of arid regions as in the highly acid Spodosols, Ultisols, and Oxisols of humid regions, where iron, aluminum, and manganese limit P availability.

Biological Influences on Inorganic P Solubility Certain bacteria and fungi enhance the solubility of both calcium and iron and aluminum phosphates by releasing various organic acids. These acids either dissolve the calcium phosphates or form metal complexes that release the P from iron and aluminum phosphates (Figure 12.37). The released P is likely used first by the microorganisms themselves, but is eventually made available to plants as well (see following).

Phosphorus-Fixation Capacity of Soils

The P-fixation capacity of a soil may be conceptualized as the total number of sites capable of reacting with phosphate ions to "fix" them in unavailable, insoluble forms. The fixation reactions include all of the sorption and precipitation reactions just discussed. The concept of a finite P-fixation capacity for a soil is illustrated schematically in Figure 12.38.

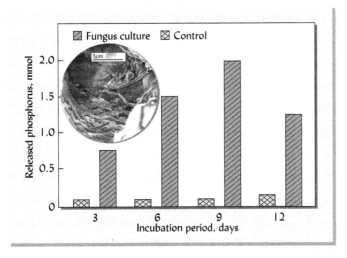

Figure 12.37 Certain soil microorganisms can increase the availability of P in minerals such as phosphate rock and aluminum phosphates that normally hold the P in very insoluble forms. (Inset) A micrograph of a fungus growing on an aluminum phosphate soil surface. The fungus is thought to produce organic acids that help solubilize P. (bar graph) In another experiment, P is released from phosphate rock by a fungus (Aspergillus niger) that had been isolated from a tropical soil. [Micrograph (inset) courtesy of Dr. Anne Taunton, University of Wisconsin; bar graph drawn from data in Goenadi et al. (2000)]

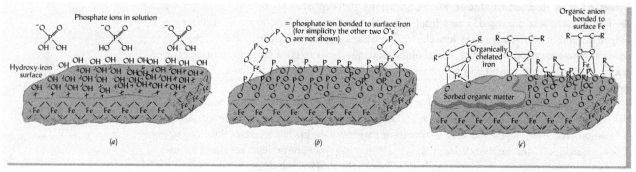

Figure 12.38 Schematic illustrations of phosphorus-fixation sites on a soil particle surface showing hydrous iron oxide as the primary fixing agent. In part (a) the sites are shown as + symbols, indicating positive charges or hydrous metal oxide sites, each capable of fixing a phosphate ion. In part (b) the fixation sites are all occupied by phosphate ions (the soil's fixation capacity is satisfied). Part (c) illustrates how organic anions, larger organic molecules, and certain strongly fixed inorganic anions can reduce the number of sites available for fixing P. Such mechanisms partially account for the reduced P fixation and greater P availability brought about when mulches and other organic materials are added to a soil. (Diagram courtesy of Ray R. Weil)

One way of determining the P-fixing capacity of a particular soil is to shake a known quantity of the soil in a solution of known P concentration. After about 24 hours an equilibrium will be approached, and the concentration of P remaining in the solution (the **equilibrium phosphorus concentration [EPC]**) can be determined. The difference between the initial and final (*equilibrium*) solution P concentrations represents the amount of P fixed by the soil. If this procedure is repeated using a series of solutions with different initial P concentrations, the results can be plotted as a P-fixation curve (sometimes referred to as a P fixation isotherm, if temperature is held constant, Figure 12.39). The maximum P-fixation capacity can be extrapolated from the value at which the curve levels off.

Phosphorus fixation by soils is not easily reversible. However, if a portion of the fixed P is present in relatively soluble forms (see Section 12.6) and most of the fixation sites are already occupied by a phosphate ion, some release of P to solution is likely to occur when the soil is exposed to water with a very low P concentration. This release (often called *desorption*) of P is indicated in Figure 12.39 where the curve for soil A crosses the zero fixation line and becomes negative (negative fixation = release). The solution EPC_0 is the

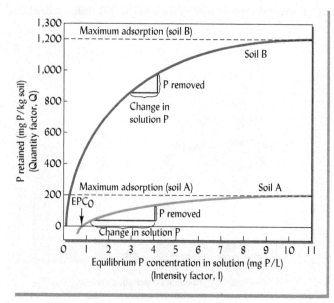

Figure 12.39 Relationships between P fixed and P remaining dissolved when soils A and B are shaken in solutions of various initial P concentrations. Each soil removes nearly all the P from dilute solutions. However, when solutions contain so much P that most P-fixation sites become occupied, a larger portion of the P remains in solution. The amount retained levels off as the maximum P-fixing capacity of the soil is reached (horizontal dashed lines: 200 or 1,200 mg P/kg of soil A or B). If the initial P concentration is equal to the equilibrium phosphorus concentration (EPC) for a particular soil, P will neither be sorbed nor released (i.e., P fixation=0 and $EPC=EPC_0$). If the solution P concentration is less than the EPC_0, the soil will release some P (i.e., the fixation will be negative). In this example, soil B has a much higher P-fixing capacity and a much lower EPC than does soil A. Soil B is highly buffered because much P must be added to achieve a small increase in the equilibrium solution P concentration. (Diagram courtesy of Ray R. Weil)

equilibrium phosphorus concentration (x-axis) at which zero fixation occurs (phosphorus is neither released nor retained). The EPC_0 is an important parameter for both soil fertility and environmental assessment because it indicates (1) the capacity to replenish the soil solution as it is depleted of P by plant roots, and (2) the rate at which the soil will release P into runoff and leaching waters.

Factors Affecting the Extent of Phosphorus Fixation in Soils Soils that remove more than 350 mg P/kg of soil (i.e., a P-fixing capacity of about 700 kg P/ha in surface soil) from solution are generally considered to be high P-fixing soils. High P-fixing soils tend to maintain low P concentrations in the soil solution and in runoff water. If soils with similar pH values and mineralogy are compared, P fixation tends to be more pronounced, and ease of P release tends to be lowest in those soils with higher clay contents.

Effect of Clay Some clay minerals are much more effective at P fixation than others. Generally, those clays that possess greater anion exchange capacity (due to positive surface charges) have a greater affinity for phosphate ions. For example, extremely high P fixation is characteristic of allophane clays typically found in Andisols and other soils associated with volcanic ash. Oxides of iron and aluminum also strongly attract and hold phosphorus ions. Among the layer silicate clays, kaolinite has a greater P-fixation capacity than the 2:1 clays of less-weathered soils. Thus, the soil components responsible for P-fixing capacity are in the order of increasing extent and degree of fixation:

2:1 clays << 1:1 clays < carbonate crystals < crystalline Al, Fe, Mn oxides < amorphous Al, Fe and Mn oxides, allophane

To some degree, the preceding P-fixing soil components are distributed among soils in relation to soil taxonomy. Vertisols and Mollisols generally are dominated by 2:1 clays and have low P-fixation capacities unless they are calcareous. Iron and aluminum oxides are prominent in Ultisols and Oxisols. Andisols, characterized by large quantities of amorphous oxides and allophane, have the greatest P-fixing capacity, and their productivity is often limited by this property.

Effect of Soil pH As a general rule in mineral soils, phosphate fixation is at its lowest (and plant availability is highest) when soil pH is maintained in the 6.0 to 7.0 range (see Figure 12.35). Even if pH ranges from 6.0 to 7.0, phosphate availability may still be very low, and added soluble phosphates will be readily fixed by soils. The low recovery by plants of phosphates added to unfertilized mineral soils in a given season is partially due to this fixation. A much higher recovery would be expected in organic soils and in many potting mixes where calcium, iron, and aluminum concentrations are not as high as in mineral soils.

Effect of Organic Matter Organic matter has little capacity to strongly fix phosphate ions. To the contrary, amending soil with organic matter, especially decomposable material, is likely to reduce P fixation by several mechanisms (see Figure 12.38). First, organic molecules adhering to sorbing surfaces can mask fixation sites and prevent them from interacting with phosphorus ions in solution. Second, organic acids produced by plant roots and microbial decay can serve as organic anions, which compete with phosphorus ions for positively charged sites on the surfaces of clays and hydrous oxides. Third, certain organic compounds can entrap reactive Al and Fe in stable organic complexes called *chelates* (see Section 12.5). Once chelated, the metals are unavailable for reaction with phosphorus ions in solution. In addition, P fixation is also likely reduced by the release of phosphorus ions by microbial mineralization of many P-rich organic materials.

How Do Plants Obtain Adequate Phosphorus?[17]

Because of the low supply of soil P and the many P-fixation reactions just discussed, terrestrial plants have had to evolve special mechanisms or strategies (see Figure 12.40) for obtaining the

---

[17]The Virtual Fertilizer Research Center has published several reviews aimed at stimulating innovations in this arena (Koele et al., 2014; Smit et al., 2013). For ways to more efficiently use P resources by applying the chemical and biological process discussed in this chapter, see also Frossard et al. (2000).

Figure 12.40 Six basic strategies that plant roots may employ to enhance their uptake of various forms of phosphorus from soils. (1) Increased root absorptive surface area. (2) Chelate iron or aluminum to release P. (3) Dissolve calcium-phosphate compounds with acid exudates. (4) Exude phosphatase enzymes to release P from organic compounds. (5) Exude substances to stimulate P-solubilizing rhizobacteria. (6) Encourage colonization by mycorrhizal fungi that help plants take up P. (Diagram courtesy of Ray R. Weil)

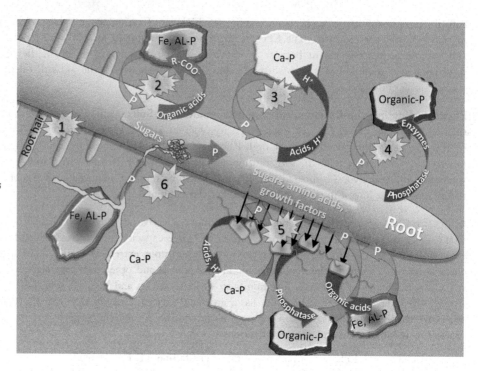

P they need. Knowledge of these strategies can help us understand ecological relationships and develop more effective plant–soil management practices.

When roots encounter low P soil they may produce more and longer fine roots and root hairs to increase the absorptive surface area available for P uptake. Roots of certain species exude specific organic compounds (e.g., citric, malic, malonic, oxalic, piscidic, and tartaric acids) that complex with iron or aluminum and thus release the P from insoluble Fe/Al-phosphates. Roots of some plants acidify their rhizosphere to speed the dissolution of sparingly soluble calcium-phosphate compounds. Roots may exude enzymes (phosphatases) that can cleave the phosphate group from organic compounds, especially in dissolved organic P, making the P available for root uptake.

Some roots exude growth-stimulating substances that attract and/or enhance the growth of rhizobacteria or fungi, which in turn excrete, organic acids, chelators, or enzymes that release soluble orthophosphate by attacking various phosphorus compounds in the soil. Many plants use root exudates as signaling compounds to attract and encourage colonization by mycorrhizal fungi. The plants then invest some of their sugars from photosynthesis as "payment" to these fungi for "phosphorus delivery services" (see Section 10.9).

Many plant species deploy a combination of several of these strategies. For example, plants in the Brassica family (cabbage, mustard, radish, etc.) compensate for their very poor to no mycorrhizal colonization by excreting citric and malic acids, forming extensive fine root hairs, and taking up high amounts of Ca^{2+} (which acidifies the rhizosphere and accelerates the dissolution of calcium-phosphates by mass action). Certain legumes, such as faba bean, in addition to forming strong symbiotic relationships with mycorrhizal fungi, also exude organic acids and phosphatase enzymes enabling access to P in Fe/Al phosphates and organic compounds. Knowledge of these plant characteristics can aid in choosing plants for ecosystem restoration, as well as for farmers who need to make more efficient use of P soil reserves and fertilizers in low-P soils.

Management Strategies for Meeting Plant Phosphorus Needs in Low-P Soils

The principles of soil phosphorus behavior discussed in this chapter suggest a number of management strategies to address the twin problems of too little and too much of this critical element. On one hand, the challenge is to maximize the efficiency by which plants obtain phosphorus from low-P soils. On the other hand, the challenge is to avoid wasting this finite and precious resource (Box 12.3) and to keep it from polluting aquatic systems.

Enhance Mycorrhizal Symbiosis Practices that enhance mycorrhizal symbiosis usually improve the utilization of soil phosphorus. One such practice is to include in the plant community or crop rotation plant species that are effective hosts for mycorrhizal fungi (Figure 12.32). It is also important to minimize tillage disruption of hyphal networks in the soil, and in some highly disturbed soils, to inoculate with appropriate fungi (see Section 11).

BOX 12.3
PEAK PHOSPHORUS?[a]

Many scientists argue that improving the efficiency of phosphorus use in farming is not only essential for profitable agriculture but is also a moral obligation to future generations that will have to depend on Earth's limited phosphorus supplies. The immediacy of this concern is not shared by everyone, but the importance of conserving phosphorus is based on two facts that are quite indisputable: (1) phosphorus has *no* substitute and (2) it is *not* a renewable resource.

All living things require phosphorus as it is *literally* in their DNA. Humans need P in their own diets. Soils need P if they are to support the plants and animals we use for food. There is no substitute for phosphorus in these roles. Economists tell us that generally goods will be replaced by something else if scarcity drives up prices. For example, if copper becomes too expensive, fiber optic cables might replace copper wires; or if fossil fuels become too expensive, people may invest in wind power to replace oil and gas in generating electricity. Since P is a basic chemical element in the structure of many essential cellular components (DNA, RNA, membranes, ATP), no such substitution will be possible.

Phosphorus is a nonrenewable resource—and one that is in quite limited supply, both in absolute global amounts and in geographic distribution. The vast majority of the world's minable phosphorus is in the North African country of Morocco. Historical examples and current resource theory suggest that as the best, easiest to mine deposits get used up first, the remaining resources get harder and more expensive to mine and refine. Thus accelerating resource exploitation to meet growing demand will eventually be limited first by escalating costs and then by dwindling absolute supply, resulting in a maximum or peak rate of production when just over half of the total resource has been used up (Figure 12.41). The remaining deposits will continue to be mined for decades beyond that time, but in ever smaller amounts and at ever greater expense. While there is considerable disagreement about the actual size of world phosphate reserves

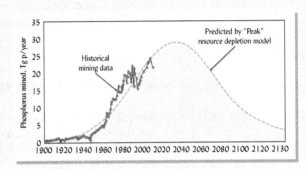

Figure 12.41 World historical and modeled trends for mined phosphorus production illustrating the concept of a peak in production occurring when somewhat more than half of the total reserves have been depleted. [Modified from Cordell and White (2011)]

and how long they will last if extraction technology improves (estimates range between 100 and 500 years to exhaustion), the data suggest peak production will come much sooner than once thought—perhaps during this century. Hence there is growing sense of urgency among many scientists and policy makers (see, e.g., the European effort at http://www.phosphorusplatform.eu/).

On the positive side, the same P atoms can be used over and over again—if they are not so carelessly dispersed as to make them virtually unrecoverable. The current model is unsustainable. We mine phosphorus and apply it as soluble fertilizer only to have much of it wash off P-saturated farmland into streams and then into lakes and oceans. Even most of the P that crops do take up and use makes its way indirectly to the oceans after a one-way trip through the food to sewage system. For these reasons—not to even mention the damage P causes in eutrophication of aquatic systems—there is a growing belief that the time has come for individuals and societies to learn more efficient and sustainable methods of using our infinitely precious but definitely finite phosphorus resource.

[a]For reviews of the complex issues and varying estimates involved with Peak Phosphorus concept, see Cordell and White (2015) and Scholz et al. (2015). For perspectives on the "trilemma" involving geopolitics, poverty, and resource limitations, see Obersteiner et al. (2013) and Wyant et al. (2013).

Grow P-Efficient Plants Some plant species require much less P in the soil solution than do others. The P requirements of native plants and agricultural species vary widely. To better utilize finite P resources, plant scientists will need to breed varieties or select species with the ability to thrive in soil with low levels of phosphorus supply and solubility.

Adjust P Application to Soil Status Where P-fixing capacity is grossly unsaturated, optimum crop yields will likely require additions that considerably exceed plant uptake. However, as the excess P begins to saturate fixation sites, rates of application should be lowered to supply no more than what plants take up so as to prevent excessive P accumulations.

Localized Placement Phosphorus fertilizer concentrated in a localized zone of soil is less likely to undergo fixation reactions than fertilizer mixed into the bulk soil. One can reduce the amount of fertilizer required by placing it in narrow bands alongside plant rows or in small holes next to individual plants and by the use of pellets instead of fine powders.

Combine Ammonium with Phosphorus When ammonium and phosphorus fertilizers are mixed in a band, the acidity produced by oxidation of ammonium ions (see reaction 12.4) and by uptake of excess cations as ammonium (see Section 9.1) keeps the phosphate in more soluble compounds and enhances plant P uptake.

Control of Soil pH Phosphorus availability can be optimized in most soils by proper liming or acidification (see Sections 9.8 and 9.10) to a pH level between 6 and 7.

Cycling of Organic Matter During the microbial breakdown of organic materials, P is released slowly and can be taken up by plants or mycorrhizae before it can be fixed by the soil. In addition, organic compounds can reduce soil P fixation capacity (Figure 12.38).

Use of P-Efficient Cover Crops and Green Manures Highly P-efficient plants solubilize and take up P from the soil profile and incorporate it into their tissues. When the cover crop is killed or tree prunings are used as a mulch, the plant tissues decompose, and the contained P is released into the surface soil, where it can be easily used by less efficient plants that follow in rotation (Figure 12.42).

Management Strategies for Controlling Over-Enrichment of Soils and Water Pollution

Avoid Excess Accumulation This goal requires carefully keeping the sum of all phosphorus inputs (deposition, fertilizers, organic amendments, plant residues, and animal feed) from consistently exceeding plant removals or accumulating beyond the lowest levels that will support near-optimum plant growth. In cases where soils are already over enriched with phosphorus, an

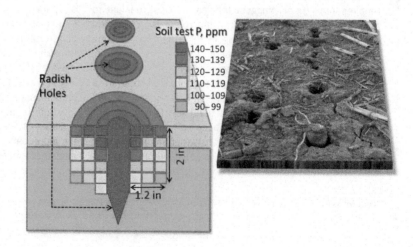

Figure 12.42 *Localized concentration and increased plant availability of soil phosphorus as a result of the growth, nutrient uptake, and subsequent decay of a tap-rooted cover crop (forage radish). (Left) Diagram of soil P concentrations as determined by Mehlich3 soil test (see Section 16.11). (Right) A photograph shows the decaying roots and root holes left by a similar forage radish cover crop after it had winter-killed.* [Diagram from data in White and Weil (2011); photo courtesy of Ray R. Weil]

effort should be made to remove the excess through intensive production and harvest export of high P-content crops without further P additions.

Minimize Loss in Runoff and Eroded Sediment Conservation tillage, cover crops, and residue mulch can minimize P loss by runoff and erosion, *especially from land already high in phosphorus.* Application of manure or fertilizer to frozen soils should be avoided, and application by injection into the soil rather than broadcasting on the surface should be implemented (see Section 13.8).

Capture P from Runoff Natural or constructed wetlands (Section 7.7) and perennially vegetated riparian (shoreline) buffer strips (Section 16.2) can tie up some P before runoff enters sensitive lakes or streams. These measures remove some dissolved P, but mainly P bound to sediment. Runoff can also be directed through structures filled with iron or calcium compounds that react with and remove dissolved P in the water.

Capture P from Drainage Water In most soils P is quite immobile and leaching losses are normally very small. However, where certain agricultural soils have been artificially drained and overfertilized with phosphorous, significant quantities of both organic and inorganic P have been observed to leach through the upper soil profile and into drain tiles or ditches. Under saturated conditions (see Section 6.8), preferential flow down macro pores may carry P in solution or fixed to suspended soil particles. One strategy for controlling such losses is to manage drainage flows so they occur only when needed. Another approach is to use iron-, aluminum-, or calcium-containing compounds as chemical filters in drainage ditches remove P from drainage water.

Tie Up P with Inorganic Compounds Various iron-, aluminum-, or calcium-containing materials can be used to react with P dissolved from soils or phosphorus-rich organic wastes. Highly insoluble compounds are formed, similar to those described in Box 12.2. Application of about 2 Mg/ha of gypsum to the surface of artificially drained cropland with certain types of soils may reduce P loss in both surface runoff and drainage water (Figure 12.43). The calcium in the gypsum reacts with dissolved P and removes it from the surface and subsurface runoff

Figure 12.43 *Gypsum applied to the soil surface reduced phosphorus loss from tile-drained croplands with fine-textured high-phosphorus soils. The photos show drainage water coming from two side-by-side fields (and bottled samples coming from similar fields), the one on the right having been treated with gypsum. This research in Ohio suggests that about 2 Mg of gypsum spread once every 2 to 3 years could reduce phosphorus losses by nearly 50% in both surface runoff and drainage water.* (Photos courtesy Warren Dick, The Ohio State University, see also Kost et al. 2018)

water. Also, increased ionic strength around dissolving gypsum particles enhances clay flocculation (see Section 10.6), reducing the dispersion and transport of clay—and its adsorbed P load—down soil macropores and across the soil surface.

12.4 POTASSIUM: NATURE AND ECOLOGICAL ROLES[18]

Unlike P (or S and, to a large extent, N), K is present in the soil solution only as a positively charged cation, K^+. Like phosphorus, potassium does not commonly form any gases that could be lost to the atmosphere. Its behavior in the soil is influenced primarily by soil cation exchange properties (see Chapter 8) and mineral weathering (Chapter 2), rather than by microbiological processes. Unlike N and P, K causes no serious off-site environmental problems when it leaves the soil system. It is not toxic and does not cause eutrophication in aquatic systems.

Potassium in Plant and Animal Nutrition

Even though most soils have large total supplies of this element, most of that present is tied up as insoluble minerals and is only slowly available for plant use. Plants require K and N in similar amounts. Potassium is the nutrient that is the third or fourth most likely (after N, P, and probably S) to limit plant productivity. For this reason it is commonly applied to soils to enhance fertility and is a component of most so-called "complete" mixed fertilizers.

Although K plays numerous roles in plant and animal nutrition, it is not actually incorporated into the structures of organic compounds. Instead, K remains in the ionic form (K^+) in solution in the cell or acts as an activator for some 80 cellular enzymes. Certain plants, many of which evolved in sodium-rich semiarid environments, can substitute sodium or other monovalent ions to carry out some, but not all, of the functions of K.

A major component of plant cytoplasmic solution, K plays a critical role in photosynthesis, protein synthesis, N fixation in legumes, starch formation, and the translocation of sugars. In most plants, healthy leaf tissue can be expected to contain 1 to 4% K, similar to the concentration of N but an order of magnitude greater than that of P or S.

Potassium is especially important in helping plants adapt to environmental stresses. Alleviation of K deficiency is commonly observed to enhance drought and cold tolerance, reduce pest damage, and lessen the incidence and severity of both fungal and bacterial plant diseases (Figure 12.44). However, application of K fertilizers beyond that needed to alleviate deficiency has not been shown to further reduce pest and disease problems.

Sufficient K also enhances the quality of flowers, fruits, and vegetables by improving flavor and color and strengthening stems (thereby reducing lodging). In many of these respects, K seems to counteract some of the detrimental effects of excess N. In animals, including humans, K counteracts effects of excessive sodium and helps regulate the nervous system and maintain healthy blood vessels. Diets that include such high-K foods as bananas, potatoes, citrus, and leafy green vegetables have been shown to lower human risk of stroke and heart disease.

Deficiency Symptoms in Plants Deficiency of K is relatively easy to recognize in most plants. Specific symptoms of deficiency usually occur earliest and most severely on the oldest leaves. In general, tips and edges of the oldest leaves begin to yellow (chlorosis) and then die (necrosis), so that the leaves appear to have been burned on the edges (Figure 12.45). On some plants the necrotic leaf edges may tear, giving the leaf a ragged appearance (Figure 12.45a,b,g). In certain forage and cover-crop legume species, K deficiency produces small, white necrotic spots that form a unique pattern along the leaflet margins; this easily recognized symptom is one that people often mistake for insect damage (see Figure 12.39e,h). Potassium deficiency should not be confused with damage from excess salinity, which can also produce brown, necrotic leaf margins. Salinity damage is more likely to affect the newer leaves (see Figures 9.25 and 9.27).

---

[18] For further information on this topic, see Mengel and Kirkby (2001) and Römheld and Kirkby (2010).

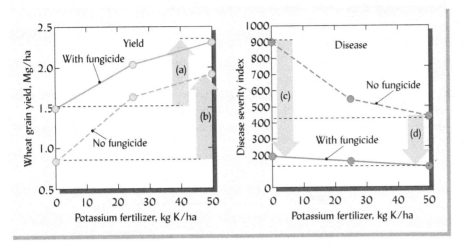

Figure 12.44 Potassium fertilization on soils with low availability of this nutrient can reduce plant disease severity and increase yields. In South Asia, Helminthosporium leaf blight causes serious loss in wheat crops. Disease severity may be aggravated by deficiencies of soil nutrients, especially K. In this experiment in Nepal, alleviation of K deficiency reduced the disease severity almost as much as six to nine applications of the fungicide propiconazole. Arrows in the left panel indicate the effect of added K on wheat yield when the crop was (a) or was not (b) protected with fungicide sprays, suggesting that about one-half of the yield increase from K fertilization was related to enhanced disease resistance. Arrows in the right panel indicate the effect of the fungicide in reducing the disease severity without (c) or with (d) K fertilization. Data shown are means of two years and three wheat cultivars. [Graphed from data in Sharma et al. (2005)]

Figure 12.45 Potassium deficiency often produces easily recognized foliar symptoms, mainly on older leaves: chlorotic margins on leaves of soybean (c), corn (d), and maple (f); ragged, necrotic margins of leaves on eggplant (a), banana (b), and piggyback plant (g); and small, white necrotic spots along edges of legume leaflets such on hairy vetch (e) and alfalfa (h). (Photo courtesy of Ray R. Weil)

The Potassium Cycle in Soil–Plant Systems

Figure 12.46 shows the major forms in which K occurs in soils and the changes it undergoes as it is cycled through the soil–plant system. The original sources of K are the primary minerals, such as micas (biotite and muscovite) and potassium feldspar (orthoclase and microcline). As these minerals weather, their rigid lattice structures become more pliable. For example, K held between the 2:1-type crystal layers of mica is in time made more available, first as non-exchangeable but slowly available forms near the weathered edges of minerals and, eventually, as readily exchangeable ions and ions dissolved in the soil solution from which it is absorbed by plant roots.

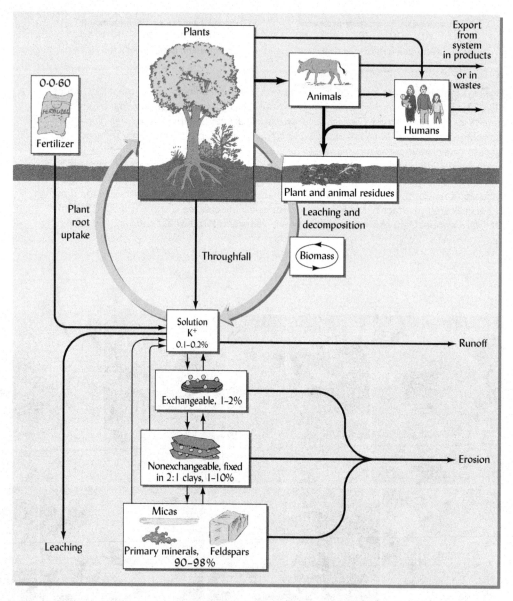

Figure 12.46 *Major components of the K cycle in soils. The large circular arrow emphasizes the biological cycling of K from the soil solution to plants and back to the soil via plant residues or animal wastes. Primary and secondary minerals are the original sources of the element. Exchangeable K may include those ions held and released by both clay and humus colloids, but K is not a structural component of soil humus. The interactions among solution, exchangeable, nonexchangeable, and structural K in primary minerals is shown. The bulk of soil K occurs in the primary and secondary minerals and is released very slowly by weathering processes.* (Diagram courtesy of Ray R. Weil)

After K is taken up by plants, a portion is leached from plant foliage by rainwater (throughfall) and returned to the soil, and a portion is returned to the soil with the plant residues. In natural ecosystems, most of the K taken up by plants is returned in these ways or as wastes (mainly urine) from animals feeding on the vegetation. Some K is lost with eroded soil particles and in runoff water, and some is lost to groundwater by leaching. In agroecosystems, from one-fifth (e.g., in cereal grains) to nearly all (e.g., in hay crops) of the K taken up by plants may be exported from the field, commonly to distant markets, from which it is unlikely to return.

At any one time, most soil K is in primary minerals and nonexchangeable forms. In relatively young, fertile soils containing substantial amounts of weatherable minerals in the soil profile, the release of K from these forms to the exchangeable and soil solution forms that plants can use directly is usually sufficiently rapid to keep plants supplied with the K needed for good growth. On the other hand, where large amounts of high K content plant biomass are repeatedly removed from the land, or where the content of weatherable K-containing minerals is low, the levels of exchangeable and solution K may have to be supplemented by outside sources, such as chemical fertilizers, poultry manure, or wood ashes. On such inherently low K soils (especially quartz sands), K additions may be necessary to prevent the K depletion and declining productivity over a period of years. An example of depletion and restoration of available soil K is illustrated in Figure 12.47.

Deep-rooted plants often act as "nutrient pumps," taking K (and other nutrients) from deep subsoil horizons into their root systems, translocating it to their leaves, and then recycling it back to the surface of the soil via leaf fall and leaching.

In most mature natural ecosystems, the small (1 to 5 kg/ha) annual losses of K by leaching and erosion are more than balanced by weathering of K from primary minerals and nonexchangeable forms in the soil profile, followed by vegetative translocation to the surface of the soil. In many agricultural annual cropping systems, leaching losses are far greater because much higher exchangeable K levels are maintained with fertilizers, while crop roots are active for only part of the year.

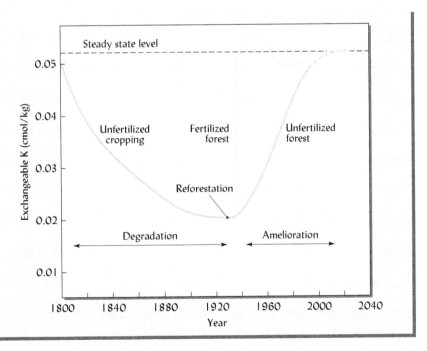

Figure 12.47 *The general pattern of depletion of A-horizon exchangeable K by decades of exploitative farming, followed by its restoration under forest vegetation. The forest consisted of red pine trees planted on a Plainfield loamy sand (Udipsamment) in New York. This soil has a very low cation exchange capacity and low levels of exchangeable K^+.* [Data from Nowak et al. (1991)]

The Potassium Problem in Soil Fertility[19]

Potassium is found in comparatively large amounts in most mineral soils, except those consisting mostly of quartz sand. In fact, the total quantity of this element is generally greater than that of any other major nutrient element. Amounts as great as 30,000 to 50,000 kg K in the upper 15 cm of 1 ha of soil are not at all uncommon (see Table 1.3), and the supplies of K in the subsoil layers are commonly manyfold greater. Yet the quantity of K held in an easily exchangeable condition at any one time may be very small.

Leaching Losses For many soils, K is quite readily lost by leaching. From representative humid-region soils growing annual crops and receiving only moderate rates of fertilizer, the annual loss of K by leaching is usually about 25 to 50 kg/ha, the greater values being typical of acid, sandy soils. Losses would undoubtedly be much larger where the leaching of K not slowed by the attraction of the positively charged K ions to the negatively charged cation exchange sites on clay and humus colloids. Liming an acid soil to raise its pH can reduce the leaching losses of K because of the *complementary ion effect* (see Section 8.10). Trivalent ions (Al^{3+}) are more tightly held than divalent ions (Ca^{2+} and Mg^{2+}), so where higher levels of exchangeable Al^{3+} saturate the exchange complex, K is less likely to replace existing exchangeable ions and become adsorbed. Liming a soil replaces the Al^{3+} with Ca^{2+} and Mg^{2+}, allowing K^+ to be more readily retained on the exchange complex, thus reducing its loss by leaching.

Plant Uptake and Removal Harvest of most or all of the aboveground plant parts, such as in whole tree harvest, hay making, or corn silage production, will seriously deplete the soil K supply by 100s of kg K/ha. While such removals are large relative to the supply of exchangeable K, they are miniscule in relation to the total supply of K that is potentially in equilibrium over time with the exchangeable pool.

Luxury Consumption Uptake and removal of K can be exaggerated by the tendency of plants to take up much more K than they need if sufficiently large quantities of easily available K are present. This tendency is termed *luxury consumption*, because the excess K absorbed does not increase plant growth or function (Figure 12.48). Applying fertilizer at rates that cause luxury consumption is decidedly wasteful. In addition, high K uptake is likely to depress the uptake of other cations, particularly calcium and magnesium, and may thereby cause nutritional imbalances both in the plants and in animals that consume them.

In summary, then, the K situation in soils is characterized by (1) a very large total supply throughout the soil profile; (2) a vastly smaller supply that is available to plants in the short term; (3) a susceptibility to substantial leaching losses; and (4) a high rate of removal in harvested agroecosystems, especially when luxury quantities of this element are supplied and consumed. With these ideas as a background, the various forms and availabilities of K in soils will now be considered.

Forms and Availability of Potassium in Soils

The total amount of K in a soil and the distribution of K among the four major pools shown in Figure 12.46 is largely a function of the kinds of clay minerals present in a soil. Generally, soils dominated by 2:1 clays contain the most K; those dominated by kaolinite contain the least (Table 12.6). Of the different forms, K^+ in solution or exchangeable on the clay and humus is most readily of available for plant uptake. Nonexchangeable K in secondary mineral is only slowly available, while K in primary minerals is slowly available or nonavailable during the current year.

All plants can easily utilize the readily available forms, but the ability to obtain K held in the slowly available and unavailable forms differs greatly among plant species. Many grass plants with fine, fibrous root systems are able to exploit K held in clay interlayers and near the

---

[19]Many of the principles discussed in this section have been studied extensively in the scientific literature, but have sometimes been ignored or downplayed in favor of greater-than-necessary use of "potash" fertilizer (Khan et al., 2014).

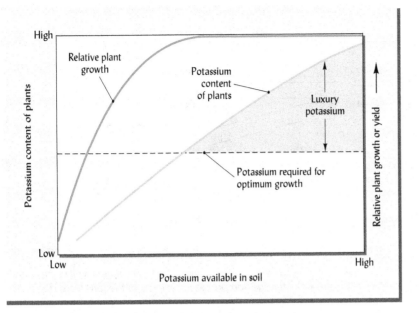

Figure 12.48 *The general relationship between available K level in soil, plant growth, and plant uptake of K. If available soil K is raised above the level needed for maximum plant growth, many plants will continue to increase their uptake of K without any corresponding increase in growth. The K taken up in excess of that needed for optimum growth is termed luxury consumption. Such luxury consumption may be wasteful, especially if the plants are completely removed from the soil. It may also cause dietary imbalance in grazing animals.* (Diagram courtesy of Ray R. Weil)

edges of mica and feldspar crystals of clay and silt size. A few plants adapted to low-fertility sandy soils, such as elephant grass (*Pennisetum purpureum* Schum.), have been shown to obtain K from even sand-sized primary minerals, a form of K usually considered to be unavailable.

Some 90 to 98% of all soil K in a mineral soil occurs in relatively unavailable forms (see Figure 12.46, Table 12.6), mostly in the crystal structure of feldspars and micas. These minerals are quite resistant to weathering, their cumulative release of K in the entire soil rooting zone and over a period of years undoubtedly is of importance. This release is enhanced by the solvent action of carbonic acid and of stronger organic and inorganic acids, as well as by the presence of acidic clays and organic colloids (see Section 2.1).

Only 1 to 2% of the total soil K is readily available, either in the soil solution or exchangeable K adsorbed on the soil colloidal surfaces. As represented in Figure 12.46, these two forms of readily available K are in dynamic equilibrium. When plants absorb K from the soil solution, some of the exchangeable K immediately moves into the soil solution until the equilibrium is

Table 12.6
THE INFLUENCE OF DOMINANT CLAY MINERALS ON THE AMOUNTS OF WATER-SOLUBLE, EXCHANGEABLE, FIXED (NONEXCHANGEABLE), AND TOTAL POTASSIUM IN SOILS

The values given are means for many soils in ten soil orders sampled in the United States.

Potassium pool	Dominant clay mineralogy of soils (mg K/kg soil)		
	Kaolinitic (26 soils)	Mixed (53 soils)	Smectitic (23 soils)
Total K	3,340	8,920	15,780
Exchangeable K	45	224	183
Water-soluble K	2	5	4

Data from Sharpley (1990).

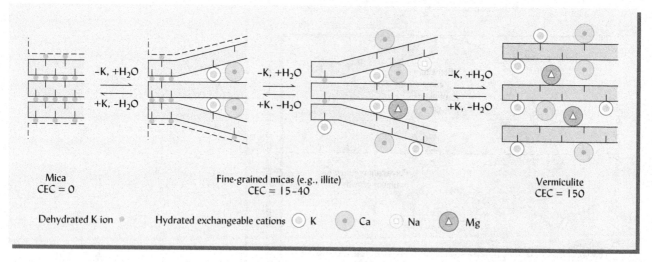

Figure 12.49 *Diagrammatic illustration of the release and fixation of K between primary micas, fine-grained mica (illite clay), and vermiculite. In the diagram, the release of K proceeds to the right, while the fixation process proceeds to the left. Note that the dehydrated K ion is much smaller than the hydrated ions of Na^+, Ca^{2+}, Mg^{2+}, etc. Thus, when K is added to a soil containing 2:1-type minerals such as vermiculite, the reaction may go to the left and K ions will be tightly held (fixed) in between layers within the crystal, producing a fine-grained mica structure. Ammonium ions (NH_4^+, not shown) are of a similar size and charge to K ions and may be fixed by similar reactions.* [Drawings based on concepts of E. O. McLean, The Ohio State University.]

again established. When water-soluble fertilizers are added, the soil solution becomes K enriched and the reverse adjustment occurs—K from soil solution moves onto the exchange complex. The exchangeable K thus *buffers* (stabilizes) the concentration of K in the soil solution.

Fixed Potassium Soil clays composed of vermiculite, smectite, and certain other 2:1-type minerals not only attract and adsorb K^+ ions in exchangeable form but may also *fix* some of the K^+ ions in tightly held, nonexchangeable form (Figure 12.49). Potassium (as well as the similar sized NH_4^+ ions) ions fit in between layers in the crystals of these normally expanding clays and become an integral part of the crystal. These ions cannot be replaced by ordinary exchange processes and consequently are referred to as *nonexchangeable ions*. As such, these ions are not readily available to most plants. This form is in equilibrium, however, with the more available forms and consequently acts as an extremely important reservoir of slowly available nutrients.

Release of Fixed Potassium Nonexchangeable or fixed K is continually released to the exchangeable form in amounts large enough to be important for plant nutrition. In some soils, the K removed by plants is supplied largely from nonexchangeable forms.

As a result of these relationships, very sandy soils with low cation exchange capacity (CEC) are poorly buffered with respect to K. In them, K^+ ion concentration may be quite high at the beginning of a growing season or just after fertilization, but these soils have little capacity to maintain the K^+ ion concentration as plants remove the dissolved K from the soil solution during the growing season. Late-season K deficiency may result. In finer-textured soils with a greater CEC (and therefore greater buffering capacity), the initial solution concentration of K may be somewhat lower, but the soil is capable of maintaining a fairly constant supply of solution K ions throughout the growing season.

Factors Affecting Potassium Fixation in Soils

Five soil conditions markedly influence the amounts of K fixed: (1) the nature of the soil colloids, (2) the levels of previous K additions and removals, (3) wetting and drying, (4) freezing and thawing, and (5) the presence of excess lime.

Type of Clay and Moisture Affect Release of Fixed Potassium Kaolinite and other 1:1-type clays fix little K, while clays of the 2:1 type, such as vermiculite, fine-grained mica (illite), and smectite, fix K very readily and in large quantities (Figure 12.50). Even silt-sized fractions of some micaceous minerals fix and subsequently release K.

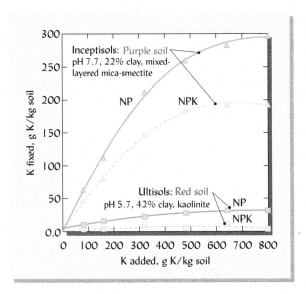

Figure 12.50 *Potassium fixation by two contrasting soils in China subjected to 15 years of cropping and fertilization with (NPK) or without (NP) the annual application of 100 kg/ha of K. The Inceptisol (known as a Purple soil in China) with mixed 2:1 clay mineralogy and an alkaline pH exhibited a much higher K fixation capacity than the Ultisol (known as a Red soil in China) with kaolinitic clays. In both soils the K fixation capacity was considerably reduced by repeated application of K fertilizers. In all cases, N and P fertilizers were applied.* [Recalculated and graphed from selected data in Zhang et al. (2009)]

Alternate wetting/drying and freezing/thawing enhance both the fixation of K in nonexchangeable forms and the release of previously fixed K to the soil solution. Although its mechanism is not well understood, it results in marked seasonal variation in the amounts of exchangeable K available for plant uptake

Some Practical Aspects of Potassium Management

Except in very sandy, or highly weathered (Oxisols) or organic (Histosols) soils, the problem of K fertility is rarely one of total supply, but rather one of adequate *rate* of transformation from nonavailable to available forms. Where little plant material is removed (e.g., in forests, pastures, rangeland, and some turfgrass), cycling between plant and soil may be adequate for continued plant growth. However, where large amounts of plant material are removed, especially if little plant residue is returned, then the plant–soil cycle may need to be supplemented by release of K from less available mineral forms and, in some cases, by fertilization.

If high yields of forage legumes such as alfalfa are to be produced, the soil may have to be capable of supplying K for very high uptake rates during certain periods, resulting in the need for high levels of fertilization even on soils well supplied with weatherable minerals. However, excessive levels of K that depress calcium and magnesium in the forage must be avoided to maintain plant and animal health (see grass tetany, Section 12.5).

Frequency of Application Although a heavy dressing applied every few years may be most convenient, more frequent light applications of K may offer the advantages of reduced luxury consumption, reduced losses by leaching, and reduced opportunity for fixation in unavailable forms before plants have had a chance to use the K applied.

The large quantities of moderately available K present in many soils means that only a part of the total amount of K removed by harvest and leaching needs to be replaced by fertilizer. Moreover, the importance of lime in reducing leaching losses of K should not be overlooked as a means of effectively utilizing the power of soils to furnish this element. For deep-rooted plants such as cotton and fruit trees, exchangeable and nonexchangeable K in deep soil layers may supply much of their needs.

Use of Potassium Fertilizers Because of the large K-supplying power of most soils, economically viable responses to K fertilization are actually quite rare and there exists little justification for the traditional viewpoint that K fertilizer should generally be applied to most soils as part of a "complete" fertilization program. Most cases in which K fertilizers have been shown to increase yield or quality of crops have occurred on soils with inherently low K-supplying power, namely, very sandy, highly organic, or highly weathered soils. Potassium fertilizers may also be called for where very large amounts of high K plant material are removed and/or the depth of plant rooting is very shallow so subsoil contributions to K nutrition are minimized. While K fertilizer has probably been over-applied in many developed countries, future high-yielding crops may require more K fertilizer than lower yielding crops of a few decades ago.

12.5 CALCIUM AS AN ESSENTIAL NUTRIENT[20]

Calcium (Ca) is a macronutrient essential for all plants. The calcium status of soils has a major influence on the species composition and productivity of terrestrial ecosystems. For animals, the calcium content of the plants they eat is important because calcium is a major component of bones and teeth and plays important roles in many physiological processes. It has even been suggested that the relatively higher calcium status of soils in Africa compared to those in South America may account for the occurrence of such large herbivores as elephants, zebras, and giraffes in semiarid savannas in Africa but not in South America.

Calcium in Plants

The concentration of Ca in plant foliage varies widely from as low as 0.1 to as high as 5% of the dry matter. Most monocots restrict are considered to be *calcifuge* (calcium avoiding) plants that grow well with 0.15 to 0.5% Ca in leaf tissues. Some dicots such as those in the Brassica family are considered to be *calcicoles* (Latin: *chalk dwelling*) and need 1 to 3% Ca in their leaves for optimal growth. Trees store a great deal of calcium in their woody tissues; the net calcium uptake by many trees is close to that of N.

Calcium is a major component of cell walls and is intimately involved with cell elongation and division, membrane permeability, and the activation of several critical enzymes. Through its role in maintaining the integrity of the cell membranes, calcium is critical for protecting the cell against toxicities of other elements. Calcium is taken up almost exclusively by young root tips and its redistribution within the plant occurs mainly with the transpiration water in the xylem, rather than in the phloem.

Since Ca^{2+} is usually the most abundant of the nonacid cations in soils, for most plants deficiencies of calcium are quite rare, except in very acid soils. When calcium deficiency does occur, it is usually associated with growing points (meristems) such as buds, unfolding leaves, fruits, and root tips (Figure 12.51). Foliar symptoms of calcium deficiency may occur in extremely acid soils where plants are likely to also suffer from aluminum toxicity and other problems. When calcium is deficient, normally harmless levels of other metals (aluminum,

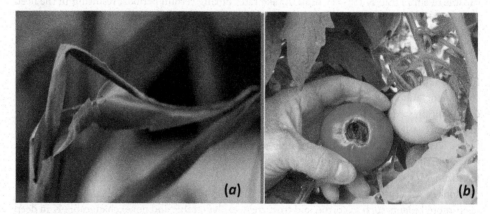

Figure 12.51 *Plant symptoms related to calcium deficiency. Plant parts not located along the transpiration stream are least likely to receive calcium.* (a) *Young leaves that stick together and fail to unfold are a symptom of Ca deficiency in monocots.* (b) *blossom-end rot (black, rotten bottom side of the tomato), associated with low calcium translocation to the fruit due to irregular soil water availability.* (Photo courtesy of Ray R. Weil)

---

[20]For a review of calcium in soils and plants, see Chapter 11 in Mengel and Kirby (2001). For a brief interview with renowned soil scientist Pedro Sanchez on the change that calcium made in the Cerrado, see Taylor (2005). For a description of long-term calcium depletion in acidified forests, see Johnson et al. (2008).

magnesium, zinc, or manganese) can become toxic to the plant. Under such conditions, calcium-deficient roots become severely stunted and gelatinous. These root effects can cause sensitive forest trees to lose branches and eventually die. In very acid soils, Ca deficiency is often accompanied by Al or Mn toxicity, and the related effects on plants are difficult to tell apart.

Certain plants show Ca deficiency related to the transport of Ca within the plant. Blossom-end rot in melons and tomatoes is caused by inadequate Ca for the cell walls of the expanding fruit and is usually associated with unevenness in the water supply that interrupts the flow of Ca.

Soil Forms and Processes

Calcium in the soil is found mainly in three pools that resupply the soil solution: (1) calcium-containing minerals (such as calcite or plagioclase), (2) calcium complexed with SOM, and (3) calcium held by cation exchange on the clay and organic colloids. The cycling of calcium among these and other soil pools, and the gains and losses of calcium by such mechanisms as plant uptake, atmospheric deposition as dust and soot, liming, and leaching, comprise the calcium cycle, as illustrated in Figure 12.52. In most soils, the principal sources supplying Ca to the soil solution for plant uptake are (1) exchangeable Ca and (2) Ca in readily weathered minerals (such as carbonates and apatite). In arid and semiarid regions, the high pH, high carbonate nature of the soil solution greatly diminishes the solubility of calcium-containing minerals.

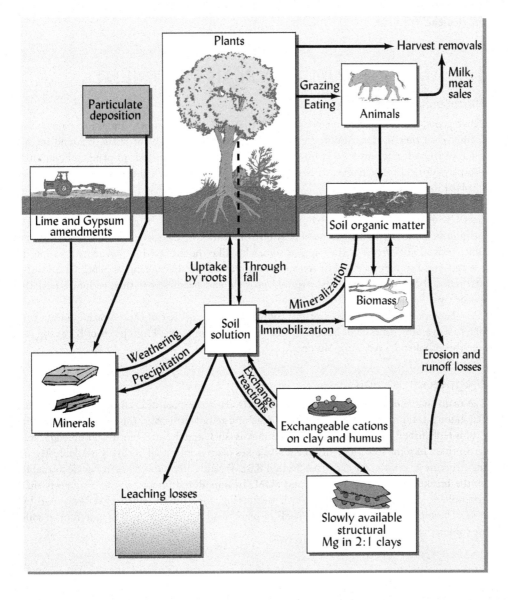

Figure 12.52 Simplified diagram of the cycling of calcium and magnesium in soils. The rectangular compartments represent pools of these elements in various forms, while the arrows represent processes by which the elements are transformed or transported from one pool to another. (Diagram courtesy of Ray R. Weil)

Downwind from "dirty" coal-burning industries without effective environmental controls, considerable calcium is deposited in particulate air pollution. Deposition of dust from wind erosion of calcareous desert soils can make substantial contributions to soil calcium thousands of kilometers downwind. Such calcium deposition can partially offset acidification caused by N and S deposition (see Sections 12.1 and 12.2).

The need for repeated applications of limestone in humid regions(Section 9.8) suggests significant losses of calcium (and magnesium) from the soil. The annual loss by leaching, erosion, and crop removal in humid-region agricultural soils is typically equivalent to about 1 Mg/ha of calcium carbonate. To maintain productivity, this loss must be compensated for by a combination of lime applications and weathering of Ca from minerals. Where Ca is inadequate, but soil pH change is not desirable, gypsum ($CaSO_4 \cdot 7H_2O$) may be applied instead of limestone.

Similarly, timber harvest methods that leave the soil open to erosion and nutrient leaching lead to rapid losses of calcium (and magnesium) from forest ecosystems. Scientists are concerned that for some soils in the humid regions, the release of Ca from mineral weathering will not be able to keep up with the losses and that acid rain combined with intensive timber harvesting may be depleting the Ca reserves in the more poorly buffered watersheds. The acidification affects a number of biogeochemical processes involving calcium. These processes in turn influence tree physiology, which ultimately impacts the ecological functioning of forests. Nonetheless, research in this area is still inconclusive and up to now forests have rarely shown a positive growth response to calcium applied as a plant nutrient supplement.

12.6 MAGNESIUM AS A PLANT NUTRIENT

Magnesium in Plants

Plants generally take up Mg in amounts (0.15 to 0.75% of dry matter) similar to or somewhat smaller than they do Ca. About one-fifth of the magnesium in plant tissue is found in the chlorophyll molecule and so is intimately involved with photosynthesis in plants. Magnesium also plays critical roles in the synthesis of oils and proteins and in the activation of enzymes involved in energy metabolism.

The most common symptom of Mg deficiency is interveinal chlorosis on the *older* leaves, which appears as a mottled green and yellow coloring in dicots (Figure 12.53) and a striping in monocots. Common on very sandy soils with low CEC, these symptoms are sometimes termed *sand drown* as they appear somewhat like those caused by oxygen starvation in a waterlogged soil. Spruce and fir trees growing on soils low in exchangeable Mg have exhibited reduced Mg in needle tissue, stunted growth, and needles that turn yellow, especially on the tips.

Forages with low contents of Mg compared to Ca and K can cause grazing animals to suffer from a sometimes-fatal Mg deficiency known as *grass tetany*. High levels of K can aggravate this problem by reducing Mg uptake.

Magnesium in Soil

The main source of plant-available Mg in most soils is the pool of exchangeable Mg on the clay-humus complex. As plant uptake and leaching remove this Mg, the easily exchangeable pool is replenished by Mg weathered from minerals (such as dolomite, biotite, hornblende, and serpentine). In some soils, replenishment also takes place from a pool of slowly available Mg in the structure of certain 2:1 clays (see Section 8.3). Variable amounts of Mg are made available by the breakdown of plant residues and SOM. In unpolluted forests, research suggests that atmospheric deposition, rather than rock weathering, may supply much of the Mg used by trees. The inputs and outputs to the pools of plant-available calcium and magnesium in soils are summarized in Figure 12.52.

Figure 12.53 *Magnesium deficiency yellows the tissue on the oldest leaves between the veins (a,b), which remain green (interveinal chlorosis). In monocots like corn, the symptom appears as yellow and green stripes (c). The raspberry plant (a) shows that application of Mg will typically stimulate increased growth and eliminate new symptoms, but will not remove the chlorosis that formed on older leaves while the plant was deficient in Mg.* (Photo courtesy of Ray R. Weil)

Ratio of Calcium to Magnesium[21]

Being less tightly held (more easily leached) than Ca, exchangeable Mg commonly saturates only 5 to 20% of the effective CEC, as compared to the 60 to 90% typical for Ca in neutral to moderately acid soils. Some agriculturists believe that optimum plant growth and soil tilth require a ratio of exchangeable Ca:Mg very near 6:1 (65% Ca and 10% Mg saturation of the CEC). This belief is based on a few out-of-date research studies from the mid-twentieth century and can lead to the wasteful use of soil amendments in an effort to achieve this so-called "ideal" Ca:Mg ratio. Numerous research studies have shown that, in fact, plants grow very well and meet their Ca and Mg needs in soils with Ca:Mg ratios anywhere from 1:1 to 15:1. Soil aggregation and biological activity in most soils are also largely unaffected by a similarly wide range in this ratio. However, in soils with easily dispersed 2:1 clays, structural stability and flocculation can be adversely affected by high levels of exchangeable Mg (see Figure 8.11), though not nearly to the degree that monovalent K or Na cause such affects. Adding Ca to such soils in the form of gypsum or (if pH also needs to be raised) lime can improve infiltration and reduce erosion and leaching loss of nutrients (P especially) associated with dispersed clay (see Figure 12.43). Also, even where plant and soil health are not likely to be affected, the ratio of Ca to Mg in plant tissue may be altered enough to influence the nutrition of grazing animals, so that Ca or Mg mineral supplements may be needed in animal diets.

Soils formed from serpentine-rich rock, which is high in Mg but contains little Ca (Ca:Mg ratio much smaller than 1.0.), offer an unusual but dramatic exception to the above statements. Serpentine-derived soils may, therefore, exhibit extreme imbalances between Ca and Mg, causing severe deficiency of Ca and toxicity of Mg for all but the few plant species that have evolved to grow in this unique soil environment. As a result, such serpentine-derived soils are characterized by natural vegetation that is sparse and stunted in comparison with that on nearby nonserpentine soils.

---

[21] For an objective review of how Ca/Mg ratios and cation balancing in general became widely, but wrongly, promoted for fertility management, see Kopittke and Menzies (2007).

12.7 SILICON IN SOIL–PLANT ECOLOGY[22]

Silicon in Plants

Silicon has been proven to be an essential nutrient(Section 1.2) for only genus of plant, Equisetum (horsetails). Yet silicon is taken up in considerable amounts by all plants and has repeatedly been shown to enhance growth and function in a wide range of plant species. Therefore, silicon is sometimes referred to as a *beneficial element* for plants or a *quasi-essential nutrient*. For humans and other animals, silicon *is* known to be an essential nutrient. However, too much Si intake along with limited water may cause kidney stones (siliceous renal calculi) in ruminant animals and humans. Silicon also plays many important, though often unrecognized, roles in soil processes, water quality, and in the biogeochemistry of such elements as carbon, calcium, and phosphorus.

Amounts and Forms Taken Up The main form of silicon present in the soil solution is H_4SiO_4 (orthosilicic acid) and this is the form taken up by plants (Figure 12.54). Unlike most plant nutrient elements, silicon is not known to comprise part of any essential plant biomolecules. However, specific silicon transporter molecules have been identified in plants. Most of the H_4SiO_4 taken up is deposited as polymerized silica in phytoliths (microscopic crystal-like bodies).

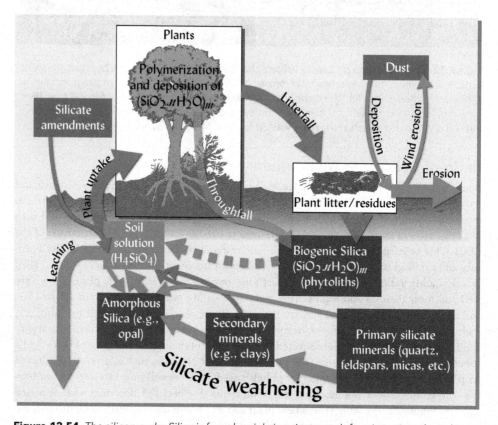

Figure 12.54 *The silicon cycle. Silica is found mainly in primary rock-forming minerals and in secondary silicate clay minerals. These minerals weather, with the aid of microbes and plant roots, releasing silicic acid (H_4SiO_4). Plants take up H_4SiO_4 and transport it into the stems and leaves where it is polymerized and deposited as silica (SiO_2). This biogenic silica solidifies into the shape of the tissue cavities in which it is deposited forming phytoliths. These silica bodies are then deposited back to the soil with litter fall or washed into the soil with throughfall. They may slowly redissolve or remain unchanged for millennia, the variable reactivity indicated by the broken arrow from this pool. Considerable silicon is leached out of the soil–plant system as silicic acid and ends up in the oceans.* (Diagram courtesy of Ray R. Weil)

---

[22]For a review of the importance of silicon in plant ecology, see Cooke and Leishman (2011); for effects of silicon on crop plants, see Guntzer et al. (2012) and for roles of Si in pedology and biogeochemistry, see Sommer et al. (2006).

The concentration of Si in plant dry matter is variable, but generally on par with the concentrations of Ca, K, or Mg. Horsetails and wetland grasses (including rice) are accumulator plants typically containing Si to levels of 5 to 10% of their dry matter. Other grasses, including crops such as wheat, seem to take up H_4SiO_4 passively and generally contain 0.5 to 1.5% Si in their dry matter. Most broadleaf (dicot) plants appear to restrict the passage of silicon into their roots and contain less than 0.25% Si in their dry matter.

Roles of Silicon in Plants Silicon strengthens and gives rigidity to plant cells, helping plants like rice and sugarcane maintain erect leaves. This rigidity also helps plants resist physical damage, insect feeding, and may allow roots to better penetrate compacted soils. Through this mechanical effect, as well as through physiological and soil chemical effects, silicon can help alleviate many types of plant stress (Table 12.7).

Silicon in Soils

Silicon is the second most abundant element (after oxygen) in the Earth's crust as well as in most soils. Exceptions include some Histosols which contain mainly organic matter and some Oxisols from which weathering and leaching have removed most of the silicon.

Cycling of Silicon in Soils As was discussed in Chapter 2, the silicate, quartz, is notoriously resistant to breakdown. However, most aluminosilicate minerals, including clay minerals and primary minerals such as Feldspars and Anorthite (shown in Eq. (12.1)), undergo weathering reactions that release silicic acid to the soil solution:

$$\underset{\text{Anorthite}}{CaAl_2Si_2O_8} + \underset{\text{Carbon dioxide}}{2CO_2} + \underset{\text{Water}}{8H_2O} \rightarrow \underset{\text{Calcium}}{Ca^{2+}} + \underset{\text{Aluminum hydroxide}}{2Al(OH)_3} + \underset{\text{Silicic acid}}{2H_4SiO_4} + \underset{\text{Bicarbonate}}{2HCO_3^{-}}$$

(12.19)

In the pH range of most soils (pH 4 to 9), silicic acid is water soluble in the nondissociated form shown in Eq. (12.17). The silicic acid is subject to removal from the soil by both leaching and plant uptake. The weathering and removal of silicon is especially rapid in soils forming from ultrabasic rocks in the humid tropics. In humid regions, the leaching process typically sends some 5 to 50 kg Si ha^{-1} yr^{-1} toward the ocean. Once in the ocean, microscopic algae termed *diatoms* use much of the silicon to form their glasslike siliceous cell walls. In this manner

Table 12.7
SILICON EFFECTS PLANT GROWTH, FUNCTION, AND STRESS TOLERANCE

	Nature and location of the silicon mechanisms		
Specific silicon effect on plant growth	Chemical effects in the soil	Physiological effects in the plant	Mechanical[a] effects in the plant
Increased resistance to herbivory and diseases		X	X
Alleviation of P deficiency	X	X	
Alleviation of drought stress		X	X
Alleviation of Al and Zn toxicity	X	X	X
Alleviation of salt stress		X	
Reduction in lodging from strong wind and rain			X
Alleviation of Mn, Cd, and As toxicity	X	X	
Reduced absorption of nutrients (P, N) in excess		X	
Enhancement of K, P, Ca intake		X	
Alleviation of Fe toxicity	X		

[a]Related to the presence of silicon phytoliths and silicon reinforced cell walls.
Based largely on concepts in Guntzer et al. (2012).

the silicon weathered and leached from soils on land plays a major role in the global carbon cycle by controlling the deposition of carbon on the ocean floor in the form of dead diatoms.

Plants take up and transport H_4SiO_4 into the stems and leaves where it is polymerized and deposited as biogenic (biologically formed) silica (SiO_2). This biogenic silica solidifies into the shape of the tissue cavities in which it is deposited, silt-sized particles termed **phytoliths** (literally *plant rocks*). These characteristically shaped and long-lasting silica bodies are then deposited back to the soil with litter fall or the death of the plant. Biogenic silica from plants is also washed into the soil with throughfall. In the case of agricultural crops, the straw may contain a large proportion of the plant's phytoliths, so the return of these crop residues may be important in replenishing the soil's supply of biogenic silica. In the soil, some phytholiths slowly redissolve, providing an important source of relatively plant-available Si.

Phytoliths also contribute, along with other processes, to silicon accumulations during soil formation that result in a number of soil characteristics and even diagnostic horizons. Eluviation (downward movement) of silica bodies during wet seasons and their desiccation during dry season can result in the formation of hardened, silica-rich subsoil layers termed duripans. Silicon accumulations may also contribute to the formation of fragipans in humid regions (see Box 3.2) and hardsetting surface crusts in semiarid regions.

Fertilization with Silicon Because Si is so abundant in most soils, it has commonly been assumed that plant growth and function must not be limited by insufficient amounts of this element. However, most of the silicon in soils is found in rather insoluble minerals such that only a miniscule fraction of the total Si present is in a form that plants can take up. Therefore, even in sandy soils made up largely of silicon (but in the inert form of quartz), plants may respond positively to the addition of relatively soluble sources of this element such as calcium silicates ($CaSiO_3$) or the mineral, wollastonite ($Ca_3(Si_3O_9)$). The effectiveness of silicon "fertilizers" is dependent on the solubility of the silicon amendment and on soil properties such as clay mineralogy, soil pH, and inherent levels of soluble silica. There are still questions about when one can expect Si applications to be effective. However, there can be little doubt that silicon plays important roles in soil–plant ecology and should be a consideration in soil fertility management.

12.8 MICRONUTRIENTS IN THE SOIL–PLANT SYSTEM[23]

Eight elements (iron, manganese, zinc, copper, boron, molybdenum, nickel, and chlorine) are essential nutrients for plant growth, but required in such small quantities that they are called *micronutrients*. This term must not be construed to imply that these nutrients are somehow less important than macronutrients. To the contrary, too little or too much of one or more of the micronutrients can stimulate dramatic effects in terms of competitive shifts in plant species, stunted growth, low yields, dieback, and even plant death.

Deficiency Versus Toxicity

At very low levels of a nutrient, reduced plant growth or function may occur because of a deficiency of the nutrient (*deficiency range*). As the level of nutrient is increased, plants respond by taking up more of the nutrient and increasing their growth. If a level of nutrient availability has been reached that is sufficient to meet the plants' needs (*sufficiency range*), raising the level further will have little effect on plant growth, although the concentration of the nutrient may continue to increase in the plant tissue. However, it is axiomatic that anything, including nutrients, can be toxic if taken in large enough amounts. At some level of availability, the plant will take up too much of the nutrient for its own good (*toxicity range*), causing adverse physiological reactions to take place. The relationship among deficient, sufficient, and toxic levels of nutrient availability is described in Figure 12.55.

---

[23]For overviews of micronutrients in global agriculture, see Alloway and Graham (2008) and Fageria (2009), the latter includes a complete chapter on each nutrient. For a detailed review of the micronutrients' physiological roles in plants, see Welch (1995).

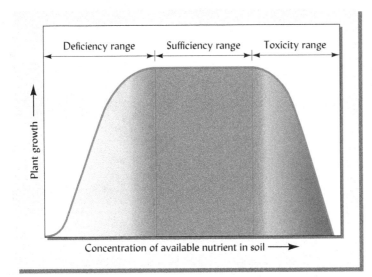

Figure 12.55 *The relationship between plant growth and the amount of a micronutrient available for plant uptake. Within the deficiency range, as nutrient availability increases, so does plant growth (and uptake, which is not shown here). Within the sufficiency range, plants can get all of the nutrient they need, and so their growth is little affected within this range. At higher levels of availability a threshold is crossed into the toxicity range, in which the amount of nutrient present is excessive and causes adverse physiological reactions that lead to reduced growth and even death of the plant.* (Diagram courtesy of Ray R. Weil).

For macronutrients, the sufficiency range is very broad and toxicity seldom occurs. However, for micronutrients the difference between deficient and toxic levels may be very narrow, making the possibility of toxicity quite real. For example, in the cases of boron and molybdenum, severe toxicity may result from applying as little as 3 to 4 kg/ha of available nutrient to a soil initially deficient in these elements. Figure 12.56 illustrates such deficiency and toxicity in the case of iron.

Figure 12.56 *Examples of deficiency (a to e) and toxicity (f) of the micronutrient iron (Fe). (a) Foliage of an Fe-deficient azalea bush, with the right side having recently received a foliar application of Fe. (b) Rose foliage exhibiting interveinal chlorosis of the newest leaves from Fe deficiency. (c) The iron status of the oak seedling is severely deficient while that of the Virginia creeper growing in the same soil is normal, illustrating the influence of plant genetics on micronutrient availability. (d) A grain sorghum plant exhibits extreme Fe deficiency due to the calcareous nature of the soil. (e) Iron-deficient basil growing in an artificial potting mix. (f) Concentrated Fe application killing weedy broadleaf plants while leaving less Fe-sensitive grasses unharmed.* (Photo courtesy of Ray R. Weil)

Figure 12.57 *Deficiency, normal, and toxicity levels in plants for seven micronutrients. Note that the range is shown on a logarithmic scale and that the upper limit for manganese is about 10,000 times the lower range for molybdenum and nickel. In using this figure, keep in mind the remarkable differences in the ability of different plant species and cultivars to accumulate and tolerate different levels of micronutrients. For example, most monocots like grasses need only about 10 ppm B in their dry matter whereas many dicots require two to three times that much. Monocots may suffer B toxicity at soil B levels that are optimal for some dicots.* (Diagram courtesy of Ray R. Weil, based on data from many sources)

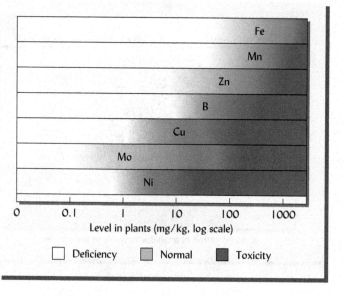

Micronutrient deficiencies are becoming increasingly commonplace in agriculture as a consequence of higher levels of their removal by ever more productive crops, but also as a result of reduced inadvertent application of these elements in fertilizers and organic amendments. Low levels of these elements in food crops can have widespread negative impacts on human health. Micronutrients are required in very small quantities, their concentrations in plant tissue being one or more orders of magnitude lower than for the macronutrients. The ranges of plant tissue concentrations considered deficient, adequate, and toxic for several micronutrients are illustrated in Figure 12.57.

Physiological Roles in Plants While most of the micronutrients participate in the functioning of a number of enzyme systems, there is considerable variation in the specific functions of the various micronutrients in plant and microbial growth processes (Table 12.8). Certain micronutrients, such as manganese, play a role in plant defenses against pests and diseases.

Table 12.8
Functions of Several Micronutrients in Higher Plants

Micronutrient	Physiological function plants
Iron	Present in several peroxidase, catalase, and cytochrome oxidase enzymes; found in ferredoxin, which participates in oxidation–reduction reactions (e.g., NO_3^- and SO_4^{2-} reduction and N fixation); important in chlorophyll formation.
Manganese	Activates decarboxylase, dehydrogenase, and oxidase enzymes; important in photosynthesis, N metabolism, and N assimilation.
Zinc	Present in several dehydrogenase, proteinase, and peptidase enzymes; promotes growth hormones and starch formation; promotes seed maturation and production.
Copper	Present in laccase and several other oxidase enzymes; important in photosynthesis, protein and carbohydrate metabolism, and probably N fixation.
Nickel	Essential for urease, hydrogenases, and methyl reductase; needed for grain filling, seed viability, iron absorption, and urea and ureide metabolism (to avoid toxic levels of these N-fixation products in legumes).
Boron	Activates certain dehydrogenase enzymes; facilitates sugar translocation and synthesis of nucleic acids and plant hormones; essential for cell division and development.
Molybdenum	Present in nitrogenase and nitrate reductase enzymes; essential for N fixation and N assimilation.
Cobalt	Essential for N fixation bacteria; found in vitamin B_{12}.
Chloride	Essential for photosynthesis and enzyme activation. Plays role in regulation of water uptake on salt-affected soils.

Figure 12.58 Foliar symptoms of manganese deficiency are similar to those caused by insufficient iron. (a) Mn-deficient ornamental oak tree growing in a sandy Ultisol. (b) A soybean field on sandy Ultisols in Maryland exhibiting large areas with Mn deficiency. (c) Deficiency of Mn in soybeans, which like that of Fe, produces interveinal chlorosis on the newest leaves. Generally the veins remain darker green in the case of Fe deficiency. (Photo courtesy of Ray R. Weil)

Deficiency Symptoms Visible plant symptoms are often helpful in diagnosing which micronutrient is deficient. The first thing to note is whether the symptoms are most pronounced on the youngest or oldest leaves on a plant. Most of the micronutrients are relatively immobile in the plant, that is, the plant cannot efficiently transfer the nutrient from older leaves to newer ones. Therefore, the concentration of the nutrient tends to be lowest, and the symptoms of deficiency most pronounced, in the younger leaves that develop after the supply of the nutrient has run low. In contrast, the macronutrients (except Ca and S) are easily translocated in plants and so become deficient first in the *older* leaves. The pattern of most pronounced *micro*nutrient deficiency symptoms on the *younger* leaves is clearly illustrated for iron in Figure 12.56b,e, for manganese in Figure 12.58, and for zinc in Figure 12.59a.

Boron deficiency generally affects plant growing points, such as buds, fruits, flowers, and root tips. Plants low in boron may produce deformed flowers (Figure 12.60c,d), aborted seeds, thickened, brittle, puckered leaves, or dead growing points. Young leaves may turn red on some plants (Figure 12.60a). In all of these cases, a small dose of boron applied at the correct time could make the difference between marketable or unmarketable plant products.

Figure 12.59 Foliar symptoms of zinc deficiency. (a) A branch of a Zn-deficient citrus tree exhibiting interveinal chlorosis with wide irregular green area along the veins and a compact leaf whirl. Young corn plants (b) develop broad chlorotic bands (sometimes striped) on either side of the midrib. (Photo courtesy of Ray R. Weil)

Figure 12.60 (a) A deficiency of boron caused the young leaves on this alfalfa plant to turn a reddish color. (b) More commonly, a lack of boron results in deformity or death of the plant growing points, as on the B starved bean plants in the right-hand pot whose terminal buds (arrow) have died and whose older leaves have thickened and crinkled in comparison to the B fertilized beans in the pot on the left. (c) A misshapen rose, or bullhead, on a boron-deficient tea rose. (d) A well-formed bloom on the same plant after a solution containing boron was applied to the leaves and young buds. (Photo courtesy of Ray R. Weil)

Micronutrient Cycles, Forms, and Reactions in the Soil

Deficiencies and toxicities of micronutrients may be related to the total amounts of these elements in the soil. More often, however, these problems result from the chemical forms present in the soil and, particularly, their solubility and availability to plants. The cycling of micronutrients through the soil–plant–animal system is illustrated in a generalized way in Figure 12.61. Although not every micronutrient will participate in every pathway shown in this figure, it can be seen that organic chelates, soil colloids, SOM, and soil minerals all contribute micronutrients to the soil solution and, in turn, to growing plants. The primary sources of micronutrient elements are the original and secondary minerals, while the breakdown of organic matter releases ions to the soil solution. Removal of nutrients in crop or timber harvest reduces the soluble ion pool. If the rate of removal exceeds the rate of replenishment by mineral weathering and atmospheric deposition, the soil supply may need to be replenished with manures or chemical fertilizers to avoid nutrient deficiencies. As we turn our attention to micronutrient availability, it will be helpful to refer back to Figure 12.61 to see the relationships among the processes involved.

Inorganic Forms Micronutrients sources vary markedly from soil to soil. All of the micronutrients can be found to some extent in igneous rocks. Iron and manganese occupy prominent structural positions in primary silicate minerals, such as biotite and hornblende. As mineral weathering and soil formation occur, oxides and, in some cases, sulfides of elements such as iron, manganese, and zinc are formed. Secondary silicates, including the clay minerals, may

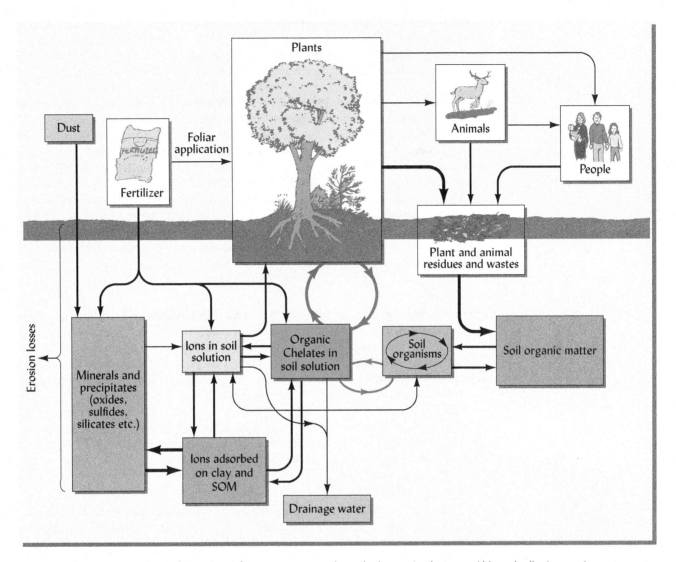

Figure 12.61 Cycling and transformations of micronutrients in the soil–plant–animal system. Although all micronutrients may not follow each of the pathways shown, most are involved in the major components of the cycle. The formation of organic chelates, which keep most of these elements in soluble forms, is a unique feature of this cycle. Also, for micronutrients, dust and other forms of atmospheric deposition can be significant additions. (Diagram courtesy of Ray R. Weil)

contain considerable quantities of iron and manganese and smaller quantities of zinc and cobalt. Ultramafic rocks, especially serpentinite, are high in nickel. Minerals in soil particles eroded from distant sites and deposited as dust can add quantities of micronutrients that are significant in relation to the small amounts used by plants.

In the same manner as calcium or aluminum ions, micronutrient cations released by mineral weathering are subject to colloidal adsorption. Micronutrient anions in soils, such as borate and molybdate, may undergo adsorption or reactions similar to those of described in Section 12.3 for phosphates. Chlorine, by far the most soluble micronutrients, is only very weakly adsorbed to soil colloids. However, it is added to soils in considerable quantities each year through rainwater and fertilizers and deficiencies of chlorine are quite rare under field conditions.

Forms Organic matter is an important secondary source of some of micronutrients and other trace elements. Several of these elements tend to be held as complex combinations by soil organic matter (humus). Copper is especially tightly held by organic matter—so much so that its availability can be very low in organic soils (Histosols). In uncultivated profiles, there is a somewhat greater concentration of micronutrients in the surface soil, much of it presumably in the organic fraction. Correlations between SOM and contents of copper, molybdenum, and zinc have been noted. Although the elements thus held are not always readily available to plants, their release through decomposition is undoubtedly an important fertility factor. Animal manures are a good source of micronutrients, much of it present in organic forms.

Trace Elements Associated with Land-Applied Wastes Trace elements are common constituents of animal, industrial, and domestic wastes that are often applied to soils. Such waste materials are usually applied to soils for their organic matter and macronutrient (particularly N, P, and calcium) contents, but we must also recognize the significance of their trace element contents. Livestock manures are generally good sources of micronutrients, both in terms of quantities and plant availability. Small quantities of trace elements applied in these wastes can help alleviate nutrient deficiencies and prevent or even reverse soil micronutrient depletion in intensive agriculture. However, in some cases repeated applications of large quantities of wastes, especially sewage and industrial by-products, may cause some trace elements to accumulate to levels that are toxic not only to plants but also to the people and other animals that may consume the plants (see Chapter 15).

Organic

Micronutrient Species in Soil Solution and Plant Uptake The dominant forms of micronutrients that occur in the soil solution and taken up by plant roots are shown in Table 1.1. For many elements, the specific forms present are determined largely by the pH and by soil aeration (i.e., redox potential). Cations are present in the form of either simple cations or hydroxy metal cations. The simple cations tend to be dominant under highly acid conditions. The more complex hydroxy metal cations become more prominent as the soil pH is increased. Molybdenum is present mainly as MoO_4^{2-}, an anionic form that reacts with soil surfaces at low pH in ways similar to those of P (see Section 12.3). Although boron also may be present in anionic form at high pH levels, undissociated boric acid (H_3BO_3) is generally dominant in the soil solution and is the form absorbed by plants.

Influence of Soil pH

The micronutrient cations (iron, manganese, zinc, copper, and nickel) are most soluble and available under acid conditions (Figure 12.62). In fact, under very acid conditions, concentrations of one or more of these elements (most commonly manganese) may become enough to cause toxicity for plants.

If the pH increases, the ionic forms of the micronutrient cations change first to the hydroxy ions and, finally, to the insoluble hydroxides or oxides of the elements. As the pH is raised, solubility and availability to plants decline. Such deficiencies associated with high pH

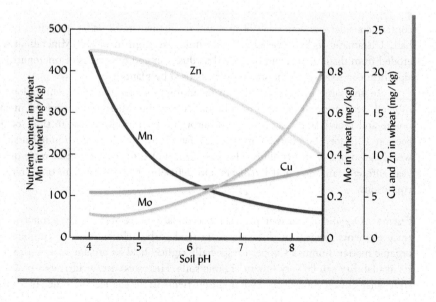

Figure 12.62 Effect of soil pH on the concentrations of manganese, zinc, copper, and molybdenum in wheat plants. The plants were grown in soils from different countries around the world. The molybdenum levels are extremely low, but increase with increasing pH. Manganese and zinc levels decrease as the pH rises, while copper is little affected. [Drawn from data in Sillanpaa (1982)]

occur naturally in many of the calcareous soils of arid regions. Overliming an acid soil may lead to a higher than desired pH at which deficiencies of iron, manganese, zinc, copper, and sometimes boron become pronounced. The general desirability of a slightly acid soil (with a pH between 6 and 7) largely stems from the fact that for most plants, this pH condition allows micronutrient cations to be soluble enough to satisfy plant needs without becoming so soluble as to be toxic (see Figure 9.18).

The effect of soil pH on the availability and plant uptake of molybdenum (Mo) is just the opposite of the cationic micronutrient just discussed. At low pH values, the molybdenum is adsorbed by silicate clays and, more especially, by oxides of iron and aluminum through *ligand exchange* with hydroxide ions on the surface of the colloidal particles. The liming of acid soils will usually increase the availability of molybdenum. Molybdenum is essential for biological N fixation and its low availability in acid soils may exert a major influence on nonsymbiotic biological fixation that provides the bulk of the N for tree growth in certain tropical rainforests (Figure 12.63).

Because molybdenum availability is high under alkaline conditions, in some arid regions molybdenum *toxicity* is a problem. The high solubility of molybdenum and low solubility of copper at high pH can combine to cause a disease known as molybdenosis, which interferes with normal copper assimilation. Ruminant animals (e.g., antelope, cattle, and sheep) grazing vegetation on high-molybdenum alkaline soils are most susceptible.

Boron (B) is one of the most commonly deficient of all the micronutrients. Boron is most available at low pH, but also most soluble and easily leached. Therefore, deficiency of B is relatively common on acid, sandy soils, but it occurs because of the low supply of total boron rather than because the boron present is not available. Boron deficiency is even more common at high pH levels—in both sandy soils (because of low boron content) and clayey soils (because the boron is tightly held by the clay). Soluble boron is present in soils mostly as boric acid [$B(OH)_3$] or as $B(OH)_4^-$. These compounds can exchange with mineral surface OH groups, forming difficult-to-reverse inner-sphere complexes (see Sections 8.7 and 8.11) on surfaces of iron and aluminum oxides and such silicate clays as kaolinite.

Oxidation State

The trace element cations iron, manganese, nickel, and copper occur in soils in more than one valence state. The lower valence states indicate that the atoms have gained electrons and the elements are considered *reduced*; in the higher valence state, they have lost electrons and are *oxidized*. Metallic cations generally become reduced when the oxygen supply is low, as occurs in wet soils containing decomposable organic matter. Reduction can also be brought about by metabolic reducing agents, such as NADPH or caffeic acid, produced by plants and

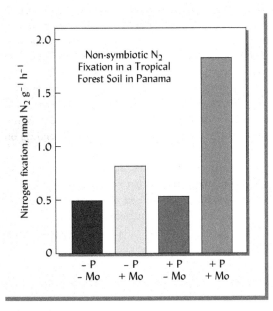

Figure 12.63 Productivity and carbon sequestration in many tropical forests are largely dependent on N from nonsymbiotic biological N_2 fixation carried out by free-living soil bacteria in the surface soil horizons. Research in lowland tropical forests of Panama has shown that P stimulates N_2 fixation only when there is sufficient amounts of the micronutrient anion, molybdenum (Mo). In fact in all the sites studied in Panama, P additions stimulated N_2 fixation only if made in conjunction with Mo additions. The main sources of naturally occurring P and Mo in these soils are weathering minerals and dust deposition. Part of the problem is that in the O horizons where much of the N_2 fixation takes place, organic matter adsorbs MoO_4 in preference to H_2PO_4. This is an example of the complex interactions that can affect the necessary balance among plant nutrients in soils. [Graphed from selected data in Wurzburger et al. (2012)]

microorganisms in the soil. Oxidation and reduction reactions in relation to soil aeration and drainage are discussed in Sections 7.4 and 7.5.

The changes from one valence state to another are, in most cases, brought about by microorganisms and organic matter.[24] In some cases, the organisms may obtain their energy directly from the inorganic reaction. For example, the oxidation of manganese from Mn(II) in manganous oxides (MnO) to Mn(IV) in manganic oxides (MnO_2) can be carried out by certain bacteria and fungi. In other cases, organic compounds formed by microbes or plant roots may be responsible for the oxidation or reduction.

At pH levels common in soils, the oxidized states of iron, manganese, and copper are much less soluble than are the reduced states. For example, the hydroxide of trivalent ferric iron precipitates at pH values of 3.0 to 4.0, whereas ferrous hydroxide does not precipitate until a pH of 6.0 or higher is reached. The interaction of soil acidity and aeration in determining micronutrient availability is of great practical importance. Iron, manganese, and copper are generally more available under conditions of restricted drainage or in flooded soils. Very acid soils that are poorly drained may supply toxic quantities of iron and manganese.

Well-oxidized calcareous soils combine high soil pH levels with good aeration. These characteristics commonly cause plant deficiencies of available iron, zinc, or manganese even though adequate total quantities of these trace elements are present in the soil. Hydroxides of the high-valence forms of these elements are too insoluble to supply the ions needed for plant growth.

Plant Genetic Differences There are marked differences among plant species and strains with regard to their ability to obtain micronutrients from low-solubility sources in soils. For example, certain iron-efficient plant varieties respond to iron stress by acidifying the immediate vicinity of the roots and by excreting compounds capable of donating electrons to reduce Fe(III) to the much more soluble Fe(II) form, with a resultant increase in its availability. Figure 12.56c illustrates the marked difference among plant species in ability to obtain iron from a particular soil.

Organic Matter and Clay

Organic matter, organic residues, and manure applications affect the immediate and potential availability of micronutrient cations. Certain water-soluble organic compounds form organometallic complexes that enhance micronutrient availability—and these are considered in the following paragraphs. Other organic compounds react with micronutrient cations to form water-insoluble complexes that protect the nutrients from interactions with mineral particles that can bind them in even more insoluble forms. Some organometallic complexes provide slowly available nutrients as they undergo microbial breakdown. Figure 12.64 illustrates the large role that sorption on SOM (and iron oxides) plays in the solubility of Cu and the much smaller role of organic matter with regard to Zn.

Deficiencies of copper and, to a lesser extent, manganese are often found on poorly drained soils high in organic matter (e.g., Histosols and other wetland soils). Zinc is also retained by organic matter, but deficiencies stemming from this retention are not common.

Boron availability can be reduced by sorption to mineral surfaces if soils have high contents of clay and Fe/Mn oxides, even if they are quite acid and sandy. Such soils cover large areas of land in India where crops suffer from yield-limiting B deficiencies. However, boron applied to these soils is of limited utility when sorption reactions reduce its availability in the soil solution. In some of these soils, organic matter can reduce the sorption by coating and masking the sorptive mineral surfaces. Thus, the role of SOM in boron sorption can vary depending on other constituents in the soil (Figure 12.65).

Microbial decomposition of organic matter in plant residues and animal manures can result in the release of micronutrients by the same mechanisms that stimulate the release of macronutrient ions. Micronutrient-rich organic products can be used as a nutrient source on soils deficient in available trace elements. As is the case for macronutrients such as N, however, temporary deficiencies of the trace elements may occur due to microbial immobilization when C/micronutrient ratio organic residues are added.

---

[24]For a review of microbial reduction of certain trace elements, see Lovley (1996).

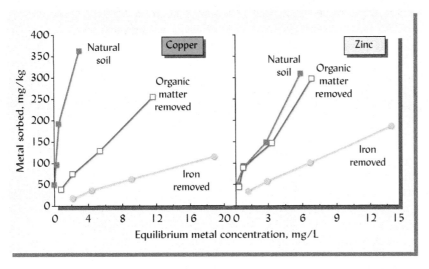

Figure 12.64 Sorption of Cu and Zn in a tropical Alfisol as affected by organic matter and iron. Removal of organic matter greatly reduced the sorption of Cu but had little effect on Zn. The subsequent removal of "free" iron from soil particle surfaces reduced the sorption of both metals. The metal binding sites on both the iron coatings and the organic matter were more selective for Cu than for Zn. The data illustrate that reactions with organic matter and amorphous iron are the major controls on Cu sorption, whereas Zn is held mainly by the iron and probably by cation exchange reactions. [Graphed from data in Agbenin and Olojo (2004)]

Role of Mycorrhizae

The roots of most plants interact symbiotically with certain fungi to form mycorrhizae (fungus roots), which are far more efficient than normal plant roots in several respects. The nature of the mycorrhizal symbiosis and its importance in phosphorus nutrition were described in Sections 10.9 and 12.3, but it is worth mentioning here that mycorrhizae can also increase plant uptake of micronutrients. Crop rotations and other practices that encourage a diversity of mycorrhizal fungi may thereby improve micronutrient nutrition utilization.

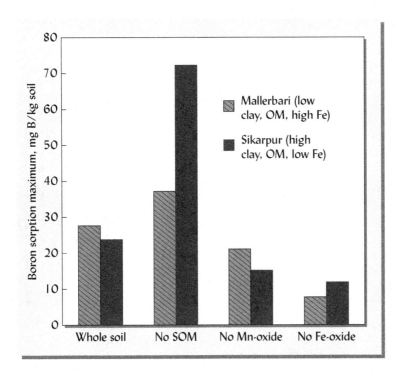

Figure 12.65 Soil organic matter and Fe- and Mn-oxides influence boron sorption from solution. The data here are for a two acidic (pH 4.8) Typic Fluvaquents in northern India. The Mallerbari soil had 12% clay, 1.2% organic C, and 1.5% free iron oxides, while the Sikarpur soil had more clay (18%) and twice as much as organic C (2.5%), but much lower free iron oxides content (0.01%). Maximum adsorption of B was measured in each soil after removal of the organic matter and the iron (and manganese) oxides and compared to B sorption in the untreated (whole) soils. Removal of organic matter, especially from the high organic C Sikarpur soil, greatly increased B sorption—probably because the organic matter had coated and masked the sorptive mineral surface of the clays and Fe- and Mn-oxides. Removal of Fe- and Mn-oxides, however, decreased the B sorption in both soils. [Graphed from selected data in Sarkar et al. (2014)]

Surprisingly, mycorrhizae also appear to protect plants from excessive uptake of micronutrients and other trace elements where these elements are present in potentially toxic concentrations (see Chapter 15). Seedlings of such trees as birch, pine, and spruce are able to grow well on sites contaminated with high levels of zinc, copper, nickel, and aluminum only if their roots are sheathed by ectomycorrhizae. The mycorrhizae apparently help exclude these metallic cations from the root stele and prevent long-distance transport of metal cations within the plant.

Organic Chelating Agents

The cationic micronutrients react with certain organic molecules to form organometallic complexes called **chelates** (from the Greek *chele*, claw). If the chelate is soluble, it increases the availability of the micronutrient and protects it from precipitation reactions. Conversely, formation of an insoluble complex will decrease the availability of the micronutrient.

A chelate is an organic compound in which two or more atoms are capable of bonding to the same metal atom, thus forming a ring. These organic molecules may be synthesized by plant roots or microorganisms and released to the surrounding soil, may be present in the SOM, or may be synthetic compounds in micronutrient fertilizers or added directly to the soil to enhance micronutrient availability. Two examples of iron chelates are shown in Figure 12.66.

In the absence of chelation, when an inorganic salt such as ferric sulfate is added to a calcareous soil, most of the iron is quickly rendered unavailable (insoluble) by reaction with hydroxide. In contrast, if the iron is chelated, it largely remains in the chelate form, which is soluble and available for uptake by plants.

Many dicots appear to remove the metallic cation from the chelate at the root surface, reducing (in the case of iron) and taking up the cation while releasing the organic chelating agent in the soil solution. Roots of certain grasses have been shown to take in the entire chelate–metal complex, reducing and removing the metallic cation inside the root cell, then releasing the organic chelating agent back to the soil solution. In both cases, it appears that the primary role of the chelate is to allow metallic cations to remain in solution so they can diffuse through the soil to the root. Once the micronutrient cations are inside the plant, other organic chelates (such as citrates) may be carriers of these cations to different parts of the plant.

Stability of Chelates Chelates vary in their stability and therefore in their suitability as sources of micronutrients for plants. The stability of a chelate is measured by its stability constant, which is related to the tenacity with which a metal ion is bound in the chelate. The stability constant is useful in predicting which chelate is best for supplying which micronutrient. An added metal chelate must be reasonably stable within the soil if it is to have lasting advantage. For example, the stability constant for EDDHA-Fe^{3+} is 33.9, but that for EDDHA-Zn^{2+}

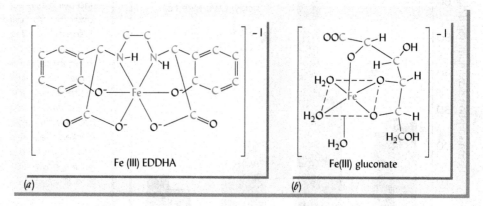

Figure 12.66 *Structural formula for two common iron chelates, ferric ethylendiamine di (o-hydroxyphenylacetic) acid (Fe-EDDHA) (a) and ferric gluconate (b). In both chelates, the iron is protected and yet can be used by plants.* (Diagram courtesy of Ray R. Weil)

is only 16.8. We can therefore predict that if EDDHA-Zn were added to a soil, the Zn in the chelate would be rapidly and almost completely replaced by Fe^{3+} from the soil.

It should not be inferred that chelates are effective only for iron. The chelates of other micronutrients, including zinc, manganese, and copper, are also used successfully to supply these nutrients. Apparently, replacement of the other micronutrients in the chelates by iron from the soil is sufficiently slow to permit plants to absorb the other added micronutrients. Also, because banded applications or foliar sprays are often used to supply zinc and manganese, the possibility of reaction of these elements in chelates with iron and calcium in the soil can be reduced or eliminated.

Chelates in Soil Ecology and Management Chelating agents play important roles in soil ecology. Some plant species or strains are better adapted and more competitive than others in a low micronutrient soil because they can produce more effective chelating agents. An important objective of SOM management is to encourage the formation of chelating agents during decomposition. Particularly intricate and interesting is the apparent ability of certain plants to preferentially stimulate certain soil bacteria which produce exceeding effective chelating agents termed siderophores. The siderophore molecules then capture and carry iron to the plant roots for uptake (Figure 12.67). A similar coordinated process also has been shown to occur with bacterial chelating agents specifically effective for zinc.

Synthetic chelates are quite expensive, so they are most economical when used to ameliorate micronutrient deficiencies of such high-value plants as vegetables, fruit trees, and ornamentals. Some of the chelators used in micronutrient fertilizers, such as gluconate, are naturally occurring and can supply certain micronutrients much more economical than the more expensive aminopolycarboxylate compounds (e.g., EDDHA).

Soil Management and Trace Element Needs

Balance among the trace elements is as essential as, but even more difficult to maintain than, macronutrient balance. Some of the plant enzyme systems that depend on micronutrients require more than one element. For example, both manganese and molybdenum are needed

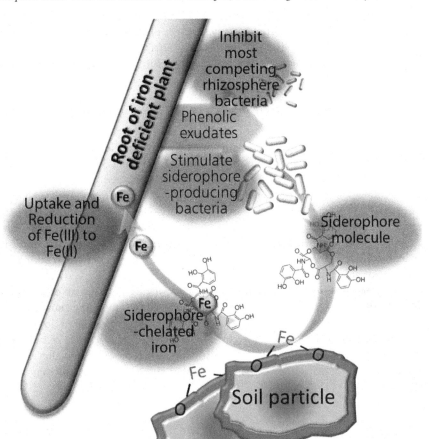

Figure 12.67 When deficient in iron, certain plants can enlist specific rhizosphere ecosystem responses to iron deficiency by increasing root exudation of specialized phenolic compounds. These root exudates inhibit most of the competing rhizosphere bacteria while stimulating the growth only of certain bacteria that produce highly effective iron chelating agents termed siderophores. The siderophore molecules released by the favored bacteria capture and bond with Fe^{3+} ions in the immediate vicinity of the root surface, greatly enhancing the amount of iron available for plant uptake. [Diagram courtesy of Ray R. Weil. Based on concepts in Jin et al. (2014)]

for the assimilation of nitrates by plants. The beneficial effects of combinations of phosphates and molybdenum have already been discussed. Apparently, some plants need zinc and phosphorus for optimum use of manganese. The use of boron and calcium depends on the proper balance between these two nutrients. A similar relationship exists between potassium and copper and between potassium and iron in the production of good-quality potatoes. Copper utilization is favored by adequate manganese, which in some plants is assimilated only if zinc is present in sufficient amounts. Ruminant animals fed plant tissue with low Cu/Mo ratios suffer molybdenum toxicity. Of course, the effects of these and other nutrients will depend on the specific plant being grown, but the complexity of the situation can be seen from the examples cited.

Fertilizer Applications

Economic responses to micronutrients are becoming more widespread as intensity of plant production and our knowledge about micronutrients increases. Yet, micronutrient deficiencies are relatively rare in most soils to which plant residues are returned and organic amendments, such as animal manure or sewage sludge, are regularly applied. Animal manures applied at normal rates sufficient to supply macronutrient needs carry enough copper, zinc, manganese, and iron to supply a major portion of micronutrient needs as well (see Chapter 13). In addition, the chelates produced during manure decomposition enhance the availability of these micronutrients.

The most common management practice to overcome micronutrient deficiencies (and some toxicities) is the use of commercial fertilizers, applied alone or in conjunction with organic amendments and manipulation of soil pH and drainage. Micronutrient fertilizers are most commonly applied to the soil, although foliar sprays and even seed treatments can be used. Foliar sprays of dilute inorganic salts or organic chelates are more effective than soil treatments where soil pH and other factors would render the soil-applied nutrients unavailable.

For soil applications, about one-half to one-fourth as much fertilizer is needed if the application is banded rather than broadcast (see Figure 13.31). About one-fifth to one-tenth as much fertilizer is needed if the material is sprayed on the plant foliage. Foliar application may, however, require repeated applications in a single year. Treating seeds with small dosages (20 to 40 g/ha) of molybdenum has had satisfactory results on molybdenum-deficient acid soils.

Fighting Micronutrient Hunger[25]

Worldwide, malnutrition is a leading contributor to human illness and micronutrient deficiencies, in particular, pose serious human health concerns. Iron deficiency anemia has been estimated to be among the top global ten health burdens, debilitating 3 to 5 billion people. Human deficiencies of zinc and copper are also widespread, afflicting perhaps a billion people. In large parts of Europe and Asia, the selenium in people's diets is also below the needed levels. In many cases these deficiencies stem from a lack of diversity—especially of animal products and green vegetables—in the diets of people who rely upon staple grain products made from rice, wheat, or corn for most of their nutrition. Evidence suggests that plant breeders have inadvertently selected for cultivars that, while high yielding, may be less effective in taking up trace elements important in human diets.

Aside from serious nutritional concerns about diets lacking in diversity or based on highly refined foods from which nutrients have been removed by processing, nutritionists have raised concerns that today's crops may be less "nutrient dense" (i.e., contain lower concentrations of nutrient elements) than crops grown in the past. Concentrations of such micronutrients as copper, zinc, and iron in many foods today have been found to be lower than those reported a half century ago. Some suggest that the cause of declining micronutrient

---

[25]For a reviews of human health impacts of micronutrient deficiencies, see Stein (2010); for assessment of global human health problems, including micronutrient deficiencies, see The World Bank (2013); for evaluation of research strategies to determine changes in nutrient density of crops, with an emphasis on vegetables, see Farnham and Grusak (2014).

Table 12.9
ZINC AND COPPER CHANGES IN SOIL AND GRAIN DURING 160 YEARS OF CONTINUOUS WHEAT FARMING IN ENGLAND[a]

In the post-1960s era, micronutrient concentrations in wheat grain declined significantly. The concentrations of micronutrients in the soil did not decline where only fertilizer was used. Where farmyard manure (a good source of micronutrients) was used, the levels in the soil were significantly higher post-1960s and increased during the period. The researchers concluded that the decline in wheat "nutrient density" was a result of genetic changes in the "green revolution era" high-yielding, short stem wheat varieties grown in the most recent four decades, rather than resulting from soil depletion.

	Concentration in wheat grain harvested				Concentration in soil sampled			
	Zinc		Copper		Zinc		Copper	
Historical period:	1845–1967	1968–2005	1845–1967	1968–2005	1845–1920	1965–2005	1845–1920	1965–2005
Soil amendments used			mg/kg					
Farm yard manure	37	27	5.4	4.0	60	90	16	22
Fertilizer (N, P, K, Mg, S)	32	22	5.0	3.6	57	70	15	17

[a] Based on data selected or extrapolated from Fan et al. (2008).

density in crops is the depletion of soil nutrients by intensive farming, especially the use of mineral fertilizers that stimulate high yields but supply only macronutrients, not micronutrients. Careful research, however, suggests that any declines that have occurred resulted more from changes in grain crop genetics (varieties) than from soil depletion (Table 12.9). Lower micronutrient concentrations in grain may owe partially to similar amounts of micronutrient taken up being diluted in a higher yield of grain, a process that may be aggravated by increasing carbon dioxide levels in the atmosphere. Lower nutrient density may also be partially a result of some loss of micronutrient accession traits through genetic selection that focused on increasing grain yield. The world may now need plant breeding and biotechnology to help select and/or create plant varieties that more efficiently absorb trace elements. The future calls for close collaboration between soil scientists, plant geneticists, and human nutritionists in exploring the long-term health and sustainable development implications of these findings.

Management Issues Trace element deficiencies and toxicities have been diagnosed in most areas of crop production and some forest areas of the United States and Europe. However, in some developing countries, particularly in the tropics, the extent of these deficiencies is much less well known. Research suggests that crop deficiencies in zinc, boron and, to a lesser degree, copper are common on highly weathered soils in the humid and subhumid tropics. Irrigation schemes that bring calcareous soils in desert areas under cultivation are often plagued with deficiencies of iron, zinc, copper, and manganese. As macronutrient deficiencies are addressed and yields are increased, more micronutrient deficiencies will undoubtedly come to the fore. One encouraging factor is the small quantity of micronutrients usually needed, making it relatively easy to add these nutrients to fertilizer materials without significantly increasing the transport costs of the macronutrient fertilizers already being shipped to remote parts of the world. The management principles established for the economically developed countries should also be helpful in alleviating micronutrient deficiencies in less-developed countries. Sustainable solutions are most likely to be achieved by simultaneous application of multiple soil and plant management strategies for enhancing plant utilization of micronutrient resources (Table 12.10).

Table 12.10
SOIL- AND PLANT-BASED STRATEGIES FOR ENHANCING MICRONUTRIENT UPTAKE[a]

Soil management	Plant management

• Root zone fertilization
  – Soil pH regulation by amendment with lime or S.
  – Regulation of rhizosphere pH by ratio of NO_3^- to NH_4^+ supplied
  – Bacterial inoculants (iron-reducing bacteria, etc.)
  – Banding inorganic salt or chelated micronutrient
• Soil water/aeration management
• Enhanced SOM by use of cover crops, reduced tillage, and organic amendments
	• Foliar spray of dilute inorganic micronutrient salts or chelates
	• Injection of micronutrient solution into plant stem/trunk
	• Genetic improvement for enhanced uptake efficiency
	• Interplanting with other species, e.g., dicot/grass intercropping
	• Choice of species or strain adapted to efficient micronutrient uptake at existing soil pH

[a]Generalized from concepts regarding iron deficiency in Zuo and Zhang (2011).

12.9 CONCLUSION

The cycles of sulfur and N have much in common, including many processes that operate in soils. Both elements are held by soil colloids in slowly available forms. In surface soil horizons, the bulk of both elements are found as constituents of SOM. Their release to inorganic ions (SO_4^{2-}, NH_4^+, and NO_3^-) is accomplished by soil microorganisms and makes them available to plants. Anaerobic soil organisms change both elements into gaseous forms, which are emitted to the atmosphere. While in the atmosphere, some of the gases accelerate climate change. Both elements are subject to deposition from the atmosphere in the form of *acid precipitation*, seriously damaging lakes, soils, and plants in susceptible ecosystems. Certain soil organisms have the ability to fix elemental N_2 gas into compounds usable by plants. No analogous process occurs for sulfur.

Plants remove about 10 to 20 times as much nitrogen as sulfur. If plant material is regularly removed from the system, it is usually more critical for N than for sulfur that the supply be regularly replenished. On the other hand, while N fertilizers are expensive and energy intensive to manufacture and are usually needed in quite large amounts (~100 kg N/ha), sulfur deficiencies can be very easily corrected with small amounts (~10 kg S/ha) of inexpensive, plentiful materials such as gypsum.

Current knowledge of the N and sulfur cycles can help us alleviate environmental degradation and enhance life-supporting productivity. Nitrogen remains the most widely limiting—but also the most widely polluting—plant nutrient. Sulfur, whose supply to the soil was taken for granted during a century of intense industrial SO_2 emissions, is now emerging as a widely limiting nutrient for plant growth. Outside the zones of heavy sulfur deposition from air pollution, plant requirements for S should be given consideration equal to P and K.

Soil P presents us with a double-edged sword. On one hand, too little P commonly limits the productivity of natural and cultivated plants and is the cause of widespread soil and environmental degradation. On the other hand, industrialized agriculture has concentrated too much P in some cases, resulting in losses from soil that cause egregious eutrophication in surface waters. Some of these situations have been caused by excessive buildup of soil P with fertilizer. Others are the result of concentration of animal production, so that P in the manure produced at many livestock facilities far exceeds that required by the crops grown on the surrounding land.

Except in cases of extreme buildup, the availability of P to plant roots has a double constraint: the low total P level in soils and the small percentage of this level that is present in available forms. Furthermore, even when soluble phosphates are added to soils, they are quickly fixed into insoluble forms that in time become quite unavailable to growing plants. In acid soils, the P is fixed primarily by iron, aluminum, and manganese; in alkaline soils, by calcium and magnesium. This fixation greatly reduces the efficiency of phosphate fertilizers, with little of the added P being taken up by plants. In time, however, this unused P can build up and serve as a reserve pool for plant absorption.

Potassium is generally abundant in soils, but it, too, is present mostly in forms that are relatively unavailable for plant absorption. Fortunately, however, most soils contain considerable nonexchangeable but slowly available forms of this element. Over time this K can be released to exchangeable and soil solution forms that can be quickly absorbed by plant roots. This is fortunate since the plant requirements for K are high—5 to 10 times that of P and similar to that of N.

Calcium and magnesium are macronutrients usually used in quantities similar to sulfur and P. However, for woody plants and for certain "calcicole" plants, calcium is used in amounts that rival N and K (1 to 3% Ca in the dry matter). Both Ca and Mg are intimately associated with the processes of soil acidification and neutralization and thus are linked to sulfur and N deposition. The cycling of these two nonacid cations is of major concern in biogeochemistry and soil Ca supplies greatly impact the nature of the ecology that will develop in an area. Calcium and magnesium also impact global carbon balance through the precipitation and dissolution of Ca and Mg carbonates.

Silicon is an important element taken up and used by most plants in macronutrient-like quantities, but not generally considered to be an essential nutrient element. Despite its occurrence in large quantities in most soils, plants often respond positively to its application in the more soluble forms. Silicon appears to help plant resist attacks by pest and pathogens and to grow better under conditions of environmental stress.

Micronutrient availability often regulates plant species competition and survival on a site. Micronutrients are becoming increasingly important to world agriculture as crop removal of these essential elements increases. Micronutrient deficiencies are due not only to low contents of these elements in soils but more often to their unavailability to growing plants. They are adsorbed by inorganic constituents such as Fe, Al oxides and form complexes with organic matter, some of which are only sparingly available to plants. Other such organic complexes, known as *chelates*, protect some of the micronutrient cations from inorganic adsorption and make them available for plant uptake.

Toxicities of micronutrients retard both plant and animal growth. Removing these elements from soil and water, or rendering them unavailable for plant uptake, is one of the challenges facing soil and plant scientists. Adequate micronutrient supply is also a potential tool for helping to manage plant diseases, to which results using Mn attest. In most cases, soil-management practices that avoid extremes in soil pH, that optimize the return of plant residues and animal manures, that encourage mycorrhizae, and that promote chelate production by actively decomposing organic matter and rhizosphere bacteria will minimize the risk of micronutrient deficiencies or toxicities

Finally, in studying the use and management of these mineral nutrients, it is important to keep in mind the sobering truth that, although accessible supplies of many elements (K, Ca, Fe) are quite large, the world's supply of usable P is predicted to become exhausted within a century or so, if today's wasteful practices continue. While substitutes can be found for many finite natural resources mined from the Earth—copper in phone lines can be replaced by fiber optics, steel in car bodies by composite plastics, and petroleum in transportation fuels by hydrogen cells—this is *not* the case for P in food production. The U.S. Geological Survey[26] lists the following under the heading *Substitutes for Phosphate Rock*: "THERE ARE NO SUBSTITUTES FOR PHOSPHORUS."

STUDY QUESTIONS

1. The manager of a landscaping company is having a bit of an argument with the landscape architect about plans to fertilize and lime the soil in a new installation before planting turf and ornamental trees. The manager planned to use mostly urea for supplying N. The landscape architect says the urea will have an alkaline reaction and raise the soil pH. The manager says that urea will have an acid reaction and lower the soil pH. Who is correct? Explain, using chemical reactions to support your argument.

2. A sandy loam soil under a golf course fairway has an organic matter content of 3% by weight. Calculate the approximate amount of N (in kg N/ha) you would expect this soil to provide for plant uptake during a typical year. Show your work and state what assumptions or estimates you made to do this calculation.

---
[26]Jasinski (2013).

3. Both sulfur and N are added to soils by atmospheric deposition. In what situations is this phenomenon beneficial and under what circumstances is it detrimental?

4. Tests showed a soil to contain 25 kg nitrate-N per ha. About 2,000 kg of wheat straw was applied to 1 ha of this land. The straw contained 0.4% N. How much N was applied in the straw? Explain why two weeks after the straw was applied, new tests showed no detectable nitrate N. Show your work and state what assumptions or estimates you made to do this calculation.

5. Why do CAFOs (concentrated animal-feeding operations) on industrial-style farms present some environmental and health problems relating to N and P? What are these problems and how can they be managed?

6. What differences would you expect in nitrate contents of streams from a forested watershed and one where agricultural crops are grown, and why?

7. Microbial transformations of N involve oxidation and reduction reactions. Explain how the N from the protein in a leaf might cycle differently in a wetland soil than in an upland soil.

8. What is *acid rain*, what are the sources of acidity in this precipitation, and how does this acidity damage natural ecosystems?

9. Why might some farmers be willing to pay more for N fertilizer that contains a nitrification inhibitor chemical in it?

10. Chemical fertilizers and manures with high N contents are commonly added to agricultural soils. Yet these soils are often lower in total N than are nearby soils under natural forest or grassland vegetation. Explain why this is the case.

11. Why are S deficiencies in agricultural crops more widespread today than 30 years ago?

12. How do riparian forests help reduce nitrate contamination of streams and rivers?

13. What are *potential acid sulfate soils*, where would you find them, and under what circumstances are they likely to cause serious problems?

14. You have learned that nitrogen, potassium, and phosphorus are all "fixed" in the soil. Compare the processes of these fixations, the benefits and constraints they each provide, and indicate the role of microbes, if any, in each process.

15. Assume you add a soluble phosphate fertilizer to an Oxisol and to an Aridisol. In each case, within a few months most of the P has been changed to insoluble forms. Indicate what these forms are and the respective compounds in each soil responsible for their formation.

16. What is meant by *eutrophication*, and how is it influenced by farm practices involving P?

17. Which is likely to have the higher buffering capacity for P and K, a loamy sand or a clay soil? Explain.

18. In the spring a certain surface soil showed the following soil test: soil solution K = 20 kg/ha; exchangeable K = 200 kg/ha. After two crops of alfalfa hay that contained 250 kg/ha of K were harvested and removed, a second soil test showed soil solution K = 15 kg/ha and exchangeable K = 200 kg/ha. Explain why there was not a greater reduction in soil solution and exchangeable K levels.

19. How does P that forms relatively insoluble inorganic compounds in soils find its way into streams and other waterways?

20. Compare the organic P levels in the upper horizons of a forested soil with those of a nearby soil that has been cultivated for 25 years. Explain the difference.

21. Discuss how the "4 Rs" concepts of nutrient management (right source, right amount, right time, and right placement) might apply to management of P and N on a specific farm of your choice.

22. What portion of the plant would you look at to find symptoms of Ca and Mg deficiencies, respectively?

23. What are phytoliths and what roles do they play in plants and soils?

24. Even though only small quantities of micronutri of the nutrentents are needed annually for normal plant growth, would it be wise to add large quantities of these elements now to satisfy future plant needs? Explain.

25. Iron deficiency is common for peaches and other fruits grown on highly alkaline irrigated soils of arid regions, even though these soils are quite high in iron. How do you account for this situation, and what would you do to alleviate the difficulty?

26. What are *chelates*, how do they function, and what are their sources?

27. The addition of only 1 kg/ha of a nutrient to an acid soil on which lime-loving cauliflower was being grown gave considerable growth response. Which of the nutrients would it likely have been? Explain.

28. Animals, both domestic and wild, are adversely affected by deficiencies and toxicities of two of the micronutrients. Which elements are these, and what are the conditions responsible for their effects?

29. Discuss the role plant breeders and geneticists might play in managing micronutrient deficiencies and toxicities.

30. Since boron is required for the production of good-quality table beets, some companies purchase only beets that have been fertilized with specified amounts of this element. Unfortunately, an oat crop following the beets does very poorly compared to oats following unfertilized beets. Give possible explanations for this situation.

REFERENCES

Agbenin, J. O., and L. A. Olojo. 2004. "Competitive adsorption of copper and zinc by a Bt horizon of a savanna Alfisol as affected by pH and selective removal of hydrous oxides and organic matter." *Geoderma* 119:85–95.

Alloway, B. J., and R. D. Graham. 2008. "Micronutrient deficiencies in crops and their global significance." *Micronutrient Deficiencies in Global Crop Production*. Springer, Amsterdam, pp. 41–61.

Anonymous http://www.soils.org/. 1974 "Cation Exchange Properties of Soils: A Slide Show" Division S-2, Soil Chemistry. S.S.S.A., Inc., Madison WI. 53711.

Bateman, E. J., and E. M. Baggs. 2005. "Contributions of nitrification and denitrification to N_2O emissions from soils at different water-filled pore space." *Biology and Fertility of Soils* 41:379–388.

Batterman, S. A., N. Wurzburger, and L. O. Hedin. 2013. "Nitrogen and phosphorus interact to control tropical symbiotic N_2 fixation: A test in Inga punctata." *Journal of Ecology* 101:1400–1408.

Bittman, S., C. G. Kowalenko, D. E. Hunt, T. A. Forge, and X. Wu. 2006. "Starter phosphorus and broadcast nutrients on corn with contrasting colonization by mycorrhizae." *Agronomy Journal* 98:394–401.

Broadbent, F. E., T. Nakashima, and G. Y. Chang. 1982. "Estimation of nitrogen fixation by isotope dilution in field and greenhouse experiments." *Agronomy Journal* 74:625–628.

Buresh, R. J., P. C. Smithson, and D. T. Hellums. 1997. "Building soil phosphorus capital in Africa." In R. J. Buresh, P. A. Sanchez, and F. Calhoun (eds.). *Replenishing Soil Fertility in Africa*. SSSA Special Publication No. 51. Soil Science Society of America, Madison, WI.

Butusov, M., and A. Jernelöv. 2012. *Phosphorus: An Element That Could Have Been Called Lucifer*. Springer, New York, p. 101.

Canfield, D. E., A. N. Glazer, and P. G. Falkowski. 2010. "The evolution and future of earth's nitrogen cycle." *Science* 330:192–196.

Churka Blum, S., J. Lehmann, D. Solomon, E. F. Caires, and L. R. F. Alleoni. 2013. "Sulfur forms in organic substrates affecting S mineralization in soil." *Geoderma* 200–201:156–164.

Cooke, J., and M. R. Leishman. 2011. "Is plant ecology more siliceous than we realise?" *Trends in Plant Science* 16:61–68.

Cordell, D., and S. White. 2011. "Peak phosphorus: Clarifying the key issues of a vigorous debate about long-term phosphorus security." *Sustainability* 3:2027–2049.

Cordell, D., and S. White. 2015. "Tracking phosphorus security: Indicators of phosphorus vulnerability in the global food system." *Food Security* 7:337–350.

Devau, N., E. L. Cadre, P. Hinsinger, B. Jaillard, and F. Gérard. 2009. "Soil pH controls the environmental availability of phosphorus: Experimental and mechanistic modelling approaches." *Applied Geochemistry* 24:2163–2174.

Dodd, R. J., and A. N. Sharpley. 2015. "Recognizing the role of soil organic phosphorus in soil fertility and water quality." *Resources, Conservation and Recycling* 105 part B:282–293.

Emsley, J. 2002. *The 13th Element: The Sordid Tale of Murder, Fire, and Phosphorus*. John Wiley & Sons, New York, p. 352.

Eriksen, J. 2009. "Soil sulfur cycling in temperate agricultural systems." *Advances in Agronomy* 102:55–89.

Fageria, N. K. 2009. *The Use of Nutrients in Crop Plants*. CRC, Boca Raton FL, USA Press.

Fan, M. -S., F. -J. Zhao, S. J. Fairweather-Tait, P. R. Poulton, S. J. Dunham, and S. P. McGrath. 2008. "Evidence of decreasing mineral density in wheat grain over the last 160 years." *Journal of Trace Elements in Medicine and Biology* 22:315–324.

Farnham, M. W., and M. A. Grusak. 2014. "Assessing nutritional changes in a vegetable over time: Issues and considerations." *HortScience* 49:128–132.

Frossard, E., L. M. Condron, A. Oberson, S. Sinaj, and J. C. Fardeau. 2000. "Processes governing phosphorus availability in temperate soils." *Journal of Environmental Quality* 29:15–23.

Goenadi, D. H., Siswanto, and Y. Sugiarto. 2000. "Bioactivation of poorly soluble phosphate rocks with a phosphorus-solubilizing fungus." *Soil Science Society of America Journal* 64:927–932.

Guntzer, F., C. Keller, and J.-D. Meunier. 2012. "Benefits of plant silicon for crops: A review." *Agronomy for Sustainable Development* 32:201–212.

Jasinski, S. M. 2013. "Phosphate rock." U.S. Department of the Interior, U.S. Geological Survey, Reston, VA. http://minerals.usgs.gov/minerals/pubs/commodity/phosphate_rock/mcs-2013-phosp.pdf.

Jez, J. (ed.). 2008. *Sulfur: A Missing Link Between Soil, Crops and Nutrition*. American Society of Agronomy, Madison, WI, pp. 1–323.

Jin, C., Y. Q. Ye, and S. J. Zheng. 2014. "An underground tale: Contribution of microbial activity to plant iron acquisition via ecological processes." *Annals of Botany* 113:7–18.

Johnson, A. H., A. Moyer, J. E. Bedison, S. L. Richter, and S. A. Willig. 2008. "Seven decades of calcium depletion in organic horizons of Adirondack forest soils." *Soil Science Society of America Journal* 72:1824–1830.

Kennedy, I. R., A. T. M. A. Choudhury, and M. L. Kecskes. 2004. "Non-symbiotic bacterial diazotrophs in crop-farming systems: Can their potential for plant growth promotion be better exploited?" *Soil Biology Biochemistry* 36:1229–1244.

Khalil, K., B. Mary, and P. Renault. 2004. "Nitrous oxide production by nitrification and denitrification in soil aggregates as affected by O2 concentration." *Soil Biology Biochemistry* 36:687–699.

Khan, S. A., R. L. Mulvaney, and T. R. Ellsworth. 2014. "The potassium paradox: Implications for soil fertility, crop production and human health." *Renewable Agriculture and Food Systems* 29:3–27.

Klimont, Z., S. J. Smith, and J. Cofala. 2013. "The last decade of global anthropogenic sulfur dioxide: 2000–2011 emissions." *Environmental Research Letters* 8:014003.

Koele, N., T. W. Kuyper1, and P. S. Bindraban. 2014. "Beneficial organisms for nutrient uptake." VFRC Report 2014/1. Virtual Fertilizer Research Center, Washington, D.C. http://www.vfrc.org.

Kopittke, P. M., and N. W. Menzies. 2007. "A review of the use of the basic cation saturation ratio and the 'ideal' soil." *Soil Science Society of America Journal* 71:259–265.

Kost, D., J. Nester, and W.A. Dick. 2018. "Gypsum as a soil amendment to enhance water quality by reducing soluble phosphorus concentrations." *Journal of Soil and Water Conservation* 73:22A–24A.

Kovar, J. L., and C. A. Grant. 2011. "Nutrient cycling in soils: Sulfur." In J. L. Hatfield, and T. J. Sauer (eds.). *Soil Management: Building a Stable Base for Agriculture.* Soil Science Society of America, Madison, WI, USA, pp. 103–112.

Leffelaar, P. A., and W. W. Wessel. 1988. "Denitrification in a homogeneous, closed system: Experimental and simulation." *Soil Science* 146:335–349.

Li, L., S. -M. Li, J. -H. Sun, L. -L. Zhou, X. -G. Bao, H. -G. Zhang, and F. -S. Zhang. 2007. "Diversity enhances agricultural productivity via rhizosphere phosphorus facilitation on phosphorus-deficient soils." *Proceedings of the National Academy of Sciences* 104:11192–11196.

Lovley, D. R. 1996. "Microbial reduction of iron, manganese and other metals." *Advances in Agronomy* 54:175–231.

Lusk, M. G., G. S. Toor, and P. W. Inglett. 2018. "Characterization of dissolved organic nitrogen in leachate from a newly established and fertilized turfgrass." *Water Research* 131:52–61.

Magill, A. H., J. D. Aber, W. S. Currie, K. J. Nadelhoffer, M. E. Martin, W. H. Mcdowell, J. M. Melillo, and P. Steudler. 2004. "Ecosystem response to 15 years of chronic nitrogen additions at the Harvard forest LTER, Massachusetts, USA." *Forest Ecology and Management* 196:7–28.

Mengel, K., and E. A. Kirkby. 2001. *Principles of Plant Nutrition.* 5th ed. Kluwer Academic Publishers, Dordrecht, Netherlands. p. 864.

Nowak, C. A., R. B. Downard, Jr., and E. H. White. 1991. "Potassium trends in red pine plantations at Pack Forest, New York." *Soil Science Society of America Journal* 55:847–850.

Nziguheba, G., E. Smolders, and R. Merckx. 2006. "Mineralization of sulfur from organic residues assessed by inverse isotope dilution." *Soil Biology and Biochemistry* 38:2278–2284.

Obersteiner, M., J. Peñuelas, P. Ciais, M. van der Velde, and I. A. Janssens. 2012. "The phosphorus trilemma." *Nature Geoscience* 6:897–898.

Patrick, W. H., Jr. 1982. "Nitrogen transformations in submerged soils." In F. J. Stevenson (ed.). *Nitrogen in Agricultural Soils.* Agronomy Series No. 27. American Society of Agronomy, Crop Science Society of America, Soil Science Society of America, Madison, WI.

Qiao, C., L. Liu, S. Hu, J. E. Compton, T. L. Greaver, and Q. Li. 2015. "How inhibiting nitrification affects nitrogen cycle and reduces environmental impacts of anthropogenic nitrogen input." *Global Change Biology* 21:1249–1257.

Römheld, V., and E. Kirkby. 2010. "Research on potassium in agriculture: Needs and prospects." *Plant and Soil* 335:155–180.

Ros, G. H., E. Hoffland, C. van Kessel, and E. J. M. Temminghoff. 2009. "Extractable and dissolved soil organic nitrogen – a quantitative assessment." *Soil Biology and Biochemistry* 41:1029–1039.

Sanchez, P. A. 2002. "Soil fertility and hunger in Africa." *Science* 295:2019–2020.

Santi, C., D. Bogusz, and C. Franche. 2012. "Biological nitrogen fixation in non-legume plants." *Annals of Botany* 111:743–767.

Sarkar, D., D. K. De, R. Das, and B. Mandal. 2014. "Removal of organic matter and oxides of iron and manganese from soil influences boron adsorption in soil." *Geoderma* 214–215:213–216.

Scheer, C., S. J. Del Grosso, W. J. Parton, D. W. Rowlings, and P. R. Grace. 2014. "Modeling nitrous oxide emissions from irrigated agriculture: Testing DayCent with high frequency measurements." *Ecological Applications* 24:528–538.

Scherer, H. W. 2009. "Sulfur in soils." *Journal of Plant Nutrition and Soil Science* 172:326–335.

Scholz, R. W., D. T. Hellums, and A. A. Roy. 2012. "Global sustainable phosphorus management: A transdisciplinary venture." *Current Science* 108:1237–1246.

Sharma, S., E. Duveiller, R. Basnet, C. B. Karki, and R. C. Sharma. 2005. "Effect of potash fertilization on Helminthosporium leaf blight severity in wheat, and associated increases in grain yield and kernel weight." *Field Crops Research* 93:142–150.

Sharpley, A. 2000. "Phosphorus availability." In M. E. Summer (ed.). *Handbook of Soil Science.* CRC Press, New York, pp. D-18–D-38.

Sharpley, A. N. 1990. "Reaction of fertilizer potassium in soils of differing mineralogy." *Soil Science* 49:44–51.

Sillanpaa, M. 1982. *Micronutrients and the Nutrient Status of Soils: A Global Study.* U.N. Food and Agricultural Organization, Rome.

Smit, A. L., M. Blom-Zands, A. van der Werf, and P. S. Bindraban. 2013. "Plant strategies and cultural practices to improve the uptake of indigenous soil P and the efficiency of fertilization." VFRC Report 2013/4. Virtual Fertilizer Research Center, Washington, D.C. http://www.vfrc.org.

Smith, S. E., I. Jakobsen, M. Grønlund, and F. A. Smith. 2011. "Roles of arbuscular mycorrhizas in plant phosphorus nutrition: Interactions between pathways of phosphorus uptake in arbuscular mycorrhizal roots have important implications for understanding and manipulating plant phosphorus acquisition." *Plant Physiology* **156**:1050–1057.

Smith, S. J., A. N. Sharpley, J. W. Naney, W. A. Berg, and O. R. Jones. 1991. "Water quality impacts associated with wheat culture in the Southern Plains." *Journal of Environmental Quality* **20**:244–249.

Sommer, M., D. Kaczorek, Y. Kuzyakov, and J. Breuer. 2006. "Silicon pools and fluxes in soils and landscapes—a review." *Journal of Plant Nutrition and Soil Science* **169**:310–329.

Stein, A. 2010. "Global impacts of human mineral malnutrition." *Plant and Soil* **335**:133–154.

Suwanwaree, P., and G. P. Robertson. 2005. "Methane oxidation in forest, successional, and no-till agricultural ecosystems: Effects of nitrogen and soil disturbance." *Soil Science Society of America Journal* **69**:1722–1729.

Taylor, M. Z. 2005. "No bad soils." AgWeb.com division of Farm Journal, Inc. http://www.agweb.com/get_article.asp?sigcat=topproducer&pageid=117257 (posted 26 August 2005; verified 26 August 2006).

Thamdrup, B. 2012. "New pathways and processes in the global nitrogen cycle." *Annual Review of Ecology, Evolution, and Systematics* **43**:407–428.

The World Bank. 2013. *The Global Burden of Disease: Generating Evidence, Guiding Policy—Sub-Saharan Africa Regional Edition*. Institute for Health Metrics and Evaluation and Human Development Network, Seattle, WA. www.healthmetricsandevaluation.org

USDA/NRCS. 2012. "Assessment of the effects of conservation practices on cultivated cropland in the Upper Mississippi River Basin." Conservation Effects Assessment Project. http://www.nrcs.usda.gov/Internet/FSE_DOCUMENTS/stelprdb1042093.pdf.

van Kessel, C., T. Clough, and J. W. van Groenigen. 2009. "Dissolved organic nitrogen: An overlooked pathway of nitrogen loss from agricultural systems?" *Journal of Environmental Quality* **38**:393–401.

van Niftrik, L., and M. S. M. Jetten. 2012. "Anaerobic ammonium-oxidizing bacteria: Unique microorganisms with exceptional properties." *Microbiology and Molecular Biology Reviews* **76**:585–596.

Welch, R. M. 1995. "Micronutrient nutrition of plants." *Critical Reviews in Plant Science* **14**(1):49–82.

White, C. M., and R. R. Weil 2011. "Forage radish cover crops increase soil test phosphorus surrounding radish taproot holes." *Soil Science Society of America Journal* **75**:121–130.

Wurzburger, N., J. P. Bellenger, A. M. L. Kraepiel, and L. O. Hedin. 2012. "Molybdenum and phosphorus interact to constrain asymbiotic nitrogen fixation in tropical forests." *PLoS ONE* **7**:e33710.

Wyant, K. A., J. E. Corman, J. R. Corman, and J. J. Elser. 2013. *Phosphorus, Food, and Our Future*. Oxford University Press, Oxford, UK, p. 224.

Young, E. O., D. S. Ross, B. J. Cade-Menun, and C. W. Liu. 2013. "Phosphorus speciation in riparian soils: A phosphorus-31 nuclear magnetic resonance spectroscopy and enzyme hydrolysis study." *Soil Science Society of America Journal* **77**:1636–1647.

Zhang, H., M. Xu, W. Zhang, and X. He. 2009. "Factors affecting potassium fixation in seven soils under 15-year long-term fertilization." *Chinese Science Bulletin* **54**:1773–1780

Zhao, F. J., J. Lehmann, D. Solomon, M. A. Fox, and S. P. McGrath. 2006. "Sulphur speciation and turnover in soils: Evidence from sulphur k-edge XANES spectroscopy and isotope dilution studies." *Soil Biology and Biochemistry* **38**:1000–1007.

Zuo, Y., and F. Zhang. 2011. "Soil and crop management strategies to prevent iron deficiency in crops." *Plant and Soil* **339**:83–95.

13 Practical Nutrient Management

Without stream fencing, cattle pollute the water with nutrients. (Photo courtesy of Ray R. Weil)

For every atom lost to the sea, the prairie pulls another out of the decaying rocks. The only certain truth is that its creatures must suck hard, live fast, and die often, lest its losses exceed its gains.
—ALDO LEOPOLD, A SAND COUNTY ALMANAC (1949)

To be good stewards of the land, soil managers must keep nutrients cycling effectively and maintain the soil's capacity to supply the nutritional needs of plants. The supply of nutrients from the soil may limit vegetative productivity to levels below what a given climate could be expected to support. Interventions to manage nutrients are typically needed for forests, farms, fairways, and flower gardens to efficiently provide us with lumber, food, biofuels, recreational opportunities, and aesthetic satisfaction.

Nutrient cycles can become unbalanced through increased removals (e.g., harvest of timber and crops), through increased system leakage (e.g., leaching and runoff), through simplification (e.g., monoculture, be it of corn, pine trees, or sugarcane), and through increased animal density (especially if imported feed brings in nutrients from outside the ecosystem). Some of the greatest impacts of land management are felt not on the land but in the water, where excess nutrients play havoc with aquatic ecosystems. Therefore, anyone attempting to manage land must also be prepared to take responsibility for nutrient and environmental management.

In this chapter, we will discuss methods of balancing and enhancing nutrient cycles, as well as sources of additional nutrients that can be applied to soils or plants. We will learn how to diagnose nutritional disorders of plants and correct soil fertility problems. Building on the principles set out in early chapters, this chapter contains practical information on profitable production of abundant, high-quality plants while maintaining the health of both the soil and the rest of the environment.

13.1 GOALS OF NUTRIENT MANAGEMENT[1]

Nutrient management is one aspect of a holistic approach to managing soils in the larger environment. It aims to achieve four broad, interrelated goals: (1) cost-effective production of high-quality plants and animals, (2) maintenance or enhancement of soil health, (3) efficient use and conservation of nutrient resources, and (4) protection of the environment beyond the soil.

---

[1] For an overview of issues and advances in nutrient management for agriculture and environmental quality, see Magdoff et al. (1997). For estimates of nutrient limitations on global vegetative productivity, see Fisher et al. (2012). For a standard textbook on management of agricultural soil fertility and fertilizers, see Havlin et al. (2014).

Plant Production

Three of the primary types of plant production in which people engage are: (1) *agriculture*, (2) *forestry*, and (3) *ornamental landscaping*. Agriculturists range from small-scale subsistence farm families or home gardeners who produce mainly for their own use to larger-scale commercial farmers and ranchers whose primary goal is to make a profit by selling the plants and animals they produce. Regardless of the scale, the main nutrient-management goal for agriculture is to increase plant yield and quality. In forestry, the principal plant product may be measured in terms of volume of lumber or paper produced. When soils are used for ornamental landscaping purposes, the principal objective is to produce quality, aesthetically pleasing plants and land surfaces. Whether the plants are produced for sale or not, relatively little attention is paid to yield of biomass produced.

Farmers usually judge the success or failure of their management schemes within the period of one or two crop-growing seasons, rather than the span of several decades or even human generations that would allow full evaluation of the practices they choose to use. The time frame in forestry is measured in decades or even centuries, and so tends to limit the intensity of nutrient management interventions that can be profitably undertaken.

Soil Health and Productivity

The concept of using nutrient management to enhance soil health goes far beyond simply supplying nutrients for the current year's plant growth. Rather, it includes the long-term nutrient-supplying and nutrient-cycling capacity of the soil, improvement of soil physical properties or tilth, maintenance of above- and belowground biological functions and diversity, and the avoidance of chemical toxicities. Likewise, the management tools employed go far beyond the application of various fertilizers (although this may be an important component of nutrient management). Nutrient management requires the integrated management of the physical, chemical, and biological processes that comprise soil health. The impact of fire on soil nutrient and water supplies (Chapter 7), the increase in nutrient availability (and sometimes losses) brought about by earthworm activity (Chapter 10), the effects of tillage on organic matter mineralization (Chapter 11), the contribution of rhizobia–legume symbiosis to nitrogen supplies (Chapter 12), and the role of mycorrhizal fungi in phosphorus uptake by plants (Chapter 12) are all examples of soil processes that need to be considered in nutrient management.

Conservation of Nutrient Resources

Two concepts that are key to the goal of conserving nutrient resources are: (1) renewal or reuse of the resources, and (2) nutrient budgeting that reflects a balance between system inputs and outputs.

The first law of thermodynamics suggests that all material resources are ultimately renewable, since the elements are not destroyed by use but are merely recombined and moved about in space. In practical terms, however, once a nutrient has been removed from a plot of land and dispersed into the larger environment, it may be difficult, if not impossible, to use it again for plant growth. For instance, phosphorus deposited in a lake bottom with eroded sediment and N buried in a landfill as a component of garbage will not likely be available for reuse. In contrast, soil application of livestock manure or composted municipal garbage and irrigation with treated sewage effluent are practical examples of nutrient reuse.

Recycling is a form of reuse in which nutrients are returned to the same land from which they were previously removed. Litter fall from perennial vegetation recycles nutrients to the soil naturally. Leaving crop residues in the field and spreading manure onto the land from which the cattle feed was harvested are both examples of managed nutrient recycling. The term *renewable resource* best applies to soil N, which can be replenished from the atmosphere by biological N fixation (see Section 12.1).

Other fertilizer nutrients, such as K and P, are mined or extracted from nonrenewable mineral deposits. The size of known global reserves varies according to the nutrient. As farmers, especially in Asia, increase their use of P fertilizer, the world's most high-quality sources of this crucial and irreplaceable element are likely to be depleted within as little as a single century (see Box 12.3). As the best, most concentrated, and most accessible sources of these nutrients are depleted, the cost of producing fertilizer will likely rise in terms of money, energy, and environmental disruption. Therefore, careful husbandry of phosphorus and other nutrient resources must be an integral part of any long-term nutrient management program.

Environmental Impact: Nutrient Budgets and Balances[2]

Examples of simplified nutrient budgets are shown in Figure 13.1. Much cultivated land receives nutrients in excess of those removed in the harvested crops and animal products. In recent years, the nearly 300 million ha of cultivated temperate-region soils had a net *positive* nutrient balance of at least 60 kg/ha N, 20 kg/ha P, and 30 kg/ha K. While nutrient deficiencies still occur in some locations, the great majority of commercially farmed soils have experienced a nutrient buildup. Some of the excess nutrients move into groundwater, surface waters, or the atmosphere, where they can contribute to environmental damage.

The separation of livestock production from crop production and the concentration of livestock production on farms that must import feed from off-farm have led to the buildup of excessive levels of nutrients in soils. The animal manure produced on these farms contains nutrients far in excess of the amounts that can be used in an efficient and environmentally safe manner by crops in nearby fields.

There are also global imbalances between the total amounts of nutrient inputs and outputs, leading to nutrient surpluses that, in the cases of N and P, result in considerable environmental damages. The problem greatly worsened during the second half of the twentieth century (Figure 13.2) with development of industrial agriculture based on inorganic fertilizers and concentrated livestock feeding, and is likely to get worse still with the spread of these practices in emerging economies around the world. In the case of P (Figure 13.2, *right*), about two-thirds of the growing surplus is expected to accumulate in agricultural soils (see Section 12.3), while about one-third will run off the land into aquatic systems. In the case of nitrogen, the fate of the surplus is more complex. About one-third of the surplus N is denitrified to N_2 gas which returns to the atmosphere with no adverse effects. However, the rest of the N surplus is either transformed into greenhouse gases that accelerate climate change or to nitrate and ammonia that can end up causing eutrophication and toxicity in aquatic and forest systems (Sections 12.1 and 12.3).

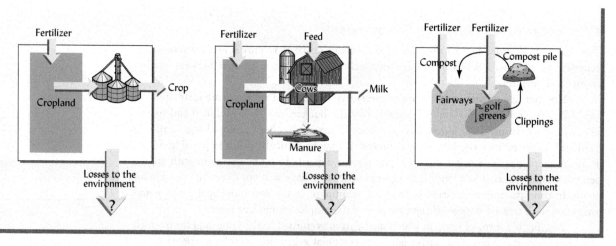

Figure 13.1 *Representative conceptual nutrient flowcharts for a cash-grain farm, a dairy farm, and a golf course. Only the managed inputs, key recycling flows, and outputs are shown. Unmanaged inputs, such as nutrient deposition in rainfall, are not shown. Outputs that are difficult to manage, such as leaching and runoff losses to the environment, are shown as being variable. Although information on unmanaged inputs and outputs is not always readily available, it must be taken into consideration in developing a complete nutrient management plan. Such flowcharts are a starting point in identifying imbalances between inputs and outputs that could lead to wasted resources, reduced profitability, and environmental damage.* (Diagram courtesy of Ray R. Weil)

---

[2]For an exploration of how global nitrogen and phosphorus budgets have changed due to changes in livestock production, see Bouwman et al. (2013); for a discussion of negative nutrient balance in African soils, see Smaling et al. (1997).

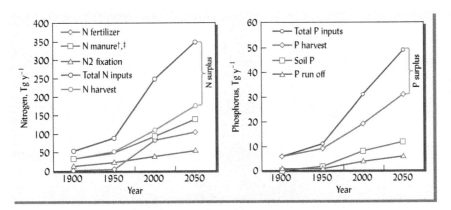

Figure 13.2 *Trends in agricultural nitrogen and phosphorus inputs and outputs. The surplus of N or P is the total inputs minus crop or animal products harvests. Projections for 2050 assume "business as usual." If better nutrient management is widely adapted, future surpluses could be smaller than shown. Total P input includes both mineral and manure forms. Surplus N or P accumulates in soil or pollutes the environment. Fluxes for N are almost 10 times greater than for P.* [Graphed from data in Bouwman et al. (2013)]

13.2 NUTRIENTS AS POLLUTANTS[3]

Nutrient management impacts the environment most directly with water-quality problems caused by N and P. Together, these two nutrients are the most widespread cause of water-quality impairment in lakes and estuaries, and are second only to sediment (see Section 14.2) among pollutants impairing the water quality of rivers and streams.

Nutrient Damage to Aquatic Ecosystems

The growth of aquatic plants (e.g., algae and phytoplankton) often explodes when concentrations of N or P exceed critical levels, leading to numerous undesirable effects in the aquatic ecosystem, including low oxygen, fish kills, and blooms of toxic algae. In most freshwater lakes and streams, P is the limiting nutrient that can set off eutrophication; as little as 0.1 mg/L of total P is thought to be potentially damaging (Box 13.1). In saltier waters (estuaries and coastal ocean waters), N is usually the nutrient most likely to cause eutrophication. Levels of total dissolved N above 2 mg/L in waters are often considered above normal and damaging to the ecosystem. In addition to stimulating eutrophication, N in the form of dissolved ammonia gas (NH_3) can be directly toxic to fish. For this reason, average concentrations of ammonium-N (which is in equilibrium with ammonia) should be kept below 2 mg/L. The nitrate form of N is also of concern, as levels above 10 mg nitrate-N/L are considered unfit for human drinking water (see Box 13.2).

Most industrialized countries have made great strides in reducing nutrient pollution from factories and municipal sewage outfalls (called **point sources** because these sources are clearly localized). However, much less progress has been made in controlling nutrients in the water coming from landscapes (called **nonpoint sources** because these sources are diffuse and not easily identified). The nature of the landscape generally dictates the impact on water quality. Streams draining forestland and rangeland are generally far lower in nutrients than those draining watersheds dominated by agricultural and urban land uses. Two examples are worth considering, as follows. Scientists estimate that agricultural activities account for some 40% of the N and P loads to the Chesapeake Bay from its 170,000 km² watershed (on the U.S. Atlantic Coast), while point sources (sewage and industrial plants) and urban runoff (including septic systems) each account for another 20%. In the more intensively farmed and less densely populated Mississippi River basin covering nearly 3 million km², some 70% of the N and 80% of the P load come from agriculture.

An understanding of the different pathways traveled by N and P is essential in designing management practices aimed at reducing nutrient pollution. Although soils and watersheds differ, we can make some general statements about N and P transport from land to water (Figure 13.3). Most commonly, the rate of loss or loading to streams is about an order of magnitude greater for N than for P. Also, the main pathways differ for the two nutrients. Nitrogen makes its way to streams mainly by the leaching of dissolved forms (nitrate, ammonium, and

---

[3] See Harmel et al. (2018) for environmentally acceptable levels of N and P in runoff and leaching water.

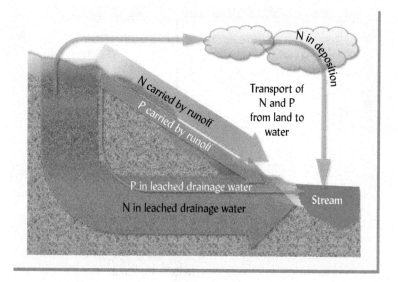

Figure 13.3 *Nitrogen moves from land to streams mainly dissolved in drainage water, while P is carried mainly in surface runoff. Therefore, control of P loading usually focuses on reduced runoff and erosion, while control of N loading usually focuses on reduced leaching. Some nitrogen leaving the soil in gaseous forms may be deposited in water from the atmosphere.* (Diagram courtesy of Ray R. Weil)

dissolved organic N) to shallow groundwater and then by subsurface flow to the streams. Phosphorus, on the other hand, tends to move off the land primarily in surface runoff, either as part of mineral or organic particles, or as dissolved inorganic and organic compounds in the runoff water. Except in certain low-lying, artificially drained soils, very little P moves by subsurface flow because of the strong attraction between mineral soil particles and dissolved inorganic P (see Section 12.3). The data in Table 13.1 describing the average N and P loading per unit land area in the Chesapeake Bay watershed is typical of many humid region watersheds and confirms the trends just described.

Nutrient Management Plans

One tool for reducing nonpoint source N and P pollution is a nutrient management plan—a document that records an integrated strategy and specific practices for how nutrients will be used in plant production. Generally, the document is prepared by a specially trained soil scientist who consults closely with the landowner to meet both environmental goals and practical needs. The plan attempts to balance the inputs of N and P (and other nutrients) with their desirable outputs (i.e., removal in harvested products) to prevent undesirable outputs (runoff, leaching) that exceed allowable **total maximum daily load** (TMDL), the largest amount of nutrient runoff and leaching (measured as $g\ ha^{-1}\ day^{-1}$) permitted from an area of land. The plan usually includes field-by-field data from soil and plant tissue tests (see Section 13.11), soil maps, location of sensitive waters, realistic yield goals, and records of all nutrients applied.

Table 13.1
AVERAGE ANNUAL LOSSES OF NUTRIENTS AND SEDIMENT BY SURFACE RUNOFF AND LEACHING FROM CROPLAND IN THE CHESAPEAKE BAY WATERSHED

Values are estimates of losses at the edge of an average crop field. Nitrogen is lost predominately by leaching and phosphorus predominately in surface runoff. Compare to Figure 13.3.

	Surface runoff	Subsurface (leaching)	Total loss
	— kg/ha —		
Nitrogen loss	9.98	36.7	46.6
Phosphorus loss	4.11	0.08	4.19
Sediment loss	1,320	0	1,320

Data selected from USDA economic research service estimates (Ribaudo et al., 2014).

BOX 13.1
PHOSPHORUS AND EUTROPHICATION

Abundant growth of plants in terrestrial systems is usually considered beneficial, but excess growth of algae (floating single-celled plants) and aquatic weeds can greatly damage aquatic ecosystems. Unpolluted water is usually clear, free of excess algae and aquatic plants, and is inhabited by diverse communities of organisms. When P is added to a P-limited lake, it stimulates a burst of algal growth (an *algal bloom*) and, often, a shift in the dominant algal species. The overfertilization is termed *eutrophication* (Greek, *eutruphos*, meaning well-nourished). Natural eutrophication—the slow accumulation of nutrients over centuries—causes lakes to slowly fill in with dead plants, eventually forming Histosols (see Figure 2.17). Excessive input of nutrients under human influence, called *cultural eutrophication*, tremendously speeds this process. Critical levels of P in water, above which eutrophication is likely to be triggered, are approximately 0.03 mg/L of dissolved P and 0.1 mg/L of total P.

Phosphorus-stimulated algae may suddenly turn water into a soupy green algal scum (Figure 13.4). When these aquatic plants die, they sink to the bottom, where their decomposition by microorganisms uses up the oxygen dissolved in the water. The process is accelerated by warm water temperatures. The decrease in oxygen (anoxic conditions) severely limits the growth of many aquatic organisms, especially fish. Such eutrophic lakes often become turbid, limiting the growth of beneficial submerged aquatic vegetation and benthic (bottom-feeding) organisms that serve as food for much of the fish community. In extreme cases, eutrophication caused algal blooms can lead to massive fish kills.

Eutrophic conditions favor the growth of *Cyanobacter*, so-called blue-green algae, at the expense of zooplankton, a major food source for fish. These *Cyanobacter* produce toxins and bad-tasting and bad-smelling compounds that make the water unsuitable for human or animal consumption. Thus, eutrophication can transform clear, oxygen-rich, good-tasting water into cloudy, greenish, oxygen-poor, foul-smelling, bad-tasting, and possibly toxic water in which a healthy aquatic community cannot survive.

Figure 13.4 In extreme cases, phosphorus runoff from overfertilized land can result in massive fish kills. Elevated levels of P in lake and river water can stimulate so much algal growth that the water turns anoxic when the algae die and decompose. The lack of oxygen, in turn, can suffocate aquatic life, including fish. (Photos courtesy of Ray R. Weil)

BOX 13.2
SOIL, NITRATE, AND YOUR HEALTH[a]

Mismanagement of soil N can result in both eutrophication of coastal waters as well as levels of nitrates in drinking water and food that may threaten human health. While nitrate itself is not directly toxic, once ingested, a portion of the nitrate is reduced by bacterial enzymes to nitrite, which is considered toxic.

The most widely known (though actually quite rare) malady caused by nitrite is **methemoglobinemia** in which nitrites decrease the ability of hemoglobin in the blood to carry oxygen to the body cells. Since inadequately oxygenated blood is blue rather than red, people with this condition take on a bluish skin color. This symptom, and the fact that infants under three months of age are much more susceptible to this illness than older individuals, accounts for the condition being commonly referred to as "blue baby syndrome." Most known deaths from this disease have been caused by infant formula made with high-nitrate water. With the aim of protecting infants, standards to limit nitrate allowed in drinking water have been set at 10 mg/L nitrate-N (= 45 mg/L nitrate) in the United States and 11 mg/L nitrate-N (= 50 mg/L nitrate) in the European Union.

Of greater potential concern is the tendency of nitrate to form N-nitroso compounds in the stomach. Certain N-nitroso compounds are highly toxic, causing cancer in some 40 species of test animals, so the threat to humans must be given serious consideration. Nitrates (or nitrites formed therefrom) have also been reported to promote certain types of diabetes, stomach cancers, thyroid gland malfunction, and birth defects. Many of these effects appear to be associated with nitrate concentrations much lower than the drinking water limits just mentioned.

On the other hand, several studies suggest that ingestion of nitrate does no harm and may actually provide protection against bacterial infections and some forms of cardiovascular diseases and stomach cancers. Therefore, while a precautionary approach is probably wise, we must conclude that the "jury is still out" on the health risks of nitrates in drinking water and vegetables.

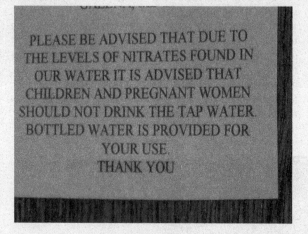

Figure 13.5 Warning on hotel room door in heavily agricultural watershed. (Photo courtesy of Ray R. Weil)

[a]For reviews of nitrate's effects on health, see Santamaria (2006) and L'hirondel and L'hirondel (2002). For a contrary view, see Addiscott (2006).

Best Management Practices (BMPs)

The primary means of preventing nutrient pollution of waters is to avoid excessive applications of N and P onto the landscape to begin with. One of the reasons nutrient balance and avoidance of surpluses on agricultural land are so important is that a large proportion of the surplus usually makes its way to water. Figure 13.6 illustrates this point. In well-drained cropland, some 80% of surplus N leaches below the crop root zone. Perennial grasslands tend not to exhibit as much N leaching until the N surplus becomes quite large (or they are plowed).

In addition to avoiding additional nutrient surpluses, it is important to manage soils and plants to reduce the transport of nutrients (and other pollutants) from soils to groundwater and surface waters. In the United States, practices officially sanctioned to implement nutrient input and transport reduction strategies are known as BMPs. Several BMPs designed to avoid additional surpluses will be discussed later in this chapter. Here we will focus on three general types of practices aimed at controlling nutrient transport to water: (1) buffer strips, (2) cover crops, and (3) conservation tillage.

Buffer Strips

Buffer strips of dense vegetation are a simple and generally cost-effective method to protect water from the polluting effects of a nutrient-generating land use. The vegetation in the buffer strips may consist of existing natural vegetation or planted species, including grasses, shrubs, trees (Figure 13.7), or a combination of these vegetation types (Figure 13.8).

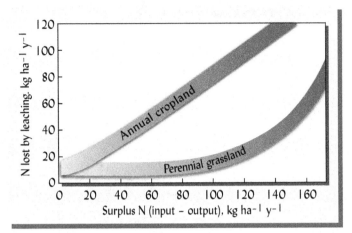

Figure 13.6 Nitrogen losses by leaching from humid region upland agricultural soils are related to the N surplus—the amount by which the N added from all sources exceeds the N removed in harvested products. While some N may accumulate temporarily in the soil organic matter and some may be lost to the atmosphere by denitrification, typically in well-drained cropland about 80% of the surplus N leaches below the crop root zone. In permanent grassland used for harvesting forage or pasturing livestock, soil storage and denitrification tend to account for most of the surplus N up to a threshold beyond which leaching losses follow a pattern similar to that exhibited by cropland. [Graph based on concepts in Billen et al. (2013)]

Riparian Buffers. Buffer strips situated along the banks (the **riparian zone**) of a stream or other water body are termed *riparian buffers*. Water running off the surface of nutrient-rich land passes through the riparian buffer before it reaches the stream. Trees (and their litter layer) or grass plants (and their thatch layer) reduce water velocity and increase the tortuosity of the water's travel paths. Under these conditions, most of the sediment and attached nutrients will settle out of the slowly flowing water. In addition, dissolved nutrients are adsorbed by the soil, immobilized by microorganisms, or are taken up by plants growing in the buffer strip. The decreased flow velocity also increases the retention time—allowing a greater length of time during which microbial action can break down pesticides before they reach the stream. Under some circumstances, buffer strips along streams can also reduce the nitrate levels in the groundwater flowing under them (although emission of nitrous oxide to the atmosphere may be a consequence of nitrate reduction; see Figure 13.17).

Design and Management of Riparian Buffers. For optimal performance, some land grading may be needed when the buffers are established to ensure that the runoff water is spread out evenly as opposed to flowing through in concentrated in rivulets. To preserve the dense vegetation and litter cover, riparian zones should be kept off limits to timber harvest machinery, and if the adjacent land is grazed, cattle should be fenced out. The width needed for optimum cleanup may vary from 6 to 60 m, although a width of 10–20 m is usually sufficient to obtain most of the nutrient- and sediment-removal benefits on slopes of less than 8% (Figure 13.8). Occasional

Figure 13.7 A forest riparian buffer separates the Connecticut River from alluvial farmland in western Massachusetts. (Photo courtesy of Ray R. Weil)

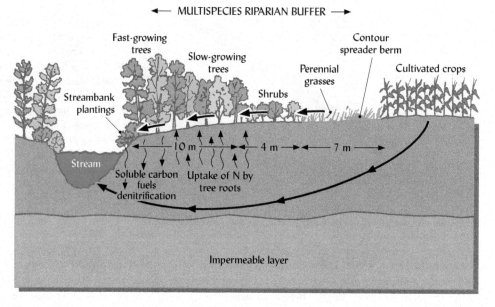

Figure 13.8 *A multispecies riparian buffer protects a stream from nutrients and sediment in cropland runoff while also providing wildlife habitat. A grassed level berm spreads runoff evenly to avoid gullies. Perennial grasses filter out sediments and take up dissolved nutrients. Deep tree roots remove nutrients from groundwater. Soluble carbon from tree litter percolates downward to provide energy for anaerobic denitrifying bacteria that remove additional N from groundwater. Woody vegetation provides wildlife habitat, shade to cool the stream, and woody debris for fish habitat. A total buffer width of 10–20 m can usually provide most of these potential benefits.* (Diagram courtesy Ray R. Weil)

mowing of the grassy zone or thinning of tree stands may help maintain high rates of nutrient uptake. Setting land aside as a buffer often represents a significant reduction in harvestable cropland or timberland (see Figure 14.27, *left*). However, a well-designed buffer strip can provide real benefits, some of which directly accrue to the landowner: turnaround space for field equipment, valuable hay from grass buffers, improved fishing by shading the stream and providing large woody debris for fish habitat, and enhanced recreational values associated with increased wildlife populations. Thus, installation of riparian buffer strips can often be a win-win situation.

In-Field Contour Buffer Strips. Small buffer areas upslope from the riparian zone, often in the agricultural fields themselves, can also be effective in preventing the loss of nutrients in runoff. Various spatial configurations and types of vegetation can be used for this purpose. Nutrient retention can be coupled with biodiversity and wildlife benefits if multiple species of native perennial grasses and forbs are used in the buffer areas. Generally, most of the nutrient and sediment removal takes place in the first 3–6 m of flow through buffer vegetation. If contour buffer strips are used, these can be as narrow as 3 m and occupy less than 10% of the field area, leaving 90% or more of the land to grow crops. However, research suggests that it is just as effective and more convenient for farming operation if the buffer area is consolidated into a single buffer area at the mouth of each small cropland watershed (Figure 13.9).

Cover Crops for Nutrient Management

Instead of being harvested, a *cover crop* is grown to provide vegetative cover for the soil and then is killed and either left on the surface as a mulch, or tilled into the soil as a *green manure*. Cover crops generally are planted so as to avoid competition with crops grown for income. Cover crops may occupy time periods when cash crops are not grown. Or they may grow simultaneously with cash crops, but occupy space between vines or fruit trees or between rows of annual crops. In climates with enough precipitation to allow for some water use by the cover crop, they can help farmers improve soil health and biodiversity. Other roles for cover crops include protecting the soil from the erosive forces of wind and rain (see Section 14.6), adding to the soil organic matter (see Section 11.7), alleviating soil compaction (see Section 4.7), and, if leguminous, increasing the available N in the soil (see Section 12.1).

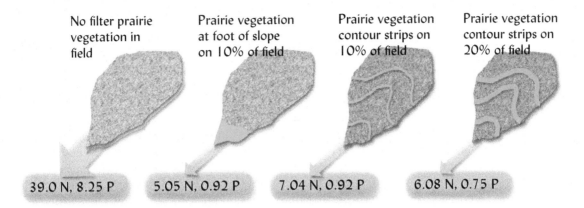

Figure 13.9 *Effectiveness of perennial herbaceous (prairie) vegetation filters for reducing nutrient loss from cropland. A diverse mixture of native prairie forbs and grasses was seeded in portions of a crop field (shown in green), either in 3–6 m wide contour strips or in the bottom of the watershed. Total N and P losses given as kg ha^{-1}. The three practices shown all reduced nutrient losses by >80%, but planting prairie vegetation in only the bottom 10% of a field proved to be the most convenient and economical practice to implement. The soils were Hapludalfs and Argiudolls with 6–10% slopes. A no-till soybean–corn rotation was grown with 135–185 kg N ha^{-1} and 50 kg P ha^{-1} applied to the corn crops. Despite initial concerns, the prairie vegetation did not contribute weeds to any of the fields.* [Diagrams courtesy of Ray R. Weil. Data and concepts selected from Zhou et al. (2014)]

Cover crops can also reduce the loss of nutrients and sediment in surface runoff. First, the protective foliage and litter prevent the formation of a crust at the soil surface, thus maintaining a high rate of infiltration (see Figure 6.5, *right*) and reducing runoff. Second, for the runoff that does occur, the cover crop helps remove both sediment and nutrients by the same mechanisms that operate in a buffer strip, as just described.

Cover Crops Reduce Leaching Losses. Cover crops can also reduce leaching losses of nutrients, principally N (Figure 13.10). In many temperate humid regions, the greatest potential for leaching of N from cropland occurs when rainfall or snowmelt is high and evapotranspiration by vegetative is low, for example, during the periods of unfrozen soil between crop maturity in early fall and the establishment of a new crop root system in spring. The main crop is not taking up soil N but soluble soil N continues to be produced by decomposition and leaches downward. During this time of vulnerability, an actively growing cover crop will reduce both

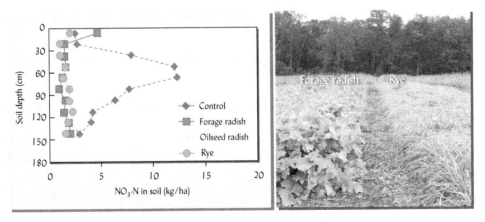

Figure 13.10 *Cover crops can capture soluble N (e.g., NO$_3$-N) left in the soil profile at the beginning of winter, thus substantially reducing N loss to groundwater before next year's crop is established. Forage radish and rye (photo) cover crops are capable of capturing >100 kg/ha of such residual N in fall, cleaning the soil profile of soluble N to considerable depths. The graph shows nitrate-N in November, expressed as kg N/ha for each 15 cm depth increment of this sandy Ultisol. The control plots had some weeds, but no cover crop.* [Data from Dean and Weil (2009); photo courtesy of Ray R. Weil]

the percolation of water and the N concentration in the water that does percolate, incorporating this nutrient into plant tissues. For this purpose, an ideal cover crop should rapidly produce an extensive and deep root system capable of catching the soluble N *before* it moves so deep into the soil that it will be subject to leaching away over winter. To accomplish this purpose, the cover crop should be established early enough to capture the soluble soil N in the fall during the weeks just before and after harvest time and not rely primarily on growth early in the following spring (Figure 13.11). If the cover crop maintains a low C/N ratio and produces easily decomposed residues, the N captured from deep in the soil profile will largely be released near the soil surface in time for use by the next spring-planted main crop. To accomplish these goals, farmers are experimenting with methods for earlier planting of cover crops. These include adopting cropping systems that allow earlier harvests with the use of winter cereals in the rotation or more rapidly maturing summer crop varieties. The use of airplanes and special high-clearance machines allow farmers to sow cover crop seeds weeks or even months before the main crop is harvested from a field. Cover crops can capture soluble nitrogen (e.g., NO_3-N) left in the soil profile at the beginning of winter, thus substantially reducing N loss to groundwater before next year's crop is established. Forage radish and rye (photo) cover crops are capable of capturing >100 kg/ha of such

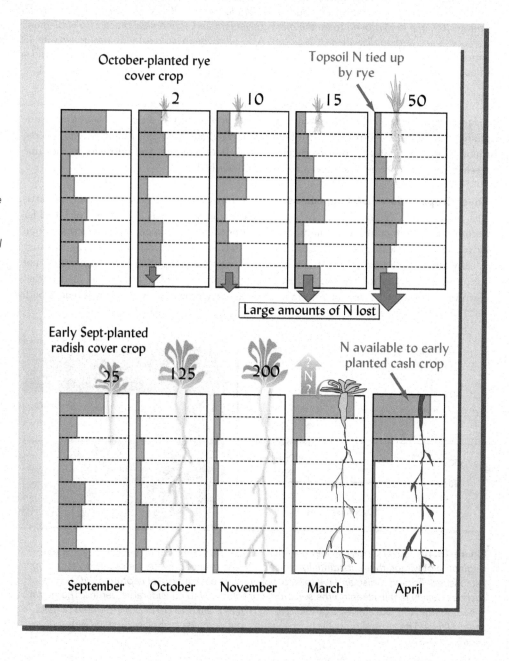

Figure 13.11 *Early sowing of fast-growing, deeply rooting cover crops in temperate humid regions may capture soluble N from deep in the soil before it escapes during the main winter leaching season. A cover crop species which is killed by extremely cold temperatures over winter and whose residues then decompose rapidly when the soil warms in the following spring may return much of the captured N to the topsoil layer where it can be used by an early spring-planted main crop, thus reducing the amount of fertilizer N needed. The scenarios in the diagram are for a climate in which N uptake by the summer crop is finished by early September, and killing cold occurs after November. The numbers indicate the expected kg/ha of N in the cover crop biomass. The blue bars indicate expected amounts of soluble N in 30 cm increments of the soil profile.* (Diagram courtesy of Ray R. Weil)

residual N in fall, cleaning the soil profile of soluble N to considerable depths. The graph shows nitrate-N in November, expressed as kg N/ha for each 15 cm depth increment of this sandy Ultisol. The control plots had some weeds, but no cover crop. Largely because of their more rapid root growth in fall, winter annual cereals (rye, wheat, oats) and Brassicas (rape, forage radish, turnips, mustards) have proven to be more efficient than legumes (vetch, clover, etc.) at mopping up leftover soluble N. Some of the most effective cover crops may be mixtures of three or more species that perform in complementary ways. A cover crop, like any plant, requires an adequate nutrient supply. If the surface soil in which the seeds germinate is so poorly supplied with nutrients (including N) that growth is retarded, subsoil N is likely to leach deeply before the cover crop roots can become established. On sandy, low organic matter soils it may even be necessary to apply a very small dressing of N in fall to promote vigorous cover crop growth that will enable the roots to catch up with N already moving deeply into the profile.

Conservation Tillage

The term **conservation tillage** applies to agricultural practices that keep at least 30% of the soil surface covered by plant residues. The effects of conservation tillage on soil properties and on the prevention of soil erosion are discussed in Sections 4.6, 6.4, and 14.6. Here, we emphasize the effects on nutrient losses.

Compared to plowed fields with little residue cover, conservation tillage reduces the total amount of water running off the land surface, whether from rainfall or snowmelt. It reduces even more the load of nutrients and sediment carried by that runoff (Figure 13.12). When combined with a cover crop, the reductions are greater still. Except, perhaps, when heavy rain closely follows the spreading of fertilizer or manure on the land surface (see Section 13.4), the less the soil surface is disturbed by tillage, the smaller are the losses of nutrients in surface runoff. The relatively small amounts of nutrients in runoff from untilled land (no-till cropland, pastures, and forests) tend to be mostly dissolved in the water rather than attached to sediment particles, while the reverse is true for tilled land. Because of large, sediment-associated nutrient losses, total nutrient loss in surface runoff from sloping plowed land generally is far greater than that from land where no-till or conservation tillage methods are used.

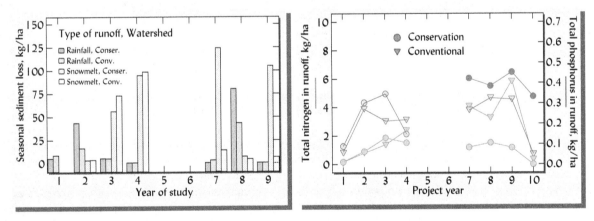

Figure 13.12 *Sediment and nutrients lost in runoff from a pair of watersheds growing a wheat, flax, and canola rotation in Manitoba, Canada. Runoff from snowmelt (light fill) and rainfall (dark fill) are shown. During years 1–4 the watersheds were managed exactly the same to collect baseline data. (Left) The data confirm that brown and yellow-coded watersheds lost very similar amounts of sediment in runoff from snowmelt and rainfall (except for a small difference in year 2). Between years 4 and 7, the watershed coded yellow was converted to conservation tillage (which left 50–60% of the surface covered with plant residues), while the watershed coded brown was managed with conventional full tillage (which left less than 20% of the surface covered with residues). During the four years after the conversion, the conservation tillage watershed lost much less sediment than did the watershed under conventional tillage. Most of the sediment loss was due to snowmelt in years 7 and 9, but due to rainfall in year 8. (Right) Likewise, the total N and P losses in runoff were very similar for the two watersheds before the tillage system was changed (years 1–4). The loss of N (green lines) was much greater under conventional tillage (triangle symbols). Phosphorus losses (blue lines) were greater for conservation tillage (circle symbols), probably because P applied in that system accumulates near the soil surface.* [Graphed from data in Tiessen et al. (2010)]

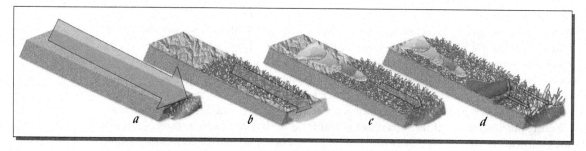

Figure 13.13 *Landscape management to moderate runoff and retain nutrients. (a) With no practices, nutrient and sediment-laden runoff flow rapidly to stream. (b) A riparian buffer zone slows the runoff and removes some nutrients. (c) Waterway check dams and contour buffers in the tilled crop field slow and partially clean runoff water before it reaches the riparian buffer. (d) Best results are obtained where the landscape management integrates many practices: conservation tillage and cover crops in the fields, buffers in the waterways, retention wetlands, and riparian buffer zones.* [For more on these concepts, see Kröger et al. (2013)]

On the other hand, the loss of nutrients by leaching can be somewhat greater with conservation tillage than conventional tillage. In conservation tillage systems, a higher percentage of the precipitation or irrigation water infiltrates into the soil, where it may carry nutrients downward. Over time, many no-till soils develop large macropores (such as worm burrows) that are open to the soil surface. Rain and irrigation water may move down rapidly through these large macropores. However, nutrients held in the finer pores of the soil matrix (as opposed to those on the soil surface) are bypassed by such flow and do not leach into the lower horizons (see Figure 6.23). In such situations, nutrient leaching may actually be less under conservation tillage.

Combining Practices on the Landscape

The most effective solutions to environmental problems often employ the concept of "many little hammers." Thus, it may be most efficient to rely on the combined effect of many nutrient management practices deployed sequentially on the landscape. We have discussed the use of nutrient budgets, nutrient cycling, buffer zones, nutrient retaining wetlands, and conservation tillage in isolation. When we put them altogether in an integrated manner, we may find that synergies are realized and resulting nutrient load reductions may exceed the sum of the individual contributions. The concept of integrating many practices sequentially on the landscape to protect water quality from N and P coming from agricultural land is illustrated in Figure 13.13.

13.3 ECOSYSTEM NUTRIENT CYCLES

Internal nutrient resources come from within the ecosystem, be it a forest, a watershed, a farm, or a home in the suburbs, and are generally preferred, since their financial and environmental costs are minimized. These resources include the process of mineral weathering within the soil profile, biological N fixation, acquisition of nutrients from atmospheric deposition, and various forms of internal recycling, such as utilization of cover crops, retention of plant residue, and application of animal manure to the fields that grew the animal feed.

If internal resources prove inadequate, nutrients must be imported from resources external to the system (see Section 13.7). Such *external* resources are usually purchased inorganic or organic fertilizers. Various types of organic residues and (so-called) wastes are available to be used on land as a source of supplemental nutrients (and organic matter). Among the most plentiful and most commonly used to enhance soil fertility are the crop residues and animal manures already mentioned. In addition, urban refuse rivals animal manure in terms of its quantity and nutrient content, but only about 10% or so is composted and applied to land in the United States and elsewhere. More than half of sewage sludge and septage (sludge from septic tanks, see Section 6.8) is used as a soil amendment, but the total quantity is much smaller than animal manures. Organic wastes from food processing and other industries as well as certain logging and wood manufacture wastes also offer some (mostly unrealized) potential as soil amendments. For most suburban homeowners, all the nutrients their landscaping and gardens might require could usually be supplied by collecting (instead of using drinking water to flush away) the residents' urine (see Box 13.3 for details).

BOX 13.3
THE LAW OF RETURN MADE EASY: USING HUMAN URINE[4]

A central tenet of organic farming—of any good farming, really—is the Law of Return which states that all the nutrients that a plant takes away from the soil should be returned to the soil to complete the cycle, preventing soil depletion and resource exhaustion. Such return of nutrients is rarely implemented. A very easy way to implement the Law of Return is by using human urine. Urine is generally free of pathogens because it has been filtered through the kidneys. Furthermore, if it is contaminated, storing it in a closed container for a week will cause it to self-sterilize via the ammonia it releases. It is easy to handle, and if kept clean, does not smell very much. But its main virtue is that it holds in dissolved, plant-available form most of the mineral nutrients that were in a person's diet. Tests have shown that food harvested from urine-treated plots does not have greater bacterial contamination than food conventionally grown. Research in Nigeria suggests that people are willing to buy urine-fertilized produce. One question that needs research is: what happens to pharmaceuticals or natural hormones excreted in the urine once they are applied to soil? For that matter, little is known what happens to these chemicals when they are flushed down the toilet.

Urine is essentially a liquid fertilizer solution that we all make from our food and drink. More than half of the N ingested is excreted in the urine. It also contains significant amounts of K, P, S, Ca, and Na and smaller amounts of Mg and micronutrients. In addition, urine is very easy to collect. Just keep a plastic jug with a suitable wide opening next to the toilet (Figure 13.14a). Or for large-scale urban collection, one can tap men's urinals or install separator toilets designed such that urine goes down one tube while feces and paper go down another. Several towns in New England, United States, and in northern Europe have organized themselves to provide urine in bulk to local farms.

So why would typical Westerners use 20–80 L of drinking-quality water to flush away every liter of urine they produce? Why would subsistence farmers waste their urine behind the bushes or in the latrine? The answer in both cases is that they generally do not know any better. Many small-scale farmers and gardeners have probably tried urinating on a tomato plant or the like and noted that the plant was injured, possibly even killed. A concentrated fertilizer solution poured onto a plant can also cause salt-burn. The problem is concentration; the osmotic potential or salinity is too high (see Sections 5.3 and 9.15). Before applying urine to plants, one needs to dilute it: one part urine in two parts water. When applied to moist soil or compost piles without young plants, a 1:1 dilution is safe.

Annually, about 2.5–4 kg of N could be collected per capita from urine as the average urination rate is about 2 L/day and the human urine contains between 1 and 8 g N/L. Urine is also an excellent source for K, S, and P. To put this in perspective, a family of six would be able to fertilize about 0.25 ha of crops if they collected and used all their urine (assuming a moderate per ha fertilization rate of about 70 kg N, 35 kg K, and 6–12 kg P and S per ha). Likewise, a suburban family of four could easily keep their lawn and garden green without purchasing any fertilizer, but using only diluted urine (Figure 13.14c). Not to mention that they would significantly reduce their water use and eliminate 60% of the N they send to the water treatment plant.

Figure 13.14 *Simple means of collecting and using human urine. (a) Everyone can have a jug with their name on it. (b) Lawn grass responds to the nitrogen (and other nutrients) in the diluted urine that was applied in the shape of the letter "N." The area around the letter received only water. (c) A backyard experiment in Africa using compost to fertilize the row of cabbage plants on the left and compost with urine on the right.* (Photos (a) and (b) courtesy of Ray R. Weil; photo (c) courtesy of Ben Waterman)

[4]The use of human urine as fertilizer has been the subject of a limited amount of research (AdeOluwa and Cofie, 2012; Heinonen-Tanski et al., 2010; Karak and Bhattacharyya, 2011; Pradhan et al., 2007; Richert et al., 2010; Shrestha et al., 2013).

Depending on the parent materials and climate, weathering of minerals (see Section 2.1) may release significant quantities of nutrients. For timber production, most nutrients are released fast enough from either parent materials or decaying organic matter to supply adequate nutrients. In agricultural systems, however, some nutrients generally must be added, since nutrients are removed from the land annually in harvested crops. Negligible amounts of N are released by weathering of mineral parent materials, so other mechanisms (including biological N fixation) are needed to resupply this important nutrient. Release of nutrients by mineralization of soil organic matter is important in short-term nutrient cycling, but in the long run, the organic matter and the nutrients it contains must be replenished or soil fertility will be depleted.

Surface Soil Enrichment. Forest trees obtain most of their nutrients from the surface horizons, but some trees are also well adapted to gathering nutrients from deep in the soil profile, where much of the nutrient release from parent material takes place. Overall nutrient-use efficiency can sometimes be increased by combining trees and annual agricultural crops into what are known as *agroforestry systems* (Figures 13.15).

Trees can improve fertility of the upper soil horizons in several ways. In Figure 2.27, we saw that trees can act as nutrient pumps, altering the course of soil formation. Tree roots can take up nutrients deep in the profile where they occur because of weathering or leaching. Then as the tree leaf and root litter decomposes, the nutrients are released into the A horizon where they can be of use to relatively shallow-rooted agricultural crops. Nitrogen-fixing trees (mostly legumes) can also add N from the atmosphere to the surface soil with their N-rich leaf litter. Trees may also enhance the fertility of the soil in their vicinity by trapping windblown dust and by providing shade and shelter for birds and animals that leave their droppings near the trees.

Nutrient Cycling in Grasslands

The burning of rangeland grasses (or crop residues) usually produces much less volatilization of N and S than do forest wildfires, as grass fires move quickly and burn at relatively low temperatures. While the loss of organic matter consumed by the fire is undeniable, the release of nutrients and the increased light penetration with the opening up of the grass canopy may stimulate plant biomass production, especially belowground. As a result, organic

Figure 13.15 *Two examples of agroforestry systems. (Left) The deep-rooted* Faidherbia (Acacia) albida *enriches the soil under its spreading branches (note man in white shirt). Conveniently, these trees lose their leaves during the rainy season when crops are grown, so the trees do not compete with crops for light. Africans traditionally leave these trees standing when land is cleared for crop production. Crops growing under the trees yield more and have higher contents of S, N, and other nutrients. (Right) Branches pruned from widely spaced rows of leguminous trees are spread as mulch in the alleys between the tree rows, thus enriching the alley soil with nutrients from the leaves as well as conserving soil moisture. Crops grown in this alley-cropping system may yield better than crops grown alone, but only if competition between trees and crop plants for light and water can be kept to a minimum.* (Photo courtesy of Ray R. Weil)

matter accumulation may be greater under grasslands that experience occasional burns than under those where fire is completely controlled. Grazing by large mammals, if not too frequent and sustained, can also stimulate increased plant production and quality (Box 13.4). Fire and grazing, both natural components in rangeland ecosystems, are important tools for managing nutrient cycles and soil productivity.

BOX 13.4
GRAZING MAMMALS AND NUTRIENT CYCLING IN SOILS

Most tourists don't travel to East Africa to see the amazing diversity of soils; most come to see wildlife in their natural habitat. However, in the vast Serengeti plains, variations in soil fertility do influence where one finds nonmigratory antelopes and gazelles—and the lions and cheetahs that prey on them. These animals graze most frequently where soils are relatively high in nitrogen and sodium, two mineral nutrients critical to health and survival, especially for pregnant or lactating females and their young (Figure 13.16). The animals not only seek out more fertile soils, they also help enhance the fertility of the soils they frequent. The animals leave manure and urine that contain most of the nutrients they consumed, but in a more easily decomposable form than in the original plant material. Also, animals' grazing seems to stimulate vigorous growth of palatable, easily decomposed plant species, thus speeding the cycling of nutrients. Nitrogen concentrations in plant regrowth after grazing generally are higher than in ungrazed plants, making the resulting plant residues more readily recyclable in the soil.

Overgrazing, however, is not such a good thing. It occurs when density and frequency of grazing exceed the *carrying capacity* of the land. Continual heavy grazing or other forms of poorly managed grazing by livestock may kill off the most palatable species, so that the vegetation becomes relatively sparse and dominated by less palatable plants. In some cases, the residues of these plants are also less decomposable. The resulting impairment of the nutrient cycling processes accelerates the deterioration of the vegetative cover and exposes the soil to the erosive action of wind and water (see Section 14.1)

The use of grasslands for grazing (pastures) contrasts sharply with the use of similar grasslands to produce hay. Hay is vegetation that has been mechanically cut, preserved by sun-drying, and then bailed and removed from the land to feed livestock in another time and place, as needed. Harvesting hay removes large amounts of nutrients that are *not* returned by grazing animals in urine and feces. Thus, hayfields usually require the addition of large amounts of N and K (and smaller amounts of other nutrients) to partially replace nutrients removed in harvest. On the other hand, a well-grazed pasture should not be fertilized as if it were a hayfield. Grazing animals typically recycle back to the soil most of the nutrients taken up by the vegetation so properly grazed pastures rarely need fertilizer. In fact, fertilization of grazed pastures may lead to excess P in runoff water and to excessive leaching of N (Figure 13.17). Of course, some fields may be both grazed and harvested for hay, and need to be managed accordingly.

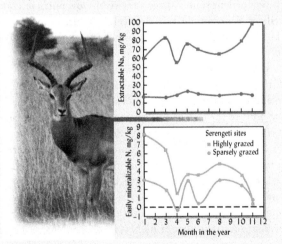

Figure 13.16 *Grazing mammals interact with soil in the east African Savannah. Soil sodium (upper) and nitrogen (lower) where gazelles habitually graze heavily or sparsely. Averages for two pairs of sites and two years in Tanzania's Serengeti savannah.* [Graphed from data in McNaughton et al. (1997). Photo courtesy of Ray R. Weil]

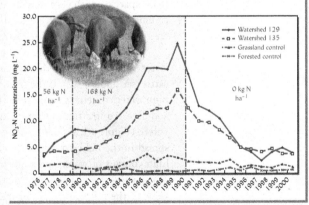

Figure 13.17 *Nitrate leaching caused by fertilizing grazed grasslands in a temperate humid region (Ohio, USA). Watersheds 129 and 135 were rotationally grazed with beef cattle. The other two watersheds (controls) were not grazed, fertilized, or harvested. When N fertilization was increased from 56 to 168 kg ha^{-1} yr^{-1}, groundwater NO$_3$-N changed little in the first few years, but then increased for 10 years. The NO$_3$-N concentrations returned to low levels within six years after discontinuation of N fertilization, even though legumes were interseeded into the grasslands.* [Data from Owens et al. (2008); photo courtesy of Ray R. Weil]

13.4 RECYCLING NUTRIENTS THROUGH ANIMAL MANURES[5]

For centuries, the use of farm manure has been synonymous with successful and stable agriculture. In this context, manure supplies organic matter and plant nutrients to the soil and is associated with the production of soil-conserving forage crops used to feed animals. About half of the solar energy captured by plants grown for animal feed ultimately is embodied in animal manure, which if returned to the soil can be a major driver of soil quality.

Huge quantities of farm manure are available each year for the recycling of essential elements to the land. For each kilogram of liveweight, farm animals produce approximately 2–4 kg dry weight of manure per year. In the United States, the farm animal population voids some 350 million Mg of manure solids per year, about 10 times as much as does the human population. Some manure is spread on pastures by grazing animals, while about 20% is excreted in a manner that allows its collection for use as a soil amendment. The U.S. Department of Agriculture estimates that about 5% of U.S. cropland is treated with livestock manure in any given year.

Nutrient Composition of Animal Manures

Generally, about 75% of the N, 80% of the P, and 90% of the K ingested by animals pass through the digestive system and appear in the manure (including urine). For this reason, animal manures are valuable sources of both macro- and micronutrients. Nutrient content varies greatly from one type of animal manure to another (e.g., poultry manure compared to horse manure). For a particular type of animal, the actual water and nutrient content of a load of manure will depend on the nutritional quality of the animals' feed, how the manure was handled, and the conditions under which it has been stored. Therefore, it is wise to regularly obtain laboratory analyses of the value of the particular manure in question rather than rely on general statements and textbook information.

Both the urine (except for poultry, which produce solid uric acid instead of urine) and feces are valuable components of animal manure. On average, a little more than *one-half of the N*, about *90% of the P*, and about *40% of the K* are found in the solid manure. Nevertheless, this higher nutrient content of the solid manure is offset by the more ready availability of the constituents carried by the urine. Effective nutrient conservation requires that manure handling and storage minimize the loss of the liquid portion.

Table 13.2 indicates that organic nutrient sources have a relatively low nutrient content in comparison with commercial fertilizer. For example, on a dry-weight basis, animal manures contain 2–5% N, 0.5–2% P, and 1–3% K. These values are one-half to one-tenth as great as are typical for commercial fertilizers.

Furthermore, manure is rarely spread in the dry form, but usually contains a great deal of water. As it comes from the animal, the water content is 30–50% for poultry to 70 or 85% for cattle (see Table 13.2). If the fresh manure is handled as a solid and spread directly on the land, high water content is a nuisance that adds to the expense of hauling. If the manure is handled and digested in a liquid form or slurry and applied to the land as such, even more water is involved (Figure 13.25, *right*). All this water dilutes the nutrient content of manure, as normally spread in the field, to values much lower than those cited for dry manure in Table 13.2. The high content of water and low content of nutrients make it difficult to economically justify transporting bulk manure to distant fields where it might do the most good. However, the value of the micronutrients in manure (Table 13.2) and the additional benefits of its organic matter (see Section 11.5) may be even greater than that of its N—P—K content, and should be included in any economic evaluation of manure transport.

Concentrated Animal-Feeding Operations (CAFOs)

In most industrialized countries, the advent of huge, CAFOs and the separation of crop production from livestock production have changed the perception of animal manure from an *opportunity for recycling nutrients and organic matter* as efficiently as possible to an *obligation for disposing of wastes*

---

[5]Some of the challenges of such recycling are discussed in Gardner (1997) and a wide range of European perspectives on manure management in modern industrial agriculture can be found in Sommer et al. (2013). Hilimire (2011) reviews livestock-cropping integration in the United States.

Table 13.2
COMMONLY USED ORGANIC NUTRIENT SOURCES: THEIR APPROXIMATE NUTRIENT CONTENTS AND OTHER CHARACTERISTICS

Along with N-fixing legumes grown in rotation and as cover crops, materials such as these (except sewage sludge and municipal solid wastes) provide the mainstay of nutrient supply in organic farming. The nutrient contents shown for animal manures are typical of well-fed livestock in confinement production systems. Manure from free-range animals not given feed supplements may be considerably lower in both N and P.

Material	%[a] Water	Percent of dry weight						g/Mg of dry weight						
		Total N	P	K	Ca	Mg	S	Fe	Mn	Zn	Cu	B	Mo	
Coffee grounds[b]	60	1.6	0.01	0.04	0.08	0.01	0.11	330	50	15	40	—	—	May acidify soil.
Cottonseed meal	<15	7	1.5	1.5	—	—	—	—	—	—	—	—	—	Acidifies soil. Commonly used as livestock feed.
Dairy cow manure[c]	75	2.4	0.7	2.1	1.4	0.8	0.3	1,800	165	165	30	20	—	May contain high-C bedding.
Dried blood	<10	13	1	1	—	—	—	—	—	—	—	—	—	Slaughterhouse by-product, N available quickly.
Dried fish meal	<15	10	3	3	—	—	—	—	—	—	—	—	—	Incorporate or compost due to bad odors. Can feed to livestock.
Feedlot cattle manure[d]	80	1.9	0.7	2.0	1.3	0.7	0.5	5,000	40	8	2	14	1	May contain soil and soluble salts.
Hardwood tree leaves[e]	20	1.0	0.1	0.4	1.6	0.2	0.1	1,500	550	80	10	38	—	High Pb for some street trees.
Horse manure[d]	65	1.4	0.4	1.0	1.6	0.6	0.3	—	200	125	25	—	—	May contain high-C bedding.
Municipal solid waste compost[f]	40	1.2	0.3	0.4	3.1	0.3	0.2	14,000	500	650	280	60	7	May have high C/N, heavy metals, plastic, and glass.
Poultry (broiler) manure[c]	35	4.4	2.1	2.6	2.3	1.0	0.6	1,000	413	480	172	40	0.7	May contain high-C bedding, high soluble salts, arsenic, and ammonia.
Sewage sludge, activated	<10	6	1.5	0.5	2.0	—	2.0	40,000	—	450	—	—	—	e.g., Milorganite®, N released 2–6 months, low solubility P, low salt index. Poor on cold soil.
Sewage sludge, digested	80	4.5	2.0	0.3	1.5g	0.2	0.2	16,000g	200	700	500	100	15	May contain high soluble salts, toxic heavy metals, pharmaceuticals.
Sheep manure[d]	68	3.5	0.6	1.0	0.5	0.2	0.2	—	150	175	30	30	—	
Spoiled legume hay	40	2.5	0.2	1.8	0.2	0.2	0.2	100	100	50	10	1,500	3	May contain weed seeds.
Swine manure[d]	72	2.1	0.8	1.2	1.6	0.3	0.3	1,100	182	500	300	75	0.6	May contain elevated Cu and Zn.
Wood wastes	—	—	0.2	0.2	0.2	1.1	0.2	2,000	8,000	500	50	30	—	Very high C/N ratio; must be supplemented by other N.
Young rye green manure	85	2.5	0.2	2.1	0.1	0.05	0.04	100	50	40	5	5	0.05	Nutrient content decreases with advancing maturity.

[a] Water content given for fresh materials. Processing and storage methods may alter water content to less than 5% (heat dried) or to more than 95% (slurry).
[b] Coffee grinds data from Krogmann et al. (2003).
[c] Broiler and dairy manure composition estimated from means of ~800 and 400 samples analyzed by the University of Maryland manure analysis program 1985–1990.
[d] Composition of swine, sheep, and horse manure calculated from Zublena et al. (1993) and Cu and Zn in swine averaged from other sources.
[e] Hardwood leaf data from Heckman and Kluchinski (1996).
[f] Composition of municipal solid waste compost based on mean values from ten composting facilities in the United States as reported by He et al. (1995). Sulfur as sulfate-S.
[g] Sludge contents of Ca and Fe may vary tenfold depending on the wastewater treatment processes used.

with as little cost and environmental damage as possible (Figure 13.18). If manure-holding and disposal practices are inadequate, waters under or near such CAFOs may become polluted with nitrates and pathogens, and water from nearby wells may be unfit to drink.

Cattle Feedlots. More than two-thirds of U.S. feedlot beef production takes place on "farms" that have almost no cropland at all. To visualize the enormity of the manure disposal problem, consider a 100,000-head beef feedlot. We can estimate that the feedlot produces 200,000 Mg of manure (dry matter) per year:

$$\frac{(4\,\text{Mg manure})}{\text{Mg liveweight}} \times \frac{0.5\,\text{Mg liveweight}}{\text{animal}} \times 100,000\,\text{animals} = 200,000\,\text{Mg manure}$$

If this manure contains 2% N and the corn grain grown to feed the cattle removes some 140 kg N/ha, then we can estimate that the manure should be applied at 7 Mg/ha (~3 tons/acre),[6]

$$\frac{1\,\text{Mg manure}}{0.02\,\text{Mg N}} \times \frac{140\,\text{kg N}}{\text{ha}} \times \frac{1\,\text{Mg}}{1,000\,\text{kg}} = \frac{7\,\text{Mg manure}}{\text{ha}}$$

and that utilization of the manure in this manner would require some 28,000 ha of land,

$$\frac{(1\,\text{ha})}{7\,\text{Mg manure}} \times 200,000\,\text{Mg manure} = 28,571\,\text{ha}$$

If corn were grown in rotation with soybean (a legume crop that does not need applied N), then the total amount of cropland required for manure utilization would double to 57,000 ha or 571 km². To find this much cropland, some of the manure would have to be hauled 20–50 km or more from the feedlot! Finally, if soil P is already at (or above) optimal levels from previous manuring, as is usually the case near CAFOs (Section 12.3), manure should be applied at a much lower rate tailored to meet the P (not N) needs of the crops, thus requiring an even larger land base.

Figure 13.18 Examples of concentrated animal feeding operations (CAFOs). Interior of broiler chicken house in Maryland with an inset showing an exterior view of a pile of manure cleaned out from such a house. (Right) An aerial view of a large cattle feedlot in Colorado. In both types of CAFOs, animals are fed on grain imported from distant farms. It is difficult for such CAFOs to recycle nutrients in the manure back to the land on which the cattle feed was grown. Instead of being seen as a valued resource, the manure in this situation may be considered a troublesome waste to be disposed of. Agriculture is challenged to structure itself in a more ecologically balanced manner that better integrates its animal and crop components. (Photo courtesy of Ray R. Weil)

---

[6]In the unlikely event that the land had not been previously manured, some N fertilizer might be needed for the corn silage in the first year or two to supplement the N released from the manure, but soon N released from previous years' applications would make supplementary fertilizer unnecessary (see Section 13.6).

In the absence of sufficient education and enforced regulation, managers unsurprisingly tend to save transportation costs and time by applying manure to nearby fields at higher-than-needed rates. Applications at such high rates typically result in the pollution of surface and groundwater by N and P (and pathogens) and may cause salinity damage to crops and soils.

Poultry and Swine Manure. Even more concentration of nutrients exists in the poultry and swine industries. Nearly all chickens and most hogs produced in the United States are grown in large CAFOs located near meat-processing plants. Not only do such CAFOs import nutrients in feed grains, but they also import calcium-P mineral feed supplements to compensate for the inability of their nonruminant animals to digest *phytic acid*, the form of P found in most seeds(see Section 12.3). The manure produced, therefore, contains more N and far more P than the local cropland base can properly utilize (see Figures 13.1 and 13.2). As a result, farm fields near large CAFOs tend to have very high levels of N and P in both the soil and in the water draining from the land.

Some Stop-Gap Measures. Although probably not long-term solutions to an unbalanced agricultural system, the public welfare may be served by such approaches as the following: (1) Discourage further manure applications to fields already saturated with nutrients; instead, facilitate transportation of manure to areas with low P soils. (2) Encourage the use of new corn varieties that contain less phytic acid P and more inorganic P, allowing better assimilation by nonruminant animals and making it less necessary to purchase P feed supplements, thereby reducing the amount of P excreted. (3) Promote manure composting to reduce the volume of material and the solubility of the nutrients in it. (4) Eliminate the overfeeding of P supplements to all types of livestock, in order to reduce the concentration of P in the manure (Figure 13.19). (5) Mix iron or aluminum compounds with the manure to reduce the solubility of its P (see Section 12.3).

Storage, Treatment, and Management of Animal Manures

Integrated Animal Production. Where animal and crop production are integrated on a farm, manure handling is not too much of a problem. The use of carefully managed pasture for cattle can be maximized so that the animals themselves spread most of the manure and urine while grazing. The total amount of nutrients in the manure produced on the farm is likely to be somewhat less than that needed to grow the crops; thus, modest amounts of inorganic fertilizers may be needed to make up the difference.

Manure Handling in Confinement Systems. Where animals are concentrated in large confinement systems, the problem of manure disposal often takes precedence over its utilization. The nutrient content of the manure can be markedly affected by the handling methods

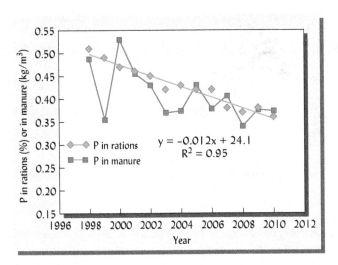

Figure 13.19 More careful formulation of animal feed can help reduce nutrient pollution from farm manure. The decline in P content of liquid dairy manure from commercial farms in Wisconsin paralleled the reduction of P in the total mixed rations fed to dairy cows on Wisconsin farms. The reduced P in the rations during a 12-year period resulted from a concerted effort to address concerns about phosphorus water pollution stemming from manure spread on dairy farmland. The graph is based on analyses of nearly 15,000 samples of liquid dairy manure. [Graphed from data of University of Wisconsin Extension (Peters, 2012)]

used. Most types of manure storage cause considerable N loss to the atmosphere, as just mentioned, while most of the P and K remain in the manure lagoon or compost pile. These changes during storage thus lower the N/P ratio in the manure, making it more difficult to utilize the manure as a N source without over applying P.

The manure can be collected and *spread daily, packed in piles* where it is allowed to partially decompose before spreading, stored in *aerated ponds* that promote oxidation of the organic materials, or stored in deep *anaerobic lagoons* (Figure 13.20, *right*) in which the manure ferments in the absence of oxygen gas. The latter method produces several environmentally damaging gases: methane (a powerful greenhouse gas that should be captured to use as fuel) and ammonia (which contributes to acrid odors and N deposition) are gaseous products of fermentation, while nitrogen gases (including NO and N_2O, which are also powerful greenhouse gasses) are produced by denitrification. Storage and microbial processing of manure in composting piles or in liquid slurry lagoons for several months before spreading on the land can greatly reduce the potential for contaminating fruits and vegetables with pathogens that can cause human illness.

Methods of manure handling that both prevent pollution and preserve nutrients, especially N, in a form that can be easily transported and sold commercially would make a major contribution to ameliorating the manure problem for concentrated animal production enterprises. Options currently being developed include the following: (1) **Heat-dry** and **pelletize** technology transforms the sloppy manure into small pellets that handle like commercial fertilizer. Although expensive in terms of energy and capital, the product is popular in the landscaping and lawn industries as a slow-release fertilizer. (2) Dry and heat the manure to high temperatures with limited oxygen to **pyrolize the manure and form a biochar** that is easy to handle, may be profitable to ship to areas where nutrients are needed, and may enhance soil properties (Box 11.2). (3) **Commercial composting** systems (see Section 11.11) represent a low-energy use, low-cost way to produce an easy-to-handle, nonodiferous, relatively high-analysis, slow-release fertilizer. Composting of livestock manure in well-managed windrows (Figure 13.20, *left*) is one way of reducing leaching and runoff losses of soluble nutrients while reducing the volume of manure that must be transported. (4) **Anaerobic digestion** (discussed previously) can be enhanced with the collection of biogas, which contains about 80% methane and 20% carbon dioxide and can be burned much like commercial natural gas. Small-scale manure digesters can supply cooking and heating fuel for remote villages and large-scale digesters can generate electricity to power modern farms.

Figure 13.20 *Contrasting methods of storing and handling manure. (Left) Composting long windrows (2) of animal manure and bedding straw. The tractor is pulling a machine (3) that stirs and turns the compost to aerate it and stimulate microbes. Microbial respiration has created so much heat that steam rises (4) from the compost as the machine stirs it. A green semipermeable fabric covering (1) prevents a second compost windrow from drying out. The finished compost will have easy to handle, slow nutrient release properties and take up about half the volume of the original manure. (Right) Liquid manure is being pumped out of a lined anaerobic storage lagoon into tractor-drawn manure wagons used to spray the manure onto crop fields on this dairy farm in the eastern United States.* [Photo courtesy of Ray R. Weil]

Methods of Manure Application

Similar to the impacts of manure handling, how farmers apply manure to their fields can have a large impact on the efficiency of nutrient use and the level of losses by surface runoff, volatilization, and leaching. Since farm animals produce manure every single day, manure handling is a constant daily process and farmers naturally tend to seek the quickest, most convenient methods of dealing with its management. The convenient thing, especially for the collect-and-spread-daily management approach, is to apply the manure to the nearest fields in all kinds of weather. Such fields will often carry a legacy of nutrient-saturated soils and continue to leak N and P into waterways for many years after this outdated practice is ceased. Another very detrimental practice is the application of manure to frozen soils in cold climates. Although convenient because the frozen soil will support heavy manure wagons without causing them to bog down in mud, the manure nutrients cannot soak into the soil and will wash off the surface with snowmelt in spring.

The advent of large-diameter drag hoses or dragline systems of manure-spreading technology has allowed more flexibility in manure application times as this method eliminates the need for driving extremely heavy tank-wagons (Figure 13.21c,d) of liquid manure across the field. In the dragline systems, liquid manure is pumped from a lagoon or large tank at the edge of a field through a long, flexible 15-cm-diameter hose to manure application implements pulled across the field by a tractor.

The various handling and storage methods discussed in the previous section can allow farmers to wait for the optimal time for applying manure, when the manure can be incorporated into the soil to avoid runoff losses and the nutrients can be immediately used by growing crops. Incorporation of manure into the soil by various tillage implements, however, can leave the soil surface exposed to erosion (see Section 14.5) and increase the loss of sediment and associated nutrients.

One solution is to inject the manure beneath the soil surface using a no-till implement with knife-like tines and rollers to leave the soil surface and residue cover almost undisturbed. Such soil injection has been widely practiced for liquid manure, but equipment that can inject solid dry manure, such as pelleted poultry house litter, is still under development. Compared to that broadcast on the soil surface, manure injected by such equipment loses far less N by ammonia volatilization and P by surface runoff. Increased adaption of such technologies, along

Figure 13.21 Examples of manure-spreading methods. Solid manure from poultry (a) and cattle (b) is spread on the surface of crop- or pastureland. Liquid manure from hog (c) and dairy (d) farms can be either injected into the soil (c) or sprayed on the surface (d). Note the crawler tread tractor and multiwheeled manure wagon in (c) designed to minimize compaction of soil by the heavy load. The liquid manure is likely greater than 95–99% water, while the solid manures may be only 50–90% water. (Photo courtesy of J. Gruver, Western Illinois University; others courtesy of Ray R. Weil)

with more closely integrated animal and crop farming systems, could help to redress the serious nutrient imbalances and pollution associated with manure production in industrialized agriculture.

13.5 INDUSTRIAL AND MUNICIPAL BY-PRODUCTS[7]

In addition to farm manures, four major types of organic wastes are of significance in land application: (1) municipal garbage, (2) sewage effluent and sludge (see also 15.7), (3) food-processing wastes, and (4) wood wastes of the lumber industry. Because of their uncertain content of toxic chemicals, these and other industrial wastes may or may not be suitable application to soils. Some of these materials are increasingly seen as sources of nutrients and organic matter that can be used beneficially to promote soil productivity in agriculture, forestry, landscaping, and disturbed-land reclamation.

Garbage and Yard Wastes

Household and municipal garbage has been used for centuries by traditional societies to enhance soil fertility. In industrial countries, most municipal solid waste (MSW) is incinerated or landfilled (see Section 15.9), but growing concerns about air quality, scarcity of landfill space, and greenhouse gas emissions are raising the level of interest in using soil application as a means of MSW disposal. About 50–60% of MSW consists of decomposable organic materials (paper, food scraps, yard waste, street tree leaves, etc.). Once the inorganic glass, metals, and so forth are removed, MSW can be *composted* (Section 11.11). Because MSW contains so much low N paper and cardboard, combining MSW with more nutrient-rich materials such as sewage sludge or animal manure greatly improves the composting process and the quality of the MSW compost that is then applied to the land. Even such mixed MSW compost is low enough in nutrient concentrations (see Table 13.2) that it is an expensive way to redistribute nutrients. However, alternative disposal options are often even more expensive. One advantage of composting MSW is the reduced production of greenhouse gases as compared to allowing the materials to decompose in anaerobic landfills or manure lagoons.

Food-Processing Wastes

Land application of food-processing wastes is being practiced in selected locations, but the practice is focused more on pollution abatement than on soil enhancement. Liquid wastes are commonly applied through sprinkle irrigation to permanently grassed fields. Kitchen food wastes, on the other hand, lend themselves to effective composting, whether at the scale of an individual household, large institution, or municipality. Food wastes usually are best composted together with another organic feed source that is lower in water and higher in N.

Wood Wastes

Sawdust, wood chips, and shredded bark from the lumber industry have long been sources of soil amendments and mulches, especially for home gardeners and landscapers. Because of their high C/N ratios and high lignin contents, these materials decompose very slowly. They make good mulching material, but do not readily supply plant nutrients. In fact, sawdust incorporated into soils to improve soil physical properties may cause plants to become N deficient unless an additional source of N is applied (see Section 11.3).

---

[7]For a collection of technical papers discussing the potential benefits and problems associated with the land application of these by-products, see Powers and Dick (2000). For an example of potentially net beneficial effects of yard waste compost application to native grasslands, see DeLonge et al. (2013).

Wastewater Treatment By-Products

For nearly two centuries, since the development of piped drinking water, flush toilets, and sewage systems, urban inhabitants have created vast quantities of sewage by using clean water as a vehicle to carry away human wastes. Although the typical system is fundamentally wasteful of water, energy, organic matter, and nutrients, at least sewage treatment has evolved to help society avoid polluting rivers and oceans with the pathogens, oxygen-demanding organic debris, and eutrophication-causing nutrients in sewage. In most advanced countries, sewage is given both primary and secondary treatment—and often advanced tertiary treatment—before the water is discharged back into a stream or river (Figure 13.22). The ever-more-stringent effort to clean up wastewater before returning it to natural waters has two basic consequences. First, the amount of material *removed* from the wastewater during the treatment process has increased tremendously. This solid material, known as sewage *sludge*, must also be disposed of safely. Second, soils can assist with the sewage problem in two ways: (1) as a system of assimilating, recycling, or disposing of the solid sludge; and (2) as a means of carrying out the final removal of nutrients and organics from the liquid **effluent** – the treated water that is returned to rivers. The final stages of effluent clean up can be accomplished safely and economically by allowing the partially treated effluent to interact with a soil–plant system.

Sewage Effluent

Some cities operate sewage farms on which they produce crops, usually animal feeds and forages, the sale of which can offset part of the expense of effluent disposal. Forest irrigation is another cost-effective method of final effluent cleanup and produces enhanced tree growth as a bonus (Figure 13.23, *left*). The rate of wood production is greatly increased as a result of both the additional water and the additional nutrients supplied therewith. This method of advanced wastewater treatment is used by a number of cities around the world.

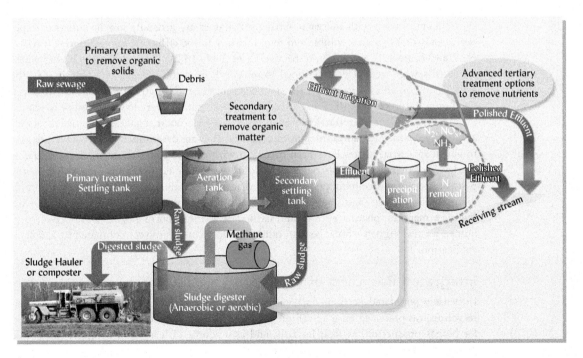

Figure 13.22 *Modern sewage treatment, which effectively cleans urban wastewater, produces two primary products: the solids removed from the wastewater, a material known as sewage sludge, and the partially cleaned up water that leaves the treatment plant, known as sewage effluent. Both products can be beneficially applied to soils if the process is properly planned and controlled.* (Diagram and photo courtesy of Ray R. Weil)

Figure 13.23 *Carefully managed application to soils has the potential to safely dispose of and beneficially use both the liquid (effluent) and solid (sludge) products of wastewater treatment. The soil–plant system may benefit from the contained water, nutrients, and organic carbon. However, a system's approach must be instituted to avoid unwanted and potentially dangerous constituents. The photos illustrate (left) irrigating with effluent in a pine plantation and (right) injecting sewage sludge into cropland soil.* (Photo courtesy of Ray R. Weil)

In a carefully planned and managed effluent irrigation system, the combination of (1) nutrient uptake by the plants, (2) adsorption of inorganic and organic constituents by soil colloids, and (3) degradation of organic compounds by soil microorganisms results in the purification of the wastewater. Percolation of the purified water eventually replenishes the groundwater supply.

Sewage Sludge or Biosolids

Sewage sludge has been spread on the land for decades and its use will likely increase in the future. If sewage sludge has been treated to meet certain standards of low pathogen and contaminant levels, the term **biosolids** may be applied. Numerous cities market dried activated (oxygenated) sludge or composted sludge for use on turfgrass, landscape plantings, and other specialty uses. However, the great bulk of sewage sludge used on land is injected into the soil as liquid slurry (Figure 13.23, *right*), or broadcast and incorporated as partially dried cake.

Composition of Sewage Sludge. As might be expected, the composition of sludge varies from one sewage treatment plant to another. The variations depend on the nature of treatment the sewage receives, especially the degree to which the organic material is allowed to digest.

In comparison with inorganic fertilizers, sludges are generally low in nutrients, especially potassium (which is soluble and found mainly in the effluent). Representative levels of N, P, and K are 4, 2, and 0.3%, respectively (see Table 13.2). The P content is higher where advanced sewage treatment is designed to remove P from the effluent and deposit it in the sludge (see Box 12.2). However, if the sewage treatment precipitates P by reactions with iron or aluminum compounds, the P in the sludge will likely have a very low availability to plants.

Levels of plant micronutrient metals (zinc, copper, iron, manganese, and nickel) as well as other heavy metals (cadmium, chromium, lead, etc.) are determined largely by the degree to which industrial wastes have been mixed in with domestic wastes. In the United States, the levels of metals and industrial chemicals in sewage are far lower than they were in the past, because of source-reduction programs that require industrial facilities to remove pollutants *before* sending their sewage to municipal treatment plants (see Section 15.7). As with livestock manures, pharmaceuticals and natural or artificial hormones are also of concern in human wastes. Vigilance must be maintained to avoid sludges too contaminated for safe land application.

Integrated Recycling of Wastes

In densely populated parts of Asia, particularly in China and Japan, complex integrated recycling was practiced long ago (Figure 13.24). Organic "wastes" were traditionally used for biogas production, as food for fish, and as a source of heat from compost piles. The plant nutrients and organic matter were recycled and returned to the soil. Despite China's increasing use of chemical fertilizers, the traditional respect for what others might see as wastes supports such complex recycling systems that continue to supply a significant proportion of the nutrients used in China's agriculture. As they look to achieve a more sustainable future, today's populations have much to learn from traditional Chinese attitudes and practices.

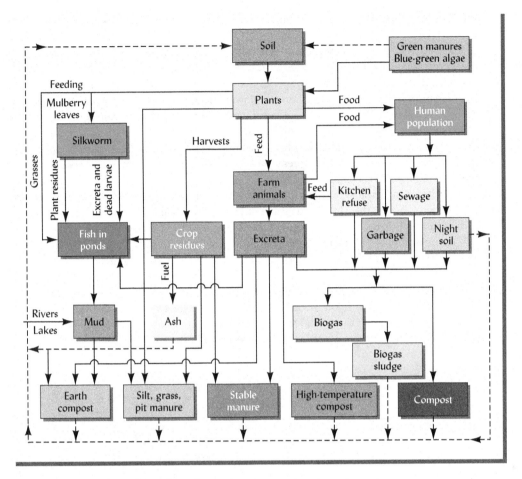

Figure 13.24 *Traditional recycling of organic wastes and nutrient elements as once commonly practiced in China. Note the degree to which the soil is involved in the recycling processes.* [Concepts from FAO (1977) and Yang (2006)]

13.6 PRACTICAL UTILIZATION OF ORGANIC NUTRIENT SOURCES[8]

In Section 11.5, we discussed the many beneficial effects on soil physical and chemical properties that can result from amendment of soils with decomposable organic materials. Here we will focus on the principles of efficient and ecologically sound management of nutrients from plant residues, sewage sludge, farm manure, MSW compost, and other organic materials.

Undoubtedly, the largest pool of organic material available to be applied to soils is the residue of plants grown on the land. Farmers are increasingly interested in using **cover crops** to produce additional nutrient-rich plant residues and to assist them in managing nutrients for their cash crops. The concept of cover crops—plants grown for soil improvement rather than for sale—was previously introduced along with the idea of using them to capture and recycle nutrients that might otherwise leach away (Section 13.2), as was the use of legume cover crops to enrich soils with N fixed from the atmosphere (Section 12.1). Cover crops may be most beneficial if nutrient release from their residues is timed to be in

---

[8]For an example of special, high rate applications of organic wastes that beg some interesting nutrient management questions, see research on a scheme to reduce greenhouse gas emissions from California agriculture while increasing the productivity and soil organic matter content of California's semiarid grasslands by making a single very large application of composted organic waste (Ryals and Silver, 2013; Wick and Creque, 2014).

synchrony with the nutrient uptake needs of the following main crop. This synchrony can be achieved with careful management of such factors as the composition of the cover crop residues and the physical contact between these residues and the soil. One way to manage cover crop nutrient composition is with the selection of species or mixture of species grown. For example, low C/N ratio legume species may be mixed with high C/N ratio grass species in different proportions to achieve faster or slower release or immobilization of N, as may be desired. The rates of decomposition and nutrient cycling can also be hastened by tillage that chops the residues into small pieces and mixes them in with moist soil. Using no-till techniques to terminate cover crops will have the opposite effect, spreading the nutrient release more evenly over the season and preserving a weed-suppressing, water-conserving surface mulch (Figure 13.25).

When using organic nutrient resources, the rate of application is generally governed by the amount of N or P that the organic material will make available to plants. Nitrogen usually is the first criterion because it is needed in the largest quantity by most plants, and because N can present a pollution problem if applied in excess (see Section 12.1). The ratio of P/N in most animal-derived organic sources is much higher than in plant tissue. Consequently, if organic materials supply sufficient N to meet plant needs, they probably supply excessive levels of P (see Section 13.2); the buildup of soil P must be taken into account in the long run. For soils already high in available P, the advisable application rate for an organic amendment may be limited by the P, rather than by the N content. Potentially, toxic-heavy metals in some materials may also limit the rate of application (see Section 15.7).

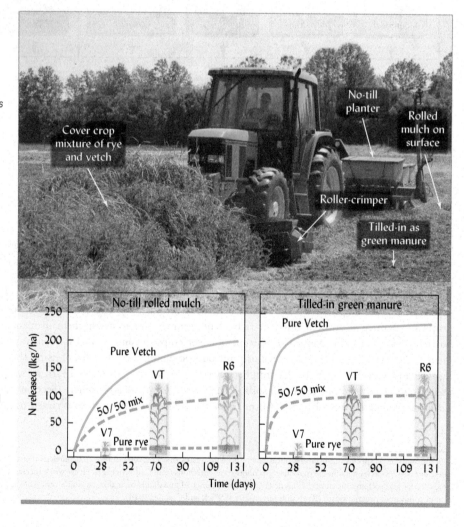

Figure 13.25 *Farmers are increasingly interested in using cover crops to assist with nutrient management. Here a mixed cover crop of vetch and rye is terminated with a roller-crimper and a corn crop is planted into the residue with a no-till planter. Nutrient release from cover crop residues in synchrony with the nutrient needs of the following main crop depends largely on the composition of the cover crop and how it is physically handled. Composition can be managed by mixing different proportions of high N legumes (e.g., vetch) with lower N grasses (e.g., rye). The rate of release is faster if the cover crop is tilled into the soil, and is more evenly spread out over the season if the cover crop is applied as surface mulch.* [Data and concepts from Poffenbarger et al. (2015). Photo courtesy of Betty H Marose]

Table 13.3
RELEASE OF MINERAL NITROGEN FROM VARIOUS ORGANIC MATERIALS APPLIED TO SOILS, AS PERCENT OF THE ORGANIC NITROGEN ORIGINALLY PRESENT[a]

For example, if 10 Mg of poultry floor litter initially contains 300 kg N in organic forms, 55% or 165 kg of N would be mineralized in year 1. Another 20% (0.20 × 300) or 60 kg of N would be released in year 2.

Organic N source	Year 1	Year 2	Year 3	Year 4
Poultry floor litter	55	20	8	3
Dairy manure (fresh solid)	25	18	9	4
Swine manure lagoon liquid	45	12	6	2
Feedlot cattle manure	30	15	6	2
Composted feedlot manure	18	18	4	1
Lime-stabilized, aerobically digested sewage sludge	40	12	5	2
Anaerobically digested sewage sludge	20	8	4	1
Composted sewage sludge	10	5	3	2
Activated, unstabilized sewage sludge	45	15	4	2
Flowering stage legume cover crop foliage	80	15	5	2

[a]These values are approximate and may need to be increased for warm climates or sandy soils and decreased for cold or dry climates or heavy clay soils. Release rates estimated from many sources.

A small fraction of the N organic amendments may be soluble (ammonium or nitrate) and immediately available, but the bulk of the N must be released by microbial mineralization. Table 13.3 estimates N mineralization rates for various organic materials. Materials partially decomposed during treatment and handling (e.g., compost or digestate) annually release a lower percentage of their organic N (Figure 13.26). However, anaerobic digestion may greatly increase the proportion of total N in soluble ammonium form.

If a field is treated repeatedly with an organic material, the application rate needed will become progressively smaller because, after the first year, the amount of N released from material applied in previous years must be subtracted from the total to be applied afresh (see Box 13.5). This is especially true for composts for which the initial availability of the N is quite low.

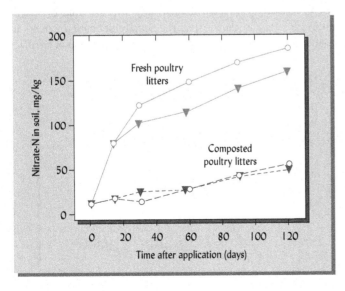

Figure 13.26 *Organic material stabilized by digestion or composting mineralizes more slowly than the raw unstabilized material. Here, nitrate nitrogen accumulates in a silt loam soil incubated with composted or fresh poultry litter from two sources. Amounts each of the litters sufficient to provide 230 mg of total N per kg of dry soil were incubated with moist soil at 25°C for 120 days.*
[Modified from Preusch et al. (2002)]

BOX 13.5
CALCULATION OF AMOUNT OF ORGANIC NUTRIENT SOURCE NEEDED TO SUPPLY A DESIRED AMOUNT OF NITROGEN

If the soil P level is not already above the optimal range, the rate of release of available N usually determines the proper amount of manure, sludge, or other organic nutrient source to apply. The amount of N made available in any year should meet, but not exceed, the amount of N that plants can use for optimum growth. Our example here is a field producing corn two years in a row. The goal is to produce 7,000 kg/ha of grain each year. For this field, this yield normally requires the application of 120 kg/ha of available nitrogen (about 58 kg of grain per kg N applied; see Section 13.13). We expect to obtain this N from a lime-stabilized sewage sludge containing 4.5% total N and 0.2% mineral N (ammonium and nitrate).

Year 1
Calculation of amount of sludge to apply per hectare:

% organic N in sludge = total N − mineral N = 4.5% − 0.2%
= 4.3%.

Organic N in 1 Mg of sludge = 0.043 × 1,000 kg = 43 kg N.

Mineral N in 1 Mg of sludge = 0.002 × 1,000 kg = 2 kg N.

Mineralization rate for lime-stabilized sludge in first year (Table 13.3) = 40% of organic N.

Available N mineralized from 1 Mg sludge in first year
= 0.40 × 43 kg N = 17.2 kg N.

Total available N from 1 Mg sludge = mineral N + mineralized N = 2.0 + 17.2 = 19.2 kg N.

Amount of (dry) sludge needed = 120 kg N/(19.2 kg available N/Mg dry sludge) = 6.25 Mg dry sludge.

Adjust for moisture content of sludge (e.g., assume sludge has 25% solids and 75% water).

Amount of wet sludge to apply:
6.25 Mg dry sludge/(0.25 Mg dry sludge/Mg wet sludge) = 6.25/0.25 = 25 Mg wet sludge.

Year 2
Calculation of amount of N mineralized in year 2 from sludge applied in year 1:

Second year mineralization rate (Table 13.3)
= 12% of original organic N.

N mineralized from sludge in year 2
= 0.12 × 43 Kg N/Mg × 6.25 Mg dry sludge
= 32.25 kg N from sludge in year 2.

Calculation of amount of sludge to apply in year 2:

N needed from sludge applied in year 2
= N needed by corn − N released from sludge applied in year 1 = 120 kg − 32.25 kg = 87.75 kg N needed/ha.

Amount of (dry) sludge needed per ha
= 87.75 kg N/(19.2 kg available N/Mg dry sludge)
= 87.75/19.2 = 4.57 Mg dry sludge/ha.

Adjust for moisture content of sludge (e.g., assume sludge has 25% solids and 75% water).

Mg of wet sludge to apply:
4.57 Mg dry sludge/(0.25 Mg dry sludge/Mg wet sludge)
= 4.57/0.25 = 18.3 Mg wet sludge.

Note that the 10.82 Mg dry sludge (6.25 + 4.57) also provided plenty of P: 216 kg P/ha (assuming 2% P; see Table 13.2), an amount that greatly exceeds the crop requirement and will soon lead to excessive buildup of P.

13.7 INORGANIC COMMERCIAL FERTILIZERS

The worldwide agricultural use of fertilizers increased dramatically after the mid-twentieth century (Figure 13.27), accounting for a significant part of the equally dramatic increases in crop yields during the same period. Improved soil fertility through the application of fertilizer

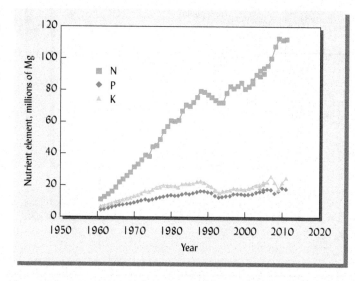

Figure 13.27 *The use of nitrogen (N) in the world has increased since 1960 much faster than that of phosphorus (P) and potassium (K). The dip in world fertilizer use in the early 1990s was due mainly to the collapse of the Soviet Union. In most industrialized countries, fertilizer use has leveled off or declined, but it continues to increase in China and India, and in much of the developing world. In much of Africa, the removal of nutrients in crop harvests still exceeds the amounts returned to the soil.* [Data selected from FAOSTAT (FAO, 2014)]

nutrients is an essential factor in enabling the world to feed the billions of people that are being added to its population. Most of the increase has been in the use of nitrogen, which has increased its share of the total N, P, and K applied from 50% in 1960 to over 70% in recent years.

Fertilizer use is also increasing in forestry as the demands for forest products increase the removal of nutrients and competing uses of land leave forestry with more infertile, marginal sites. Currently, most fertilization in forestry is concentrated on tree nurseries and on seed trees, where the benefits of fertilization are relatively short term and of high value, and the logistics of fertilizer application are not so difficult and expensive as for extensive forest stands. However, fertilization of new forest plantings and existing stands is increasingly practiced in Japan, northwestern and southeastern United States, Canada, the Scandinavian countries, and Australia, with the fertilizer spread by helicopter. Nearly all the fertilizer used in forestry has been to supply nitrogen, but in recent decades applications of phosphorus to forest have become more common.

Origin and Processing of Inorganic Fertilizers

Most fertilizers are inorganic salts containing readily soluble plant nutrient elements. Nitrogen fertilizers are manufactured using natural gas and the elemental N_2 gas in air under very high temperatures and pressures to synthesize ammonia gas, NH_3. This industrial fixation is termed the *Haber–Bosch process* and consumes tremendous quantities of energy—usually from fossil fuel. Therefore, most nitrogen fertilizer factories are located near natural gas fields in Russia, the U.S. state of Texas, China, India, and Nigeria. The industrial process of N fixation has transformed the global nitrogen cycle (see Section 12.1).

Phosphorous fertilizers are processed by dissolving phosphate mineral (apatite) bearing rocks with sulfuric acid. Phosphate rock deposits are located around the world, but the vast majority of remaining known reserves are located in Morocco and to a lesser degree in China.

Most potassium fertilizers are simply purified from natural geological deposits of salts formed from the evaporation of ancient seawater. These underground deposits are found in many locations, including sites in Canada, France, Germany, and Russia, as well as in the U.S. state of New Mexico.

When the manufactured ammonia gas is put under moderate pressure, it liquefies, forming anhydrous ammonia. Because of its low cost and ease of application, much N is applied directly to soils as anhydrous ammonia. This compound is also the starting point for the manufacture of most other N carriers, including urea, ammonium nitrate, ammonium sulfate, sodium nitrate, and aqueous nitrogen solutions.

Many commercial fertilizers contain two or more macronutrient elements. In some cases, mixtures of the primary nutrient carriers are blended together (Figure 13.28). But in fertilizers, a given compound may carry two nutrients, examples being diammonium phosphate, potassium sulfate, potassium nitrate, and ammonium sulfate. Care must be used in selecting the components of a mixed fertilizer since some compounds are not compatible with others, resulting in poor physical condition and reduced nutrient availability in the mixture.

Properties and Use of Inorganic Fertilizers

The composition of inorganic commercial fertilizers is much more precisely defined than is the case for the organic materials discussed in the previous sections. Table 13.4 lists the nutrient element contents and other properties of some of the more commonly used inorganic fertilizers. In most cases, fertilizers are used to supply plants with the macronutrients N, P, S, and/or K. Fertilizers that supply magnesium, calcium, and the micronutrients are also manufactured.

It can be seen from the data in Table 13.4 that a particular nutrient (say, N) can be supplied by many different *carriers*, or fertilizer compounds. Decisions as to which fertilizers to use must take into account not only the nutrients they contain, but also characteristics, such as the salt hazard (see also Section 9.15), acid-forming tendency (see also Section 9.1), tendency to volatilize, ease of solubility, and content of nutrients other than the principal one. Of the N

Figure 13.28 *A handful of a dry, granular fertilizer blend. Most primary fertilizer materials exhibit characteristic properties such as color, texture, particle size, and shape. In the photo, smooth, white pearl-like beads or prills of urea (1) are easily identified, as are the dark gray, rough granules of diammonium phosphate (2) and the pinkish or white crystals of potassium chloride (3). The green-blue particles are micronutrient mixtures colored by copper sulfate (4).* (Photo courtesy of Ray R. Weil)

carriers, anhydrous ammonia, nitrogen solutions, and urea are used most widely. Diammonium phosphate and potassium chloride supply the bulk of the P and K used.

Fertilizer Grade

Commercial fertilizers typically are described by their *fertilizer grade*. Every fertilizer label states the **grade** as a three-number code, such as 10-5-10 or 6-24-24. These numbers stand for percentages indicating the *total* nitrogen (N) content, the *available* (citric acid soluble) phosphate (P_2O_5) content, and the *water soluble* potash (K_2O) content. Plants do not take up phosphorus and potassium as P_2O_5 or K_2O, nor do any fertilizers actually contain these chemical forms. The oxide expressions for phosphorus (P_2O_5) and potassium (K_2O) are relics of the days when geochemists reported the contents of rocks and minerals in terms of the oxides formed upon heating. Unfortunately, these expressions found their ways into laws governing the sale of fertilizers, and there is considerable resistance to changing them, although some progress is being made. In scientific work and in this textbook, the simple elemental contents are used (P and K) wherever possible. Box 13.6 and Figure 13.29 explain how to convert between the elemental and oxide forms of expression.

The grade is important from an economic standpoint because it conveys the analysis or concentrations of the nutrients in a carrier. When properly applied, most fertilizer carriers give equally good results for a given amount of nutrient element. The more concentrated carriers are usually the most economical to use, because less weight of fertilizer must be transported to supply the needed quantity of a given nutrient. Hence, *economic comparisons among different equally suitable fertilizers should be based on the price per kilogram of nutrient, not on the price per kilogram of fertilizer.*

Fate of Fertilizer Nutrients

A common myth about fertilizers suggests that inorganic fertilizers applied to soil directly feed the plant, and that therefore the biological cycling of nutrients, such as described by Figures 12.1 for nitrogen and 12.27 for phosphorus, are of little consequence where inorganic fertilizers are used. The reality is that nutrients added by normal application of fertilizers, whether organic or inorganic, are incorporated into the complex soil nutrient cycles, and that relatively

Table 13.4
COMMONLY USED INORGANIC FERTILIZER MATERIALS: THEIR MACRONUTRIENT CONTENTS AND OTHER CHARACTERISTICS

Fertilizer	Percent by weight				Salt hazard	Acid formation[a]	Other nutrients & comments
	N	P	K	S			
Primarily sources of nitrogen							
Anhydrous ammonia (NH_3)	82				Low	−148	Needs pressurized equipment, toxic gas; inject into soil.
Urea [$CO(NH_2)_2$]	45				Moderate	−84	Soluble; hydrolyzes to ammonium forms. Volatilizes if left on soil surface.
Ammonium nitrate (NH_4NO_3)	33				High	−59	Absorbs water from air; explosive if mixed with organic dust or S.
UAN solution	30				Moderate	−52	Most commonly used liquid N.
IBDU (isobutylidene diurea)	30				Very low	—	Slowly soluble.
Ammonium sulfate [$(NH_4)_2SO_4$]	21			24	High	−110	Rapidly lowers soil pH; very easy to handle.
Sodium nitrate ($NaNO_3$)	16				Very high	+29	Hardens, disperses soil structure.
Potassium nitrate (KNO_3)	13		36	0.2	Very high	+26	Very rapid plant response.
Primarily sources of phosphorus							
Monoammonium phosphate ($NH_4H_2PO_4$)	11	21–23		1–3	Low	−65	Best as starter.
Diammonium phosphate [$(NH_4)_2HPO_4$]	18–21	20–23		0–3	Moderate	−70	Best as starter.
Triple superphosphate		19–22		1–3	Low	0	15% Ca.
Phosphate rock [$Ca_3(PO_4)_2 \cdot CaX$]		8–18[b]			Very low	Variable	Low to extremely low availability. Best as fine powder on acid soils. May contain Cd, F, radiation.
Single superphosphate		7–9		11	Low	0	Nonburning, can place with seed. 20% Ca.
Bonemeal	1–3[b]	10[b]	0.4		Very low	—	Slow availability of N, P 20% Ca.
Primarily sources of potassium							
Potassium chloride (KCl)			50		High	0	47% Cl—may reduce some diseases.
Potassium sulfate (K_2SO_4)			42	17	Moderate	0	Use where Cl not desirable.
Wood ashes		0.5–1	1–4		Moderate to high	+40	Alkaline, caustic, 10–20% Ca, 2–5% Mg, 0.2% Fe, 0.8% Mn.
Greensand		0.6	6		Very low	0	Very low availability.
Primarily sources of other nutrients							
Gypsum ($CaSO_4 \cdot 2H_2O$)				19	Low	0	Stabilizes soil structure; no effect on pH; Ca and S readily available. 23% Ca.
Calcitic limestone ($CaCO_3$)					Very low	+95	Slow availability; raises pH. 36% Ca.
Dolomitic limestone [$CaMg(CO_3)_2$]					Very low	+95	Very slow availability; raises pH. ~24% Ca, ~12% Mg.
Epsom salts ($MgSO_4 \cdot 7H_2O$)				13	Moderate	0	No pH effect; water soluble; 2% Ca, 10% Mg.
Sulfur, flowers (S)				95	—	−300	Irritates eyes; acidifying; slow acting; requires microbial oxidation.

[a] A negative number indicates that acidity is produced; a positive number indicates that alkalinity is produced; kg $CaCO_3$/100 kg material needed to neutralize acidity.
[b] Highly variable contents.

little of the fertilizer nutrient (from 10 to 60%) actually winds up in the plant being fertilized during the year of application. Even when the application of fertilizer greatly increases both plant growth and nutrient uptake, the fertilizer stimulates increased cycling of the nutrients, and the nutrient ions taken up by the plant come largely from various pools in the soil and not directly from the fertilizer. For example, some of the added N may go to satisfy the needs

> **BOX 13.6**
> **HOW MUCH N, P, AND K IS IN A BAG OF 6-24-24?**
>
> Conventional labeling of fertilizer products reports percentage N, P_2O_5, and K_2O. Thus, a fertilizer package (Figure 13.29) labeled as 6-24-24 (6% nitrogen, 24% P_2O_5, 24% K_2O) actually contains 6% N, 10.5% P, and 19.9% K (see calculations below).
>
> To determine the amount of fertilizer needed to supply the recommended amount of a given nutrient, first convert percent P_2O_5 and percent K_2O to percent P and K, by calculating the proportion of P_2O_5 that is P and the proportion of K_2O that is K. The following calculations may be used:
>
> Given that the molecular weights of P, K, and O are 31, 39, and 16 g/mol, respectively:
>
> Molecular weight of $P_2O_5 = 2(31) + 5(16) = 142\,g/mol$
>
> $$\text{Proportion P in } P_2O_5 = \frac{2P}{P_2O_5} = \frac{2(31)}{2(31)+5(16)} = 0.44$$
>
> To convert $P_2O_5 \longrightarrow P$, multiply percent P_2O_5 by 0.44
>
> Molecular weight of $K_2O = 2(39) + 16 = 94$
>
> $$\text{Proportion K in } K_2O = \frac{2K}{K_2O} = \frac{2(39)}{2(39)+16} = 0.83$$
>
> To convert $K_2O \longrightarrow K$, multiply percent K_2O by 0.83
>
> Thus, if the bag in Figure 13.29 contains 25 kg of the 6-24-24 fertilizer, it will supply 1.5 kg N (0.06 × 25); 2.6 kg P (0.24 × 0.44 × 25); and 5 kg K (0.24 × 0.83 × 25).
>
>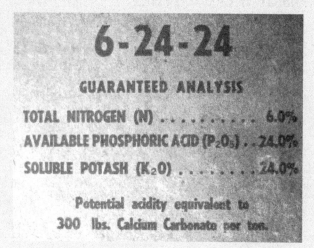
>
> **Figure 13.29** A typical commercial fertilizer label. Note that a calculation must be performed to determine the percentage of the nutrient elements P and K in the fertilizer since the contents are expressed as if the nutrients were in the forms of P_2O_5 and K_2O. Also note that after interacting with the plant and soil, this material would cause an increase in soil acidity that could be neutralized by 300 units of $CaCO_3$ per 2,000 units (1 ton = 2,000 lb) of fertilizer material. (Photo courtesy of Ray R. Weil)

of microorganisms, preventing them from competing with plants for other pools of N. This knowledge has been obtained by careful analysis of dozens of nutrient studies that used fertilizer with isotopically tagged nutrients. Results from one such study are summarized in Table 13.5, which shows somewhat more N uptake from fertilizer than is typically reported. Generally, as fertilizer rates are increased, the efficiency of fertilizer nutrient use decreases, leaving behind in the soil an increasing proportion of the added nutrient.

Table 13.5
Source of Nitrogen in Corn Plants Grown in North Carolina on an Enon Sandy Loam Soil (Ultic Hapludalf) Fertilized with Three Rates of Nitrogen as Ammonium Nitrate

The source of the N in the corn plant was determined by using fertilizer tagged with the isotope 15N. Moderate fertilizer use increased the uptake of N already in the soil system as well as that derived from the fertilizer.

Fertilizer N applied, kg/ha	Corn grain yield, Mg/ha	Total N in corn plant, kg/ha	Fertilizer-derived N in corn, kg/ha	Soil-derived N in corn, kg/ha	Fertilizer-derived N in corn as percent of total N in corn	Fertilizer-derived N in corn as percent of N applied
50	3.9	85	28	60	33	56
100	4.6	146	55	91	38	55
200	5.5	157	86	71	55	43

Calculated from Reddy and Reddy (1993).

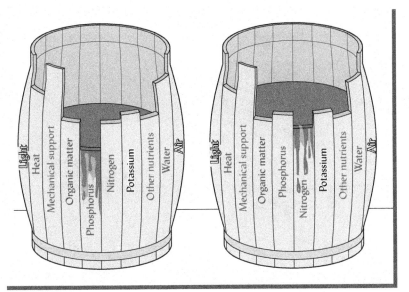

Figure 13.30 *An illustration of the law of the minimum and the concept of the limiting factor. Plant growth is constrained by the essential element (or other factor) that is most limiting. The level of liquid in the barrel represents the level of plant production. (Left) Phosphorus is represented as being the shortest barrel stave—the factor that is most limiting. Even though the other elements are present in more than adequate amounts, plant growth can be no greater than that allowed by the level of P available. (Right) When the P stave is lengthened (P is added), the level of plant production is raised until another factor becomes most limiting—in this case, nitrogen.*

The Concept of the Limiting Factor

Two German chemists (Justus von Liebig and Carl Sprengel) are credited with first publishing, in the mid-1800s, "the law of the minimum" which holds that *plant production can be no greater than that level allowed by the growth factor present in the lowest amount relative to the optimum amount for that factor.* This growth factor, be it temperature, nitrogen, sulfur, or water supply, will limit the amount of growth that can occur and is therefore called the ***limiting factor*** (Figure 13.30). This is similar to the adage that a chain is only as strong as its weakest link.

If a factor is not the limiting one, increasing it will do little or nothing to enhance plant growth. In fact, increasing the amount of a nonlimiting factor may actually reduce plant growth by throwing the system further out of balance. For example, if a plant is limited by lack of P, adding more N may only aggravate the P deficiency.

Looked at another way, applying available P (the first limiting nutrient in this example) may allow the plant to respond positively to a subsequent addition of N. Thus, there may be an *interaction* or *synergy* between two nutrients applied together such that the increased growth obtained is much greater than the sum of the growth increases obtained by applying each of the two nutrients individually.

13.8 FERTILIZER APPLICATION METHODS

As we have been discussing, wise, effective fertilizer use involves making correct decisions regarding *which* nutrient element(s) to apply, *how much* of each needed nutrient to apply, *what type* of material or carrier to use (Tables 13.2 and 13.4 list some of the choices), *in what manner* to apply the material, and, finally, *when* to apply it. We will leave information on the first two decisions until Section 13.10. Here we will discuss the alternatives available with regard to the last two decisions.

There are three general methods of applying fertilizers (Figure 13.31): (1) *broadcast application*, (2) *localized placement*, and (3) *foliar application.* Each method has some advantages and disadvantages and may be particularly suitable for different situations. Often, some combination of the three methods is used.

Broadcasting

In many instances, fertilizer is spread evenly over the entire field or area to be fertilized. This method is called *broadcasting.* Often the broadcast fertilizer is mixed into the soil layer by means of tillage, but in some situations it is left on the soil surface and allowed to be carried

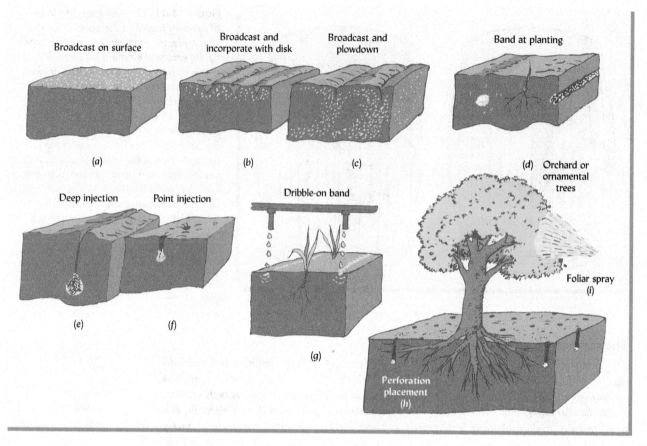

Figure 13.31 Fertilizers may be applied by many different methods, depending on the situation. Methods (a)–(c) represent broadcasting, with or without incorporation. Methods (d)–(h) are variations of localized placement. Method (i) is foliar application and has special advantages, but also limitations. Commonly, two or three of these methods may be used in sequence. For example, a field may be prepared with (c) before planting; (d) may be used during the planting operation; (g) may be used as a side dressing early in the growing season; and, finally, (i) may be used to correct a micronutrient deficiency that shows up mid-season. (Diagram courtesy of Ray R. Weil)

into the root zone by percolating rain or irrigation water. The broadcast method is most appropriate when a large amount of fertilizer is being applied with the aim of raising the fertility level of the soil over a long period of time. Broadcasting is the most economical way to spread large amounts of fertilizer over wide areas (Figure 13.32).

For close-growing vegetation, broadcasting provides an appropriate distribution of the nutrients. It is therefore the most commonly used method for rangeland, pastures, small grains, turf grass, and forests. Fertilizers are also broadcast on some row cropland, especially in the fall when it is most convenient, although certainly not most efficient. The broadcast fertilizer may or may not be incorporated into the soil. Unfortunately, for crops with wide row spacing or young tree seedlings in forest plantings, broadcasting places the fertilizer where most of it is accessible more to the *weeds* than to the target plants.

For phosphorus, zinc, manganese, and other nutrients that tend to be strongly retained by the soil, broadcast applications are usually much less efficient than localized placement. Often 2–3 kg of fertilizer must be broadcast to achieve the same response as from 1 kg that is placed in a localized area.

A one-time application of P and K fertilizer, broadcast and worked into the soil, is a good preparation, where these nutrients are needed, for establishing perennial plants such as lawns, pastures, and orchards. It may be necessary to broadcast a top dressing in subsequent years, being careful not to allow fertilizers with high salt hazards (see Table 13.4) to remain in contact with the foliage long enough to cause salt burn (Figure 13.33). Nitrogen

Figure 13.32 *Broadcast fertilization. (a) Helicopter broadcasting forestry-grade fertilizer pellets. (b) Broadcasting nitrogen solution (mixed with fungicide) on a wheat field in France. (c) Broadcasting granular phosphorus and potassium fertilizer on a hayfield.* (Helicopter photo courtesy of Westvaco, Inc.; others courtesy of Ray R. Weil)

is commonly broadcast in a liquid form (sprayed), often in a solution that also contains other nutrients or chemicals (see Figure 13.32b). Because of its mobility in the soil, nitrogen does not suffer from reduced availability when broadcast, but if left on the soil surface, much may be lost by volatilization. Volatilization losses are especially troublesome for urea and ammonium fertilizers applied to soils with a high pH. Surface broadcast nutrients are also susceptible to loss in runoff. Many studies have shown that most of the annual loss of nutrients (or herbicides, if surface broadcast) in runoff usually occurs during the first one or two heavy-rainfall events after the broadcast application. Where sprinkler irrigation is practiced, liquid fertilizers can be broadcast in the irrigation water, a practice sometimes called **fertigation**.

Figure 13.33 *Salt burn from fertilizer broadcast on plant foliage. (Left) Too much urea-ammonium nitrate solution sprayed on corn during dry weather. (Right) Dry fertilizer initially spread on home lawn at a too high rate.* (Photo courtesy of Ray R. Weil)

Localized Placement

Although it is commonly thought that nutrients must be thoroughly mixed throughout the root zone, research has clearly shown that a plant can easily obtain its entire supply of a nutrient from a concentrated localized source in contact with only a small fraction of its root system. In fact, a small portion of a plant's root system can grow and proliferate in a band of fertilizer even though the salinity level caused by the fertilizer in that small volume of soil may be very high. Such levels of salinity as found near a fertilizer granule would be fatal to a germinating seed or to a mature plant were a large part of its root system exposed to this high salinity level. This finding allowed the development of techniques for localized fertilizer placement.

Fertilizer is often more effectively used by plants if it is placed in a localized concentration rather than mixed with soil throughout the root zone. First, localized placement reduces the amount of contact between soil particles and the fertilizer nutrient, thus minimizing the opportunity for adverse fixation reactions. Second, in the fertilized zone, the concentration of the nutrient in the soil solution at the root surface will be very high, resulting in greatly enhanced uptake by the roots.

Localized placement is especially effective for young seedlings, in cool soils in early spring, and for plants that grow rapidly with a big demand for nutrients early in the season. For these reasons, **starter fertilizer** is often applied in bands on either side of the seed as the crop is planted. Since germinating seeds can be injured by fertilizer salts, and since these salts tend to move upward as water evaporates from the soil surface, the best placement for starter fertilizer is approximately 5 cm below and 5 cm off to the side from the seed row (see Figure 13.31d).

Liquid fertilizers and slurries of manure and sewage sludge can also be applied in bands rather than broadcast. Bands of these liquids are placed 10–30 cm deep in the soil by a process known as *knife injection* (Figures 13.31e and 13.34, *middle*). In addition to the advantages mentioned for banding fertilizer, injection of these organic slurries reduces runoff losses and odor problems. Anhydrous ammonia and pressurized nitrogen solutions must be injected into the soil to prevent losses by volatilization. Injecting bands at 15 and 5 cm, respectively, are considered adequate depths for these two materials.

Another approach to banding solutions (though not slurries) is to *dribble* a narrow stream of liquid fertilizer alongside the crop row as a side-dressing (Figure 13.34, *left*). The use of a stream instead of a fine spray changes the application from broadcast to banding and results in enough liquid in a narrow zone to cause the fertilizer to soak into the soil. This action greatly reduces volatilization loss of nitrogen.

Figure 13.34 (Left) *A liquid fertilizer applicator with hoses dribbling a stream of liquid between each row of corn as a side-dressing.* (Middle) *Manure slurry being knife-injected into the soil before planting crops on the land. This injection method reduces runoff losses and objectionable odors.* (Right) *A measured micro dose of solid fertilizer granules being applied by hand as a dollop in a small hole near a young crop plant in Africa.* (Photo courtesy of Ray R. Weil)

Localized placement of fertilizer by *point injection* applies small portions of fertilizer to every individual plant without significantly disturbing either the plant root or the surface residue cover left by conservation tillage. Point injection implements are modern mechanized versions of the simple dibble stick with which small-scale farmers and gardeners plant seeds and later apply a portion of fertilizer in the soil next to each plant (Figure 13.34, *right*).

The use of *drip irrigation* systems (see Section 6.9) has greatly facilitated the localized application of nutrients in irrigation water. Because drip fertigation is applied at frequent intervals, the plants are essentially spoon-fed, and the efficiency of nutrient use is quite high.

Perforation Method for Trees. Trees in orchards and ornamental plantings are best treated individually, the fertilizer being applied around each tree within the spread of the branches but beginning approximately 1 m from the trunk (see Figure 13.31h). The fertilizer is best applied by what is called the *perforation* method. Numerous small holes are dug around each tree within the outer half of the branch-spread zone and extending down into the upper subsoil where the fertilizer is placed. Special large fertilizer pellets are available for this purpose. This method of application places the nutrients within the tree root zone and avoids an undesirable stimulation of the grass or cover that may be growing around the trees. If the cover crop or lawn around the trees needs fertilization, it is treated separately, the fertilizer being drilled in at the time of seeding or broadcast later.

Foliar Application

Plants are capable of absorbing nutrients through their leaves in limited quantities. Under certain circumstances, the best way to apply a nutrient is *foliar application*—spraying a dilute nutrient solution directly onto the plant leaves (Figure 13.31i). Diluted macronutrient fertilizers, micronutrients, or small quantities of urea can be used as foliar sprays, although care must be taken to avoid significant concentrations of salts, especially Cl^- or NO_3^-, which can be toxic to some plants. Foliar fertilization may conveniently fit in with other field operations for horticultural crops, because the fertilizer is often applied simultaneously with pesticide sprays.

The amount of nutrients that can be sprayed on leaves in a single application is quite limited. Therefore, while a few spray applications may deliver the entire season's requirement for a micronutrient, only a small portion of typical macronutrient needs can be supplied in this manner. The danger of leaf injury (Figure 13.33, *left*) is especially high during dry, hot weather, when the solution quickly evaporates from the leaf surface, leaving behind the fertilizer salts. Spraying on cool, overcast days or during early morning or late evening hours reduces the risk of injury, as does the use of a dilute solution containing, for example, only 1 or 2% N.

13.9 TIMING OF NUTRIENT APPLICATION

The timing of nutrient applications in the field is governed by several basic considerations: (1) making the nutrient available when the plant needs it; (2) avoiding excess availability, especially of N, before and after the principal period of plant uptake; (3) making nutrients available when they will strengthen, not weaken, long-season and perennial plants; and (4) conducting field operations when conditions make them feasible and not damaging to the soil.

Availability When the Plants Need It

For mobile nutrients such as N (and to a lesser degree S and K), the general rule is to make applications as close as possible to the period of rapid plant nutrient uptake. For rapid-growing summer annuals, such as corn, this means making only a small starter application at planting time and applying most of the needed N as a side dressing just before the plants enter the rapid nutrient accumulation phase, usually about four to six weeks after planting. For cool-season plants, such as winter wheat or certain turf grasses, most of the N should be applied about the time of spring "green-up," when the plants resume a rapid growth rate. For trees, the best time is when new leaves are forming. With slow-release organic sources, time should be allowed for mineralization to take place prior to the plants' period of maximum uptake.

Environmentally Sensitive Periods

In temperate (**Udic** and **Xeric**) climates, most leaching takes place in the winter and early spring when precipitation is high and evapotranspiration is low. Nitrates left over or released after plant uptake has ceased have the potential for leaching during this period. In this regard, it should be noted that, for grain crops, the rate of nutrient uptake begins to decline during grain-filling stages and has largely or completely ceased long before the crop is ready for harvest. With inorganic N fertilizers, avoiding leftover nitrates is largely a matter of applying the right amount at the right time. However, for slow-release organic sources applied in late spring or early summer, mineralization is likely to continue to release nitrates after the crop has matured and ceased taking them up. To the extent that this timing of nitrate release is unavoidable, cover crops should be planted as soon as possible after the summer crop's N uptake has slowed in order to absorb the excess nitrate being released.

Split Applications. In high-rainfall conditions and on permeable soils, dividing a large dose of fertilizer into two or more split applications may avoid leaching losses prior to the crop's establishment of a deep root system. In cold climates, another environmentally sensitive period occurs during early spring, when snowmelt over frozen or saturated soils results in torrents of runoff water.

Application of urea fertilizers to mature forests is usually carried out when rains can be expected to wash the nutrients into the soil and minimize volatilization losses. Furthermore, nitrogen fertilization of forests should occur just prior to the onset of the growing season (early spring, or in warm climates, winter) so that the tree roots will have the entire growing season to utilize the N. Fertilizer applications commonly result in a pulse of nitrate leaving the watershed for several weeks following the fertilizer application. The nitrogen pollution can be reduced if a 10- to 15-m unfertilized buffer is maintained along all streams.

Physiologically Appropriate Timing

It is important to make nutrients available when they will strengthen plants and improve their quality. For example, too much N in the summer may stress cool-season turfgrasses, while high N late in the season will reduce the sugar content of root crops. A good supply of K is particularly important in the fall to enable plants to improve their winter-hardiness. Planting-hole or broadcast application of P at the time of the tree planting brings good results on phosphorus-poor sites. However, broadcast application of N to trees soon after the seedlings are planted may benefit fast-growing weeds more than the desired trees. Later in the development of a forest stand, when the tree canopy has matured, application of fertilizer, usually from a helicopter, can be quite beneficial.

Practical Field Limitations

Sometimes it is simply not possible to apply fertilizers at the ideal time of the year. For example, although a crop may respond to a late-season side dressing, such an application will be difficult if the plants are too tall to drive over without damaging them. Using an airplane or special high-clearance equipment may allow more flexibility in fertilizer timing. Early spring applications may be limited by the need to avoid compacting wet soils. Economic costs or the time demands of other activities may also require compromises in the timing of nutrient application.

13.10 DIAGNOSTIC TOOLS AND METHODS[9]

There are three basic tools available for diagnosing soil fertility problems: (1) *field observations*, (2) *plant tissue analysis*, and (3) *soil analysis* (soil testing). All three approaches should be integrated to effectively guide the application of nutrients, as well as to diagnose problems as they arise in the field. There is no substitute for careful observation and *recording* of circumstantial evidence and symptoms in the field. Effective observation and interpretation requires skill and experience, as well as an open mind. It is not uncommon for a supposed soil fertility problem

---

[9] For detailed discussions of both plant analysis and soil testing, see Jones Jr. (2012).

to actually be caused by soil compaction, weather conditions, pest damage, or human error. The task of the diagnostician is to use all the tools available in order to identify the factor that is limiting plant growth, and then devise a course of action to alleviate the limitation.

Plant Symptoms and Field Observations

This detective-like work can be one of the more exciting and challenging aspects of nutrient management. To be an effective soil fertility diagnostician, several general guidelines are helpful.

1. Develop an organized way to *record* your observations. The information you collect may be needed to properly interpret soil and plant analytical results obtained at a later date.
2. Talk to the person who owns or manages the land. Ask when the problem was first observed and if any recent changes have taken place. Obtain records on plant growth or crop yield from previous years, and ascertain the management history for as many years as possible. It is often useful to sketch a map of the site showing features you have observed and the distribution of symptoms.
3. Look for *spatial patterns*—how the problem seems to be distributed in the landscape and in individual plants. Linear patterns across a field may indicate a problem related to tillage, drain tiles, or the incorrect spreading of lime or fertilizer. Poor growth concentrated in low-lying areas may relate to the effects of soil aeration. Poor growth on the high spots in a field may reflect the effects of erosion and possibly exposure of subsoil material with an unfavorable pH. Satellite images, such as those freely available from Google Earth®, may show patterns of vegetation and soils and other patterns that help explain patterns seen on the ground and can provide an excellent map base for your observations.
4. Closely examine individual plant leaves to characterize any foliar symptoms. Nutrient deficiencies can produce characteristic symptoms on leaves and other plant parts. Examples of such symptoms are shown in several figures in Chapter 12. Determine whether the symptoms are most pronounced on the younger leaves (as is the case for S and most of the micronutrient cations) or on the older leaves (as is the case for nitrogen, potassium, and magnesium). Some nutrient deficiencies produce symptoms that may be confused with herbicide damage, insect damage, or damage from poor aeration.
5. Observe and *measure* differences in *both* aboveground and belowground plant growth that may reflect different levels of soil fertility, even though no leaf symptoms are apparent. Use a shovel to dig down to at least the depth of deepest tillage or the upper subsoil layer. Dig up a few of the ailing plants and a few of the more normal plants to compare their root systems. Are mycorrhizae associated with tree roots? Are legumes well nodulated? Is root growth restricted in any way? Are roots thin, flexible, and white (as healthy root appear), or are they thickened, stubby, brittle, and brown? Do roots grow smoothly and vertically or do they hit a soil layer and abruptly turn horizontally, indicating a compacted or cemented zone?

Plant Analysis[10]

Nutrient Concentrations. The concentration of essential elements in plant tissue is related to plant growth or crop yield, as shown in Figure 13.35. The range of tissue concentrations associated with optimal plant growth is termed the **sufficiency range**. At the upper end of the sufficiency range, plants may be participating in luxury consumption, as the additional nutrient uptake has not produced additional plant growth (see Section 12.4). At concentrations above the sufficiency range, plant growth may decline as nutrient elements reach concentrations that are toxic to plant cells or interfere with the use of other nutrients. If tissue concentrations are in the **critical range**, the supply is just marginal and growth is expected to decline if the nutrient becomes any less available, even though visible foliar symptoms may not

---

[10]Detailed information on tissue analysis for a large number of plant species can be found in Reuter and Robinson (1986).

Figure 13.35 *The relationship between plant growth or yield and the concentration of an essential element in the plant tissue. For most nutrients, there is a relatively wide range of values associated with normal, healthy plants (the sufficiency range). Beyond this range, plant growth suffers from either too little or too much of the nutrient. Nutrient concentrations below the critical range (CR) are likely to reduce plant growth. Moderate deficiency that does not produce symptoms is called hidden hunger. When extremely deficient plants are given a small dose of the limiting nutrient, the growth response may be so great that even though somewhat more of the element is taken up, it is diluted in a much greater plant mass.* (Diagram courtesy of Ray R. Weil)

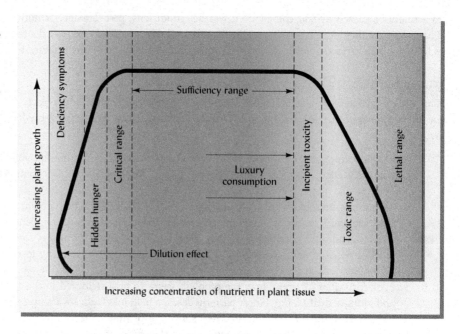

be exhibited ("hidden hunger"). Plants with tissue concentrations below the sufficiency range for a nutrient element are likely to respond to additions of that nutrient if no other factor is more limiting. The sufficiency range and critical range have been well characterized for many plants, especially for agronomic and major horticultural crops. Less is known about forest trees and ornamentals. Sufficiency ranges for 11 essential elements in a variety of plants are listed in Table 13.6.

Tissue Analysis.[11] Tissue analysis can be a powerful tool for identifying plant nutrient problems if several simple precautions are taken. First, it is critical that the correct plant part be sampled. Second, the plant part must be sampled at the specified stage of growth, because the concentrations of most nutrients decrease considerably as the plant matures. Third, it must be recognized that the concentration of one nutrient may be affected by that of another nutrient, and that sometimes the ratio of one nutrient to another (e.g., Mg/K, N/S, or Fe/Mn) may be the most reliable guide to plant nutritional status (Figure 13.36). In fact, several elaborate mathematical systems for assessing the ratios or balance among nutrients have proven useful for certain plant species. Because of the uncertainties and complexities in interpreting tissue concentration data, it is wise to sample plants from the best and worst areas in a field or stand. The difference between samples may provide valuable clues concerning the nature of the nutrient problem.

Cornstalk Nitrate. The *end-of-season cornstalk nitrate test* (CSNT) measures the nitrate content of the lower portion of mature cornstalks at harvest time. Four categories have been defined [expressed as mg of nitrate-N/kg dry cornstalk: deficient (<250), marginal (250–700), optimal (700–2,000), and excessive (>2,000)]. This is a "post-mortem" test that can be used to improve future N management for corn. This test is particularly helpful in identifying excessive levels of N in the plant at harvest time, which, in turn, is an indication of excessive levels in the soil at the end of the season. Such high cornstalk nitrate levels alert farmers to reduce their N applications to avoid excess N leaching in future years.

On-the-Spot Plant Sap Analysis. Healthy, well-fed plants will take up nutrients faster than metabolic processes can assimilate the elements into plant molecules. Therefore, certain nutrients may accumulate in the plant sap as they await assimilation. The sap can be squeezed

---

[11]The best developed of the multinutrient ratio systems is known as the Diagnostic Recommendation Integrated System (DRIS). For details, see Serra et al. (2013).

Table 13.6
A GUIDE TO SUFFICIENCY RANGES FOR TISSUE ANALYSIS OF SELECTED PLANT SPECIES
Values apply only to the indicated plant parts and stage of growth. Normally, 6–20 plants should be sampled. Leaves should be washed briefly in distilled water to remove any soil or dust and then dried before submitting for analysis.

Plant species and part to sample	Content, %						Content, µg/g				
	N	P	K	Ca	Mg	S	Fe	Mn	Zn	B	Cu
Pine trees (*Pinus* spp.) Current-year needles near terminal	1.2–1.4	0.10–0.18	0.3–0.5	0.13–0.16	0.05–0.09	0.08–0.12	20–100	50–600	20–50	3–9	2–6
Oak tree (*Quercus*) Mature leaves	1.9–3.0	0.15–0.30	1.0–1.5	0.3–0.5	0.15–0.30	0.18–0.25	50–150	35–200	15–30	15–40	6–12
Turfgrasses, warm season Clippings	2.7–3.5	0.25–0.55	1.3–3.0	0.50–1.2	0.15–0.60	0.15–0.6	35–500	25–150	15–55	6–60	5–30
Turfgrasses, cool season Clippings	3.0–5.0	0.3–0.4	2–4	0.3–0.8	0.2–0.4	0.25–0.8	40–500	20–100	20–50	5–20	6–30
Corn (*Zea mays*) Ear-leaf at tasseling	2.5–3.5	0.20–0.50	1.5–3.0	0.2–1.0	0.16–0.40	0.16–0.50	25–300	20–200	20–70	6–40	6–40
Soybean (*Glycine max*) Youngest mature leaf at flowering	4.0–5.0	0.31–0.50	2.0–3.0	0.45–2.0	0.25–0.55	0.25–0.55	50–250	30–200	25–50	25–60	8–20
Apple (*Malus* spp.) Leaf at base of nonfruiting shoots	1.8–2.4	0.15–0.30	1.2–2.0	1.0–1.5	0.25–0.50	0.13–0.30	50–250	35–100	20–50	20–50	5–20
Wheat (*Triticum* spp.) Youngest mature leaf at flowering	2.2–3.3	0.24–0.36	2.0–3.0	0.28–0.42	0.19–0.30	0.20–0.30	35–55	30–50	20–35	5–10	6–10
Rice (*Oryza sativa*) Youngest mature leaf at tillering	2.8–3.6	0.14–0.27	1.5–3.0	0.16–0.40	0.12–0.22	0.17–0.25	90–200	40–800	20–160	5–25	6–25
Tomato (*Solanum lycopersicum*) Youngest mature leaf at flowering	3.2–4.8	0.32–0.48	2.5–4.2	1.7–4.0	0.45–0.70	0.60–1.0	120–200	80–180	30–50	35–55	8–12
Alfalfa (*Medicago sativa*) Upper third of plant at first flower	3.0–4.5	0.25–0.50	2.5–3.8	1.0–2.5	0.3–0.8	0.3–0.5	50–250	25–100	25–70	6–20	30–80

Data derived from many sources.

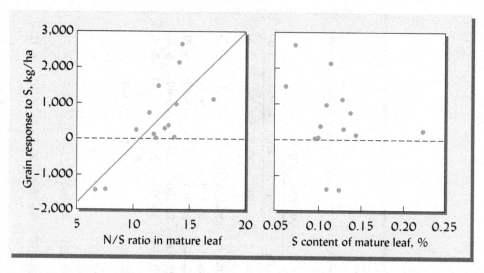

Figure 13.36 *Relationship between the corn grain response to S application and the corn leaf S content (right) or the N/S ratio (left). The graphs illustrate that the ratio of two interacting elements is often a better guide to plant nutrient status than the tissue contents of either element alone. Nitrogen and sulfur are both needed to synthesize plant proteins. As the N/S ratio of the unfertilized corn increased, so did the positive response to application of sulfur (as gypsum). There was no clear relationship between response to S application and the leaf S content by itself, even though most of the corn had S levels below the sufficiency range indicated in Table 13.6. The data are from experiments on 14 small farms in Malawi. Sulfur deficiencies are widespread in African soils.* (Graphed from data in Weil and Mughogho (2000))

from plant stems or leaf petioles and analyzed for its content of the dissolved nutrient elements to assess the sufficiency of the nutrient supplies. While nutrients in the sap can be determined in the lab, recent advances have allowed reasonably accurate measurement of sap nutrients to be made in the field immediately after picking the plant parts (Figure 13.37). Reference values for desirable concentration ranges are available online, mainly for high-value vegetable crops. Nutrient concentrations in sap can vary widely with plant growth stage and with weather conditions. Interpretation of sap data must take into account the conditions of growth during and just before the sampling.

Figure 13.37 *Analysis of plant sap in the field. Plant stems or leaf petioles are chopped into small pieces (left) and then squeezed to express a few drops of sap (middle) which is then placed on an ion selective sensor to determine the concentrations of such ions as NO_3^- or K^+. Reference values for sap NO_3^- and K, as well as tissue analysis norms for those and other elements, are given for vegetable crops in Hochmuth et al. (2012).* (Photo courtesy of Ray R. Weil)

Measurement of Plant Leaf "Greenness." Chlorophyll molecules comprise the pigments that absorb the light that plants use in photosynthesis, but do not absorb green light, hence leaves both reflect and transmit green light, giving them the green color we see (or measure with a meter). Chlorophyll concentration and therefore the greenness of a leaf are often determined by the plant's N supply, nitrogen-limited plants exhibiting somewhat lighter or yellower green colors. While other environmental stresses such as S deficiency or water stress can also make leaves less green, the nitrogen-greenness correlation has proven to be a practical indication of the nitrogen status of plants in the field.

13.11 SOIL ANALYSIS

Since the total amount of a nutrient element in a soil tells us very little about the ability of that soil to supply that element to plants, more meaningful *partial soil analyses* have been developed. *Soil testing* is the routine partial analysis of soils for the purpose of guiding nutrient management.

Soil testing consists of three critical phases: (1) *sampling* the soil, (2) chemically *analyzing* the sample, and (3) *interpreting* the analytical result to make a recommendation on the kind and amounts of nutrients to apply.

Sampling the Soil

Soil sampling is widely acknowledged to be one of the weakest links in the soil testing process. Part of the problem is that about a teaspoonful of soil (Figure 13.38) is eventually used to represent millions of kilograms of soil in the field.

Because nutrient levels vary from spot to spot, it is always advisable to divide a given field or property into as many distinct areas as practical, taking soil samples from each to determine nutrient needs. For example, suppose a 20-ha field has a 2-ha low spot in the middle, and 5 ha at one end that used to be a permanent pasture. These two areas should be sampled,

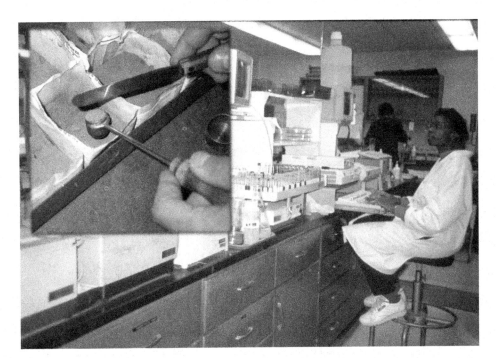

Figure 13.38 *In the testing lab, soil samples are ground and screened to make a homogenous powder, a small portion of which is scooped or weighed (inset) for analysis. This small amount of soil must represent thousands of tons of soil in the field. After a portion of the nutrients has been extracted from the soil sample, the solution containing these nutrients undergoes elemental analysis. Since soil test labs must run hundreds or thousands of samples each day, the analysis is generally automated and the results are recorded and interpreted by computer.* (Photo courtesy of Ray R. Weil)

and later managed, separately from the remainder of the field. Similarly, a homeowner should sample flower beds separately from lawn areas, low spots separately from sloping areas, and so on. On the other hand, known areas of unusual soil that are too small or irregular to be managed separately should be avoided and not included in the sample representing the whole field.

Composite Sample. Usually, a soil probe is used to remove a thin cylindrical core of soil from at least 12–15 randomly scattered places within the land area to be represented (Figure 13.39). The 12–15 subsamples are thoroughly mixed in a plastic bucket, and a **composite sample** of about 250 g of the soil is placed in a labeled container and sent to the lab. If the soil is moist, it should be air-dried without sun or heat prior to packaging for routine soil tests.

Depth to Sample. The standard depth of sampling for a plowed soil is the minimum depth of the plowed layer, about 15–20 cm, but various other depths are also used (see Figure 13.39). Because in many unplowed soils nutrients are stratified in contrasting layers, the depth of sampling can greatly alter the results obtained. Whatever depth of sampling is decided upon, it is critical that each subsample core contributing to the composite sample be of that precise depth.

Time of Year. Seasonal changes are often observed in soil test results for a given area. For example, the K level is usually highest in early spring, after freezing and thawing has released some fixed K ions from clay interlayers, and lowest in late summer, after plants have removed much of the readily available supply. The time of sampling is especially important if year-to-year comparisons are to be made. A good practice is to sample each area every year or two (always at the same time of year), so that the soil test levels can be tracked over the years to determine whether nutrient levels are being maintained, increased, or depleted.

Site-Specific Management Zones. The just-described standard procedure of mixing many small soil samples into a single composite sample to represent the "average" soil in a large field must be recognized for the compromise that it is. Fertilizer recommendations based on the *average* soil condition in the field will likely be either too high or too low for almost any

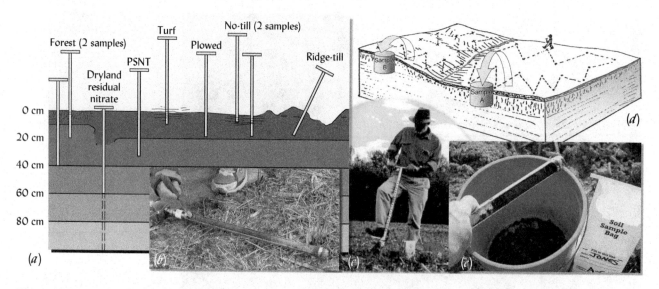

Figure 13.39 Taking a proper soil sample in the field is often the most critical and error-prone step in the soil testing process. (a) The proper depth to sample depends on the purpose of the soil test and the nature of the soil. Some suggested depths for different situations are shown. (b) An open-sided soil probe containing 40 cm long soil core is about to be cut into separate depth increments. It would not be a good idea for a sample to straddle the boundary between the A and B horizons as indicated here by the abrupt color change. (c) A soil probe with a long handle and foot step makes sampling easier, especially for deep samples or in hard soils. (d) Collecting 10–20 subsample cores for each sample sent to the lab is essential, even in a uniform-appearing area. Unusual areas (wet spots, places where manure was piled, eroded spots, etc.) should be avoided. (e) A 15 cm soil core being added to others in a plastic bucket. The cores will be crumbled and mixed in the bucket before a portion of about 250 g is placed in the sample bag for transport to the lab. (Photos and diagrams courtesy of Ray R. Weil)

particular spot in that field. Using *geographic information systems* (GIS) computer technology, fertilizer rates can be much more precisely tailored to account for soil variations within a field. However, the benefits of doing this are not always worth the costs. The costs include field labor and fees for obtaining, processing, and analyzing a large number of composite soil samples collected in a grid pattern within the field. Since each sample is geo-referenced as to its specific location (using space satellite–based *geo-positioning systems*, GPS), computer software can generate maps showing management zones with defined soil properties and fertilizer needs. The maps can then be fed into computer-controlled fertilizer-spreading equipment that can adjust the rates of application as they drive across a field to spread more or less fertilizer as called for by the map of management zones (Figure 13.40).

Chemical Analysis of the Sample

In general, soil tests attempt to extract from the soil amounts of essential elements that are correlated with the nutrients taken up by plants during a growing season. Different extraction solutions are employed by various laboratories. Buffered salt solutions, such as sodium or ammonium acetate, or mixtures of dilute acids and chelating agents are the extracting agents most commonly used. The extractions are accomplished by placing a small measured quantity of soil in a bottle with the extracting agent and shaking the mixture for a certain number of minutes. The amount of the various nutrient elements brought into solution is then determined. The whole process is usually automated so that a modern laboratory can handle hundreds or thousands of samples each day (see Figure 13.38).

The most common and reliable tests are those for soil pH, potassium, phosphorus, and magnesium. Micronutrients are sometimes extracted using synthetic chelating agents that mimic the metal-binding action of root and microbial compounds. Chelating agents are especially important for testing calcareous soils in the more arid regions.

Testing for the nitrogen and sulfur supplying capacity of soils is particularly troublesome. While the nitrate and sulfate present in the soil at the time of sampling can be measured, this "snapshot" is unlikely to predict the availability of nitrogen and sulfur during the

Figure 13.40 *Use of GPS technology in precision or site-specific farming. (Left) View from the cab of a fertilizer spreading machine that is GPS-guided and capable of changing the rates of application as required by the computer map as it drives across a field. This type of farming is also known as Prescription Farming. (Upper right) A map of the same field showing the lime required to raise the soil pH to 6.5 as determined from samples collected in 1-ha cells. (Lower right) This map of corn yields throughout a 22-ha field was generated by a harvesting machine equipped with a yield monitor and a GPS receiver. Each dot on the map was generated as the monitor recorded the yield of corn every few meters while the machine worked up and down the field. Each yield measurement was associated with a map position determined by the GPS. An agronomist can then check the red areas where yields were low to ascertain the cause and design measures to alleviate the problem in those spots.* (Photo of in-cab computer enhanced to show detail, courtesy of Ray R. Weil, maps courtesy of Hoober, Inc)

growing season because of the many weather-dependent biological processes involved in the mineralization of these elements from soil organic matter. Some progress is being made in developing laboratory tests that combine a measure of microbial activity (such as CO_2 respiration rate) with measures of organic N content to predict potential N mineralization.

Soil tests designed for use on soils or soil-based potting media generally do not give meaningful results when used on artificial peat-based soilless potting media. Special extraction procedures must be used for the latter, and the results must then be correlated to nutrient uptake and growth of plants grown in similar media. These examples should further emphasize the importance of providing the soil testing lab with complete information concerning the nature of your soil, its management history, and your plans for its future use.

Interpreting the Results to Make a Recommendation

Interpretation is, perhaps, the most controversial aspect of soil testing. The soil test values themselves are merely *indices* of nutrient-supplying power. They do not indicate the actual amount of nutrient that will be supplied. For this reason, it is best to think of soil test reports as more indicative than quantitative.

Many years of field experimentation at many sites are needed to determine which soil test levels indicate low, medium, or high capacities to supply the nutrient tested. Such categories are used to predict the likelihood of obtaining a profitable response from the application of a particular nutrient (Figure 13.41). Since the actual units of measurement (ppm, mg/L, kg/ha, or lb/acre) have little meaning with regard to evaluating the soil nutrient supplying capacity,

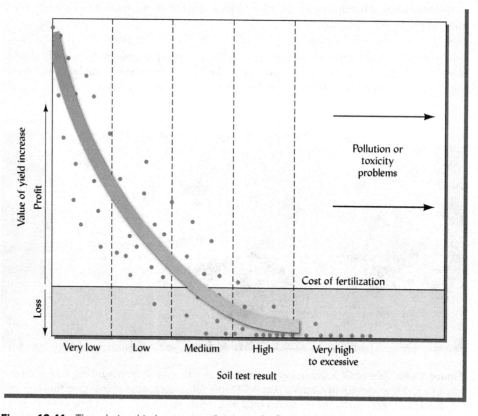

Figure 13.41 *The relationship between soil test results for a nutrient and the extra yield obtained by fertilizing with that nutrient. Each data point represents the difference in plant yield between fertilized and the unfertilized plots at one location in one year. Because many factors affect yield and because soil tests can only approximately predict nutrient availability, the relationship is not precise, but the data points are scattered about the trend line. If the point falls above the fertilizer cost line, the extra yield was worth more than the cost of fertilizing and a profit would be made. For a soil testing in the very low and low categories, a profitable response to fertilizer is very likely. For a soil testing medium, a profitable response is a 50:50 proposition. In the high category, a profitable response is unlikely.* (Diagram courtesy of Ray R. Weil)

soil test results are increasingly reported as index values on a relative scale (often with 75–100 considered optimal). The report shown in Figure 13.42 uses both interpretative categories and a relative index scale.

Recommendations for nutrient applications take into consideration practical knowledge of the plants to be grown, the characteristics of the soil under study, and other environmental conditions. Management history and field observations can help relate soil test data to fertilizer needs.

The interpretation of soil test data is best accomplished by experienced and technically trained personnel who fully understand the scientific principles underlying the common field procedures. In most commercial or university soil test laboratories, the factors to be considered in making fertilizer recommendations are programmed into a computer, and the interpretation is printed out for use by the land manager (Figure 13.42).

Merits of Soil Testing

It must not be inferred from the preceding discussion that the limitations of soil testing outweigh its advantages. When the precautions already described are observed, soil testing is an invaluable tool in making effective use of nutrients. Soil tests are most useful when they

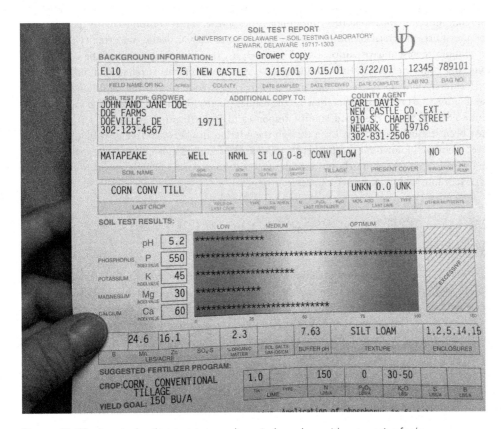

Figure 13.42 *A typical soil test giving results as index values with categories for low, medium, and optimum or high. The report includes extractable levels of certain nutrients and soil properties that were measured in the lab, and recommendations for the amounts and types of soil amendments to apply for a specific crop. For this particular field, the following interpretations can be made. The pH is low and should be raised to near 5.8–6.0. At least some of the liming material should be dolomitic limestone to add Mg, which is in low to medium supply. Moderate amounts of K may or may not give an economic response. The P level is highly excessive, so P additions should be avoided, and steps taken to limit transport of P to waterways. Sulfur and B are not part of the standard analysis package because the amounts extracted by the lab do not correlate well with plant uptake from soils. As Delaware is in a humid region, N application is recommended based on yield goal and soil properties that affect mineralization, but not on an analytical test for available N. In a dry region, test labs would also measure nitrate and electrical conductivity in the soil to support recommendations concerning N fertilization and salt management, respectively.* (Soil test report courtesy of K. L. Gartley, University of Delaware)

are correlated with the results of field fertilizer experiments. Adding amendments to achieve some "ideal" balance of nutrients in the soil is often wasteful of money and resources. Rather, soil tests used correctly in conjunction with calibration experiments can indicate what level of amendment needs to be added, if any, to allow the soil to supply sufficient nutrients for optimal plant growth.

Dependability in predicting how plants will respond to soil amendments varies with the particular soil test, the tests for some parameters being much more reliable than others because of the consistency and breadth of field correlation data. In general, there are quite reliable tests for pH (need for liming or acidification), P, K, Mg, B, and Zn. Some soil test labs report additional soil properties such as other micronutrients (Cu, Mn, Fe, Mo, etc.), "humic substances" extractions (fulvic, humic acids, etc.), nonacid cation ("base") saturation, and even levels of soil respiration or various microbial populations (fungi, bacteria, nonparasitic nematodes, etc.). However, the basis for making valid and practical recommendations for managing soil fertility from such soil test parameters is currently very scant, at best.

Testing in Forestry. Generally, soil testing has been most relied on in agricultural systems, while foliar analysis has proved more widely useful in forestry. The limited use of soil testing in forestry is partly due to the fact that trees, with their extensive perennial root systems, integrate nutrient bioavailability throughout the profile. This, combined with the typically complex nutrient stratification in forested soils, creates a great deal of uncertainty about how to obtain a representative sample of soil for analysis. In addition, because of the comparably long time frame in forestry, limited information is available on the correlation of soil test levels with timber yields in the sense that such information is widely available for agronomic crops (see Figure 13.41). An exception may be the conifer plantations on Ultisols of the southeastern United States, where the wood volume responses to P fertilization at various soil test levels have been studied extensively. Even though the relationship between tree growth and soil test level is not well known for most other forest systems, standard agronomic soil testing can still be useful in distinguishing those soils whose ability to supply P or K is adequate from those with very low supplying power for these nutrients.

13.12 SITE-INDEX APPROACH TO PHOSPHORUS MANAGEMENT

As concerns have increased about nonpoint source water pollution, research has identified phosphorus movement from land to water as a major cause of aquatic ecosystem degradation, especially for lakes. Phosphorus movement from land to water is determined by *phosphorus-transport*, *phosphorus-source*, and *phosphorus-management* parameters. Transport parameters include proximity to sensitive water bodies and how rain, snowmelt, and irrigation water cause phosphorus to move across the landscape in runoff and sediment. Source parameters include the amount and forms of phosphorus on and in the soil. Management parameters include the method of application, timing, and placement of phosphorus, tillage and establishment of buffer zones.

Overenrichment of Soils

Research in the early and mid-twentieth century suggested that farmers needed to add several times the amount of phosphorus that plants were likely to use, because soils were found to fix most of the added phosphorus in unavailable forms. Applications of 40–100 kg P/ha became standard practice, even though crop harvests removed only about one-third of that amount. The phosphorus not removed in harvest accumulated in the soil, and eventually satisfied a large fraction of the soil's phosphorus fixation capacity. It was not until the 1980s and 1990s that many farmers and researchers realized that soils with a long history of phosphorus fertilization were no longer in need of phosphorus application levels above what the crops would remove.

The second trend that contributed to the phosphorus pollution problem was the concentration of livestock and the subsequent heavy applications of animal manure to soils nearby (see Section 13.2). When manure is used as the main source of N, the amount of P applied is far more than plants can use.

Because of these trends, some farm fields have become so high in phosphorus that water interacting with the soil carries away enough phosphorus to impair the ecology of receiving waters. It is imperative that sites with such a high potential to cause phosphorus pollution be identified and appropriately managed to reduce their environmental impact. It may take decades to lower soil test P levels on these farms back to the optimum range (Figure 13.43) and phosphorus fertilizers are not likely to be needed on them for decades more.

Transport of Phosphorus from Land to Water

For a soil to cause significant phosphorus pollution, the supply of phosphorus in the soil must be relatively large and the characteristics of the site and soil must allow significant transport of P from the field to a lake or stream. A review of the phosphorus cycle (Section 12.3) reminds us that phosphorus can move from land to water by three principal pathways:

1. Attached to eroded soil particles (the main P-loss pathway from tilled soils).
2. Dissolved in water running off the surface of the land (a major pathway for pastures, woodland, and no-till cropland).
3. Dissolved or attached to suspended colloidal particles in water percolating down the profile and through the drainage pipes or groundwater aquifers that feed streams or lakes (a significant pathway in very sandy or artificially drained soils with shallow water tables).

Phosphorus Soil Test Level as Indicator of Potential Losses

As described in Section 13.11, soil testing was designed to determine how well a soil can supply nutrients to plant roots. However, routinely used phosphorus soil tests have been shown to also provide a useful (though not perfect) indication of how readily P will desorb from a soil and dissolve in runoff water. The relationship between phosphorus soil test level in the upper centimeter or two of soil and P in runoff water is not linear (Figure 13.44). Rather, the relationship exhibits a threshold effect such that little P is lost to runoff if soils are near or below the optimal range for plant growth, but the amount released from the soil to runoff water increases exponentially as P soil test levels rise much higher than this. Therefore, as suggested by Figure 13.44, it appears that soils can be maintained at phosphorus levels sufficient for optimal plant growth without undue losses of P in runoff water. As is also the case with N, significant environmental damages are associated with excess—the application of more nutrient than is really needed.

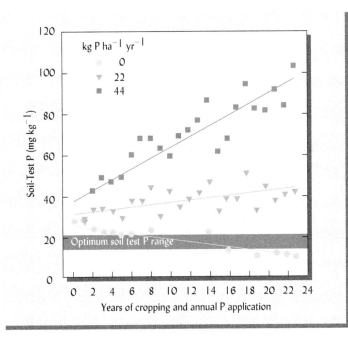

Figure 13.43 *Changes in soil test P as a result of annual applications of 0, 22, or 44 kg P ha^{-1} to a corn-soybean rotation on an Iowa Mollisol. Data from this and other sites in the study suggest that application of 12–20 kg P ha^{-1} yr^{-1} is sufficient to balance P removed in crop harvests and maintain soil test P in the optimal range for many cropping systems. The gentle slope of the zero P added line suggests that many decades of crop harvest removals would be needed to bring the most elevated P levels back to the optimum range. The soil tests were conducted using the Bray-1 method for which the optimal levels are 16–22 mg P extracted per kg soil.* [Data from Dodd and Mallarino (2005) with permission of the Soil Science Society of America]

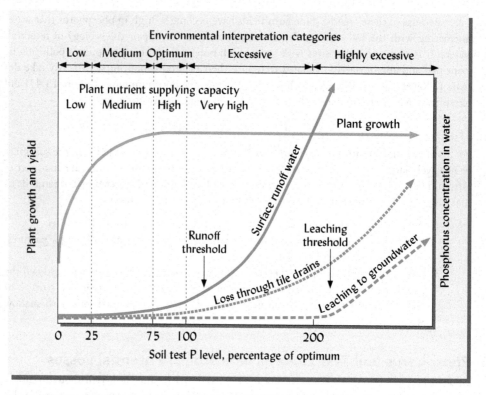

Figure 13.44 *A generalized relationship between soil test P levels and environmental losses of P dissolved in surface runoff and subsurface drainage waters. The relationship between traditional plant nutrient supply interpretation categories and environmental interpretation categories is also indicated. The diagram suggests that, fortunately, soil P levels can be achieved that are both conducive to optimum plant growth and protective of the environment. If P losses by soil erosion are controlled (although not shown, this is a big if), significant quantities of dissolved P would be lost only when soils contain P levels in excess of those needed for optimum plant growth. Losses of P by leaching in drainage water are significant only at very high soil P levels, as this pathway rapidly increases only after the P-sorption capacity of the soil is substantially saturated. As indicated by the dotted curve, short-circuiting this pathway by installing tile drainage is likely to lower the threshold for subsurface leaching P loss. The vertical axes are not to scale.* [Figure courtesy of Ray R. Weil, based on concepts discussed in Sharpley (1997), Higgs et al. (2000), and Reid et al. (2012)]

Therefore, an environmentally sound phosphorus fertility level, and one that also conserves the world's finite stocks of this essential element, should also be a profitable level of fertilization.

Phosphorus in Drainage Water. Drainage water from most soils is usually very low in phosphorus because as phosphorus-laden water percolates from the surface down through the subsurface soil horizons, dissolved $H_2PO_4^-$ is strongly sorbed by inner-sphere complexation onto Fe-, Al-, Mn-, and Ca-containing mineral surfaces (Sections 8.7 and 12.3). However, under certain circumstances, this mechanism is not so effective in removing phosphorus from drainage water. These circumstances include the following:

1. The percolating water is flowing through macropores and has little contact with soil surfaces (Section 6.6). This may occur under long-term no-till management or in pastures, especially if installation of tile drainage (see Section 6.7) has shortened the flow pathways and increased the opportunities for P loss by preferential flow output.
2. The dissolved phosphorus is not in mineral ($H_2PO_4^-$ or HPO_4^{2-}) forms, but instead is part of soluble organic molecules, which are not strongly sorbed by mineral surfaces. This situation is common under forests or heavily manured soils.
3. The phosphorus is sorbed onto the surface of dispersed colloidal particles that are so small that they remain suspended in the percolating water.

4. The phosphorus-fixation capacity of the soil is so small (as in sands, Histosols, and some waterlogged soils with reduced iron) or already so saturated with phosphorus (as in soils overloaded with phosphorus after many years of excessive manure and fertilizer applications) that little more phosphorus can be sorbed.

Phosphorus Site Index

In order to focus control measures on the sites within a watershed that account for most of the phosphorus loss, each field can be evaluated using an index of phosphorus pollution risk. Commonly referred to as a *phosphorus site index*, such a tool integrates phosphorus source, transport, and management characteristics of a site into a single number. One can then develop plans for corrective action to address problems associated with these identified parameters. An important part of such plans is the restriction of further phosphorus application, whether in manure or fertilizer.

13.13 SOME ADVANCES AND CHALLENGES IN NITROGEN MANAGEMENT[12]

The initial focus in most soil fertility plans is on N, the nutrient to which nonleguminous plants most commonly and most dramatically respond. Nitrogen is also at the forefront because of its implications for environmental quality. Applications of other nutrients are usually made to balance and supplement the N supply, whether that comes from soil organic matter, crop residues, organic wastes, or organic and inorganic fertilizers. Nonetheless, we must be mindful that the effectiveness of N applications can be limited by deficiencies of S, K, P, or other nutrients. If other nutrients are well-supplied, and soil organic matter is neither increasing nor decreasing, then the amount of N that needs to be applied should approximate the amount likely to be removed in the harvested product.

Nitrogen Credits and Predictions. This estimated fertilizer rate should be adjusted by the amount of any additional gains or losses not taken into account in the standard response curve. For example, N contributions from previous or current manure applications (see Box 13.5), legume cover crop, previous legume in the rotation, or nitrate in irrigation water should be subtracted from the amount of fertilizer recommended. In arid and semiarid regions where overwinter leaching is minimal, a N credit is often taken for the nitrate nitrogen found in the upper 60 to 120 cm of the profile in spring.

Plant Response. We can analyze the soil to estimate its nutrient supplying power and we can calculate the amount of each nutrient removed from the field by harvests, but the response of growing plant is the final arbitrator of soil fertility management. The soil–plant–weather systems that govern plant growth and crop yield are so highly complex and variable that we are challenged to be able to predict exactly what nutrient management will be optimal. Even with all the advances in sensing, analysis, and computer modeling of these systems, we still must rely on well-designed field experiments to define the relationships between the addition of nutrients and the resulting plant growth or crop yield responses. The optimal level of nitrogen fertilization is usually determined by the shape of such response curves (Figure 13.45).

Profitability. A second aspect relates to economics. Farmers do not apply nutrients just to grow big crops or to increase the nutrient content of their soils. They do so to make a living. The most profitable rate is determined by the ratio of the value of the extra yield expected to the cost of the nutrient applied. The law of diminishing returns applies. Therefore, the most profitable nutrient application rate will be somewhat less than the rate that would produce the highest yield (Figure 13.45).

---

[12]For a discussion of the various approaches to managing agricultural nitrogen application rates, see Morris et al. (2017).

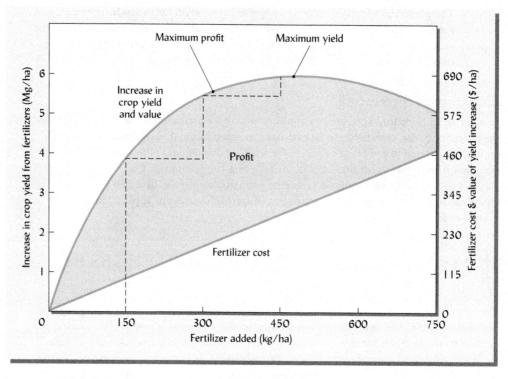

Figure 13.45 Relationships among the rate of fertilizer addition, crop yield increase, fertilizer costs, and profit from adding the fertilizer. Note that the yield increase (and profit) from the first 150 kg of fertilizer is much greater than from the second and third 150 kg. Also note that the maximum profit is obtained at a lower fertilizer rate than that needed to give a maximum yield. The calculations assume that the yield response to added fertilizer takes the form of a quadratic curve, an assumption which often leads to overestimating fertilizer needed to optimize profit.

Response Curves. Traditionally, economic analysis of optimum fertilizer rates has assumed that the plant response to fertilizer inputs was represented by a smooth curve following a quadratic function ($y = a + bx + cx^2$). In fact, actual data obtained can be just as well described by a number of other mathematical functions (Figure 13.46). This seemingly

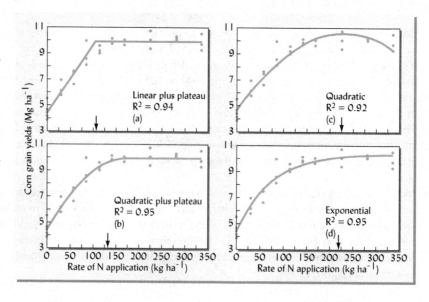

Figure 13.46 The mathematical function chosen to represent fertilizer-response data can affect the amount of fertilizer recommended. The data in all four graphs are exactly the same and represent the response of corn yields in Iowa to increasing levels of N fertilization. The vertical arrows indicate the recommended, most profitable rate of N to apply according to each mathematical model. Note that all models fit the data equally well (as indicated by the very similar R^2 values), but that the linear-plus-plateau model predicts an optimum N rate of 104 kg/ha, while the commonly used quadratic model suggests that 222 kg/ha is the optimum. The extra 118 kg/ha of N would probably have no effect on crop yield, but would cost the farmer and greatly increase the risk of environmental damage. [Redrawn from Cerrato and Blackmer (1990)]

esoteric observation can have a great effect on the amount of N recommended and, in turn, on the likelihood of environmental harm from excessive N application. Among the various models studied, the linear-plateau approach (Figure 13.46a) usually leads to the lowest fertilizer recommendation, least wasted fertilizer, least environmental damage, and often the greatest profit as well.

Both plant growth potential and soil N supply are highly influenced by temperature and rainfall, so it is difficult to predict the amount of N fertilizer needed in a given year. Differing weather patterns (a cold or warm spring, a dry or wet summer) cause the response to N and other nutrients to vary greatly from one year to the next, even in the exact same soil and location (Figure 13.47). The uncertainties related to N supply are still sufficiently high, and the cost of N fertilizer may be sufficiently low, that farmers tend to err on the side of oversupply, just for "insurance." In most years, this philosophy can be disadvantageous to both the farmer and the environment.

There are several relatively new tools that can help predict N fertilizer needs, even in humid regions with high leaching potential over winter. A pre-side-dress nitrate test (PSNT) predicts soil N availability during the season based on a single actual data point for soil nitrate in the upper 30 cm a week or so before the crop enters its exponential phase for growth and N uptake (Figure 13.48). Complex computer models are being developed to make more accurate predictions of N supply and plant N demand by integrating large amounts of weather data with information about plant and soil processes (Figure 13.48). Another advantage of the computer models is that they can be updated throughout the season. Even with these tools, it is wise to delay the main application of N so that as much of the season's weather is known before the N application rate is determined.

Trends in fertilizer use and corn yields from around the world suggest that when farmers increase N fertilizer use *without* simultaneously adopting improved crop varieties, agronomy, and soil management, the efficiency of N use declines and N surpluses and environmental impacts are worsened. For example, in the United States during the 1960–1970s and in China since 2000, fertilizer was not used very efficiently, much to the detriment of both water quality and farmer profitably. In recent years, farmers in the United States and several other industrialized countries have increased N use efficiency by adopting improved fertilizer management, improved means of accounting for weather, genetically improved seeds, and more careful crop management to produce greater yields with the same or less fertilizer.

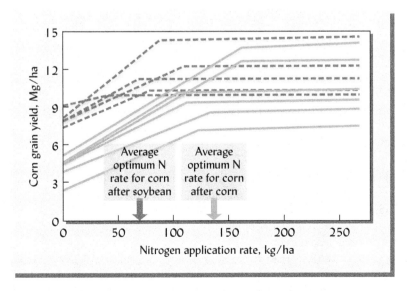

Figure 13.47 *Corn yield response to applied N in a single field varies greatly with weather and previous crop grown. Each curve is of the linear-plateau type illustrated in Figure 13.46a and represents a different year with different weather. The green dashed curves represent corn after soybean and the yellow solid curves represent corn after corn. In the same field, the optimal rate of N ranged from as low as 60 to as high as 160 kg N/ha, with rotation corn always requiring less N than continuous corn.*
[Graphed from selected data in Sawyer et al. (2006)]

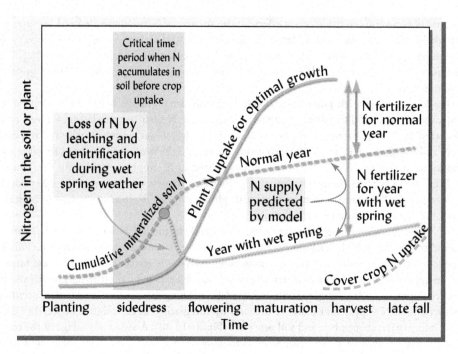

Figure 13.48 *Weather patterns during the growing season influence the amount of N fertilizer needed in a given year. The single large dot on the soil nitrate curve represents the pre-side-dress nitrate test (PSNT). Advanced, weather-driven computer models can help predict the most economic N rate to apply for a particular soil and weather combination.* [Diagram partially based on concepts in Moebius-Clune et al. (2014)]

Although some advisors continue to recommend as much as 27 kg N per Mg (1.5 lb of N/bushel) of expected corn grain, most field data suggest that 14–16 kg N/Mg grain (0.8 or 0.9 lb N/bushel) is all that is really required. Where soil health, organic matter, and nutrient cycling processes have been improved through careful management, even lower rates of fertilizer are needed. Data indicate that fertilizers are being used in most advanced and emerging economies at greater than optimal rates. In the United States, N and P balances are both positive, suggesting overapplication to soils already quite high in these nutrients. The water quality problems resulting from these overapplications are a continuing challenge. Rising K concentrations in rivers draining the main cropland areas of the United States suggest that even potassium, with a small negative balance in the United States, may be overapplied if the K-supplying capacity of the huge reserve in soil minerals is taken into consideration.

Anyone involved with the actual production of plants can testify to the enormous improvements that accrue from judicious use of organic and inorganic nutrient supplements. The preceding discussion suggests that while such supplements are often beneficial, optimizing their management presents ongoing economic, resource, and environmental challenges.

13.14 CONCLUSION

The continuous availability of plant nutrients is critical for the sustainability of most ecosystems. The challenge of nutrient management is threefold: (1) to provide adequate nutrients for plants in the system; (2) to simultaneously ensure that inputs are in balance with plant utilization of nutrients, thereby conserving nutrient resources; and (3) to prevent contamination of the environment with unutilized nutrients.

The recycling of plant nutrients must receive primary attention in any ecologically sound management system. This can be accomplished in part by returning plant residues to the soil. These residues can be supplemented by judicious application of the organic wastes that are produced in abundance by municipal, industrial, and agricultural operations worldwide. The use of cover crops grown specifically to be returned to the soil is an additional organic means of recycling nutrients.

For sites from which crops or forest products are removed, nutrient losses commonly exceed the inputs from recycling. Inorganic fertilizers will continue to supplement natural and managed recycling to replace these losses and to increase the level of soil fertility so as to enable humankind to not only survive, but to flourish on this planet as well. In extensive areas of the world, fertilizer use will have to be increased above current levels to avoid soil and ecosystem degradation, to remediate degraded soils, and to enable profitable production of food and fiber.

The use of fertilizers, both inorganic and organic, should not be done in a simply habitual manner or for so-called insurance purposes. Rather, soil testing and other diagnostic tools should be used to determine the true need for added nutrients. In managing N and phosphorus, increasing attention will have to be paid to the potential for transport of these nutrients from soils where they are applied to waterways where they can become pollutants. If soils are low in available nutrients, fertilizers often return several dollars' worth of improved yield for every dollar invested. However, where the nutrient-supplying power of the soil is already sufficient, adding fertilizers is likely to be damaging both to the bottom line and to the environment.

STUDY QUESTIONS

1. The groundwater under a heavily manured field is high in nitrates, but by the time it reaches a stream bordering the field, the nitrate concentration has declined to acceptable levels. What are likely explanations for the reduction in nitrate?
2. You want to plant a cover crop in fall to minimize nitrate leaching after the harvest of your corn crop. What characteristics would you look for in choosing a cover crop to ameliorate this situation?
3. What management practices on forested sites can lead to significant N losses, and how can the losses be prevented?
4. What effect do forest fires have on nutrient availabilities and losses to streams?
5. Compare the resource-conservation and environmental-quality issues related to each of the three so-called fertilizer elements, N, P, and K.
6. A park manager wants to fertilize an area of turfgrass with N and phosphorus at the rates of 60 kg/ha of N and 20 kg/ha of P. He has stocks of two types of fertilizers: urea (45-0-0) and diammonium phosphate (18-46-0). How much of each should he blend together to fertilize a 10-ha area of turfgrass?
7. How much phosphorus (P) is there in a 25-kg bag of fertilizer labeled "20-20-10"?
8. Compare the relative advantages and disadvantages of organic and inorganic nutrient sources.
9. A certified organic grower plans to grow a crop that requires the application of 120 kg of plant-available N/ha and 20 kg/ha of P. She has a source of compost that contains 1.5% total N (with 10% of this available in the first year) and 1.1% total P (with 80% of this available in the first year). (a) Assuming her soil has a low P soil test level, how much compost should she apply to provide the needed N and P? (b) If her soil is already optimal in P, how can she provide the needed amounts of both N and P *without* causing further P buildup?
10. Discuss the concept of the *limiting factor* and indicate its importance in enhancing or constraining plant growth.
11. Why are nutrient-cycling problems in agricultural systems more prominent than those in forested areas?
12. Discuss how GIS-based, site-specific nutrient-application technology might improve profitability and reduce environmental degradation.
13. Discuss the value and limitations of soil tests as indicators of plant nutrient needs and water pollution risks.
14. When might the use of plant tissue analyses have advantages over soil testing for correcting nutrient imbalances?
15. Compare the tools used to determine whether and how much fertilizer should be applied in the case of N and phosphorus. Why is the N requirement more difficult to predict?

REFERENCES

Addiscott, T. M. 2006. "Is it nitrate that threatens life or the scare about nitrate?" *Journal of the Science of Food and Agriculture* 86:2005–2009.

AdeOluwa, O. O., and O. Cofie. 2012. "Urine as an alternative fertilizer in agriculture: Effects in amaranths *(Amaranthus caudatus)* production." *Renewable Agriculture and Food Systems* 27:287–294.

Billen, G., J. Garnier, and L. Lassaletta. 2013. "The nitrogen cascade from agricultural soils to the sea: Modelling nitrogen transfers at regional watershed and global scales." *Philosophical Transactions of the Royal Society B: Biological Sciences* 368:20130123.

Bouwman, L., K. K. Goldewijk, K. W. Van Der Hoek, A. H. Beusen, D. P. Van Vuuren, J. Willems, M. C. Rufino, and E. Stehfest. 2013. "Exploring global changes in nitrogen and phosphorus cycles in agriculture induced by livestock production over the 1900–2050 period." *Proceedings of the National Academy of Sciences* 110:20882–20887.

Cerrato, M. E., and A. M. Blackmer. 1990. "Comparison of models from describing corn yield response to nitrogen fertilizer." *Agronomy Journal* 98:138–143.

Dean, J. E., and R. R. Weil. 2009. "Brassica cover crops for nitrogen retention in the mid-Atlantic coastal plain." *Journal of Environmental Quality* 38:520–528.

DeLonge, M., R. Ryals, and W. Silver. 2013. "A lifecycle model to evaluate carbon sequestration potential and greenhouse gas dynamics of managed grasslands." *Ecosystems* 16:962–979.

Dodd, J. R., and A. P. Mallarino. 2005. "Soil-test phosphorus and crop grain yield responses to long-term phosphorus fertilization for corn-soybean rotations." *Soil Science Society of America Journal* 69:1118–1128.

FAO. 1977. *China: Recycling of Organic Wastes in Agriculture*. FAO Soils Bulletin 40. U.N. Food and Agriculture Organization, Rome.

FAO. 2014. FAOSTAT [Online]. Food and Agriculture Organization of the United Nations. Available at http://faostat.fao.org/ (verified 20 May 2014).

Fisher, J. B., G. Badgley, and E. Blyth. 2012. "Global nutrient limitation in terrestrial vegetation." *Global Biogeochemical Cycles* 26:GB3007.

Gardner, G. 1997. *Recycling Organic Waste: From Urban Pollutant to Farm Resource*. Worldwatch Paper 135. Worldwatch Institute, Washington, D.C.

Harmel, R. D., R. A. Pampell, A. B. Leytem, D. R. Smith, and R. L. Haney. 2018. "Assessing edge-of-field nutrient runoff from agricultural lands in the United States: How clean is clean enough?" *Journal of Soil and Water Conservation* 73:9–23.

Havlin, J. L., T. S. L., W. L. Nelson, and J. D. Beaton. 2014. *Soil Fertility and Fertilizers: An Introduction to Nutrient Management*. 8th ed. Pearson, Upper Saddle River, NJ, p. 528.

He, X. -T., T. Logan, and S. Traina. 1995. "Physical and chemical characteristics of selected U.S. municipal solid waste composts." *Journal of Environmental Quality* 24:543–552.

Heckman, J. R., and D. Kluchinski. 1996. "Chemical composition of municipal leaf waste and hand-collected urban leaf litter." *Journal of Environmental Quality* 25:355–362.

Heinonen-Tanski, H., A. Sjöblom, H. Fabritius, and P. Karinen. 2007. "Pure human urine is a good fertiliser for cucumbers." *Bioresource Technology* 98:214–217.

Higgs, B., A. E. Johnston, J. L. Salter, and C. J. Dawson. 2000. "Some aspects of achieving sustainable phosphorus use in agriculture." *Journal of Environmental Quality* 29:80–87.

Hilimire, K. 2011. "Integrated crop/livestock agriculture in the United States: A review." *Journal of Sustainable Agriculture* 35:376–393.

Hochmuth, G., D. Maynard, C. Vavrina, E. Hanlon, and E. Simonne. 2012. "Plant tissue analysis and interpretation for vegetable crops in Florida." *Horticulture Science Extension HS964*. University of Florida. http://edis.ifas.ufl.edu/ep081.

Jones Jr., J. B. 2012. *Plant Nutrition and Soil Fertility Manual*. CRC Press, Boca Raton, FL, p. 285.

Karak, T., and P. Bhattacharyya. 2011. "Human urine as a source of alternative natural fertilizer in agriculture: A flight of fancy or an achievable reality." *Resources, Conservation and Recycling* 55:400–408.

Kröger, R., E. Dunne, J. Novak, K. King, E. McLellan, D. Smith, J. Strock, K. Boomer, M. Tomer, and G. Noe. 2013. "Downstream approaches to phosphorus management in agricultural landscapes: Regional applicability and use." *Science of the Total Environment* 442:263–274.

Krogmann, U., B. F. Rogers, L. S. Boyles, W. J. Bamka, and J. R. Heckman. 2003. "Guidelines for land application of non-traditional organic wastes (food processing by-products and municipal yard wastes) on farmlands in New Jersey." *Bulletin e281*. Rutgers Cooperative Extension, New Jersey Agricultural Experiment Station, Rutgers, The State University of New Jersey. http://www.rce.rutgers.edu/pubs/pdfs/e281.pdf (posted June 2003; verified 28 November 2004).

L'hirondel, J., and J. -L. L'hirondel. 2002. *Nitrate and Man: Toxic, Harmless or Beneficial?* CABI, Wallingford, UK. p. 168.

Magdoff, F., L. Lanyon, and B. Liebhardt. 1997. "Nutrient cycling, transformations, and flows: Implications for a more sustainable agriculture." *Advances in Agronomy* 60:2–73.

McNaughton, S. J., F. F. Banyikwa, and M. M. McNaughton. 1997. "Promotion of the cycling of diet-enhancing nutrients by African grazers." *Science* 278:1798–1800.

Moebius-Clune, B. N., M. Carlson, H. M. van Es, J. J. Melkonian, A. T. DeGaetano, and L. Joseph. 2014. "Adapt-n training manual: A tool for precision n management in corn." Extension Series No. E14-1, Edition 1.0. Cornell University, Ithaca, NY. http://adapt-n.cals.cornell.edu/manual/pdfs/adapt-n-manual.pdf.

Morris, T. F., T. S. Murrell, D. B. Beegle, J. J. Camberato, R. B. Ferguson, J. Grove, Q. Ketterings, P. M. Kyveryga, C. A. M. Laboski, J. M. McGrath, J. J. Meisinger, J. Melkonian, B. N. Moebius-Clune, E. D. Nafziger, D. Osmond, J. E. Sawyer, P. C. Scharf, W. Smith, J. T. Spargo, H. M. van Es, and H. Yang. 2017. "Strengths and limitations of nitrogen rate recommendations for corn and opportunities for improvement." *Agronomy Journal*. 110:1–37.

Owens, L. B., M. J. Shipitalo, and J. V. Bonta. 2008. "Water quality response times to pasture management changes in small and large watersheds." *Journal of Soil and Water Conservation* 63:292–299.

Peters, J. 2012. "Manure analysis update: 1998–2010." presentation. Department of Soil Science, University of Wisconsin, Madison, WI. http://www.soils.wisc.edu/extension/area/2010/Manure_summary_Peters.pdf.

Poffenbarger, H. J., S. B. Mirsky, R. R. Weil, M. Kramer, J. T. Spargo, and M. A. Cavigelli. 2015. "Legume proportion, poultry litter, and tillage effects on cover crop decomposition." *Agronomy Journal* 107:2083–2096.

Powers, J. F., and W. P. Dick (eds.). 2000. *Land Application of Agricultural, Industrial, and Municipal By-Products*. Soil Science Society of America Book Series No. 6. Soil Science Society of America, Madison, WI.

Pradhan, S. K., A. -M. Nerg, A. Sjöblom, J. K. Holopainen, and H. Heinonen-Tanski. 2007. "Use of human urine fertilizer in cultivation of cabbage (*Brassica oleracea*)—impacts on chemical, microbial, and flavor quality." *Journal of Agricultural and Food Chemistry* 55:8657–8663.

Preusch, P. L., P. R. Adler, L. J. Sikora, and T. J. Tworkoski. 2002. "Nitrogen and phosphorus availability in composted and uncomposted poultry litter." *Journal of Environmental Quality* 31:2051–2057.

Reddy, G. B., and K. R. Reddy. 1993. "Fate of nitrogen-15 enriched ammonium nitrate applied to corn." *Soil Science Society of America Journal* 57:111–115.

Reid, D. K., B. Ball, and T. Q. Zhang. 2012. "Accounting for the risks of phosphorus losses through tile drains in a phosphorus index." *Journal of Environmental Quality* 41:1720–1729.

Reuter, D. J., and J. B. Robinson. 1986. *Plant Analysis: An Interpretation Manual*. Inkata Press, Melbourne, Australia.

Ribaudo, M., J. Savage, and M. Aillery. 2014. "An economic assessment of policy options to reduce agricultural pollutants in the Chesapeake Bay," ERR - 166. U.S. Department of Agriculture, Economic Research Service, Washington, D.C. www.ers.usda.gov/publications/err-economic-research-report/err166.aspx.

Richert, A., R. Gensch, Håkan Jönsson, T. -A. Stenström, and L. Dagerskog. 2010. "Practical guidance on the use of urine in crop production. EcoSanRes Series 2010-1." EcoSanRes Programme, Stockholm Environment Institute, Stockholm.http://sei-international.org/mediamanager/documents/Publications/Air-land-water-resources/ecosan-urine-in-crops-100824%20web.pdf

Ryals, R., and W. L. Silver. 2013. "Effects of organic matter amendments on net primary productivity and greenhouse gas emissions in annual grasslands." *Ecological Applications* 23:46–59.

Santamaria, P. 2006. "Nitrate in vegetables: Toxicity, content, intake and ec regulation." *Journal of the Science of Food and Agriculture* 86:10–17.

Sawyer, J., E. Nafziger, G. Randall, L. Bundy, G. Rehm, and B. Joern. 2006. "Concepts and rationale for regional nitrogen rate guidelines for corn," PM 2015. Iowa State University, Ames, IA. https://store.extension.iastate.edu/Product/pm2015-pdf.

Serra, A. P., D. J. Bungenstab, M. E. Marchetti, F. C. N. Guimarães, V. D. A. Conrad, H. S. d. Morais, M. A. G. d. Silva, and R. P. Serra. 2013. "Diagnosis and recommendation integrated system (dris) to assess the nutritional state of plants." In M. D. Matovic (ed.). *Biomass Now-Sustainable Growth and Use*. InTech.

Sharpley, A. N. 1997. "Rainfall frequency and nitrogen and phosphorus runoff from soil amended with poultry litter." *Journal of Environmental Quality* 26:1127–1132.

Shrestha, D., A. Srivastava, S. M. Shakya, J. Khadka, and B. S. Acharya. 2013. "Use of compost supplemented human urine in sweet pepper (*Capsicum annuum* l.) production." *Scientia Horticulturae* 153:8–12.

Smaling, E. M. A., S. M. Nwanda, and B. H. Jensen. 1997. "Soil fertility in Africa is at stake." In R. J. Buresh, P. A. Sanchez, and F. Calhoun (eds.). *Replenishing Soil Fertility in Africa*. ASA/SSSA Publication, Soil Science Society of America, Madison, WI.

Sommer, S. G., M. L. Christensen, T. Schmidt, and L. S. Jensen (eds.). 2013. *Animal Manure Recycling: Treatment and Management*. John Wiley, New York, pp. 1–367.

Tiessen, K. H. D., J. A. Elliott, J. Yarotski, D. A. Lobb, D. N. Flaten, and N. E. Glozier. 2010. "Conventional and conservation tillage: Influence on seasonal runoff, sediment, and nutrient losses in the Canadian prairies." *Journal of Environmental Quality* 39:964–980.

Weil, R. R., and S. K. Mughogho. 2000. "Sulfur nutrition of maize in four regions of Malawi." *Agronomy Journal* 92:649–656.

Wick, J., and J. Creque. 2014. "The Marin Carbon Project: Increasing carbon capture on California's rangelands." Marin Carbon Project Washington State University YouTube. https://www.youtube.com/watch?v=bGWry8jWlmw.

Yang, H. S. 2006. "Resource management, soil fertility and sustainable crop production: Experiences of China." *Agriculture, Ecosystems and Environment* 116: 27–33.

Zhou, X., M. J. Helmers, H. Asbjornsen, R. Kolka, M. D. Tomer, and R. M. Cruse. 2014. "Nutrient removal by prairie filter strips in agricultural landscapes." *Journal of Soil and Water Conservation* 69:54–64

Zublena, J. P., J. C. Barker, and T. A. Carter. 1993. "Poultry manure as a fertilizer source." *Soil Facts*. North Carolina Cooperative Extension Service, North Carolina State University, Raleigh, NC.

14
Soil Erosion and Its Control

The wind crosses the brown land, unheard …
—T. S. Eliot, The Waste Land

Degraded land and livelihoods, Kisongo, Tanzania. (Photo courtesy of Ray R. Weil)

The first 13 chapters of this book teach us about Earth's incredibly intricate soil ecosystems, how they function to support life, and how we can best preserve and enhance their health and productivity. Now we must consider how these complex biological–physical–chemical systems are threatened with degradation and destruction by the very people they support. In a world with ever more people, but never more land, no challenge can be more central to human survival than proper care of land and soil. Soil erosion is the most destructive outcome of poor soil and land management. Throughout history people have brought the scourge of soil erosion upon themselves, suffering impoverishment and hunger in its wake. Past civilizations have disintegrated as their soils, once deep and productive, were washed or blown away, leaving only thin, rocky relics of the past. Lakes, rivers, harbors, and estuaries have been silted-in with the lost sediment and entire coastlines transformed.

From expanding cities to shrinking forests, humans acting out of ignorance or greed or just carelessness continue to expose bare soils with devastating results. As the ratio of people to land steadily rises, poor people see little choice but to clear and burn steep, forested slopes or plow up natural grasslands to plant their crops. Population pressures have also led to overgrazing of rangelands and overexploitation of timber resources. All these activities lead to a downward spiral of ecological deterioration, land degradation, and deepening poverty. The impoverished crops and rangelands leave little if any plant litter to protect the soil, leading to further erosion, driving ever-more-desperate people to clear and cultivate—and degrade—still more land. Add to this the intense, concentrated erosion on sites disturbed by urbanization, construction, or mining activity, as well as globalization pressures to expand cropping and logging to marginal lands, and it is clear that the current threat of soil erosion is more ominous than at any time in history.

The degraded productivity and loss of ecosystem services from farm, forest, range, and urban lands tells only part of the sad erosion story. Soil particles washed or blown from the eroding areas are subsequently deposited elsewhere—in nearby low-lying sites within the landscape or far away—even on other continents. Far downstream or downwind, the sediment and dust cause major pollution of water and air and bring enormous economic and social costs to society.

Combating soil erosion is everybody's responsibility and to everyone's advantage. Fortunately, much has been learned about the mechanisms of erosion and techniques have been developed that can effectively and economically control soil loss in most situations. This chapter will equip you with some of the concepts and tools you will need to do your part in solving this pressing world problem.

14.1 SIGNIFICANCE OF SOIL EROSION AND LAND DEGRADATION[1]

Land Degradation

Land degradation may be defined as a reduction in the capacity of land to provide ecosystem goods and perform functions and services that support society and nature (The Millennium Ecosystem Assessment, http://www.millenniumassessment.org/). During the past century, human land use and associated activities may have degraded almost half (some 5 billion ha) of the Earth's vegetated land to some degree. Much of this degradation (on about 2.5 billion ha) is linked to **desertification**, the spreading of desertlike conditions that disrupt semiarid and arid ecosystems (including agroecosystems). A major cause of desertification is poorly managed grazing by cattle, sheep, and goats, a process that likely accounts for about a third of all land degradation, mainly in such dry regions as northern Africa, western China, and the American Southwest. Likewise, the indiscriminate felling of rain forest trees has already degraded about 0.5 billion ha in the humid tropics. Additionally, inappropriate agricultural practices continue to degrade land in all climatic regions.

Soil-Vegetation Interdependency

On about 2 billion of the 5 billion ha of degraded lands in the world, physical soil degradation is a major part of the problem. Physical degradation of soils occurs in many ways (Table 14.1). Although the table illustrates only physical degradation, in reality soils are complex systems in which physical, chemical, and biological processes interact. About 85% of soil degradation stems from erosion—the movement of soil material by wind, water, or tillage implements. The two main components of land degradation—damage to plant communities and deterioration of soil—commonly interact to cause a downward spiral of accelerating ecosystem damage and human poverty (Figure 14.1). Improvements in both soil and vegetation management can be challenging, but have the potential to restore soil and human health by moving up rather than down the spiral.

Table 14.1
TYPES OF PHYSICAL SOIL DEGRADATION AND CONDITIONS UNDER WHICH THEY ARE FOUND

Although the table indicates only physical degradation, in reality soils are complex systems in which physical, chemical, and biological processes interact and influence one another. The first three processes listed, erosion by water, wind, and tillage, together constitute the dominant forms of soil degradation.

Specific physical degradation processes	Conditions and locations where most prominent
Soil erosion by water	Exposure of bare soils by tillage agriculture, deforestation, improper grazing on sloping lands in humid to semiarid regions.
Soil erosion by wind	Semiarid to arid regions. Disturbance of soil, vegetation, or bio-crusts by agricultural tillage and improper grazing or recreational trafficking.
Soil erosion by tillage	Summits and slopes in hilly cultivated landscapes, especially with tillage up and downslope.
Surface sealing	Hydrologic sealing of the soil surface results from pavement and compaction in urban areas, and from excessive tillage in croplands, especially those with low organic matter sandy or silty soils.
Soil compaction	Clayey or sandy soils in humid regions plowed or trafficked by heavy machinery when too wet.

---

[1] For a readable accounts of historical degradation of land and water resources, see Hillel (1991) and Montgomery (2007). For a geo-historical study of 100-fold human-accelerated erosion in the nineteenth century following deforestation for farming in the southeastern United States, see Reusser et al. (2015).

Figure 14.1 The downward spiral resulting from feedback loops between soil and vegetation. As the natural vegetation is disturbed, soil becomes exposed to raindrops and wind leading to erosion and loss of soil, organic matter, and nutrients. The now impoverished soil can support only stunted vegetation, leaving even less protective cover and root mass than before. As more soil loss occurs, the soil becomes extremely degraded and even less protected. Incapable of providing nutrients and water needed to support healthy natural vegetation or crops, the soil continues to erode, polluting rivers with sediment and impoverishing the people who attempt to grow their food and lumber on the land. (Photo courtesy of indigenous village in Mexico and diagram courtesy of Ray R. Weil)

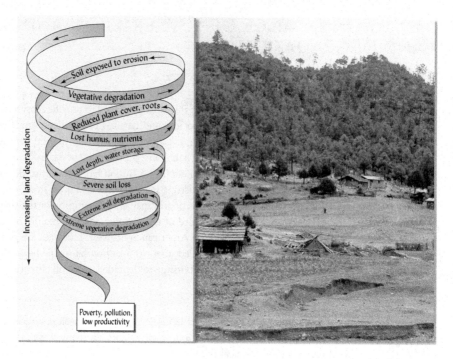

Geological Versus Accelerated Erosion

Geological Erosion Erosion is a process that transforms soil into **sediment**. Soil erosion that takes place naturally, without the influence of human activities, is termed **geological erosion**. It is a natural leveling process. It inexorably wears down hills and mountains, and through subsequent deposition of the eroded sediments, it fills in valleys, lakes, and bays. Many of the landforms we see around us—canyons, buttes, rounded mountain remnants, river valleys, deltas, plains, and pediments—are the result of geological erosion and deposition. The vast deposits that now appear as sedimentary rocks originated in this way.

In most settings, geological erosion wears down the land slowly enough that new soil forms from the underlying rock or regolith faster than the old soil is lost from the surface. The very existence of soil profiles bears witness to the net accumulation of soil and the effectiveness of undisturbed natural vegetation in protecting the land surface from erosion. The rate of geological soil erosion varies greatly with both rainfall and type of material comprising the regolith. Geological erosion by water tends to be greatest in semiarid regions where rainfall is enough to be damaging, but not enough to support dense, protective vegetation. Areas blanketed by deep deposits of silts may have exceptionally high erosion rates under such conditions.

Human-Accelerated Erosion We stand in awe at the edge of the Grand Canyon, which was formed over millennia by geologic erosion—yet few realize that humankind has become the preeminent force on the landscape, now moving nearly twice as much soil per year as global geologic processes, and two-thirds of that *unintentionally* through erosion, mainly associated with agricultural activities (Figure 14.2).

Accelerated erosion occurs when people disturb the soil or the natural vegetation by overgrazing livestock, cutting forests for agricultural use, plowing hillsides (Figure 14.3, *left*), or tearing up land for construction of roads and buildings. Accelerated erosion is often 10 to 1,000 times as destructive as geological erosion, especially on sloping lands in regions of high rainfall. Rates of erosion by wind and water on agricultural land in Africa, Asia, and South America are thought to average about 30 to 40 Mg/ha annually. In North America, the *average* erosion rates on cropland by water and wind are 7 Mg and 5 Mg/ha, respectively. Some cultivated soils are eroding at 10 times these average rates. In comparison, erosion on undisturbed humid-region grasslands and forests generally occurs at rates considerably below 0.1 Mg/ha.

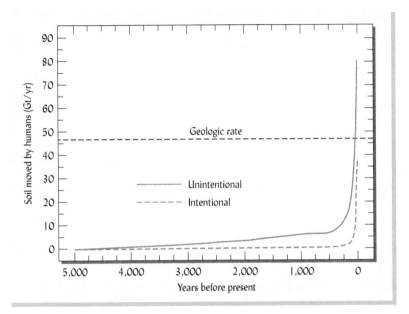

Figure 14.2 *The amount of soil moved by human beings. Intentional soil movement refers mainly to construction and excavation activities. Unintentional soil movement refers mainly to soil loss due to agricultural activities such as land clearing, tillage, overgrazing, and long periods without vegetative cover. Humans now move more soil material than all natural processes combined. This may not mean that sediment loads of major rivers have increased this dramatically, as all movement of soil within a landscape does not necessarily lead to sediment in rivers.* [For comparisons of agricultural to geologic soil movement, see Wilkinson and McElroy (2007). Graph redrawn from Hooke (2000)]

In the United States, some two-thirds of the erosion is caused by water, one-third by wind, much of which occurs on croplands that produce essential food supplies. Some cropland also suffers from soil relocation during tillage operations. Much of the remaining erosion comes from semiarid rangelands, from logging roads and timber harvest on forest lands, and from soils disturbed for highway and building construction. Although progress has been made in reducing erosion in some places, current losses are simply too great for long-term sustainability and must be further reduced.

Under the influence of accelerated erosion, soil is commonly washed, blown, or scraped away faster than new soil can form by weathering or deposition. As a result, the soil depth suitable for plant roots is often reduced. In severe cases, sloping terrain may become scarred by deep gullies, and once-forested hillsides may be stripped down to bare rock. Accelerated erosion often makes the soils in a landscape more heterogeneous. An example can be seen in

Figure 14.3 *Erosion and deposition occur simultaneously across a landscape. (Left) The soil on this hilltop was worn down by water and tillage erosion during nearly 300 years of cultivation. The surface soil exposed on the hilltop consists mainly of light-colored C horizon material. At sites lower down the slope the surface horizon shows mainly darker-colored A and B horizon materials, some of which has been deposited after eroding from locations upslope. (Right) Erosion on the sloping wheat field in the background has deposited a thick layer of sediment in the foreground, burying the plants at the foot of the hill.* (Photos courtesy of Ray R. Weil (*left*) and USDA Natural Resources Conservation Service (*right*))

the striking differences in surface soil color that develop as hilltop soils are truncated, exposing material from the B or C horizon at the land surface, while soils lower in the landscape are buried under organic-matter-enriched sediment (Figure 14.3, *right*). Much of such soil movement occurs during tillage.

14.2 ON-SITE AND OFF-SITE IMPACTS OF ACCELERATED SOIL EROSION

Erosion damages the site on which it occurs and also has undesirable effects off-site in the larger environment. The off-site costs relate to the effects of excess water, sediment, dust, and associated chemicals on downhill, downwind, and downstream environments. While the costs associated with either or both of these types of damages may not be immediately apparent, they are real and grow with time. Landowners and society as a whole eventually foot the bill.

Types of On-Site Damages

The damage done to the soil is greater than the amount of soil lost would suggest because the soil material eroded away is almost always more valuable than that left behind. Not only are surface horizons eroded while less fertile subsurface horizons remain untouched, but the quality of the remaining topsoil is also impaired. Erosion by wind and water selectively removes organic matter and fine mineral particles, while leaving behind mainly relatively less active, coarser fractions. The quantity of essential nutrients lost from the soil by erosion is quite high, although only a portion of these nutrients are lost in forms that would be available to plants in the short term. The soil left behind usually has lower water-holding and cation exchange capacities, less biological activity, and a reduced capacity to supply nutrients for plant growth.

In addition, soil movement during erosion can spread plant disease and the deterioration of soil structure often leaves a dense crust on the soil surface, which, in turn, greatly reduces water infiltration and increases water runoff. Newly planted seeds and seedlings may be washed downhill, trees may be uprooted, and small plants may be buried in sediment or by windblown soil. Finally, gullies that carve up badly eroded land may make the use of field equipment impossible, may undercut pavements and building foundations, and cause building to collapse.

Types of Off-Site Damages

Erosion moves sediment and nutrients off the land, creating the two most widespread water pollution problems in our rivers and lakes. The nutrients impact water quality largely through the process of eutrophication caused by excessive nitrogen and phosphorus, as was discussed in Sections 13.1. In addition to nutrients, sediment and runoff water may also carry toxic metals and organic compounds, such as pesticides. The sediment itself is a major water pollutant, causing a wide range of environmental damages.

Damages from Sediment Sediment deposited on the land may smother crops and other low-growing vegetation (Figure 14.3, *right*). It fills in roadside drainage ditches and creates hazardous driving conditions where mud covers the roadway. Sediment that washes into streams makes the water cloudy or turbid (Figure 14.4). High **turbidity** prevents sunlight from penetrating the water and thus reduces photosynthesis and survival of the *submerged aquatic vegetation* (SAV). The demise of the SAV, in turn, degrades the fish habitat and upsets the aquatic food chain. The muddy water also fouls the gills of some fish. Sediment deposited on the stream bottom can have a disastrous effect on many freshwater fish by burying the pebbles and rocks among which they normally spawn. The buildup of bottom sediments can actually raise the level of the river, so that flooding becomes more frequent and more severe. For example, to counter the rising river bottom, flood-control levees along the Mississippi River must be constantly enlarged.

A number of major problems occur when the sediment-laden rivers reach a lake, reservoir, or estuary. Here, the water slows down and drops its load of sediment. Eventually

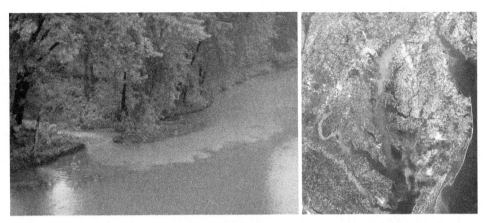

Figure 14.4 Off-site damages caused by soil erosion include the effects of sediment on aquatic systems. (Left) A sediment-laden tributary stream empties into the relatively clear waters of a larger river. The turbid water will foul fish gills, inhibit submerged aquatic vegetation, and clog water-purification systems. Part of the sediment will settle out on the river bottom, covering fish-spawning sites and raising the riverbed enough to aggravate the severity of future flooding episodes. (Right) A NASA satellite image shows the heavy sediment loads (yellow) entering the Chesapeake Bay on the US Atlantic coast from major tributary rivers such as the Potomac in the west and the Susquehanna in the north. (Photos courtesy of USDA Natural Resources Conservation Service (left), NASA (middle), and Ray R. Weil (right))

reservoirs—even those formed by giant dams—become mere mudflats, completely filled in with sediment. Prior to that, the capacity of the reservoir to store water for irrigation or municipal water systems is progressively reduced, as is the capacity for floodwater retention or hydroelectric generation. Similarly, harbors and shipping channels fill in and become impassable. The loss of function and the costs of dredging, excavation, filtering, and construction activities necessary to remedy these situations run into billions of dollars every year.

Windblown Sand and Dust Wind erosion also has its off-site effects. Blowing sands may bury roads and fill in drainage ditches, necessitating expensive maintenance. The sandblasting effect of wind-borne soil particles may damage the fruits and foliage of crops in neighboring fields, as well as the paint on vehicles and buildings many kilometers downwind from the eroding site. Finer windblown dust with clay-size particles causes the most expensive and far-reaching damages, including major human health hazards presented when people inhale these very fine windblown particles. Much of this dust—especially particulate matter (PM) with diameters between 2.5 and 10 microns (PM_{10})—arises from wind erosion on cropland, rangelands, and construction sites (as well as from traffic on unpaved roads). Even more hazardous are particles smaller than 2.5 microns ($PM_{2.5}$), which arise mainly from vehicle exhaust and smoke from fires and industrial plants. Epidemiological studies suggest that the number of deaths resulting from people inhaling this fine *fugitive dust* is in the thousands every year, and may even exceed the number of deaths from traffic accidents. The U.S. Environmental Protection Agency therefore has set standards that call for the 24-hour average concentrations in the air not to exceed 150 and 35 $\mu g/m^3$ of PM_{10} and $PM_{2.5}$, respectively. However, windblown dust is a global problem. For example, wind erosion in the Sahara desert in Africa and Gobi desert in China has been implicated in respiratory diseases in North America. The off-site damages from these dust particles, in addition to health hazards, include the aesthetics-related costs of added housecleaning, more frequent car washes, and lost tourism revenues when majestic views at recreational parks are obscured.

Maintenance of Soil Productivity[2] Although extreme soil erosion can eventually reduce soil productivity to almost zero (as when only exposed rock remains), in most cases the effect is too subtle to notice between one year and the next. Where farmers can afford to do so, they compensate for the loss of nutrients by increasing the use of fertilizer. The losses of organic matter, rooting depth, and water-holding capacity are much more difficult to overcome, though irrigation may partially do so.

---

[2]Research models can simulate how soil productivity may decline as erosion advances over time and the plow layer composition comes to include less organic matter and more and more subsoil material (Gao et al., 2015).

Ultimately, the rate of decline of soil productivity, or the cost of maintaining constant food production levels, is determined by such soil properties as *depth to a root-restricting layer* and *permeability or chemical favorability of the subsoil*. A deep, permeable, well-drained, and well-managed soil may not decline much in productivity even though it suffers some erosion. In contrast, erosion on a shallow, low-permeability soil may bring about a rapid productivity decline.

14.3 MECHANICS OF WATER EROSION

Soil erosion by water is fundamentally a three-step process (Figure 14.5):

1. *Detachment* of soil particles from the soil mass.
2. *Transportation* of the detached particles downhill by floating, rolling, dragging, and splashing.
3. *Deposition* of the transported particles at some place lower in elevation.

On comparatively smooth soil surfaces, the beating action of raindrops causes most of the detachment. Where water is concentrated into channels, the cutting action of turbulent, flowing water detaches soil particles. In some situations, freezing-thawing action also contributes to soil detachment.

Influence of Raindrops

A raindrop accelerates as it falls until it reaches *terminal velocity*—the speed at which the friction between the drop and the air balances the force of gravity. Larger raindrops fall faster, reaching a terminal velocity of about 30 km/h, or about as fast as an Olympic athlete can run. As the speeding raindrops impact the soil with explosive force, they transfer their high kinetic energy to the soil particles (see Figure 14.5).

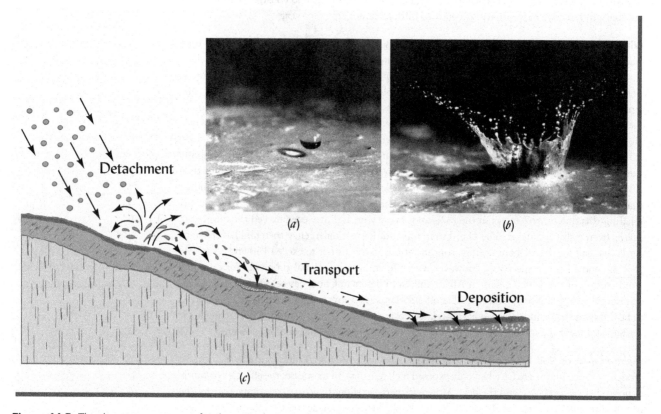

Figure 14.5 *The three-step process of soil erosion by water begins with raindrops impacting wet soil. (a) A raindrop speeding toward the ground. (b) The splash resulting when a drop strikes wet, bare soil. Raindrop impact destroys soil aggregates, encouraging sheet and interrill erosion. Considerable soil may be moved by the splashing process itself. Raindrops detach soil particles, which are then transported and eventually deposited in locations downhill (c). (Diagram courtesy of Ray R. Weil; photos by US Navy)*

Raindrop impact exerts three important detrimental effects: (1) it detaches soil; (2) it destroys granulation; and (3) its splash, under certain conditions, causes an appreciable transportation of soil. So great is the force exerted by raindrops that they not only loosen and detach soil granules but may even beat the granules to pieces. As the dispersed material dries it may develop into a hard crust, which will prevent the emergence of seedlings and will encourage runoff from subsequent precipitation (see Section 4.6).

History may someday record that one of the truly significant scientific advances of the twentieth century was the realization that most erosion is initiated by the impact of raindrops, rather than the flow of running water. For centuries prior to this realization, soil conservation efforts aimed at controlling the more visible flow of water across the land, rather than protecting the soil surface from the impact of raindrops.

Transportation of Soil

Raindrop Splash Effects When raindrops strike a wet soil surface, they detach soil particles and send them flying in all directions (see Figure 14.5). On a soil subject to easy detachment, a very heavy rain may splash as much as 225 Mg/ha of soil, some of the particles splashing as far as 0.7 m vertically and 2 m horizontally. If the land is sloping or if the wind is blowing, this splashing may be greater in one direction, leading to considerable net horizontal movement of soil.

Role of Running Water If the rate of rainfall exceeds the soil's infiltration capacity, water will pond on the surface and begin running downslope. The soil particles sent flying by raindrop impact will then land in flowing water, which will carry them down the slope. So long as the water is flowing smoothly in a thin layer (sheet flow), it has little power to detach soil. However, in most cases the water is soon channeled by irregularities in the soil surface which causes it to increase in both velocity and turbulence. The channelized flow then not only carries along soil splashed by raindrops but also begins to detach particles as it cuts into the soil mass. This is an accelerating process, for as a channel is cut deeper, it fills with greater and greater volumes of flowing water.

Types of Water Erosion

Three types of water erosion are generally recognized: (1) *sheet*, (2) *rill*, and (3) *gully* (Figure 14.6). In **sheet erosion**, splashed soil is removed more or less uniformly, except that tiny columns of soil often remain where pebbles intercept the raindrops (see Figure 14.6a). However, as the sheet flow is concentrated into tiny channels (termed **rills**), **rill erosion** becomes dominant. Rills are especially common on bare land, whether newly planted or in fallow (see Figure 14.6b). Rills are channels small enough to be smoothed by normal tillage, but the damage is already done—the soil is lost. When sheet erosion takes place primarily between irregularly spaced rills, it is called **interrill erosion**.

Where the volume of runoff is further concentrated, the rushing water cuts deeper into the soil, deepening and coalescing the rills into larger channels termed **gullies** (see Figure 14.6c). This is **gully erosion**. Gullies on cropland are obstacles for tractors and cannot be removed by ordinary tillage practices. All three types may be serious, but sheet and rill erosion, although less noticeable than gully erosion, are responsible for most of the soil moved.

Deposition of Eroded Soil

Erosion may send soil particles on a journey of a thousand kilometers or more—off the hills, into creeks, and down great muddy rivers to the ocean. On the other hand, eroded soil may travel only a meter or two before coming to rest in a slight depression on a hillside or at the foot of a slope (Figure 14.3). The amount of soil delivered to a stream, divided by the amount eroded, is termed the **delivery ratio**. As much as 60% of eroded soil may reach a stream (delivery ratio = 0.60) in certain watersheds where valley slopes are very steep. As little as 1% may reach the streams draining a gently sloping coastal plain. Typically, the delivery ratio is larger for small watersheds than for large ones, because the latter provide many more opportunities for deposition before a major stream is reached.

Figure 14.6 *Three major types of soil erosion.* (a) Sheet erosion is relatively uniform erosion from the entire soil surface. Note that the small stones have protected the soil underneath them from sheet erosion creating the soil pedestals on which they are perched. (b) Rill erosion is initiated when the water concentrates in small channels (rills) as it runs off the soil. Subsequent cultivation may erase rills, but it does not replace the lost soil. (c) Gully erosion creates deep channels that cannot be erased by cultivation. Although gully erosion looks the most catastrophic of the three, far more total soil is lost by the less obvious sheet and rill erosion. [Drawings from FAO (1987); photos courtesy of Ray R. Weil]

14.4 MODELS TO PREDICT WATER-INDUCED EROSION[3]

Land managers and policymakers need to predict the extent of soil erosion in order to plan the best management of soil resources, evaluate consequences of alternative tillage practices on a farm, determine compliance with environmental regulations, develop sediment-control plans for construction projects, and estimate the years it will take to silt-in a reservoir.

---

[3]The WEPP computer model program, supporting instructions and associated databases can be downloaded freely from: http://www.ars.usda.gov/Research/docs.htm?docid=10621. The WEPP model was developed over a 30-year period and continues to be updated and improved (Flanagan et al., 2007). The SWAT is a public domain model designed to simulate the physical processes governing water runoff and erosion from small to regional watersheds. It can be downloaded for free and is supported by extensive documentation online and has been applied to erosion and hydrological issues around the world (Gassman et al., 2014; USDA/ARS, 2014). For discussion of the original USLE, see Wischmeier and Smith (1978), and for the RUSLE, see Renard et al. (1994).

The detachment, transport, and deposition processes of soil erosion can be predicted mathematically by soil erosion *models*. These are equations—or sets of linked equations—that interrelate information about the rainfall, soil, topography, vegetation, and management of a site with the amount of soil likely to be lost by erosion. One of the most ambitious and sophisticated erosion models developed so far is a complex, process-based computer program called the Water Erosion Prediction Project (WEPP) model. It is based on an understanding of the fundamental mechanisms involved with each process leading to soil erosion.

The Soil and Water Assessment Tool (SWAT) is a widely used public-domain computer model with even greater complexity and a broader scope than WEPP. The SWAT combines many individual computer models into a large complex "super model" capable of simulating many hydrologic processes within large heterogeneous watersheds. In addition to simulating soil erosion and sediment transport, SWAT also can simulate surface and subsurface runoff, stream flows, groundwater recharge, and the movement of nutrients (N and P) in association with sediment or dissolved in water. However, when predictions from either WEPP or SWAT are compared with actual in-field measurements, it can be seen that neither model always provides a close match to reality.

The Universal Soil-Loss Equation (USLE)[4]

In contrast to the process-based operation of WEPP and SWAT, most predictions of soil erosion continue to rely on much simpler models that statistically relate soil erosion to a number of easily observed environmental factors. Scientists can make such *empirical* models if they know that certain conditions are associated with soil erosion, even if they do not understand the details of *why* this is so. At the heart of these models is the realization that water-induced erosion results from the interaction of rain and soil. More than half a century of erosion research have clearly identified the major factors affecting this interaction. These factors are quantified in the universal soil-loss equation (USLE):

$$A = R \times K \times LS \times C \times P \qquad (14.1)$$

A, the predicted annual soil loss, is the product of

R = rainfall erosivity } Rain-related factor

K = soil erodibility
L = slope length } Soil-related factors
S = slope gradient or steepness

C = cover and management } Land-management factors
P = erosion-control practices

Working together, these factors determine how much water enters the soil, how much runs off, how much soil is transported, and when and where it is redeposited. Note that because the factors are multiplied together, *if any one factor could be reduced to zero, the resulting amount of erosion (A) would also be reduced to zero*.

Unlike the WEPP and SWAT programs, the USLE was designed to predict only the amount of soil loss by sheet and rill erosion in an average year for a given location. It cannot predict erosion from a specific year or storm, nor can it predict the extent of gully erosion and sediment delivery downslope or to streams. It can, however, show how varying any combination of the soil- and land-management-related factors might be expected to influence soil erosion, and therefore can be used as a decision-making aid in choosing the most effective strategies to conserve soil.

---

[4]In this textbook, we use the scientifically acceptable SI units for the R and K factors in our discussion of these erosion equations. However, since these soil-loss equations were published for use by landowners and the general public, most maps, tables, and computer programs available in the United States supply values in customary English units, rather than in SI units. When using English units for the R and K factors, the soil loss A is expressed in tons (2,000 lb) per acre, which can be easily converted to Mg/ha by multiplying by 2.24. For details on converting the customary English units to SI units, see Foster et al. (1981).

The Revised Universal Soil-Loss Equation (RUSLE)

The USLE has been used widely since the 1970s. Beginning in the 1990s, the USLE was updated and computerized to create an erosion-prediction tool called the revised universal soil-loss equation (RUSLE). RUSLE uses the same basic factors of USLE just shown, although some are better defined, and interrelationships among them improve the accuracy of soil-loss prediction. RUSLE is a computer software package that is constantly being improved and modified as experience is gained from its use around the world. As we are about to see, the five factors that comprise the USLE (and RUSLE) provide a useful framework for understanding and controlling soil erosion by water.

14.5 FACTORS AFFECTING INTERRILL AND RILL EROSION

Rainfall Erosivity Factor, R

The rainfall **erosivity** factor, R, represents the driving force for sheet and rill erosion. It takes into consideration the total annual rainfall and, importantly, the rain intensity and seasonal distribution. An index of the kinetic energy of each storm is calculated from data related to the intensity and amount of rainfall. Then the indices for all storms occurring during a year are summed to give an annual index. An average of such indexes for many years is used as the R value in the universal soil-loss equation.

Rainfall index values for locations around the world are shown in Figure 14.7. Note that they vary from less than 100 in arid parts of the western United States, North Africa, and Central Asia to more than 50,000 in a few parts of the very humid tropics [the map gives R values in SI units of $(MJ \cdot mm)/(ha \cdot h \cdot yr)$ that can be converted to English units if divided by 17.02]. Generally, rainfall tends to be more intense and more erosive in subtropical and tropical regions than in temperate regions.

Rainfall intensity in most locations is so highly variable that actual erosivity in any one year is commonly two to five times greater or smaller than the long-term average. In fact, a few unusually intense, heavy storms often account for most of the erosion that takes place. Conservation practices based on the predictions of the USLE or RUSLE using long-term average R factors may not be sufficient to limit erosion damages from these relatively rare, but extremely damaging, storms.

Soil Erodibility Factor, K

The soil **erodibility** factor, K, indicates a soil's inherent susceptibility to erosion. The K value assigned to a particular type of soil indicates the amount of soil lost per unit of erosive energy in the rainfall, assuming a standard research plot (22 m long, 9% slope) on which the soil is kept continuously bare by tillage.

The two most significant and closely related soil characteristics influencing erodibility are (1) *infiltration capacity* and (2) *structural stability*. High infiltration means that less water will be available for runoff, and the surface is less likely to be ponded (which would make it more susceptible to splashing). Stable soil aggregates resist the beating action of rain, and resist soil detachment even though runoff may occur. Certain tropical clay soils high in hydrous oxides of iron and aluminum are known for their highly stable aggregates that resist the action of torrential rains. Downpours of a similar magnitude on swelling-type clays would be disastrous.

Topographic Factor, LS

The topographic factor, LS, reflects the influence of length and steepness of slope on soil erosion. It is expressed as a unitless ratio with soil loss from the area in question in the numerator, and that from a standard plot (9% slope, 22 m long) in the denominator. The longer the slope, the greater the opportunity for accumulation and concentration of the runoff water.

Figure 14.7 *Global distribution of R values for rainfall erosivity. Note the very high values in the humid tropics, where annual rainfall is high and intense storms are common. Similar amounts of annual rainfall in Northern Europe result in much lower R values because there the rain mostly falls gently over long periods. Values on map are in units of (MJ·mm)/(ha·h·yr). To convert to English units of 100*(ft·ton·in.)/(acre·h·yr), divide by 17.02.* [Modified from Panagos et al., 2017]

Figure 14.8 illustrates the increases in *LS* factors that occur as slope length and steepness increase for sites with low, moderate, and high ratios of rill to interrill (sheet) erosion. Most sites cultivated to row crops have moderate rill to interrill erosion ratios. On sites where this ratio is low, such as rangelands, more of the soil movement occurs by interrill erosion. On these sites, slope steepness (%) has a relatively greater influence on erosion, while the slope length has a relatively smaller influence. The opposite is true for freshly excavated construction areas and other highly disturbed sites, which have high rill to interrill erosion ratios. Here, where rill erosion predominates, slope length has a greater influence.

Table 14.2
COMPUTED *K* VALUES FOR SOILS AT DIFFERENT LOCATIONS
The values listed are in SI units. Values in the RUSLE program may differ somewhat.

Soil	Location	Compounds[a] *K*
Udalf (Dunkirk silt loam)	Geneva, NY	0.091
Udalf (Keene silt loam)	Zanesville, OH	0.063
Udult (Lodi loam)	Blacksburg, VA	0.051
Udult (Cecil sandy clay loam)	Watkinsville, GA	0.048
Udoll (Marshall silt loam)	Clarinda, IA	0.044
Udalf (Hagerstown silty clay loam)	State College, PA	0.041
Ustoll (Austin silt)	Temple, TX	0.038
Aqualf (Mexico silt loam)	McCredie, MO	0.034
Udult (Cecil sandy loam)	Watkinsville, GA	0.030
Alfisols	Indonesia	0.018
Udult (Tifton loamy sand)	Tifton, GA	0.013
Alfisols	Nigeria	0.008
Ultisols	Nigeria	0.005
Oxisols	Puerto Rico	0.001

[a]To convert *K* values from (Mg · ha · h)/(ha · MJ · mm) to English units of (ton · acre · h)/(100 acres · ft-ton · in.), multiply the values in this table by 7.6. From Wischmeier and Smith (1978); data for tropical soils cited by Cassel and Lal (1992).

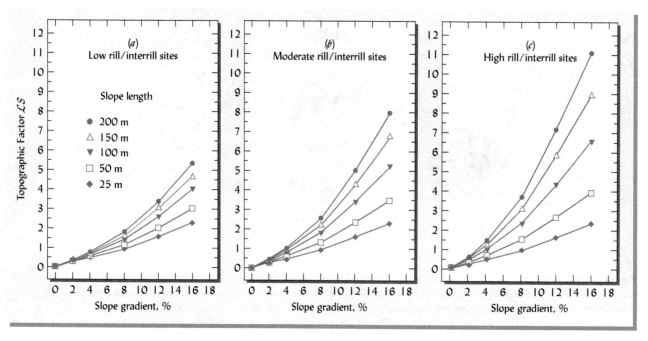

Figure 14.8 *Relationship between values of the topographic factor LS and the slope gradient for several lengths of slope on three types of sites: (a) sites with low ratios of rill to interrill erosion, such as many rangelands; (b) sites with moderate ratios of rill to interrill erosion, such as most tilled row-crop land; and (c) sites with high ratios of rill to interrill erosion, such as freshly disturbed construction sites and new seedbeds. The LS values extrapolated from these graphs can be used in the Universal Soil-Loss Equation.* [Graphs based on data in Renard et al. (1994)]

Cover and Management Factor, C

Erosion and runoff are markedly affected by different types of vegetative cover or cropping systems. Undisturbed forests and dense grass (Figure 14.9, *right*) provide the best soil protection and are about equal in their effectiveness. Even in semiarid regions unable to support dense vegetation, increasing the density of trees and shrubs to cover more than one-third to one-half of the soil surface can dramatically reduce the loss of both soil and water in runoff (Figure 14.10). On agricultural land, forage crops (both legumes and grasses) are most effective in staunching erosion because of their relatively dense cover. Small grains, such as wheat and oats, are intermediate and offer considerable obstruction to surface wash during their growing season. Row crops, such as corn, soybeans, cotton, and potatoes, offer relatively little living cover during the early growth stages and thereby leave the soil susceptible to erosion (Figure 14.9, *left*) unless residues and stubble from previous crops protect the soil surface.

Cover crops can provide soil protection during the time of year between the growing seasons for annual crops. For widely spaced perennial plantings such as orchards and vineyards, cover crops can permanently protect the soil between rows of trees or vines. A mulch of plant residue or applied materials is also effective in protecting soils. Research on all continents has shown that surface mulch does not have to be very thick or cover the soil completely to make a major contribution to soil conservation. Even small increases in surface cover result in large reductions in soil erosion, particularly interrill erosion (Figure 14.11).

Regulation of grazing to maintain a dense vegetative cover on range and pastureland and the inclusion of close-growing hay crops in rotation with row crops on arable land will help control both erosion and runoff. Likewise, the use of conservation tillage systems, which leave most of the plant residues on the surface, greatly decreases erosion hazards.

The *C* factor in the USLE or RUSLE is the ratio of soil loss under the conditions in question to that which would occur under continuously bare soil. This ratio *C* will approach 1.0 where there is little soil cover (e.g., a bare seedbed in the spring or freshly graded bare soil on a

Figure 14.9 *The importance of vegetative cover on soils. Immediately after a heavy rainstorm, runoff (left) is scouring a gully and carrying off a heavy sediment load from a cultivated soil with almost no residue cover while (right) on a soil with almost 100% cover the runoff is clear.* (Photo courtesy of Ray R. Weil)

construction site). It will be low (e.g., <0.10) where large amounts of plant residues are left on the land or in areas of dense perennial vegetation. Examples of *C* values are given in Table 14.3.

Support Practice Factor, P

On some sites with long and/or steep slopes, erosion control achieved by management of vegetative cover, residues, and tillage must be augmented by the construction of physical structures or other steps aimed at guiding and slowing the flow of runoff water. These **support practices** determine the value of the *P* factor in the USLE. The *P* factor is the ratio of soil loss with a given support practice to the corresponding loss if row crops were planted up and down the slope. If there are no support practices, the *P* factor is 1.0. The support practices include tillage on the contour, contour strip-cropping, terrace systems, and grassed waterways, all of which will tend to reduce the *P* factor. Many of the erosion-control practices or management techniques discussed with regard to the *C* and *P* factors and in later sections of this chapter are considered to be **best management practices** (BMPs).

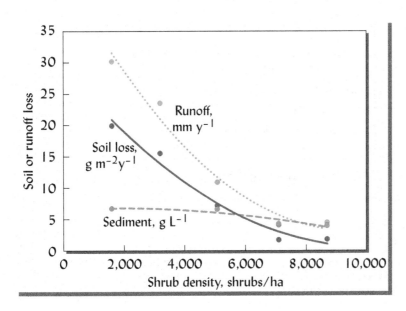

Figure 14.10 *Effect of shrub planting density on runoff and soil erosion in a hilly semiarid loess region in Ningxia province, western China. Deep-rooted, fast-growing shrubs (Caragana korshinskii Kom.) were planted for afforestation at densities that achieved canopy cover ranging from 20 to 75%. Both runoff water and soil loss were dramatically reduced by planting more shrubs, but the concentration of sediment in the runoff that did occur was only slightly affected. The Huangmian soil (Haplocalcids) is developed from loess and is highly erodible. The region receives 416 mm of rainfall annually, mainly between June and September.* [Graphed from data in Guo and Shao (2013)]

Figure 14.11 *Reduction in interrill erosion achieved by increasing ground cover percentage. The diagrams above the graph illustrate 5, 20, 40, 60, and 80% ground cover. Note that even a light covering of mulch has a major effect on soil erosion. The graph applies to interrill erosion. On steep slopes, some rill erosion may occur even if the soil is well covered. Generalized relationship based on results from many studies.* (Diagram courtesy of Ray R. Weil)

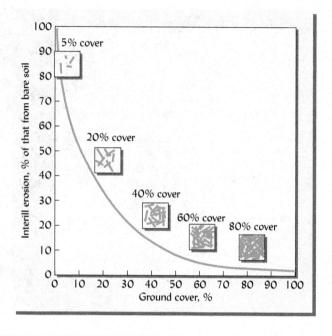

Table 14.3
EXAMPLES OF C VALUES FOR THE COVER AND VEGETATION MANAGEMENT FACTOR
The C values indicate ratios of soil eroded from a particular vegetation system to that expected if the soil were kept completely bare. Note the effects of canopy cover, surface litter (residue) cover, tillage, and crop rotation. The C values are situation specific and must be calculated from local information on plant growth habits, climate, etc. Specific values may be obtained from the RUSLE computer program or from local conservation agencies.

Vegetation	Management/condition	C value
Range grasses and low (<1 m) shrubs	75% canopy cover, no surface litter	0.17
	75% canopy cover, 60% cover with decaying litter	0.032
Scrub brush about 2 m tall	25% canopy cover, no litter	0.40
	75% canopy cover, no litter	0.28
Trees with no understory, about 4 m drop fall	75% canopy cover, no litter	0.36
	75% canopy cover, 40% leaf litter cover	0.09
	75% canopy cover, 100% leaf litter cover	0.003
Woodland with understory	90% canopy cover, 100% litter cover	0.001
Permanent pasture	Dense grass sod, properly grazed	0.003
Corn–soybean rotation	Fall plowing, conventional tillage, residues removed	0.53
	Spring chisel plow–plant, 2,500 kg/ha surface residues	0.22
	No-till planting, 5,000 kg/ha surface residues	0.06
	No-till with winter cover crops	0.02
Corn–oats–hay–hay rotation	Spring conventional plowing before planting	0.05
	No-till planting	0.03

Values typical of midwestern United States. Based on Wischmeier and Smith (1978) and Schwab et al. (1996).

Contour Cultivation Rows of plants slow the flow of runoff water if they follow the contours across the slope gradient (but the rows *encourage* channelization and gullies if they run up and down the slope; Figure 14.12). Even more effective is planting on ridges built up of soil *along the contours*. However, ridges must be designed to carry heavy runoff safely from the field, or else severe erosion may occur when the ridges overflow.

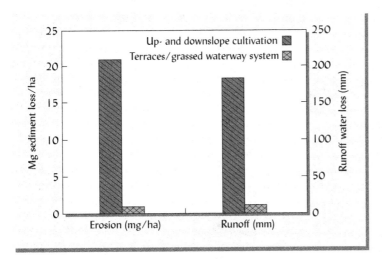

Figure 14.12 *Erosion and water runoff losses from small watersheds where potato (a row crop with high soil disturbance) was grown either up and down the slope, or on the contour in a system with diversion terraces and a grassed waterway. The contour practices provided dramatic soil and water conservation benefits. Data are annual rates averaged across three years.* [From Chow et al. (1999)]

On long slopes subject to sheet and rill erosion, fields may be laid out in narrow strips across the incline, alternating the tilled row crops, such as corn, soybean, or potato, with hay and small grains (such as wheat or barley). Water cannot achieve an undue velocity on the narrow strips of tilled land, and the hay and small grain crops check the rate of runoff. Such a layout is called **strip-cropping** and is the basis for erosion control in many hilly agricultural areas. This arrangement can be thought of as shortening the effective slope length.

When the cross strips are laid out to approximately follow the slope contours, the system is called **contour strip-cropping** (Figure 14.13). The width of the strips will depend primarily on the slope steepness and the soil permeability and erodibility. Widths of 20 to 120 m are common. Contour strip-cropping is often augmented by diversion ditches and waterways between fields. Permanent sod established in the swales produces **grassed waterways** that can safely carry water off the land without the formation of gullies (see Figure 14.14, *right*).

Terraces Construction of various types of terraces reduces the effective length and gradient of a slope (Figure 14.13, *left*). **Bench terraces** are used where nearly complete control of the water runoff must be achieved, such as in rice paddies (see Figure 6.34). Where farmers use

Figure 14.13 *Practices supporting erosion control. (Left) Contour cropping with graded terraces between crop strips and grassed waterway to safely convey excess water off a sloping field in Kansas. The broken arrows show paths taken by the runoff water. (Right) Contour strip-cropping with alternating strips of mature wheat and young green alfalfa on a farm in New York. The arrows indicate grassed waterways.* (Photos courtesy of Jeff Vanuga, USDA/NRCS (*left*) and Ray R. Weil (*right*))

Figure 14.14 Concentrated runoff erodes soil and carries away sediments on an unprotected, conventionally tilled crop field, despite planting across the slope (left). Water runs clear without scouring soil from a crop field protected by a permanent grassed waterway (right). (Photos courtesy of USDA/NRCS)

large machinery and need to farm all the land in a field, **broad-based terraces** are more common. Broad-based terraces waste little or no land and are quite effective if properly maintained. Water collected behind each terrace flows gently across (rather than down) the field in a terrace channel, which has a drop of only about 50 cm in 100 m (0.5%). The terrace channels should guide the runoff water to grassed waterways (Figures 14.13 and 14.14), so that the runoff water can move downhill off the field to a nearby stream or drainage canal.

Examples of P values for contour tillage and strip-cropping at different slope gradients are shown in Table 14.4. Note that P values increase with slope and that they are lower for strip-cropping, illustrating the importance of this practice for erosion control. Terracing also reduces the P values. Unlike the USLE, the RUSLE takes into account interactions between support practices and subfactors such as slope and soil water infiltration.

The five factors of the USLE (R, K, LS, C, and P) have suggested many approaches to the practical control of soil erosion. A sample calculation is shown in Box 14.1 to illustrate how the USLE can help evaluate erosion-control options. We will now discuss several specific erosion-control technologies appropriate for various types of land uses.

14.6 CONSERVATION TILLAGE[5]

For centuries, conventional agricultural practice around the world has encouraged extensive soil tillage that leaves the soil bare and unprotected from the ravages of erosion. Beginning in the middle of the twentieth century, two technological developments allowed many farmers to avoid this problem by managing their soils with little tillage—or no tillage at all. First came the development of herbicides that could kill weeds chemically without tillage. Second, farmers and equipment manufacturers developed machinery that could plant crop seeds even if the soil was in an unloosened state and covered by plant residues. These developments obviated two of the main reasons that farmers tilled their soils. Farmer interest in reduced tillage heightened as it was shown that these systems produced equal or even higher crop yields in many regions while saving time, fuel, money—and soil. The latter attribute earned these systems the name of **conservation tillage**. More recently, organic farmers and researchers who do not use chemical herbicides have worked to develop practical systems of conservation tillage that use cover crops to suppress weeds while reducing the amount of tillage performed (see Sections 13.2 and 13.6).

---

[5]For a review of the history of how conservation tillage transformed agriculture in humid Brazil and semi-arid regions of Canada, see Awada et al. (2014) and de Freitas and Landers (2014). To learn under what conditions notill yields better or worse than tillage, see Pittelkow et al. (2015).

Table 14.4
P Factors for Contour and Strip-Cropping at Different Slopes and the Terrace Subfactor at Different Terrace Intervals

The product of the contour or strip-cropping factors and the terrace subfactor gives the P value for terraced fields.

Slope, %	Contour P factor	Strip-cropping P factor	Terrace interval, m	Terrace subfactor Closed outlets	Open outlets
1–2	0.60	0.30	33	0.5	0.7
3–8	0.50	0.25	33–44	0.6	0.8
9–12	0.60	0.30	43–54	0.7	0.8
13–16	0.70	0.35	55–68	0.8	0.9
17–20	0.80	0.40	69–60	0.9	0.9
21–25	0.90	0.45	90	1.0	1.0

Contour and strip-cropping factors from Wischmeier and Smith (1978); terrace subfactor from Foster and Highfill (1983).

BOX 14.1
CALCULATIONS OF EXPECTED SOIL LOSS USING USLE

The principles involved in both USLE and RUSLE can be verified by making calculations using USLE and its associated factors. Note that the factors in the USLE are related to each other in a multiplicative fashion. Therefore, if any one factor can be made to be near zero, the amount of soil loss A will be near zero.

Assume, for example, a location in Iowa on a Marshall silt loam with an average slope of 6% and an average slope length of 100 m. Assume further that the land is clean-tilled and fallowed.

Figure 14.7 shows that the R factor for this location is about 2,500 (between 1,700 and 3,100) in SI units (147 in English units, 2,500/17 = 147). The K factor for a Marshall silt loam in central Iowa is 0.044 (Table 14.2) and the topographic factor LS from Figure 14.8 is 1.7 (high rill to interrill ratio on soil kept bare). The C factor is 1.0, since there is no cover or other management practice to discourage erosion. If we assume the tillage is up and down the hill, the P value is also 1.0. Thus, the anticipated soil loss can be calculated by the USLE ($A = RKLSCP$):

$$A = (2,500)(0.044)(1.7)(1.0)(1.0)$$
$$= 187 \text{ Mg/ha [or 83.5 tons/acre]}$$

If the crop rotation involved corn–soybean–wheat–hay and conservation tillage practices (e.g., spring chisel plow tillage) were used, a reasonable amount of residue would be left on the soil surface. Under these conditions, the C factor may be reduced to about 0.13 (Table 14.3). Likewise, if the tillage and planting were done on the contour, the P value would drop to about 0.5 (Table 14.4). P could be reduced further to 0.4 (0.5 × 0.8) if terraces with open outlets were installed about 40 m apart. Furthermore, with crops on the land the site would have a moderate rill to interrill erosion ratio, so the LS factor would be only 1.4 (middle of Figure 14.8). With these values the soil loss becomes as follows:

$$A = (2,500)(0.044)(1.4)(0.13)(0.4)$$
$$= 8 \text{ Mg/ha [or 3.6 tons/acre]}$$

The units for the calculation are not normally shown, but they are

$$\frac{\text{MJ} \cdot \text{mm}}{\text{ha} \cdot \text{h} \cdot \text{yr}} \times \frac{\text{Mg} \cdot \text{ha} \cdot \text{h}}{\text{ha} \cdot \text{MJ} \cdot \text{mm}}$$

$$= \frac{\text{Mg}}{\text{ha} \cdot \text{yr}} \quad (LS, C, \text{ and } P \text{ are unitless ratios.})$$

The benefits of good cover and management and support practices are obvious. The figures cited were chosen to provide an example of the utility of the universal soil-loss equation, but calculations can be made for any specific location. In the United States, pertinent factor values that can be used for erosion prediction in specific locations are generally available from state offices of the USDA Natural Resource Conservation Service. The necessary factors are also built into the RUSLE software which is freely available online.

Conservation Tillage Systems

While there are numerous conservation tillage systems in use today (Table 14.5), all have in common that they leave significant amounts of plant residues on the soil surface after a new crop is sown. Keep in mind that conventional tillage involves first moldboard plowing (Figure 14.15, *left*) to completely bury weeds and residues, followed by one to three passes with a harrow to break up large clods, then planting the crop, and subsequently several cultivations between crop rows to kill weeds. Every pass with a tillage implement bares the soil anew and also weakens the structure that helps soil resist water erosion.

Conservation tillage systems range from those that merely reduce excess tillage to the no-tillage system, which does not use any tillage beyond the slight soil disturbance that occurs as the planter cuts a planting slit through the residues to a depth of several centimeters into the soil (Figure 14.16, *inset*). The conventional moldboard plow was designed to leave the field "clean"; that is, free of surface residues. In contrast, conservation tillage systems, such as **chisel plowing** (Figure 14.15, *right*), stir the soil but only partially incorporate surface residues, ideally leaving more than 30% of the soil covered. **Stubble mulching**, whose water-conserving attributes were highlighted in Section 6.4, is another example. **Ridge tillage** is a conservation system in which crops are planted on top of permanent 15- to 20-cm-high ridges. About 30% soil coverage is maintained, even though the ridges are scraped off a bit for planting and then built up again by shallow tillage to control weeds.

With **no-till** systems (sometimes referred to as *direct seeding*) we can expect 50 to 100% of the surface to remain covered. Well-managed continuous no-till systems in humid regions include cover crops during the winter and high-residue-producing crops in the rotation. Such systems keep the soil nearly 100% covered at all times and build up organic surface layers somewhat like those found in forested soils.

Conservation tillage systems generally provide yields similar to those from conventional tillage, provided crop residues are retained as a mulch, the soil is not too poorly drained, and the climate is not too cold. In dry climates or dry years, no-till often outyields conventional tillage because water as well as soil is conserved. However, during the transition from

Table 14.5
GENERAL CLASSIFICATION OF DIFFERENT CONSERVATION TILLAGE SYSTEMS

To be considered conservation tillage, a system must maintain enough plant residues on the soil to cover at least 30% of the ground surface area after planting.

Tillage system	Operation involved
No-till	Soil undisturbed prior to planting. Seeds placed in narrow slots cut through residues and soil. Slot is closed so only 1 to 10% of soil surface exposed. Weed control usually by herbicides and/or cover crop residue mulch.
Vertical/turbo till	High-speed, shallow disturbance implements with two gangs of forward-facing rolling blades, either straight or waved, spaced about 20 to 25 cm apart set to disturb vertical slices of soil about 5 cm wide and 4 to 8 cm deep, as well as chop up surface residues.
Ridge till (till, plant)	Soil undisturbed prior to planting, which is done on ridges 10 to 15 cm higher than row middles. Residues moved aside or incorporated on about one-third of soil surface. Herbicides and cultivation to control weeds.
Strip till	Soil undisturbed prior to planting. Narrow and shallow tillage in rows using rotary tiller, in-row chisel, etc. About 25 to 50% of soil surface is tilled at planting time. Herbicides and cultivation to control weeds.
Mulch till	Soil surface disturbed by tillage prior to planting, but at least 30% of ground is covered with residue. Tools such as chisels, field cultivators, disks, and sweeps are used (e.g., stubble mulch). Herbicides and cultivation to control weeds.
Reduced till	Any other tillage and planting system that keeps at least 30% of the soil surface covered by residues.

Definitions, except vertical till, modified from Conservation Technology Information Center, West Lafayette, Indiana.

Figure 14.15 *Conventional inversion tillage and conservation tillage in action. (Left) In conventional tillage, a moldboard plow inverts the upper soil horizon, burying all plant residues and producing a bare soil surface. (Right) A chisel plow, one type of conservation tillage implement, stirs the soil but leaves a good deal of the crop residues on the soil surface.* (Photo courtesy of Ray R. Weil)

conventional tillage to no-tillage, crop yields may decline somewhat for several years for reasons associated with some of the effects outlined in the following subsections.

Adaptation by Farmers[6]

In recent years conservation tillage has become increasingly popular. No-tillage systems, especially, have spread to nearly all regions of the United States. The no-till system has been used continuously on some farms in the eastern United States since the 1970s (about half a century without any tillage). No-tillage and other conservation sys*tems are now used in all parts of the world*, but the earliest and most widespread adoption has been mainly in the Americas and

Figure 14.16 *In no-till systems, one crop is planted directly into the residue of a cover crop or of a previous cash crop, with only a narrow band of soil disturbed. No-till systems leave virtually all of the residue on the soil surface, providing up to 100% cover and nearly eliminating erosion losses. Here corn was planted into a cover crop killed with a herbicide (weed-killing chemical) to form a surface mulch. The inset shows a close-up of a no-till planter in action (direction of travel is to the right). The rolling furrow openers cut a slot through the residue and soil into which the seed is placed at a depth set by the depth wheel. Snug seed-to-soil contact is ensured by the press wheel that closes the slot.* (Photo courtesy of Ray R. Weil)

---

[6]For the inspiring story of one Chilean farmer's struggle to conquer the forces of erosion and degradation and restore the health of his soil, see Crovetto (1996).

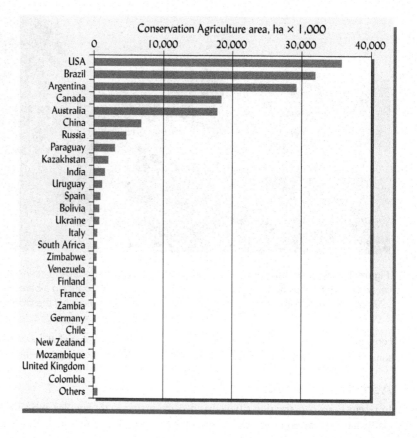

Figure 14.17 *Adoption of conservation agriculture (mainly high residue tillage or no-till) on farms in most of the world has lagged far behind than in North and South America and Australia.* [Data selected from Kassam et al. (2012)]

Australia (Figure 14.17). One of the most significant examples of no-tillage expansion has been in southern Brazil. In addition to large farms similar to those in North America, thousands of small-scale soybean and corn farmers in Brazil have successfully adapted cover-crop-based no-tillage systems using hand-pushed, animal-drawn, or small tractor-pulled equipment.

Vertical or Turbo Tillage A decade into the twenty-first century, some long-term no-till farmers began to perceive thick surface accumulations of crop residues to be an obstacle to incorporation of livestock manures and lime and to soil warming and seed placement in spring. These and other farmers have favored using new types of high-speed tillage implements designed to chop up residues and so stir, but not invert, portions of the soil to shallow depths. This new generation of machines disturbs the soil mainly in a vertical (rather than horizontal) direction, hence the process is sometimes referred to as vertical tillage. The operation cuts and stirs slots in the soil about 4 to 6 cm wide and 4 to 8 cm deep spaced every 10 to 15 cm across the width of the implement, thus disturbing about 30 to 40% of the soil surface. Turbo tilling may maintain some 30 to 80% residue coverage to protect from water erosion, but also creates a high potential for tillage erosion (see Section 14.13) because the high speed of the operation results in much soil being thrown (Figure 14.18).

Erosion Control by Conservation Tillage

Since conservation tillage systems were initiated, hundreds of field trials have demonstrated that these tillage systems allow much less soil erosion than do conventional tillage methods. Surface runoff is also decreased, although the differences are not as pronounced as with soil erosion. These differences are reflected in the much lower C factor values assigned to conservation tillage systems (see Table 14.3).

The erosion-control value of an undisturbed surface residue mulch was discussed in the previous section. The development of biopores in no-till soils promotes much more rapid infiltration and therefore much less runoff loss than is the case with conventionally tilled soils. Conservation tillage may also significantly reduce the loss of nutrients attached to sediment, although nutrients dissolved in runoff water may be increased (review Tables 12.2).

Figure 14.18 *Fast, shallow tillage operations that disturb only narrow strips of soil to less than 8 cm deep are commonly termed turbo till or vertical tillage. Turbo till implements employ sets of rolling knives to chop up crop residues and partially incorporate them into slits made in the soil about 4 to 6 cm wide and 4 to 8 cm deep. Some 30 to 80% residue coverage may be maintained to give protection from water erosion, but the potential for tillage erosion (see Section 14.13) is quite high because the high speed of the operation results in much soil being thrown.* (Photo courtesy of John Nowatzki, North Dakota State University)

Effect on Soil Properties

When soil management is converted from plow tillage to conservation tillage (especially no-tillage), numerous soil properties are affected, mostly in favorable ways. The changes are most pronounced in the upper few centimeters of soil (Figure 11.23). Generally, the changes are greatest for systems that produce the most plant residue (such as corn and small grains in humid regions, especially if accompanied by cover crops), retain the most residue coverage, and cause the least soil disturbance. Many of these changes are illustrated in other chapters of this textbook, so the discussion here will be brief.

Macroporosity and aggregation (see Sections 4.5 and 4.8) are increased as active organic matter builds up and earthworms and other organisms establish themselves. Infiltration and internal drainage are generally improved, as is soil water-holding capacity. Some of these effects are illustrated in Figure 14.19. The enhanced infiltration capacity of no-till-managed soils is generally quite desirable, but in some cases it may lead to more rapid leaching of nitrates and other water-soluble chemicals. Residue-covered soils are generally cooler and moister (see Sections 6.4 and 7.11). This is an advantage in the hot part of the year, but may be detrimental to early crop growth in the cool spring of temperate regions.

No-tillage systems significantly increase the organic matter content of the upper few centimeters of soil (see Figures 14.19 and 12.23). During the initial four to six years of no-till management, the buildup of organic matter results in the immobilization of nutrients, especially nitrogen (see Section 11.3). This is in contrast to the mineralization of nutrients that is encouraged by the decline of soil organic matter under conventional tillage. Eventually, when soil organic matter stabilizes at a new higher level, nutrient mineralization rates under no-till increase. Higher moisture and lower oxygen levels may also stimulate denitrification (see Section 12.1). These processes sometimes result in the need for greater levels of nitrogen fertilization for optimum yields during the early years of no-till management.

The abundance, activity, and diversity of soil organisms tend to be greatest in conservation tillage systems characterized by high levels of surface residue, year round diverse vegetation, and little physical soil disturbance (see Section 10.14). Earthworms and fungi, both important for soil structure, are especially favored (see Table 10.7). However, no-till residues are in less intimate contact with the soil particles so their breakdown is delayed, increasing the length of time they remain as a protective surface barrier.

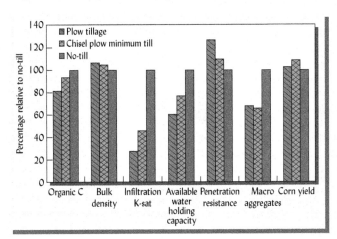

Figure 14.19 *Effects of 49 years of three tillage systems on soil organic matter (0 to 30 cm), soil physical soil properties (0 to 10 or 0 to 20 cm), and corn yields averaged across two fine-textured soils in Ohio (a Fragiudalf and an Epiaqualf) growing continuous corn and corn–soybean rotations without cover crops. Values for no-till were taken as 100, and the others are shown in comparison. Bulk density (0 to 10 cm) and corn yields (five-year average) were about the same for each tillage system, but for saturated hydraulic conductivity (0 to 10 cm), available water-holding capacity (0 to 20 cm), and macroaggregation (>2 mm) the no-till system was decidedly more beneficial than the other two systems. No-till had especially large impacts on saturated hydraulic conductivity (0–10 cm), available water holding capacity (0–20 cm), and macroaggregation (>2 mm).* [Graphed from data in Kumar et al. (2012a) and Kumar et al. (2012b)]

Figure 14.20 *Vegetative barriers create natural terraces. (Photo) Elephant grass has been vegetatively planted on the contour in a crop field. (a) Root cuttings are planted perpendicular to the slope direction. In a year or so the dense grass root and shoot growth will serve as a barrier to hold soil particles while permitting some water to pass on through. (b) Soil eventually accumulates above the grass, basically forming a terrace wall, as indicated by the different levels on which the people are standing.* (Photo and diagram courtesy of Ray R. Weil)

14.7 VEGETATIVE BARRIERS

Narrow rows of permanent vegetation (usually grasses or shrubs) planted on the contour can be used to slow down runoff, trap sediment, and eventually build up "natural" or "living" terraces (Figure 14.20). In some situations, tropical grasses (e.g., deep-rooted, drought-tolerant species like *vetiver grass or elephant grass*) have shown considerable promise as an affordable alternative to the construction of terraces.

The deep-rooted grass plants have dense, stiff stems that tend to filter out soil particles from muddy runoff and catch soil thrown downslope by tillage. This sediment and soil accumulates on the upslope side of the grass barrier and, in time, actually creates a terrace that may be more than 1 m above the soil surface on the downslope side of the plants. Narrow grass hedges have effectively reduced runoff and erosion from soils in midwestern United States (Figure 14.21). In temperate regions, conservation technologies (like the buffer strips and winter cover crops discussed in Section 13.2) can improve crop production in the short term as

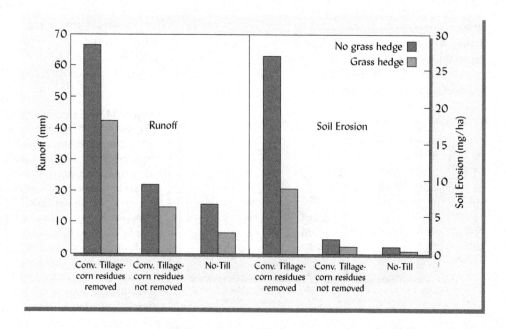

Figure 14.21 *Narrow grass hedges can be effective tools to reduce losses of runoff water and soil. Switchgrass hedges 0.72 m wide were established across the slope every 16 m in a field of continuous corn that then received a total of 120 mm of simulated rainfall. The soil was a Typic Hapludoll in Iowa with an average slope of 12%. Corn residues left on the soil surface as well as no-till systems also drastically reduced losses, especially of the soil.* [Data estimated from a figure in Gilley et al. (2000)]

Table 14.6
RANGE IN THE EFFECTS OF MULCHING, CONTOUR CULTIVATION, AND GRASS CONTOUR HEDGES ON SOIL EROSION AND CROP YIELDS

Practice	Reduction in soil erosion, %	Increase in crop yield, %
Mulching	78–98	7–188
Contour cultivation	50–86	6–66
Grass contour hedges	40–70	38–73

From Doolette and Smyle (1990), a review of more than 200 studies.

well as protect soil and water quality in the longer term. The benefits of reduced erosion and increased crop yields resulting from several vegetation-based erosion-control practices are summarized in Table 14.6.

14.8 CONTROL OF GULLY EROSION AND MASS WASTING

Gullies rarely form in soils protected by healthy, dense, forest, or sod vegetation, but are common on deserts, sparsely vegetated rangeland, and open woodland in which the soil is only partially vegetated. Gullies also readily form in soils exposed by tillage or grading if small rills are allowed to coalesce so that running water eats into the land (Figure 14.22, *left*). Water concentrated by poorly designed roads and trails may cause gullies to form even in dense forests. In many cases, neglected gullies will continue to grow and after a few years, devastate the landscape (see Figure 14.22, *right*). On the other hand, in some stony soils a layer of rock fragments left behind when finer particles are scoured away may protect the gully channel bottom from further cutting action.

Remedial Treatment of Gullies

If small enough, gullies can be filled in, shaped for smooth water flow, sown to grass, and thereafter be left undisturbed to serve as grassed waterways. When the gully erosion is too active to be checked in this manner, more extensive treatment may be required. If the gully is not

Figure 14.22 *The devastation of gully erosion. (Left) Gully erosion in action on a highly erodible soil in western Tennessee, United States. The roots of the small corn plants are powerless to prevent the cutting action of the concentrated water flow. (Right) The legacy of neglect of human-induced accelerated erosion. Tillage of sloping soils during the days of the Roman Empire began a process of accelerated erosion that eventually turned swales into jagged gullies that continue to cut into this Italian landscape with each heavy rain. For a sense of scale, note the dark olive trees and white houses on the grassy, gentle slopes of the relatively uneroded hilltops.* (Left photo courtesy of the USDA Natural Resources Conservation Service; right photo courtesy of Ray R. Weil)

too large, a series of check dams about 0.5 to 1 m high may be constructed at intervals of 4 to 9 m, depending on the slope. These small dams may be constructed from materials available on site, such as large rocks, rotted hay bales, brush, or logs. Wire netting may be used to stabilize such structures made from stones. Check dams, whether large or small, should be constructed with the general features illustrated in Figure 14.23. After a time, enough sediment may collect behind the dams to form a series of bench terraces which may be cultivated or put into permanent sod. Improperly constructed check dams, however, may make matters worse when water washes around them, enlarging the gully (Figure 14.23, *left*).

With very large gullies, it may be necessary to divert the runoff away from the head of the channel and install more permanent dams of earth, concrete, or stone in the channel itself. Again, sediment deposited above the dams will slowly fill in the gully.

Mass Wasting on Unstable Slopes

The downhill movement of large masses of unstable soil (**mass wasting**, see also Section 4.9) is quite different from the erosion of the soil surface, which is the main topic of this chapter. Mass wasting most commonly occurs on very steep slopes (usually greater than 50% slope). While this type of soil loss sometimes occurs on steep pastures, it is most common on nonagricultural land. Mass wasting can take several forms. **Soil creep** is the slow deformation (without shear failure) of the soil profile as the upper layers move imperceptibly downhill. As the soil of a forested slope deforms, trees attempting to right themselves produce curved trunks, a common sign of soil creep (see Figure 4.41). **Mud flows** involve the partial liquefaction and

Figure 14.23 *A schematic drawing of a check dam used to arrest gully erosion (upper center). Whether made from rock, brush, concrete, or other materials, a check dam should have the general features shown. The structure should be dug into the walls of the gully to prevent water from going around it. The center of the dam should be lower so that water will spill over there and not wash out the soil of the gully walls. An erosion-resistant apron made from densely bundled brush, concrete, large rocks, or similar material should be installed beneath the center of the dam to prevent the overflow from undercutting the structure. An effective check employing these features is shown at the right. In contrast to the gully-healing effect of a well-designed check dam, the haphazard dumping of rocks, brush, or junked cars into a gully will make matters worse, not better (left).* (Diagram and photo courtesy of Ray R. Weil)

Figure 14.24 While erosion by water and wind removes material from just the soil surface, mass wasting moves a much thicker layer of soil and regolith materials all at once. Mass wasting includes both gradual processes such as soil creep and moderately rapid processes such as mud flows, as well as sudden processes such as the landslide that occurred in the scene shown here (for scale, note the man standing on a sheared and partially rotated soil block). (Photo courtesy of Ray R. Weil)

moderately rapid flow of saturated soil due to loss of cohesion between particles (see Figure 4.41). **Landslides** occur when sudden shear failure, usually under very wet conditions, causes the rapid downhill movement of a mass of soil (Figure 14.24)

Mass wasting is sometimes triggered by human activities that undermine natural stabilizing forces or cause the soil to become water saturated as a result of concentrated water flow. The rotting of large soil-anchoring tree roots several years after clear-cutting a forest or excavations cutting into the toe of a steep slope are all-too-common examples.

14.9 CONTROL OF ACCELERATED EROSION ON RANGE AND FORESTLAND

Rangeland Problems

Many semiarid rangelands lose large amounts of soil under natural conditions, but accelerated erosion can lead to even greater losses if human influences are not managed carefully. Overgrazing by cattle, which leads to the deterioration of the vegetative cover on rangelands, is a prime example (Figure 14.25). Grass cover generally protects the soil better than the scattered shrubs that usually replace it under the influence of poorly managed livestock grazing. In addition, cattle congregating around poorly distributed water sources and salt licks may completely denude the soil. Cattle trails, as well as ruts from off-road vehicles, can channelize runoff water and spawn gullies that eat into the landscape. Because of the prevalence of dry conditions, wind erosion (to be discussed in Sections 14.10 and 14.11) also plays a major role in the deterioration of rangeland soils.

Erosion on Forestlands

In contrast to deserts and rangelands, land under healthy, undisturbed forests loses very small amounts of soil. However, accelerated erosion can be a serious problem on forested land, both because the rates of soil loss may be quite high and because the amount of land involved may be enormous. The main cause of accelerated erosion in forested watersheds is usually logging operations and the trampling of trails and off-trail areas by recreational users (or cattle, in some areas).

Contrary to the common perception, it is the forest floor, rather than the tree canopy or roots, that protects the soil from erosion (Figure 14.26, *left*). In fact, rainwater dripping from the leaves of tall trees forms very large drops that reach terminal velocity and impact the ground with more energy than direct rain from even intense storms. If the forest floor has been

Figure 14.25 *Overstocking and poorly managed grazing is a major cause of soil erosion and degradation worldwide. (Left) Damaged vegetative cover and eroded soil caused by too many poorly managed cattle in a semiarid rangeland in East Africa. (Right) Soil and water quality in a humid region of eastern North America damaged by overstocking with cattle (note the differences in soil erosion, vegetation, and stream configuration between the left and right sides of the fence).* (Photo courtesy of Ray R. Weil)

disturbed and mineral soil exposed, serious splash erosion can result (see Figure 14.26, *right*). Gully erosion can also occur under the forest canopy if water is concentrated, as by unplanned footpaths or poorly designed roads.

Practices to Reduce Soil Loss Caused by Timber Production[7]

The main sources of eroded soil from timber production are *logging roads* (that are built to provide access to the area by trucks), *skid trails* (the paths along which logs are dragged), and *yarding areas* (the places where collected logs are sized and loaded onto trucks). Relatively little

Figure 14.26 *The leaf mulch on the forest floor, rather than the tree roots or canopy, provides most of the protection against erosion in a wooded ecosystem. (Left) An undisturbed temperate deciduous forest floor (as seen through a rotten stump). The leafless canopy will do little to intercept rain during winter months. During the summer, rainwater dripping from the foliage of tall trees may impact the forest floor with as much energy as unimpeded rain. (Right) Severe erosion has taken place under the tree canopy in a wooded area where the protective forest floor has been destroyed by foot traffic. The exposed tree roots indicate that nearly 25 cm of the soil profile has washed away.* (Photo courtesy of Ray R. Weil)

---

[7] For general introduction to this topic, see Chapter 3 in Nyland (2016). For an analysis of erosion causes on steep forest lands, see Siddle et al. (2006) and Safari et al. (2016).

erosion results directly from the mere felling of the trees (except where large tree roots are needed to anchor the soil against mass wasting, as discussed in Section 14.8). Strategies to control erosion should include consideration of (1) intensity of timber harvest, (2) methods used to remove logs, (3) scheduling of timber harvests, and (4) design and management of roads and trails. Soil disturbance in preparation for tree regeneration (such as tillage to eliminate weed competition or provide better seed-to-soil contact) must also be limited to sites with low susceptibility to erosion.

Intensity of Timber Harvest On the steepest, most erodible sites, environmental stewardship may require that timber harvest be foregone and the land be given only protective management. On somewhat less susceptible sites, selective cutting (occasional removal of only the oldest trees) may be practiced without detrimental results. The shelterwood system (in which a substantial number of large trees are left standing after all others are harvested) is probably the next most intensive method and can be used on moderately susceptible sites. Clear-cutting (removal of all trees from large blocks of forest) should be used only on gentle slopes with stable soils.

Tree Removal The least expensive and most commonly used method of tree removal is by wheeled tractors called *skidders* (see Figure 4.32). This method generally disrupts the forest floor, exposing the mineral soil on perhaps 30 to 50% of the harvested area. In contrast, more expensive methods using cables to lift one end of the log off the ground are likely to expose mineral soil on only 15 to 25% of the area. For very sensitive sites, logs can be lifted to yarding areas by balloon or helicopter, practices which are very expensive but result in as little as 4 to 8% bare mineral soil.

Much erosion can be prevented by limiting entry into the forest by machinery to those periods when the soil is either dry or (in temperate regions) frozen and covered with snow. Damage to the forest floor (including both compaction and exposure of mineral soil) occurs much more readily if the soil is very wet. In addition, wheel ruts easily form in wet soils and channel runoff to initiate gully erosion.

Design and Management of Roads and Skid Trails Poorly built logging roads may lose as much as 100 Mg/ha of soil by erosion of the road surface, the drainage ditch walls, or the soil exposed by road cuts into the hillside. An inexpensive measure is to control water flow with cross channels (shallow ditches or **water bars**, as shown in Figure 14.27, *right*) every 25 to 100 m

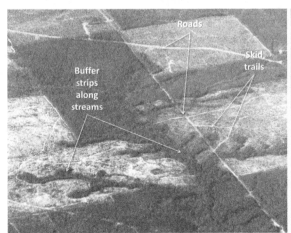 

Figure 14.27 *Forestry practices that minimize damage from erosion due to timber harvests. (Left) An aerial view of clear-cut and unharvested blocks of pine forest in Alabama, United States. Narrow skid trails can be seen leading up to a staging area located on high ground and spreading apart going downhill toward the streams. This avoids the easier practice of dragging logs downhill so that skid trails converge at a low point, inviting runoff water to concentrate into gully-cutting torrents. Also visible are several dark-green buffer strips where the trees were left undisturbed along streams to provide protection for water quality. (Right) An open-top culvert (also called a water bar) in a well-designed logging road in Montana, United States. Water bars placed at frequent intervals lead runoff water, a little at a time, off the road and into densely vegetated areas.* (Photo courtesy of Ray R. Weil)

to prevent excessive accumulation of water and safely spread it out onto areas protected by natural vegetation. After timber harvest is complete, the roads in an area should be grassed over and closed to traffic.

Skidding trails that lead runoff water downhill toward a yarding area invite the formation of gullies. Repeated trips dragging logs over the same secondary trails also greatly increase the amount of mineral soil exposed to erosive forces. Both practices should be avoided, and yarding areas should be located on the highest-elevation, most level, and well-drained areas available (Figure 14.27, *left*).

Buffer Strips Along Stream Channels When forests are harvested, buffer strips as wide as 1.5 times the height of the tallest trees should generally be left untouched along all streams (Figure 14.27, *left*). As discussed in Section 13.2, buffer strips of dense vegetation have a high capacity to remove sediment and nutrients from runoff water. Forested buffers also protect the stream from excessive logging debris. In addition, streamside trees shade the water, protecting it from the undesirable heating that would result from exposure to direct sunlight.

14.10 EROSION AND SEDIMENT CONTROL ON CONSTRUCTION SITES

Although active construction sites cover relatively little land in most watersheds, they may still be a major source of eroded sediment because the potential erosion per hectare on drastically disturbed land is commonly 100 times that on agricultural land. To prevent serious sediment pollution from construction sites, governments in the United States (e.g., through state laws and the Federal Clean Water Act of 1992) and in many other industrialized countries require that contractors develop detailed erosion- or sediment-control plans before initiating construction projects that will disturb a significant area of land. The goals of erosion control on construction sites are (1) to avoid on-site damage, such as undercutting of foundations or finished grades and loss of topsoil needed for eventual landscaping and (2) to retain eroded sediment on-site so as to avoid environmental damages (and liabilities).

Principles of Erosion Control on Construction Sites

Five basic steps are useful in developing plans to meet the aforementioned goals:

1. When possible, schedule the main excavation activities for low-rainfall periods of the year.
2. Divide the project into as many phases as possible, so that only a few small areas must be cleared of vegetation and graded at any one time.
3. Cover disturbed soils as completely as possible, using vegetation or other materials.
4. Control the flow of runoff to retain or detain most of the water and move the excess safely off the site without destructive gully formation.
5. Trap the sediment before releasing the excess runoff water off-site.

The last three steps bear further elaboration. They are best implemented as specific practices integrated into an overall erosion-control plan for the site.

Keeping the Disturbed Soil Covered

Soils freshly disturbed by excavation or grading operations are characterized by very high erodibility (K values). This is especially true for low-organic-matter subsoil materials. Potential erosion can be extremely high (200 to 400 Mg/ha is not uncommon) unless the C value is made very low by providing good soil cover. This is best accomplished by allowing the natural vegetation to remain undisturbed for as long as possible, rather than clearing and grading the entire project area at the beginning of construction (see step 1, preceding). Once a section of the site

Figure 14.28 Two means of providing rapid temporary soil cover while establishing vegetation on erodible slopes. (Left) A hydroseeder spraying a mixture of water, chopped straw, grass seed, fertilizer, and sticky polymers (inset close-up) that holds the mulch in place until the grass seed can take root. Hydroseeding is especially useful for efficiently establishing vegetative cover on difficult-to-reach areas. (Right) Erosion-control mats or blankets made of plastic netting or natural materials like wood fiber or jute (shown here) are rolled out over newly seeded grass to hold the seed and soil in place until the vegetative cover is established. In the inset, young grass seedlings are exposed as the jute erosion blanket is pulled away. (Photo courtesy of Ray R. Weil)

is graded, any sloping areas not directly involved in the construction should be sodded or sown to fast-growing grass species adapted to the soil and climatic conditions.

Seeded areas should be covered with **mulch** or specially manufactured **erosion mat or blankets** (Figure 14.28, *right*). Erosion blankets, made of various biodegradable or nonbiodegradable materials, provide instant soil cover, protect the seed from being washed away, and are highly effective at reducing soil erosion (Figure 14.29).

A commonly used technology to protect steep slopes and areas difficult to access, such as road cuts, is the **hydroseeder** (Figure 14.28, *left*) that sprays out a mixture of seed, fertilizer, lime (if needed), mulching material, and sticky polymers. Good construction-site management includes removal and stockpiling of the A horizon material before an area is graded (see Figure 1.12). This soil material is often quite high in fertility and is a potential source of

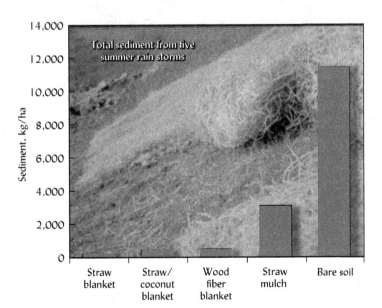

Figure 14.29 Total sediment generated by five summer storms on construction site soil left bare or protected by straw mulch or various types of commercial erosion blankets. The clayey soil met the State Department of Transportation specifications for topsoil. It was raked, fertilized, and seeded to grass mixture. The experimental plots had a 35% slope gradient and were 9.75 m long. Immediately after seeding, the soils were covered by the various erosion-control materials. The best grass vegetative cover and biomass was achieved on straw mulch plots in the first year. The best short-term sediment control was achieved by the commercial blankets, such as the wood-fiber blanket shown in the background photo. [Drawn from data in Benik et al. (2003); photo courtesy of Ray R. Weil]

sediment and nutrient pollution. The stockpile should therefore be given a grass and mulch cover to protect it from erosion until it is used to provide topsoil for landscaping around the finished structures.

Controlling the Runoff

Freshly exposed and disturbed subsoil material is highly susceptible to the cutting action of flowing water. The gullies so formed may ruin a grading job, undercut pavements and foundations, and produce enormous sediment loads. The flow of runoff water must be controlled by carefully planned grading, terracing, and channel construction. Most construction sites require a perimeter waterway to catch runoff before it leaves the site and to channel it to a retention basin.

The sides and bottom of such channels must be covered with "armor" to withstand the cutting force of flowing water. Where high water velocities are expected, the soil must be protected with **hard armor** such as **riprap** (large angular rocks, as shown in Figure 14.30), **gabions** (rectangular wire-mesh containers filled with hand-sized stone), or interlocking concrete blocks. Before installing the hard armor, the soil is first covered with a **geotextile** filter cloth (a tough nonwoven material) to prevent mixing of the soil into the rock or stone.

In smaller channels, and on more gentle slopes where relatively low water velocities will be encountered, **soft armor**, such as grass sod or erosion blankets, can be used. Generally, soft armor is cheaper and more aesthetically appealing than hard armor, and can lead to a more naturally functioning waterway once the plant community becomes fully established. Newer approaches to erosion control often involve physically reinforced vegetation such as trees or grasses planted in openings between concrete blocks or in tough erosion mats.

The term **bioengineering** describes techniques that use vegetation (locally native, noninvasive species are preferred) and natural biodegradable materials to protect channels subject to rather high water velocities. Examples include the use of **brush mattresses** to stabilize steep slopes. In this technique, live tree branches are tightly bundled together, staked down flat using long wooden pegs, and partially covered with soil. The so-called **live stake** technique (a version of which is exemplified by the willow cuttings in Figure 14.31, *right*) is another example of a bioengineering approach commonly used to stabilize soil along channels subject to high velocity water. In both cases, the soil is provided some immediate physical protection from scouring water, and eventually the dormant cuttings take root to provide permanent, deep-rooted vegetative protection.

Figure 14.30 The flow of runoff from large areas of bare soil must be carefully controlled if off-site pollution is to be avoided. Here, a carefully designed channel with a grass sod bottom and sides lined with large rocks (riprap) prevents gully erosion, reduces soil loss, and guides runoff around the perimeter of a construction site. (Photo courtesy of Ray R. Weil)

Figure 14.31 Stream bank erosion and its control with bioengineering techniques. (Left) Tumbling trees are a sign of a degraded stream with rapidly eroding banks that contribute large amounts of soils to the stream sediment load. (Right) A somewhat smaller stream that was similarly degraded, but has now been restored with several techniques that control bank erosion and improve water quality. The banks have been protected from scouring storm flows by both soft armor and hard armor. As soft armor, the stream banks have been seeded with a diversity of native grasses and forbs and the seeding protected temporarily by biodegradable erosion blankets. The live willow branches that are taking root along the base of the stream banks comprise another type of soft armor referred to as live staking. The large rocks (hard armor) protect the banks from scouring, cause water to form rapids that improve water oxygen content, and collect gravelly sediments that improve fish-spawning habitat. (Photo courtesy of Ray R. Weil)

Trapping the Sediment

For small areas of disturbed soils, several forms of sediment barriers can be used to filter the runoff before it is released. The most commonly used types of silt barriers are woven fabric silt fences and various bundles or bales of plant fiber such as straw. If installed properly, both can effectively slow the water flow so that most of the sediment is deposited on the uphill side of the barrier (Figure 14.32), while relatively clear water passes through.

On large construction sites, a system of protected slopes and channels leads storm runoff water to one or more retention and sedimentation ponds located at the lowest elevation of

Figure 14.32 Sediment-control measures used around the periphery of a construction site. (Left) A line of straw bales pegged to the ground allows water to seep through, but filters out much of the sediment and slows down the flow so that the water drops its sediment load. (Right) A properly installed silt fence effectively removes sediment from runoff water leaving the edge of a construction site. The silt fencing material is a woven plastic that allows water to flow through at a much reduced rate. Note the light-colored sediment on the inside and the undisturbed forest floor on the outside of the silt fence. A silt fence must be embedded in the soil and supported with stout stakes or posts. Improperly installed silt fencing is useless, as sediment-laden water may pass underneath or may even knock the fence over. (Photo courtesy of Ray R. Weil)

Figure 14.33 *Sediment and storm water retention. (Left) The effect of a small sediment impoundment structure on the retention of suspended soil eroded from a construction site. Note the riprap-lined channel leading runoff water into the pond and the delta-like alluvial fan of sediment deposited as the water's velocity is reduced where it enters the pond. The standpipe in the background has holes that allow the clearest water near the top to overflow and leave the construction site. (Right) A larger sediment retention pond shown several years after construction activities were completed. A well-designed sediment structure will serve as a decorative landscaping asset to the property and continue to detain storm water and retain sediment. Only rarely does enough water accumulate to reach the outflow grating, and even then, most sediment is retained on the pond bottom.* (Photo courtesy of Ray R. Weil)

the site. As the flowing water meets the still water in the pond it drops most of its sediment load (Figure 14.33), allowing the relatively clear water to be skimmed off the top and released to the next pond or off the site. Wetlands (Section 7.7) are often constructed to help detain and purify the overflow from sedimentation ponds before the water is released into a natural stream or river.

Construction site erosion-control measures are commonly designed to retain the runoff from small storms on-site. The retention ponds must also be able to deal with runoff generated from intense rainstorms—the kind that may be expected to occur on a site only once in every 10 or even 100 years. In designing the capacity of sediment-retention ponds, the erosion models discussed in Section 14.4 are used to estimate the amount of sediment that is likely to be eroded from the site. While expensive to construct, well-designed sediment-retention ponds can be incorporated as permanent aesthetic water features that enhance the value of the final project.

14.11 WIND EROSION: IMPORTANCE AND FACTORS AFFECTING IT[8]

Up to this point we have focused on soil erosion by water, but wind, too, causes much soil degradation and loss around the world. Generally wind erosion is most serious in dry regions and water erosion in wet regions. However, there are many places that suffer from both types of erosion because both wet and dry seasons occur in the same location. Regions vulnerable to both wind and water erosion include the Sahel in Africa, the Pacific coast of South America, and the Loess Plateau in China (Figure 14.34).

Overgrazing of the fragile lands of arid and semiarid areas destroys the protective biological soil crusts (see Section 10.14) and much of the native vegetation. Tillage for dryland crop production also dries out and lays bare millions of hectares, making them much more vulnerable to the wind than when the native vegetation was undisturbed. In the United States wind moves about two-thirds as much soil as is transported by water erosion. In six of the Great Plains states, annual wind erosion exceeds water erosion on cropland.

Wind erosion occurs when strong winds blow across soils with relatively dry surface layers. All kinds of soils and soil materials are affected, but silty and fine sandy soils generally

---

[8]For a clearly written and detailed explanation of wind erosion and its control, see Zobeck and Van Pelt (2014).

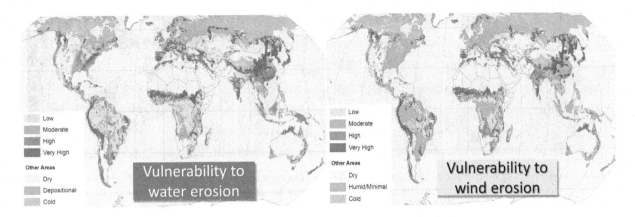

Figure 14.34 *Global distribution of soil vulnerability to erosion by water (left) and to erosion by wind (right). Climate is the main variable driving the vulnerabilities, with additional influences of topography, soil properties, and vegetation. Although generally wind erosion is most serious in dry regions and water erosion in wet regions, there are many places that suffer from both types of erosion, because both wet and dry seasons occur in the same location.* (Maps modified from USDA NRCS, Soil Survey Division, Office of World Soils)

are most quick to "blow." Sand particles may pile up in dunes, but the finer soil particles can be carried to great heights and for thousands of kilometers (Figure 14.35)—even from one continent, across the ocean, to another. Wind erosion causes widespread damage, not only to the vegetation and soils of the eroding site, but also to anything that can be damaged by the abrasiveness of soil-laden wind, and finally to the off-site area where the eroded soil material settles back to earth (Figure 14.36). The on-site and off-site damages caused by wind erosion were discussed in Section 14.2.

The erosive force of the wind varies greatly among regions of the world and among seasons of the year within a specific location. For example, the Great Plains region of North America is subject to winds with 5 to 10 times the erosive force of the winds common in the eastern parts of Canada and the United States.

Figure 14.35 *Wind erosion can have drastic off-site effects on air quality and dust deposition. Damaging dust storms arise when high winds sweep across arid regions with unprotected soils. These storms can be regional in scale and the resulting dust deposition can have intercontinental effects. (Left) A giant wall of dust swallows large parts of China and Mongolia in a typical spring dust storm spawned from severe wind erosion in the loess deserts. (Right) A smaller, but still impressively damaging, dust storm bears down upon the town of Lubbock, in arid western Texas.* (Photos courtesy of NASA (left) and Thomas Todd Lindley (right))

  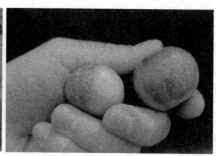

Figure 14.36 Some effects of wind erosion. (Left) Soil eroded by wind during a single dust storm has piled up to a depth of nearly 1 m along a fencerow in semiarid Idaho. The existing soil and plants are covered by deposits that are quite unproductive because the soil structure has been destroyed. Also, these deposits are subject to further movement when the wind direction changes. (Middle) Direct wind damage to a tomato crop in a sandy field in humid Delaware. The tomato plants are buried beneath windblown sandy soil. (Right) Young tomato fruits show damage incurred by sandblasting during a windstorm. (Photo courtesy of Ray R. Weil)

Mechanics of Wind Erosion

Like water erosion, wind erosion involves three processes: (1) *detachment*, (2) *transportation*, and (3) *deposition*. The moving air, itself, results in some detachment of tiny soil grains from the granules or clods of which they are a part. However, when the moving air is laden with soil particles, its abrasive power is greatly increased. The impact of these rapidly moving grains dislodges other particles from soil clods and aggregates. These dislodged particles are now ready for one of the three modes of wind-induced transportation, depending mostly on their size.

The first and most important mode of particle transportation is that of **saltation**, or the movement of soil by a series of short bounces along the ground surface (Figure 14.37). The particles remain fairly close to the ground as they bounce, seldom rising more than 30 cm or so. Depending on conditions, this process may account for 50 to 90% of the total movement of soil.

Saltation also encourages **soil creep**, or the rolling and sliding along the surface of the larger particles. The bouncing particles carried by saltation strike larger particles and aggregate and accelerate their movement along the surface. Soil creep accounts for the movement of particles up to about 1.0 mm in diameter, which may amount to 5 to 25% of the total movement.

The most visually spectacular method of transporting soil particles is by movement in **suspension**. Here, dust particles of a fine-sand size and smaller are moved parallel to the ground surface and upward. Although some of them are carried at a height no greater than a few meters, the turbulent action of the wind results in other particles being carried kilometers upward into the atmosphere and many hundreds of kilometers horizontally. These particles return to the earth only when the wind subsides and/or when precipitation washes them down.

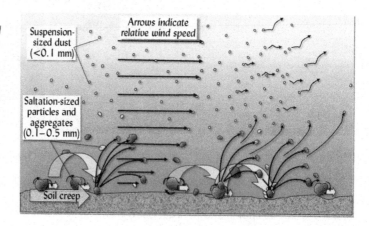

Figure 14.37 How wind moves soil. Wind is blowing from left to right and is slowed somewhat by friction and obstructions near the soil surface (straight arrows). Fine particles are picked up and carried into the atmosphere, where they remain suspended until the wind slows. Medium sized particles or aggregates, being too large to be carried up in suspension, are bounced along the soil surface. When they strike larger soil aggregates, they release particles of various sizes. This process of medium-sized particles skipping along the surface is termed saltation. In a process termed soil creep, larger particles also roll along the soil surface, kept in motion by the wind itself and by collisions with the saltating particles. (Diagram courtesy of Ray R. Weil)

Factors Affecting Wind Erosion

Wet soils do not blow because of the adhesion between water and soil particles. Dry winds generally lower the soil moisture content to below the wilting point before wind erosion takes place. Other factors that influence wind erosion are (1) wind velocity and turbulence, (2) soil surface conditions, (3) soil characteristics, and (4) the nature and orientation of the vegetation.

Wind Velocity and Turbulence The *threshold velocity*—the wind speed required to initiate soil movement—is usually about 25 km/h (7 m/s). At higher wind speeds, soil movement is proportional to the cube of the wind velocity. Thus, the quantity of soil carried by wind increases dramatically as wind speeds above 30 km/h are reached. Wind turbulence also influences the capacity of the atmosphere to transport matter. Although the wind itself has some direct influence in picking up fine soil, the impact of wind-carried particles as they strike the soil (saltation) is probably more important.

Surface Roughness Wind erosion is less severe where the soil surface is rough. This roughness can be obtained by proper tillage methods, which create large clods or ridges. Leaving a stubble mulch (see Section 6.4) is an even more effective way of reducing windborne soil losses.

Soil Properties In addition to moisture content, wind erosion is also influenced by (1) mechanical stability of soil clods and aggregates, (2) stability of soil crusts, (3) bulk density, and (4) size of erodible soil fractions. Some clods resist the abrasive action of wind-carried particles. If a soil crust resulting from a previous rain is present, it, too, may be able to withstand the wind's erosive power. The presence of clay, organic matter, and other cementing agents is also important in helping clods and aggregates resist abrasion. This is one reason why sandy soils, which are low in such agents, are so easily eroded by wind. Because they participate in saltation, soil particles or aggregates about 0.1 mm in diameter are more erodible than those larger or smaller in size.

Vegetation Vegetation or stubble mulch will reduce wind erosion hazards, especially if rows run perpendicular to the prevailing wind direction. This effectively slows wind movement near the soil surface. In addition, plant roots help bind the soil and make it less susceptible to wind damage. Biological crusts are important in protecting desert soils from erosion.

14.12 PREDICTING AND CONTROLLING WIND EROSION

As for soil erosion by water, mathematical models or equations have also been developed to predict soil loss from the action of wind. The wind erosion prediction equation (WEQ) has been in use since the late 1960s:

$$E = f(I \times C \times K \times L \times V) \tag{14.2}$$

The WEQ involves the major factors that determine the severity of the erosion, but it also considers how these factors interact with each other. Consequently, it is not as simple as is the USLE for water erosion. Evidence of the interaction among factors is seen in Figure 14.38. The **soil erodibility factor**, I, relates to the properties of the soil and to the degree of slope of the site in question. The **soil-ridge-roughness** factor, K, takes into consideration the cloddiness of the soil surface, vegetative cover, V, and ridges on the soil surface. The **climatic factor**, C, involves wind velocity, soil temperature, and precipitation (which helps control soil moisture). The **width of field factor**, L, is the width of a field in the downwind direction. Naturally, the width changes as the direction of the wind changes, so the prevailing wind direction is generally used. The **vegetative cover**, V, relates not only to the degree of soil surface covered with residues, but to the nature of the cover—whether it is living or dead, still standing, or flat on the ground. The situation may be complicated by interactions between two or more of these wind erosion factors. For instance, low rainfall may result in less vegetative cover and therefore more wind erosion in subsequent years (Figure 14.38).

Figure 14.38 Interaction of climate and vegetative cover can affect the amount of wind erosion (aeolian sediment flux). An increase in mean temperature on the Colorado plateau of 3°C resulted in greater soil drying that decreased vegetation cover from 45 to 20% of the ground surface. In turn, this climate-induced reduction in vegetative cover more than doubled the amount of erosion caused by high winds the following year. [Graph based on data selected from Munson et al. (2011)]

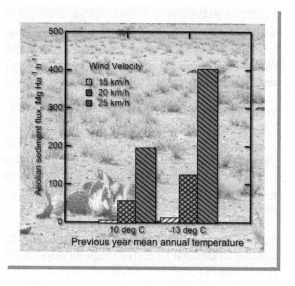

A revised, more complex, and more accurate computer-based prediction model has been developed, and is known as the **revised wind erosion equation (RWEQ)**. It is still an empirical model based on many site-years of research to characterize the relationship between observable conditions and resulting wind erosion severity. Table 14.7 outlines the factors taken into consideration by RWEQ, which (like RUSLE) calculates the erosion hazard during 15-day intervals throughout the year.

Scientists and engineers around the world are also cooperating in the continual development of a much more complex process-based model known as the **Wind Erosion Prediction System (WEPS)**. Like its water erosion sister, WEPP, this computer program simulates all the basic processes of wind interaction with soil. The USDA makes this model freely available online (https://infosys.ars.usda.gov/WindErosion/weps/wepshome.html). However, to run complex process–based models like WEPP and WEPS one must supply a great deal of information about every aspect of the site on which erosion is to be predicted. Therefore, the simpler, tried-and-true empirical models also continue to be in widespread use.

Control of Wind Erosion

Soil Moisture The factors included in the wind erosion equation suggest approaches that can be taken to reduce wind erosion. For example, since soil moisture increases cohesiveness, the wind speed required to detach soil particles increases dramatically as soil moisture increases. Therefore, where irrigation water is available, an effective protective practice is to moisten the soil surface when high winds are predicted. Unfortunately, most wind erosion occurs in dry regions without available irrigation.

Soil Cover Plant cover effectively protects soil from blowing, especially if the plant roots are well established. Crop rotations that include cover crops can greatly diminish wind erosion. In semiarid areas, however, many farmers attempt to conserve soil moisture by leaving some of their fields fallow (bare of water-using vegetation) during alternate summer seasons. Fallow, especially if practiced with tillage that maintains a smooth, bare soil surface, can greatly increase the susceptibility to wind erosion. Consequently, if farmers feel they must leave fields in fallow to conserve profile moisture, it is critical to control wind erosion by employing means other than growing vegetative cover.

Tillage Certain conservation tillage practices described in Section 14.6 were used for wind erosion control long before they became popular as water erosion control practices. No-till practices that leave the soil in a consolidated but biopore riddled mass covered by growing vegetation or anchored residues is the best protection available for cropland. On tilled cropland, a rough surface and at least partial some vegetative or residue cover can be maintained

Table 14.7
SOME FACTORS INTEGRATED IN THE REVISED WIND EROSION EQUATION (RWEQ) MODEL
The RWEQ program calculates values for each factor for each 15-day period during the year. It also calculates interactions; that is, it uses the value of one factor to modify other factors.

Model factor	Subfactor terms in the model	Comments
Weather factor	Weather factor *WF*; includes terms for wind velocity, direction, air temperature, solar radiation, rainfall, and snow cover.	Modified by other factors such as soil wetness.
Soil factors	Erodible fraction *EF*; fraction smaller than 0.84 mm diameter.	Based on sand, silt, organic matter, and rock cover.
	Soil crust factor *SCF*.	Induced by rainfall, eliminated by tillage.
	Surface roughness *SR*; a random component due to clods and/or an oriented component due to ridges.	Interacts with rainfall and tillage.
	Soil wetness *SW*.	Computed from rainfall minus evapotranspiration.
Tillage factor	Tillage factor *TF*; depends on type of implement, soil conditions, timing, and so forth.	Modifies surface roughness, crust factor, and so forth.
Hill factor	Hill factor *HF*; slope gradients and length input.	Affects wind speed (high going upslope, lower going down).
Irrigation factor	Irrigation factor *IR*.	Equivalent to added rainfall.
Crops factor	Flat residue *SLRf* factor.	Soil cover by residues lying on the surface is estimated for each crop, including changes over time due to decay, and so forth.
	Standing residue *SLRs*; depends on crop, harvest height, plant density, and so forth.	Standing residues reduce the wind speed at the soil surface. Decay is slower than for flat residues.
	Crop canopy factor *SLRc*.	Changes daily with crop growth.
Barriers	Barriers (e.g., tree windbreaks); includes orientation, density, height, spacing.	Reduce leeward wind velocity.

Based on information in Fryrear et al. (2000).

by using appropriate tillage practices. Ideally, both living and dead vegetation should be well anchored into the soil to prevent it from blowing away. Stubble mulching has proven to be an effective practice for this purpose (see Section 6.4).

The effect of tillage depends not only on the type of implement used but also on the timing of the tillage operation. Tillage can greatly reduce wind erosion if it is done while there is sufficient soil water to cause large clods to form. Tillage on a dry soil may produce a fine, dusty surface that aggravates the erosion problem. Tillage to provide for a cloddy surface condition should be at right angles to the prevailing winds. Likewise, strip-cropping and alternate strips of cropped and fallowed land should be perpendicular to the wind.

Barriers Barriers or *windbreaks* such as tree shelter belts (Figure 14.39a,b) are effective in reducing wind velocities for short distances and for trapping drifting soil. Significant protection against wind erosion extends to a distance of about 10 times the height of the barrier (Figure 14.39e). Tree windbreaks and rows of tenacious shrubs are especially effective. Rows of grasses are often used to form microwindbreaks. For example, narrow strips of cereal rye, planted in across the field perpendicular to the prevailing wind, are sometimes used on peat lands and on sandy soils (see Figure 14.39c,d). Narrow rows of perennial grasses (e.g., tall wheatgrass) have been useful for a combination of wind erosion control and capturing of winter snows in the Northern Great Plains.

Not all windbreaks are living; picket fences and burlap screens, though less efficient as windbreaks than trees, are often preferred because they can be moved from place to place as

Figure 14.39 *Wind breaks to reduce wind erosion. (a,b) Trees make good windbreaks and add beauty to a North Dakota farm homestead in winter and summer. (e) The effect of a windbreak on wind velocity. The wind is deflected upward by the trees and is slowed down even before reaching them. On the leeward side further reduction occurs; the effect being felt as far as ten times the height of the trees. (c,d) Narrow strips of cereal rye act as miniature windbreaks to protect watermelons from wind erosion on a loamy sand on the mid-Atlantic coastal plain.* [Photos courtesy of Ray R. Weil, diagram based on concepts in Tatarko (2010)]

crops and cropping practices are varied. A particularly successful nonliving microwindbreak technology has been successfully developed and deployed to control blowing sands and aid in reforestation in the arid northwest of China (Figure 14.40).

14.13 TILLAGE EROSION

Movement of Soil by Tillage

Globally, wind and water are the major agents of soil erosion. In addition, the action of tillage itself is an important third agent of erosion. Any tillage operation will loosen and move soil. Some types of tillage, such as moldboard plowing and disk harrowing pick up and invert large quantities of soil. Some, such as the chisel plow, spread it over distances of several meters. Even

Figure 14.40 *Ecological engineers use straw checkerboards for wind erosion control, reforestation, and improvement of sandy soils in arid Ningxia province, China. (a) Windblown sand has completely degraded a once forested area, but reforestation efforts can be seen as the dark areas on the sand dunes. (b) Rice straw produced in irrigated river valleys is trucked into the desert where workers use shovels to anchor it into the loose sand to form squares about 1 m on a side. (c) A tiny tree seedling will be planted in each rice straw square, which remains substantially intact and inhibits wind erosion for several decades. (d) Some 20 years after planting, the trees are growing but the straw has mostly disappeared. Dust trapped by the straw checkerboard changed soil texture from 99% sand to only 80 to 90% sand and 10 to 20% silt and clay.* [Information based on Li et al. (2006). Photos courtesy of Ray R. Weil]

Figure 14.41 *A chisel plow in action. Tillage implements such as this loosen and move large quantities of soil, some of which is thrown into the air. The amount of a certain soil that will be moved and the distance it is moved depend on the design of the implement, depth of tillage, and speed of travel. The soil moves mainly in the direction of travel, but will move much farther when travel is downslope so gravity assists the movement. When tillage is upslope, gravity hinders the forward movement of soil.* (Photo courtesy of Ray R. Weil)

field operations such as shallow cultivation to control weeds or planting seed may disturb and move substantial quantities of soil. The amount of soil moved and the distance it is moved depend on the design of the implement, the depth of tillage, and the speed of travel (as well as the type of soil). High-speed tillage especially tends to throw soil into the air as well as push it forward; chisel plowing (Figure 14.41) and vertical tillage (Figure 14.25) are examples.

Soil moves mainly in the direction of tillage. When the tillage implement is traveling upslope, gravity hinders the forward movement of soil. Soil will move much farther when the implement is traveling downslope and the force of gravity assists in moving the soil forward. Even when tillage is on the contour, soil thrown sideways will travel farther downslope than upslope. Therefore, the net movement of soil by tillage is always downslope. In this process, topsoil is removed from the hilltops and accumulates in the valleys, thus gradually leveling the landscape. Once the A horizon material is mostly scalped off the hilltops continued tillage will mix subsoil material into the tilled layer. This action often changes the color of the soil exposed on the hilltops (Figure 14.42).

Quantification of Tillage Erosion

There are several models of tillage erosion that are based on the relationship between the amounts of soil moved by tillage in the upslope and downslope directions and slope gradient. The most general of these models is the Tillage Erosion Risk Indicator model (TillERI),

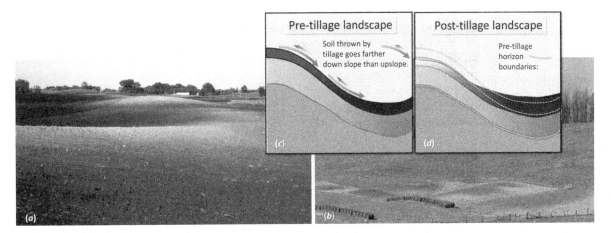

Figure 14.42 *The exposure of B horizon soil on the crests of small hills in cultivated landscapes is largely due to tillage erosion. (Left) The whitish calcareous subsoil material from Mollisols is mixed into the plow layer of a conventionally tilled field in sub-humid Minnesota. (Right) Reddish B horizon material is likewise mixed into the hilltop plow layer of Ultisols in a strip-cropped field in humid Virginia. The diagram illustrates how tillage scalps the hilltops by throwing soil farther downslope than upslope, resulting in a net movement of soil downslope and gradual leveling of the landscape.* (Diagram courtesy of Ray R. Weil. Photos courtesy of Ray R. Weil (right) and David A. Lobb, University of Manitoba (left))

which in its simplest form considers the product of a tillage implement factor and a landscape factor

$$A_t = E_t \times E_l \tag{14.3}$$

Here A_t = annual *rate of soil movement* downslope due to tillage erosion (Mg/ha/y); E_t = the *erosivity* of tillage operations (kg%/m/y^{-1}) expressed as kilograms of soil moved annually per meter of tillage width and percentage slope inclination (where a slope of 45° = 100%); and E_l = the *erodibility* of the landscape (%m/ha).

The landscape erodibility factor, E_l, is based on slope length (l) and slope gradient (s) factor values and can be determined using the topographic data from the revised universal soil loss equation (RUSLE). Tillage erosion is most active on landscapes with many small hills characterized by short, steep slopes (such as those pictured in Figure 14.42). Soil properties such as water content, texture, and structure also influence the vulnerability of a landscape to tillage erosion.

The erosivity of tillage operations, E_t, is a function of four factors (I_d, I_o, I_m, I_b) related to the tillage implement and how it is used. The first implement factor, I_d, represents the *design* of the tillage implement, i.e., the type, size, number, and angle of coulters, tines, or other steel parts that interact with the soil. The second factor, I_o, represents the mode of *operation* of the implement, namely, the speed and depth of tillage. The third factor, I_m, represents the *match* between the available power of the tractor and the power required to draw the implement. Finally, the fourth factor, I_b, represents the operator *behavior*, i.e., how steadily and in what patterns (across or up and down the slopes) the farmer drives across the field.

Compared to that devoted to water and wind erosion, very little research has been directed to determine factor values for tillage erosion. Values for E_t per tillage pass are typically in the range of 0.5 to 5 kg soil per meter of tillage width per percent slope inclination. Considering these four factors that contribute to the tillage erosivity, it is not surprising that values for E_t vary widely from one research study to another, even when the tillage operation is very similar (e.g., chisel plowing at a given speed and depth). However, the few measurements that have been made suggest that certain high speed conservation tillage operations, such as strip-till planting may cause nearly as much tillage erosion as primary tillage operations like moldboard or chisel plowing. How much and how far soil is moved is important, but it is the variation of these within the landscape (e.g., the difference between tillage moving uphill and that going down) that causes soil erosion; certain operations, particularly those carried out at high speed, tend to have much greater variation in soil movement (Table 14.8).

Annual rates of erosion by tillage have been reported in the range of 20 to 50 Mg/ha for mechanized farming on hummocky landscapes and as high as 100 to 200 Mg/ha for cultivated steep mountainsides where tillage was powered by draft animals or human being with soil pulled only in the downslope direction. Rates of any type of erosion are difficult to put in a whole landscape perspective since parts of the landscape lose soil while other parts gain, and only a small fraction (the delivery ratio, Section 14.3) of the soil moved by erosion is exported from the landscape by streamflow or in the air. The export of soil out of the watershed, or even off the field, is generally much larger for soil moved by water or wind than for soil moved by tillage. Another important difference among the agents of soil erosion is the quality of the material deposited in the downslope positions.

In our discussion of wind and water erosion, we emphasized that by the time the eroded material is deposited downwind, downslope, or downstream it has become mere dust or muddy sediment. The deposited material has little value as productive soil because transport by wind and water destroys soil structure, segregates out the various particle sizes, and often removes organic matter and nutrients. In contrast, the material moved downslope by tillage is still productive soil when it accumulates in the low-lying, concave parts of the landscape. That is, the soil eroded by tillage is not mere sediment, but generally retains most of the desirable physical, chemical, and biological properties of a productive soil. Therefore, where movement by tillage is the main erosion process, the potential may exist to remediate the erosion damage using mechanized land scrapers that can efficiently move many megagrams per hectare of eroded soil back upslope to restore the thickness of productive soil on the hilltops. In fact, studies in western Canada suggest that such soil restoration might be economically viable under certain circumstances.

Table 14.8
EROSIVITY CHARACTERISTICS OF SELECTED TILLAGE IMPLEMENTS AND OPERATIONS
Types of tillage implements are listed from most to least erosive. Note that planting and harvesting root crops like potato are highly erosive operations. The values here are for a single pass of the implement, except where indicated. The research from which these values were derived was conducted in several provinces in Canada and somewhat different values could be expected under different conditions.

Type of implement	Type of operation	T_M,[a] kg/m	E_t,[b] kg/m/%
		kg/m	kg/m/%
Potato planter + 2 passes with hiller	Conventional tillage	117	3.6
Potato harvester	Conventional tillage	70	3.0
Moldboard plow	Primary tillage	60	1.4
Chisel plow	Primary tillage	60	1.5
Tandem disk	Secondary tillage	50	2.0
Offset disk	Secondary tillage	34	1.8
Cultivator	Secondary tillage	46	0.2
Air-seeder with knives	No-till	8	0.1
Air-seeder with sweeps	No-till	30	1.0

[a] The movement of soil which would occur on a flat ground.
[b] The movement of soil due to slope gradient.
Data selected, recalculated, and rearranged from Lobb (2008).

14.14 LAND CAPABILITY CLASSIFICATION AND PROGRESS IN SOIL CONSERVATION

The land capability classification system devised by the U.S. Department of Agriculture uses eight **land capability classes** to indicate the *degree* of limitation imposed on land uses (Figure 14.43, *left*), with Class I the least limited and Class VIII the most limited. Each land-use class may have four subclasses that indicate the *type* of limitation encountered: risks of erosion (e); wetness, drainage, or flooding (w); root-zone limitations, such as acidity, density, and shallowness (s); and climatic limitations, such as a short growing season (c). The erosion (e) subclasses are the most common; for example, Class IIe land is slightly susceptible to erosion, while Class VIIe is extremely susceptible. Figure 14.43 *right* shows the appropriate intensity of use allowable for each of these land capability classes if erosion losses (or problems associated with the other subclasses) are to be avoided. In the United States erosion and sedimentation problems limit land use for nearly 60% of the land. Excessive wetness and shallowness are each problems on about 20% of the land.

Class I lands are nearly level with deep, well-drained soils and therefore little or no limitations on agricultural uses. Classes II, III, and IV can be used for agriculture, but with increasing limitations and requirements for remedial or protective measures. Class V lands are generally not suited to crop production because of factors other than erosion hazards, including (1) frequent flooding, (2) shortness of growing season, (3) stoniness, and (4) waterlogged soils where drainage is not feasible. Often, pastures can be of good quality on this class of land. Class VI and VII lands are highly susceptible to erosion or other limitations and the choices of land uses severally restricted, but often can provide good grazing or timberlands under proper management. Class VIII land is best left undisturbed in natural vegetation; it includes sand beaches, river wash, and rock outcrops and extremely steep terrain.

It wasn't until the worldwide depression and widespread droughts of the early 1930s accentuated rural poverty and displaced millions of people that governments began to pay attention to the rapid deterioration of soils. In 1930, Dr. H. H. Bennett and associates recognized the damage being done and obtained the U.S. government support for erosion-control efforts. Since then, considerable reductions in erosion have been achieved in the United States and

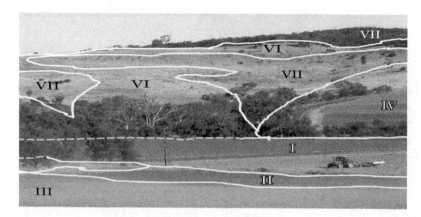 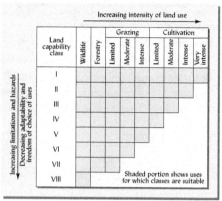

Figure 14.43 (Left) *Several land capability classes in the Lockyer Valley of Queensland, Australia. A range is shown from the nearly level land in the center (Class I), which can be cropped intensively, to the badly eroded hillsides (Classes VI and VII). Although topography and erosion hazards are emphasized here, it should be remembered that other factors—drainage, stoniness, and droughtiness—also limit soil usage and help determine the land capability class.* (Right) *Intensity with which each land capability class can be used with safety. Note the increasing limitations on the safe uses of the land as one moves from Class I to Class VIII.* [Left image courtesy of Ray R. Weil; Right diagram modified from Hockensmith and Steele (1949)]

elsewhere. During the 1940s and 1950s, such physical practices as contour strips, terraces, and windbreaks were installed with much persuasion and assistance from government agencies. Some of this progress was reversed as the terraces and windbreaks appeared to stand in the way of the "fence row to fence row" all-out crop production policies of the 1970s. But since 1982, rather remarkable progress has been achieved in reducing soil erosion, largely as a result of two factors: (1) the spread of conservation tillage (see Section 14.6), and (2) the implementation of land-use changes such as in the conservation reserve program (CRP). The CRP is basically an arrangement by which the U.S. taxpayers pay rent to farmers to forego cropping part of their farmland and instead plant grass or trees on it (more rent is paid for trees).

Nonetheless, about one-third of the cultivated cropland in the United States is still losing more than 11 Mg/ha/yr, the maximum loss that can be sustained without serious loss of productivity on most soils. After some 80 years of soil conservation efforts, soil erosion is still a major problem on about half the cropland of the United States and in much of the world the problem has actually worsened.

Conservation Management to Enhance Soil Health

In a broader sense, conservation management practices are those that improve soil health in more ways than just by protecting the soil from erosion. Properties that indicate the level of soil quality, especially those associated with soil organic matter, can be enhanced by such conservation measures as minimizing tillage, maximizing residue cover of the soil surface, providing for diversity of plant types, keeping soil under grass sod vegetation for at least part of the time, adding organic amendments where practical, and maintaining balanced soil fertility. Improved soil health, in turn, enhances the soil's capacity to support plants, resist erosion, prevent environmental contamination, and conserve water. Conversation management therefore can lead to the upward spiral of soil and environmental improvement referred to in Section 14.1.

Finding Soil Conservation Win-Win Systems

Farmers and ranchers need to make a living from their land. So it is their long-term interest to protect their soils. In much of the world, so little land is available to each farmer for food production that nations cannot afford the luxury of following the land-use capability classification recommendations as outlined in Figure 14.43. Many farmers must use *all* land capable of food production simply to stave off starvation and impoverishment. These farmers often realize that farming erodible land jeopardizes their future livelihood and that of their children, but they see no choice. It is imperative to find either nonagricultural employment for these people or farming systems that are sustainable on these erodible lands.

Fortunately, some farmers and scientists have developed, through long traditions of adaptation or through innovation and research, farming systems that *can* produce food and profits while conserving such erodible soil resources. Examples include the traditional Kandy Home Gardens of Sri Lanka's humid mountains, in which a rain forest–like mixed stand of tall fruit and nut trees is combined with an understory of pepper vines, coffee bushes, and spice plants to provide valuable harvests while keeping the soil under perennial vegetative protection. Another example comes from Central America, where farmers have learned to plant thick stands of velvet bean (*Mucuna*) or other viny legumes that can be chopped down by machete to leave a soil-protecting, water-conserving, weed-inhibiting mulch on steep farmlands. In Asia, steep lands have been carefully terraced in ways that allow production of food, even paddy rice, on very steep land without causing significant erosion. In semiarid parts of the United States and Australia, innovative rancher are using management intensive rotation grazing (see Box 13.4) to protect soil while enhancing animal production. Leading farmers in North America are using no-till and cover crops to enhance their soils' health and virtually eliminate erosion while profitably producing crops. These and similar examples show that farmers and scientists working collaboratively can develop practices that are good for the land and for profits: a win-win situation.

14.15 SUMMARY AND CONCLUSION

Accelerated soil erosion is one of the most critical environmental and social problems facing humanity today. Erosion degrades soils, making them less capable of producing the plants on which animals and people depend. Equally important, erosion causes great damage downstream in reservoirs, lakes, waterways, harbors, and municipal water supplies. Wind erosion also causes fugitive dust that may be very harmful to human health. Tillage itself is a third agent of erosion on cropland and may move more soil than wind and water combined in some cases.

Nearly 4 billion Mg of soil is eroded each year on land in the United States alone. Half of this erosion occurs on the nation's croplands, and the remainder on harvested timber areas, rangelands, and construction sites. Some one-third of the cropland still suffers from erosion that exceeds levels thought to be tolerable.

Water carries away most of the sediment in humid areas by sheet and rill erosion. Gullies created by infrequent, but violent, storms account for much of the erosion in drier areas. Wind is the primary erosion agent in many drier areas, especially where the soil is bare and low in moisture during the season when strong winds blow. Tillage erosion is most damaging on cultivated land with hummocky topography.

Protecting soil from the ravages of wind or water is by far the most effective way to constrain erosion. In croplands and forests, such protection is due mainly to the cover of plants and their residues. Conservation tillage practices maintain vegetative cover on at least 30% of the soil surface, and the widening adoption of these practices has contributed to the significant reductions in soil erosion achieved over the past two decades. Crop rotations that include sod and close-growing crops, coupled with such practices as contour tillage, strip cropping, and terracing, also help combat erosion on farmland.

In forested areas, most erosion is associated with timber-harvesting practices and forest road construction. For the sake of future forest productivity and current water quality, foresters must become more selective in their harvest practices and invest more in proper road construction.

Construction sites for roads, buildings, and other engineering projects lay bare many scattered areas of soils that add up to a serious erosion problem. Control of sediment from construction sites requires carefully phased land clearing, along with vegetative and artificial soil covers, and installation of various barriers and sediment-holding ponds. These measures may be expensive to implement, but the costs to society that result when sediment is not controlled are too high to ignore. Once construction is completed, erosion rates on urban areas are commonly as low as those on areas under undisturbed native vegetation.

Erosion-control systems must be developed in collaboration with those who use the land, and especially the poor, for whom immediate needs must overshadow concerns for the future. As put succinctly in *The River*,[9] a classic 1930s documentary film produced during the rebirth of American soil erosion consciousness, "Poor land makes poor people, and poor people make poor land."

---

[9]https://www.youtube.com/watch?v=9MRCltkSZbw—be sure to watch both parts 1 and 2.

STUDY QUESTIONS

1. Explain the distinction between *geologic erosion* and *accelerated erosion*. Is the difference between the two greater in humid or arid regions?
2. When erosion takes place by wind or water, what are the three important types of damages that result on the land whose soils are eroding? What are the five important types of damages that erosion causes in locations away from the eroding site?
3. Describe the three main steps in the water erosion process.
4. Many people assume that the amount of soil eroded on the land in a watershed (A in the universal soil-loss equation) is the same as the amount of sediment carried away by the stream draining that watershed. What factor is missing that makes this assumption incorrect? Do you think that this means the USLE should be renamed?
5. Why is the total annual rainfall in an area *not* a very good guide to the amount of erosion that will take place on a particular type of bare soil?
6. Contrast the properties you would expect in a soil with either a very high K value or a very low K value.
7. How much soil is likely to be eroded from a Keene silt loam in central Ohio, on a 12% slope, 100 m long, if it is in dense permanent pasture and has no supporting physical conservation practices like terraces applied to the land? Use the information available in this chapter to calculate an answer.
8. What type of conservation tillage leaves the greatest amount of soil cover by crop residues? What are the advantages and disadvantages of this system?
9. Why are narrow strips of grass planted on the contour sometimes called a "living terrace"?
10. In most forests, which component of the ecosystem provides the primary protection against soil erosion by water, the *tree canopy*, *tree roots*, or *leaf litter*?
11. Certain soil properties generally make land susceptible to erosion by wind or erosion by water. List four properties that characterize soils highly susceptible to wind erosion. Indicate which two of these properties should also characterize soils highly susceptible to water erosion, and which two should not.
12. Which two factors in the wind erosion prediction equation (WEQ) can be affected by tillage? Explain.
13. What type of landscape is most prone to tillage erosion?
14. Why might the approach to restoring land damaged by tillage erosion be quite different than that used to restore productivity to land damaged by water erosion? Explain.
15. Describe a soil in land capability Class IIw in comparison with one in Class IVe.

REFERENCES

Awada, L., C. W. Lindwall, and B. Sonntag. 2014. "The development and adoption of conservation tillage systems on the Canadian prairies." *International Soil and Water Conservation Research* 2:47–65.

Benik, S. R., B. N. Wilson, D. D. Biesboer, B. Hansen, and D. Stenlund. 2003. "Evaluation of erosion control products using natural rainfall events." *Journal of Soil and Water Conservation* 58:98–104.

Cassel, D. K., and R. Lal. 1992. "Soil physical properties of the tropics: Common beliefs and management constraints." In R. Lal and P. A. Sanchez (eds.). *Myths and Science of Soils of the Tropics*. SSA Special Publication No. 29. Soil Science Society of America, Madison, WI, pp. 61–89.

Chow, T. L., H. W. Rees, and J. L. Daigle. 1999. "Effectiveness of terraces/grassed waterway for soil and water conservation: A field evaluation." *Journal of Soil and Water Conservation* 54:577–583.

Crovetto, C. 1996. *Stubble Over the Soil*. American Society of Agronomy, Madison, WI.

de Freitas, P. L., and J. N. Landers. 2014. "The transformation of agriculture in Brazil through development and adoption of zero tillage conservation agriculture." *International Soil and Water Conservation Research* 2:35–46.

Doolette, J. B., and J. W. Smyle. 1990. "Soil and moisture conservation technologies: Review of literature." In J. B. Doolette and W. B. Magrath (eds.). *Watershed Development in Asia: Strategies and Technologies*. World Book Technical Paper 127, Washington, D.C.

FAO. 1987. *Protect and Produce*. U.N. Food and Agriculture Organization, Rome.

Flanagan, D. C., J. E. Gilley, and T. G. Franti. 2007. "Water erosion prediction project (WEPP): Development history, model capabilities, and future enhancements." *Transactions of the American Society of Agricultural and Biological Engineers* 50:1603–1612.

Foster, G.R., and R.E. Highfill. 1983. "Effect of terraces on soil loss: Usle P factor values for terraces." *Journal of Soil and Water Conservation* 38:48–51.

Foster, G. R., D. K. McCool, K. G. Renard, and W. C. Moldenhauer. 1981. "Conversion of the universal soil loss equation to SI metric units." *Journal of Soil and Water Conservation* 36:355–359.

Fryrear, D. W., J. D. Bilbro, A. Saleh, H. Schomberg, J. E. Stout, and T. M. Zobeck. 2000. "RWEQ: Improved wind erosion technology." *Journal of Soil and Water Conservation* 55:183–189.

Gao, X., Y. Xie, G. Liu, B. Liu, and X. Duan. 2015. "Effects of soil erosion on soybean yield as estimated by simulating gradually eroded soil profiles." *Soil and Tillage Research* 145:126–134.

Gassman, P. W., A. M. Sadeghi, and R. Srinivasan. 2014. "Applications of the SWAT model special section: Overview and insights." *Journal of Environmental Quality* 43:1–8.

Gilley, J. E., B. Eghball, L. A. Kramer, and T. B. Moorman. 2000. "Narrow grass hedge effects on runoff and soil loss." *Journal of Soil and Water Conservation* 55:190–196.

Guo, Z., and M. Shao. 2013. "Impact of afforestation density on soil and water conservation of the semiarid loess plateau, China." *Journal of Soil and Water Conservation* 68:401–410.

Hillel, D. 1991. *Out of the Earth: Civilization and the Life of the Soil*. The Free Press, New York.

Hockensmith, R. D., and J. G. Steele. 1949. "Recent trends in the use of the land-capability classification." *Soil Science Society of America Proceedings* 14:383–388.

Hooke, R. L. 2000. "On the history of humans as geomorphic agents." *Geology* 28:843–846.

Kassam, A., T. Friedrich, R. Derpsch, and J. Kienzle. 2015. "Overview of the worldwide spread of conservation agriculture". *Field Actions Science Reports*, Volume 8. Online at http://journals.openedition.org/factsreports/3966

Kumar, S., A. Kadono, R. Lal, and W. Dick. 2012a. "Long-term no-till impacts on organic carbon and properties of two contrasting soils and corn yields in Ohio." *Soil Science Society of America Journal* 76:1798–1809.

Kumar, S., A. Kadono, R. Lal, and W. Dick. 2012b. "Long-term tillage and crop rotations for 47–49 years influences hydrological properties of two soils in Ohio." *Soil Science Society of America Journal* 76:2195–2207.

Li, X. R., H. L. Xiao, M. Z. He, and J. G. Zhang. 2006. "Sand barriers of straw checkerboards for habitat restoration in extremely arid desert regions." *Ecological Engineering* 28:149–157.

Lobb, D. 2008. "Soil movement by tillage and other agricultural activities." In S. E. Jorgensen and B. D. Fath (eds.). *Encyclopedia of Ecology*. Vol. 4. Elsevier, Oxford, pp. 3295–3303.

Montgomery, D. R. 2007. "Soil erosion and agricultural sustainability." *Proceedings of the National Academy of Sciences* 104:13268–13272.

Munson, S. M., J. Belnap, and G. S. Okin. 2011. "Responses of wind erosion to climate-induced vegetation changes on the Colorado plateau." *Proceedings of the National Academy of Sciences* 108:3854–3859.

Nyland, R.D. 2016. *Silviculture: Concepts and Applications*. 3rd ed. Waveland Press, Long Grove, IL. 632 p.

Panagos, P., P. Borrelli, K. Meusburger, B. Yu, A. Klik, K. Jae Lim, J.E. Yang, J. Ni, C. Miao, N. Chattopadhyay, S.H. Sadeghi, Z. Hazbavi, M. Zabihi, G.A. Larionov, S.F. Krasnov, A.V. Gorobets, Y. Levi, G. Erpul, C. Birkel, N. Hoyos, V. Naipal, P.T.S. Oliveira, C.A. Bonilla, M. Meddi, W. Nel, H. Al Dashti, M. Boni, N. Diodato, K. Van Oost, M. Nearing, and C. Ballabio. 2017. "Global rainfall erosivity assessment based on high-temporal resolution rainfall records." *Nature (Scientific Reports)* 7:4175. DOI: 10.1038/s41598-017-04282-8

Pittelkow, C.M., B.A. Linquist, M.E. Lundy, X. Liang, K.J. van Groenigen, J. Lee, N. van Gestel, J. Six, R.T. Venterea, and C. van Kessel. 2015. "When does no-till yield more? A global meta-analysis." *Field Crops Research* 183:156–168.

Renard, K. G., G. Foster, D. Yoder, and D. McCool. 1994. "RUSLE revisited: Status, questions, answers and the future." *Journal of Soil and Water Conservation* 49:213–220.

Reusser, L., P. Bierman, and D. Rood. 2015. "Quantifying human impacts on rates of erosion and sediment transport at a landscape scale." *Geology*. doi: 10.1130/G36272.1XX:XX.

Safari, A., A. Kavian, A. Parsakhoo, I. Saleh, and A. Jordán. 2016. "Impact of different parts of skid trails on runoff and soil erosion in the Hyrcanian forest (northern Iran)." *Geoderma* 263:161–167.

Schwab, G. O., D. D. Fangmeirer, and W. J. Elliot. 1996. *Soil and Water Management Systems*, 4th ed. Wiley, New York.

Sidle, R. C., A. D. Ziegler, J. N. Negishi, A. R. Nik, R. Siew, and F. Turkelboom. 2006. "Erosion processes in steep terrain—Truths, myths, and uncertainties related to forest management in Southeast Asia." *Forest Ecology and Management* 224:199–225.

Tatarko, J. 2010. "Wind erosion: Problem, processes, and control." USDA-Agricultural Research Service, Engineering and Wind Erosion Research Unit, Manhattan, KS. http://www.nrcs.usda.gov/Internet/FSE_-DOCUMENTS/nrcs142p2_019407.pdf.

USDA/ARS. 2014. "The soil and water assessment tool (SWAT) [Online]." Available by USDA Agricultural Research Service and Texas A&M AgriLife Research (posted 30 September 2014).

WHO. 2013. "Health effects of particulate matter." World Health Organization, Copenhagen, Denmark. http://www.euro.who.int/__data/assets/pdf_file/0006/189051/Health-effects-of-particulate- matter-final-Eng.pdf.

Wilkinson, B. H., and B. J. McElroy. 2007. "The impact of humans on continental erosion and sedimentation." *Geological Society of America Bulletin* 119:140–150.

Wischmeier, W. J., and D. D. Smith. 1978. Predicting Rainfall Erosion Loss—A Guide to Conservation Planning. Agricultural Handbook No. 537. USDA, Washington, D.C.

Zobeck, T. M., and R. S. Van Pelt. 2014. "Wind erosion." USDA Agricultural Research Service, Lincoln, NE. http://digitalcommons.unl.edu/usdaarsfacpub/1409/.

15
Soils and Chemical Pollution

Polluted soil at an abandoned oil refinery. (Photo courtesy of Ray R. Weil)

Black and portentous this humor prove, unless good counsel may the cause remove ...
—W. Shakespeare, Romeo and Juliet

Every year, millions of tons of potentially toxic industrial, domestic, and agricultural products and waste materials find their way into the world's soils. Once there, they become part of biological cycles that affect all forms of life. In previous chapters, we highlighted the enormous capacity of soils to accommodate added organic and inorganic chemicals. In a typical hectare of land, tons of organic residues are broken down by soil microbes each year (Chapter 11), and large quantities of inorganic chemicals are fixed or bound tightly by soil minerals (Chapter 12). But we also learned of the limits of the soil's capacity to sorb these chemicals, and how environmental quality suffers when these limits are exceeded (Chapters 8 and 13).

We have seen some of the many ways in which soils and soil processes influence environmental quality. Soil processes affect the production and sequestering of greenhouse gases, such as nitrous oxide, methane, and carbon dioxide (see, e.g., Sections 11.10 and 12.1). Other nitrogen- and sulfur-containing gases come to earth in acid rain (see, e.g., Section 9.6). Mismanaged irrigation projects on arid-region soils result in the accumulation of naturally occurring salts (see Section 9.16), including toxic levels of sodium, selenium, and arsenic.

The brief review of soil pollution in this chapter is intended as an introduction to the nature of major pollutants, their reactions in soils, and alternative means of managing, destroying, or inactivating them. We will focus on potentially toxic chemicals that contaminate soils, threaten the health of organisms, and degrade soil quality. Some of the contaminants may escape the soil to extend their environmental damage to plants, animals, water, and air. In terms of their chemical nature, the toxic materials of concern can be viewed as being organic compounds, inorganic substances (mostly metallic elements), or radioactive isotopes, and we will consider them in that order.

15.1 TOXIC ORGANIC CHEMICALS

The quantity of synthetic organic chemicals manufactured every year is enormous—globally in 2017 more than 500 million Mg with a monetary value of some $5.2 trillion. Included are plastics and plasticizers, lubricants and refrigerants, fuels and solvents, pesticides and preservatives. Some are extremely toxic to humans and other life. Through accidental leakage and spills or through planned burial, spraying, or other treatments, synthetic organic chemicals can be found in virtually every corner of our environment—in the soil, in the groundwater, in the plants, in the ocean, and in our own bodies.

Environmental Damage from Organic Chemicals

Artificially synthesized compounds are termed **xenobiotics** because they are unfamiliar to the living world (Greek *xeno*, strange). Being nonnatural, many xenobiotics are either toxic to living organisms or resistant to biological decay, or both. The chemical structures of xenobiotic compounds may be quite similar to those of naturally occurring compounds produced by microorganisms and plants. The difference is commonly the insertion of halogen atoms (Cl, F, Br) or multivalent nonmetal atoms (such as S and N) into the structure.

Some xenobiotic compounds are relatively inert and harmless, but others are biologically damaging even in very small concentrations. Those that find their way into soils may inhibit or kill soil organisms, thereby undermining the balance of the soil community (see Section 10.14). Other chemicals may be transported from the soil to the air, water, or vegetation, where they may be contacted, inhaled, or ingested by any number of organisms, including human beings. It is imperative, therefore, that we control the release of organic chemicals and that we learn of their fate and effects once they enter the soil.

Organic chemicals may enter the soil as contaminants in industrial and municipal organic wastes applied to or spilled on soils, as components of discarded machinery, in large or small lubricant and fuel leaks, as military explosives, as deposition from the atmosphere, or as sprays applied to control pests in terrestrial ecosystems. Pesticides are probably the most widespread organic pollutants associated with soils. In the United States, pesticides are used on some 150 million ha of land, three-fourths of which is agricultural land. Soil contamination by other organic chemicals tends to be much more localized. We will therefore emphasize the pesticide problem.

The Nature of the Pesticide Problem[1]

Pesticides are chemicals that are designed to kill pests (i.e., any organism unwanted by the pesticide user). Some 700 chemicals in about 50,000 formulations are used to control various pests in all parts of the world.

Statistics on the amounts of pesticides used can be difficult to interpret. Figure 15.1 (*left*) presents trends in agricultural pesticide use in the United States across five decades. Back in 1960 when Rachel Carson wrote her famous book *Silent Spring*, most of the pesticide materials used were highly toxic insecticides, such as DDT. During the subsequent 20 years, overall pesticide use rose rapidly due mainly to the advent of chemical herbicides effective at selectively killing unwanted plants (weeds). By the 1990s, insecticides accounted for a much smaller fraction of pesticide use. Herbicide use (mainly glyphosate) increased again after 2000 with the wide adoption of corn and soybean cultivars genetically engineered to resist that broad spectrum herbicide. The total amount of insecticide used declined steadily during the entire period, mainly because of the introduction of more potent compounds requiring lower application rates and, in the late 1990s, the introduction of crops genetically engineered to produce their own insecticidal toxins (using a gene from the soil bacterium, *Bacillus thuringiensis*, or Bt).

Although the total amount of pesticides used in the United States may have dropped somewhat since the 1980s, formulations of both insecticides and herbicides in use today are generally more potent, so that smaller quantities need be applied per hectare to achieve toxicity to the pest. In general, the acute toxicity to humans and wildlife of the newer pesticides is thought to be much lower than that of those in use prior to the 1980s (Figure 15.1, *right*). However, there is concern that some of the newer, ostensibly safer compounds may have more subtle, long-term effects on humans and wildlife such as endocrine (hormone) disruption. In addition, newer pesticide compounds may be more or less persistent in the environment than the compounds they supplanted.

---

[1]Rachel Carsen's groundbreaking warning (Carson, 1962) about the ecological dangers of excessive pesticide use is still very worth reading today, as is Miguel Altieri's classic text (Altieri, 1995) on pesticide-free farming and agro-ecology. For research on glyphosate residues in genetically modified soybeans and their potential effects on animal health, see Bøhn et al. (2014) and Cuhra et al. (2015).

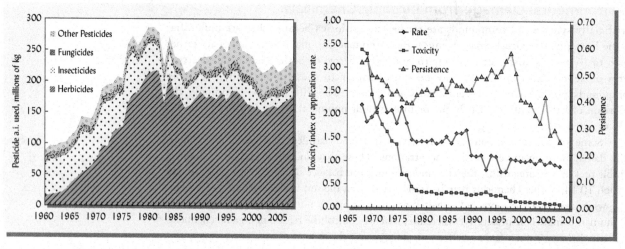

Figure 15.1 Types, amounts, and nature of pesticides used on crops in the United States across five decades. (Left) The total amount of pesticides used (kg of active ingredient, a.i.) increased dramatically between 1960 and 1980, mainly due to greater amounts of herbicides used. With the adoption of resistant crops, herbicide use (mainly glyphosate) increased again after 2000. Insecticide use decreased steadily because of more potent compounds requiring lower rates and the introduction in the late 1990s of crops genetically engineered to produce their own insecticidal Bt toxins. "Other pesticides" included soil fumigants, desiccants, harvest aids, and plant growth regulators. (Right) The index of chronic toxicity (1/threshold μg L^{-1}) of pesticides used declined dramatically during the 1970s when DDT and aldrin were phased out, and has continued to decline slowly since 1980. The pesticides in use in 2010 were far less toxic than those used several decades earlier. The persistence index (the fraction of compounds with half-lives >60 days) also declined somewhat. [Graphed from data in Fernandez-Cornejo et al. (2014)]

Benefits of Pesticides. Pesticides have provided many benefits to society. They have helped control mosquitoes and other vectors of such human diseases as malaria, yellow fever, and West Nile virus. Bed nets treated with long-lasting insecticides have recently helped turn the tide against the ravages of malaria in many tropical countries. Similar insecticides are essential in controlling the spread of tick-carried Lyme disease and bedbugs that are resurgent in many industrial countries, especially in hotels. They have protected crops and livestock against insects and diseases that could devastate food production. Without the control of weeds by chemicals called *herbicides*, conservation tillage (especially no-tillage) would be much more difficult to adopt. Without herbicides farmers would have to rely mainly on tillage to manage weeds and much of the progress made in controlling soil erosion probably would not have come about (Section 14.6). Also, pesticides reduce the spoilage of food as it moves from farm fields to distant dinner tables. In addition, long-lasting insecticides applied to soils around our houses prevent costly damages from wood eating soil-dwelling termites (Section 10.5).

Figure 15.2 Leaking underground storage tank (LUST) replacement at a gas station in California. The old rusting steel tanks have been removed and replaced by more corrosion-resistant fiberglass tanks, which are set in the ground and covered with pea gravel. The soil and groundwater aquifer beneath the tanks were cleaned up using special techniques to stimulate soil microorganisms and to pump out volatile organics such as benzene vapors. Remediation and replacement can cost about $1 million for a single large gas station. (Photo courtesy of Ray R. Weil)

Problems with Pesticides. While the benefits to society from pesticides must be acknowledged, so too must the costs (Table 15.1). Widespread and heavy use of pesticides on agricultural soils and suburban and urban landscapes has led to contamination of both surface and groundwater (Table 15.2). Therefore, when pesticides are used, they should be chosen for low toxicity to humans and wildlife, low mobility on soils, and low persistence (see Table 15.2 and Section 15.3). Even then, the use of pesticides often has wide-ranging detrimental effects on soil and aquatic microbial and faunal communities. In fact, the harm done, though not always obvious, may outweigh the benefits. Examples include insecticides that kill natural enemies of pest species as well as the target pest (sometimes creating new major pests from species formerly controlled by natural enemies) and fungicides that kill both disease-causing and beneficial fungi (see Section 10.9). Given these facts, it should not come as a surprise that despite the widespread use of pesticides, insects, diseases, and weeds still cause the loss of about one-third of today's crop production, about the same proportion of crops lost to these pests in the United States before synthetic organic pesticides were in use. Still, most economic analyses show that even when environmental and social impacts are included on the costs side, the use of pesticides brings a net benefit in countries where their use is well regulated.

Alternatives to Pesticides. Pesticides should not be seen as a panacea, or even as indispensable. Some farmers, most notably the small but increasing number who practice **organic farming**,[2] produce profitable, high-quality yields without the use of synthetic pesticides (but often with the use of pesticides based on naturally occurring toxins). In any case, chemical pesticides should be used as a *last* resort, rather than as a *first* resort, whether on a farm, an ornamental landscape, a forest, an urban neighborhood, or in a building. Before using toxic insecticides or herbicides, every effort should be made to minimize the detrimental effects of pests by managing the environment to increase biodiversity, favor natural enemies of pests, and change aspects of the environment that encourage pest outbreaks or allow pests to cause damage. Such nonchemical management requires detailed knowledge of the pest's life cycle with a view to taking advantage of its particular weaknesses.

Table 15.1
TOTAL ESTIMATED ENVIRONMENTAL AND SOCIAL COSTS FROM PESTICIDE USE IN THE UNITED STATES

Type of impacts[a]	Annual costs, millions of $
Public health impacts	1,676
Domestic animal deaths and contaminations	44
Loss of natural enemies	764
Cost of pesticide resistance	2,205
Honeybee and pollination losses	491
Crop losses	2,045
Fishery losses	147
Bird losses	3,175
Groundwater contamination	2,940
Government regulations to prevent damage	691
Total	14,178

[a]The death of ~60 million wild birds may represent additional lost revenues from hunters and bird-watchers. Data adjusted for inflation to 2018 dollars and other changes, selected from Pimentel and Burgess (2014).

---

[2]The term *organic farming* has little to do with the chemical definition of organic, which simply indicates that a compound contains carbon. Rather, it refers to a system and philosophy of farming that eschews the use of synthetic chemicals while it emphasizes soil organic matter and biological interactions to manage agroecosystems.

Table 15.2
PROPERTIES OF SELECTED PESTICIDES COMMONLY FOUND IN WATER

The distribution coefficients for organic carbon (K_{oc}), solubilities in water (S_w), and half-lives help determine how easily compounds move to ground and surface water. Detections, maximum concentrations observed, and health advisory levels should be compared to suggest the seriousness of groundwater contamination.

Pesticide compound	log K_{oc} (mL/g)	S_w (mg/L)	Half-life for transformation in aerobic soil (days)	Half-life for transformation in water (days)	Maximum concentration observed (µg/L)	Health criterion (µg/L)
Agricultural herbicides and degradates frequently detected in water						
Atrazine	2.00	30	146	742	3.8	3[a]
Metolachlor	2.26	430	26	410	5.4	100[b]
Alachlor	2.23	240	20.4	640	0.55	2[a]
Metribuzin	1.72	1,000	172	>200	0.30	200[b]
Trifluralin	4.14	0.5	169	>32	0.014	5[b]
Urban herbicides and degradates frequently detected in water						
Simazine	2.11	5	91	>32	1.3	4[a]
2,4-D	1.68	890	2.3	732	—	70[a]
Diuron	2.60	40	372	>500	—	10[b]
Insecticides frequently detected in water						
Diazinon	2.76	60	39	140	0.077	0.6[b]
Carbaryl	2.36	120	17	11	0.02	700[b]
Malathion	3.26	145	<1	6.3	0.004	100[b]
Dieldrin	4.08	0.17	NA	3,830	0.45	0.002[c]
Imidacloprid	2.2	54	200	30	10.0	0.013[a]

Data from several sources: Gilliom et al. (2006), Vijver and van den Brink (2014).
[a]Maximum contaminate level (MCL) permissible as an annual average concentration in public-use water.
[b]Health advisory level (HA-L), the concentration expected to cause health problems (noncancer) with lifetime exposure.
[c]Cancer risk concentration (CRC).

For example, instead of insecticide spraying to kill mosquitoes in an urban neighborhood, a campaign could be established to search out and remove all containers (old tires, jars, dog dishes, clogged roof gutters, etc.) that can hold 2 cm of rainwater for two days—habitat elements in which the mosquitoes can breed. The tools employed in agricultural fields may include crop diversification, establishment of habitat for beneficial insects that prey upon the pest, application of organic soil amendments to stimulate acquired immune responses, implementation of cultural practices such as planting dates, rates, patterns, and intercrops that reduce the ability of weed to compete, and the selection of pest-resistant plant cultivars. Too often, because pesticides are available as a convenient crutch, these more sophisticated ecological approaches to pest management are not thoroughly explored.

Nontarget Damages. Although some pesticides are intentionally applied to soils, most reach the soil inadvertently. When pesticides are sprayed in the field, most of the chemical misses the target organism. For pesticides aerially applied to forests, about 25% reaches the tree foliage, and far less than 1% reaches a target insect. About 30% may reach the soil, while about half of the chemical applied is likely to be lost into the atmosphere or in runoff water.

Designed to kill living things, many of these chemicals are potentially toxic to organisms other than the pests for which they are intended. Those chemicals that do not quickly break down may be biologically magnified as they move up the food chain. For example, as earthworms ingest contaminated soil, the chemicals tend to concentrate in the fatty tissues of the earthworm bodies. When birds, moles, or fish eat the earthworms, the pesticides can build up further, or biomagnify, to lethal levels. The near-extinction of the American bald eagle during the 1960s and 1970s called public attention to the sometimes-devastating environmental consequences of biomagnification of pesticides. More recently, evidence is mounting to suggest that hormone balance in humans and other animals may be disrupted by the minute traces of some pesticides found in water, air, and food. Others, such as glyphosate, may disrupt beneficial microbial communities in the human gut.

15.2 KINDS OF ORGANIC CONTAMINANTS

Industrial Organics

Industrial organics that often end up contaminating soils by accident or neglect include petroleum products used for fuel [gasoline components such as benzene, and more complex polycyclic aromatic hydrocarbons (PAHs)], solvents used in manufacturing processes [such as trichloroethylene (TCE)], and military explosives such as trinitrotoluene (TNT). Polychlorinated biphenyls (PCBs) constitute a particularly troublesome class of widely dispersed compounds. These compounds can disrupt reproduction in birds and cause cancer and hormone effects in humans and other animals. Although banned since the late twentieth century, their extreme resistance to natural decay and their ability to enter food chains means that, even today, soil and water all over the globe contain at least traces of PCBs.

The sites most intensely contaminated with organic pollutants are usually located near chemical manufacturing plants or oil storage facilities, but railway, shipping, and highway accidents also produce hot spots of contamination. Thousands of neighborhood gas stations represent potential or actual sites of soil and groundwater contamination as gasoline leaks from old, rusting underground storage tanks (Figure 15.2). However, as already mentioned, by far the most widely dispersed xenobiotics are those designed to kill unwanted organisms (i.e., pesticides).

Pesticides

Most pesticide molecules contain aromatic rings of some kind, but that there is great variability in pesticide chemical structures. Pesticides are commonly classified according to the type of pest organisms targeted: (1) *insecticides*, (2) *fungicides*, (3) *herbicides* (weed killers), (4) *rodenticides*, and (5) *nematocides*. In practice, all find their way into soils and from there into streams (Figure 15.3). Since the first three are used in the largest quantities and are therefore more likely to contaminate soils, they will be given primary consideration.

Insecticides. Most of these chemicals are included in four general groups. The *chlorinated hydrocarbons*, such as DDT, were the most extensively used until the early 1970s, when their use was banned or severely restricted in many countries due to their low biodegradability and persistence, as well as their toxicity to birds and fish. The *organophosphate* pesticides are generally biodegradable, and thus less likely to build up in soils and water. However, they are extremely toxic to humans, so great care must be used in handling and applying them. The *carbamates* are considered least dangerous because of their ready biodegradability and relatively low mammalian toxicity. However, they are highly toxic to honeybees and other beneficial insects and to earthworms. The *neonicotinoids* are a relatively new class of insecticides with chemical structures similar to that of nicotine (the toxic and addictive ingredient in tobacco). They affect the central nervous system of insects. Neonicotinoids exhibit chronic toxicity to aquatic organisms at very low concentrations; for example, imidacloprid is toxic in the range of 10–100 *nano*grams (ng) per L. Minute quantities of neonicotinoids have been found to kill brain cells in bees, causing these critical pollinators to lose their ability to navigate between their nests and flowers.

Neonicotinoid insecticides are highly soluble in water and somewhat slow to degrade in the environment; therefore, they easily leach through soils to groundwater. These highly soluble insecticides also are subject to being carried by runoff to nearby streams (Figure 15.4).

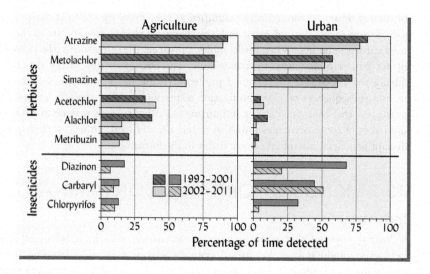

Figure 15.3 *Percentage of stream samples in the United States with detectable pesticides during two time periods in relation to land use in the watersheds. The herbicides atrazine, metolachlor, and simazine were detected in more than 50% of the stream samples. Insecticides tended to be detected more frequently in urban than in agricultural watersheds. Human-health benchmarks were rarely exceeded. During 2002–2011, only one agricultural stream and no urban or mixed-land-use streams in the study exceeded human-health benchmarks for any of the measured pesticides.* [Modified from Stone et al. (2014)]

Research suggests that dust from planting seeds coated with neonicotinoids may be partially responsible for drastic declines in honeybees and other pollinator insects. While the use of organophosphate and carbamate insecticides on crops in the United States has declined during the past two decades, the use of neonicotinoid insecticides has increased dramatically, mainly in seed coatings. Although coating seeds with neonicotinoid insecticides has become the default practice in agriculture, environmental toxicity consideration suggests that such coatings should *not* be used except in those fields where early season insect damage is *known* to be likely.

Fungicides. Fungicides are used mainly to control diseases of fruit and vegetable crops and as seed coatings to protect against seed rots. Some are also used to protect harvested fruits and vegetables from decay, to prevent wood decay, and to protect clothing from mildew. Organic materials such as the thiocarbamates and triazoles are currently in use.

Herbicides. The quantity of herbicides used in the United States exceeds that of the other types of pesticides combined (see Figure 15.1). Starting with 2,4-D (a chlorinated phenoxyalkanoic acid), dozens of chemicals in literally hundreds of formulations have been placed on the market. These include the *triazines*, used mainly for weed control in corn; *substituted ureas*; some *carbamates*; the relatively new *sulfonylureas*, which are potent at very low rates; *dinitroanilines*; and *acetanilides*, which have proved to be quite mobile in the environment. One of the most widely used herbicides, *glyphosate* (e.g., Roundup®), does not belong to any of the aforementioned chemical groups. Unlike most herbicides, it is nonselective, meaning that it will

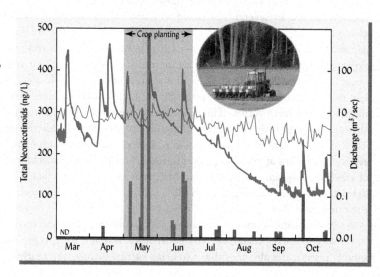

Figure 15.4 *Total streamwater neonicotinoid concentrations (brown bars) in Midwestern U.S. area of intense corn and soybean production. Stream discharge (flow) during the sampling year (heavy blue line) and average discharge are shown to highlight the wet spring and dry summer of the sampling year compared to an average year (thin blue line). The timing of corn and soybean planting is shown by the shaded rectangle. The timing of neonicotinoid pulses associated with rainfall events during crop planting suggests seed treatments as the likely source. Threshold levels of neonicotinoids that can cause acute toxicity to aquatic organisms are as low as 200 ng/L and for chronic toxicity as low as 20 ng/L.* [Modified from USGS data in Hladik et al. (2014). For perspectives on unnecessary use of seed treatment insecticides, see Tooker et al. (2017)]

kill almost any plant, including crops. However, a gene that confers resistance to its effects has been discovered and engineered into several major crops. These genetically engineered crops can then be grown with a very simple, convenient method of weed control that usually consists of one or two sprayings of glyphosate that will kill all plants other than the resistant crop. As might be expected from the interaction of human nature (overdoing a convenient thing) with Mother Nature (evolution to select the fittest), the overuse of glyphosate in this system has led to an increasing number of important weed species developing resistance to this herbicide. Its overuse has also led to the detection of potentially chronically toxic levels of glyphosate and its breakdown products in soybeans destined for human and animal foods (see footnote 1).

The wide variation in herbicide chemical makeup provides an equally wide variation in properties. Most herbicides are biodegradable, and most of them are relatively low in mammalian toxicity. However, some are quite toxic to fish, soil fauna, and perhaps to other wildlife. They can also have deleterious effects on beneficial aquatic vegetation that provides food and habitat for fish and shellfish. As for any chemicals, our knowledge about subtle chronic toxic effect of herbicides such as developmental changes, birth defects, performance changes, and degenerative diseases is much more uncertain than our knowledge of acute immediate poisoning.

Nematocides. Although nematocides are not as widely used as herbicides and insecticides, some of them are known to contaminate soils and the water draining from treated soils. For example, some carbamate nematocides dissolve readily in water, are not adsorbed onto soil surfaces, and consequently easily leach downward and into the groundwater. Other nematicidal chemicals are volatile soil fumigants that kill virtually all life in the soil, both the helpful and the harmful (Section 15.4). Methyl bromide, once the most commonly used of these fumigants, has been banned because of its adverse effects on the atmosphere and parts of the environment. Happily, the search for substitutes has led to the development of many nonchemical means to manage the pests once controlled by this highly toxic chemical (e.g., see Figure 10.17).

15.3 BEHAVIOR OF ORGANIC CHEMICALS IN SOIL[3]

Once they reach the soil, organic chemicals, such as pesticides or hydrocarbons, can follow one or more of eight pathways (Figure 15.5). They may: (1) volatilize into the atmosphere without chemical change; (2) sorb to clay or organic matter in soils; (3) leach downward through and become lost from the soil in liquid or solution form; (4) undergo chemical reactions within or on the surface of the soil; (5) decompose aerobically by the action of microorganisms in the aerated upper soil horizons; (6) decompose anaerobically by the action of microorganisms in hydric soils, saturated subsoils, or groundwater; (7) wash into streams and rivers in surface runoff; and (8) move by uptake into plants or soil animals to be metabolized or passed up the food chain. The specific environmental fate of each chemical will be determined at least in part by its particular chemical structure and properties.

Organic chemicals vary greatly in their **volatility** and subsequent susceptibility to atmospheric loss. Some soil fumigants, such as methyl bromide (now banned from most uses), were selected because of their very high vapor pressure, which permits them to penetrate soil pores to contact the target organisms. This same characteristic encourages rapid loss to the atmosphere after treatment, unless the soil is covered or sealed. A few herbicides (e.g., trifluralin) and fungicides (e.g., PCNB) are sufficiently volatile to make vaporization a primary means of their loss from soil. The lighter fractions of crude oil (e.g., gasoline and diesel) and many solvents vaporize to a large degree when spilled on the soil.

One cannot assume that disappearance of a pesticide from the soil is evidence of its breakdown into harmless products. Some chemicals lost to the atmosphere are known to return to the soil or to surface waters with the rain.

The degree to which an organic compound will be immobilized by **adsorption** to soil particles is determined largely by characteristics of both the chemical and the soil. Soil organic matter and high-surface-area clays tend to be the strongest adsorbents for some compounds

---

[3]For reviews on organic chemicals in the soil environment, see Barth et al. (2009) and Pierzynski et al. (2004).

Figure 15.5 *Eight important processes affecting the dissipation of organic chemicals (OC) in soils. Note that the OC symbol is split up by decomposition (both by light and chemical reaction) and degradation by microorganisms, indicating that these processes alter or destroy the organic chemical. In transfer processes, the OC remains intact.* [Modified from Weber and Miller (1989)]

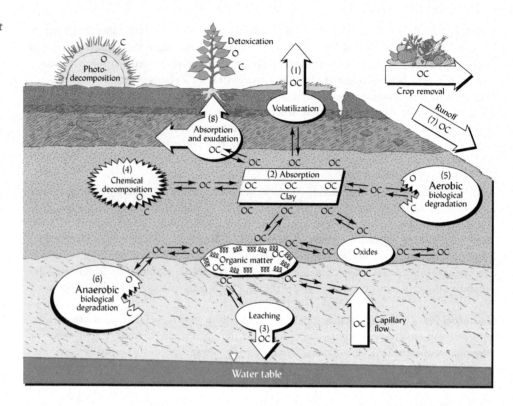

(Figure 15.6), while oxide coatings on soil particles strongly adsorb others. Everything else being equal, larger organic molecules with many charged sites adsorb more strongly.

Some organic chemicals with positively charged groups, such as the herbicides diquat and paraquat, are strongly adsorbed by silicate clays. Adsorption of certain pesticide molecules to clay particles tends to be greatest at low pH, which encourages *protonation*. Protonation, or the addition of H^+ ions to functional groups (e.g., —NH_2), yields a positive charge on the herbicide molecule, resulting in greater attraction to negatively charged soil colloids. The clay colloids in soils can thus strongly adsorb both positively charged organic compounds and other contaminants such as heavy metals (Section 15.6). On the other hand, the cations (Ca^{2+}, K^+, etc.) typically adsorbed on clay particle surfaces also make these surfaces hydrophilic and, as such, they are generally coated with adsorbed water. The hydrophilic nature of the clays greatly

Figure 15.6 *Adsorption of polychlorinated biphenyl (PCB) by different soil materials. The Lakeland sand (Typic Quartzipsamments) lost much of its adsorption capacity when treated with hydrogen peroxide H_2O_2 to remove its organic matter. The amount of soil material required to adsorb 50% of the PCB was approximately 10 times as great for montmorillonite (a 2:1 clay mineral) as for soil organic matter, and 10 times again as great for H_2O_2 treated Lakeland sand. Later tests showed that once the PCB was adsorbed, it was no longer available for uptake by plants. Note that the amount of soil material added is shown on a log scale.* [Data from Strek and Weber (1982)]

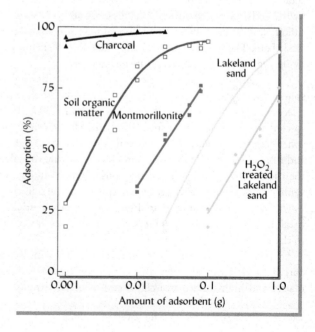

reduces their ability to adsorb many organic molecules which are hydrophobic in nature and without positively charged sites.

The tendency of organic chemicals to leach from soils is closely related to their *solubility in water* and their *potential for adsorption*. Some compounds, such as chloroform and phenoxyacetic acid, are a million times more water soluble than others, such as DDT and PCBs, which are quite soluble in oil but not in water. High water solubility favors leaching losses.

Strongly adsorbed molecules are not likely to move down the profile (Table 15.3). Leaching is apt to be favored by water movement, the greatest leaching hazard occurring in highly permeable, sandy soils that are also low in organic matter. Periods of high rainfall around the time of application of the chemical promote both leaching and runoff losses (Table 15.4). With some notable exceptions (e.g., paraquat and glyphosate), herbicides seem to be somewhat

Table 15.3
THE DEGREE OF ADSORPTION OF SELECTED HERBICIDES
Weakly adsorbed herbicides move more readily through the soil than those more tightly adsorbed.

Common name	Example trade name	Adsorptivity to soil colloids
Dalapon	Dowpon	None
Chloramben	Amiben	Weak
Bentazon	Basagran	Weak
2,4-D	Several	Moderate
Propachlor	Ramrod	Moderate
Atrazine	AAtrex	Strong
Alachlor	Lasso	Strong
EPTC	Eptam	Strong
Diuron	Karmex	Strong
Glyphosate	Roundup	Very strong
Paraquat	Paraquat	Very strong
Trifluralin	Treflan	Very strong
DCPA	Dacthal	Very strong

Table 15.4
SURFACE RUNOFF AND LEACHING LOSSES (THROUGH DRAIN TILES) OF THE HERBICIDE ATRAZINE FROM A CLAY LOAM LACUSTRINE SOIL (ALFISOLS) IN ONTARIO, CANADA
The herbicide was applied at 1,700 g/ha in late May. The data are the average of three tillage methods. Note that the rainfall for May and June is related to the amount of herbicide lost by both pathways.

Year of study	Surface runoff loss	Drainage water loss	Total dissolved loss	Percent of total applied, %	Rainfall, May–June, mm
		Atrazine loss, g/ha			
1	18	9	27	1.6	170
2	1	2	3	0.2	30
3	51	61	113	6.6	255
4	13	32	45	2.6	165

Data abstracted from Gaynor et al. (1995).

more mobile than most fungicides or insecticides, and therefore are more likely to find their way to groundwater supplies and streams, even when applied at normal rates (Figure 15.7).

Contamination of Groundwater

Experts once maintained that contamination of groundwater by pesticides occurred only from misuse or accidents such as spills, but it is now known that many pesticides reach the groundwater from normal agricultural use. Since many people (e.g., 40% of Americans) depend on groundwater for their drinking supply, leaching of pesticides is of wide concern. Table 15.2 lists some of the pesticides commonly found in well water in the United States. The concentrations are given in µg/L or parts per billion (see Box 15.1). In some cases, the amount of pesticide found in the drinking water has been high enough to raise long-term health concerns (e.g., atrazine and dieldrin exceed the health safety criteria in Table 15.2).

Chemical Reactions

Upon contacting the soil, some pesticides undergo chemical modification independent of soil organisms. For example, insecticides of the neonicotinoid class decompose within hours or days if exposed to bright sunlight. In contrast, DDT, diquat, and the triazines are subject to slow photodecomposition in sunlight. The triazine herbicides (e.g., atrazine) and organophosphate insecticides (e.g., malathion) are subject to hydrolysis and subsequent degradation. While the complexities of molecular structure of the pesticides suggest different mechanisms of breakdown, it is important to realize that degradation independent of soil organisms does in fact occur.

Microbial Metabolism

Biochemical degradation by soil organisms is the single most important method by which pesticides are removed from soils. Some microbes may metabolize a pesticide or other organic xenobiotic compound as a food source, thus deriving energy and cellular constituents from the compound. In other cases, when the xenobiotic compound is not used as food by the organism, the enzymatic breakdown of the compound is termed *cometabolism*. In some cases, microorganism can mineralize xenobiotics all the way to simple CO_2 (or CH_4) and H_2O, but in other cases more complex and possibly toxic intermediary breakdown products accumulate.

The biochemical reactions occurring in aerated environments are quite different from those in anaerobic environments. Certain soil fungi (Basidiomycetes or "white rot fungi") are

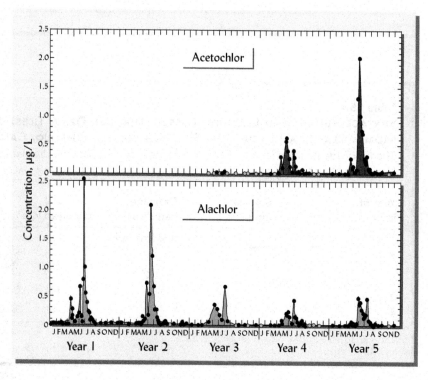

Figure 15.7 Herbicides in the main U.S. corn-growing region illustrate the direct and rapid connection between the use of a chemical on the land and its concentration in streams and rivers. The White River near Hazelton, Indiana, was monitored over a five-year period. Note that the concentrations of the herbicide alachlor peaked every year in June, about a month after most farmers in the watershed sprayed their corn and soybean fields. In year 3, a new compound, acetochlor, partially replaced the older herbicide alachlor. Within a year of the introduction of the newer compound, acetochlor concentrations increased, while alachlor concentrations decreased. [Data from Gilliom et al. (2006)]

> **BOX 15.1**
> ## CONCENTRATIONS AND TOXICITY OF CONTAMINANTS IN THE ENVIRONMENT
>
> As analytical instrumentation becomes more sophisticated, contaminants can be detected at much lower levels than was the case in the past. Since humans and other organisms can be harmed by almost any substance if large enough quantities are involved, the subject of toxicity and contamination must be looked at *quantitatively*. That is, we must ask *how much*, not simply *what*, is in the environment. Many highly toxic (meaning harmful in very small amounts) compounds are produced by natural processes and can be detected in the air, soil, and water—quite apart from any activities of humans.
>
> The mere presence of a natural toxin or a synthetic contaminant may not be a problem. Toxicity depends on: (1) the *concentration* of the contaminant, and (2) the level of *exposure* of the organism. Thus, low concentrations of certain chemicals that would cause no observable effect by a single *exposure* (e.g., one glass of drinking water) may cause harm (e.g., cancer, birth defects) to individuals exposed to these concentrations over a long period of time (e.g., three glasses of water a day for many years).
>
> Regulatory agencies attempt to estimate the effects of long-term exposure when they set standards for no-observable-effect levels (NOEL) or health-advisory levels (see Table 15.2). Some species and individuals within a species will be much more sensitive than others to any given chemical. Regulators attempt to consider the risk to the most susceptible individual in any particular case. For nitrate in groundwater, this individual might be a human infant whose entire diet consists of infant formula made with the contaminated water. For DDT, the individual at greatest risk might be a bird of prey that eats fish that eat worms that ingest lake sediment contaminated with DDT. For a pesticide taken up by plants from the soil, the individual at greatest risk might be an avid gardener who eats vegetables and fruits mainly from the treated garden over the course of a lifetime.
>
> It is important to get a feel for the meaning of the very small numbers used to express the concentration of contaminants in the environment. For instance, concentrations are often given in parts per billion (ppb). This is equivalent to micrograms per kilogram or µg/kg. In water this would be µg/L (Table 15.2). To comprehend the number 1 billion imagine a billion golf balls: lined up, they would stretch completely around the Earth. One bad ball out of a billion (1 ppb) seems like an extremely small number. On the other hand, 1 ppb can seem like a very large number. Consider water contaminated with 1 ppb of potassium cyanide, a very toxic substance consisting of a carbon, a potassium, and a nitrogen atom linked together (KCN). If you drank just one *drop* of this water, you would be ingesting almost 1 trillion (10^{12}) molecules of potassium cyanide:
>
> $$\frac{6.023 \times 10^{23} \text{ molecules}}{1 \text{ mol}} \times \frac{1 \text{ mol}}{65 \text{ g KCN}} \times \frac{1 \text{ g KCN}}{10^6 \text{ µg KCN}}$$
> $$\times \frac{1 \text{ µg KCN}}{L} \times \frac{L}{10^3 \text{ cm}^3} \times \frac{\text{cm}^3}{10 \text{ drops}}$$
> $$= \frac{9.3 \times 10^{11} \text{ molecules}}{\text{drop}}$$
>
> In the case of potassium cyanide, the molecules in this drop of water would probably not cause any observable effect. However, for other compounds, this many molecules may be enough to trigger DNA mutations, endocrine disruptions, or the beginning of cancerous growth. Assessing these risks is still an uncertain business.

especially important in the breakdown of certain recalcitrant xenobiotics that can be degraded by the same enzymes that these fungi use to metabolize natural polymers like lignin. In aerobic environments, certain polar groups on the pesticide molecules, such as —OH, —COO⁻, and —NH_2, can provide points of attack for the organisms' enzymes, but the presence of halogens may prevent this enzymatic action. For most aerobic microorganisms the oxidative removal of halogens (dehalogenation) and addition of hydroxyls is a necessary, but often very slow first enzymatic step. Anaerobic bacteria and archaea can accomplish this step much more readily using reductive dehalogenation.

Chlorinated hydrocarbons, such as DDT, aldrin, dieldrin, and heptachlor, are very slowly broken down, persisting in aerobic soils for 20 or more years. In contrast, the organophosphate insecticides, such as parathion, are degraded quite rapidly in soils, apparently by a variety of organisms. Likewise, most herbicides (e.g., 2,4-D, the phenylureas, the aliphatic acids, and the carbamates) are readily attacked by a host of organisms. However, their breakdown is often slow enough that residual herbicidal effects may be exhibited on rotation or cover crops planted 4–6 months after the initial application. The triazines are degraded even more slowly, primarily by chemical action. Most organic fungicides are also subject to microbial decomposition, although the rate of breakdown of some is slow, causing troublesome residue problems. Furthermore, it cannot be assumed that once the original compound is degraded, the problem has been removed. In many cases, the degradation products may be as toxic as the original pesticide, thus adding to the long-term environmental hazards.

Plant Absorption and Breakdown

Plants assist in the degradation of pesticides and other xenobiotic contaminants in several ways. They may degrade the compounds in their rhizosphere soil by exuding enzymes that catalyze the breakdown of the compounds or by stimulating microorganisms to do so. Pesticides, especially systemic insecticides and most herbicides, are commonly absorbed by plants via their roots or leaf surfaces. The absorbed chemicals may remain intact inside the plant, or they may be degraded. Often the pesticide is isolated by the plant inside cellular vacuoles where the degradation process takes place. Plants, being photoautotrophs, generally do not metabolize the pesticide molecule as a source of energy or carbon. Rather plant enzymes may cometabolize the pesticide into simpler breakdown products. As with microbial and chemical degradation, some plant metabolic degradation products are harmless, but others are even more toxic than the original chemical that was absorbed.

Understandably, society is quite concerned about pesticide residues found in the parts of plants that people eat, whether as fresh fruits and vegetables or as processed foods. Even foods produced without the use of synthetic pesticides (e.g., certified Organic) have been found to contain many pesticide compounds, albeit at significantly lower concentrations than found in conventionally produced foods. The use of pesticides and the amount of pesticide residues in food are strictly regulated by law to ensure human safety. Despite widespread concerns, there is little evidence that the small amounts of residues permissible in foods by law have had any ill effects on public health. However, routine testing by regulatory agencies has shown that about 1–2% of food samples tested contain pesticide residues above the levels legally permissible.

Persistence in Soils

The persistence of chemicals in the soil is the net result of all their reactions, movements, and degradations (see Figure 15.5). For example, organophosphate insecticides may last only a few days in soils. The widely used herbicide 2,4-D persists in soils for only two to four weeks. PCBs, DDT, and other chlorinated hydrocarbons may persist for 3–20 years or longer (Table 15.5). The persistence times of other pesticides and industrial organics fall generally between the extremes cited. The majority of pesticides degrade rapidly enough to prevent buildup in soils receiving normal annual applications. Those that resist degradation have a greater potential to cause environmental damage.

Table 15.5
PERSISTENCE OF SELECTED ORGANIC COMPOUNDS
Risks of environmental pollution are highest with those chemicals with greatest persistence.

Organic chemical	Persistence in soils
Chlorinated hydrocarbon insecticides (e.g., DDT, dieldrin)	3–20 yr
PCBs	2–10 yr
Triazine herbicides (e.g., atrazine and simazine)	1–2 yr
Glyphosate herbicide	6–20 mo
Benzoic acid herbicides (e.g., amiben and dicamba)	2–12 mo
Urea herbicides (e.g., monuron and diuron)	2–10 mo
Neonicotinoid insecticides (e.g., imidacloprid)	1–8 mo
Vinyl chloride	1–5 mo
Phenoxy herbicides (2,4-D and 2,4,5-T)	1–5 mo
Organophosphate insecticides (e.g., malathion and diazinon)	1–12 wk
Carbamate insecticides	1–8 wk

Continued use of the same pesticide on the same land can increase the rate of microbial breakdown of that pesticide. Apparently, having a constant food source allows a population buildup of those microbes equipped with the enzymes needed to break down the compound. This is an advantage with respect to environmental quality and is a principle sometimes applied in environmental cleanup of toxic organic compounds, but the breakdown may become sufficiently rapid to reduce a compound's effectiveness as a pesticide.

15.4 EFFECTS OF PESTICIDES ON SOIL ORGANISMS

Since pesticides are formulated to kill living things, it is not surprising that some of these compounds are also toxic to specific nontarget soil organisms. Many pesticides have been shown to affect certain soil organisms in ways that alter ecological balance and ecosystem function of soils. Fortunately, the diversity of soil organisms is so great that, most pesticides do not kill a significant proportion of the soil community. Chemicals used as soil fumigant present a major exception.

Fumigants

Fumigants are usually injected into the soil as bands of liquid or solid material. These compounds are designed to permeate or "fumigate" the soil in order to contact target pests (usually nematodes, fungi, or weed seeds) wherever they reside. As the chemical diffuses in vapor form, it moves much more rapidly through air-filled than through water-filled pores. Therefore, soil fumigation is most effective when the soil is quite *dry* and *warm*. On the other hand, nematodes live in water films around soil particles, so the chemical must also cross the partition between air and water phases to reach that target. Finally, the chemicals must also persist long enough to build up lethal concentrations that will kill the target organisms wherever they are located in the treated soil.

Most fumigant compounds have very broad spectrum toxicity and effect on both the soil fauna and microbes more drastically than do other pesticides. For example, 99% of the microarthropod population is usually killed by commercial fumigants, and it may take as long as two years for the population to fully recover. Fortunately, the recovery time for the microorganisms is generally much less. Fumigation reduces the number of species of both microbes and fauna. At the same time, the total number of bacteria may increase due to the relative absence of competitors and predators following fumigation and to the carbon and energy sources left by dead organisms for microbial utilization. The decimation of the rhizosphere community of plant-beneficial micro- and mesoorganisms generally leaves the next crop planted quite defenseless against any new infestation of pest or disease, creating a dependence on fumigants and calling into question the sustainability of the fumigation approach to soil borne pest control.

Effects on Soil Fauna

The effects of pesticides on soil animals vary greatly from chemical to chemical and from organism to organism. Nematodes are generally least affected, except by specific fumigants. Mites and springtails, two of the main groups of soil fauna at several trophic levels, vary in their sensitivity to insecticides, but many chemicals are quite toxic to these organisms.

Earthworms. Fortunately, many pesticides have only mildly depressing effects on earthworm numbers, but there are exceptions. Among insecticides, most of the carbamates (carbaryl, carbofuran, aldicarb, etc.) are highly toxic to earthworms. Among the herbicides, simazine is more toxic than most. Among the fungicides, benomyl is unusually toxic to earthworms. Toxicity may be expressed as more subtle than mortality. For example, one study showed the burrowing behavior of earthworms to be dramatically inhibited by a neonicotinoid insecticide applied to the soil (Figure 15.8).

Pesticides have significant effects on the numbers of certain predators and, in turn, on the numbers of prey organisms. For example, an insecticide that reduces the numbers of predatory mites may stimulate numbers of springtails, which serve as prey for the mites (Figure 15.9). Such organism interaction is normal in most soils.

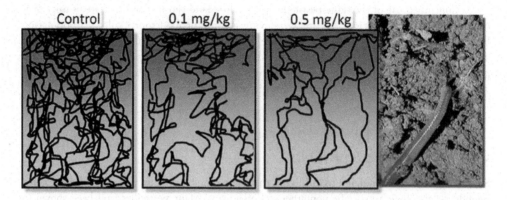

Figure 15.8 *Effects of organic pesticides on soil organisms may be more subtle than mortality. Here, the burrowing behavior of earthworms was shown to be dramatically inhibited by a neonicotinoid insecticide (imidacloprid) often used for seed treatments. The earthworms in the test columns were an endogeic (shallow burying) species (Allolobophora Icterica) collected from an apple orchard. Imidacloprid solutions were applied to give concentrations of 0.5 or 0.1 mg of the compound per kg of dry soil. The soil was a silt loam with 2.8% organic matter and pH 8.3.*
[Burrow diagrams based on concepts in Capowiez et al. (2006), photo of another earthworm species, courtesy of Ray R. Weil]

Effects on Soil Microorganisms

Overall levels of bacteria in the soil are generally not too seriously affected by pesticides. However, the organisms responsible for nitrification and nitrogen fixation are sometimes adversely affected. Insecticides and fungicides affect both processes more than do most herbicides, although some of the latter can reduce the numbers of organisms carrying out these two reactions. Certain pesticides may actually enhance biological nitrogen fixation by reducing the activity of protozoa and other organisms that are competitors or predators of the nitrogen fixing bacteria.

Fungicides, especially those used as fumigants, can have marked adverse effects on soil fungi and actinomycetes, thereby slowing down the humus formation in soils. Interestingly, however, the process of ammonification is often stimulated by pesticide use.

The negative effects of most pesticides on soil microorganisms are temporary, and after a few days or weeks, organism numbers generally recover. But exceptions are common enough to dictate caution in the use of the chemicals. Care must be taken to apply them only when alternate means of pest management are not available.

Figure 15.9 *The effects of an insecticide on the soil fauna can be both direct and indirect. Predators such as certain mites are often rapidly killed by a toxin, freeing their prey populations, such as saprophytic collembolan, from the constraints of predation. Depending on soil conditions and the persistence of the chemical, the time scale could be in days or months.* [Drawn from data and concepts in Edwards (1978) and El-Naggar and Zidan (2013). For effects of herbicides on soil organisms, refer to Rose et al. (2016)]

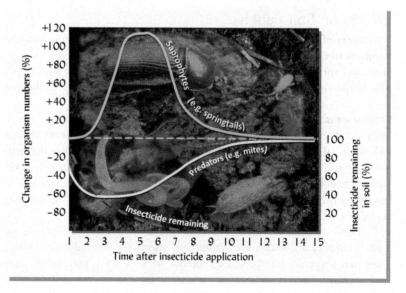

This brief overview of the behavior of organic chemicals in soils reemphasizes the complexity of the ecological changes that take place when new and exotic substances are added to our environment. Our growing knowledge of the soil processes involved certainly reaffirms the necessity for a thorough evaluation of potential soil ecological impacts prior to approval of new chemical products or the adoption of existing chemicals for use on the land.

15.5 REMEDIATION OF SOILS CONTAMINATED WITH ORGANIC CHEMICALS

Soils may become seriously contaminated where accidental spills or leakage of toxic organic materials have occurred or where, through the decades, organic wastes from industrial and domestic processes have been dumped on soils. The levels of such *acute contamination* are often sufficiently high that plant growth is restrained or even prevented. Pollutants can move into the groundwater, making it unfit for human consumption. Fish and wildlife may be decimated. The soil itself may become unsafe for human exposure. Because of public concerns, businesses and government are spending billions of dollars annually to clean up (**remediate**) these contaminated soils. We shall consider a few of the methods in use and under development in the rapidly evolving soil remediation industry.

In general, efforts to remediate polluted soils face the need to compromise between speed and certainty that cleanup standards will be met on one hand, and expense and disruption of the site on the other (Figure 15.10). In many cases of relatively mild contamination, the soil ecosystem may be able to destroy the contaminant and heal itself, unaided, by the process of **natural attenuation**. Natural attenuation may allow the soil ecosystem to recover its function and diversity over a reasonable period using any or all of the chemical, physical, and biological processes that were illustrated in Figure 15.5. However, in other cases, the contamination may be so serious that natural attenuation will occur too slowly to be a practical solution and more active technological interventions may be called for.

Physical and Chemical Methods

The earliest and still widely used methods of soil remediation involve physical and/or chemical treatment of the soil, either in place (in situ) or by moving the soil to a treatment site (ex situ). Ex situ treatment may involve excavating the soil to treatment containers where it may be incinerated to drive off volatile chemicals and to destroy other pollutants by high-temperature chemical decomposition. Water-soluble and volatile chemicals may also be removed by pushing or pulling air or water through the soil by vacuum extraction or leaching. Such treatments are usually quite effective in removing or destroying the contaminants, but are expensive, especially if large quantities of soil must be excavated and treated. And, of course,

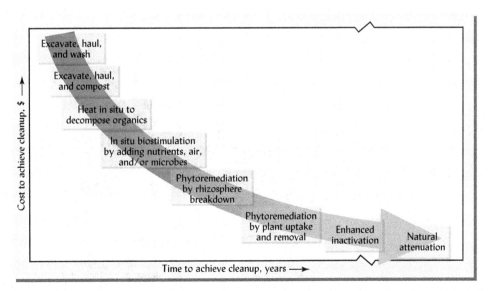

Figure 15.10 A wide range of methods is available to remediate (clean up) polluted soils. At one extreme are remediation techniques that are very expensive and disruptive, but usually quite rapid. Technologies at the other extreme may be quite inexpensive and nondisruptive, but usually take much more time to accomplish the cleanup. (Diagram courtesy of Ray R Weil.)

the site is highly disrupted and the treated soil is destroyed as a living system and must be either replaced on the site or deposited in a landfill. The costs, in terms of money and energy, for excavation and hauling large volumes of soil can be prohibitive.

In situ treatments are usually preferred if viable technologies are available. The soil is left in place, thereby reducing excavation, treatment, and disposal costs and providing greater flexibility in future land use. The contaminants are either removed from the soil (*decontamination*) or are sequestered (*bound up*) in the soil matrix (*stabilized*). Decontamination in situ involves some of the same techniques of water flushing, leaching, vacuum extraction, and heating used in ex situ processes.

Water washing treatments are not effective for oils and other nonpolar compounds that are repelled by water. To help remove such compounds, compounds called *surfactants* (similar to detergents) maybe sprayed onto the soil surface or injected into the soil. As the surfactants move downward in the soil, they dissolve organic contaminants, which can then be pumped out of the soil as in the water washing systems.

Certain surfactants may also be used to immobilize or stabilize soil contaminants. They are positively charged and through cation exchange can replace metal cations on soil clays. For example, one group of such surfactants, quaternary ammonium compounds (QACs), has the general formula $(CH_3)_3NR^+$, where R is an organic alkyl or aromatic group. The positive charges on QACs stimulate cation exchange by reactions such as the following, using a monovalent exchangeable cation such as K^+ as an example:

$$\boxed{\text{Colloid}}\ K^+ + \underset{\text{QAC}}{(CH_3)_3NR^+} \rightarrow \underset{\text{Organoclay}}{\boxed{\text{Colloid}}\ (CH_3)_3NR^+} + K^+ \quad (15.1)$$

Untreated clay

The resulting products, known as *organoclays*, have properties quite different from the untreated clays. They attract rather than repel nonpolar organic compounds. Thus, the injection of a QAC into the zone of groundwater flow can stimulate the formation of organoclays and thereby immobilize soluble organic groundwater contaminants, holding them until they can be degraded (Figure 15.11).

As we learned in Section 8.12, the degree of sorption of organic compounds by soil colloids is commonly indicated by the **coefficient of distribution** K_d, the ratio between the sorbed and solution portions of the organic compound.

$$K_d = \frac{(\text{mg contaminant}/\text{kg soil})}{(\text{mg contaminant}/\text{L solution})} \quad (15.2)$$

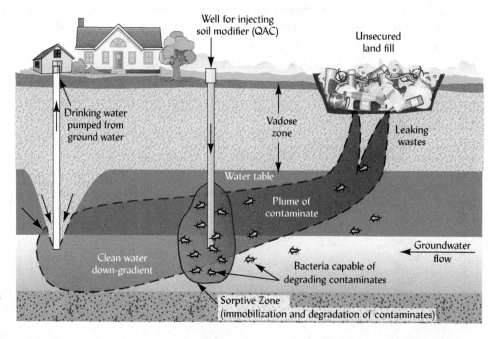

Figure 15.11 How a combination of a quaternary ammonium compound (QAC), hexadecyltrimethylammonium, and bioremediation by degrading bacteria could be used to hold and remove an organic contaminant. The pollutant is moving into groundwater from a buried waste site. The QAC reacts with soil clays to form organoclays and soil organic matter complexes that adsorb and stabilize the contaminant, giving microorganisms time to degrade or destroy it. [Diagram courtesy of Ray R. Weil using concepts from Xu et al. (1997)]

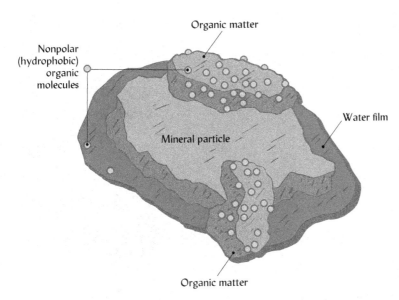

Figure 15.12 Because mineral colloids in soils are nearly always surrounded by at least a thin film of water, hydrophobic organic molecules tend to sorb onto humus more readily than clay. The nonpolar organic molecules cannot compete with the polar water molecule for a place on the charged mineral surfaces. This is one reason why, for a specific organic contaminant, the K_{oc} is more consistent than the K_d in characterizing the tendency to be held by various soils. (Diagram courtesy of Ray R. Weil)

The K_d for the adsorption of many nonpolar organic compounds on untreated clays is very low because the clays are hydrophilic (water-loving) and their adhering water films repel the hydrophobic, nonpolar organic compounds (Figure 15.12). Surface soil horizons containing significant quantities of humus often exhibit a much higher K_d because of the sorption of the organic contaminant into the organic matter coatings. This is the reason that K_{oc} is often a better measure of a compound's tendency to become immobilized in various surface soils. Deeper soil layers, especially near and below the water table, generally contain little humus, and so have limited capacity to immobilize organic contaminants.

In contrast, organoclays effectively sorb organic contaminants, leaving little in the soil solution, thereby reducing their movement into the groundwater and eventually into streams or drinking water. Consequently, the K_d values of organic contaminants on organoclays are commonly 100–200 times those measured on the untreated clays. The K_d values for common organic contaminants on organoclays tend to be very high and indicate the very high sorbing power of the newly created sorbants. Their tenacity is complemented by the very strong complexation of organic compounds by soil organic matter. Organoclays thus offer promising mechanisms for holding organic soil pollutants until they can be destroyed by biological or physicochemical processes.

Bioremediation[4]

Even for many heavily contaminated soils, there is a biological alternative to incineration, soil washing, and landfilling—namely, **bioremediation**. Simply put, this technology uses enhanced plant and/or microbial action to degrade organic contaminants into harmless metabolic products. Analysis of microbial DNA has shown that degradation of contaminants in soils is almost always the work of genetically diverse *consortia* of many organisms, rather than just one or two bacterial species. Petroleum constituents, including the more resistant PAHs, as well as several synthetic compounds, such as pentachlorophenol (PCP) and TCE, can be broken down, primarily by soil bacteria and so-called white-rot fungi. Bioremediation is usually accomplished in situ, but polluted soil may also be excavated and treated ex situ—that is, hauled to a treatment site where such techniques as high-temperature composting may be used to destroy the organic contaminants in the soil (Figure 15.13).

Bioaugmentation. In most cases, the remediation process depends on organisms native to the soil. In certain cases, the biodegradation of contaminants can be accelerated if microbes specifically selected for their ability to degrade the contaminants are introduced into the polluted soil zone to *augment* the natural microbial populations. This approach is called **bioaugmentation**. Some success has been achieved by inoculating contaminated soils with improved organisms that can degrade the pollutant more readily than can the native population. Although genetic engineering

---

[4]For reviews of the theories and technologies regarding this topic, see Hazen (2008) or Mishra and Mohan (2017). For an example of how biostimulation can be combined with bioaugmentation, view Alvarez-Cohen (2012).

Figure 15.13 Hot water vapor rises in cold winter air as windrows of high-temperature compost are mixed and aerated by a special compost-turning machine in order to accelerate the breakdown of organic compounds. The method can hasten the degradation of organic pollutants in soil material excavated from a contaminated site, mixed with decomposable organic materials, and made into windrows. (Photo courtesy of Ray R. Weil)

may prove useful in making "superbacteria" in the future, most inoculation has been achieved with naturally occurring organisms. Organisms isolated from sites with a long history of the specific contamination or grown in laboratory culture on a diet rich in the pollutant in question tend to become acclimated to metabolizing the target chemical.

For example, certain bacteria have been identified that can detoxify perchloroethene (PCE), a common, highly toxic groundwater pollutant that is suspected of being a carcinogen. Scientists can inoculate the anaerobic subsoil or groundwater zone with these organisms to expedite reductive dechlorination. This step-by-step removal of the four chlorines from the PCE produces ethylene, a gas that is relatively harmless to humans.

$$\underset{\text{(Suspected carcinogen)}}{\underset{\text{PCE}}{\text{Cl}_2\text{C}=\text{CCl}_2}} \xrightarrow[\text{4HCl}]{8\text{H}} \underset{\text{(Harmless gas)}}{\underset{\text{Ethylene}}{\text{H}_2\text{C}=\text{CH}_2}} \quad (15.3)$$

Biostimulation. The use of technology that assists the naturally occurring microbial populations in breaking down chemicals is called **biostimulation**. Usually, the soil naturally contains some bacteria or other microorganisms that can degrade the specific contaminant. Even so, the rate of natural degradation may be far too slow to be effective.

Once it is determined that organisms are present that are capable of metabolizing the contaminant, biostimulation techniques work to address one or more of four fundamental conditions that may limit the effective rate of contaminant biodegradation:

1. *Inaccessibility of the contaminant* to the organisms that could degrade it. We have already mentioned the practice of adding surfactants to make contaminants accessible to bacterial degradation by desorbing mineral-bound contaminants or by dissolving hydrophilic contaminants in a way that allows bacteria to attach themselves to the molecules.
2. *Lack of nutrients* may restrict organism growth and metabolism. This is especially likely in low organic matter sandy soils and where the contaminant is a hydrocarbon such as a petroleum product. In these cases, adding fertilizer (especially nitrogen and phosphorus) in suitable form may greatly stimulate biodegradation rates. A classic example from oil-spill cleanup technology is described below.
3. *Electron donor* substances (food) may be inadequate for rapid bacteria growth. In some cases, readily available energy sources such as molasses (for respiring organisms) or methane (for methanotrophs) have been added with success (see below).
4. *Electron acceptor* substances may be inadequate for effective metabolism. For aerobic breakdown, this situation may call for adding oxygen by pumping air into the contaminated soil or groundwater zone. For other organisms, pumping in nitrate may allow denitrifiers to degrade the compound.

Finally, the *habitat* may be inhospitable or too toxic for biodegradation to take place because either: (1) levels of the target contaminant are so high that it is toxic to even the organisms that can degrade it, or (2) the site is co-contaminated with another substance, perhaps a heavy metal (see Sections 15.6 and 15.7), in such concentrations that the potential biodegradation organisms cannot thrive, or even survive. In such cases, in situ bioremediation may not be possible and the soil may have to be excavated and removed from the site.

Oil Spill Cleanup. The 30 or more different genera of bacteria and fungi known to degrade hydrocarbons are found in almost any soil or aquatic environment. In 2012, an offshore drilling rig explosion resulted in a huge amount of oil leaking into the Gulf of Mexico and washing up the surrounding beaches. In the warm, nutrient-rich Gulf of Mexico, naturally occurring bacteria rapidly degraded most of the lighter components of the crude oil within a few weeks of the leak being sealed. The only help they got from humans was the application of dispersant chemicals that broke the oil into tiny droplets that the bacteria could better access. Fortunately, soil Bacteria and Archaea helped the beaches and wetlands recover from the oil spill much faster than scientists had expected.

The situation with the previously largest U.S. oil spill—the 1989 Exxon Valdez oil spill in Alaskan waters—was quite different. The spill took place when a supertanker broke up and sank just a short distance from the pristine shores of Prince William Sound. The pebble beaches were heavily contaminated with crude oil driven up by a large storm. The cold and nutrient-poor water and pebble beaches were not conducive to bacterial growth. The bacteria needed some help. It came in the form of the first large-scale successful bioremediation by fertilization.

The oil-eating bacteria had the carbon and oxygen they needed (electron donor and acceptor, respectively), but they were starved for nitrogen and phosphorus. The oil-soaked beaches (pebbly Entisols) were sprayed with a special fertilizer made from a biodegradable oliophilic organic and contained tri(laureth-4)-phosphate (the phosphorus source and a surfactant similar to those found in many shampoos), urea (a soluble nitrogen source), and 2-butoxyethanol (a surfactant and emulsifier that, though toxic, proved to be biodegradable).

The fertilizer was formulated to be soluble in oil but not in water so that it would stay with the oil and not only provide N and P to the bacteria but also emulsify and disperse the oil to make it more physically accessible to biodegradation (and to avoid contributing to eutrophication of Prince William Sound). Within a few weeks, and despite the cold temperatures, most of the oil in the test area was degraded (Figure 15.14).

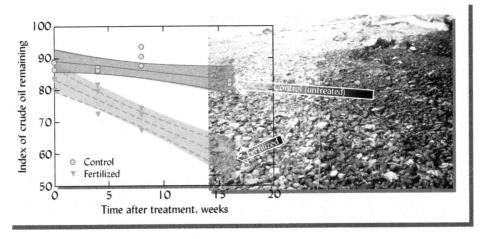

Figure 15.14 *Bioremediation of crude oil from the Exxon Valdez oil spill off the coast of Alaska. The oil contaminating the pebbly beach soils was degraded by indigenous bacteria when an oil-soluble fertilizer containing nitrogen and phosphorus was sprayed on the beach (green data points in graph on the left). The control sections of the beach (brown data points) were left unfertilized for almost 11 weeks. By then, the effect of the fertilization was so dramatic that a decision was made to treat the control sections as well. Therefore, the regression line for the control plots probably includes some remediation by fertilizer in the last set of measurements. The photo (right) shows the clear delineation between the oil-covered control section and the fertilized parts of the beach.* [Graph regressions using selected data in Bragg et al. (1994); Additional information from Pritchard, et al. (1992); Photo courtesy of P.H. Pritchard, U.S. EPA, Gulf Ecology Division, Gulf Breeze FL; from EPA]

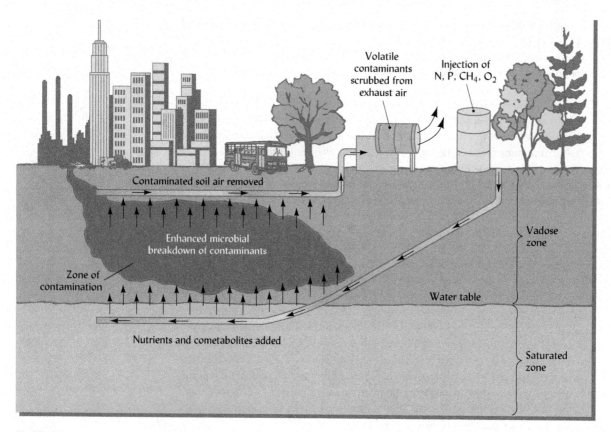

Figure 15.15 *In situ bioremediation of soil and groundwater contaminated with volatile organic solvents. The scheme illustrated is typical of the biostimulation approach to soil remediation. Breakdown of the organic contaminant is stimulated by adding such components as nutrients, oxygen, and cometabolites that improve the soil environment for the growth of native bacteria capable of metabolizing the contaminant. In this instance, methane (CH_4) is added intermittently as a substrate for certain methane-oxidizing bacteria which multiply rapidly and turn to the solvent as a carbon source whenever methane is not available. The nutrients are added and the contaminated soil air is removed by perforated pipes inserted by horizontal well-drilling techniques. Such biostimulation schemes can significantly cut the time and cost for cleanup of contaminated soils.* [Based on Hazen (1995)]

In Situ Biostimulation Techniques. Other situations call for the use of other bioremediation techniques in situ. In some cases, low soil porosity or water saturation causes oxygen deficiency that limits microbial activity. Techniques have been developed that use in situ bioremediation to clean up oxygen-deficient soils and associated groundwater contamination. For example, organic-solvent-contaminated soils have been bioremediated (Figure 15.15) by piping in a mixture of air (for oxygen), methane (to act as a carbon source to stimulate specific bacteria), and phosphorus (a nutrient that is needed for bacteria growth). The design and likely success of such schemes are very dependent on such properties as soil texture. Compared to fine-textured soil with high clay contents and low permeability, coarse-textured, sandy soils are much easier to bioremediation because their high permeability allows nutrients and other additives to be relatively easily pumped through the soil to the contaminated zone.

Phytoremediation

Plants can also participate in bioremediation, in which case the process is termed **phytoremediation**. For several decades, plant-based systems have been used for the removal of municipal wastewater contaminants, industrial soil pollutants, and organic and inorganic pollutants from shallow groundwater.

The most basic and simplest form of phytoremediation is **phytostabilization**, in which almost any kind of plant cover is grown in order to stabilize the contaminant and reduce its movement into the wider environment. In phytostabilization, plants provide several services:

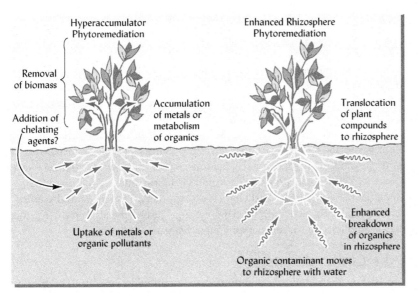

Figure 15.16 *Two approaches to phytoremediation—the use of plants to help clean up contaminated soils. (Left) Hyperaccumulating plants take up and tolerate very high concentrations of an inorganic or organic contaminant. In the case of metal contaminants, the addition of chelating agents may increase the rate of metal uptake, but can add a major expense and may allow metals to migrate below the root zone. (Right) In enhanced rhizosphere phytoremediation, the plants do not take up the contaminant. Instead, the plant roots excrete substances that stimulate the microbes in the rhizosphere soil, speeding their degradation of organic contaminants. Transpiration-driven movement of water and dissolved contaminants to the enhanced rhizosphere zone improves the system's effectiveness.* (Diagram courtesy of Ray R. Weil)

(1) they provide groundcover to reduce the movement of contaminated soil and dust by water and wind erosion; (2) their transpiration reduces the amount of rainwater moving through the contaminated soil and thus also reduces the production of contaminated leachate; (3) the plants may absorb the contaminant and sequester it in their root systems; and (4) the plant root may cause the precipitation of the contaminant in the rhizosphere. The main shortcoming of the phytostabilization approach is that the contaminant is not removed or destroyed, but remains in the soil.

At a more sophisticated level, certain plants have been found or developed through selection that can quite effectively remove, accumulate, and/or degrade specific soil contaminants. Such phytoremediation uses plants in two fundamentally different ways (Figure 15.16). In the first, plant roots take up the pollutant from the soil. The accumulation of unusually high concentrations of a contaminant in the aboveground plant biomass is called **hyperaccumulation**. Hyperaccumulating plants take up and tolerate very high concentrations of a particular contaminant, most commonly a toxic metal such as cadmium or nickel (see Section 15.9), but also certain organics such as TNT. Hyperaccumulation allows the contaminant to be removed by harvesting the plant tissue. The harvested plant material may be so concentrated in the contaminant that it must be treated as hazardous material. Fortunately, in certain organic contaminants, plants have been found that can metabolize the accumulated contaminant into harmless by-products.

A second type of cleanup using plants is called **enhanced rhizosphere phytodegradation**. In this process, the plants do not take up the contaminant. Instead, the plant roots excrete into the soil carbon compounds that serve as microbial substrates and growth regulators (see Section 10.7). These compounds stimulate the growth of the rhizosphere bacteria that, in turn, degrade the organic contaminant. The transpiration of water by the plant causes soil water, with its load of dissolved contaminant molecules, to move toward the roots, thus increasing the efficiency of the rhizosphere reactions. The process can be further enhanced by teaming the plants with appropriate symbiotic mycorrhizal fungi (see Section 10.9).

Phytoremediation is particularly advantageous where large areas of land are contaminated with only moderate concentrations of organic pollutants located at shallow depths in the soil. Phytoremediation causes little site disturbance and is relatively inexpensive, but also commonly takes a longer time to remove large quantities of contaminants than do the more costly engineering procedures.

15.6 SOIL CONTAMINATION WITH TOXIC INORGANIC SUBSTANCES[5]

The toxicity of inorganic contaminants released into the environment every year is now estimated to exceed that from organic and radioactive sources combined. A fair share of these inorganic substances ends up contaminating soils. The greatest problems most likely involve mercury, cadmium, lead, arsenic, nickel, copper, zinc, chromium, molybdenum, manganese, selenium, fluorine, and boron. To a greater or lesser degree, all of these elements are toxic to humans and other animals. Cadmium and arsenic are extremely poisonous; mercury, lead, nickel, and fluorine are moderately so; boron, copper, manganese, and zinc are relatively lower in mammalian toxicity. Table 15.6 provides background information on the uses, sources, and effects of some of these elements. Although the toxic metals and metalloid elements (see periodic table, Appendix B) are not all, strictly speaking, "heavy" metals (more dense than iron), for the sake of simplicity this term is often used in referring to them.

Sources of the Contaminants

Inorganic chemicals from many sources contaminate and accumulate in soils. The burning of fossil fuels, metal ore smelting, and many manufacturing processes release tons of these elements into the atmosphere, which can carry the contaminants and later deposit them on the

Table 15.6
SOURCES AND HEALTH EFFECTS OF SELECTED INORGANIC SOIL POLLUTANTS

Chemical	Major uses and sources of soil contamination	Organisms principally harmed[a]	Human health effects
Arsenic	Pesticides, plant desiccants, animal feed additives, coal and petroleum, mine tailings, detergents, and irrigation water	H, A, F, B	Cumulative poison, cancer, skin lesions
Cadmium	Electroplating, pigments, plastic stabilizers, batteries, and phosphate fertilizers	H, A, F, B, P	Heart and kidney disease, bone embrittlement
Chromium	Stainless steel, chrome-plated metals, pigments, refractory brick manufacture, and leather tanning	H, A, F, B	Mutagenic; also essential nutrient
Copper	Mine tailings, fly ash, fertilizers, windblown copper-containing dust, and water pipes	F, P	Rare; mental problems, fatigue; essential nutrient
Lead	Combustion of oil, gasoline, and coal; lead-acid batteries; iron and steel production; solder in water-pipes; paint pigments	H, A, F, B	Brain damage, convulsions
Mercury	Pesticides, catalysts for synthetic polymers, metallurgy, and thermometers; from coal burning	H, A, F, B	Nerve damage
Nickel	Combustion of coal, gasoline, and oil; alloy manufacture; electroplating; batteries; and mining	F, P	Lung cancer
Selenium	High Se geological formations and irrigation wastewater in which Se is concentrated	H, A, F, B	Deformities; essential nutrient
Zinc	Galvanized iron and steel, alloys, batteries, brass, rubber manufacture, mining, and old tires	F, P	Rare; essential nutrient

[a] H = humans, A = animals, F = fish, B = birds, P = plants.

---

[5] Several reviews of metal contamination and remediation of soils are worth reading (Su et al., 2014; Wu et al., 2010). Heavy metal contamination of food grown in China is discussed in Zhang et al. (2015). The basic biogeochemistry of metals is detailed by Adriano (2001).

vegetation and soil many kilometers downwind. Lead, nickel, and boron are gasoline additives that are released into the atmosphere and carried to the soil through rain and snow. Today the main use for lead is in lead-acid batteries for vehicles, but many soils continue to become contaminated with lead from the peeling of old painted surfaces. Boron as the mineral borax is used in detergents, fertilizers, and forest fire retardants, all of which commonly reach the soil. Phosphate fertilizer and limestone, two widely used soil amendments, usually contain small quantities of cadmium, copper, manganese, nickel, and zinc. Cadmium is used in plating metals and in the manufacture of rechargeable batteries. Arsenic was for many years used as a wood preservative as well as an insecticide on cotton, tobacco, fruit crops, lawns, and as a defoliant or vine killer. Some of these mentioned elements are found as constituents in specific organic substances and in domestic and industrial sewage sludge. Mercury is released from gold mining and chemical manufacturing industries, as well as from burning certain types of coal.

Accumulation in Soils

Some of these toxic metals are being released to the environment in increasing amounts, while emissions of others (most notably lead, because of changes in gasoline formulation) are decreasing. Unlike most organic contaminants, the inorganic toxins do not decompose and degrade; rather they usually remain in the soil and so accumulate from year to year. In many places, the metal contamination is historical, resulting from polluting activities that are no longer practiced. Lead in gasoline was mentioned, but lead from old water pipes, old methods of leather tanning (Cr), felt and gold processing (Hg), and old artisanal manufacture of metal alloys (Cu, Zn, Ni), etc., may leave a legacy of pollution in various soil layers. This is especially common in old urban areas where cycles of building, demolition, and rebuilding have occurred through centuries (see Figure 15.17).

Concentration in Living Tissues

Irrespective of their sources, people everywhere are exposed to these toxic elements every day, either through the air or through ingestion of food, water, and—yes—soil (see Box 15.2). Toxic metals can and do reach the soil by direct or indirect deposition, and from there they become part of the food chain: soil–plant–animal–human (Figure 15.18). Unfortunately, once the elements become part of this cycle, they may accumulate in animal and human body tissue to toxic levels.

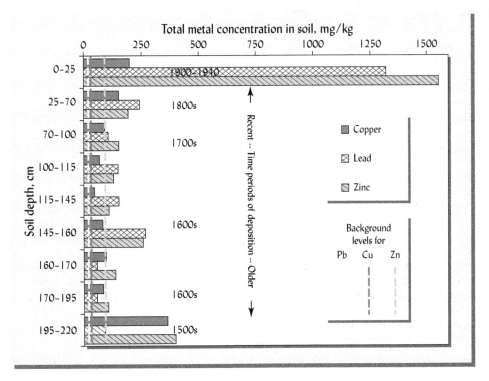

Figure 15.17 History of copper, lead, and zinc contamination of soils in central Moscow, Russia. The layers of contaminated urban soil go back to the founding of the city in the fourteenth century. The most serious contamination occurred during the industrialization of Russia following the Bolshevik revolution in the early twentieth century. However, the copper concentration is greatest in the oldest layers because of ancient artisanal metalworking. Such legacy metal contamination exists in many older cities around the world. [Graphed using data selected from Meuser (2010)]

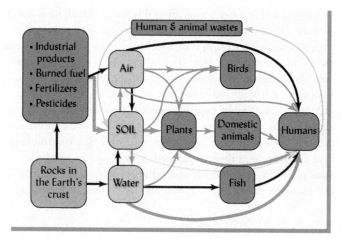

Figure 15.18 Sources of heavy metals and their cycling in the soil–water–air–organism ecosystem, with an emphasis on human exposure. The content of metals in organism tissue generally builds up from left to right, indicating the vulnerability of humans (as secondary and tertiary food web consumers) to heavy metal toxicity. Air, water, and soil may also be directly ingested by humans. (Diagram courtesy of Ray R. Weil)

BOX 15.2
LEAD POISONING AND SOIL CONTAMINATION

Lead contamination is a serious and widespread form of inorganic soil pollution. Long-term exposure to low levels of lead can profoundly affect a child's development and neurological function, including intelligence. Lead poisoning has been shown to contribute to mental retardation, poor academic performance, and juvenile delinquency. In the past (and, unfortunately, in the present in many developing countries that still use leaded gasoline), much of the lead exposure came from burning leaded fuels. The content of lead in soils commonly increases with proximity to major highways. Inner city residents generally live surrounded by lead-contaminated soils. The soil on the windward side of apartment buildings often shows the highest accumulations of lead, as it is there that the wind-carried particulates tend to settle out of the air. A second reason for high lead concentrations in urban soils is related to the lead-based pigments in paint from pre-1970 era buildings. Paint chips, flakes, and dust from sanding painted surfaces spread the lead around, and eventually much of it ends up in the soil. During dry weather, soil particles blow about, spreading the lead and contributing to the dust that settles on floors and windowsills. Although plants do not readily take up lead through their roots, lead-contaminated dust may stick to foliage and fruits.

Eating these garden products and breathing in lead-contaminated dust are two pathways for human lead exposure (see Figure 15.18). However, the most serious pathway, at least for young children, is thought to be hand-to-mouth activity—basically, eating dirt (see also Box 1.1). Anyone who has observed a toddler knows that a child's hands are continually in its mouth (Figure 15.19). Lead-contaminated dust on surfaces in the home can therefore be an important source of lead exposure for young children; so, too, can lead-contaminated soil in outdoor play areas. Having children wash their hands frequently can significantly cut down their exposure to this insidious toxin. The U.S. EPA has set standards for the cleanup of lead in soil around homes: 400 mg of lead per kg dry soil in bare soil in children's play areas

Figure 15.19 Lead from car exhaust and old paint accumulates in urban soils, and hand-to-mouth activity is a major pathway for lead poisoning in young children. (Photo courtesy of Ray R. Weil)

or 1,200 mg/kg average for bare soil in the rest of the yard. Soils with lead levels higher than these standards require some remediation.

Since 1970, major government programs have aimed to reduce lead exposure from paint (lead paints are banned and existing lead paint must be removed or sealed), drinking water, and food (lead has been banned from solder used in pipe joints and food cans). As a result, median levels of lead in blood samples from U.S. children fell from 0.18 mg/L in 1970 to 0.03 mg/L in 1994. For children under age 6, blood levels above 0.10 mg Pb/L are considered elevated and a threat to health. Unfortunately, lead in soils has not been similarly addressed. The U.S. EPA reports that nearly 1 million children in America (especially in central cities) have dangerously elevated levels of lead in their blood. Not only are the children of poor, urban families more exposed to lead-contaminated soils than most, but their diets are typically lower in phosphorus and calcium, nutrients that could make them less susceptible to lead toxicity.

(continued)

> **BOX 15.2**
> **LEAD POISONING AND SOIL CONTAMINATION** *(CONTINUED)*
>
> Current measures designed to protect children from lead in soil around the home include: (1) excavation and removal of the soil, (2) dilution by mixing in large amounts of noncontaminated soil, or (3) stabilization of the lead away from the reach of children and dust-creating winds. Excavation of soil around homes is extremely expensive and, given the low mobility of lead in soils, probably not necessary. Instead, the contaminated soil area may be covered with a thick layer of uncontaminated topsoil, a wooden deck, or pavement. Well-maintained turfgrass will prevent most dust formation and soil ingestion. Removal of lead-based paints is likewise very expensive and difficult, so isolating the lead-based paint under several coats of fresh paint can limit exposure.
>
> Effective phytoremediation techniques have not yet been developed to get the lead out of soils; however, leaving the lead in the soil but transforming it into a form that is not bioavailable might be beneficial. Research shows that phosphorus (fertilizer) can effectively transform bioavailable lead compounds into relatively inert, nonbioavailable forms. However, rates of P application needed are 500 times normal fertilization rates.

This situation is especially critical for humans, fish, and other creatures at the top of the food chain. It has already resulted in restrictions on the use of certain fish and wildlife for human consumption. Because of the globalization of our food supply, crops grown on polluted soils in countries with weak environmental regulations (especially those with a recent history of "dirty" industrialization, such as China) may threaten the safety of food consumed in importing countries around the world. By the second decade of the twenty-first century, it was clear that industrialization had severely contaminated soil in several parts of China to a degree that was affecting food safety, both in urban agriculture and in rice fields in heavily polluted regions. Protection of the environment and food supplies requires that governments in all industrial nations take seriously the need to closely regulate the release of these toxic elements and the production of food in potentially contaminated soils.

Even in American suburban home gardens, where nearby industries are not a factor, there are possible sources of contamination, including the many outdoor decks, fences, and utility poles made from wood pressure treated with chromated copper arsenate (CCA) as a preservative. Although no longer in use in the United States, most of the wood sold for outdoor use between 1975 and 2003 was treated with arsenic chemicals and remains a source of contamination. Leaching of arsenic (As) from the treated wood has been shown to contaminate the nearby soil and, in turn, the vegetables grown in it (Figure 15.20).

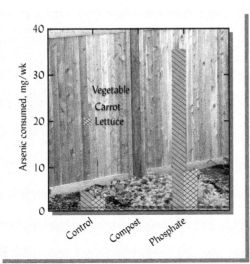

Figure 15.20 Leaching of arsenic (As) from chromated copper arsenate (CCA) treated wood may contaminate soil and, in turn, vegetables grown in it. If a gardener ate the recommended daily servings of lettuce and carrots from these gardens, the weekly As intake would be 3–4 times the provisional tolerable weekly intake (PTWI) for inorganic arsenic set at about 1 mg As per week (red dashed line). However, amending the soil with biosolids compost significantly reduced plant As compared to the untreated (control) soils. The As in water-soluble and exchangeable fractions in the soil was reduced, probably because of As adsorption by biosolids organic matter. However, adding phosphate fertilizer to this urban garden greatly increased the As content of the carrots and lettuce grown. Arsenate (AsO_4^{3-}) is an oxyanion much like phosphate (PO_4^{3-}), rather than a metallic cation like lead (Pb^{+2}). Therefore, the phosphate replaced the sorbed arsenate. Thus, the effect of adding phosphate to soil was just the opposite for arsenic as the effect discussed for lead in Box 15.2. [Graphed from tabular data in Cao and Ma (2004)]

Some Inorganic Contaminants and Their Reactions in Soils

Arsenic, as just mentioned, can leach from treated outdoor lumber. A broader problem is its accumulation in certain orchard soils following years of application of arsenic-containing pesticides. Being present in an anionic form (e.g., $H_2AsO_4^-$), this element is absorbed (as are phosphates) by hydrous iron and aluminum oxides, especially in acid soils. In spite of the capacity of most soils to tie up arsenates, long-term additions of arsenical sprays can lead to toxicities for sensitive plants and earthworms. The arsenic toxicity can be reduced by applications of sulfates of zinc, iron, and aluminum, which tie up the arsenic in insoluble forms. Toxicities from naturally occurring arsenic were discussed in Section 13.4.

Lead contaminates primarily urban soils from vehicle exhaust and from old lead-pigmented paints (paint chips and dust from painted woodwork). However, other soils at shooting ranges, hunting grounds, and battlefields become contaminated with lead from bullets and buckshot. Most of the lead is tied up in the soil as low solubility carbonates, sulfides, and in combination with iron, aluminum, and manganese oxides. Consequently, the lead is largely unavailable to plants and not mobile enough to readily leach to groundwater. However, it can be absorbed by children who put contaminated soil or dust in their mouths (Box 15.2).

Mercury is released mainly from burning coal to generate electricity. When it contaminates lake sediments and wetlands, the result is toxic levels of mercury among certain species of fish. Insoluble forms of mercury in soils, not normally available to plants or, in turn, to animals, are converted by microorganisms to an organic form, methylmercury, which is more soluble and available for plant and animal absorption. The methylmercury is concentrated in fatty tissue as it moves up the food chain, until it accumulates in some fish to levels that may be toxic to humans. This series of transformations illustrates how reactions in soil can influence human toxicities.

Chromium in trace amounts is essential for human life, but, like arsenic, it is a carcinogen when absorbed in larger doses. This element is widely used in steel, alloys, and paint pigments. Chromium is found in two major oxidation states in upland soils: a trivalent form [Cr(III)], and a hexavalent form [Cr(VI)]. In contrast to most metals, the more highly oxidized state [Cr(VI)] is the more soluble, and its solubility increases above pH 5.5. This behavior is opposite that of Cr(III), which forms insoluble oxides and hydroxides above that pH level.

To remediate Cr(VI)-contaminated soil and water, it is useful to reduce the chromium to Cr(III) (see also Section 7.5). This reduction process is enhanced by anaerobic conditions [wet soil with an abundance of decomposable organic material to provide a large, biological oxygen demand (BOD)]. The organic matter serves as an electron donor and thereby hastens the reduction of Cr(VI) to the trivalent state [Cr(III)]. Provided the pH is maintained above 5.5, chromium in this reduced state will remain relatively stable, immobile, and nontoxic.

Boron can contaminate soil via high-boron irrigation water, by excessive fertilizer application, or by the use of power plant fly ash as a soil amendment. Boron may be adsorbed by organic matter and clays but may still be available to plants, except at high soil pH. Boron is relatively soluble in soils, toxic quantities being leachable, especially from acid sandy soils. Boron toxicity in plants is usually a localized problem and is probably much less important than boron deficiency (see Section 12.8).

Fluorine toxicity is also generally localized. Drinking water for animals and fluoride fumes from industrial processes often contain toxic amounts of fluorine. The fumes can be ingested directly by animals or deposited on nearby plants. If the fluorides are adsorbed by the soil, their uptake by plants is restricted. The fluorides formed in soils are highly insoluble, the solubility being least if the soil has a pH above neutral and high calcium saturation.

Selenium, which derives mainly from certain soil parent material, can accumulate in soils and plants to toxic levels, especially in arid regions (see Section 12.8).

15.7 POTENTIAL HAZARDS OF CHEMICALS IN SEWAGE SLUDGE

The domestic and industrial sewage sludges considered as nutrient sources in Chapter 13 can be important sources of potentially toxic chemicals. About half of the municipal sewage sludge produced in the United States and in some European countries ends up applied to soils,

Figure 15.21 *Sewage sludge slurry being transferred from a tanker truck to a floatation tire spreader for injection into soils during an ecological restoration project on orphan mined land. Such sludge applications are closely regulated on the basis of allowable metal levels in soil and sludge.* (Photo courtesy of Ray R. Weil)

either on agricultural land or to remediate land disturbed by mining and industrial activities (Figure 15.21). Industrial sludges—or municipal sewage sludges from systems that treat both industrial and human wastes—commonly carry significant quantities of inorganic as well as organic chemicals that can have harmful environmental or human health effects. The sludge contaminants that have received the most attention are the heavy metals.

Heavy Metals in Sewage Sludge

Concern over the possible buildup of heavy metals in soils resulting from large land applications of sewage sludges has prompted research on the fate of these chemicals in soils. Most attention has been given to zinc, copper, nickel, cadmium, and lead, which are commonly present in significant levels in these sludges. Many studies have suggested that if only moderate amounts of sludge are added, and the soil is not very acid (pH > 6.5), these elements are generally bound by soil constituents; they do not then easily leach from the soil, nor are they readily available to plants. Only in moderately to strongly acid soils have most studies shown significant movement down the profile from the layer of application of the sludge. Monitoring soil acidity and using judicious applications of lime have been widely recommended to prevent leaching into groundwater and minimize uptake by plants.

Forms Found in Soils Treated with Sludge. By using a sequence of chemical extractants, researchers have found that heavy metals are associated with soil solids in four major ways. First, a very small proportion is held in *soluble* or *exchangeable forms*, which are available for plant uptake. Second, the elements are bound by the *soil organic matter* and by the *organic materials* in the sludge. High proportions of copper and chromium are commonly found in this form, while lead is not so highly attracted. Organically bound elements are not readily available to plants, but may be released over a period of time.

The third and fourth forms of heavy metals in soils are associations with *carbonates* and with *oxides of iron and manganese*. These forms are less available to plants than either the exchangeable or the organically bound forms, especially if the soils are not allowed to become too acid. The fifth association is commonly known as the *residual form*, which consists of sulfides and other very insoluble compounds that are less available to plants than any of the other forms.

It is fortunate that most soil-applied heavy metals are not readily absorbed by plants and that they are not easily leached from the soil. However, the immobility of the metals means that they will accumulate in soils if repeated sludge applications are made. Care must be taken not to add such large quantities that the capacity of the soil to react with a given element is exceeded. It is for this reason that regulations set maximum cumulative loading limits for each metal (see Table 15.7).

Source Reduction Programs. A great deal was learned during the 1970s and 1980s about the contents, behavior, and toxicity of metals in municipal sewage sludges. As a result of the research, source reduction programs were implemented in the United States and elsewhere,

Table 15.7
REGULATORY LIMITS ON INORGANIC POLLUTANTS (HEAVY METALS) IN SEWAGE SLUDGE APPLIED TO AGRICULTURAL LAND

Element	Maximum concentration in sludge, U.S. EPA,[a] mg/kg	Annual pollutant loading rates, U.S. EPA, kg/ha/yr	Cumulative allowable pollutant loading, kg/ha		
			U.S. EPA	Germany	Ontario, Canada
As	75	2.0	41	—	28
Cd	85	1.9	39	3.2	3.2
Cr	3,000	150.0	3,000	200	240
Cu	4,300	75.0	1,500	120	200
Hg	57	0.85	17	2	1.0
Mo	75	—	—	—	8
Ni	420	21	420	100	64
Pb	840	15	300	200	120
Se	100	5.0	100	—	3.2
Zn	7,500	140	2,800	400	440

[a]U.S. EPA (1993).

which required industries to clean pollutants out of their wastewater *before* sending it to municipal wastewater treatment plants. In many cases, the recovery of valuable metal pollutants was actually profitable for industries. Because of these programs, municipal sewage sludges are much cleaner than in the past.

Regulation of Sludge Application to Land. The lower levels of metals (and of organic pollutants) make municipal sewage sludges much more suitable for application to soils than in the past, at least in countries where this issue is effectively regulated. Today, the amount of sludge that can be applied to agricultural land is more often limited by the potential for nitrate pollution from the nitrogen or phosphorus it contains, rather than by the metal content of the sludge. Nonetheless, application of sewage sludge to farmland is closely regulated to ensure that the metal concentrations in the sludge do not exceed the standards and that the total amount of metal applied to the soil over the years does not exceed the regulatory maximum accumulative loading limit. The fact that metal-loading standards differ considerably between countries (see Table 15.7) is an indication that the nature of the metal contamination threat is still somewhat uncertain and controversial.

Toxic Effects from Sludge. The uncertainties as to the nature of many of the organic chemicals found in the sludge, as well as the cumulative nature of the metals problem, dictate continued caution in the regulations governing application of sludge to croplands. The sludge-treated soils, as well as the bodies of earthworms living in these soils, were much higher in cadmium and zinc than was the case in soils where sludge had not been applied. One would expect further concentration to take place in the tissue of birds and fish that might consume the earthworms.

Farmers must be assured that the levels of inorganic chemicals in sludge are not toxic to plants (a possibility mainly for zinc and copper) or to humans and other animals who consume the plants (a serious consideration for As, Cd, Cr, and Pb). For relatively low-metal municipal sludges, application at rates just high enough to supply needed nitrogen seems to be quite safe (Table 15.8).

Direct ingestion of soils and sludge is also an important pathway for human and animal exposure. Animals should not be allowed to graze on sludge-treated pastures until rain or irrigation has washed the sludge from the forage. Children may eat soil while they play, and a considerable amount of soil eventually becomes dust in many households. Direct ingestion of soil and dust is particularly important in lead toxicity.

Table 15.8
METAL UPTAKE BY CORN AFTER 19 YEARS OF FERTILIZING WITH LIME-STABILIZED SEWAGE SLUDGE

As is typical, metals accumulated less in grain than in leaves and stalks (stover). The annual sludge rate of about 10.5 Mg was designed to supply the nitrogen needs of the corn. The sludge had little effect on the metal content of the plants, except in the case of zinc (which increased, but not beyond the normal range for corn).

	Zn	Cu	Cd	Pb	Ni	Cr
Cumulative metal applied in sludge, kg/ha	175	135	1.2	49	4.9	1045
Treatment	Uptake in stover, mg/kg					
Fertilizer	18	8.4	0.16	0.9	0.7	0.9
Sludge	46.5	7.0	0.18	0.8	0.6	1.4
	Uptake in grain, mg/kg					
Fertilizer	20	3.2	0.29	0.4	0.4	0.2
Sludge	26	3.2	0.31	0.5	0.3	0.2

Data for Typic Hapludolls in Minnesota abstracted from Dowdy et al. (1994).

15.8 PREVENTION AND REMEDIATION OF INORGANIC SOIL CONTAMINATION[6]

Three primary methods of alleviating soil contamination by toxic inorganic compounds are: (1) to eliminate or drastically reduce the amounts of toxin applied to or allowed to reach the soil; (2) to immobilize the toxin by means of soil management, to prevent it from moving into food or water supplies; and (3) in the case of severe contamination, to remove the toxin from the soil by chemical, physical, or biological remediation.

Reducing Soil Application

The first method requires action to reduce unintentional contamination from industrial operations and from automobile, truck, and bus exhausts. Decision makers must recognize the soil as an important natural resource that can be seriously damaged if its contamination by unintended addition of inorganic toxins is not curtailed. One of the considerations in life cycle sustainability analysis should be the likelihood of soil contamination. For example, during their lifetimes, cars and trucks emit contaminants, not only in their exhaust fumes, but also from brake linings (copper) and tire wear (cadmium, zinc)—all of which can wash off paved roads onto soils. Potential contamination issues from the improper disposal of discarded electronics are of increasing concern (tin, lead, beryllium cadmium, mercury). Also, there must be judicious reductions in *intended* applications to soil of the toxins through pesticides (arsenic, copper), fertilizers (cadmium, zinc), irrigation water (selenium, boron), and composted solid wastes.

Immobilizing the Toxins

Soil and plant management can help reduce the continued cycling of these inorganic chemicals by stimulating changes that keep the chemicals in the soil rather than encouraging their uptake by plants or leaching by drainage water. By immobilizing and becoming a sink for the toxins,

---

[6]For phytoremediation of a large area of soils in Thailand too contaminated with Cd and Ni to produce rice for human consumption, see Simmons et al. (2014). For reviews of plant accumulation of metals from soils and the use of natural organic materials to enhance this process, see van der Ent et al. (2013) and Wiszniewska et al. (2016).

the soil breaks the soil–plant–animal (humans) cycle through which the toxin exerts its effect. For example, most of these elements are rendered less mobile and less available if the pH is kept near neutral or above. Liming of acid soils reduces metal mobility; hence, regulations require that the pH of sludge-treated land be maintained at 6.5 or higher. However, it is important to know which contaminants are present as certain toxic elements, such as arsenic and molybdenum, have the opposite response to pH change and become more soluble and bio-available at high pH levels.

Draining wet soils can be beneficial, since the oxidized forms of the several toxic elements are generally less soluble and less available for plant uptake than are the reduced forms. However, again, there are exceptions and the opposite is true for chromium. The oxidized Cr(VI) is mobile and highly toxic to humans (see Section 15.8).

Heavy phosphate applications may reduce the availability of metal cations such as lead that react with P to form insoluble compounds (see Box 15.2). However, again, the opposite effect may be realized, for example, with arsenic, which is found in the anionic form and is desorbed by phosphate (see Figure 15.20). Leaching may be effective in removing excess boron, although moving the toxin from the soil to water is not likely be of real benefit.

Application of organic matter, especially heavy applications of compost, generally reduces the mobility and bioavailability of many of the inorganic toxins. The stabilized organic matter seems to strongly adsorb many elements, preventing their bioconcentration in plants. That is, plants species differ in their ability to take up metals such that the contaminant may be many times more or many time less concentrated in the plant tissue than in the soil from which it was taken up.

$$\text{Bioconcentration factor}, BCF = \frac{mg/kg, \text{plant}}{mg/kg, \text{soil}} \qquad (15.4)$$

Plants with a relatively high bioconcentration factor (BCF) are likely to be more toxic to consuming organisms further up the food chain than plants with lower BCF values. For example, on contaminated urban gardens in one study (Figure 15.22), the bioconcentration factor was much higher for lead than for arsenic and about 10 times as great for carrot as for lettuce or tomato. Amending the soil with very large amounts of biosolids and sawdust-based compost reduced the BCF by a modest but significant degree in most cases.

Care should be taken in selecting food plants to be grown on metal-contaminated soil. Generally, plants translocate much larger quantities of metals to their leaves than to their fruits or seeds (see Table 15.8). Root vegetables can also be problematic. Fruits or grains only rarely accumulate high concentration of metals. The greatest risk for food-chain contamination with metals is therefore through leafy vegetables, such as lettuce and spinach, or through forage crops eaten by livestock. On the other hand, nonfood plants with a high propensity to take up metals from contaminated soil can be put to good use for remediation.

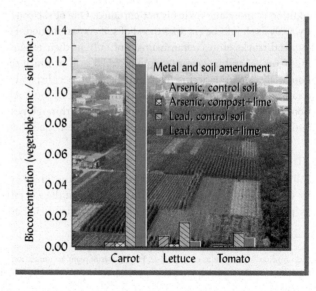

Figure 15.22 Bioconcentration factors (BFC) in three vegetable crops grown in urban lots with soil contaminated by both arsenic and lead. All three species excluded the contaminants to some extent, but for Pb carrot was much less exclusionary than the others. The soil in some plots was amended with 180 Mg/ha of biosolids + 180 Mg/ha of a sawdust/sand/lime mixture. This amendment significantly reduced the BCF for As in lettuce and for Pb in carrot and lettuce. The soil pH ranged from 5.6 to 5.9 and contained 200–300 mg/kg Pb. [Means for two years graphed from tabular data in Defoe et al. (2014), background photo of other urban vegetable farms near Beijing, China, courtesy of Ray R. Weil]

Bioremediation by Metal Hyperaccumulating Plants

As just discussed, different plant species are characterized by wide propensities to take up metals from soil. Plants also differ widely in their responses to the accumulation of high concentrations of metals in their roots—some die, some survive by sequestering the metal in their root tissue, while others thrive and translocate the metal to their shoots. Based on these characteristics, plants can be grouped into four categories: sensitive plants, indicator plants, excluder plants, or hyperaccumulator plants (Figure 15.23).

Sensitive plants tend to take up metals readily but then cannot tolerate the resulting high metal concentrations in their tissues. They become poisoned by the metals, their growth is inhibited, and they are likely to die if grown in even moderately metal-contaminated soils. With regard to most heavy metals, the majority of plants fall into this category.

Indicator plants steadily take up the metal in proportion to how much is available in the soil. This trait makes them useful as indicators of soil metal concentrations. They survive from very low to quite high soil metal concentrations, but they do not accumulate concentrations of the metal in their tissues that exceed the concentrations in the soil. They manage to tolerate moderately high levels of metal taken up by sequestering the metal nonsensitive plant parts or by binding the metals in organic compounds that reduce the toxicity.

Excluder plants can tolerate quite high levels of metal contamination, but they do so by either excluding the metal and avoiding its uptake, or by isolating the metal in the root and not translocating it to the shoot. That is, excluder plants may exhibit a high bioconcentration factor in the root ($BCF_{root} > 1$), but not in the shoot ($BCF_{shoot} = 1$). This behavior may be useful in stabilizing the metal in the soil, but does not facilitate the removal of the metal altogether.

To be useful in remediating metal-contaminated soils by *phytoextraction*, a plant must not only take up large amounts of the metal, but must also translocate the metal to the aboveground plant parts for practical metal removal with plant harvest. This is where hyperaccumulating plants come into the story. **Hyperaccumulator plants** not only tolerate extremely high metal concentrations in soils, they also efficiently extract the metal from the

Figure 15.23 *Plant responses to metal contamination in soils fall into four categories. Sensitive plants accumulate metals but cannot tolerate high levels. Indicator plants and excluder plants may be useful for bioremediation by phytostabilization. Hyperaccumulator plants present the most potential for bioremediation by phytoextraction. Examples of the wide range of crop and noncrop plants that show potential for phytoremediation: (a) wheat acts as an excluder plant for Ni; (b) collecting soil samples in a mixed stand of Brassicas; (c) sorghum accumulates high levels of Pb; (d) Thlapsi spp. hyperaccumulate Zn and Cd; (e) research plots with mustards; (f) roots and shoots of mustard plant that hyperaccumulate several metals; (g) research plots testing heavy compost application to bind toxic metals at a zinc mine site.* [Photo (g) courtesy of USDA/Agric. Research Service; other photos courtesy of Ray R. Weil; diagram redrawn with concepts from van der Ent et al. (2013)]

soil, translocate the metal to aboveground parts from their roots, and continue to grow unimpaired with extremely high metal concentrations in their shoot tissues. Researchers have found most known hyperaccumulator species among plants growing and evolving naturally in old polluted sites or soils naturally high in metals (such as serpentine-derived soils, see Section 12.7). Hundreds of hyperaccumulating plant species from dozens of plant families have been identified, but the Cruciferae, including the genera Brassica, Alyssum, and Thlaspi, stand out as supplying many hyperaccumulators.

Hyperaccumulator plants may accumulate more than 20,000 mg/kg nickel, 40,000 mg/kg zinc, and 1,000 mg/kg cadmium. While such plants would pose a serious health hazard if eaten by animals or people, they may offer one of the most practical and sustainable solutions for the remediation of large, extensive, or environmentally sensitive areas with metal-contaminated soils. With concentrations of 2–4% metal (e.g., Ni or Zn), the plant tissue is so concentrated that it could be used as an "ore" for smelting new metal. Since the total amount of metal removed is the product of tissue concentration (mg/kg) and dry matter production (kg/ha), it is not enough to find plants that accumulate 10,000–50,000 mg/kg metal in their dry matter (1–5,000 if the most highly toxic metals like Cd are considered). Up to now, the main limitation of the practicality of the phytoextraction process is the rather puny dry matter production of most hyperaccumulating plants such that site cleanup could take many decades or centuries.

Management to Enhance Phytoremediation

In addition to manipulating soil pH and redox conditions, as already mentioned, organic materials, composts, and biochars of various kinds are being experimented with to enhance phytoremediation. One way these material increase plant metal uptake is by chelation. The natural materials or synthetic chelating agents can be applied to soils to enhance the removal of lead and other metals by phytoremediation. Lead in soils is strongly bound by both mineral and organic colloids and is thus only very sparingly available to plants. In a form of enhanced rhizosphere bioremediation, added chelates solubilize the lead, and plants such as Indian mustard are used to remove it.

Encouraging an active soil faunal community may also play a role. Earthworms have been found to make lead more easily taken up by plants and leached from soils during bioremediation effort (Table 15.9).

Genetic and bioengineering techniques are being applied to develop high-yielding hyperaccumulating plants that can remove larger quantities of heavy metal contaminants from soils. For example, wide genetic variation in heavy metal accumulation by different strains of Alpine pennycress suggests the potential for breeding improved accumulating plants. Also, research to insert genes responsible for contaminant accumulation into other higher-yielding plants, such as canola and Indian mustard, is being explored.

Table 15.9
Earthworms' Effects on the Mobility of Copper, Zinc, Lead, and Arsenic in Contaminated Soil
Incubating soil with earthworms (Lumbricus terrestris) for 112 days increased both plant uptake and leaching.

	Concentration in rye plant tissue				Concentration dissolved in leachate					
	Cu	Zn	Pb	As	Cu	Zn	Pb	As	pH	DOC
	— mg kg^{-1} —				— µg L^{-1} —					mg L^{-1}
No earthworms	36	120	2.9	4.7	3.0	128	1.0	0.6	4.1	4.2
L. terrestris	48*	140*	3.5	12*	12.0*	549*	1.9*	1.6*	3.7*	3.9

*Asterisk indicates values that differ statistically between the soils that were incubated with or without earthworms. Data selected from Sizmur et al. (2011). Soil was collected (0–30 cm) from a grassed field—a former settling pond for the separation of metal from crushed ores at an abandoned copper and arsenic mine near Gunnislake, UK.

15.9 LANDFILLS[7]

A visit to the local landfill would convince anyone of the wastefulness of modern societies. And the problem is only getting worse by the day. Estimates are that 1.4 billion Mg of solid waste is generated annually and current trends suggest this amount may double before 2030. Costs for dealing with all this waste are estimated at well over $200 billion per year. In the United States, which is now second to China as the greatest waste-generating country, people generate about 260 million Mg of municipal wastes each year (Figure 15.24).

The Municipal Solid Waste Problem

Most (about 70%) municipal solid waste material is organic in nature, largely paper, cardboard, and food or yard wastes (e.g., grass clippings, leaves, and tree prunings). The other 30% consists mainly of such nonbiodegradables as glass, metals, and plastics. Environmentally speaking, the good news is that since 2000 total waste generation in the United States has not increased (it had doubled between 1965 and 1995) and the per capita generation has actually declined. Since the overall economy has grown during that period, the data indicate that the United States has been successful in decoupling waste generation from economic growth. Environmental education and municipal recycling programs have paid off with an ever-increasing proportion of wastes being recycled, composted, or burned for energy recovery (Figure 15.24). Still, the present reality is that at least half of municipal solid wastes are buried in the ground (landfilled) and will probably continue to be disposed of in this manner for some time to come. In lesser developed countries, wastes may not even be properly buried, but instead simply either dumped at the city's edge or, where municipal collection is inadequate, left to pile up in the streets, alleys, and riverbanks (Figure 15.25) where they sadly degrade the livability of the cities and release toxins and pathogens to the environment. For the poor in some of these cities, the exposed wastes represent a means of making a living by picking through the trash and selling the most valuable and easily recovered waste materials for metal recovery and recycling.

Regardless of location, it is important to understand that the entire waste disposal problem could be greatly reduced by creating less waste in the first place—implementing policies of "reduce and reuse." Second, it is possible to eliminate most problems associated with waste disposal by two simple measures: (1) keeping the metals, glass, plastics, and paper separate for easy recycling (Figure 15.25, *left*), and (2) composting the yard wastes, food wastes, and some of the paper products. Composting can be done on-site by homeowners and waste generators, or by municipalities using collected organic wastes (Section 11.10). The composted product from a number of municipalities is successfully used as a beneficial soil amendment (see Section 13.5). The small fraction of more hazardous wastes remaining can then be detoxified, concentrated and reused, or immobilized.

In the past, wastes were merely placed in open "dumps" and, often, set afire. The term *landfill* came into use because wastes were often dumped into gullies, ravines, or swampy lowland

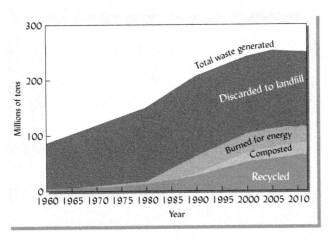

Figure 15.24 *Historical and predicted trends in municipal solid-waste management in the United States. Soils play a central role in the composting and landfilling options.* [Diagram courtesy of Ray R. Weil; Data from U.S. EPA (2014)]

---

[7] For details on the soil and geotechnical aspects of landfill design, see Qian et al. (2002) and settlement and other limitation for use after landfill closure, see Wong et al. (2013).

Figure 15.25 *Alternatives to burying wastes in the soil. (Left) Vending machines in reverse: residents of New York City bring collected wastes to be deposited in "vending" machines that pay out cash in return. (Right) Residents of a small West African city go about their business amid sea of uncollected solid wastes.* (Photo courtesy of Ray R. Weil)

areas where, eventually, their accumulation filled in the lowland, creating areas for urban parks and other facilities. Locating landfills on wetlands is no longer an acceptable practice.

Two Basic Types of Landfill Design

Although landfill designs vary with the characteristics of both the site and the wastes, two basic types of landfills can be distinguished: (1) the natural attenuation or unsecured landfill, and (2) the containment or secured landfill. We will briefly discuss the main features of each.

Natural Attenuation Landfills

The purpose of a natural attenuation landfill is to contain nonhazardous municipal wastes in a sanitary manner, protect them from animals and wind dispersal, and, finally, cover them sufficiently to allow revegetation and possible reuse of the site. Although the landfill is engineered to reduce water infiltration, some rainwater is allowed to percolate through the waste and down to the groundwater (Figure 15.26). Natural processes are relied upon to attenuate the leachate contaminants before the leachate reaches the groundwater. Soils play a major role in

Figure 15.26 *A natural attenuation landfill depends largely on soil processes to attenuate the contaminants in the leachate before they reach the groundwater. Compare to the containment landfill design illustrated in Figure 15.28.* (Diagram courtesy of Ray R. Weil)

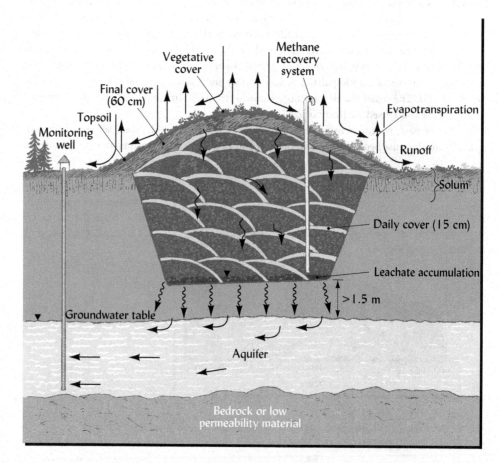

these natural attenuation processes through physical filtering, adsorption, biodegradation, and chemical precipitation (Table 15.10).

Soil Requirements. Finding a site with suitable soil characteristics is critical for a natural attenuation landfill. There must be at least 1.5 m of soil material between the bottom of the landfill and the highest groundwater level. This layer of soil should be only moderately permeable. If too permeable (sandy, gravelly, or highly structured), it will allow the leachate to pass through so quickly that little attenuation of contaminants will take place. The soil must have sufficient cation exchange capacity to adsorb NH_4^+, K^+, Na^+, Cd^{2+}, Ni^{2+}, and other cations that the wastes are expected to release. The soil should also adsorb and retard organic contaminants long enough to allow a high degree of microbial degradation. On the other hand, if the soil is too impermeable, the leachate will build up, flood the landfill, and seep out laterally.

Daily and Final Soil Cover. The site for a natural attenuation landfill should also provide soils suitable for daily and final cover materials. At the end of every workday, the waste must be covered by a layer of relatively impermeable soil material. The daily cover prevents blowing waste and odors escaping and excludes vermin to some extent.

The final cover for the landfill is much thicker than the daily covers, and includes a 60- to 100-cm-thick layer of low-permeability, clay soil material designed to minimize percolation

Table 15.10
SOME ORGANIC AND INORGANIC CONTAMINANTS IN UNTREATED LEACHATE FROM MUNICIPAL LANDFILLS

Range of concentrations and typical sources of the contaminants and mechanisms by which soils can attenuate the contaminants are also given. The ranges show that leachates vary greatly among landfills.

Chemical	Concentration, µg/L	Common sources	Mechanisms of attenuation
Organics			
Dissolved organic matter,	140,000–	Rotting yard wastes, paper, and garbage	Biological degradation
Dissolved organic matter as chemical oxygen demand (COD)	140,000–150,000,000	Rotting yard wastes, paper, garbage	Biological degradation
Benzene	0.2–1,630	Adhesives, deodorants, oven cleaner, solvents, paint thinner, and medicines	Filtration, biodegradation, and methanogenesis
Trans 1,2-dichloroethane	1.6–6,500	Adhesives and degreasers	Biodegradation and dilution
Toluene	1–12,300	Glues, paint cleaners and strippers, adhesives, paints, shampoo, and carburetor cleaners	Biodegradation and dilution
Xylene	0.8–3,500	Oil and fuel additives, paints, and carburetor cleaners	Biodegradation and dilution
Metals			
Nickel	15–1,300	Batteries, electrodes, and spark plugs	Adsorption and precipitation
Chromium	20–1,500	Cleaners, paint, linoleum, and batteries	Precipitation, adsorption, and exchange
Cadmium	0.1–40	Paint, batteries, and plastics	Precipitation and adsorption

Leachate concentration ranges from a review of hundreds of landfills built since 1965, in Kjeldsen et al. (2002).

Figure 15.27 *An engineered soil used to cap a completed section of a landfill in New Jersey. The pit excavation has revealed three basic layers of the landfill cap. A compacted clay layer seal prevents infiltration of water so as to minimize formation of landfill leachate. A layer of loose, clean, medium sand was installed above the clay as a drainage layer to guide water horizontally over the clay and off the landfill cell to a collection pond (not shown). On top, a layer of sandy loam Ap horizon material hauled in from a distant site serves as growing medium for the grassy vegetative cover that protects and stabilizes the landfill cap. This pit reveals signs of in situ soil development and acid sulfate weathering (2Bs horizon) from unfortunate use of high sulfide clay.* (Photos courtesy of Chris Smith, USDA/NRCS)

of water into the landfill. This impermeable layer of compacted clay is usually then covered with a 30- to 45-cm layer of highly permeable medium to coarse sand. This sand layer is designed to allow water to drain laterally off the landfill to a collection area.

On top of the sand, a thinner layer of loamy "topsoil" is installed. The moderately permeable topsoil layer is meant to support a vigorous plant cover that will prevent erosion and use up water by evapotranspiration. The whole system is designed to limit the amount of water percolating through the waste, so that the amounts of contaminated leachate generated will not overwhelm the attenuating capacity of the soil between the landfill bottom and the groundwater (Figure 15.27).

Containment or Secured Landfills

The second main type of landfill is much more complex and expensive to construct, but its construction and function are much less dependent on the nature of the soils at the site. The design (see Figure 15.28) is intended to contain, pump, and treat all leachate from the landfill, rather than to depend on soil processes for cleansing the leachate on its way to the groundwater. To accomplish the containment, one or more impermeable liners are set in place around the sides and bottom of the landfill. These are often made of expanding clays (e.g., bentonite) that swell to a very low permeability when wet. Plastic, watertight geomembranes are also used in making the liners. The membranes are covered with a tough, nonwoven, synthetic fabric (geotextiles) and then covered with a thick layer of fine gravel or sand to protect the liner from accidental punctures. A system of slotted pipes and pumps is installed to collect all the leachate from the bottom of the landfill (Figure 15.29, *left*). The collected leachate is then treated on or off the site. The principal soil-related concerns are the requirement for suitable sources of sand and gravel, of soil for daily cover, for clayey material to form the final cover, and for topsoil to support protective vegetation.

Environmental Impacts of Landfills

Modern regulations require that wastes be buried in carefully located and designed containment-type sanitary landfills. As a result, the number of landfill sites has been reduced; in the United States, from about 16,000 in 1970 to fewer than 2,000 in 2014. The remaining landfills are mostly very large, highly engineered containment-type systems. A major concern with regard to landfills is the potential water pollution from the rainwater that percolates through the wastes, dissolving and carrying away all manner of organic and inorganic contaminants (see Table 15.10). In addition to the general load of oxygen-demanding dissolved organic carbon compounds, many of the contaminants in landfill leachate are highly toxic and would create a serious pollution problem if they reached the groundwater under the landfill.

In addition to efficiency of resource use, avoidance of particular landfill management problems is another reason that the organic components of refuse (mainly paper, yard

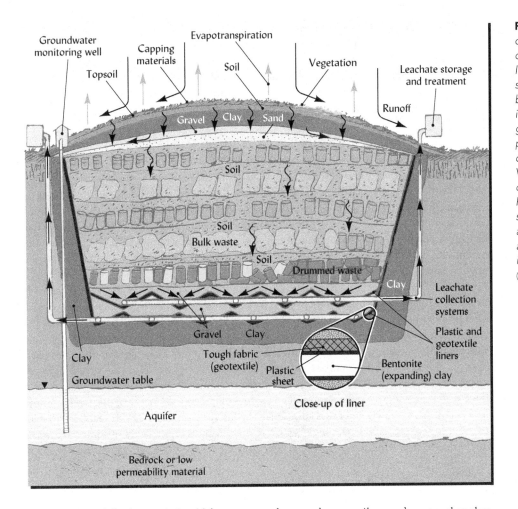

Figure 15.28 *A containment-type landfill is designed to collect all the leachate and pump it out for storage and treatment. The bottom of the landfill cell is sealed with a waterproof geomembrane which is protected by a covering of geotextile and gravel. Versions of this basic design are used for more hazardous wastes or when soil conditions on the site are unsuitable for a natural attenuation landfill design, illustrated in Figure 15.26.* (Diagram courtesy of Ray R. Weil)

trimmings, and food waste) should be composted to produce a soil amendment rather than landfilled. First, as these materials decompose in a finished landfill, they lose volume and cause the landfill to settle and the landfill surface to subside. This physical instability severely limits the uses that can be made of the land once a landfill is completed.

Second, decomposition of the organic refuse produces undesirable liquid and gaseous products. Within a few weeks, decomposition uses up the oxygen in the landfill, and the processes of anaerobic metabolism take over, changing the cellulose in paper wastes into butyric, propionic, and other volatile organic acids, as well as hydrogen and carbon dioxide. After a month or so, methane-producing bacteria become dominant, and for several years (or even decades) a gaseous mixture of about one-third carbon dioxide and two-thirds methane (known as *landfill gas*) is generated in quantity. Anaerobic decomposition in landfills also emits other harmful gases, the effects of which are less well known.

Figure 15.29 *Systems for collection of leachate and gas emissions in a containment landfill. (Left) A black geomembrane liner covered with white pea gravel and a leachate collection pipe in a new cell. The pollutant-laden leachate will be piped to a treatment facility. (Right) Gas wells collecting landfill gas (mainly methane and carbon dioxide) from anaerobic decomposition in a completed landfill cell. The methane is used to generate electric power.* (Photo courtesy of Ray R. Weil)

The production of methane gas by the anaerobic decomposition of organic wastes in a landfill can present a very serious explosion hazard if this gas is not collected and used as fuel (Figure 15.29, *right*). Where the soil is rather permeable, the gas may diffuse into basements up to several hundred meters away from the landfill. A number of fatal explosions have occurred by this process. The methane also diffuses through the final cover to the atmosphere, making landfills a significant source of greenhouse gas emissions, worldwide.

Land Use After Completion

A modern landfill may be designed to receive wastes for several decades, but eventually it will become filled to capacity and will have to be closed. The management and land use after completion and closure of a landfill is an important part of proper landfill design. This long-term management requires the application of soils' knowledge at all stages. The site will receive a final engineered soil cover and installation of landscaping vegetation, all of which will have to be maintained to prevent erosion, landslides, and exposure of wastes. Leachate treatment and monitoring of runoff and groundwater will have to continue, indefinitely. Gas collection and utilization will continue for decades.

Many landfills occupy large areas (commonly hundreds of hectares) of highly valuable land located in or near major urban centers. As such, the closed landfill sites represent opportunities for beneficial land uses. However, there are limitations. The presence of leachate and landfill gases could present hazards for buildings. But the most common hazard is the land subsidence that occurs as the wastes inside the landfill settle and decay. The settling takes place in stages. Some settlement is almost immediate as the easily compressed wastes are compacted under the weight of machines and final cover soil. Additional subsidence takes place during the next few months as wastes settle and creep downslope. The decay of organic waste components occurs more slowly and cause additional settling for a period of 25–30 years. In the end, the land surface typically settles by 25–30% of the thickness of the layers of compacted waste inside the finished landfill.

15.10 RADIONUCLIDES IN SOIL

Soils contain small quantities of ^{238}U, ^{40}K, ^{87}Rb, ^{14}C, and a number of other naturally occurring radioactive isotopes (radionuclides) that are characterized by long half-lives and give off minute amounts of radiation in the form of alpha particles (bundles of two neutrons and two protons) and beta particles (positively or negatively charged particles). As a radionuclide decays, its nucleus discharges these particles, transforming the atom into a different isotope or element with a lighter nucleus. The time it takes for one-half of the atoms of a particular radioactive isotope to undergo such decay is termed the *half-life* of the isotope. After 10 half-lives, 99.1% of the original atoms will have decayed. The intensity of radioactivity present—or, more precisely, the rate of radioactive decay—is expressed using the SI unit *becquerel* (Bq), which represents one decay per second. An older metric unit, still in wide use, is the *curie* (Ci), which equals 3.7×10^{10} Bq.[8]

Radioactivity from Nuclear Fission

The process of nuclear fission, in connection with atomic weapons testing and nuclear power generation, has contaminated soils with a number of additional radionuclides. However, only two of these are sufficiently long-lived to be of significance in soils: strontium 90 (half-life = 28 years) and cesium 137 (half-life = 30 years). The average level of ^{90}Sr in soil in the United States is about 14.4 kilobecquerels per square meter (kBq/m^2) or 388 millicuries per square kilometer (mCi/km^2). The average level for ^{137}Cs is about 22.9 kBq/m^2 (620 mCi/km^2). The levels of radioactivity caused by nuclear fallout are quite small compared to that for naturally occurring radionuclides. For example, for naturally occurring ^{40}K the average level is about 1,900 kBq/m^2 (51,800 mCi/km^2).

---

[8]The curie was named after Marie and Pierre Curie, Polish scientists who discovered radium, an element that decays at the rate of 3.7×10^{10} Bq/g. The becquerel was named after Antoine Henry Becquerel, a French scientist who discovered radioactivity in uranium.

Partly because of the cation-exchange properties of soils, the levels of these fission radionuclides found in most soils are not high enough to be hazardous. Most of the ^{90}Sr and ^{137}Cs reaching the soil is adsorbed by the soil colloids in exchange for other cations previously adsorbed, with the result that plants take up mainly the replaced cations rather than the added radionuclides (see Section 8.8). Even during the peak periods of weapons testing in the early 1960s, soils did not contribute significantly to the level of these nuclides in plants. Atmospheric fallout directly onto foliage was the primary source of radionuclides in the food chain. Consequently, only in the event of a catastrophic supply of fission products could toxic soil levels of ^{90}Sr and ^{137}Cs be expected.

Nuclear Accident at Chernobyl[9]

Such high levels of ^{90}Sr and ^{137}Cs, as well as ^{131}I, did contaminate soils in Ukraine, Scandinavia, and Eastern Europe in the wake of the 1986 reactor meltdown at Chernobyl in Ukraine (then part of the Soviet Union). While the accident deposited more than 200 kBq/m^2 of radioactivity on soils as far away as the United Kingdom, the dangerous high levels of contamination occurred closer to the reactor and resulted in the designation of a 2,600 km^2 exclusion zone from which everyone was evacuated and to which no one has been allowed to return since. This Chernobyl exclusion zone is still one of the world's most toxic environments, with ^{137}Cs levels ranging from 30 kBq/m^2 to as high as 70,000 kBq/m^2. To put these levels in perspective, 30 kBq/m^2 is the soil contamination level usually considered unsafe for use in farming.[10]

The successional vegetation has helped maintain the contamination by protecting the soil surface from erosion that might have washed or blown away some of the contaminant, while cycling of Cs and Sr by deep-rooted trees (in a manner analogous to that shown for Ca in Figure 2.21) has kept the radionuclides from leaching out of the soil profile. Scientists studying the ecosystems growing up on the abandoned farmland have found mutation-induced deformities in young trees and reduced fertility and life spans in birds. Although not officially sanctioned, some people residing nearby wander into the exclusion zone to graze cattle, although the milk from these cattle fails safety tests. Likewise, locals enter the zone to gather wild mushrooms, a local food delicacy, although those with deeper mycelia tend to be dangerously radioactive (Figure 15.30).

Figure 15.30 *Women gather mushrooms near Visokoye, Belarus, under a sign that reads: "Radiation danger! Cultivation and harvesting of agricultural crops, haymaking and cattle grazing are prohibited." The exclusion zone around Chernobyl, Ukraine, is still highly contaminated with radionuclides.* (Photo courtesy of Caroline Penn/Panos Pictures)

---

[9]For an update on the ecological consequence of the Chernobyl disaster, see Little and Bird (2013).

[10]Depending on the depth of sampling (usually 5–10 cm) and the bulk density of the soil (usually 0.8–1.5 Mg/m^3), the 30 kBq/m^2 threshold on a mass rather than area basis would be approximately 0.5 kBq/kg soil. See footnote 8 in Chapter 4.

Nuclear Accident at Fukushima

A major nuclear accident occurred in 2011 at the Fukushima Dai-ichi nuclear power plant. As a result of a large earthquake and associated tsunami that hit Japan on March 11, 2011, massive amounts of radionuclides were released into the water and air over a period of several weeks before the leakage was contained. As the reactor was on the coast, much of the radiative material fell in the ocean, but some 20% of the radioactive substances released were deposited on land. Of greatest concern was soil contamination with radioactive Cesium. Surface soil samples collected in Fukushima Prefecture some 30–40 km downwind from the stricken reactor were found to contain radiation from these two radionuclides at levels of up to 300–400 kBq/kg (equivalent to about 18,000–24,000 kBq/m^2), with many locations registering between 10 and 100 kBq/kg. The area normally produces almost half a million Mg of rice for human consumption. To avoid health risks to the rice-eating public, the government of Japan prohibited rice planting in fields with Cs levels in the soil exceeding 5.0 kBq/kg. However, even with this planting restriction in place, some rice harvested the following season from that area was soon found to contain in excess of the 0.5 kBq/kg safety limit set for rice under the Japanese Food Sanitation Law. The ensuing public panic about rice (and other produce) from Fukushima Prefecture effectively put thousands of farmers out of business and threatened to end the region's 2,000-year-old traditional lifestyle.

Fortunately, considerable research has been accomplished on the behavior of radionuclides in the soil–plant system. Research is also underway to take advantage of plant uptake of radionuclides in phytoremediation exercises. Certain plants, such as sunflowers, are being used to remove ^{90}Sr and ^{137}Cs from ponds and soils near the site of the Chernobyl nuclear disaster. Indian mustard is also being used in nearby sites to remove such nuclide contaminants.

Radioactive Wastes[11]

In addition to radionuclides added to soils because of weapons testing and nuclear power plant accidents, soils may interact with radioactive waste materials that have leaked from their holding tanks or have been intentionally buried for disposal. Plutonium, uranium, americium, neptunium, curium, and cesium are among the elements whose nuclides occur in radioactive wastes. These wastes are generated by research and medical facilities (where the radionuclides are used in cancer therapy and the like), and at power plants and weapons manufacturing sites.

Because of the secrecy and lack of regulation associated with the latter, they constitute some of the most polluted locations on Earth. For example, the U.S. Defense Department's now-abandoned plutonium-production complex at Hanford, Idaho, represents one of the biggest environmental cleanup challenges in the world. Among the hazards plaguing that site are hundreds of huge, in many cases leaking, underground tanks, in which high-level radioactive wastes have been stored for decades. Billions of cubic meters of soil and water have been contaminated with radioactive wastes at U.S. weapons manufacturing sites and at similar, equally polluted sites in the former Soviet Union.

Plutonium Toxicity. Plutonium 239, a major pollutant at these sites, is dangerous both because of its intense radioactivity and because of its high level of toxicity to humans. The ^{239}Pu itself is quite immobile in soils, having a K_d estimated at about 1,000. Nor is it taken up readily by plants, so it does not accumulate along terrestrial food chains. It does, however, accumulate in algae. Furthermore, oily liquid wastes carrying ^{239}Pu seep into the groundwater and nearby rivers, and contaminated surface soil blows in the desert wind, spreading the radionuclides for many kilometers. Cleanup may be impossible at some of these former weapons sites; the agencies responsible are struggling merely to stabilize and contain the contamination. With a half-life of 24,400 years, ^{239}Pu contamination is a problem that will not go away.

---

[11] For a description of the environmental challenges at the Hanford site, see Zorpette (1996).

15.11 RADON GAS FROM SOILS[12]

The Health Hazard

The soil is the primary source of the colorless, odorless, and tasteless radioactive gas *radon*, which has been shown to cause lung cancer. Radon is not usually considered a soil pollutant because it has not been introduced into the soil by human activity. Nonetheless, radon gas is thought to be a serious environmental health hazard when it moves from the soils and accumulates inside buildings. Deaths from breathing in radon are estimated at about 20,000 per year in the United States, some 10–50 times more numerous than deaths caused by contaminants in drinking water. The gas may be a causal factor in about 10% of lung cancer cases.

The health hazard from this gas stems from its transformation to radioactive polonium isotopes, which are solids that tend to attach to dust particles. The polonium-contaminated dust particles may lodge in the lungs, where alpha-particle radiation emitted by the polonium can penetrate the lung tissue and cause cancer. The principal concern is with radon accumulation in homes, offices, and schools where people breathe the air in basement or ground-level rooms for extended periods.

How Radon Accumulates in Buildings

Geologic Factors. Radon originates from uranium (^{238}U) found in minerals, sorbed on soil colloids, or dissolved in groundwater. Over billions of years, the uranium undergoes radioactive decay forming radium, which in turn gives off radiation over thousands of years and transforms into radon (Figure 15.31). Both uranium and radium are solids; however, radon is a gas that can diffuse through pores and cracks and emerge into the atmosphere. Soils and rocks that contain high concentrations of uranium will likely produce large amounts of radon gas. Soils formed from certain highly deformed metamorphic rocks and from marine sediments, limestones, and coal or oil-bearing shales tend to have the highest potential levels for radon production. However, nearby houses built on soils formed from the same parent material may differ widely in their indoor radon concentrations (Figure 15.32, *right*), the difference being due to variations in soil properties and/or house construction.

Soil Properties. To become a hazard, radon must travel from its source in the underlying rock or soil, up through overlying soil layers, and finally into an enclosed building where it might accumulate to unhealthful concentrations. It must also make this trip quite rapidly,

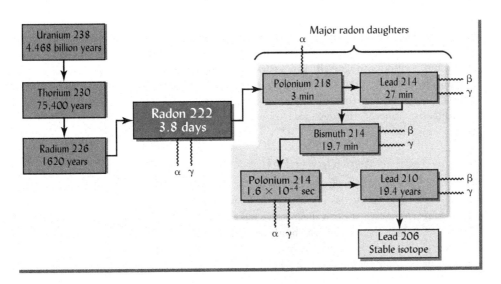

Figure 15.31 *Radon decay showing the name, atomic weight, and half-life for each isotope. Radioactive decay of uranium 238 in soils results in the formation of inert, but radioactive, radon. This gas emits alpha (α) particles and gamma (γ) rays and forms radon daughters that are capable of emitting alpha and beta (β) particles and gamma rays. The α particles damage lung tissue and cause cancer. Radon gas may cause about 20,000 deaths annually in the United States.* (Diagram courtesy of Ray R. Weil)

---

[12]For a straightforward discussion of the hazards of indoor radon and what you can do to protect yourself, see U.S. EPA (2012). For a review of the radon in soils and geologic material with a focus on mapping radon hazard areas, see Appleton (2007).

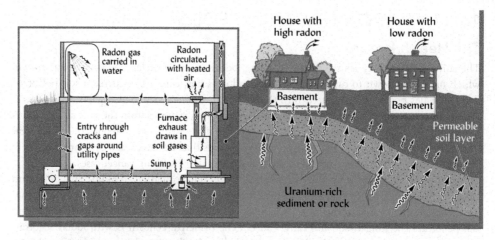

Figure 15.32 *The foundation of a house, as well as the nature of the soil and rocks in the vicinity, helps determine whether dangerous levels of radioactive radon gas are likely to accumulate inside. (Right) Levels of indoor radon can vary greatly from one house to another. On a regional basis, radon is likely to be highest where the soils are formed from uranium-bearing parent materials. The soil that underlies a house plays a role in the movement of radon gas from its source in rock and soil minerals to the air inside the house. Dry, coarse-textured permeable soil layers allow much faster diffusion of radon gas than does a wet or fine-textured soil. If soils underlying a house are relatively impermeable, radon movement will be so slow that nearly all of the radon emitted will have decayed before it can reach the house foundation. (Left) Once it arrives at a house foundation, radon may enter through a variety of openings, such as cracks in the foundation blocks, joints between the walls and concrete floor, and gaps where utility pipes enter the house. Radon risks are greatest during cold weather. Heated air escapes up the furnace chimney but windows are sealed, creating negative air pressure inside the house, which, in turn, draws in air from the soil.* (Diagram courtesy of Ray R. Weil)

because the half-life of radon is only 3.8 days. Within several weeks, radon completely decays to polonium, lead, and bismuth, radioactive solids that last for only minutes. Whether significant quantities of radon reach a building foundation depends mainly on two factors: (1) the distance that the radon must travel from its source and (2) the permeability of the soil through which it travels. Since radon is an inert gas, the soil does not react with it, but merely serves as a channel through which the gas moves.

As explained in Sections 7.1 and 7.2, gases move through soil both by diffusion and by mass flow (convection). The rate of radon diffusion through a soil depends on the total soil porosity (more so than on pore size) and the degree to which the pores are filled with water. Therefore, some of the highest indoor radon concentrations have been found in houses where only a thin, well-drained, gravelly soil separates the house foundation from uranium-rich rock. At the other extreme, a thick, wet clay layer would provide an excellent barrier against radon diffusion. Convective airflow, which is mainly stimulated by rainwater entering the soil and by changes in atmospheric pressure, may play a significant role in radon movement during stormy weather.

Radon Testing and Remediation

Testing. Since the occurrence of high radon levels cannot be accurately predicted, the only sure way to determine the risk of radon is to test for its presence. Testing is usually carried out in two stages. The first uses an inexpensive (about $15) charcoal canister, which is placed in the test area, unsealed, and left to absorb radon for the specified period (usually three days). The canister should be placed in the most high-risk location in the building—for example, in a basement bedroom without cross ventilation—during a time when the building is heated. After the test period, the canister is resealed and sent to a lab where the amount of radon absorbed can be measured and related to the radon concentration that was in the air. If the results suggest a radon level above 4 piCi/L (148 Bq/m^3), the U.S. EPA advises that a long-term test be conducted using a somewhat more expensive alpha-track detector for a period of

3–12 months. If the long-term test also suggests levels above 4 piCi/L, modifications should be made to the building to reduce the accumulation of radon inside.

Remediation. Depending on the levels of radon and the condition of the building, the modification may be as simple as caulking cracks in the floor and walls and filling gaps around utility-pipe entrances. Remediation of higher radon levels may require alterations that are more extensive. Ventilation of the room with outside air can prevent unhealthful radon buildup, but a more energy-efficient solution is a subslab ventilation system. For the latter, perforated pipes are installed in a layer of gravel under the foundation slab, and the air pressure there is lowered either by a mechanical fan or by convective draw from a special chimney. In this way, gas coming from the soil is intercepted and redirected to the atmosphere before it can enter the building. Installation of a subslab ventilation system is much less expensive during new construction than as a retrofit, and is now standard practice in many areas with high-uranium soils.

15.12 CONCLUSION

Three major conclusions may be drawn about soils in relation to environmental quality. First, since soils are valuable resources, they should be protected from environmental contamination, especially that which does permanent damage. Second, because of their vastness and remarkable capacities to absorb, bind, and break down added materials, soils offer promising mechanisms for the disposal and utilization of many wastes that otherwise may contaminate the environment. Third, soil contaminants and the products of their breakdown in soil reactions can be toxic to humans and other animals if the soil is ingested or the contaminants move from the soil into plants, soil fauna, the air, and—particularly—into water supplies.

To gain a better understanding of how soils might be used and yet protected in waste-management efforts, soil scientists devote a considerable share of their research efforts to environmental-quality problems. Furthermore, soil scientists have much to contribute to the research teams that search for better ways to clean up environmental contamination. Some of the most promising technological advances have been in the field of bioremediation, in which the biological processes of the soil are harnessed to effect soil cleanup.

STUDY QUESTIONS

1. What agricultural practices contribute to soil and water pollution, and what steps must be taken to reduce or eliminate such pollution?
2. Discuss the types of reactions pesticides undergo in soils, and indicate what we can do to encourage or prevent such reactions.
3. Discuss the environmental problems associated with the disposal of large quantities of sewage sludge on agricultural lands, and indicate how the problems could be alleviated.
4. What is *bioremediation*, and what are its advantages and disadvantages compared with physical and chemical methods of handling organic wastes?
5. Even though large quantities of the so-called heavy metals are applied to soils each year, relatively small quantities find their way into human food. Why?
6. Compare the design, operation, and management of today's containment landfills with the natural attenuation type most common 30 years ago, and indicate how the changes affect soil and water pollution.
7. What are *organoclays*, and how can they be used to help remediate soils polluted with nonpolar organic compounds?
8. Soil organic matter and some silicate clays chemically sorb some organic pollutants and protect them from microbial attack and leaching from the soil. What are the implications (positive and negative) of such protection for efforts to reduce soil and water pollution?
9. What radionuclides are of greatest concern in soil and water pollution, and why are they not more readily taken up by plants?
10. What are the comparative advantages and disadvantages of in situ and ex situ means of remediating soils polluted with organic compounds?
11. What are two approaches to *phytoremediation*, and for what kinds of pollutants are they useful? Explain.
12. Suppose a nickel-contaminated soil 15 cm deep contained 800 mg/kg Ni. Vegetation was planted to remove the nickel by phytoremediation. The aboveground plant parts average 1% Ni on a dry-weight basis and produce 4,000 kg/ha of harvestable dry matter. If two harvests are possible per year, how many years will it take to reduce the Ni level in the soil to a target of 80 mg/kg?

REFERENCES

Adriano, D. C. 2001. *Trace Elements in Terrestrial Environments: Biogeochemistry, Bioavailability, and Risks of Metals.* Springer, New York, p. 880.

Altieri, M. A. 1995. *Agroecology: The Science of Sustainable Agriculture.* 2nd ed. Westview Press, Boulder, CO, p. 448.

Appleton, J. D. 2007. "Radon: Sources, health risks, and hazard mapping." *AMBIO: A Journal of the Human Environment* 36:85–89.

Alvarez-Cohen, L. 2012. A systems approach to bioremediation. Superfund Research Program and Department of Civil and Environmental Engineering University of California Berkeley http://berkeley.edu/., https://youtu.be/YJURahqtkYY.

Barth, J. A. C., P. Grathwohl, H. J. Fowler, A. Bellin, M. H. Gerzabek, et al. 2009. "Mobility, turnover and storage of pollutants in soils, sediments and waters: Achievements and results of the EU project AquaTerra. A review." *Agronomy for Sustainable Development* 29:161–173.

Bøhn, T., M. Cuhra, T. Traavik, M. Sanden, J. Fagan, and R. Primicerio. 2014. "Compositional differences in soybeans on the market: Glyphosate accumulates in roundup ready GM soybeans." *Food Chemistry* 153:207–215.

Bragg, J. R., R. C. Prince, E. J. Harner, and R. M. Atlas. 1994. "Effectiveness of bioremediation for the Exxon Valdez oil spill." *Nature* 368:413–418.

Cao, X., and L. Q. Ma. 2004. "Effects of compost and phosphate on plant arsenic accumulation from soils near pressure-treated wood." *Environmental Pollution* 132:435–442.

Capowiez, Y., F. Bastardie, and G. Costagliola. 2006. "Sublethal effects of imidacloprid on the burrowing behaviour of two earthworm species: Modifications of the 3D burrow systems in artificial cores and consequences on gas diffusion in soil." *Soil Biology and Biochemistry* 38:285–293.

Carson, R. 1962. *Silent Spring.* Houghton Mifflin, Cambridge, MA, p. 378.

Cuhra, M., T. Traavik, M. Dando, R. Primicerio, D. F. Holderbaum, and T. Bøhn. 2015. "Glyphosate-residues in roundup-ready soybean impair Daphnia magna life-cycle." *Journal of Agricultural Chemistry and Environment* 4:24.

Defoe, P. P., G. M. Hettiarachchi, C. Benedict, and S. Martin. 2014. "Safety of gardening on lead- and arsenic-contaminated urban brownfields." *Journal of Environmental Quality* 43:2064–2078.

Dowdy, R. H., C. E. Clapp, D. R. Linden, W. E. Larson, T. R. Halbach, and R. C. Polta. 1994. "Twenty years of trace metal partitioning on the Rosemount sewage sludge watershed." In C. E. Clapp, W. E. Larson, and R. H. Dowdy (eds.). *Sewage Sludge: Land Utilization and the Environment.* Soil Science Society of America, Madison, WI, pp. 149–155.

Edwards, C. A. 1978. "Pesticides and the micro-fauna of soil and water." In I. R. Hill and S. J. Wright (eds.). *Pesticide Microbiology.* Academic Press, London, pp. 603–622.

El-Naggar, J. B., and N. E. H. A. Zidan. 2013. "Field evaluation of imidacloprid and thiamethoxam against sucking insects and their side effects on soil fauna." *Journal of Plant Protection Research* 53:375–387.

Fernandez-Cornejo, J., R. Nehring, C. Osteen, S. Wechsler, A. Martin, and A. Vialou. 2014. "Pesticide use in U.S. Agriculture: 21 selected crops, 1960-2008." Economic Information Bulletin 124. U.S. Service. www.ers.usda.gov/publications/eib-economic-information-bulletin/eib124.aspx.

Gaynor, J. D., D. C. MacTavish, and W. I. Findlay. 1995. "Atrazine and metolachlor loss in surface and subsurface runoff from three tillage treatments in corn." *Journal of Environmental Quality* 24:246–256.

Gilliom, R. J., J. E. Barbash, C. G. Crawford, P. A. Hamilton, J. D. Martin, N. Nakagaki, L. H. Nowell, J. C. Scott, P. E. Stackelberg, G. P. Thelin, and D. M. Wolock. 2006. "The quality of our nation's waters: Pesticides in the nation's streams and ground water, 1992–2001." USGS Circular 1291. U.S. Geological Survey, Reston, VA, p. 172. http://pubs.usgs.gov/circ/2005/1291/.

Hazen, T. C. 1995. "Savannah river site—a test bed for cleanup technologies." *Environmental Protection* (April):10–16.

Hazen, T. 2008. Bioremediation: Hope / hype for environmental cleanup. Video lecture 57 min. https://www.youtube.com/watch?v=MT0qY3_n1kI

Hladik, M. L., D. W. Kolpin, and K. M. Kuivila. 2014. "Widespread occurrence of neonicotinoid insecticides in streams in a high corn and soybean producing region, USA." *Environmental Pollution* 193:189–196.

Kjeldsen, P., M. Barlaz, A. Rooker, A. Baun, A. Ledin, and T. Christensen. 2002. "Present and long-term composition of MSW landfill leachate: A review." *Critical Reviews in Environmental Science and Technology* 32:297–336.

Little, J. B., and W. A. Bird. 2013. "A tale of two forests: Addressing postnuclear radiation at Chernobyl and Fukushima." *Environmental Health Perspectives.* http://ehp.niehs.nih.gov/121-a78/.

Meuser, H. 2010. *Contaminated Urban Soils.* Springer, New York, p. 340.

Mishra, M., and D. Mohan. 2017. "Bioremediation of contaminated soils: An overview." In A. Rakshit et al. (ed.). *Adaptive Soil Management: From Theory to Practices.* Springer Singapore, Singapore, pp. 323–337.

Pierzynski, G. M., J. T. Sims, and G. F. Vance. 2004. *Soils and Environmental Quality.* 3rd ed. CRC Press/Lewis Publishers, Boca Raton, FL.

Pimentel, D., and M. Burgess. 2014. "Environmental and economic costs of the application of pesticides primarily in the United States." In D. Pimentel and R. Peshin (eds.). *Integrated Pest Management*. Springer, Dordrecht, pp. 47–64.

Pritchard, P. H., J. G. Mueller, J. C. Rogers, F. V. Kremer, and J. A. Glaser. 1992. "Oil spill bioremediation: Experiences, lessons and results from the Exxon Valdez oil spill in Alaska." *Biodegradation* 3:315–335.

Qian, X., R. M. Koerner, and D. H. Gray. 2002. *Geotechnical Aspects of Landfill Design and Construction*. Prentice Hall, Upper Saddle River, NJ, p. 716.

Rose, M. T., T. R. Cavagnaro, C. A. Scanlan, T. J. Rose, T. Vancov, S. Kimber, I. R. Kennedy, R. S. Kookana, and L. Van Zwieten. 2016. "Impact of herbicides on soil biology and function." *Advances in Agronomy*. Vol. 136. Academic Press, New York, pp. 133–220.

Simmons, R. W., R. L. Chaney, J. S. Angle, M. Kruatrachue, S. Klinphoklap, R. D. Reeves, and P. Bellamy. 2014. "Towards practical cadmium phytoextraction with noccaea caerulescens." *International Journal of Phytoremediation* 17:191–199.

Sizmur, T., E. L. Tilston, J. Charnock, B. Palumbo-Roe, M. J. Watts, and M. E. Hodson. 2011. "Impacts of epigeic, anecic and endogeic earthworms on metal and metalloid mobility and availability." *Journal of Environmental Monitoring* 13:266–273.

Stone, W. W., R. J. Gilliom, and J. D. Martin. 2014. "An overview comparing results from two decades of monitoring for pesticides in the nation's streams and rivers, 1992–2001 and 2002–2011." *Scientific Investigations Report* 2014-5154. U.S. Geological Survey.

Strek, H. J., and J. B. Weber. 1982. "Adsorption and reduction in bioactivity of polychlorinated biphenyl (Aroclor 1254) to redroot pigweed by soil organic matter and montmorillonite clay." *Soil Science Society of America Journal* 46:318–322.

Su, C., L. Jiang, and W. Zhang. 2014. "A review on heavy metal contamination in the soil worldwide: Situation, impact and remediation techniques." *Environmental Skeptics and Critics* 3:24–38.

Tooker, J. F., M. R. Douglas, and C. H. Krupke. 2017. "Neonicotinoid seed treatments: Limitations and compatibility with integrated pest management." *Agricultural & Environmental Letters* 2:170026.

U.S. EPA. 1993. *Clean Water Act*. Sec. 503, Vol. 58, No. 32. U.S. Environmental Protection Agency, Washington, D.C.

U.S. EPA. 2012. "A citizen's guide to radon: The guide to protecting yourself and your family from radon." U.S. EPA 402/K-12/002. U.S. Division, Washington, D.C. http://www.epa.gov/radon/pdfs/citizensguide.pdf.

U.S. EPA. 2014. "Municipal solid waste generation, recycling, and disposal in the United States: Facts and figures for 2012," EPA-530-F-14-001. U.S. Environmental Protection Agency, Washington, D.C. http://www.epa.gov/osw/nonhaz/municipal/pubs/2012_msw_fs.pdf.

van der Ent, A., A. M. Baker, R. Reeves, A. J. Pollard, and H. Schat. 2013. "Hyperaccumulators of metal and metalloid trace elements: Facts and fiction." *Plant and Soil* 362:319–334.

Vijver, M. G., and P. J. van den Brink. 2014. "Macroinvertebrate decline in surface water polluted with imidacloprid: A rebuttal and some new analyses." *PLoS One* 9:e89837.

Weber, J. B., and C. T. Miller. 1989. "Organic chemical movement over and through soil." in B. L. Sawhney and K. Prown (eds.). *Reactions and Movement of Organic Chemicals in Soils*. SSSA Special Publication No. 22. Soil Science Society of America, Madison, WI, pp. 305–334.

Wiszniewska, A., E. Hanus-Fajerska, E. MuszyŃSka, and K. Ciarkowska. 2016. "Natural organic amendments for improved phytoremediation of polluted soils: A review of recent progress." *Pedosphere* 26:1–12.

Wong, C. T., M. K. Leung, M. K. Wong, and W. C. Tang. 2013. "Afteruse development of former landfill sites in Hong Kong." *Journal of Rock Mechanics and Geotechnical Engineering* 5:443–451.

Wu, G., H. Kang, X. Zhang, H. Shao, L. Chu, and C. Ruan. 2010. "A critical review on the bio-removal of hazardous heavy metals from contaminated soils: Issues, progress, eco-environmental concerns and opportunities." *Journal of Hazardous Materials* 174:1–8.

Xu, S., G. Sheng, and S. A. Boyd. 1997. "Use of organoclays in pollution abatement." *Advances in Agronomy* 59:25–62.

Zhang, X., T. Zhong, L. Liu, and X. Ouyang. 2015. "Impact of soil heavy metal pollution on food safety in China." *PLoS ONE* 10:e0135182.

Zorpette, G. 1996. "Hanford's nuclear wasteland." *Scientific American* (May): 88–97. Comprehensive information on cleaning contaminated soils at abandoned industrial sites: http://www.epa.gov/brownfields/.

Appendix A
World Reference Base, Canadian, and Australian Soil Classification Systems

Table A.1
SOIL REFERENCE GROUPS IN THE WORLD REFERENCE BASE (WRB) FOR SOIL RESOURCES[a]

The World Reference Base provides a global vocabulary for communicating about different kinds of soils and a reference by which various national soil classification systems (such as U.S. Soil Taxonomy and the Canadian Soil Classification System discussed in this text) can be compared and correlated. The 32 Reference Soil Groups are differentiated mainly according to the primary pedogenesis process that has produced the characteristic soil features, except where "special" soil parent materials are of overriding importance. Each Reference Soil Group can be subdivided using a unique list of possible prefix and suffix qualifiers (not shown here[b]). These qualifiers indicate secondary soil-forming processes that significantly affected the primary soil characteristics especially important to soil use. To avoid making the classification of soils dependent on the availability of climatic data, separations are not based on specific climatic characteristics (as is the case in U.S. Soil Taxonomy).

Reference Soil Group[c]	Major Soil Characteristics	Approximate Equivalents in U.S. Soil Taxonomy[d]
Organic Soils		
Histosols (HS)	Composed of organic materials	Most Histosols and Histels
Mineral Soils Dominantly Influenced by Human Activity		
Anthrosols (AT)	Soils with long and intensive agriculture use	Anthropic and Plaggic subgroups
Technosols (TC)	Soils containing many artifacts	Subgroups, mainly of Entisols, with Anthraltic, Anthroportic, or Anthropic adjectives.
Soils with Limited Rooting Due to Shallow Permafrost or Stoniness		
Cryosols (CR)	Ice-affected soils: Cryosols	Gelisols
Leptosols (LP)	Shallow or extremely gravelly soils	Lithic subgroups of Inceptisols and Entisols
Soils Influenced by Water		
Vertisols (VR)	Alternating wet-dry conditions, rich in swelling clays	Vertisols
Fluvisols (FL)	Young soils in alluvial deposits	Fluvents and Fluvaquents
Solonchaks (SC)	Strongly saline soils	Salids and salic or halic great groups of other orders
Solonetz (SN)	Soils with subsurface clay accumulation, rich in sodium	Natric great groups of Alfisols, Aridisols, and Mollisols
Gleysols (GL)	Groundwater-affected soils	Endoaquic great groups (e.g., Endoaqualfs, Endoaquolls, Endoaquults, Endoaquents, Endoaquepts)
Soils for Which Aluminium (Al) Chemistry Plays a Major Role in Their Formation		
Andosols (AN)	Young soils from volcanic ash and tuff deposits	Andisols
Podzols (PZ)	Acid soils with a subsurface accumulation of iron–aluminium–organic compounds	Spodosols

(continued)

Table A.1 (CONTINUED)

Reference Soil Group[c]	Major Soil Characteristics	Approximate Equivalents in U.S. Soil Taxonomy[d]
Plinthosols (PT)	Wet soils with an irreversibly hardening mixture of iron, clay, and quartz in the subsoil	Plinthic great groups of Aqualfs, Aquox, and Ultisols
Nitisols (NT)	Deep, dark red, brown, or yellow clayey soils having a pronounced shiny, nut-shaped structure	Parasesquic Inceptisols and some Oxisols and Ultisols
Ferralsols (FR)	Deep, strongly weathered soils with a chemically poor, but physically stable subsoil	Oxisols
Soils with Stagnant Water		
Planosols (PL)	Soils with a bleached, temporarily water-saturated topsoil on a slowly permeable subsoil	Albaqualfs and Albaquults and some albaquic subgroups of Alfisols and Ultisols
Stagnosols (ST)	Soil with temporarily water-saturated topsoil on structural or moderate textural discontinuity	Epiaquic and Anthraquic great groups
Mineral Soils with Humus-Rich Topsoils and a High Base Saturation Typically in Grasslands		
Chernozems (CH)	Soils with a thick, dark topsoil, rich in organic matter with a calcareous subsoil	Calciudolls
Kastanozems (KS)	Soils with a thick, dark brown topsoil, rich in organic matter and a calcareous or gypsum-rich subsoil	Many Calciustolls and Calcixerolls
Phaeozems (PH)	Soils with a thick, dark topsoil, rich in organic matter and evidence of removal of carbonates	Many Cryolls, Udolls, and Albolls
Soils with Accumulation of Nonsaline Substances and Influenced by Aridity		
Gypsisols (GY)	Soils with accumulation of secondary gypsum	Gypsids and some gypsic great groups of other orders
Durisols (DU)	Soils with accumulation of secondary silica	Durids and some Duric great groups of other orders
Calcisols (CL)	Soils with accumulation of secondary calcium carbonates	Calcids and Calcic great groups of Inceptisols
Mineral Soils with a Clay-Enriched Subsoil		
Albeluvisols (AB)	Acid soils with a bleached horizon penetrating into a clay-rich subsurface horizon	Some Glossudalfs
Alisols (AL)	Soils with subsurface accumulation of high activity clays, and low base saturation	Ultisols and ultic Alfisols
Acrisols (AC)	Soils with subsurface accumulation of low activity clays and low base saturation	Kandic great groups of Alfisols and Ultisols
Luvisols (LV)	Soils with subsurface accumulation of high activity clays and high base saturation	Haplo and pale great groups of Alfisols
Lixisols (LX)	Soils with subsurface accumulation of low activity clays and high base saturation	Kandic great groups of Alfisols with high base saturation
Relatively Young Soils or Soils with Little or No Profile Development		
Umbrisols (UM)	Acid soils with a thick, dark topsoil, rich in organic matter	Many Umbric great groups of Inceptisols
Arenosols (AR)	Very sandy soils featuring no or only very weak B horizon development	Psamments, grossarenic subgroups of other orders
Cambisols (CM)	Soil with only weakly to moderately developed B horizons	Cambids and many Inceptisols
Regosols (RG)	Soils with very limited soil development, often shallow to rock	Orthents, some Psamments and other Entisols

[a]Based on FAO (2006). World reference base for soil resources 2006: A framework for international classification correlation and communication. World Soil Resources Reports 103. Food and Agriculture Organization of the United Nations and United Nations Environmental Program, Rome. 128 pp. and on personal communication from Bob Engel (USDA/NRCS) and Michéli Erika (Univ. Agric. Sci., Hungary).

[b]As one example, the Kastanozems reference group can be subdivided using the prefix modifiers Vertic, Gypsic, Calcic, Luvic, Hyposodic, Siltic, Chromic, Anthric, and Haplic.

[c]Abbreviations (often used as symbols on soil maps) in parentheses.

[d]As discussed in Chapter 3 of this text.

(continued)

Table A.2
THE AUSTRALIAN SOIL CLASSIFICATION SYSTEM AND APPROXIMATE CORRELATIONS WITH U.S. SOIL TAXONOMY

Order	Main Characteristics	Soil Taxonomy Order and Suborders
Anthroposols	"Man-made" soils	Anthraltic, Anthroportic or Antrhopic subgroups
Calcarosols	B horizon calcareous and lacking a marked clay accumulation	Aridisols, Alfisols (Ustalfs, Xeralfs)
Chromosols	Strong clay accumulation and pH > 5.5 in B horizon	Alfisols, some Aridisols
Dermosols	B horizon well structured but lacking a marked clay accumulation	Mollisols, Alfisols, Ultisols
Ferrosols	B horizon high in Fe and lacking a marked clay accumulation, without vertic properties.	Oxisols, some Alfisols
Hydrosols	Prolonged seasonal water saturation	Aquic subgroups of Alfisols, Ultisols, Inceptisols, salic Aridisols, and some Histosols
Kandosols	B horizon massive or lacking strong structure, and lacking a marked clay accumulation	Alfisols, Ultisols, and Aridisols with massive B horizon structure
Kurosols	Strong clay accumulation and pH < 5.5 in B horizon; clear or abrupt B horizon boundary.	Ultisols, some Alfisols
Organosols	Organic materials; not regularly inundated by saline tidal waters.	Histosols
Podosols	Acid soils with subsurface accumulation of Fe, Al-organic compounds	Spodosols, some Entisols
Rudosols	Negligible (rudimentary) horizon differentiation	Entisols, salic Aridisols
Sodosols	Strong clay accumulation in B horizon, with high sodium saturation	Natric subgroups of Alfisols, Aridisols
Tenosols	Weak horizon differentiation	Inceptisols, Aridisols, Entisols
Vertosols	High clay (>35%), deep cracks, slickensides	Vertisols

Modified from: CSIRO Land and Water (2016). http://www.clw.csiro.au/aclep/asc_re_on_line_V2/soilhome.htm

Table A.3
SUMMARY WITH BRIEF DESCRIPTIONS OF THE SOIL ORDERS IN THE SOIL CLASSIFICATION SYSTEM OF CANADA

The Canadian Soil Classification System is one of many national soil classification systems used in various countries around the world. Of these, it is perhaps the most closely aligned with the U.S. Soil Taxonomy. It includes five hierarchical categories: order, great group, subgroup, family, and series. The system is designed to apply principally to the soils of Canada. The soil orders of the Canadian System of Soil Classification are described in this table and soil orders and some great groups are compared to the U.S. Soil Taxonomy in Table A.4.

Brunisolic	Soils with sufficient development to exclude them from the Regosolic order, but lacking the degree or kind of horizon development specified for other soil orders.
Chernozemic	Soils with high base saturation and surface horizons darkened by the accumulation of organic matter from the decomposition of plants from grassland or grassland-forest ecosystems.
Cryosolic	Soils formed in either mineral or organic materials that have permafrost either within 1 m of the surface or within 2 m if more than one-third of the pedon has been strongly cryoturbated, as indicated by disrupted, mixed, or broken horizons.

Table A.3 (Continued)

Gleysolic	Gleysolic soils have features indicative of periodic or prolonged saturation (i.e., gleying, mottling) with water and reducing conditions.
Luvisolic	Soils with light-colored, eluvial horizons that have illuvial B horizons in which silicate clay has accumulated.
Organic	Organic soils developed on well- to undecomposed peat or leaf litter.
Podzolic	Soils with a B horizon in which the dominant accumulation product is amorphous material composed mainly of humified organic matter combined in varying degrees with Al and Fe.
Regosolic	Weakly developed soils that lack development of genetic horizons.
Solonetzic	Soils that occur on saline (often high in sodium) parent materials, which have B horizons that are very hard when dry and swell to a sticky mass of very low permeability when wet. Typically the solonetzic B horizon has prismatic or columnar macrostructure that breaks to hard to extremely hard, blocky peds with dark coatings.
Vertisolic	Soils with high contents of expanding clays that have large cracks during the dry parts of the year and show evidence of swelling, such as gilgae and slickensides.

Table A.4
Comparison of U.S. Soil Taxonomy and the Canadian Soil Classification System

Note that because the boundary criteria differ between the two systems, certain U.S. Soil Taxonomy soil orders have equivalent members in more than one Canadian Soil Classification System soil order.[a]

U.S. Soil Taxonomy Soil Order	Canadian System Soil Order	Canadian System Great Group	Equivalent Lower-Level Taxa in U.S. Soil Taxonomy
Alfisols	Luvisolic	Gray Brown Luvisols	Hapludalfs
		Gray Luvisols	Haplocryalfs, Eutrocryalfs, Fragudalfs, Glossocryalfs, Palecryalfs, and some subgroups of Ustalfs and Udalfs
	Solonetzic	Solonetz	Natrudalfs and Natrustalfs
		Solod	Glossic subgroups of Natraqualfs, Natrudalfs, and Natrustalfs
Andisols	Components of Brunisolic and Cryosolic		
Aridisols	Solonetzic		Frigid families of Natrargids
Entisols	Regosolic		Cryic great groups and frigid families of Entisols, except Aquents
		Regosol	Cryic great groups and frigid families of Folists, Fluvents, Orthents, and Psamments
Gelisols	Cryosolic	Turbic Cryosol	Turbels
		Organic Cryosol	Histels
		Stagnic Cryosol	Orthels
Histosols	Organic	Fibrisol	Cryofibrists, Sphagnofibrists
		Mesisol	Cryohemists
		Humisol	Cryosaprists
Inceptisols	Brunisolic	Melanic Brunisol	Some Eutrustepts
		Eutric Brunisol	Subgroups of Cryepts; frigid and mesic families of Haplustepts
		Sombric Brunisol	Frigid and mesic families of Udepts, and Ustept and Humic Dystrudepts
		Dystric Brunisol	Frigid families of Dystrudepts and Dystrocryepts

(continued)

Table A.4 (CONTINUED)

U.S. Soil Taxonomy Soil Order	Canadian System Soil Order	Canadian System Great Group	Equivalent Lower-Level Taxa in U.S. Soil Taxonomy
	Gleysolic		Cryic subgroups and frigid families of Aqualfs, Aquolls, Aquepts, Aquents, and Aquods
		Humic Gleysol	Humaquepts
		Gleysol	Cryaquepts and frigid families of Fragaquepts, Epiaquepts, and Endoaquepts
Mollisols	Chernozemic	Brown	Xeric and Ustic subgroups of Argicryolls and Haplocryolls
		Dark Brown	Subgroups of Argicryolls and Haplocryolls
		Black	Typic subgroups of Argicryolls and Haplocryolls
		Dark Gray	Alfic subgroups of Argicryolls
	Solonetzic	Solonetz	Natricryolls and frigid families of Natraquolls and Natralbolls
		Solod	Glossic subgroups of Natricryolls
Oxisols	Not relevant in Canada		
Spodosols	Podzolic	Humic Podzol	Humicryods, Humic Placocryods, Placohumods, and frigid families of other Humods
		Ferro-Humic Podzol	Humic Haplocryods, some Placorthods, and frigid families of humic subgroups of other Orthods
		Humo-Ferric Podzol	Haplorthods, Placorthods, and frigid families of other Orthods and Cryods except humic subgroups
Ultisols	Not relevant in Canada		
Vertisols	Vertisolic	Vertisol	Haplocryerts
		Humic Vertisol	Humicryerts

[a]Based on information in Soil Classification Working Group, 1998, *The Canadian System of Soil Classification*, 3rd ed. (Ottawa: Agriculture and Agri-Food Canada). Publication No. A53-1646/1997E. http://sis.agr.gc.ca/cansis/taxa/cssc3/index.html

Appendix B
SI Units, Conversion Factors, Periodic Table of the Elements, and Plant Names

Basic SI Units of Measurement

Parameter	Basic Unit	Symbol
Amount of substance	mole	mol
Electrical current	ampere	A
Length	meter	m
Luminous intensity	candela	cd
Mass	gram (kilogram)	g (kg)
Temperature	kelvin	K
Time	second	s

Prefixes Used to Indicate Order of Magnitude

Prefix	Multiple	Abbreviation	Multiplication Factor
exa	10^{18}	E	1,000,000,000,000,000,000
peta	10^{15}	P	1,000,000,000,000,000
tera	10^{12}	T	1,000,000,000,000
giga	10^{9}	G	1,000,000,000
mega	10^{6}	M	1,000,000
kilo	10^{3}	k	1,000
hecto	10^{2}	h	100
deca	10	da	10
deci	10^{-1}	d	0.1
centi	10^{-2}	c	0.01
milli	10^{-3}	m	0.001
micro	10^{-6}	µ	0.000 001
nano	10^{-9}	n	0.000 000 001
pico	10^{-12}	p	0.000 000 000 001
femto	10^{-15}	f	0.000 000 000 000 001
atto	10^{-18}	A	0.000 000 000 000 000 001

Factors for Converting Non-SI Units to SI Units

Non-SI Unit	Multiply by[a]	To Obtain SI Unit
Length		
inch, in.	2.54	centimeters, cm (10^{-2} m)
foot, ft	0.304	meter, m
mile,	1.609	kilometer, km (10^3 m)
micron, μ	1.0	micrometer, μm (10^{-6} m)
Ångstrom unit, Å	0.1	nanometer, nm (10^{-9} m)
Area		
acre, ac	0.405	hectare, ha (10^4 m^2)
square foot, ft^2	9.29×10^{-2}	square meter, m^2
square inch, in^2	645	square millimeter, mm^2
square mile, mi^2	2.59	square kilometer, km^2
Volume		
bushel, bu	35.24	liter, L
cubic foot, ft^3	2.83×10^{-2}	cubic meter, m^3
cubic inch, in.^{3}	1.64×10^{-5}	cubic meter, m^3
gallon (U.S.), gal	3.78	liter, L
quart, qt	0.946	liter, L
acre-foot, ac-ft	12.33	hectare-centimeter, ha-cm
acre-inch, ac-in.	1.03×10^{-2}	hectare-meters, ha-m
ounce (fluid), oz	2.96×10^{-2}	liter, L
pint, pt	0.473	liter, L
Mass		
ounce (avdp), oz	28.4	gram, g
pound, lb	0.454	kilogram, kg (10^3 g)
ton (2,000 lb)	0.907	megagram, Mg (10^6 g)
tonne (metric), t	1,000	kilogram, kg
Radioactivity		
curie, Ci	3.7×10^{10}	becquerel, Bq
picocurie per gram, pCi/g	37	becquerel per kilogram, Bq/kg
Yield and Rate		
pound per acre, lb/ac	1.121	kilogram per hectare, kg/ha
pounds per 1,000 ft^2	48.8	kilogram per hectare, kg/ha
bushel per acre (60 lb), bu/ac	67.19	kilogram per hectare, kg/ha
bushel per acre (56 lb), bu/ac	62.71	kilogram per hectare, kg/ha
bushel per acre (48 lb), bu/ac	53.75	kilogram per hectare, kg/ha
gallon per acre (U.S.), gal/ac	9.35	liter per hectare, L/ha
ton (2,000 lb) per acre	2.24	megagram per hectare, Mg/ha
miles per hour, mph	0.447	meter per second, m/s
gallon per minute (U.S.), gpm	0.227	cubic meter per hour, m^3/h
cubic feet per second, cfs	101.9	cubic meter per hour, m^3/h
Pressure		
atmosphere, atm	0.101	megapascal, MPa (10^6 Pa)
bar	0.1	megapascal, MPa
pound per square foot, lb/ft^2	47.9	pascal, Pa
pound per square inch, lb/in^2	6.9×10^3	pascal, Pa
Temperature		
degrees Fahrenheit (°F − 32)	0.556	degrees, °C
degrees Celsius (°C + 273)	1	Kelvin, K
Energy		
British thermal unit, Btu	1.05×10^3	joule, J
calorie, cal	4.19	joule, J
dyne, dyn	10^{-5}	newton, N
erg	10^{-7}	joule, J
foot-pound, ft-lb	1.36	joule, J
Concentrations		
percent, %	10	gram per kilogram, g/kg
part per million, ppm	1	milligram per kilogram, mg/kg
milliequivalents per 100 grams	1	centimole per kilogram, cmol/kg

[a] To convert from SI to non-SI units, *divide* by the factor given.

PERIODIC TABLE OF THE ELEMENTS WITH NOTES CONCERNING RELEVANCE TO SOIL SCIENCE

Based on atomic mass of $^{12}C = 12.0$. Numbers in parentheses are the mass numbers of the most stable isotopes of radioactive elements.

Group IA	Group IIA		Group IIIB	Group IVB	Group VB	Group VIB	Group VIIB	Group VIIIB	Group VIIIB	Group VIIIB	Group IB	Group IIB	Group IIIA	Group IVA	Group VA	Group VIA	Group VIIA	Group VIIIA
1 H 1.01 Hydrogen																		2 He 4.00 Helium
3 Li 6.94 Lithium	4 Be 9.01 Beryllium												5 B 10.81 Boron	6 C 12.01 Carbon	7 N 14.01 Nitrogen	8 O 16.00 Oxygen	9 F 19.00 Fluorine	10 Ne 20.18 Neon
11 Na 22.99 Sodium	12 Mg 24.30 Magnesium												13 Al 26.98 Aluminum	14 Si 28.09 Silicon	15 P 30.97 Phosphorus	16 S 32.07 Sulfur	17 Cl 35.45 Chlorine	18 Ar 39.95 Argon
19 K 39.10 Potassium	20 Ca 40.08 Calcium		21 Sc 44.96 Scandium	22 Ti 47.88 Titanium	23 V 50.94 Vanadium	24 Cr 52.00 Chromium	25 Mn 54.94 Manganese	26 Fe 55.85 Iron	27 Co 58.93 Cobalt	28 Ni 58.69 Nickel	29 Cu 63.55 Copper	30 Zn 65.38 Zinc	31 Ga 69.72 Gallium	32 Ge 72.59 Germanium	33 As 74.92 Arsenic	34 Se 78.96 Selenium	35 Br 79.90 Bromine	36 Kr 83.80 Krypton
37 Rb 85.47 Rubidium	38 Sr 87.62 Strontium		39 Y 88.91 Yttrium	40 Zr 91.22 Zirconium	41 Nb 92.91 Niobium	42 Mo 95.94 Molybdenum	43 Tc (98) Technetium	44 Ru 101.07 Ruthenium	45 Rh 102.91 Rhodium	46 Pd 106.42 Palladium	47 Ag 107.87 Silver	48 Cd 112.41 Cadmium	49 In 114.82 Indium	50 Sn 118.71 Tin	51 Sb 121.75 Antimony	52 Te 127.60 Tellurium	53 I 126.90 Iodine	54 Xe 131.29 Xenon
55 Cs 132.91 Cesium	56 Ba 137.33 Barium		57 La 138.91 Lanthanum	72 Hf 178.49 Hafnium	73 Ta 180.95 Tantalum	74 W 183.85 Tungsten	75 Re 186.21 Rhenium	76 Os 190.2 Osmium	77 Ir 192.22 Iridium	78 Pt 195.08 Platinum	79 Au 196.97 Gold	80 Hg 200.59 Mercury	81 Tl 204.38 Thallium	82 Pb 207.2 Lead	83 Bi 208.98 Bismuth	84 Po (209) Polonium	85 At (210) Astatine	86 Rn (222) Radon
87 Fr (223) Francium	88 Ra (226) Radium		89 Ac (227) Actinium	104 Unq (261) Unnilquadium	105 Unp (262) Unnilpentium	106 Unh (263) Unnilhexium	107 Uns (262) Unnilseptium	108 Uno (265) Unniloctium										

58 Ce 140.12 Cerium	59 Pr 140.91 Praseodymium	60 Nd 144.24 Neodymium	61 Pm (145) Promethium	62 Sm 150.36 Samarium	63 Eu 151.96 Europium	64 Gd 157.25 Gadolinium	65 Tb 158.93 Terbium	66 Dy 162.50 Dysprosium	67 Ho 164.93 Holmium	68 Er 167.26 Erbium	69 Tm 168.93 Thulium	70 Yb 173.04 Ytterbium	71 Lu 174.97 Lutetium
90 Th (232) Thorium	91 Pa (231) Protactinium	92 U (238) Uranium	93 Np (237) Neptunium	94 Pu (244) Plutonium	95 Am (243) Americium	96 Cm (247) Curium	97 Bk (247) Berkelium	98 Cf (251) Californium	99 Es (252) Einsteinium	100 Fm (257) Fermium	101 Md (258) Mendelevium	102 No (259) Nobelium	103 Lr (260) Lawrencium

Atomic number
Symbol — 87
Fr ← (223)
Francium
Atomic mass

■ Elements known to be nutrients for animals or plants. Some are also toxic in excessive amounts.
▨ Elements toxic to organisms in small amounts, and not known to serve as nutrients.
□ Other elements commonly studied in Soil Science because of soil-environmental impacts or because of their use as tracers or electrodes. (Br is used to trace anionic solutes such as nitrate. Isotopes of Rb and Sr are used to trace K and Ca in plants and soils. Cs and Ti are used to trace geological processes such as soil erosion. Pt and Ag are used in electrodes for measuring soil redox potential and pH, respectively.)

These 22 elements are needed as essential mineral nutrients by humans: macronutrients (calcium, chloride, magnesium, phosphorus, potassium, sodium, and sulfur) and micronutrients (chromium, cobalt, copper, fluoride, iodine, iron, manganese, molybdenum, nickel, selenium, silicon, tin, vanadium, zinc).

Plants Mentioned in This Text: Their Common and Scientific Names

acacia, apple ring	*Faidherbia albida* (Del.) A. Chev. [syn. *Acacia albida*]	carrot	*Daucus carota* L. ssp. *sativus* (Hoffm.) Arcang.
acacia, catclaw	*Acacia greggii* Gray	cassava	*Manihot esculenta* Crantz
alder	*Alnus* spp. P. Mill.	casuarina (sheoak)	*Casuarina* spp. Rumph. ex L.
alder, red	*Alnus rubra* Bong.	cattail, common	*Typa latifolia* L.
alfalfa	*Medicago sativa* L.	cauliflower	*Brassica oleracea* L. (Botrytis group)
alkali grass, Nutall's	*Puccinellia nuttalliana* (J.A. Schultes) A.S. Hitchc.	ceanothus	*Ceanothus* spp. L.
alkali sacaton	*Sporobolus airoides* (Torr.) Torr.	celery	*Apium graveolens* L. var. *dulce* (Mill.) Pers.
almond	*Prunus dulcis* (P. Mill.) D.A. Webber	cherry, flowering	*Prunus serrulata* Lindl.
andromeda (bog rosemary)	*Andromeda polifolia* L.	citrus	*Citrus* spp. L.
		clover, alsike	*Trifolium hybridum* L.
apple	*Malus* spp. P. Mill.	clover, berseem	*Trifolium alexandrinum* L.
apricot	*Prunus armeniaca* L.	clover, crimson	*Trifolium incarnatum* L.
arborvitae	*Thuja occidentalis* L.	clover, ladino	*Trifolium repens* L.
ash	*Fraxinus* spp. L.	clover, red	*Trifolium pratense* L.
ash, white	*Fraxinus americana* L.	clover, strawberry	*Trifolium fragiferum* L.
asparagus	*Asparagus officinalis.* L.	clover, sweet	*Melilotus indica* All
aspen	*Populus* spp. L.	clover, white	*Trifolium repens* L.
aspen, quaking	*Populus tremuloides* Michx.	coffee	*Coffea* spp. L.
autumn olive	*Elaeagnus umbellata* Thunb.	corn	*Zea maiz* L.
azalea	*Rhododendron* spp. L.	cotton	*Gossypium hirsutum* L.
azolla	*Azolla* spp. L.	cottonwood	*Populus deltoidies* Bartr. Ex. Marsh
bahia grass	*Paspalum notatum* Flueggé		
banana	*Musa acuminata* Colla	cowpea	*Vigna unguiculata* (L.) Walp.
barley, forage	*Hordeum vulgare* L.	cranberry	*Vaccinium macrocarpon* Ait.
bean, broad (faba)	*Vicia faba* L.	cranberry, small	*Vaccinium oxycoccos* L.
bean, common	*Phaseolus vulgaris* L.	cucumber	*Cucumis sativus* L.
bean, winged	*Psophocarpus tetragonobus* L. D.C.	currant	*Ribes* spp. L.
		cypress, bald	*Taxodium distichum* (L.) L.C. Rich.
beech, American	*Fagus grandifolia* Ehrh.		
beet, garden	*Beta procumbens* L.	dallisgrass	*Paspalum dilatatum* Poir
beet, sugar	*Beta vulgaris* L.	dogwood	*Cornus* spp. L.
bentgrass	*Agrostis stolonifera* L.	dogwood, grey	*Cornus racemosa* Lam.
bermudagrass	*Cynodon dactylon* (L.) Pers.	elaeagnus	*Elaeagnus* spp. L.
birch	*Betula* spp. L.	elm	*Ulmus* spp. L.
birch, black	*Betula lenta* L.	elm, American	*Ulmus americana* L.
black cherry	*Prunus serotina* Ehrh.	eucalyptus	*Eucalyptus* spp.
black locust	*Robinia pseudoacacia* L.	eucalyptus (jarrah)	*Eucalyptus marginata* Donn ex Sm.
blackberry	*Rubus* spp. L.		
blueberry	*Vaccinium* spp. L.	fescue	*Festuca* spp. L.
bluegrass, Kentucky	*Poa pratensis* L. ssp. *pratensis*	fescue, meadow	*Festuca pratensis* Huds.
bog rosemary	*Andromeda polifolia* L.	fescue, red	*Festuca rubra* L.
bougainvillea	*Bougainvillea* spp. Comm. ex Juss.	fescue, sheep	*Festuca ovina* L.
		fescue, tall	*Festuca elatior* L.
boxwood	*Buxus* spp. L.	fig	*Ficus carica* L.
broccoli	*Brassica oleracea* L. var. *botrytis* L.	filbert	*Corylus* spp. L.
		fir, Douglas	*Pseudotsuga menziesii* (Mirbel) Franco
brome grass	*Bromus* spp. L.		
buckwheat	*Eriogonum* spp. Michx.	gamagrass, eastern	*Tripsacum dactyloides* (L.) L.
buffalo grass	*Buchloe dactyloides* (Nutt.) Engelm.	gliricidia (quickstick)	*Gliricidia sepium* (Jacq.) Kunth ex Walp.
cabbage	*Brassica oleracea* L.	grape	*Vitus* spp. L.
canola (rapeseed)	*Brassica napus* L.	grapefruit	*Citrus paradisi* Macfad. (pro sp.) [*maxima sinensis*]
cantaloupe	*Cucumis melo* L.		

(continued)

Plants Mentioned in This Text: Their Common and Scientific Names (Continued)

Common	Scientific	Common	Scientific
grevillea	*Grevillea* spp. R. Br. ex Knight	mulberry	*Morus* spp. L.
groundnut (peanut)	*Arachis hypogaea* L.	myrica	*Myrica* spp. L.
guayule	*Parthenium argentatum* Gray	nut trees (e.g., almonds, hazelnuts)	*Prunus dulcis* (P. Mill.) D.A. Webber
gunnera	*Gunnera* spp. L.		*Corylus* spp. L.
harding grass	*Phalaris tuberosa* L. var. *stenoptera* (Hack) A.S. Hitchc.	needlegrass	
hemlock, Canadian	*Tsuga canadensis* (L.) Carr.	oak	*Quercus* spp. L.
hemlock, Carolina	*Tsuga caroliniana* Engelm.	oak, blackjack	*Quercus marilandica* Muenchh.
hibiscus	*Hibiscus* spp. L.	oak, California scrub	
hickory, bitternut	*Carya cordiformis* (Wangenh.) K. Koch	oak, chestnut	*Quercus prinus* L.
		oak, northern red	*Quercus rubra* L.
hickory, shagbark	*Carya ovata* (P. Mill.) K. Koch	oak, pin	*Quercus palustrus* Muenchh.
holly, American	*Ilex opaca* Ait.	oak, southern red	*Quercus falcata* Michx.
holly, burford	*Ilex cornuta* Lindl. & Paxton	oak, swamp white	*Quercus bicolor* Wild.
honeysuckle	*Lonicera* spp. L.	oak, white	*Quercus alba* L.
hydrangea	*Hydrangea* spp. L.	oak, willow	*Quercus pellos* L.
ipil ipil tree	*Leucaena leucocephala* Benth.	oats	*Avena sativa* L.
jarrah	*Eucalyptus marginata* Donn ex Sm.	olive	*Olea europaea* L.
		onion	*Allium cepa* L.
jojoba	*Simmondsia chinensis* (Link) Schneid.	orange	*Citrus sinensis* (L.) Osbeck
		orchard grass	*Dactylis glomerata* L.
juniper	*Juniperus* spp. L.	pangola grass	*Digitaria eriantha* Steud.
kale	*Brassica oleracea* L. (Acephala group)	pueraria, kudzu	*Pueraria phaseoloides* (Roxb.) Benth.
kallargrass	*Leptochloa fusca* (L.) Kunth [syn. *Diplachne fusca* Beauv.]	pea	*Pisum sativa* L.
		pea, pigeon	*Cajanus cajan* (L.) Millsp.
kenaf	*Hibiscus cannabinus* L.	peach	*Prunus persica* (L.) Batsch
kochia, prostrate	*Kochia prostrata* (L.) Schrad.	peanut	*Arachis hypogaea* L.
kudzu	*Pueraria montana* (Lour.) Merr. Var. *lobata* (Wild.)	pear	*Pyrus communis* L.
		pecan	*Carya illinoinensis* (Wangenh.) K. Koch
larch	*Larix* spp. P. Mill.		
lemon	*Citrus limon* (L.) Burm. F.	phragmities reed	*Phragmities australis* (Cav.) Trin. Ex Steud.
lespedeza	*Lespedeza* spp. Michx.		
lettuce	*Lactuca sativa* L.	pine, loblolly	*Pinus taeda* L.
leucaena (lead tree)	*Leucaena* spp. Benth.	pine, Monterey	*Pinus radiata* D. Don
lilac	*Syringa* spp. L.	pine, ponderosa	*Pinus ponderosa* Dougl. Ex P. & C. Laws.
linden	*Tillia* spp. L.		
locust, black	*Robinia pseudoacacia* L.	pine, red	*Pinus resinosa* Ait.
locust, honey	*Gleditsia triacanthos* L.	pine, white	*Pinus strobus* L.
lovegrass, weeping	*Eragrostis curvula* (Schrad.) Nees	pine, white Scotch	*Pinus sylvestris* L.
		pineapple	*Ananas comosus* (L.) Merrill
lupine	*Lupinus* spp. L.	pitcher plant	*Sarracenia* spp. L.
magnolia	*Magnolia* spp. L.	plum (prune)	*Prunus domestica* L.
maiden cane	*Panicum hemitomon* J.A. Schultes	poinsettia	*Euphorbia pulcherrima* Willd. ex Klotzsch
maize (corn)	*Zea mays* L.	pomegranate	*Punica granatum* L.
mandarin orange	*Citrus reticulata* Blanco	poplar	*Populus* spp. L.
manzanita		potato	*Solanum tuberosum* L.
maple	*Acer* spp. L.	potato, sweet	*Ipomoea batatas* (L.) Lam.
maple, red	*Acer rubrum* L.	povertygrass	*Danthonia spicata* (L.) Beauv. Ex Roem. & Schult.
maple, sugar	*Acer saccharum* Marsh.		
marigold	*Tagetes* spp.	privet	*Ligustrum* spp. L.
milkvetch	*Astragalus* spp.	quickstick	*Gliricidia sepium* (Jacq.) Kunth ex Walp.
mosquito fern	*Azolla* spp. L.		
mountain brome grass	*Bromus breviaristatus* Buckl.	radish	*Raphanus sativus* L.
mountain laurel	*Kalmia latifolia* L.		

(continued)

PLANTS MENTIONED IN THIS TEXT: THEIR COMMON AND SCIENTIFIC NAMES (CONTINUED)

rapeseed (see also canola)	*Brassica campestris* L. [syn. *B. rapa* L.]	sumac	*Rhus* spp. L.
raspberry	*Rubus idaeus* L.	sunflower	*Helianthus annuus* L.
red top	*Agrostis alba* L.	sycamore	*Plantus occidentalis* L.
red top	*Agrostis gigantea* Roth	sweetgum	
reed canarygrass	*Phalaris arundinacea* L.	tamarix (tamarisk)	*Tamarix gallica* L.
rescuegrass	*Bromus catharticus* Vahl	tea	*Camellia sinensis* (L.) O. Kuntze
rhododendron	*Rhododendron* spp. L.	teaberry	*Gaultheria procumbens* L.
rice	*Oryza* spp. L.	timothy	*Phleum pratense* L.
rice (paddy)	*Oryza sativa* L.	tithonia	*Tithonia diversifolia* (Hemsl.) Gray
riverhemp, Egyptian	*Sesbania sesban* (L.) Merr.	tobacco	*Nicotiana* spp. L.
rose-mallow, swamp	*Hibiscus moscheutos* L.	tomato	*Solanum lycopersicum* L.
rosemary	*Rosmarinus officinalis* L.	tree marigold	*Tithonia diversifolia* (Hemsl.) Gray
roses	*Rosa* spp. L.		
rye (grain, forage)	*Secale cereale* L.	trefoil, birdsfoot	*Lotus corniculatus* L.
rye, wild	*Elymus* spp.	tulip poplar	*Liriodendron tulipifera* L.
ryegrass, perennial	*Lolium perenne* L.	turnip	*Brassica rapa* L. (Rapifera group)
safflower	*Carthamus tinctorius* L.	velvetleaf	*Abutilon theophrasti* Medik.
saltgrass, desert	*Distichlis spicta* L. var. *stricta* (Torr.) Bettle	vetch	*Vicia* spp. L.
		vetch, common	*Vicia angustifolia* L.
sesbania	*Sesbania sesban* (L.) Merr.	vetch, hairy	*Vicia villosa* Roth
skunk cabbage	*Symplocarpus foetidus* (L.) Salisb. Ex Nutt.	vetiver grass	*Vetiveria zizanioides* (L.) Nash ex Small
sorghum	*Sorghum bicolor* (L.) Moench	viburnum	*Viburnum* spp. L.
soy beans	*Glycine max* (L.) Merr.	walnut	*Juglans* spp. L.
spartina (cordgrass)	*Spartina* spp. Schreb.	water melon	*Citrullus lanatus* (Thunb.) Matsumura & Nakai
spinach	*Spinacia oleracea* L.		
spruce, black	*Picea mariana* (Mill.) B.S.P.	water tupelo	*Nyssa aquatica* L.
spruce, Norway	*Picea abies* (L.) Karst.	wheat	*Triticum aestivum* L.
spruce, red	*Picea rubens* Sarg.	wheatgrass, crested	*Agropyron sibiricum* (Willd.) Beauvois
spruce, white	*Picea glauca* (Moench) Voss		
squash	*Cucurbita pepo* L.	wheatgrass, fairway	*Agropyron cristatum* (L.) Gaertn.
squash (zucchini)	*Cucurbita pepo* L. var. *melopepo* (L.) Alef.	wheatgrass, tall	*Agropyron elongatum* (Hort) Beauvois
star jasmine	*Jasminum multiflorum* (Burm. f.) Andr	wheatgrass, western	*Pascopyrum smithii* (Rydb.) A. Löve
strawberry	*Fragaria x ananassa* Duch.	wild rye, altai	*Leymus angustus* (Trin.) Pilger
sudan grass	*Sorghum sudanense* (Piper) Stapf	wild rye, Russian	*Psathyrostachys juncea* (Fisch.) Nevski
sugar beet	*Beta vulgaris* L.	willow	*Salix* spp. L.
sugar cane	*Saccharum officinarum* L.	willow, black	*Salix nigra* L.

Glossary of Soil Science Terms[1]

A horizon The surface horizon of a mineral soil having maximum organic matter accumulation, maximum biological activity, and/or eluviation of materials such as iron and aluminum oxides and silicate clays.

abiotic Nonliving basic elements of the environment, such as rainfall, temperature, wind, and minerals.

accelerated erosion Erosion much more rapid than normal, natural, geological erosion; primarily as a result of the activities of humans or, in some cases, of animals.

acid cations Cations, principally Al^{3+}, Fe^{3+}, and H^+, that contribute to H^+ ion activity either directly or through hydrolysis reactions with water. *See also* nonacid cations.

acid rain Atmospheric precipitation with pH values less than about 5.6, the acidity being due to inorganic acids (such as nitric and sulfuric) that are formed when oxides of nitrogen and sulfur are emitted into the atmosphere.

acid saturation The proportion or percentage of a cation-exchange site occupied by acid cations.

acid soil A soil with a pH value <7.0. Usually applied to surface layer or root zone, but may be used to characterize any horizon. *See also* reaction, soil.

acid sulfate soils Soils that are potentially extremely acid (pH <3.5) because of the presence of large amounts of reduced forms of sulfur that are oxidized to sulfuric acid if the soils are exposed to oxygen when they are drained or excavated. A sulfuric horizon containing the yellow mineral jarosite is often present. *See also* cat clays.

acidity, active The activity of hydrogen ions in the aqueous phase of a soil. It is measured and expressed as a pH value.

acidity, residual Soil acidity that can be neutralized by lime or other alkaline materials but cannot be replaced by an unbuffered salt solution.

acidity, salt replaceable Exchangeable hydrogen and aluminum that can be replaced from an acid soil by an unbuffered salt solution such as KCl or NaCl.

acidity, total The total acidity in a soil. It is approximated by the sum of the salt-replaceable acidity plus the residual acidity.

Actinomycetes A group of bacteria that form branched mycelia that are thinner, but somewhat similar in appearance, to fungal hyphae. Includes many members of the order Actinomycetales.

activated sludge Sludge that has been aerated and subjected to bacterial action.

active layer The upper portion of a Gelisol that is subject to freezing and thawing and is underlain by permafrost.

active organic matter A portion of the soil organic matter that is relatively easily metabolized by microorganisms and cycles with a half-life in the soil of a few days to a few years.

adhesion Molecular attraction that holds the surfaces of two substances (e.g., water and sand particles) in contact.

adsorption The attraction of ions or compounds to the surface of a solid. Soil colloids adsorb large amounts of ions and water.

adsorption complex The group of organic and inorganic substances in soil capable of adsorbing ions and molecules.

aerate To impregnate with gas, usually air.

aeration, soil The process by which air in the soil is replaced by air from the atmosphere. In a well-aerated soil, the soil air is similar in composition to the atmosphere above the soil. Poorly aerated soils usually contain more carbon dioxide and correspondingly less oxygen than the atmosphere above the soil.

aerobic (1) Having molecular oxygen as a part of the environment. (2) Growing only in the presence of molecular oxygen, as aerobic organisms. (3) Occurring only in the presence of molecular oxygen (said of certain chemical or biochemical processes, such as aerobic decomposition).

aerosolic dust A type of eolian material that is very fine (about 1 to 10 μm) and may remain suspended in the air over distances of thousands of kilometers. Finer than most *loess*.

aggregate (soil) Many soil particles held in a single mass or cluster, such as a clod, crumb, block, or prism.

agric horizon A diagnostic subsurface horizon in which clay, silt, and humus derived from an overlying cultivated and fertilized layer have accumulated. Wormholes and illuvial clay, silt, and humus occupy at least 5% of the horizon by volume.

agroforestry Any type of multiple cropping land-use that entails complementary relations between trees and agricultural crops.

agronomy A specialization of agriculture concerned with the theory and practice of field-crop production and soil management. The scientific management of land.

air-dry (1) The state of dryness (of a soil) at equilibrium with the moisture content in the surrounding atmosphere. The actual moisture content will depend upon the relative humidity and the temperature of the surrounding atmosphere. (2) To allow to reach equilibrium in moisture content with the surrounding atmosphere.

air porosity The proportion of the bulk volume of soil that is filled with air at any given time or under a given condition, such as a specified moisture potential; usually the large pores.

albic horizon A diagnostic subsurface horizon from which clay and free iron oxides have been removed or in which the oxides have been segregated to the extent that the color of the horizon is determined primarily by the color of the primary sand and silt particles rather than by coatings on these particles.

Alfisols An order in *Soil Taxonomy*. Soils with gray to brown surface horizons, medium to high supply of bases, and B horizons of illuvial clay accumulation. These soils form mostly under forest or savanna vegetation in climates with slight to pronounced seasonal moisture deficit.

algal bloom A population explosion of algae in surface waters, such as lakes and streams, often resulting in high turbidity and green- or red-colored water, and commonly stimulated by nutrient enrichment with phosphorus and nitrogen.

alkaline soil Any soil that has pH > 7. Usually applied to the surface layer or root zone but may be used to characterize any horizon or a sample thereof. *See also* reaction, soil.

allelochemical An organic chemical by which one plant can influence another. *See* allelopathy.

allelopathy The process by which one plant may affect other plants by biologically active chemicals introduced into the soil, either directly by leaching or exudation from the source plant, or as a result of the decay of the plant residues. The effects, though usually negative, may also be positive.

allophane A poorly defined aluminosilicate mineral whose structural framework consists of short runs of three-dimensional crystals interspersed with amorphous noncrystalline materials. Along with its more weathered

---

[1] This glossary was compiled and modified from several sources, including *Online Glossary of Soil Science Terms* [Madison, Wis.: Soil Sci. Soc. Amer. (2015)] *Resource Conservation Glossary* [Anheny, Iowa: Soil Cons. Soc. Amer. (1982)], and *Soil Taxonomy* [Washington, D.C.: U.S. Department of Agriculture (1999)].

companion, it is prevalent in volcanic ash materials.

alluvial fan Fan-shaped alluvium deposited at the mouth of a canyon or ravine where debris-laden waters fan out, slow down, and deposit their burden.

alluvium A general term for all detrital material deposited or in transit by streams, including gravel, sand, silt, clay, and all variations and mixtures of these. Unless otherwise noted, alluvium is unconsolidated.

alpha particle A positively charged particle (consisting of two protons and two neutrons) that is emitted by certain radioactive compounds.

aluminosilicates Compounds containing aluminum, silicon, and oxygen as main constituents. An example is microcline, $KAlSi_3O_8$.

amendment, soil Any substance other than fertilizers, such as lime, sulfur, gypsum, and sawdust, used to alter the chemical or physical properties of a soil, generally to make it more productive.

amino acids Nitrogen-containing organic acids that couple together to form proteins. Each acid molecule contains one or more amino groups ($-NH_2$) and at least one carboxyl group ($-COOH$). In addition, some amino acids contain sulfur.

Ammanox A biochemical process in the N cycle by which certain anaerobic bacteria or archaea oxidize ammonium ions using nitrite ions as the electron acceptor, the main product being N_2 gas.

ammonification The biochemical process whereby ammoniacal nitrogen is released from nitrogen-containing organic compounds.

ammonium fixation The entrapment of ammonium ions by the mineral or organic fractions of the soil in forms that are insoluble in water and are at least temporarily nonexchangeable.

amorphous material Noncrystalline constituents of soils.

anaerobic (i) The absence of molecular oxygen. (ii) Growing or occurring in the absence of molecular oxygen (e.g., anaerobic bacteria or biochemical reduction reaction).

anaerobic respiration The metabolic process whereby electrons are transferred from a reduced compound (usually organic) to an inorganic acceptor molecule other than oxygen.

andic properties Soil properties related to volcanic origin of materials, including high organic carbon content, low bulk density, high phosphate retention, and extractable iron and aluminum.

Andisols An order in *Soil Taxonomy*. Soils developed from volcanic ejecta. The colloidal fraction is dominated by allophane and/or Al-humus compounds.

angle of repose The maximum slope steepness at which loose, cohesionless material will come to rest.

anion Negatively charged ion; during electrolysis it is attracted to the positively charged anode.

anion exchange Exchange of anions in the soil solution for anions adsorbed on the surface of clay and humus particles.

anion exchange capacity The sum total of exchangeable anions that a soil can adsorb. Expressed as centimoles of charge per kilogram ($cmol_c/kg$) of soil (or of other adsorbing material, such as clay).

anoxic *See* anaerobic.

anthropic epipedon A thick horizon that formed in material that was intentionally altered or transported by humans. It usually contains humans artifacts and/or is located in a human altered landscape feature (mounds, pit, terraces, etc.).

antibiotic A substance produced by one species of organism that, in low concentrations, will kill or inhibit growth of certain other organisms.

Ap The surface layer of a soil disturbed by cultivation or pasturing.

apatite A naturally occurring complex calcium phosphate that is the original source of most of the phosphate fertilizers. Formulas such as $[3Ca_3(PO_4)_2] \cdot CaF_2$ illustrate the complex compounds that make up apatite.

aquic conditions Continuous or periodic saturation (with water) and reduction, commonly indicated by redoximorphic features.

aquiclude A saturated body of rock or sediment that is incapable of transmitting significant quantities of water under ordinary water pressures.

aquifer A saturated, permeable layer of sediment or rock that can transmit significant quantities of water under normal pressure conditions.

arbuscular mycorrhiza A common endomycorrhizal association produced by phycomycetous fungi and characterized by the development, within root cells, of small structures known as *arbuscules*. Some also form, between root cells, storage organs known as *vesicles*. Host range includes many agricultural and horticultural crops. Formerly called vesicular arbuscular mycorrhiza (VAM). *See also* endotrophic mycorrhiza.

arbuscule Specialized branched structure formed within a root cortical cell by endotrophic mycorrhizal fungi.

Archaea One of the two domains of single-celled prokaryote microorganisms. Includes organisms adapted to extremes of salinity and heat, and those that subsist on methane. Similar appearing, but evolutionarily distinct from bacteria.

argillan A thin coating of well-oriented clay particles on the surface of a soil aggregate, particle, or pore. A clay film.

argillic horizon A diagnostic subsurface horizon characterized by the illuvial accumulation of layer-lattice silicate clays.

arid climate Climate in regions that lack sufficient moisture for crop production without irrigation. In cool regions annual precipitation is usually less than 25 cm. It may be as high as 50 cm in tropical regions. Natural vegetation is desert shrubs.

Aridisols An order in *Soil Taxonomy*. Soils of dry climates. They have pedogenic horizons, low in organic matter, that are never moist for as long as three consecutive months. They have an ochric epipedon and one or more of the following diagnostic horizons: argillic, natric, cambic, calcic, petrocalcic, gypsic, petrogypsic, salic, or a duripan.

aspect (of slopes) The direction (e.g., south or north) that a slope faces with respect to the sun.

association, soil *See* soil association.

Atterberg limits Water contents of fine-grained soils at different states of consistency.

liquid limit (LL) The water content corresponding to the arbitrary limit between the liquid and plastic states of consistency of a soil.

plastic limit (PL) The water content corresponding to an arbitrary limit between the plastic and semisolid states of consistency of a soil.

autochthonous organisms Those microorganisms thought to subsist on the more resistant soil organic matter and little affected by the addition of fresh organic materials. *Contrast with* zymogenous organisms. *See also* k-strategist.

autotroph An organism capable of utilizing carbon dioxide or carbonates as the sole source of carbon and obtaining energy for life processes from the oxidation of inorganic elements or compounds such as iron, sulfur, hydrogen, ammonium, and nitrites, or from radiant energy. *Contrast with* heterotroph.

available nutrient That portion of any element or compound in the soil that can be readily absorbed and assimilated by growing plants. ("Available" should not be confused with "exchangeable.")

available water The portion of water in a soil that can be readily absorbed by plant roots. The amount of water released between the field capacity and the permanent wilting point.

B horizon A soil horizon, usually beneath the A or E horizon, that is characterized by one or more of the following: (1) a concentration of soluble salts, silicate clays, iron and aluminum oxides, and humus, alone or in combination; (2) a blocky or prismatic structure; and (3) coatings of iron and aluminum oxides that give darker, stronger, or redder color.

Bacteria One of two domains of single-celled prokaryote microorganisms. Includes all that are not Archaea.

bar A unit of pressure equal to 1 million dynes per square centimeter (10^6 dynes/cm^2). It approximates the pressure of a standard atmosphere.

base flow The flux or flow of water in a stream fed by groundwater during periods without precipitation. Opposite of *storm flow*.

base-forming cations (Obsolete) Those cations that form strong (strongly dissociated) bases by reaction with hydroxyl; e.g., K^+ forms potassium hydroxide ($K^+ + OH$). See nonacid cations.

base saturation percentage The extent to which the adsorption complex of a soil is saturated with exchangeable cations other than hydrogen and aluminum. It is expressed as a percentage of the total cation exchange capacity. *See* nonacid saturation.

bedding (Engineering) Arranging the surface of fields by plowing and grading into a series of elevated beds separated by shallow depressions or ditches for drainage.

bedrock The solid rock underlying soils and the regolith in depths ranging from zero (where exposed by erosion) to several hundred feet.

bench terrace An embankment constructed across sloping fields with a steep drop on the downslope side.

beta particle A high-speed electron emitted in radioactive decay.

bioaccumulation A buildup within an organism of specific compounds due to biological processes. Commonly applied to heavy metals, pesticides, or metabolites.

bioaugmentation The cleanup of contaminated soils by adding exotic microorganisms that are especially efficient at breaking down an organic contaminant. A form of *bioremediation*.

biochar A black carbon condensate product purposefully made by heating organic material at 300 to 700°C under low oxygen conditions.

biodegradable Subject to degradation by biochemical processes.

biological nitrogen fixation Occurs at ordinary temperatures and pressures. It is commonly carried out by certain bacteria, algae, and actinomycetes, which may or may not be associated with higher plants.

biomass The total mass of living material of a specified type (e.g., microbial biomass) in a given environment (e.g., in a cubic meter of soil).

biopores Soil pores, usually of relatively large diameter, created by plant roots, earthworms, or other soil organisms.

bioremediation The decontamination or restoration of polluted or degraded soils by means of enhancing the chemical degradation or other activities of soil organisms.

biosequence A group of related soils that differ, one from the other, primarily because of differences in kinds and numbers of plants and soil organisms as a soil-forming factor.

biosolids Sewage sludge that meets certain regulatory standards, making it suitable for land application. *See* sewage sludge.

biostimulation The cleanup of contaminated soils through the manipulation of nutrients or other soil environmental factors to enhance the activity of naturally occurring soil microorganisms. A form of *bioremediation*.

blocky soil structure Soil aggregates with blocklike shapes; common in B horizons of soils in humid regions.

broad-base terrace A low embankment with such gentle slopes that it can be farmed, constructed across sloping fields to reduce erosion and runoff.

broadcast Scatter seed or fertilizer on the surface of the soil.

brownfields Abandoned, idled, or underused industrial and commercial facilities where expansion or redevelopment is complicated by real or perceived environmental contamination.

buffering capacity The ability of a soil to resist changes in pH. Commonly determined by presence of clay, humus, and other colloidal materials.

bulk blended fertilizers Solid fertilizer materials blended together in small blending plants, delivered to the farm in bulk, and usually spread directly on the fields by truck or other special applicator.

bulk blending Mixing dry individual granulated fertilizer materials to form a mixed fertilizer that is applied promptly to the soil.

bulk density, soil The mass of dry soil per unit of bulk volume, including the air space. The bulk volume is determined before drying to constant weight at 105°C.

buried soil Soil covered by an alluvial, loessal, or other deposit, usually to a depth greater than the thickness of the solum.

by-pass flow *See* preferential flow.

C horizon A mineral horizon, generally beneath the solum, that is relatively unaffected by biological activity and pedogenesis and is lacking properties diagnostic of an A or B horizon. It may or may not be like the material from which the A and B have formed.

calcareous soil Soil containing sufficient calcium carbonate (often with magnesium carbonate) to effervesce visibly when treated with cold 0.1 N hydrochloric acid.

calcic horizon A diagnostic subsurface horizon of secondary carbonate enrichment that is more than 15 cm thick, has a calcium carbonate equivalent of more than 15%, and has at least 5% more calcium carbonate equivalent than the underlying C horizon.

caliche A layer near the surface, more or less cemented by secondary carbonates of calcium or magnesium precipitated from the soil solution. It may occur as a soft, thin soil horizon; as a hard, thick bed just beneath the solum; or as a surface layer exposed by erosion.

cambic horizon A diagnostic subsurface horizon that has a texture of loamy very fine sand or finer, contains some weatherable minerals, and is characterized by the alteration or removal of mineral material. The cambic horizon lacks cementation or induration and has too few evidences of illuviation to meet the requirements of the argillic or spodic horizon.

capillary conductivity (Obsolete) *See* hydraulic conductivity.

capillary fringe A zone in the soil just above the plane of zero water pressure (water table) that remains saturated or almost saturated with water.

capillary water The water held in the capillary or *small* pores of a soil, usually with a tension >60 cm of water. *See also* soil water potential.

carbon cycle The sequence of transformations whereby carbon dioxide is fixed in living organisms by photosynthesis or by chemosynthesis, liberated by respiration and by the death and decomposition of the fixing organism, used by heterotrophic species, and ultimately returned to its original state.

carbon/nitrogen ratio The ratio of the weight of organic carbon (C) to the weight of total nitrogen (N) in a soil or in organic material.

carnivore An organism that feeds on animals.

casts, earthworm Rounded, water-stable aggregates of soil that have passed through the gut of an earthworm.

cat clays Wet clay soils high in reduced forms of sulfur that, upon being drained, become extremely acid because of the oxidation of the sulfur compounds and the formation of sulfuric acid. Usually found in tidal marshes. *See* acid sulfate soils.

catena A group of soils that commonly occur together in a landscape, each characterized by a different slope position and resulting set of drainage-related proprieties. *See also* toposequence.

cation A positively charged ion; during electrolysis it is attracted to the negatively charged cathode.

cation exchange The interchange between a cation in solution and another cation on the surface of any surface-active material, such as clay or organic matter.

cation exchange capacity The sum total of exchangeable cations that a soil can adsorb. Sometimes called *total-exchange capacity, base-exchange capacity*, or *cation-adsorption capacity*. Expressed in centimoles of charge per kilogram (cmol$c$/kg) of soil (or of other adsorbing material, such as clay).

cemented Indurated; having a hard, brittle consistency because the particles are held together by cementing substances, such as humus, calcium carbonate, or the oxides of silicon, iron, and aluminum.

channery Thin, flat fragments of limestone, sandstone, or schist up to 15 cm (6 in.) in major diameter.

char An important component of stable soil organic carbon created by natural fires that heat organic material under low oxygen conditions to cause charring rather than burning. Char generally has very high aromaticity, surface area, and water holding capacity.

chelate (Greek, claw) A type of chemical compound in which a metallic ion is firmly combined with an organic molecule by means of multiple chemical bonds.

chert A structureless form of silica, closely related to flint, that breaks into angular fragments.

chisel, subsoil A tillage implement with one or more cultivator-type feet to which are attached strong knifelike units used to shatter or loosen hard, compact layers, usually in the subsoil, to depths below normal plow depth. *See also* subsoiling.

chlorite A 2:1:1-type layer-structured silicate mineral having 2:1 layers alternating with a magnesium-dominated octahedral sheet.

chlorosis A condition in plants relating to the failure of chlorophyll (the green coloring matter) to develop. Chlorotic leaves range from light green through yellow to almost white.

chroma (color) *See* Munsell color system.

chronosequence A sequence of related soils that differ, one from the other, in certain properties primarily as a result of time as a soil-forming factor.

classification, soil *See* soil classification.

clay (1) A soil separate consisting of particles <0.002 mm in equivalent diameter. (2) A soil textural class containing >40% clay, <45% sand, and <40% silt.

clay mineral Naturally occurring inorganic material (usually crystalline) found in soils and other earthy deposits, the particles being of clay size, that is, <0.002 mm in diameter.

claypan A dense, compact, slowly permeable layer in the subsoil having a much higher clay content than the overlying material, from which it is separated by a sharply defined boundary. Claypans are usually hard when dry and plastic and sticky when wet. *See also* hardpan.

climosequence A group of related soils that differ, one from another, primarily because of differences in climate as a soil-forming factor.

clod A compact, coherent mass of soil produced artificially, usually by such human activities as plowing and digging, especially when these operations are performed on soils that are either too wet or too dry for normal tillage operations.

coarse fragments Mineral (rock) soil particles larger than 2 mm in diameter. *Compare to* fine earth fraction.

coarse texture The texture exhibited by sands, loamy sands, and sandy loams (except very fine sandy loam).

cobblestone Rounded or partially rounded rock or mineral fragments 7.5 to 25 cm (3 to 10 in.) in diameter.

co-composting A method of composting in which two materials of differing but complementary nature are mingled together and enhance each other's decomposition in a compost system.

cohesion Holding together: force holding a solid or liquid together, owing to attraction between like molecules. Decreases with rise in temperature.

collapsible soil Certain soil that may undergo a sudden loss in strength when wetted.

colloid, soil (Greek, gluelike) Organic and inorganic matter with very small particle size and a correspondingly large surface area per unit of mass.

colluvium A deposit of rock fragments and soil material accumulated at the base of steep slopes as a result of gravitational action.

color The property of an object that depends on the wavelength of light it reflects or emits.

columnar soil structure *See* soil structure types.

companion planting The practice of growing certain species of plants in close proximity because one species has the effect of improving the growth of the other, sometimes by positive *allelopathic* effects.

compost Organic residues, or a mixture of organic residues and soil, that have been piled, moistened, and allowed to undergo biological decomposition. Mineral fertilizers are sometimes added. Often called *artificial manure* or *synthetic manure* if produced primarily from plant residues.

concretion A local concentration of a chemical compound, such as calcium carbonate or iron oxide, in the form of grains or nodules of varying size, shape, hardness, and color.

conduction The transfer of heat by physical contact between two or more objects.

conductivity, hydraulic *See* hydraulic conductivity.

conservation tillage *See* tillage, conservation.

consistence The combination of properties of soil material that determine its resistance to crushing and its ability to be molded or changed in shape. Such terms as *loose, friable, firm, soft, plastic*, and *sticky* describe soil consistence.

consistency The interaction of adhesive and cohesive forces within a soil at various moisture contents as expressed by the relative ease with which the soil can be deformed or ruptured.

consociation *See* soil consociation.

consolidation test A laboratory test in which a soil mass is laterally confined within a ring and is compressed with a known force between two porous plates.

constant charge The net surface charge of mineral particles, the magnitude of which depends only on the chemical and structural composition of the mineral. The charge arises from isomorphous substitution and is not affected by soil pH.

consumptive use The water used by plants in transpiration and growth, plus water vapor loss from adjacent soil or snow, or from intercepted precipitation in any specified time. Usually expressed as equivalent depth of free water per unit of time.

contour An imaginary line connecting points of equal elevation on the surface of the soil. A contour terrace is laid out on a sloping soil at right angles to the direction of the slope and nearly level throughout its course.

contour strip-cropping Layout of crops in comparatively narrow strips in which the farming operations are performed approximately on the contour. Usually strips of grass, close-growing crops, or fallow are alternated with those of cultivated crops.

controlled traffic A farming system in which all wheeled traffic is confined to fixed paths so that repeated compaction of the soil does not occur outside the selected paths.

convection The transfer of heat through a gas or solution because of molecular movement.

cover crop A close-growing crop grown primarily for the purpose of protecting and improving soil between periods of regular crop production or between trees and vines in orchards and vineyards.

creep Slow mass movement of soil and soil material down relatively steep slopes, primarily under the influence of gravity, but facilitated by saturation with water and by alternate freezing and thawing.

crop rotation A planned sequence of crops growing in a regularly recurring succession on the same area of land, as contrasted to continuous culture of one crop or growing different crops in haphazard order.

crotovina A former animal burrow in one soil horizon that has been filled with organic matter or material from another horizon (also spelled *krotovina*).

crumb A soft, porous, more or less rounded natural unit of structure from 1 to 5 mm in diameter. *See also* soil structure types.

crushing strength The force required to crush a mass of dry soil or, conversely, the resistance of the dry soil mass to crushing. Expressed in units of force per unit area (pressure).

crust (soil) (i) physical A surface layer on soils, ranging in thickness from a few millimeters to as much as 3 cm, that physical-chemical processes have caused to be much more compact, hard, and brittle when dry than the material immediately beneath it.

(ii) microbiotic An assemblage of cyanobacteria, algae, lichens, liverworts, and mosses that commonly forms an irregular crust on the soil surface, especially on otherwise barren, arid-region soils. Also referred to as cryptogamic, cryptobiotic, or biological crusts.

cryophilic Pertaining to low temperatures in the range of 5 to 15°C, the range in which cryophilic organisms grow best.

cryoturbation Physical disruption and displacement of soil material within the profile by the forces of freezing and thawing. Sometimes called *frost churning*, it results in irregular, broken horizons, involutions, oriented rock fragments, and accumulation of organic matter on the permafrost table.

cryptogam *See* crust (ii) microbiotic.

crystal A homogeneous inorganic substance of definite chemical composition bounded by planar surfaces that form definite angles with each other, thus giving the substance a regular geometrical form.

crystal structure The orderly arrangement of atoms in a crystalline material.

cultivation A tillage operation used in preparing land for seeding or transplanting or later for weed control and for loosening the soil.

cutans A modification of the texture, structure, or fabric at natural surfaces in soil materials due to concentration of particular soil constituents; e.g. "clay skins."

cyanobacteria Chlorophyll-containing bacteria that accommodate both photosynthesis and nitrogen fixation. Formerly called blue-green algae.

deciduous plant A plant that sheds all its leaves every year at a certain season.

decomposition Chemical breakdown of a compound (e.g., a mineral or organic compound) into simpler compounds, often accomplished with the aid of microorganisms.

deflocculate (1) To separate the individual components of compound particles by chemical and/or physical means. (2) To cause the particles of the *disperse phase* of a colloidal system to become suspended in the *dispersion medium*.

delineation An individual polygon shown by a closed boundary on a soil map that defines the area, shape, and location of a map unit within a landscape.

delivery ratio The ratio of eroded sediment carried out of a drainage basin to the total amount of sediment moved within the basin by erosion processes.

delta An alluvial deposit formed where a stream or river drops its sediment load upon entering a quieter body of water.

denitrification The biochemical reduction of nitrate or nitrite to gaseous nitrogen, either as molecular nitrogen or as an oxide of nitrogen.

density *See* particle density; bulk density.

depletion zone The volume of soil solution adjacent to plant roots where the concentration of an element has been reduced by plant uptake that is faster than diffusion can replace the element from the bulk solution. Often used in reference to phosphorus or potassium.

desalinization Removal of salts from saline soil, usually by leaching.

desert crust A hard layer, containing calcium carbonate, gypsum, or other binding material, exposed at the surface in desert regions.

desert pavement A natural residual concentration of closely packed pebbles, boulders, and other rock fragments on a desert surface where wind and water action has removed all smaller particles.

desert varnish A thin, dark, shiny film or coating of iron oxide and lesser amounts of manganese oxide and silica formed on the surfaces of pebbles, boulders, rock fragments, and rock outcrops in arid regions.

desorption The removal of sorbed material from surfaces.

detritivore An organism that subsists on detritus.

detritus Debris from dead plants and animals.

diagnostic horizons (As used in *Soil Taxonomy*): Horizons having specific soil characteristics that are indicative of certain classes of soils. Horizons that occur at the soil surface are called *epipedons*; those below the surface, *diagnostic subsurface horizons*.

diatomaceous earth A geologic deposit of fine, grayish, siliceous material composed chiefly or wholly of the remains of diatoms. It may occur as a powder or as a porous, rigid material.

diatoms Algae having siliceous cell walls that persist as a skeleton after death; any of the microscopic unicellular or colonial algae constituting the class Bacillariaceae. They occur abundantly in fresh and salt waters and their remains are widely distributed in soils.

diffusion The movement of atoms in a gaseous mixture or of ions in a solution, primarily as a result of their own random motion.

dioctahedral sheet An octahedral sheet of silicate clays in which the sites for the six-coordinated metallic atoms are mostly filled with trivalent atoms, such as Al^{3+}.

disintegration Physical or mechanical breakup or separation of a substance into its component parts (e.g., a rock breaking into its mineral components).

disperse (1) To break up compound particles, such as aggregates, into the individual component particles. (2) To distribute or suspend fine particles, such as clay, in or throughout a dispersion medium, such as water.

dissimilatory nitrate reduction to ammonium (DNRA) A bacterial process by which nitrate is converted to ammonium under a wide range of oxygen and carbon levels. Compare to dentrification (a different type of dissimilatory nitrate reduction) which is strictly anaerobic and requires an energy source.

dissolution Process by which molecules of a gas, solid, or another liquid dissolve in a liquid, thereby becoming completely and uniformly dispersed throughout the liquid's volume.

distribution coefficient (K_d) The distribution of a chemical between soil and water.

diversion terrace *See* terrace.

drain (1) To provide channels, such as open ditches or drain tile, so that excess water can be removed by surface or by internal flow. (2) To lose water (from the soil) by percolation.

drain field, septic tank An area of soil into which the effluent from a septic tank is piped so that it will drain through the lower part of the soil profile for disposal and purification.

drainage, soil The frequency and duration of periods when the soil is free from saturation with water.

drift Material of any sort deposited by geological processes in one place after having been removed from another. Glacial drift includes material moved by the glaciers and by the streams and lakes associated with them.

drumlin Long, smooth, cigar-shaped low hills of glacial till, with their long axes parallel to the direction of ice movement.

dryland farming The practice of crop production in low-rainfall areas without irrigation.

duff The matted, partly decomposed organic surface layer of forest soils.

duripan A diagnostic subsurface horizon that is cemented by silica, to the point that air-dry fragments will not slake in water or HCL. Hardpan.

dust mulch A loose, finely granular or powdery condition on the surface of the soil, usually produced by shallow cultivation.

E horizon Horizon characterized by maximum illuviation (washing out) of silicate clays and iron and aluminum oxides; commonly occurs above the B horizon and below the A horizon.

earthworms Animals of the Lumbricidae family that burrow into and live in the soil. They mix plant residues into the soil and improve soil aeration.

ecosystem A dynamic and interacting combination of all the living organisms and nonliving elements (matter and energy) of an area.

ecosystem services Products of natural ecosystems that support and fulfill the needs of human beings. Provision of clean water and unpolluted air are examples.

ectotrophic mycorrhiza (ectomycorrhiza) A symbiotic association of the mycelium of fungi and the roots of certain plants in which the fungal hyphae form a compact mantle on the surface of the roots and extend into the surrounding soil and inward between cortical cells, but not into these cells. Associated primarily with certain trees. *See also* endotrophic mycorrhiza.

edaphology The science that deals with the influence of soils on living things, particularly plants, including human use of land for plant growth.

effective cation exchange capacity The amount of cation charges that a material (usually soil or soil colloids) can hold at the pH of the material, measured as the sum of the exchangeable Al^{3+}, Ca^{2+}, Mg^{2+}, K^+, and Na^+, and expressed as moles or cmol of charge per kg of material. *See* cation exchange capacity.

effective precipitation That portion of the total precipitation that becomes available for plant growth or for the promotion of soil formation.

E_h In soils, it is the potential created by oxidation–reduction reactions that take place on the surface of a platinum electrode measured against a reference electrode, minus the Eh of the reference electrode. This is a measure of the oxidation–reduction potential of electrode-reactive components in the soil. *See also* pe.

electrical conductivity (EC) The capacity of a substance to conduct or transmit electrical current. In soils or water, measured in siemens/meter (or often dS/m), and related to dissolved solutes.

eluviation The removal of soil material in suspension (or in solution) from a layer or layers of a soil. Usually, the loss of material in solution is described by the term "leaching." *See also* illuviation and leaching.

endoaquic (endosaturation) A condition or moisture regime in which the soil is saturated with water in all layers from the upper boundary of saturation (water table) to a depth of 200 cm or more from the mineral soil surface. *See also* epiaquic.

endotrophic mycorrhiza (endomycorrhiza) A symbiotic association of the mycelium of fungi and roots of a variety of plants in which the fungal hyphae penetrate directly into root hairs, other epidermal cells, and occasionally into cortical cells. Individual hyphae also extend from the root surface outward into the surrounding soil. *See also* arbuscular mycorrhiza.

enrichment ratio The concentration of a substance (e.g., phosphorus) in eroded sediment divided by its concentration in the source soil prior to being eroded.

Entisols An order in *Soil Taxonomy*. Soils that have no diagnostic pedogenic horizons. They may be found in virtually any climate on very recent geomorphic surfaces.

eolian soil material Soil material accumulated through wind action. The most extensive areas in the United States are silty deposits (loess), but large areas of sandy deposits also occur.

epiaquic (episaturation) A condition in which the soil is saturated with water due to a perched water table in one or more layers within 200 cm of the mineral soil surface, implying that there are also one or more unsaturated layers within 200 cm below the saturated layer. *See also* endoaquic.

epipedon A diagnostic surface horizon that includes the upper part of the soil that is darkened by organic matter, or the upper eluvial horizons, or both. (*Soil Taxonomy.*)

equilibrium phosphorus concentration The concentration of phosphorus in a solution in equilibrium with a soil, the EPC_0 being the concentration of phosphorus achieved by desorption of phosphorus from a soil to phosphorus-free distilled water.

erosion (1) The wearing away of the land surface by running water, wind, ice, or other geological agents, including such processes as gravitational creep. (2) Detachment and movement of soil or rock by water, wind, ice, or gravity.

esker A narrow ridge of gravelly or sandy glacial material deposited by a stream in an ice-walled valley or tunnel in a receding glacier.

essential element A chemical element required for the normal growth of plants.

eukaryote An organism whose cells each have a visibly evident nucleus.

eutrophic Having concentrations of nutrients optimal (or nearly so) for plant or animal growth. (Said of algal-enriched bodies of water)

eutrophication Nutrient enrichment of lakes, ponds, and other such waters that stimulates the growth of aquatic organisms, which leads to a deficiency of oxygen in the water body.

evapotranspiration The combined loss of water from a given area, and during a specified period of time, by evaporation from the soil surface and by transpiration from plants.

exchange capacity The total ionic charge of the adsorption complex active in the adsorption of ions. *See also* anion exchange capacity; cation exchange capacity.

exchangeable ions Positively or negatively charged atoms or groups of atoms that are held on or near the surface of a solid particle by attraction to charges of the opposite sign, and which may be replaced by other like-charged ions in the soil solution.

exchangeable sodium percentage The extent to which the adsorption complex of a soil is occupied by sodium. It is expressed as follows:

$$ESP = \frac{\text{exchangeable sodium (cmol}_c/\text{kg soil)}}{\text{cation exchange capacity (cmol}_c/\text{kg soil)}} \times 100$$

exfoliation Peeling away of layers of a rock from the surface inward, usually as the result of expansion and contraction that accompany changes in temperature.

expansive soil Soil that undergoes significant volume change upon wetting and drying, usually because of a high content of swelling-type clay minerals.

external surface The area of surface exposed on the top, bottom, and sides of a clay crystal.

facultative organism An organism capable of both aerobic and anaerobic metabolism.

fallow Cropland left idle in order to restore productivity, mainly through accumulation of nutrients, water, and/or organic matter. Preceding a cereal grain crop in semiarid regions, land may be left in *summer fallow* for a period during which weeds are controlled by chemicals or tillage and water is allowed to accumulate in the soil profile. In humid regions, fallow land may be allowed to grow up in natural vegetation for a period ranging from a few months to many years. *Improved fallow* involves the purposeful establishment of plant species capable of restoring soil productivity more rapidly than a natural plant succession.

family, soil In *Soil Taxonomy*, one of the categories intermediate between the great group and the soil series. Families are defined largely on the basis of physical and mineralogical properties of importance to plant growth.

fauna The animal life of a region or ecosystem.

fen A calcium-rich, peat-accumulating wetland with relatively stagnant water.

ferrihydrite, $Fe_5HO_8 \cdot 4H_2O$ A dark reddish brown poorly crystalline iron oxide that forms in wet soils.

fertigation The application of fertilizers in irrigation waters, commonly through sprinkler systems.

fertility, soil The quality of a soil that enables it to provide essential chemical elements in quantities and proportions for the growth of specified plants.

fertilizer Any organic or inorganic material of natural or synthetic origin added to a soil to supply certain elements essential to the growth of plants.

fibric materials *See* organic soil materials.

field capacity (field moisture capacity) The percentage of water remaining in a soil two or three days after its having been saturated and after free drainage has practically ceased.

fine earth fraction That portion of the soil that passes through a 2 mm diameter sieve opening. *Compare to* coarse fragments.

fine texture Consisting of or containing large quantities of the fine fractions, particularly of silt and clay. (Includes clay loam, sandy clay loam, silty clay loam, sandy clay, silty clay, and clay textural classes.)

fine-grained mica A silicate clay having a 2:1-type lattice structure with much of the silicon in the tetrahedral sheet having been replaced by aluminum and with considerable interlayer potassium, which binds the layers together, prevents interlayer expansion and swelling, and limits interlayer cation exchange capacity.

fixation (1) For other than elemental nitrogen: the process or processes in a soil by which certain chemical elements are converted from a soluble or exchangeable form to a much less soluble or to a nonexchangeable form; for example, potassium, ammonium, and phosphorus fixation. (2) For elemental nitrogen: process by which gaseous elemental nitrogen is chemically combined with hydrogen to form ammonia. *See* biological nitrogen fixation.

flagstone A relatively thin rock or mineral fragment 15 to 38 cm in length commonly composed of shale, slate, limestone, or sandstone.

flocculate To aggregate or clump together individual, tiny soil particles, especially fine clay, into small clumps or floccules. Opposite of *deflocculate* or *disperse*.

floodplain The land bordering a stream, built up of sediments from overflow of the stream and subject to inundation when the stream is at flood stage. Sometimes called *bottomland*.

flora The sum total of the kinds of plants in an area at one time. The organisms loosely considered to be of the plant kingdom.

fluorapatite A member of the apatite group of minerals containing fluorine. Most common mineral in phosphate rock.

fluvial deposits Deposits of parent materials laid down by rivers or streams.

fluvioglacial *See* glaciofluvial deposits.

foliar diagnosis An estimation of mineral nutrient deficiencies (excesses) of plants based on examination of the chemical composition of selected plant parts, and the color and growth characteristics of the foliage of the plants.

food web The community of organisms that relate to one another by sharing and passing on food substances. They are organized into trophic levels such as producers that create organic substances from sunlight and inorganic matter, to consumers and predators that eat the producers, dead organisms, waste products and each other.

forest floor The forest soil O horizons, including litter and unincorporated humus, on the mineral soil surface.

fraction A portion of a larger store of a substance operationally defined by a particular analysis or separation method. For example, the particulate organic matter fraction of soil organic matter is defined by a series of laboratory procedures by which it is separated. *Compare to* pool.

fragipan Dense and brittle pan or subsurface layer in soils that owes its hardness mainly to extreme density or compactness rather than high clay content or cementation. Removed fragments are friable, but the material in place is so dense that roots penetrate and water moves through it very slowly.

friable A soil consistency term pertaining to soils that crumble with ease.

frigid A soil temperature class with mean annual temperature below 8°C.

fritted micronutrients Sintered silicates having total guaranteed analyses of micronutrients with controlled (relatively slow) release characteristics.

fulvic acid A term of varied usage but usually referring to a mixture of organic substances soluble in both alkali and acid solution. Found in natural waters, sediments and to a limited extent in aerobic soils. Once thought to be a major constituent of soil organic matter.

functional diversity The characteristic of an ecosystem exemplified by the capacity to carry out a large number of biochemical transformations and other functions.

functional group An atom, or group of atoms, attached to a large molecule. Each functional group (e.g., —OH, —CH$_3$, —COOH) has a characteristic chemical reactivity.

fungi Eukaryote microorganisms with a rigid cell wall. Some form long filaments of cells called *hyphae* that may grow together to form a visible body.

furrow slice The uppermost layer of an arable soil to the depth of primary tillage; the layer of soil sliced away from the rest of the profile and inverted by a moldboard plow.

gabion Partitioned, wire fabric containers, filled with stone at the site of use, to form flexible, permeable, and monolithic structures for earth retention.

gamma ray A high-energy ray (photon) emitted during radioactive decay of certain elements.

Gelisols An order in *Soil Taxonomy*. Soils that have permafrost within the upper 1 m, or upper 2 m if cryoturbation is also present. They may have an ochric, histic, mollic, or other epipedon.

gellic materials Mineral or organic soil materials that have *cryoturbation* and/or ice in the form of lenses, veins, or wedges and the like.

genesis, soil The mode of origin of the soil, with special reference to the processes responsible for the development of the solum, or true soil, from the unconsolidated parent material.

genetic horizon Soil layers that resulted from soil-forming (pedogenic) processes, as opposed to sedimentation or other geologic processes.

geographic information system (GIS) A method of overlaying, statistically analyzing, and integrating large volumes of spatial data of different kinds. The data are referenced to geographical coordinates and encoded in a form suitable for handling by computer.

geological erosion Wearing away of the Earth's surface by water, ice, or other natural agents under natural environmental conditions of climate, vegetation, and so on, undisturbed by man. Synonymous with *natural erosion*.

gibbsite, Al(OH)$_3$ An aluminum trihydroxide mineral most common in highly weathered soils, such as Oxisols.

gilgai The microrelief of soils produced by expansion and contraction with changes in moisture. Found in soils that contain large amounts of clay that swells and shrinks considerably with wetting and drying. Usually a succession of microbasins and microknolls in nearly level areas or of microvalleys and microridges parallel to the direction of the slope.

glacial drift Rock debris that has been transported by glaciers and deposited, either directly from the ice or from the meltwater. The debris may or may not be heterogeneous.

glacial till *See* till.

glaciofluvial deposits Material moved by glaciers and subsequently sorted and deposited by streams flowing from the melting ice. The deposits are stratified and may occur in the form of outwash plains, deltas, kames, eskers, and kame terraces.

gleyed A soil condition resulting from prolonged saturation with water and reducing conditions that manifest themselves in greenish or bluish colors throughout the soil mass or in mottles.

glomalin A protein-sugar group of molecules secreted by certain fungi resulting in a sticky hyphal surface thought to contribute to aggregate stability.

goethite, FeOOH A yellow-brown iron oxide mineral that accounts for the brown color in many soils.

granular structure Soil structure in which the individual grains are grouped into spherical aggregates with indistinct sides. Highly porous granules are commonly called *crumbs*. A well-granulated soil has the best structure for most ordinary crop plants. *See also* soil structure types.

granulation The process of producing granular materials. Commonly used to refer to the formation of soil structural granules, but also used to refer to the processing of powdery fertilizer materials into granules.

grassed waterway Broad and shallow channel, planted with grass (usually perennial species) that is designed to move surface water downslope without causing soil erosion.

gravitational potential That portion of the total *soil water potential* due to differences in elevation of the reference pool of pure water and that of the soil water. Since the soil water elevation is usually chosen to be higher than that of the reference pool, the gravitational potential is usually positive.

gravitational water Water that moves into, through, or out of the soil under the influence of gravity.

great group A category in *Soil Taxonomy*. The classes in this category contain soils that have the same kind of horizons in the same sequence and have similar moisture and temperature regimes.

green manure Plant material incorporated with the soil while green, or soon after maturity, for improving the soil.

greenhouse effect The entrapment of heat by upper atmosphere gases, such as carbon dioxide, water vapor, and methane, just as glass traps heat for a greenhouse. Increases in the quantities of these gases in the atmosphere will likely result in global warming that may have serious consequences for humankind.

groundwater Subsurface water in the zone of saturation that is free to move under the influence of gravity, often horizontally to stream channels.

grus A sediment or soil material comprised of loose grains of coarse sand and fine gravel size composed of quartz, feldspar and rock fragments. Produced from rocks by physical weathering or selectively transported by borrowing insects.

gully erosion The erosion process whereby water accumulates in narrow channels and, over short periods, removes the soil from this narrow area to considerable depths, ranging from 1 to 2 ft to as much as 23 to 30 m (75 to 100 ft).

gypsic horizon A diagnostic subsurface horizon of secondary calcium sulfate enrichment that is more than 15 cm thick.

gypsum requirement The quantity of gypsum required to reduce the exchangeable sodium percentage in a soil to an acceptable level.

halophyte A plant that requires or tolerates a saline (high salt) environment.

hard armor Pertains to the use of hard materials (such as large stones or concrete) to prevent soil and stream bank erosion by reducing the erosive force of flowing water. *See* soft armor.

hardpan A hardened soil layer, in the lower A or in the B horizon, caused by cementation of soil particles with organic matter or with such materials as silica, sesquioxides, or calcium carbonate. The hardness does not change appreciably with changes in moisture content and pieces of the hard layer do not slake in water. *See also* caliche; claypan.

harrowing A secondary broadcast tillage operation that pulverizes, smooths, and firms the soil in seedbed preparation, controls weeds, or incorporates material spread on the surface.

heaving The partial lifting of plants, buildings, roadways, fenceposts, etc., out of the ground, as a result of freezing and thawing of the surface soil during the winter.

heavy metals Those metals that have densities of 5.0 Mg/m or greater. Elements in soils include Cd, Co, Cr, Cu, Fe, Hg, Mn, Mo, Pb, and Zn.

heavy soil (Obsolete in scientific use) A soil with a high content of clay, and a high drawbar pull, hence difficult to cultivate.

hematite, Fe_2O_3 A red iron oxide mineral that contributes red color to many soils.

hemic material *See* organic materials.

herbicide A chemical that kills plants or inhibits their growth; intended for weed control.

herbivore A plant-eating animal.

heterotroph An organism capable of deriving energy for life processes only from the decomposition of organic compounds and incapable of using inorganic compounds as sole sources of energy or for organic synthesis. *Contrast with* autotroph.

histic epipedon A diagnostic surface horizon consisting of a thin layer of organic soil material that is saturated with water at some period of the year unless artificially drained and that is at or near the surface of a mineral soil.

Histosols An order in *Soil Taxonomy*. Soils formed from materials high in organic matter. Histosols with essentially no clay must have at least 20% organic matter by weight (about 78% by volume). This minimum organic matter content rises with increasing clay content to 30% (85% by volume) in soils with at least 60% clay.

horizon, soil A layer of soil, approximately parallel to the soil surface, differing in properties and characteristics from adjacent layers below or above it. *See also* diagnostic horizons.

horticulture The art and science of growing fruits, vegetables, and ornamental plants.

hue (color) *See* Munsell color system.

humic acid A mixture of dark colored organic substances produced by extraction of soil with strong alkali followed by precipitation in acid. Once thought to be a major constituent of soil organic matter.

humic substances A series of complex, high molecular weight, brown- to black-colored organic substances that occur in natural waters, sediments, and to a small extent in aerobic soils.

humid climate Climate in regions where water, when distributed normally throughout the year, should not limit crop production. In cool climates annual precipitation may be as little as 25 cm; in hot climates, 150 cm or even more. Forest is the common natural vegetation in uncultivated areas.

humification The processes involved in the decomposition and partial stabilization of organic matter and leading to the formation of humus.

humin Colloidal sized soil organic material that is not dissolved upon extraction of the soil with strong alkali. Part of an obsolete fractionation scheme for soil organic matter.

Humus The portion of soil organic matter that is not alive or recognizable plant tissue, and is protected from rapid decomposition to some degree by the soil environment. It is generally colloidal in particle size and black in color. Previously thought to be composed mainly of very large polymer molecules operationally defined as humic acids, folic acids and humin, which see.

hydration Chemical union between an ion or compound and one or more water molecules, the reaction being stimulated by the attraction of the ion or compound for either the hydrogen or the unshared electrons of the oxygen in the water.

hydraulic conductivity An expression of the readiness with which a liquid, such as water, flows through a solid, such as soil, in response to a given potential gradient.

hydric soils Soils that are water-saturated for long enough periods to produce reduced conditions and affect the growth of plants.

hydrogen bonding Relatively low energy bonding exhibited by a hydrogen atom located between two highly electronegative atoms, such as nitrogen or oxygen.

hydrologic cycle The circuit of water movement from the atmosphere to the Earth and back to the atmosphere through various stages or processes, as precipitation, interception, runoff, infiltration, percolation, storage, evaporation, and transpiration.

hydrolysis A reaction with water that splits the water molecule into H^+ and OH^- ions. Molecules or atoms participating in such reactions are said to *hydrolyze*.

hydronium A hydrated hydrogen ion (H_3O^+), the form of the hydrogen ion usually found in an aqueous system.

hydroperiod The duration of the presence of surface water in seasonal wetlands.

hydroponics Plant-production systems that use nutrient solutions and no solid medium to grow plants.

hydrostatic potential *See* submergence potential.

hydrous mica *See* fine-grained mica.

hydroxyapatite A member of the apatite group of minerals rich in hydroxyl groups. A nearly insoluble calcium phosphate.

hygroscopic coefficient The amount of moisture in a dry soil when it is in equilibrium with some standard relative humidity near a saturated atmosphere (about 98%), expressed in terms of percentage on the basis of oven-dry soil.

hyperaccumulator A plant with unusually high capacity to take up certain elements from soil resulting in very high concentrations of these elements in the plant's tissues. Often pertaining to concentrations of heavy metals to 1% or more of the tissue dry matter.

hyperthermic A soil temperature class with mean annual temperatures >22°C.

hypha (pl. hyphae) Filament of fungal cells. Actinomycetes also produce similar, but thinner, filaments of cells.

hypoxia State of oxygen deficiency in an environment so low as to restrict biological respiration (in water, typically less than 2 to 3 mg O_2/L).

hysteresis A relationship between two variables that changes depending on the sequences or starting point. An example is the relationship between soil water content and water potential, for which different curves describe the relationship when a soil is gaining water or losing it.

igneous rock Rock formed from the cooling and solidification of magma that has not been changed appreciably since its formation.

illite *See* fine-grained mica.

illuvial horizon A soil layer or horizon in which material carried from an overlying layer has been precipitated from solution or deposited from suspension. The layer of accumulation.

illuviation The process of deposition of soil material removed from one horizon to another in the soil; usually from an upper to a lower horizon in the soil profile. *See also* eluviation.

immature soil A soil with indistinct or only slightly developed horizons because of the relatively short time it has been subjected to the various soil-forming processes. A soil that has not reached equilibrium with its environment.

immobilization The conversion of an element from the inorganic to the organic form in microbial tissues or in plant tissues, thus rendering the element not readily available to other organisms or to plants.

imogolite A poorly crystalline aluminosilicate mineral with an approximate formula $SiO_2Al_2O_3 \cdot 2.5H_2O$; occurs mostly in soils formed from volcanic ash.

impervious Resistant to penetration by fluids or by roots.

improved fallow *See* fallow.

Inceptisols An order in *Soil Taxonomy*. Soils that are usually moist with pedogenic horizons of alteration of parent materials but not of illuviation. Generally, the direction of soil development is not yet evident from the marks left by various soil-forming processes or the marks are too weak to classify in another order.

induced systemic resistance Plant defense mechanisms activated by a chemical signal produced by a rhizosphere bacteria. Although the process begins in the soil, it may confer disease resistance to leaves or other aboveground tissues.

indurated (soil) Soil material cemented into a hard mass that will not soften on wetting. *See also* consistence; hardpan.

infiltration The downward entry of water into the soil.

infiltration capacity A soil characteristic determining or describing the *maximum* rate at which water *can* enter the soil under specified conditions, including the presence of an excess of water.

inner-sphere complex A relatively strong (not easily reversed) chemical association or bonding directly between a specific ion and specific atoms or groups of atoms in the surface structure of a soil colloid.

inoculation The process of introducing pure or mixed cultures of microorganisms into natural or artificial culture media.

inorganic compounds All chemical compounds in nature except compounds of carbon other than carbon monoxide, carbon dioxide, and carbonates.

insecticide A chemical that kills insects.

intergrade A soil that possesses moderately well-developed distinguishing characteristics of two or more genetically related great soil groups.

interlayer (mineralogy) Materials between layers within a given crystal, including cations, hydrated cations, organic molecules, and hydroxide groups or sheets.

internal surface The area of surface exposed within a clay crystal between the individual crystal layers. *Compare with* external surface.

interstratification Mixing of silicate layers within the structural framework of a given silicate clay.

ionic double layer The distribution of cations in the soil solution resulting from the simultaneous attraction toward colloid particles by the particle's negative charge and the tendency of diffusion and thermal forces to move the cations away from the colloid surfaces. Also described as a diffuse double layer or a diffuse electrical double layer.

ions Atoms, groups of atoms, or compounds that are electrically charged as a result of the loss of electrons (cations) or the gain of electrons (anions).

iron-pan An indurated soil horizon in which iron oxide is the principal cementing agent.

irrigation efficiency The ratio of the water actually consumed by crops on an irrigated area to the amount of water diverted from the source onto the area.

isomorphous substitution The replacement of one atom by another of similar size in a crystal lattice without disrupting or changing the crystal structure of the mineral.

isotopes Two or more atoms of the same element that have different atomic masses because of different numbers of neutrons in the nucleus.

joule The SI energy unit defined as a force of 1 newton applied over a distance of 1 meter; 1 joule = 0.239 calorie.

K_d *See* distribution coefficient, K_d.

K_{oc} The distribution coefficient, K_d, calculated based on organic carbon content. $K_{oc}=K_d$/foc where foc is the fraction of organic carbon.

kame A conical hill or ridge of sand or gravel deposited in contact with glacial ice.

kandic horizon A subsurface diagnostic horizon having a sharp clay increase relative to overlying horizons and having low-activity clays.

kaolinite An aluminosilicate mineral of the 1:1 crystal lattice group; that is, consisting of single silicon tetrahedral sheets alternating with single aluminum octahedral sheets.

K_{sat} Hydraulic conductivity when the soil is water saturated. *See also* hydraulic conductivity.

k-strategist An organism that maintains a relatively stable population by specializing in metabolism of resistant compounds that most other organisms cannot utilize. *Contrast with* r-strategist. *See also* autochthonous organisms.

labile A substance that is readily transformed by microorganisms or is readily available for uptake by plants.

lacustrine deposit Material deposited in lake water and later exposed either by lowering of the water level or by the elevation of the land.

land A broad term embodying the total natural environment of the areas of the Earth not covered by water. In addition to soil, its attributes include other physical conditions, such as mineral deposits and water supply; location in relation to centers of commerce, populations, and other land; the size of the individual tracts or holdings; and existing plant cover, works of improvement, and the like.

land capability classification A grouping of kinds of soil into special units, subclasses, and classes according to their capability for intensive use and the treatments required for sustained use. One such system has been prepared by the USDA Natural Resources Conservation Service.

land classification The arrangement of land units into various categories based upon the properties of the land or its suitability for some particular purpose.

land forming Shaping the surface of the land by scraping off the high spots and filling in the low spots with precision grading machinery to create a uniform, smooth slope, often for irrigation purposes. Also called *land smoothing*.

land-use planning The development of plans for the uses of land that, over long periods, will best serve the general welfare, together with the formulation of ways and means for achieving such uses.

laterite An iron-rich subsoil layer found in some highly weathered humid tropical soils that, when exposed and allowed to dry, becomes very hard and will not soften when rewetted. When erosion removes the overlying layers, the laterite is exposed and a virtual pavement results. *See also* plinthite.

layer (Clay mineralogy) A combination in silicate clays of (tetrahedral and octahedral) sheets in a 1:1, 2:1, or 2:1:1 combination.

leaching The removal of materials in solution from the soil by percolating waters. *See also* eluviation.

leaching requirement The leaching fraction of irrigation water necessary to keep soil salinity from exceeding a tolerance level of the crop to be grown.

leaf area index The ratio of the area of the total upper leaf surface of a plant canopy and the unit area on which the canopy is grown.

legume A pod-bearing member of the Leguminosae family, one of the most important and widely distributed plant families. Includes many valuable food and forage species, such as peas, beans, peanuts, clovers, alfalfas, sweet clovers, lespedezas, vetches, and kudzu. Nearly all legumes are associated with nitrogen-fixing organisms.

lichen A symbiotic relationship between fungi and cyanobacteria (blue-green algae) that enhances colonization of bare minerals and rocks. The fungi supply water and nutrients, the cyanobacteria the fixed nitrogen and carbohydrates from photosynthesis.

Liebig's law The growth and reproduction of an organism are determined by the nutrient substance (oxygen, carbon dioxide, calcium, etc.) that is available in minimum quantity with respect to organic needs; the *limiting factor*. Also attributed to Sprengel.

light soil (Obsolete in scientific use) A coarse-textured soil; a soil with a low drawbar pull and hence easy to cultivate. *See also* coarse texture; soil texture.

lignin The complex organic constituent of woody fibers in plant tissue that, along with cellulose, cements the cells together and provides strength. Lignins resist microbial attack and after some modification may become part of the soil organic matter.

lime (agricultural) In strict chemical terms, calcium oxide. In practical terms, a material containing the carbonates, oxides, and/or hydroxides of calcium and/or magnesium used to neutralize soil acidity.

lime requirement The mass of agricultural limestone, or the equivalent of other specified liming material, required to raise the pH of the soil to a desired value under field conditions.

limestone A sedimentary rock composed primarily of calcite ($CaCO_3$). If dolomite ($CaCO_3 \cdot MgCO_3$) is present in appreciable quantities, it is called a *dolomitic limestone*.

limiting factor *See* Liebig's law.

liquid limit (LL) *See* Atterberg limits.

lithosequence A group of related soils that differ, one from the other, in certain properties primarily as a result of parent material as a soil-forming factor.

loam The textural-class name for soil having a moderate amount of sand, silt, and clay. Loam soils contain 7 to 27% clay, 28 to 50% silt, and 23 to 52% sand.

loamy Intermediate in texture and properties between fine-textured and coarse-textured soils. Includes all textural classes with the words *loam* or *loamy* as a part of the class name, such as clay loam or loamy sand. *See also* loam; soil texture.

lodging Falling over of plants, either by uprooting or stem breakage.

loess Material transported and deposited by wind and consisting of predominantly silt-sized particles.

luxury consumption The intake by a plant of an essential nutrient in amounts exceeding what it needs. For example, if potassium is abundant in the soil, alfalfa may take in more than it requires.

lysimeter A device for measuring percolation (leaching) and evapotranspiration losses from a column of soil under controlled conditions.

macronutrient A chemical element necessary in large amounts (usually 50 mg/kg in the plant) for the growth of plants. Includes C, H, O, N, P, K, Ca, Mg, and S. (*Macro* refers to quantity and not to the essentiality of the element.) *See also* micronutrient.

macropores Larger soil pores, generally having a diameter greater than 0.08 mm, from which water drains readily by gravity.

map unit (mapping unit), soil A conceptual group of one to many component soils, delineated or identified by the same name in a soil survey, that represent similar landscape areas. *See also* delineation, soil consociation, soil complex, soil association, and undifferentiated group.

marl Soft and unconsolidated calcium carbonate, usually mixed with varying amounts of clay or other impurities.

marsh Periodically wet or continually flooded area with the surface not deeply submerged. Covered dominantly with sedges, cattails, rushes, or other hydrophytic plants. Subclasses include freshwater and saltwater marshes.

mass flow Movement of nutrients with the flow of water to plant roots.

matric potential That portion of the total *soil water potential* due to the attractive forces between water and soil solids as represented through adsorption and capillarity. It will always be negative.

mature soil A soil with well-developed soil horizons produced by the natural processes of soil formation and essentially in equilibrium with its present environment.

maximum retentive capacity The average moisture content of a disturbed sample of soil, 1 cm high, which is at equilibrium with a water table at its lower surface.

mechanical analysis (Obsolete) *See* particle size analysis; particle size distribution.

medium texture Intermediate between fine-textured and coarse-textured (soils). It includes the following textural classes: very fine sandy loam, loam, silt loam, and silt.

melanic epipedon A diagnostic surface horizon formed in volcanic parent material that contains more than 6% organic carbon, is dark in color, and has a very low bulk density and high anion adsorption capacity.

mellow soil A very soft, very friable, porous soil without any tendency toward hardness or harshness. *See also* consistence.

mesic A soil temperature class with mean annual temperature 8 to 15°C.

mesofauna Animals of medium size, between approximately 2 and 0.2 mm in diameter.

mesophilic Pertaining to moderate temperatures in the range of 15 to 35°C, the range in which mesophilic organisms grow best and in which mesophilic composting takes place.

metamorphic rock A rock that has been greatly altered from its previous condition through the combined action of heat and pressure. For example, marble is a metamorphic rock produced from limestone, gneiss is produced from granite, and slate is produced from shale.

methane, CH_4 An odorless, colorless gas commonly produced under anaerobic conditions. When released to the upper atmosphere, methane contributes to global warming. *See also* greenhouse effect.

micas Primary aluminosilicate minerals in which two silica tetrahedral sheets alternate with one alumina/magnesia octahedral sheet with entrapped potassium atoms fitting between sheets. They separate readily into visible sheets or flakes.

microfauna That part of the animal population which consists of individuals too small to be clearly distinguished without the use of a microscope. Includes protozoans and nematodes.

microflora That part of the plant population which consists of individuals too small to be clearly distinguished without the use of a microscope. Includes actinomycetes, algae, bacteria, and fungi.

micronutrient A chemical element necessary in only extremely small amounts (<50 mg/kg in the plant) for the growth of plants. Examples are B, Cl, Cu, Fe, Mn, and Zn. (*Micro* refers to the amount used rather than to its essentiality.) *See also* macronutrient.

micropores Relatively small soil pores, generally found within structural aggregates and having a diameter less than 0.08 mm. *Contrast to* macropores.

microrelief Small-scale local differences in topography, including mounds, swales, or pits that are only 1 m or so in diameter and with elevation differences of up to 2 m. *See also* gilgai.

mineral (i) An inorganic compound of defined composition found in rocks. (ii) An adjective meaning inorganic.

mineral nutrient An element in inorganic form used by plants or animals.

mineral soil A soil consisting predominantly of, and having its properties determined predominantly by, mineral matter. Usually contains <20% organic matter, but may contain an organic surface layer up to 30 cm thick.

mineralization The conversion of an element from an organic form to an inorganic state as a result of microbial decomposition.

minimum tillage *See* tillage, conservation.

minor element (Obsolete) *See* micronutrient.

moderately coarse texture Consisting predominantly of coarse particles. In soil textural classification, it includes all the sandy loams except the very fine sandy loam. *See also* coarse texture.

moderately fine texture Consisting predominantly of intermediate-sized (soil) particles or with relatively small amounts of fine or coarse particles. In soil textural classification, it includes clay loam, sandy loam, sandy clay loam, and silty clay loam. *See also* fine texture.

moisture potential *See* soil water potential.

mole drain Unlined drain formed by pulling a bullet-shaped cylinder through the soil.

mollic epipedon A diagnostic surface horizon of mineral soil that is dark colored and relatively thick, contains at least 0.6% organic carbon, is not massive and hard when dry, has a base saturation of more than 50%, has less than 250 mg/kg P_2O_5 soluble in 1% citric acid, and is dominantly saturated with bivalent cations.

Mollisols An order in *Soil Taxonomy*. Soils with nearly black, organic-rich surface horizons and high supply of bases. They have mollic epipedons and base saturation greater than 50% in any cambic or argillic horizon. They lack the characteristics of Vertisols and must not have oxic or spodic horizons.

molybdenosis A nutritional disease of ruminant animals in which high Mo in the forage interferes with copper absorption.

montmorillonite An aluminosilicate clay mineral in the smectite group with a 2:1 expanding crystal lattice, with two silicon tetrahedral sheets enclosing an aluminum octahedral sheet. Isomorphous substitution of magnesium for some of the aluminum has occurred in the octahedral sheet. Considerable expansion may be caused by water moving between silica sheets of contiguous layers.

mor Raw humus; type of forest humus layer of unincorporated organic material, usually matted or compacted or both; distinct from the mineral soil, unless the latter has been blackened by washing in organic matter.

moraine An accumulation of drift, with an initial topographic expression of its own, built within a glaciated region chiefly by the direct action of glacial ice. Examples are ground, lateral, recessional, and terminal moraines.

morphology, soil The constitution of the soil, including the texture, structure, consistence, color, and other physical, chemical, and biological properties of the various soil horizons that make up the soil profile.

mottling Spots or blotches of different color or shades of color interspersed with the dominant color.

mucigel The gelatinous material at the surface of roots grown in unsterilized soil.

muck Highly decomposed organic material in which the original plant parts are not recognizable. Contains more mineral matter and is usually darker in color than peat. *See also* muck soil; peat.

muck soil (1) A soil containing 20 to 50% organic matter. (2) An organic soil in which the organic matter is well decomposed.

mulch Any material such as straw, sawdust, leaves, plastic film, and loose soil that is spread upon the surface of the soil to protect the soil and plant roots from the effects of raindrops, soil crusting, freezing, evaporation, etc.

mulch tillage *See* tillage, conservation.

mull A humus-rich layer of forested soils consisting of mixed organic and mineral matter. A mull blends into the upper mineral layers without an abrupt change in soil characteristics.

Munsell color system A color designation system that specifies the relative degrees of the three simple variables of color:

  chroma The relative purity, strength, or saturation of a color.

  hue The chromatic gradation (rainbow) of light that reaches the eye.

  value The degree of lightness or darkness of the color.

mycelium A stringlike mass of individual fungal or actinomycetes hyphae.

myco Prefix designating an association or relationship with a fungus (e.g., mycotoxins are toxins produced by a fungus).

mycorrhiza The association, usually symbiotic, of fungi with the roots of seed plants. *See also* ectotrophic mycorrhiza; endotrophic mycorrhiza; arbuscular mycorrhiza.

natric horizon A diagnostic subsurface horizon that satisfies the requirements of an argillic horizon, but that also has prismatic, columnar, or blocky structure and a subhorizon having more than 15% saturation with exchangeable sodium.

necrosis Death associated with discoloration and dehydration of all or parts of plant organs, such as leaves.

nematodes Very small (most are microscopic) unsegmented round worms. In soils they are abundant and perform many important functions in the soil food web. Some are plant parasites and considered pests.

neutral soil A soil in which the surface layer, at least to normal plow depth, is neither acid nor alkaline in reaction. In practice this means the soil is within the pH range of 6.6 to 7.3. *See also* acid soil; alkaline soil; pH; reaction, soil.

nitrate depression period A period of time, beginning shortly after the addition of fresh, highly carbonaceous organic materials to a soil, during which decomposer microorganisms have removed most of the soluble nitrate from the soil solution.

nitrification The biochemical oxidation of ammonium to nitrate, predominantly by autotrophic bacteria.

nitrogen assimilation The incorporation of nitrogen into organic cell substances by living organisms.

nitrogen cycle The sequence of chemical and biological changes undergone by nitrogen as it moves from the atmosphere into water, soil, and living organisms, and upon death of these organisms (plants and animals) is recycled through a part or all of the entire process.

nitrogen fixation The biological conversion of elemental nitrogen (N_2) to organic combinations or to forms readily utilized in biological processes.

nodule bacteria *See* rhizobia.

nonacid cations Those cations that do not react with water by hydrolysis to release H^+ ions to the soil solution. These cations do not remove hydroxyl ions from solution, but form strongly dissociated bases such as potassium hydroxide (K^+ + OH). Formerly called *base cations* or *base-forming cations* in soil science literature.

nonacid saturation The proportion or percentage of a cation-exchange site occupied by nonacid cations. Formerly termed *base saturation*.

nonhumic substances The portion of soil organic matter comprised of relatively low molecular weight organic substances; mostly identifiable biomolecules.

nonlimiting water range The region bounded by the upper and lower soil water content over which water, oxygen, and mechanical resistance are not limiting to plant growth. *Compare with* available water.

nonpoint source A pollution source that cannot be traced back to a single origin or source. Examples include water runoff from urban areas and leaching from croplands.

no-tillage *See* tillage, conservation.

nucleic acids Complex organic acids found in the nuclei of plant and animal cells; may be combined with proteins as nucleoproteins.

O horizon Organic horizon of mineral soils.

ochric epipedon A diagnostic surface horizon of mineral soil that is too light in color, too high in chroma, too low in organic carbon, or too thin to be a plaggen, mollic, umbric, anthropic, or histic epipedon, or that is both hard and massive when dry.

octahedral sheet Sheet of horizontally linked, octahedral-shaped units that serve as the basic structural components of silicate (clay) minerals. Each unit consists of a central, six-coordinated metallic atom (e.g., Al, Mg, or Fe) surrounded by six hydroxyl groups that, in turn, are linked with other nearby metal atoms, thereby serving as interunit linkages that hold the sheet together.

oligotrophic Environments, such as soils or lakes, which are poor in nutrients.

order, soil The category at the highest level of generalization in *Soil Taxonomy*. The properties selected to distinguish the orders are reflections of the degree of horizon development and the kinds of horizons present.

organic farming A system/philosophy of agriculture that does not allow the use of synthetic chemicals to produce plant and animal products, but instead emphasizes the management of soil organic matter and biological processes. In many countries, products are officially certified as being organic if inspections confirm that they were grown by these methods.

organic fertilizer By-product from the processing of animal or vegetable substances that contain sufficient plant nutrients to be of value as fertilizers.

organic soil A soil in which more than half of the profile thickness is comprised of organic soil materials.

organic soil materials (As used in *Soil Taxonomy*): (1) Saturated with water for prolonged periods unless artificially drained and having 18% or more organic carbon (by weight) if the mineral fraction is more than 60% clay, more than 12% organic carbon if the mineral fraction has no clay, or between 12 and 18% carbon if the clay content of the mineral fraction is between 0 and 60%. (2) Never saturated with water for more than a few days and having more than 20% organic carbon. Histosols develop on these organic soil materials. There are three kinds of organic materials:

fibric materials The least decomposed of all the organic soil materials, containing very high amounts of fiber that are well preserved and readily identifiable as to botanical origin; with very low bulk density.

hemic materials Intermediate in degree of decomposition of organic materials between the less decomposed fibric and the more decomposed sapric materials.

sapric materials The most highly decomposed of the organic materials, having the highest bulk density, least amount of plant fiber, and lowest water content at saturation.

orographic Influenced by mountains (Greek *oros*). Used in reference to increased precipitation on the windward side of a mountain range induced as clouds rise over the mountain, leaving a *rain shadow* of reduced precipitation on the leeward side.

ortstein An indurated layer in the B horizon of Spodosols in which the cementing material consists of illuviated sesquioxides (mostly iron) and organic matter.

osmotic potential That portion of the total *soil water potential* due to the presence of solutes in soil water. It will generally be negative.

osmotic pressure Pressure exerted in living bodies as a result of unequal concentrations of salts on both sides of a cell wall or membrane. Water moves from the area having the lower salt concentration through the membrane into the area having the higher salt concentration and, therefore, exerts additional pressure on the side with higher salt concentration.

outer-sphere complex A relatively weak (easily reversed) chemical association or general attraction between an ion and an oppositely charged soil colloid via mutual attraction for intervening water molecules.

outwash plain A deposit of coarse-textured materials (e.g., sands and gravels) left by streams of meltwater flowing from receding glaciers.

oven-dry soil Soil that has been dried at 105°C until it reaches constant weight.

oxic horizon A diagnostic subsurface horizon that is at least 30 cm thick and is characterized by the virtual *absence* of weatherable primary minerals or 2:1 lattice clays and the *presence* of 1:1 lattice clays and highly insoluble minerals, such as quartz sand, hydrated oxides of iron and aluminum, low cation exchange capacity, and small amounts of exchangeable bases.

oxidation The loss of electrons by a substance; therefore, a gain in positive valence charge and, in some cases, the chemical combination with oxygen gas.

oxidation ditch An artificial open channel for partial digestion of liquid organic wastes in which the wastes are circulated and aerated by a mechanical device.

oxidation–reduction potential *See* E_h and pe.

Oxisols An order in *Soil Taxonomy*. Soils with residual accumulations of low-activity clays, free oxides, kaolin, and quartz. They are mostly in tropical climates.

pans Horizons or layers in soils that are strongly compacted, indurated, or very high in clay content. *See also* caliche; claypan; fragipan; hardpan.

parent material The unconsolidated and more or less chemically weathered mineral or organic matter from which the solum of soils is developed by pedogenic processes.

particle density The mass per unit volume of the soil particles. In technical work, usually expressed as metric tons per cubic meter (Mg/m^3) or grams per cubic centimeter (g/cm^3).

particle size The effective diameter of a particle measured by sedimentation, sieving, or micrometric methods.

particle size analysis Determination of the various amounts of the different separates in a soil sample, usually by sedimentation, sieving, micrometry, or combinations of these methods.

particle size distribution The amounts of the various soil separates in a soil sample, usually expressed as weight percentages.

particulate organic matter A microbially active fraction of soil organic matter consisting largely of fine particles of partially decomposed plant tissue.

partitioning The distribution of organic chemicals (such as pollutants) into a portion that dissolves in the soil organic matter and a portion that remains undissolved in the soil solution.

pascal An SI unit of pressure equal to 1 newton per square meter.

peat Unconsolidated soil material consisting largely of undecomposed, or only slightly decomposed, organic matter accumulated under conditions of excessive moisture. *See also* organic soil materials; peat soil.

peat soil An organic soil containing more than 50% organic matter. Used in the United States to refer to the stage of decomposition of the organic matter, *peat* referring to the slightly decomposed or undecomposed deposits and *muck* to the highly decomposed materials. *See also* muck; muck soil; peat.

ped A unit of soil structure such as an aggregate, crumb, prism, block, or granule, formed by natural processes (in contrast to a *clod*, which is formed artificially).

pedology The science that deals with the formation, morphology, and classification of soil bodies as landscape components.

pedon The smallest volume that can be called *a soil*. It has three dimensions. It extends downward to the depth of plant roots or to the lower limit of the genetic soil horizons. Its lateral cross section is roughly hexagonal and ranges from 1 to 10 m^2 in size, depending on the variability in the horizons.

pedosphere The conceptual zone within the ecosystem consisting of soil bodies or directly influenced by them. A zone or sphere of activity in which mineral, water, air, and biological components come together to form soils. Usage is parallel to that for "atmosphere" or "biosphere."

pedoturbation Physical disturbance and mixing of soil horizons by such forces as burrowing animals (faunal pedoturbation) or frost churning (cryoturbation).

peneplain A once high, rugged area that has been reduced by erosion to a lower, gently rolling surface resembling a plain.

penetrability The ease with which a probe can be pushed into the soil. May be expressed in units of distance, speed, force, or work depending on the type of penetrometer used.

penetrometer An instrument consisting of a rod with a cone-shaped tip and a means of measuring the force required to push the rod into a specified increment of soil.

perc test *See* percolation test.

percolation, soil water The downward movement of water through soil. Especially, the downward flow of water in saturated or nearly saturated soil at hydraulic gradients of the order of 1.0 or less.

percolation test A measurement of the rate of percolation of water in a soil profile, usually to determine the suitability of a soil for use as a septic tank drain field.

perforated plastic pipe Pipe, sometimes flexible, with holes or slits in it that allow the entrance and exit of air and water. Used for soil drainage and for septic effluent spreading into soil.

permafrost (1) Permanently frozen material underlying the solum. (2) A perennially frozen soil horizon.

permanent charge *See* constant charge.

permanent wilting point *See* wilting point.

permeability, soil The ease with which gases, liquids, or plant roots penetrate or pass through a bulk mass of soil or a layer of soil.

petrocalcic horizon A diagnostic subsurface horizon that is a continuous, indurated calcic horizon cemented by calcium carbonate and, in some places, with magnesium carbonate. It cannot be penetrated with a spade or auger when dry; dry fragments do not slake in water; and it is impenetrable by roots.

petrogypsic horizon A diagnostic subsurface horizon that is a continuous, strongly cemented, massive gypsic horizon that is cemented by calcium sulfate. It can be chipped with a spade when dry. Dry fragments do not slake in water and it is impenetrable by roots.

pH, soil The negative logarithm of the hydrogen ion activity (concentration) of a soil. The degree of acidity (or alkalinity) of a soil as determined by means of a glass or other suitable electrode or indicator at a specified moisture content or soil-to-water ratio, and expressed in terms of the pH scale.

pH-dependent charge That portion of the total charge of the soil particles that is affected by, and varies with, changes in pH.

phase, soil A subdivision of a soil series or other unit of classification having characteristics that affect the use and management of the soil but do not vary sufficiently to differentiate it as a separate series. Included are such characteristics as degree of slope, degree of erosion, and content of stones.

photomap A mosaic map made from aerial photographs to which place names, marginal data, and other map information have been added.

phyllosphere The leaf surface.

physical properties (of soils) Those characteristics, processes, or reactions of a soil that are caused by physical forces and that can be described by, or expressed in, physical terms or equations. Examples of physical properties are bulk density, water-holding capacity, hydraulic conductivity, porosity, pore-size distribution, and so on.

physical weathering The breakdown of rock and mineral particles into smaller particles by physical forces such as frost action. *See also* weathering.

phytoliths Biologically deposited silica (SiO_2) or mineral bodies (biogenic opal) created by plants and comprising the predominant form if silica in most plant tissue.

phytotoxic substances Chemicals that are toxic to plants.

placic horizon A diagnostic subsurface horizon of a black to dark reddish mineral soil that is usually thin but that may range from 1 to 25 mm in thickness. The placic horizon is commonly cemented with iron and is slowly permeable or impenetrable to water and roots.

plaggen epipedon A diagnostic surface horizon that is human-made and more than 50 cm thick. Formed by long-continued manuring and mixing.

plant nutrients *See* essential element.

plastic limit (PL) *See* Atterberg limits.

plastic soil A soil capable of being molded or deformed continuously and permanently, by relatively moderate pressure, into various shapes. *See also* consistence.

platy Consisting of soil aggregates that are developed predominantly along the horizontal axes; laminated; flaky.

plinthite (brick) A highly weathered mixture of sesquioxides of iron and aluminum with quartz and other diluents that occurs as red mottles and that changes irreversibly to hardpan upon alternate wetting and drying.

plow layer The soil ordinarily moved when land is plowed; equivalent to *surface soil*.

plow pan A subsurface soil layer having a higher bulk density and lower total porosity than layers above or below it, as a result of pressure applied by normal plowing and other tillage operations.

plowing A primary broad-base tillage operation that is performed to shatter soil uniformly with partial to complete inversion.

point of zero charge The pH value of a solution in equilibrium with a particle whose net charge, from all sources, is zero.

point source A pollution source that can be traced back to its origin, which is usually an effluent discharge pipe. Examples are a wastewater treatment plant or a factory. *Opposite of* nonpoint source.

polypedon (As used in *Soil Taxonomy*) Two or more contiguous pedons, all of which are within the defined limits of a single soil series; commonly referred to as a *soil individual*.

pool A portion of a larger store of a substance defined by kinetic or theoretical properties. For example, the active pool of soil organic matter is defined by its rapid rate of microbial turnover. *Compare to* fraction.

pore size distribution The volume of the various sizes of pores in a soil. Expressed as percentages of the bulk volume (soil plus pore space).

porosity, soil The volume percentage of the total soil bulk not occupied by solid particles.

potential acidity The acidity that could potentially be formed if reduced sulfur compounds in a potential acid sulfate soil were to become oxidized.

precision farming The spatially variable management of a field or farm based on information specific to the soil or crop characteristics of many very small subunits of land. This technique commonly uses variable rate equipment, geo positioning systems and computer controls.

preferential flow Nonuniform movement of water and its solutes through a soil along certain pathways, which are often macropores.

primary consumer An organism that subsists on plant material.

primary mineral A mineral that has not been altered chemically since deposition and crystallization from molten lava.

primary producer An organism (usually a photosynthetic plant) that creates organic, energy-rich material from inorganic chemicals, solar energy, and water.

primary tillage *See* tillage, primary.

priming effect The increased decomposition of relatively stable, protected soil humus under the influence of much enhanced, generally biological, activity resulting from the addition of fresh organic materials to a soil.

prismatic soil structure A soil structure type with prismlike aggregates that have a vertical axis much longer than the horizontal axes.

Proctor test A laboratory procedure that indicates the maximum achievable bulk density for a soil and the optimum water content for compacting a soil.

productivity, soil The capacity of a soil for producing a specified plant or sequence of plants under a specified system of management. Productivity emphasizes the capacity of soil to produce crops and should be expressed in terms of yields.

profile, soil A vertical section of the soil through all its horizons and extending into the parent material.

prokaryote An organism whose cells do not have a distinct nucleus.

protein Any of a group of nitrogen-containing organic compounds formed by the polymerization of a large number of amino acid molecules and that, upon hydrolysis, yield these amino acids. They are essential parts of living matter and are one of the essential food substances of animals.

protonation Attachment of protons (H^+ ions) to exposed OH groups on the surface of soil particles, resulting in an overall positive charge on the particle surface.

protozoa One-celled eukaryotic organisms, such as amoeba.

puddled soil Dense, massive soil artificially compacted when wet and having no aggregated structure. The condition commonly results from the tillage of a clayey soil when it is wet.

rain, acid *See* acid rain.

reaction, soil (No longer used in soil science) The degree of acidity or alkalinity of a soil, usually expressed as a pH value or by terms ranging from extremely acid for pH values <4.5 to very strongly alkaline for pH values >9.0.

reactive nitrogen All forms of nitrogen that are readily available to biota (mainly ammonia, ammonium, and nitrate with smaller quantities of other compounds including nitrogen oxide gases) as opposed to unreactive nitrogen that exists mostly as inert N_2 gas.

recharge area A geographic area in which an otherwise confined aquifer is exposed to surficial percolation of water to recharge the groundwater in the aquifer.

redox concentrations Zones of apparent accumulations of Fe-Mn oxides in soils.

redox depletions Zones of low chroma (<2) where Fe-Mn oxides, and in some cases clay, have been stripped from the soil.

redox potential The electrical potential (measured in volts or millivolts) of a system due to the tendency of the substances in it to give up or acquire electrons.

redoximorphic features Soil properties associated with wetness that result from reduction and oxidation of iron and manganese compounds after saturation and desaturation with water. *See also* redox concentrations; redox depletions.

reduction The gain of electrons, and therefore the loss of positive valence charge, by a substance. In some cases, a loss of oxygen or a gain of hydrogen is also involved.

regolith The unconsolidated mantle of weathered rock and soil material on the Earth's surface; loose earth materials above solid rock. (Approximately equivalent to the term *soil* as used by many engineers.)

relief The relative differences in elevation between the upland summits and the lowlands or valleys of a given region.

residual material Unconsolidated and partly weathered mineral materials accumulated by disintegration of consolidated rock in place.

resilience The capacity of a soil (or other ecosystem) to return to its original state after a disturbance.

rhizobacteria Bacteria specially adapted to colonizing the surface of plant roots and the soil immediately around plant roots. Some have effects that promote plant growth, while others have effects that are deleterious to plants.

rhizobia Bacteria capable of living symbiotically with higher plants, usually in nodules on the roots of legumes, from which they receive their energy, and capable of converting atmospheric nitrogen to combined organic forms; hence the term *symbiotic nitrogen-fixing bacteria*. (Derived from the generic name *Rhizobium*.)

rhizoplane The root surface–soil interface. Used to describe the habitat of root-surface-dwelling microorganisms.

rhizosphere That portion of the soil in the immediate vicinity of plant roots in which the abundance and composition of the microbial population are influenced by the presence of roots.

rill A small, intermittent water course with steep sides; usually only a few centimeters deep and hence no obstacle to tillage operations.

rill erosion An erosion process in which numerous small channels of only several centimeters in depth are formed; occurs mainly on recently cultivated soils. *See also* rill.

riparian zone The area, both above and below the ground surface, that borders a river.

riprap Coarse rock fragments, stones, or boulders placed along a waterway or hillside to prevent erosion.

rock The material that forms the essential part of the earth's solid crust, including loose incoherent masses such as sand and gravel, as well as solid masses of granite and limestone.

root interception Acquisition of nutrients by a root as a result of the root growing into the vicinity of the nutrient source.

root nodules Swollen growths on plant roots. Often in reference to those in which symbiotic microorganisms live.

rotary tillage *See* tillage, rotary.

r-strategist Opportunistic organisms with short reproductive times that allow them to respond rapidly to the presence of easily metabolized food sources. *Contrast with* k-strategist. *See also* zymogenous organisms.

runoff The portion of the precipitation on an area that is discharged from the area through stream channels. That which is lost without entering the soil is called *surface runoff* and that which enters the soil before reaching the stream is called *groundwater runoff* or *seepage flow* from groundwater. (In soil science *runoff* usually refers to the water lost by surface flow; in geology and hydraulics *runoff* usually includes both surface and subsurface flow.)

salic horizon A diagnostic subsurface horizon of enrichment with secondary salts more soluble in cold water than gypsum. A salic horizon is 15 cm or more in thickness.

saline seep An area of land in which saline water seeps to the surface, leaving a high salt concentration behind as the water evaporates.

saline soil A nonsodic soil containing sufficient soluble salts to impair its productivity. The conductivity of a saturated extract is >4 dS/m, the exchangeable sodium adsorption ratio is less than about 13, and the pH is <8.5.

saline-sodic soil A soil containing sufficient exchangeable sodium to interfere with the growth of most crop plants and containing appreciable quantities of soluble salts. The exchangeable sodium adsorption ratio is >13, the conductivity of the saturation extract is >4 dS/m (at 25°C), and the pH is usually 8.5 or less in the saturated soil.

salinization The process of accumulation of salts in soil.

saltation Particle movement in water or wind where particles skip or bounce along the stream bed or soil surface.

sand A soil particle between 0.05 and 2.0 mm in diameter; a soil textural class.

sapric materials *See* organic soil materials.

saprolite Soft, friable, weathered bedrock that retains the fabric and structure of the parent rock but is porous and can be dug with a spade.

saprophyte An organism that lives on dead organic material.

saturated paste extract The extract from a saturated soil paste, the electrical conductivity E_c of which gives an indirect measure of salt content in a soil.

saturation extract The solution extracted from a saturated soil paste.

saturation percentage The water content of a saturated soil paste, expressed as a dry weight percentage.

savanna (savannah) A grassland with scattered trees, either as individuals or clumps. Often a transitional type between true grassland and forest.

second bottom The first terrace above the normal floodplain of a stream.

secondary mineral A mineral resulting from the decomposition of a primary mineral or from the reprecipitation of the products of decomposition of a primary mineral. *See also* primary mineral.

sediment Transported and deposited particles or aggregates derived from soils, rocks, or biological materials.

sedimentary rock A rock formed from materials deposited from suspension or precipitated from solution and usually being more or less consolidated. The principal sedimentary rocks are sandstones, shales, limestones, and conglomerates.

seedbed The soil prepared to promote the germination of seed and the growth of seedlings.

self-mulching soil A soil in which the surface layer becomes so well aggregated that it does not crust and seal under the impact of rain but instead serves as a surface mulch upon drying.

semiarid Term applied to regions or climates where moisture is more plentiful than in arid regions but still definitely limits the growth of most crop plants. Natural vegetation in uncultivated areas is short grasses.

separate, soil One of the individual-sized groups of mineral soil particles—sand, silt, or clay.

septic tank An underground tank used in the deposition of domestic wastes. Organic matter decomposes in the tank, and the effluent is drained into the surrounding soil.

series, soil The soil series is a subdivision of a family in *Soil Taxonomy* and consists of soils that are similar in all major profile characteristics.

sewage effluent The liquid part of sewage or wastewater; it is usually treated to remove some portion of the dissolved organic compounds and nutrients present from the original sewage.

sewage sludge Settled sewage solids combined with varying amounts of water and dissolved materials, removed from sewage by screening, sedimentation, chemical precipitation, or bacterial digestion. Also called *biosolids* if certain quality standards are met.

shear Force, as of a tillage implement, acting at right angles to the direction of movement.

sheet (Mineralogy) A flat array of more than one atomic thickness and composed of one or more levels of linked coordination polyhedra. A sheet is thicker than a plane and thinner than a layer. Examples: tetrahedral sheet, octahedral sheet.

sheet erosion The removal of a fairly uniform layer of soil from the land surface by runoff water.

shelterbelt A wind barrier of living trees and shrubs established and maintained for protection of farm fields. Syn. *windbreak*.

shifting cultivation A farming system in which land is cleared, the debris burned, and crops grown for 2 to 3 years. When the farmer moves on to another plot, the land is then left idle for 5 to 15 years; then the burning and planting process is repeated.

short-range order minerals Minerals, such as allophane, whose structural framework consists of short distances of well-ordered crystalline structure interspersed with distances of noncrystalline amorphous materials.

shrinkage limit (SL) The water content above which a mass of soil material will swell in volume, but below which it will shrink no further.

side-dressing The application of fertilizer alongside row-crop plants, usually on the soil surface. Nitrogen materials are most commonly side-dressed.

siderophore A nonporphyrin metabolite secreted by certain microorganisms that forms a highly stable coordination compound with iron.

silica/alumina ratio The molecules of silicon dioxide (SiO_2) per molecule of aluminum oxide (Al_2O_3) in clay minerals or in soils.

silica/sesquioxide ratio The molecules of silicon dioxide (SiO_2) per molecule of aluminum oxide (Al_2O_3) plus ferric oxide (Fe_2O_3) in clay minerals or in soils.

silt (1) A soil separate consisting of particles between 0.05 and 0.002 mm in equivalent diameter. (2) A soil textural class.

silting The deposition of waterborne sediments in stream channels, lakes, reservoirs, or on floodplains, usually resulting from a decrease in the velocity of the water.

site index A quantitative evaluation of the productivity of a soil for forest growth under the existing or specified environment.

slag A product of smelting, containing mostly silicates; the substances not sought to be produced as matte or metal and having a lower specific gravity.

slash-and-burn *See* shifting cultivation.

slick spots Small areas in a field that are slick when wet because of a high content of alkali or exchangeable sodium.

slickensides Stress surfaces that are polished and striated and are produced by one mass sliding past another.

slope The degree of deviation of a surface from horizontal, measured in a numerical ratio, percent, or degrees.

slow fraction (of soil organic matter) That portion of soil organic matter that can be metabolized with great difficulty by the microorganisms in the soil and therefore has a slow turnover rate with a half-life in the soil ranging from a few years to a few decades. Often this fraction is the product of some previous decomposition.

smectite A group of silicate clays having a 2:1-type lattice structure with sufficient isomorphous substitution in either or both the tetrahedral and octahedral sheets to give a high interlayer negative charge and high cation exchange capacity and to permit significant interlayer expansion and consequent shrinking and swelling of the clay. Montmorillonite, beidellite, and saponite are in the smectite group.

sodic soil A soil that contains sufficient sodium to interfere with the growth of most crop plants, and in which the sodium adsorption ratio is 13 or greater.

sodium adsorption ratio (SAR)

$$SAR = \frac{[Na^+]}{\sqrt{1/2([Ca^{2+}] + [Mg^{2+}])}}$$

where the cation concentrations are in millimoles of charge per liter ($mmol_c/L$).

soft armor The bioengineering use of organic and/or inorganic materials combined with plants to create a living vegetation barrier of protection against erosion.

soil (1) A dynamic natural body composed of mineral and organic solids, gases, liquids and living organisms which can serve as a medium for plant growth. (2) The collection of natural bodies occupying parts of the Earth's surface that is capable of supporting plant growth and that has properties resulting from the integrated effects of climate and living organisms acting upon parent material, as conditioned by topography, over periods of time.

soil air The soil atmosphere; the gaseous phase of the soil, being that volume not occupied by soil or liquid.

soil alkalinity The degree or intensity of alkalinity of a soil, expressed by a value >7.0 on the pH scale.

soil amendment Any material, such as lime, gypsum, sawdust, or synthetic conditioner, that is worked into the soil to make it more amenable to plant growth.

soil association A group of defined and named taxonomic soil units occurring together in an individual and characteristic pattern over a geographic region, comparable to plant associations in many ways.

soil auger A tool used to bore small holes up to several meters deep in soils in order to bring up samples of material from various soil layers. It consists of a long T-handle attached to either a cylinder with twisted teeth or a screwlike bit.

soil classification (Soil Taxonomy) The systematic arrangement of soils into groups or categories on the basis of their characteristics. See order; suborder; great group; subgroup; family; and series.

soil complex A mapping unit used in detailed soil surveys where two or more defined taxonomic units are so intimately intermixed geographically that it is undesirable or impractical, because of the scale being used, to separate them. A more intimate mixing of smaller areas of individual taxonomic units than that described under *soil association*.

soil compressibility The property of a soil pertaining to its capacity to decrease in bulk volume when subjected to a load.

soil conditioner Any material added to a soil for the purpose of improving its physical condition.

soil conservation A combination of all management and land-use methods that safeguard the soil against depletion or deterioration caused by nature and/or humans.

soil consociation A kind of soil map unit that is named for the dominant soil taxon in the delineation, and in which at least half of the pedons are of the named soil taxon, and most of the remaining pedons are so similar as to not affect most interpretations.

soil correlation The process of defining, mapping, naming, and classifying the kinds of soils in a specific soil survey area, the purpose being to ensure that soils are adequately defined, accurately mapped, and uniformly named.

soil erosion See erosion.

soil fertility See fertility, soil.

soil genesis See genesis, soil.

soil geography A subspecialization of physical geography concerned with the areal distributions of soil types.

soil health The state of self-regulation, stability, resilience, and lack of stress symptoms in a soil as an ecosystem of living organisms that supports the growth of plants. Sometimes used synonymously with soil quality, but usually focused on the plant growth supporting function of soil.

soil horizon See horizon, soil.

soil loss tolerance (T value) (i) The maximum average annual soil loss that will allow continuous cropping and maintain soil productivity without requiring additional management inputs. (ii) The maximum soil erosion loss that is offset by the theoretical maximum rate of soil development, which will maintain an equilibrium between soil losses and gains.

soil management The sum total of all tillage operations, cropping practices, fertilizer, lime, and other treatments conducted on or applied to a soil for the production of plants.

soil map A map showing the distribution of soil types or other soil mapping units in relation to the prominent physical and cultural features of the Earth's surface.

soil mechanics and engineering A subspecialization of soil science concerned with the effect of forces on the soil and the application of engineering principles to problems involving the soil.

soil moisture potential See soil water potential.

soil monolith A vertical section of a soil profile removed from the soil and mounted for display or study.

soil morphology The physical constitution, particularly the structural properties, of a soil profile as exhibited by the kinds, thicknesses, and arrangement of the horizons in the profile, and by the texture, structure, consistence, and porosity of each horizon.

soil order See order, soil.

soil organic matter The organic fraction of the soil that includes plant and animal residues at various stages of decomposition, cells and tissues of soil organisms, and substances synthesized by the soil population. Commonly determined as the amount of organic material contained in a soil sample passed through a 2-mm sieve.

Soil organic matter: The organic fraction of the soil that includes plant, animal and microbial residues in various stages of decomposition, biomass of soil microorganisms, and substances produced by plant roots and other soil organisms. It is commonly determined as the total organic (non-carbonate) carbon in a soil sample passed through a 2-mm sieve.

soil porosity See porosity, soil.

soil productivity See productivity, soil.

soil profile See profile, soil.

soil quality The capacity of a specific kind of soil to function physically, chemically and biologically, within natural or managed ecosystem boundaries, so as to maximize provisioning and regulatory ecosystem services. Often considered in relative to this capacity in the undisturbed, natural state.

soil reaction See reaction, soil; pH, soil.

soil salinity The amount of soluble salts in a soil, expressed in terms of percentage, milligrams per kilogram, parts per million (ppm), or other convenient ratios.

soil separates See separate, soil.

soil series See series, soil.

soil solution The aqueous liquid phase of the soil and its solutes, consisting of ions dissociated from the surfaces of the soil particles and of other soluble materials.

soil strength A transient soil property related to the soil's solid phase cohesion and adhesion.

soil structure The combination or arrangement of primary soil particles into secondary particles, units, or peds. These secondary units may be, but usually are not, arranged in the profile in such a manner as to give a distinctive characteristic pattern. The secondary units are characterized and classified on the basis of size, shape, and degree of distinctness into classes, types, and grades, respectively.

soil structure classes A grouping of soil structural units or peds on the basis of size from the very fine to very coarse.

soil structure grades A grouping or classification of soil structure on the basis of inter- and intraaggregate adhesion, cohesion, or stability within the profile. Four grades of structure, designated from 0 to 3, are recognized: *structureless, weak, moderate,* and *strong*.

soil structure types A classification of soil structure based on the shape of the aggregates or peds and their arrangement in the profile, including platy, prismatic, columnar, blocky, subangular blocky, granulated, and crumb.

soil survey The systematic examination, description, classification, and mapping of soils in an area. Soil surveys are classified according to the kind and intensity of field examination.

soil temperature classes A criterion used to differentiate soil in *Soil Taxonomy*, mainly at the family level. Classes are based on mean annual soil temperature and on differences between summer and winter temperatures at a depth of 50 cm.

soil textural class A grouping of soil textural units based on the relative proportions of the various soil separates (sand, silt, and clay). These textural classes, listed from the coarsest to the finest in texture, are sand, loamy sand, sandy loam, loam, silt loam, silt, sandy clay loam, clay loam, silty clay loam, sandy clay, silty clay, and clay. There are several subclasses of the sand, loamy sand, and sandy loam classes based on the dominant particle size of the sand fraction (e.g., loamy fine sand, coarse sandy loam).

soil texture The relative proportions of the various soil separates in a soil.

soil water deficit The difference between PET and ET, representing the gap between the amount of evapotranspiration water atmospheric conditions "demand" and the amount the soil can actually supply. A measure of the limitation that water supply places on plant productivity.

soil water potential (total) A measure of the difference between the free energy state of soil water and that of pure water. Technically it is defined as "that amount of work that must be done per unit quantity of pure water in order to transport reversibly and isothermically an infinitesimal quantity of water from a pool of pure water, at a specified elevation and at atmospheric pressure, to the soil water (at the point under consideration)." This *total* potential consists of *gravitational*, *matric*, and *osmotic* potentials.

solarization The process of heating a soil in the field by covering it with clear plastic sheeting during sunny conditions. The heat is meant to partially sterilize the upper 5 to 15 cm of soil to reduce pest and pathogen populations.

solum (pl. sola) The upper and most weathered part of the soil profile; the A, E, and B horizons.

sombric horizon A diagnostic subsurface horizon that contains illuvial humus but has a low cation exchange capacity and low percentage base saturation. Mostly restricted to cool, moist soils of high plateaus and mountainous areas of tropical and subtropical regions.

sorption The removal from the soil solution of an ion or molecule by adsorption and absorption. This term is often used when the exact mechanism of removal is not known.

species diversity The variety of different biological species present in an ecosystem. Generally, high diversity is marked by many species with few individuals in each.

species richness The number of different species present in an ecosystem, without regard to the distribution of individuals among those species.

specific gravity The ratio of the density of a mineral to the density of water at standard temperature and pressure.

specific heat capacity The amount of kinetic (heat) energy required to raise the temperature of 1 g of a substance (usually in reference to soil or soil components).

specific surface The solid particle surface area per unit mass or volume of the solid particles.

splash erosion The spattering of small soil particles caused by the impact of raindrops on very wet soils. The loosened and separated particles may or may not be subsequently removed by surface runoff.

spodic horizon A diagnostic subsurface horizon characterized by the illuvial accumulation of amorphous materials composed of aluminum and organic carbon with or without iron.

Spodosols An order in *Soil Taxonomy*. Soils with subsurface illuvial accumulations of organic matter and compounds of aluminum and usually iron. These soils are formed in acid, mainly coarse-textured materials in humid and mostly cool or temperate climates.

stem flow The process by which rain or irrigation water is directed by a plant canopy toward the plant stem so as to wet the soil unevenly under the plant canopy.

storm flow The folks or flow of water in a stream coming from surface runoff during or just after rainfall or snow melt. Opposite of *base flow*.

stratified Arranged in or composed of strata or layers.

strip-cropping The practice of growing crops that require different types of tillage, such as row and sod, in alternate strips along contours or across the prevailing direction of wind.

structure, soil *See* soil structure.

stubble mulch The stubble of crops or crop residues left essentially in place on the land as a surface cover before and during the preparation of the seedbed and at least partly during the growing of a succeeding crop.

subgroup, soil The *great groups* in *Soil Taxonomy* are subdivided into central concept subgroups that show the central properties of the great group, intergrade subgroups that show properties of more than one great group, and other subgroups for soils with atypical properties that are not characteristic of any great group.

submergence potential The positive hydrostatic pressure that occurs below the water table.

suborder, soil A category in *Soil Taxonomy* that narrows the ranges in soil moisture and temperature regimes, kinds of horizons, and composition, according to which of these is most important.

subsoil That part of the soil below the plow layer.

subsoiling Breaking of compact subsoils, without inverting them, with a special knifelike instrument (chisel), which is pulled through the soil at depths usually of 30 to 60 cm and at spacings usually of 1 to 2 m.

sulfidic Adjective used to describe sulfide-containing soil materials that initially have a pH > 4.0 and exhibit a drop of at least 0.5 pH unit within 8 weeks of aerated, moist incubation. Found in potential acid sulfate soils.

sulfuric horizon A diagnostic subsurface horizon in either mineral or organic soils that has a pH < 3.5 and fresh straw-colored mottles (called *jarosite mottles*). Forms by oxidation of sulfide-rich materials and is highly toxic to plants.

summer fallow *See* fallow.

surface runoff *See* runoff.

surface seal A thin layer of fine particles deposited on the surface of a soil that greatly reduces the permeability of the soil surface to water.

surface soil The uppermost part of the soil, ordinarily moved in tillage, or its equivalent in uncultivated soils. Ranges in depth from 7 to 25 cm. Frequently designated as the *plow layer*, the *Ap layer*, or the *Ap horizon*.

surface tension The elasticlike phenomenon resulting from the unbalanced attractions among liquid molecules (usually water) and between liquid and gaseous molecules (usually air) at the liquid–gas interface.

swamp An area of land that is usually wet or submerged under shallow fresh water and typically supports hydrophilic trees and shrubs.

symbiosis The living together in intimate association of two dissimilar organisms, the cohabitation being mutually beneficial.

synergism (i) The nonobligatory association between organisms that is mutually beneficial. Both populations can survive in their natural environment on their own, although, when formed, the association offers mutual advantages. (ii) The simultaneous actions of two or more factors that have a greater total effect together than the sum of their individual effects.

talus Fragments of rock and other soil material accumulated by gravity at the foot of cliffs or steep slopes.

taxonomy, soil The science of classification of soils; laws and principles governing the classifying of soil. Also a specific *soil* *classification* system developed by the U.S. Department of Agriculture.

tensiometer A device for measuring the negative pressure (or tension) of water in soil *in situ*; a porous, permeable ceramic cup connected through a tube to a manometer or vacuum gauge.

tension, soil-moisture *See* soil water potential.

terrace (1) A level, usually narrow, plain bordering a river, lake, or the sea. Rivers sometimes are bordered by terraces at different levels. (2) A raised, more or less level or horizontal strip of earth usually constructed on or nearly on a contour and designed to make the land suitable for tillage and to prevent accelerated erosion by diverting water from undesirable channels of concentration; sometimes called *diversion terrace*.

tetrahedral sheet Sheet of horizontally linked, tetrahedron-shaped units that serve as one of the basic structural components of silicate (clay) minerals. Each unit consists of a central four-coordinated atom (e.g., Si, Al, Fe) surrounded by four oxygen atoms that, in turn, are linked with other nearby atoms (e.g., Si, Al, Fe), thereby serving as interunit linkages to hold the sheet together.

texture *See* soil texture.

thermal analysis (differential thermal analysis) A method of analyzing a soil sample for constituents, based on a differential rate of heating of the unknown and standard samples when a uniform source of heat is applied.

thermic A soil temperature class with mean annual temperature 15 to 22°C.

thermophilic Pertaining to temperatures in the range of 45 to 90°C, the range in which thermophilic organisms grow best and in which thermophilic composting takes place.

thermophilic organisms Organisms that grow readily at temperatures above 45°C.

thixotrophy The property of certain clay soils of becoming fluid when jarred or agitated and then setting again when at rest. Similar to *quick*, as in quick clays or quicksand.

tile, drain Pipe made of burned clay, concrete, or ceramic material, in short lengths, usually laid with open joints to collect and carry excess water from the soil.

till (1) Unstratified glacial drift deposited directly by the ice and consisting of clay, sand, gravel, and boulders intermingled in any proportion. (2) To plow and prepare for seeding; to seed or cultivate the soil.

tillage The mechanical manipulation of soil for any purpose; but in agriculture it is usually restricted to the modifying of soil conditions for crop production.

tillage, conservation Any tillage sequence that reduces loss of soil or water relative to conventional tillage, which generally leaves at least 30% of the soil surface covered by residues, including the following systems:

minimum tillage The minimum soil manipulation necessary for crop production or meeting tillage requirements under the existing soil and climatic conditions.

mulch tillage Tillage or preparation of the soil in such a way that plant residues or other materials are left to cover the surface; also called *mulch farming, trash farming, stubble mulch tillage,* and *plowless farming*.

no-tillage system A procedure whereby a crop is planted directly into a seedbed not tilled since harvest of the previous crop; also called *zero tillage*.

ridge till Planting on ridges formed by cultivation during the previous growing period.

strip till Planting is done in a narrow strip that has been tilled and mixed, leaving the remainder of the soil surface undisturbed.

tillage, conventional The combined primary and secondary tillage operations normally performed in preparing a seedbed for a given crop grown in a given geographic area. Usually said of non-conservation tillage.

tillage, primary Tillage that contributes to the major soil manipulation, commonly with a plow.

tillage, rotary An operation using a power-driven rotary tillage tool to loosen and mix soil.

tillage, secondary Any tillage operations following primary tillage designed to prepare a satisfactory seedbed for planting.

tilth The physical condition of soil as related to its ease of tillage, fitness as a seedbed, and its impedance to seedling emergence and root penetration.

topdressing An application of fertilizer to a soil after the crop stand has been established.

toposequence A sequence of related soils that differ, one from the other, primarily because of *topography* as a soil-formation factor, with other factors constant.

topsoil (1) The layer of soil moved in cultivation. *See also* surface soil. (2) Presumably fertile soil material used to top-dress roadbanks, gardens, and lawns.

trace elements Elements present in the Earth's crust in concentrations less than 1,000 mg/kg. When referring to plant nutrients, the term *micronutrients* is preferred.

trioctahedral An octahedral sheet of silicate clays in which the sites for the six-coordinated metallic atoms are mostly filled with divalent cations, such as Mg^{2+}.

trophic level Levels in a food chain that pass nutrients and energy from one group of organisms to another.

truncated Having lost all or part of the upper soil horizon or horizons.

tuff Volcanic ash usually more or less stratified and in various states of consolidation.

tundra A level or undulating treeless plain characteristic of arctic regions.

Ultisols An order in *Soil Taxonomy*. Soils that are low in bases and have subsurface horizons of illuvial clay accumulations. They are usually moist, but during the warm season of the year some are dry part of the time.

umbric epipedon A diagnostic surface horizon of mineral soil that has the same requirements as the mollic epipedon with respect to color, thickness, organic carbon content, consistence, structure, and P_2O_5 content, but that has a base saturation of less than 50%.

universal soil loss equation (USLE) An equation for predicting the average annual soil loss per unit area per year; $A = RKLSPC$, where R is the climatic erosivity factor (rainfall plus runoff), K is the soil erodibility factor, L is the length of slope, S is the percent slope, P is the soil erosion practice factor, and C is the cropping and management factor.

unsaturated flow The movement of water in a soil that is not filled to capacity with water.

vadose zone The aerated region of soil above the permanent water table.

value (color) *See* Munsell color system.

variable charge *See* pH-dependent charge.

varnish, desert A glossy sheen or coating on stones and gravel in arid regions.

vermicompost Compost made by earthworms eating raw organic materials in moist aerated piles, which are kept shallow to avoid heat buildup that could kill the worms.

vermiculite A 2:1-type silicate clay, usually formed from mica, that has a high net negative charge stemming mostly from extensive isomorphous substitution of aluminum for silicon in the tetrahedral sheet.

Vertisols An order in *Soil Taxonomy*. Clayey soils with high shrink–swell potential that have wide, deep cracks when dry. Most of these soils have distinct wet and dry periods throughout the year.

vesicles (1) Unconnected voids with smooth walls. (2) Spherical structures formed inside root cortical cells by vesicular arbuscular mycorrhizal fungi.

virgin soil A soil that has not been significantly disturbed from its natural condition.

water deficit (soil) The amount of available water removed from the soil within the vegetation's active rooting depth, or the amount of water required to bring the soil to field capacity.

water potential, soil *See* soil water potential.

water table The upper surface of groundwater or that level below which the soil is saturated with water.

water table, perched The surface of a local zone of saturation held above the main body of groundwater by an impermeable layer of stratum, usually clay, and separated from the main body of groundwater by an unsaturated zone.

water use efficiency Dry matter or harvested portion of crop produced per unit of water consumed.

waterlogged Saturated with water.

watershed All the land and water within the geographical confines of a drainage divide or surrounding ridges that separate the area from neighboring watersheds.

water-stable aggregate A soil aggregate stable to the action of water, such as falling drops or agitation, as in wet-sieving analysis.

weathering All physical and chemical changes produced in rocks, at or near the Earth's surface, by atmospheric agents.

wetland An area of land that has hydric soil and hydrophytic vegetation, typically flooded for part of the year, and forming a transition zone between aquatic and terrestrial systems.

wetting front The boundary between the wetted soil and dry soil during infiltration of water.

wilting point (permanent wilting point) The moisture content of soil, on an oven-dry basis, at which plants wilt and fail to recover their turgidity when placed in a dark, humid atmosphere.

windbreak Planting of trees, shrubs, or other vegetation perpendicular, or nearly so, to the principal wind direction to protect soils, crops, homesteads, etc., from wind and snow.

xenobiotic Compounds foreign to biological systems. Often refers to compounds resistant to decomposition.

xerophytes Plants that grow in or on extremely dry soils or soil materials.

yield gap The difference between actual crop production levels at a given location with current management and the potential productivity of the soils and climate with best available management.

zero tillage *See* tillage, conservation.

zymogenous organisms So-called opportunist organisms found in soils in large numbers immediately following addition of readily decomposable organic materials. *Contrast with* autochthonous organisms. *See also* r-strategist.

Index

A

A horizons, 13, 63
Abrasion, rock and mineral weathering and, 33
Absorption
  of methane, 7
  of oxygen, 7
  of rainwater, 3
  sodium absorption ratio, 350
  of sulfates, 495
Abundance, of soil organisms, 378–379
Accelerated erosion, 608–610
  on-site and off-site effects of, 610–612
  on range- and forestland, 631–632
  timber production, 632–634
Acetanilides, 658
Acid cations, 315, 318
Acid mine drainage, 329
Acid rain
  acidity falls in, 327
  aquatic ecosystem effect of, 328
  forest effect of, 326–328
  overview of, 326
  sensitive soils of, 328
  soil acidification with, 326
Acid saturation, 320
Acid-forming tendency, of inorganic fertilizers, 579
Acidic soils, 9
  aluminum role in, 317–318
  buffering of, 321–323
  gypsum for, 341
  human-influenced, 325–330
  introduction to, 312
  liming for, 336–341
  nutrient availability in, 332–333
  organic matter for, 341–343
  pools of soil acidity, 318–321
  process for, 313–317
  Spodosols, 101–102
  without lime, 341–343
Acidification, 312
Actinomycetes, 403
Active acid sulfate soils, 328
Active acidity, 318
Activity
  of earthworms, 381–383
  of fungi, 395–396
  of soil organisms, 378–379
Adenosine triphosphate (ATP), 496
Adhesion, in water, 166, 168
Adsorption
  of anions and cations, 276–277, 293–295
  clay formation and, 132–133
  on colloids of cations and anions, 277
  of essential elements, 22
  of organic compound, 659
  water retention and movement, 166
AEC. *See* Anion exchange capacity
Aeration
  characterization of, 241
  compaction and, 190–191
  ecological effects of, 247–250
  nitrification and, 474
  oxidation-reduction potential, 242–245
  in urban landscapes, 251–253
  process of, 239–241
  respiration rates and, 245
  for septic drain field, 228
  seasonal differences of, 247
  small-scale soil heterogeneity and, 246
  soil compaction and, 250
  vegetation effects of, 247
  water drainage and, 246
  of wetlands, 253–258
Aerenchyma tissues, 240
Aerobic microorganisms, 405
Aerobic soils, decomposition in, 424
Aerosolic dust, 43, 44
Aggregates
  description of, 127–128
  formation and stabilization of, 131–137
  maintenance of, 140–141
  pore space and, 150
  tillage and, 137–141
Aging, of inorganic phosphorus, 505
Agricultural limes, 336
Agriculture, 1–2
  1:1 silicate clays in, 281–282
  Andisols and, 87
  animal manures in, 564–567
  Aridisols and, 92
  conventional tillage and, 138–139
  conversation tillage and, 139
  Histosols and, 91
  irrigation for, 229
  soil density and, 146–147
  soil formation and, 51–53
  soil–plant–atmosphere continuum, 205–209
  SOM influence of, 453
  Spodosols, 102
  ET control in, 209–213
  Verisols and, 94
Agroecosystems
  carbon balance in, 444–446
  potassium in, 516
Air. *See also* Soil air
  atmospheric air v., 21–22
Air injection, for saline-sodic and sodic soils reclamation, 363
Air-filled porosity, 242
Albedo, 264
Albic horizon, 77, 83, 98, 101
Albolls, 96–98
Alfisols
  profile of, 81, 98–99
  distribution and use, 99
  profile, 98
Algae, as soil conditioner, 140, 393
Alkaline soils
  introduction to, 312
  nutrient availability in, 332–333
  process for, 313–317
Alkalinity
  irrigation-induced, 347–348
Allelochemical effects, 439–440
Allelopathy, 439
Allophane, 44, 86
  charge of, 290–293
  genesis of, 289
  properties of, 278, 286–287
Alluvial deposits, 39–41
Alluvial fans, 40–41
Aluminum
  exchangeable, 338–339
  phosphorus precipitation by, 505
  soil acidity role of, 317–318
  soil aggregation and, 135
  soil buffering and, 321–322
  in soil colloids, 279–281
  toxicity of, 330–331
Aluminum oxide clays
  characteristics of, 286–287
  charge of, 290–293
  soil colloids and, 278–279
  Spodosols, 102
  sulfur in, 490
  weathering and, 33–34
Aluminum Toxicity
  symptoms of, 330
  tolerance of, 330
Aluminum-bound inorganic phosphorus, 500
AM. *See* Arbuscular mycorrhizae
Ammonia, volatilization of, 472–473
Ammonification, 474
Ammonium
  fixation, 471
  in nitrogen cycle, 468
Anaerobic digestion, of animal manures, 568
Anaerobic microorganisms, 405
Anaerobic soils
  decomposition in, 426
  description of, 241
  microorganisms in, 243
Anammox, 476
Andic properties, of Andisols, 87
Andisols, profile of, 81, 86–87
Anecic earthworms, 379
Angle of repose, 155
Animal manures
  nutrient composition of, 564
  in nutrient management, 564–570
Animals
  potassium and, 514–515
  rock abrasion by, 33–34
  soil formation and, 51–53
  sulfur in, 487–488

Anion effects, on mass action, 297
Anion exchange, 296, 304–306. *See also* Exchangeable anions
Anion exchange capacity (AEC), 305
Anions
 role of carbonate and bicarbonate, 316–317
 soil colloid adsorption of, 276–277, 293–295
Anoxic conditions, 9
Anthrepts, 85–86
Antibiotics, 308, 403
Ants, 52, 384
Apatite, 496, 504
Aqualfs, 98–99
Aquands, 86–87
Aquatic ecosystem, acid rain effect on, 328
Aquents, 83–85
Aquepts, 85–86
Aquerts, 94–96
Aquifers, 216
Aquods, 101–102
Aquolls, 79, 96
Aquox, 102–103
Aquults, 99–101
Arbuscular mycorrhizae (AM), 397–398
Arbuscules, 398
Archaea, 371, 400–403, 426
Archeology and soil classification, 107
Arents, 83
Argiaquolls, 80
Argids, 92–94
Argillic horizon, 77, 93, 99–101
Argillans, 77
Aridisols
 characteristics, 92–93
 distribution and use, 93
 profile of, 81–82, 92–94
 sulfur in, 489
Arsenic
 in drainage water, 358
 in orchard soils, 678
Artificial drainage, benefits and detriments of, 221–222
Atmosphere
 carbon cycle and, 455–456
 nitrogen deposition from, 484
 soil interface with, 10
 soil modification of, 7
 soil pH from, 326–328
 soil-plant-atmosphere continuum, 205–209
 as sulfur source, 488–491
Atmospheric pollution, denitrification and, 476
ATP. *See* Adenosine triphosphate
Atrazine, soil pH and, 336
Atterberg limits, 157–158
Australian soil classification system, 700
Autotrophs, 371, 373, 473, 484
Available water-holding capacity, 189–190
 compaction, 190
 osmotic potential, 190
 root distribution, 193
 soil depth and layering, 190–191

B

B horizons, 13–15, 62–64
Bacteria, 135, 371, 400–403

Bacterivores, in soil, 371
Barriers, wind erosion and, 643
Basalt, Vertisols from, 94
"Base" cations, 320
Base saturation percentage, 303
Basicity, of soil, 9
Bearing strength, 154–156
Bedrock, 12
Bench terraces, 621
Beneficial soil organic matter, 7
Bicarbonate, pH and, 316–317
Bioaugmentation, 670
Bioconcentration factor (BCF), 682
Biocrusts, 412–413
Bioengineering, 636–637
Biofuel production, greenhouse effect and, 456–457
Biogeochemical weathering, 34
Biological nitrogen fixation, 478–479
Biomass, 378–379
 in soil, 19
 in SOM, 432–433
Biomolecules, clay and humus binding of, 308–309
Biopores, 151, 179
Bioremediation, 669–672, 683
Biosequence, 36
Biosolids, 572
Biosphere 2, 420
Biosphere, soil interface with, 10
Biostimulation, 670–671
Black alkali, 352
Black belt, 96
Block-like structure, 129–130
Blue-green algae, 393
Border irrigation, 233
Boron
 contaminants, 678
 deficiency v. toxicity of, 528–530
 in drainage water, 358
Boulders, 123
Broad-based terraces, 622
Broadcasting, of fertilizer, 581–582
Brush mattresses, 636
Bt, 308
Buffer strips, along stream channels, 634
Buffering. *See also* Soil buffering capacity
 of acidic soils, 321–323
 lime needs based on, 339
 mechanisms of, 322
Buffering capacity
 potassium and, 520
 of soil solution, 21
Building foundations
 1:1 silicate clays, 281–282
 bulk density and, 145
 of clay in soil, 19–20
 enhancing soil drainage, 220–221
 in Oxisols, 102
 regolith properties and, 16
 soils and, 9–10
Building materials, soils as, 9
Bulk density
 description of, 141
 factors affecting, 141–144
 management practices for, 145–150
Buried perforated pipes, for drainage, 226, 228

Burrows, of earthworms, 379–380
Bypass flow, 219

C

C. *See* Climatic factor; Cover and management factor
C horizons, 13–15, 62–64
CAFOs. *See* Concentrated Animal-Feeding Operations
Calcicoles, 522
Calcids, 92–94
Calcifuge, 522
Calcium
 available quantities of, 23
 leaching of, 524
 magnesium ratio to, 525
 in Mollisols, 96
 in plants, 522
 soil forms and processes, 523–524
 soil pH and, 334–335
Calcium saturation percentage, 303
Calcium-bound inorganic phosphorus, 500
Cambids, 92–94
Camping, soil density and, 145
Canadian soil classification system, 700–701
 U.S. Soil classification taxonomy *vs.*, 701–702
Capacitance, for water content measurement, 175
Capillary forces, 166–167
Capillary movement, 193
Capillary water, 187, 200
Carbamate pesticides, 657
Carbohydrates, in plants, 423
Carbon
 in organic matter, 419
 sources of, 373–375, 421–423
Carbon balance
 in agroecosystems, 444–446
 in natural ecosystems, 447
 in SPAC, 444–447
Carbon cycle
 atmosphere and, 455–456
 in Biosphere 2, 420
 carbon sources, 421–423
 basic process of, 420–421
Carbon dioxide
 in carbon cycle, 420–423
 soils' role in climate change, 454–461
 microbial respiration and, 425
 oxygen v., in soil, 21–22
 pH and, 317
 production of, 248–249
 release of, 7, 19
 soil aeration of, 239
 in soil air, 241
Carbonaceous remains, in soil, 19
Carbonates
 in carbon cycle, 421
 pH and, 316–317
 soil buffering and, 321
Carbonic acid, 314
 in carbon cycle, 421
Carbon/nitrogen (C/N) ratio
 in composting, 461–463
 of organic residue, 428–429
Carnivores, 375
Casts, of earthworms, 380–381

Catchment, 200
Catena, 54
Cation exchange, 276, 321–322. *See also* Exchangeable cations
Cation exchange capacity (CEC), 298–302
   aluminum and, 317–318
   method of measuring, 320–321
   potassium and, 520
Cation exchange reactions, 295–298
Cation saturation, 303, 320–321
Cation selectivity, with cation exchange reactions, 297
Cations. *See also* Acid cations; Nonacidcations
   adsorbed in clays, 132–133
   micronutrient, 534–540
   plant uptake of, 315
   soil colloid adsorption of, 276–277, 293–295
   tree cycling of, 50
Cattle manure, nutrient recycling and, 564–565
CEC. *See* Cation exchange capacity
Cellulose, decomposition of, 424
Channers, 123
Charge equivalence, with cation exchange reactions, 296
Chelates, 538
Chemical analysis, of soil sample, 593–594
Chemical fertilizers, soil pH with, 325–326
Chemical remediation methods, 667–669
Chemical weathering, of rocks and minerals, 29–36
Chemicals
   drainage movement of, 218
   macropores movement of, 218–219
Chisel plowing, 624
Chlorites, 286
   charge of, 290–293
Chlorophyll, magnesium in, 524
Chlorosis, 335, 467, 487, 524
Chroma, soil color, 118
Chromium, contaminants, 678
Chronosequence, 36
Clay
   ammonium fixation by, 471
   biomolecule binding to, 308–309
   capillary action in, 166–167
   description of, 18, 120–121
   flocculation of, 132–133
   geographic distribution of, 289–290
   humus and, 20, 438
   organic matter, 536
   in Oxisols, 102
   soil aggregation and, 135
   soil behavior and, 19–20
   soil texture and, 123
   in Ultisols, 99–101
   in Vertisols, 94
   volume changes in, 134
Clay domain, 133
Clay loam, 18, 123, 179
Clay particles, 18
Clay skins, 77
Clayey soils
   Atterberg limits and, 157–158
   engineering considerations with, 156
   for landfill, 688
   particle size class, 108
   tillage of, 137–138

Claypans
   septic drain field and, 228
Climate
   parent materials and, 47–49
   SOM influence of, 449–450
   topography and, 53–55
Climate change
   Gelisols and, 89
   soils' role in, 454–461
Climatic factor (C), 641
Climosequence, 36
Clods, 128
C/N ratio. *See* Carbon/nitrogen ratio
Coarse fragments, 120, 123
Coastal sediments, 41–42
Cobbles, 123
Coefficient of linear extensibility (COLE), 158
Cohesion, in water, 166
Cohesive soils, 154
COLE. *See* Coefficient of linear extensibility
Collapsible soils, 155
Colloidal fraction, 276
Colloidal properties
   of clay, 18, 122
Colloids
   description of, 275
   effect of, 304
   pH and, 320
   phosphorus with, 597
Colluvial debris, 38–39
Color
   albedo and, 264
   of hydric soils, 256
   of soil, 117–118
   of soil horizons, 15
Common names, of plants in text, 706–708
Compaction
   plant available water and, 190–191
   soil density and, 141–150
   water infiltration and, 202
Comparative organism activity, 378–379
Complementary cations
   with cation exchange reactions, 298
   influence of, 303–304
Composite sample, for soil analysis, 591–592
Compost, 410, 461–463
   disadvantages of, 462–463
   nature of, 462
Composting, 461–463
   benefits of, 462
   thermophilic, 461
Compressibility, 156
Concentrated Animal-Feeding Operations (CAFOs), 564–567
Conservation management, for soil quality, 648
Conservation Reserve Program (CRP), 648
Conservation tillage
   for erosion, 559, 622–628
   ET control with, 210–211
   hydraulic conductivity with, 179–180
   in nutrient management, 559–560
   soil temperature and, 270
   soil tilth and, 139
   vapor loss control with, 213
Consociations, 113
Consolidation test, 156
Constant charge, 290
Constructed wetlands, 257–258

Construction sites, 634–638
Consumption of soil, 5
Container-grown plants
   aeration for, 251
   salt problems with, 355
Containment landfills, 688
Contaminants. *See also* Inorganic chemicals; Organic chemicals
   concentrations and toxicity of, 663
   inorganic substances, 674–681
   lead poisoning, 676–677
Contour cultivation, 620
Contour strip-cropping, 619
Controlled traffic, soil compaction and, 148
Conventional tillage
   crop production and, 138–139
   hydraulic conductivity with, 179–180
Conversion factors, for SI units, 704
Cool climates, organic mulches in, 271
Copper, deficiency *vs.* toxicity of, 528
Core cultivation, 253
Cornstalk nitrate. *See* Cornstalk nitrate test (CSNT)
Cornstalk nitrate test (CSNT), 588
Cover and management factor (C), 618–619
Cover crops, 618
   for acidic soils, 341–343
   for nutrient management, 556–557
   for soil tilth, 140
   for water infiltration, 201–202
Cover, for disturbed soil, 634
Critical range, of nutrient concentrations, 588
Crop production, conventional tillage and, 138–139
Crop residue
   for acidic soils, 341–343
   sulfur and, 495
   for ET control, 213
Crotovinas, 51–52
CRP. *See* Conservation Reserve Program
Crustal warping, 37
Crusting, water infiltration and, 139–140
Crusts, microbiotic, 393
Cryalfs, 98–99
Cryands, 86–87
Cryepts, 85–86
Cryic regimes, 108
Cryids, 92–94
Cryods, 101–102
Cryolls, 96–98
Cryoturbation, 87–89
Cryptobiotic state, 386
Cryptopores, 152
Crystalline silicate clays. *See* Silicate clays
Cultivation, contour, 620
Cultural eutrophication, 553
Curing stage, in composting, 462
Cyanobacteria, 403
Cryerts, 94–96
Cysts, 388

D

Darcy's law, 177–178
DDT, 662–664
Decomposition
   inorganic nitrogen release during, 430–431
   of organic material, 425

phosphorus and, 500
process of, 424
rates of, 427–432
in soils, 423–427
Deep open-ditch drainage, 223–226
Degradation
of land, 607
soil (*see* Soil degradation)
Deicing salts, 355
Deleterious rhizobacteria, 408
Delivery ratio, 613
Delta deposits, 41
Denitrification, 474, 475–478
air pollution with, 476
description of, 475
in flooded soils, 477
in groundwater, 477
quantity of nitrogen lost with, 477
Deoxyribonucleic acid (DNA), 496
Deposition, in erosion, 612, 613, 640
Depth ratio, 173
Desert pavement, 93
Desert varnish, 93
Detachment, in erosion, 612, 640
Detritivores, 371, 375
Detritus, 375, 432–433
Diagnostic soil horizons, 75
Diagnostic subsurface horizons, 75–79
Dictyostelium, 389
Diffusion, 25–26, 241
Dinitroanilines, 658
Diquat, 660
Disease-suppressive soils, 409–411
Dispersion
causes of, 353–354
of sodic soils, 352
Dissolved organic nitrogen, 471–472
Dissolved organic phosphorus (DOP), 503
Distribution coefficient, soil, 307–308
Disturbed soil
cover on, 634
DNA. *See* Deoxyribonucleic acid
DNRA. *See* Dissimilatory nitrate reduction to ammonium
Dokuchaev, V. V., 72, 74
Dolomitic limestone, 337
DOP. *See* Dissolved organic phosphorus
Double disk opener, 136
Drain field
for septic system, 227–228
soil properties for, 228–229
Drainage
chemical movement in, 218
enhancing soil, 220–226
groundwater and, 216–217
hydrologic cycle and, 200
topiary garden design for, 224–225
phosphorus in, 598–599
salinity of, 357
septic tank drain fields, 226–229
soil aeration and, 246
SOM influence of, 450–451
subsurface systems for, 222–226
surface systems for, 222
toxic elements in, 358
Drainage ditches
deep open, 223–226
surface, 222

Dribble, for fertilizer, 584
Drift, glacial, 42
Drip irrigation, 235–236, 584–585
of fertilizer, 585
for saline soils reclamation, 361–362
for saline-sodic and sodic soils reclamation, 363
Dry well, 252
Dune sand, 44
Dung beetles, 377
Durids, 92–94
Duripan, 78
Dust
aerosolic, 44
in atmosphere, 7
windblown, 611

E
E horizons, 63
Earthworms, 379–383
deleterious effects of, 381–382
effects of pesticides, soil organisms, 665–667
factors affecting activity of, 382–383
nutrients in, 380–381
physical effects of, 381
soil fertility, productivity, and environmental quality and, 380–381
soil formation and, 52
EC. *See* Electrical conductivity
ECEC. *See* Effective cation exchange capacity
Effects of nitrate, 554
Enchytraeid worms, 379
Ecological relationships, among soil organisms, 412–414
Ecology of soil, 369–415
Ecosystem engineers, 377–378
Ecosystems
acid rain effect on, 328
carbon balance in, 447
decline of, 25–26
dynamics of, 372
soils' role in, 1–2
Ectomycorrhiza, 397–398
Edaphologists, 12
Effects of pesticides, soil organisms, 665–667
fumigants, 665
on soil fauna, 665
earthworms, 665
soil microorganisms, 666–667
Effective cation exchange capacity (ECEC), 299, 319
Effective precipitation, 48
Effluent, 228, 571–572
Electrical conductivity (EC), 348–350
Electrical resistance blocks, 176–177
Electromagnetic induction (EM), 349
Electron acceptors, 242–243
Elemental sulfur, cycle of, 491–492
Elements, periodic table of, 705
Eluviation, 63
EM. *See* Electromagnetic induction
Endogeic earthworms, 379
Endomycorrhiza, 398–399
Endorhizosphere, 391
Energy curves, water, 171–172
Energy potential, soil water, 171–177
Energy production, on Aridisols, 92–94

Energy, sources of, 373–375, 402
Engineering considerations
1:1 silicate clays, 281–282
bulk density and, 145
of clay in soil, 19–20
enhancing soil drainage, 220–221
Gelisols, 87–88
Histosols, 89–92
matric potential and, 169–170
in Oxisols, 102
regolith and, 16
soil properties and, 9–10, 153–159
unified classification system, 158–159
Vertisols, 94
Engineers, ecosystem, 377–378
Enhanced rhizosphere phytoremediation, 673
Entisols, profile of, 83–85
Environmental impacts
of landfills, 688–690
of organic chemicals, 653
Environmental quality
earthworms' influence on, 380–381
Environmental uses, of smectite-type expanding clay, 284–285
Eolian deposits, 43
EPC. *See* Equilibrium phosphorus concentration
Epigeic earthworms, 379
Epipedons, 75
Equilibrium phosphorus concentration (EPC), 508
Eroded soil, deposition of, 613–614
Erodibility, 616–617, 641
Erosion, 203–204, 606–649
conservation tillage for, 559, 622–628
on construction sites, 634–638
control of accelerated erosion on range- and forestland, 631–634
factors affecting interrill and rill erosion, 616–622
geological *vs.* accelerated, 608–610
gully erosion and mass wasting, 629–631
models for predicting water-induced erosion, 614–616
on-site and off-site effects of, 610–612
phosphorus with, 597
prediction and control of wind erosion, 642–644
progress in soil conservation, 647–649
significance of, 607–610
tillage, 644–647
vegetative barriers for, 628–629
water, 612–614
by wind, 638–641
Erosion blankets, 635
Erosivity, 616–617
Essential elements
availability of, 22
for plant growth, 5
ET. *See* Evapotranspiration
ET, control of, 209–213
unwanted vegetation, 209
alternative weed controls, 210
Ethnopedology, 74
Ethylene
soil aeration of, 239
in soil air, 241

Eukarya, 371
Eutrophication, 218
  phosphorus and, 496, 553
Evaporation, 198–199
  control of, 211–213
  percolation balance with, 214–215
  salt-affected soils and, 346
Evapotranspiration (ET), 199, 205–209
  efficiency of, 208
  hydraulic redistribution, 208–209
  measurement of, 205–206
  plant characteristics, 207
  plant water stress, 207
  salinity and, 357
  soil moisture, 206
  water use efficiency, 207–208
Exchangeable acidity, 318
Exchangeable aluminum, 338–339
Exchangeable anions, soil colloids and, 277, 296
Exchangeable cations
  in field soils, 302–304
  soil colloids and, 277, 296
  soil microorganisms requirement of, 405
Exchangeable ions, soil solution and, 22
Exfoliation, of rocks, 33
Expansive soils, 157
External surface, of colloids, 276

F

Facultative bacteria, 405
Fairy rings, 395
Fallow, 210
Families, in *Soil Taxonomy*, 80, 108
Farmers, and soil conservation, 648
Fats, in plants, 423–424
Fauna, 408
"Feel" method, for soil textural class, 124–127
Ferrihydrite, 86
Fertigation, 583
Fertility. *See also* Soil fertility
  and control of plant disease, 408–409
  earthworms' influence on, 379–380
  soil fungi and, 395
Fertilization
  application methods for, 581–585
  organic application rates of, 574–576
  timing of, 585–586
Fertilizer grade, 578
Fertilizers
  inorganic commercial, 576–581
  liquid, 583
  nitrogen, 325–326
  starter, 584
Fiber production. *See* Agriculture
Fibric material, 47
Fibrists, 90
Field capacity, 186–187, 190
Field observations, nutrient management and, 587
Field soils
  aeration in, 240
  exchangeable cations in, 302–304
  pH in, 323–325
Field water efficiency, 232
Fine earth fraction, 120

Fine-grained mica
  charge of, 290–293
  properties of, 282, 285–286
Fire
  in grassland carbon balance, 447
  soil heating by, 263
Fixation
  of ammonium, 471
  of nitrogen, 406–407, 470, 478–483
  of phosphorus, 503, 507–509
  of potassium, 520
Flocculation
  of clay, 132–133
  gypsum and, 140
Floccules, 353
Flooded soils, denitrification in, 477–478
Floodplains, 39–40
Fluorine, contaminant, 678
Fluvents, 83
Foliar application, of fertilizer, 585
Folists, 90
Food chain, 373
Food production. *See* Agriculture
Food web, 373
Food-processing wastes, nutrient recycling of, 570
Forest ecosystems, earthworms in, 383
Forest floor. *See* O horizons
Forestlands, accelerated erosion on, 631–634
Forestry, soil density and, 145
  testing, 595
Forests
  acid rain effect on, 326
  Alfisols in, 98
  calcium and, 522–524
  carbon balance in, 444
  decline of, 25–26
  effects, 485
  fire and, 263
  liming of, 340
  microorganisms in, 49–50
  soil density and, 145
  soil temperature in, 270
  Spodosols, 101–102
  urea fertilizer application in, 586
Fossil fuels, carbon dioxide and, 423, 455
Foundation drains, 226
Fragipan, 78
Fragipan horizons, 107
  septic drain field and, 228
  water infiltration and, 204
Frankia, 482
Freeze-thaw cycles, 262
  clay volume changes and, 134
  nitrification and, 474
Friability, 137
Frigid regimes, 108
Frost churning, 87–89
Frost heaving, 262
Functional diversity, of soil organisms, 372
Functional redundancy, of ecosystems, 372
Fungal hyphae, soil aggregation and, 134–135
Fungi, 393–400
  activity of, 395–396
  mycorrhizae, 397–400
Fungicides, 658
Fungivores, in soil, 371
Furrows, for irrigation, 232–233

G

Garbage, nutrient recycling and, 570
Gas exchange, in soil aeration, 240–241
Gelands, 86–87
Gelepts, 85–86
Gelisols
  microbial activity in, 261
  profile of, 82–83, 87–88
Gelods, 101
Gelolls, 82
General disease suppression, 409–410
Genetic horizon, 67
Genetic resources, of ecosystems, 373
Genetically modified organisms (GMOs), clay and humus binding of, 308
Geographic information systems (GIS), soil analysis with, 593
Geological erosion, 608–610
Geophagy, 4
Geosmins, 403
Geotextile, 636
Gibbsite, 286
  charge of, 290–293
Gilgai, 94–95
GIS. *See* Geographic information systems
Glacial drift, 42
Glacial ice, parent materials transported by, 42–43
Glacial outwash, 43
Glacial till, 42–43
Glauconite, 285
Gleying, 119, 228
Global biodiversity, 373
Global climate change, 419
Global hydrologic cycle, 198–200
Global warming, 7. *See also* Greenhouse effect
Global-positioning system (GPS), soil analysis with, 593
Glomalin, 134
Glyphosate, 658
GMOs. *See* Genetically modified organisms
Goethite, 278, 287, 290–293
Golf courses
  irrigation for, 230–231
  soil water movement in, 184
GPS. *See* Global-positioning system
Grade, of inorganic fertilizers, 578
Grass tetany, 524
Grassed waterways, 619
Grassland
  fire and carbon balance in, 447
  microorganisms in, 49–50
  Mollisols, 96
Gravel, 123
Gravimetric method, for water content measurement, 174–175
Gravitational potential, 169
Gravitational water, 186–187
Gravity, soil water and, 168
Grazing, on Aridisols, 93
Great groups, in *Soil Taxonomy*, 79, 104–105
Green manure crops
  lignin and polyphenols in, 431–432
  for soil tilth, 140–141
  for symbiotic nitrogen fixation, 483
Green roofs, 146
Greenhouse effect

biofuel production and, 456–457
description of, 454
methane and, 458–460
organic matter role in, 419
soils' role in, 7, 454
wetlands and, 457–458
Greenhouse gases
carbon cycle and, 422–423
description of, 454
nitrous oxide, 460–461
production of, 248–249
Groundwater
contamination of, 662
denitrification in, 477–478
description of, 216
percolation and, 216–220
pesticides and, 664
resources of, 216–217
septic drain fields and, 228
Groundwater pollution
gravitational water, 187
preferential flow and, 179–180
Growth, of soil microorganisms, 405
Gullies, 613
remedial treatment of, 629–630
Gully erosion, 613, 629–631
Gypsids, 92–94
Gypsum
for saline-sodic and sodic soils reclamation, 362–364
as soil conditioner, 140

H

Hard armor, 636
Hapludalf, 14
Health hazards, from particulate matter, 611
Heat of vaporization, 267–268
Heavy metals, in sewage sludge, 679–681
HEL. *See* Highly erodible land
Hemists, 89–90
Herbaceous peat, 46
Herbicides, 209–210, 658
Herbivores, in soil, 371, 375
Hermaphrodites, 379
Heterotrophs, 371–373
denitrification, 475, 484
nonsymbiotic fixation, 484
Hierarchy of aggregation, 131, 133
Histels, 89
Histosols
distribution and use, 91
formation process, 89–90
greenhouse effect and, 456–457
profile of, 82–83, 90
Hue, soil color, 118
Human-accelerated erosion, 608
Human influences and urban soils, 52–53
Humans, soil formation and, 52–53
Humic materials
from microbial activity, 434–436
as soil conditioner, 438
Hummocks, 88
Humods, 101
Humults, 99–100
Humus
biomolecule binding to, 308–309
charge of, 290–293
clay and, 20, 438

description of, 20
in forests, 447
genesis of, 432–438
plant influence of, 438–439
production of, 7, 426
soil colloids and, 287
in SOM, 433
sorption of, 306–309
stability of, 438
Hurricane Katrina, 160
Hydrated radius, 297
Hydration, 35
Hydraulic conductivity
percolation and, 182–183
in saturated soils, 179–180
in unsaturated soils, 180–181
Hydraulic gradient, 177–178
Hydraulic lift, 209
Hydraulic redistribution, 208–209
Hydric soils, 255
Hydrogen bonding
soil colloids, 282
in water, 165
Hydrogen sulfide, in soil air, 241
Hydrologic cycle
global, 198–200
introduction to, 197–198
irrigation principles and practices, 229–236
percolation and groundwater, 216–220
septic tank drain fields, 226–229
soil-plant-atmosphere continuum, 205–209
water loss from soil, 213–216
Hydrology, wetland, 254
Hydroperiod, 255
Hydrophobic substances, 307
Hydrophytes, 240, 255–257
Hydroponics, 5, 438
Hydroseeder, 635
Hydrosphere, soil interface with, 10
Hydrostatic potential, 168–169
Hydrous oxides, phosphorus reactions with, 505–506
Hydroxylated surface, 281
Hydroxyl ions, production of, 313–317
Hygroscopic coefficient, 188–189
Hyperaccumulation, 673
Hyperaccumulator plants, bioremediation by, 684
Hyphae, 393

I

I. See Soil erodibility factor
Ice, 33, 42–43
Ice lenses, 262
Igneous rock, 30–31
Illite, 285
Illuviation, 63
Immobilization
description of, 470
of nitrogen, 468–470
of sulfur, 492–493
Imogolite, 86, 278, 289
Inceptisols, profile of, 82–83, 85–86
Incoming water
fate of, 200–205
Induced systemic resistance, 411

Industrial by-products, nutrient management of, 570–573
Industrial organics, 657
Infiltrability, 182
Infiltration of water, 182, 200
fires and, 263
plastic mulch and, 212–213
precipitation timing and, 200
soil management for, 201–202
soil properties and, 203–205
stem flow, 200–201
in urban watersheds, 203–205
vegetation type, 200
Inner-sphere complex, 293, 294–295, 304
Inorganic chemicals
contamination with, 674–678
in landfill, 687
for lowering soil pH, 343–345
prevention and remediation of, 681–684
Inorganic commercial fertilizers, 576–581
fate of, 578–581
grade of, 578
limiting factor, 581
properties of, 577–578
use of, 577
Inorganic matters
compositions, 17–18
production of, 426
soil minerals, 18–19
soil structure, 18, 19
Inorganic micronutrients, 532–534
Inorganic phosphorus, 498–499
in soils, 503–507
solubility of, 505
Inorganic transformations, by soil organisms, 406
Insecticides, 657–658
Interception of water by plants, 200
Internal surface, of colloids, 276
International System of Units (SI)
conversion factors for, 704
prefixes in, 703
units of measure, 703
Interped pores, 151
Interrill erosion, 613, 616–622
Invertebrate Animals
role in soil formation, 52
Ion exchange, of essential elements, 22
Ionic double layer, 277
Iron
in hydric soils, 255
phosphorus precipitation by, 505–507
soil aggregation and, 135
in soil colloids, 279–281
soil color and, 118
soil pH and, 334
Iron oxide clays
characteristics of, 286–288
charge of, 290–293
pseudosand, 133
Spodosols, 101
sulfur in, 490
weathering and, 33–34
Iron-bound inorganic phosphorus, 499–500
Irrigation
Aridisols and, 93
fate of, 200–205
microirrigation, 232–236

Irrigation (*Continued*)
  modern, 230–231
  nitrification and, 474
  percolation with, 182–183
  salt accumulation with, 346
  soil formation and, 52
  soil salinity with, 359
  sprinkler systems, 232–234
  surface, 232–233
  water content after, 185
  water-quality considerations for, 356–358
  water-use efficiency, 231–232
Isomorphous substitution, 280

K

K. See Soil erodibility factor; Soil-ridge-roughness factor
Kandic horizon, 77, 98
Kaolinite clays, 281
  charge of, 290–293
  liquid limits of, 158
  mineralogical organization of, 281, 283
  potassium in, 518
  sulfur in, 490
Keystone species, 372
Knife injection, of fertilizer, 584
K-strategist, 425–426

L

L. See Width of field factor
Labile soil organic matter
  description of, 442
  soil management and, 443–444
Lacustrine deposits, 43
LAI. *See* Leaf area index
Land capability classification, 647–649
Land degradation, 607–608
  soil phosphorus and, 496
Land smoothing, 222
Landfills, 685–690
  containment, or secured, 688
  environmental impacts of, 688
  Land Use, Completion, 690
  natural attenuation, 686–688
  Solid Waste Problem, 685
Landscaping
  drainage design for, 224–225
  irrigation for, 230–231
Landslides, 631
Lawn management, aeration and, 252–253
Layering, available water-holding capacity and, 192
Leachate, from landfill, 688
Leaching
  in Alfisols, 98
  of calcium, 524
  of chemicals, 218
  cover crops and, 556–559
  for inorganic contaminants, 681
  of nitrogen, 470
  of organic chemicals, 661
  in Oxisols, 102–103
  of potassium, 518
  in Spodosols, 101–102
  of sulfates, 495
  in Ultisols, 99
Leaching requirement (LR)
  limitations of, 361
  for saline soils reclamation, 358–359

Lead poisoning, 676, 678
Leaf area index (LAI), 207, 209
Least limiting water range, 190
Legumes
  symbiotic nitrogen fixation with, 479–481
Level basin technique, for irrigation, 233
Lichens, 393
Lignin
  decomposition of, 424
  in plants, 423
Lime requirement
  influence of, 339
Limestone, Vertisols from, 94
Liming
  application of, 340
  calculating needs for, 339
  chemical reactions of, 337
  overliming, 340
  for soil pH, 336–341
Limiting factor, concept of, 581
Liquid fertilizers, 583
Liquid limit, 158–160
Lithosequence, 36
Lithosphere, soil interface with, 10
Live stake technique, 636–637
Loam, 122–124
Loam surface soil, components of, 17
Loamy sand, 123
Loamy soil, particle size class, 108
Localized Placement, of fertilizer, 584–585
Loess, 43–44
Losses, in soil formation, 59–60
Low-impact urban design, 203
LR. *See* Leaching requirement
LS. *See* Topographic factor
Luxury consumption, of potassium, 518

M

Macroaggregates, 131
Macrofauna, 371
Macronutrients, 5, 529
Macropores, 150–151
  hydraulic conductivity with, 179–180
  infiltration into, 181
  water movement through, 183–185
Magnesium
  available quantities of, 22
  calcium ratio to, 525
  inorganic fertilizers for, 579
  in plants, 524
  in soil, 524
  in soil colloids, 279–281
  soil pH and, 336–341
Manganese
  in hydric soils, 255
  phosphorus precipitation by, 505
  soil color and, 118
  toxicity of, 331–332
Maple seedlings, acidity and survival of, 333
Maps. *See* Soil maps
Marbut, C. F., 73
Marine deposits, 41–42
Mass action, with cation exchange reactions, 297
Mass flow, 24–25, 240
Mass wasting, 629–631
Mass water content, 173
Massive structural condition, 128

Master horizons
  overview of, 62–64
  subdivisions within, 64–65
  transition layers between, 65
Matric force, 168
Matric potential, 168–171
  compaction and, 190
  field capacity and, 186–187
  hydraulic conductivity and, 181
Matric potential gradient, 181
Maximum allowable daily loadings (MDL), 552
Maximum retentive capacity, 186
MDL. *See* Maximum allowable daily loadings
Meltwaters, parent materials transported by, 42–43
Meniscus, 166
Mercury, contaminants, 678
Mesic regimes, 108
Mesofauna, 371
Mesophilic stage, in composting, 461
Metabolic activity, of soil organisms, 378–379
Metamorphic rock, 30–32
Metamorphism, 31
Methane
  absorption of, 7
  greenhouse effect and, 458–460
  production of, 248–249
  soil aeration of, 239
  in soil air, 241
Methanogenic bacteria, 426
Methanotrophic bacteria, 459
Methemoglobinemia, 554
Microaggregates, 131
Microanimals, 386–390
  nematodes, 386–388
  protozoa, 388–389
Microbial biomass, 425–426
Microbial decomposers, 376
Microbial Transformations of, organic matter, 434–435
Microbiotic crusts, 393, 412–413
Microbiovorous feeders, 376
Microfauna, 371
Microirrigation, 232–236
Micronutrients, 5
  deficiency symptoms of, 531
  deficiency v. toxicity of, 528–532
  inorganic fertilizers for, 579
  in plants, 530
  forms and reactions, 532–534
Microorganisms. *See also* Bacteria; Fungi
  in anaerobic conditions, 243
  conditions affecting growth of, 405
  effects of pesticides, soil organisms, 665–667
  nitrogen fixation with, 478–479
  organic decay and, 425–426
  phosphorus solubility and, 507–509
  in soil formation, 44
  soil pH and, 334
  soil temperature and, 261
  sulfur cycle and, 492
Micropores, 151–152
  in subsoil, 17
Midden, 379

Mineral soils
    pore space of, 150–153
    potassium in, 518
    structure of, 127–131
Mineralization, 426
    of nitrogen, 429, 468–470
    of organic phosphorus, 503
    rates of, 427–432
    of sulfur, 491–493
Minerals
    characteristics of, 30–32
    chemical weathering of, 32–36
    in composting, 461–463
    as electron acceptors, 243
    organic matter and, 19
    pH and, 320
    potassium in, 516
    in soil, 17–18
    as sulfur source, 488–491
    weathering of, 29–36
Moisture content
    aeration and excess, 240
    evapotranspiration and, 206
    heat capacity, 267
    oxidation and, 242–245
    soil water potential v., 171–172
    SOM influence of, 450
Moisture control
    for soil temperature, 272
Moisture, soil microorganisms requirement
    of, 405
Molds, 393
Mole Drain, 226
Mollicepipedon, 75
Mollisols
    profile of, 82–83, 96–98
    sulfur in, 488–491
Molybdenum
    deficiency v. toxicity of, 528–532
    distribution and use, 97–98
    formation process, 96–97
    in drainage water, 357–358
Montmorillonite, 283
Moraines, 42
Moss peat, 46
MSW. See Municipal solid waste
Mucigel, 392
Muck, 47
    in Histosols, 89–90
Mud flows, 630
Mulching
    on construction sites, 635
    for soil tilth, 140
    for vapor loss control, 212–213
Municipal by-products, nutrient management
    of, 570–573
Municipal solid waste (MSW), nutrient
    recycling of, 570–571
Munsell color charts, 117
Mushrooms, 393
Mutualistic associations, among soil
    organisms, 412
Mycelia, 393
Mycorrhizae, 134, 397–400
    micronutrient cation availability
        and, 536
    phosphorus and, 498–499
Mycotoxins, 395

N
Natric horizon, 98
Natural attenuation, 667
Natural attenuation landfills, 686–688
Natural eutrophication, 553
Nematicides, 388, 659
Nematodes, 386–388
Nitrate
    in drinking water, 551–554
    in nitrogen cycle, 468
Nitrate Leaching Problem, 486
Nitrification, 473–474
    acidity and, 314
Nitrogen
    available quantities of, 22
    distribution and cycling of, 468–469
    fixation of, 406–407
    inorganic fertilizers for, 579
    in organic farming, 574–576
    from organic matter, 19
    oxidation of, 314
    plant growth and development and,
        467–468
    for SOM management, 453–454
Nitrogen cycle, 468–469
    ammonia volatilization, 472–473
    ammonium fixation, 471
    biological fixation, 478–479
    denitrification in, 475–477
    deposition from atmosphere, 484–485
    ecosystem, 560–563
    immobilization and mineralization,
        468–471
    nitrification, 473–474
    nonsymbiotic fixation, 484
    SON, 471
    symbiotic fixation, 479–482
Nitrogen deposition, 484–485
Nitrogen fertilization
    soil pH with, 325
Nitrogen fixation, 469
    biological, 478–479
    nonsymbiotic, 484
    quantity of, 480
    with root nodules, 479
    symbiotic with legumes, 479–480
    symbiotic with nonlegumes, 482–483
    symbiotic without nodules, 483–484
Nitrogen management, 599–602
    challenges in, 599
    credits and predictions, 600
    plant response, 600
    profitability, 600
    response curves, 600
Nitrogen mineralization, 468–471
    calculation of, 470
    decomposition release of, 430–41
    in organic materials, 575–576
    soil ecology and, 431
Nitrogen pollution, 553–554
"Nitrogen solutions," 577
Nitrogenase, 478–479
Nitrobacter, 473
Nitrosomonas, 473
Nitrous oxide
    production of, 248–249
    release of, 7
    soil production of, 460

Nonacidcations
    accumulation of, 316
    leaching of, 312, 315
    saturation of, 303
    weathering of, 316
Nonacid saturation, 320–321
Noncohesive soils, 155
Noncrystalline silicate clays. See Silicate clays
Nonirrigated soils, salt accumulation in,
    346–347
Nonpoint sources, of pollution, 552
No-till, 139, 624–625
    water conservation with, 213
Nuclear Accident
    at Chernobyl, 691
    at Fukushima, 692
Nuclear waste storage, 284
Nutrient application timing, 585–586
    split applications, 586
    physiologically appropriate Timing, 586
    practical field limitations, 586
Nutrient availability
    in acidic soils, 332–333
    in alkaline soils, 332–333
    of boron, 535
    of essential elements, 22–24
    fertilizer timing, 585–586
    of potassium, 517–520
Nutrient composition, of animal
    manures, 564
Nutrient concentration, 587–591
Nutrient cycling, process of, 51
    organic material formation and, 406
Nutrient management
    animal manures, 564–570
    diagnostic tools and methods, 586–591
    pollutant, 551–560
    fertilizer application methods, 581–585
    fertilizer application timing, 585–586
    goals of, 548–551
    industrial and municipal by-products,
        570–573
    inorganic commercial fertilizers, 576–581
    organic nutrient sources, 573–576
    overview of, 548
    plans for, 552
    site-index approach for phosphorus,
        596–599
    soil analysis, 591–596
Nutrients
    cation saturation and, 303
    humus and, 20
    from organic matter, 20
    oxidation and reduction of, 248
    plant availability of, 332–333
    root metabolism and, 25
    soil aeration and, 349
Nutrient Resources, conservation, 549–550

O
O horizons, 13, 62
Ochricepipedon
    in Aridisols, 92
    in Entisols, 83
    in Oxisols, 102
Off-site damage, of soil erosion, 610–612
Oil Spill Cleanup, 671
Oils, in plants, 423

1:1 Silicate clays, 281, 284
　charge of, 290–293
On-site damage, of soil erosion, 610
Organic acids, 314
Organic chemicals
　absorption and breakdown, plant, 664
　behavior in soil, 659–665
　chemical reaction, 662
　environmental damage from, 653
　groundwater, contamination, 662
　kinds of, 657–659
　in landfill, 687
　microbial metabolism, 662
　persistence, soils, 664
　remediation for, 667–673
　oil spill Cleanup, 671
Organic compounds
　decomposition of, 424–425
　micronutrient cation availability and, 538
　soil colloid sorption of, 306
Organic farming
　description, 19–20
　nitrogen application, 574–575
　pesticides and, 655
Organic matter. See also Humus; Soil organic matter
　for acidic soils, 341–343
　acidity and, 314
　available water-holding capacity and, 190
　balance of, 444–446
　in carbon cycle, 420–421
　decomposition of, 424
　deposits of, 44–47
　factors and practices influencing, 447–451
　friability and, 137
　in Histosols, 89–90
　lignin and polyphenol content of, 431–432
　for lowering soil pH, 343
　microbial transformations of, 434–436
　micronutrient cation availability and, 536
　in Mollisols, 96–98
　phosphorus fixation and, 509
　plant growth and soils influence of, 438–442
　pore space and, 150
　in soil, 19–20
　soil aggregation and, 135
　soil color and, 117–120
　soil microorganisms requirement of, 405–406
　soil pH with, 325–326
　soil property effect of, 440–442
　soil quality and, 26
　soil tilth and, 453
　for SOM management, 455
　in subsoils, 17
　as sulfur source, 489
Organic micronutrients, 533–534
Organic molecules, soil pH and, 335–336
Organic mulch
　for soil temperature control, 270–271, 272
　for vapor loss control, 211
Organic nutrient sources, 573–576
Organic phosphorus, 498–503
Organic residue. See also Crop residue; Plant residue
　aeration and, 247–248
　carbon/nitrogen ratio of, 428–429
　physical factors of, 427

Organic toxins, and control of plant disease, 409
Organisms. See also Soil organisms
　abundance, biomass, and metabolic activity of, 378–379
　actions of, 373–378
　ants, 384
　beneficial effects of, 406–407
　conditions affecting growth of, 405–406
　and damage to higher plants, 407–411
　diversity of, 370–373
　earthworms, 379–383
　ecological relationships among, 412–414
　effects of agricultural practices on, 413–414
　metabolic grouping of, 373
　plants, 390–393
　prokaryotes: bacteria and archaea, 400–404
　sizes of, 371
　soil algae, 393
　soil ecology and, 369–415
　soil formation and, 49–53
　soil fungi, 393–400
　soil microanimals, 386–390
　termites, 384–386
Organoclays, 307, 668–669
Organophosphate pesticides, 657
Orthels, 87–89
Orthents, 85
Orthods, 101–102
Osmosis, 170
Osmotic force, 168
Osmotic potential, 168–169, 190, 354
Outer-sphere complex, 293, 294–295
Outwash plains, 43
Overenrichment, of phosphorus, 596–597
Overliming, 340
Oxic horizon, 102–103
Oxidation
　aeration and, 248
　of nitrogen, 469
　Oxidation-reduction (redox) potential, 242–245
　in wetlands, 257
Oxidized root zones, 256
Oxidizing agent, 243
Oxisols
　distribution and use, 102–103
　formation, 102
　nitrate leaching in, 486
　profile of, 81–83, 102–103
　sulfur in, 489–490
Oxygen
　absorption of, 7
　carbon dioxide v., in soil, 21–22
　denitrification and, 475
　as electron acceptor, 242–243
　in hydric soils, 255
　for nutrient absorption, 22
　for root supply, 240
　in soil air, 239, 241
　soil microorganisms requirement of, 405

P

PAM. See Polyacrylamide
Pans, 78
Paramecium, 8
Paraquat, 660

Parasites
　of plants, 408
　in soil, 371
Parent materials
　alluvial deposits, 39–41
　classification of, 37
　climate acting on, 47–49
　coastal sediments, 41–42
　colluvial debris, 38–39
　description of, 13
　factors influencing, 36
　glacial ice and meltwater transport of, 42–43
　organic deposits, 44–47
　residual, 37–38
　rock and mineral weathering, 19–36
　time and, 55–58
　topography and, 53–55
　wind transport of, 43–44
Partial pressure gradients, 240–241
Particle density
　definition of, 141
　pore space and, 150
Particulate matter (PM), wind and, 611
Partitioning, 307
　PCB, 664
Peat, 44, 47
　in Histosols, 90
　types of, 46–47
　pots and power, 459
Pebbles, 123
Pedogenic, 58
Pedology, 12
Pedon, 73
Pedosphere, 10–11. See also Soils
Pedoturbation, 52
Peds, 128–129
Perc test, septic drain fields and, 228–229
Percent "base" cations, 320
Perched water table, 220
Percolation
　evaporation balance with, 214–216
　groundwater and, 216–220
　septic drain fields and, 228–229
　of water, 182–183
Perforation, of fertilizer, 585
Periodic table of the elements, 705
Permafrost, 87–89, 262–263
Permanent charge, 290
Permanent wilting percentage, 187–188
Perox, 102–103
Persistence, of organic chemicals, 664–665
Pesticides, 413–414, 657–659
　alternatives to, 655
　benefits of, 654
　nontarget damages of, 656–657
　preferential flow and, 179–181, 219
　problems with, 653–655
　on soil organisms, 665–667
Pests, of plants, 408
PET. See Potential evapotranspiration
Petrocalcic horizons, 93
pH meter, 323
pH, soil microorganisms requirement of, 405–406
Phosphates, for inorganic contaminants, 682
Phospholipids, 496
Phosphorus

available quantities of, 22–23
eutrophication and, 551
hydrous oxide and silicate clay reaction, 505–506
inorganic fertilizers for, 579
iron reduction of, 507
organic, 498–500
from organic matter, 20
plant growth and, 496
plant nutrition and soil fertility, 496–514
precipitation of, 505
site-index approach to management of, 596–599
soil fertility and, 497
Phosphorus cycle, 498–501
Phosphorus fixation, 503–504, 507–509
Phosphorus pollution, 553–554
Phosphorus site index, 599
Phyllosilicates, 279
Physical disintegration, of rocks and minerals, 32–34
Physical remediation methods, 667–669
Phytoremediation, 672–673, 684
  management, 684
Phytotoxic substances, 3–4
Piping, silty soil, 121
Placic Horizons, 78
Plant available water, 186–194
  compaction effects on, 190
Plant communities, soil organisms' beneficial effects on, 406–407
Plant disease, soil management for control of, 408–409
Plant growth
  enhancing soil drainage for, 220–222
  nitrogen and, 467–468
  organic matter and, 438–442
  phosphorus and, 496
  pore space and, 17
  problems with, 16
  soil aeration and, 249
  soil pH and, 335–336
  soil water and, 21
  plant growth-promoting rhizobacteria, 407
Plant nutrients. *See also* Nutrients
  in soil solution, 24–25
  supply of, 22–24
  uptake of, 24–25
Plant residue
  phosphorus and, 500
  for soil temperature control, 272–271, 272
  for SOM management, 453–454
Plant roots
  aeration and, 245–247
  aluminum toxicity and, 330–331
  anion uptake by, 317
  bulk density and, 149–150
  cation uptake by, 314
  field capacity, 186–187
  in hydric soils, 256
  hygroscopic coefficient, 188
  matric potential and, 169–170
  nutrient uptake by, 24–25
  osmotic potential and, 170
  permanent wilting percentage, 187–188
  phosphorus and, 499–500
  soil aggregation and, 134–135

soil consistence and, 154
water depth ratio, 173–174
Plant analysis, nutrient management and, 587–591
  greenness, 591
  on-the-spot plant sap, 588
Plant water stress, 207
Plants
  for acidic soils, 343
  aluminum toxicity and, 330–331
  anion uptake by, 317
  bioremediation by, 683–684
  boron in, 535
  calcium in, 522
  cation uptake by, 314
  container-grown, 251, 355
  evapotranspiration of, 205–209
  humus influence on, 438–439
  magnesium in, 524
  manganese toxicity and, 331
  mentioned in text, 706–708
  molybdenum in, 535
  nitrogen forms taken up by, 468
  nutrient availability to, 332–333
  organic compounds in, 423
  organic phosphorus for, 503
  pesticide uptake by, 663–664
  pests and parasites of, 408
  potassium and, 514–515
  salt affect on, 354
  sodic soils and, 352, 355
  soil aeration and, 249
  in soil ecosystems, 390–393
  soil organisms' damage to, 407–411
  soil temperature and, 259–261
  soil–plant–atmosphere continuum, 205–209
  sulfur in, 487–488
  water potential and, 205
  Uptake and Removal, 518
Plastic limit, 157–158
  field capacity and, 187
Plastic mulch
  for soil temperature control, 271–272
  for vapor loss control, 212–213
Plasticity, of silt, 121
Plate-like structure, 128–130
Plinthite, 100
Plow layer, 15
Plow pans, 147
Plowing, 15
Plutonium Toxicity, 692
PM. *See* Particulate matter
PM2.5, 611
PM$_{10}$, 611
Point injection, of fertilizer, 585
Point sources, of pollution, 551–552
Pollutants
  concentration in soil, 16
  pH and mobility of, 312
  preferential flow of, 179–180
Pollution. *See also* Atmospheric pollution; Contaminants; Groundwater pollution; Toxic compounds
  atmospheric, 476–479
  groundwater, 179–180, 187
  landfills, 685–690
  nitrogen, 553–554

nonpoint sources of, 552
and nutrient management, 551–560
organic chemicals, 652–657
overview of, 652
phosphorus, 553–554
point sources of, 551–552
soil temperature and, 263
Polyacrylamide (PAM), as soil conditioner, 140
Polypedon, 73
Polyphenols
  in organic matter, 431–432
  in plants, 423
POM. *See* Particulate organic matter
Poor soil aeration, 240
Pore space
  aeration and, 245–247
  air-filled, 242
  bulk density and, 141
  calculation of, 151
  capillary action and, 166–167
  field capacity and, 187
  least limiting water range and, 190
  of mineral soils, 150–153
  plant growth and, 17
  water movement and, 183–185
Potassium
  available quantities of, 23
  forms and availability of, 518–520
  inorganic fertilizers for, 579
  management of, 521–522
  in plant and animal nutrition, 514–515
  soil fertility and, 517–518
Potassium cycle, 516–517
Potassium fixation, 520
Potential acid sulfate soils, 328
Potential acidity, 328, 330
Potential evapotranspiration (PET), 205
Poultry manure, nutrient recycling and, 564
Precipitation. *See also* Rainwater
  in acids, 315
  chemical leaching and, 219
  effective, 48
  percolation with, 182–183
  water content after, 186
Precision agriculture, for saline soils, 348–350
Predators, 371, 373
Preferential flow, 179–180
Primary minerals, 18
Primary producers, 373–374
Priming effect, 425
Prism-like structure, 129, 130
Proctor test, 155–156
Productivity
  earthworms' influence on, 379–380
  maintenance of, 611–612
Prokaryotes
  actinomycetes, 403
  cyanobacteria, 403
  energy sources of, 401
  importance of, 402–403
Protein
  decomposition of, 424
  in plants, 424
Protonation, 292, 335
Protozoa, 388–389
Psamments, 85

Pseudosand, 133
Purity, of water, 7
Pyrite-containing materials, excavation of, 329

Q

Quality, soil. See Soil quality
Quartz
   in sandy soil, 120–121
   weathering and, 33–34

R

R. See Rainfall erosivity factor
R layers, 64
Radionuclides, 690–692
Radon gas, 693–695
   accumulation in buildings, 693–695
   hazard of, 693–694
   testing and remediation for, 694–695
Raindrop splash effects, 613
Raindrops, influence of, 612–613
Rainfall erosivity factor (R), 616–617
Rainwater
   acidity and, 314
   soil temperature and, 265
Rangeland
   accelerated erosion on, 631
   evaporation in, 211
   heterogeneity in, 51
Ratio law, with cation exchange reactions, 296
Reactive nitrogen, 468
Reclamation
   of saline soils, 357, 358–362
   of saline-sodic and sodic soils, 362–364
   spatial variability, 361
   management of, 363–364
Recycling of nutrient resources, 7
   through animal manure, 564–570
Redox depletions, 256
Redox potential. See Oxidation-reduction potential
Redoximorphic features, 256
Reduced sulfur, acidity from, 328
Reducing agent, 243
Reduction
   aeration and, 248
   in hydric soils, 256
   of sulfur, 493
Redundancy, of ecosystems, 372
Regolith
   building foundations and, 14
   description of, 12–13
Remediation
   bioremediation, 669–672
   of gullies, 629–630
   for organic chemicals, 667–673
   physical and chemical methods for, 667–669
   phytoremediation, 672–673
   for radon gas, 695
Residence time, in wetlands, 255
Residual acidity, 318
Resilience
   of ecosystems, 372
   soil. See Soil resilience
Respiration
   aeration and, 245
   Restoration ecology, 27

Retention
   of sulfur, 494–495
Reusable resource, soil as, 25–26
Reversibility, of cation exchange reactions, 296
Revised universal soil-loss equation (RUSLE), 616–622
Revised wind erosion equation (RWEQ), 642
Rhizobacteria, 407, 410–411
   deleterious, 408
Rhizobium, 480–483
Rhizodeposition, 391–393
Rhizosphere, 391
Ribonucleic acid (RNA), 496
Ridge tillage, 624
Rill erosion, 613
   factors affecting, 616–622
Rills, 613
Riprap, 636
RNA. See Ribonucleic acid
Roadbeds
   design and management of, 633
   soils and, 9–10
Rocks
   characteristics of, 30–32
   weathering of, 30–36
Root distribution, 193–194
Root extension, 193
Root growth
   bulk density and, 149–150 compaction and, 190
   soil consistence and, 154
   soil temperature and, 259, 261
Root hairs, 24
Root interception, 24
Root metabolism, nutrient uptake and, 25
Root nodules, for symbiotic nitrogen fixation, 479
Root–soil contact, 194
Root zone, salts in, 460
Roots
   morphology of, 390–391
   rhizosphere, 391
   rhizodeposition, 391–393
   in soil ecosystems, 390–393
r-strategist, 425–426
Running water, erosion and, 613
Runoff, 199–200
   with conservation tillage, 559
   control of, 636–638
   erosion with, 203–204
   phosphorus in, 597
   with sprinkler systems, 233–235
   surface, 200
RUSLE. See Revised universal soil-loss equation
RWEQ. See Revised wind erosion equation

S

Salids, 92–94
Saline seeps, 346–347
Saline soils
   overview of, 350–352
   plant tolerance to, 355
   precision agriculture for, 348–350
   reclamation of, 358–362
Saline-sodic soils
   overview of, 351–352

   physical degradation in, 352–354
   reclamation of, 362–364
Salinity
   of drainage water, 357
   irrigation-induced, 347–348
   measurement of, 348–350
Salinization, 348, 350
Salt balance, 356
Salt concentration, soil dispersion and, 354
Salt hazard, of inorganic fertilizers, 579
Salt-affected soils
   biological impacts of, 354
   classes of, 350–352
   development of, 346–348
   ion effects, 354
   management of, 348–350, 363–364
   osmotic effects, 354
   osmotic potential, 190
   physical degradation of, 352–354
   physical effects of, 354
   reclamation of, 358–362
   topography and, 54–55
Saltation, 640
Saltwater intrusion, 217
Sand
   capillary action in, 166–167
   description of, 18, 120–121
   soil texture and, 123–124
   windblown, 611
Sandstone, 31
Sandy clay, 18, 123–124
Sandy clay loams, 123–124
Sandy loam, 123–124
   water movement in, 178
Sandy soils
   aggregation in, 132
   engineering considerations with, 156
   particle size class, 108
   Spodosols, 101–102
   tillage of, 137–138
Sapric, 47
Saprists, 90
Saprolite, 11
Saprotrophic, in soil, 375
SAR. See Sodium absorption ratio
Saturated hydraulic conductivity, 177–180
   for septic drain field, 228
Saturated soils
   hydraulic conductivity in, 179–180
   water flow through, 177–179
Saturation percentage, 303
SAV. See Submerged aquatic vegetation
Scientific names, of plants in text, 706–708
Secondary minerals, 18
Secured landfills, 688
Sediment, 608
   damages from, 610–611
   with runoff, 200
   trapping of, 637–638
Sediment control, on construction sites, 634–638
Sedimentary peat, 46–47
Sedimentary rock, 30–31
Seed germination
   fires and, 263
   heat of vaporization and, 267–268
   soil temperature and, 260

INDEX

Selenium
    contaminants, 678
    in drainage water, 357–358
Separates, soil, 120–121
Septic tank drain fields, 226–229
    alternative systems, 229
    septic system operation, 227–228
    soil properties for, 228–229
Series, in *Soil Taxonomy*, 81, 110
Settlement of soil, 156
Sewage sludge. *See* Sludge
Sewage systems, regolith properties and, 16
Sheet erosion, 613
Shrinkage limit, 157
SI. *See* International System of Units
Silicate clays
    in Alfisols, 98
    constant charges on, 290
    genesis of, 288–290
    mineralogical organization of, 281–286
    phosphorus reactions with, 505–506
    soil colloids and, 278–279
    structure of, 279–281
    in Ultisols, 99–101
    weathering and, 32
Silicon, in soil colloids, 279–281
Silt
    capillary action in, 166–167
    description of, 18, 121
    soil texture and, 123–124
Silt loam, 123–124
Silt particles, 18
Silty clay, 18, 123–124
Silty clay loams, 123–124
Site-specific management zones, with soil analysis, 592–593
Skidding trails, design of, 634
Slaking, 353
Slickensides, 94–95
Slope angle and aspect, of solar radiation, 265
Slopes
    mass wasting on, 630–631
    septic drain fields and, 228
    solar radiation and, 55
    water infiltration and, 204
Slow soil organic matter, 442–443
Sludge, 228, 678–681
    composition of, 572
    heavy metals in, 679–681
    nutrient recycling of, 571–572
Smectite clays
    charge of, 290–293
    expansion of, 157
    liquid limits of, 158
    properties of, 278–279, 281–283
SMR. *See* Soil moisture regimes
SOC. *See* Soil organic carbon
Sodic soils, 348
    measurement of, 350
    overview of, 350–352
    reclamation of, 362–364
Sodium absorption ratio (SAR), sodic soil measurement with, 350
Sodium, soil dispersion and, 352–354
Soft armor, 636
Soil
    concept of, 72–75
    conservation tillage effects on, 626–628
    cover on, 634
    definition of, 11, 72, 74
    disease-suppressive, 409–411
    earthworms' exposure of, 381
    grouping of, 74
    magnesium in, 524
    minerals in, 17–19
    organic chemical behavior in, 659–665
    organic matter in, 16–19
    pedon, polypedon, and series, 59–60
    persistence in, 664
    phosphorus and, 498
    radionuclides in, 690–692
    soil–plant–atmosphere continuum, 205–209
    texture and structure influence, 18
Soil air
    composition of, 241
    interface of, 17
    overview of, 11, 21–22
    in pore spaces, 150–153
    soil water v., 21
    texture and structure influence on, 18–19
Soil analysis, 591–596
    of phosphorus, 596–603
Soil and Water Assessment Tool (SWAT), 615
Soil associations, 112
Soil biomass, 19, 378
Soil boundaries, delineating, 112–113
Soil buffering capacity
    importance of, 322–323
Soil classification, 72–115
    color's role in, 118
    comprehensive system for, 75–79
    individual soils, 72–74
    introduction to, 72
    mapping techniques, 111–113
    nomenclature of, 79
    orders, 83–103
    separates, 120
    soil surveys, 113–115
Soil colloids, 275
    biomolecule binding, 308–309
    cation and anion adsorption, 276–277, 293–295
    cation exchange reactions, 295–298
    charges on, 290–293
    genesis of, 288–290
    geographic distribution of, 288–290
    humus, 287–288
    iron and aluminum oxides, 286–287
    organic compound molecules, 306
    pH and, 320
    properties of, 276–279
    silicate clay structure, 279–281
    types of, 278–279
    water adsorption, 277
Soil color, 117–118
    oxidation and reduction of, 248
Soil structure, 131
Soil conditioners, 140
Soil conservation
    progress in, 641–649
    win-win system, 648
Soil consistency, 153–154
Soil cover, soil temperature and, 266
Soil creep, 630, 640
Soil degradation, 1–2, 25–27

Soil density
    bulk density, 141–142
    management practices and, 145–150
    particle density, 141
Soil depth
    available water-holding capacity, 190–192
    pH with, 324
    ratio, 173
    for soil analysis, 591–592
    soil temperature and, 268–270
Soil distribution coefficient, 307
Soil ecology, nitrogen mineralization and, 431–432
Soil erodibility factor (K, I), 616–618, 641
Soil fertility
    phosphorus and, 496
    potassium and, 518
Soil formation, 21, 29–70
    basic processes of, 58–61
    climate and, 47–49
    effects of time, 55–58
    parent materials for, 36–47
    rock and mineral weathering, 29–34
    time and, 55–58
Soil genesis, 67
Soil horizons. *See also specific horizons*
    pH with, 324
    regolith properties, 12
    soil profile and, 13–15, 62–63
Soil individual, 74
Soil Loss, Reduction
    by Timber Production, 632–634
Soil management
    aeration and, 251–253
    infiltration and, 201–202
    inorganic contaminants and, 681
    plant disease control through, 408–409
    labile and human pools SOM changes with, 443–444
Soil maps, 111–113
Soil mineral constituents, 17–19
    effect of clay, 18
Soil moisture, and wind erosion, 641–644
Soil moisture regimes (SMR), 104–110
Soil orders
    description of, 81–83
    families, 79, 108
    great groups, 79, 104–106
    key for, 84
    names of, 79–81
    overview of, 83–103
    pesticides on, 665–667
    series, 80, 108
    subgroups, 105
    suborders, 79, 104
Soil organic carbon (SOC), 432–434
Soil organic matter (SOM)
    genesis of, 432–438
    management of, 451–452
    nitrogen in, 468
Soil organisms
    habitat for, 7–8
    interface of, 17
    organic matter and, 19–20
    soil aggregation and, 134–135

Soil particles
　classification of, 120–123
　laboratory analysis of, 125–127
　soil texture and, 123–124
　surface area and, 121
Soil pH
　buffering of, 321–323
　CEC and, 300
　description of, 21
　ECEC and, 319
　in field soils, 323–325
　influence of, 534–535
　liming for, 336–341
　living things, affect of, 330–336
　lowering of, 343–345
　mobility of pollutants and, 312
　pH-dependent charge, 291–292
　potassium fixation and, 520
　in potentiometric methods, 323–324
　rangeland v. forests, 51
　season variation in, 325
　soil colloid charge, 290–293
Soil phases, 81
Soil pit, 13, 111
Soil plant ecology, silicon, 526–528
Soil pores
　crusting and, 139
　infiltration into, 181
　oxygen v. carbon dioxide in, 21
　and soil structure, 127–131
　water and, 21
Soil productivity, soil water and, 21
Soil profile
　available water-holding capacity of, 192
　horizons of, 62–64
　infiltration of water and, 201–205
　master horizons subdivisions, 64
　overview of, 13–15, 58
　using information from, 16
Soil properties, 409
　color, 117
　effect on, 627–628
　for engineering uses, 153–160
　infiltration of water and, 201–205
　for landfill, 687–688
　observation of, 88
　organic matter effect on, 440–442
　for radon gas, 693–694
　for septic tank drain fields, 228–229
　soil density, 141–150
　soil moisture regimes, 104–110
　Soil Taxonomy and, 75
　soil temperature regimes, 108
　soil texture, 120–123
　subsurface horizons, 75–76
　surface horizons, 75
　and wind erosion, 641
Soil quality
　conservation management for, 648
　description of, 26
　organic matter in, 419
　productivity and, 549
Soil resilience, 27
Soil respiration, 425
Soil restoration, 27
Soil sampling, for soil analysis, 591–593
Soil series, 73

Soil solarization, 261
Soil solution
　description of, 21
Soil storage water, 200
Soil strength, bulk density and, 149–150
Soil structure
　description of, 19, 117, 127
　mineral soils, 127–131
　silicate clays and, 279–281, 285
　soil quality and, 26
　tillage and, 137–141
　types of, 131
　water infiltration and, 201
Soil Survey Staff of the U.S. Department of Agriculture, 60
Soil Taxonomy, 75–79
　wetland, 457
Soil temperature
　control of, 270–272
　moderation of, 3–6
　soil microorganisms requirement of, 405
　in soil orders, 82
　soil processes and, 259–264
　soil thermal properties, 266–270
　solar energy and, 264–266
Soil temperature regimes, 64
Soil texture
　bulk density and, 141
　classes of, 124–127
　classification of, 120–123
　description of, 18, 117
　hydraulic conductivity and, 181
　soil quality and, 26
　SOM influence of, 452–453
Soil tilth
　conservation tillage and, 139
　description of, 137
　management of, 140–141
　soil fungi and, 395–396
　tillage and, 137–138
Soil-Vegetation Interdependency, 607
Soil water
　Atterberg limits, 157–160
　capacity for, 3–4
　capillary forces and, 166–167
　capillary movement, 193
　complex and dynamic nature, 21
　compression and, 154–155
　consistence and, 153
　content v. potential, 171–177
　energy concepts of, 168–171
　humus and, 20
　infiltration, 181
　interface of, 15
　loss of, 213–216
　organic matter and, 19
　overview of, 21, 164–165
　percolation, 182
　plant-available, 186–194
　soil air v., 21–22
　in soil orders, 82
　in stratified soils, 183–185
　root distribution, 193–194
　root extension rate, 193
　root-soil contact, 194
　strength and, 155–156
　texture and structure influence on, 18–19

　vapor movement, 185
　water structure and, 165–166
　wetness description of, 185–189
Soil water deficit, 211
Soil–plant–atmosphere continuum (SPAC), 205–209
　carbon balance in, 444–447
Soil-ridge-roughness factor (K), 641
Soils
　carbon/nitrogen ratio of, 428–429
　component interfaces in, 17
　concluding remarks on, 27
　decomposition in, 423–427
　degradation of, 1–2
　ecological functions of, 27
　engineering of, 9–10, 52
　human consumption of, 4–5
　inorganic phosphorus in, 503–504
　as natural body, 11–12
　organic matter and, 438–442
　organic phosphorus in, 501–503
　overview of, 1–2, 10
　parent material of, 13–15
　phosphorus fixation in, 507–508
　potassium in, 519–520
　thermal conductivity of, 268–270
Solar radiation
　description of, 264
　hydrologic cycle and, 198–200
　slopes and, 55
　soil temperature and, 264–266
Solarization, 409
Solubility
　of calcium, 522–524
　of inorganic phosphorus, 505
Solum, 12
SOM. *See* Soil organic matter
SON. *See* Soluble organic nitrogen
SPAC. *See* Soil–plant–atmosphere continuum
Species diversity, of soil organisms, 372
Specific disease suppression, 409
Specific gravity, 141
Specific heat of soils, 266
Specific surface area, 98, 277, 283
Spheroidal structure, 128–129
Splash effects, of raindrops, 613
Spodic horizon, 77, 101
Spodosols
　distribution and use, 102
　formation process, 102
　profile of, 81–84, 101–102
　sulfur in, 490
Sprinkler systems
　water control, 234
　suitable soils of, 234
Sprinkler systems, for irrigation, 233–236
Stability
　of ecosystems, 372
　micronutrient cation availability and, 538–539
Starch, decomposition of, 424
Starter fertilizer, 584
Stickiness, of silt, 121
Stomata, 199
Stones, 123
Stratified soils, soil water movement in, 183–185
Stream channels, buffer strips along, 634

Strengite, 505
Strip-cropping, 619
Stubble mulch tillage, 213, 624
Subgroups, in *Soil Taxonomy*, 105
Sublimation, 200
Submerged aquatic vegetation (SAV), 610
Suborders, in *Soil Taxonomy*, 79, 104
Subsoil, 15. *See also* B horizons
    compaction of, 144
    organic matter in, 17
Subsoiling, 146–147
Sufficiency range, 528
    of nutrient concentrations, 587–591
Sulfates
    absorption and leaching of, 495
    cycle of, 491–492
Sulfides, cycle of, 491–492
Sulfidic materials, 328
Sulfonylureas, 658
Sulfur
    agronomic deficiencies, 487–488
    available quantities of, 22
    immobilization of, 492
    inorganic fertilizers for, 579
    for lowering soil pH, 343–345
    from organic matter, 19
    oxidation of, 314, 328–330, 492–493
    in plants and animals, 487–488
    reduction process, 493
    retention and exchange of, 494–495
    for saline-sodic and sodic soils reclamation, 362
    soil fertility and, 495
    soil sources of, 488–491
    soil system in, 487–495
Sulfur cycle, of soil compounds, 491–492
Sulfur mineralization, 492
Support practice factor (*P*), 619–622
Surface area
    colloids and, 275
    other soil properties and, 122
    soil particles and, 121
Surface charge, colloids and, 276
Surface drainage Swales, 222
Surface evaporation, control of, 211–212
Surface horizons, 75
Surface irrigation, 232–233
Surface roughness, and wind erosion, 641
Surface runoff, 200
Surface seal, 139
Surface tension, in water, 166
Surfactants, 668
Suspension, of soil particles, 648
Swelling, of sodic soils, 353
Swine manure, nutrient recycling and, 567
Symbiosis, 397
Symbiotic nitrogen fixation
    with legumes, 479–482
    with nonlegumes, 482–483
    without nodules, 483

T

TDR. *See* Time domain reflectometry
TDS. *See* Total dissolved solids
Temperature, atmospheric. *See also* Soil temperature
    parent materials and, 39–41

rock and mineral weathering and, 3
    SOM influence of, 447
Tensile strength, of soil aggregates, 137
Tensiometers, 175–177
Termites, 384
    soil formation and, 51
Terra firma, 9
Terraces, 39, 621
    in irrigation, 233
Testing, for radon gas, 694
Tetrahedral sheets, 279–281, 285
Textural classes
    alteration of, 124
    determination of, 124–125
    overview of, 123
    triangle, 124
Thermal conductivity of soils, 268–270
Thermic regimes, 108
Thermophilic composting, 461
Thixotropy, 155
Threshold velocity, of wind erosion, 641
Till, glacial, 42
Tillage, 413
    aeration and, 245
    conservation, 139, 211, 212–213, 270, 559–560, 622–628
    conventional, 138–139, 179–180
    erosion, 644–647
    ET control with, 209–213
    hydraulic conductivity with, 179–180
    liming and, 340
    pore space and, 150
    soil aggregation and, 135
    soil structure and, 137–141, 146–147
    soil tilth and, 137–138, 140
    SOM and, 453
    stubble mulch, 213
    ET control with, 209–213
    Vertical or Turbo, 626
    wind erosion and, 642
Tillage erosion, 644–647
    quantification of, 645–647
Timber harvesting
    calcium and, 524
    soil density and, 145–146
    intensity of, 633
Time
    soil formation and, 55–58
    soil temperature with, 268–270
Time domain reflectometry (TDR), for water content measurement, 175–176
Timing
    of nutrient, 585–586
Tires, soil compaction and, 146–149
Tissue analysis, nutrient management and, 587–591
Titration curves, 321–322
Topographic factor (*LS*), 616–618
Topography
    interaction with vegetation, 55
    salt builtup, 55
    slope aspect, 55
    soil formation and, 53–55
Toposequence, 36, 54
Topsoil, 15. *See also* A horizons
Torrands, 86–87
Torrerts, 94–96
Torrox, 102–103

Total acidity, 318
Total dissolved solids (TDS), salinity measurement, 348
Total porosity, 150
Total soil water potential, 168–169
Toxic compounds
    aluminum, 330–331
    and control of plant disease, 408–409
    in drainage water, 357–358
    inorganic substances, 674–681
    lead poisoning, 676–677
    manganese, 331–333
    organic chemicals, 652–657
    oxidation and reduction of, 248
    salts, 354–355
    soil as protection from, 3–4
    soil organisms' breakdown of, 406
    soil temperature and, 263
Toxicity range, 528
Traffic pans, 147
Transformations, in soil formation, 58–61
Transition layers, 65
Translocations, in soil formation, 58–61
Transpiration, 199, 206
    control of, 209–211
Transportation
    in erosion, 612, 640
    of phosphorus, 597–599
    of soil, 613
Trees
    aeration and, 252–253
    cation cycling by, 50
Tree removal, 633
Triazines, 658
Trickle irrigation, 235–236
    for saline soils reclamation, 359–362
    for saline-sodic and sodic soils reclamation, 362
Trophic level, 373
Turbels, 87–89
Turbidity, 610
2:1 Silicate clays, 283–286
    charge of, 290–293
    potassium in, 519

U

Udalfs, 98–99
Udands, 86–87
Udepts, 85
Uderts, 94–96
Udic moisture regime, 104
Udolls, 96–98
Udox, 102–103
Udults, 99–101
Ultisols
    distribution and use, 100–101
    nitrate leaching in, 486
    profile of, 81–83, 99–101
Ultramicropores, 151–152, 283
Umbricepipedon
    description of, 75
    in Oxisols, 102–103
Unconfined aquifer, 216
Unconfined compression test, 154–155
Unified classification system, for soil materials, 158
Universal soil-loss equation (USLE), 615–622
Unsaturated soil, water flow in, 180–181

Unstable slopes, mass wasting on, 630–631
Untilled soils, liming of, 340
Urban landscapes, aeration and, 251–253
Urban soils
 biological Properties, 69
 chemical Properties, 68–69
 human influences, 52–53
 pedological Properties, 67–68
 physical Properties, 68
 water infiltration and, 202–205
Urbanization, influence on soils, 1–2, 52, 146
USLE. *See* Universal soil-loss equation
Ustalfs, 98–99
Ustands, 86–87
Ustepts, 85–86
Usterts, 94–96
Ustic moisture regime, 104
Ustolls, 96
Ustox, 102–103
Ustults, 99–101

V

V. See Vegetative cover
Vadose zone, 216–217
Value, soil color, 118
Vapor movement, 177, 185
Vapor pressure gradient, 205
Variscite, 505
Vegetation
 albedo of, 264
 infiltration and, 201
 for saline-sodic and sodic soils reclamation, 362
 soil formation and, 49–50
 SOM and, 449–450, 454
 temperature and, 48–49
 topography and, 54
 and wind erosion, 641
Vegetative barriers, 628–629
Vegetative cover (*V*), 641
Vegetative mulch, for vapor loss control, 212
Velocity, of wind, 641
Ventilation, in soil, 3
Vermiculite, 283, 285
 charge of, 290–293
Vernalization, 260
Vertisols
 distribution and use, 95–96
 profile of, 81–84, 94–96
Vesicles, 398
Vitrands, 86–87
Volatilization, of nitrogen, 469
Volcanic ash, 43–44
 Andisols, 86–87
Volumetric water content, 173

W

Warm climates, organic mulches in, 270–271
Wastes, integrated recycling of, 572–573
Wastewater
 nutrient recycling of, 571
 phosphorus removal from, 504
 in septic system, 227–228
Water
 in cation and anion adsorption, 293–295
 crusting and, 139
 field capacity and, 187
 flow in soil, 177–181
 global stocks of, 198, 213
 nitrate impact on, 486
 pesticides in, 662
 in pore spaces, 150–153
 rock abrasion by, 33
 soil aeration and, 251
 soil colloid adsorption of, 278
 soil color and, 117
 soil consistence and, 153
 soil formation and, 21
 soil loss of, 213–214
 soil strength and, 155
 soils' capacity for, 3–4, 6–7
 structure of, 165–166
Water balance, salt-affected soils and, 346
Water bars, 633
Water characteristic curves, 171–172
Water deficit, 207
Water erosion
 mechanics of, 612–614
 models for predicting, 614–616
 types of, 613
Water potential, 168–169, 171–177, 205–206
Water potential gradient, 177–179
Water saturated soil, 240
Water table, 216
Water use efficiency, 207–208
 in irrigation, 231–232
Water-balance equation, 199
Waterlogged soil
 in container-grown plants, 251–252
 description of, 240
 salt-affected soils and, 346
Water-quality considerations, for irrigation, 356–358
Watershed, 199
Waxes, in plants, 423
Weathering
 in Alfisols, 98–99
 biogeochemical, 34
 calcium and, 524
 CEC and AEC levels and, 305
 chemical, 30–32
 description of, 29
 general case of, 32–33
 of nonacidcations, 315–316
 organic phosphorus and, 503
 in Oxisols, 102–103
 physical, 33–34
 rates of, 56
 of rocks and minerals, 29–36
 soil colloids and, 288–290
 in soil orders, 82
 in Ultisols, 99–101
Weeds, ET and, 209
WEPS. *See* Wind Erosion Prediction System
WEQ. *See* Wind erosion equation
Wetland delineation, 254
Wetland mitigation, 257
Wetlands
 aeration and, 253–258
 carbon balance in, 447
 chemistry of, 257
 constructed, 257–258
 description of, 253–254
 drainage of, 328–329
 greenhouse effect and, 457
 greenhouse gas production by, 248–249
 hydrology, 254
 moisture content of, 240
 selenium in, 678
 vegetation of, 249, 255–257
Wetlands functions, 258
Wetting front, 183
White alkali, 352
White ants, 384
Width of field factor (*L*), 641
Wilting coefficient, 187–188
Wind
 parent materials transport by, 43–44
 rock abrasion by, 33
Wind erosion
 factors affecting, 641
 importance of, 638–641
 mechanics of, 640
 prediction and control of, 641–644
Wind erosion equation (WEQ), 641
Wind Erosion Prediction System (WEPS), 642
Wind velocity, 641
 and turbulence, 641
Windblown dust, 611
Windblown sand, 611
Winter burn, 260–261
Wood wastes, nutrient recycling of, 570
Woody peat, 46–47
World Reference Base (WRB), 698–699
WRB. *See* World Reference Base

X

Xenobiotics, 406, 635
Xeralfs, 98–99
Xerands, 86–87
Xerepts, 85–86
Xererts, 94–96
Xeric moisture regime, 104
Xerolls, 96–97
Xerults, 99–100

Z

Zinc, deficiency v. toxicity of, 530